Epigenetics in Aquaculture

Epigenetics in Aquaculture

Edited by

Francesc Piferrer
Institute of Marine Sciences (ICM)
Spanish National Research Council (CSIC)
Barcelona, Spain

Han-Ping Wang
Ohio Center for Aquaculture Research and Development
The Ohio State University
Piketon, Ohio, USA

WILEY

This edition first published 2023

Registered Office(s)
John Wiley & Sons, Inc., 111 River Street, Hoboken, NJ 07030, USA
John Wiley & Sons Ltd, The Atrium, Southern Gate, Chichester, West Sussex, PO19 8SQ, UK

For details of our global editorial offices, customer services, and more information about Wiley products visit us at www.wiley.com.

Wiley also publishes its books in a variety of electronic formats and by print-on-demand. Some content that appears in standard print versions of this book may not be available in other formats.

Library of Congress Cataloging-in-Publication Data
Names: Piferrer, Francesc, 1960– editor. | Wang, Han-Ping, 1958– editor.
Title: Epigenetics in aquaculture / [edited by] Francesc Piferrer, CSIC, Institute of Marine Sciences, Barcelona, Spain, Han-Ping Wang, Ohio Center for Aquaculture Research and Development, The Ohio State University, Piketon, Ohio, USA.
Description: First edition. | Chichester, West Sussex, UK ; Hoboken : John Wiley & Sons Ltd., 2023. | Includes bibliographical references and index.
Identifiers: LCCN 2023000224 (print) | LCCN 2023000225 (ebook) | ISBN 9781119821915 (Hardback) | ISBN 9781119821922 (adobe pdf) | ISBN 9781119821939 (epub)
Subjects: LCSH: Aquacultural biotechnology. | Epigenetics.
Classification: LCC SH136.B56 P54 2023 (print) | LCC SH136.B56 (ebook) | DDC 639.8–dc23/eng/20230215
LC record available at https://lccn.loc.gov/2023000224
LC ebook record available at https://lccn.loc.gov/2023000225

Cover Design: Wiley
Cover Image: © Alisles/Adobe Stock Photos; Adobe Express; phonlamaiphoto/Adobe Stock Photos; issaronow/Adobe Stock Photos; Ahmed.yosri/Wikimedia Common

Set in 9.5/12.5pt STIXTwoText by Straive, Pondicherry, India

Dedication

Dedicated to Marta Burcet, for her affection and love – Francesc Piferrer.

To my grandboy Harrison Wang, who arrived in the world with this book and brings us so much laughter and energy, and his parents, Alan Wang and Kathy Zhang, for their understanding of the influences of parental care, nutrition, and other environmental factors on offspring's early development and beyond – Han-ping Wang.

Contents

About the Editors

Francesc Piferrer obtained his PhD in biology in 1990 and currently is research professor and deputy-director at the Institute of Marine Sciences (CSIC) in Barcelona. His traditional field of research is sexual development in fish. He has made relevant contributions to our understanding of how genetic, epigenetic, and environmental factors determine whether an animal will differentiate as male or female. In the last years, he has been studying environmental influences on the establishment of epimutations and the contribution of epigenetic inheritance to enable animals to cope with a changing environment. He is significantly contributing to the integration of epigenetics in aquaculture research. His lab developed the world's first epigenetic test to predict sex and the first epigenetic clock in fish. He organized the first epigenetics session in an international aquaculture conference. Dr. Piferrer has been the principal investigator in numerous research projects and contracts with companies, having signed two patents. He is the editor of 2 previous books and the author of more than 160 articles in indexed journals and has directed 14 doctoral theses. He has carried out management tasks at the Catalan Society of Biology and the Spanish Ministry of Science and Innovation and provided advice to the European Commission and the Council of Europe. He was awarded the XII Jacumar Prize for the Best Aquaculture Research in 2013 and the Research Award of the Official College of Biologists in 2020. In 2019, he became a full member of the Royal Academy of Sciences and Arts of Barcelona.

Dr. Han-Ping Wang is principal scientist/research professor and director of the Ohio Center for Aquaculture Research and Development at The Ohio State University (OSU). He has provided leadership as project director for about 75 research projects in the areas of breeding, genetics, sex control, and epigenetics in fish and aquaculture. He was the first to achieve success in controlled breeding and culture of Reeves shad, and in developing large-scale populations of all-male bluegill, all-female yellow perch, and superior perch strains. He also completed whole genome sequencing of these two species. Dr. Wang has published more than 170 papers in prestigious international journals and proceedings as a principal author or corresponding author and three books, including this book and *Sex Control in Aquaculture*. Dr. Wang has advised more than 30 PhD students and post-doctoral researchers and organized and chaired 2 international conferences. He has served as a member of the US Department of Agriculture (USDA)-NCRAC Research and Technical Committee and a research review panellist of the USDA and the US National Oceanic and Atmospheric Administration (NOAA). Dr. Wang has been awarded approximately $11 million grants for his research and outreach. He has won 6 S&T Achievement Awards, and 10 "best paper" and other professional awards from national and international agencies.

List of Contributors

Anne-Catrin Adam
Feed and Nutrition, Institute of Marine Research (IMR)
Bergen, Norway

Dafni Anastasiadi
The New Zealand Institute for Plant and Food Research
Limited, Nelson Research Centre, Nelson, New Zealand

Ishrat Anka
Biosciences Department, Faculty of Science and
Engineering, Swansea University, Swansea, UK

Jana Asselman
Laboratory for Environmental Toxicology and Aquatic
Ecology, Ghent University, Ghent, Belgium

Anne Beemelmanns
Institut de Biologie Intégrative et des Systèmes (IBIS)
Université Laval, Québec City, Quebec, Canada

Ramji K. Bhandari
Department of Biology, University of North Carolina
Greensboro, Greensboro, NC, USA

Susanne M. Brander
Coastal Oregon Marine Experiment Station, Department
of Fisheries, Wildlife, and Conservation Sciences
Oregon State University, Newport, OR, USA

Sofia Consuegra
Biosciences Department, Faculty of Science and
Engineering, Swansea University, Swansea, UK

Marit Espe
Feed and Nutrition, Institute of Marine Research (IMR)
Bergen, Norway

Manon Fallet
Man-Technology-Environment Research Centre (MTM)
School of Science and Technology, Örebro University
Örebro, Sweden

Bo Feng
Yellow Sea Fisheries Research Institute, Chinese Academy
of Fishery Sciences, Qingdao, China

Jorge M.O. Fernandes
Faculty of Biosciences and Aquaculture, Nord University
Bodø, Norway

Ignacio Fernández
Centro Oceanográfico de Vigo, Instituto Español de
Oceanografía (IEO-CSIC), Spanish National Research
Council, Vigo, Spain

Paulo Gavaia
Centro de Ciências do Mar (CCMAR) and Department
of Biomedical Sciences and Medicine, Universidade do
Algarve, Faro, Portugal

Jian-Fang Gui
State Key Laboratory of Freshwater Ecology and
Biotechnology, Institute of Hydrobiology, Wuhan, China

Bruno Guinand
Institut des Sciences de l'Evolution de Montpellier
University of Montpellier, CNRS, IRD, EPHE
Montpellier, France

María Paz Herráez
Department of Molecular Biology, Faculty of Biology
and Environmental Sciences, Universidad de León
León, Spain

Timothy A. Hore
Department of Anatomy, University of Otago
Dunedin, New Zealand

Sara J. Hutton
Department of Environmental and Molecular Toxicology
Oregon State University, Corvallis, OR, USA

Ioannis Konstantinidis
Faculty of Biosciences and Aquaculture, Nord University
Bodø, Norway

Catherine Labbé
INRAE, Department of Fish Physiology and Genomics
Campus de Beaulieu, Rennes, France

Audrey Laurent
INRAE, Department of Fish Physiology and Genomics
Campus de Beaulieu, Rennes, France

Qian Liu
Yellow Sea Fisheries Research Institute, Chinese Academy of Fishery Sciences, Qingdao, China

Marta Lombó Alonso
Department of Molecular Biology, Faculty of Biology and Environmental Sciences, Universidad de León
León, Spain
Dipartimento Scienze della Vita e dell'Ambiente, Università Politecnica delle Marche, Ancona, Italy

Wenxiu Ma
Yellow Sea Fisheries Research Institute, Chinese Academy of Fishery Sciences, Qingdao, China

Shokouoh Makvandi-Nejad
Department of Immunology and Virology, Norwegian Veterinary Institute, Ås, Norway

Jan A. Mennigen
Department of Biology, University of Ottawa, Ottawa
Ontario, Canada

Hooman Moghadam
Breeding and Genetics, Benchmark Genetics Norway AS
Bergen, Norway

Laia Navarro-Martín
Institute of Environmental Assessment and Water Research, IDAEA-CSIC, Spanish National Research Council (CSIC), Barcelona, Spain

Artem V. Nedoluzhko
Faculty of Biosciences and Aquaculture, Nord University
Bodø, Norway
Paleogenomics Laboratory, European University at Saint Petersburg, Saint Petersburg, Russia

Oscar Ortega-Recalde
Department of Anatomy, University of Otago, Dunedin
New Zealand

Francesc Piferrer
Institute of Marine Sciences, Spanish National Research Council (CSIC), Barcelona, Spain

Takaya Saito
Feed and Nutrition, Institute of Marine Research (IMR)
Bergen, Norway

Athanasios Samaras
Department of Biology, University of Crete, Heraklion
Crete, Greece

Elena Sarropoulou
Institute of Marine Biology, Biotechnology, and Aquaculture, Hellenic Centre for Marine Research
Hraklion, Crete, Greece

Changwei Shao
Yellow Sea Fisheries Research Institute, Chinese Academy of Fishery Sciences, Qingdao, China

Zhi-Gang Shen
College of Fisheries, Huazhong Agricultural University
Wuhan, China

Kaja H. Skjærven
Feed and Nutrition, Institute of Marine Research (IMR)
Bergen, Norway

Christina R. Steadman
Earth & Environmental Sciences Division, Climate, Ecology & Environment Group, Los Alamos National laboratory, Los Alamos, NM, USA

Lili Tang
Yellow Sea Fisheries Research Institute, Chinese Academy of Fishery Sciences, Qingdao, China

Günter Vogt
Faculty of Biosciences, University of Heidelberg
Heidelberg, Germany

Rune Waagbø
Feed and Nutrition, Institute of Marine Research (IMR)
Bergen, Norway

Han-Ping Wang
Ohio Center for Aquaculture Research and Development
The Ohio State University South Centers
Piketon, OH, USA

Qian Wang
Yellow Sea Fisheries Research Institute, Chinese Academy of Fishery Sciences, Qingdao, China

Tamsyn Uren Webster
Biosciences Department, Faculty of Science and Engineering, Swansea University, Swansea, UK

Yue Yu
College of Fisheries, Huazhong Agricultural University, Wuhan, China

Xiaona Zhao
Yellow Sea Fisheries Research Institute, Chinese Academy of Fishery Sciences, Qingdao, China

Li Zhou
State Key Laboratory of Freshwater Ecology and Biotechnology, Institute of Hydrobiology, Wuhan, China

Preface

Aquaculture is the fastest food production sector in the world and prospects are that this position will be maintained for years to come. According to data from the Food and Agriculture Organization (FAO) of the United Nations, in 2020 global aquaculture production reached a record of 122.6 million tons worth USD 281.5 billion. Animals accounted for 87.5 million tons while algae comprised 35.1 million tons. However, aquaculture must become more sustainable to meet the growing demand for aquatic foods of an ever-increasing human population. Thus, improved aquaculture production requires further technical innovations, including more focus on breeding programs, feed utilization, well-being, and disease control. Similar to any other food production system, aquaculture is about producing the phenotypes with superior value. In this endeavor, new advances on our understanding of the epigenetic regulation of the phenotype have the potential to play an increasing role in achieving aquaculture production sustainability.

The term "epigenetics" was coined by Conrad Waddington in the 1940s, but with a meaning different from how it is understood today. Initially, it was essentially related to what today is understood as the field of developmental biology and how the phenotype comes into being. However, the modern concept of epigenetics, i.e., "the study of mitotically and/or meiotically heritable changes in gene function that cannot be explained by changes in DNA sequence," arose in the mid-1990s and around the turn of this century. The field has largely benefited especially from the advancements made after the sequence of the human genome, the characterization of the regulatory elements, and all emerging technologies to interrogate different aspects of the genome and epigenome.

Epigenetics is now considered one of the "hot topics" in biology. Epigenetic modifications or "marks" can be easily identified, and they constitute therapeutic approaches for the treatment of an increasing number of diseases. Thus, there is a lot of research ongoing in the epigenetics of cancer, for example.

There are three very important aspects to take into account when dealing with epigenetics. First, epigenetics integrates genomic and environmental influences to bring about the phenotype. Second, there is a fraction of the phenotypic variance that cannot be explained solely on genetic variation, but that can be explained by taking into account epigenetic variation. Third, epigenetic changes can be inherited and thus passed from parents to offspring into the following generations. Combined, this has prompted the implementation of epigenetic research, not only in ecology and evolution for its contribution to adaptation to new environments, but also into agriculture and livestock for improved food production. Consequently, recently there has been both a clear interest in marine epigenetics and in the application of epigenetics in aquaculture. One of the main reasons is that aquatic organisms are quite susceptible to environmental cues since, for example, temperature in a cold-blooded animal influences growth rates more strongly than in a warm-blooded animal. Further, in contrast to mammals, fishes seem to have less reprogramming and erasing of epigenetic marks after fertilization, thus facilitating epigenetic transmission of environmental influences to the next generation. Thus, there is a lot of interest for the application of epigenetics in aquaculture. However, and to the best of our knowledge, there are currently no books that address this need.

"Epigenetics in Aquaculture" consists of 20 chapters and is arranged into three parts: Part I: Theoretical and practical bases of epigenetics in aquaculture; Part II: Epigenetics insights from major aquatic groups; and Part III: Implementation of epigenetics in aquaculture. All chapters are written by top specialists with ample experience and at the forefront in their respective research fields.

Part I contains six chapters (Chapters 1–6) and provides the necessary background to understand what epigenetics is about and what are the major mechanisms and phenomena. The first chapter covers the overall roles and the diversity of epigenetic mechanisms across major taxa and provides insights into their potential applications in aquaculture and aquatic animals. The following two chapters are devoted to the three main epigenetic mechanisms regulating gene expression, namely, DNA methylation, histone modifications, and non-coding RNAs. The next two chapters are devoted to two key aspects of epigenetics. One explains how epigenetic modifications can be inherited across different generations, a hot topic in different areas of biology, and the other elucidates the role of epigenetics in integrating environmental cues as a powerful mechanism in the adaptation and the basis of organismal plastic responses to rapid environmental change. The last chapter of Part I presents the currently available methods to analyze the epigenetic modifications including the latest developments, as well as some basic resources for the bioinformatics analysis of the data. It also explains how to choose among the different approaches based on the type of question that one aims to answer.

Part II contains 11 chapters (Chapters 7–17) and constitutes the bulk of the book. These chapters explore the roles of epigenetic regulatory mechanisms in key biological process and their relevance for aquatic production. The first two chapters deal with epigenetic sex determination and differentiation as well as the dynamics of epigenetic marks during gametogenesis and early development. The following two chapters are devoted to growth, with one focusing on skeletal muscle and the other emphasizing nutritional programming. The next two chapters of Part II are devoted to the epigenetics of stress response, immune response, and the emerging topic of the role of the microbiome in shaping epigenetic responses of the host. Additionally, one chapter is focused on epigenetics in hybridization and polyploidy and another on how epigenetics can contribute to explain organismal responses to toxins present in the aquatic environment. Many of the above-cited chapters of Part II focus on fish, where considerable work has been carried out so far. Thus, this part ends with three chapters dedicated to the epigenetics of other taxa that are also very important for aquaculture production, namely mollusks, crustaceans, and algae, where interesting discoveries related to similarities and differences with the situation in vertebrates are being made.

Finally, Part III includes the final three chapters (Chapters 18–20), dealing with the actual integration of epigenetics into aquaculture practice. For this, the development of biomarkers and their applications in aquaculture is discussed. Particular attention is then paid on the integration of epigenetic selection into current genetic breeding programs. The final chapter identifies knowledge gaps, discusses challenges that must be overcome, and outlines future prospects on the application of epigenetics in aquaculture. In addition to tables, figures, and abundant bibliography, each chapter contains a glossary of terms used with pertinent definitions.

Thus, this book provides an update on the state-of-the-art on the knowledge of epigenetic regulatory mechanisms in major taxa of aquatic organisms including algae, crustaceans, mollusks, and fish and how this new knowledge can be applied to increase aquaculture production. It covers both basic and applied aspects of epigenetics related to reproduction, development, growth, nutrition, and disease of aquatic species, which we hope will benefit the aquatic scientific community and the aquaculture sector.

This book will be appealing to anyone interested in knowing all major aspects related to epigenetics, including mechanisms, inheritance, methodology, etc. Information contained within will be particularly useful to researchers working on epigenetics in aquatic animals and aquaculture, including basic aspects of fish and shellfish epigenetics, reproductive endocrinology, genetics, and evolutionary and environmental biology. It will also appeal to PhD and MSc students and biologists working in hatcheries or in breeding companies, who will all benefit from reading about epigenetics and the opportunities it can provide. More broadly, aquatic biologists, including fisheries managers and conservation biologists, will also benefit from clear and practical information in epigenetics. The epigenetic insights from fish, shellfish, and aquatic model species will attract readers from other disciplines as well, who might find inspiration in findings made on epigenetics of aquatic organisms.

Our previous book, "Sex Control in Aquaculture," was based on knowledge accumulated after decades of applying sex control techniques to improve aquatic productivity. In contrast, the present book is based on research that is just a few years old because the study of epigenetics in aquatic organisms and its application in aquaculture is still in its infancy. Thus, a lot remains to be done. We hope that our efforts in providing a comprehensive picture of the current situation will be of much help and foster future research. The well-known British zoologist D'Arcy Thompson (1860–1948) wrote in the preface of his most famous book "On Growth and Form" that "this book of mine has little need of preface, for indeed it is 'all preface' from beginning to end." Given that the application to epigenetics into aquaculture is still in the first steps of a hopefully long and profitable journey, it could be stated that in a good way the same applies to the present book.

Francesc Piferrer and Han-Ping Wang

Acknowledgments

We thank Bradford Sherman at The Ohio State University for the English editing of all chapters of this book. Thanks also go to Sarah Swanson and Hong Yao at The Ohio State University for their assistance in chapter coordination, format review, and reference/citation editing. We thank Hong Yao at The Ohio State University for designing the front cover of the book and Zh-Gang Shen at the Huazhong Agricultural University for helping with the indexes of this book. Special thanks go to all anonymous reviewers for their efforts in the peer review of the chapters and for their constructive comments that helped improve the quality of the book. We thank Rebecca Ralf, Stacey Woods and Vinitha Kannaperan at Wiley for their guidance throughout the production of this book.

Part I

Theoretical and Practical Bases of Epigenetics in Aquaculture

1

The Potential Role of Epigenetics in Aquaculture: Insights from Different Taxa to Diverse Teleosts

Han-Ping Wang[1] and Zhi-Gang Shen[2]

[1] *Ohio Center for Aquaculture Research and Development, The Ohio State University South Centers, Piketon, OH, USA*
[2] *College of Fisheries, Huazhong Agricultural University, Wuhan, China*

1.1 Introduction

1.1.1 Concepts and Terminology

As aquaculture and livestock production and human demands increase, recent efforts have been focused on improving the production efficiency and health of agricultural animals and fish and their sustainability using advanced approaches. One of the recent approaches has been the application of sequencing technologies, genotype analysis, and genome editing for genetic improvement programs. Although great advancement has been achieved through genetic selection in animals, including those used in aquaculture, genetics can only elucidate part of the phenotypic variability of economic traits to researchers and breeders. Recent research strongly supports the view that epigenetic modifications contribute to additional layers of variation that could facilitate the improvement of different aspects of production, including reproduction, health, growth and nutrition, and overall sustainability of agricultural and aquacultural animals.

The concept of "epigenetics" was originally proposed based on advances in plant research in 1942 by Conrad H. Waddington, indicating the effect of the environment on the development of phenotypes at that time [1, 2]. Epigenetics has evolved considerably since then and is now implicated as the event of molecular modifications that are heritable and in control of the genome activity and gene expression modulation leading to phenotypic variations without changing the DNA base sequence [3, 4]. In other words, epigenetics is the set of heritable marks (including DNA methylation, chromatin remodeling, posttranslational histone changes, noncoding RNAs, and other molecules) on the genome that can modify gene expression without changing the DNA sequence content [5]. These epigenetic marks are able to transmit messages through mitosis by mediating gene expression [6]. The epigenome, or the set of epigenetic modifications of the genome of a given cell, has a dynamic role throughout the lifetime as it mediates genetic and environmental interaction [7, 8]. The general terminology of epigenetics is listed in Box 1.1.

1.1.2 Epigenetic Mechanisms and Phenomena

Genetics deals with the inheritance of information encoded by the DNA nucleotide sequence and its variants, which can have severe consequences as seen in genetically inherited diseases [9]. In contrast, epigenetic mechanisms modulate chromatin packaging and gene expression by adding or removing chemical tags to/from DNA and histones without altering the DNA sequence. Environmental exposures acting on enzymes can alter the epigenetic signatures, which can be transmitted to the next generation, without directly changing epigenetic signatures. In the process of DNA methylation and DNA demethylation, DNMTs (DNA methyltransferases) add methyl groups to cytosine residues, and TET (ten-eleven-translocation) proteins actively catalyze the oxidation of 5mC (5-methylcytosine) to 5hmC (5-hydroxymethylcytosine) for demethylation; demethylation also occurs passively through cell division without de novo methylation; DNA hypomethylation and hypermethylation play an essential role in gene transcription and transcriptional repression respectively [10, 11].

Epigenetics in Aquaculture, First Edition. Edited by Francesc Piferrer and Han-Ping Wang.

Box 1.1 General Glossary of Epigenetics

CpG sites: Regions of DNA where guanine follows a cytosine nucleotide. In vertebrates, cytosines at CpG dinucleotides constitute the principal target of DNA methylation, while cytosine methylation takes place in other sequence contexts in invertebrates. CpG sites occur with high frequency in genomic regions called CpG islands.
DNA methylation: The adding of methyl groups to the DNA molecule, usually on a cytosine of cytosine-guanine dinucleotides, as a means of chemical DNA modification.
Epigenetic modifications: Chromatin and DNA modifications that affect genome function without alterations in the underlying DNA sequence.
Epimutation: A heritable change resulting in an alteration in gene expression but not affecting the DNA-coding sequence. Epimutations take place when methyl groups are added or removed from DNA or when changes are made to histones that bind to the DNA in chromosomes.
Histone modification: A covalent posttranslational alteration of histone proteins such as lysine acetylation, methylation, serine phosphorylation, arginine methylation, etc.
Noncoding RNA: A generic term for functional RNA molecules including miRNA, siRNA, piRNA, and lncRNA, which are not translated into proteins that can regulate gene expression at both transcriptional and posttranscriptional levels.
Phenotypic plasticity: The capacity of a genotype to generate various phenotypes in response to different environmental changes.
Transgenerational epigenetic inheritance: The epigenetic effect on the phenotype that can be transmitted to subsequent generations of cells or organisms without leading to changes in DNA base sequences.

For histone methylation, HATs (histone acetyltransferases) add acetyl groups to lysine residues within the N-terminal end of histones, causing local chromatin decondensation that allows transcriptional activation to occur [10, 12, 13]. HDACs (histone deacetylases) remove the acetyl group's chromatin condensation and repress gene regions transcriptionally [10, 14], whereas HMTs (histone methyltransferases) add methyl groups to lysine and arginine residues and HDMs (histone demethylases) remove methyl groups. Different from DNA methylation, whether histone methylation activity is associated with activation or repression of gene expression is dependent on the amino acid methylation position and extent [10, 15, 16]. The detailed roles of these key players in epigenetics are described in the section later.

It is understood that ncRNAs also possess vital roles in the regulation of gene expression, particularly micro- (miRs) and long noncoding RNAs, which regulate important genetic programs through interaction with the coding genome. For example, interfering RNAs can assist in the recruitment of enzymes modifying chromatin to mediate transcription or promote the posttranscriptional degradation of transcripts [9].

1.2 Key Players of Epigenetics

1.2.1 DNMTs

In animals, DNA cytosine methylation takes place mainly in the CpG dinucleotides in the genome [17, 18], while in humans, approximately 70–80% of CpG dinucleotides are methylated, with 5mC enriched on transposons, satellite repeats, and intergenic regions [19]. 5mC discovered initially in tubercle bacillus and calf thymus DNA [20, 21] is now the main epigenetic modification identified in the genomes of plants and animals [19]. This modification is implicated in the suppressing transcription of transposable elements (TEs) [22] and maintaining chromosomal integrity [22, 23] with start site DNA methylation of promoters, enhancers, and transcription associated with coding regions' gene repression [24].

In mammals, three key members (DNMT1, DNMT2, and DNMT3) of DNA-cytosine-5-methyltransferase enzymes, which catalyze the transfer of methyl groups to the 5-position of cytosines, are responsible for establishing and maintaining the methylation patterns as key "writers" [25]. The DNMT1 preferentially methylates hemimethylated DNA [26, 27] to ensure the maintenance and propagation of methylation patterns through cell division [28]. Knocking out DNMT1 in mice leads to embryonic lethality during the developmental stage and dilution of DNA methylation during cell cycles [29]. During cell differentiation processes, the DNMT3a and DNMT3b are actively involved in the de novo methylation with different functions – the former is contributory to establishing the patterns of DNA methylation at maternally imprinted loci [30] and the latter is involved in methylation of CpG islands and pericentromeric repeats during inactivation of X-chromosome [31]. Loss of germline function of DNMT3a and DNMT3b leads to an absence of methylation in spermatogonia, resulting in arrest in meiosis and infertility [10, 32–35] and embryonic lethality in mice [36]. In addition, the DNMT family has a cofactor, DNMT3l, which is instrumental in recruiting de novo methyltransferases to target sequences [30, 37]. A study shows that DNMT3l can serve as both a positive modulator of DNA methylation at gene bodies of housekeeping genes and a negative mediator at promoters of bivalent genes, during embryonic cell differentiation in a mouse [17, 38]. Like DNMTs, DNMT3l involvement in de novo DNA methylation is so important that a lack of DNMT3l causes infertility [39–43].

In contrast, DNMTs are not founded to be involved in DNA methylation in plants, and their non-CpG methylation is typically targeted to TEs with modulation by siRNAs. The 5mC modification of plants takes place in multiple dinucleotide contexts, including CpG, CHH, and CHG. Instead, three other DNA methyltransferase enzymes, methyltransferase 1 (MET1) and chromomethylase 2 and 3 (CMT2 and CMT3) are responsible for these modifications. The enzyme DNA methylation "readers" that mediate gene expression include SUVH1 and SUVH3, which are the SU(VAR)3-9 protein homologs [44, chapter 17]. Similarly, algae do not have mammalian DNMTs for DNA methylation [45, 46]. However, like animals, CpG methylation of TEs and gene bodies have also been reported in plants [17, 47–49].

Overall, DNA methylation distribution in fish is similar to mammals, with Dnmt1, Dnmt3a, and Dnmt3b involvement in CpG methylation and very little in non-CpG sites. However, DNA methyltransferase isoforms are more expanded and DNA methylation machinery is more diversified in teleosts than mammals due to whole genome duplication and locus-specific rearrangements [50–52]. Different mammalian paralogues of DNA methylation enzymes, Dnmt1, Dnmt3a, and Dnmt3b are identified in fish. For example, *dnmt3aa* and *dnmt3ab* have been found to mediate sex differentiation of zebrafish related to temperature [53] and gonadal *dntm3aa* plays an important role in gamete development through gonadal DNA methylation in tilapia [54]. Dnmts are also involved in the methylation related to muscle growth as evidenced in Atlantic cod (*Gadus morhua*), in which *dnmt1* and *dnmt3a* are found to be associated with improving photoperiod-stimulated growth [55]. In addition, higher percentages of CpG methylation were observed, and PGCs are predetermined by germplasm and not elicited via epigenetic mechanisms in fish when compared to mammals [56]. Furthermore, studies show that Dnmt3l is absent from fish genomes, which is also the case in bird and amphibian genomes. DNMT3L is a key cofactor for the relatively high methylation of mammalian sperm and the lack of Dnmt3l in fish suggests their low levels of methylation in sperm.

In summary, DNMTs play a crucial "writer" role in maintaining and propagating 5mC/cytosine methylation patterns through early development across taxa. Other than DNMT1, DNMT3a, and DNMT3b, mammals have a cofactor, DNMT3L. Knockout or deletion of DNMT1, DNMT3b, and DNMT3L in mice leads to embryonic lethality. DNMTs have not been found to be involved in DNA methylation in plants and algae. DNA methylation distribution in fish is similar to mammals but more diversified with more and different mammalian paralogues of DNA methylation enzymes. However, fish lack a Dnmt3l gene and the global hypomethylation of sperm DNA [57], the same as birds and amphibia. These specific features of the DNA methylation landscape in fish make them unique as model species for studying comparative DNA methylation patterns and early evolutionary fates of duplicated genes, including the determination of the mechanisms for *dnmt3aa* and *dnmt3ab* regulation of sex differentiation with temperature and identification of basic DNA methylation patterns in somatic and germ cells within lifecycle and between generations of aquatic species (Chapter 2). The related epigenetic research will benefit fish, the aquatic field, and the entire taxa in general.

1.2.2 TETs

In mammals, TETs (ten-eleven translocations), including TET1, TET2, and TET3, play an important role as an "eraser" in DNA demethylation during early development and multiple adult somatic tissues [58]. The TET proteins contain a C-terminal catalytic dioxygenase domain that can recognize CpG dinucleotides and bind to 5mC to perform oxidation reactions, producing 5hmC, 5fC (5-formylcytosine), and 5caC (5-carboxylcytosine) [29, 36, 59, 60]. These oxidized derivatives are then diluted during replication or replaced by unmethylated cytosines by the DNA repair machinery [61]. Thus, while the old 5mC patterns were being erased and new ones established right after fertilization during organogenesis and early development, TETs performed two waves of erasure process in mammalian genomes [62–64]. In mouse preimplantation embryos, demethylation of both parental genomes is likely to follow a TDG (thymine DNA glycosylase)-independent pathway in spite of involving 5mC oxidation to 5fC/5caC [17, 65]. Depleting the three TET enzymes together leads to incorrect differentiation of embryonic stem cells [66], and gastrulation defects in mice highlight the importance of the 5mC derivatives for development and differentiation [67]. However, studies show that the experimental mice with knockout of individual *TET1* or *TET2* can survive [68–70], with *TET1*-deleted mice showing a neurogenesis-related phenotype [71, 72] and *TET2* deficiency affecting hematopoietic stem cell (HSC) differentiation [70, 73, 74], and combined deletion unchanging organogenesis [75]. *TET3* knockout leads only to neonatal lethality during mouse embryonic development [17, 76].

TETs possess an important role in response to numerous environmental stressors. Recent evidence shows that restricting maternal food during late gestation causes a reduced fetal and liver weight and fetal heart in goats, with a significant increase of *TET1* and *MBD2* (methyl-CpG-binding domain protein 2) expression in fetal heart and liver [77, 78]. Another study suggests that an adverse maternal environment downregulated *TET1-3* expression in the heart of male offspring of pregnant mice [77, 79]. Linking vitamins and TET function has also been evidenced. A study conducted by Chen et al. [80] revealed that *TET1* effect on the reprogramming of embryonic fibroblast was mediated by vitamin C, leading to the induction of pluripotent stem cells in the mouse. These results suggest the possibility of applying nutritional intervention to control certain diseases. In terms of chemicals and pollutants, Feng et al. [81] found that repeated exposure of both mice and humans to cocaine-recessed TET1 in the Nucleus accumbens and TET1 deficiency promoted the preference for cocaine in mice. Other studies suggested that global increases of 5hmC levels and *TET1* activity were elevated by exposure to toxicants benzene and its metabolites, which can be found in cigarette smoke and petroleum products [77, 82, 83]. TET1 can be affected by radiation in different types of cells and tissues. Coulter et al. [84] and Kuhns et al. [85] found that *TET1* depletion promoted ionizing radiation-induced aberrant cell cycles and apoptosis processes. Another study showed that loss of *TET1* resulted in augmented X-ray-induced deterioration and accumulated DNA damage and genomic instability of embryonic cells in mice. These results indicate that *TETs* are required for genomic stability, DNA damage repair, and cell cycle maintenance after radiation.

In plants, mammalian TET proteins and homologs have not been identified [86], although 5hmC and 5fC can be discernable in the DNA of several plants [87–89]. Some studies show that plants use a demethylation mechanism without 5mC enzymatic oxidation, and it is moved from the DNA by specific DNA glycosylases directly [17, 48, 90]; however, the roles of oxidized forms in 5mC are not clear yet in plants.

However, TET homologs have been identified in algae and fungi [91–93]. Recently, two studies show that a novel DNA modification C5-glyceryl-methylcytosine (5gmC) was catalyzed by algal TET homologue CMD1 from *Chlamydomonas reinhardtii*, using vitamin C as a co-substrate [94, 95]. Also, a mushroom TET homolog CcTET from *Coprinopsis cinerea* was found to preferentially oxidize 5mC to 5fC, but not to 5hmC [17, 92, 95].

In teleosts, a few studies in zebrafish (*Danio rerio*) suggest that both the canonical TET-TDG pathway [96, 97] and the AICDA/APOBEC (activation-induced cytosine deaminases/apolipoprotein B mRNA editing complex) pathway are involved in demethylation activity and there is only one copy of tet1-3 encoded in their genome as mammals (Chapter 2). There are seven *tet* and four *tdg* found in salmonids with duplicates of all *tet* genes following Ss4R [50, 52]. Overall, there is limited knowledge of the demethylation activity of both pathways in teleosts and more related research is needed.

In summary, evidence suggests that TET proteins play an important role in DNA demethylation during early development for differentiation and embryogenesis, and in responding to environmental stresses in vertebrates, including fish. *TET1* absence causes DNA repair deficiency, severe DNA damage, and increased genomic instability. Although TET proteins and homologs have been found in algae and fungi [17, 91–93], this is not evidenced in plants [17, 86]. The *tet1, tet2, tet3,* and *tet4* are identified in fish genomes due to specific whole genome duplication and locus-specific rearrangements [51, 98, 99]. This specific feature of genome duplication is worth much attention for future research. The application of nutritional intervention could be a potential way to control certain diseases in aquatic species and aquaculture.

The epigenetic similarities and differences in the above key regulators and DNA methylation among terrestrial vertebrates, fish, and plants are listed in Table 1.1.

1.2.3 KMTs/KDMs

KMTs (histone lysine methyltransferases) and KDMs (histone lysine demethylases) are histone code writers and erasers that play a critical role in operating epigenetic control through histone modification [109–113]. In the past two decades, there have been many KMTs and KDMs and related biochemical properties and their regulation identified and investigated [114]. The activity of these histone modifiers needs to be in dynamic steadiness to balance epigenetic homeostasis. Evidence shows that alterations in histone lysine methylation possess an important role in the determination of neuronal development and differentiation, neuropsychiatric disorders, mental retardation, and behavioral changes in humans and

Table 1.1 Differences and similarities in DNA methylation and related key payers among fish (exemplified by zebrafish), terrestrial vertebrates, and plants.

	Terrestrial vertebrates	Fish	Plants	References
Presence of CpG methylation	CpG methylation	CpG methylation; very little in non-CpG sites	Non-CpG methylation targets to TEs and is commonly modulated by siRNAs	[100–103]
Timing of germline separation from somatic tissues	PGCs are obtained from the epiblast and appear in the posterior primitive streak during gastrulation; there is limited time for epigenetic alterations to be transmitted into the germline cells	PGCs are predetermined by germplasm and are not elicited via epigenetic mechanisms	There is no early separation of germline and soma, and the gametes obtained from vegetative tissue right before completing development	[104]
Targets of DNA methylation	In general, gene bodies are methylated, whereas CpG islands are often unmethylated	Same as other vertebrates	Methylation generally takes place on repetitive DNA elements and TEs	[105]
Methylation patterns	Typically, DNA methylation takes place globally in vertebrates and around 70–80% of cytosines in CpG dinucleotides are methylated	Same as other vertebrates	In general, there are mosaic DNA methylation patterns, which are characterized by domains of massively methylated DNA scattered with domains of free methylation	[105–107]
Genes involved in the cytosine methylation and oxidation pathway	*Dnmt1, Dnmt3a, Dnmt3b, Dnmt3l, Uhrf1, Tet1, Tet2, Tet2, Tdg*	*dnmt1, dnmt3aa, dnmt3ab, dnmt3ba, dnmt3bb.1, dnmt3bb.2, dnmt3bb.3, Tet1, Tet2, Tet3,* Tet4, *Tdg.1, Tdg.2*		[108]
Proteins involved in the cytosine methylation and oxidation pathway	Dnmt1, Dnmt1s, Dnmt1o, Dnmt3a, Dnmt3b, Uhrf1, Tet1, Tet2, Tet2, Tdg	dnmt1, dnmt3aa, dnmt3ab, dnmt3ba, dnmt3bb.1, dnmt3bb.2, dnmt3bb.3, Tet1, Tet2, Tet3, Tet4, Tdg.1, Tdg.2		[108]

Source: Data from Youngson and Whitelaw [104] and Jessop et al. [108].

animals [114] and lysine methylation balance is instrumental in the maintenance of genome integrity, gene regulation, and disease evasion [114–116].

Many diseases including aging, neurological pathologies, and cancer are correlated with KMT/KDM amplifications, mutations, deletions, and misexpression [114]. In humans, extreme and epigenome-wide histone hypo- or hypermethylation have been observed in cancer, and deregulation of KMTs and KDMs can turn epigenetic master modulators into cancer drivers [117], which means hyperactive KMT or inactive KDM can contribute to the repression of tumor suppressor genes due to accumulation of histone lysine methylation marks [114, 118], and inactive KMTs or hyperactive KDMs can result in transcriptional activation of oncogenes. Knockout or loss of the H3K4me3 methyltransferase MLL1/KMT2A in mice resulted in reduced response to fear stimulation, indicating that MLL1/KMT2A is essential for forming proper memory [119]. Another study found that feeding methyl-deficient diets resulted in increased liver cancer [120]. These results suggest that KMT inhibitors could be important in the therapy of cancer and mental disorders.

KMTs and KDMs have been involved in the differentiation of mesenchymal stem cells into various cell lineages. In mammals, skeletal myogenesis and myoblast differentiation take place during early embryonic development and in response to damaged mature muscle. The process is modulated by key transcription factors, including MEF and MyoD. The KMTs G9a mediate both MyoD and MEF transcription during myogenesis [121, 122], and the KDM LSD1 is required for myogenesis [122]. Additionally, a recent study [121] specifically identified KDM3B, KDM6A, and KDM8 as modulators of osteogenic differentiation with only KDM LSD1 possessing a role in myotube differentiation.

Evidence suggests that, in the case of hypoxia, KDM4A stabilization enhances the amplification of epidermal growth factor receptor (EGFR) [123]. In that experiment, cells were exposed to hypoxia for an initial 24 hours, then followed by supplementation with EGF under hypoxia conditions for an additional 24 hours; afterward, cells were harvested, and their EGFR copy numbers were evaluated. Results showed that, in combination with EGF, hypoxia caused an increased EGFR DNA copy number, indicating that KDM4A stabilization induced by hypoxia in accordance with increases in H3K4 methylation stimulated by EGF promotes EGFR locus plasticity. This study suggests that KMTs/KDMs-targeting inhibitors can increase or inhibit extrachromosomal EGFR amplification, which uncovers potential therapeutic opportunities for dealing with EGFR copy-number heterogeneity and the related drug response [123]. KDM also plays a critical role in the mediation of stress hematopoiesis and early B-cell differentiation. A recent study revealed that loss of *KDM6A* affected the function of early B cells and erythroid and myeloid progenitor cells through global gene expression analyses in animals [124].

In fish and aquatic species, KMT- and KDM-related studies and information are relatively limited. In zebrafish, as shown in mammals, SMCX/KDM5C was found to be mutated with cognitive impairments [125] and the mutations were shown to impact demethylation, resulting in aberrant neural branching [114, 126]. KMTs are found both as paralogues and as single loci being involved in histone methylation dynamics in zebrafish [98]. In salmoniform, however, it appears that duplicates are retained in genomes. Fellous et al. [127] studied KMTs and KDMs in a self-fertilizing hermaphroditic fish species, the mangrove rivulus fish (*Kryptolebias marmoratus*), and discovered 25 Kdm orthologues. By examining expression patterns of both Kdm and Kmt, they found them at peak level in the gastrula stage and a reduction in later embryogenesis, and higher in male brains compared to hermaphrodite brains. They also identified the Kdm and Kmt expression patterns in the male testes and hermaphrodite ovotestes. Observed in Pacific oyster (*Crassostrea gigas*), a few studies suggest that methylation of Histone H3 lysine K4 (H3K4), which is regulated by Kdm and Kmt, modulates some gene expression in the early embryo [128–130], and abnormal histone methylation patterns were associated with mortalities of embryos under thermal stress [131]. Evidence shows that *ehmt2* (autosomal lysine *N*-methyltransferase 2) is involved in the regulation of hypoxia in marine medaka (*Oryzias melastigma*) via inducing dimethylation of the lysine residue of histone H3 (H3K9me2). In that study, the hypoxia-induced reductions in sperm quality and quantity took place through histone modification mediated by EHMT2 [132]. These findings suggest that, similar to mammals, Kdm and Kmt play important roles during early development, neurogenesis, and stress in fish and aquatic animals.

In summary, KMTs and KDMs play critical roles in multiple tissues with Yin-Yang opposing mechanisms in maintaining genome integrity, gamete development, and disease control in mammals and aquatic species. However, the retention patterns of KTMs in zebrafish and salmoniform genomes are different from mammals. Developing tools and strategies that can inhibit or activate them with drugs have great potential for disease therapy research and practice. Designing small molecules and peptides that can block active sites has been suggested to disrupt protein–protein interactions [114]. This area and related approach would have great potential for aquaculture research and practice.

1.2.4 HATs/KATs and HDACs

Organism cells can turn genes on or off by adding or removing acetyl moieties (CH_3CO) to histones. The addition of acetyl groups to histones allows the DNA to be more accessible for transcription "turning on" genes of interest [133]. This biological process of histone acetylation and deacetylation is modulated by two major enzyme families: HATs or KATs (lysine acetyltransferases) and HDACs or KDACs (lysine deacetylases). As "writers," HATs transfer the acetyl group to a target histone from acetyl-CoA (Ac-CoA), while HDACs function as "erasers," removing the acetyl group of a target histone [134–136]. HATs are divided into two major families: type-A and type-B, with type-A further being divided into GNAT, MYST, and CBP/p300 families [98]. Thus, their balance and performance play crucial roles in the regulation of gene expression and metabolic/physiological function [133]. Currently, there are at least 18 HDAC proteins identified since the first HDAC1 in mammals was isolated in 1996 [137]. HDACs are divided into four classes [138, 139]: the Class I Rpd3-like proteins including HDAC1, HDAC2, HDAC3, and HDAC8; the Class II Hda1-like proteins with HDAC4, HDAC5, HDAC6, HDAC7, HDAC9, and HDAC10; the Class III Sir2-like proteins including SIRT1, SIRT2, SIRT3, SIRT4, SIRT5, SIRT6, and SIRT7; and the Class IV protein with HDAC11.

It is understandable that HATs and HDACs play a crucial role in stem cell maintenance and differentiation because cell identity is determined by the pattern of gene expression. Several HATs and HDACs have been identified to possess an essential role in activating genes that allow stem cells to self-renew. The deficiency of the activity results in decreased expression and/or differentiation [133]. Within the MYST family of HATs, the MOZ complex has a major function in the maintenance of HSCs. In mice, loss of MOZ significantly reduces the HSC level in the fetal liver and causes embryonic death [133, 140]. MOZ is also essential for the reconstitution of hematopoiesis for HSCs that are transplanted into mice [133, 140, 141].

HATs and HDACs play important roles in the regulation of growth, development, and reproduction through mediating juvenile hormones as shown in *Drosophila melanogaster* and other insects [142]. Evidence suggests that exposure of

Tribolium castaneum cells (TcA cells) to juvenile hormone or Trichostatin A (TSA, HDAC inhibitor) promotes expression of Kr-h1 (a known JH-response gene) and knockdown or loss of the gene coding for CREB-binding protein (CBP, contains HAT domain) leads to a reduction in the Kr-h1 expression [142].

HATs and HDACs are necessary for skeletal muscle development and normality through maintaining and moderating metabolic homeostasis, motor adaptation, and exercise capacity of skeletal muscles. The two enzymes may be involved in turning on and off metabolic fuel switching and controlling skeletal muscle homeostasis from the myogenesis process [137]. The master transcription factor MyoD A is a driver for satellite cells differentiating to myotubes [143]. By targeting MyoD gene regulatory elements, the HAT p300 possesses an important role in myotube differentiation [144]. In humans, evidence shows that the HAT GCN5 drives the osteogenic differentiation of mesenchymal stem cells (MSC) via an acetyltransferase-independent mechanism [145]. A study shows that class I HDACs are increased in models of disuse atrophy and Hdac2, HDAC4, HDAC6, and SIRT1, which belong to class I–III HDACs, are promoted in multiple models of muscle atrophy, suggesting HATs and HDACs serve as essential elements for normal muscle atrophy in response to pathophysiological conditions [146].

HATs and HDACs are key regulators of fat formation, lipid metabolism, and energy balance. For example, the HAT enzyme PCAF possesses an important role in the modulation of the myogenic program and adipocyte proliferation [147, 148]. Accumulating evidence suggests, as shown in mice, that loss of HDAC3 and HDAC9 affects the development and accumulation of fat-laden cells as adipose tissue at various sites in the body [147, 149–151]. Through a feeding experiment using a high-fat diet, it was found that the knockout of HAT p/CIP and SRC-1 suppressed the development of brown adipose tissue (BAT) and energy balance, suggesting the essential role of SRC-1 in the regulation of energy balance [152, 153].

HATs and HDACs possess pivotal roles in mounting an effective immune response against pathogens and controlling the disease. Research data shows that HATs and HDACs are both required for mycelium growth and pathogenicity [154]. In humans, accumulating data reveals that HDAC activity and/or expression upgraded in numerous disease states [155] and histone acetylation, which is regulated by HATs and HDACs, in host cells is dynamically mediated during infection [156]. In addition, numerous studies provide evidence that HAT MOF expression is downregulated in numberers of cancers and mutation residues on HATs in certain cancer [134, 157, 158], but a higher level of HDACs is found in cancer cells [134, 158–161]. Because virulence factors and metabolic products of pathogenic microorganisms change the activity of HATs and HDACs to restrain transcription of host defense genes, this new avenue of pathogen–host interactions has significant applications by introducing HDAC inhibitors (HDACi) into clinical practice [156].

HATs and HDACs are essential elements in the regulation of hypoxia. This process is mainly through targeting hypoxia-inducible factors (HIFs) because the foundation of hypoxia mainly depends on the stability of HIF 1 alpha (HIF1α). In mammals, it has been reported that HIF1α stability can be reduced by protein FIH1 (factor-inhibiting HIF1) under normoxic conditions and inhibition of HAT activity leads to decreased viability and invasiveness of germ cell cancers during hypoxia [138, 162]. Meanwhile, strong evidence suggests that HDAC1, 2, 3, 4, and 6 enhance HIF1α stabilization by removing the inhibitory acetyl groups [138, 163, 164]. It was reported that HDAC7 translocates with HIF1α into the nucleus, leading to the amplification of HIF1α transcriptional activity by interacting with p300 during hypoxia [165]. Another study showed that the expression of EMT (epithelial–mesenchymal transition) markers in cancer cells was modulated by hypoxia-induced HDAC3, and HDAC3 deficiency caused ablation of the hypoxia-induced EMT process [166]. In addition, there is evidence that class III HDAC sirtuins (SIRTs) play dynamic roles during hypoxia [167]. These findings indicate that HDACs could be potential targets for developing therapeutic tools for treating diseases.

There are only a few studies found related to the roles of HAT and HDAC families in teleosts. A recent study in marine medaka suggests that a reduced HAT and HDAC expression is correlated with increased EHMT2 dimethylation and activation, which is involved in the regulation of hypoxia [132]. Accumulating evidence shows that both HAT kat5a and HAT ep300 have been found as paralogues in zebrafish, and additional duplication and differential retention of these paralogues are present in Salmoniformes. Furthermore, *hdac3*, *hdac5*, *hdac11*, and *sirt1* genes are identified in single loci in zebrafish, with duplication for *hdac3*, *hdac5*, and *sirt1* in genomes of salmoniform [98]. These results suggest that the retention patterns in zebrafish and salmoniform genomes are different from mammals and Ts3R and Ss4R genome duplication events differentially affected components of histone acetylation [98].

In summary, HATs and HDACs play important roles in the modulation of stem cell differentiation, early development, reproduction, skeletal muscle growth, fat formation, energy balance, disease control, and hypoxia. Although limited results have been found in the roles of HAT and HDAC families in teleosts, studies indicate there are different retention patterns in zebrafish and salmoniform genomes relative to mammals, and the genome duplication events affected components of histone acetylation differentially. The mounting research findings in mammals and teleosts should have potential applications by introducing HDACi into aquaculture practice.

1.3 Divergent Epigenetic Mechanisms from Different Taxa to Diverse Teleosts

Epigenetic modifications induced by the environment can be transmitted across generations. Although this transgenerational epigenetic inheritance takes place in all taxa, the loyalty of this transmission is conditional, variable, and taxon-dependent [168], resulting in conserved and species-specific aspects of the epigenetic machinery and dynamics [169]. Figure 1.1 shows the common and divergent epigenetic mechanisms in animal taxa [169]. For instance, genome cytosine

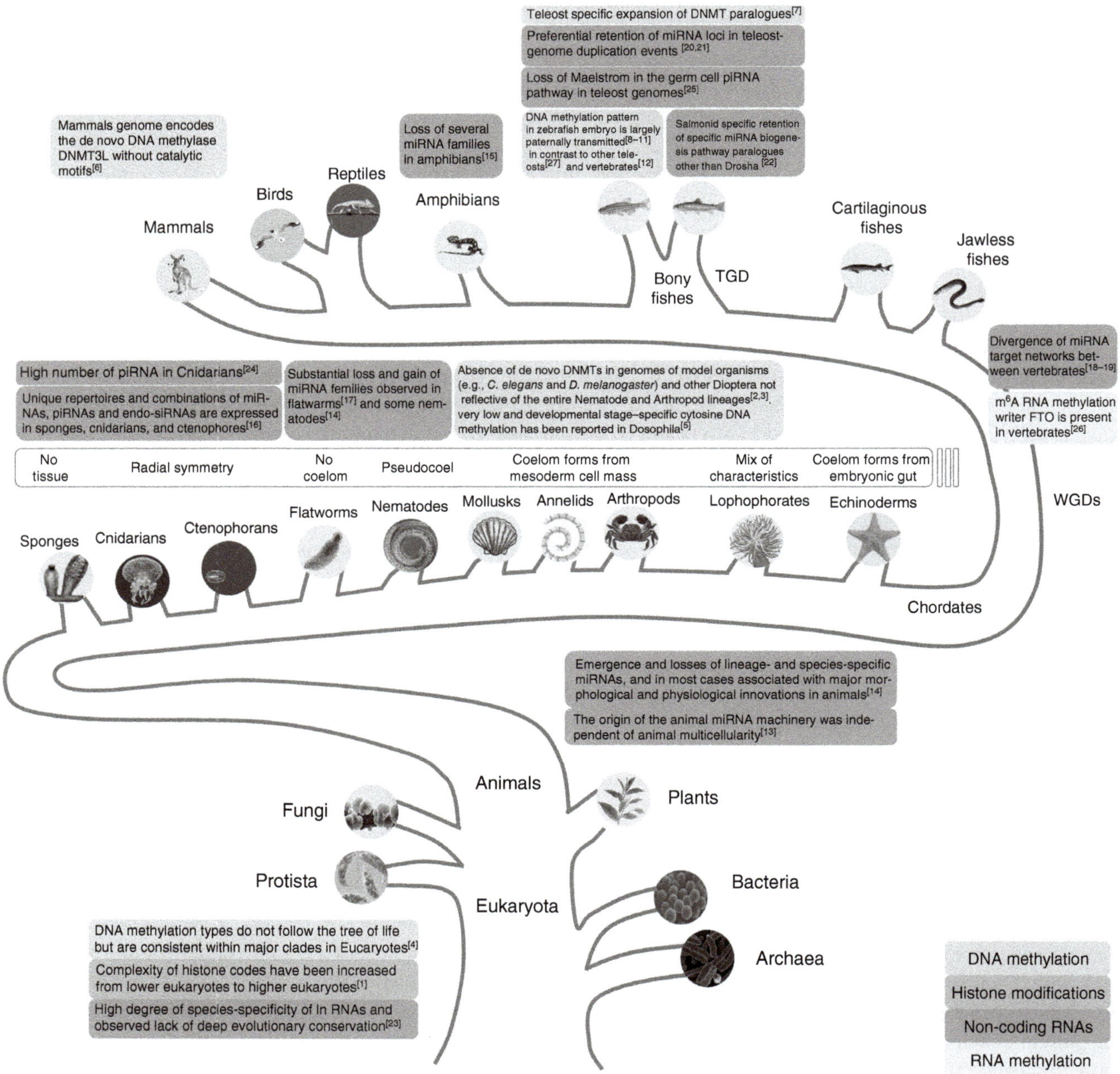

Figure 1.1 Phylogenetic epigenetics showing the common and divergent epigenetic molecular mechanisms in animal taxa. *Source:* Modified from Navarro-Martín et al. [169]. (a) Bradleyblackburn/Adobe Stock, (b) Vadim_petrakov/Adobe Stock, (c) RooM The Agency/Adobe Stock, (d) Brian Gratwicke/Wikimedia Commons/CC BY 2.0, (e) Mirko Rosenau/Adobe Stock, (f) U.S. Fish & Wildlife Service/Wikimedia Commons/Public Domain, (g) U.S. Department of Commerce, (h) Andrei Nekrassov/Adobe Stock, (i) Thanunchakorn/Adobe Stock, (j) Devin/Adobe Stock, (k) Sinhyu/Adobe Stock, (l) Uckyo/Adobe Stock, (m) Siloto/Adobe Stock, (n) Peteri/Adobe Stock, (q) Artphotoclub/Adobe Stock, (s) National Institute of Allergy and Infectious Diseases (NIAID)/CDC Public Health Image Library. References: [1][170]; [2][171]; [3][172]; [4][173]; [5][174]; [6][175]; [7][176]; [8][177]; [9][178]; [10][179]; [11][180]; [12][181]; [13][182]; [14][183]; [15][184]; [16][185]; [17][186]; [18][187]; [19][188]; [20][189]; [21][190]; [22][191]; [23][192]; [24][193]; [25][194]; [26][195]; [27][196]. (*See colour plate section for colour representation of this figure.*)

DNA methylation was found not correlated with the presence of de novo DNA-methyltransferase encoding loci in some insect genomes [175], indicating the presence of potentially undiscovered pathways controlling DNA cytosine methylation in these species. Meanwhile, cytosine DNA methylation was missing in a fruit fly (*Drosphila melanogaster*) and perhaps the entire clade of Diptera [169, 175]. This phenomenon was also discovered in *Caenorhabditis elegans* [197]. One or more genome duplication events with protein expansions have been identified in teleosts that were contributory to DNA methylation and miRNA biogenesis [169]. Moreover, piwi-RNAs were identified to be involved in germline development in zebrafish despite the absence of maelstrom, a protein essential for piRNA-mediated transcriptional transposon silencing in various species such as in the fruit fly and mice [194, 198], which indicates that some teleosts have different molecular mechanisms in epigenetic transgenerational inheritance than mammals. In addition, Dnmt3l is absent in fish genomes [57], the same as in bird and amphibian genomes. DNMT3L is a key cofactor for the relatively high methylation of mammalian sperm, and the loss of Dnmt3l in teleosts suggests their low levels of methylation in sperm. Collectively, these specific features of the DNA methylation landscape in some teleosts make them unique as model species for investigating patterns of comparative DNA methylation and early evolutionary fates of duplicated genes.

1.4 The Roles and Applications of Epigenetics

Mounting evidence from epigenetic research in humans and animals has revealed that epigenetics plays a complementary role to genetics, touching many aspects of biological processes such as reproduction, growth, development, health, and nutrition [199]. Since the role of epigenetics in aquaculture or aquatic animals is similar to other vertebrates in principle, and epigenetics in aquaculture is behind other livestock animals, in the later sections, we will briefly review the roles of epigenetic processes in reproduction and early development, health, and well-being management, nutrition and growth advancement, and sustainability enhancement in different taxa and discuss the potential role in aquaculture based on the most recent research progress.

1.4.1 Reproduction and Early Development

1.4.1.1 The Roles of Epigenetics in Early Development

Epigenetic regulation, including DNA methylation and histone modification, plays a crucial role in early development and reproduction. This is because the latter requires the formation of gametes, which develop from PGCs, and the modulation of gene expression during the development significantly relies on epigenetic mechanisms [108]. During the embryonic development of vertebrates, epigenetic mechanisms act on cis-regulatory elements to regulate gene expression during tissue differentiation, with DNA methylation usually leading to the repression of gene expression. Studies show that mouse and zebrafish establish, maintain, regulate, and remove methylation marks using essentially the same enzymatic machinery, but the way they use this machinery early in embryogenesis is different. In most terrestrial animals, PGCs, obtained from the epiblast, arise in the posterior primitive streak during gastrulation (inductive species) and there is limited time for transmitting epigenetic changes into the germline cell [104]. In teleosts, PGCs, predetermined by germplasm (preformative species), are not induced via epigenetic mechanisms; early separation of germline and soma, and the gametes obtained from vegetative tissue right before completion of early development, is not a case [104] (Table 1.1).

DNMTs and TETs play essential roles in maintaining and propagating DNA methylation patterns during early development across taxa. Since discovering a novel function of TETs, the active demethylation and the 5mC's role in early development have been better understood [108, 200, 201]. To maintain good order of the expression of germline differentiation genes and the progression of gametogenesis, it is essential to have a process of losing 5mC and then repressing histone marks, and gaining 5hmC [6, 202]. During the process, DNA re-methylation of GCs occurs through using the de novo methylation enzymes DNMT3A, DNMT3B, and DNMT3L. Specifically, as shown in mice, the DNMT1 ensures the maintenance and propagation of methylation patterns through cell division [6, 28]; DNMT3a and DNMT3b are in control of de novo methylation of DNA, which takes place during cell differentiation processes with different functions. DNA methylation distribution in teleost is like mammals but more diversified with more diverse mammalian paralogues of DNA methylation enzymes. In addition, there is no *dnmt3l* gene and the global hypomethylation of sperm DNA is found in fish genomes [57]. In both mammals and teleosts, the TETs play an opposite role as an "eraser" in DNA demethylation during early development (see details in Section 1.2 and Chapter 7). However, *et1*, *tet2*, *tet3*, and *tet4* are identified in fish genomes due to specific whole genome duplication [51, 98, 99]. This specific feature of genome duplication is worth much attention for future research.

In mice, knockout of germline function of DNMT3A and DNMT3B leads to a loss of methylation in spermatogonia, subsequent arrest in meiosis, and infertility [10, 32–35] and embryonic lethality [36], and depletion of the three TET enzymes together leads to incorrect differentiation of embryonic stem cells [75] and gastrulation defects [17, 67]. In fish (zebrafish), although they use similar enzymatic machinery establishing, maintaining, modifying, and removing these methylation marks, their way of using this machinery in early development and embryogenesis is quite different [108]. Therefore, experimental results from the homozygous knockout of cytosine methylation and oxidation pathway genes of zebrafish are considerably different from those in the mouse. The detailed results of comparison for early development and beyond are listed in Table 1.2. Briefly, the mammalian embryo needs to produce extraembryonic membranes for its survival and utilize imprinting to prevent parthenogenesis. In addition, producing naïve pluripotent cells in the blastocyst for the germline is essential for early development in mammals. In zebrafish, both fertilization and early development occur externally with PGCs predetermined by germplasm and nutrition supplied by yolk. Thus, the roles of TETs are limited to the mediation of specific processes during early development in zebrafish [108].

Development of the embryo is a process of the paternal and maternal genome reprogramming, which is characterized by a series of epigenetic modifications starting right after fertilization [223]. A study conducted by Jiang et al. shows that levels of global 5meC decrease significantly during preimplantation stage in mice embryos, which is linked to successive waves of demethylation and de novo methylation [224]. Requiring the expression of TET enzymes, demethylation is more important in the paternal genome as shown in mice. In addition, DNMT3B seems primarily in charge of mediating 5meC levels in embryos, with the presence of DNMT1 and DNMT3A. These epigenetic alterations are involved in triggering embryonic genome activation. Thereafter, embryo development relies heavily on the stock of RNA and proteins accumulated in the oocyte. When the first cell differentiation occurs, the global levels of repressive histone marks arrive at a minimum level at embryonic genome activation and recover to the blastocyst stage [6]. For example, the H3K27me3 mark is removed by HMT KDM6B during the first embryonic divisions [6, 225]. When the KDM6B expression is repressed, preventing the decrease of H3K27me3, this results in a change to embryonic genome activation and a reduction in blastocyst rate [6, 226]. Camargo et al. [6, 226] have shown that an H3K9me3 accumulation and excessive heterochromatin formation in embryos resulted from the heat shock exposure of oocytes in mammals, indicating the sensitivity of oocytes to culture conditions. Moreover, evidence suggests that the different incorporation of the histone variant macroH2A1 may be responsible for the sexual dimorphism of the steatotic phenotype in males and females of mice. MacroH2A1 has been found to be enriched on the inactive X chromosome in females, indicating it plays a role in sex chromosome dosage compensation through its ability of regulating gene expression [227].

In teleosts, the promoters of genes involved in early development and germline formation, such as *hox*, *piwil1*, *dazl*, and *vasa*, selectively demethylate in the zygote at zygotic genome activation in females [228]. In males, it has been reported that several permissive histone marks such as H3k4me3, H3k4me2, and H4k16ac, and multiple hypomethylated regions that correspond to promoters of genes during early development or meiosis are persistent in zebrafish sperm [229]. There are several best-characterized epigenetic writers found in the fish gonads due to specific whole genome duplication [230–232]. These results from both mammals and teleosts highlight the important roles of histone marks KMTs/KDMs and HATs/HDACs in the regulation of early development as described in Section 1.2. Detailed information about epigenetics in sexual maturation and gametes of fish is described in Chapter 7.

It is well recognized that there is an inverse relationship between DNA methylation and expression levels of the female-promoting gene *cyp19a1a* and the male-promoting gene *dmrt1* in many fish species [233]. For example, knockout of cyp19a1a by TALEN and CRISPR/Cas9 results in all-male progeny in zebrafish and further knockout of male promoting gene *dmrt1* in the *cyp19a1a* mutant can save the all-male phenotype by overcoming estrogen deficiency [234]. Recently, our group found that the downregulation of *cyp19a1* and *dmrt1* genes affected the gonadal development, resulting in certain percentages of sex reversal in yellow perch (*Perca flavescens*) through the interference of siRNA (Xie and Wang, unpublished data). In addition, high levels of DNA methylation in females and low levels of DNA methylation in males are correlated with the expression of *dmrt1* [233, 235].

Mounting evidence in teleosts also shows that culture conditions or environment affect embryogenesis and early development including sex determination and differentiation, gamete quality, sex ratio, egg size, larvae survival, growth, etc., through epigenetic mechanisms [94]. Temperature can impact the sex determination and differentiation of many heat-sensitive teleost species by affecting the methylation level of typical sex-specific molecules and interfering with the interaction between transcriptional elements [236–241]. Studies also show that temperature affects the sex ratio of some fish species in natural

Table 1.2 Comparative outcomes resulting from the gene knockout of cytosine methylation and oxidation pathway in the mouse and zebrafish.

	Homozygous knockout phenotypes			
Gene knockout	**Mouse**	**Mouse germline**	**Zebrafish**	**References**
Tet1	Smaller size, reduced oocyte numbers – meiotic gene expression reduced	Male and female germline → smaller litter size when bred to WT	No phenotype	[68, 203, 204]
Tet2	2–4 months – increased white cell count; Adult – predisposition to myeloid leukemia;	No phenotype	Adults – myelodysplastic syndrome susceptibility	[70, 205, 206]
Tet3	Parental knockout → neo-natal lethality in heterozygous pups	No phenotype	No phenotype	[65, 67, 76, 203, 207]
Tet1 + *Tet2*	Smaller ovaries and reduced fertility; mid-gestational lethality of some embryos.		No phenotype in embryo	[68, 203]
Tet1 + *Tet3*			No phenotype in embryos	[203]
Tet2 + *Tet3*			Lethality – larval period. Reduction of HSC formation	[203]
Tet1 + *Tet2* + *Tet3*	Lethality at E6.5 – gastrulation failure		Phenocopy of tet2 + tet3	[67, 203]
Tdg	Lethality at E11.5 – hemorrhage			[208, 209]
tdg1 + *tdg2*			Unknown	
Dnmt1	Lethality at E11	Infertility, premature meiotic gene expression and premature imprint erasure	Lethality at 8dpf	[29, 210–213]
Dnmt1o	Lethality at mid-gestation; imprinting largely maintained			[214, 215]
Dnmt1s	Non-lethal, partial loss of maternal imprints			[216]
dnmt3bb.1			Defects in hematopoietic progenitor maintenance	[217]
Uhrf1	Lethality at E11, phenocopy of Dnmt1 knockout		Defects in liver outgrowth and regeneration, lens defects, defects in intestinal barrier function and hematopoietic progenitor maintenance	[36, 218–220]
Dnmt3a	Lethality – post-natal. Stunted growth, imprinting defects	Male germline → gametogenic defects Female germline → mid-gestational lethality of heterozygous offspring		[34, 36, 221]
Dnmt3b	Lethality – post-E9.5	No phenotype		[34, 36]
dnmt3bb.2			Defects in brain and retinal neurogenesis	[222]
Dnmt3l		Phenocopy of Dnmt3a		[222]
Dnmt3a + *Dnmt3b*	Lethality – earlier than Dnmt3b			[36]

Source: Modified from Jessop et al. [108].

environments via epigenetic mechanisms (Chapter 5). More interestingly, a recent study found that five days of starvation right after hatching resulted in sex reversal of female-to-male in medaka due to low levels of lipids caused by a limited amount of acetyl-CoA, which is produced by the pantothenate pathway and is an essential metabolite involved in histone acetylation [228]. In *Daphnia*, additional administration of pantothenate induced the production of males. These findings suggest that the pantothenate pathway may play an essential role in sex differentiation by regulating the acetyl-CoA supply. Another study in marine medaka (*Oryzias melastigma*) suggests that hypoxic stress can damage normal gonad function, leading to reproductive disorders through the regulation of miRNAs [242]. The hypoxia-induced miRNA activity can also affect fish testis function and signal pathways, resulting in testicular dysfunction [243]. In addition, a study conducted by Wang et al. [132] shows that the hypoxia exposure of medaka across their lifecycle causes consistent intra- (F0), multi- (F1), and transgenerational (F2) impairment of sperm quality and quantity of the fish through histone modification regulated by EHMT2. Furthermore, exposure to toxicants during early development or life stages can also negatively affect reproduction and early development through an epigenetic mechanism. The effects of EDCs on the global DNA methylation of fish gonads have been well reported [244–250]. Zebrafish exposed to both hypoxia and venlafaxine (a drug used to treat depression) was found to have miRNA expression changed through the F1 generation but not the F2 [251–253]. Other chemicals have also been shown to affect the sperm DNA methylation pattern in teleosts [254]. Zebrafish males exposed to methyl mercury were found to have a change of sperm DNA methylation in both F0 and F2 fish [255]. BPA exposure in zebrafish leads to an increase in acetylation levels of activating marks (H3k9ac, H3k14ac, and H4k12ac) and a decrease in the levels of the repressive mark (H3k27me3) [248]. Collectively, these findings further highlight that DNA methylation, histone modifications, and ncRNAs play crucial roles in the regulation of ovarian and testicular metabolism and early development in teleosts under different stress environments. More details can be found in Chapters 2, 5, 7, 8, and 14 of this book.

1.4.1.2 The Potential Applications of Epigenetics in Reproduction and Breeding

During gametogenesis and early development, reprogramming of epigenetic marks takes place. This is particularly achieved by changes in global DNA methylation of the gametes and through waves of demethylation and methylation in the early embryo, with different outcomes depending on whether distinct tissues or germ cells are being specified. In animal breeding and reproduction, the application of epigenetic manipulations to favor a desired phenotype in many cases relies on internal or external manipulation of fertilization and early development, when epigenetic marks are most easily established or can be influenced. In mammals, both fertilization and early development are internal in the sense that they occur within the gestating mother, and related reprogramming of the epigenome has been well studied [256]. In poultry, or birds in general, fertilization is internal, while embryonic development of the fertilized eggs is external. In contrast, in most farmed aquatic animals including mollusks, crustaceans, and fish, both fertilization and early development are external (Figure 1.2).

Figure 1.2 Environment (internal vs. external) where fertilization and early development events take place in mammals, birds, and aquatic animals. Because both events take place externally, aquatic animals in principle offer more opportunities for influencing the epigenetic reprogramming events that take place during these early stages. *Source:* F. Piferrer. (*See colour plate section for colour representation of this figure.*)

Consequently, a conserved feature across species is that gametes and early embryos are susceptible to environmental changes and this fact can perhaps be exploited to manipulate the epigenome in such a way that it retains marks associated with a phenotype of interest. In this regard, chances to manipulate the environment are scarce in farmed mammals with internal fertilization and early development. Thus, it has been argued that the chicken is a good model for performing epigenetic studies and manipulations, with its embryo being easily manipulated in vivo and in vitro [257]. Although the information is available only in a few species of cultured aquatic animals, the external fertilization and early development make epigenetic modifications of them theoretically more feasible, either by manipulation of gametes or by manipulation of environmental conditions during early embryonic development. By this principle, we argue that aquatic animals are excellent models for epigenetic research and application.

The knowledge of epigenetics from mammals, birds/poultry, and aquatic animals has been applied to medicine and other aspects of humans and agricultural animals and will have great potential for aquaculture practice. Below we will summarize the current and potential application of epigenetics in the reproduction and early development of different taxa, attempting to gain insights into aquaculture.

In mammals, as described earlier, the sperm RNA content has long been regarded as a remnant of spermatogenesis. Transcriptional profiling has great potential for the discovery of fertility biomarkers in disease treatments. Defining a panel of male (in)fertility biomarkers with improved discriminatory power relative to individual biomarkers has been attempted recently [258–260] and several miRNA pairs with potential use as fertility biomarkers have been validated to be correlated with infertility conditions in human sperm [258, 259, 261]. In agricultural animals, the effective biomarkers for predicting male fertility in the sperm methylome are being isolated using artificial intelligence tools and the genetic value of sires has been established at an early stage based on genotypes for genomic selection [258, 262]. Another example is to obtain cloned animals, which are built on epigenetic reprogramming of the somatic nucleus by the oocyte machinery, and many strategies are being used to improve cloning efficiency [263]. One example is to reduce the hypermethylation of cloned embryos and improve cleavage and blastocyst rates through the overexpression of AID encoding an enzyme associated with active DNA demethylation in cattle [263].

The development and application of epigenetic methods to improve gamete quality and larvae survival have considerable potential for aquaculture research and practice. A recent study in common carp (*Cyprinus carpio*) shows that oocytes aged in vitro exhibit time-dependent increases in H4k12ac, suggesting a good possibility of epigenetic marker development for improving egg quality [254]. In addition, the findings that parental nutrition regulates nutritional programming of changing egg composition and improving sperm production [264, 265] provide an opportunity for aquaculture to improve larvae survival. Moreover, the research results that parental exposure of shellfish to low pH during crucial life stages improves fitness and growth rate of offspring could be applied to improving larvae/fingerling quality and production through epigenetic approaches [245, 266].

Controlling the maturation and onset of ovulation of broodstock by manipulation of temperature and photoperiod for out-of-season spawning have been applied in many aquaculture fish species for producing fry and fingerling in different seasons through the mediation of the endocrine kinetic and DNA methylation profile of the broodstock [267, 268]. Grafting spermatogonia or gametogenesis into the recipient breeder to improve the production of ova and sperm has been used in aquaculture research and practice [269, 270]. This rises to the bipotential of germinal stem cells and the establishment of the gametic epigenetic profile during gametogenesis for application (Chapter 7).

Sex control techniques for producing monosex populations have been widely used in aquaculture research and practice [271]. The current epigenetic knowledge of modulation of gene expression in this field has provided technical feasibility for the development of epigenetic tools for sex control in aquaculture. The epigenetic editing technology developed recently, e.g., CRISPRoff and CRISPR/dCas9 systems, can establish DNA methylation and repressive histone modifications in mammals and should have good potential for sex manipulation techniques in aquaculture. A recent study demonstrates that DNA hypermethylation can promote the silencing of the female-determining *CFSH* gene in decapod males [272], suggesting the potential of applying this knowledge to produce monosex fish for aquaculture. Currently, sex-linked DNA markers have been used in sex manipulation breeding for genotypic sex identification and progeny testing. Developing and applying epi-markers could have the potential to contribute to significant improvements in the accuracy and success of sex identification for sex control in aquaculture (Chapter 8). For example, the finding of the methylation state of individual CpG sites from a few candidate genes such as *cyp19a1a* and *dmrt1* in the European sea bass (*Dicentrarchus labrax*) and other species can be used as a tool to predict the sex of a fish species [273]. The finding that knockout of macroH2A1 leads to dimorphism in mice could have an application in studying dimorphism in teleosts and other aquaculture species.

Although epigenetic variations might not significantly affect the phenotypic diversity among or between populations, emerging evidence shows that epigenetic mechanisms may play an important role in the selective breeding of animals through shaping the epigenome in early development, persisting during the lifetime, and transferred to the next and multiple generations. Such interindividual differences in the epigenetic signatures can significantly contribute to the phenotypic variations through gene expression or patterns of alternative splicing [274], potentially making use of the epigenetic marks as biomarkers for selective breeding. With sequencing technology development and related costs decreasing, it is possible to use Epigenome-Wide Association Study (EWAS) to identify epigenetic marks with significantly different frequencies between groups and link them to the traits of interest [275, 276]. EWAS could be implemented for intergenerational analysis to evaluate and determine the transgenerational transmission of epigenetic signatures (Chapter 19).

1.4.2 Health and Well-Being Management

1.4.2.1 The Roles of Epigenetics in Controlling Stress and Disease

Epigenetic changes in response to environmental stressors and the essential roles of DNA and histone methylation in stress responses have been well documented in different farm animals [277, 278]. For example, epigenetic regulation is found to be involved in heat stress processes in pigs through differentially methylated regions (DMRs) and associated genes [277, 279]; the adaptation of chickens to thermal manipulation of embryos was partially mediated by DNA methylation and histone H3K27me3 and H3K4me3 [277, 280–282]; exposure of sheep to heat stress led to increased m6A RNA methylation level and related enzymes in the liver [277, 283]. In addition, Zhang et al. [284] found some differentially methylated genes (DMGs) being significantly enriched in HIF 1 in heart tissues of Tibetan pigs from high- and low-altitude regions. These findings indicate the involvement of epigenetic mechanisms in the resistance to heat stress and hypoxia in animals.

DNA methylation and its catalyzing enzymes play an essential role in stress response and disease control. Recent evidence suggests that exposure of rats to chronic stress conditions results in a substantial reduction of DNMT3A in the prefrontal cortex (PFC), a key target region of stress, indicating the functional roles of DNMT3A in stress-related behaviors and neurotransmission through epigenetics mechanisms [285]. In addition, HATs and HDACs are involved in the resistance to multiple stresses such as oxidative stress, osmotic stress, cell wall-targeting agents, and fungicide [286]. Similarly, KDM6A was found to mediate stress through regulating gene expression programs in hematopoiesis and early B-cell differentiation [124].

Epigenetic processes and related specific transcription factors mediate complex cell differentiation processes associated with a succession of gene expression profiles, which shape innate and adaptive immune responses, and disease control. There are numerous studies showing DNA methylation and the immune response to different disease pathogens. For instance, in cattle, the differential expression of genes that encode enzymes of the epigenetic machinery in circulating leukocytes was reduced by exposure to bacterial lipopolysaccharide (LPS) in vivo [287]. In vitro, an increased expression of proinflammatory cytokine genes was induced by exposing fibroblasts to LPS. The presence of an inhibitor of HATs can reduce the response [288]; suppressing *DNMT1* expression in the kidney of cattle resulted in decreased methylation and upregulated expression of miR-29b promoter, confirming the interaction between DNA methylation and miRNA in the regulation of animal health [277, 289].

In humans, accumulating evidence reveals that HAT/HDAC-regulated histone acetylation in host cells is dynamically modulated during infection [156] and HDAC expression is upgraded in numerous disease conditions [155]. Similarly, KMTs'/KDMs' amplifications, deletions, mutations, and misexpression correlate with numerous diseases [114]. Deregulation of KMTs and KDMs can turn epigenetic master mediators into cancer drivers and hyperactive KMT or inactive KDM can contribute to the repression of tumor suppressor genes resulting from the accumulation of histone lysine methylation marks [114, 118]. These results suggest that epigenetic processes and HATs/HDACs and KMTs/KDMs are actively involved in the immune response against pathogens and controlling the disease.

In fish and aquatic animals, recent evidence suggests that epigenetics can contribute to increased immune responses and improved disease and stress tolerance. For example, the exposure of parents to a synthetic dsRNA resulted in increased expression of an antiviral gene in larvae in oysters, indicating a potential role of epigenetic mechanism in modulating increased resilience to the Ostreid herpesvirus [290]; stress experience in the early development of Atlantic salmon (*Salmo salar*) can increase immune responses of adult [291, 292]; and exposure of brine shrimp to heat shock during early development improved pathogen resistance and heat stress tolerance [293]. This improved resistance lasted for several generations with the involvement of increased histones H3 and H4 acetylation, which are regulated by HATs/HDACs. Other epigenetic mechanisms such as DNA methylation, m6A RNA methylation, and active chromatin markers H3K4Me3, H3K4me1, H3K27me1, H3K14ac, and repression marker H3K27me3 might also play a role in modulation of gene expression, resulting in the observed transgenerational inheritance of the resistant brine shrimp progenies [293, 294]. In addition, the role of DNA methylation and its catalyzing enzymes in handling stress and anoxia has also been recognized. Exposing a turtle to anoxia for 5 and 20 hours causes an overall increase in DNA methylation, DNMT protein expression, and enzymatic activity in the liver and white muscle, suggesting an epigenetics-induced downregulation of gene expression during oxygen deprivation [295]. These findings suggest that the innate immune responses can be trained for increasing resistance against pathogens with the involvement of epigenetic mechanisms in this process, and epigenetic conditioning of the immune responses to stress or pathogens could be a potential approach to control diseases and improve well-being of aquaculture species.

1.4.2.2 The Potential Applications of Epigenetics in Health and Well-Being Management

As diverse epigenetic markers have been evidenced to be linked with human and animal health, epigenetic markers could be promising tools for the diagnosis and therapy monitoring of diseases. For farm animals, environmental factors such as living conditions, feed, chemicals, pathogens, and parental stress have direct effects on production and productivity. These effects can be captured via epigenetic markers and included in animal health management, especially for the detection and management of chronic diseases [277]. However, in farm animals including fish, epigenetic studies are in the early stages compared to epigenetic research and development in humans.

In humans, different epigenetic biomarkers have been recognized for various diseases and applied in various disease diagnoses and treatments, which is supported by growing evidence of EWAS [215, 231–233]. To predict cancer evaluation, some epigenetic drugs have been discovered, and sensitive epigenetic biomarkers have been developed [277, 296]. Most of the epidrugs act by inhibiting the enzyme machinery responsible for transferring methyl, acetyl, or alkyl groups, either to DNA or to histones. Furthermore, DNMT inhibitors (DNMTi) are used in treating the disease [297]. For example, highly efficient DNMTi were approved by the FDA to treat hematological malignancies [298, 299] and some DNMTi were recognized to have significant efficacy in improving the survival of myelodysplastic syndrome patients. To treat the silent stage of diseases, predictive epigenetic biomarkers could have great potential to decrease the possible effects of ineffective treatments for clinical trials and practice [300].

HAT and HDACi are also used in controlling the disease [297]. Acetylation of histones by HATs and initiation of an open-chromatin structure leading to its active transcription is counteracted by HDACs, which triggers a compact nucleosome structure preventing active transcription. It is assumed that HDACi can decrease the expression of pro- and anti-inflammatory mediators, which blocks a serious condition resulting from the presence of harmful microorganisms in the blood or other tissues. Therefore, restraining the activity of HDACs by HDACi can be used to restore the expression profile of the cells [301]. More recently, increased HDAC expression has been discovered to be associated with numerous disease states in many studies, with HDAC upregulation leading to dysregulation of genes and proteins involved in cell proliferation, cell cycle regulation, and apoptosis. These findings have resulted in the therapeutic development of using HDACi, such as cyclin D1 and HIF1α, to treat diseases. [155].

Thus, epigenetic modification by inhibiting HDAC activity is an emerging approach in cancer and other disease treatment. Currently, many active small molecules that target histone acetylation regulatory enzymes such as HDACs, HATs, and bromodomains (BRDs) have been developed to restore abnormal histone acetylation levels to normal in humans. Since HDACs, HATs, and BRDs regulate dynamic histone acetylation/deacetylation, many research/development activities are focused on small molecules and novel synthesized compounds targeting enzyme catalytic activity, with the clinical therapeutic scope of such agents emerging in treating diseases [134]. To date, at least 700 clinical trials are registered and more than 200 are currently recruiting participants, among which more than 30 HDACi have been extensively studied in clinical trials and several of them have been approved by regulatory agencies for treating a broad range of diseases [302–304].

Moreover, HDACs regulate many myogenic and metabolic factors, which are considered therapeutic targets for treating diseases related to inflammation, neurology, and metabolism in clinical research and practice [137]. In addition, some RNA therapies can specifically inhibit a certain protein in skeletal muscle tissue as the biotechnology that could target HDACs and prevent related diseases further develops [137, 305, 306].

Like HATs/HDACs, KMTs and KDMs also possess essential roles in multiple tissues with opposing mechanisms for disease prevention. Efforts have also been focused on developing drugs to inhibit or activate KMTs/KDMs as potential therapies. The design and development of small molecules and peptides are under consideration with an assumption that they can close active sites, interfere with cofactor binding, modify cofactors, or potentially interfere with protein–protein interactions. For instance, disrupting EZH2/KMT6 in response to UTX/KDM6A or JMJD3/KDM6B amplification might reduce off-target effects; inhibiting LSD1/KDM1A interaction with CoREST may effectively block LSD1/KDM1A function in vivo; and identifying molecules that block the recruitment of KMTs and KDMs could provide a potentially better function [114, 307]. For these, designing enzyme-specific inhibitors, as well as specific targeting mechanisms, may be a good approach [114].

It is well accepted that noncoding RNAs (miRNA, lncRNA, and circular RNA) are involved in the regulation of hypoxia through epigenetic mechanisms and using lncRNAs as biomarkers or therapeutic tools is a promising approach to control hypoxia-induced stress and diseases. Studies on epigenetic regulators targeting HIFs show diverse results in controlling signaling from hypoxia, some of which are conflicting for HDAC methylations and HIF1α. Future research is needed for this at the molecular level [138].

As discussed in the previous section, epigenetics can contribute to increased immune responses and improved disease and stress tolerance since the immune responses can be epigenetically trained to increase resistance against pathogens with long-lasting effects in fish and aquatic animals. Aquaculture species regularly experience multiple stressors and pathogen challenges; thus, these findings will have potential application for health and well-being management in the aquaculture industry through epigenetic training of the immune responses to stress or pathogens in farmed species. In addition, exposure of sea lice, which is a parasite affecting salmon aquaculture, to azamethiphos, cypermethrin, and deltamethrin led to a significant correlation between gene expression and phenotypic variability in sea lice, as well as a linkage to lncRNA expression, suggesting the potential application of epigenetic markers to determine sea lice resistance for salmon aquaculture [308]. Exposing Japanese medaka fish to glyphosate during embryonic development resulted in a decrease in transcription of *dmnt1* in males but not females [309], indicating the possibility of using the DMGs and DHRs as biomarkers for transgenerational disease inheritance [310].

Although studies on the epigenetic application in aquaculture species are very limited, the current mechanisms and strategies used in humans and other animals could be a potential strategy to control stress and diseases in aquaculture. For example, the CRISPR-Cas technique has already been implemented in crustaceans and proved effective in producing heritable genetic knockouts and knockins in *Daphnia pulex* [311, 312]. This technique could be used in other aquatic species. Numerous studies have indicated that increased HDAC expression is associated with many disease conditions, resulting in the therapeutic development of using HDACi, such as cyclin D1 and HIF1α, to control stress and disease [155]. HDACi could be used to prevent hypoxia-induced stress and disease in aquaculture species. The development of drugs to inhibit or activate KMTs/KDMs as potential therapies in humans is underway. A similar approach with small molecules that inhibit certain KMTs and KDMs to treat a disease could be considered in aquaculture. Noncoding RNAs are involved in the regulation of hypoxia and using lncRNAs as biomarkers or therapeutic tools is a promising approach to controlling hypoxia-induced stress and diseases [138]. This would enable aquaculture researchers to develop novel therapeutic strategies targeting HIFs in hypoxia-driven diseases in aquatic species.

1.4.3 Nutrition and Growth Advancement

1.4.3.1 The Roles of Epigenetics in Nutrition and Growth

Nutrition

DNA methylation modifications may result in changes in growth performance or disease susceptibility in response to nutritional or environmental stimulus [277, 313–315]. The programming effect of altered feed type and components is more significant during early development than in other life stages [316]. In farm animals, there are several reports of the interaction of changes in feed composition by epigenetic mechanisms. Li et al. [317] showed that the methylation abundance of corresponding DMRs was associated with differential expression of *ADAMTS3* and *ENPP3* genes between animals fed grass and grain. These genes are involved in the biosynthesis and modulation of glycosyltransferase activity. In addition, the changed gene expression mediated by DNA methylation caused the downregulation of immune-related genes in mammary and liver tissues of cows fed with high-concentrate diets [277, 318–320].

Responses of histone modifications to the stimulus of nutritional factors have also been reported. The expression of HDAC2, HDAC3, SIRT2, KAT2A, and EHMT2 (HMT) was significantly repressed by the supplementation of linseed oil to cows in mid-lactation, which caused a 30% decrease in milk fat production. This finding indicates potential epigenetic modulation of milk fatty acid synthesis [277, 321]. Fed with a high-concentrate corn straw diet, cow's LPS concentrations in the blood were negatively correlated with Histone H3 acetylation [318]. In addition, an increase in genome coverage of CTCF, H3K27me3, and H3K4me3, and a decrease in coverage of H3K27ac and H3K4me1 resulted from butyrate treating of rumen epithelial cells [277, 322].

HATs, HDACs, DNMTs, and MBDs are also actively involved in the fat metabolism process to determine gene expression patterns related to the metabolic synthesis of fats. Growing evidence suggests that methylating dietary micronutrients induce differential expressions of genes involved in lipid metabolism [23]. For example, DNMT3A inhibits the differentiation of intramuscular preadipocytes through reducing the expression of cyclin-dependent kinase inhibitor 1A [323, 324]. An increased transcript level of *DNMT1* gene, followed by a decrease, resulted from the onset of adipogenesis, with an increase of *DNMT3A* and *DNMT3B* gene transcripts during the in vitro differentiation [323, 325].

Nutritional supplementation to parents during early development can have long-term effects on the development and productivity of the offspring as evidenced in farm animals. Supplementation of maternal methyl donor was found to significantly change the methylome of offspring in bovine animals [277, 326]. Additionally, diet changes during pregnancy can affect the reproduction ability of female offspring, which may be mediated by epigenetic modifications [277, 327].

Moreover, the high-fat diet of the mother affected the weight of the offspring, which was potentially regulated by epigenetic modifications [277, 328, 329]. Furthermore, the body weight of offspring was found to be negatively correlated with 5hmC but positively correlated with 5mC [330]. In farm pigs, changed patterns of global DNA methylation resulted from the supplementation of prenatal and postnatal dietary omega-3 fatty acid, suggesting probable implications in the growth and inflammatory processes [331]; methyl donor supplementation during gestation could increase offspring intestinal digestion and the growth rate, which was associated with modifications of DNA methylation in specific genes and related expression [277, 332, 333]. In poultry, maternal betaine supplementation induced a change of expression of the DNA methylation-regulated gene related to corticosteroid and cholesterol synthesis in offspring [334–336]; H3K36me3 and H4K12ac in the promoter region of *PPARδ* gene were found to be involved in mediating altered lipid metabolism and growth performance following supplementation of maternal genistein [337]. Additionally, mediating transcriptional regulatory network was found to be associated with DNA methylation in geese fed a diet with methionine and betaine supplementation [338]. The effect of betaine, which is an often used supplement to chicken by industry, on intercellular metabolism may be shaped by DNA methylation [339]. Furthermore, the content and concentration of methionine, a key nutritional factor limiting protein synthesis, in diet can alter the expression of genes involved in the health and growth of animals through DNA methylations [340].

Worthy to note, the activity of DNMTs, HATs, and HDACs requires zinc, as these epigenetic enzymes have several zinc-binding sites. Similarly, it is expected that several other epigenetic enzymes may also possess zinc-binding sites, suggesting the importance of zinc supplementation in diets. In addition, the activity of TET and JHDM enzymes needs vitamin C involvement for DNA and histone demethylation, thus vitamin C deficiency is likely to lead to the impairment of cellular programming [341, 342]. DNA methylation modifications during the development of porcine germline cells and embryos are elicited by vitamin C supplemental and feed restrictions [277, 343, 344].

In teleost, although epigenetic research in nutrition is still in the infant stage compared to other farm animals, there are several studies regarding the nutritional effect of broodstock programming on offspring and beyond through mechanisms of DNA methylation and histone modifications, exampled by a few species. In zebrafish, epigenetic programming of broodstock was evidenced through DNA methylation modulating impacts on offspring by parent fish fed plant-based diets enriched in arachidonic acid or with low concentrations of vitamin B, folate, choline, and methionine [345, 346]. In Atlantic salmon, fatty liver and downregulated mRNA levels of lipid-related genes were found in the offspring from parents fed low concentrations of the selected micronutrients [346]; increased levels of DNA methylation on the acetyl-CoA carboxylase alpha promoter were induced by feeding different concentrations of micronutrient feed to the salmon [347]; the supplementation of micronutrient to feed also impacted the DNA methylation of the promoter regions of genes associated with cell signaling and embryonic development [348]. In rainbow trout (*Oncorhynchus mykiss*), changed DNA methylation of specific genes was identified in the progeny from broodfish fed with a low methionine diet [349] and increased feed carbohydrate concentrations affected DNA hypomethylation and hypoacetylation of histone H3K9 liver [350]. Additionally, increased global histone H3 acetylation and global hypomethylation in the larvae of Gilthead sea bream (*Sparus aurata*) resulted from feeding larvae with a soybean meal diet [351]. Details on epigenetics in fish nutrition are described by Skjærven et al. in Chapter 10 of this book.

Finally, several hypermethylated CpG sites are found to be associated with nutrition/fat composition-related genes. The gene *capn1* encoding Calpain 1 related to a hypomethylated CpG was identified to be involved with the nutritional state of halibut [352, 353] and Gilthead sea bream [352, 354]. Obesity-induced metabolic complications linked to *Mlkl* codes for mixed lineage kinase domain-like protein [352, 355]. Three exonic DM CpG sites were related to *irx3*, encoding for Iroquois homeobox 3, and *fto*, a gene associated with obese phenotype and obesity [352, 356]. High-fat diet upregulates *Acot1* coding for Acyl-coenzyme A thioesterase 1 in animals [352, 357]. Finally, the gene *elovl6* encoding a fatty acid elongase involves a QTL effect on fatty composition [352, 358], as well as insulin resistance induced by obesity in vertebrates [352, 359].

Growth

Other than genetics, the growth of organisms can be significantly affected by nutritional and environmental factors. Muscular and skeletal are two parallel and interlaced systems to support and affect the development and growth of organisms, with epigenetics playing crucial roles in that process.

For muscular development and growth, dynamic changes in DNA modification modulate myogenic commitment and differentiation [360, 361]. The epigenetic changes mediate DNA binding of MyoD and induction of muscle-determining genes to binding sites in the promoters of the target genes [360, 362]. For example, Hu et al. identified global changes in DNA methylation patterns in fast- and slow-growing strains of farm chicken, suggesting a direct epigenetic effect on muscle growth [360, 363]. The intramuscular fat deposition was shaped by DNA methylation through modulating the expression of *ABCA1*, *COL6A1*, and *GSTT1L* [277, 284, 364]. Meanwhile, the methylation levels of *UCP3* and *FATP1* genes were affected in muscle by different breeds or feed conditions of chicken [277, 365, 366]. Moreover, differential expression of *DNMT3A*, *DNMT3B*, and *DNMT1* was found to be significantly associated with body weight and growth in cattle [212, 277, 367]. The differential muscle development of several cattle breeds has been linked to some key genes with DMRs [277, 368, 369]. Furthermore, feed restriction resulted in more change in the methylation level of IGF2 in the longissimus dorsi muscle than in the semitendinosus muscle [277, 370]. These findings highlight the significant involvement of DNA modification and related genes in the controls of muscular development and growth.

In addition to DNA modification, mounting evidence has demonstrated that histone modification key regulators, HATs and HDACs, play an essential role in the control of muscle differentiation and the transcription factors involved in muscle gene activation. During development, HATs/HDACs provide a mechanism for activating and repressing muscle-specific genes in the myogenic pathway through a myogenic partnership of MyoD/bHLH and MEF2 proteins. While evidence shows that Class II HDACs (HDAC4 and HDAC5) can impact MyoD activity indirectly via MEF2 [371, 372], some reports suggest that Class I HDACs, HDAC1 in particular, target MyoD directly [371, 373, 374]. Thus, myocytes may have complex regulatory mechanisms to master the HAT effects on MyoD and MEF2 and block precocious activation of downstream target genes. Meanwhile, HDAC1, HDAC2, HDAC4, and HDAC6 were found to be required for normal muscle atrophy in response to various conditions. Therefore, selecting chemical probes for these specific HDACs could be a potential strategy to inhibit muscle atrophy [146]. Moreover, histone H4 and E2F2 potentially regulated DNA methylation of the key promoter region of the SIX1 gene in muscle tissues as shown in farm animals [277, 375]. Besides, extensive evidence reveals that KMTs and KDMs have also been involved in the differentiation of myotubes, a developmental stage of a muscle fiber, and there are dynamic changes in histone modification states during myotube differentiation [121, 376, 377]. Specifically, LSD1 is required for myogenic differentiation; KDM3B, KDM6A, and KDM8 serve as the candidate KDMs essential for osteoblast differentiation, and MLL4 is needed for efficient osteoblast differentiation [121]. Collectively, these studies demonstrate the importance of histone modifications in the regulation of muscular development and growth, with HATs and HDACs acting as mediators between the environmental cues that impact myogenesis and the transcriptional circuitry for muscle-specific gene expression, and with KMTs and KDMs involved in myotube differentiation.

Furthermore, long noncoding RNAs are actively involved in the modulation of muscle growth through the controlling activity of muscle-specific promoters [360, 378]. It is worth mentioning that mitochondrial activity also affects myogenesis, and thus, muscle development and growth. Inhibiting mitochondrial translation of myoblast cells results in the complete stop of the differentiation of cells [379]. In addition, the mutation accumulation related to muscle tissue degeneration is associated with age-related changes in mitochondria and a decrease in effectiveness [380].

For the skeletal development of the organisms, the major epigenetic players that are involved in the formation and differentiation of the skeleton are among the families of DNA methylation and histone modifications. Disrupting mitochondrial NADP+-dependent isocitrate dehydrogenase (IDH2) in mammals causes high bone mass through decreased osteoclast number and resorption activity. Decreased serum markers of osteoclast activity and bone resorption have been observed in IDH2-deficient animals. Knockout of IDH2 resulted in bone marrow stromal cells defective in promoting osteoclastogenesis due to the downregulation of RANKL in osteoblasts.

Similar to other organisms, DNA modifications, histone modifications, and noncoding RNAs play an essential role in muscle development and growth in teleosts. For instance, in Nile tilapia (*Oreochromis niloticus*), increased growth has been found to be associated with differential methylation of key myogenic genes and biological processes, such as the TGF-β1 signaling pathway; the activity of *sirt3*, a gene contributing to metabolic flexibility of skeletal muscle, has been identified to be associated with a hypomethylated CpG site [352, 381]. The gene activity has also been linked to the fast-growing trait of Gilthead sea bream [352, 382]. HATs/HDACs also play an important role in the control of muscle differentiation and growth in teleost. Hdac1 has been found to suppress cartilage ossification and mediate neural crest formation during early development in zebrafish [383]. Knockdown of *hdac1* resulted in a total loss of cartilage formation in the cranium and pectoral fins, suggesting HDAC-1's essential role in the regulation of the formation of craniofacial cartilages and pectoral fins in zebrafish [384]. Details on epigenetics in fish growth are described by Fernandes et al. in Chapter 9 of this book.

Nutrition Effect on Muscle Growth

For skeletal muscle development and maintenance, diet protein is a key factor for protein synthesis across the lifespan of organisms. Mounting evidence has shown that protein-restricted maternal diets affect the muscle development of offspring. In mammals, feeding pregnant mothers with dietary components with 60% of the ad libitum level, but supplemented with protein to the same level as controls during gestation, resulted in the same level of muscle fiber number in the offspring as the ad libitum-fed controls [385, 386]. Mallinson et al. showed that offspring muscle fiber number and density were reduced by protein intake restriction during mid-pregnancy in rats [387]. Another study revealed that the reduced size of neuromuscular junctions resulted from protein restriction during gestation, with age-related decline in muscle associated with denervation [385, 388]. These findings suggest a crucial role of early-life nutrition in muscle development and function over lifespan. Meanwhile, reduced *Glut4* transcription was elicited by early high-carbohydrate diets in the skeletal muscle of the adult offspring through epigenetic modifications associated with an increase in the methylation level of the Glut4 promoter and acetylation of H2 and H4 [385, 389]. Changes in gene expression of the ligand, insulin-like growth factor II (IGF-II), were induced by restricted global maternal nutrition in the skeletal muscle of sheep [385, 390]. In addition, the protein restriction led to reduced protein synthetic signaling through p70S6 kinase 1 and 4E binding protein-1 phosphorylation, as well as manipulating FOXO and NF-jB action [385, 386]. Muscle sizes are also reported to be negatively affected by high-fat diets through reduced myogenic cues, decreased protein synthetic signaling, and increased protein degradation [385, 386]. These findings support the involvement of epigenetic mechanisms in regulating muscle size following maternal nutrient restriction or high-fat diet intake.

In human aging cohort studies, the decline in skeletal muscle mass and low strength into old age have been found to be associated with low birth weight and malnutrition [385, 391–394], indicating that loss of muscle tissue as a natural part of the aging process might have origins in fetal and early life. Additionally, only adult body mass index and fat-free mass have shown a positive association with birth weight [385, 391], suggesting that early-life environmental conditions affect lean body mass in old age. Furthermore, higher lean mass and lower gene expression of interleukin 6 (IL-6) and interleukin-1 receptor (IL1R1) linked to lower transcript expression of vitamin D receptor and interferon gamma in skeletal muscle tissue, and higher strength was associated with lower myostatin in older men [394]. Moreover, recent evidence suggests that skeletal muscle cells can be trained to morphologically memorize catabolic stress in myoblasts. A study by Sharples et al. showed that exposure of skeletal muscle cells to acute cytokine stress in early life increased their susceptibility to impaired differentiation after encountering the TNF-α stress in later proliferative life, with the muscle cells retaining elevated MyoD methylation even after 30 population doublings [395]. These findings reveal that skeletal muscle can memorize environmental stimuli and respond to them in an adaptive manner if the stimuli have been previously encountered, and epigenetic modifications play an essential role in modulating and maintaining the memory (epi-memory) and process.

In teleost, the vitamin A (retinol) metabolite and all-trans retinoic acid (RA) play crucial roles in the development of the body plan and the differentiation of many types of cells. RA signaling is essential for cell positional memory, including muscle memory, through combinatorial Hox gene expression in the early life and adult stages [396]. In addition, a few hypermethylated CpG sites are found to be associated with genes related to nutrition/muscle interaction. Overfeeding leads to increased expression of *adcy3* in skeletal muscle in zebrafish [396]. Regenerating heart, adult fin, and larval fin after amputation induced Aldh1a2, an enzyme that catalyzes the synthesis of RA from retinaldehyde, in zebrafish, suggesting local generation of RA plays an important role in fin formation during early development and adult fin regeneration [397]. In addition, HMTs are involved in the regulation of muscle cell differentiation in teleost [398].

1.4.3.2 Potential Applications of Epigenetics in Nutrition and Growth

As discussed in the previous section, maternal nutritional and methyl donor supplementation to parents during early development can significantly change the methylome of their offspring, thus shaping phenotypic traits in farm animals. This forms the concept of nutriepigenomics and nutrition-epigenetic-phenotype relationships in farm animals [399, 400]. Meanwhile, considerable evidence suggests that the effects of the early-life environment on the epigenome and phenotype can be reversed and/or prevented [401–403]. Additionally, feeding diets with enzyme-inhibiting substances, e.g., DNMT, led to significant changes in the DNA methylation pattern and related production traits [404–406]. These concepts and findings provide a new perspective on developing and utilizing epigenetic intervention strategies to manipulate growth and/or reverse/prevent unfavorable epigenetic changes in aquaculture species using epigenetic biomarkers early in life.

Epigenetic events can change heritable phenotypes and epigenetic enzymes' activity requires zinc's involvement. For example, DNMTs, HATs, HDACs, and HDMs have several zinc-binding sites. Therefore, epigenetic changes can result from the dysregulation of zinc homeostasis, which is modulated by zinc transporters (ZnTs), Zrt- and IRT-like proteins (ZIPs), and the zinc storage protein metallothionein (MT) [407]. Loss of ZIP10 decreases HATs activity, indicating its involvement in histone acetylation for rigid skin barrier formation. ZIP13 deficiency leads to increased DNMT, leading to dysgenesis of the dermis via improper gene expressions [407]. Like zinc, vitamin C also plays an essential role in epigenetic regulation through its function as a cofactor for maintaining optimal TET and JHDM activities [36]. For DNA demethylation, TETs catalyze the oxidation of 5mC to 5hmC [200, 341], and then iteratively oxidize them to 5fC and 5caC, and 5mC and 5hmC are removed and replaced by unmodified cytosine through base-excision repair [341, 408]. Similarly, JHDM catalyzes the oxidation of methyl groups on arginine and lysine residues on histone tails. This reaction generates a hydroxymethyl intermediate that is spontaneously removed [341, 409]. Thus, vitamin C is so important that its deficiency could result in the impairment of cellular programming that may lead to neoplasia [341, 342]. Therefore, Z- and C-drugs/diets that control zinc/C level and their transporters represent a potential therapeutic approach for controlling diseases related to epigenetics in farm animals including aquaculture species.

HATs and HDACs are two key enzymes that mediate the biological process of histone acetylation and deacetylation, and their balance possesses a crucial role in controlling gene expression and metabolic/physiological function. HDACs regulate numerous myogenic and metabolic factors and are essential for maintaining metabolic homeostasis of skeletal muscles via modulating motor adaptation and exercise capacity of skeletal muscles. HDACs may be associated with mitochondrial remodeling and physiological homeostasis orchestrating skeletal muscles from the process of myogenesis [137]. Thus, HDACs could be potential therapeutic targets in clinical trials and applications for the treatment of metabolic diseases related to nutrition and muscle growth. Some of the following approaches are being considered: HDACs interacting with proteins/complex co-IP, histone acetylation target gene ChIP, high throughput proteomics/acetylome with specific HDACs inhibitor, etc. [137]. Meanwhile, some potential RNA therapies presented by adeno-associated virus (AAV) could be applied to inhibit certain proteins in skeletal muscle tissue targeting HDACs and related diseases [137, 305, 306]. Furthermore, IDH2 plays a key role in the modulation of bone mass through osteoblastic regulation of osteoclast activity [410]. This provides the possibility of applying small molecule IDH2 inhibitors for the treatment of bone loss and osteoporosis. Some of these strategies could be considered for the treatment of metabolic diseases in aquatic animals in the future.

Diverse nutrition/muscle-related epigenetic markers and their alteration and relation to phenotypic traits of farm animals have been identified [277] (Table 1.3). For farm animals, living/farm environment, feed, parental stress, and other factors directly affect production traits. These effects can be captured through epigenetic markers and are included in animal nutrition and growth enhancement [277]. Therefore, some of these identified epigenetic markers could be potential targets or tools for the development of strategies and methods for nutrition management and growth enhancement in farm animals including aquaculture species.

1.4.4 Sustainability Enhancement

1.4.4.1 The Roles of Epigenetics in Adaptation and Sustainability

Environment induces epigenetic alteration and subsequent phenotypic variations that can be transmitted across generations. This phenomenon has often been referred to as "rapid evolution," as it increases the resilience of an organism's offspring and subsequent generations. It is recognized that epigenetic mechanisms play an important role in the rapid evolution of organisms, e.g., the local fitness effects of the epigenome can result in the changes or rapid evolution of these patterns, thus impacting sustainability [421–424] (Figure 1.3). In this context, much interest has been focused on potential contributions and fitness consequences of transgenerational effects of epigenetic alterations to adaptive responses to environmental change and global warming, aimed to prevent farm populations from degradation and rescue natural populations from extinction [421–424].

Genetic and epigenetic effects of parents are present in a broad range of traits and taxa [425–428] and transgenerational effects have been recognized as evidenced by not only parental but also grandparental effects [429–431]. The trait phenotypic expression can be shaped by parental genetics and the correlation between direct and parental genetic effects [432] and their subsequent adaptive alterations to environmental change [433–435], which will be dependent on genetic and nongenetic factors that affect both within and across generations [432, 436] (Figure 1.3).

For natural populations, epigenetic mechanisms may play a role in the speciation process by increasing the range of phenotypes available for the action of natural selection and further facilitating the speed of speciation [294, 437–439]. For example, epigenetic mechanisms potentially contributed to the establishment of the marbled crayfish as a new species [294] and increased rapid phenotypic plasticity of the species [440]. Considerable evidence shows that the short-term adaptation of species to environmental changes is partly modulated through epigenetic mechanisms. For instance, exposing zebrafish to crude oil led to adaptive responses in the F1 generation, which was found to perform better in crude oil-contaminated water than in clean water [441]. Exposure of F0 adult wild guinea pigs (*Cavia aperea*) to an increased ambient temperature resulted in differentially methylated patterns and then transferred to the subsequent F1 generation [294, 440]. Additionally, increased levels of the heat shock protein 70 and heat stress tolerance resulted from the exposure of a parthenogenetic

Table 1.3 Change of nutrition-/muscle-related epigenetic markers in relation to phenotypic traits of farm animals.

Parameter	Organ/tissue	Epigenetic change	Phenotypic trait	References
Feed restriction Cow	*Longissimus dorsi* and *semitendinosus* muscle	One differentially methylated (DM) region in *IGF2*	Muscle function	[370]
Fetal and adult cattle	*Longissimus dorsi* muscles	Three DM CpGs in the core promoter region of *SIX1*, histone H4 and E2F2 bind to *SLX1*	Muscle development	[375]
Obese, lean, and miniature pig breeds	Blood leukocytes	2807, 2969, and 5547 DM genes in the Tongcheng vs Landrace, Tongcheng vs Wuzhishan, and Landrace vs Wuzhishan comparisons, respectively	Fat-related phenotype variance	[411]
Obese and lean-type pig breeds	Backfat	483 DM regions in the promoter regions	Fat deposition and fatty acid composition	[412]
Castrated and non-castrated pigs	Liver and adipose tissues	*GHR* methylation rates in the liver of castrated and noncastrated pigs were 93.33% and 0, respectively	Castration-induced fat deposition	[413]
Pig breeds differing in metabolic characters	*Longissimus dorsi* muscles	More than 2000 DM CpG s	Muscle metabolism	[414]
Three pig breeds	Subcutaneous fat,	H3K9ac and H3K4me3	Body fat	[415]
Differing in fatness traits	Visceral fat and *longissimus dorsi* muscle	Correlated to the expression level of selected genes	Accumulation	
Highest and lowest pH among littermates pig	*Longissimus dorsi* muscle	3468 DM regions, including 44 and 21 protein-coding genes with hyper- and hypomethylation regions in their gene bodies	Postmortem energy metabolism and pH	[416]
Different growth stages Chickens	Breast muscle	2714 DM regions and 378 DM genes	Intramuscular fat deposition and water-holding capacity	[417]
Different feed conditions and breeds Chickens	Breast muscle	46 CpG sites and 3 CpG islands in *UCP3*, different methylation levels of *UCP* and *FATP1* between groups	Intramuscular fat content	[365, 366]
Two sheep breeds differing in meat production ability	*Longissimus dorsi* muscles	808 DM regions and global loss of DNA methylation in the DM regions in the crossbred sheep, 12 potential DM genes	Meat production	[418]
Two cattle breeds exhibiting different meat production ability	*Longissimus dorsi* muscles	23,150 DM regions identified; 331 DM regions correlated negatively with expression of differentially expressed (DE) genes, 21 DM regions located in promoter regions	Muscle and related meat quality traits	[419]
Divergent beef tenderness	*Longissimus dorsi* muscles	DNA methylation profiles related to beef tenderness, and 7215 DM regions between tender and tough beef	Beef tenderness	[420]
Three cattle breeds differing in meat production abilities	*Longissimus dorsi* muscles	18 DM and DE genes between Simmental and Wenshan cattle, 14 DM and DE genes between Simmental and Yunling cattle, 28 DM genes between Wenshan and Yunling cattle	Meat quality	[368]
Three growth stages Polled yak	*Longissimus dorsi* muscles	1344, 822, and 420 genes with DM CCGG sites and 2282, 3056, and 537 genes with DM CCWGG sites in early life	Muscle development and growth	[369]

Source: Modified from Wang and Ibeagha [277].

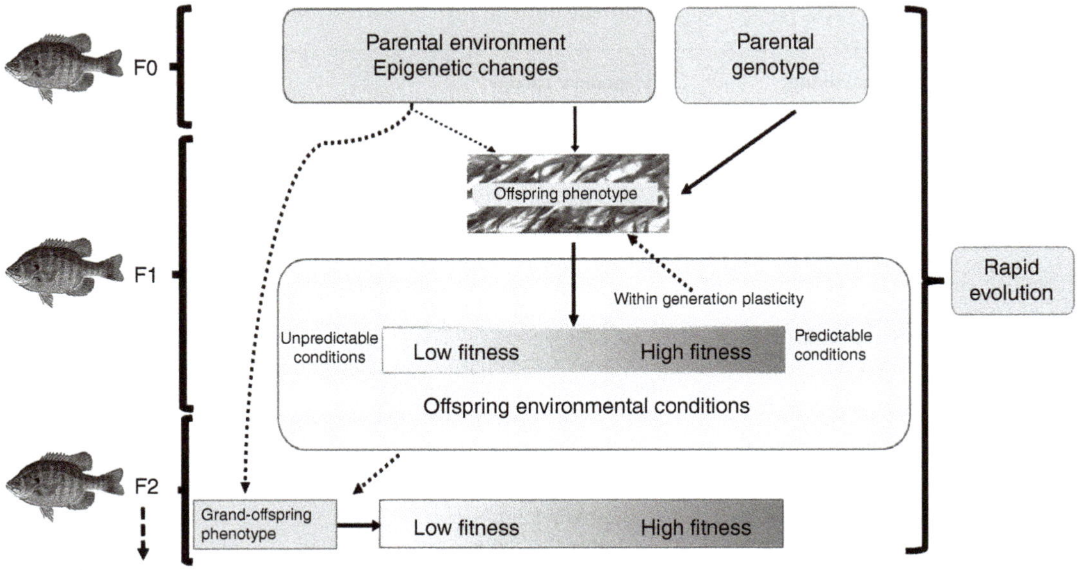

Figure 1.3 Genetic and epigenetic effects of parents across generations. Offspring phenotype can be affected by the parental environment directly through body condition or indirectly via epigenetic changes (dashed arrows). The impact of the environment-induced parental effects on offspring fitness is dependent on genetic and nongenetic factors that interact both within and across generations. This phenomenon can result in "rapid evolution," as it increases the resilience of an organism's offspring and subsequent generations. *Source:* Modified from McGuigan et al.

population of Artemia to a nonlethal heat shock. The effects were transmitted to three successive nonexposed generations with a correlation with changed DNA methylation and acetylated H3 and H4 [293, 294].

However, despite results like the ones just discussed earlier, firm knowledge of the contribution of epigenetic mechanisms to long-term responses to global changes is still poorly documented since such studies require the monitoring of multiple generations. Kronholm et al. [442] monitored the level of epigenetic variation generated in populations adapting to three different challenging environments for 200 asexual generations in unicellular green algae (*Chlamydomonas reinhardtii*) and found that reducing the amount of epigenetic variation available to populations could decrease adaptation to the stressing environments while the opposite would occur when levels of epigenetic variation were kept unchanged. These findings confirm that the effects of transgenerational epigenetics possess an important role in adaptive evolution.

For farm populations, domestication is a regular process of modifying a population phenotype through artificial culture and selection [352, 443]. Phenotypic changes over generations are induced by farm environmental factors and often promoted by strong artificial selection [444]. Domestication can result in unintentional phenotypic traits, which could jeopardize fish or animal farming sustainability of the target species. For example, representation profiling of methylated cytosines in fast muscle was reduced in farm-born and raised female offspring, i.e., subjected to one generation of domestication, compared to their wild mothers in Nile tilapia [352]. The presence of unwelcome traits in an artificially created phenotype could result from low genetic variation due to a limited size founder population [445]. Epigenetic mechanisms such as DNA methylation can lead to rapid changes in gene expression through mediating the response to the new environment at the beginning and over generations. Higher rates of spontaneous mutation could result from epigenetic variation than genetic variation [446] to mediate a more sensitive reaction to environmental changes [447]. Thus, epigenetic mechanisms' role can be essential in the creation of phenotypic variants for the process of adaptation or artificial selection in the new culture environment, while regulating genetics-driven phenotypic plasticity.

In this context, it has been documented that epigenetic regulatory mechanisms are involved in the acquisition of numerous phenotypic traits including disease resistance, growth, and response to diet [347, 365, 448, 449] in several domesticates. Moreover, rearing conditions or environmental factors have stronger effects on the epigenome than on fish origin, as shown in a study performed on coho salmon (*Oncorhynchus kisutch*) [450]. The onset of specific traits, such as lower jaw malformations has been associated with epigenetic modification in European sea bass under the early stages of domestication [451]. These findings highlight that epigenetic mechanisms possess a broad role in mediating environment-induced phenotypic traits in farm animals. Thus, understanding the mechanisms involved in the domestication process could facilitate the consideration of the contribution of epigenetics in the development of breeding programs, and thus aquaculture sustainability.

1.4.4.2 The Potential Applications of Epigenetics in Sustainability Enhancement

For natural populations, the summarized findings earlier on phenotypes modified by epigenetic mechanisms and the epi-mark transmission across generations contribute to shedding new light on species adaptations to global change. Thus, empirical data deriving from such efforts are important for better management of natural populations [294]. In addition, there are growing numbers of examples of transgenerational epigenetic inheritance in model systems [448, 449, 452], which can also be used as a reference for other species and future studies regarding long-term epigenetic responses to environmental change. The expansion of invasive species worldwide is being expedited by global environmental change, threatening freshwater biodiversity [453]. A better understanding of the role of epigenetic mechanisms in invasive species expansion will enable us to develop more efficient management strategies to control invasive events [294].

For farmed populations, knowledge of the role of epigenetic regulatory mechanisms that act under diverse farm environments can allow for more effective control and management of such effects and, thus, provide new management and nutritional tools to enhance productivity and sustainability. This can be achieved by manipulating the environment and management practices to elicit specific epigenetic modifications that result in the desired phenotype. This sort of intervention is aligned with the concept of "flash evolution," defined as fast phenotypic changes, achieved through epigenetic modifications [352]. While interest in epigenetics studies applied to aquaculture grows, the findings on the effects of domestication and culture environments through epigenetic mechanisms can be used to improve breeding programs. Although some fish species have been domesticated for a long time, many new aquatic species are still in the very early stages of domestication [454]. Thus, comprehensive knowledge of epigenetic mechanisms associated with the domestication process of aquaculture species can have a broad application for the development and sustainability of the aquaculture industry.

1.5 Conclusion and Perspectives

This review summarized some of the recent epigenetic studies and findings in different taxa and discussed the potential roles and applications of epigenetics in different aspects of aquaculture production, which will be described in more detail in the following chapters. New findings regarding epigenetics being made in Metazoa have the potential to contribute to future applications related to the well-being and productivity of farm animals and aquaculture species with minimal environmental impacts [277]. However, the roles of epigenetic mechanisms and their potential in the aforementioned aspects are far from being fully understood. Thus, more studies are needed to generate/identify comprehensive data and fill the knowledge gaps concerning the contributions of epigenetics to reproduction, growth, nutrition, health, and the immune response of farm animals and aquaculture species.

To generate comprehensive and meaningful data, eventually EWAS will need to be implemented in farm animals and aquatic species (279). For instance, array-based DNA methylation has been used for EWAS in humans [455–457]. Meanwhile, identifying epigenetic effects on production traits under the impact of multiple factors is also worth investigating. This will facilitate epigenetic biomarker development and utilization in farm animal management. In addition, a comprehensive exploration of the impacts of transgenerational epigenetic inheritance on domestication and selective breeding is needed. This will further facilitate understanding of currently identified effects and applications of epigenetics in sustainable production [279]. Moreover, there is a need for studying the correlation between epigenetics and related physiologic responses. This could enhance our knowledge of the regulation of key production traits of farm animals and aquaculture species. For detailed information on knowledge gaps, challenges, and prospects in epigenetics with aquaculture and aquatic species, please see Chapter 20 of this book.

Acknowledgments

This work was financially supported by the National Institute of Food and Agriculture (NIFA), U.S. Department of Agriculture (No. 2021-70007-34785) and U.S. National Oceanic & Atmospheric Administration (No. 2023-1270-04). Salaries and research support were provided by state and federal funds appropriated to The Ohio State University. The authors wish to thank Francesc Piferrer and Bradford Sherman for their comments on the manuscript.

References

1 Takatsuka, H., Umeda-Hara, C., and Umeda, M. (2015). Cyclin-dependent kinase-activating kinases CDKD;1 and CDKD;3 are essential for preserving mitotic activity in *Arabidopsis thaliana*. *Plant J.* **82** (6): 1004–1017.

2 Pikaard, C.S. and Mittelsten Scheid, O. (2014). Epigenetic regulation in plants. *Cold Spring Harb. Perspect. Biol.* **6** (12): a019315.

3 Nicoglou, A. and Merlin, F. (2017). Epigenetics: a way to bridge the gap between biological fields. *Stud. Hist. Phil. Biol. Biomed. Sci.* **66**: 73–82.

4 Greally, J.M. (2018). A user's guide to the ambiguous word "epigenetics". *Nat. Rev. Mol. Cell Biol.* **19** (4): 207–208.

5 Berger, S.L., Kouzarides, T., Shiekhattar, R., and Shilatifard, A. (2009). An operational definition of epigenetics. *Genes Dev.* **23** (7): 781–783.

6 Beaujean, N., Boutinaud, M., Devinoy, È. et al. (2020). Epigenetics and the elaboration of phenotype in cattle. *Epigenetics Chromatin* **9**: 31.

7 Kouzarides, T. (2007). Chromatin modifications and their function. *Cell* **128** (4): 693–705.

8 Monk, C., Lugo-Candelas, C., and Trumpff, C. (2019). Prenatal developmental origins of future psychopathology: mechanisms and pathways. *Annu. Rev. Clin. Psychol.* **15**: 317–344.

9 Kornak, U., Bischof, O., Ebert, R. et al. (2021). Epigenetics and noncoding RNA – principles and clinical impact. *Osteologie* **30** (3): 201–210.

10 Schagdarsurengin, U., Western, P., Steger, K., and Meinhardt, A. (2016). Developmental origins of male subfertility: role of infection, inflammation, and environmental factors. *Semin. Immunopathol.* **38** (6): 765–781.

11 Baylin, S.B. (2005). DNA methylation and gene silencing in cancer. *Nat. Clin. Pract. Oncol.* **2** (1): S4–S11.

12 Rice, J.C. and Allis, C.D. (2001). Histone methylation versus histone acetylation: new insights into epigenetic regulation. *Curr. Opin. Cell Biol.* **13** (3): 263–273.

13 Roth, S.Y., Denu, J.M., and Allis, C.D. (2001). Histone acetyltransferases. *Annu. Rev. Biochem.* **70**: 81–120.

14 Thiagalingam, S., Cheng, K.-H., Lee, H.J. et al. (2003). Histone deacetylases: unique players in shaping the epigenetic histone code. *Ann. N. Y. Acad. Sci.* **983**: 84–100.

15 Fischle, W., Wang, Y., Jacobs, S.A. et al. (2003). Molecular basis for the discrimination of repressive methyl-lysine marks in histone H3 by Polycomb and HP1 chromodomains. *Genes Dev.* **17** (15): 1870–1881.

16 Santos-Rosa, H., Schneider, R., Bannister, A.J. et al. (2002). Active genes are tri-methylated at K4 of histone H3. *Nature* **419** (6905): 407–411.

17 Eleftheriou, M. and Ruzov, A. (2021). Modified forms of cytosine in eukaryotes: DNA (de)methylation and beyond. In: *DNA Modifications: Methods and Protocols* (ed. A. Ruzov and M. Gering), 3–13. New York: Springer US.

18 Kumar, S., Chinnusamy, V., and Mohapatra, T. (2018). Epigenetics of modified DNA bases: 5-methylcytosine and beyond. *Front. Genet.* **9**: 640.

19 Li, E. and Zhang, Y. (2014). DNA methylation in mammals. *Cold Spring Harb. Perspect. Biol.* **6** (5): a019133.

20 Johnson, T.B. and Coghill, R.D. (2002). Researches on pyrimidines. C111. The discovery of 5-methyl-cytosine in tuberculinic acid, the nucleic acid of the tubercle bacillus1. *J. Am. Chem. Soc.* **47**: 2838–2844.

21 Bickle, T.A. and Krüger, D.H. (1993). Biology of DNA restriction. *Microbiol. Rev.* **57** (2): 434–450.

22 Moarefi, A.H. and Chédin, F. (2011). ICF syndrome mutations cause a broad spectrum of biochemical defects in DNMT3B-mediated de novo DNA methylation. *J. Mol. Biol.* **409** (5): 758–772.

23 Traynor, S., Møllegaard, N.E., Jørgensen, M.G. et al. (2019). Remodeling and destabilization of chromosome 1 pericentromeric heterochromatin by SSX proteins. *Nucleic Acids Res.* **47** (13): 6668–6684.

24 Charlet, J., Duymich, C.E., Lay, F.D. et al. (2016). Bivalent regions of cytosine methylation and H3K27 acetylation suggest an active role for DNA methylation at enhancers. *Mol. Cell* **62** (3): 422–431.

25 Vassena, R., Schramm, R.D., and Latham, K.E. (2005). Species-dependent expression patterns of DNA methyltransferase genes in mammalian oocytes and preimplantation embryos. *Mol. Reprod. Dev.* **72** (4): 430–436.

26 Ramsahoye, B.H., Biniszkiewicz, D., Lyko, F. et al. (2000). Non-CpG methylation is prevalent in embryonic stem cells and may be mediated by DNA methyltransferase 3a. *Proc. Natl. Acad. Sci.* **97** (10): 5237–5242.

27 Bacolla, A., Pradhan, S., Roberts, R.J., and Wells, R.D. (1999). Recombinant human DNA (cytosine-5) methyltransferase: II. Steady-state kinetics reveal allosteric activation by methylated DNA. *J. Biol. Chem.* **274** (46): 33011–33019.

28 Lyko, F. (2018). The DNA methyltransferase family: a versatile toolkit for epigenetic regulation. *Nat. Rev. Genet.* **19** (2): 81–92.

29 Li, E., Bestor, T.H., and Jaenisch, R. (1992). Targeted mutation of the DNA methyltransferase gene results in embryonic lethality. *Cell* **69** (6): 915–926.

30 Hata, K., Okano, M., Lei, H., and Li, E. (2002). Dnmt3L cooperates with the Dnmt3 family of de novo DNA methyltransferases to establish maternal imprints in mice. *Dev. Camb. Engl.* **129** (8): 1983–1993.

31 Kim, J.K., Samaranayake, M., and Pradhan, S. (2009). Epigenetic mechanisms in mammals. *Cell. Mol. Life Sci. CMLS* **66** (4): 596–612.

32 Kato, Y., Kaneda, M., Hata, K. et al. (2007). Role of the Dnmt3 family in de novo methylation of imprinted and repetitive sequences during male germ cell development in the mouse. *Hum. Mol. Genet.* **16** (19): 2272–2280.

33 Bourc'his, D. and Bestor, T.H. (2004). Meiotic catastrophe and retrotransposon reactivation in male germ cells lacking Dnmt3L. *Nature* **431** (7004): 96–99.

34 Kaneda, M., Okano, M., Hata, K. et al. (2004). Essential role for de novo DNA methyltransferase Dnmt3a in paternal and maternal imprinting. *Nature* **429** (6994): 900–903.

35 La Salle, S., Oakes, C.C., Neaga, O.R. et al. (2007). Loss of spermatogonia and wide-spread DNA methylation defects in newborn male mice deficient in DNMT3L. *BMC Dev. Biol.* **7**: 104.

36 Okano, M., Bell, D.W., Haber, D.A., and Li, E. (1999). DNA methyltransferases Dnmt3a and Dnmt3b are essential for de novo methylation and mammalian development. *Cell* **99** (3): 247–257.

37 Aapola, U., Kawasaki, K., Scott, H.S. et al. (2000). Isolation and initial characterization of a novel zinc finger gene, DNMT3L, on 21q22.3, related to the cytosine-5-methyltransferase 3 gene family. *Genomics* **65** (3): 293–298.

38 Neri, F., Krepelova, A., Incarnato, D. et al. (2013). Dnmt3L antagonizes DNA methylation at bivalent promoters and favors DNA methylation at gene bodies in ESCs. *Cell* **155** (1): 121–134.

39 Bhushan, S., Schuppe, H.-C., Fijak, M., and Meinhardt, A. (2009). Testicular infection: microorganisms, clinical implications and host–pathogen interaction. *J. Reprod. Immunol.* **83** (1–2): 164–167.

40 Hedger, M.P. (2002). Macrophages and the immune responsiveness of the testis. *J. Reprod. Immunol.* **57** (1–2): 19–34.

41 Nashan, D., Malorny, U., Sorg, C. et al. (1989). Immuno-competent cells in the murine epididymis. *Int. J. Androl.* **12** (1): 85–94.

42 Nashan, D., Cooper, T.G., Knuth, U.A. et al. (1990). Presence and distribution of leucocyte subsets in the murine epididymis after vasectomy. *Int. J. Androl.* **13** (1): 39–49.

43 Wheeler, K., Tardif, S., Rival, C. et al. (2011). Regulatory T cells control tolerogenic versus autoimmune response to sperm in vasectomy. *Proc. Natl. Acad. Sci. U. S. A.* **108** (18): 7511–7516.

44 Gallego-Bartolomé, J. (2020). DNA methylation in plants: mechanisms and tools for targeted manipulation. *New Phytol.* **227** (1): 38–44.

45 Hovde, B.T., Hanschen, E.R., Steadman Tyler, C.R. et al. (2018). Genomic characterization reveals significant divergence within *Chlorella sorokiniana* (Chlorellales, Trebouxiophyceae). *Algal Res.* **35**: 449–461.

46 Huff, J.T. and Zilberman, D. (2014). Dnmt1-independent CG methylation contributes to nucleosome positioning in diverse eukaryotes. *Cell* **156** (6): 1286–1297.

47 Finnegan, E.J., Peacock, W.J., and Dennis, E.S. (2000). DNA methylation, a key regulator of plant development and other processes. *Curr. Opin. Genet. Dev.* **10** (2): 217–223.

48 Bartels, A., Han, Q., Nair, P. et al. (2018). Dynamic DNA methylation in plant growth and development. *Int. J. Mol. Sci.* **19** (7): 2144.

49 Feng, S., Jacobsen, S.E., and Reik, W. (2010). Epigenetic reprogramming in plant and animal development. *Science* **330** (6004): 622–627.

50 Best, C., Ikert, H., Kostyniuk, D.J. et al. (2018). Epigenetics in teleost fish: from molecular mechanisms to physiological phenotypes. *Comp. Biochem. Physiol. B Biochem. Mol. Biol.* **224**: 210–244.

51 Campos, C., Valente, L.M.P., and Fernandes, J.M.O. (2012). Molecular evolution of zebrafish dnmt3 genes and thermal plasticity of their expression during embryonic development. *Gene* **500** (1): 93–100.

52 Liu, J., Hu, H., Panserat, S., and Marandel, L. (2020). Evolutionary history of DNA methylation related genes in chordates: new insights from multiple whole genome duplications. *Sci. Rep.* **10** (1): 970.

53 Loughland, I., Little, A., and Seebacher, F. (2021). DNA methyltransferase 3a mediates developmental thermal plasticity. *BMC Biol.* **19** (1): 11.

54 Wang, F., Qin, Z., Li, Z. et al. (2021). Dnmt3aa but not Dnmt3ab is required for maintenance of gametogenesis in Nile tilapia (*Oreochromis niloticus*). *Int. J. Mol. Sci.* **22** (18): 10170.

55 Giannetto, A., Nagasawa, K., Fasulo, S., and Fernandes, J.M.O. (2013). Influence of photoperiod on expression of DNA (cytosine-5) methyltransferases in Atlantic cod (*Gadus morhua*). *Gene* **519** (2): 222–230.

56 Head, J.A. (2014). Patterns of DNA methylation in animals: an ecotoxicological perspective. *Integr. Comp. Biol.* **54** (1): 77–86.

57 Chédin, F. (2011). The DNMT3 family of mammalian de novo DNA methyltransferases. *Prog. Mol. Biol. Transl. Sci.* **101**: 255–285.

58 Veland, N. and Chen, T. (2017). Mechanisms of DNA methylation and demethylation during mammalian development. In: *Handbook of Epigenetics* (ed. T.O. Tollefsbol), 11–24. Elsevier.

59 Laurent, L., Wong, E., Li, G. et al. (2010). Dynamic changes in the human methylome during differentiation. *Genome Res.* **20** (3): 320–331.

60 Bostick, M., Kim, J.K., Estève, P.-O. et al. (2007). UHRF1 plays a role in maintaining DNA methylation in mammalian cells. *Science* **317** (5845): 1760–1764.

61 Wu, X. and Zhang, Y. (2017). TET-mediated active DNA demethylation: mechanism, function and beyond. *Nat. Rev. Genet.* **18** (9): 517–534.

62 Saitou, M., Kagiwada, S., and Kurimoto, K. (2012). Epigenetic reprogramming in mouse pre-implantation development and primordial germ cells. *Dev. Camb. Engl.* **139** (1): 15–31.

63 Lee, H.J., Hore, T.A., and Reik, W. (2014). Reprogramming the methylome: erasing memory and creating diversity. *Cell Stem Cell* **14** (6): 710–719.

64 Bagci, H. and Fisher, A.G. (2013). DNA demethylation in pluripotency and reprogramming: the role of tet proteins and cell division. *Cell Stem Cell* **13** (3): 265–269.

65 Guo, F., Li, X., Liang, D. et al. (2014). Active and passive demethylation of male and female pronuclear DNA in the mammalian zygote. *Cell Stem Cell* **15** (4): 447–459.

66 Dawlaty, M.M., Breiling, A., Le, T. et al. (2014). Loss of Tet enzymes compromises proper differentiation of embryonic stem cells. *Dev. Cell* **29** (1): 102–111.

67 Dai, H.-Q., Wang, B.-A., Yang, L. et al. (2016). TET-mediated DNA demethylation controls gastrulation by regulating Lefty-Nodal signalling. *Nature* **538** (7626): 528–532.

68 Dawlaty, M.M., Ganz, K., Powell, B.E. et al. (2011). Tet1 is dispensable for maintaining pluripotency and its loss is compatible with embryonic and postnatal development. *Cell Stem Cell* **9** (2): 166–175.

69 Li, Z., Lu, X., Gao, Y. et al. (2011). Polyploidization and epigenetics. *Chin. Sci. Bull.* **56** (3): 245–252.

70 Moran-Crusio, K., Reavie, L., Shih, A. et al. (2011). Tet2 loss leads to increased hematopoietic stem cell self-renewal and myeloid transformation. *Cancer Cell* **20** (1): 11–24.

71 Zhang, Y., Zhang, S., Liu, Z. et al. (2013). Epigenetic modifications during sex change repress gonadotropin stimulation of Cyp19a1a in a teleost ricefield eel (*Monopterus albus*). *Endocrinology* **154** (8): 2881–2890.

72 Rudenko, A., Dawlaty, M.M., Seo, J. et al. (2013). Tet1 is critical for neuronal activity-regulated gene expression and memory extinction. *Neuron* **79** (6): 1109–1122.

73 Ko, M., Bandukwala, H.S., An, J. et al. (2011). Ten-Eleven-Translocation 2 (TET2) negatively regulates homeostasis and differentiation of hematopoietic stem cells in mice. *Proc. Natl. Acad. Sci. U. S. A.* **108** (35): 14566–14571.

74 Quivoron, C., Couronné, L., Della Valle, V. et al. (2011). TET2 inactivation results in pleiotropic hematopoietic abnormalities in mouse and is a recurrent event during human lymphomagenesis. *Cancer Cell* **20** (1): 25–38.

75 Dawlaty, M.M., Breiling, A., Le, T. et al. (2013). Combined deficiency of Tet1 and Tet2 causes epigenetic abnormalities but is compatible with postnatal development. *Dev. Cell* **24** (3): 310–323.

76 Gu, T.-P., Guo, F., Yang, H. et al. (2011). The role of Tet3 DNA dioxygenase in epigenetic reprogramming by oocytes. *Nature* **477** (7366): 606–610.

77 Zhu, T., Brown, A.P., and Ji, H. (2020). The emerging role of Ten-Eleven Translocation 1 in epigenetic responses to environmental exposures. *Epigenetics Insights* **13**: 251686572091015.

78 Li, Y.-H., Wang, H.-P., Yao, H. et al. (2017). De novo transcriptome sequencing and analysis of male, pseudo-male and female yellow perch, *Perca flavescens*. *PLoS One* **12** (2): e0171187.

79 Spearman, A.D., Ke, X., Fu, Q. et al. (2018). Adverse maternal environment leads to cardiac fibrosis in adult male mice. *Birth Defects Res.* **110** (20): 1551–1555.

80 Chen, J., Guo, L., Zhang, L. et al. (2013). Vitamin C modulates TET1 function during somatic cell reprogramming. *Nat. Genet.* **45** (12): 1504–1509.

81 Feng, J., Shao, N., Szulwach, K.E. et al. (2015). Role of Tet1 and 5-hydroxymethylcytosine in cocaine action. *Nat. Neurosci.* **18** (4): 536–544.

82 Bollati, V., Baccarelli, A., Hou, L. et al. (2007). Changes in DNA methylation patterns in subjects exposed to low-dose benzene. *Cancer Res.* **67** (3): 876–880.

83 Coulter, J.B., O'Driscoll, C.M., and Bressler, J.P. (2013). Hydroquinone increases 5-hydroxymethylcytosine formation through Ten Eleven Translocation 1 (TET1) 5-methylcytosine dioxygenase. *J. Biol. Chem.* **288** (40): 28792–28800.

84 Coulter, J.B., Lopez-Bertoni, H., Kuhns, K.J. et al. (2017). TET1 deficiency attenuates the DNA damage response and promotes resistance to DNA damaging agents. *Epigenetics* **12** (10): 854–864.

85 Kuhns, K.J., Lopez-Bertoni, H., Coulter, J.B., and Bressler, J.P. (2019). TET1 regulates DNA repair in human glial cells. *Toxicol. Appl. Pharmacol.* **380**: 114646.

86 Mahmood, A.M. and Dunwell, J.M. (2019). Evidence for novel epigenetic marks within plants. *AIMS Genet.* **6** (4): 70–87.

87 Tang, Y., Xiong, J., Jiang, H.-P. et al. (2014). Determination of oxidation products of 5-methylcytosine in plants by chemical derivatization coupled with liquid chromatography/tandem mass spectrometry analysis. *Anal. Chem.* **86** (15): 7764–7772.

88 Wang, X., Song, S., Wu, Y.-S. et al. (2015). Genome-wide mapping of 5-hydroxymethylcytosine in three rice cultivars reveals its preferential localization in transcriptionally silent transposable element genes. *J. Exp. Bot.* **66** (21): 6651–6663.

89 Yakovlev, I.A., Gackowski, D., Abakir, A. et al. (2019). Mass spectrometry reveals the presence of specific set of epigenetic DNA modifications in the Norway spruce genome. *Sci. Rep.* **9** (1): 19314.

90 Jang, H., Shin, H., Eichman, B.F., and Huh, J.H. (2014). Excision of 5-hydroxymethylcytosine by DEMETER family DNA glycosylases. *Biochem. Biophys. Res. Commun.* **446** (4): 1067–1072.

91 Nabel, C.S., Manning, S.A., and Kohli, R.M. (2012). The curious chemical biology of cytosine: deamination, methylation, and oxidation as modulators of genomic potential. *ACS Chem. Biol.* **7** (1): 20–30.

92 Zhang, L., Chen, W., Iyer, L.M. et al. (2014). A TET homologue protein from *Coprinopsis cinerea* (CcTET) that biochemically converts 5-methylcytosine to 5-hydroxymethylcytosine, 5-formylcytosine, and 5-carboxylcytosine. *J. Am. Chem. Soc.* **136** (13): 4801–4804.

93 Iyer, L.M., Zhang, D., de Souza, R.F. et al. (2014). Lineage-specific expansions of TET/JBP genes and a new class of DNA transposons shape fungal genomic and epigenetic landscapes. *Proc. Natl. Acad. Sci. U. S. A.* **111** (5): 1676–1683.

94 Xue, J.-H., Chen, G.-D., Hao, F. et al. (2019). A vitamin-C-derived DNA modification catalysed by an algal TET homologue. *Nature* **569** (7757): 581–585.

95 Li, W., Zhang, T., Sun, M. et al. (2021). Molecular mechanism for vitamin C-derived C5-glyceryl-methylcytosine DNA modification catalyzed by algal TET homologue CMD1. *Nat. Commun.* **12** (1): 744.

96 Almeida, R.D., Loose, M., Sottile, V. et al. (2012). 5-Hydroxymethyl-cytosine enrichment of non-committed cells is not a universal feature of vertebrate development. *Epigenetics* **7** (4): 383–389.

97 Ge, L., Zhang, R., Wan, F. et al. (2014). TET2 plays an essential role in erythropoiesis by regulating lineage-specific genes via DNA oxidative demethylation in a zebrafish model. *Mol. Cell. Biol.* **34** (6): 989–1002.

98 Best, C., Ikert, H., Kostyniuk, D.J.D.J. et al. (2018). Epigenetics in teleost fish: from molecular mechanisms to physiological phenotypes. *Comp. Biochem. Physiol. Part – B Biochem. Mol. Biol.* **224**: 210–244.

99 Liu, J., Hu, H., Panserat, S., and Marandel, L. (2020). Evolutionary history of DNA methylation related genes in chordates: new insights from multiple whole genome duplications. *Sci. Rep.* **10** (1): 1–14.

100 Mette, M.F., Aufsatz, W., van der Winden, J. et al. (2000). Transcriptional silencing and promoter methylation triggered by double-stranded RNA. *EMBO J.* **19** (19): 5194–5201.

101 Chan, S.W.-L., Zilberman, D., Xie, Z. et al. (2004). RNA silencing genes control de novo DNA methylation. *Science* **303** (5662): 1336–1336.

102 Chan, S.W.-L., Henderson, I.R., and Jacobsen, S.E. (2005). Gardening the genome: DNA methylation in *Arabidopsis thaliana*. *Nat. Rev. Genet.* **6** (5): 351–360.

103 Potok, M.E., Nix, D.A., Parnell, T.J., and Cairns, B.R. (2013). Reprogramming the maternal zebrafish genome after fertilization to match the paternal methylation pattern. *Cell* **153** (4): 759–772.

104 Youngson, N.A. and Whitelaw, E. (2008). Transgenerational epigenetic effects. *Annu. Rev. Genomics Hum. Genet.* **9** (1): 233–257.

105 Suzuki, M.M. and Bird, A. (2008). DNA methylation landscapes: provocative insights from epigenomics. *Nat. Rev. Genet.* **9** (6): 465–476.

106 Bird, A.P. and Taggart, M.H. (1980). Variable patterns of total DNA and rDNA methylation in animals. *Nucleic Acids Res.* **8** (7): 1485–1497.

107 Tweedie, S., Charlton, J., Clark, V., and Bird, A. (1997). Methylation of genomes and genes at the invertebrate–vertebrate boundary. *Mol. Cell. Biol.* **17** (3): 1469–1475.

108 Jessop, P., Ruzov, A., and Gering, M. (2018). Developmental functions of the dynamic DNA methylome and hydroxymethylome in the mouse and zebrafish: similarities and differences. *Front. Cell Dev. Biol.* **6**: 27.

109 Filipp, F.V. (2017). Crosstalk between epigenetics and metabolism-Yin and Yang of histone demethylases and methyltransferases in cancer. *Brief. Funct. Genomics* **16** (6): 320–325.

110 Rea, S., Eisenhaber, F., O'Carroll, D. et al. (2000). Regulation of chromatin structure by site-specific histone H3 methyltransferases. *Nature* **406** (6796): 593–599.

111 Shi, Y., Lan, F., Matson, C. et al. (2004). Histone demethylation mediated by the nuclear amine oxidase homolog LSD1. *Cell* **119** (7): 941–953.

112 Yamane, K., Toumazou, C., Tsukada, Y. et al. (2006). JHDM2A, a JmjC-containing H3K9 demethylase, facilitates transcription activation by androgen receptor. *Cell* **125** (3): 483–495.

113 Whetstine, J.R., Nottke, A., Lan, F. et al. (2006). Reversal of histone lysine trimethylation by the JMJD2 family of histone demethylases. *Cell* **125** (3): 467–481.

114 Black, J.C. and Whetstine, J.R. (2013). Tipping the lysine methylation balance in disease. *Biopolymers* **99** (2): 127–135.

115 Chi, P., Allis, C.D., and Wang, G.G. (2010). Covalent histone modifications – miswritten, misinterpreted and mis-erased in human cancers. *Nat. Rev. Cancer* **10** (7): 457–469.

116 Lim, S., Metzger, E., Schüle, R. et al. (2010). Epigenetic regulation of cancer growth by histone demethylases. *Int. J. Cancer* **127** (9): 1991–1998.

117 Hayami, S., Kelly, J.D., Cho, H.-S. et al. (2011). Overexpression of LSD1 contributes to human carcinogenesis through chromatin regulation in various cancers. *Int. J. Cancer* **128** (3): 574–586.

118 Huang, H.-S., Matevossian, A., Whittle, C. et al. (2007). Prefrontal dysfunction in schizophrenia involves mixed-lineage leukemia 1-regulated histone methylation at GABAergic gene promoters. *J. Neurosci.* **27** (42): 11254–11262.

119 Gupta, S., Kim, S.Y., Artis, S. et al. (2010). Histone methylation regulates memory formation. *J. Neurosci.* **30** (10): 3589–3599.

120 Henning, S.M., Swendseid, M.E., and Coulson, W.F. (1997). Male rats fed methyl- and folate-deficient diets with or without niacin develop hepatic carcinomas associated with decreased tissue NAD concentrations and altered poly(ADP-ribose) polymerase activity. *J. Nutr.* **127** (1): 30–36.

121 Munehira, Y., Yang, Z., and Gozani, O. (2017). Systematic analysis of known and candidate lysine demethylases in the regulation of myoblast differentiation. *J. Mol. Biol.* **429** (13): 2055–2065.

122 Yahi, H., Philipot, O., Guasconi, V. et al. (2006). Chromatin modification and muscle differentiation. *Expert Opin. Ther. Targets* **10** (6): 923–934.

123 Clarke, T.L., Tang, R., Chakraborty, D. et al. (2020). Histone lysine methylation dynamics control *EGFR* DNA copy-number amplification. *Cancer Discov.* **10** (2): 306–325.

124 Huppertz, S., Senger, K., Brown, A. et al. (2021). KDM6A, a histone demethylase, regulates stress hematopoiesis and early B-cell differentiation. *Exp. Hematol.* **99**: 32–43. e13.

125 Jensen, L.R., Amende, M., Gurok, U. et al. (2005). Mutations in the JARID1C gene, which is involved in transcriptional regulation and chromatin remodeling, cause X-linked mental retardation. *Am. J. Hum. Genet.* **76** (2): 227–236.

126 Iwase, S., Lan, F., Bayliss, P. et al. (2007). The X-linked mental retardation gene SMCX/JARID1C defines a family of histone H3 lysine 4 demethylases. *Cell* **128** (6): 1077–1088.

127 Fellous, A., Earley, R.L., and Silvestre, F. (2019). The Kdm/Kmt gene families in the self-fertilizing mangrove rivulus fish, *Kryptolebias marmoratus*, suggest involvement of histone methylation machinery in development and reproduction. *Gene* **687**: 173–187.

128 Fellous, A. and Shama, L.N.S. (2019). Genome survey of chromatin-modifying enzymes in threespine stickleback: a crucial epigenetic toolkit for adaptation? *Front. Mar. Sci.* **6**: 721.

129 Jia, D., Jurkowska, R.Z., Zhang, X. et al. (2007). Structure of Dnmt3a bound to Dnmt3L suggests a model for de novo DNA methylation. *Nature* **449** (7159): 248–251.

130 Ooi, S.K.T., Qiu, C., Bernstein, E. et al. (2007). DNMT3L connects unmethylated lysine 4 of histone H3 to de novo methylation of DNA. *Nature* **448** (7154): 714–717.

131 Fellous, A., Favrel, P., and Riviere, G. (2015). Temperature influences histone methylation and mRNA expression of the Jmj-C histone-demethylase orthologues during the early development of the oyster *Crassostrea gigas*. *Mar. Genomics* **19**: 23–30.

132 Wang, S.Y., Lau, K., Lai, K.-P. et al. (2016). Hypoxia causes transgenerational impairments in reproduction of fish. *Nat. Commun.* **7** (1): 12114.

133 He, R., Dantas, A., and Riabowol, K. (2021). Histone acetyltransferases and stem cell identity. *Cancers* **13** (10): 2407.

134 Wu, D., Qiu, Y., Jiao, Y. et al. (2020). Small molecules targeting HATs, HDACs, and BRDs in cancer therapy. *Front. Oncol.* **10**: 560487.

135 Iizuka, M. and Smith, M.M. (2003). Functional consequences of histone modifications. *Curr. Opin. Genet. Dev.* **13** (2): 154–160.

136 Johnstone, R.W. (2002). Histone-deacetylase inhibitors: novel drugs for the treatment of cancer. *Nat. Rev. Drug Discov.* **1** (4): 287–299.

137 Tian, H., Liu, S., Ren, J. et al. (2020). Role of histone deacetylases in skeletal muscle physiology and systemic energy homeostasis: implications for metabolic diseases and therapy. *Front. Physiol.* **11**: 949.

138 Islam, S. and Mukherjee, C. (2022). Molecular regulation of hypoxia through the lenses of noncoding RNAs and epitranscriptome. *WIREs RNA* e1750.

139 Luo, L., Martin, S.C., Parkington, J. et al. (2019). HDAC4 controls muscle homeostasis through deacetylation of myosin heavy chain, PGC-1α, and Hsc70. *Cell Rep.* **29** (3): 749–763. e12.

140 Katsumoto, T., Aikawa, Y., Iwama, A. et al. (2006). MOZ is essential for maintenance of hematopoietic stem cells. *Genes Dev.* **20** (10): 1321–1330.

141 Thomas, T., Corcoran, L.M., Gugasyan, R. et al. (2006). Monocytic leukemia zinc finger protein is essential for the development of long-term reconstituting hematopoietic stem cells. *Genes Dev.* **20** (9): 1175–1186.

142 Roy, A. and Palli, S.R. (2018). Epigenetic modifications acetylation and deacetylation play important roles in juvenile hormone action. *BMC Genomics* **19** (1): 934.

143 Yin, H., Price, F., and Rudnicki, M.A. (2013). Satellite cells and the muscle stem cell niche. *Physiol. Rev.* **93** (1): 23–67.

144 Hamed, M., Khilji, S., Chen, J., and Li, Q. (2013). Stepwise acetyltransferase association and histone acetylation at the Myod1 locus during myogenic differentiation. *Sci. Rep.* **3**: 2390.

145 Zhang, P., Liu, Y., Jin, C. et al. (2016). Histone acetyltransferase GCN5 regulates osteogenic differentiation of mesenchymal stem cells by inhibiting NF-κB. *J. Bone Miner. Res. Off. J. Am. Soc. Bone Miner. Res.* **31** (2): 391–402.

146 Beharry, A.W. and Judge, A.R. (2015). Differential expression of *HDAC* and *HAT* genes in atrophying skeletal muscle: HDACs and HAT genes and muscle atrophy. *Muscle Nerve* **52** (6): 1098–1101.

147 Zhou, Y., Peng, J., and Jiang, S. (2014). Role of histone acetyltransferases and histone deacetylases in adipocyte differentiation and adipogenesis. *Eur. J. Cell Biol.* **93** (4): 170–177.

148 Chérasse, Y., Maurin, A.-C., Chaveroux, C. et al. (2007). The p300/CBP-associated factor (PCAF) is a cofactor of ATF4 for amino acid-regulated transcription of CHOP. *Nucleic Acids Res.* **35** (17): 5954–5965.

149 Knutson, S.K., Chyla, B.J., Amann, J.M. et al. (2008). Liver-specific deletion of histone deacetylase 3 disrupts metabolic transcriptional networks. *EMBO J.* **27** (7): 1017–1028.

150 Montgomery, R.L., Potthoff, M.J., Haberland, M. et al. (2008). Maintenance of cardiac energy metabolism by histone deacetylase 3 in mice. *J. Clin. Invest.* **118** (11): 3588–3597.

151 Chatterjee, T.K., Idelman, G., Blanco, V. et al. (2011). Histone deacetylase 9 is a negative regulator of adipogenic differentiation. *J. Biol. Chem.* **286** (31): 27836–27847.

152 Ong, B.X., Brunmeir, R., Zhang, Q. et al. (2020). Regulation of thermogenic adipocyte differentiation and adaptive thermogenesis through histone acetylation. *Front. Endocrinol.* **11**: 95.

153 Wang, Z., Qi, C., Krones, A. et al. (2006). Critical roles of the p160 transcriptional coactivators p/CIP and SRC-1 in energy balance. *Cell Metab.* **3** (2): 111–122.

154 Ma, H., Li, L., Gai, Y. et al. (2021). Histone acetyltransferases and deacetylases are required for virulence, conidiation, DNA damage repair, and multiple stresses resistance of *Alternaria alternata*. *Front. Microbiol.* **12**: 783633.

155 Hutt, D.M., Roth, D.M., Marchal, C., and Bouchecareilh, M. (2017). Using histone deacetylase inhibitors to analyze the relevance of HDACs for translation. *Methods Mol. Biol.* **1510**: 77–91.

156 Grabiec, A.M. and Potempa, J. (2018). Epigenetic regulation in bacterial infections: targeting histone deacetylases. *Crit. Rev. Microbiol.* **44** (3): 336–350.

157 Di Cerbo, V. and Schneider, R. (2013). Cancers with wrong HATs: the impact of acetylation. *Brief. Funct. Genomics* **12** (3): 231–243.

158 Sauer, T., Arteaga, M.F., Isken, F. et al. (2015). MYST2 acetyltransferase expression and histone H4 lysine acetylation are suppressed in AML. *Exp. Hematol.* **43** (9): 794–802. e4.

159 Singh, P.B., Shloma, V.V., and Belyakin, S.N. (2019). Maternal regulation of chromosomal imprinting in animals. *Chromosoma* **128** (2): 69–80.

160 Cha, E.J., Noh, S.J., Kwon, K.S. et al. (2009). Expression of DBC1 and SIRT1 is associated with poor prognosis of gastric carcinoma. *Clin. Cancer Res. Off. J. Am. Assoc. Cancer Res.* **15** (13): 4453–4459.

161 Kim, J.K., Noh, J.H., Jung, K.H. et al. (2013). Sirtuin7 oncogenic potential in human hepatocellular carcinoma and its regulation by the tumor suppressors MiR-125a-5p and MiR-125b. *Hepatol. Baltim. Md* **57** (3): 1055–1067.

162 Rath, S., Das, L., Kokate, S.B. et al. (2017). Inhibition of histone/lysine acetyltransferase activity kills CoCl2-treated and hypoxia-exposed gastric cancer cells and reduces their invasiveness. *Int. J. Biochem. Cell Biol.* **82**: 28–40.

163 Luo, W. and Wang, Y. (2018). Epigenetic regulators: multifunctional proteins modulating hypoxia-inducible factor-α protein stability and activity. *Cell. Mol. Life Sci. CMLS* **75** (6): 1043–1056.

164 Qian, D.Z., Kachhap, S.K., Collis, S.J. et al. (2006). Class II histone deacetylases are associated with VHL-independent regulation of hypoxia-inducible factor 1 alpha. *Cancer Res.* **66** (17): 8814–8821.

165 Kato, H., Tamamizu-Kato, S., and Shibasaki, F. (2004). Histone deacetylase 7 associates with hypoxia-inducible factor 1alpha and increases transcriptional activity. *J. Biol. Chem.* **279** (40): 41966–41974.

166 Nakazawa, M.S., Eisinger-Mathason, T.S.K., Sadri, N. et al. (2016). Epigenetic re-expression of HIF-2α suppresses soft tissue sarcoma growth. *Nat. Commun.* **7**: 10539.

167 German, N.J. and Haigis, M.C. (2015). Sirtuins and the metabolic hurdles in cancer. *Curr. Biol. CB* **25** (13): R569–R583.

168 Santourlidis, S. (2022). Phyloepigenetics. *Biol.-Basel* **11** (5): 754.

169 Navarro-Martín, L., Martyniuk, C.J., and Mennigen, J.A. (2020). Comparative epigenetics in animal physiology: an emerging frontier. *Comp. Biochem. Physiol. Part D Genomics Proteom.* **36**: 100745.

170 Willbanks, A., Leary, M., Greenshields, M. et al. (2016). The evolution of epigenetics: from prokaryotes to humans and its biological consequences. *Genet. Epigenetics* **8**: GEG.S31863.

171 Zemach, A., McDaniel, I.E., Silva, P., and Zilberman, D. (2010). Genome-wide evolutionary analysis of eukaryotic DNA methylation. *Science* **328** (5980): 916–919.

172 Yi, S.V. (2012). Birds do it, bees do it, worms and ciliates do it too: DNA methylation from unexpected corners of the tree of life. *Genome Biol.* **13** (10): 174.

173 Aliaga, B., Bulla, I., Mouahid, G. et al. (2019). Universality of the DNA methylation codes in eucaryotes. *Sci. Rep.* **9** (1): 173.

174 Dunwell, T.L. and Pfeifer, G.P. (2014). Drosophila genomic methylation: new evidence and new questions. *Epigenomics* **6** (5): 459–461.

175 Schmitz, R.J., Lewis, Z.A., and Goll, M.G. (2019). DNA methylation: shared and divergent features across eukaryotes. *Trends Genet.* **35** (11): 818–827.

176 Liu, J., Hu, H., Panserat, S., and Marandel, L. (2020). Evolutionary history of DNA methylation related genes in chordates: new insights from multiple whole genome duplications. *Sci. Rep.* **10** (1): 970.

177 Skvortsova, K., Tarbashevich, K., Stehling, M. et al. (2019). Retention of paternal DNA methylome in the developing zebrafish germline. *Nat. Commun.* **10** (1): 3054.

178 Ortega-Recalde, O., Day, R.C., Gemmell, N.J., and Hore, T.A. (2019). Zebrafish preserve global germline DNA methylation while sex-linked rDNA is amplified and demethylated during feminisation. *Nat. Commun.* **10** (1): 3053.

179 Potok, M.E., Nix, D.A., Parnell, T.J., and Cairns, B.R. (2013). Reprogramming the maternal zebrafish genome after fertilization to match the paternal methylation pattern. *Cell* **153** (4): 759–772.

180 Jiang, L., Zhang, J., Wang, J.-J. et al. (2013). Sperm, but not oocyte, DNA methylome is inherited by zebrafish early embryos. *Cell* **153** (4): 773–784.

181 Zeng, Y. and Chen, T. (2019). DNA methylation reprogramming during mammalian development. *Genes* **10** (4): 257.

182 Bråte, J., Neumann, R.S., Fromm, B. et al. (2018). Unicellular origin of the animal microRNA machinery. *Curr. Biol.* **28** (20): 3288–3295. e5.

183 Hertel, J. and Stadler, P.F. (2015). The expansion of animal microRNA families revisited. *Life* **5** (1): 905–920.

184 Guerra-Assunção, J.A. and Enright, A.J. (2012). Large-scale analysis of microRNA evolution. *BMC Genomics* **13** (1): 218.

185 Calcino, A.D., Fernandez-Valverde, S.L., Taft, R.J., and Degnan, B.M. (2018). Diverse RNA interference strategies in early-branching metazoans. *BMC Evol. Biol.* **18** (1): 160.

186 Fromm, B., Worren, M.M., Hahn, C. et al. (2013). Substantial loss of conserved and gain of novel microRNA families in flatworms. *Mol. Biol. Evol.* **30** (12): 2619–2628.

187 Simkin, A., Geissler, R., McIntyre, A.B.R., and Grimson, A. (2020). Evolutionary dynamics of microRNA target sites across vertebrate evolution. *PLoS Genet.* **16** (2): e1008285.

188 Xu, J., Zhang, R., Shen, Y. et al. (2013). The evolution of evolvability in microRNA target sites in vertebrates. *Genome Res.* **23** (11): 1810–1816.

189 Berthelot, C., Brunet, F., Chalopin, D. et al. (2014). The rainbow trout genome provides novel insights into evolution after whole-genome duplication in vertebrates. *Nat. Commun.* **5** (1): 3657.

190 Mennigen, J.A. (2016). Micromanaging metabolism – a role for miRNAs in teleost energy metabolism. *Comp. Biochem. Physiol. B Biochem. Mol. Biol.* **199**: 115–125.

191 Best, C., Ikert, H., Kostyniuk, D.J. et al. (2018). Epigenetics in teleost fish: from molecular mechanisms to physiological phenotypes. *Comp. Biochem. Physiol. B Biochem. Mol. Biol.* **224**: 210–244.

192 Ulitsky, I. (2016). Evolution to the rescue: using comparative genomics to understand long non-coding RNAs. *Nat. Rev. Genet.* **17** (10): 601–614.

193 Praher, D., Zimmermann, B., Genikhovich, G. et al. (2017). Characterization of the piRNA pathway during development of the sea anemone *Nematostella vectensis*. *RNA Biol.* **14** (12): 1727–1741.

194 Zhang, D., Xiong, H., Shan, J. et al. (2008). Functional insight into Maelstrom in the germline piRNA pathway: a unique domain homologous to the DnaQ-H 3′–5′ exonuclease, its lineage-specific expansion/loss and evolutionarily active site switch. *Biol. Direct* **3** (1): 48.

195 Robbens, S., Rouzé, P., Cock, J.M. et al. (2008). The FTO gene, implicated in human obesity, is found only in vertebrates and marine algae. *J. Mol. Evol.* **66** (1): 80–84.

196 Wang, X. and Bhandari, R.K. (2019). DNA methylation dynamics during epigenetic reprogramming of medaka embryo. *Epigenetics* **14** (6): 611–622.

197 Lee, R.C., Feinbaum, R.L., and Ambros, V. (1993). The *C. elegans* heterochronic gene lin-4 encodes small RNAs with antisense complementarity to lin-14. *Cell* **75** (5): 843–854.

198 Houwing, S., Kamminga, L.M., Berezikov, E. et al. (2007). A role for Piwi and piRNAs in germ cell maintenance and transposon silencing in zebrafish. *Cell* **129** (1): 69–82.

199 Paiva, S.-L. and Crews, C.M. (2019). Targeted protein degradation: elements of PROTAC design. *Curr. Opin. Chem. Biol.* **50**: 111–119.

200 Tahiliani, M., Koh, K.P., Shen, Y. et al. (2009). Conversion of 5-methylcytosine to 5-hydroxymethylcytosine in mammalian DNA by MLL partner TET1. *Science* **324** (5929): 930–935.

201 Ito, S., Shen, L., Dai, Q. et al. (2011). Tet proteins can convert 5-methylcytosine to 5-formylcytosine and 5-carboxylcytosine. *Science* **333** (6047): 1300–1303.

202 Hill, W.D., Arslan, R.C., Xia, C. et al. (2018). Genomic analysis of family data reveals additional genetic effects on intelligence and personality. *Mol. Psychiatry* **23** (12): 2347–2362.

203 Li, C., Lan, Y., Schwartz-Orbach, L. et al. (2015). Overlapping requirements for Tet2 and Tet3 in normal development and hematopoietic stem cell emergence. *Cell Rep.* **12** (7): 1133–1143.

204 Yamaguchi, S., Hong, K., Liu, R. et al. (2012). Tet1 controls meiosis by regulating meiotic gene expression. *Nature* **492** (7429): 443–447.

205 Li, Z., Cai, X., Cai, C.-L. et al. (2011). Deletion of Tet2 in mice leads to dysregulated hematopoietic stem cells and subsequent development of myeloid malignancies. *Blood* **118** (17): 4509–4518.

206 Gjini, E., Mansour, M.R., Sander, J.D. et al. (2015). A zebrafish model of myelodysplastic syndrome produced through tet2 genomic editing. *Mol. Cell. Biol.* **35** (5): 789–804.

207 Inoue, A., Shen, L., Matoba, S., and Zhang, Y. (2015). Haploinsufficiency, but not defective paternal 5mC oxidation, accounts for the developmental defects of maternal Tet3 knockouts. *Cell Rep.* **10** (4): 463–470.

208 Cortázar, D., Kunz, C., Selfridge, J. et al. (2011). Embryonic lethal phenotype reveals a function of TDG in maintaining epigenetic stability. *Nature* **470** (7334): 419–423.

209 Cortellino, S., Xu, J., Sannai, M. et al. (2011). Thymine DNA glycosylase is essential for active DNA demethylation by linked deamination-base excision repair. *Cell* **146** (1): 67–79.

210 Anderson, R.M., Bosch, J.A., Goll, M.G. et al. (2009). Loss of Dnmt1 catalytic activity reveals multiple roles for DNA methylation during pancreas development and regeneration. *Dev. Biol.* **334** (1): 213–223.

211 Tittle, R.K., Sze, R., Ng, A. et al. (2011). Uhrf1 and Dnmt1 are required for development and maintenance of the zebrafish lens. *Dev. Biol.* **350** (1): 50–63.

212 Liu, X., Usman, T., Wang, Y. et al. (2015). Polymorphisms in epigenetic and meat quality related genes in fourteen cattle breeds and association with beef quality and carcass traits. *Asian-Australas. J. Anim. Sci.* **28** (4): 467–475.

213 Hargan-Calvopina, J., Taylor, S., Cook, H. et al. (2016). Stage-specific demethylation in primordial germ cells safeguards against precocious differentiation. *Dev. Cell* **39** (1): 75–86.

214 Howell, C.Y., Bestor, T.H., Ding, F. et al. (2001). Genomic imprinting disrupted by a maternal effect mutation in the Dnmt1 gene. *Cell* **104** (6): 829–838.

215 Hirasawa, R., Chiba, H., Kaneda, M. et al. (2008). Maternal and zygotic Dnmt1 are necessary and sufficient for the maintenance of DNA methylation imprints during preimplantation development. *Genes Dev.* **22** (12): 1607–1616.

216 Kurihara, Y., Kawamura, Y., Uchijima, Y. et al. (2008). Maintenance of genomic methylation patterns during preimplantation development requires the somatic form of DNA methyltransferase 1. *Dev. Biol.* **313** (1): 335–346.

217 Gore, A.V., Athans, B., Iben, J.R. et al. (2016). Epigenetic regulation of hematopoiesis by DNA methylation. *elife* **5**: e11813.

218 Sharif, J., Muto, M., Takebayashi, S. et al. (2007). The SRA protein Np95 mediates epigenetic inheritance by recruiting Dnmt1 to methylated DNA. *Nature* **450** (7171): 908–912.

219 Sadler, K.C., Krahn, K.N., Gaur, N.A., and Ukomadu, C. (2007). Liver growth in the embryo and during liver regeneration in zebrafish requires the cell cycle regulator, uhrf1. *Proc. Natl. Acad. Sci. U. S. A.* **104** (5): 1570–1575.

220 Jacob, V., Chernyavskaya, Y., Chen, X. et al. (2015). DNA hypomethylation induces a DNA replication-associated cell cycle arrest to block hepatic outgrowth in uhrf1 mutant zebrafish embryos. *Dev. Camb. Engl.* **142** (3): 510–521.

221 Bourc'his, D., Xu, G.L., Lin, C.S. et al. (2001). Dnmt3L and the establishment of maternal genomic imprints. *Science* **294** (5551): 2536–2539.

222 Rai, K., Jafri, I.F., Chidester, S. et al. (2010). Dnmt3 and G9a cooperate for tissue-specific development in zebrafish. *J. Biol. Chem.* **285** (6): 4110–4121.

223 Ross, P.J. and Sampaio, R.V. (2018). Epigenetic remodeling in preimplantation embryos: cows are not big mice. *Anim. Reprod.* **15** (3): 204–214.

224 Jiang, B.-C., Zhang, W.-W., Yang, T. et al. (2018). Demethylation of G-protein-coupled receptor 151 promoter facilitates the binding of Krüppel-like factor 5 and enhances neuropathic pain after nerve injury in mice. *J. Neurosci.* **38** (49): 10535–10551.

225 Chung, N., Bogliotti, Y.S., Ding, W. et al. (2017). Active H3K27me3 demethylation by KDM6B is required for normal development of bovine preimplantation embryos. *Epigenetics* **12** (12): 1048–1056.

226 Camargo, L.S.A., Aguirre-Lavin, T., Adenot, P. et al. (2019). Heat shock during in vitro maturation induces chromatin modifications in the bovine embryo. *Reproduction* **158** (4): 313–322.

227 Boulard, M., Storck, S., Cong, R. et al. (2010). Histone variant macroH2A1 deletion in mice causes female-specific steatosis. *Epigenetics Chromatin* **3** (1): 8.

228 Sakae, Y. and Tanaka, M. (2021). Metabolism and sex differentiation in animals from a starvation perspective. *Sex. Dev.* **15** (1–3): 168–178.

229 Wu, S.-F., Zhang, H., and Cairns, B.R. (2011). Genes for embryo development are packaged in blocks of multivalent chromatin in zebrafish sperm. *Genome Res.* **21** (4): 578–589.

230 Zhang, Y., Sun, X., Zhang, L., and Zhang, W. (2017). Testicular Dnmt3 expression and global DNA methylation are down-regulated by gonadotropin releasing hormones in the ricefield eel *Monopterus albus*. *Sci. Rep.* **7**: 43158.

231 Wang, F.-L., Yan, L.-X., Shi, H.-J. et al. (2018). Genome-wide identification, evolution of DNA methyltransferases and their expression during gonadal development in Nile tilapia. *Comp. Biochem. Physiol. B Biochem. Mol. Biol.* **226**: 73–84.

232 Liu, J., Hu, H., Panserat, S., and Marandel, L. (2020). Evolutionary history of DNA methylation related genes in chordates: new insights from multiple whole genome duplications. *Sci. Rep.* **10** (1): 970.

233 Piferrer, F., Anastasiadi, D., Valdivieso, A. et al. (2019). The model of the conserved epigenetic regulation of sex. *Front. Genet.* **10**: 857.

234 Lau, E.S.-W., Zhang, Z., Qin, M., and Ge, W. (2016). Knockout of zebrafish ovarian aromatase gene (cyp19a1a) by TALEN and CRISPR/Cas9 leads to all-male offspring due to failed ovarian differentiation. *Sci. Rep.* **6**: 37357.

235 Jia, Y., Zheng, J., Chi, M. et al. (2019). Molecular identification of dmrt1 and its promoter CpG methylation in correlation with gene expression during gonad development in *Culter alburnus*. *Fish Physiol. Biochem.* **45** (1): 245–252.

236 Anastasiadi, D., Díaz, N., and Piferrer, F. (2017). Small ocean temperature increases elicit stage-dependent changes in DNA methylation and gene expression in a fish, the European sea bass. *Sci. Rep.* **7** (1): 12401.

237 Metzger, D.C.H. and Schulte, P.M. (2017). Persistent and plastic effects of temperature on DNA methylation across the genome of threespine stickleback (*Gasterosteus aculeatus*). *Proc. R. Soc. B Biol. Sci.* **284** (1864): 20171667.

238 Han, B., Li, W., Chen, Z. et al. (2016). Variation of DNA methylome of zebrafish cells under cold pressure. *PLoS One* **11** (8): e0160358.

239 Varriale, A. (2014). DNA methylation, epigenetics, and evolution in vertebrates: facts and challenges. *Int. J. Evol. Biol.* **2014**: 475981.

240 Sun, L.-X., Wang, Y.-Y., Zhao, Y. et al. (2016). Global DNA methylation changes in Nile tilapia gonads during high temperature-induced masculinization. *PLoS One* **11** (8): e0158483.

241 He, K., Cao, X., and Deng, X. (2021). Histone methylation in epigenetic regulation and temperature responses. *Curr. Opin. Plant Biol.* **61**: 102001.

242 Lai, K.P., Li, J.-W., Tse, A.C.-K. et al. (2016). Hypoxia alters steroidogenesis in female marine medaka through miRNAs regulation. *Aquat. Toxicol.* **172**: 1–8.

243 Tse, A.C.-K., Li, J.-W., Wang, S.Y. et al. (2016). Hypoxia alters testicular functions of marine medaka through microRNAs regulation. *Aquat. Toxicol.* **180**: 266–273.

244 Aniagu, S.O., Williams, T.D., Allen, Y. et al. (2008). Global genomic methylation levels in the liver and gonads of the three-spine stickleback (*Gasterosteus aculeatus*) after exposure to hexabromocyclododecane and 17-beta oestradiol. *Environ. Int.* **34** (3): 310–317.

245 Yuan, C., Zhang, Y., Liu, Y. et al. (2016). Enhanced GSH synthesis by bisphenol A exposure promoted DNA methylation process in the testes of adult rare minnow *Gobiocypris rarus*. *Aquat. Toxicol.* **178**: 99–105.

246 Laing, L.V., Viana, J., Dempster, E.L. et al. (2016). Bisphenol A causes reproductive toxicity, decreases dnmt1 transcription, and reduces global DNA methylation in breeding zebrafish (*Danio rerio*). *Epigenetics* **11** (7): 526–538.

247 Liu, Y., Zhang, Y., Tao, S. et al. (2016). Global DNA methylation in gonads of adult zebrafish *Danio rerio* under bisphenol A exposure. *Ecotoxicol. Environ. Saf.* **130**: 124–132.

248 González-Rojo, S., Lombó, M., Fernández-Díez, C., and Herráez, M.P. (2019). Male exposure to bisphenol A impairs spermatogenesis and triggers histone hyperacetylation in zebrafish testes. *Environ. Pollut. 1987* (248): 368–379.

249 Lombó, M., Fernández-Díez, C., González-Rojo, S., and Herráez, M.P. (2019). Genetic and epigenetic alterations induced by bisphenol A exposure during different periods of spermatogenesis: from spermatozoa to the progeny. *Sci. Rep.* **9** (1): 18029.

250 Lombó, M., Getino-Álvarez, L., Depincé, A. et al. (2019). Embryonic exposure to bisphenol A impairs primordial germ cell migration without jeopardizing male breeding capacity. *Biomolecules* **9** (8): E307.

251 Wu, R.S.S., Zhou, B.S., Randall, D.J. et al. (2003). Aquatic hypoxia is an disrupter and impairs fish reproduction. *Environ. Sci. Technol.* **37** (6): 1137–1141.

252 Vera-Chang, M.N., Moon, T.W., and Trudeau, V.L. (2019). Ancestral fluoxetine exposure sensitizes zebrafish to venlafaxine-induced reductions in cortisol and spawning. *Endocrinology* **160** (9): 2137–2142.

253 Thompson, W.A. and Vijayan, M.M. (2021). Zygotic exposure to venlafaxine disrupts cortisol stress axis activity in multiple generations of zebrafish. *Environ. Pollut.* **1987** (274): 116535.

254 Waghmare, S.G., Samarin, A.M., Samarin, A.M. et al. (2021). Histone acetylation dynamics during in vivo and in vitro oocyte aging in common carp *Cyprinus carpio*. *Int. J. Mol. Sci.* **22** (11): 6036.

255 Iii, M.J.C., Kalluvila, T.A., Klingler, R.H. et al. (2017). Mercury-induced epigenetic transgenerational inheritance of abnormal neurobehavior is correlated with sperm epimutations in zebrafish. *PLoS One* **12** (5): e0176155.

256 Mann, M.R.W. and Bartolomei, M.S. (2002). Epigenetic reprogramming in the mammalian embryo: struggle of the clones. *Genome Biol.* **3** (2): REVIEWS 1003.

257 Bednarczyk, M., Dunislawska, A., Stadnicka, K., and Grochowska, E. (2021). Chicken embryo as a model in epigenetic research. *Poult. Sci.* **100** (7): 101164.

258 Santiago, J., Silva, J.V., Howl, J. et al. (2021). All you need to know about sperm RNAs. *Hum. Reprod. Update* **28** (1): 67–91.

259 Corral-Vazquez, C., Salas-Huetos, A., Blanco, J. et al. (2019). Sperm microRNA pairs: new perspectives in the search for male fertility biomarkers. *Fertil. Steril.* **112** (5): 831–841.

260 Hua, M., Liu, W., Chen, Y. et al. (2019). Identification of small non-coding RNAs as sperm quality biomarkers for in vitro fertilization. *Cell Discov.* **5**: 20.

261 Abu-Halima, M., Hammadeh, M., Schmitt, J. et al. (2013). Altered microRNA expression profiles of human spermatozoa in patients with different spermatogenic impairments. *Fertil. Steril.* **99** (5): 1249–1255. e16.

262 Byrne, P.F., Volk, G.M., Gardner, C. et al. (2018). Sustaining the future of plant breeding: the critical role of the USDA-ARS national plant germplasm system. *Crop Sci.* **58** (2): 451–468.

263 Ao, X., Sa, R., Wang, J. et al. (2016). Activation-induced cytidine deaminase selectively catalyzed active DNA demethylation in pluripotency gene and improved cell reprogramming in bovine SCNT embryo. *Cytotechnology* **68** (6): 2637–2648.

264 Hou, Z. and Fuiman, L.A. (2020). Nutritional programming in fishes: insights from mammalian studies. *Rev. Fish Biol. Fish.* **30**: 67–92.

265 Cabrita, E., Martínez-Páramo, S., Gavaia, P.J. et al. (2014). Factors enhancing fish sperm quality and emerging tools for sperm analysis. *Aquaculture* **432**: 389–401.

266 Parker, L.M., O'Connor, W.A., Raftos, D.A. et al. (2015). Persistence of positive carryover effects in the oyster, *Saccostrea glomerata*, following transgenerational exposure to ocean acidification. *PLoS One* **10** (7): e0132276.

267 Migaud, H., Davie, A., and Taylor, J.F. (2010). Current knowledge on the photoneuroendocrine regulation of reproduction in temperate fish species. *J. Fish Biol.* **76** (1): 27–68.

268 Migaud, H., Bell, G., Cabrita, E. et al. (2013). Gamete quality and broodstock management in temperate fish. *Rev. Aquac.* **5** (s1): S194–S223.

269 Okutsu, T., Suzuki, K., Takeuchi, Y. et al. (2006). Testicular germ cells can colonize sexually undifferentiated embryonic gonad and produce functional eggs in fish. *Proc. Natl. Acad. Sci. U. S. A.* **103** (8): 2725–2729.

270 Lee, S., Katayama, N., and Yoshizaki, G. (2016). Generation of juvenile rainbow trout derived from cryopreserved whole ovaries by intraperitoneal transplantation of ovarian germ cells. *Biochem. Biophys. Res. Commun.* **478** (3): 1478–1483.

271 Wang, H.-P. and Shen, Z.-G. (2018). *Sex Control in Aquaculture*, 1–34. Wiley.

272 Jiang, Q., Lin, D., Huang, H. et al. (2020). DNA methylation inhibits the expression of CFSH in mud crab. *Front. Endocrinol.* **11**: 163.

273 Anastasiadi, D., Vandeputte, M., Sánchez-Baizán, N. et al. (2018). Dynamic epimarks in sex-related genes predict gonad phenotype in the European sea bass, a fish with mixed genetic and environmental sex determination. *Epigenetics* **13** (9): 988–1011.

274 Lyko, F., Foret, S., Kucharski, R. et al. (2010). The honey bee epigenomes: differential methylation of brain DNA in queens and workers. *PLoS Biol.* **8** (11): e1000506.

275 Kang, H.M., Sul, J.H., Service, S.K. et al. (2010). Variance component model to account for sample structure in genome-wide association studies. *Nat. Genet.* **42** (4): 348–354.

276 Sul, J.H., Martin, L.S., and Eskin, E. (2018). Population structure in genetic studies: confounding factors and mixed models. *PLoS Genet.* **14** (12): e1007309.

277 Wang, M. and Ibeagha-Awemu, E.M. (2020). Impacts of epigenetic processes on the health and productivity of livestock. *Front. Genet.* **11**: 613636.

278 Wu, J., Zhang, W., and Li, C. (2020). Recent advances in genetic and epigenetic modulation of animal exposure to high temperature. *Front. Genet.* **11**: 653.

279 Hao, Y., Cui, Y., and Gu, X. (2016). Genome-wide DNA methylation profiles changes associated with constant heat stress in pigs as measured by bisulfite sequencing. *Sci. Rep.* **6**: 27507.

280 Kisliouk, T., Cramer, T., and Meiri, N. (2017). Methyl CpG level at distal part of heat-shock protein promoter HSP70 exhibits epigenetic memory for heat stress by modulating recruitment of POU2F1-associated nucleosome-remodeling deacetylase (NuRD) complex. *J. Neurochem.* **141** (3): 358–372.

281 Vinoth, A., Thirunalasundari, T., Shanmugam, M. et al. (2018). Evaluation of DNA methylation and mRNA expression of heat shock proteins in thermal manipulated chicken. *Cell Stress Chaperones* **23** (2): 235–252.

282 David, S.-A., Vitorino Carvalho, A., Gimonnet, C. et al. (2019). Thermal manipulation during embryogenesis impacts H3K4me3 and H3K27me3 histone marks in chicken hypothalamus. *Front. Genet.* **10**: 1207.

283 Lu, Z., Ma, Y., Li, Q. et al. (2019). The role of N6-methyladenosine RNA methylation in the heat stress response of sheep (*Ovis aries*). *Cell Stress Chaperones* **24** (2): 333–342.

284 Zhang, Y., Li, F., Feng, X. et al. (2017). Genome-wide analysis of DNA Methylation profiles on sheep ovaries associated with prolificacy using whole-genome bisulfite sequencing. *BMC Genomics* **18** (1): 759.

285 Wei, J., Cheng, J., Waddell, N.J. et al. (2021). DNA methyltransferase 3A is involved in the sustained effects of chronic stress on synaptic functions and behaviors. *Cereb. Cortex* **31** (4): 1998–2012.

286 Sanaei, M. and Kavoosi, F. (2021). Histone deacetylase inhibitors, intrinsic and extrinsic apoptotic pathways, and epigenetic alterations of histone deacetylases (HDACs) in hepatocellular carcinoma. *Iran. J. Pharm. Res. IJPR* **20** (3): 324–336.

287 Doherty, R., Farrelly, C.O., and Meade, K.G. (2014). Comparative epigenetics: relevance to the regulation of production and health traits in cattle. *Anim. Genet.* **45** (s1): 3–14.

288 Green, B.B. and Kerr, D.E. (2014). Epigenetic contribution to individual variation in response to lipopolysaccharide in bovine dermal fibroblasts. *Vet. Immunol. Immunopathol.* **157** (1): 49–58.

289 Fu, Q., Shi, H., and Chen, C. (2017). Roles of bta-miR-29b promoter regions DNA methylation in regulating miR-29b expression and bovine viral diarrhea virus NADL replication in MDBK cells. *Arch. Virol.* **162** (2): 401–408.

290 Green, T.J., Helbig, K., Speck, P., and Raftos, D.A. (2016). Primed for success: oyster parents treated with poly(I:C) produce offspring with enhanced protection against ostreid herpesvirus type I infection. *Mol. Immunol.* **78**: 113–120.

291 Uren Webster, T.M., Rodriguez-Barreto, D., Martin, S.A.M. et al. (2018). Contrasting effects of acute and chronic stress on the transcriptome, epigenome, and immune response of Atlantic salmon. *Epigenetics* **13** (12): 1191–1207.

292 Robinson, N.A., Johnsen, H., Moghadam, H. et al. (2019). Early developmental stress affects subsequent gene expression response to an acute stress in atlantic salmon: an approach for creating robust fish for aquaculture? *G3 Bethesda Md.* **9** (5): 1597–1611.

293 Norouzitallab, P., Baruah, K., Vandegehuchte, M. et al. (2014). Environmental heat stress induces epigenetic transgenerational inheritance of robustness in parthenogenetic *Artemia* model. *FASEB J.* **28** (8): 3552–3563.

294 Jeremias, G., Barbosa, J., Marques, S.M. et al. (2018). Synthesizing the role of epigenetics in the response and adaptation of species to climate change in freshwater ecosystems. *Mol. Ecol.* **27** (13): 2790–2806.

295 Wijenayake, S. and Storey, K.B. (2016). The role of DNA methylation during anoxia tolerance in a freshwater turtle (*Trachemys scripta elegans*). *J. Comp. Physiol. B.* **186** (3): 333–342.

296 Sistare, F.D. and DeGeorge, J.J. (2007). Preclinical predictors of clinical safety: opportunities for improvement. *Clin. Pharmacol. Ther.* **82** (2): 210–214.

297 Poças-Fonseca, M.J., Cabral, C.G., and Manfrão-Netto, J.H.C. (2020). Epigenetic manipulation of filamentous fungi for biotechnological applications: a systematic review. *Biotechnol. Lett.* **42** (6): 885–904.

298 Kantarjian, H.M., Thomas, X.G., Dmoszynska, A. et al. (2012). Multicenter, randomized, open-label, phase III trial of decitabine versus patient choice, with physician advice, of either supportive care or low-dose cytarabine for the treatment of older patients with newly diagnosed acute myeloid leukemia. *J. Clin. Oncol. Off. J. Am. Soc. Clin. Oncol.* **30** (21): 2670–2677.

299 Adès, L., Sekeres, M.A., Wolfromm, A. et al. (2013). Predictive factors of response and survival among chronic myelomonocytic leukemia patients treated with azacitidine. *Leuk. Res.* **37** (6): 609–613.

300 Treppendahl, M.B., Kristensen, L.S., and Grønbæk, K. (2016). Chapter 5 – biomarkers and methodologies for monitoring epigenetic drug effects in cancer. In: *Epigenetic Biomarkers and Diagnostics* (ed. J.L. García-Giménez), 91–118. Boston: Academic Press.

301 von Knethen, A. and Brüne, B. (2019). Histone deacetylation inhibitors as therapy concept in sepsis. *Int. J. Mol. Sci.* **20** (2): E346.

302 Gomez-Sanchez, J.A., Patel, N., Martirena, F. et al. (2022). Emerging role of HDACs in regeneration and ageing in the peripheral nervous system: repair Schwann cells as pivotal targets. *Int. J. Mol. Sci.* **23** (6): 2996.

303 Montgomery, R.L., Davis, C.A., Potthoff, M.J. et al. (2007). Histone deacetylases 1 and 2 redundantly regulate cardiac morphogenesis, growth, and contractility. *Genes Dev.* **21** (14): 1790–1802.

304 Haberland, M., Montgomery, R.L., and Olson, E.N. (2009). The many roles of histone deacetylases in development and physiology: implications for disease and therapy. *Nat. Rev. Genet.* **10** (1): 32–42.

305 Wang, Z., Zhu, T., Qiao, C. et al. (2005). Adeno-associated virus serotype 8 efficiently delivers genes to muscle and heart. *Nat. Biotechnol.* **23** (3): 321–328.

306 Long, C., Amoasii, L., Mireault, A.A. et al. (2016). Postnatal genome editing partially restores dystrophin expression in a mouse model of muscular dystrophy. *Science* **351** (6271): 400–403.

307 Filippakopoulos, P., Qi, J., Picaud, S. et al. (2010). Selective inhibition of BET bromodomains. *Nature* **468** (7327): 1067–1073.

308 Xing, H., Wang, C., Wu, H. et al. (2015). Effects of atrazine and chlorpyrifos on DNA methylation in the brain and gonad of the common carp. *Comp. Biochem. Physiol. Toxicol. Pharmacol. CBP* **168**: 11–19.

309 Smith, C.M., Vera, M.K.M., and Bhandari, R.K. (2019). Developmental and epigenetic effects of Roundup and glyphosate exposure on Japanese medaka (*Oryzias latipes*). *Aquat. Toxicol. Amst. Neth.* **210**: 215–226.

310 Dimond, J.L. and Roberts, S.B. (2016). Germline DNA methylation in reef corals: patterns and potential roles in response to environmental change. *Mol. Ecol.* **25** (8): 1895–1904.

311 Nakanishi, T., Kato, Y., Matsuura, T., and Watanabe, H. (2014). CRISPR/Cas-mediated targeted mutagenesis in *Daphnia magna*. *PLoS One* **9** (5): e98363.

312 Xu, Y., Xu, C., Kato, A. et al. (2012). Tet3 CXXC domain and dioxygenase activity cooperatively regulate key genes for xenopus eye and neural development. *Cell* **151** (6): 1200–1213.

313 Jang, H. and Serra, C. (2014). Nutrition, epigenetics, and diseases. *Clin. Nutr. Res.* **3** (1): 1–8.

314 Block, T. and El-Osta, A. (2017). Epigenetic programming, early life nutrition and the risk of metabolic disease. *Atherosclerosis* **266**: 31–40.

315 Maugeri, A. and Barchitta, M. (2020). How dietary factors affect DNA methylation: lesson from epidemiological studies. *Med. Kaunas Lith.* **56** (8): E374.

316 Burggren, W.W. (2019). Phenotypic switching resulting from developmental plasticity: fixed or reversible? *Front. Physiol.* **10**: 1634.

317 Li, Y., Carrillo, J.A., Ding, Y. et al. (2019). DNA methylation, microRNA expression profiles and their relationships with transcriptome in grass-fed and grain-fed Angus cattle rumen tissue. *PLoS One* **14** (10): e0214559.

318 Dong, G., Qiu, M., Ao, C. et al. (2014). Feeding a high-concentrate corn straw diet induced epigenetic alterations in the mammary tissue of dairy cows. *PLoS One* **9** (9): e107659.

319 Chang, G., Zhang, K., Xu, T. et al. (2015). Epigenetic mechanisms contribute to the expression of immune related genes in the livers of dairy cows fed a high concentrate diet. *PLoS One* **10** (4): e0123942.

320 Xu, T.L., Seyfert, H.M., and Shen, X.Z. (2018). Epigenetic mechanisms contribute to decrease stearoyl-CoA desaturase 1 expression in the liver of dairy cows after prolonged feeding of high-concentrate diet. *J. Dairy Sci.* **101** (3): 2506–2518.

321 Li, R. and Ibeagha-Awemu, E.M. (2017). Altered gene expression of epigenetic modifying enzymes in response to dietary supplementation with linseed oil. *J. Dairy Res.* **84** (2): 119–123.

322 Fang, L., Liu, S., Liu, M. et al. (2019). Functional annotation of the cattle genome through systematic discovery and characterization of chromatin states and butyrate-induced variations. *BMC Biol.* **17** (1): 68.

323 Malgwi, I.H., Halas, V., Grünvald, P. et al. (2022). Genes related to fat metabolism in pigs and intramuscular fat content of pork: a focus on nutrigenetics and nutrigenomics. *Anim. Open Access J. MDPI* **12** (2): 150.

324 Qimuge, N., He, Z., Qin, J. et al. (2019). Overexpression of DNMT3A promotes proliferation and inhibits differentiation of porcine intramuscular preadipocytes by methylating p21 and PPARg promoters. *Gene* **696**: 54–62.

325 Stachecka, J., Lemanska, W., Noak, M., and Szczerbal, I. (2020). Expression of key genes involved in DNA methylation during in vitro differentiation of porcine mesenchymal stem cells (MSCs) into adipocytes. *Biochem. Biophys. Res. Commun.* **522** (3): 811–818.

326 Bach, A., Aris, A., and Guasch, I. (2017). Consequences of supplying methyl donors during pregnancy on the methylome of the offspring from lactating and non-lactating dairy cattle. *PLoS One* **12** (12): e0189581.

327 Noya, A., Casasús, I., Ferrer, J., and Sanz, A. (2019). Effects of developmental programming caused by maternal nutrient intake on postnatal performance of beef heifers and their calves. *Anim. Open Access J. MDPI* **9** (12): E1072.

328 Keleher, M.R., Zaidi, R., Shah, S. et al. (2018). Maternal high-fat diet associated with altered gene expression, DNA methylation, and obesity risk in mouse offspring. *PLoS One* **13** (2): e0192606.

329 Glendining, K.A. and Jasoni, C.L. (2019). Maternal high fat diet-induced obesity modifies histone binding and expression of Oxtr in offspring hippocampus in a sex-specific manner. *Int. J. Mol. Sci.* **20** (2): E329.

330 Marco, A., Kisliouk, T., Tabachnik, T. et al. (2016). DNA CpG methylation (5-methylcytosine) and its derivative (5-hydroxymethylcytosine) alter histone posttranslational modifications at the Pomc promoter, affecting the impact of perinatal diet on leanness and obesity of the offspring. *Diabetes* **65** (8): 2258–2267.

331 Boddicker, R.L., Koltes, J.E., Fritz-Waters, E.R. et al. (2016). Genome-wide methylation profile following prenatal and postnatal dietary omega-3 fatty acid supplementation in pigs. *Anim. Genet.* **47** (6): 658–671.

332 Liu, H., Wang, J., Mou, D. et al. (2017). Maternal methyl donor supplementation during gestation counteracts the bisphenol A-induced impairment of intestinal morphology, disaccharidase activity, and nutrient transporters gene expression in newborn and weaning pigs. *Nutrients* **9** (5): E423.

333 Jin, C., Zhuo, Y., Wang, J. et al. (2018). Methyl donors dietary supplementation to gestating sows diet improves the growth rate of offspring and is associating with changes in expression and DNA methylation of insulin-like growth factor-1 gene. *J. Anim. Physiol. Anim. Nutr.* **102** (5): 1340–1350.

334 Hu, Y., Sun, Q., Zong, Y. et al. (2017). Prenatal betaine exposure alleviates corticosterone-induced inhibition of CYP27A1 expression in the liver of juvenile chickens associated with its promoter DNA methylation. *Gen. Comp. Endocrinol.* **246**: 241–248.

335 Idriss, A.A., Hu, Y., Hou, Z. et al. (2018). Dietary betaine supplementation in hens modulates hypothalamic expression of cholesterol metabolic genes in F1 cockerels through modification of DNA methylation. *Comp. Biochem. Physiol. B Biochem. Mol. Biol.* **217**: 14–20.

336 Abobaker, H., Hu, Y., Omer, N.A. et al. (2019). Maternal betaine suppresses adrenal expression of cholesterol trafficking genes and decreases plasma corticosterone concentration in offspring pullets. *J. Anim. Sci. Biotechnol.* **10**: 87.

337 Lv, Z., Fan, H., Song, B. et al. (2019). Supplementing genistein for breeder hens alters the fatty acid metabolism and growth performance of offsprings by epigenetic modification. *Oxidative Med. Cell. Longev.* **2019**: 9214209.

338 Yang, Z., Yang, H.M., Gong, D.Q. et al. (2018). Transcriptome analysis of hepatic gene expression and DNA methylation in methionine- and betaine-supplemented geese (*Anser cygnoides domesticus*). *Poult. Sci.* **97** (10): 3463–3477.

339 Hu, Y., Sun, Q., Liu, J. et al. (2017). In ovo injection of betaine alleviates corticosterone-induced fatty liver in chickens through epigenetic modifications. *Sci. Rep.* **7**: 40251.

340 Zhang, N. (2018). Role of methionine on epigenetic modification of DNA methylation and gene expression in animals. *Anim. Nutr. Zhongguo Xu Mu Shou Yi Xue Hui* **4** (1): 11–16.

341 Mikkelsen, S.U., Gillberg, L., Lykkesfeldt, J., and Grønbæk, K. (2021). The role of vitamin C in epigenetic cancer therapy. *Free Radic. Biol. Med.* **170**: 179–193.

342 Gillberg, L., Ørskov, A.D., Liu, M. et al. (2018). Vitamin C – a new player in regulation of the cancer epigenome. *Semin. Cancer Biol.* **51**: 59–67.

343 Yu, X.-X., Liu, Y.-H., Liu, X.-M. et al. (2018). Ascorbic acid induces global epigenetic reprogramming to promote meiotic maturation and developmental competence of porcine oocytes. *Sci. Rep.* **8** (1): 6132.

344 Zglejc-Waszak, K., Waszkiewicz, E.M., and Franczak, A. (2019). Periconceptional undernutrition affects the levels of DNA methylation in the peri-implantation pig endometrium and in embryos. *Theriogenology* **123**: 185–193.

345 Adam, A.-C., Lie, K.K., Whatmore, P. et al. (2019). Profiling DNA methylation patterns of zebrafish liver associated with parental high dietary arachidonic acid. *PLoS One* **14** (8): e0220934.

346 Skjærven, K.H., Jakt, L.M., Fernandes, J.M.O. et al. (2018). Parental micronutrient deficiency distorts liver DNA methylation and expression of lipid genes associated with a fatty-liver-like phenotype in offspring. *Sci. Rep.* **8** (1): 3055.

347 Saito, T., Whatmore, P., Taylor, J.F. et al. (2021). Micronutrient supplementation affects transcriptional and epigenetic regulation of lipid metabolism in a dose-dependent manner. *Epigenetics* **16** (11): 1217–1234.

348 Saito, T., Whatmore, P., Taylor, J.F. et al. (2022). Micronutrient supplementation affects DNA methylation in male gonads with potential intergenerational epigenetic inheritance involving the embryonic development through glutamate receptor-associated genes. *BMC Genomics* **23** (1): 115.

349 Veron, V., Marandel, L., Liu, J. et al. (2018). DNA methylation of the promoter region of bnip3 and bnip3l genes induced by metabolic programming. *BMC Genomics* **19** (1): 677.

350 Marandel, L., Lepais, O., Arbenoits, E. et al. (2016). Remodelling of the hepatic epigenetic landscape of glucose-intolerant rainbow trout (*Oncorhynchus mykiss*) by nutritional status and dietary carbohydrates. *Sci. Rep.* **6**: 32187.

351 Perera, E. and Yúfera, M. (2017). Effects of soybean meal on digestive enzymes activity, expression of inflammation-related genes, and chromatin modifications in marine fish (*Sparus aurata* L.) larvae. *Fish Physiol. Biochem.* **43** (2): 563–578.

352 Podgorniak, T., Dhanasiri, A., Chen, X. et al. (2022). Early fish domestication affects methylation of key genes involved in the rapid onset of the farmed phenotype. *Epigenetics* **17**: 1–18.

353 Macqueen, D.J., Meischke, L., Manthri, S. et al. (2010). Characterisation of capn1, capn2-like, capn3 and capn11 genes in Atlantic halibut (*Hippoglossus hippoglossus* L.): transcriptional regulation across tissues and in skeletal muscle at distinct nutritional states. *Gene* **453** (1–2): 45–58.

354 Vélez, E.J., Azizi, S., Verheyden, D. et al. (2017). Proteolytic systems' expression during myogenesis and transcriptional regulation by amino acids in Gilthead sea bream cultured muscle cells. *PLoS One* **12** (12): e0187339.

355 Xu, H., Du, X., Liu, G. et al. (2019). The pseudokinase MLKL regulates hepatic insulin sensitivity independently of inflammation. *Mol. Metab.* **23**: 14–23.

356 Schneeberger, M. (2019). Irx3, a new leader on obesity genetics. *EBioMedicine* **39**: 19–20.

357 Fujita, M., Momose, A., Ohtomo, T. et al. (2011). Upregulation of fatty acyl-CoA thioesterases in the heart and skeletal muscle of rats fed a high-fat diet. *Biol. Pharm. Bull.* **34** (1): 87–91.

358 Corominas, J., Ramayo-Caldas, Y., Puig-Oliveras, A. et al. (2013). Polymorphism in the ELOVL6 gene is associated with a major QTL effect on fatty acid composition in pigs. *PLoS One* **8** (1): e53687.

359 Matsuzaka, T., Shimano, H., Yahagi, N. et al. (2007). Crucial role of a long-chain fatty acid elongase, Elovl6, in obesity-induced insulin resistance. *Nat. Med.* **13** (10): 1193–1202.

360 Sinclair, K.D., Rutherford, K.M.D., Wallace, J.M. et al. (2016). Epigenetics and developmental programming of welfare and production traits in farm animals. *Reprod. Fertil. Dev.* **28** (10): 1443.

361 Tsumagari, K., Baribault, C., Terragni, J. et al. (2013). Early de novo DNA methylation and prolonged demethylation in the muscle lineage. *Epigenetics* **8** (3): 317–332.

362 Fong, A.P., Yao, Z., Zhong, J.W. et al. (2012). Genetic and epigenetic determinants of neurogenesis and myogenesis. *Dev. Cell* **22** (4): 721–735.

363 Hu, Y., Xu, H., Li, Z. et al. (2013). Comparison of the genome-wide DNA methylation profiles between fast-growing and slow-growing broilers. *PLoS One* **8** (2): e56411.

364 Zhang, M., Li, D., Zhai, Y. et al. (2020). The landscape of DNA methylation associated with the transcriptomic network of intramuscular adipocytes generates insight into intramuscular fat deposition in chicken. *Front. Cell Dev. Biol.* **8**: 206.

365 Gao, G.L., Wang, H.W., Zhao, X.Z. et al. (2015). Feeding conditions and breed affect the level of DNA methylation of the mitochondrial uncoupling protein 3 gene in chicken breast muscle. *J. Anim. Sci.* **93** (4): 1522–1534.

366 Gao, G., Wang, H., Zhao, X. et al. (2017). Effect of feeding conditions on the methylation status of Fatp1 gene in chicken breast muscle. *R. Bras. Ci Avic* 55–58.

367 Guo, X., Liu, X., Xu, X. et al. (2012). The expression levels of DNMT3a/3b and their relationship with meat quality in beef cattle. *Mol. Biol. Rep.* **39** (5): 5473–5479.

368 Chen, Z., Chu, S., Xu, X. et al. (2019). Analysis of longissimus muscle quality characteristics and associations with DNA methylation status in cattle. *Genes Genomics* **41** (10): 1147–1163.

369 Ma, X., Jia, C., Chu, M. et al. (2019). Transcriptome and DNA methylation analyses of the molecular mechanisms underlying with longissimus dorsi muscles at different stages of development in the Polled yak. *Genes* **10** (12): E970.

370 Paradis, F., Wood, K.M., Swanson, K.C. et al. (2017). Maternal nutrient restriction in mid-to-late gestation influences fetal mRNA expression in muscle tissues in beef cattle. *BMC Genomics* **18** (1): 632.

371 McKinsey, T.A., Zhang, C.L., and Olson, E.N. (2001). Control of muscle development by dueling HATs and HDACs. *Curr. Opin. Genet. Dev.* **11** (5): 497–504.

372 Lu, J., McKinsey, T.A., Zhang, C.-L., and Olson, E.N. (2000). Regulation of skeletal myogenesis by association of the MEF2 transcription factor with Class II histone deacetylases. *Mol. Cell* **6** (2): 233–244.

373 Bailey, P., Downes, M., Lau, P. et al. (1999). The nuclear receptor corepressor N-CoR regulates differentiation: N-CoR directly interacts with MyoD. *Mol. Endocrinol.* **13** (7): 1155–1168.

374 Mal, A., Sturniolo, M., Schiltz, R.L. et al. (2001). A role for histone deacetylase HDAC1 in modulating the transcriptional activity of MyoD: inhibition of the myogenic program. *EMBO J.* **20** (7): 1739–1753.

375 Wei, D., Li, A., Zhao, C. et al. (2018). Transcriptional regulation by CpG sites methylation in the core promoter region of the bovine SIX1 gene: roles of histone H4 and E2F2. *Int. J. Mol. Sci.* **19** (1): E213.

376 Blum, R., Vethantham, V., Bowman, C. et al. (2012). Genome-wide identification of enhancers in skeletal muscle: the role of MyoD1. *Genes Dev.* **26** (24): 2763–2779.

377 Asp, P., Blum, R., Vethantham, V. et al. (2011). Genome-wide remodeling of the epigenetic landscape during myogenic differentiation. *Proc. Natl. Acad. Sci.* **108** (22): E149–E158.

378 Mousavi, K., Zare, H., Dell'orso, S. et al. (2013). eRNAs promote transcription by establishing chromatin accessibility at defined genomic loci. *Mol. Cell* **51** (5): 606–617.

379 Rochard, P., Rodier, A., Casas, F. et al. (2000). Mitochondrial activity is involved in the regulation of myoblast differentiation through myogenin expression and activity of myogenic factors. *J. Biol. Chem.* **275** (4): 2733–2744.

380 Hartmann, N., Reichwald, K., Wittig, I. et al. (2011). Mitochondrial DNA copy number and function decrease with age in the short-lived fish *Nothobranchius furzeri*. *Aging Cell* **10** (5): 824–831.

381 Jing, E., O'Neill, B.T., Rardin, M.J. et al. (2013). Sirt3 regulates metabolic flexibility of skeletal muscle through reversible enzymatic deacetylation. *Diabetes* **62** (10): 3404–3417.

382 Simó-Mirabet, P., Perera, E., Calduch-Giner, J.A. et al. (2018). Co-expression analysis of sirtuins and related metabolic biomarkers in juveniles of Gilthead sea bream (*Sparus aurata*) with differences in growth performance. *Front. Physiol.* **9**: 608.

383 DeLaurier, A., Nakamura, Y., Braasch, I. et al. (2012). Histone deacetylase-4 is required during early cranial neural crest development for generation of the zebrafish palatal skeleton. *BMC Dev. Biol.* **12**: 16.

384 Pillai, R., Coverdale, L.E., Dubey, G., and Martin, C.C. (2004). Histone deacetylase 1 (HDAC-1) required for the normal formation of craniofacial cartilage and pectoral fins of the zebrafish. *Dev. Dyn. Off. Publ. Am. Assoc. Anat.* **231** (3): 647–654.

385 Sharples, A.P., Stewart, C.E., and Seaborne, R.A. (2016). Does skeletal muscle have an 'epi'-memory? The role of epigenetics in nutritional programming, metabolic disease, aging and exercise. *Aging Cell* **15** (4): 603–616.

386 Dwyer, C.M. and Stickland, N.C. (1994). Supplementation of a restricted maternal diet with protein or carbohydrate alone prevents a reduction in fetal muscle fibre number in the guinea-pig. *Br. J. Nutr.* **72** (2): 173–180.

387 Mallinson, J.E., Sculley, D.V., Craigon, J. et al. (2007). Fetal exposure to a maternal low-protein diet during mid-gestation results in muscle-specific effects on fibre type composition in young rats. *Br. J. Nutr.* **98** (2): 292–299.

388 Confortim, H.D., Jerônimo, L.C., Centenaro, L.A. et al. (2015). Effects of aging and maternal protein restriction on the muscle fibers morphology and neuromuscular junctions of rats after nutritional recovery. *Micron Oxf. Engl.* **1993** (71): 7–13.

389 Raychaudhuri, N., Thamotharan, S., Srinivasan, M. et al. (2014). Postnatal exposure to a high-carbohydrate diet interferes epigenetically with thyroid hormone receptor induction of the adult male rat skeletal muscle glucose transporter isoform 4 expression. *J. Nutr. Biochem.* **25** (10): 1066–1076.

390 Brameld, J.M., Mostyn, A., Dandrea, J. et al. (2000). Maternal nutrition alters the expression of insulin-like growth factors in fetal sheep liver and skeletal muscle. *J. Endocrinol.* **167** (3): 429–437.

391 Sayer, A.A., Syddall, H.E., Dennison, E.M. et al. (2004). Birth weight, weight at 1 y of age, and body composition in older men: findings from the Hertfordshire Cohort Study. *Am. J. Clin. Nutr.* **80** (1): 199–203.

392 Sayer, A.A. and Cooper, C. (2005). Fetal programming of body composition and musculoskeletal development. *Early Hum. Dev.* **81** (9): 735–744.

393 Sayer, A.A., Dennison, E.M., Syddall, H.E. et al. (2008). The developmental origins of sarcopenia: using peripheral quantitative computed tomography to assess muscle size in older people. *J. Gerontol. A Biol. Sci. Med. Sci.* **63** (8): 835–840.

394 Patel, H.P., Al-Shanti, N., Davies, L.C. et al. (2014). Lean mass, muscle strength and gene expression in community dwelling older men: findings from the Hertfordshire Sarcopenia Study (HSS). *Calcif. Tissue Int.* **95** (4): 308–316.

395 Sharples, A.P., Polydorou, I., Hughes, D.C. et al. (2016). Skeletal muscle cells possess a "memory" of acute early life TNF-α exposure: role of epigenetic adaptation. *Biogerontology* **17** (3): 603–617.

396 Tong, T., Shen, Y., Lee, H.-W. et al. (2016). Adenylyl cyclase 3 haploinsufficiency confers susceptibility to diet-induced obesity and insulin resistance in mice. *Sci. Rep.* **6**: 34179.

397 Gudas, L.J. (2012). Emerging roles for retinoids in regeneration and differentiation in normal and disease states. *Biochim. Biophys. Acta BBA – Mol. Cell Biol. Lipids* **1821** (1): 213–221.

398 Adhikari, A. and Davie, J. (2018). JARID2 and the PRC2 complex regulate skeletal muscle differentiation through regulation of canonical Wnt signaling. *Epigenetics Chromatin* **11** (1): 46.

399 Scholtz, M.M., Maiwashe, A., Neser, F.W.C. et al. (2013). Livestock breeding for sustainability to mitigate global warming, with the emphasis on developing countries. *South Afr. J. Anim. Sci.* **43** (3): 269–281.

400 Neibergs, H.L. and Johnson, K.A. (2012). Alpharma beef cattle nutrition symposium: nutrition and the genome. *J. Anim. Sci.* **90** (7): 2308–2316.

401 Godfrey, K.M., Costello, P.M., and Lillycrop, K.A. (2015). The developmental environment, epigenetic biomarkers and long-term health. *J. Dev. Orig. Health Dis.* **6** (5): 399–406.

402 Gluckman, P.D., Lillycrop, K.A., Vickers, M.H. et al. (2007). Metabolic plasticity during mammalian development is directionally dependent on early nutritional status. *Proc. Natl. Acad. Sci. U. S. A.* **104** (31): 12796–12800.

403 Lillycrop, K.A., Phillips, E.S., Jackson, A.A. et al. (2005). Dietary protein restriction of pregnant rats induces and folic acid supplementation prevents epigenetic modification of hepatic gene expression in the offspring. *J. Nutr.* **135** (6): 1382–1386.

404 Kvist, J., Gonçalves Athanàsio, C., Shams Solari, O. et al. (2018). Pattern of DNA methylation in *Daphnia*: evolutionary perspective. *Genome Biol. Evol.* **10** (8): 1988–2007.

405 Lindeman, L.C., Thaulow, J., Song, Y. et al. (2019). Epigenetic, transcriptional and phenotypic responses in two generations of *Daphnia magna* exposed to the DNA methylation inhibitor 5-azacytidine. *Environ. Epigenetics* **5** (3): dvz016.

406 Pizzorusso, T. and Tognini, P. (2020). Interplay between metabolism, nutrition and epigenetics in shaping brain DNA methylation, neural function and behavior. *Genes* **11** (7): E742.

407 Brito, S., Lee, M.-G., Bin, B.-H., and Lee, J.-S. Zinc and its transporters in epigenetics. *Mol. Cell* **43**: 323–330.

408 Kohli, R.M. and Zhang, Y. (2013). TET enzymes, TDG and the dynamics of DNA demethylation. *Nature* **502** (7472): 472–479.

409 Tsukada, Y., Fang, J., Erdjument-Bromage, H. et al. (2006). Histone demethylation by a family of JmjC domain-containing proteins. *Nature* **439** (7078): 811–816.

410 Lee, S.H., Lee, S.-H., Lee, J.H. et al. (2019). IDH2 deficiency increases bone mass with reduced osteoclastogenesis by limiting RANKL expression in osteoblasts. *Bone* **129**: 115056.

411 Yang, Y., Zhou, R., Mu, Y. et al. (2016). Genome-wide analysis of DNA methylation in obese, lean, and miniature pig breeds. *Sci. Rep.* **6**: 30160.

412 Zhang, S., Shen, L., Xia, Y. et al. (2016). DNA methylation landscape of fat deposits and fatty acid composition in obese and lean pigs. *Sci. Rep.* **6**: 35063.

413 Wang, J., Chen, J., Zhang, J. et al. (2017). Castration-induced changes in the expression profiles and promoter methylation of the GHR gene in Huainan male pigs. *Anim. Sci. J. Nihon Chikusan Gakkaiho* **88** (8): 1113–1119.

414 Ponsuksili, S., Trakooljul, N., Basavaraj, S. et al. (2019). Epigenome-wide skeletal muscle DNA methylation profiles at the background of distinct metabolic types and ryanodine receptor variation in pigs. *BMC Genomics* **20** (1): 492.

415 Kociucka, B., Stachecka, J., Szydlowski, M., and Szczerbal, I. (2017). Rapid communication: the correlation between histone modifications and expression of key genes involved in accumulation of adipose tissue in the pig. *J. Anim. Sci.* **95** (10): 4514–4519.

416 Park, H., Seo, K.-S., Lee, M., and Seo, S. (2020). Identification of meat quality-related differentially methylated regions in the DNA of the longissimus dorsi muscle in pig. *Anim. Biotechnol.* **31** (3): 189–194.

417 Zhang, M., Yan, F.-B., Li, F. et al. (2017). Genome-wide DNA methylation profiles reveal novel candidate genes associated with meat quality at different age stages in hens. *Sci. Rep.* **7**: 45564.

418 Cao, Y., Jin, H.-G., Ma, H.-H., and Zhao, Z.-H. (2017). Comparative analysis on genome-wide DNA methylation in longissimus dorsi muscle between Small Tailed Han and Dorper×Small Tailed Han crossbred sheep. *Asian-Australas. J. Anim. Sci.* **30** (11): 1529–1539.

419 Fang, X., Zhao, Z., Yu, H. et al. (2017). Comparative genome-wide methylation analysis of longissimus dorsi muscles between Japanese black (Wagyu) and Chinese Red Steppes cattle. *PLoS One* **12** (8): e0182492.

420 Zhao, C., Ji, G., Carrillo, J.A. et al. (2020). The profiling of DNA methylation and its regulation on divergent tenderness in angus beef cattle. *Front. Genet.* **11**: 939.

421 McGuigan, K., Hoffmann, A.A., and Sgrò, C.M. (2021). How is epigenetics predicted to contribute to climate change adaptation? What evidence do we need? *Philos. Trans. R. Soc. Lond. Ser. B Biol. Sci.* **376** (1826): 20200119.

422 Eirin-Lopez, J.M. and Putnam, H.M. (2018). Marine environmental epigenetics. *Annu. Rev. Mar. Sci.* **11**: 335–368.

423 Torda, G., Donelson, J.M., Aranda, M. et al. (2017). Rapid adaptive responses to climate change in corals. *Nat. Clim. Chang.* **7** (9): 627–636.

424 Donelson, J.M., Salinas, S., Munday, P.L., and Shama, L.N.S. (2018). Transgenerational plasticity and climate change experiments: Where do we go from here? *Glob. Chang. Biol.* **24** (1): 13–34.

425 Mousseau, T.A. and Fox, C.W. (1998). *Maternal Effects as Adaptations*. Oxford University Press.

426 Roach, D.A. and Wulff, R.D. (1987). Maternal effects in plants. *Annu. Rev. Ecol. Syst.* **18**: 209–235.

427 Räsänen, K. and Kruuk, L.E.B. (2007). Maternal effects and evolution at ecological time-scales. *Funct. Ecol.* **21** (3): 408–421.

428 Wolf, J.B. and Wade, M.J. (2009). What are maternal effects (and what are they not)? *Philos. Trans. R. Soc. Lond. Ser. B Biol. Sci.* **364** (1520): 1107–1115.

429 Bell, A.M. and Hellmann, J.K. (2019). An integrative framework for understanding the mechanisms and multigenerational consequences of transgenerational plasticity. *Annu. Rev. Ecol. Evol. Syst.* **50** (1): 97–118.

430 Donelan, S.C., Hellmann, J.K., Bell, A.M. et al. (2020). Transgenerational plasticity in human-altered environments. *Trends Ecol. Evol.* **35** (2): 115–124.

431 Uller, T. (2008). Developmental plasticity and the evolution of parental effects. *Trends Ecol. Evol.* **23** (8): 432–438.

432 Galloway, L.F. and Etterson, J.R. (2009). Plasticity to canopy shade in a monocarpic herb: within- and between-generation effects. *New Phytol.* **182** (4): 1003–1012.

433 Ezard, T.H.G., Prizak, R., and Hoyle, R.B. (2014). The fitness costs of adaptation via phenotypic plasticity and maternal effects. *Funct. Ecol.* **28** (3): 693–701.

434 McGlothlin, J.W. and Galloway, L.F. (2014). The contribution of maternal effects to selection response: an empirical test of competing models. *Evol. Int. J. Org. Evol.* **68** (2): 549–558.

435 Kuijper, B. and Hoyle, R.B. (2015). When to rely on maternal effects and when on phenotypic plasticity? *Evol. Int. J. Org. Evol.* **69** (4): 950–968.

436 Day, T. and Bonduriansky, R. (2011). A unified approach to the evolutionary consequences of genetic and nongenetic inheritance. *Am. Nat.* **178** (2): E18–E36.

437 Laland, K., Uller, T., Feldman, M. et al. (2014). Does evolutionary theory need a rethink? *Nature* **514** (7521): 161–164.

438 Mendizabal, I., Keller, T.E., Zeng, J., and Yi, S.V. (2014). Epigenetics and evolution. *Integr. Comp. Biol.* **54** (1): 31–42.

439 Pu, C. and Zhan, A. (2017). Epigenetic divergence of key genes associated with water temperature and salinity in a highly invasive model ascidian. *Biol. Invasions* **19** (7): 2015–2028.

440 Weyrich, A., Lenz, D., Jeschek, M. et al. (2016). Paternal intergenerational epigenetic response to heat exposure in male Wild guinea pigs. *Mol. Ecol.* **25** (8): 1729–1740.

441 Bautista, N.M. and Burggren, W.W. (2019). Parental stressor exposure simultaneously conveys both adaptive and maladaptive larval phenotypes through epigenetic inheritance in the zebrafish (*Danio rerio*). *J. Exp. Biol.* **222** (Pt 17): jeb208918.

442 Kronholm, I., Bassett, A., Baulcombe, D., and Collins, S. (2017). Epigenetic and genetic contributions to adaptation in chlamydomonas. *Mol. Biol. Evol.* **34** (9): 2285–2306.

443 Price, E.O. (2002). *Animal Domestication and Behavior*. Wallingford, Oxon; New York: CABI Pub.

444 Trut, L., Oskina, I., and Kharlamova, A. (2009). Animal evolution during domestication: the domesticated fox as a model. *BioEssays News Rev. Mol. Cell. Dev. Biol.* **31** (3): 349–360.

445 Selechnik, D., Richardson, M.F., Shine, R. et al. (2019). Increased adaptive variation despite reduced overall genetic diversity in a rapidly adapting invader. *Front. Genet.* **10**: 1221.

446 Schmitz, R.J., Schultz, M.D., Lewsey, M.G. et al. (2011). Transgenerational epigenetic instability is a source of novel methylation variants. *Science* **334** (6054): 369–373.

447 Feil, R. and Fraga, M.F. (2012). Epigenetics and the environment: emerging patterns and implications. *Nat. Rev. Genet.* **13** (2): 97–109.

448 Jablonka, E. and Raz, G. (2009). Transgenerational epigenetic inheritance: prevalence, mechanisms, and implications for the study of heredity and evolution. *Q. Rev. Biol.* **84** (2): 131–176.

449 Daxinger, L. and Whitelaw, E. (2012). Understanding transgenerational epigenetic inheritance via the gametes in mammals. *Nat. Rev. Genet.* **13** (3): 153–162.

450 Le Luyer, J., Laporte, M., Beacham, T.D. et al. (2017). Parallel epigenetic modifications induced by hatchery rearing in a Pacific salmon. *Proc. Natl. Acad. Sci. U. S. A.* **114** (49): 12964–12969.

451 Anastasiadi, D. and Piferrer, F. (2019). Epimutations in developmental genes underlie the onset of domestication in farmed European sea bass. *Mol. Biol. Evol.* **36** (10): 2252–2264.

452 Lim, J.P. and Brunet, A. (2013). Bridging the transgenerational gap with epigenetic memory. *Trends Genet. TIG* **29** (3): 176–186.

453 Friberg, N. (2014). Impacts and indicators of change in lotic ecosystems. *WIREs Water* **1** (6): 513–531.

454 Teletchea, F. and Fontaine, P. (2014). Levels of domestication in fish: implications for the sustainable future of aquaculture. *Fish Fish.* **15** (2): 181–195.

455 Birney, E., Smith, G.D., and Greally, J.M. (2016). Epigenome-wide association studies and the interpretation of disease -omics. *PLoS Genet.* **12** (6): e1006105.

456 Carnero-Montoro, E. and Alarcón-Riquelme, M.E. (2018). Epigenome-wide association studies for systemic autoimmune diseases: the road behind and the road ahead. *Clin. Immunol.* **196**: 21–33.

457 Feeney, A., Nilsson, E., and Skinner, M.K. (2014). Epigenetics and transgenerational inheritance in domesticated farm animals. *J. Anim. Sci. Biotechnol.* **5** (1): 48.

2

Transcriptional Epigenetic Mechanisms in Aquatic Species

Laia Navarro-Martín[1], Jan A. Mennigen[2], and Jana Asselman[3]

[1] *Institute of Environmental Assessment and Water Research, IDAEA-CSIC, Spanish National Research Council (CSIC), Barcelona, Spain*
[2] *Department of Biology, University of Ottawa, Ottawa, Ontario, Canada*
[3] *Laboratory for Environmental Toxicology and Aquatic Ecology, Ghent University, Ghent, Belgium*

2.1 Epigenetic Mechanisms as Modulators of Transcription

Mechanisms regulating gene transcription are diverse among organisms. In unicellular organisms, most genes are active and only a small number are repressed. In contrast, in pluricellular organisms, repression is by default the dominant regulator of gene expression and more than 50% of the genome is silenced in all cell types [1]. Two different strategies are used for gene long-term repression: (i) genes of almost all somatic cell types are active during early embryogenesis and then silenced at late development and adult life; and (ii) genes remain silent during embryonic development and then become activated during late development or adult life in a time- and tissue-specific manner [1]. Epigenetic modifications are known to be responsible for this maintenance of long-term silencing, regulating temporal and spatial gene transcription, and thus ensuring correct expression patterns during development and controlling transcriptomic tissue specificity. In eukaryotes, epigenetics has been linked to the regulation of biological processes such as stem cell differentiation and development [2], embryogenesis [3], aging [4], neuronal function [5], and reproduction and nongenetic inheritance [6], among many others.

In past decades, studies have also demonstrated that epigenetics not only is highly important for the proper development and differentiation of organisms in general, but also acts as a direct link between the genome and the environment. Environmental factors such as diet, exposure to abiotic factors (e.g., temperature, oxygen concentration, and salinity), and environmental pollutants (e.g., heavy metals and endocrine disruptors) are known to alter epigenetic mark dynamics, allowing the organism to respond to environmental changes through changes in gene expression [7], which translates into changes in phenotypes [8]. This chain of events can occur both within (context-dependent) and between (germline-dependent) generations [9, 10]. From an environmental/ecological point of view, it has been suggested that epigenetic mechanisms are a driving force in responding and adapting to rapidly changing environmental conditions [11], linked to increased phenotypic plasticity [12]. In addition, it is well known that stress caused by changes in the environmental conditions during development can lead to effects at the physiological level later in life, which are, at least in part, also mediated by epigenetic modifications in somatic cells. Furthermore, epigenetics has been found to be related to inter- and transgenerational inheritance, in which early life experiences impact phenotypes of subsequent generations via molecular epigenetic marks in germ cells [13]. At the population and ecological level, epigenetics is also related to the variation of diversity, enabling individuals to respond in real-time to environmental stressors and selection pressures, resulting in the ability of species to respond to environmental stress [14]. Epigenetics must thus also be considered in an evolutionary context [15]. Herein, we review the importance of molecular epigenetic mechanisms to better understand the role of epigenetics in mediating developmental processes and to characterize their role in responding to environmental cues. Given the reality of global climate change, the study of environmental epigenetics holds particular relevance for aquaculture species and their aquatic environment, which increasingly contribute to the global food system [16]. In this chapter, we briefly describe epigenetic mechanisms that are capable of modulating mRNA abundance at the transcriptional level (DNA methylation

Epigenetics in Aquaculture, First Edition. Edited by Francesc Piferrer and Han-Ping Wang.

Box 2.1 Key Molecular Epigenetic Concept Mechanisms

DNA methylation: Reversible cytosine 5 methylation in specific genomic contexts, often CpG-rich. DNA methylation can be distinguished into maintenance and de novo methylation, mediated by different writers, Dnmt1 and Dnmt3s, respectively. DNA methylation, especially in promoter regions and first intron regions, is generally linked to transcriptional repression.
Epigenetic mark: A reversible, but mitotically and meiotically stable, molecular modification resulting in changes in gene expression without alteration of the coding sequence (mutation). As such, epigenetic marks play crucial roles in cellular differentiation, plasticity, and nongenetic inheritance.
Histone modifications: Reversible posttranslational modification of N-terminal residues of core histones or their variants. While many modifications exist, methylation, acetylation, and phosphorylation are the most studied with regard to transcriptional control. Histone modifications alter chromatin states and depending on the specific location and type of PTM, render DNA accessible or inaccessible for transcription machinery.
Parental imprinting: Silencing of one of the paternal alleles that occur to express only one of the gene copies during embryonic development.
Phenotypic plasticity: Producing different phenotypes from the same genotype.
Transgenerational inheritance: Germline transmission of epigenetic information between generations in the absence of direct environmental exposures.
Writer, Reader, Eraser: Proteins able to create, read, and erase specific epigenetic marks.

and chromatin remodeling through histone modifications, summarized in Table 2.1) with special attention to those that modulate key biological functions in aquatic species of interest in aquaculture.

2.1.1 DNA Methylation

DNA methylation is present in a variety of organisms ranging from animals (invertebrates and vertebrates) to plants and fungi, and it is one of the most studied epigenetic mechanisms in eukaryotes [6]. It consists of the addition of a methyl group of a cytosine (5mC) located commonly in CpG dinucleotides, and its function is to regulate chromatin structure and the activation/repression of transcription. In vertebrates, genomes are globally methylated, with higher percentages observed in fish and frogs compared to others such as mammals or birds [17]. In these species, 5mC can be generally found in repetitive DNA sequences that are often heavily methylated. On the other hand, 1–2 kilobases of GC-rich regions, the so-called CpG Islands (CGIs), are found to be mostly demethylated especially when located in the promoters of genes with active transcription. In contrast to vertebrate genomes, invertebrate genomes tend to be only sparsely methylated, referred to as mosaic methylation, with large differences between species [18]. Methylated DNA is often targeted to gene bodies interspersed with intergenic sections devoid of methylation. So far, transposable elements and repetitive sequences seem to lack methylation in most invertebrate species [19, 20]. This leads to the typical binomial distribution of DNA methylation levels observed in invertebrate genomes [21]. Figure 2.1 represents a simplified diagram of the DNA methylation landscape in both vertebrate and invertebrate genomes.

2.1.1.1 Regulation of DNA Methylation Status by Key Enzymes

Regulation of DNA methylation is mediated by several proteins found to be responsible to establish, maintain, read, and erase methylation patterns. DNA methylation patterns are established by the activity of opposite reactions mediated by the methylation and demethylation machineries, which have been mostly investigated and well characterized in mammals, but those seem to be conserved across the vertebrate-invertebrate boundary [22]. DNA-cytosine-5-methyltransferases (DNMTs) act as "writers" and are the enzymes responsible for establishing and maintaining the methylation patterns [23]. DNMTs can be classified depending on their function as de novo methylases that add new methyl groups to previously unmethylated DNA, or maintenance methylases that maintain methylation patterns during cell division, contributing to genome stability. Five mammalian DNMTs have been identified and characterized: DNMT1, DNMT2, DNMT3a, DNMT3b, and DNMT3L (reviewed by Bestor [24]). DNMT1 is the predominant enzyme in somatic tissues, and it has been demonstrated to have a preference for hemimethylated DNA (where only one of the two strands is methylated), being critical for the maintenance of methylation patterns during the replication of DNA [25, 26]. DNMT1 is then commonly referred to as

Table 2.1 A simplified overview of key molecular epigenetic mechanisms and their role on gene expression via reversible modification of key genomic features. Please refer to the text for detailed explanations.

Molecular epigenetic mechanism	Writer	Reader	Eraser	Reversible mark	Relationship between epimarks and transcription
DNA methylation (transcriptional)	Dnmt 1 (maintenance) Dnmt3 (de novo)	Methyl-binding domain (MBD) proteins	Passive loss Active loss via TET enzymes Other implicated enzymes: apolipoprotein B mRNA editing complex (APOBEC), base excision repair (BEC), etc.	Cysteine methylation Cysteine-hydroxy methylation	Vertebrates Active transcription related to: Unmethylated enhancers and promoters (especially if they are CpG-rich), first intron Methylated gene bodies
Histone modification (transcriptional)	Histone methyl transferases (HMTs) Histone acetylases (HATs) Protein kinases (PTK)	Chromo domain proteins (methylation) Tudor domain proteins (methylation) etc. Bromo domain proteins (acetylation) etc. BRCT domain of MDC1 (phosphorylation) etc.	Histone demethylases (KDMs) Histone deacetylases (HDACS) Protein phosphatases (PP)	Posttranslational modification (PTM) such as methylation, acetylation, phosphorylation of core histones H2A, H2B, H3, and H4 residues, especially N-terminal lysins (K)	Depends on histone-specific modification. Histone acetylation and phosphorylation generally promote euchromatin states to stimulate transcription. Examples: H3K27ac Histone methylation can promote both eu- or heterochromatin states depending on the modified residues and thus stimulate or inhibit transcription. Examples: H3K4me2 and 3 (+) H3K9me2 and 3 (−) Open and inaccessible chromatin states can exist through co-existing (+/−) marks -> bivalent or poised chromatin

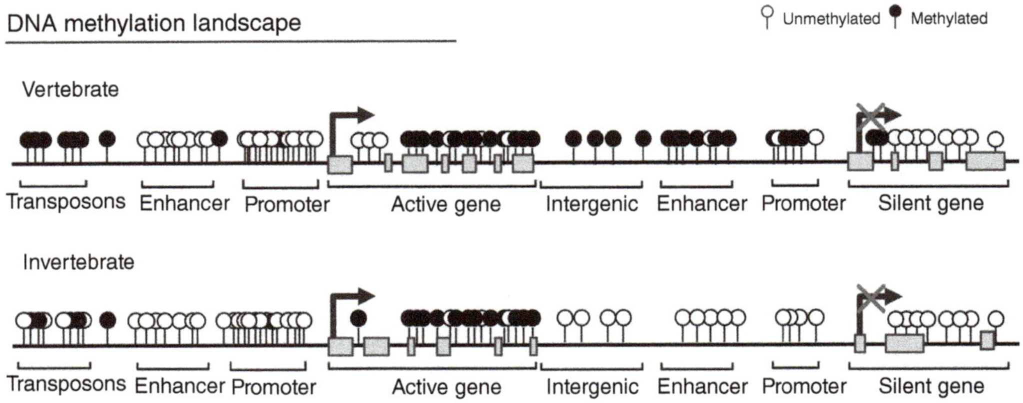

Figure 2.1 DNA methylation landscape of vertebrate and invertebrate genomes. The genomic regions represented include transposon enhancers, promoters, and genic and intergenic regions. The most common DNA methylation states are shown as examples.

the maintenance DNA methyltransferase. In contrast, DNMT3a and DNMT3b have been found to function preferentially as de novo methyltransferases [27, 28], and it has been observed that both have roles in the methylation of previously unmethylated DNA as well as demethylation [29]. In contrast, DNMT3L does not possess DNA methyltransferase activity, but seems to cooperate with DNMT3 methyltransferase family members to carry out de novo methylation of maternally imprinted genes in mouse oocytes [30]. Dnmt3 family members show gene-specific and germline-specific patterns of expression. An interaction of DNMT3A and DNMT3L in the de novo methylation in the male germline has been reported, whereas Dnmt3b is involved in the maintenance of DNA methylation in spermatogonia [31]. Despite earlier studies suggesting that DNMT2 is not essential for de novo methylation [32], other studies have demonstrated that DNMT2 also possesses methyltransferase activity [33, 34].

DNA demethylation can be attained passively or actively by two different mechanisms. Passive demethylation occurs through cell division and the absence of DNA methylation maintenance [35–39]. On the other hand, active demethylation takes place by the action of enzymes that act as “erasers” and play key roles in establishing methylation patterns. These enzymes are responsible for performing enzymatic reactions to remove or modify 5mC, oxidizing it to 5-hydroxymethylcytosine (5hmC), 5-formylcytosine (5fC), and 5-carboxylcytosine (5caC). The ten-eleven translocation (TET) family of dioxygenases is capable of recognizing CpG dinucleotides and binding to 5mC or its derivatives to perform oxidation reactions. Mammalian TETs are expressed in primordial germ cells, oocytes, and zygotes, but also play an important role in DNA methylation during early development and multiple adult somatic tissues, for example, neuronal cells [40]. Moreover, other enzymes have been identified and found in many animal species that are capable of producing deamination reactions and repair mechanisms, which in turn contribute actively to DNA demethylation [39–41]. These include thymine DNA glycosylases (TDG), activation-induced cytosine deaminases (AICDA), the apolipoprotein B mRNA editing complex (APOBEC), single-strand-selective monofunctional uracil-DNA glycosylase 1 (SMUG1), Nei-Like DNA Glycosylase 1 (NEIL1), methyl-CpG-binding protein 4 (MBD4), and base excision repair (BER) enzymes.

2.1.1.2 Methylation Changes Translated into Functional States in the Genome

Changes in methylation patterns are known to be linked to key events occurring during development and are associated with tissue-specific activation/repression of transcription of genes crucially involved in different biological functions. In vertebrates, promoter methylation has been often linked to transcription repression, while gene body methylation usually appears to be related to transcriptional activation [42]. For many years, DNA methylation was thought to be inversely correlated with transcription levels when located in gene promoters. However, this inverse correlation has more recently been demonstrated in the first intron regions also [43]. Therefore, traditionally DNA methylation has been associated with gene silencing. In this regard, some CGIs located in the promoter of key genes become methylated during development and the associated gene becomes silenced in a stable way, demonstrating that DNA methylation acts somewhat like a system of cellular memory [44]. However, a significant number of CGIs located in the promoter of genes have a tissue-restricted expression and can remain unmethylated even if the associated gene is silent. Interestingly, DNA methylation may also act context-dependently to stimulate transcription [45], which denotes the complexity of understanding the interplay between DNA methylation and transcription. In this regard, it was discovered that in the human methylome areas with low CpG concentration located in the vicinity of CGIs, also known as CGI *shores*, mediated tissue-dependent expression is reported to be more dynamic than CGI itself [46]. Several studies have demonstrated the existence of other human methylome domains including the so-called *shelves*, *open sea*, *valleys*, *canyons*, and *ravines*, some associated with high expression and others found to be highly dynamic [46]. In the same manner, evidence suggests that DNA methylation may also play other roles besides transcriptional repression in invertebrate genomes [47].

Repression of transcription can be achieved by the action of a conserved family of proteins, the methyl-CpG-binding domain (MBD) proteins, also known as "readers," that selectively bind to methylated CpG dinucleotides. From their discovery, MBD proteins have been associated with the translation of regulatory information provided by DNA methylation into functional states [48]. A major role of MBD proteins is to coordinate the crosstalk between DNA methylation, histone modifications, and chromatin organization to regulate proper transcriptional programs [49, 50]. So far, 11 proteins have been identified that contain an MBD domain, with the methyl-CpG-binding protein 2 (MeCP2) being the first to be discovered and subsequently reviewed by Du et al. [51]. It is possible that during evolution, MBPs have emerged to allow for transcriptional control by DNA methylation-mediated silencing by protecting DNA against mutation, thus promoting the appearance of diverse MBPs in protovertebrate genomes [52]. Although some MBD proteins seem to have lost their MBD, and consequently their methyl-binding function, several MBD protein homologs and orthologues have been identified in vertebrate and invertebrate genomes, with a putative ancestor in the MBD2/3 protein encoded by a single gene in invertebrate genomes [52].

A recent comparative review on epigenetic mark dynamics and animal physiology showed that animals have a diverse array of strategies to epigenetically modify physiological responses [6]. Therefore, despite the conserved structure and function of the main enzymes involved in the establishment and interpretation of the methylation patterns in plants, fungi, and animals, the exact mechanisms by which locus- and cell-specific DNA methylation patterns are established are divergent between taxa [53]. Current evidence highlights the complex role of DNA methylation in both repressing and activating gene transcription, but requires further investigations from a comparative point of view.

2.1.2 Chromatin Remodeling Through Histone Modifications

Chromatin is a dynamic structure that allows packing of the genome into the nucleus at the same time that regulates gene transcription. DNA compaction is achieved by the presence of nucleosomes, key subunits composed of wrapped DNA surrounding a histone octamer made up of a set of four histone protein (H) pairs: H2A, H2B, H3, and H4, which are linked along the DNA strand by the action of the H1 histone. Although less studied, histone variants and posttranslational modifications are also key epigenetic events capable of performing chromatin remodeling. These changes in chromatin compaction (conversion of accessible open euchromatin to tightly packed heterochromatin associated with gene silencing and vice versa) result in changes in DNA accessibility to transcription factors, RNA polymerase II, and other DNA-binding proteins that ultimately regulate transcription.

Histone modifications occur on the aminoterminal region (also called "tail") by histone-modifying enzymes, altering the overall arrangement of the chromatin. The most common histone tail modifications (Figure 2.2a) are acetylation, methylation, phosphorylation, and ubiquitylation, and are often located in various lysine or arginine residues, among others [54].

The best-studied modifications are acetylation and methylation, both of which were discovered by Allfrey et al. [55]. These modifications change histone-DNA and histone-histone interactions and serve as a docking site for protein effectors, the "readers" [56]. The addition or removal of chemical groups into the histone tails is performed by different "writers" and "erasers": histone acetyltransferases (HATs) and histone deacetylases (HDACs) for acetyl groups; histone methyltransferases (HMTs) and histone demethylases for methyl groups; phosphatases and kinases for phosphate groups; and E1, E2, and E3 ligases and deubiquitinating enzymes (DUBs) for ubiquitin groups [57–59]. As seen in Figure 2.2b, acetylation and phosphorylation act to incorporate a negative charge that is associated with a more relaxed and open form of chromatin and an increase in DNA accessibility promoting transcription [57, 58]. Although histone methylation has been associated with promoting heterochromatin remodeling and gene silencing, as is the case of methylation of lysine 9 (K9) of histone H3, it has also been found that methylation of lysine 4 (K4) of histone H3 is associated with transcriptional activation [58]. Comparative studies found that although histone variants are mostly universal for all eukaryotes [60], sufficient plasticity exists across the vertebrate-invertebrate evolutionary boundary, particularly for histone lysine methylation [61]. Although some histone variants appear to be unique to fish [62], conservation of histone modifications has also been observed in insects [63], suggesting a conserved and integrative role of these modifications across the animal kingdom. Posttranslational modifications are also mostly conserved among vertebrates, as demonstrated in zebrafish and various bivalve and annelid species [64, 65]. An interesting aspect to highlight is the existence of crosstalk between DNA methylation and histone modifications, for example, as seen by the fact that ubiquitin marks are involved in the regulation of DNA methylation [59]

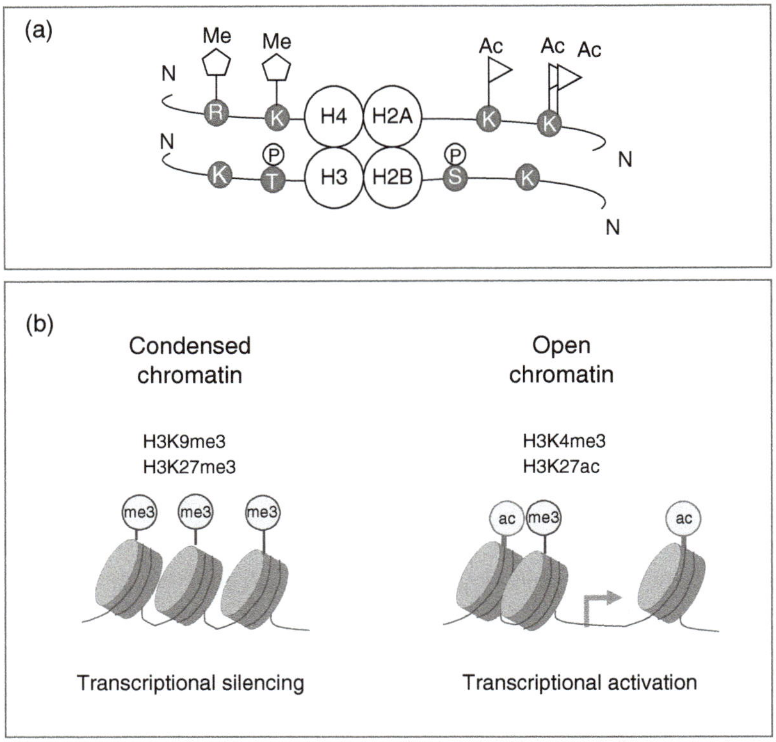

Figure 2.2 Histone modifications (a) and simplified diagram of chromatin state and DNA accessibility linked to transcription (b). (*See colour plate section for colour representation of this figure.*)

or that H3K9 histone methylation can increase the likelihood of DNA cytosine methylation and vice versa [66]. In addition, the interaction between histone "readers" and "writers/erasers" is frequently found [67], which further highlights the complexity and stringent regulation of the epigenetic machinery.

In summary, Figure 2.3 shows an overview of the key players involved in DNA methylation and histone modifications.

Function	Writers	Erasers	Readers
DNA methylation	• DNA methyltransferases (Dnmts) Dnmt1s Maintenance methylation Dnmt3s *de novo* methylation	Active demethylation by • Ten-eleven translocation Dioxygenases (Tets 1–4) • Apolipoprotein B mRNA editing complex (APOBEC), base excision repair (BER), etc Passive demethylation by dilution through cell division	Methyl binding domain (Mbd) proteins (Mecp2, MBD1, Mbd2)
Histone modifications	Histone methyltransferases (HMTs) Histone acetyltransferases (HATs) Protein kinases (PTKs)	Chromo-domain proteins etc. Bromo-domain proteins etc. BRCT-domain of MDC1 etc.	Histone demethylases (KMDs) Histone deacetylases (HDACs) Protein phosphatases (PP)

Figure 2.3 Key players that contribute to create, read, and erase specific epigenetic marks.

2.2 Transcriptional Epigenetic Mechanisms in Aquatic Species

Among aquatic species, the focus of this chapter is placed on teleost (bony) fish, the largest vertebrate group characterized by a high degree of phenotypic variation, and aquatic invertebrates including mollusks, arthropods, and sponges.

2.2.1 Teleost Fish

The unique evolutionary history of teleost- and lineage-specific genome duplications and differential retention of duplicated genetic material are generally considered to have provided the molecular substrates for such diversity. It is thus not surprising that the molecular epigenetic toolbox has equally experienced a comparatively high degree of diversification [6, 41]. While epigenetic mechanisms have, or are beginning to be characterized in, only a small number of fish species (Table 2.2), we herein summarize emerging evidence of epigenetic diversity in both laboratory model fish species such as zebrafish (*Danio rerio*) and Japanese medaka (*Oryzas latipes*), as well as relevant aquaculture species, especially the salmonids rainbow trout (*Oncorhynchus mykiss*) and Atlantic salmon (*Salmo salar*), but also the European sea bass (*Dicentrarchus labrax*), the Nile tilapia (*Oreochromis niloticus*), the grass carp (*Ctenopharyngodon idella*), the common carp (*Cyprinus carpio*), and the Senegalese sole (*Solea senegalensis*).

An overview of the evolutionary history of "writer" (*dnmt*) and "eraser" (*tet*) genes involved in DNA methylation dynamics in key teleost fish species has been described in detail by Best and collaborators [41]. Briefly, a different number of mammalian paralogues of DNA methylation "writers," the maintenance DNA methyltransferase 1 (*dnmt1*) and the de novo DNA methyltransferase 3 (*dnmt3a* and *dnmt3b*), and "erasers," the ten-eleven translocases 1 to 4 (*tet1, tet2, tet3,* and *tet4*), are found in fish genomes due to teleost and/or lineage-specific whole genome duplication and locus-specific rearrangements [41, 83, 84]. Context-dependent regulation of *dnmt* gene paralogue transcripts, as well as global DNA methylation and genome-wide or locus-specific differentially methylated regions in response to biotic and abiotic factors, have been well described in both laboratory models such as zebrafish and aquaculture species such as rainbow trout [41]. Together, these findings point to a functional role of DNA methylation dynamics in intragenerational phenotypic plasticity in fish. Direct evidence for functional roles of *dnmt* genes in mediating phenotypic plasticity remains limited in fishes to date, but CRISPR/Cas9-based studies in genetically tractable laboratory model fish species, but also in aquatic species, are emerging. In zebrafish, *dnmt3aa* and *dnmt3ab* paralogues have been shown to additively mediate developmental thermal plasticity in offspring [85], although the specificity and mechanistic basis for this effect remain to be elucidated. In Nile tilapia, expression of *dntm3aa* in gonadal tissues, but not the *dnmt3ab* paralogue, contributes to female and male

Table 2.2 Fish species for which epigenetic mechanisms have been studied.

Species	Family	Molecular epigenetic marks studied
Zebrafish	Cyprinidae	DNA methylation [68, 69] Histone modification [70]
Japanese medaka	Adrianichthyidae	DNA methylation [71] Histone modification [72]
Rainbow trout	Salmonidae	DNA methylation [73] Histone modification [73]
Atlantic salmon	Salmonidae	DNA methylation [74] Histone modification [75]
Grass carp/Common Carp	Cyprinidae	DNA methylation [76] Histone modification [77]
Tilapia	Cichlidae	DNA methylation [78] Histone modification [79]
European sea bass	Moronidae	DNA methylation [80] Histone modification [81]
Senegalese sole	Soleidae	DNA methylation [82]

gametogenesis via direct action on gonadal DNA methylation [86]. Evidence suggests that teleost-specific diversification of DNA methylation machinery accompanied genome duplication and paralogue retention events [41, 84]. Interestingly, in rainbow trout, it is suggested that *dnmt* gene paralogues might play a sex-specific subfunctionalization role in gametogenesis [84]. Similarly, evidence also suggests increased retention of critical enzymes involved in demethylation processes, notably seven *tet* and four thymine DNA glycosylase (*tdg*) in salmonids [41, 84]. As reported for *dnmt* paralogues, differential expression of *tet* and *tdg* paralogues was found to be largely concordant across ontogeny, with a general induction around the time of organ differentiation (Vernier stages; [22, 23]), suggesting a global role in mediating DNA turnover during tissue specification. Again, as reported for *dnmt* gene paralogues, *tet* and *tdg* gene paralogues revealed sex- and season-specific regulation when comparing male and female gonads [84]. In addition to (sex-specific) developmental and seasonal changes, the expression of *dnmt* gene paralogues has recently been investigated in heterozygous isogenic trout lines characterized by differential sensitivity to acute developmental temperature. This study revealed a moderate but complex interplay between genetic background and temperature on the differential regulation of *dnmt3* paralogues, which was furthermore associated with highly lineage-specific DMRs [87].

With regard to inter- and transgenerational inheritance, orthologues of the mammalian *DNMT3L* gene, which in eutherian mammals function to ensure epigenomic imprinting, are missing in teleost fishes [88]. Although parental imprinting has been found to be essential in mammals [89, 90], it remains unclear if it exists in other organisms such as teleost fish, and has thus been subject to debate [91]. Indeed, increasing evidence suggests a variety in DNA methylation dynamics between generations, even between the few model teleost species investigated to date. For example, it has been shown in zebrafish that offspring somatic cells [68, 92], and also developing germ cells [93, 94], largely inherit the paternal DNA methylation patterns. Conversely, DNA methylation dynamics in another teleost model species, the Japanese medaka, for example, have been reported to follow a pattern similar to mammalians, characterized by demethylation and erasure of paternal and maternal DNA methylation in the zygote followed by differential de novo methylation in female and male primordial germ cells during early sexual differentiation [71, 95]. Regardless of the distinct DNA methylation dynamics during early development, a recent investigation has concluded that methylome inheritance and enhancer dememorization (full methylation of adult enhancers) have a key role in safeguarding embryonic programming [96]. Resetting adult and gametic epigenetic memories to a ground state prevents precocious activation of adult enhancers in early embryos. Interestingly, and contrary to adult enhancers, which are CG-rich and are bound by TFs with motifs containing CGs, embryonic enhancers are mostly CG-poor and are less affected by DNA methylation. Altogether, differences in sensitivity to DNA methylation between adult and embryonic enhancers, coupled with the inheritance of global DNA methylome, generate an epigenetic gate that separates embryonic and adult programs, essential to ensure ordered spatiotemporal gene expression during early development [96]. Thus, these findings reconcile the reported difference in DNA methylation dynamics during the parental-to-embryonic transition and reveal that, albeit through distinct mechanisms, both serve to reset inherited DNA methylation marks to "ground" states [97]. While studies in genetically tractable fish laboratory models clearly indicate differences in DNA methylation dynamics between generations, such dynamics remain to date largely unknown or only partially understood in aquaculture species such as salmonid rainbow trout or Atlantic salmon [84]. It will thus be important to determine basic DNA methylation profiles in somatic and (primordial) germ cells within an aquaculture species' lifecycle and between generations.

Histone modifications, while evolutionarily conserved [6], are comparatively less studied than DNA methylation or non-coding RNAs in aquatic organisms in general and aquaculture species in particular. In teleost fishes, genome retention (paralogues) of key histone-modifying enzymes attests to a conserved histone modification via acetylation/deacetylation, methylation/demethylation, and phosphorylation/dephosphorylation in zebrafish, rainbow trout, and Atlantic salmon gene expression regulation [41]. Profiling of histone modification in relation to chromatin states and gene expression changes in teleost fishes are best studied in classically, genetically tractable lab model species. In zebrafish, histone modifications have been shown to play critical developmental roles, controlling spatiotemporal gene expression and tissue differentiation [70, 98, 99]. Importantly, environmental factors have been linked to mediating zebrafish plasticity through histone modifications, exemplified by the demonstration that pharmacologically inhibited HDAC-1- and HDAC-2-dependent histone deacetylations globally affect physiological acclimation processes such as stroke volume changes in response to cold temperature [100]. While more detailed studies regarding the genomic location of affected marks are clearly warranted, such studies demonstrate that histone modifications may mediate intergenerational effects of the environment on fish physiology. In aquaculture species, evidence for histone modification-dependent physiological responses to abiotic and biotic factors relevant to aquaculture is equally emerging. For example, global and gene-specific responses to dietary modulation have been demonstrated in rainbow trout [73]. Specifically, the liver of fed compared to fasted trout exhibited increased global H3K9me3 marks, while a high protein diet resulted in increased H3K9ac/H3 abundance compared to both fasted and high carbohydrate diet-fed fish. Regarding specific gene loci, the H3K36me3 marks were generally more enriched at putative gluconeogenic gene TSS in fasted and high protein-diet fed trout liver compared to high carbohydrate diet-fed trout liver.

2.2.2 Aquatic Invertebrates

As far as we know, transcriptional epigenetic mechanisms in aquatic invertebrates have been studied in a few species from different phyla including Arthropoda, Cnidaria, Ctenophora, Echinodermata, Mollusca, and Porifera (Table 2.3). At present, it is clear that DNA methylation in aquatic invertebrates plays both a role in stochastic variation and targeted regulation [102]. It has been postulated that limited DNA methylation in invertebrates may allow for large phenotypic plasticity. At present, less than a dozen aquatic invertebrates have genome-wide methylation data available, making it difficult to draw conclusions on the functional role of DNA methylation throughout this group. Nevertheless, key observations in different aquatic invertebrate species seem to suggest that the lack of germline methylation in gene bodies allows for more phenotypic plasticity as it facilitates random genetic variation and thus may increase the adaptive potential in a population [103]. This is particularly important for aquatic invertebrates that tend to live in unpredictable habitats and experience a wide range of stressors, including anthropogenic pollution [102]. Functional analysis in oysters, for example, has indicated that genes with low methylation are involved in tissue-specific expression or inducible expression. In contrast, the heavy germline methylation of conserved or "housekeeping genes/critically expressed genes" would then protect these genes from genetic variation and mutations, given their essential role [21, 103]. More specifically, the lack of methylation facilitates transcriptional opportunities, including access to transcription start sites [103]. Research in insects has indicated that DNA methylation may play an additional role in sparsely methylated gene bodies that are linked to alternative splicing [104]. In particular, DNA methylation of exons may positively influence exon inclusion during transcription through interactions with DNA-binding proteins and subsequent effects on DNA polymerase II [104, 105]. Both in oysters and waterfleas, similar results indicate that DNA methylation in sparsely methylated genes may impact alternative splicing

Table 2.3 Invertebrate species for which epigenetic mechanisms have been studied.

Species	Phylum	Epigenetics
Daphnia sp.	Arthropoda	DNA methylation[a], histone modification
Crassostrea gigas	Mollusca	DNA methylation[a]
Chlamys farreri	Mollusca	DNA methylation
Donax trunculus	Mollusca	DNA methylation
Nematostella vectensis	Cnidaria	DNA methylation[a]
Stylphora pistillata	Cnidaria	DNA methylation[a]
Mnemiopsis leidyi	Ctenophora	DNA methylation[a]
Apostichopus japonicus	Echinodermata	DNA methylation[a]
Amphimedon queenslandica	Porifera	DNA methylation[a]
Sycon ciliatum	Porifera	DNA methylation[a]
Artemia franciscana	Arthropoda	DNA methylation, histone modification
Gammarus fossarum	Arthropoda	DNA methylation
Cantareus aspersus	Mollusca	DNA methylation
Mytilus sp.,	Mollusca	DNA methylation, histone modification
Xenostrobus securis	Mollusca	DNA methylation
Physella acuta	Mollusca	DNA methylation
Octopus vulgaris	Mollusca	DNA methylation

[a] Denotes the use of sequencing technologies.
Source: Adapted from Šrut [101].

and splice variants [103, 106]. Recently, high levels of DNA methylation have been reported in sponges, challenging the previously conceived notions of the unique aspects of hypermethylation in vertebrate genomes [107]. For these species, global methylC levels have been reported to range between 50 and 80%, contrasting the low methylation levels in other invertebrates (0–10%). The genomes of these sponges seem to show increasingly striking similarities with methylation patterns in vertebrate genomes, with the highly methylated *Amphimedon queenslandica* showing the highest similarities [107]. This seems to suggest a gradual evolutionary transition from sparsely methylated genomes to hypermethylated genomes [107]. As genomic technologies become more readily available, we can expect to see further epigenomic evidence in other aquatic invertebrates that will contribute to elucidating the roles of DNA methylation in invertebrates. Histone modifications have been reported in *Daphnia* to be present in embryos, blastula, and gastrula cells, but specific modifications were reduced in oocytes and nurse cells [108]. Additionally, differences in histone modifications were observed in males versus females, suggesting a role for epigenetics in the parthenogenic and sexual reproduction cycle of daphnia [109]. Similarly, specific histone modifications were also reported in sperm cells of the Japanese mantis shrimp (*Oratosquilla oratoria*) [110] and during early development in the Pacific oyster (*Crassostrea gigas*) [111].

At the same time, an increasing number of laboratory studies have reported changes in DNA methylation, and to a lesser extent histone modifications, in response to changing environmental and ecological conditions [8, 101]. Causal links between changes in methylation and increased phenotypic plasticity remain to be established, particularly the underlying mechanisms on how methylation affects plasticity through subsequent generations remain unclear. Nevertheless, a recent number of experimental studies have observed transgenerational inheritance of methylation after exposure to environmental stressors. Effects have been observed in a wide range of invertebrates including mollusks, crustaceans, and nematodes [112]. Depending on the reproduction cycle of the organisms, this implies methylation transfer to at least either F2 or F3 generations to exclude confounding effects of prenatal exposure during reproduction. This has often not been taken into consideration in the past, making it difficult to identify true transgenerational effects in studies only focusing on F1 or F2 generations [112]. As such, many reported transgenerational effects are only of limited duration and refer to intergenerational inheritance rather than transgenerational inheritance [113]. Overall, it is clear that epigenetics plays a crucial role in aquatic invertebrates and their interactions with environmental stress, albeit the exact mechanisms of response leading to inheritance, phenotypic plasticity, or even adaptation remain to be unraveled.

2.3 Modulation of Biological Functions by Transcriptional Epigenetic Mechanisms in Aquaculture Species of Interest

A key aspect within aquaculture is the selection of phenotypic characteristics resulting in optimal product yield. This is often a combination not only of traits involving development, growth reproduction, immunology, and disease leading to a high yield in terms of biomass, but also of traits that lead to healthy organisms that are less susceptible to diseases. In this regard, recent advances within the field of epigenetics have indicated that a deeper understanding of epigenetic mechanisms and their role in physiological traits can provide not only improvements on broodstock selection but also a measurable link between environment and phenotype within aquaculture practices (Figure 2.4). In this regard, epigenetics has been proposed to have the potential to promote more sustainable aquaculture while still maintaining economic benefits [64]. In the following sections, we briefly introduce and give some examples of the involvement of epigenetics in key

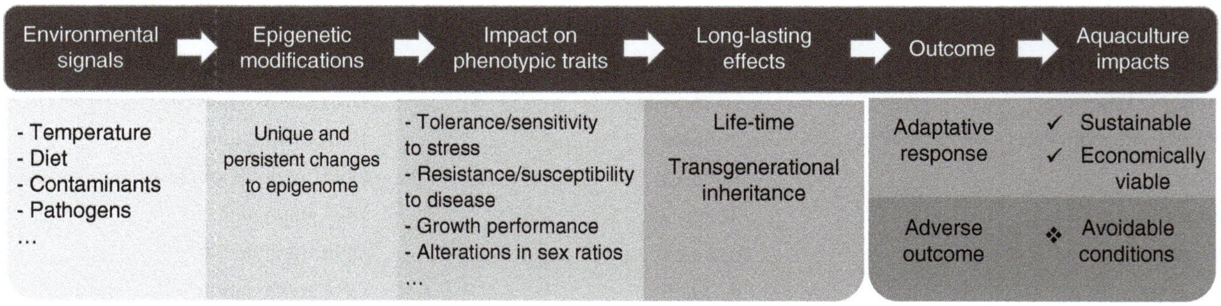

Figure 2.4 Epigenetic modifications linking environmental cues with physiological traits can be key for the production of high-quality aquaculture products.

biological functions (i.e., development, growth, reproduction, nutrition and metabolism, and disease) of importance for the production of high-quality aquaculture products.

2.3.1 Growth and Development

Within the production cycle of farmed aquatic organisms, larvae development is a particularly crucial life stage in which animals are both sensitive and susceptible to environmental manipulations. Indeed, it has been reported that favorable conditions during fish embryogenesis and larval development (e.g., optimal temperature) may have long-lasting positive effects related to sex ratio, egg size, growth, adult body size, and lifespan [114]. One of the first studies to show a link between environmental factors and epigenetics in an aquaculture fish species demonstrated that during development, the temperature was capable of modifying DNA methylation levels of a specific gene involved in the sexual differentiation of European sea bass [115]. In particular, exposures to high temperatures during early larval development caused long-term changes in DNA methylation of the gonadal aromatase (*cyp19a1a*) gene promoter that was associated with changes in the proportion of sexes [115]. Similarly, high-temperature regimes during sensitive development windows in the half-smooth tongue sole (*Cynoglossus semilaevis*) lead to pseudomales mediated through epigenetic inheritance of environmentally induced sex reversal [116]. For Senegalese sole, exposure to higher temperatures during larval metamorphosis led to increased thermal plasticity of muscle growth and was associated with cytosine methylation of the *myog* gene promoter [82]. A link between development and growth and epigenetics was also observed when exposing Atlantic cod (*Gadus morhua*) to different light regimes. Results demonstrated that juveniles reared under continuous light grew larger than juveniles reared under natural light conditions due to increased expression of DNA methyl transferases 1 and 3a [117]. In aquaculture settings, fish and bivalves are generally reared in different environmental conditions in larvae versus adult life stages. Furthermore, each species and life stage have different sensitive windows. Therefore, knowledge of how these environmental conditions may interact with sensitive windows in development and growth processes through epigenetic changes may significantly improve animal husbandry practices.

2.3.2 Nutrition and Metabolism

Nutrition in aquaculture is a major challenge. It is not only linked to key economic drivers such as biomass and health of the animal, but also to unsustainable practices, as nutrition often includes fish oil and fish feed from capture fisheries [118]. On the other hand, challenges such as the poor utilization of dietary carbohydrate macronutrients in several fish species [119] represent a major bottleneck toward the utilization of economically and ecologically sustainable protein-sparing, carbohydrate-rich diets, and increasing efforts have been directed at potential direct and indirect programming effects across generations and in early development [120]. The understanding of epigenetic mechanisms may provide crucial insight into how animals respond to sustainable changes in diet and how these responses may affect important phenotypic traits. We here highlight recent advances made in both freshwater and marine aquaculture species, as well as lab model fish species such as zebrafish, which is considered to be a genetically tractable fish model for aquaculture research [121].

Nutrient intake during early life stages, which in fish includes both endogenous (maternally derived nutrients in the yolk) and exogenous (larval feeding), is known to influence growth, survival, brain development, stress tolerance/sensitivity, intestinal inflammation, and lipid and carbohydrate metabolism later in life [122]. For example, studies have demonstrated that early nutrition challenges with high carbohydrate diets resulted in long-term changes in genes related to digestive enzymes or carbohydrate metabolism [123, 124]. Nutrients and bioactives in food can directly influence epigenetic patterns through their interactions with the enzymatic reactions catalyzing epigenetic modifications (e.g., inhibition of enzymes and altering substrate availability) [125]. Although studies on nutritional programming in fish are still scarce, there is growing evidence of epigenetic regulation of metabolism-related genes associated with early nutrition [122]. The introduction of a fully plant-based diet at first feeding in carnivorous rainbow trout led to higher growth rate, feed intake, and feed efficiency upon later life-stage challenges, a phenotype that has been suggested to be mediated epigenetically [126, 127]. On the other hand, changes in broodstock diets and/or early nutrition challenges result in long-term, intergenerational consequences in macronutrient-dependent nutritional programming in fish that have been linked to changes in offspring gene expression of genes related to digestive enzymes, energy and carbohydrate metabolism, or lipid metabolism [123, 124, 128–130]. While these studies reveal that juvenile gene expression, as well as metabolic physiology, is sensitive to developmental nutritional stimuli, very few have investigated the role of epigenetics. Recent studies investigating detailed epigenetic and posttranscriptional epigenetic mechanisms in response to high dietary carbohydrates in adult trout uncovered additional complexity in epigenetic regulation in the complex rainbow trout genome. Indeed, among gluconeogenic enzyme coding gene paralogues in the liver, several transcripts (*pck2b, g6pcb*) show an, at least in part, epigenetically regulated atypical induction in response to high dietary carbohydrates. This regulation is in contrast to the "typical" inhibition of gluconeogenic enzyme coding gene paralogues and is believed to contribute to glucose intolerance in this species [73, 120]. Future work is thus needed to identify specific genome-wide epigenetic marks and longer-term physiological consequences and phenotypic traits which, in addition to current steady-state approaches, should also include integrative analysis such as metabolic fluxes [131]. Nutritional programming of metabolic gene expression has recently been studied in invertebrate species, most notably the whiteleg shrimp (*Litopenaeus vannamei*). Feeding restriction studies revealed that both the development stage as well as the degree of feeding restriction caused long-term effects on growth performance and metabolic gene expression. While dietary history and challenge interacted to affect several metabolic gene expression levels [132], feeding restriction provoked long-term increases in glycolytic and ATP synthesis genes, again suggesting possible long-term programming effects on gene expression [133]. Further studies in invertebrates are required to determine the potential epigenetic regulation and functional relevance to the growth phenotype.

In addition to macronutrient-dependent regulation of molecular epigenetic marks, the role of micronutrients in epigenetic regulation is equally beginning to be investigated in aquaculture fish species. For example, parental dietary selenium supplementation resulted in reduced free methionine and the methyl group donor S-adenosyl methionine (SAM), as well as induction of methionine synthase mRNA on the methionine cycle in rainbow trout swim-up fry stage [134]. Furthermore, it was demonstrated widespread differential methylation in hepatic gene-enriched signal transduction and immune function [135]. Another study has demonstrated that micronutrient supplementation quantities of several B vitamins and microminerals were sufficient to improve the growth and welfare of Atlantic salmon, which was linked, at least in part, to gene-specific DNA methylation changes [136]. In particular, DNA methylation levels on the acetyl-CoA carboxylase alpha (*acca*) promoter increased in response to dietary micronutrient supplementation in a concentration-dependent manner, suggesting that epigenetic silencing of *acca* expression is involved in the observed inhibition of downstream lipid biosynthesis activities [136].

One of the challenges associated with studying molecular epigenetic effects with relevance to aquaculture nutrition is related to the existence of complex genomes, and that comparative baseline information on paternal and maternal epigenetic mark dynamics within and across generations is lacking [6]. To overcome and mitigate this challenge, we suggest the use of established fish lab model species, such as zebrafish, with relevance to aquaculture nutrition [41, 121], which may provide useful complementary insight when investigating molecular epigenetic mechanisms involved in (nutritional) metabolic programming [137–140]. Notwithstanding, all the available information and tools on genome-wide specific analysis of epigenetic marks and/or epimutations in gametes are currently lacking, even in zebrafish, and should be a focus in future studies. It is, however, important to keep in mind that as a phylogenetically highly diverse group, dietary and metabolic phenotypes, genomic context, as well as epigenetic mark dynamics are highly divergent, and thus care needs to be exerted when making inferences.

2.3.3 Reproduction and Broodstock Selection

In the marine medaka, a role for histone modifications in mediating inter- and transgenerational effects of environmental factors has been demonstrated in the context of hypoxia-mediated inhibition of reproductive function [141]. Indeed, fish exposed to hypoxia across their lifecycle exhibited consistent intra- (F0), multi- (F1), and transgenerational (F2) impairment of sperm quality and quantity [141]. These effects were linked to consistent transgenerational gene expression changes in testes of the ancestral hypoxia-lineage animals, which notably included the induction of euchromatic histone-lysine *N*-methyltransferase 2 (*ehmt2*) with a vital role in DNA methylation and chromatin modification during germ cell development and sperm maturation. While the F2 sperm was globally hypermethylated, the promoter region of the *ehmt2* was hypomethylated, providing a plausible mechanistic epigenetic basis for the observed paternally mediated transgenerational reduction of sperm development in the hypoxic lineage. Further evidence supporting this mode of action is the finding that this transgenerational upregulation of *ehmt2* in testes of the ancestrally hypoxia-exposed lineage was associated with the elevation in histone H3 lysine 9 dimethylation (H3K9me2), an important epigenetic marker for gene silencing. Thus, it appears likely that, at least in part, ancestral hypoxia exposure affects male reproductive function via sperm methylation of genes that include histone-modifying enzymes to regulate spermatogenesis in unexposed F1 and F2 offspring.

The development of methods to evaluate gamete quality also has great potential for future aquaculture applications. For example, broodstock nutrition and epigenetics have been identified as important factors for enhancing sperm production [142]. Moreover, it is known that maternal nutrition mediates nutritional programming by altering egg composition [122], which has consequences in fish offspring growth rates (see Section 2.3.2). A recent finding that carp oocytes aged in vitro exhibit time-dependent increases in H4K12ac, raising the interesting possibility of developing epigenetic markers of egg quality in the context of aquaculture broodstock [77]. It has also been demonstrated in shellfish that parental exposure to low pH during crucial life stages can lead to better offspring in terms of fitness and growth rate [143, 144], suggesting that epigenetic inheritance might also be involved in invertebrates. Selective breeding can play a major role to improve key species' characteristics such as growth and survival rates. While currently breeding efforts have been almost exclusively focused on genetic variation, increasing evidence emerges that epigenetic variation may also play an important role in selective breeding and could have the potential to contribute to significant improvements in growth and survival performance [68, 145].

2.3.4 Stress and Immune Responses

In aquaculture, diseases are a major economic and sustainability challenge leading to reduced production due to mortality. Research in Atlantic salmon has indicated that early life stress can contribute to better stress resilience or improved immune responses in the adult stages [146, 147]. More specifically, changes in promoter and gene body methylation and associated changes in the expression of immune-related genes after chronic stress resulted in a suppressed inflammatory immune response when exposed to a model pathogen later in life [146]. Other studies have provided evidence of epigenetic modifications in response to infection. Trimethylation of H3K4 (H3K4me3) in gene promoters of over 200 immune-related genes was observed in adult zebrafish after virus infection [148]. More evidence was provided where dynamic changes in DNA methylation occurred in the guppy (*Poecilia reticulata*) skin when infected with its ectoparasitic monogenean *Gyrodactylus turnbulli* [149]. All these supporting lines of evidence demonstrate the interrelationship between the epigenome and the immune system in fish.

In contrast to vertebrates, invertebrates are a heterogeneous group in terms of immune regulation and immune responses and lack an adaptive immune system [150]. This has raised questions on how invertebrates are able to cope with pathogens, and research indicates that invertebrates used a variety of different strategies, some of which are epigenetically regulated [150, 151]. Indeed, Norouzitallab and collaborators [152] suggested a role for epigenetics in trained immunity in the brine shrimp (*Artemia* sp.): heat shock exposure during early life stages resulted in increased tolerance to lethal heat stress and resistance against the pathogen *Vibrio campbellii*. This phenotype was furthermore inherited for three subsequent generations and was associated with changes in DNA methylation and increased acetylation of histones H3 and H4 [152]. Green et al. [153] suggest a potential epigenetic mechanism for mediating increased tolerance to the Ostreid herpesvirus in oysters by the increased expression of an antiviral gene in larvae of parents exposed to poly(I:C), a synthetic dsRNA. For shrimps, different studies have also observed a role for miRNAs in immunoregulation and immune responses [151]. To summarize, epigenetic mechanisms have been shown to influence disease susceptibility and drive increased immune responses, but studies are scarce and further investigations are required.

2.4 Conclusions and Perspectives

The available literature reviewed here denotes how the study of different epigenetic mechanisms regulating gene transcription is unbalanced. While studies on DNA methylation are dominant, how histone modifications mediate environmental influences in phenotypical traits, especially for nonmammalian species, are less studied and require further investigation. We have also highlighted the complexity of epigenetic regulation of gene transcription, which may contribute to the lack of correlation between epigenetic modifications and changes in gene expression due to environmental factors [6]. Moreover, although epigenetic mechanisms are often studied individually, evidence highlights the existence of crosstalk between the different mechanisms. In this regard, with the use of technological advances, we highlight the need for global multi-omic studies at higher resolution (including single-cell studies) when trying to link epigenetics to gene expression profiles at a global scale, which needs to include posttranscriptional and noncoding RNA mechanisms. On the other hand, it has been highlighted that comparative epigenetic studies are needed to reveal and account for the existing differences in how phylogenetically related or distinct animal taxa respond to environmental challenges via different epigenetic differences [154]. Another level of complexity is that under aquaculture husbandry practices (some of them even carried in open waters with a lower possibility to control environmental conditions), multiple relevant factors (stocking density, nutrients, hypoxia, temperature, etc.) can be simultaneously acting to modulate epigenetic mechanisms. This level of complexity is often disregarded since studies usually focus on a single environmental factor at a time. Therefore, to better understand the role of epigenetics mediating the appearance of different phenotypes, it is necessary to use holistic approaches that simultaneously study multiple environmental cues and epigenetic mechanisms to link changes in gene transcription to phenotypic outcomes using distinct animal taxa.

Advancing in the knowledge of how the environment interacts with the epigenetic machinery to modulate responses and phenotypic plasticity is key for aquaculture practices. Herein, we have provided examples of how environmental signals are capable of causing epigenetic modifications that impact phenotypic traits related to metabolism, growth, development, reproduction, and immune response in many different aquaculture species of interest. Moreover, we have reviewed several lines of evidence of how, in some cases, long-lasting effects on these phenotypic traits are observed within a lifetime or even through several generations. This opens the door to developing aquaculture practices that select environmental conditions that provoke unique and persistent changes to the epigenome, or even perform "epigenetic selection" (in addition to traditional genetic selection) that, for example, favor higher growth rates, tolerance to stress, resistance to disease, or monosex culture, not only within the lifespan of the organism, but also in subsequent generations. In this context, more comparative work regarding epigenetic mark dynamics in commercial aquaculture species is clearly warranted, as even between the few studied lab model fish species, such as zebrafish and Japanese medaka, significant differences exist [6]. Recent evidence suggests that epigenetic mark dynamics are further dependent on (complex) genomic contexts [75] and domestication [78], demonstrating the need for a synthetic integration of the field of aquaculture epigenetics into traditional breeding strategies. Of note, the state of the art of the increasingly important invertebrate aquaculture sector appears to be less developed and research in comparative epigenetic mechanisms in species such as shrimp is thus clearly warranted.

We envision that the application of such future advances in aquatic species epigenetics will significantly benefit aquaculture by exploring a largely untapped mechanistic basis to make it more sustainable and economically viable.

Acknowledgments

This work was supported by Grants RyC2019-026426-I awarded to LNM and CEX2018-000794-S, both funded by MCIN/AEI/10.13039/501100011033. Grant 3G003622 was awarded by Science Foundation Flanders (FWO) to JA and Grant BOF/24J/2021/164 funded by UGent Research Fund (BOF). National Science and Engineering Research Council (NSERC) Discovery Grant 2114456-2017 awarded to Jan Alexander Mennigen (JAM).

References

1 Lande-Diner, L. and Cedar, H. (2005). Silence of the genes – mechanisms of long-term repression. *Nature Reviews Genetics* **6**: 648–654.

2 Atlasi, Y. and Stunnenberg, H.G. (2017). The interplay of epigenetic marks during stem cell differentiation and development. *Nature Reviews Genetics* **18** (11): 643–658.

3 O'Neill, C. (2015). The epigenetics of embryo development. *Animal Frontiers* **5** (1): 42–49.

4 Soriano-Tárraga, C., Jiménez-Conde, J., and Roquer, J. (2019). Epigenetics and aging. *Handbook of Nutrition, Diet, and Epigenetics* **2** (July): 1413–1433.

5 Feng, J., Fouse, S., and Fan, G. (2007). Epigenetic regulation of neural gene expression and neuronal function. *Pediatric Research* **61** (5 PART 2 SUPPL): 58–63.

6 Navarro-Martín, L., Martyniuk, C.J., and Mennigen, J.A. (2020). Comparative epigenetics in animal physiology: an emerging frontier. *Comparative Biochemistry and Physiology – Part D: Genomics and Proteomics* **36** (June): 100745. https://doi.org/10.1016/j.cbd.2020.100745.

7 Jaenisch, R. and Bird, A. (2003). Epigenetic regulation of gene expression: how the genome integrates intrinsic and environmental signals. *Nature Genetics* **33** (3): 245–254.

8 Vandegehuchte, M.B. and Janssen, C.R. (2014). Epigenetics in an ecotoxicological context. *Mutation Research, Genetic Toxicology and Environmental Mutagenesis* **764–765**: 36–45.

9 Burggren, W.W. (2015). Dynamics of epigenetic phenomena: intergenerational and intragenerational phenotype 'washout'. Hoppeler HH, editor. *Journal of Experimental Biology* **218** (1): 80–87.

10 Burggren, W.W. and Crews, D. (2014). Epigenetics in comparative biology: why we should pay attention. *Integrative and Comparative Biology* **54** (1): 7–20.

11 Bossdorf, O., Richards, C.L., and Pigliucci, M. (2008). Epigenetics for ecologists. *Ecology Letters* **11** (2): 106–115.

12 Jeremias, G., Barbosa, J., Marques, S.M. et al. (2018). Transgenerational inheritance of DNA hypomethylation in *Daphnia magna* in response to salinity stress. *Environmental Science and Technology* **52** (17): 10114–10123.

13 Norouzitallab, P., Baruah, K., Vanrompay, D., and Bossier, P. (2019). Can epigenetics translate environmental cues into phenotypes? *Science of the Total Environment* **647**: 1281–1293.

14 Rey, O., Eizaguirre, C., Angers, B. et al. (2020). Linking epigenetics and biological conservation: towards a conservation epigenetics perspective. *Functional Ecology* **34** (2): 414–427.

15 Ashe, A., Colot, V., and Oldroyd, B.P. (2021). How does epigenetics influence the course of evolution? *Philosophical Transactions of the Royal Society, B: Biological Sciences* **376** (1826): https://doi.org/10.1098/rstb.2020.0111.

16 Naylor, R.L., Hardy, R.W., Buschmann, A.H. et al. (2021). A 20-year retrospective review of global aquaculture. *Nature* **591** (7851): 551–563.

17 Head, J.A. (2014). Patterns of DNA methylation in animals: an ecotoxicological perspective. *Integrative and Comparative Biology* **54** (1): 77–86.

18 Tweedie, S., Charlton, J., Clark, V., and Bird, A. (1997). Methylation of genomes and genes at the invertebrate–vertebrate boundary. *Molecular and Cellular Biology* **17** (3): 1469–1475.

19 Glastad, K.M., Hunt, B.G., Yi, S.V., and Goodisman, M.A.D. (2011). DNA methylation in insects: on the brink of the epigenomic era. *Insect Molecular Biology* **20** (5): 553–565.

20 Glastad, K.M., Hunt, B.G., and Goodisman, M.A. (2014). Evolutionary insights into DNA methylation in insects. *Current Opinion in Insect Science* **1**: 25–30.

21 Sarda, S., Zeng, J., Hunt, B.G., and Yi, S.V. (2012). The evolution of invertebrate gene body methylation. *Molecular Biology and Evolution* **29** (8): 1907–1916.

22 Zemach, A. and Zilberman, D. (2010). Evolution of eukaryotic DNA methylation and the pursuit of safer sex. *Current Biology* **20** (17): R780–R785.

23 Vassena, R., Schramm, R.D., and Latham, K.E. (2005). Species-dependent expression patterns of DNA methyltransferase genes in mammalian oocytes and preimplantation embryos. *Molecular Reproduction and Development* **72** (4): 430–436.

24 Bestor, T.H. (2000). The DNA methyltransferases of mammals. *Human Molecular Genetics* **9**: 2395–2402.

25 Bestor, T.H. (1992). Activation of mammalian DNA methyltransferase by cleavage of a Zn binding regulatory domain. *The EMBO Journal* **11** (7): 2611–2617.

26 Yoder, J.A., Soman, N.S., Verdine, G.L., and Bestor, T.H. (1997). DNA (cytosine-5)-methyltransferases in mouse cells and tissues. Studies with a mechanism-based probe. *Journal of Molecular Biology* **270** (3): 385–395.

27 Okano, M., Bell, D.W., Haber, D.A., and Li, E. (1999). DNA methyltransferases Dnmt3a and Dnmt3b are essential for de novo methylation and mammalian development. *Cell* **99** (3): 247–257.

28 Chen, T., Ueda, Y., Xie, S., and Li, E. (2002). A novel Dnmt3a isoform produced from an alternative promoter localizes to euchromatin and its expression correlates with active de novo methylation. *Journal of Biological Chemistry* **277** (41): 38746–38754.

29 Métivier, R., Gallais, R., Tiffoche, C. et al. (2008). Cyclical DNA methylation of a transcriptionally active promoter. *Nature* **452** (7183): 45–50.

30 Hata, K., Okano, M., Lei, H., and Li, E. (2002). Dnmt3L cooperates with the Dnmt3 family of de novo DNA methyltransferases to establish maternal imprints in mice. *Development* **129**: 1983–1993.

31 La Salle, S., Mertineit, C., Taketo, T. et al. (2004). Windows for sex-specific methylation marked by DNA methyltransferase expression profiles in mouse germ cells. *Developmental Biology* **268** (2): 403–415.

32 Okano, M., Xie, S., and Li, E. (1998). Dnmt2 is not required for de novo and maintenance methylation of viral DNA in embryonic stem cells. *Nucleic Acids Research* **26** (11): 2536–2540.

33 Hermann, A., Schmitt, S., and Jeltsch, A. (2003). The human Dnmt2 has residual DNA-(cytosine-C5) methyltransferase activity. *Journal of Biological Chemistry.* **278** (34): 31717–31721.

34 Tang, L.Y., Reddy, M.N., Rasheva, V. et al. (2003). The eukaryotic DNMT2 genes encode a new class of cytosine-5 DNA methyltransferases. *Journal of Biological Chemistry* **278** (36): 33613–33616.

35 Bhutani, N., Burns, D.M., and Blau, H.M. (2011). DNA demethylation dynamics. *Cell* **146**: 866–872.

36 Inoue, A. and Zhang, Y. (2011). Replication-dependent loss of 5-hydroxymethylcytosine in mouse preimplantation embryos. *Science* **334**: 194.

37 Smith, Z.D. and Meissner, A. (2013). DNA methylation: roles in mammalian development. *Nature Reviews Genetics* **14**: 204–220.

38 Zhao, H. and Chen, T. (2013). Tet family of 5-methylcytosine dioxygenases in mammalian development. *Journal of Human Genetics* **58**: 421–427.

39 de Mendoza, A., Lister, R., and Bogdanovic, O. (2020). Evolution of DNA methylome diversity in eukaryotes. *Journal of Molecular Biology* **432** (6): 1687–1705.

40 Veland, N. and Chen, T. (2017). Mechanisms of DNA methylation and demethylation during mammalian development. In: *Handbook of Epigenetics* (ed. T. Tollefsbol), 11–24. Elsevier.

41 Best, C., Ikert, H., Kostyniuk, D.J.D.J. et al. (2018). Epigenetics in teleost fish: from molecular mechanisms to physiological phenotypes. *Comparative Biochemistry and Physiology Part – B: Biochemistry and Molecular Biology* **224**: 210–244.

42 Jones, P.A. (2012). Functions of DNA methylation: islands, start sites, gene bodies and beyond. *Nature Reviews Genetics* **13**: 484–492.

43 Anastasiadi, D., Vandeputte, M., Sánchez-Baizán, N. et al. (2018). Dynamic epimarks in sex-related genes predict gonad phenotype in the European sea bass, a fish with mixed genetic and environmental sex determination. *Epigenetics* **13** (9): 988–1011.

44 Bird, A. (2002). DNA methylation patterns and epigenetic memory. *Genes & Development* **16**: 6–21.

45 Ambrosi, C., Manzo, M., and Baubec, T. (2017). Dynamics and context-dependent roles of DNA methylation. *Journal of Molecular Biology* **429**: 1459–1475.

46 Edgar, R., Tan, P.P.C., Portales-Casamar, E., and Pavlidis, P. (2014). Meta-analysis of human methylomes reveals stably methylated sequences surrounding CpG islands associated with high gene expression. *Epigenetics & Chromatin* **7** (1): 1–3.

47 Suzuki, M.M. and Bird, A. (2008). DNA methylation landscapes: provocative insights from epigenomics. *Nature Reviews Genetics* **9** (6): 465–476.

48 Wade, P.A. (2001). Methyl CpG binding proteins: coupling chromatin architecture to gene regulation. *Oncogene* **20** (24 REV. ISS. 3): 3166–3173.

49 Bogdanović, O. and Veenstra, G.J.C. (2009). DNA methylation and methyl-CpG binding proteins: developmental requirements and function. *Chromosoma* **118** (5): 549–565.

50 Hashimoto, H., Vertino, P.M., and Cheng, X. (2010). Molecular coupling of DNA methylation and histone methylation. *Epigenomics* **2** (5): 657–669.

51 Du, Q., Luu, P.L., Stirzaker, C., and Clark, S.J. (2015). Methyl-CpG-binding domain proteins: readers of the epigenome. *Epigenomics* **7** (6): 1051–1073.

52 Hendrich, B. and Tweedie, S. (2003). The methyl-CpG binding domain and the evolving role of DNA methylation in animals. *Trends in Genetics* **19** (5): 269–277.

53 Zhong, X. (2016). Comparative epigenomics: a powerful tool to understand the evolution of DNA methylation. *New Phytologist* **210** (1): 76–80.

54 Lawrence, M., Daujat, S., and Schneider, R. (2016). Lateral thinking: how histone modifications regulate gene expression. *Trends in Genetics* **32**: 42–56.

55 Allfrey, V.G., Faulkner, R., and Mirsky, A.E. (1964). Acetylation and methylation of histones and their possible role in the regulation of RNA synthesis. *Proceedings of the National Academy of Sciences USA* **51**: 786–794.

56 Musselman, C.A., Lalonde, M.E., Côté, J., and Kutateladze, T.G. (2012). Perceiving the epigenetic landscape through histone readers. *Nature Structural & Molecular Biology* **19**: 1218–1227.

57 Bannister, A.J. and Kouzarides, T. (2011). Regulation of chromatin by histone modifications. *Cell Research* **21**: 381–395.

58 Singh, D., Nishi, K., Khambata, K., and Balasinor, N.H. (2020). Introduction to epigenetics: basic concepts and advancements in the field. In: *Epigenetics and Reproductive Health*. Academic Press.

59 Vaughan, R.M., Kupai, A., and Rothbart, S.B. (2021). Chromatin regulation through ubiquitin and ubiquitin-like histone modifications. *Trends in Biochemical Sciences* **46** (4): 258–269.

60 Talbert, P.B. and Henikoff, S. (2014). Environmental responses mediated by histone variants. *Trends in Cell Biology* **24**: 642–650.

61 Spada, F., Vincent, M., and Thompson, E.M. (2005). Plasticity of histone modifications across the invertebrate to vertebrate transition: histone H3 lysine 4 trimethylation in heterochromatin. *Chromosome Research* **13** (1): 57–72.

62 Wu, N., Yue, H.M., Chen, B., and Gui, J.F. (2009). Histone H2A has a novel variant in fish oocytes. *Biology of Reproduction* **81** (2): 275–283.

63 Hunt, B.G., Glastad, K.M., Yi, S.V., and Goodisman, M.A.D. (2013). Patterning and regulatory associations of DNA methylation are mirrored by histone modifications in insects. *Genome Biology and Evolution* **5** (3): 591–598.

64 Gavery, M.R. and Roberts, S.B. (2017). Epigenetic considerations in aquaculture. *PeerJ* **2017** (12): e4147.

65 Gibson, G., Hart, C., Pierce, R., and Lloyd, V. (2012). Ontogenetic survey of histone modifications in an annelid. *Genetics Research International.* **2012**: 1–11.

66 Fuks, F., Burgers, W.A., Brehm, A. et al. (2000). DNA methyltransferase Dnmt1 associates with histone deacetylase activity. *Nature Genetics* **24** (1): 88–91.

67 Torres, I. and Fujimori, D. (2016). Functional coupling between writers, readers and erasers of the histone code. *Current Opinion in Structural Biology* **33**: 68–75.

68 Jiang, L., Zhang, J., Wang, J.J. et al. (2013). Sperm, but not oocyte, DNA methylome is inherited by zebrafish early embryos. *Cell* **153** (4): 773–784.

69 Cavalieri, V. and Spinelli, G. (2017). Environmental epigenetics in zebrafish. *Epigenetics and Chromatin* **10**: 46. https://doi.org/10.1186/s13072-017-0154-0.

70 Cavalieri, V. (2020). Histones, their variants and post-translational modifications in zebrafish development. *Frontiers in Cell and Development Biology* **8**: 456.

71 Wang, X. and Bhandari, R. (2020). DNA methylation reprogramming in medaka fish, a promising animal model for environmental epigenetics research. *Environmental Epigenetics* **6** (1): 1–4.

72 Fukushima, H.S., Takeda, H., and Nakamura, R. (2022). Histone marks retained during epigenetic reprogramming and their roles essential for fish early development. *bioRxiv* **129**: 1983–1993.

73 Marandel, L., Lepais, O., Arbenoits, E. et al. (2016). Remodelling of the hepatic epigenetic landscape of glucose-intolerant rainbow trout (*Oncorhynchus mykiss*) by nutritional status and dietary carbohydrates. *Scientific Reports* **6**: 1–12.

74 Beemelmanns, A., Ribas, L., Anastasiadi, D. et al. (2021). DNA methylation dynamics in Atlantic salmon (*Salmo salar*) challenged with high temperature and moderate hypoxia. *Frontiers in Marine Science* **7** (January): 3.

75 Verta, J.P., Barton, H.J., Pritchard, V., and Primmer, C.R. (2021). Genetic drift dominates genome-wide regulatory evolution following an ancient whole-genome duplication in Atlantic salmon. *Genome Biology and Evolution* **13** (5): 1–13.

76 Cai, W.J., Liang, X.F., Yuan, X.C. et al. (2020). Changes of DNA methylation pattern in metabolic pathways induced by high-carbohydrate diet contribute to hyperglycemia and fat deposition in grass carp (*Ctenopharyngodon idella*). *Frontiers in Endocrinology* **11** (July): 1–18.

77 Waghmare, S.G., Samarin, A.M., Samarin, A.M. et al. (2021). Histone acetylation dynamics during in vivo and in vitro oocyte aging in common carp cyprinus carpio. *International Journal of Molecular Sciences* **22** (11): 6036.

78 Konstantinidis, I., Sætrom, P., Mjelle, R. et al. (2020). Major gene expression changes and epigenetic remodelling in Nile tilapia muscle after just one generation of domestication. *Epigenetics [Internet]* **15** (10): 1052–1067. https://doi.org/10.1080/15592294.2020.1748914.

79 Kratochwil, C.F. and Meyer, A. (2015). Mapping active promoters by ChIP-seq profiling of H3K4me3 in cichlid fish – a first step to uncover cis-regulatory elements in ecological model teleosts. *Molecular Ecology Resources* **15** (4): 761–771.

80 Anastasiadi, D., and Piferrer, F., P. (2019). Epimutations in developmental genes underlie the onset of domestication in farmed European sea bass. *Molecular Biology and Evolution* **36** (10): 2252–2264.

81 Terova, G., Díaz, N., Rimoldi, S. et al. (2016). Effects of sodium butyrate treatment on histone modifications and the expression of genes related to epigenetic regulatory mechanisms and immune response in European sea bass (*Dicentrarchus Labrax*) fed a plant-based diet. *PLoS one* **11** (7): 1–20.

82 Campos, C., Valente, L.M.P., Conceição, L.E.C. et al. (2013). Temperature affects methylation of the myogenin putative promoter, its expression and muscle cellularity in *Senegalese sole* larvae. *Epigenetics* **8** (4): 389–397.

83 Campos, C., Valente, L.M.P., and Fernandes, J.M.O. (2012). Molecular evolution of zebrafish dnmt3 genes and thermal plasticity of their expression during embryonic development. *Gene* **500** (1): 93–100.

84 Liu, J., Hu, H., Panserat, S., and Marandel, L. (2020). Evolutionary history of DNA methylation related genes in chordates: new insights from multiple whole genome duplications. *Scientific Reports* **10** (1): 1–14.

85 Loughland, I., Little, A., and Seebacher, F. (2021). DNA methyltransferase 3a mediates developmental thermal plasticity. *BMC Biology* **19** (1): 1–11.

86 Wang, F., Qin, Z., Li, Z. et al. (2021). Dnmt3aa but not dnmt3ab is required for maintenance of gametogenesis in nile tilapia (*Oreochromis niloticus*). *International Journal of Molecular Sciences* **22** (18): 10170.

87 Lallias, D., Bernard, M., Ciobotaru, C. et al. (2021). Sources of variation of DNA methylation in rainbow trout: combined effects of temperature and genetic background. *Epigenetics* **16** (9): 1031–1052.

88 Chédin, F. (2011). The DNMT3 family of mammalian de novo DNA methyltransferases. *Progress in Molecular Biology and Translational Science* **101**: 255–285.

89 Inoue, A., Jiang, L., Lu, F. et al. (2017). Maternal H3K27me3 controls DNA methylation-independent imprinting. *Nature* **547** (7664): 419–424.

90 Gallou-Kabani, C., Gabory, A., Tost, J. et al. (2010). Sex- and diet-specific changes of imprinted gene expression and DNA methylation in mouse placenta under a high-fat diet. *PLoS one* **5** (12): e14398.

91 Labbé, C., Robles, V., and Herraez, M.P. (2017). Epigenetics in fish gametes and early embryo. *Aquaculture* **472**: 93–106.

92 Potok, M.E., Nix, D.A., Parnell, T.J., and Cairns, B.R. (2013). Reprogramming the maternal zebrafish genome after fertilization to match the paternal methylation pattern. *Cell* **153** (4): 759–772.

93 Skvortsova, K., Tarbashevich, K., Stehling, M. et al. (2019). Retention of paternal DNA methylome in the developing zebrafish germline. *Nature Communications* **10** (1): 1–13.

94 Ortega-Recalde, O., Day, R.C., Gemmell, N.J., and Hore, T.A. (2019). Zebrafish preserve global germline DNA methylation while sex-linked rDNA is amplified and demethylated during feminisation. *Nature Communications* **10** (1): 1–10.

95 Wang, X. and Bhandari, R.K. (2020). The dynamics of DNA methylation during epigenetic reprogramming of primordial germ cells in medaka (Oryzias latipes). *Epigenetics* **15** (5): 483–498.

96 Wu, X., Zhang, H., Zhang, B. et al. (2021). Methylome inheritance and enhancer dememorization reset an epigenetic gate safeguarding embryonic programs. *Science Advances* **7** (52): 3858. www.science.org.

97 Wu, X., Zhang, H., Zhang, B. et al. (2021). Methylome inheritance and enhancer dememorization reset an epigenetic gate safeguarding embryonic programs. *bioRxiv* **2021**: 10.1126/sciadv.abl3858.

98 Horsfield JA. Packaging development: how chromatin controls transcription in zebrafish embryogenesis. Vol. **47**, *Biochemical Society Transactions* 2019 (2):713–724. doi: https://doi.org/10.1042/BST20180617.

99 Marchione, A.D., Thompson, Z., and Kathrein, K.L. (2021). DNA methylation and histone modifications are essential for regulation of stem cell formation and differentiation in zebrafish development. *Briefings in Functional Genomics* **47**: 713–724.

100 Seebacher, F. and Simmonds, A.I.M. (2019). Histone deacetylase activity mediates thermal plasticity in zebrafish (*Danio rerio*). *Scientific Reports* **9** (1): 1–3.

101 Šrut, M. (2021). Ecotoxicological epigenetics in invertebrates: emerging tool for the evaluation of present and past pollution burden. *Chemosphere* **282** (June): 131026.

102 Gavery, M.R. and Roberts, S.B. (2014). A context dependent role for DNA methylation in bivalves. *Briefings in Functional Genomics* **13** (3): 217–222.

103 Roberts, S.B. and Gavery, M.R. (2012). Is there a relationship between DNA methylation and phenotypic plasticity in invertebrates? *Frontiers in Physiology* **2** (January): 1–5.

104 Flores, K., Wolschin, F., Corneveaux, J.J. et al. (2012). Genome-wide association between DNA methylation and alternative splicing in an invertebrate. *BMC Genomics* **13** (1): 1–9.

105 Shukla, S., Kavak, E., Gregory, M. et al. (2011). CTCF-promoted RNA polymerase II pausing links DNA methylation to splicing. *Nature* **479** (7371): 74–79.

106 Asselman, J., De Coninck, D.I.M., Beert, E. et al. (2017). Bisulfite sequencing with *Daphnia* highlights a role for epigenetics in regulating stress response to microcystis through preferential differential methylation of serine and threonine amino acids. *Environmental Science and Technology* **51** (2): 924–931.

107 de Mendoza, A., Hatleberg, W.L., Pang, K. et al. (2019). Convergent evolution of a vertebrate-like methylome in a marine sponge. *Nature Ecology and Evolution* **3** (10): 1464–1473.

108 Robichaud, N.F., Sassine, J., Beaton, M.J., and Lloyd, V.K. (2012). The epigenetic repertoire of *Daphnia magna* includes modified histones. *Genetics Research International* **2012**: 1–7.

109 Kvist, J., Athanàsio, C.G., Pfrender, M.E. et al. (2020). A comprehensive epigenomic analysis of phenotypically distinguishable, genetically identical female and male *Daphnia pulex*. *BMC Genomics* **21** (1): 1–17.

110 Chen, T., Sun, Z., Mu, S. et al. (2019). Ultrastructure of spermiogenesis and the distribution of spermatozoal nuclear histones in the Japanese mantis shrimp, *Oratosquilla oratoria* (Crustacea: Stomatopoda). *Journal of Morphology* **280** (8): 1170–1184.

111 Fellous, A., Lefranc, L., Jouaux, A. et al. (2019). Histone methylation participates in gene expression control during the early development of the Pacific oyster *Crassostrea gigas*. *Genes (Basel)* **10** (9): 695.

112 Shaw, J.L.A., Judy, J.D., Kumar, A. et al. (2017). Incorporating transgenerational epigenetic inheritance into ecological risk assessment frameworks. *Environmental Science and Technology* **51** (17): 9433–9445.

113 Perez, M.F. and Lehner, B. (2019). Intergenerational and transgenerational epigenetic inheritance in animals. *Nature Cell Biology* **21**: 143–151. https://doi.org/10.1038/s41556-018-0242-9.

114 Jonsson, B. and Jonsson, N. (2014). Early environment influences later performance in fishes. *Journal of Fish Biology* **85** (2): 151–188.

115 Navarro-Martín, L., Viñas, J., Ribas, L. et al. (2011). DNA methylation of the gonadal aromatase (cyp19a) promoter is involved in temperature-dependent sex ratio shifts in the European sea bass. Whitelaw E, editor. *PLoS Genetics* **7** (12): e1002447.

116 Shao, C., Li, Q., Chen, S. et al. (2014). Epigenetic modification and inheritance in sexual reversal of fish. *Genome Research* **24** (4): 604–615.

117 Giannetto, A., Nagasawa, K., Fasulo, S., and Fernandes, J.M.O. (2013). Influence of photoperiod on expression of DNA (cytosine-5) methyltransferases in Atlantic cod (*Gadus morhua*). *Gene* **519** (2): 222–230.

118 Naylor, R.L., Kishore, A., Sumaila, U.R. et al. (2021). Blue food demand across geographic and temporal scales. *Nature Communications* **12** (1): 1–14.

119 Polakof, S., Panserat, S., Soengas, J.L., and Moon, T.W. (2012). Glucose metabolism in fish: a review. *Journal of Comparative Physiology B* **182** (8): 1015–1045. https://link.springer.com/article/10.1007/s00360-012-0658-7.

120 Kostyniuk, D.J., Marandel, L., Jubouri, M. et al. (2019). Profiling the rainbow trout hepatic miRNAome under diet-induced hyperglycemia. *Physiological Genomics* **51** (9): 411–431.

121 Ulloa, P.E., Medrano, J.F., and Feijo, C.G. (2014). Zebrafish as animal model for aquaculture nutrition research. *Frontiers in Genetics* **5** (SEP): 1–6.

122 Hou, Z. and Fuiman, L.A. (2020). Nutritional programming in fishes: insights from mammalian studies. *Reviews in Fish Biology and Fisheries* **30** (1): 67–92.

123 Geurden, I., Aramendi, M., Zambonino-Infante, J., and Panserat, S. (2007). Early feeding of carnivorous rainbow trout (*Oncorhynchus mykiss*) with a hyperglucidic diet during a short period: effect on dietary glucose utilization in juveniles. *American Journal of Physiology – Regulatory Integrative and Comparative Physiology* **292** (6): 2275–2283.

124 Geurden, I., Mennigen, J., Plagnes-Juan, E. et al. (2014). High or low dietary carbohydrate: protein ratios during firstfeeding affect glucose metabolism and intestinal microbiota in juvenile rainbow trout. *Journal of Experimental Biology* **217** (19): 3396–3406.

125 Moghadam, H., Mørkøre, T., and Robinson, N. (2015). Epigenetics-potential for programming fish for aquaculture? *Journal of Marine Science and Engineering* **3** (2): 175–192.

126 Geurden, I., Borchert, P., Balasubramanian, M.N. et al. (2013). The positive impact of the early-feeding of a plant-based diet on its future acceptance and utilisation in rainbow trout. *PLoS one* **8** (12): e83162.

127 Balasubramanian, M.N., Panserat, S., Dupont-Nivet, M. et al. (2016). Molecular pathways associated with the nutritional programming of plant-based diet acceptance in rainbow trout following an early feeding exposure. *BMC Genomics* **17** (1): 1–20.

128 Turkmen, S., Zamorano, M.J., Fernández-Palacios, H. et al. (2017). Parental nutritional programming and a reminder during juvenile stage affect growth, lipid metabolism and utilisation in later developmental stages of a marine teleost, the gilthead sea bream (*Sparus aurata*). *British Journal of Nutrition* **118** (7): 500–512.

129 Turkmen, S., Zamorano, M.J., Xu, H. et al. (2020). Parental LC-PUFA biosynthesis capacity and nutritional intervention with alpha-linolenic acid affect performance of *Sparus aurata* progeny. *Journal of Experimental Biology* **223** (23): 1–12.

130 Lazzarotto, V., Corraze, G., Larroquet, L. et al. (2016). Does broodstock nutritional history affect the response of progeny to different first-feeding diets? A whole-body transcriptomic study of rainbow trout alevins. *British Journal of Nutrition* **115** (12): 2079–2092.

131 Forbes, J.L.I., Kostyniuk, D.J., Mennigen, J.A., and Weber, J.M. (2019). Glucagon regulation of carbohydrate metabolism in rainbow trout: in vivo glucose fluxes and gene expression. *Journal of Experimental Biology* **222** (24): jeb211730.

132 Lage, L.P.A., Serusier, M., Weissman, D. et al. (2018). Metabolic programming in juveniles of the whiteleg shrimp (*Litopenaeus vannamei*) linked to an early feed restriction at the post-larval stage. *Aquaculture* **495**: 328–338.

133 Lage, L.P.A., Weissman, D., Serusier, M. et al. (2020). Long-term impact of a 4-day feed restriction at the protozoea stage on metabolic gene expressions of whiteleg shrimp (*Litopenaeus vannamei*). *PeerJ* **2020** (3): e8715.

134 Wischhusen, P., Saito, T., Heraud, C. et al. (2020). Parental selenium nutrition affects the one-carbon metabolism and the hepatic dna methylation pattern of rainbow trout (*Oncorhynchus mykiss*) in the progeny. *Lifestyles* **10** (8): 1–25.

135 Saito, T., Whatmore, P., Taylor, J.F. et al. (2021). Micronutrient supplementation affects transcriptional and epigenetic regulation of lipid metabolism in a dose-dependent manner. *Epigenetics* **16** (11): 1217–1234.

136 Saito, T., Whatmore, P., Taylor, J.F. et al. (2022). Micronutrient supplementation affects DNA methylation in male gonads with potential intergenerational epigenetic inheritance involving the embryonic development through glutamate receptor-associated genes. *BMC Genomics* **23** (1): 1–18.

137 Fang, L., Liang, X.F., Zhou, Y. et al. (2014). Programming effects of high-carbohydrate feeding of larvae on adult glucose metabolism in zebrafish, Danio rerio. *British Journal of Nutrition* **111** (5): 808–818.

138 Rocha, F., Dias, J., Engrola, S. et al. (2015). Glucose metabolism and gene expression in juvenile zebrafish (*Danio rerio*) challenged with a high carbohydrate diet: effects of an acute glucose stimulus during late embryonic life. *British Journal of Nutrition* **113** (3): 403–413.

139 Zhu, Q.S., Wang, J., He, S. et al. (2020). Early leucine programming on protein utilization and mTOR signaling by DNA methylation in zebrafish (*Danio rerio*). *Nutrition and Metabolism* **17** (1): 1–13.

140 Zarantoniello, M., Randazzo, B., Cardinaletti, G. et al. (2021). Possible dietary effects of insect-based diets across zebrafish (*Danio rerio*) generations: a multidisciplinary study on the larval phase. *Animals* **11** (3): 1–18.

141 Wang, S.Y., Lau, K., Lai, K.P. et al. (2016). Hypoxia causes transgenerational impairments in reproduction of fish. *Nature Communications* **7**: 1–9.

142 Cabrita, E., Martínez-Páramo, S., Gavaia, P.J. et al. (2014). Factors enhancing fish sperm quality and emerging tools for sperm analysis. *Aquaculture* **432**: 389–401.

143 Zhao, L., Yang, F., Milano, S. et al. (2018). Transgenerational acclimation to seawater acidification in the Manila clam *Ruditapes philippinarum*: preferential uptake of metabolic carbon. *Science of the Total Environment* **627**: 95–103.

144 Parker, L.M., O'Connor, W.A., Raftos, D.A. et al. (2015). Persistence of positive carryover effects in the oyster, Saccostrea glomerata, following transgenerational exposure to ocean acidification. *PLoS one* **10** (7): 1–19.

145 Houston, R.D., Bean, T.P., Macqueen, D.J. et al. (2020). Harnessing genomics to fast-track genetic improvement in aquaculture. *Nature Reviews Genetics* **21** (7): 389–409.

146 Uren Webster, T.M., Rodriguez-Barreto, D., Martin, S.A.M. et al. (2018). Contrasting effects of acute and chronic stress on the transcriptome, epigenome, and immune response of Atlantic salmon. *Epigenetics* **13** (12): 1191–1207.

147 Robinson, N.A., Johnsen, H., Moghadam, H. et al. (2019). Early developmental stress affects subsequent gene expression response to an acute stress in Atlantic salmon: an approach for creating robust fish for aquaculture? *G3: Genes, Genomes, Genetics.* **9** (5): 1597–1611.

148 Medina-Gali, R., Belló-Pérez, M., Martínez-López, A. et al. (2018). Chromatin immunoprecipitation and high throughput sequencing of SVCV-infected zebrafish reveals novel epigenetic histone methylation patterns involved in antiviral immune response. *Fish & Shellfish Immunology* **82** (May): 514–521.

149 Hu, J., Pérez-Jvostov, F., Blondel, L., and Barrett, R.D.H. (2018). Genome-wide DNA methylation signatures of infection status in Trinidadian guppies (*Poecilia reticulata*). *Molecular Ecology* **27** (15): 3087–3102.

150 Loker, E.S., Adema, C.M., Zhang, S.M., and Kepler, T.B. (2004). Invertebrate immune systems – not homogeneous, not simple, not well understood. *Immunological Reviews* **198**: 10–24.

151 Roy, S., Bossier, P., Norouzitallab, P., and Vanrompay, D. (2020). Trained immunity and perspectives for shrimp aquaculture. *Reviews in Aquaculture* **12** (4): 2351–2370.

152 Norouzitallab, P., Baruah, K., Vandegehuchte, M. et al. (2014). Environmental heat stress induces epigenetic transgenerational inheritance of robustness in parthenogenetic Artemia model. *The FASEB Journal* **28** (8): 3552–3563.

153 Green, T.J., Helbig, K., Speck, P., and Raftos, D.A. (2016). Primed for success: oyster parents treated with poly(I:C) produce offspring with enhanced protection against Ostreid herpesvirus type I infection. *Molecular Immunology* **78**: 113–120.

154 Navarro-Martín, L., Martyniuk, C.J., and Mennigen, J.A. (2020). Comparative epigenetics in animal physiology: An emerging frontier. *Comparative. Biochemistry and Physiology Part D: Genomics and Proteomics* **36**: 100745.

3

Epigenetic Regulation of Gene Expression by Noncoding RNAs

Elena Sarropoulou[1] and Ignacio Fernández[2]

[1] *Institute of Marine Biology, Biotechnology, and Aquaculture, Hellenic Centre for Marine Research, Heraklion, Crete, Greece*
[2] *Centro Oceanográfico de Vigo, Instituto Español de Oceanografía (IEO-CSIC), Spanish National Research Council, Vigo, Spain*

3.1 General Introduction

The regulation of gene expression can occur by various mechanisms at different levels and may affect numerous biological processes (Figure 3.1). As a consequence, gene regulation mechanisms require high accuracy as well as precise and tight control. Therefore, comprehensive investigations of the transcriptome are of utmost importance to improve our understanding of the diverse physiological processes from cell activities to tissue-specific peculiarities within organisms. Per definition, an organism's transcriptome is a snapshot from a specific tissue or cell at a particular functional state and comprises all types of RNAs – coding (messenger RNA, mRNA) and noncoding RNA (ncRNA). The sequencing of the transcriptome, also known as RNA-Seq, is a relatively recently developed technique that describes the transcriptome in a given sample at a specific time-point and may also provide an insight into the relative transcript abundances among tissues, treatments, and conditions. While mRNAs in terms of gene expression in teleosts are widely examined, the importance of ncRNAs as regulatory molecules during gene expression has become apparent only over the past two decades. Particularly, the advent of novel, high-throughput sequencing methodologies have significantly enhanced the research of teleost ncRNAs (Figure 3.2).

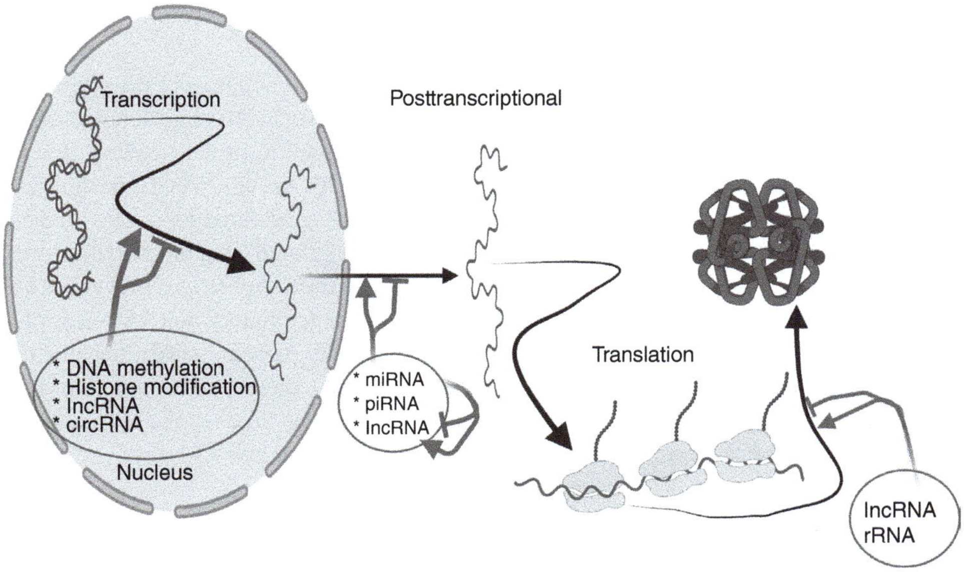

Figure 3.1 Overview of epigenetic regulation of gene expression. Arrows and blocked lines indicate up- and downregulation of the indicated processes, respectively.

Epigenetics in Aquaculture, First Edition. Edited by Francesc Piferrer and Han-Ping Wang.

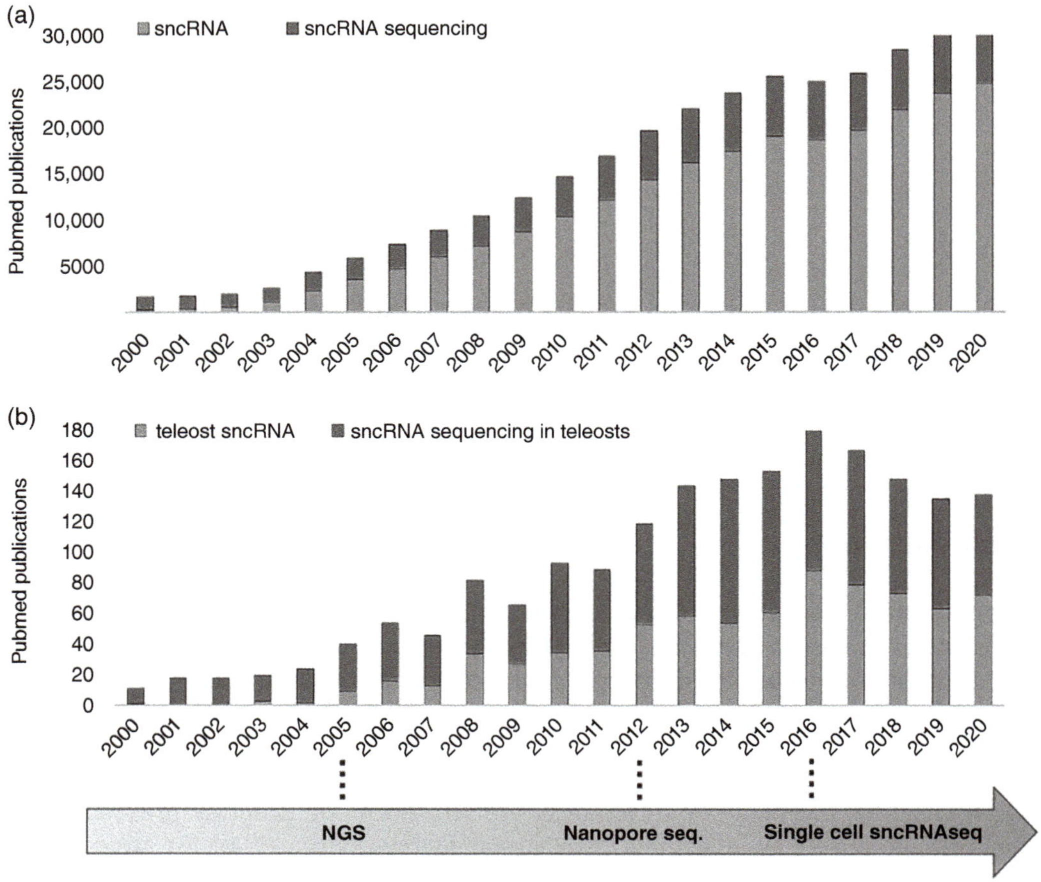

Figure 3.2 Overview of PubMed publications comprising sncRNA work and sncRNA sequencing. (a) Overall search for publication applying search query for sncRNA work: "small noncoding RNA OR miRNA OR piRNA OR snoRNA OR snRNA" (Title/Abstract), and for sncRNA sequencing search query: "RNA sequencing" (Title/Abstract); (b) Teleost-specific publications with: search query: "noncoding RNA sequencing OR noncoding RNA sequencing OR ncRNA sequencing OR noncoding RNAseq OR noncoding RNAseq" (Title/Abstract) Search query: "(Teleost OR zebrafish OR medaka OR tilapia) AND (noncoding RNA sequencing OR ncRNA sequencing OR ncRNA sequencing OR noncoding RNAseq OR noncoding RNAseq [Title/Abstract]), respectively.

These rapid advancements in sequencing technologies are today facing the challenge of discovering, quantifying, and correctly annotating transcripts encoded by the genome. Within the human ENCODE project, it was found that 76% of the genome is transcribed into RNA, but only about 3% of the human genome are protein-coding genes. Subsequently, a plethora of novel RNA transcripts were reported [1] and named noncoding RNAs (ncRNAs) since most of those transcripts were revealed not to have a clear potential to encode for a protein and were considered for a long time as "transcriptional noise." Today, ncRNAs are known to act not only mainly at the posttranscriptional level but also posttranslational [2].

Regarding classification, ncRNAs are arbitrarily grouped based on their transcript length into small ncRNAs (sncRNA) and long (lncRNA) [3], or based on their functionality as housekeeping or regulatory ncRNAs. This chapter follows the ncRNA definition by length since distinguishing categories per functionality presents difficulties due to the crossover of properties. Thus, sncRNAs are less than 200 nucleotides (nt) and are further categorized again according to their length, while long lncRNAs are noncoding transcripts with a length of more than 200 nt. Some of the best-known and well-studied lncRNA are shown in Figure 3.3. Major terms and ncRNA types are described in Box 3.1.

3.2 Major Types of ncRNAs

3.2.1 Small Noncoding RNA (sncRNA)

The main role of sncRNAs is the regulation of gene expression through either RNA interference, RNA modification, or spliceosomal involvement. Consequently, their expression can change according to the individual's particular condition.

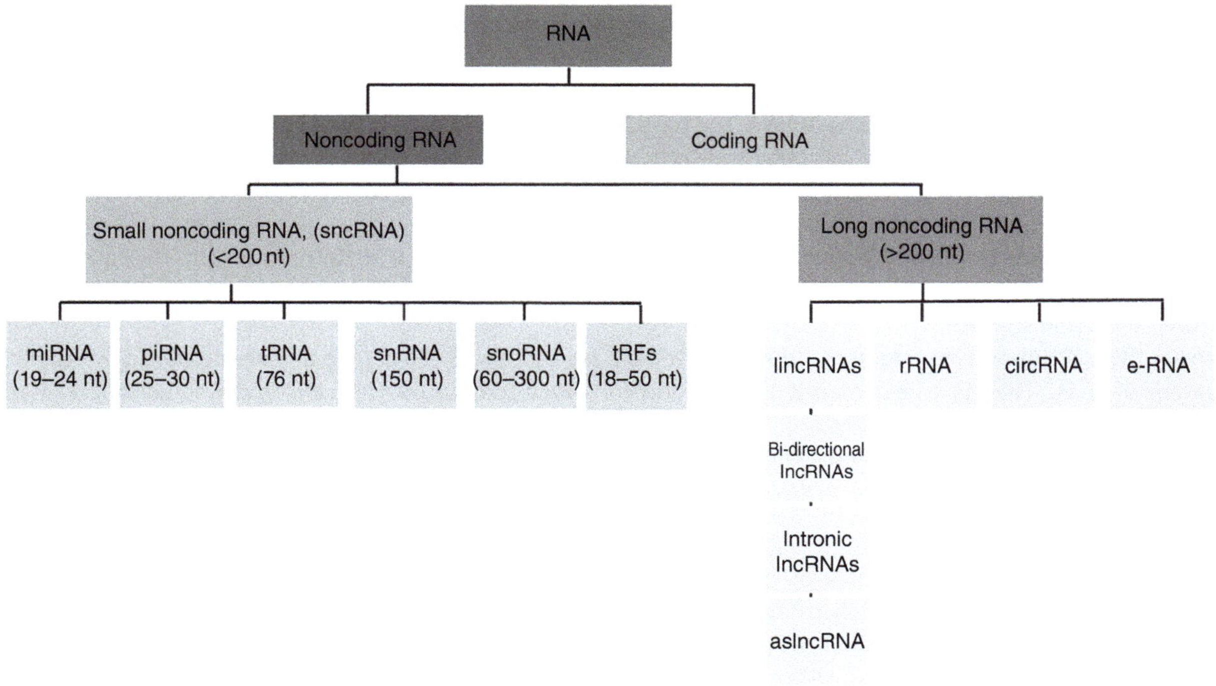

Figure 3.3 Schematic overview of RNA types divided into noncoding and coding RNAs. Noncoding RNA is further grouped into small noncoding RNA (sncRNA), including microRNA (miRNA), P-element-induced wimpy testis (Piwi)-interacting RNA (piRNA), transfer RNA (tRNA), small nuclear RNA (snRNA), small nucleolar RNA (snoRNA), as well as transfer RNA-derived RNA fragments (tRFs), and long noncoding RNA (lncRNA), comprising ribosomal RNA (rRNA), circular RNA (circRNA), intergenic lncRNAs (lincRNAs), intronic lncRNAs, antisense lncRNAs (aslncRNA), bidirectional lncRNAs, and enhancer RNAs (eRNAs).

Box 3.1 Major ncRNA Terms

Circular RNA (circRNA): A unique type of noncoding RNA molecule. Unlike conventional linear RNA, circRNA is a covalently closed loop formed by a process called back splicing. CircRNAs are highly abundant in many cells and have multiple regulatory functions.

Enhancer RNAs (eRNAs): A type of comparatively lncRNA molecules of 50–2000 nt. They appear to be important for the function of enhancers.

Long intergenic noncoding RNAs (lincRNAs): LncRNAs located in the intergenic regions of the genome.

Long noncoding RNAs (lncRNAs): A type of noncoding RNA defined as transcripts longer than 200 nt. This arbitrarily chosen length criterion differentiates lncRNAs from sncRNAs. It is reported that lncRNAs are involved in transcriptional regulation.

(Mature) microRNAs (miRNAs): Single-stranded RNAs with an average length of 20–22 nt. Their function is mostly repressive of gene expression by binding to the mRNA 3′ UTR. However, some examples have been reported with enhancing functions binding to the mRNA 5′ UTR.

miRNA recognition element (MRE): Also called miRNA response element, is the complementary sequence on the target mRNA interacting with the miRNA.

P-element-induced wimpy testis (Piwi)-interacting RNAs (piRNAs): Class of small, noncoding RNAs with a length of 25–30 nt. They are involved in the transcriptional regulation of genes. piRNAs may also function in the epigenetic regulation of DNA methylation, transposable silencing, and chromatin modification.

Precursor-miRNAs (pre-miRNAs): Pre-miRNAs are formed by the cleavage of primary miRNAs by DCGR8 and Drosha and also have the typical hairpin structure. They are transported from the nucleus to the cytoplasm where they are further processed to generate the mature miRNA.

Primary miRNAs (Pri-miRNAs): Pri-miRNAs are transcribed from their respective genes and can be more than 1000 nt in length. The pri-miRNA hairpin is defined by a terminal loop and basal unpaired flanking sequences, with the upper stem comprising the mature miRNA sequence. The hairpin, which typically comprises 60–120 nt, is further processed in the nucleus to generate the pre-miRNA.

Ribosomal RNAs (rRNAs): A type of ncRNA that is the major component of ribosomes, vital to all cells. It is the predominant form of RNA found in most cells.
Seed region: Also known as the seed sequence, is a conserved heptametric sequence between the second and the seventh nt of the 5′ end of the mature miRNA. The base pairing of miRNA and its target mRNA may not always match completely, but the "seed sequence" has been shown to be perfectly complementary in most of the cases.
Small nuclear RNAs (snRNAs): A type of very abundant sncRNA with an average size of 150 nt. They are localized in the nucleus and have important functions during the RNA processing steps.
Small nucleolar RNAs (snoRNAs): A type of small RNA molecule between 60 and 300 nt in length. They are predominantly found in the nucleolus and mainly control chemical modifications of other RNAs, especially ribosomal RNAs, transfer RNAs, and snRNAs.
Transfer RNA (tRNA)-derived fragments (tRFs): Active tRFs are usually 18–50 nt long and are produced by enzymatic cleavage of tRNAs. They have been reported to be highly conserved across eukaryotes and to play important regulatory functions similar to those of microRNAs.
Transfer RNAs (tRNAs): RNA molecules with an average length of 76 nt. They assist as a physical link between the mRNA and the amino acid sequence. tRNAs are essential for translation.

The most extensively studied sncRNA class is microRNAs (miRNAs) and will therefore be described in more detail within this chapter. Another upcoming popular sncRNA group discussed in more detail here is the class of P-element-induced wimpy testis (Piwi)-interacting RNA (piRNA). Other sncRNA classes are the small nuclear RNA (snRNA), small nucleolar RNA (snoRNA), transfer RNA (tRNA), and transfer RNA (tRNA)-derived fragments (tRFs), which are shortly discussed hereafter.

3.2.1.1 MicroRNA (miRNA)

The first miRNA discovered, lin-4 of the nematode *Caenorhabditis elegans*, dates almost 30 years back [4, 5]. It was stated that lin-4 suppresses the lin-14 protein levels important for the proper development of *C. elegans*. Subsequently, it was noted that lin-4 does regulate not only lin-14 but also other target genes (e.g., LIN-28) [6]. The appearance of the miRNA lin-4 was thought to be an isolated case of *C. elegans*. The discovery of other regulating miRNAs [7, 8], as well as their presence in other annelids, in the fruit fly (*Drosophila melanogaster*) and zebrafish (*Danio rerio*) [9], promoted their intensive investigations to better understand their role in gene regulation. Within the genome, miRNAs are organized as a single gene or gene clusters, and miRNA genes are transcribed either as monocistronic or polycistronic transcripts [10]. miRNA genes or gene clusters are either located in the intergenic region, in transcription units, or in exons of noncoding (and more rarely of coding) regions [11] (Figure 3.4, Box 3.1). To note, the miRNA gene organization may indicate the possible miRNA function, i.e., two or more miRNA genes in one cluster and their subsequent co-transcription indicates synergistic miRNAs that are involved in the same biological process.

The mature miRNA is single-stranded with an average length of 22 nt comprising the seed region (between the second and the seventh nt of the 5′ end). Mature miRNAs are generated through either canonical or noncanonical miRNA biogenesis pathways, which have been thoroughly reviewed by O'Brien and colleagues [13]. Characteristic of miRNAs and important for annotation purposes is the classical hairpin structure formed by the Pri-miRNA, maintained through the Pre-miRNA, and cleaved at the final step to generate the mature miRNA duplex. miRNAs are frequently annotated as e.g., miR-202-3p or miR-202-5p, which is the result of which strand is maintained through the biogenesis process. The maintained strand is called the guide strand, whereas the other strand is called the passenger strand and gets degraded. Emerging knowledge will help us understand the rules governing miRNA strand selection (reviewed in [14]) critical to specifying which target genes (molecular pathways) are regulated. The main function of miRNAs is the negative regulation of gene expression by binding to the 3′ untranslated region (UTR) of the genes [15]. Some studies have shown that miRNA may also have an enhancing role [16], which seems to be dominant when binding to the 5′ UTR [17, 18]. The negative gene regulation by miRNA is known to act either by direct cleavage of the target mRNA or by modifying the target mRNA, thus impeding the translation [19].

miRNA binding to the UTR is based on sequence complementarity, which in most cases is not complete. However, the alignment of the seed region to the target UTR region perfectly matches in the majority of cases. miRNA investigations in medical research have been significantly advanced due to their role in human diseases and particularly in cancer. miRNAs are found, in many cases, to be the causal factor in disease progression, which, on the other hand, renders them a powerful

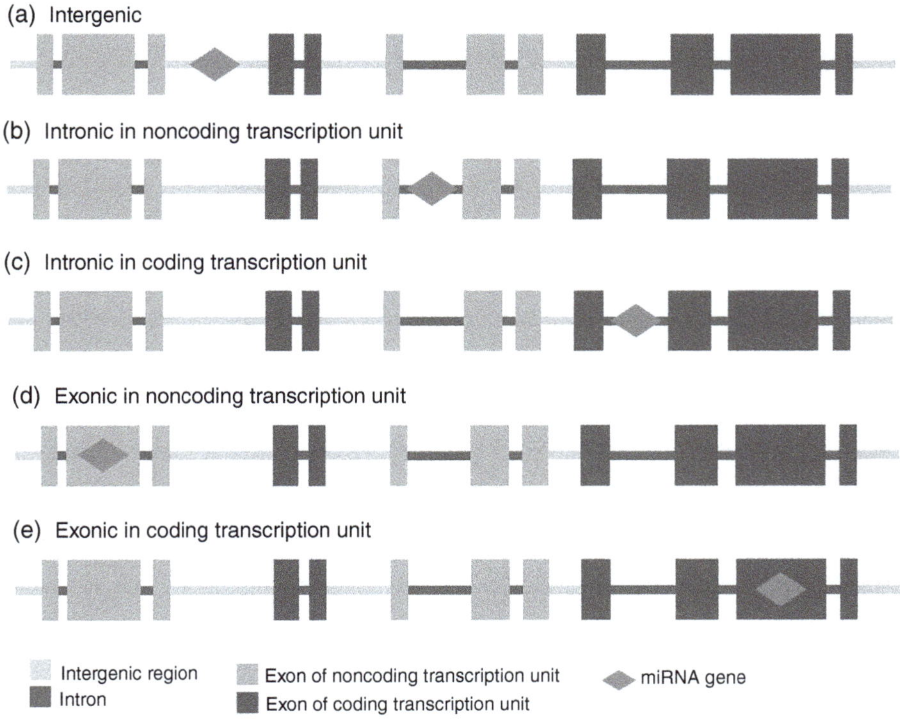

Figure 3.4 Five different types of the genomic organization of miRNA genes. *Source:* Based on Melo and Melo (2014); modified by Elisavet Kaitetzidou [12]. (a) miRNA gene located in the intergenic region, (b) miRNA gene located in the noncoding transcription unit, (c) miRNA gene located in the coding transcription unit, (d) miRNA gene located in noncoding transcription unit, and (e) miRNA located in the coding region of the transcription unit.

diagnostic marker as well as a promising candidate to develop therapeutic drugs. One popular example is breast cancer known to be linked with the expression of BRCA genes. During the last decade, multiple miRNAs have been identified to interact with BRCA genes, but still, further validation and standardization are needed before their use in routine clinical practice [20, 21]. Regarding teleosts, early miRNA studies were carried out on zebrafish development with the most popular one being the miR-430 (see Section 3.3.1). miRNA investigations in teleosts started very hesitantly with 1250 miRNA in eight teleost species in 2014 (miRbase v.20, [22]). Today (miRbase v.22), 3687 mature miRNAs are reported in 16 teleost species, and a teleost-specific miRNA database comprised of 3967 mature miRNAs has been established [23]. Detection of different miRNA abundancies according to tissues, treatments, developmental stages, also organisms, has been carried out to contribute to our understanding of posttranscriptional regulation. Regarding mature miRNAs, it has been shown that their sequence is highly conserved across species [24]. To unravel their mode of action, miRNA target sites have to be identified. Since the target sites are either in the 5′ UTR and, most frequently, in the 3′ UTR, they may not display the same level of conservation as the mature miRNAs, and 3′ UTRs are not conserved among species and not even among different isoforms. UTRs, and particularly 3′ UTRs, are known to be unique tags for the coding gene.

Today, several advanced bioinformatic approaches are available to effectively perform *in silico* miRNA target site predictions to identify miRNA recognition elements (MRE) and are reviewed by Akhatar and colleagues [25]. Nevertheless, this is a very dynamic field and new or updated algorithms are published consistently. Most of the tools base their target recognition on the seed region, which consists of perfect sequence matches between the 2nd to the 8th nt of the miRNA 5′ end. In contrast to plant miRNAs, animal miRNAs have only partial complementarity [26, 27], which broadens the determination of potential miRNA targets and increases the probability of false discovery [28]. The major seed types are (i) 6mer, (ii) 7mer, and (iii) 8mer perfect Watson–Crick base pair matches. Regarding the miRNA 3′ end, it has been proposed to compensate for imperfect seed matches [29]. Loosened target prediction criteria may increase false positives, while on the other hand, too strong criteria may filter out important functional miRNAs (e.g., let-7/lin-41 regulation) [30]. In the event that the miRNA is perfectly complementary to its target, like in plants, it will act as a siRNA, thus initiating target cleavage and not repression. Besides the criteria of the seed match, free energy (also called Gibbs free energy) may be applied to measure the stability of the miRNA-mRNA binding. Since free energy is difficult to be measured directly, the change in free energy

(ΔG) during a reaction is considered. Consequently, lower ΔGs indicate increased stability and a higher probability to have identified true miRNA targets. Therefore, most target prediction tools use the minimum free energy (mfe) as a hybridization indicator for each miRNA-mRNA pair.

Teleosts, being by far the most diverse vertebrate group, are occupying a wide range of several habitats, and thus present different adaptation capabilities. Since miRNA expression is highly influenced by environmental parameters, investigating miRNA targets in teleost is of particular interest. In zebrafish, for example, *cyp19a1a*, an important sex differentiation and reproduction gene, TargetScan (TargetScanFish, vs 6.2) predict miR-181 to regulate its gene expression, while in the protogynous hermaphroditic fish, *Epinephelus coioides*, miR-26a-5p has been reported to downregulate *cyp19a1a* in vitro and in vivo [31]. Similarly, in the sharp snout sea bream (*Puntazzo puntazzo*), differences in the 3′ UTR of another important gene in fish reproduction, the *follicle stimulator hormone receptor gene* (*fshr*), have been shown [32] to result in different target site predictions. While for the two protandrous hermaphrodite species, i.e., the sharp snout sea bream and the gilthead sea bream (*Sparus aurata*), mfe values for miR-125c as putative *fshr* regulators were higher than the set threshold of −25; for the other species examined, this could not be confirmed. In mammals, on the other hand, miR-143 has been shown to regulate *fshr* [33]. Further, in-deep experimental studies are necessary to better understand miRNA gene regulation and miRNA target validation by applying the classical in vitro approach by a reporter gene assay (e.g., luciferase) and then measuring the reporter intensity in the presence and absence of the candidate. The drawbacks of these approaches, apart from the labor-intensive process, also include the interpretation of negative results (this might also be due to experimental failure) and artificial in vitro miRNA concentrations, which may result in not-biological-relevant conclusions.

3.2.1.2 P-Element-Induced Wimpy Testis (Piwi)-Interacting RNA (piRNA)

The first piRNA was described in the fruit fly, *D. melanogaster*, testis as a novel group of small interfering RNA (siRNAs), silencing a multicopy gene on the X chromosome and influencing male fertility [34]. Although this novel group was of larger size and in contrast to siRNA, single-stranded, its recognition as a separate group was only after the findings of an association with Piwi proteins [35]. Studies in mice, rats, and human germ cells followed, reporting small RNA sequences binding to murine Piwi protein (MIWI) [36]. Today, piRNAs are recognized as the most recent class of sncRNAs associated with the Piwi subclade of the Argonaute protein family. They are single-stranded, small RNAs of a length typically between 26 and 31 nt. Further characteristic features include their modifications on the 3′ and 5′ ends of the nt sequence. Studies in zebrafish have shown that piRNAs are Dicer independent and that the 5′ end of piRNAs are comprising a monophosphate group, while the 3′ end carries a 2′ *O*-methyl modification. These modifications have been suggested to increase the stability of piRNAs [39]. In addition, piRNAs are found all over the genome in clusters, from as few as 10 up to thousands of piRNAs. This clustering, as well as the Piwi-piRNA pathway, has been shown to be highly conserved across species [40]. On the other hand, unlike miRNAs, piRNA sequences are not strongly conserved among organisms [41, 42]. piRNAs have a wide range of functions from the transcriptional level inside the nucleus to the posttranscriptional level in the cytoplasm by forming RNA–protein complexes over interactions with Piwi proteins. Their major role, however, lies in the regulation of genetic elements such as transposable elements and certain gene transcripts, especially in the germline of animals [43, 44]. This regulation, mainly by silencing of transposons in the germline, is sought to be the ancestral function, although mosquitos have been reported to use piRNAs in their immune defense against viruses. Nevertheless, several piRNAs identified are unrelated to transposable elements and their proof of function in mRNA regulation remains challenging (reviewed in [45, 46]). Further mechanisms of piRNA action, i.e., target mRNA deadenylation, have also been reported, and are also implied in the regulation of protein-coding gene expression rather than transposons [47]. Today, piRNA information is available for 21 organisms in searchable databases like piRBase (http://www.regulatoryrna.org/database/piRNA/) and has been mainly detected in high abundance in male and female gonads. Important functional roles of gonad piRNAs guiding Piwi proteins have been reported mainly in insects [34, 41] and mammals [48–50]. Regarding teleosts, only recently piRNAs have been reported, including the first steps toward understanding their functional role. Besides investigations in the model fish zebrafish [43], piRNAs have been identified in the gonads of the Japanese flounder (*Paralichthys olivaceous*), presenting candidate piRNAs with significantly different expression patterns between male and female gonads [51]. Association of piRNA expression in male gonads with dietary supplementation has been shown to play an important role in the reproductive performance of the Senegalese sole (*Solea senegalensis*) [52], while it has been hypothesized that piRNAs may play an important role in ovary maturation in the common carp (*Cyprinus carpio*) and the European sea bass (*Dicentrarchus labrax*) [32, 53] (see also Section 3.3.2). The importance of further piRNA research in farmed fish has been reviewed by Lite et al. [54].

3.2.1.3 Small Nuclear RNA (snRNA) and Small Nucleolar RNA (snoRNA)

The third group among the major sncRNA is snRNA. Historically, snRNAs are referred to as U-RNAs due to their uridylate-rich sequence [55]. Synonyms of snRNA are also U-sn RNAs or splicing snRNAs [55]. The first evidence was reported comparing RNA migration of agarose gel electrophoresis to sucrose density gradient electrophoresis [56]. They are known to be localized in the nucleus, extremely conserved in both primary and secondary structures [57], very abundant, and transcribed by either RNA polymerase II or RNA polymerase III. snRNAs have an average size of 150 nt and carry out important functions such as intron splicing during RNA processing [48]. Therefore, snRNAs form an RNA–protein complex termed small nuclear ribonucleoproteins (snRNPs), and together with additional proteins, they are part of the spliceosomes, which in turn bind to the unspliced primary RNA transcript. The binding is reached by first recognizing the critical sequence elements by base-pairing interactions between snRNAs and pre-mRNA. In addition to their role during the splicing process, snRNPs mediate the nuclear maturation of primary transcripts in mRNAs and gene expression regulation.

3.2.1.4 Transfer RNA (tRNA)-Derived Fragments (tRFs)

Transfer RNA (tRNA)-derived fragments (tRFs) were first reported 40 years ago in cancer patients. Nevertheless, they belong to a new emerging group of sncRNAs that play important roles in posttranscriptional gene regulation. The generation of high-throughput, small RNA data applying next-generation sequencing (NGS) technology permitted the detection and characterization of tRFs, which had been neglected until recently. Active tRFs are found to be typically 18–50-nt long and have been reported to be highly conserved across eukaryotes. They are a ubiquitously expressed class representing about 10% of the cellular RNA content. They are derived from different regions of tRNAs and can be classified based on their size. Not all tRNAs are fragmented into tRFs and thus their biogenesis may be selective [reviewed in [35]]. tRFs have been mainly studied in cancer research, but up until today, no studies have been published reporting the roles of tRFs in teleosts.

3.2.2 Measurement of sncRNAs

The analysis pipeline from data acquisition to sncRNA expression analysis is summarized in Figure 3.5. Subsequent paragraphs explain in more detail each of the steps and possible pitfalls.

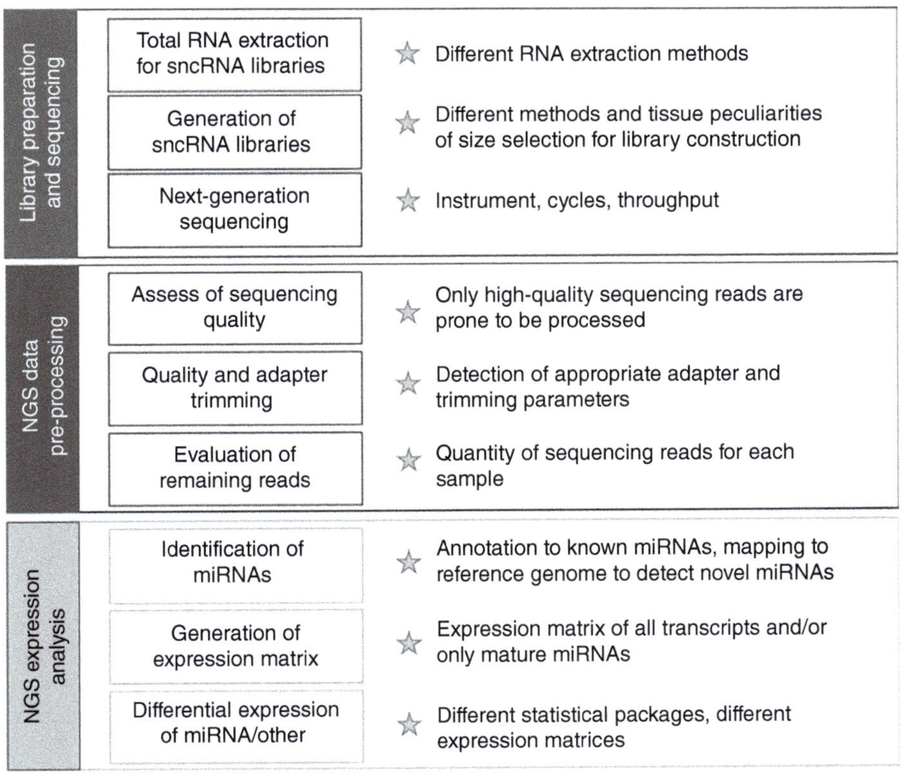

Figure 3.5 Summary of analysis pipeline from data acquisition to sncRNA expression analysis.

3.2.2.1 Methods for sncRNA Detection

A successful RNA study starts with the extraction of high-quality RNA. The RNA integrity is evaluated by the 28S–18S ratio, also known as the RNA integrity number (RIN value). As a general rule, high total RNA quality is assured with a RIN value of 8 or higher. However, this is not always valid. Exceptions of this rule are: total RNA extracted out of biofluids like blood plasma, where only one peak is visible (Figure 3.6a); immature and mature female gonads, where the 5S RNA peak is dominant (Figures 3.6b and 3.9); fixed tissue samples (Figure 3.6c); and mollusks, where the hidden break in 28S rRNA (like for most protostomes) causes 18S and 28S rRNA fragments to co-migrate during electrophoresis [49] (Figure 3.6d). In these cases, different evaluation methods are applied, like agarose gel runs to inspect mRNA integrity, and exclusion of 5S rRNA to determine the 28S : 18S ratio.

Since the accumulation of different RNA datasets is rapidly progressing, the upcoming challenge is to perform meta-analyses across a broad range of already generated datasets. It is therefore of great importance to consider possible technical variations that may result in batch effects. One of these variations may be due to the choice of the RNA extraction method. The possible effects of five different total RNA extraction kits have been investigated by Brown and colleagues [50]. The group showed a significant difference by comparing the expression values of six key miRNAs. Having successfully extracted total RNAs for small RNA library construction, several commercial kits are available based on similar protocols. One of the important steps though is correct size selection, which may vary across tissue samples and organisms. Indeed, different percentages of sncRNA types were detected when different RNA-Seq library kits were tested [61]. Obtained sequence reads are further submitted to common preanalysis procedures comprising adapter and low-quality trimming, sequence size distribution, as well as to the number or sequence reads evaluation, which should be at least five million reads per sample [62]. Annotation of sncRNAs is carried out either by aligning to a reference genome or by annotating to already known miRNA by blasting to databases such as miRbase [63], MirGeneDB [64], RefSeq, and Rfam databases, or the most global one, the RNA Central website (https://rnacentral.org/), where up to 44 different resources are integrated [65] (Table 3.1). Identifying novel miRNAs turns out to face several bottlenecks that have been reviewed by Ambros and colleagues [66], where guidelines for correct annotation have also been reported.

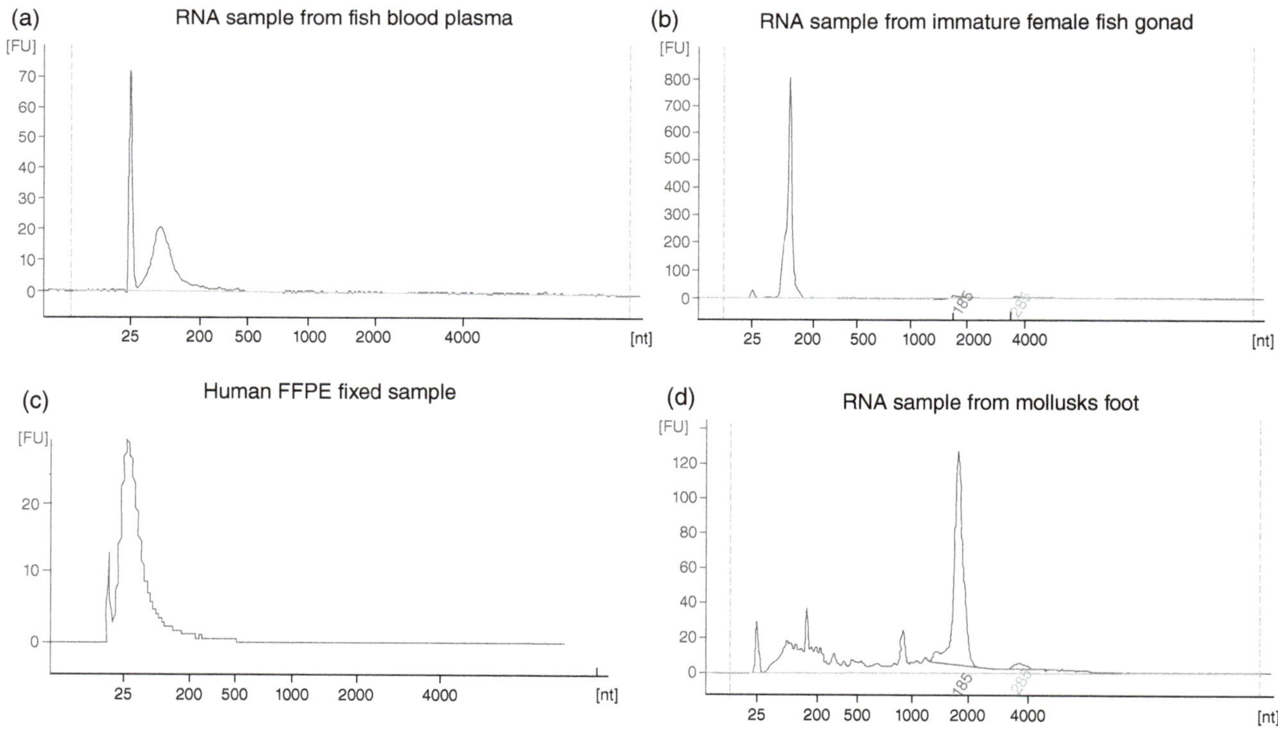

Figure 3.6 Bioanalyzer files presenting total RNA extractions from four different tissue samples: (a) RNA sample from fish blood plasma, (b) RNA sample from immature female fish gonad, (c) human FFPE fixed sample, and (d) RNA sample from mollusk foot.

Table 3.1 Most commonly used databases and software programs.

Databases	Website
MirGeneDB	https://mirgenedb.org/
miRbase	https://www.mirbase.org/
RNA Central	https://rnacentral.org/
Tools4miRs	https://tools4mirs.org/software/mirna_databases/
TargetScan	http://www.targetscan.org/vert_80/
FishmiRNA	http://www.fishmirna.org/
Software programs	Website
Mireap	https://github.com/liqb/mireap
miRDeep2	https://www.mdc-berlin.de/content/mirdeep2-documentation
Prost!	https://prost.readthedocs.io/en/latest/ https://github.com/uoregon-postlethwait/prost
CLC Genomic workbench	Commercially available software tool
miRGalaxy	https://usegalaxy.org/u/desl/w/analysis https://github.com/Glogobyte/miRGalaxy.

3.2.2.2 sncRNA Expression

To assess sncRNA differential expression from RNA-Seq data, similar software analysis programs for mRNA expression analysis are used, based on the generated expression matrices. The most commonly used are DESeq2 [67] and edgeR [68], which use raw read counts as input files, and differences between the two methods were reviewed by Maza and colleagues in 2016 [69]. Regardless of the software applied to detect differential expression, the generation of the applied read count matrix is crucial to obtaining robust and reproducible results, including similar read counts among biological replicates as well as count thresholds [70]. In this line, one of the most common queries in sncRNA sequencing design is how deep miRNA sequencing needs to be for each biological replicate. Since the amount of sequencing for each biological replicate is also dependent on the posed research question and studied tissues/stages, only rough recommendations can be made. The comparison of only two profiles certainly does not require extensive sequencing. Most sequencing providers recommend 5–10 million reads per sample. Assuming that most of the reads are annotated as miRNAs, and considering an miRNA as expressed with a read count of ≥2 reads, a simulation study by Sun and colleagues [62] showed that a sequencing depth of 5 million reads would spot around 72%, while a sequencing depth of 10 million reads around 82% of the expressed miRNAs. To detect miRNAs with low-copy numbers, thus ~100% of expressed miRNAs, a sequence depth of at least 24 million reads would be needed. Consequently, to determine low-copy number miRNAs, novel transcripts and their isoforms, higher sequencing depths are recommended, while for more classical miRNA expression approaches, a sequencing depth of around 10 million reads may capture a reasonable number of expressed miRNAs. Most of the sncRNA studies are focusing on the differential expression of miRNAs and are therefore extracting read counts only for known and novel miRNAs, neglecting the rest of obtained transcript data. Differential expression analysis may therefore also vary across publications according to their approaches of either applying all quality-trimmed sequencing reads or extracting only miRNA expression data to generate the expression matrix. Besides NGS to detect differentially expressed miRNAs, the use of microarray has also been reported. For more targeted approaches, but also to confirm results obtained from high-throughput methods, qPCR and droplet digital PCR (ddPCR) are typically carried out. Regarding the quantification of single or low-copy number miRNAs, it is recommended to carry out ddPCR, as it is the most sensitive one among today's available molecular tools.

3.2.3 Long Noncoding RNA (lncRNA)

lncRNAs represent not only the largest but also the less-studied group of ncRNAs [71–73]. Growing evidence shows that lncRNAs are major regulators of a broad range of physiological and pathological processes such as cell proliferation, differentiation, metastasis, and apoptosis. In teleosts, in contrast to mammals, investigations of lncRNA are still at their beginning and a degree of lncRNA-mediated regulation remains to be explored. The categorization of lncRNA has not been uniformly defined and, thus, no golden standard has been established, even though several databases like GENCODE,

NONCODE, and RNAcentral are available and specific lncRNA criteria such as transcript length, tissue-specific expression, low expression, less conservation, and particularly secondary structures have been set. Still, their correct identification and annotation remain challenging due to their structural similarity to mRNA. An additional difficulty is the fact that some functional open reading frames (ORFs) may also comprise noncoding properties. Up until today, lncRNAs are categorized based on:

1) length, i.e., (i) the long and large noncoding lncRNA sequence (lncRNA), (ii) very long intergenic RNA (vlincRNA), and (iii) macro RNA lncRNA.
2) according to their proximity to protein-coding genes, i.e., (i) overlaps with the antisense strand of a protein-coding gene (antisense lncRNAs), (ii) the lncRNA sequence is not located near any other protein-coding loci (intergenic lncRNAs, lincRNAs), (iii) the lncRNA sequence is derived completely from within an intron of another transcript (intronic lncRNAs, e.g., circRNA), (iv) the lncRNA sequence is located on the opposite strand from a protein-coding gene (bidirectional lncRNAs), and (v) the lncRNA sequence overlaps with the sense strand of a protein-coding gene (sense RNA) (Figure 3.7).
3) lncRNA residence within specific DNA regulatory elements and loci (e.g., enhancer lncRNA eRNAs).
4) lncRNA biogenesis pathway.
5) function (e.g., architectural lncRNA, arcRNA; competing endogenous lncRNA, ceRNAs).
6) subcellular localization (e.g., nuclear lncRNA).

The first eukaryotic lncRNA, H19, was discovered during the pregenomic era, followed by Xist, both of which became stepping stones of dosage compensation in mammals. H19 was identified in the early 1980s [75] and was falsely classified as mRNA. Intensive research followed to unravel the functions of this multitasking lncRNA, up until the second lncRNA, Xist, was discovered and reported to be involved in dosage compensation in mammals. With the advent of high-throughput sequencing technologies, including NGS, a plethora of novel noncoding transcripts were reported. Today, ~270,000 human lncRNAs retrieved not only from publications, but also novel RNA assemblies, are reported in the LncBook database (https://ngdc.cncb.ac.cn/lncbook/index) [76], and ~174,000 are compiled in the Noncode database (http://www.noncode.org/analysis.php) [77].

Generally, lncRNAs have a length from 200 bp to 100 kb and represent a very large and heterogeneous group. They are expressed in a tissue-specific (~78%) manner, and in most cases, they present about 10-fold lower abundance than those of protein-coding genes [78]. This has been explained by the higher cell-to-cell variation of lncRNA expression levels in the individual cells when compared to protein-coding genes [79]. The regulation of gene expression by lncRNAs occurs either by modulating the chromatin structure, binding transcription factors, controlling RNA splicing, or modifying mRNA stability as well as miRNA sponges [2, 80, 81]. Furthermore, lncRNAs have been reported to encode micropeptides and

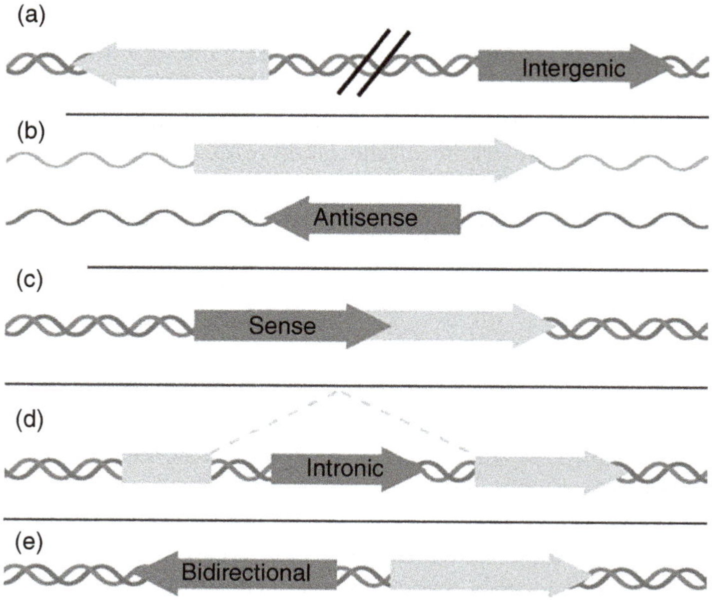

Figure 3.7 Categories of lncRNA. lncRNAs are classified according to their genomic position, orientation, and their relative location to nearby protein-coding genes. (a) Intergenic, (b) antisense, (c) sense, (d) intronic, and (e) bidirectional. lncRNA genes are marked in purple and protein-coding sequences in yellow. *Source:* Tsagakis et al. [74]/John Wiley & Sons/CC BY 4.0.

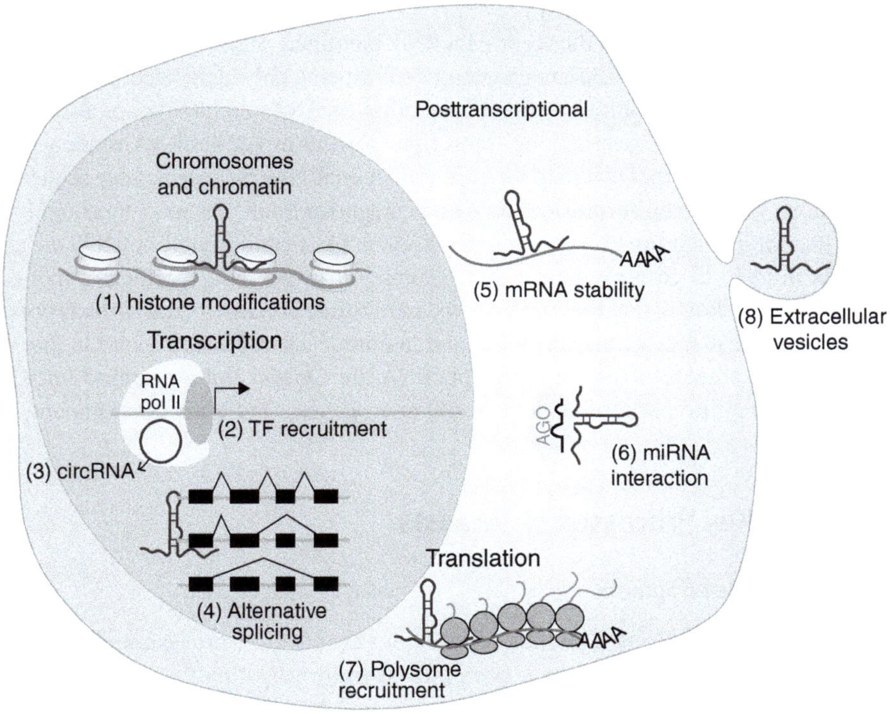

Figure 3.8 Different types of lncRNA regulatory roles during gene expression. *Source:* Fernandes et al. [83]/MDPI/CC BY 4.0. (1) Chromatin and chromosome condensation through histone modifications, (2) transcription factors (TF) direct recruitment, (3) binding to RNA polymerase (pol) II, (4) alternative splicing, (5) mRNA stability, (6) miRNA availability, (7) polysomes recruitment, and (8) modulation of gene expression in neighbor cells through packaging into extracellular vesicles.

control several distinct cellular functions comprising the regulation of transcription [82] (Figure 3.8). For those readers who wish to read more about the history, discovery, and classification of lncRNAs, an extensive review of lncRNAs has been published by Jarroux and colleagues [84].

Still, most of the work in lncRNAs is carried out in mammals, with the best-known examples being the transcripts encoding for the sex-determining region Y, and oskar RNAs [85], both of which are imperative for reproduction. Regarding teleosts, tissue-specific expression patterns have been demonstrated [86] and their participation in numerous biological processes such as immune response [87–89], sex differentiation [90, 91], sex-related differences [92], smoltification process [93], skin color regulation [94], intestinal homeostasis [95], and lipid metabolism [96] has been reported. More recently, lncRNAs in the gilthead seabream fast skeletal muscle have been investigated and provided the first indications concerning their involvement in muscle development [97]. The best-known lncRNAs are shown in Figure 3.3. These types of lncRNAs for which investigations in teleosts have been carried out are briefly outlined here.

3.2.3.1 circRNAs

circRNAs were first identified in the late 1970s in plant viroids and hepatitis delta virus, but until 2010, they were thought to be derivatives of endogenous RNA splicing errors or the result of experimental flaws. Improved protocols alongside advanced computational methods enabled their detection *ab initio* investigating RNA sequencing data across model and nonmodel organisms. CircRNAs are further categorized into three groups, i.e., exonic, intronic, and exon-intronic circRNAs, all of which are characterized by a covalently closed-loop structure. The most commonly known role of circRNAs is their function as miRNA sponges. Another mode of action is the capability of circRNA to reduce the amount of pre-mRNA and the interaction with RNA polymerase II and U1 small nuclear ribonucleoprotein at the promoter of their parental genes (reviewed in [98, 99]). One of the first and most prominent examples of circRNA in mammals is the circular isoform of human antisense to cerebellar degeneration-related protein 1 RNA (CDR1as), which was shown to be functional in neural development [100]. In teleosts, the first studies were carried out on the grass carp, where circRNAs were shown to be associated with grass carp reovirus (GCRV) infection [101, 102]. Increasing interest in this particular type of lncRNA will certainly expand to further different biological processes such as reproduction [103, 104].

3.2.3.2 Large Intergenic Noncoding RNAs (lincRNAs)

Xist was among the first lincRNAs studied and the second lncRNA identified. Other well-studied and prominent lincRNAs are Malat1, Neat1, and Miat [105–107]. lincRNAs are characterized by being polyadenylated, capped, and products of stable RNA polymerase II. Additionally, and in contrast to that of the other lncRNAs, lincRNAs are also defined based on exclusion, i.e., not coding for proteins and not overlapping transcripts. Furthermore, lincRNA sequences have been found to evolve very rapidly. It has been reported that less than 6% of zebrafish lincRNAs have detectable sequence conservation with humans or mice [108]. Efforts made to uniformly annotate and categorize lincRNAs were thoroughly reviewed by Ulitsky and Bartel in 2013 [109]. Regarding their function, not much is known. Most commonly, lincRNAs are acting in concert with other ncRNAs, as well as mRNA, to either enhance their expression or act as an artificial "sponge," as circRNAs do. Investigations into lincRNAs have shown, that they are involved in multiple processes such as the procession of p53 response pathways, regulation of epigenetic marks, gene expression, and maintenance of pluripotency. On the other hand, they may also act as "enhancer RNAs." Examples of the conserved lincRNA are Cyrano and Megamind lincRNAs, which perform important functions in zebrafish for proper morphogenesis and neurogenesis and brain development, respectively [108].

3.3 Roles of ncRNA in Key Processes of Teleosts

3.3.1 Roles of ncRNA During Development

Development is a complex process requiring a plethora of different events to be orchestrated accurately, temporally, and spatially. Regarding lncRNAs, only a few studies have been published investigating their regulative roles during development. Moreover, having investigated in depth 25 selected zebrafish lncRNAs, Goudarzi and colleagues [110] hypothesized that lncRNAs do not play any evident role in embryogenesis. The importance of miRNA during teleost development, however, has been shown in a myriad of studies comprised of model and nonmodel fish species. In this sense, among ncRNAs, miRNAs were reported as chief regulatory elements for proper organism development based on their capability to modulate gene expression, including the expression of transcriptional factors [111]. Consequently, investigations of ncRNAs during teleost development have been focused on miRNAs. The first miRNAs were reported in zebrafish [112], applying a newly developed computational approach (named MiRscan) to identify miRNA genes relying heavily on the high conservation of the stem-loop precursor RNAs among vertebrates. This approach revealed 38 miRNAs obtained out of zebrafish cDNA libraries comprising 8 embryonic stages (i.e., zygote, 256 cells, germ ring, 75% epiboly, tailbud, 11 somite, 24 hours, and 48 hours). With the rapid advances in sequencing techniques, further studies investigating gene regulation by miRNAs during zebrafish development followed, identifying not only new miRNAs for zebrafish [113] but also novel, not-yet-identified miRNAs [114]. Among the earliest miRNAs (in double-stranded form) was let-7, for which its deletion resulted in specific defects in the vertebrate embryo [115]. Subsequent studies proposed that let-7 miRNAs have significant roles in the regulation of fish growth development [116]. An initial high-throughput study investigating miRNAs in early teleost development was carried out in rainbow trout (*Oncorhynchus mykiss*) by cloning concatermized small RNAs using a TOPO TA cloning kit [117]. Authors identified, out of 150 sequenced clones, 14 unique miRNAs that showed a high conservation rate across numerous species, and described four distinct miRNA expression patterns supporting the stage-specific miRNA function in teleosts.

With the new possibilities of the NGS technologies, several studies including other model fish species like the medaka (*Oryzias latipes*) [118] and the three-spined stickleback (*Gasterosteus aculeatus*) [119], but also nonmodel species like the Atlantic halibut (*Hippoglossus hippoglossus*) [59] and the European sea bass [58], have been carried out, generating high-throughput sncRNA data in early teleost development. The studies mentioned here also showed these development-specific patterns reported by Ramachandra and colleagues [117]. High-throughput investigations of the three-spined stickleback also reported an increase in total read count at later developmental stages. Thus, in the early developmental stages, miRNA representation might be poor in terms of diversity and abundance. In contrast, at a later stage (24 hours post-hatch), both miRNA diversity and miRNA number of reads are high (Figure 3.9). Similar results are documented in the Atlantic halibut, in which, at the first stages, poor miRNA expression has been identified [59]; in the Atlantic salmon (*Salmo salar*) [119], reporting an increase of miRNA diversity across developmental stages; and the Arctic charr (*Salvelinus alpinus*), with less number of differently expressed miRNA at the earlier stages [120].

All these studies reported distinct miRNA functions during teleost development, applying different total RNA isolation, miRNA identification, and expression methods. The role of miRNAs during development can therefore not be doubted.

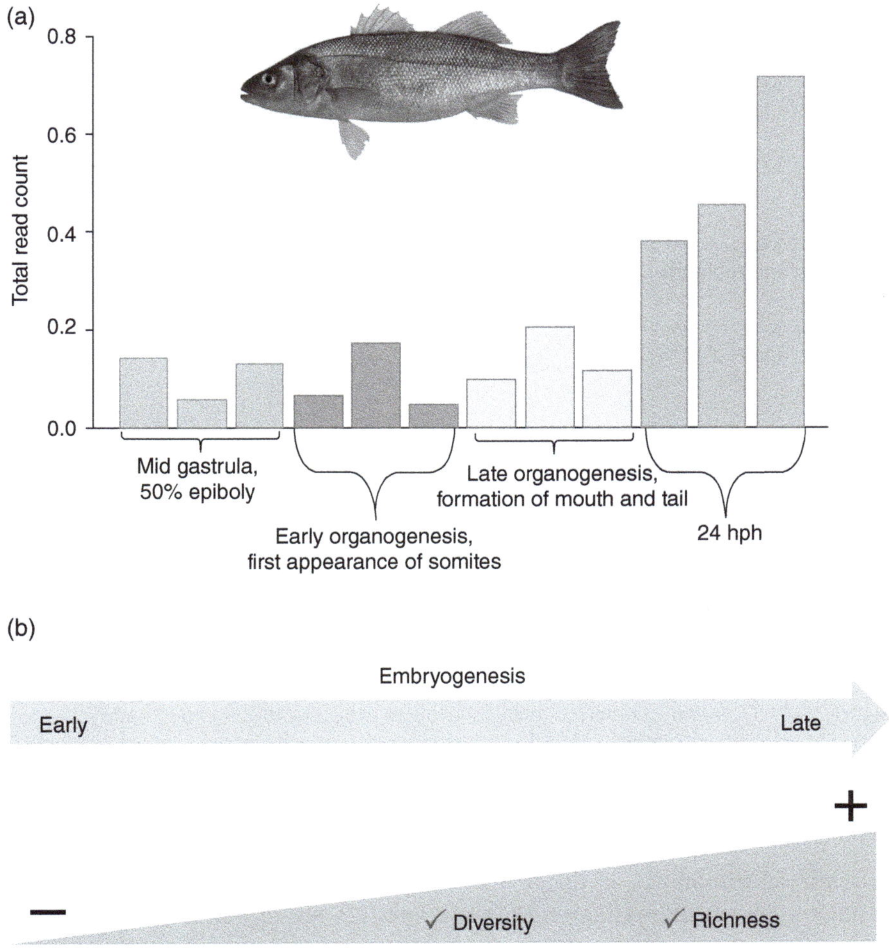

Figure 3.9 miRNA abundance during early teleost development. (a) Total read counts per sample extracted out of the European sea bass miRNA RNA-Seq results, (b) increased miRNA diversity and richness along later developmental stages reported in four different species, i.e., European sea bass. *Source:* Sergey Goruppa/Adobe Stock; Modified from Sarropoulou et al. [58], Kaitetzidou [12], Bizuayehu et al. [59], Robledo et al. [60]. (*See colour plate section for colour representation of this figure.*)

Certain mature miRNAs may have even been conserved across species, even though it remains to unravel whether or not they also share common target genes. Besides, it has been shown that miRNAs may add to phenotypic canalization [121–125] since they are assumed to be robust to genetic noise, and thus provide stability to embryonic development [126–130].

3.3.2 Evaluated miRNA Functions During Teleost Development

Apart from expression analysis, important miRNA functions in teleost development have been investigated, e.g., miR-430, highly expressed in the zebrafish blastula stage, is involved in maternal transcript clearance [131]. Similar to the situation in zebrafish, miR-430 was found to be expressed at the morula stage of the European sea bass [58], in the blastula and epiboly stage of the Atlantic halibut [59], and in the mid-gastrula and early organogenesis stages, the three-spined stickleback [12]. The presence of miR-430a was also shown from the late blastula to early neurula stages in medaka [118], supporting the importance of miR-430 function during early development. miRNAs have also been shown to hold important functions concerning cell movements, particularly during gastrulation. Pioneering work investigating the regulating role of miR-206 during development in zebrafish has been carried out, demonstrating that miR-206 is essential for normal gastrulation and cell movements, and is maternally expressed [132]. In other teleosts, miR-206 is detected throughout development, except for the morula stage, and is essential for proper muscle development [58, 133]. Among the most prominent genes in development are the well-known *homeobox* (*hox*) genes, which determine the organism's body plan. *Hox* genes encode a family of transcriptional regulators that function either by enhancing or by repressing genes key for distinct developmental programs pattering the anterior–posterior axis. First, in the fruit fly drosophila, miRNA-mediated regulation of *hox* genes has been reported. Since several hox-regulating vertebrate miRNAs are encoded with the Hox cluster, i.e., miR-10 and miR-196, miRNAs and their *hox* mRNA targets are co-expressed, facilitating the modulation of *hox* expression during

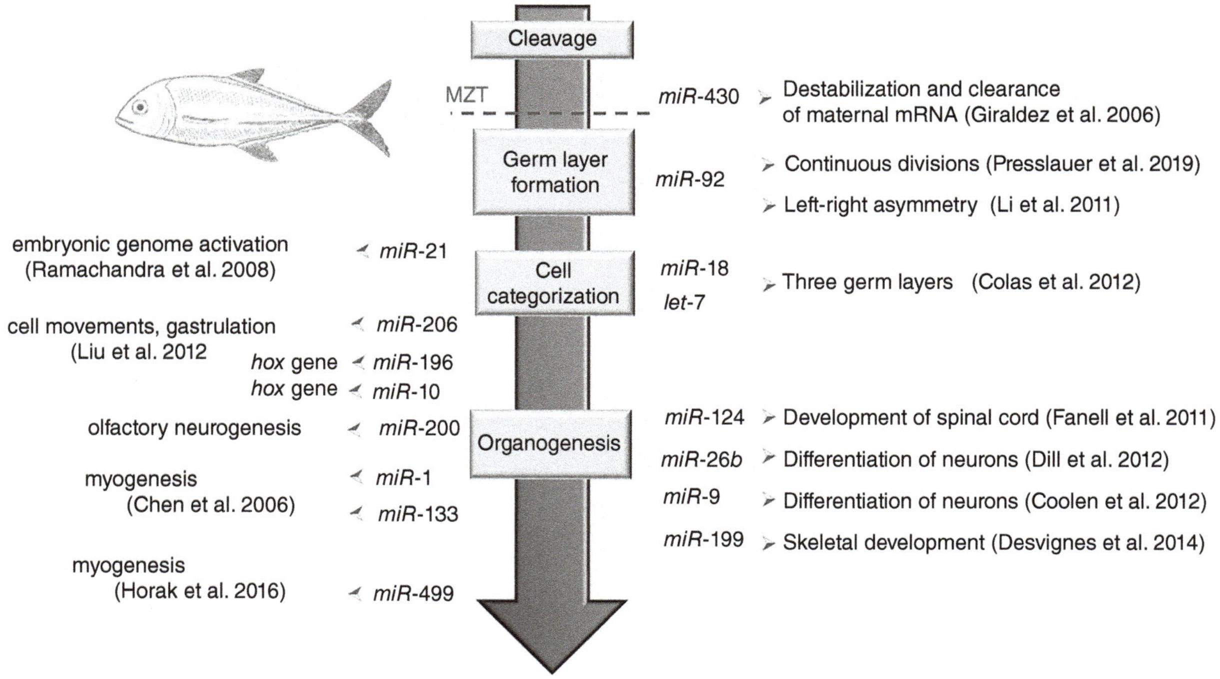

Figure 3.10 Overview of reported miRNAs involved in teleost development embryonic genome activation [117], cell movements, gastrulation [132], myogenesis [137, 138], destabiliatization and clearance of maternal mRNA [131], continuous cell division [139], left-right asymmetry [140], three germ layers [141], development of spinal cord [142], differentiation of neurons [139, 143], skeletal development [144]. Abbreviations: MZT, mid zygote transition.

vertebrate development. In teleosts, the role of the miR-10 family has been investigated in Nile tilapia (*Oreochromis niloticus*), demonstrating that *hoxB3a* and *hoxD10a* are regulated by miR-10b-5p by applying luciferase target reporter assays for target validation [134]. In turbot (*Scophthalmus maximus*), miR-10b was higher expressed at five days post-fertilization (dpf). This stage is characterized by endogenous feeding and low motor activity [135]. The role of miR-196 during teleost development has been investigated in zebrafish, with research proving that miR-196 is involved in the regulation of axial patterning and pectoral appendage initiation [136]. In the European sea bass, miR-196b was reported to be expressed from half epiboly onward, while in Atlantic halibut miR-196 was relatively highly expressed during somitogenesis and hatching [59] (Figure 3.10).

3.3.3 Roles of ncRNA During Reproduction

A few attempts have been made to investigate the roles of ncRNA in reproduction. In 1975, Mazabraud and colleagues [145] reported high quantities of tRNA and 5S rRNA in the ovaries of nine teleosts. They further showed that in four of the species being studied, more than 90% of the RNA belonged to tRNA and 5S rRNA, suggesting the latter as a sensitive and universal marker of oogenesis and oocyte differentiation in fish. In amphibian oocytes, it has been shown that ribosomal DNA (rDNA) is amplified about 1000-fold higher during early oogenesis. Additionally, 5S rRNA expression changes during oogenesis, with the peak activity coinciding with early vitellogenesis. The transcription rate in later stages is slowed down. In Xenopus, two 5S rDNA types have been isolated within the gonads, one being massively expressed during oocyte maturation, but not during male maturation. Over three decades later, Cancio and colleagues documented that the ratio of the 5S rRNA to the 18S peaks correlates to the oocyte developmental stage [146] (Figure 3.11).

Subsequently, other studies followed [148, 149], confirming this correlation and underlining the significance of RNA profile examination before RNA-Seq analysis, and representing an effective maturity index to estimate fish fecundity.

Regarding sequencing of sncRNA libraries generated out of male and female gonads, and in contrast to other tissues, read length distribution reveals two main peaks for both mature and immature fish, at 19–24 nt and 25–30 nt (Figure 3.12). While the first peak is characteristic of miRNAs, the second one is presumed to comprise piRNAs.

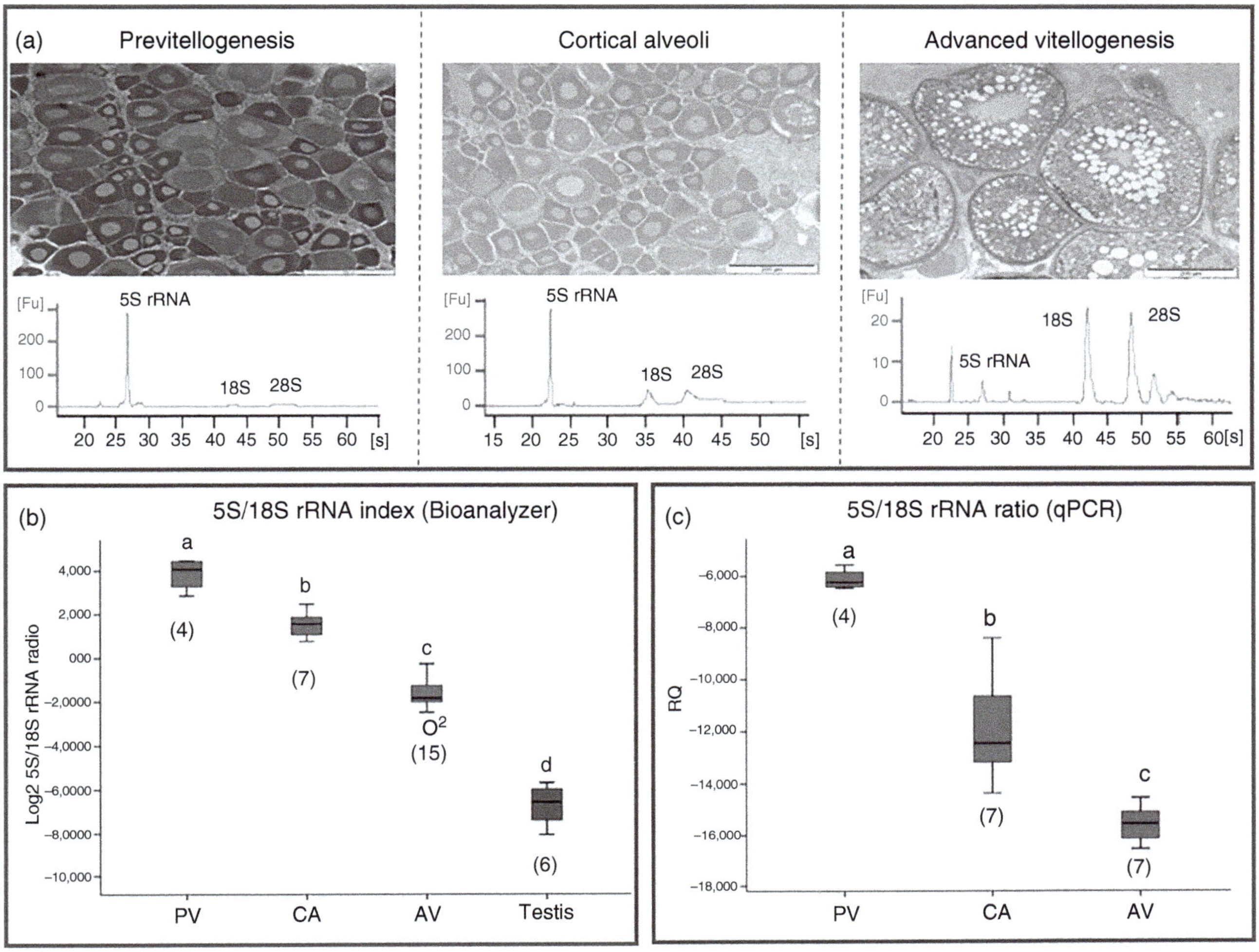

Figure 3.11 5S/18S abundance in the gonads. (a) Total RNA electropherograms and 5S/18S rRNA index in the gonads of megrims as shown in gonad histology and electropherograms of total RNA distribution in megrim (*Lepidorhombus whiffiagonis*) ovaries at different stages of oogenesis (advanced vitellogenesis -AV-, cortical alveoli -CA-, and previtellogenesis -PV-). [FU] = Fluorescence, [s] = time in seconds. (b) 5S/18S rRNA index in ovaries at different developmental stages and in testis as quantified from electropherograms. (c) 5S/18S rRNA index after qPCR analysis of the levels of transcription of both genes in ovaries at different developmental stages. Numbers between parentheses indicate the number of individuals analyzed per group. Letters show significant differences among pairs of means (one-way ANOVA, $p < 0{,}05$ in B, Jonckheere-Terpstra, $p < 0{,}05$ in C), *Source:* Rojo-Bartolomé et al. [147], PUBLIC LIBRARY OF SCIENCE (PLOS), CC BY 4.0.

Figure 3.12 sncRNA size distribution from teleost developmental stages and gonads. *Source:* Data retrieved from Papadaki et al. [32] (Sharpsnout sea bream); Lau et al. [150] (Medaka); Kaitetzidou et al. [151] (Stickleback), and Sarropoulou et al. [58] (European sea bass development).

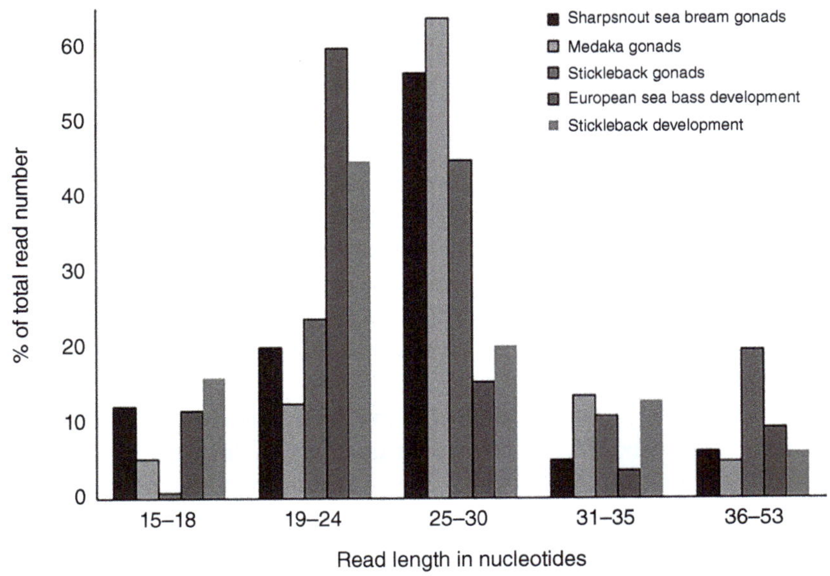

piRNAs have been reported to play a crucial role in sex determination [152] and germline development [153]. Mutations within Piwi have been shown to lead to gametogenic defects and, thus, sterility [154]. piRNAs are found in both male and female gonads [154, 155]. While the role in female gonads has not been entirely studied, deletions in piRNA of the male gonads significantly interfere with spermatogenesis, resulting in male infertility [156]. Concerning teleosts, most piRNA studies have been carried out in zebrafish, where piRNAs most likely have an important role in the silencing of repetitive elements in gonads [43]. Nevertheless, along with the rapid progress of NGS technologies, more sncRNA-Seq studies have been carried out that identify putative teleost piRNAs, mostly by size distribution in combination with excluding other sncRNAs by annotation (e.g., [58, 59]). First investigations to identify candidate piRNAs have been performed in the Japanese flounder [51]. Illumina RNA-Seq was used to isolate sncRNAs, and then size selection combined with blast search against the zebrafish piRNA database allowed for the isolation of 5181 and 3915 piRNA sequences, which clustered into 289 piRNAs. Differential expression analysis revealed 38 ovary-specific and 24 testis-specific piRNAs. Only recently, piRNAs have also been investigated in the seminal plasma exosomes in the half-smooth tongue sole (*Cynoglossus semilaevis*) [157], where piRNAs were identified first by length and exclusion of other sncRNAs (tRNA, snRNA, snoRNA, miRNA, and rRNA), blasting against the piRBase. To predict unannotated reads, the software program Piano [158] was applied. The study identified and characterized 44 piRNAs in total with differential expression profiles between males and pseudomales. Putative target genes were also predicted, among which several genes were found to be involved in methylation or transposition. Based on the data, it was concluded that piRNA might play an important role in the regulation of sex development, and thus it may provide the basis for the development of biomarkers to identify the sex of fish. The functional role of piRNAs and their prospects in aquaculture have recently been reviewed by Lite and colleagues [54]. Sex-specific differences in miRNA expression have been reported in a broad range of teleost species (e.g., [159–161]) and reviewed by Bhat et al. [162]. Therefore, tissues known to have gender-specific peculiarities such as the gonads, brain (but also adipose tissue), and liver have been the main targets. The main function of miRNA in reproduction has been hypothesized to be during sex steroid hormone synthesis [163]. Successfully annotated miRNAs in the gonads ranged from 2.8% in the Atlantic halibut [164] to 20% in the marine medaka (*Oryzias melastigma*) [150]. Besides the total number of annotated miRNAs, the numbers of differentially expressed miRNAs between sexes also varied among the species. This variation may be explained by the maturation stage of the examined gonads, the significant threshold set by different researchers, the applied algorithm for differential expression, and the pursued analysis pipeline. Most of the studies investigated differential expression only on the subset of identified miRNAs and thus, applied statistics may result in different significantly expressed miRNAs compared to those studies applying differential expression analysis for the whole set of sequenced reads after quality and adaptor trimming. In the discus fish (*Symphysodon aequifasciatus*), for example, 551 miRNAs were described, out of which 47 (~9%) were differentially expressed [161]. In the Atlantic halibut gonads, 31 miRNAs (16%) with at least a twofold difference in expression between sexes were reported out of 188 miRNAs identified. A major issue in some farmed fish is their inability for gonad maturation. Therefore, several studies examined sncRNAs, and in particular miRNAs, at different time stages of gonad maturation [165], or by comparing mature and immature individuals [32]. Based on the different approaches to detect differentially expressed miRNAs not only between ovary and testis, but also between mature and immature gonads, overall comparisons of the various studies are not feasible. Nevertheless, looking at the chief miRNAs with either the highest abundances or the most significant log2fold changes may allow for the identification of the most important miRNAs in reproduction. One of the most prominent miRNAs predominantly expressed in the gonads currently is miR-202, which has been detected in mammals [62, 166–168], frogs (*Xenopus tropicalis*) [169], and birds (*Gallus gallus domesticus*) [170, 171], as well as in teleosts [32, 160, 164, 172]. Of note, miR-202-3p has been reported to be higher expressed in the testis of Atlantic halibut and in the three-spined stickleback. In the marine medaka, on the other hand, miR-202-5p was found to be gonad-enriched in both sexes, regulating the early steps of oogenesis [150]. Studies in mice reported that both miR-202-3p and -5p are highly expressed in testis, specifically in the spermatogonial stem cells [173]. Additionally, cellular and molecular analyses in medaka showed that the lack of the miR-202 gene (miR-202 knockdown in female medaka by CRISPR/Cas genome editing) negatively affects oogenesis/folliculogenesis, resulting in reduced female fecundity [172]. The possible altered function of miR-202-3p and -5p has been investigated during zebrafish gonad development and adult gonads. During development, the predominant arm (5p/3p) has been reported to be miR-202-3p, while in adult gonads and mature oocytes, it is miR-202-5p [165, 174].

Another broadly discussed miRNA is miR-143. In mammals, it has been found to hinder the formation of primordial follicles, negatively affecting the expression of cell cycle genes and the inhabitation of the proliferation of pregranulosa cells [175]. In teleosts, miR-143 has been found in the testis and in the ovaries of the marine medaka [150], in sexually mature and immature Atlantic halibut males [164], and in a plethora of other teleosts. Interestingly, miR-143 has been found in all relevant gonad tissues, e.g., ovaries, testis, and vitellogenic follicles. Bizuyehu and colleagues [164] proposed that miR-143, together with let-7a and miR-202, are upregulated to induce testes differentiation. In stickleback and yellowfin seabream (*Acanthopagrus latus*) gonads, miR-143 has also been found to be more abundant in testis [126, 155]. On the other hand, reports have also been published describing high abundance in ovaries [53, 176, 177]. In humans, miR-143 alterations have been associated with tumor cells and cardiogenesis [178, 179], but no record has been published yet about miR-143 function during reproduction, although it has been named as one of the most important and conserved sex-biased miRNAs based on the findings in teleosts. Other frequently discussed miRNAs being differentially expressed, either during gonad maturation, between male and female gonads, or mature and immature gonads, are let-7a, miR-7a, miR-146a, miR-181a, miR-455, miR-727, and miR-2188. let-7a and miR-7a [180] are involved in the regulatory functions of mice spermatogonial differentiation, with both showing higher expression in testis compared to ovaries [151].

miRNAs are known to have a broad range of functions since they may regulate multiple genes across several pathways and conditions. Specific database search methods like text mining, also referred to as text analytics, may help extract their main functions, if they exist. Text mining is the exploratory process of large collections of documents to support answering specific research questions or to discover new information and interconnections. In this respect, miR-146a and miR-455 are of special interest since the text mining of the alleged function of both miRNAs reveals that their main functions are in immune and disease response. miR-146a, in particular, has gained importance in medical research as a regulator of the innate and adaptive immune response (reviewed in [181]). Similarly, also in teleost fish, miR-146a and miR-455 seem to be involved in an immune response. Finding miR-146a and miR-455 being differentially expressed not only between ovary and testis but also during gonad maturation [151, 161, 182, 183] may indicate the importance of a strong immune system for successful reproduction. In medicine, the field of research known as "reproductive immunology" aims to elucidate the interactions (or the absence of them) between the immune system and components linked to the reproductive system. In fish, this approach has been investigated in the gilthead sea bream, where it was demonstrated that immune-relevant molecules such as Il1b and Mcsfr are generated by germ cells (Il1b) and ovarian and testicular somatic cells (Mcsfr). Thus, it is proposed that, as for mammals, immune molecules may play an essential role in the regulation of gametogenesis and the maintenance of gonad tissue homeostasis in fish [184]. Independently, studies in two teleost species also identified two miRNAs, miR-451 and miR-190, being expressed in the small vitellogenesis stage of oocyte maturation [32, 185]. The former, however, has been reported to be mainly involved in erythrocyte maturation and consequently needs further investigation to understand its role in reproduction. Several miRNAs have been detected among various fish species with different sex-determining systems and under different conditions. Consequently, miRNAs play an important role in sex determination, differentiation, and gonad maturation. Further research implementing miRNA knockouts, or at least, luciferase reporter gene assays, will provide the ultimate proof of the functions of distinct miRNAs in these processes.

3.3.4 Roles of ncRNA in Immune and Stress Response

The immune response to infectious diseases in fish consists of a complex interaction between nonspecific and specific defense mechanisms. The prevention of disease and adequate management are essential for the well-being of teleosts, and in particular, to maintain the sustainability of the modern aquaculture industry. Along with the high diversity of teleosts, a broad range of pathogens also exists, and major losses have been documented in some fish farms. The main pathogens of concern for fish farming are: (i) *Vibrio anguillarum*, (ii) *Aeromonas salmonicida*, (iii) *Renibacterium salmoninarum*, (iv) *Yersinia ruckeri*, (v) *Edwardsiella 22ctalurid*, and (vi) *Pasteurella piscicida* [186]. Several studies at not only the physiological, but also the transcriptomic, level have been carried out to investigate fish immune response. Accordingly, a comprehensive number of lncRNAs and sncRNAs have been identified to exhibit altered expression during pathogen infection [87, 187]. The importance of lncRNA and its regulatory roles in the innate human immune system has been reviewed by Xie and colleagues [188], and for teleosts by Wang and colleagues [87]. In teleosts, most of the work is still in the early stages, and published work mainly describes the identification of lncRNA in various tissues after infection. Nevertheless, specific lncRNA with antimicrobial and antiviral functions have also recently been described [189, 190]. The importance of the specific class of lncRNA, the circRNA in immune response, as well as the first approaches to investigate these mechanisms in teleost fish species, have been reported in the grass carp [101] and zebrafish [191]. Only recently in the miiuy croaker (*Miichthys miiuy*), an important circRNA, circBCL2L1, has been reported with the putative interplay between miRNA and mRNA, regulating TRAF6-mediated innate responses [192]. The use of circRNAs in fish as a potential biomarker application has been recently reviewed extensively by Rbbani and colleagues [193].

Regarding sncRNA studies related to immune and stress response in teleosts, the focus has been given to the investigation of miRNA occurrence as a possible regulator of important genes involved in immune response. Since candidate miRNAs have to be detected first for further downstream experiments, the majority of the studies, like in reproduction and development, are investigating the general miRNA or lncRNA expression patterns by sncRNA-Seq between control and challenged groups. Nonetheless, several publications report the function of specific miRNAs after pathogen infection [194–200]. Thus, apart from the diverse miRNA lists in several fish species, alleged functional roles are assigned to miRNAs following viral or bacterial challenges. It should be noted that some miRNAs are found to be dysregulated after either viral or bacterial challenges. This group of miRNAs includes, for example, miR-462, miR-731, miR-181, miR-223, and miR-146. The latter miRNA has two paralogs, miR-146a and -b, and has been extensively studied in several species. In mammals, miR-146a is considered to be a negative regulator of immune response and can be induced by viral infection [178, 179]. The first studies in zebrafish found that both miR-146-a and -b are upregulated in embryos infected with *Salmonella typhimurium* as well as in adults infected with *Mycobacterium marinum* [201]. In humans, the focus has been particularly placed on miR-146a. For more than a decade, miR-146a has been known to have important functions in the adaptive immune response by regulating INFγ-producing T-cells, and innate immunity by targeting *IRAK1* and *TRAF6* [181]. Studies in the half-smooth tongue sole predicted eight possible miR-146a target genes (*protein C [activated], nucleolin, kynureninase, beta BTB/POZ domain-containing protein KCTD12-like, protein FAM102A-like, CD83, hypothetical protein LOC100712247*, and *interleukin*) [202]. Further searches for negative correlations revealed that downregulation of cse-miR-146a expression enhances the expression of the target gene *il-1b*. Similarly, in the orange-spotted grouper (*E. coioides*), the regulatory roles of miR-146a previously found during red-spotted grouper nervous necrosis virus (RGNNV) infection were investigated. The authors highlighted that miR-146a overexpression facilitated viral infection, while miR-146a knockdown by specific inhibitors decreased RGNNV replication. Like in mammals, it was found that miR-146a targets the 3′ UTR region of grouper *traf6* [203]. While both miR-146a and *traf6* were upregulated during RGNNV infection, miR-146a suppressed *traf6*-mediated cellular immune response. Repression of *traf6*, in turn, influences the NF-kappaB-mediated inflammatory response through the expression patterns of inflammatory signaling molecules like interleukins IL-6, -8, -1β, and TNF alpha (TNF-α) [204]. The detection of miR-146a involvement not only in immune response in several fish species, but also in higher vertebrates, suggests that the regulation by this miRNA has been evolutionarily conserved across vertebrates. Other miRNAs with functions similar to miR-146a include miR-21, miR-181, miR-155, miR-223, miR-100, and miR-99, but still, more work needs to be carried out [187]. One miRNA can act on several different target genes involved not only in one pathway and/or physiological condition. On the other hand, several miRNAs can also regulate the same target gene. This has been shown through studying specific miRNAs after bacterial or viral challenges by functional assays. For instance, it was shown that after bacterial challenges, miR-19a, miR-148, miR-214, as well as miR-3570 are targeting the 3′ UTR of the *myeloid differentiation primary response protein* (*myD88*) in the miiuy croaker [194–196, 200]. All four miRNAs show different mature and seed sequences (Figure 3.13). Instead, in human cancer cells, another miRNA, miR-155, has been reported to target *myD88* [205]. Furthermore, miR-155 is also associated with immune response genes after viral challenge in the Japanese flounder, rainbow trout, Atlantic salmon, and miiuy croaker.

MyD88 belongs to the toll-like receptor (TLR) signaling pathway, for which also other genes have been found to be regulated by miRNAs (reviewed by Chu and Xu [204]) (Figure 3.14).

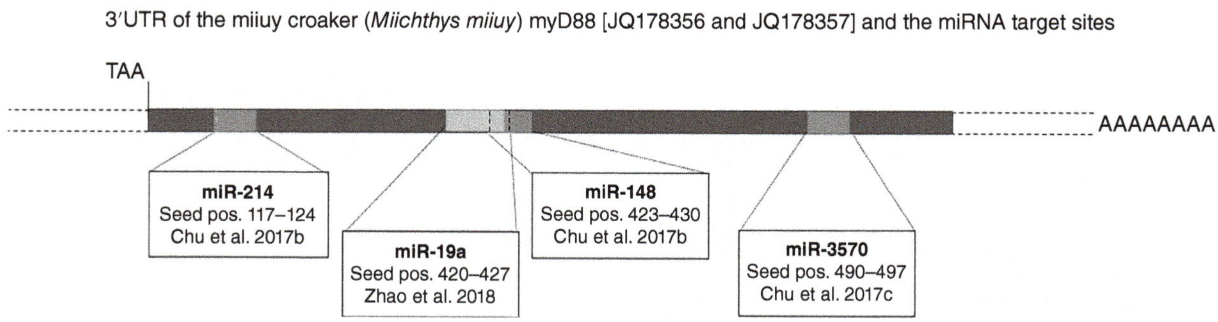

Figure 3.13 Target sites of miR-19a Zhao et al. [200], miR-148 Chu et al. [194], miR-214 Chu et al. [195], and miR-3570 Chu et al. [196] in the 3′ UTR of the myeloid differentiation primary response protein (MyD88) in the miiuy croaker (*Miichthys miiuy*). *Source:* Adopted from Tang et al. [205].

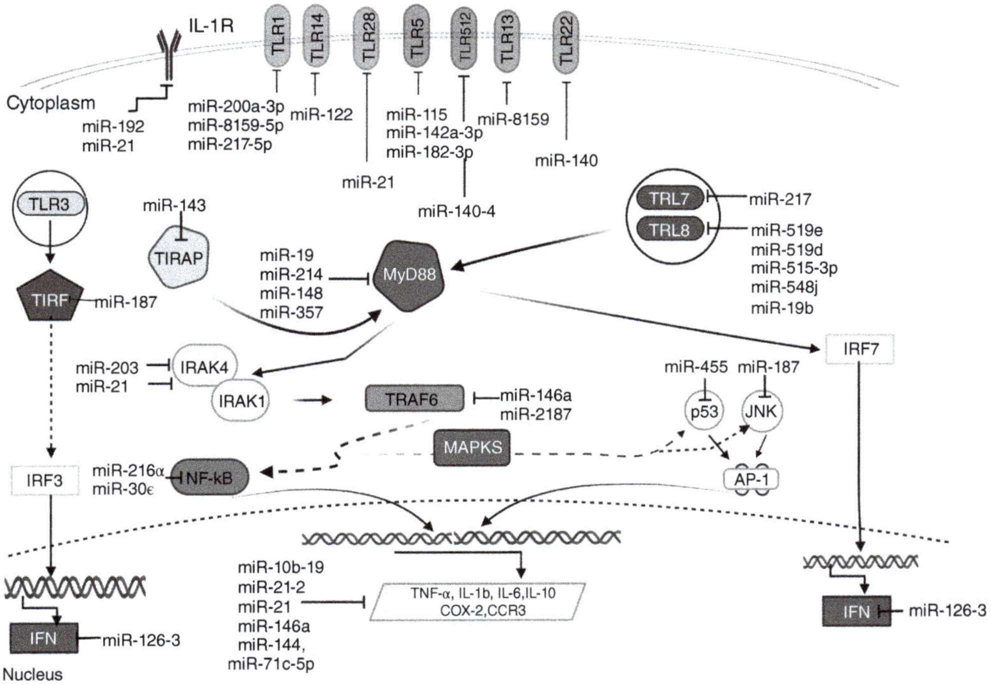

Figure 3.14 Schematic overview of miRs acting in toll-like receptor (TLR) signaling pathway. miRs are set according to published work on the miiuy croaker and orange-spotted grouper as described within the text. AP-1, activator protein; CCR3, CC chemokine receptor 3; COX2, cyclooxygenase-2 enzyme; IFN, interferon; IRAK, IL-1 receptor-associated protein kinase; IRF, activate interferon regulatory factor; JNK, c-Jun N-terminal kinases; MAPKs, mitogen-activated protein kinases; MyD88, myeloid differentiation primary response protein 88; NF-κB, nuclear factor κB; P53, tumor protein P53, also known as p53, cellular tumor antigen p53; TIRAP, TIR domain-containing adaptor protein; TLR, Toll-like receptor; TNF, tumor necrosis factor; TRAF, TNFR-associated factor; TRIF, TIR domain-containing adaptor molecule 1. *Source:* Modified form Chu and Xu [204], Cui et al. [206]. *(See colour plate section for colour representation of this figure.)*

Besides the TLR signaling pathway, miRNAs are also identified in the NOD-like receptor (NLR) (Figure 3.15) and RIG-I-like (RLR) (Figure 3.16) signaling pathways. The former is of great significance in pathogen recognition once bacteria or viruses have invaded the organism. The NLR family is a large intracellular receptor family, with NOD1 and NOD2 being the most studied cytosolic NLRs that can activate a sequence of signal cascades through the key adaptor protein receptor interacting serine/threonine kinase 2 (RIPK2), initiating NF-κB transcription factors and hence supporting the progress of inflammatory factors production. In the miiuy croaker, two miRNA (miR-210 and miR-3570) were identified as possible regulators of RIPK2 [207]. Another miRNA-inhibiting NOD1 expression was reported to be miR-217-5p [190]. In the same study, it was shown that distinct lncRNAs may have important regulatory roles by competing with miR-217-5p and thus relieving the NOD1 repression (see also below the function of lncRNA as competing endogenous RNA).

Regarding the RLR signaling pathway, two important miRNAs found to be upregulated after viral or bacterial infections in several fish species are miR-731 and miR-462. In rainbow trout, for example, both miRNAs were found to be upregulated in the liver of *viral hemorrhagic septicaemia virus* (VHSV)-infected fish [208]. miR-731 has been assigned to four target genes: *ATP-dependent RNA helicase* DHX58 (*RIG-I-like 3*), *C–C motif chemokine 19*, *macrosialin precursor CD68*, and *Interferon regulatory factor 7* (*irf7*). *irf7* was also identified as a potential miR-731 target gene in the Japanese flounder, where *irf7* suppression by miR-731 was detected after a megalocytivirus infection [211].

Another study in Atlantic cod (*Gadus morhua*) found miR-731-3p at higher abundances after treatment with the viral mimicking polyriboinosinic polyribocytidylic acid (polyI:C) [209], but no particular target was reported. Since in two fish species, the same target transcript was identified, it can be assumed that this site may be evolutionary conserved. Another example of miRNA involved in the RLR signaling pathway is miR-122, which was upregulated after infection with the *Singapore grouper iridovirus* (SGIV) in the grouper, which in turn may influence the activation of inflammation factors, AP-1 activity, and NF-κB and cell apoptosis [199]. Similar to this study, miR-122 has been detected to be upregulated in the

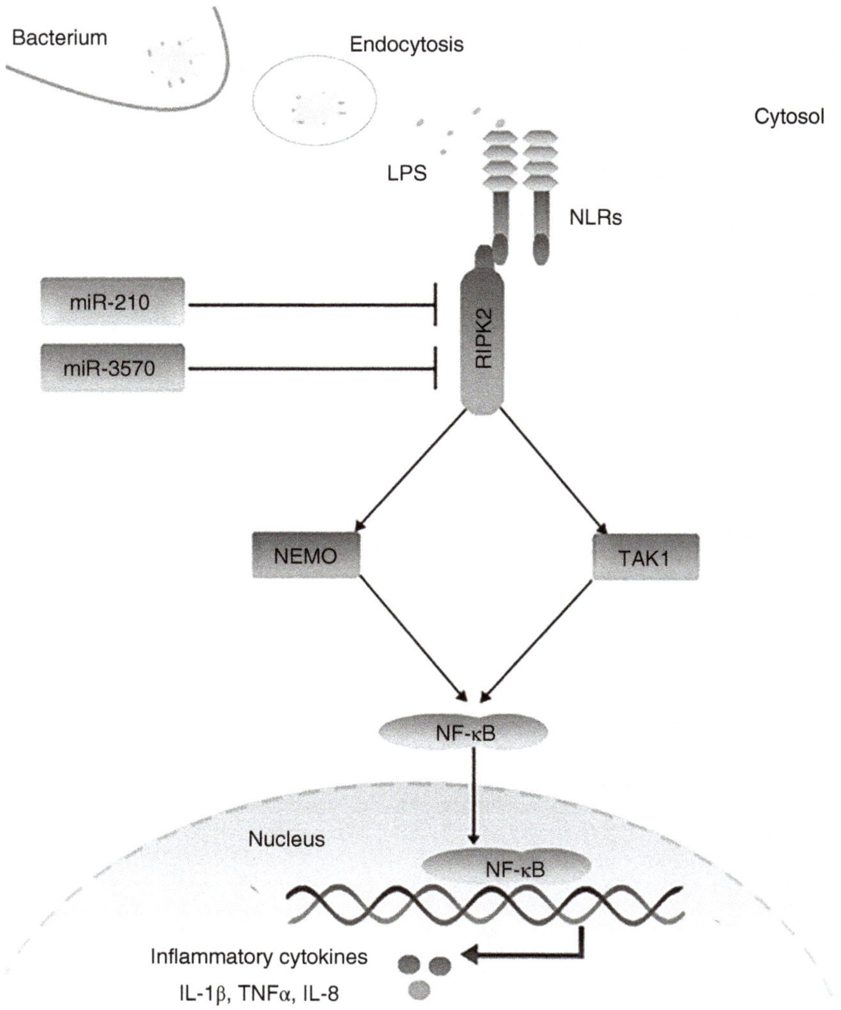

Figure 3.15 Schematic overview of miRs acting on the NOD-like receptor (NLR) signaling pathway. miRs are set according to published work on the miiuy croaker and orange-spotted grouper as described within the text. IL-1β, interleukin-1β; interleukin 8, IL-8; LPS, lipopolysaccharide; NEMO, NF-kappa-B essential modulator (NEMO) or inhibitor of nuclear factor kappa-B kinase subunit gamma (IKK-γ); NF-κB, nuclear factor κB; NLR, nod-like receptors; RIPK2, receptor-interacting serine/threonine kinase 2; TAK1, transforming growth factor b-activated kinase 1; TNF, tumor necrosis factor-alpha. *Source:* Adopted from Su et al. [207].

miiuy croaker after Poly(I:C) induction [210]. Han and colleagues showed by *in silico* target prediction, qPCR, and luciferase reporter gene assay that miR-122 most probably targets the gene *dihydroxyacetone kinase* (*dak*). However, miR-122 was also shown to be involved in the TLR signaling pathway after *V. anguillarum* induction in the miiuy croaker [206]. Moreover, as a famous physicist, Freeman J. Dyson once noted: "Things are always more complicated than most people believe.", it has been shown that miR-122 is regulated by a distinct lncRNA, adding a level of posttranscriptional regulation [212].

As shown earlier, numerous fish miRNAs control host-immune responses by targeting immune-related genes. Nevertheless, to understand the overall mechanism of the fish immune system, analysis of the miRNA regulatory machinery alone is not sufficient. Knowledge gained from mammalian studies showed that there is a complex multigene regulatory network comprising lncRNAs and circular RNAs (circRNA) with essential roles in such a network. These lncRNAs and circRNAs competing for endogenous RNAs (ceRNA) may counterbalance miRNA-mediated gene silencing.

3.4 ncRNAs as Biomarkers and Future Perspectives

The study of sncRNA in teleosts has gained increasing interest during the last two decades, particularly since its importance has been highlighted by the growing evidence of involvement in human diseases. Comparing different studies is rather challenging since different methods, sample numbers, and, most importantly, statistical approaches/thresholds are applied. Several studies refer to miRNA genes and do not refer to the family (a, b, c, etc.) or arm (5p/3p) specific expression. On the other hand, although miRNAs are known to be highly conserved across species, their target sites and genes might

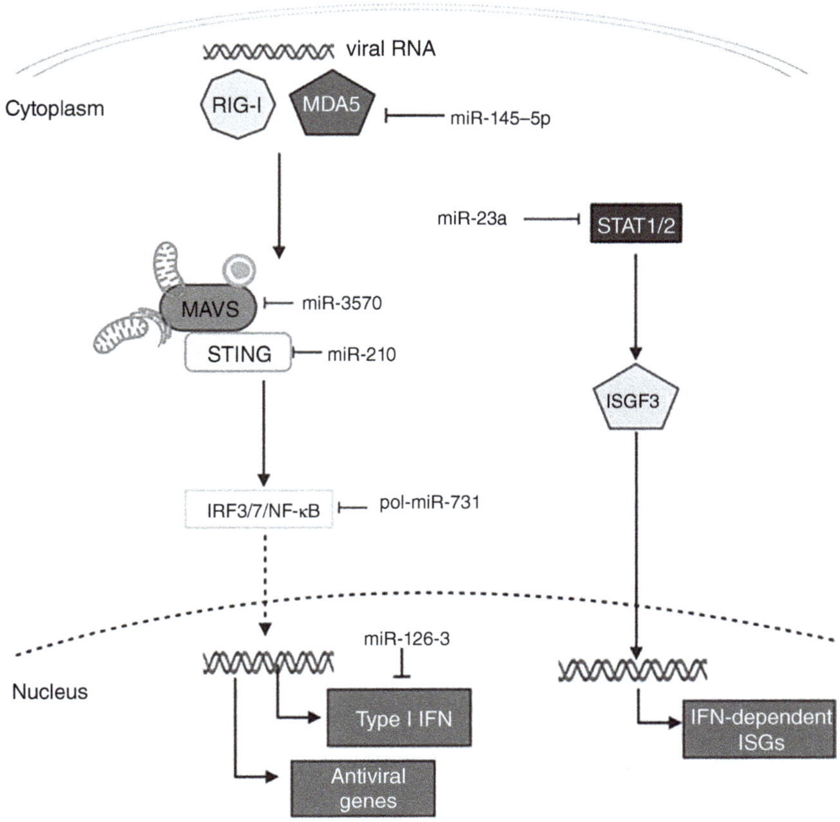

Figure 3.16 Schematic overview of miRs acting in the RLR signaling pathway. miRs are set according to published work on the miiuy croaker and orange-spotted grouper as described within the text. IFN, interferon; IRF, interferon regulatory factors; ISG, interferon stimulated gene; ISGF3, interferon stimulated gene factor 3; MAVS, mitochondrial antiviral signaling; MDA5, melanoma differentiation-associated gene 5; NFκB, nuclear factor κB; RIG-I, retinoid acid-inducible gene-I; STAT, signal transducer and activator of transcription; STING, stimulator of IFN genes. *Source:* Modified from Zheng et al. [190], Sun et al. [199], Schyth et al. [208], Eslamloo et al. [209], Han et al. [210].

not be, and thus they may be differentially regulated according to the species under study. Investigating the involvement of sncRNAs, e.g., in the immune response, certainly will have an impact on our understanding of fish defense mechanisms and can serve as a stepping stone to further investigate more circulating miRNAs. These miRNAs found in biofluids like blood, seminal plasma and urine hold promise to be the basis for a still-lacking nonlethal method for early disease detection in teleosts. In medical research, accumulating evidence demonstrates that miRNAs can serve as critical biomarkers in various complex human diseases. Besides miRNA, lncRNAs and circRNAs are also becoming a focus of important research areas for their possible application as biomarkers in fish aquaculture [193]. Thus, research on teleost ncRNA in association with their regulatory roles, not only in the immune response but also in connection with reproduction and feed intake/metabolism, may become a hot research topic in fish for providing not only a better understanding of the onset of disorders and/or diseases, but also a new method to prevent abnormal growth and development, reproductive dysfunctions, and/or disease outbreaks.

Acknowledgments

E.S. would like to acknowledge the funding from the Ministry of Education and Religious Affairs, under the Call “ARISTEIA I” of the National Strategic Reference Framework 2007–2013 (ANnOTATE), co-funded by the European Union and the Hellenic Republic through the European Social Fund; the European Union Seventh Framework Program (FP7 2010–2014) under grant agreement no. 265957 (CopeWell), and MedAID (Mediterranean Aquaculture Integrated Development) funded by the European Union in the frame of Horizon 2020, grant agreement number 727315. I.F. acknowledges funding by the European Regional Development Fund (ERDF) – Ministerio de Ciencia, Innovación y Universidades (MICIU) – Agencia Estatal de Investigación of the Spanish Government through MET2VI project (Ref. RTI2018-099029-A-I00), and from the MICIU and the European Social Fund, “The European Social Fund invests in your future” through the Ramón y Cajal (Ref. RYC2018-025337-I) contract from the Plan Estatal de Investigación Científica y Técnica e Innovación 2017–2020.

References

1 Encode Consortium, Carolina, N., and Hill, C. (2013). For junk DNA. *Nature* **489**: 57–74.
2 Statello, L., Guo, C.-J., Chen, L.-L., and Huarte, M. (2021). Gene regulation by long non-coding RNAs and its biological functions. *Nat. Rev. Mol. Cell Biol.* **22**: 96–118.
3 Mattick, J.S. (2001). Non-coding RNAs: the architects of eukaryotic complexity. *EMBO Rep.* **2**: 986–991.
4 Lee, R., Feinbaum, R., and Ambros, V. (2004). The *C. elegans* heterochronic gene lin-4 encodes small RNAs with antisense complementarity to lin-14. *Cell* **116**: 843–854.
5 Wightman, B., Ha, I., and Ruvkun, G. (1993). Posttranscriptional regulation of the heterochronic gene lin-14 by lin-4 mediates temporal pattern formation in *C. elegans*. *Cell* **75**: 855–862.
6 Moss, E., Lee, R., and Ambros, V. (1997). Control of developmental timing by the cold shock domain protein Lin-28 and its regulation by the lin-4 RNA. *Cell* **88**: 637–646.
7 Reinhart, B.J., Slack, F.J., Basson, M. et al. (2000). The 21-nucleotide let-7 RNA regulates developmental timing in *Caenorhabditis elegans*. *Nature* **403**: 901–906.
8 Lee, R.C. and Ambros, V. (2001). An extensive class of small RNAs in *Caenorhabditis elegans*. *Science* **294**: 862–864.
9 Pasquinelli, A.E., Reinhart, B.J., Slack, F. et al. (2000). Conservation of the sequence and temporal expression of let-7 heterochronic regulatory RNA. *Nature* **408** (6808): 86–89.
10 Lagos-Quintana, M., Rauhut, R., Lendeckel, W., and Tuschl, T. (2001). Identification of novel genes coding for small expressed RNAs. *Science* **294**: 853–858.
11 Olena, F.O. and Patton, G.J. (2010). Genomic organization of microRNAs. *J. Cell. Physiol.* **222**: 540–545.
12 Kaitetzidou, E. (2019). Annotation, molecular characterization and functional study of orthologous and paralogous genes in different fish species. PhD thesis. Aristotle University of Thessaloniki.
13 O'Brien, J., Hayder, H., Zayed, Y., and Peng, C. (2018). Overview of microRNA biogenesis, mechanisms of actions, and circulation. *Front. endocrinol. (Lausanne)* **9**: 1–12.
14 Medley, J.C., Panzade, G., and Zinovyeva, A.Y. (2021). MicroRNA strand selection: unwinding the rules. *Wiley Interdiscip. Rev. RNA* **12**: 1–22.
15 Pasquinelli, A.E. (2012). MicroRNAs and their targets: recognition, regulation and an emerging reciprocal relationship. *Nat. Rev. Genet.* **13**: 271–282.
16 Ørom, U.A., Nielsen, F.C., and Lund, A.H. (2008). MicroRNA-10a binds the 5′ UTR of ribosomal protein mRNAs and enhances their translation. *Mol. Cell* **30**: 460–471.
17 Wilkie, G.S., Dickson, K.S., and Gray, N.K. (2003). Regulation of mRNA translation by 5′- and 3′-UTR-binding factors. *Trends Biochem. Sci.* **28**: 182–188.
18 Lytle, J.R., Yario, T.A., and Steitz, J.A. (2007). Target mRNAs are repressed as efficiently by microRNA-binding sites in the 5′ UTR as in the 3′ UTR. *Proc. Natl. Acad. Sci.* https://doi.org/10.1073/pnas.0703820104.
19 Bartel, D.P., Lee, R., and Feinbaum, R. (2004). MicroRNAs: genomics, biogenesis, mechanism, and function genomics: the miRNA genes. *Cell* **116**: 281–297.
20 Tommasi, C., Pellegrino, B., Boggiani, D. et al. (2021). Biological role and clinical implications of microRNAs in BRCA mutation carriers. *Front. Oncol.* **11**: 1–10.
21 Ardekani, A.M. and Naeini, M.M. (2010). The role of microRNAs in human diseases. *Avicenna J. Med. Biotechnol.* **2**: 161–179.
22 Bizuayehu, T.T. and Babiak, I. (2014). MicroRNA in teleost fish. *Genome Biol. Evol.* **6**: 1911–1937.
23 Desvignes, T., Bardou, P., Montfort, J. et al. (2022). FishmiRNA: an evolutionarily supported microRNA annotation and expression database for ray-finned fishes. *Mol. Biol. Evol.* **39**: 1–7.
24 Bartel, D.P. (2009). MicroRNAs: target recognition and regulatory functions. *Cell* **136**: 215–233.
25 Akhtar, M.M., Micolucci, L., Islam, M.S. et al. (2019). A practical guide to miRNA target prediction. *Methods Mol. Biol.* https://doi.org/10.1007/978-1-4939-9207-2_1.
26 Zhang, Y. (2005). miRU: an automated plant miRNA target prediction server. *Nucleic Acids Res.* **33**: 701–704.
27 Ambros, V. (2004). The functions of animal microRNAs. *Nature* **431**: 350–355.
28 Peterson, S.M., Thompson, J.A., Ufkin, M.L. et al. (2014). Common features of microRNA target prediction tools. *Front. Genet.* https://doi.org/10.3389/fgene.2014.00023.
29 Bartel, D.P. and Chen, C.Z. (2004). Micromanagers of gene expression: the potentially widespread influence of metazoan microRNAs. *Nat. Rev. Genet.* https://doi.org/10.1038/nrg1328.

30 Brennecke, J., Stark, A., Russell, R.B., and Cohen, S.M. (2005). Principles of microRNA-target recognition. *PLoS Biol.* **3**: 0404–0418.

31 Yu, Q., Peng, C., Ye, Z. et al. (2020). An estradiol-17β/miRNA-26a/cyp19a1a regulatory feedback loop in the protogynous hermaphroditic fish, *Epinephelus coioides*. *Mol. Cell. Endocrinol.* https://doi.org/10.1016/j.mce.2019.110689.

32 Papadaki, M., Kaitetzidou, E., Mylonas, C.C., and Sarropoulou, E. (2020). Non-coding RNA expression patterns of two different teleost gonad maturation stages. *Mar. Biotechnol.* https://doi.org/10.1007/s10126-020-09991-2.

33 Du, X., Zhang, L., Li, X. et al. (2016). Tgf-β signaling controls FSHR signaling-reduced ovarian granulosa cell apoptosis through the SMAD4/miR-143 axis. *Cell Death Dis.* **7**: 1–12.

34 Aravin, A.A., Naumova, N.M., Tulin, A.V. et al. (2001). Double-stranded RNA-mediated silencing of genomic tandem repeats and transposable elements in the *D. melanogaster* germline. *Curr. Biol.* **11**: 1017–1027.

35 Krishna, S., Raghavan, S., DasGupta, R., and Palakodeti D.(2021). tRNA-derived fragments (tRFs): establishing their turf in post-transcriptional gene regulation. Cell Mol Life Sci [Internet].**78**(6):2607–19. Available from: https://doi.org/10.1007/s00018-020-03720-7

36 Grivna, S.T., Beyret, E., Wang, Z., and Lin, H. (2006). A novel class of small RNAs in mouse spermatogenic cells. *Genes Dev.* **20**: 1709–1714.

37 Girard, A., Sachidanandam, R., Hannon, G.J., and Carmell, M.A. (2006). A germline-specific class of small RNAs binds mammalian Piwi proteins. *Nature* **442**: 199–202.

38 Lau, N.C., Seto, A.G., Kim, J. et al. (2006). Characterization of the piRNA complex from rat testes. *Science* **313**: 363–367.

39 Angelica, M.D. and Fong, Y. (2008). Argonautes confront new small RNAs. *Curr. Opin. Chem. Biol.* **141**: 520–529.

40 Malone, C.D. and Hannon, G.J. (2009). Small RNAs as guardians of the genome. *Cell* **136**: 656–668.

41 Vagin, V.V., Sigova, A., Li, C. et al. (2006). A distinct small RNA pathway silences selfish genetic elements in the germline. *Science (80-)* **313**: 320–324.

42 Parhad, S.S. and Theurkauf, W.E. (2019). Rapid evolution and conserved function of the piRNA pathway. *Open Biol.* https://doi.org/10.1098/rsob.18.0181.

43 Houwing, S., Kamminga, L.M., Berezikov, E. et al. (2007). A role for Piwi and piRNAs in germ cell maintenance and transposon silencing in Zebrafish. *Cell* **129**: 69–82.

44 Sarot, E., Payen-Groschêne, G., Bucheton, A., and Pélisson, A. (2004). Evidence for a Piwi-dependent RNA silencing of the gypsy endogenous retrovirus by the *Drosophila melanogaster* flamenco gene. *Genetics* **166**: 1313–1321.

45 Ozata, D.M., Gainetdinov, I., Zoch, A. et al. (2019). Piwi-interacting RNAs: small RNAs with big functions. *Nat. Rev. Genet.* **20**: 89–108.

46 Ramat, A. and Simonelig, M. (2021). Functions of Piwi proteins in gene regulation: new arrows added to the piRNA quiver. *Trends Genet.* **37**: 188–200.

47 Weick, E.M. and Miska, E.A. (2014). piRNAs: from biogenesis to function. *Dev. Dent.* **141**: 3458–3471.

48 Maniatis, T. and Reed, R. (1987). The role of snRNS in pre-mRNA splicing. *Nature* **325**: 673–678.

49 Adema, C.M. (2021). Sticky problems: extraction of nucleic acids from molluscs. *Philos. Trans. R. Soc. B* **376**: 20200162.

50 Brown, R.A.M., Epis, M.R., Horsham, J.L. et al. (2018). Total RNA extraction from tissues for microRNA and target gene expression analysis: not all kits are created equal. *BMC Biotechnol.* **18**: 1–11.

51 Wang, C.L., Wang, Z.P., Wang, J.Q. et al. (2016). Identification of candidate piRNAs in the gonads of *Paralichthys olivaceus* (Japanese flounder). *Zool. Res.* **37**: 301–306.

52 Fernández, I., Fernandes, J.M.O., Roberto, V.P. et al. (2019). Circulating small non-coding RNAs provide new insights into vitamin K nutrition and reproductive physiology in teleost fish. *Biochim. Biophys. – Acta Gen. Subj.* **1863**: 39–51.

53 Wang, F., Jia, Y., Wang, P. et al. (2017). Identification and profiling of *Cyprinus carpio* microRNAs during ovary differentiation by deep sequencing. *BMC Genomics* **18**: 1–17.

54 Lite, C., Sridhar, V.V., Sriram, S. et al. (2021). Functional role of piRNAs in animal models and its prospects in aquaculture. *Rev. Aquac.* **13**: 2038–2052.

55 Busch, H., Ramachandra, R., Rothblum, L., and Choi, Y.C. (1982). SnRNAs, SnRNPs and RNA processing. *Annu. Rev. Biochem.* **51**: 617–654.

56 Hadjiolov, A., Venkov, P., and Tsanev, R. (1966). Ribonucleic acids fractionation by density-gradient centrifugation and by agar gel electrophoresis: a comparison. *Anal. Biochem.* **17**: 263–267.

57 Adams, R., Knowler, J., and Leader, D. (1997). *Biochemistry of the Nucleic Acids*, 11e. Chapman & Hall Publishing.

58 Sarropoulou, E., Kaitetzidou, E., Papandroulaki, N. et al. (2019). Inventory of European sea bass (*Dicentrarchus labrax*) sncRNAs vital during early teleost development. *Front. Genet.* **10**: 1–11.

59 Bizuayehu, T.T., Lanes, C.F.C., Furmanek, T. et al. (2012). Differential expression patterns of conserved miRNAs and IsomiRs during Atlantic halibut development. *BMC Genomics* **13**: 11.

60 Robledo, D., Martin, A.P., Álvarez-Dios, J.A. et al. (2017). First characterization and validation of turbot microRNAs. *Aquaculture* **472**: 76–83.

61 Coenen-Stass, A.M.L., Magen, I., Brooks, T. et al. (2018). Evaluation of methodologies for microRNA biomarker detection by next generation sequencing. *RNA Biol.* **15**: 1133–1145.

62 Sun, Z., Evans, J., Bhagwate, A. et al. (2014). CAP-miRSeq: a comprehensive analysis pipeline for microRNA sequencing data. *BMC Genomics* **15**: 423.

63 Kozomara, A. and Griffiths-Jones, S. (2014). MiRBase: annotating high confidence microRNAs using deep sequencing data. *Nucleic Acids Res.* **42**: 68–73.

64 Fromm, B., Domanska, D., Høye, E. et al. (2020). MirGeneDB 2.0: the metazoan microRNA complement. *Nucleic Acids Res.* **48**: D132–D141.

65 Sweeney, B.A., Petrov, A.I., Ribas, C.E. et al. (2021). RNAcentral 2021: secondary structure integration, improved sequence search and new member databases. *Nucleic Acids Res.* **49**: D212–D220.

66 Ambros, V., Bartel, B., Bartel, D.P. et al. (2003). A uniform system for microRNA annotation. *RNA* **9**: 277–279.

67 Love, M.I., Huber, W., and Anders, S. (2014). Moderated estimation of fold change and dispersion for RNA-seq data with DESeq2. *Genome Biol.* **15**: 550.

68 Varet, H., Brillet-Guéguen, L., Coppée, J.-Y., and Dillies, M.-A. (2016). SARTools: a DESeq2- and EdgeR-based R pipeline for comprehensive differential analysis of RNA-Seq data. *PLoS One* **11**: e0157022.

69 Maza, E. (2016). In papyro comparison of TMM (edgeR), RLE (DESeq2), and MRN normalization methods for a simple two-conditions-without-replicates RNA-seq experimental design. *Front. Genet.* **7**: 1–8.

70 The Encode Consortium (2016). Standards, guidelines and best practices for RNA-Seq. *Vasa* https://doi.org/10.1073/pnas.0703993104.

71 Bhartiya, D., Pal, K., Ghosh, S. et al. (2013). LncRNome: a comprehensive knowledgebase of human long noncoding RNAs. *Database* **2013**: 1–6.

72 Liao, Q., Liu, C., Yuan, X. et al. (2011). Large-scale prediction of long non-coding RNA functions in a coding-non-coding gene co-expression network. *Nucleic Acids Res.* **39**: 3864–3878.

73 Lipovich, L., Johnson, R., and Lin, C.Y. (2010). MacroRNA underdogs in a microRNA world: evolutionary, regulatory, and biomedical significance of mammalian long non-protein-coding RNA. *Biochim. Biophys. Acta – Gene Regul. Mech.* **1799**: 597–615.

74 Tsagakis, I., Douka, K., Birds, I., and Aspden, J.L. (2020). Long non-coding RNAs in development and disease: conservation to mechanisms. *J. Pathol.* https://doi.org/10.1002/path.5405.

75 Pachnis, V., Belayew, A., and Tilghman, S.M. (1984). Locus unlinked to α-fetoprotein under the control of the murine raf and Rif genes. *Proc. Natl. Acad. Sci. U. S. A.* **81**: 5523–5527.

76 Ma, L., Cao, J., Liu, L. et al. (2019). Lncbook: a curated knowledgebase of human long non-coding RNAs. *Nucleic Acids Res.* **47**: D128–D134.

77 Fang, S., Zhang, L., Guo, J. et al. (2018). NONCODEV5: a comprehensive annotation database for long non-coding RNAs. *Nucleic Acids Res.* **46**: D308–D314.

78 Bhat, S.A., Ahmad, S.M., Mumtaz, P.T. et al. (2016). Long non-coding RNAs: mechanism of action and functional utility. *Non-coding RNA Res.* **1**: 43–50.

79 Yunusov, D., Anderson, L., Dasilva, L.F. et al. (2016). HIPSTR and thousands of lncRNAs are heterogeneously expressed in human embryos, primordial germ cells and stable cell lines. *Sci. Rep.* https://doi.org/10.1038/srep32753.

80 Long, Y., Wang, X., Youmans, D.T., and Cech, T.R. (2017). How do lncRNAs regulate transcription? *Sci. Adv.* **3**: eaao2110.

81 Neguembor, M.V., Jothi, M., and Gabellini, D. (2014). Long noncoding RNAs, emerging players in muscle differentiation and disease. *Skelet. Muscle* **4**: 8.

82 Khaitovich, P., Kelso, J., Franz, H. et al. (2006). Functionality of intergenic transcription: an evolutionary comparison. *PLoS Genet.* **2**: 1590–1598.

83 Fernandes, J.C.R., Acuña, S.M., Aoki, J.I. et al. (2019). Long non-coding RNAs in the regulation of gene expression: physiology and disease. *Non-coding RNA* https://doi.org/10.3390/ncrna5010017.

84 Jarroux, J., Morillon, A., Pinskaya, M. et al. (2017). History, discovery, and classification of lncRNAs. *Adv. Exp. Med. Biol.* https://doi.org/10.1007/978-981-10-5203-3_1.

85 Taylor, D.H., Chu, E.T.J., Spektor, R., and Soloway, P.D. (2015). Long non-coding RNA regulation of reproduction and development. *Mol. Reprod. Dev.* https://doi.org/10.1002/mrd.22581.

86 Kaushik, K., Leonard, V.E., Shamsudheen, K.V. et al. (2013). Dynamic expression of long non-coding RNAs (lncRNAs) in adult zebrafish. *PLoS One* https://doi.org/10.1371/journal.pone.0083616.

87 Wang, M., Jiang, S., Wu, W. et al. (2018). Non-coding RNAs function as immune regulators in teleost fish. *Front. Immunol.* **9**: 1–15.

88 Wei, S., Chen, Y., Huang, L. et al. (2021). Analysis of lncRNA and mRNA expression profiles in peripheral blood leukocytes of the half-smooth tongue sole (*Cynoglossus semilaevis*) treated with chitosan oligosaccharide. *Dev. Comp. Immunol.* **120**: 104043.

89 Jiang, L., Liu, W., Zhu, A. et al. (2016). Transcriptome analysis demonstrate widespread differential expression of long noncoding RNAs involve in *Larimichthys crocea* immune response. *Fish Shellfish Immunol.* **51**: 1–8.

90 Yuan, W., Jiang, S., Sun, D. et al. (2019). Transcriptome profiling analysis of sex-based differentially expressed mRNAs and lncRNAs in the brains of mature zebrafish (*Danio rerio*). *BMC Genomics* https://doi.org/10.1186/S12864-019-6197-9.

91 Cai, J., Li, L., Song, L. et al. (2019). Effects of long term antiprogestine mifepristone (RU486) exposure on sexually dimorphic lncRNA expression and gonadal masculinization in Nile tilapia (*Oreochromis niloticus*). *Aquat. Toxicol.* https://doi.org/10.1016/J.AQUATOX.2019.105289.

92 Zhang, S., Wang, Y., Jia, L. et al. (2019). Profiling the long noncoding RNA interaction network in the regulatory elements of target genes by chromatin in situ reverse transcription sequencing. *Genome Res.* **29** (9): 1521–1532.

93 Valenzuela-Muñoz, V., Váldes, J.A., and Gallardo-Escárate, C. (2021). Transcriptome profiling of long non-coding RNAs during the Atlantic salmon smoltification process. *Mar. Biotechnol.* **23** (2): 308–320.

94 Luo, M., Wang, L., Yin, H. et al. (2019). Integrated analysis of long non-coding RNA and mRNA expression in different colored skin of koi carp. *BMC Genomics* **20**: 1–14.

95 Núñez-Acuña, G., Détrée, C., Gallardo-Escárate, C., and Gonçalves, A.T. (2017). Functional diets modulate lncRNA-coding RNAs and gene interactions in the intestine of rainbow trout *Oncorhynchus mykiss*. *Mar. Biotechnol.* **19**: 287–300.

96 Xu, H., Cao, L., Sun, B. et al. (2019). Transcriptomic analysis of potential "lncRNA-mRNA" interactions in liver of the marine teleost *Cynoglossus semilaevis* fed diets with different DHA/EPA ratios. *Front. Physiol.* **10**: 1–16.

97 García-Pérez I, Molsosa-Solanas A, Perelló-Amorós M, Sarropoulou E, Blasco J, Gutiérrez J (2022) The emerging role of long non-coding RNAs in development and function of gilthead sea Bream (*Sparus aurata*) fast skeletal muscle. *Cell*https://doi.org/10.3390/cells11030428

98 Patop, I.L., Wüst, S., and Kadener, S. (2019). Past, present, and future of circRNAs. *EMBO J.* **38**: 1–13.

99 Li, X., Yang, L., and Chen, L.L. (2018). The biogenesis, functions, and challenges of circular RNAs. *Mol. Cell* **71**: 428–442.

100 Yu, C.Y. and Kuo, H.C. (2019). The emerging roles and functions of circular RNAs and their generation. *J. Biomed. Sci.* **26**: 1–12.

101 He, L., Zhang, A., Xiong, L. et al. (2017). Deep circular RNA sequencing provides insights into the mechanism underlying grass carp reovirus infection. *Int. J. Mol. Sci.* https://doi.org/10.3390/ijms18091977.

102 Liu, B., Yuan, R., Liang, Z. et al. (2019). Comprehensive analysis of circRNA expression pattern and circRNA–mRNA–miRNA network in *Ctenopharyngodon idella* kidney (CIK) cells after grass carp reovirus (GCRV) infection. *Aquaculture* **512**: 734349.

103 Li, J., Shi, B.S., Wang, C. et al. (2021). Comprehensive CircRNA profiling and selection of key CircRNAs reveal the potential regulatory roles of CircRNAs throughout ovarian development and maturation in cynoglossus semilaevis. *Biology* **10** (9): 830.

104 Zhang, Y., Zhong, Y., Guo, S. et al. (2021). CircRNA profiling reveals circ880 functions as miR-375-3p sponge in medaka gonads. *Comp. Biochem. Physiol. – Part D Genomics Proteomics* **38**: 100797.

105 Brown, C.J., Hendrich, B.D., Rupert, J.L. et al. (1992). The human XIST gene: analysis of a 17 kb inactive X-specific RNA that contains conserved repeats and is highly localized within the nucleus. *Cell* **71**: 527–542.

106 Hutchinson, J.N., Ensminger, A.W., Clemson, C.M. et al. (2007). A screen for nuclear transcripts identifies two linked noncoding RNAs associated with SC35 splicing domains. *BMC Genomics* **8**: 1–16.

107 Sone, M., Hayashi, T., Tarui, H. et al. (2007). The mRNA-like noncoding RNA Gomafu constitutes a novel nuclear domain in a subset of neurons. *J. Cell Sci.* **120**: 2498–2506.

108 Ulitsky, I., Shkumatava, A., Jan, C.H. et al. (2011). Conserved function of lincRNAs in vertebrate embryonic development despite rapid sequence evolution. *Cell* **147**: 1537–1550.

109 Ulitsky, I. and Bartel, D.P. (2013). lincRNAs: genomics, evolution, and mechanisms. *Cell* **154**: 26–46.

110 Goudarzi, M., Berg, K., Pieper, L.M., and Schier, A.F. (2019). Individual long non-coding RNAs have no overt functions in zebrafish embryogenesis, viability and fertility. *Elife* **8**: 1–17.

111 Mallo, M. and Alonso, C.R. (2013). The regulation of Hox gene expression during animal development. *Development* **140**: 3951–3963.
112 Lim, L.P., Glasner, M.E., Yekta, S. et al. (2003). Vertebrate microRNA genes. *Science* **299**: 1540.
113 Chen, P.Y., Manninga, H., Slanchev, K. et al. (2005). The developmental miRNA profiles of zebrafish as determined by small RNA cloning. *Genes Dev.* **19**: 1288–1293.
114 Kloosterman, W.P., Steiner, F.A., Berezikov, E. et al. (2006). Cloning and expression of new microRNAs from zebrafish. *Nucleic Acids Res.* **34**: 2558–2569.
115 Kloosterman, W.P., Wienholds, E., Ketting, R.F., and Plasterk, R.H.A. (2004). Substrate requirements for let-7 function in the developing zebrafish embryo. *Nucleic Acids Res.* https://doi.org/10.1093/nar/gkh968.
116 Zhao, B.W., Zhou, L.F., Liu, Y.L. et al. (2017). Evolution of fish let-7 microRNAs and their expression correlated to growth development in blunt snout bream. *Int. J. Mol. Sci.* **18**: 3.
117 Ramachandra, R.K., Salem, M., Gahr, S. et al. (2008). Cloning and characterization of microRNAs from rainbow trout (*Oncorhynchus mykiss*): their expression during early embryonic development. *BMC Dev. Biol.* **8**: 1–11.
118 Wang, X., Song, X., and Bhandari, R.K. (2020). Distinct expression patterns of seven crucial microRNAs during early embryonic development in medaka (*Oryzias latipes*). *Gene Expr. Patterns* **37**: 119133.
119 Woldemariam, N.T., Agafonov, O., Høyheim, B. et al. (2019). Expanding the miRNA repertoire in Atlantic salmon; discovery of IsomiRs and miRNAs highly expressed in different tissues and developmental stages. *MDPI Cells* **8** (1): 1–23.
120 Snorrason SS, Franzdo R, Kapralova KH (2014) Patterns of MiRNA expression in Arctic Charr development. *PLoS One*.https://doi.org/10.1371/journal.pone.0106084
121 Ebert, M.S. and Sharp, P.A. (2012). Roles for microRNAs in conferring robustness to biological processes. *Cell* **149**: 515–524.
122 Posadas, M.D. and Carthew, W.R. (2014). MicroRNAs and their roles in developmental canalization. *Curr. Opin. Genet. Dev.* **27**: 1–6.
123 Alberti, C. and Cochella, L. (2017). A framework for understanding the roles of miRNAs in animal development. *Dev. Dent.* **144**: 2548–2559.
124 Hornstein, E. and Shomron, N. (2006). Canalization of development by microRNAs. *Nat. Genet.* **38**: S20.
125 Wu, C.I., Shen, Y., and Tang, T. (2009). Evolution under canalization and the dual roles of microRNAs – a hypothesis. *Genome Res.* **19**: 734–743.
126 Mendell, J.T. and Olson, E.N. (2012). MicroRNAs in stress signaling and human disease. *Cell* **148**: 1172–1187.
127 Kasper, D.M., Moro, A., Ristori, E. et al. (2017). MicroRNAs establish uniform traits during the architecture of vertebrate embryos. *Dev. Cell.* **40** (6): 552–565.
128 Liufu, Z., Zhao, Y., Guo, L. et al. (2017). Redundant and incoherent regulations of multiple phenotypes suggest microRNAs' role in stability control. *Genome Res.* **27**: 1665–1673.
129 Cassidy, J.J., Jha, A.R., Posadas, D.M. et al. (2013). MiR-9a minimizes the phenotypic impact of genomic diversity by buffering a transcription factor. *Cell* **155**: 1556–1567.
130 Schmiedel, J.M., Klemm, S.L., Zheng, Y. et al. (2015). MicroRNA control of protein expression noise. *Science* **348**: 128–131.
131 Giraldez, A.J., Mishima, Y., Rihel, J. et al. (2006). Zebrafish MiR-430 promotes deadenylation and clearance of maternal mRNAs. *Science* **312**: 75–79.
132 Liu, X., Ning, G., Meng, A., and Wang, Q. (2012). MicroRNA-206 regulates cell movements during zebrafish gastrulation by targeting prickle1a and regulating c-Jun N-terminal kinase 2 phosphorylation. *Mol. Cell. Biol.* **32**: 2934–2942.
133 Campos, C., Sundaram, A.Y.M., Valente, L.M.P. et al. (2014). Thermal plasticity of the miRNA transcriptome during Senegalese sole development. *BMC Genomics* **15**: 525.
134 Giusti, J., Pinhal, D., Moxon, S. et al. (2016). MicroRNA-10 modulates Hox genes expression during Nile tilapia embryonic development. *Mech. Dev.* **140**: 12–18.
135 Robledo, D., Martin, A.P., Álvarez-Dios, J.A. et al. (2017). First characterization and validation of turbot microRNAs. *Aquaculture* **472**: 76–83.
136 He, X., Yan, Y.L., Eberhart, J.K. et al. (2011). MiR-196 regulates axial patterning and pectoral appendage initiation. *Dev. Biol.* **357**: 463–477.
137 Chen, J.F., Mandel, E.M., Thomson, J.M. et al. (2006). The role of microRNA-1 and microRNA-133 in skeletal muscle proliferation and differentiation. *Nat. Genet.* **38**: 228–233.
138 Horak, M., Novak, J., and Bienertova-Vasku, J. (2016). Muscle-specific microRNAs in skeletal muscle development. *Dev. Biol.* **410**: 1–13.

139 Coolen, M., Thieffry, D., Drivenes, Ø. et al. (2012). MiR-9 controls the timing of neurogenesis through the direct inhibition of antagonistic factors. *Dev. Cell* **22**: 1052–1064.

140 Li, N., Wei, C., Olena, A.F., and Patton, J.G. (2011). Regulation of endoderm formation and left-right asymmetry by miR-92 during early zebrafish development. *Development* **138**: 1817–1826.

141 Colas, A.R., Mckeithan, W.L., Cunningham, T.J. et al. (2012). Whole-genome microRNA screening identifies let-7 and miR-18 as regulators of germ layer formation during early embryogenesis. *Genes Dev.* **26**: 2567–2579.

142 Farrell, B.C., Power, E.M., and Dermott, K.W.M. (2011). Developmentally regulated expression of Sox9 and microRNAs 124, 128 and 23 in neuroepithelial stem cells in the developing spinal cord. *Int. J. Dev. Neurosci.* **29**: 31–36.

143 Dill, H., Linder, B., Fehr, A., and Fischer, U. (2012). Intronic miR-26b controls neuronal differentiation by repressing its host transcript, ctdsp2. *Genes Dev.* **26**: 25–30.

144 Desvignes, T., Contreras, A., and Postlethwait, J.H. (2014). Evolution of the miR199-214 cluster and vertebrate skeletal development. *RNA Biol.* **11**: 281–294.

145 Mazabraud, A., Wegnez, M., and Denis, H. (1975). Biochemical RNA accumulation research on oogenesis. In the oocytes of teleosts. *Dev. Biol.* **332**: 326–332.

146 Diaz De Cerio, O., Rojo-Bartolomé, I., Bizarro, C. et al. (2012). 5S rRNA and accompanying proteins in gonads: powerful markers to identify sex and reproductive endocrine disruption in fish. *Environ. Sci. Technol.* **46**: 7763–7771.

147 Rojo-Bartolomé, I., De Cerio, O.D., Diez, G., and Cancio, I. (2016). Identification of sex and female's reproductive stage in commercial fish species through the quantification of ribosomal transcripts in gonads. *PLoS One* https://doi.org/10.1371/journal.pone.0149711.

148 Rojo-Bartolomé, I., Martínez-Miguel, L., Lafont, A.G. et al. (2017). Molecular markers of oocyte differentiation in European eel during hormonally induced oogenesis. *Comp. Biochem. Physiol. - Part A Mol. Integr. Physiol.* **211**: 17–25.

149 Rojo-Bartolomé, I., Valencia, A., and Cancio, I. (2017). Transcription of *ribogenesis* genes in fish gonads: applications in the identification of stages of oogenesis and in environmental monitoring of intersex condition. *Mar. Pollut. Bull.* **121**: 292–301.

150 Lau, K., Lai, K.P., Bao, J.Y.J. et al. (2014). Identification and expression profiling of microRNAs in the brain, liver and gonads of marine medaka (*Oryzias melastigma*) and in response to hypoxia. *PLoS One* https://doi.org/10.1371/journal.pone.0110698.

151 Kaitetzidou, E., Gilfillan, G.D., Antonopoulou, E., and Sarropoulou, E. (2022). Sex-biased dynamics of three-spined stickleback (*Gasterosteus aculeatus*) gene expression patterns. *Genomics* **114**: 266–277.

152 Dai, X., Shu, Y., Lou, Q. et al. (2017). Tdrd12 is essential for germ cell development and maintenance in zebrafish. *Int. J. Mol. Sci.* https://doi.org/10.3390/ijms18061127.

153 Kamalidehghan, B., Habibi, M., Afjeh, S.S. et al. (2020). The importance of small non-coding RNAs in human reproduction: a review article. *Appl. Clin. Genet.* **13**: 1–11.

154 Gong, J., Zhang, Q., Wang, Q. et al. (2018). Identification and verification of potential piRNAs from domesticated yak testis. *Reproduction* **155**: 117–127.

155 Waiho, K., Fazhan, H., Zhang, Y. et al. (2020). Comparative profiling of ovarian and testicular piRNAs in the mud crab *Scylla paramamosain*. *Genomics* **112**: 323–331.

156 Lempradl, A. (2020). Germ cell-mediated mechanisms of epigenetic inheritance. *Semin. Cell Dev. Biol.* **97**: 116–122.

157 Zhang, B., Zhao, N., Jia, L. et al. (2020). Identification and application of Piwi-interacting RNAs from seminal plasma exosomes in *Cynoglossus semilaevis*. *BMC Genomics* **21**: 1–11.

158 Wang, K., Liang, C., Liu, J. et al. (2014). Prediction of piRNAs using transposon interaction and a support vector machine. *BMC Bioinform.* https://doi.org/10.1186/s12859-014-0419-6.

159 Li, S., Lin, G., Fang, W. et al. (2020). Identification and comparison of microRNAs in the gonad of the yellowfin seabream (*Acanthopagrus latus*). *Int. J. Mol. Sci.* **21**: 5690.

160 Xiao, J., Zhong, H., Zhou, Y. et al. (2014). Identification and characterization of microRNAs in ovary and testis of Nile tilapia (*Oreochromis niloticus*) by using Solexa sequencing technology. *PLoS One* https://doi.org/10.1371/journal.pone.0086821.

161 Fu, Y., Xu, Z., Wen, B. et al. (2020). Gonad-specific transcriptomes reveal differential expression of gene and miRNA between male and female of the discus fish (*Symphysodon aequifasciatus*). *Front. Physiol.* **11**: 1–12.

162 Bhat, R.A., Priyam, M., Foysal, M.J. et al. (2021). Role of sex-biased miRNAs in teleosts – a review. *Rev. Aquac.* **13**: 269–281.

163 Azhar, S., Dong, S., Shen, W. et al. (2020). The role of miRNAs in regulating adrenal and gonadal steroidogenesis. *J. Mol. Endocrinol.* **64**: R21–R43.

164 Bizuayehu, T.T., Babiak, J., Norberg, B. et al. (2012). Sex-biased miRNA expression in atlantic halibut (*Hippoglossus hippoglossus*) brain and gonads. *Sex Dev.* **6**: 257–266.

165 Presslauer, C., Tilahun Bizuayehu, T., Kopp, M. et al. (2017). Dynamics of miRNA transcriptome during gonadal development of zebrafish. *Sci. Rep.* **7**: 1–16.

166 Tang, X., Feng, D., Li, M. et al. (2019). Transcriptomic analysis of mRNA–lncRNA–miRNA interactions in hepatocellular carcinoma. *Sci. Rep.* **9**: 1–12.

167 Todd, E.V., Ortega-Recalde, O., Liu, H. et al. (2019). Stress, novel sex genes, and epigenetic reprogramming orchestrate socially controlled sex change. *Sci. Adv.* **5**: 1–15.

168 Ding, Q., Jin, M., Wang, Y. et al. (2020). Transactivation of miR-202-5p by steroidogenic factor 1 (SF1) induces apoptosis in goat granulosa cells by targeting TGFβR2. *Cell* **9**: 1–16.

169 Armisen, J., Gilchrist, M.J., Wilczynska, A. et al. (2009). Abundant and dynamically expressed miRNAs, piRNAs, and other small RNAs in the vertebrate *Xenopus tropicalis*. *Genome Res.* **19**: 1766–1775.

170 Xu, L., Guo, Q., Chang, G. et al. (2017). Discovery of microRNAs during early spermatogenesis in chicken. *PLoS One* https://doi.org/10.1371/journal.pone.0177098.

171 Bannister, S.C., Smith, C.A., Roeszler, K.N. et al. (2011). Manipulation of estrogen synthesis alters MIR202* expression in embryonic chicken gonads. *Biol. Reprod.* **85**: 22–30.

172 Gay, S., Bugeon, J., Bouchareb, A. et al. (2018). MiR-202 controls female fecundity by regulating medaka oogenesis. *PLoS Genet.* https://doi.org/10.1371/journal.pgen.1007593.

173 Chen, J., Cai, T., Zheng, C. et al. (2017). MicroRNA-202 maintains spermatogonial stem cells by inhibiting cell cycle regulators and RNA binding proteins. *Nucleic Acids Res.* **45**: 4142–4157.

174 Zhang, J., Liu, W., Jin, Y. et al. (2017). MiR-202-5p is a novel germ plasm-specific microRNA in zebrafish. *Sci. Rep.* **7**: 1–7.

175 Zhang, J., Ji, X., Zhou, D. et al. (2013). miR-143 is critical for the formation of primordial follicles in mice. *Front. Biosci. (Landmark Ed)* **2**: 588–597.

176 Jing, J., Wu, J., Liu, W. et al. (2014). Sex-biased miRNAs in gonad and their potential roles for testis development in yellow catfish. *PLoS One* **9**: e107946.

177 Wang, F., Yang, Q.W., Zhao, W.J. et al. (2019). Effects of short-time exposure to atrazine on miRNA expression profiles in the gonad of common carp (*Cyprinus carpio*). *BMC Genomics* **20**: 1–15.

178 Cordes, K.R., Sheehy, N.T., White, M.P. et al. (2009). MiR-145 and miR-143 regulate smooth muscle cell fate and plasticity. *Nature* **460**: 705–710.

179 Kitade, Y. and Akao, Y. (2010). MicroRNAs and their therapeutic potential for human diseases: microRNAs, miR-143 and -145, function as anti-oncomirs and the application of chemically modified mir-143 as an anti-cancer drug. *J. Pharmacol. Sci.* **114**: 276–280.

180 Tong, M., Mitchell, D., Evanoff, R., and Griswold, M.D. (2011). Expression of Mirlet7 family microRNAs in response to retinoic acid-induced spermatogonial differentiation in mice. *Biol. Reprod.* 1 from In Situ miRNA Hybridization with Locked Nucleic **197**: 189–197.

181 Rusca, N. and Monticelli, S. (2011). MiR-146a in immunity and disease. *Mol. Biol. Int.* https://doi.org/10.4061/2011/437301.

182 Tao, W., Sun, L., Shi, H. et al. (2016). Integrated analysis of miRNA and mRNA expression profiles in tilapia gonads at an early stage of sex differentiation. *BMC Genomics* **17**: 328.

183 Wang, W., Liu, W., Liu, Q. et al. (2016). Coordinated microRNA and messenger RNA expression profiles for understanding sexual dimorphism of gonads and the potential roles of microRNA in the steroidogenesis pathway in Nile tilapia (*Oreochromis niloticus*). *Theriogenology* **85**: 970–978.

184 Chaves-pozo, E., Liarte, S., Fern, L., and Abell, E. (2008). Pattern of expression of immune-relevant genes in the gonad of a teleost, the gilthead seabream (*Sparus aurata* L.). *Mol. Immunol.* **45**: 2998–3011.

185 Zayed, Y., Qi, X., and Peng, C. (2019). Identification of novel microRNAs and characterization of microRNA expression profiles in zebrafish ovarian follicular cells. *Front. Endocrinol. (Lausanne)* **10**: 1–12.

186 Iwama, G. and Nakanishi, T. (1997). *The Fish Immune System: Organism, Pathogen, and Environment*. Academic Publishing.

187 Andreassen, R. and Høyheim, B. (2017). miRNAs associated with immune response in teleost fish. *Dev. Comp. Immunol.* https://doi.org/10.1016/j.dci.2017.02.023.

188 Xie, Y. and Wei, Y. (2021). A novel regulatory player in the innate immune system: long non-coding RNAs. *Int. J. Mol. Sci.* https://doi.org/10.3390/ijms22179535.

189 Zheng, W., Chu, Q., and Xu, T. (2021). Long noncoding RNA IRL regulates NF-κB-mediated immune responses through suppression of miR-27c-3p-dependent IRAK4 downregulation in teleost fish. *J. Biol. Chem.* **296**: 100304.

190 Zheng, W., Chu, Q., and Xu, T. (2021). The long noncoding RNA NARL regulates immune responses via microRNA-mediated NOD1 downregulation in teleost fish. *J. Biol. Chem.* **296**: 100414.

191 Shen, Y., Guo, X., and Wang, W. (2017). Identification and characterization of circular RNAs in zebrafish. *FEBS Lett.* **591**: 213–220.

192 Zheng, W., Sun, L., Yang, L., and Xu, T. (2021). The circular RNA circBCL2L1 regulates innate immune responses via microRNA-mediated downregulation of TRAF6 in teleost fish. *J. Biol. Chem.* **279**: 101199.

193 Rbbani, G., Nedoluzhko, A., Galindo-villegas, J., and Fernandes, J.M.O. (2021). Function of circular RNAs in fish and their potential application as biomarkers. *Int. J. Mol. Sci.* https://doi.org/10.3390/ijms22137119.

194 Chu, Q., Gao, Y., Bi, D., and Xu, T. (2017a). MicroRNA-148 as a negative regulator of the common TLR adaptor mediates inflammatory response in teleost fish. *Sci. Rep.* **7**: 1–9.

195 Chu, Q., Sun, Y., Cui, J., and Xu, T. (2017b). Inducible microRNA-214 contributes to the suppression of NF-κB-mediated inflammatory response via targeting myd88 gene in fish. *J. Biol. Chem.* **292**: 5282–5290.

196 Chu, Q., Sun, Y., Cui, J., and Xu, T. (2017c). MicroRNA-3570 modulates the NF-κB pathway in teleost fish by targeting MyD88. *J. Immunol.* **198**: 3274–3282.

197 Chu, Q. and Xu, T. (2016). miR-192 targeting IL-1RI regulates the immune response in miiuy croaker after pathogen infection in vitro and in vivo. *Fish Shellfish Immunol.* **54**: 537–543.

198 Gao, W., Chang, R., Sun, Y., and Xu, T. (2021). MicroRNA-2187 modulates the NF-κB and IRF3 pathway in teleost fish by targeting TRAF6. *Front. Immunol.* **12**: 1–14.

199 Sun, H.Y., Su, Y.L., Li, P.H. et al. (2021). The roles of *Epinephelus coioides* miR-122 in SGIV infection and replication. *Mar. Biotechnol.* **23**: 294–307.

200 Zhao, X., Chu, Q., Cui, J., and Xu, T. (2018). microRNA-19a as a negative regulator in TLR signaling pathway by direct targeting myeloid differentiation factor 88 in miiuy croaker. *Dev. Comp. Immunol.* **87**: 171–175.

201 Ordas, A., Kanwal, Z., Lindenberg, V. et al. (2013). MicroRNA-146 function in the innate immune transcriptome response of zebrafish embryos to *Salmonella typhimurium* infection. *BMC Genomics* **14**: 1.

202 Hu, Y.H., Zhang, B.C., Zhou, H.Z. et al. (2017). *Edwardsiella tarda*-induced miRNAs in a teleost host: global profile and role in bacterial infection as revealed by integrative miRNA–mRNA analysis. *Virulence* **8**: 1457–1464.

203 Ni, S., Yu, Y., Wei, J. et al. (2018). MicroRNA-146a promotes red spotted grouper nervous necrosis virus (RGNNV) replication by targeting TRAF6 in orange spotted grouper, *Epinephelus coioides*. *Fish Shellfish Immunol.* **72**: 9–13.

204 Chu, Q. and Xu, T. (2020). MicroRNA regulation of toll-like receptor, RIG-I-like receptor and nod-like receptor pathways in teleost fish. *Rev. Aquac.* **12**: 2177–2193.

205 Tang, B., Xiao, B., Liu, Z. et al. (2010). Identification of MyD88 as a novel target of miR-155, involved in negative regulation of *Helicobacter pylori*-induced inflammation. *FEBS Lett.* **584**: 1481–1486.

206 Cui, J., Chu, Q., and Xu, T. (2016). miR-122 involved in the regulation of toll-like receptor signaling pathway after *Vibrio anguillarum* infection by targeting TLR14 in miiuy croaker. *Fish Shellfish Immunol.* **58**: 67–72.

207 Su, H., Chang, R., Zheng, W. et al. (2021). microRNA-210 and microRNA-3570 negatively regulate NF-κB-mediated inflammatory responses by targeting RIPK2 in teleost fish. *Front. Immunol.* **12**: 1–13.

208 Schyth, B.D., Bela-Ong, D.B., Jalali, S.A.H. et al. (2015). Two virus-induced MicroRNAs known only from teleost fishes are orthologues of MicroRNAs involved in cell cycle control in humans. *PLoS One* **10**: 1–22.

209 Eslamloo, K., Inkpen, S.M., Rise, M.L., and Andreassen, R. (2018). Discovery of microRNAs associated with the antiviral immune response of Atlantic cod (*Gadus morhua*) macrophages. *Mol. Immunol.* **93**: 152–161.

210 Han, J., Chu, Q., Huo, R., and Xu, T. (2018). Inducible microRNA-122 modulates RIG-I signaling pathway via targeting DAK in miiuy croaker after poly(I:C) stimulation. *Dev. Comp. Immunol.* **78**: 52–60.

211 Zhang, B.C., Zhou, Z.J., and Sun, L. (2016). Pol-miR-731, a teleost miRNA upregulated by megalocytivirus, negatively regulates virus-induced type I interferon response, apoptosis, and cell cycle arrest. *Sci. Rep.* **6**: 1–14.

212 Chu, Q., Xu, T., Zheng, W. et al. (2020). Long noncoding RNA MARL regulates antiviral responses through suppression miR-122-dependent MAVS downregulation in lower vertebrates. *PLoS Pathog.* **16**: 1–27.

4

Epigenetic Inheritance in Aquatic Organisms

Ramji K. Bhandari

Department of Biology, University of North Carolina Greensboro, Greensboro, NC, USA

4.1 Introduction

4.1.1 Gene–Environment Interaction and Epigenetic Inheritance

Diverse phenotypic traits seen in organisms may have numerous origins. Some traits are strongly influenced by DNA sequence and genes, while others can be regulated by the environment. Interactions of the environment with the genome thus influence the expression of a trait resulting from the interplay between genetics and the environment. Genetic alterations can predispose organisms to develop a trait, but often it would not be complete unless environmental influences trigger the event. It is, therefore, essential to understand how genes and the environment work together and how alterations in environmental influences can change the developmental trajectories of phenotypic traits. Such phenotypic traits are tightly associated with molecular processes, particularly epigenetic mechanisms (Box 4.1), governing altered gene expression patterns [1, 2].

Biological embedding occurs when early experience changes an individual's biological processes, affecting their subsequent development, fitness, and behavior [15, 16]. In a more straightforward form, it implies that adverse events very early in development can exert disproportionately potent effects on health and disease later in life and can influence the fitness of an organism. The genetic makeup of the organism profoundly controls such effects; however, the epigenetic embedding (Box 4.1) of genetic elements can also result in similar outcomes as demonstrated in diverse phyla. Environmentally induced heritable effects include developmental origins of adult-onset traits with altered fitness and transmissible phenotypic traits with altered fitness developing in a non-Mendelian inheritance fashion [17–25]. These observations strongly suggest that such heritable effects observed in experimental animals can have seriously alarming consequences if this happens, as predicted, in the natural population. Phenotypic outcomes, particularly susceptibility to pathogens, reproductive impairment, behavioral alterations, and physiological anomalies, can be concerning. Such adverse outcomes can lead to population loss and gradual extinction of the species over time. On the other hand, some phenotypic outcomes can be beneficial for the organism, as past stress experiences and lifestyles may promote adaptive phenotypes in offspring encountering similar stresses [17, 26–28]. Although there is no empirical data demonstrating the degree of positive and negative outcomes of environmentally induced, heritable, phenotypic traits in the natural population of a species, mounting laboratory and field experiments suggest that negative outcomes such as impaired behavioral and physiological anomalies may outweigh positive outcomes such as physiological resilience and adaptive traits [21, 22, 28–31].

For a long time, genetic forces and the environment have been considered two separate aspects in which their interactions can lead to a phenotype [32]. Recent advances in gene–environment interaction research have led to a conclusion that the stimuli exerted by the environment are mediated by nongenetic mechanisms, which affect the physiology of the individual experiencing stimuli and their cells that give rise to future generations. The mechanisms by which such environmentally induced heritable traits can develop have been defined as epigenetic inheritance, where DNA sequence is not altered, but the epigenetic profiles, particularly DNA methylation/demethylation and histone modifications (Box 4.1) that constantly affect how the genes are expressed by maintaining their profile both mitotically and meiotically, are altered. Environmental cues, such as chemicals, nutrition, pathogens, hypoxia, temperature, photoperiod, etc., can impart

Epigenetics in Aquaculture, First Edition. Edited by Francesc Piferrer and Han-Ping Wang.

Box 4.1 Terminologies frequently used in Chapter 4

DNA methylation: DNA methylation is an enzymatically controlled process that incorporates a methyl group, as a covalent attachment, onto the 5-carbon position of the nucleotide base cytosine. Among many other means, this chemical modification reduces the expression of a gene either by directly blocking the access of regulatory proteins to the gene promoter/enhancer or by recruiting other inhibitory proteins that wrap around the DNA and make the promoter/enhancer inaccessible to transcription factor machinery [2].
Epigenetic assimilation: Method by which environmentally induced epigenetic traits become fixed after a few generations, such that the original stimulus that induced the phenotype is no longer required to maintain the phenotypic trait.
Endocrine-disrupting chemicals: The chemicals that mimic endogenous hormones and can induce the physiological events that are supposed to be induced by endogenous hormones.
Epigenetic embedding: The epigenetic process by which life experiences alter biological functions, leading to later life health and well-being of the organism [15].
Epigenetic inheritance: Mechanism underlying the transfer of nongenetic signatures from one stage of development to another, from embryo to adulthood, from one generation to the other; it can be via somatic cells or germline cells.
Epigenetic mechanisms: Processes by which heritable and stable changes in gene expression occur through modifications in the DNA bases and changes to chromosomal superstructure without altering DNA nucleotide sequences [1].
Germline cells: The cells that give rise to gametes (eggs or sperm).
Histone modifications: Histones are highly alkaline (positively charged) scaffolding proteins that package DNA (negatively charged) into smaller structural units. Posttranslational modifications to histone proteins include enzyme-catalyzed acetylation, methylation, phosphorylation, and ubiquitylation, each of which alters the DNA-histone interactions in nucleosomes. These histone modifications regulate the transcription of a gene by making a portion of the chromatin accessible or open to allow for the binding of essential transcription factors. For example, histone acetylation occurs at lysine of the tail (positively charged), weakening the binding of DNA histones and opening chromatin [3]. In an open chromatin state, DNA methylation can still block the transcription of a gene. In contrast, the closed chromatin state completely blocks the access of transcription factors to DNA-binding sites, resulting in the silencing of a gene or entire set of genes, depending on how interactive the blocked site is. If inappropriate blocking of gene expression occurs through this mechanism, a failure of protein synthesis results, leading to physiological imbalance(s) in the cell/organ/organism, ultimately leading to adverse fitness outcomes [4]. Such modifications in histones are often mitotically and meiotically stable, allowing their passage through cell proliferation or to subsequent generations via germline transmission [5–9]. Apart from the role of histone acetylation in gene transcription, the roles of other histone modifications in gene expression and genomic stability are complex, as they are influenced by intrinsic and extrinsic factors.
Small RNAs: Small noncoding RNAs, such as micro RNAs (miRNAs), control gene expression posttranscriptionally. When an miRNA binds its target with complete complementarity, it leads to the target mRNA cleavage. At the same time, it can also bind to its target with incomplete complementarity, leading to translational repression [10–13]. One miRNA can target several genes, and one gene can similarly be targeted by several miRNAs [14]. Several miRNAs have well-defined promoters for which they have a complementary sequence, but the regulatory elements are not as clear for many miRNAs. miRNAs are often present in introns of protein-coding genes. The promoters for the genes can also act as promoters for the intragenic miRNAs. All miRNAs can act at the posttranscriptional level. In addition to the post-transcriptional inhibition of gene function to encode proteins, miRNAs can also directly or indirectly target key effectors of the epigenetic machinery.
Transgenerational inheritance: Mechanism by which phenotypic traits or induced memories are transferred from the stressed parents to offspring in subsequent generations who did not experience it directly. The stress can be abiotic or biotic in nature.

epigenetic changes in the cells undergoing organizational profiling, differentiation, or proliferation (Figure 4.1). Such environmentally established epigenetic alterations, which are different from naturally occurring epigenetic profiles, are often called epigenetic memories or epimutations. Epimutations can stably be passed on to daughter cells during mitotic cell division. If this happens, the newly developed tissue can harbor epimutations on regulatory regions of the genome, which can eventually lead to an imbalance in gene expression patterns [33].

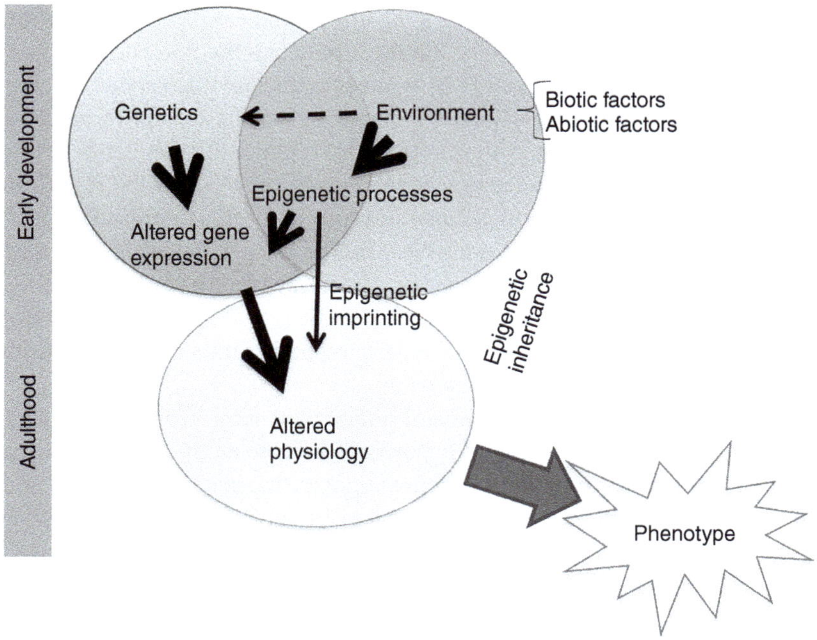

Figure 4.1 A schematic showing gene–environment interactions and emergence of phenotypes. External stimuli can induce epigenetic alterations in the cells of developing organisms, which result in a shift in the expression pattern of genes altering physiological responses. These altered physiological responses may cause a temporary or permanent expression of phenotypic traits, which may be adaptive or maladaptive in nature.

Although underlying mechanisms could be similar, epigenetic inheritance has been found to occur within, between, or across generations [34]. The inheritance that occurs during mitotic divisions of somatic cells within the same generation, mainly from early development to the adult stage, is called within-generation epigenetic inheritance. The within-generation inheritance can take place in germline cells (Box 4.1) as well. The traits associated with within-generation inheritance are associated with reproductive fitness. The paternal or maternal contribution to offspring fitness is mediated by intergenerational inheritance through the transmission of parental epigenetic marks contained in their gametes [35, 36]. Transgenerational epigenetic inheritance involves the passage of epigenetic information from grandparents to a grandchild via germline transmission at multigenerational levels [5, 6]. Since epigenetic inheritance of phenotypic traits in aquatic organisms, including fish, is comparatively understudied, this chapter will discuss epigenetic inheritance caused by environmental cues, referencing comparative examples in mammals, select fish, and other organismal models.

Chemicals used in everyday life are often released into the environment, leading to widespread exposure of aquatic and terrestrial species. These chemicals have been detected in the air, natural waters, and human consumables. At very low concentrations, they can interfere with biological processes [37–41]. The degree of biological impact caused by environmental stressors depends on the type (biotic or abiotic), the nature (lipophilic vs. hydrophilic, persistent vs. nonpersistent), and the life-history stage (particularly if exposure is during a "critical window" of germ tissue differentiation) of the organism at the time of exposure. Environmental abiotic stressors, such as hypoxia [42, 43] and rearing conditions [44–48], can impart similar phenotypic traits upon exposure during the embryonic stages, suggesting that early developmental timelines are vectors for later life phenotypic traits development in organisms across phyla.

Endocrine-disrupting chemicals (EDCs) (Box 4.1) are a class of chemical constituents of diet (natural or commercial) or various consumer products that affect the normal endocrine function and cause many adverse outcomes in development, reproduction, immunity, and other endocrine-related functions in fish, wildlife, and humans [25]. Susceptibility to endocrine disruption varies temporally, increasing the risk occurring during critical developmental windows. Accumulating evidence suggests that the effects of early-life exposure to chemicals may appear immediately or remain undetected for a long time until their manifestation as adult-onset phenotypes in the same generation or subsequent generations [18, 49–56]. The mechanisms underlying epigenetic inheritance, particularly the environmentally induced developmental origins of adult disease and adverse health outcomes in offspring, including transgenerational effects, are uncharacterized. A transgenerational effect is an abnormal outcome that develops in organisms that were never directly exposed, but the effects are inherited because prior generations were exposed to stressful conditions [5, 6]. Multigenerational and transgenerational observations take a significant amount of time to understand the risk in a natural population. It may be too late until risk assessment guidelines focus on the epigenetic inheritance of stressor effects in organisms, including fish.

Environmentally induced effects in organisms that are passed on to offspring are often mediated by nongenetic (epigenetic) mechanisms of inheritance that can influence the rate and direction of adaptation [57]. Epigenetic processes are chemical modifications around DNA sequences that are often stable and heritable. Epigenetically induced environmental effects have been described in plants, laboratory animals, and field populations of several animal species. Studies involving laboratory animals provide evidence for the inheritance of phenotypes by offspring several generations down the line, with some contributions to a rapid emergence of phenotypes [7, 18, 49, 52, 58–66]. With the advent of tools such as high-throughput next-generation sequencing of genome and epigenome and improved bioinformatics, quantification of epigenomic changes has been made possible. Questions regarding what specific epigenetic changes are transmitted to offspring and whether they are causative for the emergence of phenotypes are yet to be answered. Several review papers describe the effects of environmental contaminants on fish and wildlife [25, 49, 61, 67–72]; however, only a few studies have discussed transgenerational fitness changes caused by environmental and dietary stress during early embryonic development when epigenetic changes in germ cells are occurring.

The concept that gene–environment interactions may lead to the emergence of phenotypes has been widely accepted recently. Environmentally induced epigenetic links to adult-onset phenotypes often have their origin in embryonic development [73–75]. Accumulating evidence suggests that gene–environment interactions result in a phenotype that can either adapt to the changing environment or fail to adapt due to multiple health and fitness complications [68, 73, 76, 77]. The phenotypes that adapt to the environmental stressor may develop a certain degree of insensitivity to concurrent environmental changes [78–81]. Genetic mutations play a significant role in the development of insensitivity to intrinsic or extrinsic signals [82, 83]; however, whether epigenetic modifications, such as differential DNA methylation profiles or altered histone marks, play a role in developmentally related insensitivity to intrinsic or extrinsic signals is currently unknown. Although small RNAs (Box 4.1) have been implicated in environmentally induced health disparities [84–86], how they regulate the entire process is still unknown.

Stressful environmental conditions can be defined as those that lead to a sharp reduction in fitness in populations. When altered environmental conditions result in a sudden decline in reproductive output, leading to permanent damage, these conditions constitute environmental stress [87]. Species specificity exists with regard to the degree of sensitivity toward stressors. Some species respond to stressors immediately through the development of phenotypes, whereas other species may not have immediate responses based on lower susceptibility during the specific timing of exposure. Increased fitness in an unfavorable environment can be associated with decreased fitness in favorable environments. For instance, the same environmental exposure to ubiquitous chemical bisphenol A (BPA) caused developmental defects in zebrafish (*Danio rerio*) [88], whereas, in medaka (*Oryzias latipes*), it did not cause apparent phenotypic changes in the exposed generation, but led to transgenerational developmental and reproductive defects in subsequent generations [18, 89, 90]. Such discrepancies in sensitivity to the same chemical depend on gene–environment interactions, epigenetic susceptibility, genetic makeup, the ability of the embryo to reprogram the epigenome, and the life history characteristics of the organism. Also, it seems that environmental and dietary stressors during early developmental stages, which are also called the window of susceptibility, induce latent effects that result in physiological and behavioral traits later in life or in subsequent generations. It is not unusual to have different effects in the exposed generation that are mediated by changes in somatic cells and transgenerational effects that are mediated by changes in germ cells. To fully understand how environmental stress (biotic and abiotic) causes phenotypic alterations within, between, or across generations, it is important to study the ability of organisms to process epigenome and underlying mechanisms of epigenetic inheritance of environmentally induced signals in each generation.

In summary, epigenetic inheritance can occur at three levels: within, inter-, and transgenerational. Although they may be controlled by similar mechanisms, the resulting phenotypes can differ. Gene–environment interactions can negatively regulate physiological processes leading to maladaptive phenotypes, whereas some effects may allow the exposed individuals and their descendants to develop adaptive traits against parental or ancestral stress conditions.

4.1.2 Key Mechanisms Underlying Epigenetic Inheritance

Genetics has been an extensively studied field of science for over a century. Although gene dysregulation was found to be associated with altered phenotypic traits, including diseases, genetic mutations had long been considered a major cause for the onset of diseases and the emergence of phenotypic changes in a population. This concept was challenged when monozygotic twins were found to develop adult-onset phenotypic traits independent of their identical genetic constitution [91, 92]. This further validated the role of differential gene regulation in the development of phenotypic traits. Apart from the genetic basis for trait development, anthropological stressors, including EDCs, at levels of exposure experienced by the general public, have been shown not to cause genetic mutations, but rather more often to induce phenotypic changes via epigenetic modifications [44, 45]. The global understanding of environmental health has been revised further, as accumulating evidence has confirmed the role of epigenetic modifications in disease onset and phenotypic trait development [8, 93]. Hence, both genetic and epigenetic factors seem to contribute to the onset of disease and the emergence of altered phenotypes [57, 94] in organisms ranging from worms to mammals in the phylogenetic tree.

Epigenetic changes are chemical modifications in chromatin or the addition of chemical tags on the DNA nucleotide that influence the manner in which a gene is regulated, often locking a gene in the "ON" or "OFF" position at the appropriate times, although the result can include modulating the level of activity of genes and gene networks. Some epigenetic changes are mitotically stable and heritable within somatic tissues throughout the life of an organism, which is separate from epigenetic modifications of germ cells that can be transmitted across generations. Epigenetic modifications and their roles in the well-being of humans and wildlife have been recently described [19, 30, 48, 72, 95, 96], many of which are in relation to human health. Many human diseases, including diabetes and certain cancers, have been linked to the environment as a major contributor, and epigenetic modifications are considered to be one of the main mechanisms promoting their onset and progression [8, 97]. These studies provide insights into our understanding of how epigenetics serve as a mechanism through which environmental stressors induce phenotypic traits, particularly maladaptive in nature [49, 52, 98–102]. Since little is understood about how epigenetic inheritance occurs in fish and other aquatic organisms, most of the discussions in this chapter will include mechanistic information from mammals and other model organisms. They will be brought into the perspective of the growth, behavior, reproduction, and well-being of fish in captive and natural populations.

Within the last two decades, research on environmental health and epigenetic control of development and disease has gained significant momentum. A simple PubMed search (March 2022) with the keyword "epigenetic" yielded 118,069 publications, while "gene–environment interactions" yielded 9613 publications, "epigenetic inheritance" yielded 580 publications, and "epigenetic inheritance and fish" yielded 114 publications. In the year 2021 alone, 179 articles on epigenetic inheritance were published; out of these, 36 were related to epigenetic inheritance in fish. The growing number of publications on environmental epigenetics and epigenetic inheritance in fish is indicative of the importance of epigenetic inheritance in the well-being of aquatic organisms in the context of past, present, and future environmental changes.

The three different epigenetic modifications, i.e., DNA methylation, histone modifications, and noncoding RNAs, interact and influence each other [14, 103]. For example, the expression of miRNAs can be regulated by DNA methylation in their respective promoter regions [104], whereas miRNAs (such as miR-34b, miR-101, and miR-499a) can directly target chromatin-remodeling enzymes and DNA methyltransferases (DNMTs) and modulate the downstream effects [10–13, 105, 106]. Recent studies have found miRNA freely circulating in the blood, which is associated with diseases [107, 108] and is a biomarker of metabolic and reproductive states in fish [109]. These findings open a whole new arena for research to screen for miRNA and exosome biomarkers that can potentially serve as biomarkers of fitness and as sensitive and specific biological indicators of environmental stress and associated phenotypic outcomes [110–112]. PIWI-interacting RNAs (piRNAs), a class of small noncoding RNAs that tightly repress transposable element (TE) activity in the germline, can mediate stressor-induced epigenetic memory through methylation of transposons [113, 114]. Identification of such miRNAs in the natural population and their relevance to the phenotypic traits that they are developing would be an essential tool to detect fitness, adaptation, and population dynamics. Long noncoding RNAs (lncRNAs) are usually >200 nucleotides long and located in the nucleus or cytoplasm. A large body of evidence suggests that male germ cells at mitotic, meiotic, and postmeiotic stages and female gametes have been found to express thousands of lncRNAs that can be passed to the zygote upon fertilization [115]. They can participate in the epigenetic regulation of embryonic genes in trans across numerous loci or in cis at selected maternally or paternally imprinted loci [116]. They are involved in genomic imprinting, X chromosome inactivation [117], and intergenerational inheritance [115]. Topologically associated domains (TADs) on the genome correlate well with functional epigenetic domains defined by chromatin marks and CCCTC-binding factor (CTCF)-binding sites and regulate enhancer binding [118]. The TADs have an epigenetic association and are flanked by CpG islands, short interspersed nuclear elements (SINEs), and enrichment in CTCF and H3K4me3 binding [119, 120], suggesting their potential roles in imprinting and intergenerational epigenetic inheritance of environmentally induced traits. Posttranslational modifications of histones are associated with chromosome silencing and chromatin compaction and can persist through several rounds of mitotic cell divisions [3, 4]. They form a memory, often called a histone memory, in developing germ cells and permit the epigenetic transmission of an established memory from one stage of germ cell development to another during gametogenesis [5–9, 121, 122]. They are also believed to be carrying a stressor-specific memory in the germline during intergenerational epigenetic inheritance [123]. The majority of studies that concern epigenetic inheritance are focused on DNA methylation as a mechanism. The scope of this chapter is to provide DNA methylation as an endpoint of gene–environment interactions and an epigenetic driver for the inheritance of phenotypic traits.

4.1.3 Epigenetic Inheritance of Traits

Epigenetic inheritance in aquatic organisms can occur due to gene–environment interaction, which can occur at any stage of their life history. The environment represents all biotic and abiotic factors grossly present in the aquatic ecosystem. However, for the environmentally induced effects to be inherited within a generation level, the effects must be established in the cells that divide or differentiate into subtypes as organisms grow. For inter- and transgenerational effects, the stressor should be gametic and affect germ cells that give rise to eggs and sperm. There might be many steps, but the following two steps seem crucial. First, the gene–environment interaction should establish nongenetic memories (e.g., differential DNA methylation, histone modifications, etc.) in the genome that are mitotically or meiotically stable. It can be during early embryogenesis, juvenile, adolescent, or adult stages. Second, these memories should alter the transcription of the target gene or genes in the specific tissues that inherited them and this should result in an observable phenotypic change, or, alternatively, these alterations should take place with little or no change in the phenotype. Local and Ventura [32] beautifully portrayed an overall picture of epigenetic inheritance and its importance in phenotype development. According to them, parental exposure to external stimuli can influence an offspring's vulnerability to many pathological conditions, which can endure for several generations. They further highlighted that direct epigenetic effects occur in the lifespan of an individual due to direct experience with their environment, whereas indirect epigenetic effects occur preconceptionally or embryonically. Ample studies provide experimental evidence for the inheritance of induced traits that develop in subsequent generations due to ancestral stress experiences. DNA methylation and histone modifications and small noncoding RNAs (miRNAs/piRNAs) that play essential roles in genomic stability gene expression in diverse phyla [4, 97, 124] have been found to be associated with within, inter-, and transgenerational inheritance [5–7, 86, 112, 125] (Box 4.1). Inheritance of epigenetically induced traits can take place in individuals within the same generation. This phenomenon became evident when David Barker first described it in his human case studies [126]. Barker's hypothesis of prenatal programming and developmental origins of health and disease implies that environmental or endogenous influence in embryonic development can lead to the deviation of developmental trajectories, which often result in the development of altered phenotype and behavioral conditions in adulthood (Figure 4.2). This concept has been found to be applicable to many vertebrate species, including fish [127], where early-life stressful experiences can either make them more vulnerable or resilient to similar stress later in life [128, 129]. There are several windows of susceptibility during which the stress can induce long-term physiological consequences, leading to adaptive or maladaptive traits (Figure 4.2).

In fish, early-life environmental manipulation leading to altered phenotypic outcomes in adulthood has been previously demonstrated in salmon [130, 131]. Such growth acceleration effects on increased fitness of the exposed individuals have been extensively reported since the 1980s and have been in practice in aquaculture industries, including federal fisheries bureaus. Although the mechanistic view as to what causes enhanced growth and adaptive phenotypes in salmonid species after manipulation of neonatal growth in the hatchery has not been studied, based on the information from other relevant studies in mammalian models, observations clearly suggest that altered environmental cues and diet provided at the time of their growth phase can impart molecular memories that are inherited by somatic cells of the body, leading to altered physiological conditions in adulthood. Such a within-generation pattern of epigenetic inheritance in the somatic tissues may also have long-term consequences if developing germ cells that give rise to eggs and sperm are impacted.

Environmental manipulations in hatcheries have been common since the late twentieth century without studying the long-term consequences on the population of salmon. In the twenty-first century, after a few decades of such practices, reports are frequently coming out demonstrating that there is decreased fitness of salmon in the ocean and a significant reduction in the return of wild species to their natal river for spawning [132]. Epigenetic inheritance, particularly transgenerational epigenetic inheritance, could be one of the mechanisms driving these population-level effects in the salmon, as evidence suggests that environmental exposure can alter the osmoregulatory ability of gills in the subsequent generations without affecting the first two generations [44, 89]. It is imperative to understand if early-life exposure to environmental or climatic factors in their natural environment is causing phenotypic alterations within the generation or across generations. Transgenerational negative effects on fitness may have serious consequences at the population level. To study the inheritance of environmental influence on the growth and performance of fish in aquaculture (captive or wild), an understanding

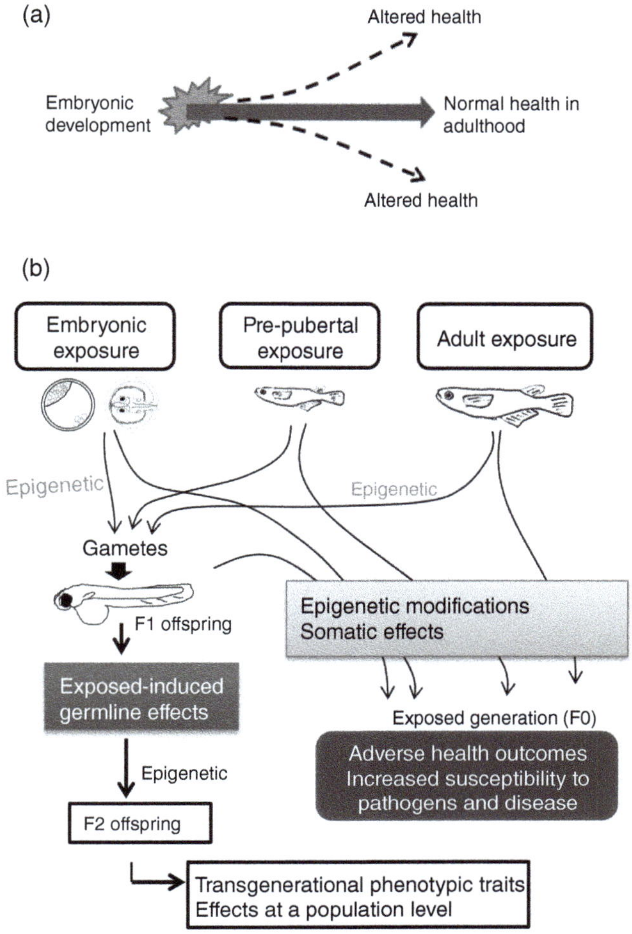

Figure 4.2 Schematic showing life history stages sensitive to external stressors and associated phenotypic traits. (a) A small change caused by external stimuli during the critical phase of embryonic development when tissues are actively differentiating and can alter the developmental trajectory away from the standard developmental path, causing altered phenotypic outcomes later in life, which may include phenotypes of super performing, adaptive, or maladaptive in nature. (b) Environmental stressors can cause immediate or long-term alterations in physiological processes leading to within, inter-, or transgenerational phenotypic traits. Effects mediated via somatic cell inheritance can produce a phenotype within the generation, while effects mediated by germline cells can produce phenotypes in multiple generations. The mechanisms underlying the inheritance of environmentally induced phenotypic traits are mostly epigenetic.

of epigenetic reprogramming of embryo and germline cells is imperative since all inter- and transgenerational traits are mediated by germline transmission.

4.2 Epigenetic Reprogramming of Embryo and Germline Cells

4.2.1 Reprogramming of the Embryo

To understand how environmental signals are processed in developing organisms and their roles in developing phenotypic traits, it is imperative to understand how the epigenome is processed during early life history stages in the embryo and how it is reprogrammed in the germline stem cells. Although several epigenetic players control genomic stability and gene expression, how cells regain pluripotency has been studied at the level of DNA methylation in humans [133, 134], mice [135–137], medaka [138–140], and zebrafish [141–143]. Reprogramming of histone modifications during early embryogenesis and the development of primordial germ cells (PGCs) has also been studied, but not at the resolution of DNA methylation. Histone profiles are dynamic, and there are several forms. It is not within the scope of this chapter to cover them; however, it is demonstrated that histone marks of transcriptional repression and activation follow the pattern exhibited by global DNA methylation and demethylation and have been reviewed elsewhere [144, 145].

There are two critical windows of embryo reprogramming of epigenetic marks – the first one takes place in postfertilization stage embryos and the second takes place in PGCs undergoing sexual differentiation. In medaka fish and mammals, fertilization leads to the loss of both paternal and maternal epigenetic modifications, which reset during the morula stage of cleavage and are reestablished at the time of the blastula stage (Figure 4.3). Aberrant epigenetic alterations, if occurred at this early stage of embryo reprogramming, can result in differential gene expression across many tissue types. This can potentially lead to altered physiological states in adulthood, and this concept has been described as the developmental

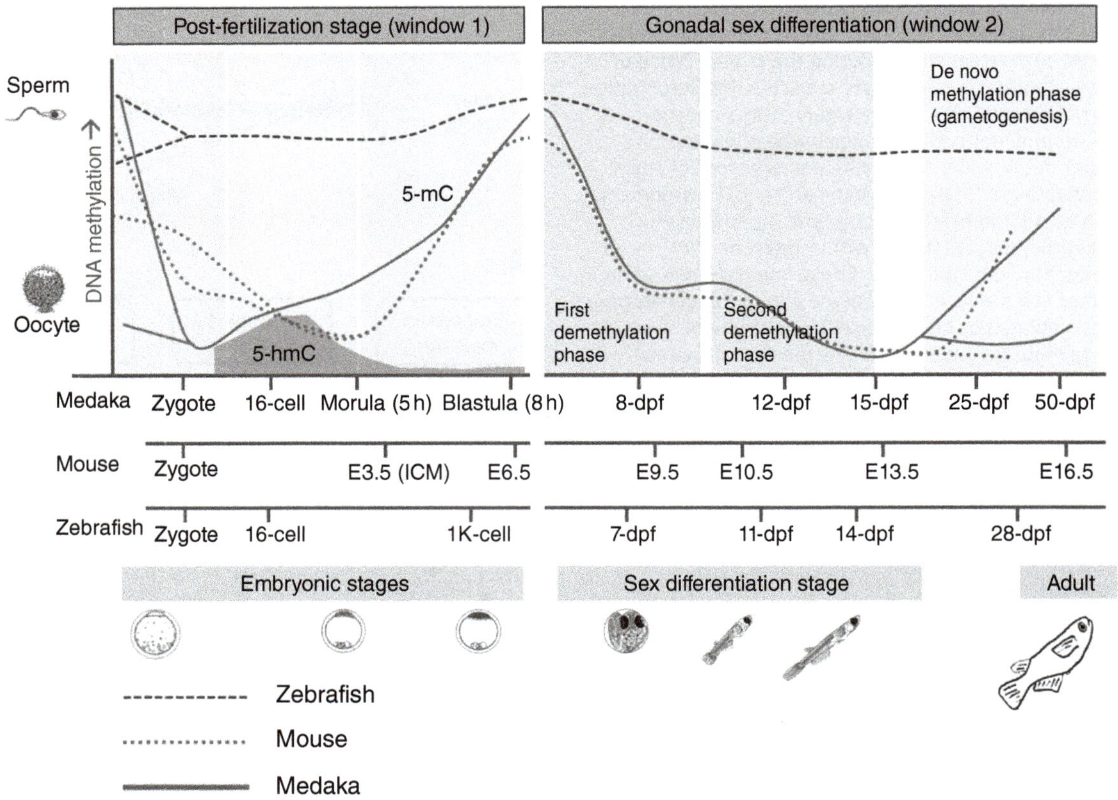

Figure 4.3 DNA methylation reprogramming model during embryogenesis (left panel) and PGC specification (right panel) in medaka. Left panel: Sperm is hypermethylated, and oocyte is hypomethylated. Paternal genomic methylation is erased in the zygote. Embryos maintain genome hypomethylation during the first several cell cycles. Global DNA methylation levels increase from the 16-cell stage to the Blastula stage. Right panel: In PGCs, the demethylation occurs in two phases: the first phase occurs before the 8-dpf stage (first demethylation phase), and the second phase occurs between the 10-dpf and the 12-dpf stages (second demethylation phase). Global DNA methylation level increases from 15-dpf to 25-dpf in male PGCs, whereas female PGCs remain hypomethylated. Data set *sources:* medaka embryogenesis, medaka PGCs, mouse embryogenesis, mouse PGCs, zebrafish embryogenesis, and zebrafish PGCs. *Source:* Wang and Bhandari [139]/Oxford University Press/public domain CC BY 4.0. (*See colour plate section for colour representation of this figure.*)

origins of health and disease [139, 146, 147]. During this stage, the paternal genome undergoes rapid demethylation during the first cell cycle and continues throughout the cleavage, and remains the lowest at the morula stage, whereas the maternal genome undergoes the process of gradual demethylation (passive), and the parental nucleus gains a methylation pattern similar to the maternal one by blastula stage. In medaka, sperm undergoes rapid demethylation (60% drop) upon fertilization, whereas eggs maintain a low level of methylation [140]. This suggests that the paternal DNA methylation memories in the sperm can be massively reprogrammed upon fertilization, whereas a low level of genome methylation in the egg suggests limited or no reprogramming of maternal memories. In contrast, the dynamics of DNA methylation in zebrafish embryos maintain a slightly different profile. The methylation of the paternal genome is maintained high, whereas maternal genome methylation gradually increases during cleavage and reaches a level comparable to the paternal level by the blastula stage, suggesting that the genome exclusively acquires methylation marks in embryos undergoing cleavage and blastulation. There are events involving demethylation or remethylation of specific genomic loci globally; the maternal genome gains 6% methylation from 80 to 86% during this process [142]. These observations suggest that zebrafish maintain genome DNA methylation patterns different from that of medaka and mammals during embryogenesis [139] and the filtration of epigenetic memories and the inheritance of traits in zebrafish could be maintained by multiple mechanisms, which need to be experimentally verified. In another fish, the mangrove killifish (*Kryptolebias maroratus*), although sperm and egg methylome profiles were not characterized, the global CpG methylation levels in the cleavage stage embryos drop to 15.8% by the gastrula stage and gained to 70% by stage 26 when organogenesis continued [148], suggesting that there is a clear pattern of reprogramming matching somewhat like that of medaka and mammals. According to Ortega-Recalde and Hore, epigenetic erasure is a strong barrier to intergenerational and transgenerational epigenetic inheritance in mammals. In those species where epigenetic erasure does not occur, increased opportunity for epigenetic inheritance between generations may exist [149]. In such circumstances, parental stress experience-induced epigenetic profiles may promote the offspring's ability to cope with the same environmental stress. The evolutionary mechanisms driving the divergence in epigenetic programming are unknown, nevertheless, a direct consequence of retaining epigenetic memory in the form of DNA methylation (5mC) is the enhanced potential for transgenerational epigenetic inheritance. Given that DNA methylation dynamics remains underexplored in most vertebrate lineages, the extent of information transferred to offspring by epigenetic modification might be underestimated [149].

Why is the reprogramming of embryo epigenome essential since all the species thrive with varying mechanisms of embryonic epigenome programming? It is for understanding epigenetic inheritance of effects established during early-life history stages in the well-being of the same individual in adulthood. Epigenetic modifications can act as drivers of welcomed or unwelcomed passengers created by environmental cues. Suppose such modifications are not reset during embryo reprogramming. In that case, the cells and tissues that result from the gastrulation of the embryo will harbor multitudes of defects, resulting in enhanced or physiologically defective phenotypes in adulthood. For instance, alterations in DNA methylation or histone modifications during the two critical windows of embryo reprogramming can be either completely erased and reestablished during de novo establishment of cell-specific epigenetic modifications occurring at the time of cell lineage specification stages, or they can be resistant to reprogramming and remain allele-specific (Figure 4.3, left panel). Such differential epigenetic alterations, if mitotically stable, can be transmitted to daughter cells, causing transcriptional alterations in the somatic tissues that develop. Such embryonically established alterations can be maintained in the somatic cells throughout the life of the organism, which may lead to adverse health outcomes immediately or later in life.

4.2.2 Reprogramming of Primordial Germ Cells

PGCs are the only embryonic germ cells that can transmit genetic and epigenetic information to the next generation. In medaka and mice, PGCs undergo a similar pattern of reprogramming. Genome DNA demethylation takes place in PGCs in two phases. During phase 1, PGCs undergo active demethylation, which erases DNA methylation marks established in epiblast during embryo reprogramming (embryonic day 6.5 in mice and 10 hours postfertilization [hpf], in medaka, Figure 4.3). During phase II, locus-specific DNA demethylation takes place, leading to the ground state of reprogramming, during which more than 80% of the PGC genome remains hypomethylated. In mammals, a small portion of the PGC genome, comprised of imprinted genes and small interspersed DNA repeat sequences, is resistant to global DNA demethylation [122, 141], suggesting that these genomic regions might be responsible for the inheritance of parental phenotype. In zebrafish, PGCs do not undergo such massive demethylation, suggesting that zebrafish maintain a different level of genome methylation in the PGCs during this phase. Since mice, humans, and the Hd-rR strain of medaka exhibit genetic sex determination under the control of the Y chromosome and maintain a similar pattern of PGC epigenetic reprogramming, it raises a possibility that the reprogramming of the PGC epigenome might be linked to genetic sex determination in sex cells. Interestingly, mice, medaka, and zebrafish demonstrate within, inter-, and transgenerational inheritance of phenotypic traits caused by environmental influences, despite zebrafish maintaining a genome DNA methylation profile different from mice and medaka [139, 141, 143]. This raises the question of whether the reprogramming of PGCs is dispensable for the transgenerational inheritance of phenotypes or whether a very small portion of PGC genome DNA methylation is sufficient to trigger this process, or whether there are secondary processes such as miRNA, piRNAs, RNA modifications, or histones playing greater roles than DNA methylation in the inheritance of transgenerational phenotypes. Although DNA epimutations have been shown in the sperm and eggs of zebrafish in response to environmental stressors and their persistence in gametes across subsequent generations [150], their roles in the development of phenotypic traits have, however, not yet been verified. It is essential to trace the dynamics of DNA methylation in PGCs at various stages of germ cell development after the environmental insult of developing germ cells. This would produce a comprehensive map of DNA epimutations that were established by exposure, their resetting, and heritability in the sperm/eggs. Preliminary studies suggest that environmentally established epimutations in the PGCs in medaka can resist epigenome reprogramming and remain heritable in the sperm (Wang et al., unpublished), which are extensively reprogrammed by the embryo prior to the blastula stage, leading to the inheritance of less than 5% of epimutations that were present in the founder sperm. The dynamics of epimutation resetting in mammalian PGCs following environmental stress are currently unknown, although inheritance of exposure-specific epimutations has been reported in the sperm and eggs across generations. Since sperm and eggs harbor environmentally established DNA epimutations, their resetting during embryo reprogramming is essential to avoid deleterious effects in the offspring across subsequent generations. Transgenerational passage of epimutations across subsequent generations leads to several somatic cell defects in adulthood. In some instances, they can convert into genetic mutations [33, 151]. If environmentally induced germline epimutations are converted into genetic mutations in the somatic/germ cells after certain generations of inheritance, different physiological and behavioral consequences can be anticipated at the population level of that species [7, 151].

4.3 Heritable Effects of Environmental Stress

4.3.1 Developmental Effects

As briefly described in the earlier section, developmental exposures to environmental stressors (chemical or nonchemical, biotic or abiotic) at the sublethal level may impose long-term phenotypic effects that can appear later in life or in subsequent generations. Environmental stress during embryo development and perinatal tissue differentiation can lead to immediate or adult-onset phenotypic traits. If developing germ cells are not affected, the effects can only be observed indirectly in exposed individuals. For effects to be transmitted to offspring, an exposure of the germ cells is required, mainly when the germ cells are at the specific stage of reprogramming or cellular differentiation [25, 131, 137, 139, 152–154]. Environmental perturbations during epigenetic reprogramming events can establish abnormal gene expression programs in the cells that affect the timely activation or repression of genes in the perturbed cells. Such alterations are believed to be initial triggers for the development of diseases, which then appear at a specific time in life, especially when the affected genes are typically required to be expressed or repressed. Effects of chronic or acute exposure to environmental or exogenous stressors lead to toxicological endpoints where physiological manifestations, structural deformities, reproductive impairment, behavioral alterations, and mortalities can be expected in the generation of the exposed [19, 25, 30]. Such effects are due to the inheritance of exposure-induced alterations through within-generation inheritance (Figure 4.2). Fish living in ecologically diverse environments can serve as excellent examples to understand the epigenomic contribution to adaptation [155]. For example, in Cichlid fish in Lake Malawi with extremely low genetic diversity (sequence divergence), differentially methylated regions (DMRs), enriched in evolutionary young transposons, are associated with transcription changes of ecologically relevant genes related to energy expenditure and lipid metabolism, pointing to a link between dietary ecology and methylome divergence [155]. Environmentally or parental stress-established effects cannot be transmitted to subsequent generations via somatic cells; however, somatic cell effects, e.g., traumatic experience, can be inherited by subsequent generations via germline transmission of epigenetic memory [156, 157]. In such cases, not DNA methylation, but small or noncoding RNAs are involved.

4.3.2 Postnatal or Parental Effects

Postnatal exposure effects can be somatic or germline. Epigenetic modifications in somatic cells can also be propagated through mitotic cell divisions, a process called somatic epigenetic inheritance, resulting in somatic memories [158]. Somatic epigenetic memories cannot be inherited by offspring in the subsequent generation unless they are transferred to germline cells. According to Anastasiadi et al. [159], soma-to-germline communication may be possible through extracellular RNA; however, it is unclear to what extent the germline absorbs somatic epigenetic changes after segregation. Therefore, the timing of germline segregation may influence epigenetic inheritance due to the potential for whole-epigenome inheritance upon germline formation, which is unlikely to occur after segregation. In animals where the germline segregates and differentiates early in development, the timing of gametogenesis and mode of reproduction (oviparity vs. viviparity) are expected to impact epigenetic inheritance [159]. It appears that for an epigenetic inheritance to occur via somatic cells (within generation inheritance), the stressor-affected somatic cells must maintain a distinct epigenetic profile (e.g., DNA methylation) and undergo cell proliferation and differentiation. When cells undergo proliferation, the somatic epigenetic memory will be maintained mitotically. As they differentiate into a specific cell type (e.g., from mesenchymal cells to precursor cells or from precursor cells to specialized cells), they undergo dynamic changes in epigenetic profiles, such as DNA methylation or histone modification profiles. During the process of differentiation, they tend to turn off the genes that are required for the maintenance of the undifferentiated state and, at the same time, turn on other genes required for the maintenance of the differentiated state. External stressors, including endogenous hormonal surges (e.g., corticosteroids) occurring during these sensitive stages of tissue differentiation, can alter the epigenetic profiling of the differentiated somatic cells, leading to altered gene expression patterns of target genes [160, 161], which eventually results in the disruption of tissue function and phenotypic trait (physiological, morphological, or behavioral). This can happen immediately or later in life, depending on the rate of the mitotic cycle and the type of differentiated tissue. Germline cells undergo mitosis and meiosis in postnatal species specifically. The epigenetic inheritance process varies significantly between male and female germlines. How female germ cells process their epigenetic profiles, particularly DNA methylation since they maintain a low level of genome methylation profile, is not well known; however, male germ cells undergoing spermatogenesis have a distinct methylome profile that includes loss of methylation from PGCs to spermatogonia, again a gain of methylation from spermatogonia to spermatocyte following parallel maintenance of methylation profile until the spermatozoa stage, in which methylation profiles remain the highest [162]. Paternal stress effects on PGCs or spermatogonia may have possibilities for establishing more germline methylome effects than other stages, as DNA methylation profiles remain dynamic at these stages. Although the spermatozoa stage has a higher level of DNA methylation, the nucleus in the sperm head is inaccessible to external stressor effects. Histone modifications control transcription in somatic tissues and germ cells. Protamines replace them in many higher vertebrates, including many fish. Some environmentally or endogenously established histone marks can be nonrandomly retained by the sperm genome, affecting embryo development in the following generation due to alteration in the pattern of embryonic gene expression [36, 163–166]. The somatic tissues developed with altered DNA methylation profiles can give rise to distinct phenotypic traits in the F1 offspring, which may be deleterious or adaptive. Although the mechanisms underlying intergenerational and transgenerational inheritance of epigenetic memories are well studied in males, the process in female germline cells is yet to be discovered [167].

4.3.3 Germline Transmission of Epigenetic Alterations: Experimental Evidence

The period of sex determination, when PGCs differentiate into spermatogonia or oogonia, is known as a susceptible window for environment–gene interactions via epigenetic mechanisms. Environmental stressors such as certain metals, components of plastics, some pesticides, and abiotic factors such as hypoxia, ocean acidification, etc., can alter the epigenetic status of the PGCs by creating exposure-specific DNA methylation marks [43, 49, 52, 55, 101, 102, 150, 153, 168–173]. The altered DNA methylation marks, which are involved with the epigenetic transmission, are believed to be resistant to the reprogramming process and are thus passed to subsequent generations via either sperm or eggs. Such environmentally induced epigenetic alterations in genomic regions are associated with a variety of adaptive and maladaptive phenotypic traits [97, 99, 152, 174–176]. When oocytes (F0) with altered epigenomes can be fertilized by similarly altered sperm, the resulting embryo (F1) acquires an already programmed epigenetic profile from the parental germline. If not reprogrammed during F1 embryonic development, every cell that develops from the zygote then carries the altered landscape of inherited epigenetic information, thereby causing transcriptional dysregulation in germ and somatic cells of the derived offspring (F1). Since germ cells are reproductive cells, the germ cell effects are confined to reproductive fitness only. If the genes that were epigenetically modified in the F1 germ cells have no role in reproduction, the F1 adult will not demonstrate reproductive defects. In contrast, the same modification may have several functional roles in somatic tissues and cause them to produce physiological and morphological responses. Therefore, epigenetic alterations in germ cells may have far-reaching and unanticipated outcomes in subsequent generations. Similarly, as F1 germ cells turn into gametes, they will transmit an F1-altered epigenetic profile to F2 germ, and somatic cells with better transmission rates as reprogramming-resistant environmentally established profiles are undergoing a second round of reprogramming again [25, 177]. A resilience in reprogramming may develop reprogramming resistant loci, leading to epigenetic assimilation (Box 4.1).

4.3.4 Multigenerational Versus Transgenerational Phenotypes

Transgenerational studies were first carried out in mammalian models. So, the terminologies were defined for the development of mammalian embryos. Multiple generations are directly exposed when environmental exposure occurs during gestation in mammals. In the case of exposure to a pregnant mammal, simultaneous exposure occurs in three generations: the pregnant mother herself (F0), her embryo (F1), and the embryo's germ cells that will produce the F2-exposed generation (Figure. 4.4a). In the case of a postnatal exposure of organisms other than a gestating female, simultaneous exposure occurs in two generations, i.e., the exposed individual itself and its germ cells (sperm or eggs or their precursor cells). Similarly, when the embryo of oviparous species is directly exposed, two generations are exposed. These direct exposures cause effects that are generally defined as multigenerational effects, and the phenotypes resulting from such exposures are called multigenerational phenotypes. Other multigenerational exposures could be the exposure of an F1 generation for a length of time and reexposure of their offspring to the same stressor condition.

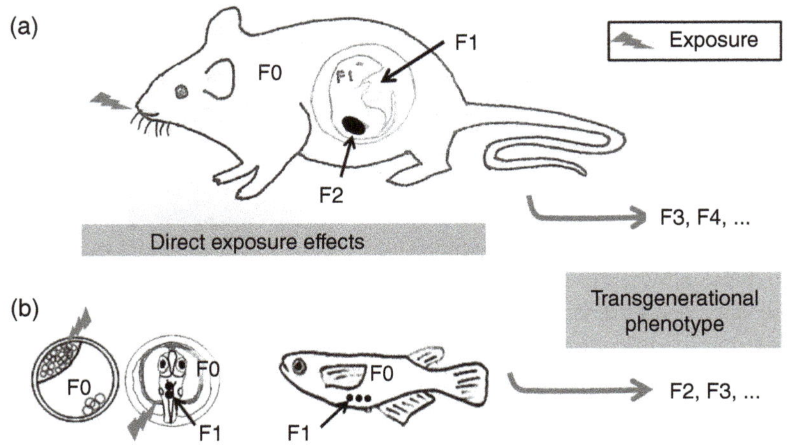

Figure 4.4 Initial exposure and generations of the exposed organisms in mice (exposure of a pregnant dam-F0) and fish (exposure of the externally fertilized embryo-F0). PGCs in the gonads of F1 mouse embryo (a) and F0 fish embryo (b) are directly exposed. Transgenerational effects are seen in mice at F3 and in fish at F2 generations.

In contrast, transgenerational phenotypes develop in organisms that were not directly exposed to chemicals or other stressors as described earlier, but due to exposure to their ancestors. In fish, truly transgenerational phenotypes can be measured at the F2 generation and thereafter, whereas gestational exposures in mammals promote transgenerational phenotypes at the F3 generation or later (Figure 4.4a, b). In many instances, multigenerational studies have been mistakenly termed transgenerational.

As mentioned in the earlier sections, embryonic development is one of the critical periods at which exposure to environmental chemicals potentially induces phenotypic effects. Lineage specification and tissue differentiation processes are in progress during this period and are epigenetically regulated [97, 178]. Each cell type acquires a distinct epigenetic landscape that is maintained throughout the individual's life. Environmental chemical/nonchemical insults can alter the epigenetic profile of differentiating cells. Consequently, an alteration in the epigenetic state may result in unprecedented gene expression profiles leading to the onset of diseases and altered behavioral phenotypes. These effects can be transmitted to subsequent generations via the affected germline, i.e., eggs or sperm (Figure 4.5).

4.3.5 Parent-of-Origin and Transgenerational Phenotypes

The parental line (eggs or sperm) responsible for transmitting altered phenotypes has become a major concern in epigenetics research [152]. Initially, the research focused on the heritability of epigenetic signatures in the somatic cell lineage. Following the observations that environmental factors (i.e., stressors) can also alter the DNA methylation profile of the germline during the embryonic period of gonadal sex determination [49], scientific communities started focusing on the intergenerational and transgenerational inheritance phenotypes. If this occurred due to exposure to a rapidly cleared (nonpersistent) chemical, this might not result in corresponding changes in somatic cell lineages that differentiate at later times in development, leading to the possibility that there could be little or no effect on the exposed (F0) organism, but there could be effects transmitted to offspring. Many studies have pointed out that sperm is a major source of transmission [24, 52]. In medaka fish, BPA exposure caused transgenerational inheritance of fertility impairment via the male germline [90]. Paternal exposure to low folate supplements causes faulty sperm formation and transmission of altered phenotype to offspring via epigenetic modifications [179]. Experiments with rats showed that exposure to an EDC during this critical window in germ cell development created epigenetic signatures that can be transmitted through sperm to subsequent generations [49, 180]. Offspring born with the inherited altered epigenome have developed a host of adult-onset diseases. Environmental toxicants (plastics, pesticides, dioxin [TCDD]), hydrocarbons, or their mixtures have been found to promote the epigenetic transgenerational inheritance of adult-onset disease phenotypes in laboratory rodents [8, 49, 52, 63, 99–102, 180–189]. Similar observations of transgenerational inheritance of altered phenotypes, which do not reflect the parent of origin, have been shown in worms [64], flies [190], plants [191], fish [51, 54, 192], rodents [24, 52, 55], domesticated farm animals [58, 59], and humans [193]. A recent study has demonstrated a paternal transmission of transgenerationally acquired traumatic memory to the offspring in mice, opening a new arena for research, especially how the acquired memory in the brain translates into an imprint in germ cells [194]. Sperm or sperm-mediated transgenerational effects have rapidly accumulated information on the paternal transmission of epigenetically acquired phenotypes because these studies are more convenient to conduct [35].

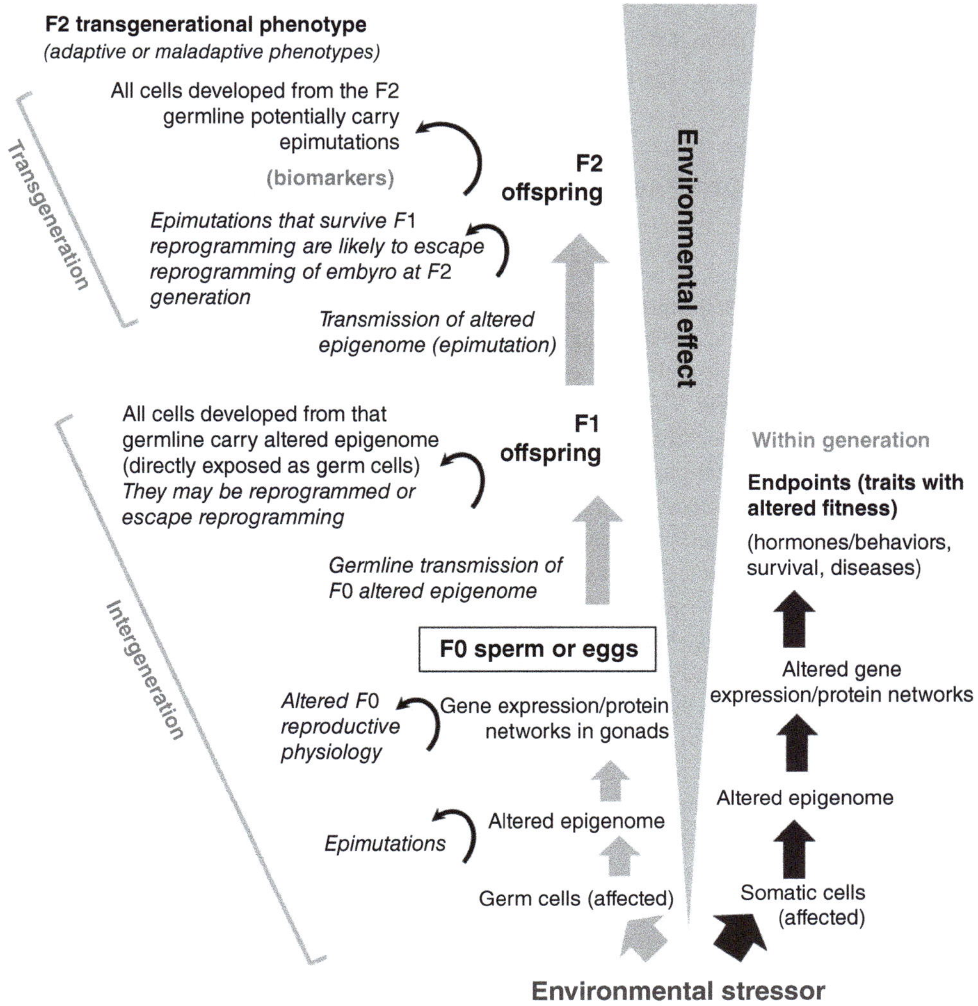

Figure 4.5 Environmental stressors cause somatic effects in the exposed organism via within-generation inheritance, whereas the exposure effects can be transmitted to subsequent generations via eggs or sperm (via inter- and transgenerational inheritance). Environmentally induced epimutations may escape the defined reprogramming processes during embryogenesis and germ cell sex determination. It is currently unknown how mitotically stable these epimutations are, but studies have demonstrated that they are heritable and can pass to the offspring via germline transmission. Upon fertilization of eggs by the sperm with established epimutation or vice versa, the resulting embryo acquires the epimutation in each cell type, just like genetic mutations, if they are not reprogrammed. The depth of the triangle shows environmental effects across generations that become profound in subsequent generations.

Maternal transmission is relatively more challenging to study due to complications in female reproductive processes, pregnancy, and egg availability. However, there is evidence for transmission of inter- or transgenerational phenotypes via the female germline. For maternal gametes to mediate inter- or transgenerational inheritance, stressor-induced effects should either affect the oocyte nuclear genome or mitochondrial genome [195]. In shortfin molly (*Poecilia mexicana*), the epigenome was stably inherited generationally through the female germline after the removal of an environmental stressor (hydrogen sulfide) [196]. In contrast, in silverside (*Menidia beryllina*), ancestral exposure to synthetic sex hormones induced maternal transfer of exposure effects to F2 and F3 generations [197]. Maternal tobacco exposure was found to affect offspring lungs due to *in utero* exposure in humans [198]. Ancestral DDT exposure during fetal development in rats caused the development of diseases such as ovarian cyst formation, prostate hyperplasia, and obesity [187], and the grandchildren of diethylstilbestrol (DES) exposed women developed a variety of health problems, including cancer [199]. The DES finding concerns a multigenerational effect of *in utero* exposure in humans, and it remains to be determined if great-grandchildren (F3 generation) are similarly affected (https://www.cancer.gov/about-cancer/causes-prevention/risk/hormones/des-fact-sheet). Literature strongly suggests that parental gametes, both eggs and sperm, are capable of harboring environmental chemical-induced epigenetic changes since imprinting of genes occurs in both sperm and eggs [200]. However, due to technical limitations and the lack of appropriate animal models, the mechanisms underlying oocyte contributions to intergenerational and transgenerational phenotypic traits have remained largely unexplored.

4.4 Past Exposure and Future Phenotypic Consequences in Aquatic Species

4.4.1 Effects on Fish

The technical challenges of studying mammalian oocytes as the source of transgenerational epigenetic phenotype transmission can be overcome if egg-laying vertebrates, such as fish that have already been extensively studied, are developed as models. External fertilization, which occurs in fish, allows experimental designs that are simply not possible with internal fertilization in mammalian reproduction. The epigenome of eggs, sperm, and developing embryos can be studied in fish without sacrificing the founder broods. Additionally, advantageous within fish models is the ability to trace the parental lineage of the epigenome through each germline, while potentially confounding background effects originating from mixed founders can be eliminated. For example, in a study with medaka discussed later, a cohort of oocytes from a female was fertilized by a male and then assigned to different treatment conditions, thus allowing one to trace the same parental lineage through multiple generations under various environmental conditions.

Fish are thus excellent models for environmental toxicology and epigenetic transgenerational inheritance research [6]. Extensive literature is available regarding the direct deleterious or advantageous effects of environmental stressors in fish and aquatic wildlife; however, as yet, relatively few studies have examined transgenerational effects of exposure using fish species (see Table 4.1). Previous reviews by Bhandari et al. [25], Schwindt [218], Mirbahai and Chapman [219], Robaire et al. [30], Best et al. [19], and Anastasiadi et al. [48] discussed environmental stressor effects on aquatic wildlife and the implications of epigenetic memory for chemical safety assessment and environmental biomonitoring. Because of the homology of the genome sequence between fish and higher vertebrates, including humans, health issues common between these two classes of vertebrates can be addressed using fish models. In addition to being widely known as model organisms for ecotoxicological research, zebrafish and medaka serve as excellent models for genetic and epigenetic studies relevant to human diseases, such as neurological disorders, melanoma, reproductive and ocular diseases, etc. [220–224]. Scientific reagents and molecular tools for genome/epigenome manipulation are available. Scientific information such as gene ontology and key pathways have already been characterized and are publicly available (GEO, http://www.ncbi.nlm.nih.gov/geoprofiles and KEGG, http://www.genome.jp/kegg). Zebrafish have been used to study the short- and long-term effects of environmentally toxic substances. Medaka, another biomedical and environmental research model [220–224], can also be a valuable model organism to study these effects. Medaka has several advantages over zebrafish [6] relevant to the study of transgenerational transmission of acquired phenotype, namely similarity in mechanisms underlying epigenetic reprogramming of the embryonic genome and the presence of an XX/XY sex-determination system [140, 222, 225]. Medaka sex determination occurs between days 5 and 8 after fertilization. Germ cell reprogramming events complement mice and humans [138–140, 226].

Transgenerational phenotypes were first demonstrated in medaka and zebrafish [18, 61]. The first medaka experiment was designed to examine the effects of exposure only during the critical window of differentiation of PGCs, precursors of eggs and sperm. Relatively higher doses of BPA or EE2 than the current environmental levels were used. Still, quantities were within the level of seasonal hit-and-run situations (i.e., up to 120 μg/L of BPA and up to 100 ng/L of EE2) [25, 227, 228]. The uptake study suggested that only up to 80 pg of BPA and 4 pg of EE2 were taken up by the eggs in seven days. For those studies, exposure occurred only during embryonic development, continuously for seven days, and never after. No noticeable phenotypic differences in egg production, survival, and fertilization were observed for the first two generations (F0 and F1).

Although the fecundity was unaffected in F2 and F3 generations, the fertilization capacity was significantly reduced in both EE2 and BPA lineages, and the survival of embryos that developed from fertilized eggs was compromised in the F3 and F4 generations [18]. In a separate study, BPA exposure during the first phase of demethylation in medaka resulted in subfertility in males in F0, F1, and F2 generations and severe steatosis of the liver, dysbiosis of gut microbiota, and ovarian atresia in females at F2 through F5 generations. The BPA-induced DNA epimutations predicting such phenotypic alterations in various somatic tissues of the offspring were found to be resistant to epigenetic reprogramming of embryos and PGCs [229, 230]. In zebrafish, exposure of adult females to BPA (20 μg/L) affected adult female fertility, promoter

Table 4.1 Epigenetic inheritance of environmentally induced effects in fish.

Stressor	Concentration	Developmental stage during exposure	Exposure duration	Generation examined, endpoints measured	References
Atrazine	5 μg/L	Medaka embryos, d-rR	0–12 days postfertilization	F2 generation, reduced fertilization rate, reduced sperm count, reduced sperm motility	Cleary et al. [56]
B[a]P	1 μg/L	Medaka (d-rR), adults	21 days	Decrease in bone thickness and the occurrence of microcracks	Seemann et al. [201, 202]
B[a]P	5 and 10 μM	Zebrafish, embryos	6–120 h postfertilization	F2 generation, hyper locomotor activity, decreased heartbeat, and mitochondrial function	Knecht et al. [203]
BPA	0.1 mg/g diet, mixed with food grain	Zebrafish, adults	6-months	F3 generation, reduced fecundity, and embryo survival rate	Akhter et al. [204]
BPA	100 μg/L	Medaka embryo, d-rR	0–7 days postfertilization	F2 and F3 generation, reduced fertilization success. F3 and F4 generation, increased embryo mortality	Bhandari et al. [18]
BPA	100 and 2000 μg/L BPA (0.44 and 8.8 μM)	Zebrafish males, adults	14 days	F2 generation, increased cardiac malformations, disrupted insulin pathway	Lombo et al. [205]
BPA	20 μg/L	Zebrafish	Adult females exposed and mated with unexposed males	Affected adult fertility, Amh mRNA decereased, and promoter hypermethylated, H3K4me3/H3K27me3 enriched.	Santangeli et al. [206]
EE2	0.05 μg/L	Medaka embryos, d-rR	0–7 days postfertilization	F2 generation, reduced fertilization success. F3 and F4 generation, increased embryo mortality	Bhandari et al. [18]
EE2	0.002 μg/L	Medaka embryo, d-rR	0–12 days postfertilization	F2 generation, reduced fertilization rate, reduced sperm count, reduced sperm motility	Cleary et al. [56]
Ethinylestradiol	0.02 μg/L	Guppies, adults, maternal	28 days	Increased anxiety behavior	Volkova et al. [207]
Floxetine	Fetus-relevant concentration, 0.54 and 54 μg/L	Zebrafish, embryos	6-day	F2 generation, Suppressed cortisol levels, pathway associated with cortisol synthesis altered significantly	Vera-Chang et al. [208, 209]
Hypoxia	Chronic, lifelong, DO 1.4 ± 0.2 mg/L	Marine medaka, adults	1 month	F2 generation: retarded gonad development, decrease in sperm count, and sperm motility	Wang et al. [173]
Hypoxia	1.5 ± 0.2 mg O_2/L	Medaka	One month, exposed at every generation until F3	Female reproducion impaired, mainly ollicle atresia and retarded oocyte development, leading to a drastic reduction in hatching success in the F2 generation	Lai et al. [172]
Methylmercury	0, 1, 3, 10, 30, and 100 nM	Zebrafish embryos	24 hours	F2 generation, abnormal neurobehavior, hyperactivity, and visual deficit	Carvan et al. [150]

(*Continued*)

Table 4.1 (Continued)

Stressor	Concentration	Developmental stage during exposure	Exposure duration	Generation examined, endpoints measured	References
PAH mixture	Environmentally relevant	Sheepshead minnow, adults	21-day?	F2 generation, deficits in prey capture,	Jasperse et al. [210]
PAH mixtures	N/A, environmentally relevant	Zebrafish, adults	21 days	F2 generation, disrupted behavior in juveniles	Vignet et al. [211]
PCB and PBDEs mixtures	ΣPCBs = 1991 ng/g diet ΣPBDEs = 411 ng/g diet	Zebrafish	Chronic, from day 5 to 180 postfertilization	Disruption of larval behavior at F2–F4 generation – mostly hypactive after light to dark transition, Dnmt3ba modified in all generations	Alfonso et al. [212]
Perfluorobutanesulfonate (PFBS)	0, 1, 2.9, 9.5 μg/L	Zebrafish	Life long, fertilization to adulthood	Skewed sex ratio toward males, small ovaries, blocked oocyte development, decreased egg production	Chen et al. [213]
PFOS alternative, 6:2 chlorinated polyfluorinated ether sulfonate	0, 5, 50, or 500 μg/L	Zebrafish	180 days, embryo to adulthood	No direct phenotye at F2 generation but transcription levels of several genes along the hypothalamus–pituitary–thyroid axis were significantly modified	Shi et al. [214]
Phthalates, MEHP	30 μM	Zebrafish, embryos	0–6 days	F2 generation, no phenotype reported, osteogenic pathways activated	Kamstra et al. [215]
Rearing condition in the hatchery	Temperature, food, photoperiod	Atlantic Salmon	Embryonic period	Sperm DNA methylation profile changed	Rodriguez Barreto et al. [46]
TCDD	50 pg/mL	Zebrafish, embryos	1 h at 3 and 7 weeks postfertiization	F2 generation, scoliosis-like axial skeleton abnormalities, reduced egg release, and fertilization success	Baker et al. [61]
ZnO, and dissolved organic matter	1–100 ppm, DOM 0.4 and 40 mg-C/L	Zebrafish, embryos	72–96 hours postfertilization	F2 generation, abnormal secondary motoneuron (SMN) axonal projections, abnormal dorsal root ganglion development, and abnormal blood vessel development	Kteeba et al. [216]
Levonorgestrel	10 ng/L	Silversides	8 hpf to 21 dph	Increased incidence of larval deformities, reduced larval growth and survival, impaired immune function, skewed sex ratios, ovarian atresia, reduced egg production, and altered gene expression.	DeCourten et al. [217]

hypermethylation, and histone modification, mainly enrichment of H3K4me3/H3K27me3 marks [206]. The observed transgenerational reproductive impairments in this study were associated with a differential methylation pattern of specific genes in sperm of both F0 and F2, coupled with relevant transcriptomic and proteomic alterations to impair spermatogenesis.

Additionally, a subsequent study demonstrated that atrazine exposure (5 μg/L) resulted in reduced fertilization rate, sperm count, and motility in adult medaka at the third generation after exposure [56]. A study by Wang et al. [173] demonstrated transgenerational abnormalities, such as delayed gonad development, decrease in sperm count, and sperm motility in F1 and F2 generation medaka caused by lifelong chronic exposure to hypoxia in the F0 generation. Additionally, female reproduction was impaired and oocyte development was delayed, leading to a drastic reduction in hatching success in the F2 generation [172]. In the past four years, hundreds of studies have examined transgenerational effects of environmental EDCs in fish, but no mechanistic explanations, especially epigenetic changes that are causative for phenotype development, have been reported with the exception of one study concerning the effects of methylmercury administered to zebrafish embryos on subsequent spatial navigation [150]. DMRs were identified in the sperm of the founder fathers, who produced offspring with navigational and visual impairments in the third and fourth generations after exposure.

Ethynylestradiol-17α (EE2) exposure (20 ng/L) induced anxiety and nonreproductive behavior transgenerationally in guppies [207]. The persistent environmental pollutant, dioxin, caused transgenerational structural abnormalities. More transgenerational effects were reported in zebrafish and medaka. Baker et al. [61] observed a reduced rate of egg release and fertilization success and scoliosis-like axial skeleton abnormalities in the F2 generation of the dioxin (TCDD)-exposed zebrafish, evident signs of transgenerational responses persisting into the generations that were never directly exposed to TCDD. Lifelong exposure to perfluorobutane sulfonate (PFBS) skewed the sex ratio toward males, small ovaries, blocked oocyte development, and decreased egg production in the F2 generation [213]. Exposures of adult parents to Benzo[a]pyrene (BaP) caused premature hatching and increased mortalities in F1 offspring; body shape, tail, and fin abnormalities in F1 and F2 offspring; and jaw, eye, and brain deformities in F2 offspring [51], as well as hyperlocomotor activity, decreased heartbeat, and mitochondrial function in the F2 generation [203].

In the medaka, Benzo[a]pyrene (BaP) exposure resulted in decreased bone thickness and increased incidences of microcracks [201–203]. White et al. [54] investigated the survivorship of F2 fathead minnow larvae whose F0 ancestors were exposed to BaP. The F2 broods with a history of grandparental exposure showed a marked decrease in survival and significantly reduced reproductive capacity and larval survivorship in the highest-exposure group. Additional transgenerational phenotypic traits have been found, such as increased cardiac malformations [231], reproductive behavior [213], immune status, osmoregulation [232], larval behavior [212], abnormal blood vessel development [216], suppressed cortisol levels [208], deficits in prey capture [210], navigational capacity [150], etc. Environmental stressors are not always present at the same concentration or levels. Because of their variability over time, the exposure of organisms to these stressors results in a variety of inheritance patterns and a multitude of adverse fitness outcomes across generations. Since the environmentally established epigenetic alterations can be heritable and stable, how the offspring with a transgenerational phenotype would respond to stressors of emerging concern (chemical or nonchemical) is currently a subject of interest. Further experimentation is required to address these issues and to examine the ability of other aquatic pollutants of concern to induce phenotypic effects in the present and future generations. The transgenerational phenotypes pose ecological risks and might potentially have unfavorable evolutionary consequences.

4.4.2 Transgenerational Fish Phenotype and Population Effects: A Perspective

Most published information on transgenerational inheritance is dominated by data from studies involving laboratory rodent models. Given that molecular mechanisms of gene regulation, including epigenetics, are conserved in eukaryotic organisms, those operating in environmentally induced transgenerational inheritance of phenotype are also believed to be conserved. Organisms living in the aquatic environment are exposed to chemical contaminants to varying degrees. It is challenging to evaluate endocrine disruption as a mechanism of deteriorating ecosystem health due to the complexity of biotic and abiotic interactions that exist in nature. Chemical contaminants are typically present in the environment as complex mixtures of hundreds or thousands of chemicals, of which their effects on organisms may not represent the simple sum of individual chemical effects; therefore, laboratory studies should be based on environmentally relevant mixtures and concentrations to establish and characterize relationships. Chemicals that share common properties and a common mode of action may alter common molecular pathways. For example, estrogenic chemicals may activate estrogen receptors or inhibit enzyme pathways, including DNA methyltransferases, resulting in specific phenotypic endpoints such as infertility in males. A suite of common biomarkers thus can be developed to detect the phenotype caused by those groups of chemicals. This may apply to every group of chemicals, although some variations may exist; for example, different environmental chemicals classified as having estrogenic activity are typically classified as "Selective Estrogen Receptor Modulators (SERMs)" that interact with estrogen receptors, but do not result in identical effects. In a transgenerational study that utilized two manufactured estrogenic chemicals present in aquatic environments, BPA and EE2, which come from the urine of women taking oral contraceptives using the same route and time of exposure (but a lower dose of EE2 than BPA), transgenerational phenotypes were common between the two [18]. Taking into account the fertilization success and embryo survival in each generation of medaka in controlled laboratory conditions, an overall population loss of 40% was

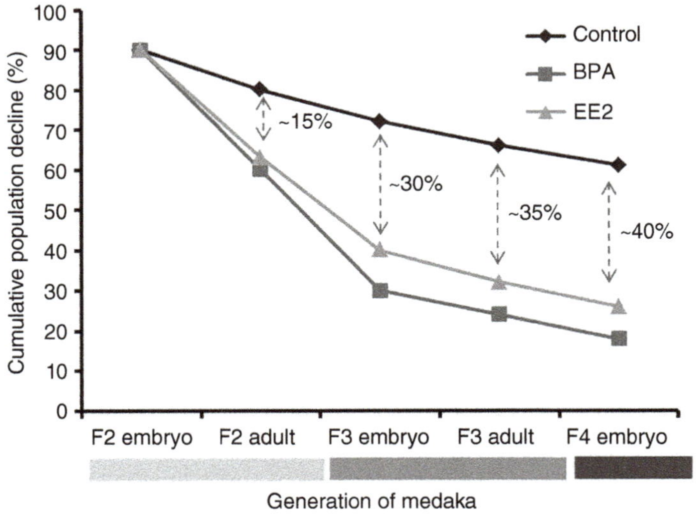

Figure 4.6 Hypothetical cumulative transgenerational effects of BPA or EE2 exposure in medaka populations. Developmental exposure to these two chemicals in controlled laboratory conditions leads to more than 40% reduction in population due to transgenerational effects on the reproduction and survival of embryos. *Source:* Adapted from Bhandari et al. [18].

observed (Figure 4.6). This picture may not represent the trend of population loss in a natural situation, given the complexity of nature; however, the findings provide insights into the ecological consequences estrogenic chemicals could cause in future generations. A prior multiyear experimental study in a Canadian lake that involved chronic administration of a low dose of EE2 led to the feminization of male fathead minnows and the near extinction of this species in the lake [233].

Earlier sections briefly described the evidence for the parental ability to generate phenotypic effects in offspring on a nongenetic basis. The parental response to these stressors may be transient, and yet the acquired effects may exert long-lasting effects on offspring if it is imposed during the critical sensitive window in germline development. The offspring then transmit these effects to their offspring and so on, generating long-term transgenerational effects. In order for the evolution of a phenotype, there are time lags between when molecular information is imprinted in the first offspring phenotype and when natural selection operates in the offspring and subsequent generations; this is the traditional "gradualism" view of evolution occurring at a slow rate. Bonduriansky et al. [69] suggest that "nongenetic inheritance mechanisms mediate the transmission to offspring of phenotypic variants that are immediately or at least potentially adaptive." Although the possibility of such effects is supported by empirical evidence, empirical studies also show that nongenetic inheritance can allow for the transmission of acquired pathological states that are likely to be detrimental in any environment [69]. Given the potential for anthropogenic changes such as pollution to engender transgenerational pathologies, theories must also address the consequences of such phenomena for population persistence and adaptation [69]. Recently, Adrian-Kalchhauser et al. further elucidated nongenetic inheritance as inherited gene regulation (IGR) [28]. Accordingly, regulation may be functional or adaptive, but (as exemplified by negative effects of "nongenetic" inheritance) this is not necessarily the case. IGR accommodates genome-associated mechanisms as well as non-nuclear factors such as maternally provided RNA, proteins, hormones, or nutrients (all of which can be potent gene expression regulators) and encompasses the broad concept of transgenerational plasticity. Contextual specification through prefixes (e.g., gamete-mediated IGR, DNA methylation-mediated IGR, hormone-mediated IGR, or RNA-mediated IGR) permits descriptive precision while maintaining conceptual unity [28]. Accumulating evidence and concepts make the process of inheritance of traits complex and multifactorial.

Compared to terrestrial organisms, environmental stressor effects are profound in aquatic organisms. Fish can be exposed to environmental stressors, including chemical contaminants, at various stages of their life history. Exposure to fish during early developmental stages may alter the epigenetic profiles of cells in both somatic and germline tissues. Somatic tissue effects manifest as immediate phenotypic changes such as notable morphological and physiological defects, whereas germline effects can only be observed in germ cells. The somatic cells in the offspring that develop from the affected germline carry epigenetic signatures, which may turn into various diseases, including reproductive impairment. Examples of somatic cell abnormalities include alteration in the function of endocrine glands, alteration of brain function, and defective morphological features such as fin, gill, eyes, cartilages, development of tumors, etc. The environmentally altered epigenetic imprints of the germline cells are transmitted to offspring. Environmentally exposed, developing germ cells may show immediate effects, such as germ cell death or their transdifferentiation into the germ cell of the opposite sex. These exposed germ cells may harbor latent long-term effects as heritable epigenetic signatures, ultimately resulting in reproductive defects.

4.4.3 Designing Transgenerational Laboratory Experiments

Studies suggest that embryonic exposures cause epigenetic changes in germ cells, but exposure during gametogenesis can induce similar changes causing transgenerational effects [204, 211, 231]. To study the transgenerational impacts of environmental stressors, experiments should be designed to cover the epigenetic programming window of embryonic development and the weeks before mating and fertilization, particularly the period spanning their spermatogenesis and oogenesis (Figure 4.7). Exposed individuals (F0) can be separated and mated with unexposed F0 of the opposite sex to avoid inbreeding and determine parental line effects. Similarly, outcrossing the exposed line with the wild-type line can be continued until the desired generation is achieved to demonstrate differences between multigenerational and transgenerational effects. Mating between exposed individuals increases the possibility of having inbreeding effects. Still, it would best reflect a more natural situation observed in the wild, where both exposed males and females freely breed.

The epigenome varies drastically among different cell populations. So, it is advisable to focus on homogeneous cell populations, such as isolated germ cells, either sperm or eggs, or somatic cells, such as peripheral blood monocytes, hepatocytes of the liver, radial cells of the brain, and intestinal epithelial cells. As a lower vertebrate, fish fall under the Institutional Animal Care and Use Committee (IACUC) procedures and guidelines for the humane treatment of research animals. One fish can serve as a biological replicate, just like rodents. So, space issues can be overcome by keeping all lineage groups in a single or two separate tanks. At least 10–15 biological replicates are desired, as transgenerational effects are variable and larger sample sizes provide greater statistical power.

4.4.4 Transgenerational Effects on Hatchery-Raised Fish

Well-functioning physiological processes such as growth, stress response, ion regulation, and energetics are critical for survival, particularly for migratory species. Environmental factors such as climate change, water temperature, food source, stress, hatchery practices, dams, and pollution may all affect these complex physiological systems [234]. State and private hatcheries in the United States have long been producing fish for stock enhancement programs, where growth acceleration methods may include extended photoperiod, elevated temperature, and high-calorie fish feed. Although the consequence

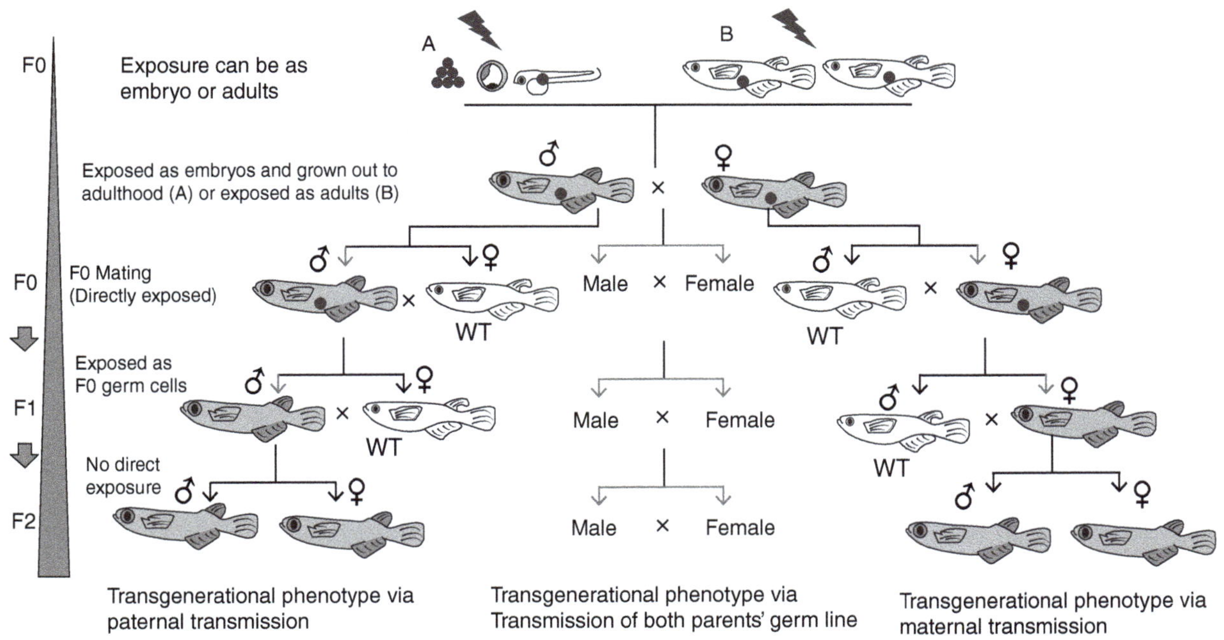

Figure 4.7 Design of transgenerational laboratory experiments in fish. Exposure to F0 generation can be carried out at embryonic stages or in adulthood. Parent-of-origin effects can be studied by mating exposed individuals with the unexposed across generations. To obtain a transgenerational phenotype, the F2 or subsequent generations must be screened. For subsequent generations beyond the F2, the same strategy of breeding should be applied, as shown for the breeding of the F0 and F1 generation fish. The shaded fish shows the lineage of the fish exposed to the stressor at F0 generation (as embryo or adult). The arrows on the top of shaded fish indicate the parent of origin, × indicates crossbreeding, and the line immediately below the fish indicates the founder fish (parent) for the next generation. Dark spots on the abdomen of the fish indicate direct exposure effects on gonad. (*See colour plate section for colour representation of this figure.*)

of growth enhancement techniques on adult physiological processes is unknown, reproductive competence, disease resistance, and accelerated gill performance of the offspring of these growth-enhanced fish may be susceptible to epigenetically based alterations. Chamberlin et al. [235] reported that hatchery-reared salmon from Puget Sound did not undergo a complete parr-smolt transformation, and homing migration rates and seawater tolerance of these fish were reduced. Changes in rearing conditions, such as pH, temperature, and carbon dioxide concentration, can significantly alter the epigenome of developing cells, and the tissues that differentiate from these stem cells can develop with altered physiological capabilities [236]. The environmental manipulations adopted by fish hatcheries during the early development of fish are likely to cause shifts in the epigenome of the exposed generation. As a result, the offspring of the exposed fish may be expected to develop transgenerational phenotypes. Phenotypes that would potentially affect population dynamics include reduced navigation and osmoregulatory ability and increased susceptibility to pathogens. If such epigenetically transmitted outcomes were consistently observed in the enhanced population of salmonids, further experimentation would be required to ensure growth acceleration enhancement programs were safe for the development of long-term sustainable populations of fish. Exposure- and species-specific biomarkers would need to be developed to examine populations of fish to evaluate the natural populations of the migratory fish. In a study on domestication syndrome, Anastasiadi and Piferrer [151] reported an association of DNA methylation with the process of domestication of European sea bass (*Dicentrarchus labrax* L.). Epimutations observed in adulthood were established during gastrulation, predominantly in genes involved in developmental processes and embryonic structures such as jaws. Such epimutations significantly overlapped with cytosine to thymine polymorphisms after 25 years of selective breeding, suggesting a shift occurring from epimutation to genetic polymorphism over time [151]. Collectively, research suggests that environmental stress, embryonic growth manipulation, and domestication may have long-term ecological consequences in farmed fish.

4.4.5 Potential for the Mitigation of Epigenetically Inherited Harmful Effects on Fish

A stable form of epigenetic inheritance can be found in the transgenerational mode of transmission. Unlike genetic mutations, which can potentially contribute to the recurrence of risks across generations, inherited transgenerational epigenetic modifications can be lost at some point [237]. Given that epigenetic alterations are reversible chemical modifications, DNA methylation and histone modifications can be erased or reestablished nonspecifically by applying epigenetic modifiers. Studies have demonstrated altered DNA methylation profiles in germ cells and somatic cells following treatment with epigenetic erasers such as vitamin C [238, 239] or writers such as folate compounds [240]. Chromatin state and nucleosome positions primarily determine the success or failure of epigenome manipulation. Epigenome remains dynamic when cells are differentiating. Nucleosomes remain open for the genes that are going to be transcribed, and some nucleosomes that were open are required to be closed to block transcription in the newly formed tissue. Taking advantage of such a dynamic state of the tissue, the application of epigenetic modifiers can reestablish epigenetic profiles that were lost in the particular somatic or germline cells if they are precisely applied at the time of tissue differentiation. For example, in medaka fish, the application of vitamin C at the time of liver differentiation (two days' postfertilization through day 12, when the liver is completely differentiated) can bring the transgenerational nonalcoholic fatty liver disease back to the normal condition [230]. In environmental exposure studies, vitamin C or melatonin co-exposure has been found to alleviate chemical-induced toxicity in the liver [241] and reproductive cells [242], suggesting that proper treatment of embryos or adult fish with epigenetic erasers can alleviate environmentally induced phenotypic alterations at the intergenerational and transgenerational levels. Such insights may provide opportunities for hatchery management and sustainable aquaculture.

4.5 Conclusions and Perspectives

Environmental chemicals can act in several ways, such as acting as a direct ligand for a receptor, bringing about a rapid physiological response, as a potent modulator of gene networks, and as epigenetic modifiers by interacting with cofactors that are required for epigenetic modifications. The first two stated here may have short-term consequences, but epigenetic modifications that occur during the critical window of development may be long-lasting. When offspring develop with an altered epigenome, they carry those epigenetic marks in the affected site of the genome, resulting in abnormal transcription of genes. Depending on what gene is affected, phenotypic effects will appear and, via the germline, continue to subsequent generations as a mild or strong phenotypic trait. Exposure effects that are not producing morphologically distinct

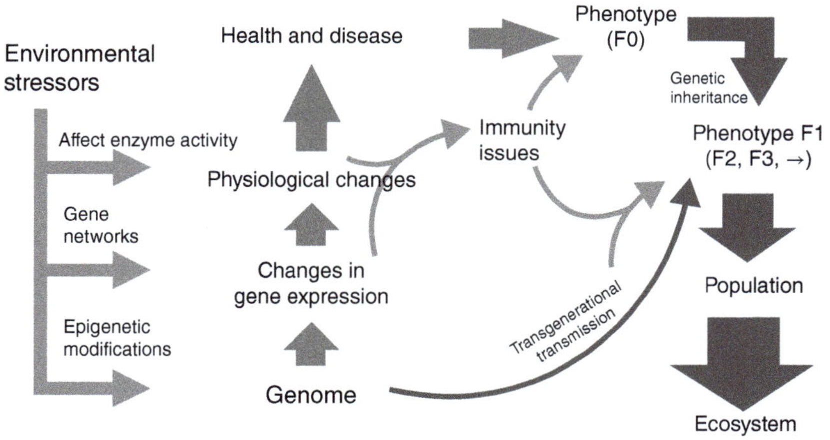

Figure 4.8 Environmental stressors (biotic and abiotic) can induce adaptive or maladaptive phenotypic outcomes due to direct exposure. The stressors can alter enzymatic activity, causing abnormal physiological changes, altering gene expression in a network by altering intracellular networks or induce epigenetic modifications, causing genome instability and abnormal gene expression patterns. All of these effects are linked to the development of robust or adverse health outcomes in exposed organisms. Additionally, environmental stressors affect the immune system of an organism via the modulation of epigenetic, transcriptional, and physiological processes. Some stressors can cause genetic changes in the germline, while others can be mediated epigenetically. The effects that pass through within, inter-, and transgenerational inheritance may alter the health and fitness of organisms at the population or ecosystem level.

phenotypic traits can still show a behavior phenotype [181]. For example, alterations in the epigenome of neurons associated with reproductive behavior may cause a shift in the mating behavior of exposed individuals. In contrast, the same modification in the parental germ cell may cause mating behavior problems in offspring [181]. In both cases, it is harmful to the success of reproduction, and the reproduction failure will cause a population crash (Figure 4.8). In addition to reproductive defects, ancestral exposure may also lead to other health issues, such as energy deficiency, compromised immune system, and increased susceptibility to microbial pathogens and parasites, ultimately impacting the survival of the species within a population.

Epigenetic alterations on germ cells may have far-reaching, unexpected outcomes and should be prevented. Currently, several phenotype-associated biological indicators (adverse outcome pathways, AOP) have been developed to assess endocrine and reproductive damage; however, the biomarkers (genetic and epigenetic) to identify the history of environmental exposure (past exposure) and to predict future phenotypes and the degree of an organism's fitness in a population are not as yet available. Biomarkers would allow restoration agencies to detect chemical-specific exposures and may effectively predict later life consequences.

Environmental conditions are known to affect phenotypic development in many organisms through genotype-by-environment (G × E) interaction, which may be adaptive or maladaptive [243, 244]. Environmental stressors can produce both short- and long-term adverse physiological and behavioral outcomes in all vertebrate organisms, including fish. The period of embryonic development acts as a susceptible window for epigenetic effects, although gametogenesis can also provide a susceptible window. In addition to exposure-related somatic effects in the same generation, subsequent generations may inherit epigenetic alterations via germline transmission. The individuals that develop from an epigenetically altered germline may maintain the altered epigenome in their somatic and germline cells, consequently affecting transcriptional networks in all cells. There is need for the identification of biomarkers capable of distinguishing ancestral exposure and predict future phenotypic traits reliably. Identification of such biomarkers will further strengthen the understanding of the transgenerational inheritance of phenotypes.

References

1 Berger, S.L., Kouzarides, T., Shiekhattar, R. et al. (2009). An operational definition of epigenetics. *Genes Dev.* **23** (7): 781–783.

2 Moore, L.D., Le, T., and Fan, G. (2013). DNA methylation and its basic function. *Neuropsychopharmacology* **38** (1): 23–38.

3 Bannister, A.J. and Kouzarides, T. (2011). Regulation of chromatin by histone modifications. *Cell Res.* **21** (3): 381–395.

4 Russo, V.E., Martienssen, R.A., and Riggs, A.D. (1996). *Epigenetic Mechanisms of Gene Regulation*. Cold Spring Harbor Laboratory Press.
5 Skinner, M.K. (2008). What is an epigenetic transgenerational phenotype? F3 or F2. *Reprod. Toxicol.* **25** (1): 2–6.
6 Bhandari, R.K. (2016). Medaka as a model for studying environmentally induced epigenetic transgenerational inheritance of phenotypes. *Environ. Epigenet.* **2** (1): dvv010.
7 Skinner, M.K. (2015). Environmental epigenetics and a unified theory of the molecular aspects of evolution: a neo-Lamarckian concept that facilitates neo-Darwinian evolution. *Genome Biol. Evol.* **7** (5): 1296–1302.
8 Jirtle, R.L. and Skinner, M.K. (2007). Environmental epigenomics and disease susceptibility. *Nat. Rev. Genet.* **8** (4): 253–262.
9 D'Urso, A. and Brickner, J.H. (2014). Mechanisms of epigenetic memory. *Trends Genet. TIG* **30** (6): 230–236.
10 Singh, P.K. and Campbell, M.J. (2013). The interactions of microRNA and epigenetic modifications in prostate cancer. *Cancers* **5** (3): 998–1019.
11 Chuang, J.C. and Jones, P.A. (2007). Epigenetics and microRNAs. *Pediatr. Res.* **61** (7): 24–29.
12 Bartel, D.P. (2009). MicroRNAs: target recognition and regulatory functions. *Cell* **136** (2): 215–233.
13 Yao, Q., Chen, Y., and Zhou, X. (2019). The roles of microRNAs in epigenetic regulation. *Curr. Opin. Chem. Biol.* **51**: 11–17.
14 Filipowicz, W., Bhattacharyya, S.N., and Sonenberg, N. (2008). Mechanisms of post-transcriptional regulation by microRNAs: are the answers in sight? *Nat. Rev. Genet.* **9** (2): 102–114.
15 Aristizabal, M.J., Anreiter, I., Halldorsdottir, T. et al. (2020). Biological embedding of experience: a primer on epigenetics. *Proc. Natl. Acad. Sci.* **117** (38): 23261–23269.
16 Boyce, W.T., Sokolowski, M.B., and Robinson, G.E. (2020). Genes and environments, development and time. *Proc. Natl. Acad. Sci.* **117** (38): 23235–23241.
17 Shama, L.N., Strobel, A., Mark, F.C. et al. (2014). Transgenerational plasticity in marine sticklebacks: maternal effects mediate impacts of a warming ocean. *Funct. Ecol.* **28** (6): 1482–1493.
18 Bhandari, R.K., vom Saal, F.S., and Tillitt, D.E. (2015). Transgenerational effects from early developmental exposures to bisphenol a or 17alpha-ethinylestradiol in medaka, Oryzias latipes. *Sci. Rep.* **5**: 9303.
19 Best, C., Ikert, H., Kostyniuk, D.J. et al. (2018). Epigenetics in teleost fish: from molecular mechanisms to physiological phenotypes. *Comp. Biochem. Physiol. B: Biochem. Mol. Biol.* **224**: 210–244.
20 Ryu, T., Veilleux, H.D., Donelson, J.M. et al. (2018). The epigenetic landscape of transgenerational acclimation to ocean warming. *Nat. Clim. Chang.* **8** (6): 504–509.
21 Salinas, S., Brown, S.C., Mangel, M. et al. (2013). Non-genetic inheritance and changing environments. *Non-Genet. Inherit.* **1**: 1.
22 Fellous, A., Wegner, K.M., John, U. et al. (2022). Windows of opportunity: ocean warming shapes temperature-sensitive epigenetic reprogramming and gene expression across gametogenesis and embryogenesis in marine stickleback. *Glob. Chang. Biol.* **28** (1): 54–71.
23 Petitjean, Q., Jean, S., Gandar, A. et al. (2019). Stress responses in fish: from molecular to evolutionary processes. *Sci. Total Environ.* **684**: 371–380.
24 Nilsson, E.E. and Skinner, M.K. (2014). Environmentally induced epigenetic transgenerational inheritance of disease susceptibility. *Transl. Res.* **93**: 145.
25 Bhandari, R.K., Deem, S.L., Holliday, D.K. et al. (2015). Effects of the environmental estrogenic contaminants bisphenol A and 17alpha-ethinyl estradiol on sexual development and adult behaviors in aquatic wildlife species. *Gen. Comp. Endocrinol.* **214**: 195–219.
26 Jeremias, G., Barbosa, J., Marques, S.M. et al. (2018). Transgenerational inheritance of DNA hypomethylation in *Daphnia magna* in response to salinity stress. *Environ. Sci. Technol.* **52** (17): 10114–10123.
27 Spinks, R.K., Donelson, J.M., Bonzi, L.C. et al. (2022). Parents exposed to warming produce offspring lower in weight and condition. *Ecol. Evol.* **12** (7): e9044.
28 Adrian-Kalchhauser, I., Sultan, S.E., Shama, L.N.S. et al. (2020). Understanding 'non-genetic' inheritance: insights from molecular-evolutionary crosstalk. *Trends Ecol. Evol.* **35** (12): 1078–1089.
29 Bautista, N.M. and Burggren, W.W. (2019). Parental stressor exposure simultaneously conveys both adaptive and maladaptive larval phenotypes through epigenetic inheritance in the zebrafish (*Danio rerio*). *J. Exp. Biol.* **222** (Pt 17).
30 Robaire, B., Delbes, G., Head, J.A. et al. (2022). A cross-species comparative approach to assessing multi- and transgenerational effects of endocrine disrupting chemicals. *Environ. Res.* **204** (Pt B): 112063.
31 Jablonka, E. and Lamb, M.J. (1995). *Epigenetic Inheritance and Evolution: The Lamarckian Dimension*. UK: Oxford University Press.

32 Lacal, I. and Ventura, R. (2018). Epigenetic inheritance: concepts, mechanisms and perspectives. *Front. Mol. Neurosci.* **11**: 292.
33 Skinner, M.K., Guerrero-Bosagna, C.M., and Haque, M.M. (2015). Environmentally induced epigenetic transgenerational inheritance of sperm epimutations promote genetic mutations. *Epigenetics* **10**: 762–771.
34 Auge, G.A., Leverett, L.D., Edwards, B.R. et al. (2017). Adjusting phenotypes via within- and a cross-generational plasticity. *N. Phytol.* **216** (2): 343–349.
35 Zheng, X., Li, Z., Wang, G. et al. (2021). Sperm epigenetic alterations contribute to inter- and transgenerational effects of paternal exposure to long-term psychological stress via evading offspring embryonic reprogramming. *Cell Discov.* **7** (1): 101.
36 Ben Maamar, M., Beck, D., Nilsson, E. et al. (2020). Developmental origins of transgenerational sperm histone retention following ancestral exposures. *Dev. Biol.* **465** (1): 31–45.
37 Kolpin, D.W., Furlong, E.T., Meyer, M.T. et al. (2002). Pharmaceuticals, hormones, and other organic wastewater contaminants in US streams, 1999–2000: a national reconnaissance. *Environ. Sci. Technol.* **36** (6): 1202–1211.
38 Clayton, R., Erin, M., Todd, M. et al. (2010). The impact of bisphenol A and triclosan on immune parameters in the U.S. population, NHANES 2003–2006. *Environ. Health Perspect.* **119** (3): 390–396.
39 Melzer, D., Rice, N.E., Lewis, C. et al. (2010). Association of urinary bisphenol A concentration with heart disease: evidence from NHANES 2003/06. *PLoS One* **5** (1): e8673.
40 Castorina, R., Bradman, A., Fenster, L. et al. (2010). Comparison of current-use pesticide and other toxicant urinary metabolite levels among pregnant women in the CHAMACOS cohort and NHANES. *Environ. Health Perspect.* **118**: 856–863.
41 Vandenberg, L.N., T. Colborn, T.B. Hayes, sssssset al., Hormones and endocrine-disrupting chemicals: low-dose effects and nonmonotonic dose responses. *Endocr. Rev.*, 2012. **33**:378–455
42 Wang, S.Y., Lau, K., Lai, K.P. et al. (2016). Hypoxia causes transgenerational impairments in reproduction of fish. *Nature Commun.* **7** (1): 1–9.
43 Crispo, E., Suman, P.D., and Chapman, L.J. (2020). Quantifying genome-wide cytosine methylation in response to hypoxia in the gills, muscle, and brain of an African cichlid fish. *Environ. Biol. Fish* **103** (3): 223–232.
44 Nilsson, E., Sadler-Riggleman, I., Beck, D. et al. (2021). Differential DNA methylation in somatic and sperm cells of hatchery vs wild (natural-origin) steelhead trout populations. *Environ. Epigenetics* **7** (1): dvab002.
45 Gavery, M.R., Nichols, K.M., Berejikian, B.A. et al. (2019). Temporal dynamics of DNA methylation patterns in response to rearing juvenile steelhead (*Oncorhynchus mykiss*) in a hatchery versus simulated stream environment. *Genes* **10** (5): 356.
46 Rodriguez Barreto, D., Garcia de Leaniz, C., Verspoor, E. et al. (2019). DNA methylation changes in the sperm of captive-reared fish: a route to epigenetic introgression in wild populations. *Mol. Biol. Evol.* **36** (10): 2205–2211.
47 Navarro-Martín, L., Viñas, J., Ribas, L. et al. (2011). DNA methylation of the gonadal aromatase (cyp19a) promoter is involved in temperature-dependent sex ratio shifts in the European sea bass. *PLoS Genet.* **7** (12): e1002447.
48 Anastasiadi, D., Venney, C.J., Bernatchez, L. et al. (2021). Footprints of global change in marine life: inferring past environment based on DNA methylation and gene expression marks. *Mol. Ecol.* **30** (3): 747–760.
49 Anway, M.D., Cupp, A.S., Uzumcu, M. et al. (2005). Epigenetic transgenerational actions of endocrine disruptors and male fertility. *Science* **308** (5727): 1466–1469.
50 Baker, T.R., King-Heiden, T.C., Peterson, R.E. et al. (2014). Dioxin induction of transgenerational inheritance of disease in zebrafish. *Mol. Cell. Endocrinol.* **398** (1–2): 36–41.
51 Corrales, J., Thornton, C., White, M. et al. (2014). Multigenerational effects of benzo[a]pyrene exposure on survival and developmental deformities in zebrafish larvae. *Aquat. Toxicol.* **148**: 16–26.
52 Doyle, T.J., Bowman, J.L., Windell, V.L. et al. (2013). Transgenerational effects of di-(2-ethylhexyl) phthalate on testicular germ cell associations and spermatogonial stem cells in mice. *Biol. Reprod.* **88** (5): 1–15.
53 Quinnies, K.M., Doyle, T.J., Hee Kim, K. et al. (2015). Transgenerational effects of di-(2-ethylhexyl) phthalate, DEHP, on stress hormones and behavior. *Endocrinology* **156**: EN20151326.
54 White, P.A., Robitaille, S., and Rasmussen, J.B. (1999). Heritable reproductive effects of benzo[a]pyrene on the fathead minnow (*Pimephales promelas*). *Environ. Toxicol. Chem.* **18** (8): 1843–1847.
55 Wolstenholme, J.T., Edwards, M., Shetty, S.R. et al. (2012). Gestational exposure to bisphenol A produces transgenerational changes in behaviors and gene expression. *Endocrinology* **153** (8): 3828–3838.
56 Cleary, J.A., Tillitt, D.E., Vom Saal, F.S. et al. (2019). Atrazine induced transgenerational reproductive effects in medaka (*Oryzias latipes*). *Environ. Pollut.* **251**: 639–650.
57 Varriale, A. (2014). DNA methylation, epigenetics, and evolution in vertebrates: facts and challenges. *Int. J. Evol. Biol.* **2014**: 7.

58 Feeney, A., Nilsson, E., and Skinner, M.K. (2014). Epigenetics and transgenerational inheritance in domesticated farm animals. *J. Anim. Sci. Biotechnol.* **5** (1): 48.

59 Braunschweig, M., Jagannathan, V., Gutzwiller, A. et al. (2012). Investigations on transgenerational epigenetic response down the male line in F2 pigs. *PLoS One* **7** (2): e30583.

60 Aldrich, J.C. and Maggert, K.A. (2015). Transgenerational inheritance of diet-induced genome rearrangements in *Drosophila*. *PLoS Genet.* **11** (4): e1005148.

61 Baker, T.R., Peterson, R.E., and Heideman, W. (2014). Using zebrafish as a model system for studying the transgenerational effects of dioxin. *Toxicol. Sci.* **138** (2): 403–411.

62 Buescher, J.L., Musselman, L.P., Wilson, C.A. et al. (2013). Evidence for transgenerational metabolic programming in *Drosophila*. *Dis. Model. Mech.* **6** (5): 1123–1132.

63 Crews, D., Gillette, R., Scarpino, S.V. et al. (2012). Epigenetic transgenerational inheritance of altered stress responses. *Proc. Natl. Acad. Sci. U. S. A.* **109** (23): 9143–9148.

64 Greer, E.L., Maures, T.J., Ucar, D. et al. (2011). Transgenerational epigenetic inheritance of longevity in *Caenorhabditis elegans*. *Nature* **479** (7373): 365–371.

65 Guerrero-Bosagna, C. and Jensen, P. (2015). Globalization, climate change, and transgenerational epigenetic inheritance: will our descendants be at risk? *Clin. Epigenet.* **7** (1): 8.

66 Skinner, M.K., Gurerrero-Bosagna, C., Haque, M.M. et al. (2014). Epigenetics and the evolution of Darwin's finches. *Genome Biol. Evol.* **6** (8): 1972–1989.

67 Bateson, P., Barker, D., Clutton-Brock, T. et al. (2004). Developmental plasticity and human health. *Nature* **430** (6998): 419–421.

68 Bernanke, J. and Kohler, H.R. (2009). The impact of environmental chemicals on wildlife vertebrates. *Rev. Environ. Contam. Toxicol.* **198**: 1–47.

69 Bonduriansky, R., Crean, A.J., and Day, T. (2012). The implications of nongenetic inheritance for evolution in changing environments. *Evol. Appl.* **5** (2): 192–201.

70 Colborn, T., vom Saal, F.S., and Soto, A.M. (1993). Developmental effects of endocrine-disrupting chemicals in wildlife and humans. *Environ. Health Perspect.* **101** (5): 378–384.

71 Flint, S., Markle, T., Thompson, S. et al. (2012). Bisphenol A exposure, effects, and policy: a wildlife perspective. *J. Environ. Manag.* **104** (2012): 19–34.

72 Vandegehuchte, M.B. and Janssen, C.R. (2011). Epigenetics and its implications for ecotoxicology. *Ecotoxicology* **20** (3): 607–624.

73 Jablonka, E. and Raz, G. (2009). Transgenerational epigenetic inheritance: prevalence, mechanisms, and implications for the study of heredity and evolution. *Q. Rev. Biol.* **84** (2): 131–176.

74 Gillman, M.W. (2005). Developmental origins of health and disease. *N. Engl. J. Med.* **353** (17): 1848.

75 Godfrey, K.M., K.A. Lillycrop, G.C. Burdge, set al., Epigenetic mechanisms and the mismatch concept of the developmental origins of health and disease. *Pediatr. Res.*, 2007. **61**: p. 5R–10R.

76 Feinberg, A.P. and Irizarry, R.A. (2010). Stochastic epigenetic variation as a driving force of development, evolutionary adaptation, and disease. *Proc. Natl. Acad. Sci.* **107** (suppl 1): 1757–1764.

77 Lachmann, M. and Jablonka, E. (1996). The inheritance of phenotypes: an adaptation to fluctuating environments. *J. Theor. Biol.* **181** (1): 1–9.

78 Hengen, K.B., Nelson, N.R., Stang, K.M. et al. (2015). Daily isoflurane exposure increases barbiturate insensitivity in medullary respiratory and cortical neurons via expression of epsilon-subunit containing GABA ARs. *PLoS One* **10** (3): e0119351.

79 Nkya, T.E., Akhouayri, I., Kisinza, W. et al. (2013). Impact of environment on mosquito response to pyrethroid insecticides: facts, evidences and prospects. *Insect Biochem. Mol. Biol.* **43** (4): 407–416.

80 Donelson, J.M. and Munday, P.L. (2015). Transgenerational plasticity mitigates the impact of global warming to offspring sex ratios. *Glob. Chang. Biol.* **21** (8): 2954–2962.

81 Donelson, J.M., Wong, M., Booth, D.J. et al. (2016). Transgenerational plasticity of reproduction depends on rate of warming across generations. *Evol. Appl.* **9** (9): 1072–1081.

82 Zuloaga, D.G., Puts, D.A., Jordan, C.L. et al. (2008). The role of androgen receptors in the masculinization of brain and behavior: what we've learned from the testicular feminization mutation. *Horm. Behav.* **53** (5): 613–626.

83 Mongan, N.P., Tadokoro-Cuccaro, R., Bunch, T. et al. (2015). Androgen insensitivity syndrome. *Best Pract. Res. Clin. Endocrinol. Metab.* **29** (4): 569–580.

84 Klastrup, L.K., Bak, S.T., and Nielsen, A.L. (2019). The influence of paternal diet on sncRNA-mediated epigenetic inheritance. *Mol. Gen. Genomics.* **294** (1): 1–11.

85 Fitz-James, M.H. and Cavalli, G. (2022). Molecular mechanisms of transgenerational epigenetic inheritance. *Nat. Rev. Genet.* **23**: 325–341.

86 Gapp, K., Jawaid, A., Sarkies, P. et al. (2014). Implication of sperm RNAs in transgenerational inheritance of the effects of early trauma in mice. *Nat. Neurosci.* **17** (5): 667–669.

87 Hoffmann, A.A. and Hercus, M.J. (2000). Environmental stress as an evolutionary force. *Bioscience* **50** (3): 217–226.

88 Lam, S.H., Hlaing, M.M., Zhang, X. et al. (2011). Toxicogenomic and phenotypic analyses of bisphenol-A early-life exposure toxicity in zebrafish. *PLoS One* **6** (12): e28273.

89 Wang, X., Hill, D., Tillitt, D.E. et al. (2019). Bisphenol A and 17alpha-ethinylestradiol-induced transgenerational differences in expression of osmoregulatory genes in the gill of medaka (*Oryzias latipes*). *Aquat. Toxicol.* **211**: 227–234.

90 Thayil, A.J., Wang, X., Bhandari, P. et al. (2020). Bisphenol A and 17alpha-ethinylestradiol-induced transgenerational gene expression differences in the brain–pituitary–testis axis of medaka, Oryzias latipes. *Biol. Reprod.* **103** (6): 1324–1335.

91 Fraga, M.F., Ballestar, E., Paz, M.F. et al. (2005). Epigenetic differences arise during the lifetime of monozygotic twins. *Proc. Natl. Acad. Sci. U. S. A.* **102** (30): 10604–10609.

92 Wong, C.C., Caspi, A., Williams, B. et al. (2010). A longitudinal study of epigenetic variation in twins. *Epigenetics* **5** (6): 516–526.

93 Simmons, D. (2008). Epigenetic influence and disease. *Nature Educ.* **1** (1): 6.

94 Penny, D. (2015). Epigenetics, Darwin, and Lamarck. *Genome Biol. Evol.* **7** (6): 1758–1760.

95 Fleisch, A.F., Wright, R.O., and Baccarelli, A.A. (2012). Environmental epigenetics: a role in endocrine disease? *J. Mol. Endocrinol.* **49** (2): R61–R67.

96 Laird, P.W. and Jaenisch, R. (1996). The role of DNA methylation in cancer genetic and epigenetics. *Annu. Rev. Genet.* **30**: 441–464.

97 Jaenisch, R. and Bird, A. (2003). Epigenetic regulation of gene expression: how the genome integrates intrinsic and environmental signals. *Nat. Genet.* **33**: 245–254.

98 Skinner, M.K., Manikkam, M., and Guerrero-Bosagna, C. (2011). Epigenetic transgenerational actions of endocrine disruptors. *Reprod. Toxicol.* **31** (3): 337–343.

99 Manikkam, M., Guerrero-Bosagna, C., Tracey, R. et al. (2012). Transgenerational actions of environmental compounds on reproductive disease and identification of epigenetic biomarkers of ancestral exposures. *PLoS One* **7** (2): e31901.

100 Manikkam, M., Tracey, R., Guerrero-Bosagna, C. et al. (2012). Dioxin (TCDD) induces epigenetic transgenerational inheritance of adult onset disease and sperm epimutations. *PLoS One* **7** (9): e46249.

101 Manikkam, M., Tracey, R., Guerrero-Bosagna, C. et al. (2012). Pesticide and insect repellent mixture (permethrin and DEET) induces epigenetic transgenerational inheritance of disease and sperm epimutations. *Reprod. Toxicol.* **34** (4): 708–719.

102 Manikkam, M., Tracey, R., Guerrero-Bosagna, C. et al. (2013). Plastics derived endocrine disruptors (BPA, DEHP and DBP) induce epigenetic transgenerational inheritance of obesity, reproductive disease and sperm epimutations. *PLoS One* **8** (1): e55387.

103 Liep, J., Rabien, A., and Jung, K. (2012). Feedback networks between microRNAs and epigenetic modifications in urological tumors. *Epigenetics* **7** (4): 315–325.

104 Saito, Y. and Jones, P.M. (2006). Epigenetic activation of tumor suppressor microRNAs in human cancer cells. *Cell Cycle* **5** (19): 2220–2222.

105 Chen, J.-F., Mandel, E.M., Thomson, J.M. et al. (2006). The role of microRNA-1 and microRNA-133 in skeletal muscle proliferation and differentiation. *Nat. Genet.* **38** (2): 228–233.

106 Duursma, A.M., Kedde, M., Schrier, M. et al. (2008). miR-148 targets human DNMT3b protein coding region. *RNA* **14** (5): 872–877.

107 Schraml, E., Hackl, M., and Grillari, J. (2017). MicroRNAs and toxicology: a love marriage. *Toxicol. Rep.* **4**: 634–636.

108 Mitchell, P.S., Parkin, R.K., Kroh, E.M. et al. (2008). Circulating microRNAs as stable blood-based markers for cancer detection. *Proc. Natl. Acad. Sci. U. S. A.* **105** (30): 10513–10518.

109 Cardona, E., Guyomar, C., Desvignes, T. et al. (2021). Circulating miRNA repertoire as a biomarker of metabolic and reproductive states in rainbow trout. *BMC Biol.* **19** (1): 235.

110 Song, M.K. and Ryu, J.C. (2015). Blood miRNAs as sensitive and specific biological indicators of environmental and occupational exposure to volatile organic compound (VOC). *Int. J. Hyg. Environ. Health* **218**: 590–602.

111 Inns, J. and James, V. (2015). Circulating microRNAs for the prediction of metastasis in breast cancer patients diagnosed with early stage disease. *Breast* **24** (4): 364–369.

112 Sharma, A. (2014). Novel transcriptome data analysis implicates circulating microRNAs in epigenetic inheritance in mammals. *Gene* **538** (2): 366–372.

113 Casier, K., Boivin, A., Carré, C. et al. (2019). Environmentally-induced transgenerational epigenetic inheritance: implication of PIWI interacting RNAs. *Cells* **8** (9): 1108.

114 Kuramochi-Miyagawa, S., Watanabe, T., Gotoh, K. et al. (2008). DNA methylation of retrotransposon genes is regulated by Piwi family members MILI and MIWI2 in murine fetal testes. *Genes Dev.* **22** (7): 908–917.

115 Legoff, L., D'Cruz, S.C., Tevosian, S. et al. (2019). Chapter 10 – long noncoding RNA and epigenetic inheritance. In: *Transgenerational Epigenetics*, 2e (ed. T.O. Tollefsbol), 213–248. Academic Press.

116 Mercer, T.R. and Mattick, J.S. (2013). Structure and function of long noncoding RNAs in epigenetic regulation. *Nat. Struct. Mol. Biol.* **20** (3): 300–307.

117 Wei, J.-W., Huang, K., Yang, C. et al. (2017). Non-coding RNAs as regulators in epigenetics (review). *Oncol. Rep.* **37** (1): 3–9.

118 Szabo, Q., Jost, D., Chang, J.M. et al. (2018). TADs are 3D structural units of higher-order chromosome organization in *Drosophila*. *Sci. Adv.* **4** (2): eaar8082.

119 Lazar, N.H., Nevonen, K.A., O'Connell, B. et al. (2018). Epigenetic maintenance of topological domains in the highly rearranged gibbon genome. *Genome Res.* **28** (7): 983–997.

120 Kaaij, L.J.T., van der Weide, R.H., Ketting, R.F. et al. (2018). Systemic loss and gain of chromatin architecture throughout zebrafish development. *Cell Rep.* **24** (1): 1–10.e4.

121 Siklenka, K., Erkek, S., Godmann, M. et al. (2015). Disruption of histone methylation in developing sperm impairs offspring health transgenerationally. *Science* **350** (6261): aab2006.

122 Teperek, M., Simeone, A., Gaggioli, V. et al. (2016). Sperm is epigenetically programmed to regulate gene transcription in embryos. *Genome Res.* **26** (8): 1034–1046.

123 Hammond, C.M., Strømme, C.B., Huang, H. et al. (2017). Histone chaperone networks shaping chromatin function. *Nat. Rev. Mol. Cell Biol.* **18** (3): 141–158.

124 Guil, S. and Esteller, M. (2009). DNA methylomes, histone codes and miRNAs: tying it all together. *Int. J. Biochem. Cell Biol.* **41** (1): 87–95.

125 Stuwe, E., Toth, K.F., and Aravin, A.A. (2014). Small but sturdy: small RNAs in cellular memory and epigenetics. *Genes Dev.* **28** (5): 423–431.

126 Barker, D.J. (2002). Fetal programming of coronary heart disease. *Trends Endocrinol. Metab.* **13** (9): 364–368.

127 Edwards, M. (2017). The Barker hypothesis. In: *Handbook of Famine, Starvation, and Nutrient Deprivation: From Biology to Policy* (ed. V. Preedy and V. Patel), 191–211. Cham: Springer.

128 Vindas, M.A., Fokos, S., Pavlidis, M. et al. (2018). Early life stress induces long-term changes in limbic areas of a teleost fish: the role of catecholamine systems in stress coping. *Sci. Rep.* **8** (1): 5638.

129 Fontana, B.D., Gibbon, A.J., Cleal, M. et al. (2021). Moderate early life stress improves adult zebrafish (*Danio rerio*) working memory but does not affect social and anxiety-like responses. *Dev. Psychobiol.* **63** (1): 54–64.

130 Beckman, B.R., Dickhoff, W.W., Zaugg, W.S. et al. (1999). Growth, smoltification, and smolt-to-adult return of spring chinook salmon from hatcheries on the Deschutes River, Oregon. *Trans. Am. Fish. Soc.* **128** (6): 1125–1150.

131 Wedemeyer, G.A., Saunders, R.L., and Clarke, W.C. (1980). Environmental factors affecting smoltification and early marine survival of anadromous salmonids. *Mar. Fisheries Rev. 42* (6): 1–14.

132 Oke, K.B., Cunningham, C.J., Westley, P.A.H. et al. (2020). Recent declines in salmon body size impact ecosystems and fisheries. *Nat. Commun.* **11** (1): 4155.

133 Guo, H., Zhu, P., Yan, L. et al. (2014). The DNA methylation landscape of human early embryos. *Nature* **511** (7511): 606–610.

134 Guo, F., Yan, L., Guo, H. et al. (2015). The transcriptome and DNA methylome landscapes of human primordial germ cells. *Cell* **161** (6): 1437–1452.

135 Wang, L., Zhang, J., Duan, J. et al. (2014). Programming and inheritance of parental DNA methylomes in mammals. *Cell* **157** (4): 979–991.

136 Guo, F., Li, X., Liang, D. et al. (2014). Active and passive demethylation of male and female pronuclear DNA in the mammalian zygote. *Cell Stem Cell* **15** (4): 447–459.

137 Seisenberger, S., Andrews, S., Krueger, F. et al. (2012). The dynamics of genome-wide DNA methylation reprogramming in mouse primordial germ cells. *Mol. Cell* **48** (6): 849–862.

138 Wang, X. and Bhandari, R.K. (2020). The dynamics of DNA methylation during epigenetic reprogramming of primordial germ cells in medaka (*Oryzias latipes*). *Epigenetics* **15** (5): 483–498.

139 Wang, X. and Bhandari, R.K. (2020). DNA methylation reprogramming in medaka fish, a promising animal model for environmental epigenetics research. *Environ. Epigenet.* **6** (1): dvaa008. https://doi.org/10.1093/eep/dvaa008.

140 Wang, X. and Bhandari, R.K. (2019). DNA methylation dynamics during epigenetic reprogramming of medaka embryo. *Epigenetics* **14** (6): 611–622.

141 Ortega-Recalde, O., Day, R.C., Gemmell, N.J. et al. (2019). Zebrafish preserve global germline DNA methylation while sex-linked rDNA is amplified and demethylated during feminisation. *Nat. Commun.* **10** (1): 3053.

142 Jiang, L., Zhang, J., Wang, J.J. et al. (2013). Sperm, but not oocyte, DNA methylome is inherited by zebrafish early embryos. *Cell* **153** (4): 773–784.

143 Skvortsova, K., Tarbashevich, K., Stehling, M. et al. (2019). Retention of paternal DNA methylome in the developing zebrafish germline. *Nat. Commun.* **10** (1): 3054.

144 von Meyenn, F. and Reik, W. (2015). Forget the parents: epigenetic reprogramming in human germ cells. *Cell* **161** (6): 1248–1251.

145 Saitou, M., Kagiwada, S., and Kurimoto, K. (2012). Epigenetic reprogramming in mouse pre-implantation development and primordial germ cells. *Development* **139** (1): 15–31.

146 Felix, J.F. and Cecil, C.A.M. (2019). Population DNA methylation studies in the developmental origins of health and disease (DOHaD) framework. *J. Dev. Orig. Health Dis.* **10** (3): 306–313.

147 Hanson, M.A. and Skinner, M.K. (2016). Developmental origins of epigenetic transgenerational inheritance. *Environ. Epigenet.* **2** (1): dvw002.

148 Fellous, A., Labed-Veydert, T., Locrel, M. et al. (2018). DNA methylation in adults and during development of the self-fertilizing mangrove rivulus, *Kryptolebias marmoratus*. *Ecol. Evol.* **8** (12): 6016–6033.

149 Ortega-Recalde, O. and Timothy, A. (2019). Hore, DNA methylation in the vertebrate germline: balancing memory and erasure. *Essays Biochem.* **63** (6): 649–661.

150 Carvan, M.J. 3rd, Kalluvila, T.A., Klingler, R.H. et al. (2017). Mercury-induced epigenetic transgenerational inheritance of abnormal neurobehavior is correlated with sperm epimutations in zebrafish. *PLoS One* **12** (5): e0176155.

151 Anastasiadi, D. and Piferrer, F. (2019). Epimutations in developmental genes underlie the onset of domestication in farmed European sea bass. *Mol. Biol. Evol.* **36** (10): 2252–2264.

152 Soubry, A., Hoyo, C., Jirtle, R.L. et al. (2014). A paternal environmental legacy: evidence for epigenetic inheritance through the male germ line. *BioEssays* **36** (4): 351–371.

153 Guerrero-Bosagna, C. and Skinner, M.K. (2012). Environmentally induced epigenetic transgenerational inheritance of phenotype and disease. *Mol. Cell. Endocrinol.* **354** (1–2): 3–8.

154 Hajkova, P., Erhardt, S., Lane, N. et al. (2002). Epigenetic reprogramming in mouse primordial germ cells. *Mech. Dev.* **117** (1–2): 15–23.

155 Vernaz, G., Malinsky, M., Svardal, H. et al. (2021). Mapping epigenetic divergence in the massive radiation of Lake Malawi cichlid fishes. *Nat. Commun.* **12** (1): 5870.

156 Houri-Zeevi, L., Korem Kohanim, Y., Antonova, O. et al. (2020). Three rules explain transgenerational small RNA inheritance in *C. elegans*. *Cell* **182** (5): 1186–1197. e12.

157 Gapp, K., van Steenwyk, G., Germain, P.L. et al. (2020). Alterations in sperm long RNA contribute to the epigenetic inheritance of the effects of postnatal trauma. *Mol. Psychiatry* **25** (9): 2162–2174.

158 Migicovsky, Z. and Kovalchuk, I. (2011). Epigenetic memory in mammals. *Front. Genet.* **2**: 28.

159 Anastasiadi, D., Venney, C.J., Bernatchez, L. et al. (2021). Epigenetic inheritance and reproductive mode in plants and animals. *Trends Ecol. Evol.* **36** (12): 1124–1140.

160 Zhang, X. and Ho, S.M. (2011). Epigenetics meets endocrinology. *J. Mol. Endocrinol.* **46** (1): R11–R32.

161 Johansson, A., Palli, D., Masala, G. et al. (2019). Epigenome-wide association study for lifetime estrogen exposure identifies an epigenetic signature associated with breast cancer risk. *Clin. Epigenet.* **11** (1): 66.

162 Ben Maamar, M., Beck, D., Nilsson, E. et al. (2022). Developmental alterations in DNA methylation during gametogenesis from primordial germ cells to sperm. *iScience* **25** (2): 103786.

163 Beck, D., Ben Maamar, M., and Skinner, M.K. (2021). Integration of sperm ncRNA-directed DNA methylation and DNA methylation-directed histone retention in epigenetic transgenerational inheritance. *Epigenet. Chromatin* **14** (1): 6.

164 Ben Maamar, M., Sadler-Riggleman, I., Beck, D. et al. (2018). Epigenetic transgenerational inheritance of altered sperm histone retention sites. *Sci. Rep.* **8** (1): 5308.

165 Skinner, M.K., Ben Maamar, M., Sadler-Riggleman, I. et al. (2018). Alterations in sperm DNA methylation, non-coding RNA and histone retention associate with DDT-induced epigenetic transgenerational inheritance of disease. *Epigenet. Chromatin* **11** (1): 8.

166 Thorson, J.L.M., Beck, D., Ben Maamar, M. et al. (2020). Epigenome-wide association study for atrazine induced transgenerational DNA methylation and histone retention sperm epigenetic biomarkers for disease. *PLoS One* **15** (12): e0239380.

167 Clarke, H.J. and Vieux, K.F. (2015). Epigenetic inheritance through the female germ-line: the known, the unknown, and the possible. *Semin. Cell Dev. Biol.* **43**: 106–116.

168 Head, J.A., Dolinoy, D.C., and Basu, N. (2012). Epigenetics for ecotoxicologists. *Environ. Toxicol. Chem.* **31** (2): 221–227.

169 Manikkam, M., Haque, M.M., Guerrero-Bosagna, C. et al. (2014). Pesticide methoxychlor promotes the epigenetic transgenerational inheritance of adult-onset disease through the female germline. *PLoS One* **9** (7): e102091.

170 McBirney, M., King, S.E., Pappalardo, M. et al. (2017). Atrazine induced epigenetic transgenerational inheritance of disease, lean phenotype and sperm epimutation pathology biomarkers. *PLoS One* **12** (9): e0184306.

171 Hao, C., Gely-Pernot, A., Kervarrec, C. et al. (2016). Exposure to the widely used herbicide atrazine results in deregulation of global tissue-specific RNA transcription in the third generation and is associated with a global decrease of histone trimethylation in mice. *Nucl. Acids Res.* **44** (20): 9784–9802.

172 Lai, K.P., Wang, S.Y., Li, J.W. et al. (2019). Hypoxia causes transgenerational impairment of ovarian development and hatching success in fish. *Environ. Sci. Technol.* **53** (7): 3917–3928.

173 Wang, S.Y., Lau, K., Lai, K.-P. et al. (2016). Hypoxia causes transgenerational impairments in reproduction of fish. *Nat. Commun.* **7**: 12114.

174 Daxinger, L. and Whitelaw, E. (2012). Understanding transgenerational epigenetic inheritance via the gametes in mammals. *Nat. Rev. Genet.* **13** (3): 153–162.

175 Guerrero-Bosagna, C.M. and Skinner, M.K. (2014). Environmental epigenetics and phytoestrogen/phytochemical exposures. *J. Steroid Biochem. Mol. Biol.* **139**: 270–276.

176 Skinner, M.K. and Guerrero-Bosagna, C. (2014). Role of CpG deserts in the epigenetic transgenerational inheritance of differential DNA methylation regions. *BMC Genom.* **15** (1): 692.

177 Curley, J.P., Mashoodh, R., and Champagne, F.A. (2011). Epigenetics and the origins of paternal effects. *Horm. Behav.* **59** (3): 306–314.

178 Kelsey, G., Stegle, O., and Reik, W. (2017). Single-cell epigenomics: recording the past and predicting the future. *Science* **358** (6359): 69–75.

179 Lambrot, R., Xu, C., Saint-Phar, S. et al. (2013). Low paternal dietary folate alters the mouse sperm epigenome and is associated with negative pregnancy outcomes. *Nat. Commun.* **4**: 2889.

180 Guerrero-Bosagna, C., Settles, M., Lucker, B. et al. (2010). Epigenetic transgenerational actions of vinclozolin on promoter regions of the sperm epigenome. *PLoS One* **5** (9): e13100.

181 Crews, D., Gore, A.C., Hsu, T.S. et al. (2007). Transgenerational epigenetic imprints on mate preference. *Proc. Natl. Acad. Sci. U. S. A.* **104** (14): 5942–5946.

182 Guerrero-Bosagna, C., Covert, T.R., Haque, M.M. et al. (2012). Epigenetic transgenerational inheritance of vinclozolin induced mouse adult onset disease and associated sperm epigenome biomarkers. *Reprod. Toxicol.* **34** (4): 694–707.

183 Nilsson, E., Larsen, G., Manikkam, M. et al. (2012). Environmentally induced epigenetic transgenerational inheritance of ovarian disease. *PLoS One* **7** (5): e36129.

184 Nilsson, E.E., Anway, M.D., Stanfield, J. et al. (2008). Transgenerational epigenetic effects of the endocrine disruptor vinclozolin on pregnancies and female adult onset disease. *Reproduction* **135** (5): 713–721.

185 Skinner, M.K., Anway, M.D., Savenkova, M.I. et al. (2008). Transgenerational epigenetic programming of the brain transcriptome and anxiety behavior. *PLoS One* **3** (11): e3745.

186 Skinner, M.K., Haque, C.G., Nilsson, E. et al. (2013). Environmentally induced transgenerational epigenetic reprogramming of primordial germ cells and the subsequent germ line. *PLoS One* **8** (7): e66318.

187 Skinner, M.K., Manikkam, M., Tracey, R. et al. (2013). Ancestral dichlorodiphenyltrichloroethane (DDT) exposure promotes epigenetic transgenerational inheritance of obesity. *BMC Med.* **11**: 228.

188 Skinner, M.K., Mohan, M., Haque, M.M. et al. (2012). Epigenetic transgenerational inheritance of somatic transcriptomes and epigenetic control regions. *Genome Biol.* **13** (10): R91.

189 Tracey, R., Manikkam, M., Guerrero-Bosagna, C. et al. (2013). Hydrocarbons (jet fuel JP-8) induce epigenetic transgenerational inheritance of obesity, reproductive disease and sperm epimutations. *Reprod. Toxicol.* **36**: 104–116.

190 Ruden, D.M. and Lu, X. (2008). Hsp90 affecting chromatin remodeling might explain transgenerational epigenetic inheritance in *Drosophila*. *Curr. Genom.* **9** (7): 500–508.

191 Hauser, M.T., Aufsatz, W., Jonak, C. et al. (2011). Transgenerational epigenetic inheritance in plants. *Biochim. Biophys. Acta* **1809** (8): 459–468.

192 Schwindt, A.R., Winkelman, D.L., Keteles, K. et al. (2014). An environmental oestrogen disrupts fish population dynamics through direct and transgenerational effects on survival and fecundity. *J. Appl. Ecol.* **51**: 582–591.

193 Pembrey, M.E. (2010). Male-line transgenerational responses in humans. *Hum. Fertil.* **13** (4): 268–271.

194 Dias, B.G. and Ressler, K.J. (2014). Experimental evidence needed to demonstrate inter- and trans-generational effects of ancestral experiences in mammals. *BioEssays* **36** (10): 919–923.

195 Ge, Z.-J. and Sun, Q.-Y. (2019). Chapter 5 – maternal epigenetic inheritance. In: *Transgenerational Epigenetics*, 2e (ed. T.O. Tollefsbol), 75–105. Academic Press.

196 Kelley, J.L., Tobler, M., Beck, D. et al. (2021). Epigenetic inheritance of DNA methylation changes in fish living in hydrogen sulfide-rich springs. *Proc. Natl. Acad. Sci. U. S. A.* **118** (26): e2014929118.

197 Major, K.M., DeCourten, B.M., Li, J. et al. (2020). Early life exposure to environmentally relevant levels of endocrine disruptors drive multigenerational and transgenerational epigenetic changes in a fish model. *Front. Mar. Sci.* **7**: 1–17.

198 Zakarya, R., Adcock, I., and Oliver, B.G. (2019). Epigenetic impacts of maternal tobacco and e-vapour exposure on the offspring lung. *Clin. Epigenet.* **11** (1): 32.

199 Kalfa, N., Paris, F., Soyer-Gobillard, M.-O. et al. (2011). Prevalence of hypospadias in grandsons of women exposed to diethylstilbestrol during pregnancy: a multigenerational national cohort study. *Fertil. Steril.* **95** (8): 2574–2577.

200 Bajrami, E. and Spiroski, M. (2016). Genomic imprinting. *Open Access Maced. J. Med. Sci.* **4** (1): 181–184.

201 Seemann, F., Peterson, D.R., Witten, P.E. et al. (2015). Insight into the transgenerational effect of benzo[a]pyrene on bone formation in a teleost fish (*Oryzias latipes*). *Comp. Biochem. Physiol. Part C: Toxicol. Pharmacol.* **178**: 60–67.

202 Seemann, F., Jeong, C.-B., Zhang, G. et al. (2017). Ancestral benzo[a]pyrene exposure affects bone integrity in F3 adult fish (*Oryzias latipes*). *Aquat. Toxicol.* **183**: 127–134.

203 Knecht, A.L., Truong, L., Marvel, S.W. et al. (2017). Transgenerational inheritance of neurobehavioral and physiological deficits from developmental exposure to benzo[a]pyrene in zebrafish. *Toxicol. Appl. Pharmacol.* **329**: 148–157.

204 Akhter, A., Rahaman, M., Suzuki, R.T. et al. (2018). Next-generation and further transgenerational effects of bisphenol A on zebrafish reproductive tissues. *Heliyon* **4** (9): e00788.

205 Lombo M., Fernandez-Diez C., Gonzalez-Rojo S. et al. 2015. *Environ. sssssPollut.* 2015. **206**:667–678.

206 Santangeli, S., Consales, C., Pacchierotti, F. et al. (2019). Transgenerational effects of BPA on female reproduction. *Sci. Total Environ.* **685**: 1294–1305.

207 Volkova, K., Reyhanian Caspillo, N., Porseryd, T. et al. (2015). Transgenerational effects of 17alpha-ethinyl estradiol on anxiety behavior in the guppy, Poecilia reticulata. *Gen. Comp. Endocrinol.* **223**: 66–72.

208 Vera-Chang, M.N., Moon, T.W., and Trudeau, V.L. (2019). Cortisol disruption and transgenerational alterations in the expression of stress-related genes in zebrafish larvae following fluoxetine exposure. *Toxicol. Appl. Pharmacol.* **382**: 114742.

209 Vera-Chang, M.N., St-Jacques, A.D., Gagné, R. et al. (2018). *Proc. Natl. Acad. Sci.* **115** (52): E12435–E12442.

210 Jasperse, L., Levin, M., Rogers, K. et al. (2019). Transgenerational effects of polycyclic aromatic hydrocarbon exposure on sheepshead minnows (*Cyprinodon variegatus*). *Environ. Toxicol. Chem.* **38** (3): 638–649.

211 Vignet, C., Joassard, L., Lyphout, L. et al. (2015). Exposures of zebrafish through diet to three environmentally relevant mixtures of PAHs produce behavioral disruptions in unexposed F1 and F2 descendant. *Environ. Sci. Pollut. Res.* **22** (21): 16371–16383.

212 Alfonso, S., Blanc, M., Joassard, L. et al. (2019). Examining multi-and transgenerational behavioral and molecular alterations resulting from parental exposure to an environmental PCB and PBDE mixture. *Aquat. Toxicol.* **208**: 29–38.

213 Chen, L., Lam, J.C.-W., Hu, C. et al. (2019). Perfluorobutanesulfonate exposure skews sex ratio in fish and transgenerationally impairs reproduction. *Environ. Sci. Technol.* **53**: 8389–8397.

214 Shi, G., Guo, H., Sheng, N. et al. (2018). Two-generational reproductive toxicity assessment of 6:2 chlorinated polyfluorinated ether sulfonate (F-53B, a novel alternative to perfluorooctane sulfonate) in zebrafish. *Environ Pollut.* **243** (Pt B): 1517–1527.

215 Kamstra, J.H., Sales, L.B., Aleström, P. et al. (2017). Differential DNA methylation at conserved non-genic elements and evidence for transgenerational inheritance following developmental exposure to mono(2-ethylhexyl) phthalate and 5-azacytidine in zebrafish. *Epigenetics & Chromatin* **10** (20): 1–18.

216 Kteeba, S.M., El-Ghobashy, A.E., El-Adawi, H.I. et al. (2018). Exposure to ZnO nanoparticles alters neuronal and vascular development in zebrafish: acute and transgenerational effects mitigated with dissolved organic matter. *Environ. Pollut.* **242**: 433–448.

217 DeCourten, B.M., Forbes, J.P., Roark, H.K. et al. (2020). Multigenerational and Transgenerational Effects of Environmentally Relevant Concentrations of Endocrine Disruptors in an Estuarine Fish Model. *Environ. Sci. Technol.* **54** (21): 13849–13860.

218 Schwindt, A.R. (2015). Parental effects of endocrine disrupting compounds in aquatic wildlife: is there evidence of transgenerational inheritance? *Gen. Comp. Endocrinol.* **219**: 152–164.

219 Mirbahai, L. and Chipman, J.K. (2014). Epigenetic memory of environmental organisms: a reflection of lifetime stressor exposures. *Mutat. Res. Genet. Toxicol. Environ. Mutagen.* **764–765**: 10–17.

220 Papoulias, D.M., Noltie, D.B., and Tillitt, D.E. (2000). An in vivo model fish system to test chemical effects on sexual differentiation and development: exposure to ethinyl estradiol. *Aquat. Toxicol.* **48** (1): 37–50.

221 Papoulias, D.M., Villalobos, S.A., Meadows, J. et al. (2003). In ovo exposure to o,p-DDE affects sexual development but not sexual differentiation in Japanese medaka (Oryzias latipes). *Environ. Health Perspect.* **111** (1): 29–32.

222 Hyodo-Taguchi, Y. and Egami, N. (1989). Use of small fish in biomedical research, with special reference to in-bred strains of medaka. In: *Nonmammalian Animal Models for Biomedical Research*, vol. **1** (ed. A.D. Woodhead and K. Vivirito), 30. Boca Raton, FL: CRC Press.

223 Ishikawa, Y. (2000). Medakafish as a model system for vertebrate developmental genetics. *BioEssays* **22** (5): 487–495.

224 Wittbrodt, J., Shima, A., and Schartl, M. (2002). Medaka – a model organism from the far East. *Nat. Rev. Genet.* **3** (1): 53–64.

225 Hyodo-Taguchi, Y. and Egami, N. (1985). Establishment of inbred strains of the medaka *Oryzias latipes* and the usefulness of the strains for biomedical research. *Zool. Sci.* **2**: 12.

226 Shinomiya, A., Tanaka, M., Kobayashi, T. et al. (2000). The vasa-like gene, olvas, identifies the migration path of primordial germ cells during embryonic body formation stage in the medaka, Oryzias latipes. *Develop. Growth Differ.* **42** (4): 317–326.

227 Crain, D.A., Eriksen, M., Iguchi, T. et al. (2007). An ecological assessment of bisphenol-A: evidence from comparative biology. *Reprod. Toxicol.* **24** (2): 225–239.

228 Kolpin, D.D., Furlong, E.T., Meyer, M.T. et al. (2002). Pharmaceuticals, hormones, and other organic wastewater contaminants in U.S. streams, 1999–2000: a national reconnaissance. *Environ. Sci. Technol.* **36** (6): 1202–1211.

229 Guzman, E. (2021). *Transgenerational Differences in Gut Microbiota Population and Epigenetic Responses of Host Intestinal Epithelial Cells*. The University of North Carolina at Greensboro.

230 Chakraborty, S., Dissanayake, M., Godwin, J. et al. (2022). Ancestral BPA exposure caused defects in the liver of medaka for four generations. *Sci. Total Environ.* **856**: 159067.

231 Lombó, M., Fernández-Díez, C., González-Rojo, S. et al. (2015). Transgenerational inheritance of heart disorders caused by paternal bisphenol A exposure. *Environ. Pollut.* **206**: 667–678.

232 Wang, X., Hill, D., Tillitt, D.E. et al. (2019). Bisphenol A and 17α-ethinylestradiol-induced transgenerational differences in expression of osmoregulatory genes in the gill of medaka (*Oryzias latipes*). *Aquat. Toxicol.* **211**: 227–234.

233 Kidd, K.A., Blanchfield, P.J., Mills, K.H. et al. (2007). Collapse of a fish population after exposure to a synthetic estrogen. *Proc. Natl. Acad. Sci.* **104** (21): 8897–8901.

234 Bjornsson, B.T., Stefansson, S.O., and McCormick, S.D. (2011). Environmental endocrinology of salmon smoltification. *Gen. Comp. Endocrinol.* **170** (2): 290–298.

235 Chamberlin, J.W., Essington, T.E., Ferguson, J.W. et al. (2011). The influence of hatchery rearing practices on salmon migratory behavior: is the tendency of Chinook salmon to remain within Puget Sound affected by size and date of release? *Trans. Am. Fish. Soc.* **140** (5): 1398–1408.

236 Le Luyer, J., Laporte, M., Beacham, T.D. et al. (2017). Parallel epigenetic modifications induced by hatchery rearing in a Pacific salmon. *Proc. Natl. Acad. Sci.* **114** (49): 12964–12969.

237 Spencer, H.G. (2016). Epigenetic inheritance. In: *Encyclopedia of Evolutionary Biology* (ed. R.M. Kliman), 1–5. Oxford: Academic Press.

238 Lee Chong, T., Ahearn, E.L., and Cimmino, L. (2019). Reprogramming the epigenome with vitamin C. *Front. Cell Dev. Biol.* **7**: 128.

239 DiTroia, S.P., Percharde, M., Guerquin, M.-J. et al. (2019). Maternal vitamin C regulates reprogramming of DNA methylation and germline development. *Nature* **573** (7773): 271–275.

240 Crider, K.S., Yang, T.P., Berry, R.J. et al. (2012). Folate and DNA methylation: a review of molecular mechanisms and the evidence for folate's role. *Adv. Nutr. (Bethesda, Md.)* **3** (1): 21–38.

241 Mozhdeganloo, Z., Jafari, A.M., Koohi, M.K. et al. (2015). Methylmercury-induced oxidative stress in rainbow trout (*Oncorhynchus mykiss*) liver: ameliorating effect of vitamin C. *Biol. Trace Elem. Res.* **165** (1): 103–109.

242 Alagbonsi, A.I. and Olayaki, L.A. (2020). Vitamin C ameliorates tetrahydrocannabinol-induced spermatotoxicity in-vitro. *BMC Nutr.* **6** (1): 59.

243 Falconer, D.S. (1996). *Introduction to Quantitative Genetics*. Pearson Education Limited.

244 Sundstrom, L.F., Lohmus, M., and Devlin, R.H. (2016). Gene–environment interactions influence feeding and anti-predator behavior in wild and transgenic coho salmon. *Ecol. Appl.* **26** (1): 67–76.

5

Environmental Epigenetics in Fish: Response to Climate Change Stressors

Zhi-Gang Shen[1], *Yue Yu*[1], *and Han-Ping Wang*[2]

[1] College of Fisheries, Huazhong Agricultural University, Wuhan, China
[2] Ohio Center for Aquaculture Research and Development, The Ohio State University South Centers, Piketon, OH, USA

5.1 Overview of Climate Change and Environmental Stressors

5.1.1 Temperature Rise and Extreme Weather Events

The acceleration of anthropogenic global temperature rise has become indisputable. The latest IPCC Sixth Assessment Report (AR6) indicated that, since 1850, each of the last four decades has been successively warmer than any decade that preceded it. Global surface temperatures (refers to both global mean surface temperature and global surface air temperature) between 2011 and 2020 are 1.09 [0.95–1.20] °C warmer than 1850–1900, with larger increases over land (1.59 [1.34–1.83] °C) than over the ocean (0.88 [0.68–1.01] °C), which may lead to the extinction of thousands of species in the next 100 years [1–7].

Researchers have found a well-documented correlation between global climate and the atmospheric partial pressure of carbon dioxide (CO_2), whereby the concentration of CO_2 in the atmosphere varies in accordance with the glacial–interglacial cycle [8, 9], likely triggered by the "biological pump" action of the deep sea [9–11]. Greenhouse gases are likely to be an intermediary between the orbitally driven cycle of solar radiation and climatic cyclicity change. However, human activities, mainly rapid industrial development, have begun to stir up this natural pattern. Since the eighteenth century, humans have enjoyed the benefits created by the industrial revolution while bringing various risks and challenges to ecosystems. Predictably, the abnormal accelerated rise in anthropogenic atmospheric greenhouse gas levels will be accompanied by atypical elevated global surface temperatures, extreme weather, ice layer collapse, sea-level rise, and pathogen flooding. In the absence of concerted mitigation measures to limit greenhouse gas emissions, these stressors will accelerate to exacerbate species extinctions in the future. More importantly, there is an urgent requirement to explore the interaction mechanism and feedback between different species and environmental stresses, which can provide theoretical and practical guidance for the maintenance and protection of biodiversity in time.

Among many environmental stressors, rising temperature provides the most compelling evidence of the out-of-balance global climate, while also directly affecting species' life-history strategies. If the global warming rate is maintained, tens of thousands of species' persistence will be threatened in a short time. Thus, researchers have reasonably focused on the interaction (physiology or behavior) between organisms and high temperature, including physiological tolerance to high-temperature stress, regulation of body temperature and related hormones, phenological adaptability, high-temperature evolutionary adaptability, etc. [12–16]. Cahill et al. [17] made detailed statistics on reports about local species extinction in 136 relevant studies related to climate change up to 2012. Although none of these showed a direct relationship between local extinction and limited tolerances to high temperature, almost all extinctions were attributed to other proximate causes of climate change, such as limited ground activity time during biological reproductive periods caused by high temperatures, food shortage caused by changes in precipitation, and loss of aquatic habitats due to drought [18–24].

Based on simple physical considerations, the elevated temperature will increase the number of heat records, decrease cold extremes, lead to more evaporations and surface drying, and increase the intensity and duration of drought [25]. Even

Epigenetics in Aquaculture, First Edition. Edited by Francesc Piferrer and Han-Ping Wang.

though some key findings on the ecological effects of extreme weather events, including diversity, reproduction, phenology, nutrient cycling, and resistance to invasion, have been documented [26], it has long been recognized as one of the main gaps in knowledge in ecology.

Animals can make adaptive adjustments according to the surrounding environment, e.g., increasing temperatures promote reproductive capacity in some spring-spawning species [15], and global warming will induce earlier spawning [16]. However, unsteady heat waves still increase the risk that fertilized eggs or larvae will be exposed to high temperatures during the critical period of sex determination (Box 5.1). In species with temperature-dependent sex determination (TSD) (mainly reptiles, amphibians, and fish), small changes in environmental temperature can affect the population sex ratio. For fish with genotypic sex determination plus temperature effects (GSD + TE), a male bias sex ratio under heat stress appears to be a general rule, although different species have different temperature sensitivity and masculinization patterns. The sex switch at the initial stage of growth may not affect growth and development, but the continuous mismatch between genotypic sex and phenotypic sex may drive the disappearance of the Y chromosome in male heterogametic species (XX/XY) [27–29]. Sudden termination of high temperature will lead to species extinction to a large extent, but some scholars believe that this disappearance of the Y chromosome induced by high temperature may be the mechanism of adaptive evolution of GSD to environmental sex determination (ESD [28]). Under certain conditions, some fish species may display thermotropism during sex differentiation, which may result in active masculinization and thus increase the risk of imbalance in population dynamics [30, 31].

5.1.2 Acidification

The CO_2 concentration in the atmosphere has ranged from 180 to 300 ppm during the last 650,000 years [32], rose to ~390 ppm in 2011, and reached 410 ppm in 2021 [33]. The increase of CO_2 in the atmosphere causes acidification of the ocean. It is predicted that atmospheric CO_2 concentration will rise to ~790 ppm by the end of the century. Consequently, the ocean mean surface pH is expected to decrease by 0.3 units from the current 8.1 as the alternation of the seawater carbonate chemistry [34]. The acidification caused by excess CO_2 can have significant effects on marine organisms, including impacts on physiology, growth performance, reproduction, behavior, habitat degradation, and changes in skeletal and otolith development [35]. Non-neutral pH has also been found to affect sex differentiation in fish and lead to masculinization or an increase in the male ratio [36], and may consequently disturb the operational sex ratio. Moreover, the growth of calcifying organisms will be hampered by the acidified aquatic environment [37–39].

Box 5.1 Glossary

Acclimation: Physiological, anatomical, or morphological adjustment of an organism in response to environmental changes to improve performance or survival. Acclimation occurs within the lifetime of an organism.
Adaptation: The process by which an organism becomes fitted to its environment through heritable variations over several generations. Adaptation produces a novel phenotypic trait that has a functional role in the environment.
Environmental stress: Beyond the optima of environmental conditions, either manmade or natural, including temperature, oxygen, pH, salinity, population density, competition, etc. Global climate change causes a series of environmental stresses.
Phenology: The study of recurring seasonal events in biological life cycles and how these are influenced by periodic and interannual biotic and abiotic forces, e.g., spawning period and periodic migrations.
Sex determination: The genetic or environmental process that establishes the sex of an organism. In gonochoristic fish, sex is determined by genetics (genotypic sex determination, GSD), environmental factors (environmental sex determination, ESD), or the combination of both (GSD plus environmental effects, GSD + EE).
Sex differentiation: The process by which an undifferentiated gonad is transformed into an ovary or a testis. Sex differentiation can be influenced by almost all environmental factors, including stress temperature, non-neutral pH, hypoxia, food availability and quality, background color, and social interaction, which will result in the mismatch between genotypic sex and phenotypic sex. This provides a new path for environment-friendly mono-sex production in aquaculture.

5.1.3 Hypoxia

Hypoxia is primarily a consequence of ocean warming. Oceans continue to absorb heat from atmospheric warming and their oxygen (O_2) concentration is expected to decrease because the gas solubility reduces with the increase in water temperature. Further, elevated temperatures increase the rate of oxygen consumption by organisms through respiration [40]. Besides, surface heating reduces the mixing of O_2-rich surface water with the deeper ocean waters and further aggravates the overall hypoxia condition [41]. Previous work estimated that between 100 and 1000 m in depth, O_2 amount lost by 550 ± 130 teramoles (10^{12} mol) per decade based on the data from the 1970s and 1990s [42]. Further research found that the global O_2 concentration of 227.4 ± 1.1 petamoles (10^{15} mol) has decreased by more than 2% (4.8 ± 2.1 petamoles) since 1960, with significant fluctuation in oxygen loss in different ocean basins and at different depths [43]. Hypoxia has negative effects on fish in various aspects, including weakening appetite, slowing growth and development, changing normal physiological behavior, damaging reproduction, and increasing mortality [44–48].

5.1.4 Phenology and Distribution

Temperature is the most important abiotic factor affecting phenology and global warming is increasingly disrupting natural phenological patterns [49, 50]. The annual activities of many organisms in the aquatic environment are temperature-dependent, including seasonal mating, spawning, and periodic migrations [51–56]. Studies have indicated that many of today's spring and summer phenological events occur earlier in the year than ever before [16, 57–62]. For example, a study of fish phenology in the waters off Southern California found that 39% of the seasonal peaks in juvenile populations of 43 species occurred earlier in the year, associated with changes in ocean surface temperature and mesozooplankton displacement [50]. Overall, varying degrees of changes in the timing of interspecific phenological events can decouple biological interactions. In other words, peak resource demand periods for predators (usually breeding periods) and peak supply periods for prey cannot overlap with each other because of the differential sensitivity of different species to climate change [63]. Using a meta-analysis of 10,003 terrestrial and aquatic organisms, Stephen et al. [63] found that secondary consumers exhibited consistently lower climate sensitivity than other groups, and the time of phenological events of primary consumers could change more than for species in other trophic levels, which means that a significant proportion of predators have declining population abundance as a result of misalignment with prey's phenology. All components of the ecosystem are affected by varying degrees of phenological changes, including intraspecific and interspecific, resulting in unpredictable complex dynamics [64]. Population dynamics, community structure, ecosystem processes, and ecosystem stability will be disrupted by the imbalance of phenological synchronization [64–66].

The natural distribution of a species is at least partially restricted by climate, which is a core premise of biogeography [67–69]. Thus, the most significant environmental impact of global warming at the biogeographic scale is the shift of species ranges to higher latitudes, higher altitudes, or deeper water layers to escape excessive warm temperatures [57, 70, 71]. For species with limited dispersal ability or unable to find suitable habitats, a change in distribution is likely to result in widespread extinction [7]. To date, several reports have documented shifts in the distribution of fish and aquatic invertebrates [72–74]. It is important to note that the latitudinal response to warming is uneven across species. Smaller, shorter-lived, faster-reproducing species are likely to have a faster demographic response to temperature change, resulting in a stronger response to the distribution of global warming and faster adaptation to changes in environmental temperature. In contrast, larger species have a weaker ability to adapt to variable temperatures [75, 76]. This pattern may be similar to non-synchronization phenological changes, leading to the dynamic decoupling of interacting species [64]. Eventually, the ecological and economic impact will be far-reaching.

5.2 Epigenetic Response to Climate Change

Anthropogenic global climate change means, for many of Earth's organisms, significant changes in their habitats. In this context, it is believed that epigenetics, as a mechanism that promotes heritable phenotypic variation through genetic–environmental interactions, can help organisms to respond rapidly to changes in their surroundings, leading to physiological adjustments across generations that have important implications for ecological and evolutionary studies [77]. Varriale and Bernardi [78] found that polar fish adapted to cold temperatures had higher levels of 5-methylcytosine (5mC) than tropical fish adapted to warmer temperatures. Under laboratory simulations of future climate changes, a growing number of reports provide strong evidence that epigenetics can link global climate change with environmentally dependent plastic phenotypic adaptation in fish (Table 5.1).

Table 5.1 Effects of environmental stressors and epigenetic/phenotypic changes in fish.

Stressor	Species	Exposure conditions	Epigenetic changes	Phenotypic changes	Reference
Temperature	European sea bass *Dicentrarchus labrax*	High temperature during thermosensitive period at larvae and juvenile stages	Increase of DNA methylation of the promoter region of *cyp19a1a* at the male-producing temperature	Male-biased sex ratio	[79]
	Half-smooth tongue sole *Cynoglossus semilaevis*	High temperature during thermosensitive period at larvae and juvenile stages	Genomic methylation patterns of high temperature-induced pseudo-male testes were similar to those of normal male testes	Masculinization	[80, 81]
	Senegalese sole *Solea senegalensis*	High temperature during metamorphosis	Increased cytosine methylation of the *myog* promoter in skeletal muscle at a low temperature	Increased thermal plasticity of myogenesis of larvae reared at higher temperatures	[82]
	Zebrafish *Danio rerio*	High temperature during thermosensitive period	Significant decrease of global DNA methylation in the testis of the 35 °C-derived males in the F1 generation but not of the P and F2 generations	Family-dependent increase in the male percentage	[83]
	Brook charr *Salvelinus fontinalis*	Warm and cold temperatures during sexual maturation of parents	188 differentially methylated regions were observed due to parental maturation temperature	Improved offspring adaptation to a warming environment	[84]
	Nile tilapia *Oreochromis niloticus*	High temperature during thermosensitive period	Differentially methylated regions of gene bodies and promoters among high temperature-induced males, high temperature-treated females, control females, and control males	Masculinization	[85]
	Nile Tilapia *Oreochromis niloticus*	High temperature during thermosensitive period	Increased DNA methylation of *cyp19a1a* promoter in high temperatures	Masculinization	[86]

	Olive flounder *Paralichthys olivaceus*	High temperature during thermosensitive period	Hypermethylation maintenance in *cyp19a* promoter in high temperatures	Masculinization	[87]
	Tiger pufferfish *Takifugu rubripes*	Low temperature during thermosensitive period	Low methylation level in normal males and pseudo-male, and high methylation level in normal females of *amhr2*	Low-temperature induced masculinization	[88]
Photoperiod	Atlantic cod *Gadus morhua*	During the early juvenile stages	Increased expression of *dnmt1* and *dnmt3a* in fast muscle exposed to continuous illumination relative to fish reared under a natural photoperiod	Fish reared under continuous light during the early juvenile stages grow faster than that reared at natural photoperiod	[89]
Salinity, nutrition	Brown trout *Salmo trutta*	Exposure of freshwater fish to seawater conditions, after salt-enriched diet intake	Reversible and temporary salt-induced changes in genome-wide methylation changes of hatchery-reared trout	Hatchery-reared trout physiologically adapted to seawater	[90]
Salinity	Half-smooth tongue sole *Cynoglossus semilaevis*	Low salinity (15‰) exposure of 10-month-old fry for 0, 7, and 60 days	Three exons in the *igf1* gene had significantly different methylation levels under salt stress	Reduced growth performance	[91]
Dissolved oxygen	Medaka *Oryzias melastigma*	Hypoxia exposure of sexually mature (F0) fish	Global DNA hypermethylation in sperm of F2 fish; hypomethylation of promoter regions of *ehmt2* and *ptk2b*; hypermethylation of exonic regions of *foxp2*	Transgenerational negative impacts of hypoxia on reproductive performance	[92]

5.2.1 Sex Determination and Differentiation

Fish are the most speciose and diverse group of vertebrates and accordingly also exhibit complex sex determination strategies [93]. In general, sex in gonochorism fish is determined by genetics (i.e., GSD), environment (i.e., ESD), or a combination of both (i.e., GSD + EE) [36, 94]. The orientation of sexual differentiation is liable to be distorted by the tilting force of external environmental stresses, usually resulting in masculinization [36].

Examples of environmental stress impacting the sex ratio of fish populations are systematically summarized and collected (Figure 5.1). Firstly, we rechecked the original cases of three review articles [141, 159, 160]. We then screened all the cited references of these reviews with a high degree of correlation. Secondly, we conducted an extensive search in the database Web of Science, using the following keyword: "((TS = (environment OR temperature OR pH OR photoperiod OR light OR illumination OR density OR crowding stress OR food OR hypoxia)) AND TS = (sex determination OR sex differentiation OR sex ratio OR masculinization OR sex reversal)) AND TS = (fish OR teleost)" with a date up to June 2022 and obtained 6211 results. We also searched the CNKI, the Chinese database, using a similar keyword and obtained 95 results. Thirdly, based on the relevant reports obtained by the above two methods, we then checked the cited reference of all

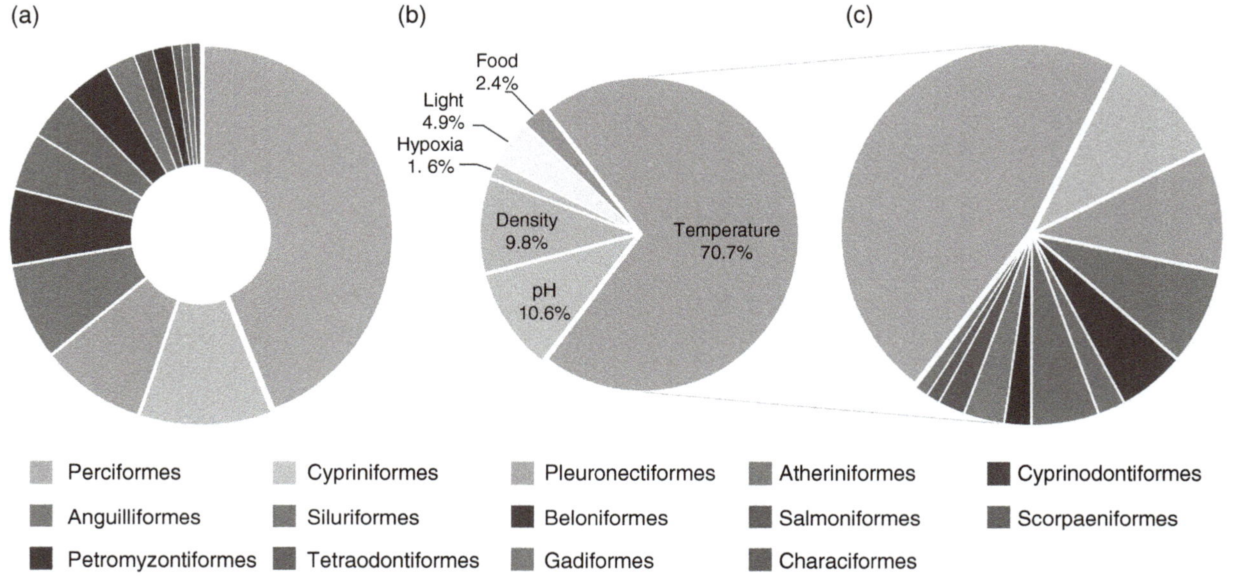

Figure 5.1 A summary of environmental stress on sex differentiation in fish. This survey covers 100 fish species and 123 cases belonging to 14 orders, which is from 65 published papers. A case represents the corresponding relationship between a fish species and a stressor. (a) Distribution of the 123 cases according to the order in which the target species belongs. (b) Percentage of type of environmental stress in the entire 123 cases. (c) Percentage of 14 orders in the 87 cases in temperature-dependent sex differentiation. Temperature, high or low temperature, or stress temperature; pH, acidic or alkali water; density, including stress density and high levels of hierarchy; hypoxia, low dissolved oxygen; light, stress illumination condition, such as a long light period or continuous green light; food, including low food availability and low food quality. References of each order: Petromyzontiformes, *Lampetra aepyptera* [95], *Petromyzon marinus* [96]; Anguilliformes, *Anguilla anguilla* [97], *Anguilla rostrata* [97], *Anguilla japonica* [97], *Anguilla dieffenbachii* [97]; Cypriniformes, *Carassius auratus* [98], *Carassius Carassius* [99, 100], *Carassius gibelio* [101], *Cyprinus carpio* [102], *Danio rerio* [103], *Gnathopogon caerulescens* [104], *Misgurnus anguillicaudatus* [105], *Paramisgurnus dabryanus* [105], *Sabanejewia balcanica* [106]; Siluriformes, *Ictalurus punctatus* [107], *Tachysurus fulvidraco* [108], *Tachysurus vachellii* [109], *Hoplosternum littorale* [110], *Clarias gariepinus* [31, 111]; Characiformes, *Nematobrycon palmeri* [112]; Salmoniformes, *Oncorhynchus nerka* [113], *Oncorhynchus mykiss* [114], *Thymallus thymallus* [115]; Gadiformes, *Gadus chalcogrammus* [116]; Perciformes, *Apistogramma* spp. [117], *Apistogramma caetei* [117], *Pelvicachromis pulcher* [118], *Pelvicachromis subocellatus* [117], *Pelvicachromis taeniatus* [117], *Pelmatochromis* spp. [119], *Oreochromis niloticus* [120], *Oreochromis aureus* [120], *Oreochromis mosambicus* [120], *Dicentrarchus labrax* [121], *Macropodus opercularis* [122], *Trichogaster lalius* [123], *Betta splendens* [124, 125], *Lepomis macrochirus* [126], *Acanthochromis polyacanthus* [127]; Scorpaeniformes, *Sebastes schlegeli* [128, 129], *Anoplopoma fimbria* [130]; Atheriniformes, *Menidia menidia* [131], *Menidia peninsulae* [132], *Odontesthes argentinensis* [133], *Odontesthes bonariensis* [134], *Odontesthes hatcheri* [135], *Chirostoma estor* [136], *Leuresthes tenuis* [137], *Hypoatherina tsurugae* [138]; Tetraodontiformes, *Takifugu rubripes* [139, 140]; Cyprinodontiformes, *Limia melanogaster* [141], *Poeciliopsis lucida* [141], *Poecilia sphenops* [142], *Poecilia reticulata* [143], *Xiphophorus helleri* [144]; Beloniformes, *Oryzias latipes* [145–147], *Oryzias sakaizumii* [148]; Pleuronectiformes, *Paralichthys olivaceus* [149], *Paralichthys dentatus* [150], *Paralichthys lethostigma* [151], *Pseudopleuronectes yokohamae* [152], *Pseudopleuronectes americanus* [153], *Verasper moseri* [154], *Verasper variegatus* [155], *Hippoglossus hippoglossus* [156], *Scophthalmus maximus* [157], *Cynoglossus semilaevis* [158]. (*See colour plate section for colour representation of this figure.*)

obtained reports and screened related articles, and sorted obtained results by species. We finally found that the correlation between environmental stress and alternation of sex ratio in fish was reported in 123 cases belonging to 14 orders and 100 species (Figure 5.1). Stressors such as stress temperature, non-neutral pH, and hypoxia, which account for about 80% of all cases, are closely related to global warming. Among the 123 cases of ESD/GSD + EE in fish identified to date, reports of extreme temperature as the environmental stressor account for the largest proportion (Figure 5.1), and the mechanism of TSD or GSD + TE has received the most attention.

For most of these cases, the results have been derived from studies under experimental controlled conditions. Thus, sex reversal induced by the temperature rise, subsequently the mismatch between phenotypic sex and genotypic sex, and the unbalanced operational sex ratio should be given attention. Importantly, concerns have increased about the correlation between global warming and disruption of operational sex ratio, but little attention has been paid to the reproductivity of sex-reversed fish [161].

For GSD + EE fish, negative environmental pressures triggered by global climate change may reverse the native sex fate, causing genotype–phenotype mismatches of sex, i.e., sex reversal, in natural populations. The mechanism of this sex differentiation strategy of environmental participation is still ambiguous, and it appears to be achieved by the coordinated interaction of different mechanisms. In addition to molecular biology [94], physiology [108, 162–164], and cytology [165, 166], epigenetics is widely recognized as a crucial player in sex differentiation in ESD/GSD + EE species [167–170].

Global methylation levels are highly temperature-sensitive [85, 171–175]. High temperature can affect the methylation level of typical sex-specific molecules and hinder the interaction between transcriptional elements, including DNA methylation and histone modification, which is the commonality of all epigenetic mechanisms involved in TSD/GSD + TE identified so far [169].

The pro-female gene *cyp19a1a* and the pro-male gene *dmrt1* consistently show an inverse relationship between their DNA methylation and expression levels across many species, an observation that led to the model of the conserved epigenetic regulation of sex (CERS) [169]. For example, in European sea bass (*Dicentrarchus labrax*), high temperatures consistently increase DNA methylation levels in the promoter region of *cyp19a1a*, thereby blocking transcription of *cyp19a1a*, while at the same time decreasing DNA methylation in the promoter of *dmrt1* [79, 176] (Table 5.1). Ultimately, the lack of estrogen leads to the masculinization of female individuals.

Studies have proved that the male-specific gene *dmrt1* showed a sexual dimorphic DNA methylation pattern in many fish species, high in females and low in males [169, 177, 178]. Similarly, high-temperature stress reduces the DNA methylation level of *dmrt1* (i.e., DNA demethylation) in the female half-smooth tongue sole and induces female-to-male sex reversal in genotypic ZW females, and parents can allele-specifically transfer *dmrt1* methylation levels across generations to their offspring [81]. The stable CERS model based on *cyp19a1a* and/or *dmrt1* genes has been well established in numerous TSD or GSD + TE teleost fishes [169], including European sea bass [79, 176], zebrafish (*Danio rerio*) [83, 179], half-smooth tongue sole (*Cynoglossus semilaevis*) [80, 180], Japanese flounder (*Paralichthys olivaceus*) [177], Nile tilapia (*Oreochromis niloticus*) [86, 181], etc. Thus, mounting evidence shows that a high-temperature environment can affect the population sex ratio of heat-sensitive teleost through an epigenetic mechanism (Table 5.1).

It is important to note that gene expression is not necessarily negatively correlated with DNA methylation levels. In the half-smooth tongue sole, the *amh* gene is always expressed in males at higher levels than in females. Correspondingly, the DNA methylation level of *amh* was higher in males than in females [80, 81]. Thus, the higher the DNA methylation level of *amh*, the higher the expression level, which seems to be fairly conserved across fish [169].

In addition, the temperature can also influence the expression levels of sex-specific genes by regulating histone microstructure (histone modification). Under high-temperature conditions, compared with females, the methylation level of histone H3K9 in the promoter region of *cyp19a1a* gene in male Nile tilapia was significantly increased, and the amount of transcriptional binding RNA polymerase and the transcriptional level of *cyp19a1a* gene was significantly decreased [86]. The pattern is also seen in European sea bass [79]. This evidence suggests that histone methylation is also involved in the mechanism of TSD/GSD + TE.

The acquired sex bias mediated by an epigenetic mechanism, particularly DNA methylation, has the potential for intergenerational inheritance. Genotypic ZW pseudo-male of tongue sole offspring induced by high temperature (28 °C) retained and enhanced masculinization sensitivity and showed a high masculinization rate (up to 94% male in the population) at medium temperature (22 °C). Further investigation revealed that ZW offspring maintained a similar low *dmrt1* DNA methylation pattern compared to their ZW pseudo-male parents [81]. However, this intergenerational inheritance of temperature-sensitive sex bias does not have a persistent effect and gradually fades away (reversibility of intergenerational inheritance phenomena). For example, in zebrafish, F1 generation populations produced by parents exposed to higher temperatures still exhibit male-biased sex ratios in normothermic environments, whereas the sex ratios of unexposed F2 generations are unaffected by heat exposure [83]. Furthermore, global DNA methylation is affected in the testis of the F1 generation of exposed parents [182]. Epigenetic pathways may represent a conservative adaptation mechanism that integrates genomic and temperature change information to control the ultimate fate of fish sex [183]. Current investigations of epigenetic mechanisms involved in sex determination/differentiation have focused on high temperatures, but gonadal plasticity as influenced by environmental conditions is considered to be a general rule rather than an exception in fish [184]. Therefore, other environmental stressors accompanying global climate change may also trigger ESD/GSD + EE mechanisms in environmentally sensitive fish, while the links between environmental stress and epigenetics need to be further explored.

5.2.2 Gonadal Development and Reproduction

Gonadal development and reproduction in fish are regulated by the endocrine system and environmental conditions [185]. Although most fish can actively migrate in search of suitable environments, rising ocean surface temperatures, decreasing salinity, and low dissolved oxygen due to global warming may be potential drivers of gonadal dysgenesis for fish with poor motility and limited habitat range. Several reports have summarized that climate warming-induced changes in the physicochemical properties of the aquatic environment interfere with normal reproduction in fish by affecting the hypothalamic–pituitary–gonadal (HPG) axis [186, 187]. In addition, some neo-males derived from ESD/GSD + EE fish have low or little fertility, mainly due to slow and incomplete development of sex-reversed gonads or mismatch between the gonads and genital opening [188, 189], which further reduces the reproductive potential [161].

Rozenfeld et al. [190] found that sustained cold stress stimulated a significant increase in androgen plasma levels, A-type spermatogonia differentiation, and differentially expressed genes in the HPG axis in European eels, accompanied by an enrichment of histone H3-K9 methylation and histone H3 deacetylation genes, confirming that temperature-dependent epigenetics is involved in the phenotype of gonad development and related gene expression.

Evidence of hypoxia causing impairment of gonadal function in fish through epigenetic mechanisms has been reported. Using the marine medaka (*Oryzias melastigma*) as a model, it was found that hypoxic stress can impair normal gonad function and cause reproductive disorders in fish by regulating miRNAs (a collection of short-stranded noncoding RNAs). Lai et al. [191] found that a large number of hypoxia-inhibited miRNAs targeted a variety of steroid-producing enzymes, including steroidogenic acute regulatory proteins, aromatase, and 17-α-monooxygenase. Furthermore, estrogen receptor 2 and androgen receptor was also found to be downstream targets of hypoxia-responsive miRNAs. Thus, the activity of miRNAs influenced by hypoxia stress can also interfere with fish testis function (e.g., cell movement, ATP production, ATP enzyme activity, etc.) and signal pathways (e.g., testicular cell cycle, GnRH signal pathway, glucose metabolism, etc.), leading to testicular dysfunction [192]. This evidence suggests that miRNA may be a major mediator of ovarian steroid synthesis and testicular metabolism in fish under hypoxic stress conditions.

DNA methylation is also reported to be involved in the reproductive dysfunction of fish due to hypoxic stress. In marine medaka, DNA methylation marks on the genome were found to be dependent on hypoxia and induced dysregulation of a series of gene pathways related to cell cycle, apoptosis, and autophagy, represented by the TP53-TERT-ZEB1 complex, which induced follicular atresia and inhibits oocyte development to hinder the normal spawning behavior of fish [193].

In testis, hypoxia induces upregulation of genes related to male reproductive function, including *ptk2b*, *mapk9*, and *ehmt2* [92]. Among them, *ehmt2* (autosomal lysine *N*-methyltransferase 2) can induce dimethylation of the lysine residue of histone H3 (H3K9me2), which is an important epigenetic marker of hypoxia-induced gene silencing. This may lead to the blocking of the molecular pathway of sperm proliferation and capacitation [92]. Interestingly, *miR-125-5p* has also been proven to be an upstream regulator of *ehmt2* [192]. This evidence strongly suggests that reproductive impairments (e.g., decrease in sperm motility, testis weight, percentage of spermatids, and the number of spermatids) in fish are mediated by the synergistic interplay of epigenetic mechanisms. Notably, these studies on DNA methylation have found intergenerational reproductive impairments. Hypoxia-induced dysregulations of some typical pathways related to cell proliferation and cell death in the ovary of the F0 generation were also observed in the F2 generation, including autophagy, protein kinase A signaling, and telomerase signaling [193]. In the latter study on testes, it was also found that F2 offspring not exposed to hypoxic conditions also exhibited similar reproductive impairments as their parents that were exposed to hypoxia [92]. This intergenerational hypoxic reproductive damage is based on a similar DNA methylation pattern between offspring and parents.

Global climate change has led to the emergence of over 400 hypoxic "dead zones" (dissolved oxygen [DO] <2 mg/L) in coastal waters worldwide [194]. Therefore, an in-depth investigation into the mechanisms of hypoxia-dependent epigenetic damage to reproductive function is needed for the protection of fish in a global change scenario.

5.2.3 Growth, Size, and Morphology

Several studies have established a link between global climate change and aquatic animal growth rate [195, 196]. Martin et al. [195] suggest that organisms shrinking in size is the third general response to global warming, in addition to changes in phenology and distribution. Changing body size appears to be an adaptive evolution for organisms in the face of climate change, as the size of living organisms directly affects thermoregulation, energy acquisition and utilization, and life-history traits [196]. For example, the winter skate (*Leucoraja ocellata*), much smaller than other species of the same genus, has greatly reduced its body size, by about 45%, to adapt to warmer habitats [197]. For water-breathing ectotherms, oxygen limits their growth ability and it determines the basic biological responses to environmental change [198]. The decrease of oxygen concentration in the atmosphere and ocean with global warming is an important factor to induce the reduction of the body size of organisms. It is predicted that with the highest CO_2 emissions [199], assemblage-averaged maximum body weight is expected to reduce by 14–24% globally between 2000 and 2050, and it will be dramatically influenced in the tropical and intermediate latitudinal areas (>20%).

Lighten et al. [197] propose that epigenetic-driven phenotypic plasticity (adaptation through regulation of gene expression) may provide a physiological buffer for species to overcome environmental change. This buffer leaves a gap period for the population to adapt or acclimate under environmental stress. The winter skate (*Leucoraja ocellata*), which settled in the warmer Northumberland Strait within the southern Gulf of St Lawrence, has differentiated into many phenotypic changes different from other marine populations in just 7000 years, including a 45% reduction in body size, a reduction in the tooth row and precaudal vertebrae, and a reduction in maximum lifespan and age at sexual maturity. Notably, the two morphologically significantly different populations of winter skate are genetically indistinguishable, which reinforces the contribution of epigenetics in morphological adaptation [197]. Light stress also has been demonstrated to affect the growth phenotype of fish through DNA methylation. The average weight of Atlantic cod (*Gadus morhua*) larvae reared under continuous light was 13% greater than of those reared under natural light, and transcript levels of *dnmt1* and *dnmt3a* were significantly increased at key time points in the fast muscle of individuals exposed to continuous light [89]. In Senegalese sole (*Solea senegalensis*), embryos exposed to different temperatures displayed differences in growth performance and muscle phenotype, and DNA methylation was found to be involved in the phenotypic plasticity [82]. In Atlantic cod (*Gadus morhua*), several *dnmts* (encode DNA methyltransferases) were found to be associated with increased growth performance promoted by photoperiod. It appears that compared to warming, illumination is not in the first rank of climate change-induced environmental stresses, but is still a potential threat affecting aquatic animal physiology. The average daily hours of illumination for some aquatic species will certainly be altered as a result of the poleward shift in global distribution.

Myogenesis is inextricably linked to the growth of body size. A conserved epigenetic mechanism regulating muscle growth in fish has recently been identified. Myogenin (Myog), a highly conserved regulator of myogenesis in vertebrates, is highly sensitive to temperature. Low temperature significantly increases the methylation of the CpG island in the *myog* promoter in skeletal muscle, and *myog* transcription was downregulated, accompanied by upregulation of *dnmt1* and *dnmt3b* expression. Thus, the Senegalese sole produces larger muscle fibers at higher temperatures [82]. A negative correlation between *myog* expression and DNA methylation levels was also found in Atlantic salmon (*Salmo salar*), and warmer embryo incubation temperatures resulted in larvae with almost twice as high body weight and more white fibers when compared to those cultured in cold environments [200]. Johnston et al. [201] found that zebrafish incubated at 26 °C produced the longest fast muscle fibers and the largest number of final fibers when compared to fish incubated at lower (22 °C) or higher (31 °C) temperatures. Besides, miRNAs were found to be involved in regulating muscle phenotypic transitions in this species.

A few studies have also traced environment-dependent epigenetic influences on the growth of aquatic species. For example, two populations of migratory wild Atlantic salmon spawned in the colder lowland (Sheeoch) tributary and the warmer upland (Baddoch) tributary of the River Dee in Aberdeenshire, Scotland shows different patterns of early myogenesis [202]. High-temperature stress significantly inhibits the growth of *Didemnum vexillum*, a marine colonial tunicate, and is accompanied by significant changes in global DNA methylation [203]. Environmentally sensitive coral (*Pocillopora damicornis*) exhibits reduced growth, increased DNA methylation levels, and segregated metabolome profiles after six weeks of simulated ocean acidification [204]. Unfortunately, the specific mechanisms reported here have not been explored in more detail.

There is currently very little research linking aquatic animal growth, size, and morphology to environment-dependent epigenetics, and further work is urgently needed.

5.2.4 Nutrition

As global climate change intensifies in the next century, the advancement of phenology will likely further desynchronize the interactions of species [205], including predator–prey interaction. Importantly, as primary producers and food for organisms at higher trophic levels, phytoplankton respond quickly to changes in environmental conditions and therefore will have a non-negligible influence on the water ecosystem. Therefore, climate change will likely influence the nutrient availability and composition of consumers of water ecosystems.

Nutrients are thought to be epigenetic regulators that can influence fish physiology by altering or rescuing epigenetic states. Canada et al. [206] demonstrated that different degrees of hydrolysis and complexity of dietary protein-induced different expression patterns of genes encoding myogenic regulatory factors and influenced muscle growth dynamics in juvenile Senegalese sole. Moreover, gene transcript levels of DNA methyltransferases were found to be upregulated in the high degree of the hydrolysis protein group. In addition, it is well known that parents' dietary and nutritional patterns, such as starvation or over-nutrition, can impact the normal physiological activities of the offspring by epigenetic modification, which may be transmitted across multiple generations [207]. This phenomenon has been fully explored in mammals. For example, pregnant female rats on a protein-restricted diet produce offspring that exhibit high levels of cholesterol because of reduced expression of *cholesterol 7α-hydroxylase* (*cyp7a1*) [208]. Similarly, low-protein diets of fathers lead to changes in promoter methylation status and expression levels of genes involved in fat and cholesterol biosynthesis in the offspring [209].

In fish, few studies have explored nutrient-dependent transgenerational epigenetic effects. Reduced levels of B vitamins in the feed consumed by parental zebrafish resulted in high levels of lipids in the liver cells of the offspring [210]. Unfortunately, no phenotype testing across generations was provided. A small number of reports unrelated to nutrient metabolism but associated with stress tolerance have linked parental diet to offspring epigenetic patterns in fish. In brown trout (*Salmo trutta*), increasing the salinity of the parental diet helped the offspring increase adaptability to a high osmotic pressure environment, most likely through salt-induced changes in global DNA methylation patterns [90]. Offspring zebrafish whose parents underwent exposure to a diet containing water accommodated fractions of crude oil for three weeks improved survival in oiled conditions, and they retained similar low levels of DNA methylation in the heart as their parents [211].

In the background of climate change, phenological alterations and species distribution shifts can force many aquatic animals to face the risk of food shortage in a period of high nutritional demand. Starvation due to food shortages will likely impact the physiology of some fish and can subsequently affect the life activities of offspring through transgenerational epigenetic effects. Importantly, the programming of nutritional epigenetics in early life may have a significantly positive influence on the long-term performance of natural populations.

5.2.5 Stress Response and Survival

At present, organisms around the world are becoming extinct at a rate more than 100 times the normal background extinction rate due to a synergy of unusual climate dynamics, extreme atmosphere, and high levels of environmental stress [212–214]. High temperatures will threaten the sustainability of organisms in numerous ways, but most crucially, the key to species' survival is whether they can evolve the adaptability to break through the thermal upper limit before reaching the peak of heat tolerance. Therefore, researchers have focused their attention on the physiological tolerances or stress responses of species [215–217]. The most pervasive threat posed by global climate change, particularly rising temperatures, is probably the sublethal effect, in which the diminished use of oxygen transport is considered to be the most critical factor [198, 215, 218, 219]. Tropical coral reef fishes lose almost 50% of their aerobic range when the water temperature increases by 2 °C above the current average summer temperature [220]. When sockeye salmon (*O. nerka*) are exposed to temperatures above their optimal swimming performance, their skeletal muscles increase their extraction of blood oxygen, thereby reducing venous oxygen levels and compromising the internal oxygen supply to tissues [221]. When elevated metabolic oxygen demand and reduced oxygen transport potential of fish meet with low dissolved oxygen in the aquatic environment, undesirable physiological performance is provoked, and local extinction of aquatic species is highly likely to occur under extreme conditions.

It is also interesting to note that in fish (also in annelids, arthropods, amphibians, reptiles, and mammals), females generally exhibit a longer lifespan and/or greater stress resistance [222–227]. For example, female mosquitofish (*Gambusia affinis*) and guppy (*Poecilia reticulata*) adapt better to high temperatures and low oxygen conditions than males [228–230]. Such male-biased sex-specific mortality may interact with stress-induced masculinization and discounted reproductive performance under environmental stress [161], and threaten the sustainability of sensitive species.

Much evidence suggests that epigenetic mechanisms are important players in the molecular mechanisms of the stress response [231]. A well-recognized and typical feature of heat-adaptation equilibrium is the establishment of enhanced heat shock protein (HSP) reserves [232], in which epigenetic mechanism inevitably plays a role. Several studies found that short-term heat exposure induced a decrease in *miR-1* and *miR-206* and promoted the synthesis of HSP70 [233–235]. In addition, using mammalian HSP72/HSP90 as a vector, Horowitz et al. [231] developed a model of heat adaptation for epigenetic mechanisms to enhance HSP storage, i.e., short-term heat stress induces histone H3ser10 phosphorylation and histone H4 acetylation on the promoters of *HSP72* and *HSP90* [236], and these processes remodeled chromatin to euchromatin (open chromatin), enabling the transcription factor HSF1 to bind to specific loci and initiate HSP translation. This model has been validated in aquatic invertebrates. Exposed to nonlethal heat shock, *Artemia* exhibits increased levels of HSP70, accompanied by an increased level of global DNA methylation and histone H3 and H4 acetylation [237]. However, the model does not appear to apply to short-term heat stress in fish [238]. Adult coral reef fish (*Acanthochromis polyacanthus*) were exposed to simulated future ocean temperatures and results found that, among all key genes involved in aerobically competent thermal adaptation, genes associated with metabolic, immune, and stress responses were upregulated across generations, which conferred on second-generation fish a better ability to cope with elevated temperatures. However, the expression of HSP genes was not significantly altered [238]. Therefore, the generality of the HSP model in fish needs to be validated by further studies.

Indeed, response to environmental stress should be the result of coordinated interactions of multiple molecular pathways rather than a single phenotypic shift. Similarly, in a heat stress experiment simulating climate change in a coral reef fish, 2467 differentially methylated DNA regions (DMRs) and 1870 associated genes were identified [232]. Among these genes, 193 were significantly correlated to the transgenerationally acclimating phenotypic trait, aerobic scope, with functions in insulin response, energy homeostasis, mitochondrial activity, oxygen consumption, and angiogenesis [239]. These genes may play a key role in restoring transgenerational performance in fish exposed to the elevated temperatures associated with climate change.

There is still a paucity of research into epigenetic mechanisms of stress response in fish, but epigenetics certainly has a prominent contribution to the response of fish to global climate change. This indispensable adaptive trait can amplify tolerance intervals to the same environmental stresses in several generations [231], which has broad implications not only ecologically, but also evolutionarily.

5.3 Conclusions and Future Perspectives

Here we have provided some examples of the impact of different environmental cues on various aspects of the biology of fish in which epigenetic mechanisms play a role. Furthermore, far from being structurally and functionally isolated, different types of epigenetic modifications act in concert to shape the epigenome and thus bring about phenotypes with the interactions of the genome. Based on what has been presented in this chapter, it seems clear that epigenetic inheritance has a significant contribution to fish adaptation to climate change and in assisting offspring to cope with environmental stresses across generations.

There are several knowledge gaps in the connection between epigenetics and fish response to climate change. Firstly, to date, epigenetic connections among different responses to various environmental stressors are unclear. Unlike in physiology or molecular biology, where the networks have been established, epigenetic networks of organisms' responses to environmental stressors are far from being decoded. Secondly, there is still a challenge to distinguish causal from consequential changes in epigenetic changes, and the essential ecological and evolutionary functions of these epigenetic changes. Thirdly, epigenetic markers that can represent a specific, measurable phenotype urgently need to be explored. Last but not least, how many generations of intergenerational/transgenerational epigenetic inheritance can be passed on from parents to offspring is unclear.

Acknowledgments

This research was supported by the National Natural Science Foundation of China (32273122), Fundamental Research Funds for the Central Universities (2662021SCPY001), and the Natural Science Foundation of Hubei Province (2018CFB215) awarded to ZG Shen. We thank Francesc Piferrer for his critical comments on this chapter.

References

1 Barbarossa, V., Bosmans, J., Wanders, N. et al. (2021). Threats of global warming to the world's freshwater fishes. *Nature Communications* **12** (1): 1701.

2 Bellard, C., Bertelsmeier, C., Leadley, P. et al. (2012). Impacts of climate change on the future of biodiversity. *Ecology Letters* **15** (4): 365–377.

3 Dawson, T.P., Jackson, S.T., House, J.I. et al. (2011). Beyond predictions: biodiversity conservation in a changing climate. *Science* **332** (6025): 53–58.

4 Hof, C., Araújo, M.B., Jetz, W., and Rahbek, C. (2011). Additive threats from pathogens, climate and land-use change for global amphibian diversity. *Nature* **480** (7378): 516–519.

5 Jetz, W., Wilcove, D.S., and Dobson, A.P. (2007). Projected impacts of climate and land-use change on the global diversity of birds. *PLoS Biology* **5** (6): e157.

6 Pereira, H.M., Leadley, P.W., Proença, V. et al. (2010). Scenarios for global biodiversity in the 21st century. *Science* **330** (6010): 1496–1501.

7 Thomas, C.D., Cameron, A., Green, R.E. et al. (2004). Extinction risk from climate change. *Nature* **427** (6970): 145–148.

8 Petit, J.R., Jouzel, J., Raynaud, D. et al. (1999). Climate and atmospheric history of the past 420,000 years from the Vostok ice core, Antarctica. *Nature* **399** (6735): 429–436.

9 Sigman, D.M., Hain, M.P., and Haug, G.H. (2010). The polar ocean and glacial cycles in atmospheric CO(2) concentration. *Nature* **466** (7302): 47–55.

10 Kohfeld, K.E., Le Quéré, C., Harrison, S.P., and Anderson, R.F. (2005). Role of marine biology in glacial–interglacial CO_2 cycles. *Science* **308** (5718): 74–78.

11 Sigman, D.M. and Boyle, E.A. (2000). Glacial/interglacial variations in atmospheric carbon dioxide. *Nature* **407** (6806): 859–869.

12 González-Tokman, D., Córdoba-Aguilar, A., Dáttilo, W. et al. (2020). Insect responses to heat: physiological mechanisms, evolution and ecological implications in a warming world. *Biological Reviews of the Cambridge Philosophical Society* **95** (3): 802–821.

13 Lockley, E.C. and Eizaguirre, C. (2021). Effects of global warming on species with temperature-dependent sex determination: bridging the gap between empirical research and management. *Evolutionary Applications* **14** (10): 2361–2377.

14 Morgan, R., Finnøen, M.H., Jensen, H. et al. (2020). Low potential for evolutionary rescue from climate change in a tropical fish. *Proceedings of the National Academy of Sciences of the United States of America* **117** (52): 33365–33372.

15 Pankhurst, N.W., Munday, P.L., Pankhurst, N.W., and Munday, P.L. (2011). Effects of climate change on fish reproduction and early life history stages. *Marine and Freshwater Research* **62** (9): 1015–1026.

16 Thackeray, S.J., Sparks, T.H., Frederiksen, M. et al. (2010). Trophic level asynchrony in rates of phenological change for marine, freshwater and terrestrial environments. *Global Change Biology* **16** (12): 3304–3313.

17 Cahill, A.E., Aiello-Lammens, M.E., Fisher-Reid, M.C. et al. (2013). How does climate change cause extinction? *Proceedings of the Biological Sciences* **280** (1750): 20121890.

18 Durance, I. and Ormerod, S.J. (2010). Evidence for the role of climate in the local extinction of a cool-water triclad. *Journal of the North American Benthological Society* **29** (4): 1367–1378.

19 Epps, C.W., Mccullough, D.R., Wehausen, J.D. et al. (2004). Effects of climate change on population persistence of desert-dwelling mountain sheep in California. *Conservation Biology* **18** (1): 102–113.

20 McLaughlin, J.F., Hellmann, J.J., Boggs, C.L., and Ehrlich, P.R. (2002). Climate change hastens population extinctions. *Proceedings of the National Academy of Sciences of the United States of America* **99** (9): 6070–6074.

21 Melamed, Y., Kislev, M., Weiss, E., and Simchoni, O. (2011). Extinction of water plants in the Hula Valley: evidence for climate change. *Journal of Human Evolution* **60** (4): 320–327.

22 Munday, P.L. (2004). Habitat loss, resource specialization, and extinction on coral reefs. *Global Change Biology* **10** (10): 1642–1647.

23 Sinervo, B., Méndez-de-la-Cruz, F., Miles, D.B. et al. (2010). Erosion of lizard diversity by climate change and altered thermal niches. *Science* **328** (5980): 894–899.

24 Trape, S. (2009). Impact of climate change on the relict tropical fish fauna of Central Sahara: threat for the survival of Adrar mountains fishes, Mauritania. *PLoS One* **4** (2): e4400.

25 Coumou, D. and Rahmstorf, S. (2012). A decade of weather extremes. *Nature Climate Change* **2** (7): 491–496.

26 Jentsch, A. and Beierkuhnlein, C. (2008). Research frontiers in climate change: effects of extreme meteorological events on ecosystems. *Comptes Rendus Geoscience* **340** (9): 621–628.

27 Geffroy, B. and Wedekind, C. (2020). Effects of global warming on sex ratios in fishes. *Journal of Fish Biology* **97** (3): 596–606.

28 Grossen, C., Neuenschwander, S., and Perrin, N. (2011). Temperature-dependent turnovers in sex-determination mechanisms: a quantitative model. *Evolution* **65** (1): 64–78.

29 Kanaiwa, M. and Harada, Y. (2008). Collapse of one-locus two-allele sex determining system by releasing sex-reversed hatchery fish. *Reviews in Fisheries Science* **16** (1–3): 296–302.

30 Nivelle, R., Gennotte, V., Kalala, E.J.K. et al. (2019). Temperature preference of Nile tilapia (*Oreochromis niloticus*) juveniles induces spontaneous sex reversal. *PLoS One* **14** (2): e0212504.

31 Santi, S., Rougeot, C., Toguyeni, A. et al. (2017). Temperature preference and sex differentiation in African catfish, *Clarias gariepinus*. *Journal of Experimental Zoology Part A: Ecological and Integratice Physiology* **327** (1): 28–37.

32 Siegenthaler, U., Stocker, T.F., Monnin, E. et al. (2005). Stable carbon cycle–climate relationship during the late Pleistocene. *Science* **310**: 1313–1317.

33 IPCC (2021). Summary for policymakers. In: *Climate Change 2021: The Physical Science Basis. Contribution of Working Group I to the Sixth Assessment Report of the Intergovernmental Panel on Climate Change* (ed. V. Masson-Delmotte, P. Zhai, A. Pirani, et al.), 5–23. Cambridge University Press, SPM.

34 Gattuso, J.-P. and Hansson, L. (ed.) (2011). Ocean acidification: background and history. In: *Ocean Acidification*, 1–20. Oxford University Press.

35 Noor, N.M. and Das, S.K. (2019). Effects of elevated carbon dioxide on marine ecosystem and associated fishes. *Thalassas* **35** (2): 421–429.

36 Shen, Z.G. and Wang, H.P. (2018). Environmental sex determination and sex differentiation in teleosts – how sex is established. In: *Sex Control in Aquaculture*, 85–115. Wiley.

37 Talmage, S.C. and Gobler, C.J. (2009). The effects of elevated carbon dioxide concentrations on the metamorphosis, size, and survival of larval hard clams (*Mercenaria mercenaria*), bay scallops (*Argopecten irradians*), and Eastern oysters (*Crassostrea virginica*). *Limnology and Oceanography* **54** (6): 2072–2080.

38 Waldbusser, G.G. and Salisbury, J.E. (2014). Ocean acidification in the coastal zone from an organism's perspective: multiple system parameters, frequency domains, and habitats. *Annual Review of Marine Science* **6** (1): 221–247.

39 Doney, S.C., Ruckelshaus, M., Emmett Duffy, J. et al. (2012). Climate change impacts on marine ecosystems. *Annual Review of Marine Science* **4** (1): 11–37.

40 Breitburg, D., Levin, L.A., Oschlies, A. et al. (2018). Declining oxygen in the global ocean and coastal waters. *Science* **359** (6371): eaam7240.

41 Deutsch, C., Brix, H., Ito, T. et al. (2011). Climate-forced variability of ocean hypoxia. *Science* **333** (6040): 336–339.

42 Helm, K.P., Bindoff, N.L., and Church, J.A. (2011). Observed decreases in oxygen content of the global ocean. *Geophysical Research Letters* **38** (23): 23602.

43 Schmidtko, S., Stramma, L., and Visbeck, M. (2017). Decline in global oceanic oxygen content during the past five decades. *Nature* **542** (7641): 335–339.

44 Bell, G. and Eggleston, D. (2005). Species-specific avoidance responses by blue crabs and fish to chronic and episodic hypoxia. *Marine Biology* **146** (4): 761–770.

45 Goodman, L.R. and Campbell, J.G. (2007). Lethal levels of hypoxia for gulf coast estuarine animals. *Marine Biology* **152** (1): 37–42.

46 Wu, R.S.S., Zhou, B.S., Randall, D.J. et al. (2003). Aquatic hypoxia is an endocrine disruptor and impairs fish reproduction. *Environmental Science & Technology* **37** (6): 1137–1141.

47 Shang, E.H.H. and Wu, R.S.S. (2004). Aquatic hypoxia is a teratogen and affects fish embryonic development. *Environmental Science & Technology* **38** (18): 4763–4767.

48 Tan, W. and Thomas, P. (2011). Exposure to chronic hypoxia in the Gulf of Mexico significantly decreases the reproductive potential of male Atlantic Croaker. *Biology of Reproduction* **85** (Suppl_1): 593–599.

49 Cleland, E.E., Chuine, I., Menzel, A. et al. (2007). Shifting plant phenology in response to global change. *Trends in Ecology & Evolution* **22** (7): 357–365.

50 Asch, R.G. (2015). Climate change and decadal shifts in the phenology of larval fishes in the California Current ecosystem. *PNAS* **112** (30): E4065–E4074.

51 Fitchett, J.M., Grab, S.W., and Portwig, H. (2019). Progressive delays in the timing of sardine migration in the Southwest Indian Ocean. *South African Journal of Science* **115** (7–8): 1–6.

52 Asch, R.G., Stock, C.A., and Sarmiento, J.L. (2019). Climate change impacts on mismatches between phytoplankton blooms and fish spawning phenology. *Global Change Biology* **25** (8): 2544–2559.
53 Krabbenhoft, T.J., Platania, S.P., and Turner, T.F. (2014). Interannual variation in reproductive phenology in a riverine fish assemblage: implications for predicting the effects of climate change and altered flow regimes. *Freshwater Biology* **59** (8): 1744–1754.
54 Peer, A.C. and Miller, T.J. (2014). Climate change, migration phenology, and fisheries management interact with unanticipated consequences. *North American Journal of Fisheries Management* **34** (1): 94–110.
55 Manhard, C.V., Joyce, J.E., and Gharrett, A.J. (2017). Evolution of phenology in a salmonid population: a potential adaptive response to climate change. *Canadian Journal of Fisheries and Aquatic Sciences* **74** (10): 1519–1527.
56 Kanamori, Y., Takasuka, A., Nishijima, S., and Okamura, H. (2019). Climate change shifts the spawning ground northward and extends the spawning period of chub mackerel in the Western North Pacific. *Marine Ecology Progress Series* **624**: 155–166.
57 Parmesan, C. and Yohe, G. (2003). A globally coherent fingerprint of climate change impacts across natural systems. *Nature* **421** (6918): 37–42.
58 Root, T.L., Price, J.T., Hall, K.R. et al. (2003). Fingerprints of global warming on wild animals and plants. *Nature* **421** (6918): 57–60.
59 Both, C., Van Asch, M., Bijlsma, R.G. et al. (2009). Climate change and unequal phenological changes across four trophic levels: constraints or adaptations? *Journal of Animal Ecology* **78** (1): 73–83.
60 Visser, M.E., Holleman, L.J.M., and Gienapp, P. (2006). Shifts in caterpillar biomass phenology due to climate change and its impact on the breeding biology of an insectivorous bird. *Oecologia* **147** (1): 164–172.
61 Burthe, S., Daunt, F., Butler, A. et al. (2012). Phenological trends and trophic mismatch across multiple levels of a North Sea pelagic food web. *Marine Ecology Progress Series* **454**: 119–133.
62 Jonsson, T. and Setzer, M. (2015). A freshwater predator hit twice by the effects of warming across trophic levels. *Nature Communications* **6** (1): 5992.
63 Thackeray, S.J., Henrys, P.A., Hemming, D. et al. (2016). Phenological sensitivity to climate across taxa and trophic levels. *Nature* **535** (7611): 241–245.
64 Durant, J.M., Hjermann, D.Ø., Ottersen, G., and Stenseth, N.C. (2007). Climate and the match or mismatch between predator requirements and resource availability. *Climate Research* **33** (3): 271–283.
65 Miller-Rushing, A.J., Høye, T.T., Inouye, D.W., and Post, E. (2010). The effects of phenological mismatches on demography. *Philosophical Transactions of the Royal Society, B: Biological Sciences* **365** (1555): 3177–3186.
66 Edwards, M. and Richardson, A.J. (2004). Impact of climate change on marine pelagic phenology and trophic mismatch. *Nature* **430** (7002): 881–884.
67 Parmesan, C., Gaines, S., Gonzalez, L. et al. (2005). Empirical perspectives on species borders: from traditional biogeography to global change. *Oikos* **108** (1): 58–75.
68 Guisan, A. and Zimmermann, N.E. (2000). Predictive habitat distribution models in ecology. *Ecological Modelling* **135** (2): 147–186.
69 Sunday, J.M., Bates, A.E., and Dulvy, N.K. (2012). Thermal tolerance and the global redistribution of animals. *Nature Climate Change* **2** (9): 686–690.
70 Hoegh-Guldberg, O. and Bruno, J.F. (2010). The impact of climate change on the world's marine ecosystems. *Science* **328** (5985): 1523–1528.
71 Walther, G.-R., Post, E., Convey, P. et al. (2002). Ecological responses to recent climate change. *Nature* **416** (6879): 389–395.
72 Dulvy, N.K., Rogers, S.I., Jennings, S. et al. (2008). Climate change and deepening of the North Sea fish assemblage: a biotic indicator of warming seas. *Journal of Applied Ecology* **45** (4): 1029–1039.
73 Nye, J.A., Link, J.S., Hare, J.A., and Overholtz, W.J. (2009). Changing spatial distribution of fish stocks in relation to climate and population size on the Northeast United States continental shelf. *Marine Ecology Progress Series* **393**: 111–129.
74 Stefansdottir, L., Solmundsson, J., Marteinsdottir, G. et al. (2010). Groundfish species diversity and assemblage structure in Icelandic waters during recent years of warming. *Fisheries Oceanography* **19** (1): 42–62.
75 Perry, A.L., Low, P.J., Ellis, J.R., and Reynolds, J.D. (2005). Climate change and distribution shifts in marine fishes. *Science* **308** (5730): 1912–1915.
76 Feary, D.A., Pratchett, M.S., Emslie, M.J. et al. (2014). Latitudinal shifts in coral reef fishes: why some species do and others do not shift. *Fish and Fisheries* **15** (4): 593–615.
77 Eirin-Lopez, J.M. and Putnam, H.M. (2019). Marine environmental epigenetics. *Annual Review of Marine Science* **11** (1): 335–368.

78 Varriale, A. and Bernardi, G. (2006). DNA methylation and body temperature in fishes. *Gene* **385**: 111–121.

79 Navarro-Martín, L., Viñas, J., Ribas, L. et al. (2011). DNA methylation of the gonadal aromatase (*cyp19a*) promoter is involved in temperature-dependent sex ratio shifts in the European sea bass. *PLoS Genetics* **7** (12): e1002447.

80 Shao, C., Li, Q., Chen, S. et al. (2014). Epigenetic modification and inheritance in sexual reversal of fish. *Genome Research* **24** (4): 604–615.

81 Chen, S., Zhang, G., Shao, C. et al. (2014). Whole-genome sequence of a flatfish provides insights into ZW sex chromosome evolution and adaptation to a benthic lifestyle. *Nature Genetics* **46** (3): 253–260.

82 Campos, C., Valente, L.M.P., Conceição, L.E.C. et al. (2013). Temperature affects methylation of the myogenin putative promoter, its expression and muscle cellularity in *Senegalese sole* larvae. *Epigenetics* **8** (4): 389–397.

83 Valdivieso, A., Ribas, L., Monleón-Getino, A. et al. (2020). Exposure of zebrafish to elevated temperature induces sex ratio shifts and alterations in the testicular epigenome of unexposed offspring. *Environmental Research* **186**: 109601.

84 Venney, C.J., Wellband, K.W., Normandeau, E. et al. (2022). Thermal regime during parental sexual maturation, but not during offspring rearing, modulates DNA methylation in brook charr (*Salvelinus fontinalis*). *Proceedings of the Royal Society B: Biological Sciences* **289** (1974): 20220670.

85 Sun, L.-X., Wang, Y.-Y., Zhao, Y. et al. (2016). Global DNA methylation changes in Nile tilapia gonads during high temperature-induced masculinization. *PLoS One* **11** (8): e0158483.

86 Wang, Y.Y., Sun, L.X., Zhu, J.J. et al. (2017). Epigenetic control of *cyp19a1a* expression is critical for high temperature induced Nile tilapia masculinization. *Journal of Thermal Biology* **69**: 76–84.

87 Fan, Z., Zou, Y., Jiao, S. et al. (2017). Significant association of cyp19a promoter methylation with environmental factors and gonadal differentiation in olive flounder *Paralichthys olivaceus. Comparative Biochemistry and Physiology. Part A, Molecular & Integrative Physiology* **208**: 70–79.

88 Zhou, H., Zhuang, Z.-X., Sun, Y.-Q. et al. (2019). Changes in DNA methylation during epigenetic-associated sex reversal under low temperature in *Takifugu rubripes. PLoS One* **14** (8): e0221641.

89 Giannetto, A., Nagasawa, K., Fasulo, S., and Fernandes, J.M.O. (2013). Influence of photoperiod on expression of DNA (cytosine-5) methyltransferases in Atlantic cod (*Gadus morhua*). *Gene* **519** (2): 222–230.

90 Morán, P., Marco-Rius, F., Megías, M. et al. (2013). Environmental induced methylation changes associated with seawater adaptation in brown trout. *Aquaculture* **392–395**: 77–83.

91 Li, S., He, F., Wen, H. et al. (2017). Low salinity affects cellularity, DNA methylation, and mRNA expression of igf1 in the liver of half smooth tongue sole (*Cynoglossus semilaevis*). *Fish Physiology and Biochemistry* **43** (6): 1587–1602.

92 Wang, S.Y., Lau, K., Lai, K.-P. et al. (2016). Hypoxia causes transgenerational impairments in reproduction of fish. *Nature Communications* **7** (1): 12114.

93 Wang, H.-P. and Shen, Z.-G. (2018). Sex control in aquaculture. In: *Sex Control in Aquaculture*, 1–34. Wiley.

94 Shen, Z.-G. and Wang, H.-P. (2014). Molecular players involved in temperature-dependent sex determination and sex differentiation in teleost fish. *Genetics Selection Evolution* **46** (1): 26.

95 Docker, M.F., William, F., and Beamish, H. (1994). Age, growth, and sex ratio among populations of least brook lamprey, *Lampetra aepyptera*, larvae: an argument for environmental sex determination. *Environmental Biology of Fishes* **41** (1): 191–205.

96 Johnson, N.S., Swink, W.D., and Brenden, T.O. (2017). Field study suggests that sex determination in sea lamprey is directly influenced by larval growth rate. *Proceedings of the Biological Sciences* **284** (1851): 20170262.

97 Qu, X.C. (2018). Sex determination and control in eels. In: *Sex Control in Aquaculture*, 775–792. Wiley.

98 Goto-Kazeto, R., Abe, Y., Masai, K. et al. (2006). Temperature-dependent sex differentiation in goldfish: establishing the temperature-sensitive period and effect of constant and fluctuating water temperatures. *Aquaculture* **254** (1): 617–624.

99 Fujioka, Y. (2002). Effects of hormone treatments and temperature on sex-reversal of Nigorobuna *Carassius carassius grandoculis. Fisheries Science* **68** (4): 889–893.

100 Zheng, Y., Qu, J., Bing, X. et al. (2015). Molecular mechanism and the cause for high ratio of male Pengze crucian carp under high-density culture condition. *Journal of Shanghai Ocean University* **24** (06): 826–833.

101 Li, X.Y., Liu, X.L., Zhu, Y.J. et al. (2018). Origin and transition of sex determination mechanisms in a gynogenetic hexaploid fish. *Heredity* **121** (1): 64–74.

102 Biswas, C., Chakraborty, S., Munilkumar, S. et al. (2021). Effect of high temperature during larval and juvenile stages on masculinization of common carp (*Cyprinus carpio*, L). *Aquaculture* **530**: 735803.

103 Santos, D., Luzio, A., and Coimbra, A.M. (2017). Zebrafish sex differentiation and gonad development: a review on the impact of environmental factors. *Aquatic Toxicology* **191**: 141–163.

104 Fujioka, Y. (2006). Patterns of sex ratio response to water temperature during sex determination in honmoroko *Gnathopogon caerulescens*. *Fisheries Science* **72** (5): 1034–1041.

105 Nan, P., Du, Q.Y., Yan, S.G., and Chang, Z.J. (2005). Effects of temperature on sex differentiation of gonads and the cloning and expression of CYP19a in two species of loaches. *Journal of Fishery Sciences of China* **04**: 407–413.

106 Bohlen, J., Freyhof, J., and Nolte, A. (2008). Sex ratio and body size in *Cobitis elongatoides* and *Sabanejewia balcanica* (Cypriniformes, Cobitidae) from a thermal spring. *Folia Zoologica* **57**: 191–197.

107 Patiño, R., Davis, K.B., Schoore, J.E. et al. (1996). Sex differentiation of channel catfish gonads: normal development and effects of temperature. *Journal of Experimental Zoology* **276** (3): 209–218.

108 Yu, Y., Chen, M., Qi, P.-P. et al. (2021). High temperature-induced masculinization in yellow catfish *Tachysurus fulvidraco*: a potential approach for environmental-friendly mono-sex production. *Aquaculture* **534**: 736263.

109 Cheng, X., Lin, D., and You, Y. (2007). Influence of temperature on sex differentiation of teleost, *Pseudobagrus vachelli*. *Zoological Research* **28**: 73–80.

110 Hostache, G., Pascal, M., and Tessier, C. (1995). Influence de la température d'incubation sur le rapport mâle: femelle chez l'atipa, *Hoplosternum littorale* Hancock (1828). *Canadian Journal of Zoology* **73** (7): 1239–1246.

111 Santi, S., Gennotte, V., Muller, M. et al. (2019). Sex-ratio, early sex steroid profiles and *cyp19a1b*, *dmrt1* and *foxl2* gene expressions upon high temperature treatment of undifferentiated African catfish juveniles (*Clarias gariepinus*). *Aquaculture* **499**: 140–148.

112 García, L.N., Marín, A.F., and Chapman, F.A. (2020). Effects of different color artificial lights on growth, survival, and sex ratio on an experimental population of freshwater ornamental emperor tetra fish *Nematobrycon palmeri*. *Aquaculture, Aquarium, Conservation & Legislation Bioflux* **13** (2): 1048–1054.

113 Azuma, T., Takeda, K., Doi, T. et al. (2004). The influence of temperature on sex determination in sockeye salmon *Oncorhynchus nerka*. *Aquaculture* **234** (1): 461–473.

114 Magerhans, A. and Hörstgen-Schwark, G. (2010). Selection experiments to alter the sex ratio in rainbow trout (*Oncorhynchus mykiss*) by means of temperature treatment. *Aquaculture* **306** (1): 63–67.

115 Wedekind, C., Evanno, G., Székely, T. et al. (2013). Persistent unequal sex ratio in a population of Grayling (Salmonidae) and possible role of temperature increase. *Conservation Biology* **27** (1): 229–234.

116 Kim, S.-S., Kim, D.N.-J., Lee, C.-J. et al. (2020). The potential sex determination genes, *Sox9a* and *Cyp19a*, in walleye Pollock (*Gadus Chalcogrammus*) are influenced by water temperature. *Journal of Marine Science and Engineering* **8** (7): 501.

117 Rubin, D.A. (1985). Effect of ph on sex ratio in cichlids and a Poecilliid (Teleostei). *Copeia* **1**: 233–235.

118 Reddon, A.R. and Hurd, P.L. (2013). Water pH during early development influences sex ratio and male morph in a West African cichlid fish, Pelvicachromis pulcher. *Zoology* **116** (3): 139–143.

119 Heiligenberg, W. (1965). Colour polymorphism in the males of an African cichlid fish. *Proceedings of the Zoological Society of London* **146** (1): 95–97.

120 Baroiller, J.F., D'Cotta, H., Bezault, E. et al. (2009). Tilapia sex determination: where temperature and genetics meet. *Comparative Biochemistry and Physiology. Part A, Molecular & Integrative Physiology* **153** (1): 30–38.

121 Piferrer, F., Blázquez, M., Navarro, L., and González, A. (2005). Genetic, endocrine, and environmental components of sex determination and differentiation in the European sea bass (*Dicentrarchus labrax* L.). *General and Comparative Endocrinology* **142** (1): 102–110.

122 Baroiller, J.F., D'Cotta, H., and Saillant, E. (2009). Environmental effects on fish sex determination and differentiation. *Sexual Development* **3** (2–3): 118–135.

123 Ramee, S.W., Lipscomb, T.N., and DiMaggio, M.A. (2020). Evaluation of the effect of larval stocking density, salinity, and temperature on stress response and sex differentiation in the Dwarf Gourami and Rosy Barb. *Aquaculture Reports* **16**: 100287.

124 Pandian, T.J. and Koteeswaran, R. (1999). Lability of sex differentiation in fish. *Current Science* **76** (4): 580–583.

125 Francis, R.C. (1992). Sexual lability in teleosts: developmental factors. *The Quarterly Review of Biology* **67** (1): 1–18.

126 Shen, Z.G., Wang, H.P., Yao, H. et al. (2016). Sex determination in bluegill sunfish *Lepomis macrochirus*: effect of temperature on sex ratio of four geographic strains. *The Biological Bulletin* **230** (3): 197–208.

127 Rodgers, G.G., Donelson, J.M., and Munday, P.L. (2017). Thermosensitive period of sex determination in the coral-reef damselfish *Acanthochromis polyacanthus* and the implications of projected ocean warming. *Coral Reefs* **36** (1): 131–138.

128 Omoto, N., Koya, Y., Chin, B. et al. (2010). Gonadal sex differentiation and effect of rearing temperature on sex ratio in black rockfish (*Sebastes schlegeli*). *Ichthyological Research* **57** (2): 133–138.

129 Hoon, L.C., Soo, N.O., Kn, Y.I. et al. (2000). Effects of sex steroid hormones and high temperature on sex differentiation in black rockfish, *Sebastes schlegeli*. *Korean Journal of Fisheries and Aquatic Sciences* **33** (5): 373–377.

130 Huynh, T.B., Fairgrieve, W.T., Hayman, E.S. et al. (2019). Inhibition of ovarian development and instances of sex reversal in genotypic female sablefish (*Anoplopoma fimbria*) exposed to elevated water temperature. *General and Comparative Endocrinology* **279**: 88–98.

131 Duffy, T.A., Hice, L.A., and Conover, D.O. (2015). Pattern and scale of geographic variation in environmental sex determination in the Atlantic silverside, Menidia menidia. *Evolution* **69** (8): 2187–2195.

132 Yamahira, K. and Conover, D.O. (2003). Interpopulation variability in temperature-dependent sex determination of the Tidewater silverside *Menidia peninsulae* (Pisces: Atherinidae). *Copeia* **2003** (1): 155–159.

133 Strüssmann, C.A., Calsina Cota, J.C., Phonlor, G. et al. (1996). Temperature effects on sex differentiation of two south American atherinids, *Odontesthes argentinensis* and *Patagonina hatcheri*. *Environmental Biology of Fishes* **47** (2): 143–154.

134 García-Cruz, E.L., Yamamoto, Y., Hattori, R.S. et al. (2020). Crowding stress during the period of sex determination causes masculinization in pejerrey *Odontesthes bonariensis*, a fish with temperature-dependent sex determination. *Comparative Biochemistry and Physiology. Part A, Molecular & Integrative Physiology* **245**: 110701.

135 Strüssmann, C.A., Saito, T., Usui, M. et al. (1997). Thermal thresholds and critical period of thermolabile sex determination in two atherinid fishes, *Odontesthes bonariensis* and *Patagonina hatcheri*. *Journal of Experimental Zoology* **278** (3): 167–177.

136 Corona-Herrera, G.A., Arranz, S.E., Martínez-Palacios, C.A. et al. (2018). Experimental evidence of masculinization by continuous illumination in a temperature sex determination teleost (*Atherinopsidae*) model: is oxidative stress involved? *Journal of Fish Biology* **93** (2): 229–237.

137 Brown, E.E., Baumann, H., and Conover, D.O. (2014). Temperature and photoperiod effects on sex determination in a fish. *Journal of Experimental Marine Biology and Ecology* **461**: 39–43.

138 Miyoshi, K., Hattori, R.S., Strüssmann, C.A. et al. (2020). Phenotypic/genotypic sex mismatches and temperature-dependent sex determination in a wild population of an Old World atherinid, the cobaltcap silverside *Hypoatherina tsurugae*. *Molecular Ecology* **29** (13): 2349–2358.

139 Zhou, H., Zhuang, Z., Zhang, R. et al. (2019). Temperature-control-induced masculinization in tiger puffer *Takifugu rubripes*. *Journal of Oceanology and Limnology* **37** (3): 1125–1135.

140 Lee, K.H., Yamaguchi, A., Rashid, H. et al. (2009). Germ cell degeneration in high-temperature treated pufferfish, *Takifugu rubripes*. *Sexual Development* **3** (4): 225–232.

141 Devlin, R.H. and Nagahama, Y. (2002). Sex determination and sex differentiation in fish: an overview of genetic, physiological, and environmental influences. *Aquaculture* **208** (3): 191–364.

142 Barón, S.B., Bückle, R.F., and Espina, S. (2002). Environmental factors and sexual differentiation in *Poecilia sphenops Valenciennes* (Pisces: Poeciliidae). *Aquaculture Research* **33** (8): 615–619.

143 Karayücel, İ., Ak, O., and Karayücel, S. (2006). Effect of temperature on sex ratio in guppy *Poecilia reticulata* (Peters 1860). *Aquaculture Research* **37** (2): 139–150.

144 Vallowe, H.H. (1957). Sexual differentiation in the teleost fish *Xiphophorus hellerii*, as modified by experimental treatment. *The Biological Bulletin* **112** (3): 422–429.

145 Hayasaka, O., Takeuchi, Y., Shiozaki, K. et al. (2019). Green light irradiation during sex differentiation induces female-to-male sex reversal in the medaka *Oryzias latipes*. *Scientific Reports* **9** (1): 2383.

146 Cheung, C.H.Y., Chiu, J.M.Y., and Wu, R.S.S. (2014). Hypoxia turns genotypic female medaka fish into phenotypic males. *Ecotoxicology* **23** (7): 1260–1269.

147 Hara, S., Furukawa, F., Mukai, K. et al. (2020). Peroxisome proliferator-activated receptor alpha is involved in the temperature-induced sex differentiation of a vertebrate. *Scientific Reports* **10** (1): 11672.

148 Horie, Y., Myosho, T., Sato, T. et al. (2016). Androgen induces gonadal soma-derived factor, Gsdf, in XX gonads correlated to sex-reversal but not Dmrt1 directly, in the teleost fish, northern medaka (*Oryzias sakaizumii*). *Molecular and Cellular Endocrinology* **436**: 141–149.

149 Yang, Y., Liu, Q., Xiao, Y. et al. (2020). Effects of environmental stress (sex steroids and heat) during sex differentiation in Japanese flounder (*Paralichthys olivaceus*): insight from germ cell proliferation and *gsdf-amh-cyp19a1a* expression. *Aquaculture* **515**: 734536.

150 Caruso, C.C., Breton, T.S., and Berlinsky, D.L. (2016). The effects of temperature on ovarian aromatase (*cyp19a1a*) expression and sex differentiation in summer flounder (*Paralichthys dentatus*). *Fish Physiology and Biochemistry* **42** (2): 795–805.

151 Luckenbach, J.A., Godwin, J., Daniels, H.V., and Borski, R.J. (2003). Gonadal differentiation and effects of temperature on sex determination in southern flounder (*Paralichthys lethostigma*). *Aquaculture* **216** (1): 315–327.

152 Goto, R.I., Kayaba, T.A., Adachi, S.H., and Yamauchi, K.O. (2000). Effects of temperature on sex determination in marbled sole *Limanda yokohamae*. *Fisheries Science* **66** (2): 400–402.

153 Fairchild, E.A., Rennels, N., Howell, W.H., and Wells, R.E. (2007). Gonadal development and differentiation in cultured juvenile winter flounder, *Pseudopleuronectes americanus*. *Journal of the World Aquaculture Society* **38** (1): 114–121.

154 Goto, R., Mori, T., Kawamata, K. et al. (1999). Effects of temperature on gonadal sex determination in barfin flounder *Verasper moseri*. *Fisheries Science* **65** (6): 884–887.

155 Tsukiyama, Y. and Soyano, K. (2013). Effects of water temperature on survival and sex differentiation in hatchery-reared spotted halibut *Verasper variegatus*. *Aquaculture Science* **61** (1): 103–111.

156 van Nes, S. and Andersen, Ø. (2006). Temperature effects on sex determination and ontogenetic gene expression of the aromatases *cyp19a* and *cyp19b*, and the estrogen receptors *esr1* and *esr2* in Atlantic halibut (*Hippoglossus hippoglossus*). *Molecular Reproduction and Development* **73** (12): 1481–1490.

157 Taboada, X., Robledo, D., Bouza, C. et al. (2018). Reproduction and sex control in turbot. In: *Sex Control in Aquaculture* (ed. H.-P. Wang, F. Piferrer, S.-L. Chen, and Z.-G. Shen), 565–582. Wiley.

158 Wang, Q., Liu, K., Feng, B. et al. (2019). Transcriptome of gonads from high temperature induced sex reversal during sex determination and differentiation in Chinese tongue sole, Cynoglossus semilaevis. *Frontiers in Genetics* **10**: 1128.

159 Ospina-Álvarez, N. and Piferrer, F. (2008). Temperature-dependent sex determination in fish revisited: prevalence, a single sex ratio response pattern, and possible effects of climate change. *PLoS One* **3** (7): e2837.

160 Wedekind, C. (2018). Population consequences of releasing sex-reversed fish. In: *Sex Control in Aquaculture*, 179–188. Wiley.

161 Yu, Y., Chen, M., Lu, Z.-Y. et al. (2022). High-temperature stress will put the thermo-sensitive teleost yellow catfish (*Tachysurus fulvidraco*) in danger through reducing reproductivity. *Ecotoxicology and Environmental Safety* **239**: 113638.

162 Duarte-Guterman, P., Navarro-Martín, L., and Trudeau, V.L. (2014). Mechanisms of crosstalk between endocrine systems: regulation of sex steroid hormone synthesis and action by thyroid hormones. *General and Comparative Endocrinology* **203**: 69–85.

163 Flood, D.E.K., Fernandino, J.I., and Langlois, V.S. (2013). Thyroid hormones in male reproductive development: evidence for direct crosstalk between the androgen and thyroid hormone axes. *General and Comparative Endocrinology* **192** (2013): 2–14.

164 Davey, A.J.H. and Jellyman, D.J. (2005). Sex determination in freshwater eels and management options for manipulation of sex. *Reviews in Fish Biology and Fisheries* **15** (1–2): 37–52.

165 Kondo, M., Nanda, I., Hornung, U. et al. (2004). Evolutionary origin of the Medaka Y chromosome. *Current Biology* **14** (18): 1664–1669.

166 Quirk, J.G. and Hamilton, J.B. (1973). Number of germ cells in known male and known female genotypes of vertebrate embryos (*Oryzias latipes*). *Science* **180** (4089): 963–964.

167 Ge, C., Ye, J., Weber, C. et al. (2018). The histone demethylase KDM6B regulates temperature-dependent sex determination in a turtle species. *Ence* **360** (6389): 645–648.

168 Ge, C., Ye, J., Zhang, H. et al. (2017). Dmrt1 induces the male pathway in a turtle species with temperature-dependent sex determination. *Development (Cambridge, England)* **144** (12): 2222–2233.

169 Piferrer, F., Anastasiadi, D., Valdivieso, A. et al. (2019). The model of the conserved epigenetic regulation of sex. *Frontiers in Genetics* **10**: 857.

170 Berger, S.L., Kouzarides, T., Shiekhattar, R., and Shilatifard, A. (2009). An operational definition of epigenetics. *Genes & Development* **23** (7): 781–783.

171 Anastasiadi, D., Díaz, N., and Piferrer, F. (2017). Small ocean temperature increases elicit stage-dependent changes in DNA methylation and gene expression in a fish, the European sea bass. *Scientific Reports* **7** (1): 12401.

172 Metzger, D.C.H. and Schulte, P.M. (2017). Persistent and plastic effects of temperature on DNA methylation across the genome of threespine stickleback (*Gasterosteus aculeatus*). *Proceedings of the Royal Society B: Biological Sciences* **284** (1864): 20171667.

173 Han, B., Li, W., Chen, Z. et al. (2016). Variation of DNA methylome of zebrafish cells under cold pressure. *PLoS One* **11** (8): e0160358.

174 Varriale, A. (2014). DNA methylation, epigenetics, and evolution in vertebrates: facts and challenges. *International Journal of Evolutionary Biology* **2014**: 475981.

175 He, K., Cao, X., and Deng, X. (2021). Histone methylation in epigenetic regulation and temperature responses. *Current Opinion in Plant Biology* **61**: 102001.

176 Anastasiadi, D., Vandeputte, M., Sánchez-Baizán, N. et al. (2018). Dynamic epimarks in sex-related genes predict gonad phenotype in the European sea bass, a fish with mixed genetic and environmental sex determination. *Epigenetics* **13** (9): 988–1011.

177 Wen, A.Y., You, F., Sun, P. et al. (2014). CpG methylation of *dmrt1* and *cyp19a* promoters in relation to their sexual dimorphic expression in the Japanese flounder *Paralichthys olivaceus*. *Journal of Fish Biology* **84** (1): 193–205.

178 Jia, Y., Zheng, J., Chi, M. et al. (2019). Molecular identification of dmrt1 and its promoter CpG methylation in correlation with gene expression during gonad development in *Culter alburnus*. *Fish Physiology and Biochemistry* **45** (1): 245–252.

179 Pierron, F., Lorioux, S., Héroin, D. et al. (2021). Transgenerational epigenetic sex determination: environment experienced by female fish affects offspring sex ratio. *Environmental Pollution* **277**: 116864.

180 Liu, J., Liu, X., Jin, C. et al. (2019). Transcriptome profiling insights the feature of sex reversal induced by high temperature in tongue sole *Cynoglossus semilaevis*. *Frontiers in Genetics* **10**: 522.

181 Wang, J., Liu, Y., Jiang, S. et al. (2019). Transcriptomic and epigenomic alterations of Nile tilapia gonads sexually reversed by high temperature. *Aquaculture* **508**: 167–177.

182 Franke, A. and Clemmesen, C. (2011). Effect of ocean acidification on early life stages of Atlantic herring (*Clupea harengus* L.). *Biogeosciences Discussions* **8** (4): 3697–3707.

183 Hattori, R.S., Castañeda-Cortés, D.C., Arias Padilla, L.F. et al. (2020). Activation of stress response axis as a key process in environment-induced sex plasticity in fish. *Cellular and Molecular Life Sciences* **77** (21): 4223–4236.

184 Nagahama, Y., Chakraborty, T., Paul-Prasanth, B. et al. (2021). Sex determination, gonadal sex differentiation, and plasticity in vertebrate species. *Physiological Reviews* **101** (3): 1237–1308.

185 Pankhurst, N.W. and Porter, M.J.R. (2003). Cold and dark or warm and light: variations on the theme of environmental control of reproduction. *Fish Physiology and Biochemistry* **28** (1): 385–389.

186 Servili, A., Canario, A.V.M., Mouchel, O., and Muñoz-Cueto, J.A. (2020). Climate change impacts on fish reproduction are mediated at multiple levels of the brain–pituitary–gonad axis. *General and Comparative Endocrinology* **291**: 113439.

187 Miranda, L.A., Chalde, T., Elisio, M., and Strüssmann, C.A. (2013). Effects of global warming on fish reproductive endocrine axis, with special emphasis in pejerrey *Odontesthes bonariensis*. *General and Comparative Endocrinology* **192**: 45–54.

188 Woodling, J.D., Lopez, E.M., Maldonado, T.A. et al. (2006). Intersex and other reproductive disruption of fish in wastewater effluent dominated Colorado streams. *Comparative Biochemistry and Physiology Part C: Toxicology & Pharmacology* **144** (1): 10–15.

189 Tetreault, G.R., Bennett, C.J., Shires, K. et al. (2011). Intersex and reproductive impairment of wild fish exposed to multiple municipal wastewater discharges. *Aquatic Toxicology* **104** (3): 278–290.

190 Rozenfeld, C., García-Carpintero, V., Pérez, L. et al. (2019). Cold seawater induces early sexual developmental stages in the BPG axis of European eel males. *BMC Genomics* **20** (1): 597.

191 Lai, K.P., Li, J.-W., Tse, A.C.-K. et al. (2016). Hypoxia alters steroidogenesis in female marine medaka through miRNAs regulation. *Aquatic Toxicology* **172**: 1–8.

192 Tse, A.C.-K., Li, J.-W., Wang, S.Y. et al. (2016). Hypoxia alters testicular functions of marine medaka through microRNAs regulation. *Aquatic Toxicology* **180**: 266–273.

193 Lai, K.P., Wang, S.Y., Li, J.W. et al. (2019). Hypoxia causes transgenerational impairment of ovarian development and hatching success in fish. *Environmental Science & Technology* **53** (7): 3917–3928.

194 Diaz, R.J. and Rosenberg, R. (2008). Spreading dead zones and consequences for marine ecosystems. *Science* **321** (5891): 926–929.

195 Daufresne, M., Lengfellner, K., and Sommer, U. (2009). Global warming benefits the small in aquatic ecosystems. *PNAS* **106** (31): 12788–12793.

196 Gardner, J.L., Peters, A., Kearney, M.R. et al. (2011). Declining body size: a third universal response to warming? *Trends in Ecology & Evolution* **26** (6): 285–291.

197 Lighten, J., Incarnato, D., Ward, B.J. et al. Adaptive phenotypic response to climate enabled by epigenetics in a K-strategy species, the fish Leucoraja ocellata (Rajidae). *Royal Society Open Science* **3** (10): 160299.

198 Pörtner, H.-O. (2010). Oxygen- and capacity-limitation of thermal tolerance: a matrix for integrating climate-related stressor effects in marine ecosystems. *Journal of Experimental Biology* **213** (6): 881–893.

199 Cheung, W.W.L., Sarmiento, J.L., Dunne, J. et al. (2013). Shrinking of fishes exacerbates impacts of global ocean changes on marine ecosystems. *Nature Climate Change* **3** (3): 254–258.

200 Burgerhout, E., Mommens, M., Johnsen, H. et al. (2017). Genetic background and embryonic temperature affect DNA methylation and expression of myogenin and muscle development in Atlantic salmon (*Salmo salar*). *PLoS One* **12**: e0179918.

201 Johnston, I.A., Lee, H.-T., Macqueen, D.J. et al. (2009). Embryonic temperature affects muscle fibre recruitment in adult zebrafish: genome-wide changes in gene and microRNA expression associated with the transition from hyperplastic to hypertrophic growth phenotypes. *Journal of Experimental Biology* **212** (12): 1781–1793.

202 Johnston, I.A., McLay, H.A., Abercromby, M., and Robins, D. (2000). Phenotypic plasticity of early myogenesis and satellite cell numbers in Atlantic salmon spawning in upland and lowland tributaries of a river system. *Journal of Experimental Biology* **203** (17): 2539–2552.

203 Hawes, N.A., Tremblay, L.A., Pochon, X. et al. (2018). Effects of temperature and salinity stress on DNA methylation in a highly invasive marine invertebrate, the colonial ascidian *Didemnum vexillum*. *PeerJ* **6**: e5003.

204 Putnam, H.M., Davidson, J.M., and Gates, R.D. (2016). Ocean acidification influences host DNA methylation and phenotypic plasticity in environmentally susceptible corals. *Evolutionary Applications* **9** (9): 1165–1178.

205 Cohen, J.M., Lajeunesse, M.J., and Rohr, J.R. (2018). A global synthesis of animal phenological responses to climate change. *Nature Climate Change* **8**: 224–228.

206 Canada, P., Engrola, S., Mira, S. et al. (2018). Larval dietary protein complexity affects the regulation of muscle growth and the expression of DNA methyltransferases in Senegalese sole. *Aquaculture* **491**: 28–38.

207 Susiarjo, M. and Bartolomei, M.S. (2014). Epigenetics. You are what you eat, but what about your DNA? *Science* **345** (6198): 733–734.

208 Sohi, G., Marchand, K., Revesz, A. et al. (2011). Maternal protein restriction elevates cholesterol in adult rat offspring due to repressive changes in histone modifications at the *cholesterol 7α-hydroxylase* promoter. *Molecular Endocrinology* **25** (5): 785–798.

209 Carone, B.R., Fauquier, L., Habib, N. et al. (2010). Paternally induced transgenerational environmental reprogramming of metabolic gene expression in mammals. *Cell* **143** (7): 1084–1096.

210 Labbé, C., Robles, V., and Herraez, M.P. (2017). Epigenetics in fish gametes and early embryo. *Aquaculture* **472**: 93–106.

211 Bautista, N.M., Crespel, A., Crossley, J. et al. (2020). Parental transgenerational epigenetic inheritance related to dietary crude oil exposure in *Danio rerio*. *Journal of Experimental Biology* **223** (16): jeb222224.

212 De Vos, J.M., Joppa, L.N., Gittleman, J.L. et al. (2015). Estimating the normal background rate of species extinction. *Conservation Biology* **29** (2): 452–462.

213 Pimm, S.L., Russell, G.J., Gittleman, J.L., and Brooks, T.M. (1995). The future of biodiversity. *Science* **269** (5222): 347–350.

214 Barnosky, A.D., Matzke, N., Tomiya, S. et al. (2011). Has the Earth's sixth mass extinction already arrived? *Nature* **471** (7336): 51–57.

215 Portner, H.O. and Farrell, A.P. (2008). Physiology and climate change. *Science* **322** (5902): 690–692.

216 Somero, G.N. (2011). Comparative physiology: a "crystal ball" for predicting consequences of global change. *American Journal of Physiology - Regulatory, Integrative and Comparative Physiology* **301** (1): R1–R14.

217 Huey, R.B., Deutsch, C.A., Tewksbury, J.J. et al. (2009). Why tropical forest lizards are vulnerable to climate warming. *Proceedings of the Biological Sciences* **276** (1664): 1939–1948.

218 Pörtner, H.O. and Knust, R. (2007). Climate change affects marine fishes through the oxygen limitation of thermal tolerance. *Science* **315** (5808): 95–97.

219 Pörtner, H.O., Peck, L., and Somero, G. (2007). Thermal limits and adaptation in marine Antarctic ectotherms: an integrative view. *Philosophical Transactions of the Royal Society, B: Biological Sciences* **362** (1488): 2233–2258.

220 Nilsson, G.E., Crawley, N., Lunde, I.G., and Munday, P.L. (2010). Elevated temperature reduces the respiratory scope of coral reef fishes. *Global Change Biology* **15**: 1405–1412.

221 Farrell, A.P. (2002). Cardiorespiratory performance in salmonids during exercise at high temperature: insights into cardiovascular design limitations in fishes. *Comparative Biochemistry and Physiology Part A: Molecular & Integrative Physiology* **132** (4): 797–810.

222 Baxter, E.M., Jarvis, S., Palarea-Albaladejo, J., and Edwards, S.A. (2012). The weaker sex? The propensity for male-biased piglet mortality. *PLoS One* **7** (1): e30318.

223 Blanckenhorn, W.U., Gautier, R., Nick, M. et al. (2014). Stage- and sex-specific heat tolerance in the yellow dung fly *Scathophaga stercoraria*. *Journal of Thermal Biology* **46**: 1–9.

224 Foley, H.B., Sun, P.Y., Ramirez, R. et al. (2019). Sex-specific stress tolerance, proteolysis, and lifespan in the invertebrate *Tigriopus californicus*. *Experimental Gerontology* **119**: 146–156.

225 Winne, C.T. and Keck, M.B. (2005). Intraspecific differences in thermal tolerance of the diamondback watersnake (*Nerodia rhombifer*): effects of ontogeny, latitude, and sex. *Comparative Biochemistry and Physiology. Part A, Molecular & Integrative Physiology* **140** (1): 141–149.

226 Liwanag, H.E.M., Haro, D., Callejas, B. et al. (2018). Thermal tolerance varies with age and sex for the nonnative Italian wall lizard (*Podarcis siculus*) in Southern California. *Journal of Thermal Biology* **78**: 263–269.

227 Gopalakriswhnan, S., Cheung, N.K., Yip, B.W., and Au, D.W. (2013). Medaka fish exhibits longevity gender gap, a natural drop in estrogen and telomere shortening during aging: a unique model for studying sex-dependent longevity. *Frontiers in Zoology* **10** (1): 78.

228 Johnson, C.R. (1976). Diel variation in the thermal tolerance of *Gambusia affinis affinis* (Pisces: Poeciliidae). *Comparative Biochemistry and Physiology. A, Comparative Physiology* **55** (4A): 337–340.

229 Nakajima, M., Fujisawa, K., and Taniguchi, N. (2009). Estimation of the mode of inheritance of thermal tolerance in the guppy *Poecilia reticulata*. *Fisheries Science* **75** (3): 683–687.

230 Cech, J.J., Massingill, M.J., Vondracek, B., and Linden, A.L. (1985). Respiratory metabolism of mosquitofish, *Gambusia affinis*: effects of temperature, dissolved oxygen, and sex difference. *Environmental Biology of Fishes* **13** (4): 297–307.

231 Horowitz, M. (2016). Epigenetics and cytoprotection with heat acclimation. *Journal of Applied Physiology* **120** (6): 702–710.

232 Maloyan, A., Palmon, A., and Horowitz, M. (1999). Heat acclimation increases the basal HSP72 level and alters its production dynamics during heat stress. *American Journal of Physiology - Regulatory, Integrative and Comparative Physiology* **276** (5): R1506–R1515.

233 Kukreti, H., Amuthavalli, K., Harikumar, A. et al. (2013). Muscle-specific microRNA1 (miR1) targets heat shock protein 70 (HSP70) during dexamethasone-mediated atrophy. *Journal of Biological Chemistry* **288** (9): 6663–6678.

234 Kwon, C., Han, Z., Olson, E.N., and Srivastava, D. (2005). MicroRNA1 influences cardiac differentiation in *Drosophila* and regulates notch signaling. *PNAS* **102** (52): 18986–18991.

235 Radom-Aizik, S., Zaldivar, F.P., Nance, D.M. et al. (2013). Growth inhibition and compensation in response to neonatal hypoxia in rats. *Pediatric Research* **74** (2): 111–120.

236 Tetievsky, A. and Horowitz, M. (2010). Posttranslational modifications in histones underlie heat acclimation-mediated cytoprotective memory. *Journal of Applied Physiology* **109** (5): 1552–1561.

237 Norouzitallab, P., Baruah, K., Vandegehuchte, M. et al. (2014). Environmental heat stress induces epigenetic transgenerational inheritance of robustness in parthenogenetic *Artemia* model. *The FASEB Journal* **28** (8): 3552–3563.

238 Veilleux, H.D., Ryu, T., Donelson, J.M. et al. (2015). Molecular processes of transgenerational acclimation to a warming ocean. *Nature Climate Change* **5** (12): 1074–1078.

239 Ryu, T., Veilleux, H.D., Donelson, J.M. et al. (2018). The epigenetic landscape of transgenerational acclimation to ocean warming. *Nature Climate Change* **8** (6): 504–509.

6

Analytical Methods and Tools to Study the Epigenome

Oscar Ortega-Recalde and Timothy A. Hore

Department of Anatomy, University of Otago, Dunedin, New Zealand

6.1 Introduction

The epigenome (Box 6.1) constitutes a biological information module critical for many cellular and organismal processes. The most notable of these is during development where epigenetics allows, at least in part, the formation and maintenance of cellular identity despite the vast array of divergent cell types sharing essentially the same genetic information [1]. Epigenetic information is generally encoded as chemical modifications either to DNA (such as cytosine methylation) or modifications to the histone proteins associated with DNA. In contrast to genomic information, the epigenome is highly dynamic and responsive to environmental and physiological stimuli. The stable, yet dynamic, nature of this information layer and its important role in several biological processes has attracted growing attention in diverse fields and led to the development of new techniques. The development of new analytical methods and tools, in turn, has accelerated our understanding of epigenome architecture and function.

Modern techniques based on high-throughput sequencing have revolutionized epigenomic research and offer the unprecedented opportunity to study epigenetic modification in an unbiased, genome-wide manner. Although robust, these techniques rely on careful experimental design, as well as bioinformatics and biostatistics tools for obtaining accurate and reproducible results. Recent studies in aquaculture have shown their utility to explore phenomena such as domestication, adaptation, and inbreeding depression in hatchery fish [2–6]. On the other hand, techniques targeted to specific genes and loci are cost-efficient and reliable tools that are widely used in epigenetic research. Given the plasticity and extensive phenotypic diversity of fish compared to other vertebrates, it is expected the study of epigenetic modifications will shed light on important processes of fish biology. To illustrate this point, the application of these methods in fish research has been useful to study biological phenomena such as sex determination, sex change, transgenerational epigenetic inheritance, and adaptation to environmental changes, including global warming [7–10]. Importantly, to promote collaborative and transparent research, epigenomic data has become more available and accessible through databases. The future in this field is promising, galvanized by new technologies, advances in bioinformatic tools, and exciting biological questions to be explored. Similarly, these techniques may be used in different fields of aquaculture research and development, including broodstock management, sex control, environmental manipulation, and epigenetic selection [11].

Currently, numerous techniques are available to study the epigenome. Nevertheless, selecting the best strategy for a given situation and research question might be challenging. This review will describe different assays to examine the epigenome. Initially, some recommendations for choosing the right method to study the epigenome are presented. Next, different techniques, classified according to methodological characteristics and the modification type, are discussed. Special focus is given to accessible and broadly used methods and tools in aquaculture. Subsequent sections include a brief overview of bioinformatics tools and resources, particularly useful for the analysis of large datasets. Finally, a conclusion and outlook of current challenges and opportunities in this evolving field are given.

Epigenetics in Aquaculture, First Edition. Edited by Francesc Piferrer and Han-Ping Wang.

Box 6.1 Definitions of Terms Used in the Text

Bisulfite sequencing: Analytical method that uses sodium bisulfite treatment of DNA before sequencing to study cytosine methylation.
Chromatin accessibility: Level of physical compaction of chromatinized DNA in the nucleus.
Chromatin conformation: Spatial organization of chromatin in the nucleus, determined by the physical interactions between genomic elements that are nearby in 3D space.
Coverage level: Percentage of the epigenome that is examined by the analytical method (e.g., 1%).
Cytosine methylation: Archetypal form of epigenetic modification involving the transfer of a methyl group onto the C5 position of the cytosine to form 5-methylcytosine.
Detection strategy: Technological approach or platform used to assess the modification of interest (e.g., Bisulfite conversion).
Epigenome: The complete set of epigenetic modifications possessed by an organism or an individual cell.
Methylation-sensitive restriction enzymes: Endonuclease enzymes able to selectively recognize and cleave specific DNA recognition sites based on their methylation status.
non-CpG methylation: DNA methylation that occurs at sites other than CpG dinucleotide sequences (e.g., CHG, CHH, where H represents any nucleotide but not guanine).
Resolution level: Degree of detail provided by the analytical method (e.g., single-nucleotide resolution).
Sodium bisulfite treatment: Chemical conversion method of DNA that modifies unmethylated cytosines to uracils, whereas methylated cytosines remain unchanged.
Ten-eleven translocation (TET) enzymes: Family of dioxygenase enzymes that catalyze the oxidation of 5-methylcytosine to 5-hydroxymethylcytosine and further oxidation products during the active DNA demethylation process.

6.2 Recommendations for Choosing a Method to Study the Epigenome

There is no optimal method for all the variety of applications and scenarios in which epigenomic techniques can be used, and it is suggested this should be carefully chosen depending on the research goals. The most important criterion for this selection is its feasibility to answer the research question. Therefore, it is advised the technique is adapted to the study goals and several other critical factors should be considered:

- Study scale. The study might require the analysis of just a few genes or a broader-scale technique.
- Availability of a reference genome. Several techniques rely on a reference genome to extract critical information (e.g., species-specific microarrays) or to analyze the data acquired (e.g., most sequencing-based methods). If this is not available, the methodological repertoire is considerably reduced.
- Sample quality and quantity. Although less common, sample collection and processing could be a potential limitation. It is important to consider these aspects to ensure the minimum requirements of any technique are met.
- Technological platform. Accessibility and availability of equipment, reagents, and computational and human resources. Most robust techniques require a dedicated platform for sample processing and data analysis.
- Complementary approaches. Often, an orthogonal (i.e., independent) technique is used to validate and contrast meaningful results.
- Financial resources. According to the research group budget, the method should be cost-efficient.

An additional perspective on epigenetic method selection is provided in an excellent 2018 review by Kurdyukov and Bullow [12], whereby the authors classify the methods to assess DNA methylation into two groups according to the knowledge of the regions/genes to study and include several criteria including throughput, coverage along the genome, and technical requirements to choose the most-suited technique based on the research needs.

6.3 Methods and Tools to Analyze Epigenetic Modifications

Epigenetic alterations include DNA modification, histone posttranslational modifications (PTMs), and noncoding RNA (ncRNA). Cytosine methylation has been described as an archetypal form of epigenetic modification [13]. 5-methylcytosine (5mC) was first described as a constituent of nucleic acids by Johnson and Coghill in 1925. However, it was not until 1975 that two groups independently suggested that this nucleotide could regulate stable changes in gene expression and cell fate [14–16]. This modification has been studied longer than any other and has the most methods available for its analysis. Methods to assess 5mC can be classified according to several parameters, including detection strategy (Section 6.3.1), resolution level (Section 6.3.2), and coverage (Section 6.3.3). Detection strategy refers to the detection technology used to assess the modification of interest, for example, methylation-sensitive enzymes or bisulfite conversion. The resolution level refers to the amount of detail provided by the methodology. For example, low-resolution techniques permit the exploration of epigenetic changes at a global level, specifically genome-wide, without providing details about specific regions or positions. In contrast, high-resolution methods provide detailed information at single-nucleotide resolution. At an intermediate level, medium-resolution techniques give information about specific locations through the genome (e.g., CpG islands), without single-base discrimination. Finally, coverage refers to the percentage of the epigenome that is covered by the analysis. According to this classification, methods can be categorized as: gene- or locus-specific, when the coverage target is limited to one or a few regions, typically less than 50 per run [17, 18]; genome-wide, when numerous regions along the genome are analyzed, typically thousands of specific loci or enriched genome fractions, and less than 5% of genomic content [19]; and whole genome, when the vast majority of the genome is covered to assess the modification of interest (note: some regions cannot be covered due to limitations in sequencing, for example, GC bias, mapping issues, or highly repetitive regions). Histones, on the other hand, are basic proteins around which DNA winds. Histones are key chromatin components and modulate DNA accessibility, regulating gene expression [20, 21]. Importantly, numerous PTMs in the histone tails that affect their binding with the DNA have been described [22]. Methods to analyze these modifications are described in Section 6.3.4. Finally, ncRNA constitutes a vast number of nonprotein coding transcripts with numerous roles in cellular functions and gene regulation [23, 24]. Several studies have highlighted the importance of these molecules in epigenetic inheritance. Assays to examine these molecules are presented in Section 6.3.5.

6.3.1 DNA Methylation Methods According to Detection Strategy

Different detection strategies have been developed to assess cytosine methylation. A graphic summary of the most common techniques used for this purpose is presented in Figure 6.1.

6.3.1.1 Enzyme-Based Methods

Methylation-sensitive enzymes (Box 6.1) are a useful tool to assess cytosine methylation on a local or global scale. Techniques based on this approach use restriction enzymes that recognize specific motifs and are sensitive to cytosine methylation (Figure 6.1a). Historically, enzyme-based methods were the first techniques used for assessing methylation. Currently, several methods apply this approach coupled with next-generation sequencing (NGS) (e.g., reduced representation bisulfite sequencing – RRBS) [25–27]. Two commonly used restriction enzymes to assess DNA methylation are *HpaII* and *MspI*. Both enzymes recognize the motif 5′-CCGG-3′; however, while *HpaII* is sensitive to methylation in the second C, *MspI* activity is not inhibited by methylation at this position. Overall, enzyme-based methods are reliable and scalable, ranging from locus-specific to whole genome scale.

The methylation-sensitive amplification polymorphism (MSAP) assay is a method based on digestion with a methylation-sensitive enzyme, followed by selective amplification of DNA and gel-based visualization [28]. Differential cleavage by these enzymes produces a distinctive global cytosine methylation pattern. Because this technique does not require a reference genome, it has been particularly used in nonmodel species, for example, the European sea bass (*Dicentrarchus labrax*) and the Staghorn Coral (*Acropora cervicornis*) [29, 30]. Restriction landmark genomic scanning (RLGS) is a method to detect a large number of restriction landmarks across the genome. In this method, digestion by restriction enzymes is used to create genomic landmarks, followed by end-labeling of the fragments and visualization in high-resolution, two-dimensional electrophoresis [31]. Both MSAP and RLGS can be applied to any species, do not require a reference genome, and have similar limitations. Nevertheless, both techniques have largely been superseded by sequencing-based methods.

LUMA (LUminometric Methylation Assay) is a genome-wide technique that uses a combination of two DNA restriction enzymes and polymerase extension assay by pyrosequencing [32]. Two reactions are performed in parallel. One uses the methylation-sensitive enzyme *HpaII* while the second uses the methylation-insensitive enzyme *MspI*. Both enzymes generate 5′-CG overhangs after cleavage. In addition, the enzyme *ecoRI* is included in both reactions as an internal control. *EcoRI* generates 5′-AATT overhangs. Resulting overhangs are extended and quantified by pyrosequencing. The *HpaII*/*MspI* ratio obtained is proportional to the amount of methylated/unmethylated DNA. Although this technique is accurate and reproducible, it is limited to the motif 5′-CCGG-3′ and has low resolution.

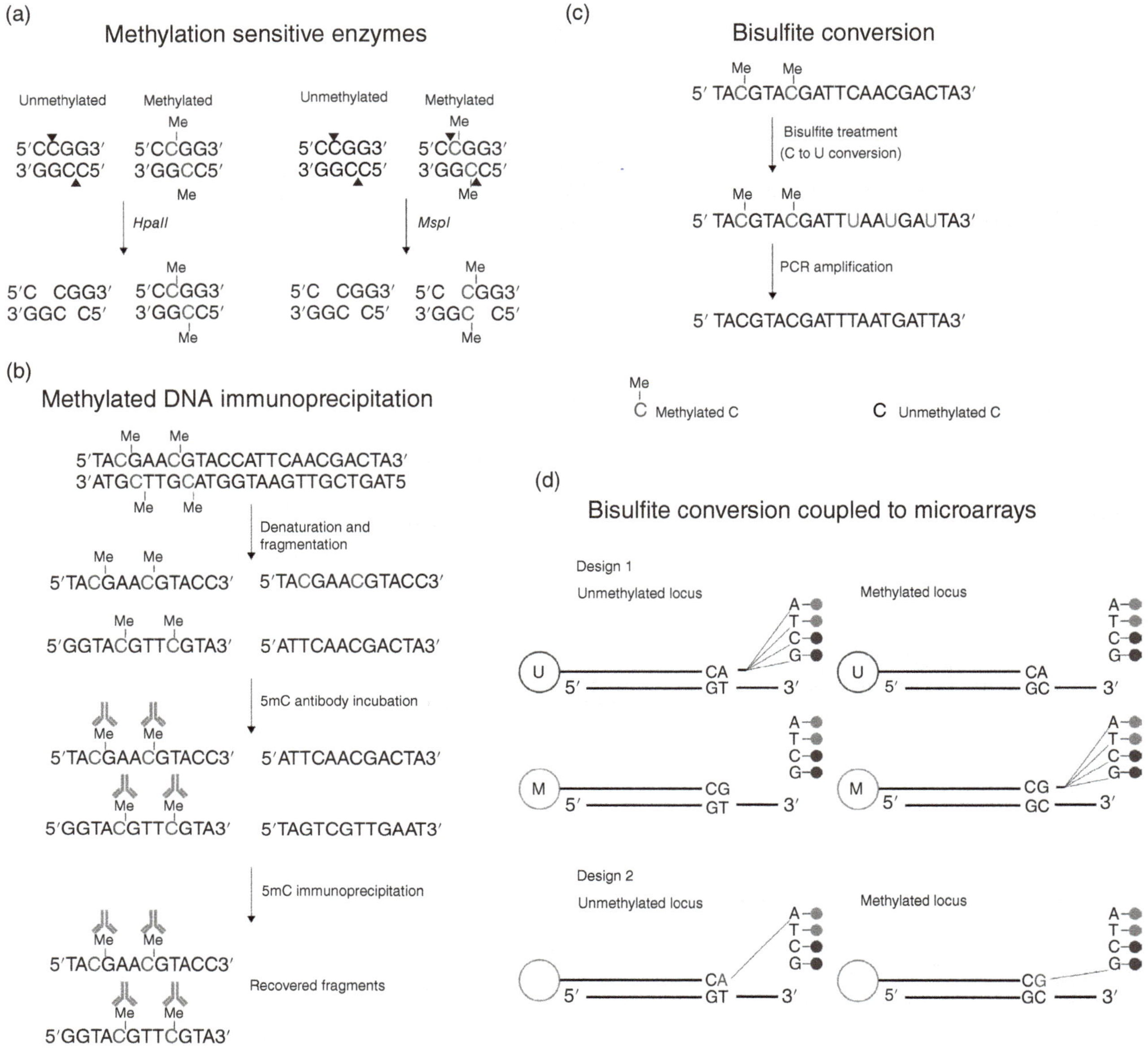

Figure 6.1 Common detection methods used to assess cytosine methylation. (a) Methylation-sensitive enzymes. (b) Bisulfite conversion. (c) Methylated DNA immunoprecipitation. (d) Bisulfite coupled to microarrays. In the Infinium assay design 2, two color channels are used simultaneously, specific incorporation of labeled nucleotides allows methylation status discrimination. Arrows indicate subsequent steps in each strategy. (*See colour plate section for colour representation of this figure.*)

New methods based on conventional methylation-sensitive restriction enzymes coupled with NGS (e.g., MSAP-seq, Methyl-seq, methyl-sensitive restriction enzymes sequencing – MRE-seq) have been developed [33–35]. In these methods, genomic DNA is digested with different methylation-sensitive enzymes and then used to construct a library and sequenced. Whereas feasible and cost-efficient, sensitivity and coverage rely on enzyme activity and specificity. In addition, individual CpGs cannot be analyzed, and the quantitation is not absolute but relative to the abundance of fragments.

Furthermore, a novel and more robust enzymatic-based approach involves genome-wide sequencing of single molecules, and with base pair resolution. This method called Enzymatic methyl-seq (EM-seq) uses two sequential enzymatic steps to detect 5mC and 5-hydroxymethylcytosine (5hmC) [36]. 5hmC is a stable epigenetic mark generated from 5mC by TET enzymes [37, 38]. Importantly, this epigenetic modification has been shown to play a role in the regulation of genomic function and several cellular and developmental processes, including pluripotency and tissue-specific development [39, 40]. In the first step of EM-seq, TET2 and T4-BGT oxidize methylated and hydroxymethylated cytosines. In the second step,

APOBEC3A deaminates unmethylated cytosines to uracils. Subsequent PCR replaces uracil by thymine in newly synthesized DNA strands. Another technique, TET-assisted pyridine borane sequencing (TAPS), uses a similar approach [41]. TAPS combines oxidation of 5mC and 5hmC by TET and reduction to dihydrouracil (DHU) by pyridine borane. DHU is converted to thymine during PCR. Importantly, these methods are not destructive, preserving long fragments and allowing higher mapping rates and accuracy.

6.3.1.2 Affinity-Based Methods

This group of techniques relies on the use of antibodies to either detect or enrich DNA methylation sequences. There are three strategies in this group: quantitation using enzyme-linked immunosorbent assay (ELISA), methylated DNA immunoprecipitation (MeDIP), and hybridization based on methyl-CpG-binding domain (MBD).

Global DNA methylation can be quantified using the ELISA method [42]. This technique employs the DNA as the antigen, which is incubated in a microplate for adhesion. Subsequently, a primary anti-5mC antibody is added, enabling antigen–antibody binding. Finally, a secondary antibody conjugated with a colorimetric substrate allows detection through color change. This parameter is measured and compared to standards to determine global methylation levels. While this technique is simple, fast, and cost-efficient, the resolution level is low and is less robust than other methods.

Methylated DNA immunoprecipitation (MeDIP) is an affinity-based method for detecting methylated DNA [43]. Initially, genomic DNA is randomly sheared by sonication, resulting in fragments mostly between 400 and 600 bp. Fragments are denatured to produce single-stranded DNA, improving the binding activity of the antibodies. Then, DNA is incubated with monoclonal 5mC antibodies and a classical immunoprecipitation technique is applied to pull down methylated sequences. The resulting enriched fraction can be assessed by PCR to study individual locus or combined with microarray hybridization or NGS (Figure 6.1b). This technique has been used to study numerous plants and animals, including several fish species [44–48].

Enrichment of methylated DNA can be achieved with anti-methylcytosine-binding proteins, specifically the MBD [49]. In nature, these proteins are important readers of methylated CpGs and interact with both DNA strands [50]. Examples of this family include MeCP2 (methyl-CpG binding protein 2) and MBD1-4 (methyl-CpG binding domain protein 1–4). Similar to MeDIP, MBD protocol starts with DNA fragmentation. The capture of methylated DNA sequences is achieved through MBD-tagged antibodies. Finally, PCR or sequencing is conducted to quantify the enriched fraction. MeDIP and MBD are robust and precise methods to study DNA methylation. A comparison of these two enrichment methods showed a 99% concordance when coupled to NGS [51].

6.3.1.3 Bisulfite-Based Methods

The bisulfite-based methods are primarily based on the fact that cytosine modifications respond differently to the sodium bisulfite chemical treatment (Box 6.1) [52, 53]. Unmodified cytosine undergoes rapid deamination to uracil in response to bisulfite treatment, whereas methylated and hydroxymethylated cytosine is somewhat resistant to this process (Figure 6.1c). 5-formylcytosine (5fC) and 5-carboxylcytosine (5caC), which are unstable and extremely rare within the genome, are also rapidly deaminated to uracil following bisulfite treatment. Together, this leads to a recognizable difference between the major modified bases (5mC and 5hmC) and unmodified cytosine. In addition, bisulfite treatment is a harsh biochemical reaction, leading to substantial DNA fragmentation.

One limitation of conventional bisulfite sequencing is that 5mC and 5hmC cannot be distinguished; however, further bisulfite-based sequencing techniques such as oxidative bisulfite sequencing [54] and TET-assisted bisulfite sequencing [55] have been developed to overcome this obstacle. On the other hand, given that these last methods are technically challenging, they require more than double the amount of sequencing and 5hmC is at least an order of magnitude less common in the genome than 5mC, thus they are often not undertaken.

Methods based on bisulfite include a large number of techniques, ranging in coverage and resolution level. Bisulfite-based methods to analyze methylation in specific loci include methylation-specific PCR (MSP), bisulfite sequencing PCR (BSP), multiplex bisulfite sequencing (MBS), combined bisulfite restriction analysis (COBRA), and multilocus methylation-specific single nucleotide primer extension (MS-SNuPE) [18, 56–59]. In MSP, genomic DNA is purified and treated with sodium bisulfite. Then, two parallel PCRs are conducted, each using a different set of primers. The first set is specific for methylated DNA (M) and the second set is specific for unmethylated DNA (U). The assessed region should contain at least one CpG to discriminate between a methylated and an unmethylated state. Finally, the region is amplified and evaluated. This technique is very sensitive to methylated alleles and uses very low quantities of DNA, as well as low-quality DNA. Nevertheless, their main limitations are restricted resolution and low throughput. In BSP, genomic DNA is denatured and undergoes bisulfite treatment. Converted DNA is amplified by PCR using primers without CpGs in their sequences flanking the region to be assessed. Finally, the PCR product is sequenced directly or cloned and sequenced. Using a similar approach, multiple-locus and samples can be assessed in a single reaction using a technique called MBS [60]. In this technique, genomic DNA is first treated with sodium bisulfite. Later, target regions are amplified through PCR and sample-specific indices are incorporated into the amplicons. Finally, multiplexed libraries are sequenced using NGS. This technique has several advantages including low cost and the capacity to assess several loci and samples in the same experiment. COBRA is a technique that relies on the selective cleavage of bisulfite-converted fragments. First, genomic DNA is bisulfite converted and specific loci are amplified through PCR. Importantly, primers do not contain CpGs to avoid enzymatic digestion. Next, the PCR product is digested with a restriction enzyme in which the recognition sequence is affected by methylation in the original DNA. Lastly, digested products are analyzed through gel electrophoresis. Similarly, MS-SNuPE is a technique based on bisulfite conversion, where genomic DNA is first treated with this compound, and the regions of interest are amplified and isolated using PCR and agarose electrophoresis, respectively [59]. Next, the SNuPE reaction is performed using oligonucleotides that hybridize immediately upstream of the CpG site being analyzed. The incorporation of a labeled dNTP using a single-nucleotide extension reaction allows discrimination of the methylation state using polyacrylamide gel electrophoresis. Overall, bisulfite-based methods in this group are cost-efficient and reliable techniques that can be used to assess methylation in a limited number of candidate regions.

Several bisulfite-based methods allow the enrichment of regions of interest. Among these techniques, RRBS is one of the most commonly used [61]. RRBS combines specific enzymatic cleavage, bisulfite conversion, and high-throughput sequencing to enrich CpG-rich regions across the genome. These regions are particularly important because they overlap with regulatory regions such as promoters and CpG islands. A conventional RRBS protocol starts with the digestion of genomic DNA with one or multiple restriction enzymes to produce selective fragmentation. Often the methylation-insensitive enzyme *MspI* is used for this purpose. After library preparation, short fragments corresponding to CpG-rich regions are size selected using methods such as columns and preparative gel electrophoresis. Selected fragments undergo bisulfite conversion, PCR amplification, and NGS. This technique is particularly effective to evaluate areas dense in CpG and provides single-base resolution level at a cost-effective manner. RRBS has been used to study fish species of aquaculture interest [3, 6]. Despite its advantages, RRBS has several limitations including bias in sequence selection and limited coverage across the genome. A graphic summary of this and other common bisulfite-based techniques coupled to NGS is presented in Figure 6.2.

Another reduced representation technique has recently been introduced, bisulfite-converted restriction site associated DNA sequencing (bsRADseq) [62]. This technique uses a restriction enzyme to digest genomic DNA samples into fragments. Fragments are ligated to methylated adapters and treated with sodium bisulfite. Library preparation continues with the addition of barcodes and the samples are sequenced. Interestingly, when a reference genome is not available, bsRAD-seq allows the construction of a synthetic reference using standard RAD-seq markers. Similarly, Tagmentation-based Indexing for Methylation Sequencing (TIME-seq), another bisulfite-based technique that enables the enrichment of regions of interest, has been recently described by Griffin et al. [63]. This method involves initial treatment with a Tn5 transposase that simultaneously cuts DNA and inserts sample indexes. Samples are then pooled and methylated end-repair is performed. Enrichment by hybridization is performed using biotinylated-RNA baits produced in-house from oligonucleotide libraries. Finally, captured DNA is bisulfite-treated, amplified using indexed-PCR for each pool, and sequenced. Multiplexing capacity, targeted sequencing, and reduced costs constitute important advantages of this technique.

Bisulfite treatment has been combined with downstream microarray analysis in the Illumina GoldenGate and Infinium platforms [64–66]. Different chemistry technologies are available for this purpose (Figure 6.1d). For the Infinium I assay chemistry technology, two probes are employed per CpG locus. Each probe has a different 3′ terminus designed to match either a cytosine, resulting from an unmodified methylated position, or a thymine, a product of C-to-U-converted unmethylated cytosine. For the Infinium II assay, just one probe per locus is required. In this case, the 3′ terminus is complementary to the upstream nucleotide to interrogate. Discrimination between methylated or unmethylated cytosines occurs by extension using labeled nucleotides: addition of a labeled G/C, detected in a green color channel when the cytosine was protected from bisulfite action by methylation, or an A/T, detected in a red color channel when the nucleotide is uracil resulting from cytosine deamination or its complementary nucleotide. The GoldenGate assay is based on initial conversion by bisulfite treatment, followed by detection of the methylation status of individual CpGs by two probe pairs: allele-specific oligonucleotides (ASO) and a locus-specific oligonucleotides (LSO) probe pair for the unmethylated CpG state and a corresponding ASO/LSO probe pair for the methylated CpG state. Extension and ligation of probe pairs are followed by PCR amplification with fluorescently labeled primers and

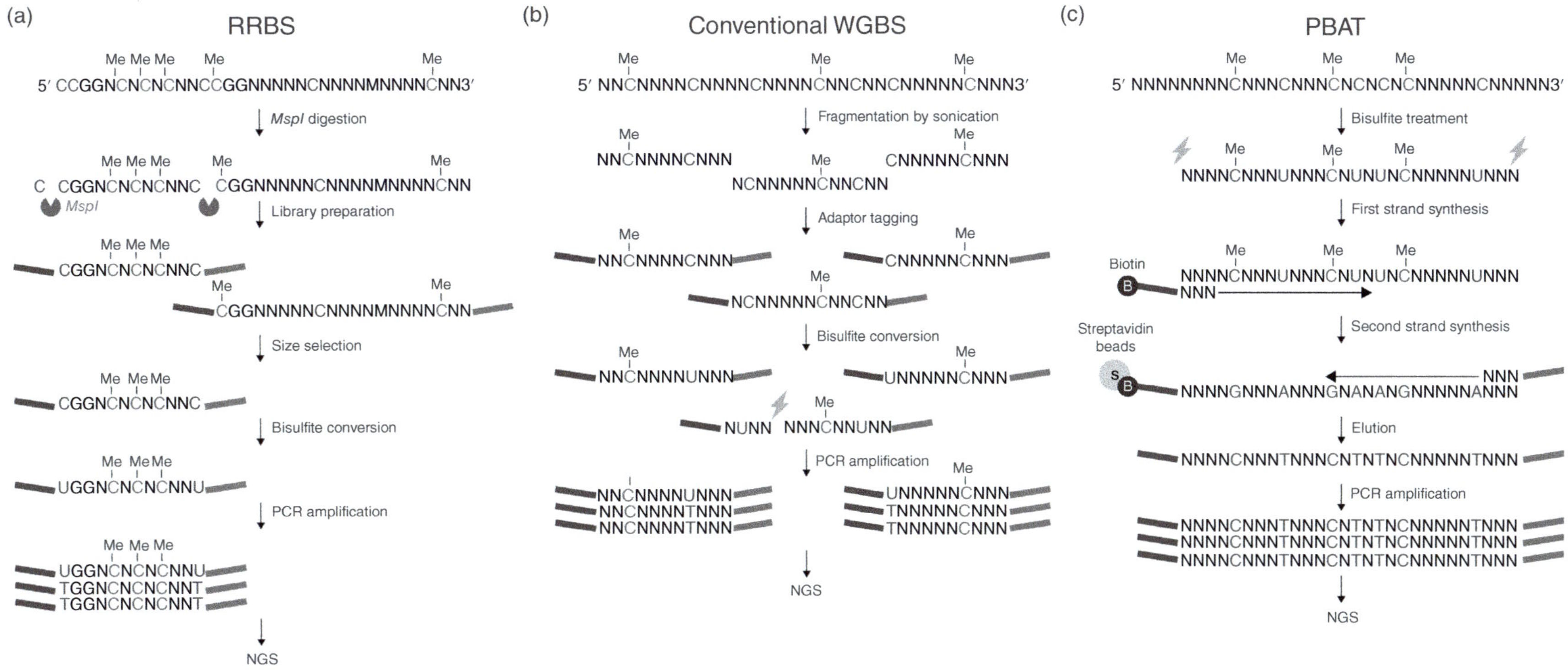

Figure 6.2 Common bisulfite-based techniques coupled to NGS. (a) Reduced representation bisulfite sequencing (RRBS). (b) Conventional Whole Genome Bisulfite Sequencing (WGBS). (c) Post-bisulfite Adaptor Tagging (PBAT). (*See colour plate section for colour representation of this figure.*)

hybridization to a complementary sequence in an array. Differential labeling with two fluorophores allows the discrimination between unmethylated, semi-methylated, and methylated states. Overall, these techniques enable to measure quantitative methylation in a large number of loci genome-wide in a cost-efficient manner. Infinium methylation platforms are available for human and mice research; interestingly, they have been used in other mammalian species with moderated success [67, 68]. Despite there being no currently available methods to assess DNA methylation in species of aquaculture interest using microarray-based techniques, it may constitute an interesting area of research and development. Recently, universal epigenetic clocks have been developed using conserved regions among mammalian species [69]. Although more genetically diverse, the development of similar approaches in farmed fish species targeting phenotypic traits of interest might result in interest for commercial and research purposes.

Advances in sequencing technology and significant cost reductions have made whole genome bisulfite sequencing (WGBS) an increasingly accessible method to assess the methylome [70–72]. WGBS can detect changes in DNA methylation in the whole genome at a single-base resolution level and is fully quantitative. For these reasons, it is now commonly used and also seen as the gold standard for methylation analysis [73]. However, these may soon be replaced by EM-seq, which is becoming increasingly popular due to its increased performance to detect 5mC and 5hmC modifications and more uniform coverage [36]. The conventional WGBS protocols consist of the following steps. First, genomic DNA is fragmented and adaptors are ligated (Figure 6.2b). Next, fragments are bisulfite treated and an additional PCR amplification step is conducted to enrich sequences with both adaptors. Finally, libraries are sequenced using NGS. An important consideration is that bisulfite treatment is harsh and induces DNA breakage. Treatment with bisulfite after adaptor ligation induces degradation of DNA templates, resulting in loss of material and decreasing library complexity. For this reason, a large amount of starting material (>5 ug) is required for this procedure.

To circumvent the degradative effects of bisulfite on library generation, Miura et al. developed post-bisulfite adaptor tagging (PBAT) [74]. This method consists of initial conversion and fragmentation by sodium bisulfite, followed by two sequential strand synthesis steps (Figure 6.2c). The first strand synthesis is performed with biotin-labeled adaptors possessing random nucleotides at their 3′ end. The products are captured using streptavidin-coated coats and magnetic immobilization. Second-strand synthesis is achieved using the immobilized first strand as a template and adaptors that also possess random nucleotides at their 3′ ends. Barcodes and additional sequences necessary for sequencing can be added using PCR. One of the main advantages of this last approach is that ultra-low quantities of DNA can be analyzed. A modification of PBAT has even been used for single-cell WGBS [75].

Despite its advantages, WGBS has several limitations. First, as mentioned previously, bisulfite-based methods cannot differentiate between 5mC and 5hmC. Second, although increasingly affordable, this technique is relatively expensive. In this regard, low-coverage WGBS, a technique where the whole genome is sequenced commonly with a coverage depth of less than 5X, has shown to be a cost-efficient method to study global methylation patterns in a cost-efficient manner [76]. Third, mapping bisulfite short reads to a reference genome is a challenging task. Although several algorithms have been developed for this task, sequence length, DNA methylation status, and sequencing error considerably affect this procedure [77]. Lastly, bisulfite conversion can result in sequencing bias. Although this drawback can be alleviated by using amplification-free library preparation, it should be taken into account during the analysis of WGBS data [78].

6.3.1.4 Direct Detection Methods

Direct methods of detection of 5mC are available for research. These include high-performance liquid chromatography-ultraviolet (HPLC/UV), liquid chromatography coupled with tandem mass spectrometry (LC-MS/MS), single-molecule real-time (SMRT) sequencing, and Nanopore sequencing. HPLC/UV allows quantification of the global methylation level. In this method, genomic DNA is digested to deoxynucleotide monophosphates, purified, and quantified using HPLC. Methylation levels are measured by using nucleotide standards [79]. Similarly, LC-MS/MS has been adapted to incorporate analysis of 5mC [80]. Despite its accuracy, the widespread use of these techniques is limited due to the specialized equipment required. Thus, these methods are not usually used in aquaculture-related research.

More recently, the increasingly available third-generation sequencing techniques offer an exciting avenue to assess DNA methylation using direct detection. SMRT sequencing, by Pacific Biosciences (PacBio), is a technology based on small reaction chambers called zero-mode waveguide (ZMW) where a unique DNA polymerase is affixed at the bottom [81]. After library preparation, a single DNA molecule is incorporated into the reaction chamber. As fluorescently labeled nucleotides are incorporated in the growing DNA strand by the DNA polymerase, fluorescence is measured at the bottom of the ZMW. Base modifications are analyzed by using kinetic signals of the DNA polymerase. In this case, the incorporation time of a nucleotide in the growing strand is altered according to the modified nucleotide in template

DNA produced by the DNA polymerase. Despite the subtle signal, this approach has been tested in several biological samples, reaching a high correlation level with WGBS [82].

Nanopore sequencing by Oxford Nanopore is a technique that allows single-molecule sequencing by passing a DNA sequence through a nanoscale pore [83]. Protein nanopores are embedded into a synthetic membrane with high electric resistance bathed in an electrolyte solution. Ionic flux through nanopores provides a baseline current signal detected by an electrometer. When a DNA molecule passes through, the nanopore modifies the current in pattern or magnitude, producing different signals. Finally, these signals are decoded to provide specific nucleotide sequences and are able to distinguish between cytosine and 5mC. The computational challenge that this task represents has led to the development of several algorithms [84]. SMRT and nanopore sequencing offer considerable benefits compared to more traditional methods, including the capacity to analyze native DNA, real-time sequencing, and longer reads that are easier to map and allow haplotype reconstruction. These approaches have shown to be useful to analyze DNA methylation in different biological systems and will probably be more commonly used in the future [85, 86].

6.3.2 DNA Methylation Methods According to Resolution Level

6.3.2.1 Low Resolution

Methods of quantifying DNA methylation can be classified according to their resolution level, as mentioned previously. Techniques such as HPLC/UV, LC-MS/MS, ELISA, and LUMA are highly reproducible and can quantify 5mC global levels within an entire genome [42, 79, 80]. Despite being much less common in the genome, 5hmC, 5fC, and 5caC can also be detected [87]. While these techniques are informative, they do not resolve the context in which cytosine modification is occurring, nor their relationship to genes and other genomic sequences. Furthermore, they can provide biased results in the event a DNA sequence is particularly enriched. Given that DNA methylation is a key regulator of transcription on a gene-by-gene basis, understanding where DNA methylation occurs within specific genomic loci is often considered important.

6.3.2.2 Medium Resolution

On an intermediate level, DNA methylation can be identified with medium resolution level. In contrast to single-nucleotide resolution techniques, these methods analyze several CpG positions at once, providing an average methylation level for the loci but no individual CpGs. Examples of these methods are the pull-down of methylated DNA by antibodies against 5mC-binding proteins (MBD) or 5mC (MeDIP) used to enrich CpG-rich sequences and downstream PCR, microarray hybridization, or NGS [43, 88]. Alternatively, other sequencing-based techniques, such as low-coverage WGBS, can discriminate different regions in the genome after mapping. These methods are particularly useful to study regions of interest, for example, CpG islands.

6.3.2.3 Single-Nucleotide Resolution

Finally, single-nucleotide resolution techniques are robust methods that can interrogate methylation status at unique positions. In this group, we find several techniques such as enzymatic and bisulfite-based methods. Among the enzymatic methods, for example, bisulfite-free C-to-U conversion by TET and subsequent sequencing by NGS allows discrimination of methylation levels across the genome at single-nucleotide detail. One of the most widely used techniques in this group is WGBS. Reduction in the cost of NGS and increased accessibility have made WGBS a fundamental part of several epigenetic analyses. Reduced representation-based techniques such as RRBS, which uses a combination of restriction enzymes and bisulfite treatment to enrich sequences with high CpG content, are also in this group [27]. RRBS displays considerable sensitivity and enables interrogation of the genomic regions, most likely to hold differentially methylated regions in a cost-effective manner. Finally, third-generation sequencing technologies have shown to be promissory tools to detect DNA methylated and other epigenetic modifications in native DNA. These technologies will considerably increase our arsenal of tools to study the epigenome.

6.3.3 DNA Methylation Methods According to Genome Coverage

6.3.3.1 Targeted Approaches

On a small scale, gene- or locus-specific targeted approaches are useful and cost-efficient techniques to analyze epigenetic modifications. Several enzymatic, affinity, and bisulfite methods are in this category. Importantly, most of these techniques are accessible and do not require highly specialized equipment or materials. These methods can be used in two scenarios: first, as a primary tools to analyze a subset of regions of interest, and second, as a validation tool for genome-wide or whole genome techniques. In the first scenario, the regions are analyzed, aiming to characterize or quantify methylation levels for one or a few loci that may be part of one or different genes of interest. In the second scenario, the analysis aims to contrast previous results obtained by genome-wide or whole-genome methods, such as microarrays or NGS. Overall, these techniques are reliable, highly reproducible, and versatile; nevertheless, their main setback is the limited number of individual CpG or loci able to be analyzed.

6.3.3.2 Genome-Wide

Genome-wide approaches include techniques able to interrogate epigenetic modifications in a representative fraction of the genome. These techniques include reduced representation bisulfite methods, microarrays, high-throughput affinity-based approaches, and most enzymatic-based approaches. By assessing a large number of regions, these methods are considerably more robust than targeted approaches and have the main advantage of being unbiased. Other advantages of these techniques are the significantly lower price than whole genome approaches, and once standardized, they are highly reproducible. In addition, because enriched regions usually correspond to regulatory DNA, the results are highly informative. Among the disadvantages of these techniques are the limited coverage across the genome and the increasingly specialized equipment and personnel necessary for sample processing and data analysis.

6.3.3.3 Whole Genome

Whole genome techniques assess epigenetic modifications in the full genome, being particularly robust for in-depth analysis when coupled to medium or high resolution. Similar to genome-wide methods, whole genome techniques are "hypothesis-free" and often are able to identify a large number of potential candidates for downstream analysis. These techniques include WGBS, bisulfite-free enzymatic sequencing methods, and third-generation sequencing methods used to directly sequence native DNA. Whereas WGBS is currently the standard method of choice for studies that generate reference methylomes, bisulfite-free enzymatic and third-generation sequencing methods are also increasingly used in basic and applied research. The principal advantage of these techniques is the ability to determine quantitively DNA methylation levels all across the genome, including low-CpG-density regions and non-CpG methylation (Box 6.1). Despite the increased accessibility and use of these methods, the cost may be a constraint, especially for large numbers of samples. In addition, a highly specialized technological platform for sample processing and trained personnel are required for data analysis.

6.3.4 Histone Modifications

Histones are positively charged proteins that bind to the negatively charged DNA molecule [21]. In eukaryotic cells, a histone octamer wraps ~147 bp of DNA around its surface, helping condense DNA and forming structures called nucleosomes, the functional unit of chromatin. Histone proteins undergo distinct PTMs, controlling the structure and function of chromatin and gene expression. Among the most-studied histone PTM are methylation, acetylation, phosphorylation, and ubiquitylation in specific residues. Histone PTMs are catalyzed by specific enzymes, including histone acetyl-transferases (HATs), histone deacetylases (HDACs), histone methyl-transferases (HMTs), kinases, and phosphatases. Importantly, these enzymes and their related PTM have been linked to chromatin remodeling, acting as fine tuners of gene expression [89]. Methods to analyze histone PTM are mainly based on antibody tagging (Figure 6.3).

6.3.4.1 Chromatin Immunoprecipitation

Chromatin immunoprecipitation (ChIP) is an extensively used technique to study gene–protein interactions [90]. ChIP uses antibodies that selectively bind and isolate proteins such as histones and histones with specific PTM (Figure 6.3a). This technique is based on four major steps: fixation, fragmentation, immunoprecipitation, and analysis of immunoprecipitated DNA. During the first step, the DNA-protein complexes are stabilized by cross-linking using fixation compounds such as formaldehyde. Next, DNA-protein complexes are fragmented into small pieces, aiming to separate the protein or modification of interest, along with any bound DNA. Usually, this step is achieved using sonication or enzymatic digestion. Fragmentation is followed by immunoprecipitation, using specific antibodies directed against histones with particular PTM. The DNA-protein complexes are then purified by affinity eliminating other DNA fragments. Finally, for the analysis of the immunoprecipitated DNA, cross-links between DNA and protein are reversed. This procedure is done through proteases or thermal treatment. The resulting DNA material can be analyzed through different platforms including ChIP-PCR,

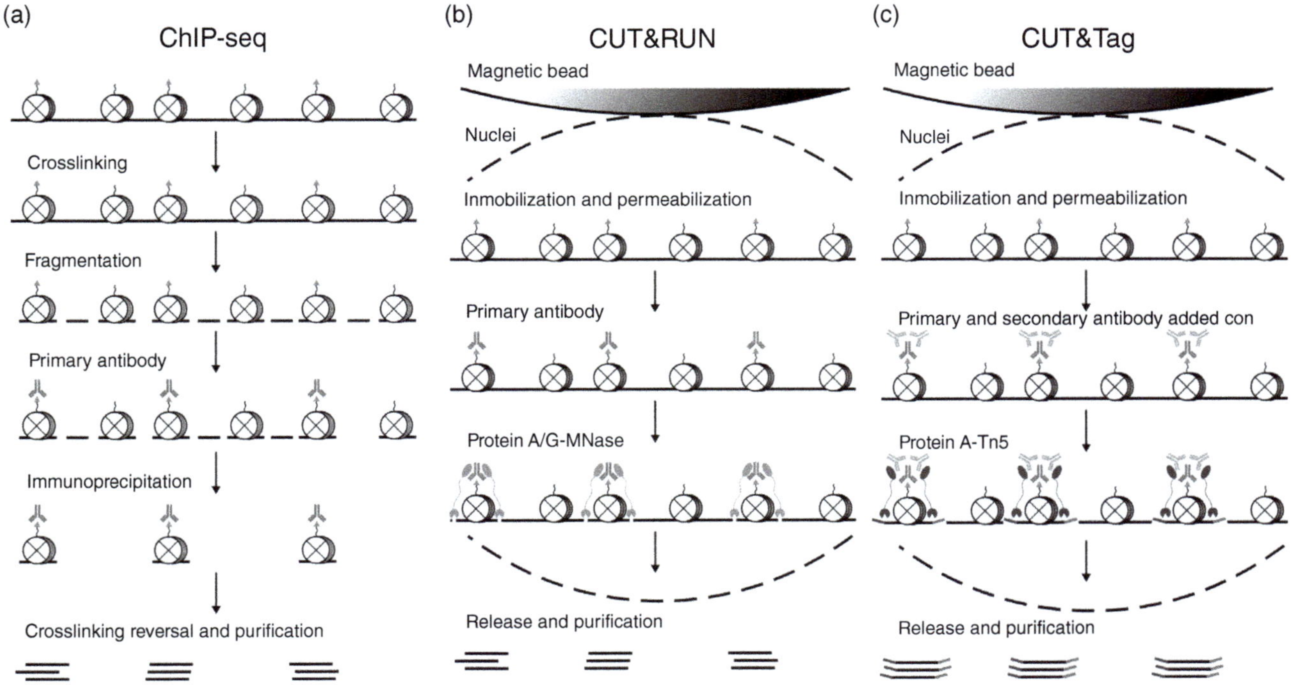

Figure 6.3 Common detection methods used to study histone PTMs. (a) Chromatin immunoprecipitation (ChIP). (b) Cleavage Under Targets and Release Using Nuclease (CUT&RUN). (c) Cleavage Under Targets and Tagmentation (CUT&Tag). Primary and secondary antibodies in CUT&Tag are added in consecutive steps, the secondary antibody aids in signal amplification by facilitating A-Tn5 activity.

ChIP-chip, and ChIP-seq. Briefly, these methods aim to quantify the enrichment or depletion of DNA sequences that were captured during ChIP.

ChIP-PCR is a cost-efficient technique to analyze histone PTM in a subset of known loci [91]. PCR or qPCR can be used to identify or quantify differences in these targeted loci. Although versatile and feasible, each region to be analyzed requires individualized primer designs and standardization, making the analysis of many regions impractical. ChIP-chip is a genome-wide method able to interrogate a large number of regions in a single assay [92]. This technique uses a DNA microarray chip to analyze the enriched fragments obtained by ChIP (ChIP-on-chip). DNA microarrays contain oligonucleotide probes designed to assess sequences across the genome, particular interest is given to regulatory regions. Competitive hybridization between ChIP-enriched DNA and input DNA allows the identification of regions that interact with a particular protein. Whereas reliable and reproducible, this technique requires large amounts of DNA and a dedicated array, which is not available for any farmed fish species and just a few model species. Finally, ChIP-seq is a high-throughput method based on NGS [93]. After library preparation and NGS, the reads obtained are mapped against the whole genome, allowing to assess in an unbiased manner signal enrichment or depletion in relation to the background noise. This technique is robust and recent advances in library preparation are compatible with low amounts of input DNA. Despite considerable reductions in NGS prices, the main limitations of this method are the associated cost compared to other techniques and the antibodies' availability and specificity. On the other hand, access to specialized equipment and bioinformatic expertise to analyze the data should be considered.

6.3.4.2 CUT&RUN and CUT&Tag

Two new methods have been recently developed to assess nucleosome modifications and other DNA-protein interactions: cleavage under targets and release using nuclease (CUT&RUN) and cleavage under targets and tagmentation (CUT&Tag) [94, 95]. During CUT&RUN, samples initially undergo nuclei extraction and immobilization by beads (Figure 6.3b). Then, the nuclei are permeabilized and incubated with a specific primary antibody against the protein of interest. Next, the fusion protein A/G-MNase is added to the mixture. A secondary antibody might be added to increase the amount of protein A-binding sites. Protein A/G-MNase targets antibodies and cleavages DNA on both sides of the complex DNA protein of interest. These complexes are released from the nuclei and collected for subsequent analysis, for example, NGS. Similarly,

CUT&Tag is a technique based on nuclei extraction, immobilization, permeabilization, and incubation with specific antibodies against the protein of interest (Figure 6.3c). Incubation with a secondary antibody is required in the CUT&Tag protocol to increase the signal. Next, the fusion protein A-Tn5 transposase is added to the reaction. Protein A-Tn5 targets an antibody and has tagmentase activity, cuts DNA, and inserts adaptors or any other sequence into the target DNA. Potential limitations of these techniques include antibody specificity, amount of input material, low reproducibility, and complexity of the isolation method. In addition, for CUT&Tag, nonoptimal salt conditions lead to unspecific Tn5 activity. In comparison to ChIP, these techniques have several advantages including better signal-to-noise ratios and lower quantities of starting material required, costs per sample, and protocol length. In aquaculture, these advantages may be particularly useful when a limited amount of input sample is available, for example, to assess nucleosome modifications during early development. Similarly, the reduced cost and protocol length might facilitate the analysis of more samples compared to ChIP-based techniques, increasing the study power and thus increasing the likelihood of finding epigenetic markers or epigenotype/phenotype associations.

6.3.5 Assessment of Other Epigenetic Modifications

The study of epigenetic regulation in gene expression is an active area of research, and growing evidence supports the active role of several mechanisms in this field [96]. Interesting discussions have taken place over the last few years about this topic and the scope and limitations of the term "epigenetics," and which mechanisms can be classified as epigenetic [97, 98]. While not considered a classical epigenetic modification, noncoding RNAs (ncRNA) have emerged as an important regulator of gene expression in fish and other eukaryotes [99, 100]. Exposure to environmental factors has been shown to induce long-standing changes in ncRNA expression, indicating they may play a role in phenotypic plasticity and act as a reservoir of epigenetic information [99, 101]. Noncoding RNA constitutes a large family of RNAs with different biogenesis and functions. Several techniques are available to evaluate different types of ncRNAs [102, 103].

Traditional methods including northern blotting, RT-qPCR, microarrays, and RNA-seq can be used to characterize and quantify ncRNAs. According to the RNA size, sequence homology and relative abundance ncRNA may be challenging to work with. Alternatively, specific techniques have been developed to study RNA-protein interactions. RNA immunoprecipitation is a technique used to assess RNA interaction with proteins [104]. Similar to ChIP, the RNA-protein complexes are immunoprecipitated with specific antibodies targeting the protein of interest. Complexes are then digested with RNAses to remove sequences not protected by protein binding. Next, RNA is purified by RNA extraction and reverse transcribed to cDNA. cDNA can be analyzed through microarrays (RIP-chip) or sequencing (RIP-seq). Another technique to assess RNA-protein interactions is cross-linking and immunoprecipitation (CLIP) [105]. This technique relies on the formation of irreversible cross-links between RNA and protein using UV light. Complexes are then immunoprecipitated with a specific antibody and purified RNA is obtained through protein digestion by proteases. Finally, DNA is synthesized from an mRNA template and sequenced by NGS (HITS-CLIP and CLIP-seq). These methods have different strengths and limitations, making it important to select an optimal approach according to the research goals and biological questions [102, 103, 106].

Methods to assess transcriptional regulation, including chromatin accessibility and chromatin conformation (Box 6.1), are useful for providing information about active/inactive regions along the genome as well as chromosome organization within the nucleus. Among the techniques to assess chromatin accessibility are: assay for transposase-accessible chromatin using sequencing (ATAC-seq), DNase I hypersensitive sites sequencing (DNAse-seq), micrococcal nuclease digestion sequencing (MNase-seq), and nucleosome occupancy and methylome sequencing (NOMe-seq). These techniques are increasingly available for research purposes and have been revised recently in previous reviews [107, 108]. ATAC-seq, one of the most widely used techniques to assess genome-wide chromatin accessibility, uses a hyperactive Tn5 transposase to simultaneously cleave and insert sequencing adaptors into open regions of the genome [109]. These regions correspond to accessible chromatin and correlate with transcriptional regulation on a fine scale [110, 111]. Similarly, different techniques to examine chromatin conformation have been described. These include chromosome conformation capture (3-C), chromosome conformation capture-on-chip (4-C), chromosome conformation capture carbon copy (5-C), high throughput chromosome conformation capture (Hi-C), and Micro-C and Micro-Capture-C (MCC), among others [112–114]. These techniques aim to explore the 3D organization of the genome as an additional layer of information to understand gene regulation in physiological and perturbed states.

A general overview of the methods presented in this review is compiled in Table 6.1 and a summary of the characteristics of common methods is presented in Table 6.2.

Table 6.1 Classification of epigenetic methods according to the type of modification, detection method, coverage, and resolution level.

		Genome coverage				
Epigenetic modification	**Detection strategy**	**Gene- or locus-specific**	**Genome-wide**	**Whole genome**	**Resolution level**	**References**
DNA methylation	Enzymatic		LUMA, MSAP, RLGS		Low	[31, 32, 115]
			MSAP-seq, MRE-seq, methyl-seq		Medium	[33–35]
		HpaII/*MspI*-PCR		EM-seq, TAPS	High	[36, 41, 116]
	Affinity			ELISA	Low	[42]
		MeDIP-PCR, MBD-PCR	MeDIP-chip, MeDIP-seq, MBD-seq		Medium	[43]
	Bisulfite	MSP			Low	[56]
				Low-coverage WGBS	Medium	[76]
		BSP, COBRA, MBS, Ms-SNuPE	RRBS, bsRAD-seq, TIME-seq, microarrays	WGBS, PBAT	High	[18, 27, 57–59, 62–66, 72]
	Direct detection			HPLC-UV, LC-MS/MS	Low	[79, 80]
				SMRT-seq, nanopore	High	[82, 117]
Histone modifications				LC-MS/MS	Low	[118]
		ChIP-PCR	ChIP-chip	ChIP-seq, CUT&RUN, CUT&Tag	High	[91–95]
Noncoding RNA		Northern blotting, PCR, qPCR	Microarrays, RIP-chip	ncRNA-seq, RIP-seq, HITS-CLIP	High	[102–105].

Table 6.2 Comparison of common methods for assessing epigenetic modifications.

Technique	Resolution	Coverage	Input material	Cost	Reference genome needed	Identification of differences at specific loci	Advantages	Considerations/limitations	References
Enzymatic methods to assess DNA methylation									
LUMA	Low	Genome-wide	250–500 ng gDNA	$	No	No	Short time around, high reproducibility	Low resolution (global methylation levels)	[32]
MSAP, RLGS	Low	Genome-wide	200–1,000 ng gDNA	$	No	No	Cost, no special equipment necessary	Low resolution, prone to false negatives, large DNA quantity required	[29, 31]
MSAP-seq	Medium	Genome-wide	500 ng gDNA	$$	Yes	Yes	High-throughput, multiplexing capacity, intermediate resolution	Laborious protocol, NGS equipment and expertise required	[33]
EM-seq	High	Whole genome	10–200 ng gDNA (low-input)	$$$	Yes	Yes	Single-base resolution, discrimination between 5mC and 5hmC, uniform coverage	Laborious protocol, NGS equipment and expertise required	[36]
TAPS	High	Whole genome	500 ng gDNA	$$$	Yes	Yes	Single-base resolution, discrimination between 5mC and 5hmC, uniform coverage	Laborious protocol, NGS equipment and expertise required	[41]
Affinity-based methods to assess DNA methylation									
ELISA	Low	Whole genome	100 ng gDNA	$	No	No	Short time around, simple procedure	Low resolution (global methylation levels)	[42]
MeDIP-seq	Medium	Genome-wide	200 ng gDNA	$$	Yes	Yes	High-throughput, specificity for 5mC, enrichment capacity	Laborious protocol, NGS equipment and expertise required, low base-pair resolution	[119]
MBD-seq	Medium	Genome-wide	15 ng gDNA (low-input protocol)	$$	Yes	Yes	High-throughput, specificity for 5mC, enrichment capacity	Laborious protocol, NGS equipment and expertise required, low base-pair resolution	[120]
Bisulfite-based methods to assess DNA methylation									
BSP, COBRA	High	Gene- or locus-specific	100 ng bisulfite treated DNA	$	No whole genome but sequence needed for primer design	Yes	High resolution, multiplexing capacity	Low-throughput, target sequence required	[58, 121]
MBS	High	Gene- or locus-specific	100 ng bisulfite-treated DNA	$	No whole genome but sequence needed for primer design	Yes	High resolution, quantitation of multiple loci, multiplexing capacity	NGS equipment and expertise required, target sequence required	[18]

Microarrays	High	Genome-wide	200 ng bisulfite-treated DNA	$$	Yes	Yes	High reproducibility, cost-efficient	Target sequences required, microarrays equipment required, not available for fish species	[122]
RRBS	High	Genome-wide	2.5 μg gDNA	$$	Yes	Yes	High reproducibility, multiplexing capacity, enrichment capacity	Laborious protocol, limited coverage, NGS equipment and expertise required, quantity of input material required	[123]
bsRAD-seq	High	Genome-wide	1 μg gDNA	$$	No	Yes	Useful for nonmodel species, multiplexing and enrichment capacity	Laborious protocol, limited coverage, NGS equipment and expertise required, low sensitivity to identify differentially methylated loci	[62]
TIME-seq	High	Genome-wide	100 ng gDNA	$	Yes	Yes	Cost-efficient, multiplexing and enrichment capacity	Laborious protocol, target sequences required, NGS equipment and expertise required	[63]
Low-coverage WGBS	Medium	Whole genome	0.6–30 ng gDNA (low-input)	$	Yes	Yes	Price, single-base resolution, ultra-low input	Laborious protocol, limited coverage, NGS equipment, and expertise required	[76, 124]
WGBS	High	Whole genome	5 μg gDNA	$$$	Yes	Yes	Gold-standard, single-base resolution, coverage	Laborious protocol, NGS equipment and expertise required, quantity of input material required, price	[125]
PBAT and similar	High	Whole genome	0.6–30 ng gDNA (low-input)	$$$	Yes	Yes	Gold-standard, single-base resolution, coverage, low-input, compatible with single-cell analysis	Laborious protocol, NGS equipment and expertise required, price	[74, 124]
Direct methods to assess DNA methylation									
HPLC-UV	Low	Whole genome	3–10 μg gDNA	$$	No	No	High sensitivity, discrimination between 5mC and derivatives	Specialized equipment and expertise required, low resolution, quantity of input material required	[79, 126]
LC-MS/MS	Low	Whole genome	50–100 ng gDNA	$$	No	No	High sensitivity, discrimination between 5mC and derivatives	Specialized equipment and expertise required, low resolution	[80]

(Continued)

Table 6.2 (Continued)

Technique	Resolution	Coverage	Input material	Cost	Reference genome needed	Identification of differences at specific loci	Advantages	Considerations/limitations	References
SMRT-seq	High	Whole genome	100 ng gDNA (low-input)	$$$	Yes	Yes	Single-base resolution, discrimination between 5mC and derivatives, uniform coverage, long-reads	Technology in development (incipient), NGS equipment, and expertise required	[127]
Nanopore	High	Whole genome	2 μg gDNA	$$	Yes	Yes	Single-base resolution, discrimination between 5mC and derivatives, uniform coverage, long-reads, portability	NGS equipment and expertise required, technology in development, quantity of input material required	[128]
Methods to assess histone modifications									
ChIP-seq	High	Whole genome	1–10 million cells	$$$	Yes	Yes	High resolution, versatility	Laborious protocol, NGS equipment and expertise required, quality of antibodies is critical, higher number of reads required	[129]
CUT&Run	High	Whole genome	500,000 cells	$$$	Yes	Yes	Lower number of reads required, high-throughput	Laborious protocol, NGS equipment and expertise required, quality of antibodies is critical	[130, 131]
CUT&Tag	High	Whole genome	5000–500,000 cells	$$$	Yes	Yes	Lower number of reads required, high-throughput, compatible with single-cell analysis	Laborious protocol, NGS equipment and expertise required, quality of antibodies is critical	[95]
Methods to assess other epigenetic modifications/transcriptional regulation									
ncRNA-seq	High	Whole genome	1–200 ng total RNA	$$$	No	Yes	Discovery of novel transcripts, high sensitivity, useful for nonmodel species	Laborious protocol, NGS equipment and expertise required	[132]
ATAC-seq	High	Whole genome	50,000 cells	$$$	Yes	Yes	Lower number of reads required compared to other methods, high resolution	Laborious protocol, NGS equipment and expertise required, quality of antibodies is critical, amount of input material, reproducibility	[133]

6.4 Bioinformatics Analysis

Advances in sequencing methods and genomic tools are rapidly transforming our ability to study the epigenome and solve important biological questions. In practice, large datasets obtained by these technologies are increasingly common and require specialized analysis using robust bioinformatics and biostatistics tools. On the bright side, these methods constitute an active area of research and offer a powerful approach to studying the epigenome with unprecedented resolution. On the other hand, the vast quantity of information obtained may be overwhelming and often requires specialized personnel for data analysis. Importantly, there is an increasing necessity for interdisciplinary teams, including bioinformaticians and wet lab and field scientists, to strengthen research capacity and capabilities in epigenomics. Furthermore, several data standards and processing pipelines have been proposed by recognized groups and consortia, for example, the Encyclopedia of DNA Elements (ENCODE) consortium has proposed a pipeline to analyze histone ChIP-seq data and several reviews discuss computational methods and tools for analyzing and interpreting DNA methylation data [134–136]. Here we will briefly discuss the general steps to analyze data obtained by high-throughput methods.

There are different kinds of data that can be obtained from DNA methylation studies. The output for microarray-based techniques, for example, is fluorescence intensity at specific probes, quantifying relative abundance of methylated and unmethylated in specific loci. The analysis for this kind of data includes filtering problematic probes, within-array normalization, and between-array normalization [64, 137]. There are commercial and public tools to perform this analysis (e.g., GenomeStudio software by Illumina; BMIQ) [122, 138]. The output for NGS-based techniques, in contrast, is a FASTQ file, a text format that contains the sequence data and its corresponding quality. The general pipeline for this data includes quality control and trimming, alignment to a reference genome, quantification of DNA methylation, and interpretation. Quality control identifies potential problems in the raw data, such as low-quality libraries and sequencing problems. FASTQC and MultiQC are useful tools for this purpose [139, 140]. Next, quality trimming removes low-quality bases and adaptors from sequence reads, as well as very short reads. Packages such as TrimGalore, Cutadapt, and Trimmomatic are examples of trimming tools [141–143]. Processed reads are mapped to the genome using different tools. For non-bisulfite converted reads, for example, conventional alignment tools such as BWA or Bowtie2 can be used [144, 145]. Likewise, several tools have been developed for bisulfite-converted reads mapping, including Bismark, BSMAP, and BS-Seeker [146–148]. Although these algorithms have similar performance, Bismark has higher mapping efficiency and is generally considered a robust method [77, 149]. The sequence alignment information is stored as SAM (Sequence Alignment Map) files or its compressed version BAM (Binary Alignment Map). There are different tools to quantify DNA methylation and interpret the results obtained according to the technique used. For restriction enzyme and enrichment affinity-based methods (e.g., MRE-seq, MeDIP-seq, and MBD-seq), DNA methylation relies on the number of reads mapped to the corresponding genomic locations. Several specific tools have been described for this purpose (e.g., msgbsR, MEDIPS, and MEDUSA). These have been described in previous reviews [150, 151]. For bisulfite sequencing-based methods, methylation calls at individual positions are determined using software such as Bismark, Bis-SNP, and MethylDackel [148, 152, 153]. Downstream analysis to identify differentially methylated loci (DML), also referred to as differentially methylated cytosines (DMC) or differentially methylated regions (DMR), can be performed using software such as SeqMonk, BSmooth, DSS, or methylKit, among others [154–157]. Some packages such as Bismark and BSmooth allow a comprehensive analysis of DNA methylation including several tools for different purposes. Lastly, data visualization tools including SeqMonk, Integrative Genome Viewer, and UCSC Genome Browser are useful resources to interpret and contextualize epigenetic information [157–159].

Bioinformatics analysis of sequencing-based methods to analyze histone modifications such as ChIP-seq and CUT&RUN/CUT&Tag include quality control and trimming, alignment to a reference genome, peak calling, and downstream analysis such as detection of loci with differences in binding events and motif analysis [93]. After read mapping, peak calling identifies significantly enriched loci across the genome. Several tools have been developed for this purpose; among these, MACS2 and HOMER are commonly used [160–162]. In addition, a specific peak caller for CUT&RUN datasets named and Sparse Enrichment Analysis for CUT&RUN (SEACR) has been recently described [163]. Importantly, statistical methods for data normalization are recommended to improve the signal-to-noise ratio [164]. Furthermore, visualization of peaks utilizing tools such as SeqMonk, UCSC Genome Browser, and WashU Epigenome Browser is a useful approach to identifying global patterns, and the software is compatible with different datasets and annotation tracks [157, 158, 165]. Several bioinformatic tools for differential binding analysis have been developed, some of them specifically designed for histone modification data [166–168]. Motif analysis identifies potentially enriched short sequences in the peaks called, allowing to detect known binding motifs, for example, for transcription factors or novel patterns [164]. Examples of motif analysis software include the MEME suite and HOMER [162, 169].

Lastly, bioinformatics analysis of ncRNA includes conventional pipelines for RNA-seq and tools to detect and predict new ncRNA [170, 171]. In addition, given their importance as regulatory mechanisms, several tools have been developed to predict gene targets and identify functional enrichment [172, 173]. Noteworthy, integrative approaches using multiple epigenomic data and other datasets (e.g., RNA-seq, chromatin accessibility data) have shown to be particularly powerful and useful to dissect the molecular mechanisms underlying complex biological phenomena [9, 174].

6.5 Databases and Other Public Resources

Genomic and epigenomic databases allow the storing, sharing, and comparison of data improving transparency, reproducibility, and strengthening research activities [175]. Given the complexity and diversity of biological systems, these tools are critical for epigenomic research. Currently, there is a growing ecosystem of general and dedicated databases, including repositories for raw and processed data. Similarly, tools for data analysis and visualization are increasingly available and facilitate bioinformatic workflows and interpretation.

International public repositories such as Gene Expression Omnibus (GEO), ArrayExpress, Sequence Read Archive (SRA), and the European Nucleotide Archive (ENA) constitute large databases of high-throughput data [176–179]. In addition, an interesting effort in fish research has been done by the DANIO-CODE consortium to annotate functional elements in the zebrafish (*Danio rerio*) genome using a similar approach to ENCODE for humans. Raw and processed information, as well as experimental details, are contained in these repositories, constituting a valuable resource for comparison and new analyses. Importantly, any researcher can submit their data, after which a specialized team validates and curates the data, and finally, the information is publicly accessible worldwide. Alternatively, data can remain private until it is published or released by the submitter. Meanwhile, the owner can share a token that allows confidential sharing, which is useful, for example, for reviewers or collaborators. A search of specific datasets is facilitated by structured data entry forms, access codes, and cross-references with papers and other databases.

Another important resource in epigenomics is the development of accessible tools for bioinformatics analysis and code repositories. Galaxy, for instance, is a scientific analysis platform that provides a user-friendly and versatile interface for computationally intensive tasks including epigenomic data processing [180]. Galaxy is distributed both as a publicly available web-based service or can be installed on a local server. Noteworthy, Galaxy's developers and collaborators have made publicly accessible training material for epigenomics and other data analyses (https://training.galaxyproject.org/training-material). Likewise, source code used for specific analysis and an increasing number of bioinformatic packages can also be shared by GitHub or other code repositories [181]. Overall, data and code sharing are currently considered essential and pervasive in scientific research and facilitate more transparent and reproducible research practices. Several practical guidelines and recommendations have been suggested by the research community to facilitate these good research practices [182–184].

International consortia aiming to generate high-quality genomes and comprehensive information about fish and other species are valuable initiatives with a broad impact on biological sciences and aquaculture. The G10K consortium, for example, is constituted by a large community of scientists including tissue curators, biologists, bioinformaticians, and computer scientists aiming to sequence, assemble, annotate, and analyze the genome of 10,000 vertebrate species [185]. Although still in development, this initiative has already sequenced the genome of the southern bluefin tuna (*Thunnus maccoyii*) and brown trout (*Salmo trutta*), two species of aquaculture interest [186]. This data is particularly valuable to understand several biological processes, including epigenetic regulation. Data visualization tools, on the other hand, are useful resources to explore genomic information and integrate epigenomic data. The UCSC Genome Browser, for example, has incorporated genome assemblies for numerous fish species [158]. Similar to desktop visualization tools, it is possible to upload custom tracks facilitating data analysis and interpretation. Ensembl, another genome browser, is a practical tool to explore genomic and epigenomic information about vertebrate species, particularly fitted for comparative genomics, evolution, sequence variation, and transcriptional regulation [187].

6.6 Conclusions and Outlook

In this review, we have presented several analytical methods to study the epigenome. These include a wide spectrum of techniques to explore different types of epigenetic modifications, including cytosine methylation, histone modifications, and noncoding RNA. Importantly, there is not a unique method fitted for all situations, yet this should be carefully chosen by the research team, taking into consideration the research question and the pros and cons of each technique. Over the last decade, the development of high-throughput methods, particularly NGS, has revolutionized the way how epigenomics is studied in aquatic animals. While NGS-based techniques constitute robust and sensible tools to analyze epigenetic modifications, special attention should be taken to obtain high-quality genomes from species of aquaculture interest allowing further exploration and careful experimental design and bioinformatics analysis following existing guidelines and protocols. These suggestions are made with the aim of obtaining reproducible and meaningful biological results and making progress in this promising area. Furthermore, the development of new experimental methods and biological tools to analyze the epigenome is an active area of study, making the future bright for epigenomics research in aquatic species.

In comparison to other vertebrate systems, fish are more plastic and physiologically diverse, meaning epigenetic studies are more relevant to understand fish biology, including important phenomena such as sex determination, development, and adaptation to a changing environment. These studies are possible thanks to a confluence of technologies, and among these, integrative approaches including phenotypic, genetic, epigenetic, transcriptomic, and ecological traits are particularly useful and have shed light on different aspects of fish biology and fishery research. These approaches require an interdisciplinary team to address the challenges of a rapidly evolving field and complex biological questions. Lastly, a growing network of collaborators and shared resources will be critical to accelerate research in this area. International consortia and initiatives have shown the utility of such approximation. The increasing integration of epigenomic methods into aquaculture will not only be interesting for the research community but also useful for practical purposes, including environmental manipulation and epigenetic selection.

Acknowledgments

The Hore laboratory is supported by funding from the University of Otago (111899.01.R.LA), the Marsden Fund (18-UOO-200), and the Ministry for Business, Innovation, and Enterprise (UOOX1909).

References

1 Reik, W. (2007). Stability and flexibility of epigenetic gene regulation in mammalian development. *Nature* **447** (7143): 425–432.

2 Le Luyer, J., Laporte, M., Beacham, T.D. et al. (2017). Parallel epigenetic modifications induced by hatchery rearing in a Pacific salmon. *Proc. Natl. Acad. Sci.* **114** (49): 12964–12969.

3 Anastasiadi, D. and Piferrer, F. (2019). Epimutations in developmental genes underlie the onset of domestication in farmed European sea bass. *Mol. Biol. Evol.* **36** (10): 2252–2264.

4 Podgorniak, T., Brockmann, S., Konstantinidis, I., and Fernandes, J.M.O. (2019). Differences in the fast muscle methylome provide insight into sex-specific epigenetic regulation of growth in Nile tilapia during early stages of domestication. *Epigenetics* **14** (8): 818–836.

5 Konstantinidis, I., Sætrom, P., Mjelle, R. et al. (2020). Major gene expression changes and epigenetic remodelling in Nile tilapia muscle after just one generation of domestication. *Epigenetics* **15** (10): 1052–1067.

6 Gavery, M.R., Nichols, K.M., Goetz, G.W. et al. (2018). Characterization of genetic and epigenetic variation in sperm and red blood cells from adult hatchery and natural-origin steelhead, Oncorhynchus mykiss. *G3 (Bethesda)* **8** (11): 3723–3736.

7 Heckwolf, M.J., Meyer, B.S., Häsler, R. et al. (2020). Two different epigenetic information channels in wild three-spined sticklebacks are involved in salinity adaptation. *Sci. Adv.* **6** (12): eaaz1138.

8 Ortega-Recalde, O., Day, R.C., Gemmell, N.J., and Hore, T.A. (2019). Zebrafish preserve global germline DNA methylation while sex-linked rDNA is amplified and demethylated during feminisation. *Nat. Commun.* **10** (1): 3053.

9 Todd, E.V., Ortega-Recalde, O., Liu, H. et al. (2019). Stress, novel sex genes, and epigenetic reprogramming orchestrate socially controlled sex change. *Sci. Adv.* **5** (7): eaaw7006.

10 Ryu, T., Veilleux, H.D., Donelson, J.M. et al. (2018). The epigenetic landscape of transgenerational acclimation to ocean warming. *Nat. Clim. Chang.* **8** (6): 504–509.

11 Gavery, M.R. and Roberts, S.B. (2017). Epigenetic considerations in aquaculture. *PeerJ* **5**: e4147.

12 Kurdyukov, S. and Bullock, M. (2016). DNA methylation analysis: choosing the right method. *Biology (Basel)* **5** (1): 3.

13 Goldberg, A.D., Allis, C.D., and Bernstein, E. (2007). Epigenetics: a landscape takes shape. *Cell* **128** (4): 635–638.

14 Johnson, T.B. and Coghill, R.D. (1925). Reseraches on pyrimidines. C111. The dicovery of 5-methyl-cytosine in tuberculinic acid, the nucleic acid of the tubercle bacillus 1. *J. Am. Chem. Soc.* **47** (11): 2838–2844.
15 Holliday, R. and Pugh, J.E. (1975). DNA modification mechanisms and gene activity during development. *Science* **187** (4173): 226–232.
16 Riggs, A.D. (1975). X inactivation, differentiation, and DNA methylation. *Cytogenet. Cell Genet.* **14** (1): 9–25.
17 Lam, D., Luu, P.-L., Song, J.Z. et al. (2020). Comprehensive evaluation of targeted multiplex bisulphite PCR sequencing for validation of DNA methylation biomarker panels. *Clin. Epigenetics* **12** (1): 90.
18 Anastasiadi, D., Vandeputte, M., Sánchez-Baizán, N. et al. (2018). Dynamic epimarks in sex-related genes predict gonad phenotype in the European sea bass, a fish with mixed genetic and environmental sex determination. *Epigenetics* **13** (9): 988–1011.
19 Bock, C., Tomazou, E.M., Brinkman, A.B. et al. (2010). Quantitative comparison of genome-wide DNA methylation mapping technologies. *Nat. Biotechnol.* **28** (10): 1106–1114.
20 Tessarz, P. and Kouzarides, T. (2014). Histone core modifications regulating nucleosome structure and dynamics. *Nat. Rev. Mol. Cell Biol.* **15** (11): 703–708.
21 Bannister, A.J. and Kouzarides, T. (2011). Regulation of chromatin by histone modifications. *Cell Res.* **21** (3): 381–395.
22 Zhang, T., Cooper, S., and Brockdorff, N. (2015). The interplay of histone modifications – writers that read. *EMBO Rep.* **16** (11): 1467–1481.
23 Lee, J.T. (2012). Epigenetic regulation by long noncoding RNAs. *Science* **338** (6113): 1435–1439.
24 Wei, J.-W., Huang, K., Yang, C., and Kang, C.-S. (2017). Non-coding RNAs as regulators in epigenetics. *Oncol. Rep.* **37** (1): 3–9.
25 Cedar, H., Solage, A., Glaser, G., and Razin, A. (1979). Direct detection of methylated cytosine in DNA by use of the restriction enzyme Mspl. *Nucleic Acids Res.* **6** (6): 2125–2132.
26 Bird, A.P. and Southern, E.M. (1978). Use of restriction enzymes to study eukaryotic DNA methylation. *J. Mol. Biol.* **118** (1): 27–47.
27 Meissner, A., Gnirke, A., Bell, G.W. et al. (2005). Reduced representation bisulfite sequencing for comparative high-resolution DNA methylation analysis. *Nucleic Acids Res.* **33** (18): 5868–5877.
28 Reyna-López, G.E., Simpson, J., and Ruiz-Herrera, J. (1997). Differences in DNA methylation patterns are detectable during the dimorphic transition of fungi by amplification of restriction polymorphisms. *Mol. Gen. Genet. MGG* **253** (6): 703–710.
29 Anastasiadi, D., Díaz, N., and Piferrer, F. (2017). Small ocean temperature increases elicit stage-dependent changes in DNA methylation and gene expression in a fish, the European sea bass. *Sci. Rep.* **7** (1): 12401.
30 Rodríguez-Casariego, J.A., Mercado-Molina, A.E., Garcia-Souto, D. et al. (2020). Genome-wide DNA methylation analysis reveals a conserved epigenetic response to seasonal environmental variation in the staghorn coral Acropora cervicornis. *Front. Mar. Sci.* **7**: 560424.
31 Ando, Y. and Hayashizaki, Y. (2006). Restriction landmark genomic scanning. *Nat. Protoc.* **1** (6): 2774–2783.
32 Mohsen, K., Johansson, S., and Ekström, T.J. (2006). Using LUMA: a luminometric-based assay for global DNA-methylation. *Epigenetics* **1** (1): 46–49.
33 Chwialkowska, K., Korotko, U., Kosinska, J. et al. (2017). Methylation sensitive amplification polymorphism sequencing (MSAP-seq)-a method for high-throughput analysis of differentially methylated CCGG sites in plants with large genomes. *Front. Plant Sci.* **8**: 2056.
34 Maunakea, A.K., Nagarajan, R.P., Bilenky, M. et al. (2010). Conserved role of intragenic DNA methylation in regulating alternative promoters. *Nature* **466** (7303): 253–257.
35 Brunner, A.L., Johnson, D.S., Kim, S.W. et al. (2009). Distinct DNA methylation patterns characterize differentiated human embryonic stem cells and developing human fetal liver. *Genome Res.* **19** (6): 1044–1056.
36 Vaisvila, R., Ponnaluri, V.K.C., Sun, Z. et al. (2021). Enzymatic methyl sequencing detects DNA methylation at single-base resolution from picograms of DNA. *Genome Res.* **31** (7): 1280–1289.
37 Ficz, G., Branco, M.R., Seisenberger, S. et al. (2011). Dynamic regulation of 5-hydroxymethylcytosine in mouse ES cells and during differentiation. *Nature* **473** (7347): 398–402.
38 Shi, D.-Q., Ali, I., Tang, J., and Yang, W.-C. (2017). New insights into 5hmC DNA modification: generation, distribution and function. *Front. Genet.* **8**: 100.
39 Ito, S., D'Alessio, A.C., Taranova, O.V. et al. (2010). Role of Tet proteins in 5mC to 5hmC conversion, ES-cell self-renewal and inner cell mass specification. *Nature* **466** (7310): 1129–1133.

40 Mellén, M., Ayata, P., Dewell, S. et al. (2012). MeCP2 binds to 5hmC enriched within active genes and accessible chromatin in the nervous system. *Cell* **151** (7): 1417–1430.

41 Liu, Y., Siejka-Zielińska, P., Velikova, G. et al. (2019). Bisulfite-free direct detection of 5-methylcytosine and 5-hydroxymethylcytosine at base resolution. *Nat. Biotechnol.* **37** (4): 424–429.

42 de Oliveira, N.F.P., de Castro Coêlho, M., and Viana Filho, J.M.C. (2020). ELISA analysis of global methylation levels. In: *Epigenetics Methods*, Translational Epigenetics, vol. **18** (ed. T. Tollefsbol), 83–92. Academic Press.

43 Mohn, F., Weber, M., Schübeler, D., and Roloff, T.-C. (2009). Methylated DNA immunoprecipitation (MeDIP). In: *DNA Methylation*, Methods in Molecular Biology, 2ee, vol. **507** (ed. J. Tost), 55–64. Humana Press.

44 Sun, L.-X., Wang, Y.-Y., Zhao, Y. et al. (2016). Global DNA methylation changes in Nile tilapia gonads during high temperature-induced masculinization. *PLoS one* **11** (8): e0158483.

45 Mirbahai, L., Yin, G., Bignell, J.P. et al. (2011). DNA methylation in liver tumorigenesis in fish from the environment. *Epigenetics* **6** (11): 1319–1333.

46 Han, B., Li, W., Chen, Z. et al. (2016). Variation of DNA methylome of zebrafish cells under cold pressure. *PLoS one* **11** (8): e0160358.

47 Gim, J.-A., Hong, C.P., Kim, D.-S. et al. (2015). Genome-wide analysis of DNA methylation before-and after exercise in the thoroughbred horse with MeDIP-seq. *Mol. Cells* **38** (3): 210–220.

48 Cortijo, S., Wardenaar, R., Colomé-Tatché, M. et al. (2014). Genome-wide analysis of DNA methylation in *Arabidopsis* using MeDIP-chip. In: *Plant Epigenetics and Epigenomics*, Methods in Molecular Biology, vol. **1112** (ed. C. Spillane and P.C. McKeown), 125–149. Humana Press.

49 Aberg, K.A., Chan, R.F., Xie, L. et al. (2018). Methyl-CpG-binding domain sequencing: MBD-seq. In: *DNA Methylation Protocols*, Methods in Molecular Biology, 3ee, vol. **1708** (ed. s.), 171–189. Humana Press.

50 Jeltsch, A., Broche, J., Lungu, C., and Bashtrykov, P. (2020). Biotechnological applications of MBD domain proteins for DNA methylation analysis. *J. Mol. Biol.* **432** (6): 1816–1823.

51 Harris, R.A., Wang, T., Coarfa, C. et al. (2010). Comparison of sequencing-based methods to profile DNA methylation and identification of monoallelic epigenetic modifications. *Nat. Biotechnol.* **28** (10): 1097–1105.

52 Frommer, M., McDonald, L.E., Millar, D.S. et al. (1992). A genomic sequencing protocol that yields a positive display of 5-methylcytosine residues in individual DNA strands. *Proc. Natl. Acad. Sci. U. S. A.* **89** (5): 1827–1831.

53 Huang, Y., Pastor, W.A., Shen, Y. et al. (2010). The behaviour of 5-hydroxymethylcytosine in bisulfite sequencing. *PLoS one* **5** (1): e8888.

54 Booth, M.J., Branco, M.R., Ficz, G. et al. (2012). Quantitative sequencing of 5-methylcytosine and 5-hydroxymethylcytosine at single-base resolution. *Science* **336** (6083): 934–937.

55 Yu, M., Hon, G.C., Szulwach, K.E. et al. (2012). Tet-assisted bisulfite sequencing of 5-hydroxymethylcytosine. *Nat. Protoc.* **7** (12): 2159–2170.

56 Ramalho-Carvalho, J., Henrique, R., and Jerónimo, C. (2018). Methylation-specific PCR. In: *DNA Methylation Protocols*, Methods in Molecular Biology, 3ee, vol. **1708** (ed. J. Tost), 447–472. Humana Press.

57 Darst, R.P., Pardo, C.E., Ai, L. et al. (2010). Bisulfite sequencing of DNA. *Curr. Protoc. Mol. Biol.* **91** (1): 7–9.

58 Eads, C.A. and Laird, P.W. (2002). Combined bisulfite restriction analysis (COBRA). *Methods Mol. Biol.* **200**: 71–85.

59 Gonzalgo, M.L. and Liang, G. (2007). Methylation-sensitive single-nucleotide primer extension (Ms-SNuPE) for quantitative measurement of DNA methylation. *Nat. Protoc.* **2** (8): 1931–1936.

60 Anastasiadi, D. and Piferrer, F. (2020). A clockwork fish: age prediction using DNA methylation-based biomarkers in the European sea bass. *Mol. Ecol. Resour.* **20** (2): 387–397.

61 Gu, H., Smith, Z.D., Bock, C. et al. (2011). Preparation of reduced representation bisulfite sequencing libraries for genome-scale DNA methylation profiling. *Nat. Protoc.* **6** (4): 468–481.

62 Trucchi, E., Mazzarella, A.B., Gilfillan, G.D. et al. (2016). BsRADseq: screening DNA methylation in natural populations of non-model species. *Mol. Ecol.* **25** (8): 1697–1713.

63 Griffin, P.T., Kane, A.E., Trapp, A. et al. (2021). Ultra-cheap and scalable epigenetic age predictions with TIME-seq. *bioRxiv* 10.1101/2021.10.25.465725.

64 Nakabayashi, K. (2020). The Illumina Infinium methylation assay for genome-wide methylation analyses. In: *Epigenetics Methods*, Translational Epigenetics, vol. **18** (ed. T. Tollefsbol), 117–140. Academic Press.

65 Schumacher, A., Kapranov, P., Kaminsky, Z. et al. (2006). Microarray-based DNA methylation profiling: technology and applications. *Nucleic Acids Res.* **34** (2): 528–542.

66 Bibikova, M. and Fan, J.-B. (2009). GoldenGate® assay for DNA methylation profiling. In: *DNA Methylation*, Methods in Molecular Biology, 2ee, vol. **507** (ed. J. Tost), 149–163. Humana Press.

67 Epiphanio, T.M.F., de Fernandes, N.C.C., A., de Oliveira, T.F. et al. (2019). Global DNA methylation of peripheral blood leukocytes from dogs bearing multicentric non-Hodgkin lymphomas and healthy dogs: a comparative study. *PLoS one* **14** (3): e0211898.

68 Pichon, F., Shen, Y., Busato, F. et al. (2021). Analysis and annotation of DNA methylation in two nonhuman primate species using the Infinium human methylation 450K and EPIC beadchips. *Epigenomics* **13** (3): 169–186.

69 Consortium, M.M., Lu, A.T., Fei, Z. et al. (2021). Universal DNA methylation age across mammalian tissues. *bioRxiv* 10.1093/geroni/igab046.1588.

70 Laird, P.W. (2010). Principles and challenges of genomewide DNA methylation analysis. *Nat. Rev. Genet.* **11** (3): 191–203.

71 Lister, R., O'Malley, R.C., Tonti-Filippini, J. et al. (2008). Highly integrated single-base resolution maps of the epigenome in *Arabidopsis*. *Cell* **133** (3): 523–536.

72 Grehl, C., Kuhlmann, M., Becker, C. et al. (2018). How to design a whole-genome bisulfite sequencing experiment. *Epigenomes* **2** (4): 21.

73 Li, N., Ye, M., Li, Y. et al. (2010). Whole genome DNA methylation analysis based on high throughput sequencing technology. *Methods* **52** (3): 203–212.

74 Miura, F., Enomoto, Y., Dairiki, R., and Ito, T. (2012). Amplification-free whole-genome bisulfite sequencing by post-bisulfite adaptor tagging. *Nucleic Acids Res.* **40** (17): e136.

75 Smallwood, S.A., Lee, H.J., Angermueller, C. et al. (2014). Single-cell genome-wide bisulfite sequencing for assessing epigenetic heterogeneity. *Nat. Methods* **11** (8): 817–820.

76 Ortega-Recalde, O., Peat, J.R., Bond, D.M., and Hore, T.A. (2021). Estimating global methylation and erasure using low-coverage whole-genome bisulfite sequencing (WGBS). *Methods Mol. Biol.* **2272**: 29–44.

77 Tran, H., Porter, J., Sun, M. et al. (2014). Objective and comprehensive evaluation of bisulfite short read mapping tools. *Adv. Bioinforma.* **2014**: 1–11.

78 Olova, N., Krueger, F., Andrews, S. et al. (2018). Comparison of whole-genome bisulfite sequencing library preparation strategies identifies sources of biases affecting DNA methylation data. *Genome Biol.* **19** (1): 33.

79 Armstrong, K.M., Bermingham, E.N., Bassett, S.A. et al. (2011). Global DNA methylation measurement by HPLC using low amounts of DNA. *Biotechnol. J.* **6** (1): 113–117.

80 Le, T., Kim, K.-P., Fan, G., and Faull, K.F. (2011). A sensitive mass spectrometry method for simultaneous quantification of DNA methylation and hydroxymethylation levels in biological samples. *Anal. Biochem.* **412** (2): 203–209.

81 Flusberg, B.A., Webster, D.R., Lee, J.H. et al. (2010). Direct detection of DNA methylation during single-molecule, real-time sequencing. *Nat. Methods* **7** (6): 461–465.

82 Tse, O.Y.O., Jiang, P., Cheng, S.H. et al. (2021). Genome-wide detection of cytosine methylation by single molecule real-time sequencing. *Proc. Natl. Acad. Sci. U. S. A.* **118** (5): e2019768118.

83 Goodwin, S., McPherson, J.D., and McCombie, W.R. (2016). Coming of age: ten years of next-generation sequencing technologies. *Nat. Rev. Genet.* **17** (6): 333–351.

84 Yuen, Z.W.-S., Srivastava, A., Daniel, R. et al. (2021). Systematic benchmarking of tools for CpG methylation detection from nanopore sequencing. *Nat. Commun.* **12** (1): 3438.

85 Jain, M., Koren, S., Miga, K.H. et al. (2018). Nanopore sequencing and assembly of a human genome with ultra-long reads. *Nat. Biotechnol.* **36** (4): 338–345.

86 Gigante, S., Gouil, Q., Lucattini, A. et al. (2019). Using long-read sequencing to detect imprinted DNA methylation. *Nucleic Acids Res.* **47** (8): e46–e46.

87 Tang, Y., Zheng, S.-J., Qi, C.-B. et al. (2015). Sensitive and simultaneous determination of 5-methylcytosine and its oxidation products in genomic DNA by chemical derivatization coupled with liquid chromatography-tandem mass spectrometry analysis. *Anal. Chem.* **87** (6): 3445–3452.

88 Nair, S.S., Coolen, M.W., Stirzaker, C. et al. (2011). Comparison of methyl-DNA immunoprecipitation (MeDIP) and methyl-CpG binding domain (MBD) protein capture for genome-wide DNA methylation analysis reveal CpG sequence coverage bias. *Epigenetics* **6** (1): 34–44.

89 Lawrence, M., Daujat, S., and Schneider, R. (2016). Lateral thinking: how histone modifications regulate gene expression. *Trends Genet.* **32** (1): 42–56.

90 Carey, M.F., Peterson, C.L., and Smale, S.T. (2009). Chromatin immunoprecipitation (ChIP). *Cold Spring Harb Protoc* **2009** (9): pdb.prot5279.

91 Migeot, V. and Hermand, D. (2018). Chromatin immunoprecipitation-polymerase chain reaction (ChIP-PCR) detects methylation, acetylation, and ubiquitylation in *S. pombe*. In: *Schizosaccharomyces Pombe*, Methods in Molecular Biology, vol. **1721** (ed. T.L. Singleton), 25–34. Humana Press.

92 Kim, T.H. and Dekker, J. (2018). ChIP-chip. *Cold Spring Harb. Protoc.* **2018** (5): pdb.prot082636.

93 Nakato, R. and Sakata, T. (2021). Methods for ChIP-seq analysis: a practical workflow and advanced applications. *Methods* **187**: 44–53.

94 Skene, P.J. and Henikoff, S. (2017). An efficient targeted nuclease strategy for high-resolution mapping of DNA binding sites. *Elife* **6**: e21856.

95 Kaya-Okur, H.S., Wu, S.J., Codomo, C.A. et al. (2019). CUT&Tag for efficient epigenomic profiling of small samples and single cells. *Nat. Commun.* **10** (1): 1930.

96 Deans, C. and Maggert, K.A. (2015). What do you mean, "epigenetic"? *Genetics* **199** (4): 887–896.

97 Greally, J.M. (2018). A user's guide to the ambiguous word "epigenetics". *Nat. Rev. Mol. Cell Biol.* **19** (4): 207–208.

98 Deichmann, U. (2016). Epigenetics: the origins and evolution of a fashionable topic. *Dev. Biol.* **416** (1): 249–254.

99 Peschansky, V.J. and Wahlestedt, C. (2014). Non-coding RNAs as direct and indirect modulators of epigenetic regulation. *Epigenetics* **9** (1): 3–12.

100 Frías-Lasserre, D. and Villagra, C.A. (2017). The importance of ncRNAs as epigenetic mechanisms in phenotypic variation and organic evolution. *Front. Microbiol.* **8**: 2483.

101 Bizuayehu, T.T., Johansen, S.D., Puvanendran, V. et al. (2015). Temperature during early development has long-term effects on microRNA expression in Atlantic cod (*Gadus morhua*). *BMC Genomics* **16** (1): 305.

102 Ye, J., Xu, M., Tian, X. et al. (2019). Research advances in the detection of miRNA. *J. Pharm. Anal.* **9** (4): 217–226.

103 Feng, Y., Hu, X., Zhang, Y. et al. (2014). Methods for the study of long noncoding RNA in cancer cell Signaling. In: *Cancer Cell Signaling*, Mehtods in Molecular Biology, 2ee, vol. **1165** (ed. M. Robles-Flores), 115–143.

104 Gagliardi, M. and Matarazzo, M.R. (2016). RIP: RNA immunoprecipitation. *Methods Mol. Biol.* **1480**: 73–86.

105 Hafner, M., Katsantoni, M., Köster, T. et al. (2021). CLIP and complementary methods. *Nat. Rev. Methods Prim.* **1** (1): 20.

106 Ramanathan, M., Porter, D.F., and Khavari, P.A. (2019). Methods to study RNA–protein interactions. *Nat. Methods* **16** (3): 225–234.

107 Klemm, S.L., Shipony, Z., and Greenleaf, W.J. (2019). Chromatin accessibility and the regulatory epigenome. *Nat. Rev. Genet.* **20** (4): 207–220.

108 Minnoye, L., Marinov, G.K., Krausgruber, T. et al. (2021). Chromatin accessibility profiling methods. *Nat. Rev. Methods Prim.* **1** (1): 10.

109 Bogdanović, O., Smits, A.H., de la Calle Mustienes, E. et al. (2016). Active DNA demethylation at enhancers during the vertebrate phylotypic period. *Nat. Genet.* **48** (4): 417–426.

110 Ranzoni, A.M., Tangherloni, A., Berest, I. et al. (2021). Integrative single-cell RNA-seq and ATAC-seq analysis of human developmental hematopoiesis. *Cell Stem Cell* **28** (3): 472–487.e7.

111 Wu, X., Yang, Y., Zhong, C. et al. (2020). Integration of ATAC-seq and RNA-seq unravels chromatin accessibility during sex reversal in orange-spotted grouper (*Epinephelus coioides*). *Int. J. Mol. Sci.* **21** (8): 2800.

112 McCord, R.P., Kaplan, N., and Giorgetti, L. (2020). Chromosome conformation capture and beyond: toward an integrative view of chromosome structure and function. *Mol. Cell* **77** (4): 688–708.

113 Hua, P., Badat, M., Hanssen, L.L.P. et al. (2021). Defining genome architecture at base-pair resolution. *Nature* **595** (7865): 125–129.

114 Goel, V.Y. and Hansen, A.S. (2021). The macro and micro of chromosome conformation capture. *WIREs Dev. Biol.* **10** (6): e395.

115 Guevara, M.Á., de María, N., Sáez-Laguna, E. et al. (2017). Analysis of DNA cytosine methylation patterns using methylation-sensitive amplification polymorphism (MSAP). In: *Plant Epigenetics*, Methods in Molecular Biology, 2ee, vol. **1456** (ed. I. Kovalchuk), 99–112. Humana Press.

116 Wang, C.-H., Lai, H.-C., Liou, T.-M. et al. (2013). A DNA methylation assay for detection of ovarian cancer cells using a HpaII/MspI digestion-based PCR assay in an integrated microfluidic system. *Microfluid Nanofluidics* **15** (5): 575–585.

117 Simpson, J.T., Workman, R.E., Zuzarte, P.C. et al. (2017). Detecting DNA cytosine methylation using nanopore sequencing. *Nat. Methods* **14** (4): 407–410.

118 Sidoli, S., Bhanu, N.V., Karch, K.R. et al. (2016). Complete workflow for analysis of histone post-translational modifications using bottom-up mass spectrometry: from histone extraction to data analysis. *J. Vis. Exp.* (111): e54112. https://doi.org/10.3791/54112.

119 Taiwo, O., Wilson, G.A., Morris, T. et al. (2012). Methylome analysis using MeDIP-seq with low DNA concentrations. *Nat. Protoc.* **7** (4): 617–636.

120 Lan, X., Adams, C., Landers, M. et al. (2011). High resolution detection and analysis of CpG dinucleotides methylation using MBD-seq technology. *PLoS one* **6** (7): e22226.

121 Jiang, M., Zhang, Y., Fei, J. et al. (2010). Rapid quantification of DNA methylation by measuring relative peak heights in direct bisulfite-PCR sequencing traces. *Lab. Investig.* **90** (2): 282–290.

122 Dedeurwaerder, S., Defrance, M., Calonne, E. et al. (2011). Evaluation of the Infinium methylation 450K technology. *Epigenomics* **3** (6): 771–784.

123 Chatterjee, A., Rodger, E.J., Stockwell, P.A. et al. (2012). Technical considerations for reduced representation bisulfite sequencing with multiplexed libraries. *J. Biomed. Biotechnol.* **2012**: 1–8.

124 Peat, J.R. and Smallwood, S.A. (2018). Low input whole-genome bisulfite sequencing using a post-bisulfite adapter tagging approach. *Methods Mol. Biol.* **1708**: 161–169.

125 Lister, R., Pelizzola, M., Dowen, R.H. et al. (2009). Human DNA methylomes at base resolution show widespread epigenomic differences. *Nature* **462** (7271): 315–322.

126 Kuo, K.C., McCune, R.A., Gehrke, C.W. et al. (1980). Quantitative reversed-phase high performance liquid chromatographic determination of major and modified deoxyribonucleosides in DNA. *Nucleic Acids Res.* **8** (20): 4763–4776.

127 Kingan, S., Heaton, H., Cudini, J. et al. (2019). A high-quality De novo genome assembly from a single mosquito using PacBio sequencing. *Genes (Basel)* **10** (1): 62.

128 Davenport, C.F., Scheithauer, T., Dunst, A. et al. (2021). Genome-wide methylation mapping using nanopore sequencing technology identifies novel tumor suppressor genes in hepatocellular carcinoma. *Int. J. Mol. Sci.* **22** (8): 3937.

129 Kidder, B.L., Hu, G., and Zhao, K. (2011). ChIP-seq: technical considerations for obtaining high-quality data. *Nat. Immunol.* **12** (10): 918–922.

130 Kong, N.R., Chai, L., Tenen, D.G., and Bassal, M.A. (2021). A modified CUT&RUN protocol and analysis pipeline to identify transcription factor binding sites in human cell lines. *STAR Protoc.* **2** (3): 100750.

131 Hainer, S.J. and Fazzio, T.G. (2019). High-resolution chromatin profiling using CUT&RUN. *Curr. Protoc. Mol. Biol.* **126** (1): e85.

132 Heinicke, F., Zhong, X., Zucknick, M. et al. (2020). Systematic assessment of commercially available low-input miRNA library preparation kits. *RNA Biol.* **17** (1): 75–86.

133 Buenrostro, J.D., Wu, B., Chang, H.Y., and Greenleaf, W.J. (2015). ATAC-seq: a method for assaying chromatin accessibility genome-wide. *Curr. Protoc. Mol. Biol.* **109** (1): 21–29.

134 Landt, S.G., Marinov, G.K., Kundaje, A. et al. (2012). ChIP-seq guidelines and practices of the ENCODE and modENCODE consortia. *Genome Res.* **22** (9): 1813–1831.

135 Bock, C. (2012). Analysing and interpreting DNA methylation data. *Nat. Rev. Genet.* **13** (10): 705–719.

136 Wreczycka, K., Gosdschan, A., Yusuf, D. et al. (2017). Strategies for analyzing bisulfite sequencing data. *J. Biotechnol.* **261**: 105–115.

137 Wu, M.C. and Kuan, P.-F. (2018). A guide to illumina BeadChip data analysis. In: *DNA Methylation Protocols*, Methods in Molecular Biology, 3ee, vol. **1708** (ed. J. Tost), 303–330. Humana Press.

138 Teschendorff, A.E., Marabita, F., Lechner, M. et al. (2013). A beta-mixture quantile normalization method for correcting probe design bias in Illumina Infinium 450 k DNA methylation data. *Bioinformatics* **29** (2): 189–196.

139 Andrews, S. (2018). FastQC. https://www.bioinformatics.babraham.ac.uk/projects/fastqc/ (accessed 9 February 2023).

140 Ewels, P., Magnusson, M., Lundin, S., and Käller, M. (2016). MultiQC: summarize analysis results for multiple tools and samples in a single report. *Bioinformatics* **32** (19): 3047–3048.

141 Bolger, A.M., Lohse, M., and Usadel, B. (2014). Trimmomatic: a flexible trimmer for Illumina sequence data. *Bioinformatics* **30** (15): 2114–2120.

142 Martin, M. (2011). Cutadapt removes adapter sequences from high-throughput sequencing reads. *EMBnet.Journal* **17** (1): 10.

143 Krueger, F. (2019). Trim Galore! https://www.bioinformatics.babraham.ac.uk/projects/fastqc/ (accessed 9 February 2023).

144 Langmead, B. and Salzberg, S.L. (2012). Fast gapped-read alignment with Bowtie 2. *Nat. Methods* **9** (4): 357–359.

145 Li, H. and Durbin, R. (2009). Fast and accurate short read alignment with Burrows–Wheeler transform. *Bioinformatics* **25**: 1754–1760.

146 Chen, P.-Y., Cokus, S.J., and Pellegrini, M. (2010). BS seeker: precise mapping for bisulfite sequencing. *BMC Bioinforma.* **11** (1): 203.

147 Xi, Y. and Li, W. (2009). BSMAP: whole genome bisulfite sequence MAPping program. *BMC Bioinforma.* **10** (1): 232.

148 Krueger, F. and Andrews, S.R. (2011). Bismark: a flexible aligner and methylation caller for bisulfite-seq applications. *Bioinformatics* **27** (11): 1571–1572.

149 Tsuji, J. and Weng, Z. (2015). Evaluation of preprocessing, mapping and postprocessing algorithms for analyzing whole genome bisulfite sequencing data. *Brief. Bioinform.* **17**: 938–952.

150 Rauluseviciute, I., Drabløs, F., and Rye, M.B. (2019). DNA methylation data by sequencing: experimental approaches and recommendations for tools and pipelines for data analysis. *Clin. Epigenetics* **11** (1): 193.

151 Wilson, G.A. and Beck, S. (2016). Computational analysis and integration of MeDIP-seq methylome data. In: *Next Generation Sequencing – Advances, Applications and Challenges*, 1ee (ed. J. Kulski). InTech.

152 Liu, Y., Siegmund, K.D., Laird, P.W., and Berman, B.P. (2012). Bis-SNP: combined DNA methylation and SNP calling for bisulfite-seq data. *Genome Biol.* **13** (7): R61.

153 Ryan, D. (2021). MethylDackel. https://github.com/dpryan79/MethylDackel (accessed 9 February 2023).

154 Hansen, K.D., Langmead, B., and Irizarry, R.A. (2012). BSmooth: from whole genome bisulfite sequencing reads to differentially methylated regions. *Genome Biol.* **13** (10): R83.

155 Wu, H., Xu, T., Feng, H. et al. (2015). Detection of differentially methylated regions from whole-genome bisulfite sequencing data without replicates. *Nucleic Acids Res.* **43** (21): 1–9.

156 Akalin, A., Kormaksson, M., Li, S. et al. (2012). methylKit: a comprehensive R package for the analysis of genome-wide DNA methylation profiles. *Genome Biol.* **13** (10): R87.

157 Andrews, S. (2021). SeqMonk. https://www.bioinformatics.babraham.ac.uk/projects/seqmonk/ (accessed 9 February 2023).

158 Kuhn, R.M., Haussler, D., and Kent, W.J. (2013). The UCSC genome browser and associated tools. *Brief. Bioinform.* **14** (2): 144–161.

159 Robinson, J.T., Thorvaldsdóttir, H., Winckler, W. et al. (2011). Integrative genomics viewer. *Nat. Biotechnol.* **29** (1): 24–26.

160 Zhang, Y., Liu, T., Meyer, C.A. et al. (2008). Model-based analysis of ChIP-seq (MACS). *Genome Biol.* **9** (9): R137.

161 Nakato, R. and Shirahige, K. (2016). Recent advances in ChIP-seq analysis: from quality management to whole-genome annotation. *Brief. Bioinform.* **18**: 279–290.

162 Heinz, S., Benner, C., Spann, N. et al. (2010). Simple combinations of lineage-determining transcription factors prime cis-regulatory elements required for macrophage and B cell identities. *Mol. Cell* **38** (4): 576–589.

163 Meers, M.P., Tenenbaum, D., and Henikoff, S. (2019). Peak calling by sparse enrichment analysis for CUT RUN chromatin profiling. *Epigenetics Chromatin* **12** (1): 42.

164 Bailey, T., Krajewski, P., Ladunga, I. et al. (2013). Practical guidelines for the comprehensive analysis of ChIP-seq data. *PLoS Comput. Biol.* **9** (11): e1003326.

165 Li, D., Hsu, S., Purushotham, D. et al. (2019). WashU epigenome browser update 2019. *Nucleic Acids Res.* **47** (W1): W158–W165.

166 Steinhauser, S., Kurzawa, N., Eils, R., and Herrmann, C. (2016). A comprehensive comparison of tools for differential ChIP-seq analysis. *Brief. Bioinform.* **17**: 953–966.

167 Ashoor, H., Louis-Brennetot, C., Janoueix-Lerosey, I. et al. (2017). HMCan-diff: a method to detect changes in histone modifications in cells with different genetic characteristics. *Nucleic Acids Res.* **45**: e58.

168 Heinig, M., Colomé-Tatché, M., Taudt, A. et al. (2015). histoneHMM: differential analysis of histone modifications with broad genomic footprints. *BMC Bioinforma.* **16** (1): 60.

169 Bailey, T.L., Johnson, J., Grant, C.E., and Noble, W.S. (2015). The MEME suite. *Nucleic Acids Res.* **43** (W1): W39–W49.

170 Veneziano, D., Nigita, G., and Ferro, A. (2015). Computational approaches for the analysis of ncRNA through deep sequencing techniques. *Front. Bioeng. Biotechnol.* **3**: 77.

171 Veneziano, D., Di Bella, S., Nigita, G. et al. (2016). Noncoding RNA: current deep sequencing data analysis approaches and challenges. *Hum. Mutat.* **37** (12): 1283–1298.

172 Riffo-Campos, Á., Riquelme, I., and Brebi-Mieville, P. (2016). Tools for sequence-based miRNA target prediction: what to choose? *Int. J. Mol. Sci.* **17** (12): 1987.

173 Garcia-Moreno, A. and Carmona-Saez, P. (2020). Computational methods and software tools for functional analysis of miRNA data. *Biomolecules* **10** (9): 1252.

174 Zhang, H., Xu, P., Jiang, Y. et al. (2019). Genomic, transcriptomic, and epigenomic features differentiate genes that are relevant for muscular polyunsaturated fatty acids in the common carp. *Front. Genet.* **10**: 217.

175 Baxevanis, A.D. (2011). The importance of biological databases in biological discovery. *Curr. Protoc. Bioinforma.* **50**: 1.

176 Barrett, T., Wilhite, S.E., Ledoux, P. et al. (2012). NCBI GEO: archive for functional genomics data sets – update. *Nucleic Acids Res.* **41** (D1): D991–D995.

177 Harrison, P.W., Ahamed, A., Aslam, R. et al. (2021). The European nucleotide archive in 2020. *Nucleic Acids Res.* **49** (D1): D82–D85.

178 Agarwala, R., Barrett, T., Beck, J. et al. (2018). Database resources of the National Center for Biotechnology Information. *Nucleic Acids Res.* **46** (D1): D8–D13.

179 Brazma, A. (2003). ArrayExpress – a public repository for microarray gene expression data at the EBI. *Nucleic Acids Res.* **31** (1): 68–71.

180 Afgan, E., Baker, D., Batut, B. et al. (2018). The galaxy platform for accessible, reproducible and collaborative biomedical analyses: 2018 update. *Nucleic Acids Res.* **46** (W1): W537–W544.

181 Perkel, J. (2016). Democratic databases: science on GitHub. *Nature* **538** (7623): 127–128.

182 Byrd, J.B., Greene, A.C., Prasad, D.V. et al. (2020). Responsible, practical genomic data sharing that accelerates research. *Nat. Rev. Genet.* **21** (10): 615–629.

183 Brown, A.V., Campbell, J.D., Assefa, T. et al. (2018). Ten quick tips for sharing open genomic data. *PLoS Comput. Biol.* **14** (12): e1006472.

184 Whitlock, M.C. (2011). Data archiving in ecology and evolution: best practices. *Trends Ecol. Evol.* **26** (2): 61–65.

185 Koepfli, K.-P., Paten, B., and O'Brien, S.J. (2015). The genome 10K project: a way forward. *Annu. Rev. Anim. Biosci.* **3** (1): 57–111.

186 Rhie, A., McCarthy, S.A., Fedrigo, O. et al. (2021). Towards complete and error-free genome assemblies of all vertebrate species. *Nature* **592** (7856): 737–746.

187 Howe, K.L., Achuthan, P., Allen, J. et al. (2021). Ensembl 2021. *Nucleic Acids Res.* **49** (D1): D884–D891.

Part II

Epigenetics Insights from Major Aquatic Groups

7

Epigenetics in Sexual Maturation and Gametes of Fish

Marta Lombó Alonso[1,2], Audrey Laurent[3], María Paz Herráez[1], and Catherine Labbé[3]

[1] *Department of Molecular Biology, Faculty of Biology and Environmental Sciences, Universidad de León, León, Spain*
[2] *Dipartimento Scienze della Vita e dell'Ambiente, Università Politecnica delle Marche, Ancona, Italy*
[3] *Department of Fish Physiology and Genomics, INRAE, Rennes, Campus de Beaulieu, France*

7.1 Introduction

Research on epigenetics has been developing rapidly for the last 15 years in all areas of life science. In gamete physiology, a considerable breakthrough occurred as a result of a better understanding of chromatin organization in human spermatozoa. It was known that not all DNA was coiled into toroids (doughnut-like shape) by protamines, as about 4% of the genome was retained in nucleosomes. Hammoud et al. [1] showed that permissive or poising histone modifications in these sperm nucleosomes were enriched at developmental promoters known to be important during early development. This was the first hint that spermatozoa's epigenetic profile could directly influence embryo development. It had been known for decades that the environment affects gene expression in organisms but the possibility that this influence could be transmitted to following generations by way of gamete epigenetic profile boosted research in this area, beginning with human and mouse gametes. Because among fish species, the zebrafish (*Danio rerio*) was a very popular animal model for basic science, it rapidly benefited from the attention of the most advanced groups in the field of epigenetics [2–4]. Therefore, today, most of our knowledge on gamete epigenetics in fish is about zebrafish, although publications focusing on species of broader interest for aquaculture are now arising.

The objective of this chapter is to gather the current knowledge on gamete epigenetics in fish, from zebrafish to aquaculture species. How and which epigenetic marks are established during gametogenesis, from PGCs' determination to highly differentiated spermatozoa and ova, will be described with a parallel to the more extensive knowledge available in humans and mice. To this end, we will focus on the epigenetic marks at the chromatin level (DNA methylation and histone modifications). Epigenetic marks are sensitive to the environment and are transmittable from cell to cell, and even from generation to generation. Fish are readily exposed to environmental and water quality variations, or reproduction technologies, which raise concern and draw interest from fish farmers. In aquaculture, the best understanding of the molecular mechanisms behind the epigenetic setup in the germline is therefore a key issue toward both: (i) the management of the risk of intergenerational transmission of environmental detrimental effects and (ii) the exploitation of accrued epigenetic variations that might be beneficial for offspring performance. This environmental issue will be developed in the second section of this chapter. Although more and more research is being published on this topic, the present review highlights the areas where a deeper investigation is still needed, so that the cascade between broodstock environment during gametogenesis, or gamete-handling conditions, and transmission of changes to the offspring can be better understood.

7.2 Epigenetics During Spermatogenesis and Oogenesis

7.2.1 PGCs' Epigenetic Remodeling During Embryo Life

The cell differentiation process occurring during embryo development entails the acquisition of specific epigenetic signatures by the cells, depending on their lineage and fate. Among the embryonic cell populations, it is considered in humans and mice that the primordial germ cells (PGCs) undergo the most intense epigenetic changes [5]. PGCs, also

Epigenetics in Aquaculture, First Edition. Edited by Francesc Piferrer and Han-Ping Wang.

evidenced in fish, are the progenitors of germ cells (GCs) that give rise to both gametes: spermatozoa and ova [6]. There are two basic mechanisms of germline specification: induction and preformation. In mammals, turtles, and urodeles (inductive species), the determination of PGCs is induced by cellular interactions occurring during gastrulation [7]. However, in fish, frogs, Drosophila, and other animals (preformative species), this specification results from the asymmetric distribution of maternally inherited cytoplasmic determinants called the germ plasm [8]. During early embryo development, the embryonic cells that have sequestered the germ plasm acquire a germline fate, and these PGCs migrate and colonize the gonadal primordium. From then onwards, they remain exposed to the specific microenvironment of the gonad. Under the effect of cytokines, metabolites, and hormones, the GC epigenetic profile will differentially progress during spermatogenesis and oogenesis to achieve the unique epigenetic profile of the mature spermatozoa and oocytes.

In mice and humans, PGCs can epigenetically be programmed before but mainly after, they have reached the genital ridge. This involves a global erasure of the embryonic epigenetic marks and a progressive repositioning of new ones [9]. DNA global demethylation occurs via two mechanisms. First, the repression of the DNA methyltransferases DNMT1, DNMT3a, and DNMT3b leads to passive demethylation, allowing the acquisition of an opened configuration of chromatin and the restoration of pluripotency [10]. In the second step, the action of ten-eleven translocation (TET) enzymes promotes an active and locus-specific DNA demethylation, affecting, among others, the imprinted regions of the genome [11, 12]. The process of PGC remethylation starts soon after sex determination in males and later in female development; GCs thereby lose their pluripotent epigenetic profile. In addition to DNA methylation, the epigenetic remodeling of PGCs also implies changes in several histone repressive (H3K9me3, H3K27me3) and activating marks (H3K9ac) [13].

In fish, the epigenetic reprogramming of PGCs has just begun to be studied in model species such as zebrafish and medaka (*Oryzias latipes*), as summarized in Figure 7.1. In zebrafish, a study of DNA methylome in PGCs, from 4 to 24 hours post fertilization (hpf), has demonstrated the retention of paternal epigenetic configuration in these cells [14]. This indicates that genome-wide DNA methylation erasure does not occur in zebrafish germline from the blastula stages to genital ridge colonization. Such stability of DNA methylation was also described to extend from PGCs of 24 hpf embryos to sexual maturation in zebrafish spermatozoa and ova [15]. In contrast, the dynamics of genome DNA demethylation and

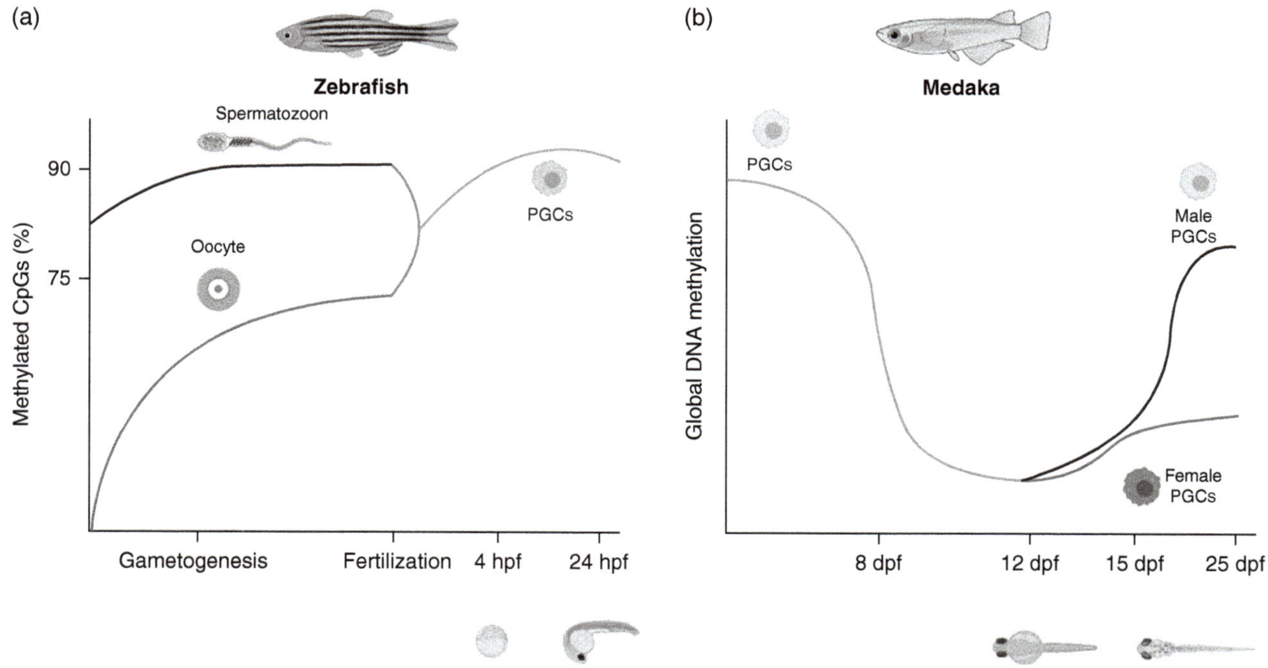

Figure 7.1 DNA methylation dynamics in fish PGCs and gametes. (a) Percentage of CpG methylation in zebrafish gametes and PGCs. The sperm cells have around 91–95% of methylated CpGs, whereas, in the oocyte, the percentage of CpG methylation is 75%. After fertilization, the DNA methylome of PGCs from 4 to 24 hpf is similar to that of spermatozoa. (b) Percentage of CpG methylation in embryonic cells at 4 hpf and in PGCs of medaka. A first round of demethylation would take place before or at 8 dpf (the first PGCs sampling point), reaching the epigenetic ground state in a second round occurring at 12 dpf. At 25 dpf, the male PGCs suffer a process of *de novo* methylation but the female PGCs retain the same pattern as at 15 dpf. (*See colour plate section for colour representation of this figure.*)

remethylation during PGC specification in medaka are more similar to those of mice and humans [16]. Although the study began after hatching, 8 days post fertilization (dpf), these authors proposed two phases of demethylation in medaka PGCs: the first was deduced from a global methylation that is much lower in PGCs 8 days post fertilization (dpf) than in the whole embryonic cells at 4 hpf; while the second, which leads to the epigenetic ground state, takes place in PGCs between 10 and 12 dpf. By 25 dpf, the male PGCs undergo a process of *de novo* methylation, whereas in the female PGCs the level of hypomethylation remains the same as that at 15 dpf. The difference between sexes is also reflected in the methylation of transposable elements: retrotransposable elements and Chapaev transposons remain demethylated in female PGCs, whereas these elements undergo a remethylation in male PGCs [16, 17]. Therefore, major differences exist between species and challenge the concept of pluripotency and its molecular basis mainly studied in mouse.

7.2.2 Establishing the Epigenetic Profile of Eggs and Sperm During Gametogenesis

The epigenetic profile acquired by GCs during embryo life will then be modified in adulthood during gametogenesis when GCs activate their proliferative and/or differentiation programs under hormonal stimulation. In females, the meiotic process, which was arrested at the pachytene stage during the embryonic period, is resumed; conversely, in males, the activation of gametogenesis starts with an active mitotic period. These processes are accompanied by an intense epigenetic remodeling that involves the establishment of sex-specific patterns in both the spermatozoa and the oocyte.

At puberty, in the females, oocytes grow, mature, and resume the arrested meiosis. A study carried out in cyprinid hybrid fish showed that meiosis arrest at pachytene is a result of histone methylation (mainly H3K4me3, H3K9me3, H3K79me, and H3K79me3) since it induces DNA repair and the transcription of transposons [18]. The initial maturation implies a DNA remethylation process that, according to the available data in fish, as it occurs in mice, is less intense than in spermatogenesis, with the oocyte reaching lower DNA methylation levels than those of sperm [19]. In zebrafish oocytes, the abundant mitochondrial DNA is almost entirely hypomethylated, whereas the nuclear DNA has 75% of methylated CpGs [4, 20]. These low levels in mitochondrial DNA necessitate that the DNA methylation assessment methods clearly identify the origin of the DNA (nuclear or mitochondrial) so that no error is made about which DNA is really assessed. This condition is not met, for example, with DNA methylation ELISA tests, leading to a risk of misinterpretation of the true nuclear oocyte values [21].

The fish oocyte nuclear genome is not homogeneous and, as in mice [9], it displays hypermethylated blocks of genes, including some involved in development (*hox*) and germline formation (*piwil1*, *dazl*, and *vasa*). These gene promoters selectively demethylate in the zygote at zygotic genome activation [4]. Moreover, zebrafish oocyte maturation entails the transcription of meiosis-related genes whose tightly regulated expression [22] is related to histone posttranslational modifications that have been claimed to play an essential role in this process in mammals [9]. The activity of histone deacetylases has been shown to be crucial in female mice to maintain the appropriate levels of histone acetylation in the oocytes. Its inhibition, which increases the level of histone acetylation, does not affect fertilization ability but strongly reduces embryo survival by increasing the rate of embryonic aneuploidy. This has been related to the observed increase of histone acetylation in oocytes from aged mice and with the increased rates of chromosome segregation failures of the progenies from aged females [23].

In males, three different periods can be distinguished during spermatogenesis: mitotic proliferation, meiotic division, and final differentiation or spermiogenesis. In mammals, it has been described that the mitotic period is characterized by a decrease in repressive marks (DNA methylation, H3K27me3, and H3K9me3) and an increase in permissive marks (H3ac and H4ac, as well as H3K4me2/3) [1, 24]. During the meiotic phase of spermatogenesis, progressive gene silencing has been reported in humans [25] and fish [26]; the permissive mark H3K4me3 decreases and the silencing marks, H3K27me3 and H3K9me3, increase in spermatocytes. As a result, spermatozoa, especially those of zebrafish which have 91–95% of methylated CpGs [4], arise as highly methylated and transcriptionally repressed cells. Still, several hypomethylated regions, as well as permissive histone marks (H3K4me3, H3K4me2, and H4K16ac), corresponding to promoters of genes expressed during early development or meiosis, have been reported to persist in zebrafish sperm [3]. Moreover, they demonstrated that, as it occurs in mammals, key developmental genes are packaged in bivalent or multivalent marks, in which activating and repressing histone marks and DNA hypomethylation coexist [3]. A similar conclusion was reached for H3K4me3/H3K27me3 modifications in a more recent study [27]. In summary, the repressive marks might well avoid the pointless expression of certain genes in the male germline, whereas the presence of permissive marks may elude DNA methylation in the promoters of some genes, allowing their further activation in the embryos when the expression is required [26]. Interestingly, Murphy and colleagues [28] have described in zebrafish the existence of "placeholder" nucleosomes, containing histone H2A.Z and H3K4me1, which occupy all hypomethylated DNA regions in both sperm and early embryos. Upon zygotic genome activation, genes carrying placeholder nucleosomes become either active (housekeeping), repressed (late development), or primed (early development) by the gain of either H3K4me3, H3K27me3, or both marks, respectively.

The enzymes responsible for all these epigenetic modifications in fish are well described. DNA methyltransferase isoforms are more expanded in teleost fish compared to mammals because of several rounds of whole genome duplication [29, 30]. They are the best-characterized epigenetic writers in the gonads of different fish species [31–33]. Histone modifiers and histone variants were also described in mangrove rivulus (*Kryptolebias marmoratus*) [34] and recently reviewed for zebrafish [35]. To the best of our knowledge, however, no isoforms specific to gonads and GCs were identified in fish.

7.2.3 The Different Actors of Chromatin Packaging During Spermatogenesis

At early stages of spermatogenesis in mice and humans, H2A and H2B are replaced by testis-specific histone variants TH2A and TH2B, at the mitotic phase, encompassing the transition to meiosis-specific open chromatin loci from mitotic-specific open chromatin [9]. During meiosis, H1 is replaced by the H1t variant and H3 by H3t, generating a less compacted chromatin, which is more permissive to meiotic transcription and recombination. Later, during spermiogenesis, a particularly intense reorganization and compaction of the chromatin fiber lead to the formation of a condensed and transcriptionally silent nucleus. This involves the histone-to-protamine transition, a massive replacement of histones by the more positively charged protamines PRM1 and PRM2. These generate toroids [36], a highly compacted and unique form of chromatin involved in transcription silencing [37]. Moreover, the substitution of histones by protamines has long been believed to erase almost all histone epigenetic codes, until it was shown that nucleosomes are preserved at some specific areas of the genome, associated with genes that are transcribed early after fertilization [1] (Box 7.1).

In fish, the diversity of SNBPs is much higher, allowing a similarly diverse protein transition during spermiogenesis. Some species such as rainbow trout (*Oncorhynchus mykiss*), sea bass (*Dicentrarchus labrax*), and the red mullet (*Mullus surmuletus*) totally replace histones by protamines, whereas others such as zebrafish, gilthead seabream (*Sparus aurata*), and goldfish (*Carassius auratus*) maintain somatic-like histones with no traces of protamines [38–40]. Additionally, some species such as the winter flounder (*Pseudopleuronectes americanus*) show SNBPs with intermediate characteristics between protamines and histones, known as protamine-like [41].

As summarized by Herráez et al. [38], the protein exchange starts with the acetylation of H3 and H4 and the replacement of the acetylated histones by the final SNBPs in a direct process, or through an intermediate replacement by transition proteins as occurs in some chondrichthyes. During this process, different condensation states (granules, fibers, or lamellae) can be observed. The condensation model is dependent on the type of SNBPs and their posttranslational modifications, mainly acetylation and phosphorylation [42]. In species with spermatic histones such as zebrafish, the compaction degree is reached thanks to a higher expression of linker histone H1 [3]. Despite its high level of compaction, zebrafish sperm chromatin exhibits differential packaging, which is represented by the distribution of epigenetic marks. Specific blocks of chromatin, with a low DNA methylation status and particular histone modifications, harbor early transcribed genes required for embryo development with similar features to those recognized in the nucleosomal regions of mammalian sperm.

Box 7.1 Chromatin Organization in Somatic Cells and Spermatozoa

The basic unit of the chromatin organization in somatic cells is the nucleosome, consisting of a histone octamer comprised of two copies of each core histone (H2A, H2B, H3, and H4), highly conserved during evolution. An additional linker histone (H1) binds to DNA and the nucleosome, thus stabilizing the two DNA loops wrapped around the octamer. The replacement of somatic histones by sperm nuclear basic proteins (SNBPs) is a key event during spermiogenesis (postmeiotic phase of spermatogenesis). Fish display different patterns of replacement; in some species, the core histones are sequentially replaced by specific classes of testicular histones, whereas in others, the somatic histones are totally or partially replaced by protamines or by protamine-like proteins throughout the postmeiotic steps.

Box 7.2 Parental Imprinting

Among monoallelically expressed genes [43], one category refers to genes where the parental origin of the alleles (either paternal or maternal) determines which one is expressed or repressed in the individual. For some genes, the allele transmitted by the spermatozoa (paternal chromosome) is expressed, while for other genes, the maternal allele is expressed, leading to parent-specific expression of these genes (imprinted from the parents). Silencing of the repressed allele is mediated by epigenetic mechanisms, namely DNA methylation, and, to some extent, the H3K27me3 modification. In mammals, the epigenetic marks conveying parental imprinting are established during gametogenesis in a gamete-specific fashion (ova/spermatozoon) [44] and they are maintained after fertilization, despite the thorough epigenetic reprogramming taking place in the embryos. In humans, more than 100 imprinted genes have been characterized. They have important roles in embryo development, and any alteration of the monoallelic expression will induce developmental defects or have a lifelong impact on health.

Parental imprinting was reported not only in eutherian mammals and marsupials [45] but also in flowering plants [46] and some insects [47]. It was therefore generalized that parental imprinting is present in species with placental, or endosperm in the case of plants [46]. The question is still open in fish.

7.2.4 Parental Imprinting in Fish Gametes

In fish, parental imprinting (Box 7.2) was discussed from an evolutionary perspective more than 20 years ago [48], and it was sought whether some genes known to be parentally imprinted in humans would bear differential epigenetic marks in fish. For example, a region in the goldfish *igf2* gene is hypermethylated in sperm and hypomethylated in the oocyte. However, the parent-specific methylation pattern was not maintained in the embryo [49]. In the same course, the zebrafish ortholog of the PEG1-/MEST-imprinted gene in mammals showed biallelic expression in zebrafish larvae [50], therefore refuting the presence of a parental-of-origin dosage. Paulsen et al. [51] also demonstrated that some imprinting control elements identified in mammals could not be identified in fugu or in zebrafish, and imprinted gene clusters in mammals were poorly conserved in fish [52]. The absence of a DNMT3L ortholog (a key enzyme in methylating imprinted genes in mammals) has been considered a hint of the lack of an imprinting mechanism in zebrafish [3, 53]. On the contrary, Ma et al. [54] showed that the goldfish *ntl* gene is differentially methylated between sperm and eggs and that the paternal allele is transcribed earlier than the maternal one in the embryo. Some hints of a possible imprinting mechanism were also provided in a study exploring transgene expression in zebrafish [55]. From these divergent studies, it is still impossible to make a definitive statement about the existence of parental imprinting in fish, thus more evolutionary-based studies are still necessary.

7.2.5 Fate of the Gamete Epigenome

The epigenome acquired by the mature GCs withstands a thorough remodeling after fertilization, with the exception of imprinted loci in mammals, to achieve a naïve state compatible with the differentiation of the new embryonic tissues. This remodeling includes passive and active DNA demethylation, and this is believed to erase most of the DNA methylation-related epimutations acquired by the germline cells during adult life, thus avoiding their transmission to following generations [56]. Thus, two major waves of epigenetic erasure and reprogramming are taking place during one's life cycle: (i) after gamete fusion in preparation for zygotic genome activation and (ii) in embryonic germinal stem cells, named PGCs, after their induction and migration in the genital ridge, as developed earlier. Depending on species, the occurrence of two of these events represents a safeguarding barrier to the intergenerational transmission of environment-induced epigenetic alterations, although it is now determined that this barrier is not thorough.

7.3 Epigenetic Changes in the Gametes Triggered by Environmental Constraints

The term "environmental epigenetics" refers to how environmental factors (diet, habits, exposure to pollutants, abiotic factors, etc.) can modulate the epigenetic pattern of a cell. Changes in the epigenetic profile promoted by a variety of factors may result in alterations of gene expression, which can in turn induce changes in the phenotype and/or several diseases [57]. Here, we address how the environment affects GC epigenetics while questioning whether these changes will influence progeny fitness. By extension, reproductive biotechnologies such as *in vitro* fertilization, hormonal induction, or gamete manipulation (ageing, storage, and cryopreservation) are also at risk to induce a change in the GCs' environment, leading to modifications of their basal epigenetic pattern. This issue, reviewed in mammals [58, 59] and, in aquatic species [60], is the object of growing attention in fish. Here, we address how environmental changes could directly affect the epigenetic marks (DNA methylation and histone modifications) of the GCs. It is known that other vectors are transmitting information to the progeny via the gametes, including extracellular vesicles (for example, in chicken [61]), or noncoding RNAs (for example, in zebrafish [62]), maternal mRNA cargo and reserves, etc. These vectors may influence the development of the offspring, including the epigenetic profile of some tissues, although not necessarily via epigenetic mechanisms. The complexity of the situation where some environmental clues are transmitted to the offspring led us to focus our review on the research where some mediation via gamete epigenetics was identified (Table 7.1).

7.3.1 Environmental Contaminants

Since 2000, more than 300 million tons of anthropogenic chemicals have been produced every year in Europe, with more than 50% of them considered environmentally harmful and around 10% reported to have a severe chronic impact on the environment [84]. Nowadays, research is mainly focused on compounds described as biologically active or even toxic, especially on those affecting the endocrine system since it plays a crucial role in animal homeostasis, reproduction, development, and/or behavior [85]. These agents being able to interfere with the synthesis, secretion, transport, binding, or elimination of endogenous hormones are known as endocrine-disrupting chemicals (EDCs) or simply as endocrine disruptors [86], and they can affect the epigenome of exposed fish [87]. As it has been reviewed by Carnevali et al. [88], the EDCs (including dioxins, phthalates, and polychlorinated biphenyls) can directly affect the epigenetic pattern of fish gonads. However, few studies are focused on the epigenetic alterations of the fish gametes and their transmission to the following generations.

7.3.1.1 DNA Methylation

The detrimental effects of EDCs on global DNA methylation of fish gonads were described a long time ago when the exposure of three-spine stickleback (*Gasterosteus aculeatus*) to hexabromocyclododecane (HBCD) and to oestradiol-17ß (E2) caused a significant increase in global DNA methylation of the testes [63]. Exposure to 225 µg/L bisphenol A (BPA) in adult male rare minnow (*Gobiocypris rarus*) also led to global DNA hypermethylation in the testes [64]. Nevertheless, in zebrafish, exposure to 1 mg/mL BPA has been reported to decrease DNA methylation of both ovaries and testes, and decrease the expression of *dnmt1* in the ovaries [65]. Likewise, exposure of zebrafish to a low dose of BPA (15 µg/L) resulted in lower global genome methylation of male and female gonads and dysregulated the expression of *dnmts*, *gnmts*, and *tets* [66]. More specifically, zebrafish exposure to 2000 µg/L BPA increased the levels of 5-methylcytosine (5mC) of testicular cells [67] but no specific changes were reported in DNA methylation of spermatozoa after males were exposed to BPA during either embryonic life or adulthood [72, 89]. An altered expression of *dnmts* and *tets* was also noticed in PGCs from medaka exposed to different concentrations of triclosan, which is suggestive of epigenetic effects on GCs [75]. Chemicals other than EDCs were also reported to directly affect sperm DNA methylation patterns, such as copper in zebrafish [68], with some consequences on the nervous system, retina, and digestive system of the offspring. Methyl mercury exposure in zebrafish males was also shown to alter sperm DNA methylation, both in F0 and F2 males [69].

7.3.1.2 Histone Modifications

In female zebrafish, Santangeli et al. [70] reported that exposure to 5 µg/L BPA triggered a decrease in H3K4me3 and H3K27me3 of *star* and *fshr* genes, which play a crucial role in female reproduction. In males, zebrafish exposure to 100 and 2000 µg/L BPA increased the global H3 acetylation level of testicular cells. The acetylation levels of activating marks (H3K9ac, H3K14ac, and H4K12ac) rose with BPA exposure, and the levels of the repressive mark (H3K27me3) were decreased [67]. In rare minnows, exposure to 15 µg/L BPA increased the level of H3K4me3 in the testes, while it decreased in males exposed to 225 µg/L BPA [71]. Interestingly, exposure of zebrafish embryos to 4000 µg/L BPA decreased levels of H3K9ac [72], whereas exposure to 2000 µg/L BPA during the mitotic phase of spermatogenesis increased the levels of H3K9ac and H3k27ac. Lastly, exposure during the whole spermatogenesis increased levels of H3K9ac and H4K12ac [89]. Altogether, these results suggest that epigenetic toxicity depends on the species, the sex, the doses of the toxicants, as well as the window during which the exposure takes place.

Table 7.1 Impact of external factors on the epigenetic profile of fish gonads and gametes.

	External factor	Epigenetic modifications	Affected organs/ cells	Species	References
Environmental contaminants	HBCD	Increase in global DNA methylation	Testes	Three-spine stickleback	[63]
	BPA	Increase in global DNA methylation	Testes	Rare minnow	[64]
	BPA	Decrease in DNA methylation	Testes and ovaries	Zebrafish	[65]
	BPA	Decrease in DNA methylation	Testes and ovaries	Zebrafish	[66]
	BPA	Increased levels of 5mC	Testes	Zebrafish	[67]
	Copper	Loci-specific changes in methylation level	Sperm	Zebrafish	[68]
	Methylmercury	Loci-specific changes in methylation level	Sperm	Zebrafish	[69]
	BPA	Decrease in H3K4me3 and H3K27me3 in specific genes	Ovary	Zebrafish	[70]
	BPA	Increase in H3K9, H3K27, H4K12 and decrease in H3K27me3	Testes	Zebrafish	[67]
	BPA	Alterations in H3K4me3 levels	Testes	Rare minnow	[71]
	BPA	Alterations in H3K9ac, H3K27ac, and H4K12ac	Spermatozoa	Zebrafish	[72]
	BPA	Loci-specific changes in methylation level	Ovary	Zebrafish	[73]
	BPA	Alterations in DNA methylation and histone marks	Ovary	Rare minnow	[74]
	Triclosan	Dysregulation of DNA methylation enzymes	PGCs	Medaka	[75]
Domestication	One year of life in a hatchery	108 differentially methylated regions	Spermatozoa	Steelhead trout	[76]
	Hatchery	183 differentially methylated regions	Spermatozoa	Coho salmon	[77]
	Growth and maturation in rearing tanks from smolt stage to adult	Loci-specific changes in methylation level	Spermatozoa	Atlantic salmon	[78]
ARTs	Temperature increase	Alterations in DNA methylation	Spermatozoa	Salmonids	(Laurent et al., in preparation)
	Sperm storage	Alterations in DNA methylation	Spermatozoa	Common carp	[79]
	Egg storage	Increase in H4K12ac	Egg	Common carp	[80]
	Sperm cryopreservation	Alterations in DNA methylation	Spermatozoa	European eel, tambaqui, goldfish, and zebrafish	[81], [82], [83]

7.3.2 Domestication

A direct effect of the environment on the epigenetic profile of fish gametes is still little studied, although it has resulted in increased attention with regards to domestication. Indeed, domestication induces dramatic environmental disruption for fish [90], known to affect all physiological functions including gamete production and quality. At the epigenetic level, the effect of rearing wild fish in a hatchery was first explored in rainbow trout, from a perspective of restocking. A single year of life in hatchery conditions, prior to releasing the males in the wild, was shown to induce some DNA methylation changes, with 108 differentially methylated regions (DMRs) between spermatozoa of hatchery- versus wild-reared fish [91]. These results were later confirmed at a larger genomic scale in the same species [76]. This indicates that although this hatchery period was far ahead of the gamete production, it imprinted to some extent the DNA methylation pattern of the sperm produced later. A similar response to hatchery conditions was reported in Coho salmon (*Oncorhynchus kisutch*) sperm DNA methylation profile [77]. In Atlantic salmon (*Salmo salar*), DMRs were identified in spermatozoa of males that underwent growth and maturation in rearing tanks from smolt stage to adult, compared to males that underwent a natural wild growth and maturation cycle [78], thus confirming a similar hatchery effect on sperm DNA methylation. In this species, some DMRs observed in spermatozoa were additionally maintained in the progeny after the embryonic genome activation [92]. Overall, these studies on the onset of domestication show that both short- and long-term exposure of fish to environmental changes can alter sperm DNA methylation, and this may drive fitness changes in the progeny. It should be kept in mind that domestication also induces changes in the epigenetic profile of the different organs of the fish, including the testis [93] but we restricted this section to effects reported on the gametes.

7.3.3 Reproductive Biotechnologies

As mentioned earlier, it was proposed that the DNA methylation status of some genomic regions in spermatozoa serves as a blueprint for early genome expression in the embryo [3, 4], whereas maternal chromatin would only be reprogrammed according to the paternally inherited pattern [4, 20]. DNA methylation information is then used by embryonic actors such as specific nucleosomes [28], or transcription factors [94] to pass on repression or activation clues to the gene expression machinery at early developmental stages. Although little studied, it is likely that the oocyte epigenetic pattern also needs to fulfil some requisites for proper embryonic development. It is therefore of paramount importance to ensure that reproductive biotechnologies and gamete manipulation are not at risk of inducing gametic epigenetic changes that may alter proper gene expression in embryos and, at later stages, during the life of the fish.

7.3.3.1 Biotechnologies Targeting the Gametogenesis Stages

In many farmed fish species, the maturation and onset of ovulation are controlled by a photoperiodic treatment of the broodstock that is applied several weeks or months before spawning [95, 96]. This allows delaying or advancing gamete production compared to the normal fish reproductive cycle, by way of altering the endocrine kinetic of the broodstock. The question as to whether these practices bear any epigenetic risk is still to be answered, and this issue is made even more complex by the fact that off-season gamete production is sometimes associated with inadequate water temperatures (i.e., reproduction in summer instead of winter for salmonids, for example), which may induce additional perturbation. Indeed, we recently showed that changing water temperature from 12 °C (winter temperature) to 16 °C (summer temperature) during the six months preceding reproduction had little effect on sperm quality, whereas it induced significant changes in sperm DNA methylation profile (Laurent et al., in preparation). Also, hormonal induction of gamete release is often necessary to compensate for too little gamete production, or for broodstock synchronization [97], and it was shown in humans and mice that hormonal induction of ovulation alters the DNA methylation of several developmental genes [98]. Consequences of hormonal induction on gamete epigenetics were studied only at a global DNA methylation level in goldfish sperm [83]: it was shown that after a single dose of GnRH analogue, DNA methylation levels were unchanged between spermatozoa released immediately under the pressure of hormonal treatment and those released in a more natural manner four days after the treatment. More studies in different species with different treatments (hormone type and concentrations), and by way of a genome-wide analysis, are still missing, however, before conclusive positions can be reached. Additionally, data on the consequences of ovulation induction on the oocyte profile are still missing.

Today, the production of gametes from surrogate broodstock is increasingly studied, in line with the pioneering work of Yoshizaki's group in salmonids [99]: germinal stem cells are collected from donor fish bearing a valuable genetic background, and they are grafted into recipient fish that take up these cells in their gonads and ensure the production of functional gamete. Again, how germinal stem cell isolation, manipulation, grafting, and gametogenesis into a surrogate breeder may affect the epigenetic profile of the produced gametes is yet to be studied. This is especially interesting with regard to the bipotential of germinal stem cells and the establishment of the gametic epigenetic profile during gametogenesis. Indeed, the grafting of spermatogonia into female recipients induced the production of ova [100], and vice versa [101]. These manipulations raise an interesting question about the epigenetic changes taking place in the spermatogonia after they are injected into female recipients, where they will undergo normal oogenesis in the ovaries to finally result in functional eggs.

7.3.3.2 Biotechnologies Targeting the Gametes

After collection, gametes storage is often required prior to fertilization, leading to some possible ageing process. Gametes ageing during *in vitro* storage has been recently studied in carp. Several differentially methylated regions in sperm DNA were identified between 1-hour and 24–96-hour storage duration and between 24 and 96 hours' sperm storage duration [79], although the identity of the altered regions remains to be characterized. In eggs, histone marks proved to be very stable over a 28-hour storage duration, except for a sharp increase in H4K12ac [80]. The functional connections with the observed offspring phenotypes (survival rates and malformations) are however complicated to achieve in these pioneer studies on gamete handling in fish.

Cryopreservation of spermatozoa is the most widespread technology used for the preservation of genetic resources in fish (reviewed in [102], and the question of the epigenetic risk of cell cryopreservation was raised previously [60, 103]. The permeating cryoprotectants used in fish (DMSO, MeOH, and DMF) often bear reactive methyl groups, and some chemical interaction with DNA during the straining phases of freezing and thawing may induce some changes in DNA methylation. Indeed, depending on the cryoprotectant, some global changes in DNA methylation were reported after sperm cryopreservation in European eel [81], tambaqui (*Colossoma macropomum*) [82], goldfish, and zebrafish [83]. However, the genomic regions affected by these changes remain to be characterized by whole genome approaches before the functional consequences of these changes can be assessed.

7.3.4 Transmission of Gamete Epimutations to the Following Generations

When changes in some DNA methylation features (discrete sites or entire genomic regions) are found to be maintained between several generations, and when they are altering noticeably and durably the expression of some genes, these DNA methylation changes are often referred to as epimutation, inherited from one generation to another. The latter would then suffer the consequences even if they have never been in contact with the disrupting agent [104].

Maternal transmission of epigenetic alterations induced by the mother's lifestyle and/or by environmental exposure during pregnancy has been widely studied in mammals [73, 107–109]. Recently, much attention has been paid to the paternal inheritance of epigenetic alterations since it can also have an impact on offspring development. As a result, the paternal lifestyle has also become a subject of research, as reviewed by Lombó and Herráez [110]. In all, transgenerational epigenetic inheritance requires that germline cells are affected by the history of the animal's ancestors. This raises the question of the vectors ensuring the soma-to-germline transmission of this information, an issue that is still under study in vertebrates, and poorly addressed in fish.

7.3.4.1 Transmission of Epigenotoxic Effect

Zebrafish is an appropriate species for the study of inter- and transgenerational inheritance (Box 7.3) because of its rapid and visually accessible development [111]. The transmission of altered phenotypes up to the F3 generation has been confirmed in this species by several studies [73, 74, 112–114], and has been linked to epigenotoxic effects after maternal, paternal, and embryonic exposure. Concerning maternal inheritance, Santangeli et al. [73] revealed that female exposure to 20 μg/L BPA triggers DNA hypermethylation in the promoter of *amh* (encoding the anti-Mullerian hormone), leading to repression in this transcript up to F3. As for the paternal inheritance, it has been reported that adult male exposure to 2000 μg/L BPA promotes heart malformations in the progeny up to the F2 [112]. The same authors proved that these malformations were caused by an increase in histone acetylation of spermatozoa that is transmitted to the F1, specifically affecting some genes involved in cardiogenesis (*hand2*, *esr2b*, and *kat6a*) [114]. Furthermore, embryonic exposure to 30 μM mono (2-ethylhexyl) phthalate (MEHP) and 10 μM of the DNA methyltransferase 1 inhibitor 5-azacytidine (5-Aza) has also been reported to trigger changes in DNA methylation of specific sequences in three generations. Moreover, treatment with 5-Aza did promote phenotypic alterations up to the F2, such as a decrease in larval length, whereas no phenotypic

Box 7.3 Intergenerational and Transgenerational Inheritance

Maternal exposure to environmental constraints during pregnancy in mammals results in exposure of this F0 female, her F1 fetus, and as a consequence, the GCs of the fetus that will generate F2. Therefore, any effects shown in F0, F1, and F2 may be attributed to direct exposure to the toxic substance, which is called intergenerational inheritance [105]. An assessment of F3 would be needed to determine the epigenetic transgenerational inheritance of such effects. On the contrary, in paternal exposure, only the F0 male and its GCs (that potentially represent the F1) are affected, so the study of F2 generation would be enough to prove a transgenerational transmission of the effects [106]. In oviparous fish, paternal and maternal exposure can have consequences on the F1 generation, so the changes in the F2 generation will sign transgenerational inheritance. In viviparous fish, they will follow the mammalian number of generations to state about inter- or transgenerational inheritance.

alterations were observed in the offspring after MEHP exposure [115]. In rare minnows, parental exposure to BPA affected DNA methylation and histone marks in the ovaries of the offspring, leading to ovary development inhibition [74].

7.3.4.2 Transmission of Biotechnological Clues

The few works reporting alterations of offspring in response to the application of gamete biotechnologies are focused on intergenerational inheritance (transmission from the exposed gametes to the resulting offspring). The expression of several marker genes was affected in rainbow trout embryos sired with cryopreserved spermatozoa [116], together with stress resistance changes at later stages [117]. In a recent study of rainbow trout, exposure of the males to a high temperature (19 °C) at juvenile stages induced better survival and growth performances of their progeny at warm temperatures [118]. These scarce studies are indicating some possible transmission of environmental clues via the spermatozoa but the extent of an epigenetic mediation of this transmission is still to be deciphered.

7.4 Conclusion

The setting up of the gamete epigenetic features in fish during gametogenesis and epigenetic profile in the terminally differentiated spermatozoa and ova is increasingly studied today. Although a restricted number of model and farmed species have been investigated thus far, the diversity of the reported profiles demonstrates that the generalized "fish" quoting is groundless. Moreover, in contrast with the extensive knowledge now available on the epigenetics of terminally differentiated gametes, information on the epigenetic set-up in PGCs and during gametogenesis is very recent in fish, and focused on two model species only, namely zebrafish and medaka. These few publications demonstrated, however, that the ground state of DNA methylation in PGCs does not operate the same in all fish species as it does in mice or humans, and more studies in other teleost groups are still needed to identify the physiological reasons for this variability. Among other missing information in fish, monoallelic gene expression and its regulation by epigenetic mechanisms, including parental imprinting clues in fish gametes, is still lacking convincing evidence, despite its evolutionary interest regarding the different rounds of genome duplication in teleost.

There are two complementary ways of addressing the animal epigenetic response to environmental constraints. One way is that the effect of an environmental constraint is analyzed on the gamete epigenetic marks of the exposed animal. If the consequences on the offspring are unknown, the gamete alteration raises the question of the epigenetic risk for the progeny. Another way is that the environmental constraint is shown to have an impact on the epigenetic profile of some tissues in the progeny but the mechanisms of epigenetic transmission by the gametes are not known (not treated in this review). We showed here that the epigenetic effect of endocrine disruptors on gametes and gametogenesis is better described today, together with the effect of a short-term rearing of wild fish in a hatchery on their gamete epigenetic profile. There is, however, a major lack of knowledge in gametes in regards to the effect of altered or improved rearing conditions in aquaculture (water quality, temperature, and stressors), and the effect of reproductive technologies, for which whole genome studies need to be more developed. In all, it should not be overlooked that inter- or transgenerational transmission of environmental clues by way of gamete epigenetics is often difficult to characterize, especially when genetic parameters or rearing conditions of the offspring overflow the more cryptic effects linked to gametic changes. Therefore, studies at the gamete level are important: they are the beacon of the risk taken, and the identification of altered epigenetic profiles in the gametes is the first step before reparation, or lack of it, by the embryo can be explored.

Acknowledgments

The author's work (CL) was partially funded by the French CRB Anim project (ANR-11-INBS-0003). The author's interactions fully benefited from the COST Actions FA1205 AQUAGAMETE and FA1201 EPICONCEPT.

References

1 Hammoud, S.S., Nix, D.A., Zhang, H. et al. (2009). Distinctive chromatin in human sperm packages genes for embryo development. *Nature* **460** (7254): 473–478.
2 Lindeman, L.C., Winata, C.L., Aanes, H. et al. (2010). Chromatin states of developmentally-regulated genes revealed by DNA and histone methylation patterns in zebrafish embryos. *The International Journal of Developmental Biology* **54** (5): 803–813.
3 Wu, S.F., Zhang, H., and Cairns, B.R. (2011). Genes for embryo development are packaged in blocks of multivalent chromatin in zebrafish sperm. *Genome Research* **21** (4): 578–589.
4 Potok, M.E., Nix, D.A., Parnell, T.J. et al. (2013). Reprogramming the maternal zebrafish genome after fertilization to match the paternal methylation pattern. *Cell* **153** (4): 759–772.
5 Tang, W.W.C., Kobayashi, T., Irie, N. et al. (2016). Specification and epigenetic programming of the human germ line. *Nature Reviews Genetics* **17** (10): 585–600.
6 Schulz, R.W., de França, L.R., Lareyre, J.J. et al. (2010). Spermatogenesis in fish. *General and Comparative Endocrinology* **165** (3): 390–411.
7 Houston, D.W. and King, M.L. (2000). Germ plasm and molecular determinants of germ cell fate. *Current Topics in Developmental Biology* **50**: 155–181.
8 Hansen, C.L. and Pelegri, F. (2021). Primordial germ cell specification in vertebrate embryos: phylogenetic distribution and conserved molecular features of preformation and induction. *Frontiers in Cell and Development Biology* **9**: 730332.
9 Wasserzug-Pash, P. and Klutstein, M. (2019). Epigenetic changes in mammalian gametes throughout their lifetime: the four seasons metaphor. *Chromosoma* **128** (3): 423–441.
10 Hajkova, P. (2011). Epigenetic reprogramming in the germline: towards the ground state of the epigenome. *Philosophical Transactions of the Royal Society, B: Biological Sciences*. The Royal Society **366**: 2266–2273.
11 Yamaguchi, S., Hong, K., Liu, R. et al. (2012). Tet1 controls meiosis by regulating meiotic gene expression. *Nature* **492** (7429): 443–447.
12 Guibert, S. and Weber, M. (2012). Epigenetics erase for a new start. *Nature* **492** (7429): 363–364.
13 Hajkova, P., Ancelin, K., Waldmann, T. et al. (2008). Chromatin dynamics during epigenetic reprogramming in the mouse germ line. *Nature* **452** (7189): 877–881.
14 Skvortsova, K., Tarbashevich, K., Stehling, M. et al. (2019). Retention of paternal DNA methylome in the developing zebrafish germline. *Nature Communications* **10** (1): 3054.
15 Ortega-Recalde, O., Day, R.C., Gemmell, N.J. et al. (2019). Zebrafish preserve global germline DNA methylation while sex-linked rDNA is amplified and demethylated during feminisation. *Nature Communications* **10** (1): 3053.
16 Wang, X. and Bhandari, R.K. (2020). DNA methylation reprogramming in medaka fish, a promising animal model for environmental epigenetics research. *Environmental Epigenetics* **6** (1): 1–4.
17 Wang, X. and Bhandari, R.K. (2020). The dynamics of DNA methylation during epigenetic reprogramming of primordial germ cells in medaka (*Oryzias latipes*). *Epigenetics* **15** (5): 483–498.
18 Zhou, R., Shang, R., Gong, D. et al. (2019). Characterization of H3 methylation in regulating oocyte development in cyprinid fish. *Science China Life Sciences* **62** (6): 829–837.
19 Kobayashi, H., Sakurai, T., Imai, M. et al. (2012). Contribution of intragenic DNA methylation in mouse gametic DNA methylomes to establish oocyte-specific heritable marks. *PLoS Genetics* **8** (1): e1002440.
20 Jiang, L., Zhang, J., Wang, J.J. et al. (2013). Sperm, but not oocyte, DNA methylome is inherited by zebrafish early embryos. *Cell* **153** (4): 773–784.
21 Wang, X. and Bhandari, R.K. (2019). DNA methylation dynamics during epigenetic reprogramming of medaka embryo. *Epigenetics* **14** (6): 611–622.
22 Handel, M.A. and Schimenti, J.C. (2010). Genetics of mammalian meiosis: regulation, dynamics and impact on fertility. *Nature Reviews Genetics* **11** (2): 124–136.

23 Akiyama, T., Nagata, M., and Aoki, F. (2006). Inadequate histone deacetylation during oocyte meiosis causes aneuploidy and embryo death in mice. *Proceedings of the National Academy of Sciences* **103** (19): 7339–7344.

24 Dada, R., Kumar, M., Jesudasan, R. et al. (2012). Epigenetics and its role in male infertility. *Journal of Assisted Reproduction and Genetics* **29** (3): 213–223.

25 Carrell, D.T. and Hammoud, S.S. (2009). The human sperm epigenome and its potential role in embryonic development. *Molecular Human Reproduction* **16** (1): 37–47.

26 Carrell, D.T. (2011). Epigenetic marks in zebrafish sperm: insights into chromatin compaction, maintenance of pluripotency, and the role of the paternal genome after fertilization. *Asian Journal of Andrology* **13** (4): 620–621.

27 Zhu, W., Xu, X., Wang, X. et al. (2019). Reprogramming histone modification patterns to coordinate gene expression in early zebrafish embryos. *BMC Genomics* **20**: 248.

28 Murphy, P.J., Wu, S.F., James, C.R. et al. (2018). Placeholder nucleosomes underlie germline-to-embryo DNA methylation reprogramming. *Cell* **172** (5): 993–1006. e13.

29 Glasauer, S.M.K. and Neuhauss, S.C.F. (2014). Whole-genome duplication in teleost fishes and its evolutionary consequences. *Molecular Genetics and Genomics* **289** (6): 1045–1060.

30 Zhou, R., Cheng, H., and Tiersch, T.R. (2002). Differential genome duplication and fish diversity. *Reviews in Fish Biology and Fisheries* **11** (4): 331–337.

31 Zhang, Y., Sun, X., Zhang, L. et al. (2017). Testicular Dnmt3 expression and global DNA methylation are down-regulated by gonadotropin releasing hormones in the ricefield eel monopterus albus. *Scientific Reports* **7**: 43158.

32 Wang, F.L., Yan, L.X., Shi, H.J. et al. (2018). Genome-wide identification, evolution of DNA methyltransferases and their expression during gonadal development in Nile tilapia. *Comparative Biochemistry and Physiology Part B: Biochemistry and Molecular Biology* **226**: 73–84.

33 Liu, J., Hu, H., Panserat, S. et al. (2020). Evolutionary history of DNA methylation related genes in chordates: new insights from multiple whole genome duplications. *Scientific Reports* **10** (1): 970.

34 Fellous, A., Earley, R.L., and Silvestre, F. (2019). Identification and expression of mangrove rivulus (*Kryptolebias marmoratus*) histone deacetylase (HDAC) and lysine acetyltransferase (KAT) genes. *Gene* **691**: 56–69.

35 Cavalieri, V. (2020). Histones, their variants and post-translational modifications in zebrafish development. *Frontiers in Cell and Development Biology* **8**: 456.

36 Brewer, L.R. (2011). Deciphering the structure of DNA toroids. *Integrative Biology* **3** (5): 540–547.

37 Ward, W.S. (2010). Function of sperm chromatin structural elements in fertilization and development. *Molecular Human Reproduction* **16** (1): 30–36.

38 Herráez, M.P., Ausió, J., Devaux, A. et al. (2017). Paternal contribution to development: sperm genetic damage and repair in fish. *Aquaculture* **472**: 45–59.

39 González-Rojo, S., Fernández-Díez, C., Lombó, M. et al. (2018). Distribution of DNA damage in the sperm nucleus: a study of zebrafish as a model of histone-packaged chromatin. *Theriogenology* **122**: 109–115.

40 Barrachina, F., Anastasiadi, D., Jodar, M. et al. (2018). Identification of a complex population of chromatin-associated proteins in the European sea bass (*Dicentrarchus labrax*) sperm. *Systems Biology in Reproductive Medicine* **64** (6): 502–517.

41 Saperas, N., Chiva, M., Casas, M.T. et al. (2006). A unique vertebrate histone H1-related protamine-like protein results in an unusual sperm chromatin organization. *The FEBS Journal* **273** (19): 4548–4561.

42 Kasinsky, H.E., Maria Eirin-Lopez, J., and Ausio, J. (2012). Protamines: structural complexity, evolution and chromatin patterning. *Protein & Peptide Letters* **18** (8): 755–771.

43 Khamlichi, A.A. and Feil, R. (2018). Parallels between mammalian mechanisms of monoallelic gene expression. *Trends in Genetics* **34** (12): 954–971.

44 Farhadova, S., Gomez-Velazquez, M., and Feil, R. (2019). Stability and lability of parental methylation imprints in development and disease. *Genes* **10** (12): 999.

45 Renfree, M.B., Hore, T.A., Shaw, G. et al. (2009). Evolution of genomic imprinting: insights from marsupials and monotremes. *Annual Review of Genomics and Human Genetics* **10**: 241–262.

46 Pires, N.D. and Grossniklaus, U. (2014). Different yet similar: evolution of imprinting in flowering plants and mammals. *F1000Prime Reports* **6**: 63.

47 Marshall, H., van Zweden, J.S., Van Geystelen, A. et al. (2020). Parent of origin gene expression in the bumblebee, *Bombus terrestris*, supports Haig's kinship theory for the evolution of genomic imprinting. *Evolution Letters* **4** (6): 479–490.

48 McGowan, R.A. and Martin, C.C. (1997). DNA methylation and genome imprinting in the zebrafish, *Danio rerio*: some evolutionary ramifications. *Biochemistry and Cell Biology* **75** (5): 499–506.

49 Xie, B., Zhang, L., Zheng, K. et al. (2009). The evolutionary foundation of genomic imprinting in lower vertebrates. *Chinese Science Bulletin* **54** (8): 1354–1360.

50 Hahn, Y., Yang, S.K., and Chung, J.H. (2005). Structure and expression of the zebrafish mest gene, an ortholog of mammalian imprinted gene PEG1/MEST. *Biochimica et Biophysica Acta (BBA) – Gene Structure and Expression* **1731** (2): 125–132.

51 Paulsen, M., Khare, T., Burgard, C. et al. (2005). Evolution of the Beckwith-Wiedemann syndrome region in vertebrates. *Genome Research* **15** (1): 146–153.

52 Dunzinger, U., Haaf, T., and Zechner, U. (2007). Conserved synteny of mammalian imprinted genes in chicken, frog, and fish genomes. *Cytogenetic and Genome Research* **117** (1–4): 78–85.

53 Smith, T.H.L., Collins, T.M., and McGowan, R.A. (2011). Expression of the dnmt3 genes in zebrafish development: similarity to Dnmt3a and Dnmt3b. *Development Genes and Evolution* **220** (11): 347–353.

54 Ma, C.G., Fan, L.C., Ganassin, R. et al. (2001). Production of zebrafish germ-line chimeras from embryo cell cultures. *Proceedings of the National Academy of Sciences of the United States of America* **98** (5): 2461–2466.

55 Martin, C.C. and McGowan, R. (1995). Parent-of-origin specific effects on the methylation of a transgene in the zebrafish *Danio rerio*. *Developmental Genetics* **17** (3): 233–239.

56 Matsui, Y. and Mochizuki, K. (2014). A current view of the epigenome in mouse primordial germ cells. *Molecular Reproduction and Development* **81** (2): 160–170.

57 Tiffon, C. (2018). The impact of nutrition and environmental epigenetics on human health and disease. *International Journal of Molecular Sciences* **19** (11): 3425.

58 Donkin, I. and Barres, R. (2018). Sperm epigenetics and influence of environmental factors. *Molecular Metabolism* **14**: 1–11.

59 Jenkins, T.G. and Carrell, D.T. (2012). The sperm epigenome and potential implications for the developing embryo. *Reproduction* **143** (6): 727–734.

60 Labbé, C., Robles, V., and Herraez, M.P. (2017). Epigenetics in fish gametes and early embryo. *Aquaculture* **472**: 93–106.

61 Cordeiro, L., Riou, C., Uzbekov, R. et al. (2021). Avian sperm increase in vitro the release of exosomes from SST-enriched organoids. *Reproduction (Cambridge, England)* **161** (4): 375–384.

62 Sarangdhar, M.A., Chaubey, D., Srikakulam, N. et al. (2018). Parentally inherited long non-coding RNA Cyrano is involved in zebrafish neurodevelopment. *Nucleic Acids Research* **46** (18): 9726–9735.

63 Aniagu, S.O., Williams, T.D., Allen, Y. et al. (2008). Global genomic methylation levels in the liver and gonads of the three-spine stickleback (*Gasterosteus aculeatus*) after exposure to hexabromocyclododecane and 17-β oestradiol. *Environment International* **34** (3): 310–317.

64 Yuan, C., Zhang, Y., Liu, Y. et al. (2016). Enhanced GSH synthesis by bisphenol A exposure promoted DNA methylation process in the testes of adult rare minnow *Gobiocypris rarus*. *Aquatic Toxicology* **178**: 99–105.

65 Laing, L.V., Viana, J., Dempster, E.L. et al. (2016). Bisphenol A causes reproductive toxicity, decreases dnmt1 transcription, and reduces global DNA methylation in breeding zebrafish (*Danio rerio*). *Epigenetics* **11** (7): 526–538.

66 Liu, Y., Zhang, Y., Tao, S. et al. (2016). Global DNA methylation in gonads of adult zebrafish *Danio rerio* under bisphenol A exposure. *Ecotoxicology and Environmental Safety* **130**: 124–132.

67 González-Rojo, S., Lombó, M., Fernández-Díez, C. et al. (2019). Male exposure to bisphenol A impairs spermatogenesis and triggers histone hyperacetylation in zebrafish testes. *Environmental Pollution* **248**: 368–379.

68 Tai, Z., Guan, P., Zhang, T. et al. (2022). Effects of parental environmental copper stress on offspring development: DNA methylation modification and responses of differentially methylated region-related genes in transcriptional expression. *Journal of Hazardous Materials* **424**: 127600.

69 Iii, M.J.C., Kalluvila, T.A., Klingler, R.H. et al. (2017). Mercury-induced epigenetic transgenerational inheritance of abnormal neurobehavior is correlated with sperm epimutations in zebrafish. *PLoS one* **12** (5): e0176155.

70 Santangeli, S., Maradonna, F., Gioacchini, G. et al. (2016). BPA-induced deregulation of epigenetic patterns: effects on female zebrafish reproduction. *Scientific Reports* **6** (1): 21982.

71 Zhu, L., Wang, L., Fan, X. et al. (2021). Chronic exposure to bisphenol A resulted in alterations of reproductive functions via immune defense, oxidative damage and disruption DNA/histone methylation in male rare minnow *Gobiocypris rarus*. *Aquatic Toxicology* **236**: 105849.

72 Lombó, M., Getino-álvarez, L., Depincé, A. et al. (2019). Embryonic exposure to bisphenol A impairs primordial germ cell migration without jeopardizing male breeding capacity. *Biomolecules* **9** (8): 307.

73 Santangeli, S., Consales, C., Pacchierotti, F. et al. (2019). Transgenerational effects of BPA on female reproduction. *Science of the Total Environment* **685**: 1294–1305.

74 Zhu, L., Liu, Y., Xue, X. et al. (2021). BPA's transgenerational disturbance to transcription of ovarian steroidogenic genes in rare minnow *Gobiocypris rarus* via DNA and histone methylation. *Science of the Total Environment* **762**: 143055.

75 Song, X., Wang, X., and Bhandari, R.K. (2020). Developmental abnormalities and epigenetic alterations in medaka (*Oryzias latipes*) embryos induced by triclosan exposure. *Chemosphere* **261**: 127613.

76 Nilsson, E., Sadler-Riggleman, I., Beck, D. et al. (2021). Differential DNA methylation in somatic and sperm cells of hatchery vs wild (natural-origin) steelhead trout populations. *Environmental Epigenetics* **7** (1): dvab002.

77 Leitwein, M., Laporte, M., Le Luyer, J. et al. (2021). Epigenomic modifications induced by hatchery rearing persist in germ line cells of adult salmon after their oceanic migration. *Evolutionary Applications* **14** (10): 2402–2413.

78 Wellband, K., Roth, D., Linnansaari, T. et al. (2021). Environment-driven reprogramming of gamete DNA methylation occurs during maturation and is transmitted intergenerationally in Atlantic salmon. *G3 Genes|Genomes|Genetics* **11** (12): jkab353.

79 Cheng, Y., Vechtova, P., Fussy, Z. et al. (2021). Changes in phenotypes and DNA methylation of in vitro aging sperm in common carp *cyprinus carpio*. *International Journal of Molecular Sciences* **22** (11): 5925.

80 Waghmare, S.G., Samarin, A.M., Samarin, A.M. et al. (2021). Histone acetylation dynamics during in vivo and in vitro oocyte aging in common carp *cyprinus carpio*. *International Journal of Molecular Sciences* **22** (11): 6036.

81 Herranz-Jusdado, J.G., Gallego, V., Morini, M. et al. (2019). Comparison of European eel sperm cryopreservation protocols with standardization as a target. *Aquaculture* **498**: 539–544.

82 de Mello, F., Garcia, J.S., Godoy, L.C. et al. (2017). The effect of cryoprotectant agents on DNA methylation patterns and progeny development in the spermatozoa of *Colossoma macropomum*. *General and Comparative Endocrinology* **245**: 94–101.

83 Depince, A., Gabory, A., Dziewulska, K. et al. (2019). DNA methylation stability in fish spermatozoa upon external constraint: Impact of fish hormonal stimulation and sperm cryopreservation. *Molecular Reproduction and Development* **87** (1): 124–134.

84 Gavrilescu, M., Demnerová, K., Aamand, J. et al. (2015). Emerging pollutants in the environment: present and future challenges in biomonitoring, ecological risks and bioremediation. *New Biotechnology* **32** (1): 147–156.

85 Thomaidis, N.S., Asimakopoulos, A.G., and Bletsou, A.A. (2012). Emerging contaminants: a tutorial mini-review. *Global NEST Journal* **14** (1): 72–79.

86 Kavlock, R.J., Daston, G.P., DeRosa, C. et al. (1996). Research needs for the risk assessment of health and environmental effects of endocrine disruptiors. *Environmental Health Perspectives* **104** (4): 715–740.

87 Cavalieri, V. and Spinelli, G. (2017). Environmental epigenetics in zebrafish. *Epigenetics & Chromatin* **10** (1): 46.

88 Carnevali, O., Santangeli, S., Forner-Piquer, I. et al. (2018). Endocrine-disrupting chemicals in aquatic environment: what are the risks for fish gametes? *Fish Physiology and Biochemistry* **44** (6): 1561–1576.

89 Lombó, M., Fernández-díez, C., González-rojo, S. et al. (2019). Genetic and epigenetic alterations induced by bisphenol A exposure during different periods of spermatogenesis: from spermatozoa to the progeny. *Scientific Reports* **9**: 1–13.

90 Teletchea, F. and Fontaine, P. (2014). Levels of domestication in fish: implications for the sustainable future of aquaculture. *Fish and Fisheries* **15** (2): 181–195.

91 Gavery, M.R., Nichols, K.M., Goetz, G.W. et al. (2018). Characterization of genetic and epigenetic variation in sperm and red blood cells from adult hatchery and natural-origin steelhead, *Oncorhynchus mykiss*. *G3-Genes Genomes Genetics* **8** (11): 3723–3736.

92 Rodriguez Barreto, D., Garcia de Leaniz, C., Verspoor, E. et al. (2019). DNA methylation changes in the sperm of captive-reared fish: a route to epigenetic introgression in wild populations. Mulligan C, editor. *Molecular Biology and Evolution* **36** (10): 2205–2211.

93 Anastasiadi, D. and Piferrer, F. (2019). Epimutations in developmental genes underlie the onset of domestication in farmed european sea bass. *Molecular Biology and Evolution* **36** (10): 2252–2264.

94 Liu, G., Wang, W., Hu, S. et al. (2018). Inherited DNA methylation primes the establishment of accessible chromatin during genome activation. *Genome Research* **28** (7): 998–1007.

95 Migaud, H., Davie, A., and Taylor, J.F. (2010). Current knowledge on the photoneuroendocrine regulation of reproduction in temperate fish species. *Journal of Fish Biology* **76** (1): 27–68.

96 Migaud, H., Bell, G., Cabrita, E. et al. (2013). Gamete quality and broodstock management in temperate fish. *Reviews in Aquaculture* **5** (s1): S194–S223.

97 Mylonas, C.C., Duncan, N.J., and Asturiano, J.F. (2017). Hormonal manipulations for the enhancement of sperm production in cultured fish and evaluation of sperm quality. *Aquaculture* **472**: 21–44.

98 Fauque, P. (2013). Ovulation induction and epigenetic anomalies. *Fertility and Sterility* **99** (3): 616–623.

99 Yoshizaki, G. and Yazawa, R. (2019). Application of surrogate broodstock technology in aquaculture. *Fisheries Science* **85** (3): 429–437.

100 Okutsu, T., Suzuki, K., Takeuchi, Y. et al. (2006). Testicular germ cells can colonize sexually undifferentiated embryonic gonad and produce functional eggs in fish. *Proccedings of the National Academy of Sciences of the United States of America* **103** (8): 2725–2729.

101 Lee, S., Katayama, N., and Yoshizaki, G. (2016). Generation of juvenile rainbow trout derived from cryopreserved whole ovaries by intraperitoneal transplantation of ovarian germ cells. *Biochemical and Biophysical Research Communications* **478** (3): 1478–1483.

102 Martinez-Paramo, S., Horvath, A., Labbe, C. et al. (2017). Cryobanking of aquatic species. *Aquaculture* **472**: 156–177.

103 Chatterjee, A., Saha, D., Niemann, H. et al. (2017). Effects of cryopreservation on the epigenetic profile of cells. *Cryobiology* **74** (Supplement C): 1–7.

104 McCarrey, J.R. (2014). Distinctions between transgenerational and non-transgenerational epimutations. *Molecular and Cellular Endocrinology* **398** (1–2): 13–23.

105 Mørkve Knudsen, T., Rezwan, F.I., Jiang, Y. et al. (2018). Transgenerational and intergenerational epigenetic inheritance in allergic diseases. *Journal of Allergy and Clinical Immunology* **142** (3): 765–772.

106 Nilsson, E.E. and Skinner, M.K. (2015). Environmentally induced epigenetic transgenerational inheritance of disease susceptibility. *Translational Research* **165**: 12–17.

107 Manikkam, M., Tracey, R., Guerrero-Bosagna, C. et al. (2013). Plastics derived endocrine disruptors (BPA, DEHP and DBP) induce epigenetic transgenerational inheritance of obesity, reproductive disease and sperm epimutations. *PLoS one* **8** (1): e55387.

108 Dolinoy, D.C., Huang, D., and Jirtle, R.L. (2007). Maternal nutrient supplementation counteracts bisphenol A-induced DNA hypomethylation in early development. *Proceedings of the National Academy of Sciences* **104** (32): 13056–13061.

109 Bansal, A., Li, C., Xin, F. et al. (2019). Transgenerational effects of maternal bisphenol: a exposure on offspring metabolic health. *Journal of Developmental Origins of Health and Disease* **10** (2): 164–175.

110 Lombó, M. and Herráez, P. (2021). The effects of endocrine disruptors on the male germline: an intergenerational health risk. *Biological Reviews* **96** (4): 1243–1262.

111 Singleman, C. and Holtzman, N.G. (2012). Analysis of postembryonic heart development and maturation in the zebrafish, *Danio reri*. *Developmental Dynamics* **241** (12): 1993–2004.

112 Lombo, M., Fernandez-Diez, C., Gonzalez-Rojo, S. et al. (2015). Transgenerational inheritance of heart disorders caused by paternal bisphenol A exposure. *Environmental Pollution* **206**: 667–678.

113 Blanc, M., Cormier, B., Hyötyläinen, T. et al. (2020). Multi- and transgenerational effects following early-life exposure of zebrafish to permethrin and coumarin 47: impact on growth, fertility, behavior and lipid metabolism. *Ecotoxicology and Environmental Safety* **205**: 111348.

114 Lombo, M. and Paz Herraez, M. (2021). Paternal inheritance of bisphenol A cardiotoxic effects: the implications of sperm epigenome. *International Journal of Molecular Sciences* **22** (4): 2125.

115 Kamstra, J.H., Sales, L.B., Aleström, P. et al. (2017). Differential DNA methylation at conserved non-genic elements and evidence for transgenerational inheritance following developmental exposure to mono(2-ethylhexyl) phthalate and 5-azacytidine in zebrafish. *Epigenetics & Chromatin* **10**: 20.

116 Pérez-Cerezales, S., Gutiérrez-Adán, A., Martínez-Páramo, S. et al. (2011). Altered gene transcription and telomere length in trout embryo and larvae obtained with DNA cryodamaged sperm. *Theriogenology* **76** (7): 1234–1245.

117 Hayes, M.C., Rubin, S.P., Hensleigh, J.E. et al. (2005). Performance of juvenile steelhead trout (*Oncorhynchus mykiss*) produced from untreated and cryopreserved milt. *Aquaculture* **249** (1): 291–302.

118 Butzge, A.J., Yoshinaga, T.T., Acosta, O.D.M. et al. (2021). Early warming stress on rainbow trout juveniles impairs male reproduction but contrastingly elicits intergenerational thermotolerance. *Scientific Reports* **11** (1): 17053.

8

Epigenetics in Sex Determination and Differentiation of Fish

Qian Wang, Qian Liu, Xiaona Zhao, Wenxiu Ma, Lili Tang, Bo Feng, and Changwei Shao

Yellow Sea Fisheries Research Institute, Chinese Academy of Fishery Sciences, Qingdao, China

8.1 Introduction

Sex determination and differentiation is a universal biological phenomenon in animals. Sex determination is the designation for the whole development stage that leads the undifferentiated gonads to the differentiated reproductive organs, either male testes or female ovaries. Sex differentiation is the pathway or the procedure that ultimately produces the male or female sexual phenotype [1]. With more than 30,000 species, teleost fishes represent the most diverse group of vertebrates [2]. Different from mammals and birds, which possess highly conserved genetic networks and master genes responsible for sex determination and differentiation, fish exhibit a large diversity of sex-determining mechanisms, including gonochorism or hermaphroditism [3]. Thus, fish can be used as an ideal model organism to study sex determination and differentiation.

8.1.1 Sex Chromosome in Fish

The evolution of the mechanism of sex determination has been linked to a particular genomic process, the formation of sex chromosomes. Sex chromosomes have evolved in different ways from a pair of autosomal ancestors [4]. As the largest group of vertebrates, fish have all the sex chromosome types found in animals, although a vast majority of fishes have not been identified to have sex chromosomes. Types of sex chromosomes in fish species are mainly divided into the following categories: (i) male heterogamy type: XX/XY – this type, similar to mammals, has been found in most fish, including Nile tilapia (*Oreochromis niloticus*); (ii) female heterogamy type: ZW/ZZ – contrary to XX/XY type, ZW is female and ZZ is male, such as in Chinese tongue sole (*Cynoglossus semilaevis*); (iii) no sex chromosomes – multiple genes on the autosomes jointly determine the sex formation, such as in European sea bass (*Dicentrarchus labrax*); (iv) sex chromosome number type: include the Y or W chromosome-lacking system, i.e., XX/X0, ZZ/Z0, as well as multiple sex chromosome system, i.e., $X_1X_1X_2X_2/X_1X_2Y$, XX/XY_1Y_2, $X_1X_1X_2X_2/X_1Y_1X_2Y_2$, ZZ/ZW_1W_2, $Z_1Z_1Z_2Z_2/Z_1W_1Z_2W_2$ [5–8]. Generally speaking, fishes are considered to be in the primitive stage of sex chromosome evolution.

8.1.2 Sex Determination and Differentiation in Fish

Sex determination mechanisms in fish are strongly diversified among species and can be mainly divided into two categories, genetic sex determination (GSD) and environmental sex determination (ESD) (Box 8.1). In species with GSD, the sex is genetically determined by chromosomes (especially the sex chromosomes) and sex-determining genes, while in species with ESD, the sex is controlled by environmental factors, such as temperature, pH, hormones, and social interactions [3, 4]. The involvement of epigenetic mechanisms in sex determination and corresponding differentiation is now supported by studies in both GSD and ESD species across different taxa. Especially in ESD, which is not considered to rely on genetic sequence, the epigenetic mechanisms therefore show their importance during sexual fate formation [9, 10]. Many fish have the features of sexual size dimorphism, meaning sex-related differences in growth, and artificial methods have been used

Epigenetics in Aquaculture, First Edition. Edited by Francesc Piferrer and Han-Ping Wang.

Box 8.1 Glossary of Terms Related to Sex Determination and Sex Differentiation

ESD: Environmental sex determination is a mechanism by which the sex of an individual is decided after conception, according to its immediate environment within a discrete period after fertilization.
Gonochoristic fish: In gonochoristic fishes, sex is determined genetically or by environmental factors. After sex determination, the gonads are differentiated into ovary or testis and retain the same sex throughout the fish's life span.
GSD: Genetic sex determination, with which the primary sex of an individual is determined at the moment of fertilization by heritable genetic elements that differ between the sexes, often based on sex chromosomes (chromosomal sex determination) or distributed among the autosomes (polygenic sex determination).
Hermaphroditic fish: In hermaphroditic fishes, the individual has both male and female reproductive organs and can perform both the male and female parts of reproduction. Sex can change sequentially from male to female (protandrous), female to male (protogynous), or serially (bidirectional sex change) during the life cycle of a hermaphroditic individual. Also, simultaneous hermaphroditism exists so that the individual can produce sperm and eggs at the same time.
Sex determination: In bisexual organisms, sex determination is the biological process of how sex is established and the determination of whether the gonads will develop as testes or ovaries.
Sex differentiation: Sex differentiation is a series of development events that allow an undifferentiated gonad to be transformed into either a testis or ovary after the sex has been determined. This produces the phenotypic sex on an individual level and the sex ratio on a population level.
Sex reversal: In some species, abnormal conditions during early development cause the individuals to develop a phenotypic sex different from their genotypic sex.

for sex control in aquaculture [11, 12]. Therefore, the epigenetic mechanism of sex determination and differentiation in fish is an attractive area of both great theoretical and economic significance.

8.1.3 Sexual Plasticity of Fish – Gonochoristic and Hermaphroditic Species

Teleost fish exhibit remarkably diverse and plastic patterns of sexual development, including gonochorism and hermaphroditism [13]. Hermaphroditism, where an individual can change from one sex to the other during the adult stage naturally, is divided into synchronous hermaphroditism and sequential hermaphroditism. The latter is subdivided into three types: protogynous hermaphroditism (female-to-male, such as three-spotted wrasse, *Halichoeres trimaculatus*), protandrous hermaphroditism (male-to-female, such as yellowtail clownfishes, *Amphiprion clarkii*), and bidirectional sex change (change sex in both directions, such as orange-red pygmy goby, *Trimma okinawae*) [14, 15]. The physical, morphological, and gonadal changes involved in the sexual fate transition process are well described, yet the molecular mechanisms underpinning those transitions are still limited. Epigenetic modifications, including DNA methylation status, associated histone proteins, and chromatin accessibility appear to be major enactors of the onset of sex change and subsequent maintenance of sexual phenotype [16, 17]. The influence of environmental stimuli such as temperature, hormones, and social cues on epigenetic modifications of key genes in sex determination will be described in detail in related sections of this chapter.

8.1.4 Phenomenon of Sex Reversal

Sexual plasticity of fish is also widespread in gonochoristic fish. In many GSD fish species, even if the sex system is inherited, the primary genetic sex can be turned over to the opposite direction by environmental factors during a critical window period in early development, and this phenomenon is known as sex reversal [18, 19]. Environmental factors can influence the sex ratio in many fish species; they can either determine sex or affect sex differentiation. Fish are ectotherm species and may face rapid environmental changes such as elevated seawater temperature caused by global warming, with embryos and spawning adults being more susceptible [20, 21]. Epigenetic regulation plays multiple crucial roles in the adaptation of sex ratios in fish, and is the key to understanding the mechanisms by which genotype and environment interact to control early sex determination and differentiation.

In this chapter, we focus on epigenetics in sex determination and differentiation of fish. Firstly, we briefly describe the role of epigenetics in sex chromosome evolution, including the contribution of the different types of epigenetic mechanisms (DNA methylation, histone modification, and chromatin structure remodeling). Secondly, the role of epigenetics in sex determination is summarized, with the concern of the sex determination regulation network, how epigenetics regulates key sex-related genes, and subsequently introduces recently developed epigenetic markers. Thirdly, epigenetic regulation of sex differentiation is discussed for both gonochoristic and hermaphroditic species. Afterward, the transgenerational epigenetic regulation mechanism is discussed with regard to sex reversal, which focuses on the inheritance of epigenetic patterns across one or more generations associated with the effect of environmental factors such as water temperature. At the end of this chapter, we will summarize and provide some insights into the future perspectives of research on epigenetics in sex determination and differentiation of fish.

8.2 Epigenetics and Sex Chromosome Evolution

The chromosome is the carrier of genetic information and the material basis of the genetic variation of bisexual organisms. Chromosomes can be divided into autosomes and sex chromosomes, and sex chromosomes have evolved in different ways from a pair of autosomal ancestors [5]. Since the first study used cytology to discover fish sex chromosomes, many scientists have conducted karyotype analysis and chromosome-banning research on fish, but only in about 10% of fish species have sex chromosomes been found [22]. This may be due to the young sex chromosomes in fish species, which have a recent origin. Therefore, fish species are excellent models for studying the early stages of sex chromosome evolution.

The formation of recombination suppression is the result of autosomal inversion, which may be the main driving force for the evolution of sex chromosomes (Figure 8.1) [23, 24]. DNA methylation and histone modification are the main epigenetic processes involved in gene expression regulation and chromatin structure dynamics. DNA hypermethylation and H3K9me3 trimethylation are usually related to the transcriptional inhibition of chromatin and can reduce the recombination of large regions on chromosomes, which indicates that epigenetics plays an important role in recombination suppression [25].

In addition to the earlier mentioned influences on the evolution of sex chromosomes, DNA methylation, histone modification, chromatin structure, and other effects also exert certain influences on the evolution of sex chromosomes and warrant further investigation.

8.2.1 The Role of DNA Methylation in the Evolution of Sex Chromosomes

The role of epigenetics on sex chromosome evolution was proposed that initial differences between females and males are determined by differential methylated states in nuclear DNA between sexes, whereafter driving Muller's ratchet. In evolutionary genetics, Muller's ratchet refers to the irreversible deleterious mutation accumulation caused by the absence of recombination. According to this hypothesis, differential methylation of sex chromosomes results in recombination suppression and increased mutation rate, thereby dramatically accelerating the speed of Muller's ratchet [26]. Another role of DNA methylation in regulating sex chromosome evolution is the dose compensation in the sex chromosome system. Classic theory predicts that sex chromosomes originate from a pair of homologous autosomes and recombination between them is suppressed, thus resulting in degradation of the Y chromosome gene content. This process creates the need for dosage compensation in the heterogametic sex [23, 25]. Although theoretically well understood, the evolution of sex chromosomes via methylation is challenging to test. The reasons have been summarized as follows: (i) the correlation between DNA methylation and corresponding gene transcription varies with genome contents (e.g., in general, methylation in the

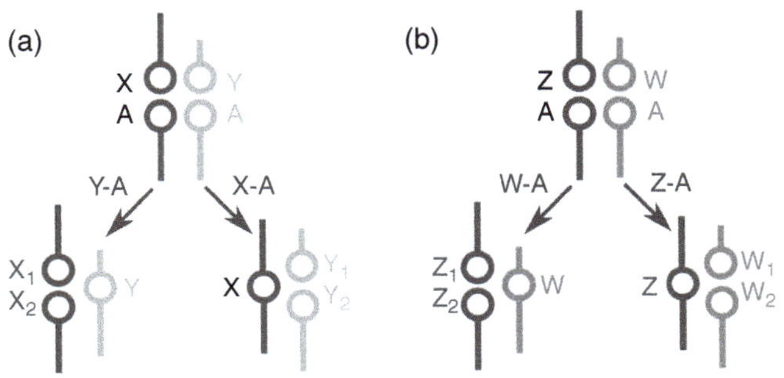

Figure 8.1 Sex chromosome and autosome fusions create multiple sex chromosome systems. (a) In XY systems, X-autosome (X-A) and Y-autosome (Y-A) fusions make XY_1Y_2 and X_1X_2Y systems, respectively. (b) In ZW systems, Z-autosome (Z-A) and W-autosome (W-A) fusions make ZW_1W_2 and Z_1Z_2W systems, respectively [24].

promoter region, especially the CpG islands silences gene expression, while in the gene body region activates gene expression); (ii) DNA methylation in distal regions has a causal effect on enhancer or repressor activity; and (iii) that methylation of specific loci may be erased and reprogramed during gamete formation [23, 27].

Along with the whole genomic bisulfite sequencing (WGBS) technique, different DNA methylation patterns between the sex chromosomes have been found in fish. In the Chinese tongue sole, the reconstructed evolutionary history of the chromosomal-level genome showed that Z and W chromosomes evolved from one duplicate descendant of the proto-chromosome. The Z and W chromosomes shared a small 640 kb pseudoautosomal region (PAR) at their distal telomeric ends with no relevant sequence divergence, and the remaining non-PAR region showed reduced or even absent recombination [28]. Gene number comparison revealed degeneration on the W chromosome had not progressed to a stage where almost its entire original gene content had disappeared as in mammals and birds, indicating a relatively recent evolutionary origin for the tongue sole sex chromosome pair. Moreover, a compensatory mechanism working through the upregulation of female genes was found by global gene expression pattern analyses. By looking into gonadal DNA methylomes of female, normal male, and sex-reversed neomales (ZW genetic females sex reversed into phenotypic males), the dosage compensation was found in a restricted, methylated cytosine-enriched Z chromosomal region in neomale testes, which helped them achieve equal expression level in normal male testes [28, 29]. In threespine stickleback (*Gasterosteus aculeatus*, XX/XY system), WGBS also found apparent hypermethylation of the younger evolutionary stratum of the stickleback X chromosome in females relative to males, which suggested a potential role of DNA methylation in the evolution of heteromorphic sex chromosomes [30].

8.2.2 The Role of Histone Modifications on Sex Chromosome Evolution

The origin of sex chromosomes requires the establishment of recombination suppression between the proto-sex chromosomes. The chromosomes in the nucleus are highly compressed. When folded, the DNA is wrapped with histones. Histones are octamers formed by the polymerization of H2A, H2B, H3, and H4. They are the structural proteins of chromosomes [31]. All core histones are intracellular, but histones H3 and H4 modify more extensively than H2A and H2B. Histones undergo different kinds of posttranslational modifications (PTMs), including methylation, acetylation, ubiquitination, phosphorylation, etc. Key sites for acetylation include at least four highly conserved lysine residues (K5, K8, K12, and K16) in histone H4 and two highly conserved lysine residues (K9 and K14) in histone H3. Histone methylation usually occurs in the arginine residues at the N terminus of histone H3, H4, such as H3R3, H3R17, or lysine residues, such as H3K4, H3K9, and H3K79 [32]. All core histones contain phosphorylation sites in their N terminal domain: serine 1 on H2A and H4, serine 14/32 on H2B, and serine 10 and threonine 3 on H3 (Figure 8.2). These PTMs can change chromatin conformation, regulate differential recruitment of transcription factors and binding proteins to DNA, regulate transcription, silence or activate genes, affect biological functions, and change biological phenotypes.

Recent research on zig-zag eel (*Mastacembelus armatus*) produced a haplotype-resolved assembled genome to explore the origin of a young sex chromosome pair and how recombination suppression evolved in the initial stages of sex chromosome differentiation. The study found that the sex-linked region (SLR) in zig-zag eel spanned the centromere, which might explain the lack of recombination of the SLR. The entire short arm of the X and Y chromosomes exhibited features typical of heterochromatin, including an average repeat content [above 50% in contrast to the long arm (~20%)], low gene density, extensive H3K9me3 modifications, and widespread inactive compartments. Because of its physical location near the centromere, it was suspected that the short-arm heterochromatin likely originated from the pericentromeric heterochromatin. On the other hand, the pericentromeric region on the autosomes was examined, and large regions (~4 Mb) around the centromeres have elevated repeat contents, lower gene density, lower recombination rate, and more frequent H3k9me3 modifications (53.7% of the genome-wide H3K9me3 peaks, whereas versus only 17.7% of the genome). These results suggest that histone methylation may have a limited role in gene repression in pericentromeric heterochromatin, and other epigenetic modifications may exist [33].

8.2.3 The Role of Chromatin Structure on Sex Chromosome Evolution

In the process of sex chromosome evolution, sex determination function is often affected by chromosome rearrangements, such as fusion or translocation, transferred from one pair of chromosomes to another pair of chromosomes to form sex chromosomes. In *Drosophila simulans*, Helleu et al. [34] have identified an evolutionarily young and dynamic X-linked gene, heterochromatin protein 1 D2 (*HP1D2*), which plays a key role in the Paris type X-linked distorters that lead to Y sperm

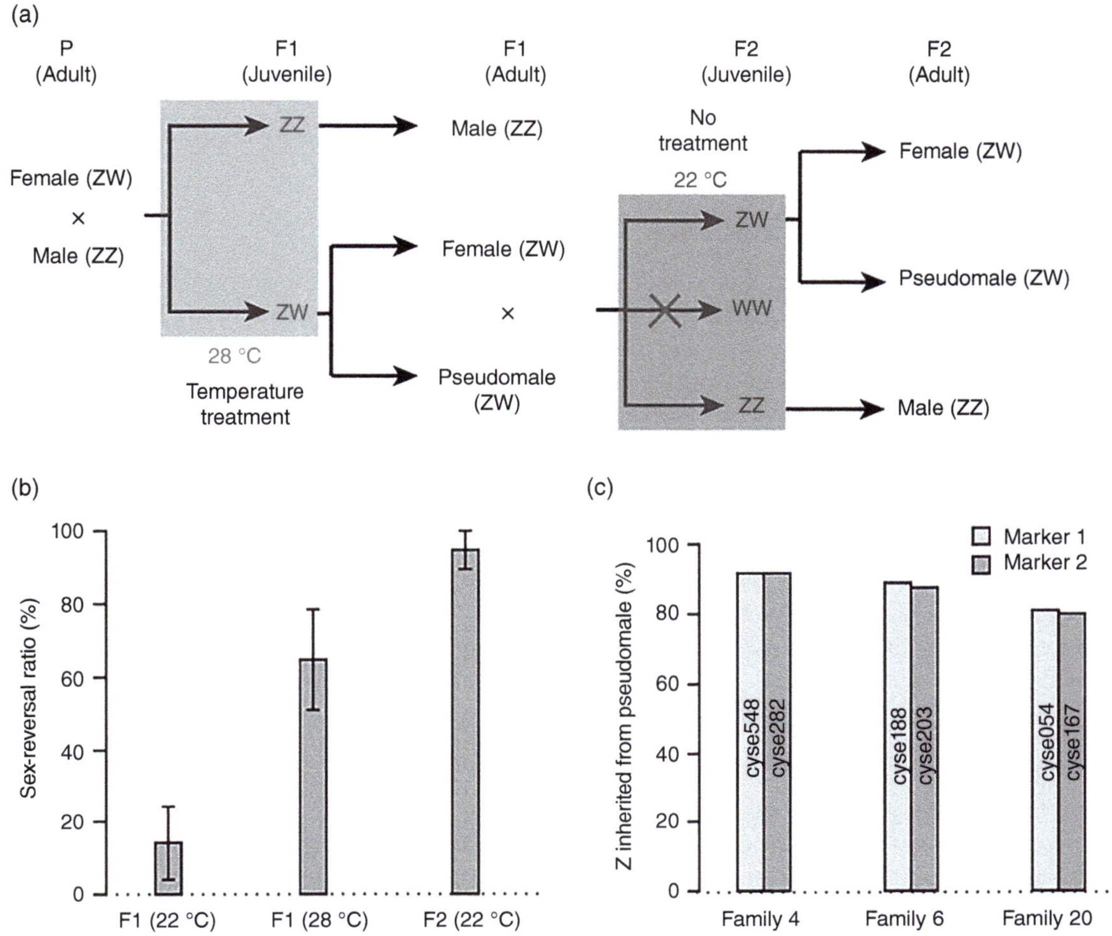

Figure 8.2 Sex reversal in Chinese tongue sole (*Cynoglossus semilaevis*). (a) A schematic flow chart of the production of sex-reversed tongue sole over two generations. (b) Ratio of sex reversal neomales in families raised at normal temperature (22 °C), high temperature (28 °C), and offspring of high temperature-induced neomale raised at normal temperature (22 °C). (c) Paternal inheritance of the Z chromosome in neomale families. The Z chromosome (identified by microsatellite markers 1 and 2) of about 84–90% of ZW individuals is inherited from the sex-reversed paternal chromosome in three neomale families tested. *Source:* Chen et al. [28]/ Springer Nature.

shortage in anaphase II. *HP1D2* alleles can function as a driver to inhibit the isolation of Y chromatids at the second meiotic division stage, resulting in female-dominated offspring. In male germ cells, HP1D2 enriches the heterochromatin Y chromosome, indicating that it can control the separation of sister chromatids through heterochromatin modification. Furthermore, *HP1D2* is discretely distributed on the Y chromosome. Considering the Y chromosome is rich in repetitive sequences, HP1D2 may affect the chromatin organization of the repeated sequences. Thus, these results provide definitive evidence that the co-evolution of highly repetitive sequences and heterochromatin-interacting proteins can produce unstable interactions and ultimately lead to genetic conflicts and promote the evolution of sex chromosomes [34].

Heteromorphic chromosomes, such as X/Y chromosomes in mammals, are different in size and genomic sequence, but play a role as homologs in meiosis. This divalent asymmetry poses a special challenge to the completion of meiosis. In the study of asymmetric chromosomes during meiosis of nematode worm (*Caenorhabditis elegans*), Henzel et al. [35] found that male worms carry the translocation chromosome mnT12, which contains the fusion site of chromosomes X and chromosomes IV. The mnT12 and *IV* can form an asymmetric bivalent, of which chromosome IV parts have the ability to perform pairing and synapsis, with the adjacent chromosome X parts of mnT12 massing the other half of the homologous pair. The unpaired *X* axis has a distinctly coiled shape, while the synaptic axis is linear and extended. Therefore, it can be concluded that bivalent asymmetry can directly inhibit cross-pairing in meiosis through synaptic regulation, thereby increasing the diversity of neo-sex chromosomes. In other words, chromosomal fusion events that produce asymmetric bivalents can cause fundamental changes in the chromosomal recombination landscape and may have an important role in the evolution of sex chromosomes [35]. Nevertheless, studies in fish species are still limited, studies in these model animals will provide valuable information and strategies for the young sex chromosomes in fish.

8.3 Epigenetics and Sex Determination

8.3.1 Regulation Network of Sex Determination

The sex determination system of fish is complex, containing the GSD and ESD mechanisms and the combination of both. External environmental factors including water temperature, pH, population density, and social cues can modify or determine the sex in several fish species. There are also many kinds of sex chromosome types in fish. Tiger puffer (*Takifugu rubripes*), medaka (*Oryzias latipes*), rainbow trout (*Oncorhynchus mykiss*), and pejerrey (*Odontesthes bonariensis*) have the XX/XY male heterogametic system, and some others such as *Cynoglossus semilaevis* have a ZZ/ZW female heterogametic system. Interestingly, many closely related species show different types of heterogamety. In tilapia for example, Nile tilapia (*Oreochromis niloticus*) and blue tilapia (*O. aureus*) have the XX/XY system, while the Mozambique tilapia (*O. mossambicus*) has the ZZ/ZW system. Even within the same sex-determining system, the sex-determining genes may vary. The *dmy* (also reported as *dmrt1y* or *dmrt1by*) gene can determine the male sex in *O. latipes* and *O. curvinotus*, whereas in *O. luzonensis*, the male-determining gene is *gsdf*, and *sox3* can determine male formation in *O. dancena* [36].

In the field of aquaculture, there has been detailed evidence of how epigenetic mechanisms function in the initiation, determination, transformation, and maintenance of the sex formation of fish. We will discuss the examples of epigenetic regulation on the sex-determining network and its applications in the following sections.

8.3.2 Epigenetic Regulation of Sex Determination on Sex-Related Genes

8.3.2.1 DNA Methylation Regulate Sex Determination

It is generally considered that the DNA methylation level of the gene promoter and the first intron is negatively correlated with its expression level [10, 37]. Recent reports have focused on the DNA methylation levels of sex-related genes between different sexes in both GSD and ESD species of fish. The first evidence coupling environmental temperature to sex-related gene expression via epigenetics is in European sea bass. The sex determination in this species is a combination of GSD and ESD, and high temperature in the early stage of sexual development can lead to a predominately male population. Navarro-Martín et al. [38] have illustrated that there are sex differences in the DNA methylation level of *cyp19a* promoter, and the CpG island methylation level in the male phenotype was significantly higher than that in the female phenotype. High temperature can reduce the expression level of *cyp19a* mRNA, the female-related aromatase gene that participates in estrogen synthesis, in the female European sea bass gonads. This is associated with the hypermethylation level of the *cyp19a* promoter [38]. Subsequent work found that epigenetic regulators such as euchromatic histone-lysine *N*-methyltransferase 2 (*ehmt2*), histone demethylase Jumonji (*jarid2a*), and polycomb group ring finger 2 (*pcgf2*) have expression level changes under high-temperature treatment in the early stages of life [39]. In Chinese tongue sole, high temperature can lead to the masculinization of ZW females, which can be inherited by the ZW offspring without exposure to heat. The methylation level of *dmrt1* gene in ZW females is significantly higher than that in ZZ males, and high temperature causes the demethylation of *dmrt1* in females to produce fertile ZW neomales [29]. In tiger pufferfish, low temperature can trigger the transition from genetic female to phenotypic male [40]. The most differentially methylated regions (DMRs) are enriched on chromosome 19, which is the sex chromosome in tiger pufferfish. The differentially methylated genes in chromosome 19 between each pair-wise comparison of male, female, and neomale have enriched gonad development pathways such as Ovarian steroidogenesis. Furthermore, the *amhr2* gene has shown lower methylation levels in males and neomales than in females, while *pfcyp19a* gene has shown an opposite trend [40]. Taken together, these results provide evidence of how DNA methylation regulates the sex determination of GSD and ESD fish species by regulating the sex-determining genes.

8.3.2.2 Histone Modification Regulates Sex Determination

The possible linkage between sex determination and histone modification has been reported in the red-eared slider turtle (*Trachemys scripta elegans*) [41, 42]. In female turtles, the elimination of the methylation marker *kdm6b*, the histone demethylase of H3K27me3 on male-determining gene *dmrt1* promoter, will produce the sex reversal from female to male. In male turtles, the increase of H3K27 methylation by knocking down *kdm6b* demethylase will change the expression level of *dmrt1*, causing the reverse from male to female. This suggests that histone modification regulates temperature-dependent sex determination in a red-eared slider turtle [42]. Similarly, research found that in the American alligator, histone modification may also regulate sex determination [43]. The *kdm6b* was identified as an early responder to male incubation temperature and could provide the molecular basis for research on the role of histone modification in sex determination [43]. However, information on the histone modification regulating sex determination in fish is still limited.

8.3.2.3 ncRNA Regulates Sex Determination

Noncoding RNA (ncRNA) is an RNA molecule transcribed from the genome that does not encode protein. It can play an indispensable role at the transcriptional and posttranscriptional levels, and in the epigenetic regulation of gene expression [44]. The classification of ncRNAs is based on their transcript size, folding structure, and function [45]. Among ncRNAs, the most illustrated type are microRNAs (miRNAs), which have a length range of 19–25 nt and can complement mRNA to silence or degrade the target mRNA. Long noncoding RNA (lncRNA) is an RNA molecule with a length of 200–100,000 nt. It has low sequence conservation and plays an important role in the process of tissue differentiation and development. The difference in ncRNA distribution between females and males has been illustrated in several fish species, suggesting candidate ncRNAs are regulators of critical sex-related genes and have the ability to respond to hormone signaling [46, 47]. In Chinese tongue sole, the dynamic expression pattern of ncRNAs and mRNA in gonads during sex determination and differentiation was assessed [48]. A circRNA from the sex-determining gene *dmrt1* (circular RNA *dmrt1*) and an lncRNA called *AMSDT* (which stands for associated with male sex differentiation of tongue sole) were found significantly upregulated in the testis. These two ncRNAs both had a higher testis-specific expression during the critical stage of sex differentiation, shared the same miRNA response elements with *gsdf*, and acted as miRNA cse-miR-196 sponges to increase the *gsdf* expression. Thus, the expression of female-related genes was inhibited by ceRNA-mediated upregulation of *gsdf* and, in this way, triggered male development. This is the first demonstration in fish that ceRNA crosstalk mediated by ncRNAs modulates sexual development and unveils a novel regulatory mechanism for sex determination and differentiation [48].

These findings provide an insight into the epigenetic regulatory mechanism for the molecular regulatory network of fish sex determination and provide a new avenue for us to better understand the pathway of fish sex determination.

8.3.3 Epigenetic Markers of Sex Determination in Fish

In the late 1980s, researchers found that DNA methylation patterns were different between female and male gametes (sperm and oocyte), which was detected by the restriction enzymes that can distinguish the methylated from unmethylated cytosines, thus exploring the epigenetic markers. Subsequent studies continue to demonstrate the nature of epigenetic markers [49, 50]. Epigenetic markers include DNA methylation markers, histone modification markers, and ncRNAs, and can be heritable between generations without changing genomic sequences. In European sea bass, the families with predictable sex ratios were used as experimental materials. DNA methylation profiles of sex-related genes were pictured by a targeted sequencing approach on the juvenile gonads. Methylation of *cyp19a1a*, *dmrt1*, and *amhr2* were different between females and males, as well as masculinizing temperatures and normal temperatures. The application of these selected methylation sites can be used as epigenetic markers to predict sex with approximately 90% accuracy [51]. This was the first report on the development of epigenetic markers in fish, and further evidence of epigenetic markers of sex determination and differentiation is still needed.

8.4 Epigenetic Regulation of Sex Differentiation in Gonochoristic Species and Sex Change in Hermaphrodites

Teleost fish are unique in sex differentiation among vertebrates because of the diversity of their sex determination system, and more intriguing is that many fish species have the capacity to change sex from one direction to the other [52]. Thus, the variety of different sexual systems and mechanisms of sex determination and differentiation between fish and other vertebrates underpins the high sexual plasticity in fish. This may be related to the lower evolutionary status of fish. Another explanation may be that fish brain is not permanently sexually differentiated, as it is in mammals or birds, and has the properties of obvious plasticity between sexes. There is a growing view that sex determination depends not only on inducing the deduced sex-specific networks but also on the continuous inhibition of the relative regulatory network and maintaining this regulation throughout life [53]. This renders sex easily manipulated by various factors; these factors will interfere with the role of the sex regulatory network, resulting in sex reversal in gonochoristic species such as Chinese tongue sole, Nile tilapia, and common carp (*Cyprinus carpio*), or in sex change in hermaphroditic species such as ricefield eel (*Monopterus albus*), black porgy (*Acanthopagrus schlegelii*), and blue-head wrasses (*Thalassoma bifasciatum*).

8.4.1 Epigenetic Regulation of Sex Differentiation in Gonochoristic Species

Sex differentiation cannot be easily considered as a fixed commitment during early gonadal development but as a competition for priority between male and female developmental trajectories [53]. As far as we know, like vertebrates, fish possess highly variable major sex-determining genes, but they possess a conserved sex regulatory network. Sexual differentiation is initiated in undifferentiated germ cells by instructive signals provided by surrounding somatic cells. Several lines of evidence suggest that epigenetic regulation of key genes may affect sexual differentiation in fishes and may cause sex reversal.

In Chinese tongue sole, integrative analysis of whole genome methylation and gene expression revealed that the sex-related genes were differentially expressed between sexes, and the methylation patterns of males were similar to those of sex-reversed neomales. Specifically, in sex-reversed ZW neomales, genes located on the female-specific W chromosome were suppressed through methylation. DNA methylation and sex-specific epigenetic reprogramming may suppress the female sexual differentiation pathway, and thus participate in environment-induced sex reversal [29]. Genome-wide methylation of male and female tilapia shows that females and males have very large differences in the regulatory regions of sex chromosomes, and the MAPK/ERK pathway was upregulated in male tilapia [54]. All these results indicate that DNA methylation has a huge impact on the sex determination and differentiation of fish. In recent years, there have been preliminary studies on the sex regulation of ncRNA in fish. For example, in common carp, by sequencing the gonadal tissue before and after interference with *igf3*, 124 differentially expressed long noncoding RNAs (lncRNAs) that may play an important role in the sex differentiation and gonadal development of carp were obtained [55]. In tilapia, an lncRNA (TCONS_02477925) has also been identified to be related to oocyte development [56]. In Chinese tongue sole, an lncRNA DMRT2-AS (referred to as *dmrt2* antisense) could promote the expression of *dmrt2* at the transcriptional level, thus playing an important role in testis differentiation [57].

8.4.2 Epigenetic Regulation of Sex Change in Hermaphroditic Species

Sex change in sequential hermaphrodites involves a thorough restructuring of the gonad, where the original gonadal system undergoes apoptosis and the opposite gonadal system begins to be created. Histological observations of the gonads of many hermaphrodite fish found that they all go through the bisexual phase when they undergo gonadal transitions [14, 15]. The balance between estrogen and androgen synthesis in the gonad plays an important role during the transitions to different sexes. In the protogynous hermaphrodite ricefield eel, which requires a continuous decrease in gonadal *cyp19a1a* expression for a female-to-male sex change, the expression of *cyp19a1a* was stimulated by gonadotropin via the cAMP signaling in the ovary, but not in the ovotestis in transition or testis [58]. This was related to the hypermethylation and deacetylation of putative promoter-binding sites for cAMP response elements (CREs) and *nr5a1* in ovotestis and testis. *in vitro* experiment found DNA methylation decreased the basal and forskolin-induced activities of the *cyp19a1a* promoter, and DNA methylation inhibitor 5-aza-2′-deoxycytidine reversed the natural female-to-male sex change of ricefield eels. These results suggested that epigenetic modification of the *cyp19a1a* promoter may abrogate the stimulation of *cyp19a1a* by gonadotropins in a male-specific pattern. This may be a mechanism that drives the natural sex change in ricefield eel [58]. In protandrous black porgy, the *cyp19a1a* promoter was hypomethylated (<20%) in the follicle cells in the ovary, and ovarian *cyp19a1a* expression was regulated by the testis through epigenetic modifications [59].

The balance between estrogen and androgens is known to be regulated not only through the classic hypothalamic-pituitary-gonadal (HPG) axis but also its neighboring hypothalamic-pituitary-interregnal (HPI) axis. The HPI axis plays a critical role in responding to stimuli and regulating the biological processes during important life-history transitions, including changes in behavior, metabolism, and development. The HPI axis responds to environmental stressors and transmits the signal into the internal system mainly through the activation of corticotropic-releasing hormone (CRH) and glucocorticoid steroids (GCs), which may have potential effects on gonadal formation [60]. More and more evidence has shown that in sequential hermaphrodites, cortisol plays a role in regulating natural sex change by affecting the expression of steroid-related genes and transmitting environmental information such as social status along the HPI axis. In protogynous blue-head wrasses, the environmental stimulus is exerted via the stress axis and that repression of the aromatase gene (encoding the enzyme converting androgens to estrogens) triggers a cascaded collapse of feminizing gene expression [61]. The DNA methylation reprogramming and trimethylation of histone H3 lysine 27 (H3K27me3) were involved in the sex change process. Dynamic cortisol production and signaling at early sex change may be a key factor triggering female-to-male transition. A mechanistic hypothesis showed that cortisol may regulate blue-head wrasses sex change in three ways: (i) it preferentially binds to the glucocorticoid response element in the *cyp19a1a* promoter to inhibit aromatase expression; (ii) upregulate *amh* to induce germ cell apoptosis and promote masculinization; and (iii) stimulate the expression of *hsd11b2* (11β-HSD) to promote 11-KT synthesis [61]. In protogynous orange-spotted grouper (*Epinephelus coioides*), transposase-accessible chromatin with high-throughput sequencing (ATAC-seq) and RNA-seq analyses identified that numerous enriched transcription factor motifs had differences in chromatin accessibility in the intersex group, and found that their target genes (including *dmrt1*, *sox11*, and *hsd17b7*) were related to sex reversal [62]. Besides, core genes (*gnas*, *ccnb2*, and *cyp21a*) in the estrogen-signaling pathway, oocyte meiosis pathway, and steroid hormone biosynthesis pathway were found to interact with certain accessible chromatin regions during the early stage of sex reversal [62].

Although many aspects of histone modifications and nucleosome structure are still unclear, they are increasingly recognized as key factors in hermaphroditic species sexual change. These data indicate that epigenetic regulation plays a critical role in shaping the phenotype of fish gonads.

8.5 Transgenerational Epigenetic Sex Reversal

8.5.1 Transgenerational Epigenetic Inheritance in Fish

The higher sensitivity to environmental stimulations on sex ratio of offspring of sex-reversed parents indicates that the perception of environmental factors, or even the ability of sex reversal, can transmit over generations and that there are heritable components of this process. In Nile tilapia, the ability of sex reversal in males under temperature treatment can be selected as a quantitative trait, and the expression profiles of genes on sex determination pathway (*cyp19a*, *foxl2*, *amh*, *sox9a*, *sox9b*, and *dmrt1*) from the gonad and brain were found to have relationship between environmental factors and sex differentiation [63]. Also, family-specific QTLs (quantitative trait loci) were associated with temperature-dependent sex reversal [64]. In Chinese tongue sole, the Z chromosome-linked *dmrt1* gene is a male determination gene [65]. Locus of single nucleotide polymorphism (SNP) associated with sex reversal was identified and located in the third intron of the *dmrt1* gene and has a genetic interaction with another SNP located on the Z chromosome [66, 67].

In Nile tilapia, an XX/XY sex determination system species without morphologically differentiated sex chromosomes, GSD polygenic factors can be overridden by a masculinizing effect such as high temperature (above 32 °C), and androgen treatment during the sensitive period of perceiving environmental factors (not later than 13 days after fertilization) can also masculinize the genetic females [63, 68]. Monosex genotypic females can be produced by mating normal females, and high temperature (36 °C) treatment on the offspring of neomales can increase the proportion of males from 33 to 81%, depending on different families [69]. Furthermore, the offspring of XX sex-reversed males produced by hormone treatment were found to have spontaneous neomales. In 35 families tested, only five families had 100% female progeny, and the remaining 30 families produced spontaneous neomales with the proportion ranging between 1 and 35% among different families [70]. A similar situation was found in sex-reversed males under high temperature induction. The effect of high temperature stimulation on sex ratio was tested in a two-generation selection experiment. Around 14% of the tested families can produce divergent lines of which the sex ratio was higher or lower under the influence of environmental temperature, which provided evidence that the sensitivity of sex determination to temperature treatment in Nile tilapia is a heritable trait that shows a strong response to selection [71].

In yellow catfish (*Pseudobagrus fulvidraco*), sex-reversed XX males can be created by aromatase inhibitor letrozole and confirmed with a sex-specific molecular marker [72]. The offspring of normal XX female and sex-reversed XX male were all females when raised at normal temperature (26–28.5 °C), showing the sex was determined by both sex chromosomes and genetic factors at normal conditions. However, when treated with high temperature (32 and 34 °C), a high ratio of XX male offspring occurred, and the ratio gradually increased when the temperature was elevated, or the incubation time was extended. Under 32 °C, a proportion of 40–47% of the XX genotype fish were sex reversed into XX males, while the other fish became intersexual. Moreover, when treated with 34 °C for 53 days, the percentage of XX male can reach 99% [73]. These results indicated offspring of sex-reversed males by hormone treatment were more sensitive to environmental high temperature, and artificially induced sex reversal can lead to the transition from GSD to temperature-dependent sex determination.

In Chinese tongue sole with a ZW/ZZ sex determination system, the females can be sex reversed into phenotypic males with a proportion of approximately 14% under normal incubation temperature (22 °C). When under high-temperature (28 °C) treatment during early development, the sex reversal rate of genetic females can rise to approximately 73% [28]. The neomales are fertile with complete male reproductive functions and can mate with normal females to produce offspring. Interestingly, even reared under normal temperature, the offspring of neomales and females can still exhibit an extremely high sex reversal rate of 94% [28].

It has been reported that offspring of sex-reversed individuals can continue to exhibit a higher sensitivity to environmental factors than the offspring of the normal parents, and in some species, the offspring have extremely high sex reversal rates even without environment-affecting factors [74]. In chinook salmon (*Oncorhynchus tshawytscha*), which has an XX/XY sex determination system, the genetic female larva treated with aromatase inhibitor for two hours of immersion at three days after 50% hatching (the critical period of bipotent gonad differentiation) can develop into phenotypic males with the ability of successful spermatogenesis. The offspring of sex-reversed neomales were more sensitive to non-aromatizable synthetic androgen 17α-methyldihydrotestosterone compared to the offspring of untreated normal XX females [75, 76].

In these examples, offspring of sex-reversed neomales tended to have higher sensitivity to environmental signals such as high temperature and some species even produce more neomales spontaneously. These observations demonstrate that sensitivity to environmental factors has a genetic component, and the ability of sex reversal can be inherited transgenerational. Thus, the underlying mechanism of epigenetic regulation, such as DNA methylation in this process, needs further exploration.

8.5.2 DNA Methylation Reprogramming Associated with Transgenerational Inheritance

In Chinese tongue sole, the offspring of sex-reversed males can spontaneously develop into functional neomales with a fairly high proportion of approximately 94% without environmental stimuli [28]. Shao et al. [29] illustrated the role of DNA methylation during this process. Comparative analysis of the gonadal DNA methylomes of normal female, normal male, and neomale Chinese tongue sole showed that genes in the sex determination pathways are the major targets of substantial methylation modification during high temperature-induced sexual reversal, and approximately 28% of sex determination-related genes exhibited remarkable methylation contrast between ZW female ovary versus normal male and neomale testes (Figure 8.3). Methylation modification changes in sex-reversed males caused by high temperature are globally inherited by their ZW offspring, which may be the reason that they can naturally develop into neomales without high temperature

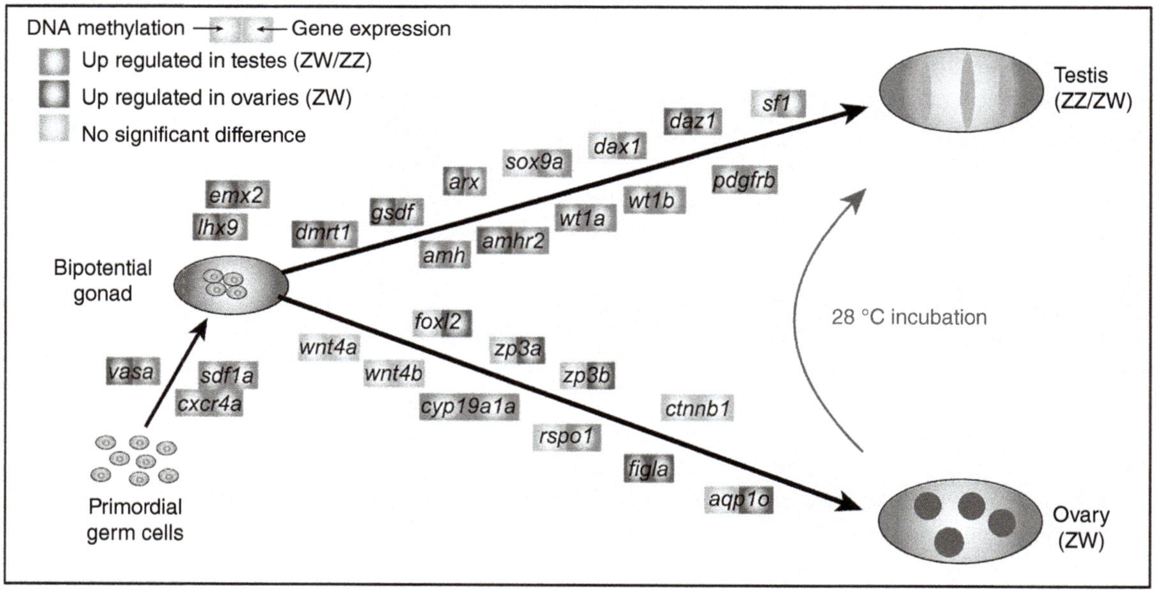

Figure 8.3 Differentially methylated and differentially expressed genes in the putative sex determination pathway of Chinese tongue sole. The compared methylation level (left square) and expression level (right square) between testes and ovaries of each gene are shown by different colors. Dark gray indicates upregulated in testes, light gray indicates upregulated in ovaries, and medium gray indicates no significant difference between testes and ovaries. *Source:* Shao et al. [29]/Cold Spring Harbor Laboratory Press. (*See colour plate section for colour representation of this figure.*)

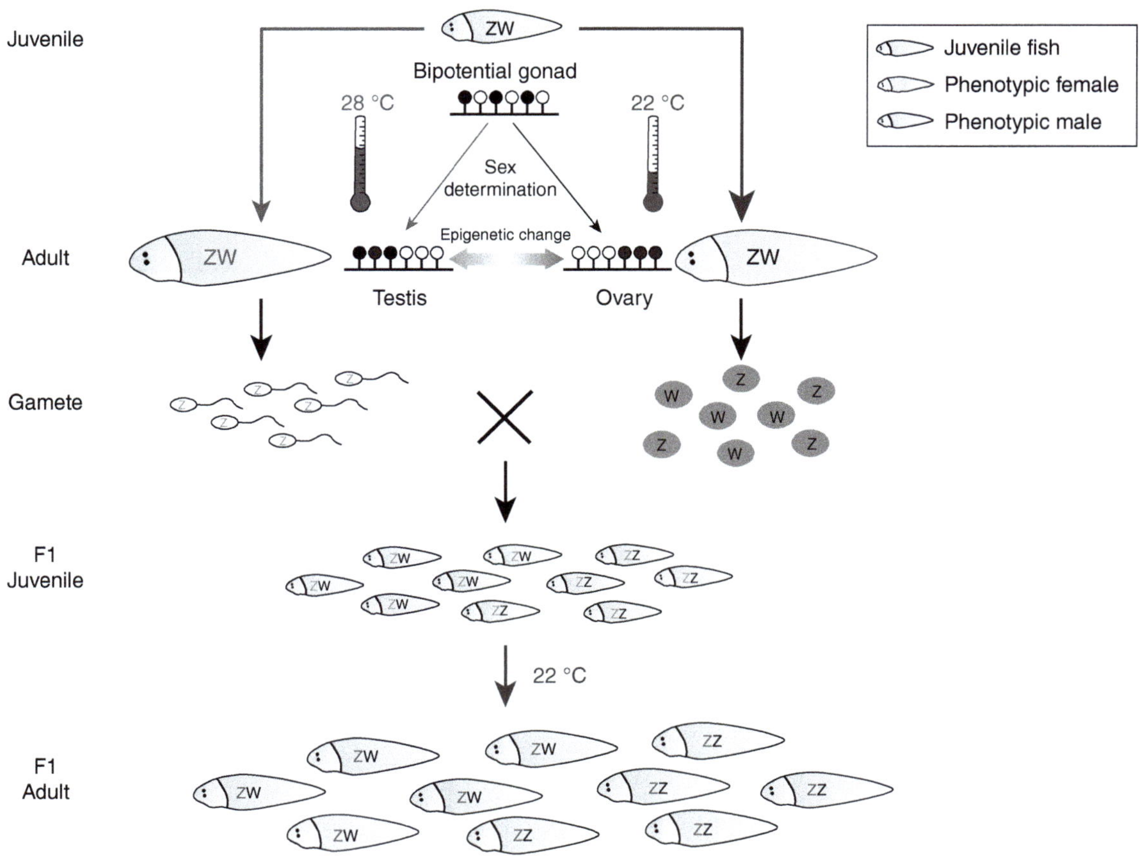

Figure 8.4 Transgenerational epigenetic inheritance of sexual reversal in Chinese tongue sole. High temperature treatment of juvenile ZW fish can change the epigenetic modification of the Z chromosome to develop into neomales. The neomales can only produce Z-type sperm, and some of these alterations are transferred to the next generation. Thus, most of the genetic female offspring form into neomales spontaneously. *Source:* Adapted from Shao et al. [29]/Cold Spring Harbor Laboratory Press. (*See colour plate section for colour representation of this figure.*)

incubation [29]. In mammals mainly with a GSD system, there always exists an epigenetic reprogramming event that erases and resets most of the epigenetic marks between generations [77, 78]. However, the erasing of epigenetic marks in teleost is not a universal rule [79, 80], which provides the possibilities for transgenerational epigenetic inheritance in fishes. Shao et al. [29] found that when the potential bidirectional undifferenced juvenile Chinese tongue sole were treated with high temperature, novel DNA methylation patterns were established in bipotential germ cells to cause the fish develop into the opposite sex; these new methylation patterns were then imprinted in the genome and appear in the offspring. Furthermore, offspring produced with the neomale paternal W chromosome cannot survive, thus the well-developed offspring of neomales only inherit the paternal Z allele, of which the methylation patterns have been altered by environmental stimulation, and spontaneously transferred into neomales reared under normal conditions (Figure 8.4) [29]. By calculating the gene expression ratio between ZZ testis to ZW testis, a specific dosage compensation mechanism on the Z chromosome was investigated that enriched with Z-chromosomal DMRs between neomale and normal male. The results can help further explain how the neomale offspring balance gene expression, and the role of DNA methylation in regulating dosage compensation (Figure 8.5) [29].

8.6 Conclusions and Future Perspectives

The role of epigenetics has been elaborated in many fields such as growth, development, immunity, and aging. Sex determination and differentiation is an important life process and is closely related to the reproduction and continuation of biological populations. Fish have a lower evolutionary status among vertebrates and exhibit multiple sex determination types including GSD, ESD, and the interaction between GSD and ESD. The external environment, especially the water

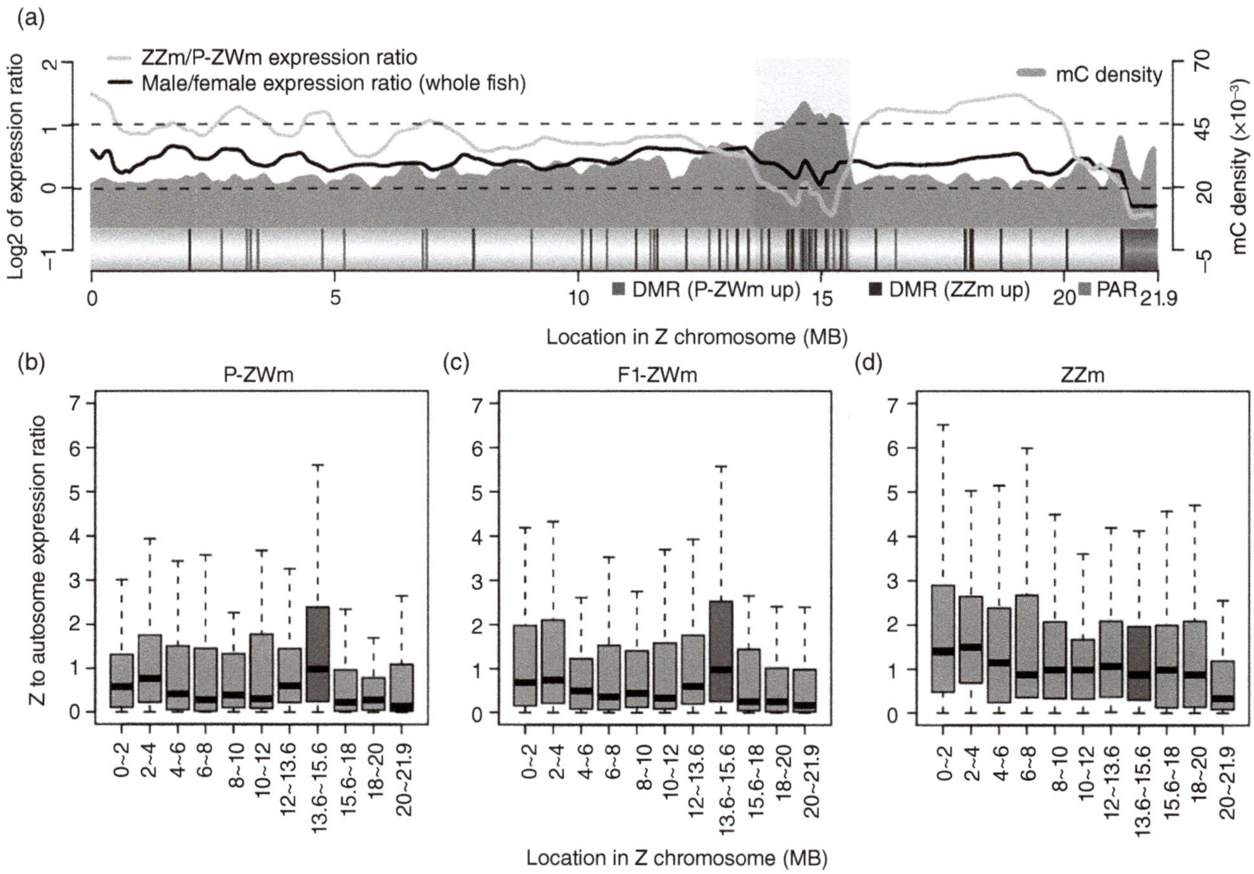

Figure 8.5 Dosage compensation of the Z chromosome in neomale testes. (a) Methylated cytosine (mC) density, expression ratios, and DMR profiles of the Z chromosome. The light gray box indicates the outstanding dosage-compensated region where DMRs were concentrated. (b–d), Z-chromosomal to autosomal gene expression ratios (Z : A) in paternal neomale (b), offspring neomale (c), and normal male (d), respectively. The dosage compensation region (box in a) is plotted in dark gray. *Source:* Shao et al. [29]/Cold Spring Harbor Laboratory Press.

temperature, can affect the sex in a variety of fish, causing sexual reversal. Epigenetic mechanisms affect phenotypic variation through the environment without changing genomic DNA sequence, exhibiting great potential to improve fish traits. Later we will look forward to the research on epigenetics in fish sex determination and differentiation:

1) Application of epigenetic editing technology based on CRISPR/Cas9
   Regulating gene expression, only by modifying genes without changing the sequence, is one of the current hotspots in the study of gene expression regulation at the transcriptional level. Due to the limitation of verification technology, epigenetics has been greatly restricted on the phenotypic research of sex determination and differentiation. Fortunately, recently developed epigenetic editing technology such as CRISPRoff and CRISPR/dCas9 systems can establish DNA methylation and repressive histone modifications in mammals. The regulation of gene expression at the epigenetic level provides technical feasibility for the interaction between genetics and environment during sex determination and differentiation in fish.
2) Development of sex-reversal epi-marker for sex control
   In many fish species, one sex may grow faster or mature earlier than the other sex, and these differences can be more pronounced under aquaculture conditions. Currently, sex-linked DNA markers are widely used in genotypic sex identification, identifying the sex of offspring in genetic breeding, and conducting sex control in selected populations. Sex-linked molecular markers have been applied in molecular marker-assisted breeding to accelerate the acquisition of unisex populations and promote aquaculture development. The stability and effectiveness of sex-linked DNA markers are largely dependent on the complexity of the genetic background of sex chromosomes and complex sex determination among different populations in a single species. Moreover, phenotype is the result of the interaction between

environment and genetics, and epi-markers are important for the identification of individuals with epigenetic information about sex reversal. The development and utilization of epi-markers in aquaculture species, when taking environmental factors into account, can help explain the genotype–phenotype gap, improve the accuracy and precision of sex screening, and develop environment-friendly sex control.

In general, the development of epigenetic studies on sex determination and differentiation has significance for both basic theoretical research and industrial development of fish.

Acknowledgments

This work was supported by the National Key R&D Program of China (2022YFD2400100, 2022YFD2401500); the AoShan Talents Cultivation Program Supported by Qingdao National Laboratory for Marine Science and Technology (2017ASTCP-ES06); the Taishan Scholar Program (tstp20221149) to C.S.; the National Ten-Thousands Talents Special Support Program to C.S.; the Central Public-interest Scientific Institution Basal Research Fund, CAFS (2020TD19); the China Agriculture Research System (CARS-47-G03), and the project by the National Marine Genetic Resource Center.

References

1 Beukeboom, L.W. and Perrin, N. (2014). *The Evolution of Sex Determination*. Oxford: Oxford University Press.

2 Nelson, J.S., Grande, T.C., and Wilson, M.V.H. (ed.) (2016). *Fishes of the World*, 5e. Chichester: Wiley.

3 Kobayashi, Y., Nagahama, Y., and Nakamura, M. (2013). Diversity and plasticity of sex determination and differentiation in fishes. *Sexual Development* **7**: 115–125.

4 Wang, H.P., Piferrer, F., Chen, S.L. et al. (2018). *Sex Control in Aquaculture*. Chichester: Wiley.

5 Zhou, Q. and Bachtrog, D. (2015). Ancestral chromatin configuration constrains chromatin evolution on differentiating sex chromosomes in *Drosophila*. *PLoS Genetics* **11** (6): e1005331.

6 Araya-Jaime, C., Mateussi, N.T.B., Utsunomia, R. et al. (2017). ZZ/Z0: the new system of sex chromosomes in *Eigenmannia* aff. *trilineata* (Teleostei: Gymnotiformes: Sternopygidae) characterized by molecular cytogenetics and DNA barcoding. *Zebrafish* **14** (5): 464–470.

7 Sember, A., Nguyen, P., Perez, M.F. et al. (2021). Multiple sex chromosomes in teleost fishes from a cytogenetic perspective: state of the art and future challenges. *Philosophical Transactions of the Royal Society B* **376**: 20200098.

8 Ferreira, M., Garcia, C., Matoso, D.A. et al. (2016). A new multiple sex chromosome system $X_1X_1X_2X_2/X_1Y_1X_2Y_2$ in Siluriformes: cytogenetic characterization of *Bunocephalus coracoideus* (Aspredinidae). *Genetica* **144**: 591–599.

9 Vandeputte, M., Dupont-Nivet, M., Chavanne, H. et al. (2007). A polygenic hypothesis for sex determination in the European sea bass *Dicentrarchus labrax*. *Genetics* **176** (2): 1049–1057.

10 Piferrer, F., Anastasiadi, D., Valdivieso, A. et al. (2019). The model of the conserved epigenetic regulation of sex. *Frontiers in Genetics* **10**: 857.

11 Xu, K., Duan, W., Xiao, J. et al. (2015). Development and application of biological technologies in fish genetic breeding. *Science China Life Sciences* **58** (2): 187–201.

12 Martinez, P., Vinas, A.M., Sanchez, L. et al. (2014). Genetic architecture of sex determination in fish: applications to sex ratio control in aquaculture. *Frontiers in Genetics* **5**: 340.

13 Kuwamura, T., Sunobe, T., Sakai, Y. et al. (2020). Hermaphroditism in fishes: an annotated list of species, phylogeny, and mating system. *Ichthyological Research* **67**: 341–360.

14 Frisch, A. (2004). Sex-change and gonadal steroids in sequentially-hermaphroditic teleost fish. *Reviews in Fish Biology and Fisheries* **14**: 481–499.

15 de Mitcheson, Y.S. and Liu, M. (2008). Functional hermaphroditism in teleosts. *Fish and Fisheries* **9**: 1–43.

16 Domingos, J.A., Budd, A.M., Banh, Q.Q. et al. (2018). Sex-specific *dmrt1* and *cyp19a1* methylation and alternative splicing in gonads of the protandrous hermaphrodite barramundi. *PLoS one* **13** (9): e0204182.

17 Zhang, Y., Zhang, S., Liu, Z. et al. (2013). Epigenetic modifications during sex change repress gonadotropin stimulation of *cyp19a1a* in a teleost ricefield eel (*Monopterus albus*). *Endocrinology* **154** (8): 2881–2890.

18 Stelkens, R.B. and Wedekind, C. (2010). Environmental sex reversal, Trojan sex genes, and sex ratio adjustment: conditions and population consequences. *Molecular Ecology* **19** (4): 627–646.

19 Jablonka, E. and Raz, G. (2009). Transgenerational epigenetic inheritance: prevalence, mechanisms, and implications for the study of heredity and evolution. *The Quarterly Review of Biology* **84** (2): 131–176.

20 Thomas, C.D., Cameron, A., Green, R.E. et al. (2004). Extinction risk from climate change. *Nature* **427**: 145–148.

21 Sunde, J., Larsson, P., and Forsman, A. (2019). Adaptations of early development to local spawning temperature in anadromous populations of pike (*Esox lucius*). *BMC Evolutionary Biology* **19**: 148.

22 Kitano, J. and Peichel, C.L. (2012). Turnover of sex chromosomes and speciation in fishes. *Environmental Biology of Fishes* **94** (3): 549–558.

23 Furman, B.L.S., Metzger, D.C.H., Darolti, I. et al. (2020). Sex chromosome evolution: so many exceptions to the rules. *Genome Biology and Evolution* **12** (6): 750–763.

24 Pennell, M.W., Kirkpatrick, M., Otto, S.P. et al. (2015). Y fuse? Sex chromosome fusions in fishes and reptiles. *PLoS Genetics* **11** (5): e1005237.

25 Graves, J.A. (2016). Evolution of vertebrate sex chromosomes and dosage compensation. *Nature Reviews Genetics* **17** (1): 33–46.

26 Gorelick, R. (2003). Evolution of dioecy and sex chromosomes via methylation driving Muller's ratchet. *Biological Journal of the Linnean Society* **80**: 353–368.

27 Piferrer, F. (2021). Epigenetic mechanisms in sex determination and in the evolutionary transitions between sexual systems. *Philosophical Transactions of the Royal Society, B: Biological Sciences* **376** (1832): 20200110.

28 Chen, S., Zhang, G., Shao, C. et al. (2014). Whole-genome sequence of a flatfish provides insights into ZW sex chromosome evolution and adaptation to a benthic lifestyle. *Nature Genetics* **46** (3): 253–260.

29 Shao, C., Li, Q., Chen, S. et al. (2014). Epigenetic modification and inheritance in sexual reversal of fish. *Genome Research* **24** (4): 604–615.

30 Metzger, D.C.H. and Schulte, P.M. (2018). The DNA methylation landscape of stickleback reveals patterns of sex chromosome evolution and effects of environmental salinity. *Genome Biology and Evolution* **10** (3): 775–785.

31 Allis, C.D. and Jenuwein, T. (2016). The molecular hallmarks of epigenetic control. *Nature Reviews Genetics* **17** (8): 487–500.

32 Sendzikaite, G. and Kelsey, G. (2019). The role and mechanisms of DNA methylation in the oocyte. *Essays in Biochemistry* **63** (6): 691–705.

33 Xue, L., Gao, Y., Wu, M. et al. (2021). Telomere-to-telomere assembly of a fish Y chromosome reveals the origin of a young sex chromosome pair. *Genome Biology* **22**: 203.

34 Helleu, Q., Gerard, P.R., Dubruille, R. et al. (2016). Rapid evolution of a Y-chromosome heterochromatin protein underlies sex chromosome meiotic drive. *Proceedings of the National Academy of Sciences of the USA* **113** (15): 4110–4115.

35 Henzel, J.V., Nabeshima, K., Schvarzstein, M. et al. (2011). An asymmetric chromosome pair undergoes synaptic adjustment and crossover redistribution during *Caenorhabditis elegans* meiosis: implications for sex chromosome evolution. *Genetics* **187** (3): 685–699.

36 Capel, B. (2017). Vertebrate sex determination: evolutionary plasticity of a fundamental switch. *Nature Reviews Genetics* **18**: 675–689.

37 Anastasiadi, D., Esteve-Codina, A., and Piferrer, F. (2018). Consistent inverse correlation between DNA methylation of the first intron and gene expression across tissues and species. *Epigenetics & Chromatin* **11**: 37.

38 Navarro-Martín, L., Viñas, J., Ribas, L. et al. (2011). DNA methylation of the gonadal aromatase (*cyp19a*) promoter is involved in temperature-dependent sex ratio shifts in the European sea bass. *PLoS Genetics* **7** (12): e1002447.

39 Diaz, N. and Piferrer, F. (2015). Lasting effects of early exposure to temperature on the gonadal transcriptome at the time of sex differentiation in the European sea bass, a fish with mixed genetic and environmental sex determination. *BMC Genomics* **16**: 679.

40 Zhou, H., Zhuang, Z.X., Sun, Y.Q. et al. (2019). Changes in DNA methylation during epigenetic-associated sex reversal under low temperature in *Takifugu rubripes*. *PLoS one* **14** (8): e0221641.

41 Czerwinski, M., Natarajan, A., Barske, L. et al. (2016). A timecourse analysis of systemic and gonadal effects of temperature on sexual development of the red-eared slider turtle *Trachemys scripta elegans*. *Developmental Biology* **420** (1): 166–177.

42 Ge, C., Ye, J., Weber, C. et al. (2018). The histone demethylase KDM6B regulates temperature-dependent sex determination in a turtle species. *Science* **360** (6389): 645–648.

43 Yatsu, R., Miyagawa, S., Kohno, S. et al. (2016). RNA-seq analysis of the gonadal transcriptome during *Alligator mississippiensis* temperature-dependent sex determination and differentiation. *BMC Genomics* **17**: 77.

44 Costa, F.F. (2008). Non-coding RNAs, epigenetics and complexity. *Gene* **410** (1): 9–12.

45 Panwar, B., Arora, A., and Raghava, G.P. (2014). Prediction and classification of ncRNAs using structural information. *BMC Genomics* **15**: 127.

46 Bizuayehu, T.T., Lanes, C.F., Furmanek, T. et al. (2012). Differential expression patterns of conserved miRNAs and isomiRs during Atlantic halibut development. *BMC Genomics* **13**: 11.

47 Yan, H., Liu, Q., Jiang, J. et al. (2021). Identification of sex differentiation-related microRNA and long non-coding RNA in *Takifugu rubripes* gonads. *Scientific Reports* **11** (1): 7459.

48 Tang, L., Huang, F., You, W. et al. (2022). ceRNA crosstalk mediated by ncRNAs is a novel regulatory mechanism in fish sex determination and differentiation. *Genome Research* https://doi.org/10.1101/gr.275962.121.

49 Reik, W. and Walter, J. (2001). Genomic imprinting: parental influence on the genome. *Nature Reviews Genetics* **2**: 21–32.

50 Wood, A.J. and Oakey, R.J. (2006). Genomic imprinting in mammals: emerging themes and established theories. *PLoS Genetics* **2** (11): e147.

51 Anastasiadi, D., Vandeputte, M., Sanchez-Baizan, N. et al. (2018). Dynamic epimarks in sex-related genes predict gonad phenotype in the European sea bass, a fish with mixed genetic and environmental sex determination. *Epigenetics* **13** (9): 988–1011.

52 Todd, E.V., Liu, H., Muncaster, S. et al. (2016). Bending genders: the biology of natural sex change in fish. *Sexual Development* **10** (5–6): 223–241.

53 Nagahama, Y., Chakraborty, T., Paul-Prasanth, B. et al. (2021). Sex determination, gonadal sex differentiation, and plasticity in vertebrate species. *Physiological Reviews* **101**: 1237–1308.

54 Wan, Z.Y., Xia, J.H., Lin, G. et al. (2016). Genome-wide methylation analysis identified sexually dimorphic methylated regions in hybrid tilapia. *Scientific Reports* **6** (1): 1–14.

55 Song, F., Wang, L., Zhu, W. et al. (2019). Long noncoding RNA and mRNA expression profiles following *igf3* knockdown in common carp, *Cyprinus carpio*. *Scientific Data* **6**: 190024.

56 Kun, P., Lingyun, S., Xiwu, Q. et al. (2019). Identification and expression analysis of a key lncRNA in the gonads of female Nile tilapia (*Oreochromis niloticus*). *Journal of Southwest China Normal University (Natural Science Edition)* **44** (6): 40–45.

57 Feng, B., Li, S., Wang, Q. et al. (2021). lncRNA DMRT2-AS acts as a transcriptional regulator of *dmrt2* involving in sex differentiation in the Chinese tongue sole (*Cynoglossus semilaevis*). *Comparative Biochemistry and Physiology Part B: Biochemistry and Molecular Biology* **253**: 110542.

58 Davey, A.J.H. and Jellyman, D.J. (2005). Sex determination in freshwater eels and management options for manipulation of sex. *Reviews in Fish Biology and Fisheries* **15** (1): 37–52.

59 Wu, G.C., Li, H.W., Huang, C.H. et al. (2016). The testis is a primary factor that contributes to epigenetic modifications in the ovaries of the protandrous black porgy, *Acanthopagrus schlegelii*. *Biology of Reproduction* **94** (6): 132.

60 Herman, J.P., McKlveen, J.M., Ghosal, S. et al. (2016). Regulation of the hypothalamic-pituitary-adrenocortical stress response. *Comprehensive Physiology* **6** (2): 603–621.

61 Todd, E.V., Ortega-Recalde, O., Liu, H. et al. (2019). Stress, novel sex genes, and epigenetic reprogramming orchestrate socially controlled sex change. *Science Advances* **5** (7): eaaw7006.

62 Wu, X., Yang, Y., Zhong, C. et al. (2020). Integration of ATAC-seq and RNA-seq unravels chromatin accessibility during sex reversal in orange-spotted grouper (*Epinephelus coioides*). *International Journal of Molecular Sciences* **21** (8): 2800.

63 Baroiller, J.F., D'Cotta, H., Bezault, E. et al. (2009). Tilapia sex determination: where temperature and genetics meet. *Comparative Biochemistry and Physiology Part A: Molecular & Integrative Physiology* **153** (1): 30–38.

64 Luhmann, L.M., Knorr, C., Horstgen-Schwark, G. et al. (2012). First evidence for family-specific QTL for temperature-dependent sex reversal in Nile tilapia (*Oreochromis niloticus*). *Sexual Development* **6** (5): 247–256.

65 Cui, Z., Liu, Y., Wang, W. et al. (2017). Genome editing reveals *dmrt1* as an essential male sex-determining gene in Chinese tongue sole (*Cynoglossus semilaevis*). *Scientific Reports* **7**: 42213.

66 Cui, Y., Wang, W., Ma, L. et al. (2018). New locus reveals the genetic architecture of sex reversal in the Chinese tongue sole (*Cynoglossus semilaevis*). *Heredity* **121** (4): 319–326.

67 Jiang, L. and Li, H. (2017). Single locus maintains large variation of sex reversal in half-smooth tongue sole (*Cynoglossus semilaevis*). *G3: Genes|Genomes|Genetics* **7** (2): 583.

68 Mair, G.C., Scott, A.G., Penman, D.J. et al. (1991). Sex determination in the genus *Oreochromis*. *Theoretical and Applied Genetics* **82** (2): 153–160.

69 Baroiller, J.F., Chourrout, D., Fostier, A. et al. (1995). Temperature and sex chromosomes govern sex ratios of the mouthbrooding Cichlid fish *Oreochromis niloticus*. *Journal of Experimental Zoology* **273** (3): 216–223.

70 Baroiller, J.F. (1996). Significant proportions of unexpected males in progenies from single pair matings with sibling sex-reversed males of Oreochromis niloticus. *The third International Symposium on Tilapia Aquaculture, ICLARM Conference Proceedings*, Abidjan, Côte d'Ivoire, 41, 229–238.

71 Wessels, S. and Hörstgen-Schwark, G. (2007). Selection experiments to increase the proportion of males in Nile tilapia (*Oreochromis niloticus*) by means of temperature treatment. *Aquaculture* **272**: S80–S87.

72 Yang, T., Xiong, Y., Dan, C. et al. (2018). Production of XX male yellow catfish by sex-reversal technology. *Acta Hydrobiologica Sinica* **42**: 871–878.

73 Xiong, Y., Wang, S., Gui, J.F. et al. (2020). Artificially induced sex-reversal leads to transition from genetic to temperature-dependent sex determination in fish species. *Science China Life Sciences* **63** (1): 157–159.

74 Piferrer, F. and Anastasiadi, D. (2021). Do the offspring of sex reversals have higher sensitivity to environmental perturbations? *Sexual Development* **15** (1–3): 134–147.

75 Piferrer, F., Zanuy, S., Carrillo, M. et al. (1994). Brief treatment with an aromatase inhibitor during sex differentiation causes chromosomally female salmon to develop as normal, functional males. *Journal of Experimental Zoology* **270** (3): 255–262.

76 Piferrer, F., Baker, I.J., and Donaldson, E.M. (1993). Effects of natural, synthetic, aromatizable, and nonaromatizable androgens in inducing male sex differentiation in genotypic female chinook salmon (*Oncorhynchus tshawytscha*). *General and Comparative Endocrinology* **91** (1): 59–65.

77 Daxinger, L. and Whitelaw, E. (2010). Transgenerational epigenetic inheritance: more questions than answers. *Genome Research* **20** (12): 1623–1628.

78 Daxinger, L. and Whitelaw, E. (2012). Understanding transgenerational epigenetic inheritance via the gametes in mammals. *Nature Reviews Genetics* **13** (3): 153–162.

79 Jiang, L., Zhang, J., Wang, J.J. et al. (2013). Sperm, but not oocyte, DNA methylome is inherited by zebrafish early embryos. *Cell* **153** (4): 773–784.

80 Potok, M.E., Nix, D.A., Parnell, T.J. et al. (2013). Reprogramming the maternal zebrafish genome after fertilization to match the paternal methylation pattern. *Cell* **153** (4): 759–772.

9

Epigenetics in Fish Growth

Jorge M.O. Fernandes[1], *Artem V. Nedoluzhko*[1,2], *Ioannis Konstantinidis*[1], *and Paulo Gavaia*[3]

[1] *Faculty of Biosciences and Aquaculture, Nord University, Bodø, Norway*
[2] *Paleogenomics Laboratory, European University at Saint Petersburg, Saint Petersburg, Russia*
[3] *Centro de Ciências do Mar (CCMAR) and Department of Biomedical Sciences and Medicine, Universidade do Algarve, Faro, Portugal*

9.1 Myogenesis in Teleosts

9.1.1 Introduction to Myogenesis, Highlighting the Peculiarities of Fish Muscle

Myogenesis is a term that originated from the Greek "*μύς*" for muscle, and "*γένεση*" for birth, and describes the earliest commitment of progenitor cells to the myogenic lineage and the formation of the skeletal muscle. For over three decades, it has been known that myogenic determination has a genetic basis [1–3]. Critical factors in this process include the gene coding for myoblast determination protein (*MyoD*) [3], myogenic factor 5 (*Myf5*) [4, 5], myogenic regulatory factor 4 (*Mrf4*) [6, 7], and myogenin (*MyoG*) [1, 8]. In particular, overexpression of *MyoD* in nonmuscle cells was proven to be sufficient for the determination of cell fate, activation of the myogenic program, and the formation of differentiated muscle [9]. However, parallel research showed that either MyoD, Myf5, or Mrf4 could drive myogenic determination through their basic helix-loop-helix (bHLH) protein structure that binds to the E-box CANNTG sequence [10] and induce myogenic conversion [11]. In zebrafish (*Danio rerio*), MyoD is critical for fast muscle differentiation; however, Myf5 can effectively replace MyoD in slow muscle differentiation [12, 13]. This was also shown by knocking down Pbx homeodomain proteins, which contribute to fast muscle differentiation and bind exclusively to MyoD [14]. Nevertheless, in the absence of all master muscle regulatory factors, skeletal muscle is unable to form.

At the onset of myogenesis, progenitor cells are being committed to the myogenic lineage; they proliferate, migrate, exit the cell cycle, differentiate, and fuse to form mature muscle fibers (Figure 9.1). This process, including stage-specific functions of the transcriptional factors mentioned earlier, is highly conserved between lower and higher vertebrates (for review, see Rescan [15]). However, teleost muscle development has a higher number of interconnected myogenic differentiation phases, different timings, cell coordination, and specification compared to mammals.

In zebrafish, one of the most significant differences compared to mammals is the initial induction of slow muscle precursor cells that do not require hedgehog (HH) signaling [16]. Although HH signaling plays a significant role in the development of slow muscle in fish and mammals, the HH effector Gli2 has been characterized as a prerequisite for the expression of MyoD during skeletal myogenesis, and the timepoints of HH activation are vastly different. In zebrafish, slow muscle precursor cells, also known as adaxial cells, originate from a mesoderm *myoD*- and *myf5*-positive cell population located adjacent to the notochord [17, 18] (Figure 9.2). In contrast, adjacent cells to the notochord in mammals exclusively give rise to the sclerotome, and in amniotes, both slow and fast muscle fibers arise from the same myogenic population. Due to this fundamental difference in slow muscle formation, teleosts have anatomically well-separated slow- and fast muscle fibers, which are admixed in amniotes. Furthermore, slow muscle progenitors in teleosts are unable to fuse into multinucleated myofibers, whereas mammalian slow muscle is similar to fast muscle, but differs in size [19].

Genes that code for Pax3 and Pax7 are critical to the formation of the teleost dermomyotome (Box 9.1). However, adaxial and posterior cells (fast muscle precursors) express myogenic regulatory factors without expressing Pax3 and Pax7. As these cells precede somite formation [21], it is now evident that in teleosts, the primary myotome develops independently and precedes the formation of the dermomyotome [22], as opposed to mammals where the dermomyotome is a

Epigenetics in Aquaculture, First Edition. Edited by Francesc Piferrer and Han-Ping Wang.

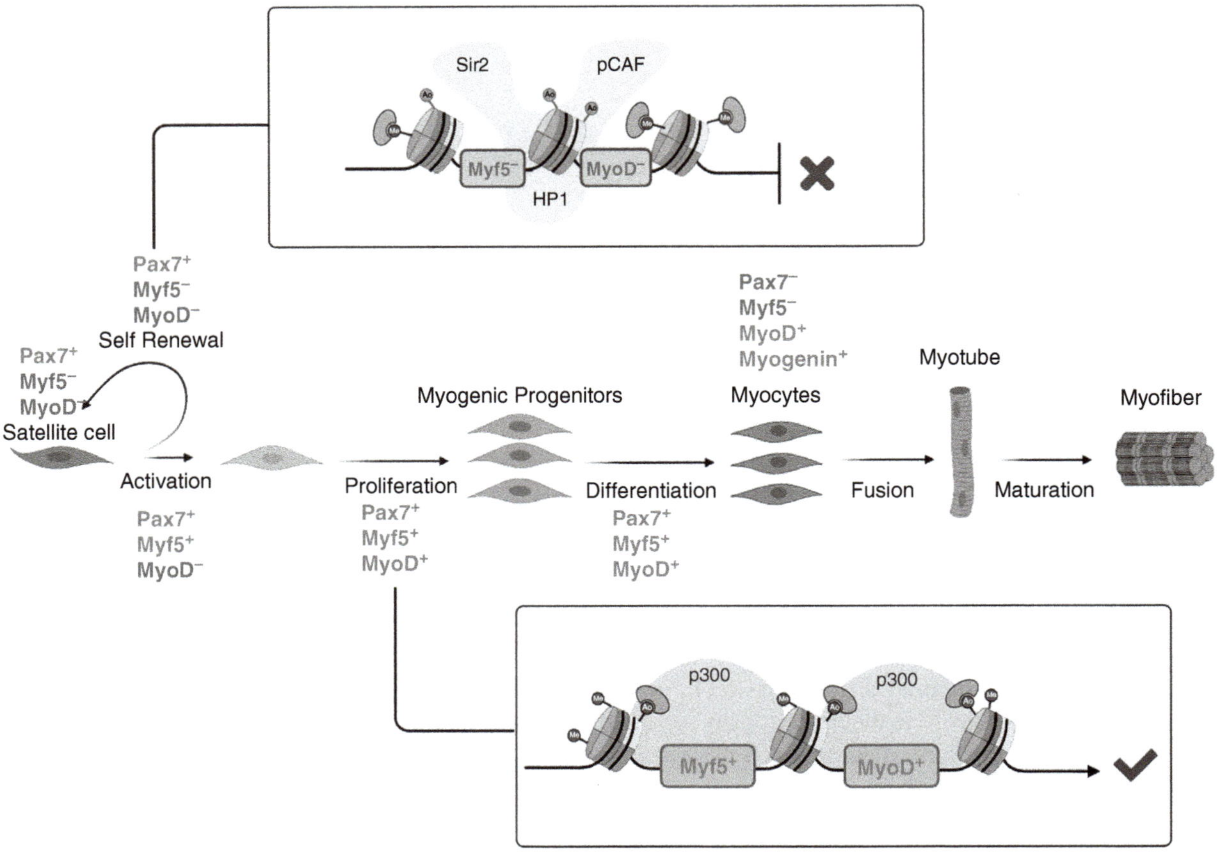

Figure 9.1 The process of myofiber formation in vertebrates. The muscle regulatory factors MyoD, Myf5, and Myogenin are expressed at different stages. The paired box protein Pax 7 plays a key role in myogenesis and is a marker of quiescent satellite cells. Repressive methylation signatures associated with the action of *Sir2*, *pCAF*, and HP1 lead to repression of MyoD and satellite cell self-renewal, whereas phosphorylation and acetylation signals associated with histone acetyltransferase p300 cause differentiation of myogenic progenitor cells. Gene name abbreviations: HP1, Heterochromatin Protein 1-Alpha; Myf5, Myogenic Factor 5; MyoD, Myoblast Determination Protein; MyoG, Myogenin; Pax7, Paired Box Protein 7; p300, E1A Binding Protein P300; pCAF, P300/CBP-Associated Factor; and Sir2, Sirtuin 2. *Source:* Created with BioRender.com. (*See colour plate section for colour representation of this figure.*)

prerequisite for the formation of the myotome. This peculiar difference between teleosts and mammals indicates that HH signaling has a similar function among vertebrates only after the formation of the dermomyotome.

The secondary myotome in zebrafish originates from the anterior border cells of the somite [23], which remain *myoD*-negative until late somite segmentation and early larval stages. When these cells are activated, they proliferate, contribute to the formation of the dorsal fin and dermis, and migrate from the myotomal to the inner surface of the slow muscle. Although, limb formation occurs only after cells have migrated outside the somite, in zebrafish, both pelvic and pectoral fins develop from fin buds adjacent to the somites. Lastly, in teleost species, the secondary myotome retains its hyperplastic activity throughout life and the addition of continuous fiber formation results in indeterminate growth and size [24].

9.1.2 Myogenesis During Early Development

The molecular machinery responsible for cell fate and muscle development is far more intricate than the function of several genes. In the early 1940s, Conrad H. Waddington introduced the term epigenetics to describe how physiological and environmental factors can affect gene expression and ultimately the phenotype. In 1957, he illustrated how pluripotent cells can differentiate following different cell fates during development in his famous epigenetic landscape, which was featured in the book "The Strategy of the Genes" [209].

In 1975, DNA methylation (i.e., the addition of a methyl group into the fifth carbon of the cytosine DNA nucleotide; see Box 9.2 for details) was the first epigenetic modification to be discovered [25, 26]. By the late 1970s and early 1980s, it was known that DNA demethylating agents, 5-azacytidine in particular, could convert embryonic mouse fibroblasts into an adipogenic, chondrogenic or myogenic lineage [27, 28]. Evidence that epigenetic modifications such as DNA methylation

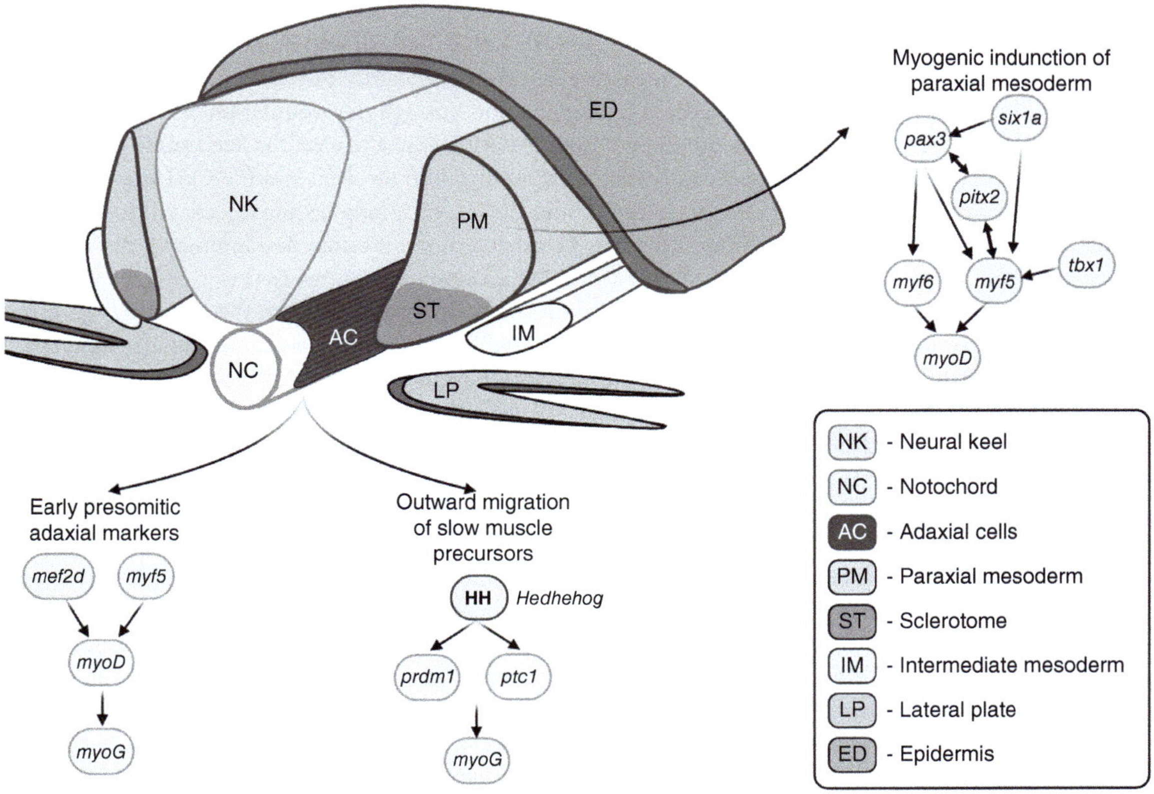

Figure 9.2 Transverse 3D section of the mid trunk during somitogenesis in fish, depicting an overview of the mesodermal structures: (i) the early presomitic phase in which *mef2d* is an adaxial marker; (ii) the outward migration of slow muscle precursors through the activation of HH signaling and action of *prdm1* and *ptc1*; and (iii) the myogenic induction of the paraxial mesoderm, where *pax3*, *six1a*, *pitx2*, and *tbx1* promote the expression of *myf5* and *myf6*. Gene name abbreviations: *Mef2d*, Myocyte Enhancer Factor 2D; *Myf5*, Myogenic Factor 5; *Myf6*, Myogenic Factor 6; *MyoD*, Myoblast Determination Protein; *MyoG*, Myogenin; *Pax3*, Paired Box Protein 3; *Pitx2*, Paired Like Homeodomain 2; *Prdm1*, PR/SET Domain 1; *Ptc1*, Patched 1; *Six1a*, SIX Homeobox 1a; *Tbx1*, T-Box Transcription Factor 1. *Source:* Created with BioRender.com. (*See colour plate section for colour representation of this figure.*)

Box 9.1 The Dermomyotome in Teleosts

During segmentation, the paraxial mesoderm is subdivided into the sclerotome, myotome, and dermomyotome. Although the sclerotome and myotome are known to be responsible for the formation of the axial skeleton and musculature, respectively, the teleost dermomyotome has been described for several decades as the outermost epithelial layer of the lateral surface of the myotome. Its distinction from the dermomyotome of amniotes is primarily attributed to the lack of immunolabeling experiments. However, the dermomyotome in fish resembles that of amniotes and is a major source of myogenic precursors that contribute significantly to myotome growth. Source: Adapted from Devoto et al. [20].

Box 9.2 Epigenetic Mechanisms

Epigenetics is a word of Greek origin with the prefix "epi-" meaning "above" and the suffix "-genetics" referring to the genetic code of DNA. Epigenetic modifications are chemical alterations that occur in the DNA without changing the sequence itself. DNA methylation is the addition of a methyl group (—CH_3) into the fifth carbon of the cytosine DNA nucleotide. Similarly, DNA hydroxymethylation is formed when an existing methyl group is oxidized (—CH_2OH) by ten-eleven translocation enzymes (TET). Consecutive oxidations of the same group can lead to formyl- (—CHO) and carboxyl- (—COOH) cytosines. Due to the different molecular geometry, composition, stability, and accumulation (based on TET expressional dynamics), these DNA modifications play major roles in transcription binding efficiency, gene expression, and silencing. Histone modifications, such as histone methylation and acetylation, are critical for the regulation of chromatin accessibility by targeting either lysine or arginine amino acids. Post transcriptionally, the epigenetic function of noncoding RNAs such as micro- and circular-RNAs can regulate translational dynamics by mRNA degradation and microRNA/protein absorption, respectively.

were modulating gene expression by chromatin inactivation [29, 30] led to the discovery that a single or a few methylated DNA loci were controlling the myogenic lineage [31]. Indeed, Brunk et al. [32] showed that the distal *myoD* enhancer was demethylated only in myogenic cells, whereas nonmuscle cells were significantly methylated. These findings suggested that DNA methylation is responsible for tissue-specific gene regulation through the modulation of DNA accessibility. Our lab provided the first evidence of epigenetic modulation of muscle plasticity in a teleost, the Senegalese sole (*Solea senegalensis*) [33]. A lower temperature (15 °C) significantly increased methylation levels at specific CpG sites within the *myogenin* promoter, relative to higher rearing temperatures (18 and 21 °C), with long-lasting effects on growth and muscle cellularity composition. These methylation differences may be related to the thermal developmental plasticity of *dnmt3* paralogues, which code for the DNA methyltransferase involved in *de novo* methylation [34].

Early embryonic development in zebrafish is separated into seven main stages. The first three stages (zygote, cleavage, and blastula) are characterized by a one-dimensional layer of epithelial cells that are precursors to all embryonic tissues. In the fourth stage, gastrulation, these cells reorganize into a multilayered and multidimensional structure with axial orientation forming the ectoderm, mesoderm, and endoderm. Myogenesis starts from the axial mesoderm, which gives rise to the notochord, a structure that extends along the animal body and is critical for developmental synchronization and signaling. Adjacent cells to the notochord and derived from the paraxial mesoderm express myogenic regulatory factors, including MyoD, which commits them to the myogenic lineage (Figure 9.2). The factors that are translated into proteins belong only to a small fraction of the eukaryotic transcripts. The vast majority of transcripts are noncoding and include microRNAs (miRNAs), circular RNAs (circRNAs), and long noncoding RNAs (lncRNAs) among others. MicroRNAs are short noncoding RNAs that regulate gene expression posttranscriptionally and are known to control myogenesis in vertebrates. In mammals, one of the miRNAs involved in muscle cell differentiation is miR-181, which is responsible for the downregulation of the differentiation inhibitor Hox-A11 [35]. In addition, miR-1, miR-133, and miR-206 are all highly conserved miRNAs across vertebrates and have been associated with the regulation of myogenesis. Interestingly, MyoG and MyoD are most likely the triggers for the expression of these muscle-specific microRNAs since they have a high binding affinity to regions upstream of their genomic sequences [36]. In teleosts, a large number of miRNAs are associated with embryonic development, while some of them (miR-17a, miR181-5p, and miR-206) have also been positively associated with muscle growth [37, 38].

CircRNAs are circular, long noncoding RNAs that are generated by pre-mRNA back-splicing. They consist either exclusively of exonic or intronic sequences, or they can also contain both exonic–intronic pre-mRNA sequences by intron retention between exons. Functionally, circRNAs act primarily as miRNA sponges that sequester their target miRNAs with complementary sequences and therefore control mRNA degradation and posttranscriptional silencing. Although circRNAs play an important role during transcription, splicing events, translation, and protein function, and are present in the skeletal muscle of many taxa including teleosts, there are currently no functional studies demonstrating a relationship between circRNAs and myogenesis in teleosts. In mammals, however, several circular RNAs have been found to regulate myoblast proliferation, differentiation, and apoptosis (see Zhou et al. [39], Das et al. [40], and Rbbani et al. [41] for comprehensive reviews). Therefore, it is likely that circRNAs share similar functionality among vertebrates.

Recent research in mammals has also revealed a strong link between long noncoding RNAs and the regulation of myogenesis. In mice, a myogenesis-associated lncRNA (*lnc-mg*) promotes myogenesis by binding to microRNA-125b, which is responsible for the control of insulin-like growth factor 2 abundance [42]. The lncRNA *SYISL* repressed muscle development by recruiting the polycomb repressive complex 2 and leading to H3K27 trimethylation of muscle-specific genes such as *MyoG*, muscle creatine kinase (*MCK*), and myosin heavy chain 4 (*Myh4*) [43]. Furthermore, the presence of *lrm* in C2C12 cells enhanced the transcriptional activity of MyoD/MEF2D complex, which is involved in skeletal muscle development of [44]. Another lncRNA (*lncMyoD*) was found to modulate the accessibility of *MyoD* to chromatin, its binding affinity, and the transactivation of target genes [45].

Segmentation is the fifth stage of embryonic development in teleosts such as zebrafish, and corresponds to the organization of paraxial mesodermal cells into whorls, forming the somitomeres. With an anterior-to-posterior direction, the paraxial mesoderm progressively segments into well-defined epithelial structures, the somites, which are chevron-shaped and bilaterally symmetrical. During the pharyngula stage (sixth developmental stage), the spinal cord strengthens and straightens the body axis. Many muscle-specific and collagen-related genes display a peak of expression during this stage, which persists even in post-embryonic stages as they are involved in the formation of the skeletal system [46]. During hatching (seventh developmental stage), most of the primary organs are complete and morphogenesis slows down, except for the formation of the gastrointestinal tract and gut, which undergo a developmental remodeling. Muscle development in many teleost species is both hyperplastic (formation of new muscle fibers) and hypertrophic (increase in mass of existing muscle fibers) during the hatching period [47].

9.1.3 Post-Embryonic Muscle Growth

Post-embryonic development in teleosts includes extensive somatic growth, patterning, and metamorphosis from the larval to the adult stage. As described earlier, the first distinct myogenic wave in teleosts stops at the segmentation period, while the second myogenic wave is characterized by stratified hyperplasia. Stratified hyperplasia is a common feature across many teleost species and can be observed from early developmental until late larval stages and even after metamorphosis is complete. The formation of new muscle fibers originates from dermomyotomal cells (Box 9.1) that express *Pax7* and migrate from the outer to the inner surface of the slow muscle [48]. These cells maintain their ability to proliferate undifferentiated and contribute to a slow increase of growth throughout life [49]. Stratified hyperplasia takes place in four well-separated growth zones. The main growth zone of the fast muscle is the superficial monolayer of external cells that cover the entire lateral surface as well as the slow and pink muscle layers. The admixture of new and old muscle fibers is termed "mosaic hyperplasia," which is morphologically distinct from stratified hyperplasia and variable with ontogeny across teleost species [50]. In teleosts, muscle growth has remarkable plasticity, modulated by nutritional [51–54] and environmental factors such as temperature [55] and photoperiod. For example, the mean weight of Atlantic cod (*Gadus morhua*) juveniles reared under continuous illumination was 13% greater than those kept under natural photoperiod conditions for 120 days, concomitantly with a decrease in expression of histone methyltransferase genes (*mll1*, *mll4b*, and *setd1a*) and an increase in transcript levels of *myogenin* [56]. Moreover, *dnmt1* and *dnmt3a* transcript levels were also significantly different between light groups [57], indicating that both DNA and histone methylation may be involved in the epigenetic regulation of muscle growth by photoperiod in Atlantic cod (*Gadus morhua*).

Several epigenetic mechanisms including DNA methylation, DNA hydroxymethylation, miRNAs, circRNAs, and long noncoding RNAs (Box 9.2) are inextricably linked to the process of muscle growth and repair through physiological (genotypic) [58] and environmental stimuli such as temperature [33, 59–61], nutrition [62], photoperiod [57], and domestication [63, 64]. We have recently shown that a circRNA derived from the *mef2c* gene has binding sites for several myomiRs and is differentially expressed with growth rate in Nile tilapia. Hence, it is likely that it forms multiple competing endogenous RNA networks that regulate muscle growth in this fish species [210]. In a study comparing the DNA methylomes of fast- and slow-growing male Nile tilapia (*Oreochromis niloticus*), we have identified several genes being differentially methylated, including the autophagy-related gene *atg14*, which is involved in muscle protein degradation [65]. DNA hydroxymethylation, which is an oxidized derivative of DNA methylation, has received little attention in epigenetic studies. Our lab has reported, for the first time, that the muscle hydroxymethylome is very different between wild Nile tilapia and their progeny reared in captivity [63]. Several differentially hydroxymethylated cytosines between groups were associated with genes involved in growth-related pathways. We have also observed an abundance of this epigenetic modification across major tissue types that comprise the somatotropic axis, including skeletal muscle [66].

9.2 Skeletogenesis in Teleosts

9.2.1 Mechanisms of Skeletal Formation – The Origin of Skeletal Tissues

The fish skeleton shares similarities with all other vertebrates, since it is composed of bone, cartilage, notochord, connective tissues, scales, and tooth components such as enamel and dentin [67, 68]. However, some particularities are present in teleosts that can be explained by a large number of species, with adaptations to a wide range of habitats and lifestyles, and to the long evolutionary history of this group of animals. In fact, across the diversity of species, their skeletons have undergone exceptional adaptations, including the presence or absence of osteocytes (a type of bone cell) contained in the matrix of bone. Osteocytes are supposed to play a central role in the homeostasis and physiology of bone. Nevertheless, this cell type has disappeared from the bone of most extant teleost species, a phylogenetic group that comprises the majority of bony fishes and represents more than half of vertebrate species. This fact has been studied since the mid-nineteenth century with the pioneering work of Kölliker [70], who made observations of bone preparations from 200 fish species and for the first time proposed a separation of the osseous fishes in two groups, species with bone corpuscles and species without. This terminology was later improved by the modern nomenclature of fishes with cellular or osteocytic bones, and fishes with acellular or anosteocytic bones, a more precise terminology since this tissue still has osteoblasts and osteoclasts on its surface [69, 71–75].

The origin of these two types of bony tissue is linked to the origin of skeletal tissues and how they developed in the evolutionary history of vertebrates. Bone is specific to vertebrates and the first bony skeletal elements appeared in ancestral fish as a mineralized tissue around the basal membranes of the skin. Analysis of the fossil record provides evidence that vertebrates arose from amphioxus-like forms with only a notochord as internal axial support that evolved into jawless fish with a cartilaginous endoskeleton, ancestors to the lampreys and hagfishes of our days. These ancestral jawless fish later developed an external dermal skeleton, or exoskeleton, for protection that was composed of mineralized tissues [72, 76–78]. This first mineralized tissue of early vertebrate's exoskeleton was similar to skeletal tissues occurring in modern vertebrates. The first manifestation of bone was as an acellular, matrix-rich mineralized tissue termed "aspidin" that was found in pteraspidomorphs, whereas cellular bone appeared only in Osteostracans [79–81]. Both the endo- and exoskeleton of osteostracans were made up of cellular bone. It appears that acellularity constitutes the primitive condition for bone, and that the presence of acellular bone in modern teleosts seems to have been independently derived at a later stage of evolution [72, 77, 82].

There are two different types of bones recognizable in vertebrate skeletons, and for a long time, it was assumed that they reflect embryonic origin, or more specifically, whether these bones developed from a cartilaginous precursor or not. The bones that develop from precursor cartilages will form in two different ways, either by deposition on the surface of the cartilage by perichondral ossification or by deposition within the cartilaginous matrix, gradually replacing the cartilage template or endochondral ossification [83]. It has been considered that these two types of bone formation, developed from a cartilaginous precursor or not, were a reflection of evolution, as mentioned by Huxley [84], who considered that, in vertebrates, certain bones are always of cartilaginous origin, while others always have a membranous origin. It was subsequently discriminated that not all bones are maintained in this manner, as some animals possess intramembranous bones that are homologous to endochondral bones of other species. This is the case of the vertebral bodies found in all vertebrates that are thought to have evolved independently in different lineages, including tetrapods, cartilaginous fishes, and teleosts [85, 86]. It is known that in tetrapods, the elements of the vertebral skeleton are derived from somitic paraxial mesoderm, which will give rise to cartilaginous tissues that form arches and centra in the embryonic vertebral skeleton and will later be substituted by bony tissue, therefore having an endochondral origin. However, in teleosts, the vertebral centra derive from cells of the notochord, the chondroblasts, which secrete matrix proteins and give rise to a mineral layer known as a chordacentrum, which develops around the notochord sheath, forming an inner layer of the vertebra. Later, it will be covered by an outer layer formed by intramembranous ossification [87, 88]. It is known from zebrafish mutants with alterations in somite patterning that the vertebral skeleton will form normally, while the vertebral arches will form with deformities, demonstrating that the endochondral forming arches have a paraxial mesoderm origin [89]. Hence, it has been postulated that the somitic origin of the vertebrae is specific to teleosts and has evolved separately in this taxon. Thus, teleosts are the only group of vertebrates that have a vertebral skeleton with a notochord contribution to the formation of bony centra [85, 90]. Unlike most bone tissues that are mesodermal in origin, the craniofacial bones (and cartilage) are derived from the neural crest, a transient and vertebrate-specific group of cells that arise from the embryonic ectoderm germ layer [91]. Anastasiadi et al. [92] have recently reported persistent changes in DNA methylation (epimutations) established by gastrulation and affect genes expressed in the neural crest of hatcher-reared European sea bass (*Dicentrarchus labrax*). This epigenetic regulation could have an impact on the normal development of the jaw and maxilla and may be one of the mechanisms underlying the emergence of Darwin's domestication syndrome in vertebrates.

9.2.2 Bone

The anatomic and developmental features found in teleosts and mammalian skeletons share a high similarity, with most bones forming the skull, axial, and appendicular skeletons being identical. The teleost bones are comprised of the same basic building blocks as bones of tetrapods, consisting of collagen type I, fibers, and hydroxyapatite crystals, providing a similarly light, yet stiff material. There is a high degree of conservation of developmental events of skeletogenesis and the underlying mechanisms controlling bone formation, including the early formation of a cartilaginous anlage, followed by bone formation through endochondral and perichondral ossification of endoskeletal tissues and the intramembranous and dermal ossification of flat bones and exoskeletal elements [67, 93, 94].

In general, vertebrate bones develop by two distinct processes known as intramembranous (including dermal) ossification, which occurs through direct differentiation of mesenchymal precursor cells into osteoblasts, giving rise to a large portion of the skull and other flat bones and to the exoskeleton (scales and rays). On the other hand, in endochondral ossification, there is the formation of an embryonic cartilage template that is progressively remodeled and replaced by bone, giving rise to the majority of the large bones in the axial skeleton. Besides endochondral bone, in perichondral ossification, mesenchymal cells mature into periosteal progenitors that form a bone collar surrounding cartilage [95, 96]. Another particularity of the teleost skeleton is that it has continuous growth throughout life, with interruptions or lower rates depending on the feeding and reproductive period [97].

The teleost skeleton contains the same cell types as other vertebrates, with cartilages formed by chondroblasts and containing chondrocytes in its matrix and a bone composed of three main cell types: osteoblasts, osteoclasts, and osteocytes. In most vertebrates, osteocytes are the most numerous cell type, representing 90–95% of all bone cells, and are widely studied since they play a key role in the maintenance of bone physiology and homeostasis, mechanosensing, and are part of the endocrine regulatory mechanism of mineral homeostasis [67, 98]. However, one of the most prominent differences in teleosts is the occurrence of bone devoid of osteocytes (acellular or anosteocytic) while tetrapods have only cellular bone. Furthermore, osteoclasts are mononucleated in most species of teleosts, in contrast with mammals where most of the osteoclasts are multinucleated. Nevertheless, cellular bone (osteocytic) is present in many of the so-called primitive teleosts, such as chondrosteans and dipnoans, but also salmonids or cyprinids that include the zebrafish, where multinucleated osteoclasts are also present [71, 72, 99]. However, despite the absence of osteocytes, teleost bone still requires resorption and remodeling for maintenance [99] and is still responsive to mechanical loading [100]. Bone formation by the osteoblastic lineage and the regulation of its functions involve multiple regulators, like morphogens and developmental signals including BMPs, Wnt ligands, hormones and vitamins, growth factors like Runx2 and Osx, and cytokines, which exert their action at multiple stages for the control of osteogenesis [101, 102].

9.2.3 Cartilage

The cells that form cartilage, the chondroblasts, are derived from mesenchymal progenitor cells. These mesenchymal cells proliferate during embryonic development and aggregate forming mesenchymal condensations. Once these mesenchymal cell condensations receive signaling from chondrogenic transcription factors like Sox9 and Ihh, they differentiate into chondroblasts that start secreting an extracellular matrix rich in type II collagen and proteoglycans, and they eventually become entrapped and transform into chondrocytes. This process forms the cartilage templates for future bones and cartilage [103, 104]. Chondrocytes secrete unique extracellular matrix proteins, such as type II collagen and aggrecan, to form the cartilaginous matrix. In vertebrates, three main types of cartilage can be found: (i) hyalin cartilage, containing chondrocytes well separated by water-rich extracellular matrix; (ii) elastic cartilage, with an extracellular matrix rich in elastin; and (iii) fibrous cartilage, with a dense web of collagen type I as the major extracellular matrix component [104–107]. However, it is noteworthy that teleosts also have a much greater diversity of cartilage types and intermediate tissues than what is found in mammals [72, 108], resulting from and representing their large diversity of sizes, habitats, and living habitats.

9.3 Epigenetic Regulation of Sexually Dimorphic Growth

9.3.1 Ecological and Physiological Relevance of Sexual Dimorphism

Sexual dimorphism refers to the phenotypic differentiation between sexes of the same species and occurs across many taxa, including mammals, birds, amphibians, reptiles, insects, and fish [109–116]. Sexual dimorphism displays a broad phenotypic variability with differences in size, shape, coloration, and behavior between males and females.

One of the most common sexually dimorphic traits is sexual size dimorphism (SSD), with either males being larger than females or vice versa. SSD can be the result of physiological differences whose origin may be genetic (sex chromosomes) and hormonal, as well as behavioral, i.e., male–male competition for reproductive success or positive correlation between female fecundity and size. Although the impact of the genotype and gene expression between sexes can be in many cases quantitative, the evolution of behavioral characteristics can be attributed to more than one causative factors, namely sexual selection, fecundity selection, and ecological divergence [117]. Sexual selection is a concept that was introduced as early as the eighteenth century by Charles Darwin and describes the sex-dependent selection of adult body sizes based on reproductive success (i.e., male-to-male combat applying an intersexual selective pressure toward larger body sizes) [118]. Fecundity selection follows the same principle (increasing fecundity with increasing body size of certain species). However, it does not involve inter- or intrasexual competition [117]. To that extent, males and females are diverging toward a specific phenotype independently of each other, and therefore, both processes result in sexual dimorphism. Although these two types of selection are considered the main causative factors for sexual dimorphism, new research reveals that sex-dependent ecological divergence is possible. For example, the freshwater blenny (*Salaria fluviatilis*) has an increased SSD and shape differentiation (i.e., larger males than females and more distinct body shapes between sexes) in populations inhabiting rivers compared to lakes [119].

It is the environmental conditions that play a significant role in repressing or promoting differentiation of existing sexually dependent characteristics, likely acting in concert or sequentially with other sexually antagonistic selection pressures [120]. An extreme example of ecological sexual dimorphism in teleost fishes is the evolution of some deep-sea ceratioid anglerfishes. Due to the vast space of the deep sea, males and females rarely meet. This environmental pressure led toward the evolution of sexual dimorphism and parasitism, where males are dwarfed compared to females and multiple structures are enhanced on males (i.e., eyes and nostrils) for the identification of females in the dark deep sea. In some cases, males permanently attach to or degenerate leaving only a functional (parasitic) testis attached to females through the fusion of their tissues, a process that requires major flexibility of molecular networks and immunity [121].

9.3.2 Relationship Between Sex and Growth with an Overview of Key Molecular Networks

As discussed earlier, SSD is the evolutionary result of a very complex relationship between sex and size across the animal kingdom. In aquaculture, a large number of economically important fish species display SSD, and therefore sex control is one important strategy for maximizing production. Although the primary goal is to achieve growth superiority and reach the marketable size in a short time, monosex cultivation systems in favor of either males or females provide additional benefits such as prevention of gonadal development and courtship behavior, reduction of unwanted offspring recruitment and aggression, avoidance of escapees, and flesh quality deterioration due to sexual maturation [122]. In Nile tilapia, the monosex culture of males is desirable, since they grow faster and can be much larger than females [123]. Although body weight is largely controlled by the same genes in both males and females [124], SSD is largely attributed to sex-determining genes rather than puberty [125]. Moreover, sex-specific growth in Nile tilapia may be finely tuned by epigenetic modifications across the genome [65]. We have identified sex-specific DNA methylation markers in fast muscle autophagy-related processes of Nile tilapia, with males having significantly higher methylation levels in *atg14* and potentially repressed autophagic function compared to females. Other important fish species where males have significantly higher growth rates compared to females include several catfish species such as the yellow (*Pelteobagrus fulvidraco*) [126] and channel catfish (*Ictalurus punctatus*). Male yellow catfish outgrow females by approximately 30% within a single year posthatching and SSD is evident primarily due to higher expression of growth hormone (GH) as well as insulin growth factors I and II (*igf1* and *igf2*) in males compared to females. Interestingly, differences in gene expression were observed as early as the first-week posthatching (larval stage), which reveals that SSD occurs independently of the reproduction onset [127]. In channel catfish, SSD is evident after a total weight of 50 g between males and females and becomes increasingly apparent within two years of age, reaching up to 42% difference in terms of the total weight in favor of males. Interestingly, a recent study revealed a putative epigenetically regulated sex determination region in channel catfish, which is heavily methylated in females compared to males during early development. However, a single biological replicate per developmental stage per sex limits the scope of any conclusions from this study [128].

SSD in favor of females is observed in the majority of important aquaculture fish species, including the common carp (*Cyprinus carpio*), Atlantic and Chinook salmon (*Salmo salar* and *Oncorhynchus tshawytscha*, respectively), European sea bass, rainbow trout (*Oncorhynchus mykiss*), Southern flounder (*Paralichthys lethostigma*), and half-smooth tongue sole (*Cynoglossus semilaevis*). Refer to Mei and Gui [129] for a review of important sexually size-dimorphic fish species. A recent study using genetically modified *cyp17a1*-deficient common carp neomales (*cyp17a1*$^{-/-}$ XX genotype) revealed that crosses with wild-type females (*cyp17a1*$^{+/+}$ XX genotype) resulted in all-female offspring (*cyp17a1*$^{+/-}$ XX genotype), showing the importance of single-locus manipulations for the species in question [130]. In European sea bass, significant size differences between sexes (female-biased) are observed within three months posthatching under normal rearing conditions [131]. To date, the link between SSD and epigenetics remains elusive, but environmentally induced and sex-specific DNA methylation changes in genes such as *cyp19a1a* and *dmrt1* unveil epigenetic mechanisms of sex determination [132–134], which have an indirect impact on growth potential.

9.3.3 Implications of DNA Methylation and Hydroxymethylation in Growth Differences Between Males and Females

DNA methylation is one of the most studied epigenetic modifications. Research of DNA methylation patterns demonstrating SSD in fish has revealed key genes and pathways under epigenetic control. Wan et al. [135] identified over 4700 differentially methylated regions (DMRs) in the skeletal muscle of male and female hybrid (*O. mossambicus* X *O. niloticus*) tilapia. Interestingly, several DMRs were located within the linkage groups LG1 and LG23, which contain sex-determination loci in Nile tilapia, blue tilapia (*O. aureus*), and Mozambique tilapia (*O. mossambicus*). Furthermore, several differentially methylated genes coding for transcription factors were involved in the development of the skeletal muscle and the nervous system as well as the overexpression of the MAPK/ERK signaling pathway in males, pinpointing toward a functional distinction between male and female growth rates.

Our lab has identified differentially methylated cytosine-guanine dinucleotides (DM-CpGs) between fast- and slow-growing male and female Nile tilapia. Some DM-CpGs are related to size differences (i.e., large males versus small males and females) rather than the sex (male versus female) [65]. By adopting a multifactorial analysis, we managed to separate size- and sex-dependent DM-CpGs. Sexually dimorphic DM-CpGs were associated with the epigenetic histone modification factor *hdac11*, which can affect histone acetylation of *mef2c* and *myoG*, and the leptin receptor *lepr*, which is involved in myoblast proliferation and differentiation. The integrin receptor of fibronectin *itga4*, which plays a role in myogenesis, and the transforming growth factor *tgfb2* were also differentially methylated between sexes. Lastly, the autophagy-related key regulator *atg14* was hypomethylated in large males compared to females, which suggests that the regulation of muscle mass and rate of muscle degradation is sexually dimorphic in tilapia.

DNA hydroxymethylation, an oxidation derivative of DNA methylation, has received scant attention. It has been recently acknowledged as a predominantly stable epigenetic modification, but most epigenetics studies using bisulfite-sequencing techniques report both DNA methylation and hydroxymethylation residues as one. The inability to distinguish them and the inverse nature between the two modifications adds another layer of complexity toward the analysis and validity of their functionality. DNA hydroxymethylation has been found to accumulate in the brain more than in any other tissue across several species [136, 137], including teleost fishes [138]. The brain is a critical organ for the perception and translation of external signals, especially for fish whose sexual dimorphism is linked to environmental stimuli. To the best of our knowledge, there is currently no information available on sex-dependent DNA hydroxymethylation levels and patterns in fish. Recent research in mouse brain provided limited insight into sex-dependent DNA hydroxymethylation differences but reported a negative correlation of expression between the ten-eleven translocation enzymes 2 and 3 (*tet2* and *tet3*) (Box 9.2) and the DNA (cytosine-5) methyltransferase 1 (D*nmt1*) in the preoptic area of the hypothalamus of females compared to males [139].

9.3.4 miRNAs Differentially Expressed with Sex and Their Role in Muscle Growth

As a functional component of cells, miRNAs are involved in a wide range of biological pathways including sex determination and growth. Xiao et al. [140] showed that several miRNAs (miR-181a, miR-181a-5p, miR-143, and miR-143-3p) serve a tissue-specific role in the gonads (testis and ovaries) of Nile tilapia. Another gonad-specific miRNA is miR-202, which is exclusively present in the gonads of rainbow trout [141], Atlantic halibut [142], and medaka [143].

Sexually dimorphic expression of miRNAs has been reported in many organisms, including teleost fishes. In ovaries, miR-29 and miR-129 have significant overexpression compared to testis and are involved in oogenesis and ovary maturation, respectively [140]. In contrast, miR-33a, miR-132, miR-135b, and miR-212 were overexpressed in Nile tilapia testes. Insulin receptor substrate 2 is a target for miR-33a [144], and its repression may contribute to testis maturation via the regulation of the insulin-signaling pathway [140]. Although the function of miR-132 in fish has not been investigated to date, research on miR-132 expression and its transcriptional network in adrenal cells suggests that it regulates adrenal and ovarian steroidogenesis [145]. In addition, Wang et al. [146] have identified several inverse correlated miRNA/mRNA pairs (miRNAs and their target genes) by sequencing and RT-PCR (real-time polymerase chain reaction) in Nile tilapia. In particular, miR-17-5p and miR-20a were associated with *dmrt1*, miR-456, and miR-138 with *amh* and miR-138, miR-338, and miR-200a with *cyp17a2* mRNA, respectively. Both *dmrt1* and *amh* are master sex-determining genes and their expression levels determine the ovarian or testicular fate. The cytochrome P450 aromatase Cyp17a2 is an enzyme that catalyzes the 17a-hydroxylation of progesterone and pregnenolone to glucocorticoids and androgens in fish [147]. Similar results were shown by Tao et al. [148], who identified a large regulatory network of miRNAs involved in steroid synthesis and, consequently, early sex differentiation. Interestingly, another aromatase (Cyp19a1a) that is responsible for ovarian differentiation and development in fish [149] is regulated via the activity of miR-4585 as part of the oxidoreductase pathway for oocyte maturation [150].

Recent research adds additional layers of complexity in sex-specificity by the regulation of miRNA arm switching. During the last stages of miRNA biogenesis, precursor-miRNAs are exported to the cytoplasm and the terminal loop is cleaved by the RNase III endonuclease Dicer. Although both 3′ and 5′ arms are functional and they originate from the 3p and 5p RNA duplex strands respectively, the evidence pinpoints that one arm is preferentially accumulated while the other is degraded [151]. Thus, arm switching involves changes in the selection of the 5p or 3p arm, which is shown to be tissue and developmental stage-specific. Pinhal et al. [152] showed that sexual dimorphism in Nile tilapia is accompanied by sex-specific arm switching of miR-145 in the gonads. In particular, miR-145-5p had a 2.5-fold higher expression in testis compared to ovaries and miR-145-3p had a 3.4-fold higher expression in ovaries compared to testis. Controlling sex determination alone can lead to large differences in growth due to sexual dimorphism. Although some miRNAs have been associated with fast growth (e.g., let-7b, let-7c, and let-7j) [153] and several miRNA/mRNA pairs (miR-1/*hdac4*, miR-133-a/*srf*, miR-133-b/*srf*, miR-206/*pax7*, and miR-499/*sox6*) have been reported to play a critical role in myoblast differentiation and proliferation throughout development [154], sex-specific miRNAs involved in muscle growth have yet to be found.

9.4 Epigenetic Control of the Skeleton in Teleosts

The molecular control of the processes of skeletal formation is highly conserved from teleosts to higher vertebrates, and both cartilage and bone formation and maintenance are under tight regulation by osteogenic-signaling pathways and by specific transcription factors that will induce cartilage and bone-specific transactivation genes [93, 155]. In the last few decades, a number of articles have been published providing new insights on the posttranscriptional regulation of bone and cartilage formation [156–158] homeostasis and differentiation [159–161] by noncoding RNAs, of which the most studied have been miRNAs.

It has been shown that miRNAs have an important role in regulating bone development. Lian et al. [102] have identified 11 miRNAs that inhibit Runx2 expression (miR-133, miR-23, miR-205, miR-135, miR-137, miR-338, miR-218, miR-34, miR-30, miR-217, and miR-204). Osterix (Osx/Sp7), another key expression factor for osteoblastic differentiation, is inhibited by miR-125b, miR-138, and miR-637. Additionally, miR-93 was described to regulate osteoblastic differentiation via Osx feedback, and miR-23a-27a-24-2 inhibits the process of transition from osteoblast to osteocyte [102]. Additionally, miR-181a targets Tgf-beta-induced (Tgfbi) and TGF-β type I receptor (TβR-I) promoting osteoblast differentiation [162].

A study by Wan et al. [163] specifically identified miRNAs from intermuscular bones during the development of six-month-old blunt snout bream (*Megalobrama amblycephala*). These authors have identified 13 miRNAs exclusively expressed in intermuscular bones (let-7d-3p, miR-106a, miR-153a, miR-15c, miR-190a, miR-196d, miR-218a, miR-301, miR-454b, miR-457b, miR-460-5p, miR-96, and miR-9a-3p) when subtracted from an expression library of connective tissue. A total of 20 differentially expressed miRNAs were common in the two libraries and exhibited higher expression in intermuscular bones. The unique pattern of expression of microRNAs in this bone indicates that they can be involved in the formation and differentiation of these skeletal structures.

In another study aimed at identifying new miRNAs in zebrafish, it was found that miR-455 is associated with skeletal tissues in zebrafish larvae, with its expression restricted to the cartilages of the pharyngeal arches and in the head skeleton, starting only after 48 hours post fertilization [164]. This pattern of expression indicates a role in the regulation of the genes controlling pharyngeal arches development and/or chondrocytes differentiation and homeostasis in the Zellknorpel cartilage during larval development. In fact, it has been recently shown that increased miR-455 and miR-140 expression will promote the TGF-β Smad 1/5/8 pathway through the suppression of the Smad 2/3 pathway, therefore stimulating a degradative response by the chondrocytes [165]. This is a regulatory mechanism that might be involved in the morphogenesis of the cartilaginous structure during early development stages.

The function and differentiation of osteoclasts were observed to be regulated by miR-21. It was first reported through *in vitro* experiments that miR-21 promotes osteogenesis, and later confirmed *in vivo* that there was an increment of trabecular bone in miR-22$^{-/-}$ mice [145]. Other authors have conducted studies in cells that revealed miR-21 to be a regulator of osteoclastogenesis and a promoter of osteoclast differentiation [166]. It was found that the action of miR-21 is made through an increased expression of receptor activator of nuclear factor κB ligand (RANKL) and a decreased expression of osteoprotegerin (Opg) through miR-21 targeting of Sprouty 1 (Spry1). Another known regulator of osteoclastogenesis in mammals is miR-223, and its ortholog has been described in zebrafish, showing that there is a conservation of its function. In fact, the identification of known miR-223 targets in zebrafish showed that osteoclast regulatory genes like Lmos2 and Nfia were also targeted, indicating that in fish MiR-223 expression is correlated with hematopoiesis and osteoclastogenesis [167].

The miR-214 family is known to play important roles in vertebrate skeletogenesis, particularly acting on chondrogenesis, being highly conserved among vertebrates [157, 168]. It was observed that miR-214 is highly expressed during early stages of cartilage formation on larval zebrafish, and when miR-214 is overexpressed, it impedes the normal formation of cranial cartilages and affects the formation of otoliths [157]. Therefore, it is assumed that miR-214 is an inhibitor of chondrogenesis as was evidenced by the downregulation of chondrogenic marker genes, and by phenotypic alterations in cartilage formation when zebrafish embryos were injected with miR-214. It was also shown that the action of miR-214 in chondrogenesis is accomplished in an Atf4-dependent manner, with the downregulation of this gene upon miR-214 overexpression [157].

Another miRNA described to be involved in embryonic bone development is miRNA-140-5p [169], which negatively regulates the expression of bone morphogenetic protein 2 (Bmp-2), thus preventing bone formation in zebrafish. Bmp-2 is a member of the transforming growth factor (TGF-betta) superfamily, which is predominantly expressed in the osteoblastic lineage, inducing the differentiation of osteoblasts and regulating osteogenesis and cartilage formation during development [170]. Injecting miRNA-140-5p-induced alterations in cranial cartilage induced deformities and shortened tails. However, this phenotype could be successfully rescued by microinjecting Bmp2 mRNA. Additionally, genes directly implicated in cartilage development like *ADAMTS-5* and *FGFRL1*, and *Dnpep*, which is involved in endochondral ossification, are targeted by miRNA-140-5p. This indicates that miR-140-5p is a key player in the regulation of bone and cartilage formation during fish development [169]. miR-92a has been implicated in the regulation of pharyngeal cartilage formation by targeting the Bmp antagonist *noggin3*, a negative regulator of Bmp2 signaling. Inactivation of miR-92a led to inactivation of Bmp signaling, causing loss of pharyngeal cartilage elements due to impairments in proliferation, differentiation, and survival of chondrogenic progenitors. Conversely, miR92a ectopic expression led to a decrease in Nog3 mRNA levels, derepressing Bmp signaling and promoting cell apoptosis [171].

It has also been shown that mir-20a is capable of regulating in vitro mineralization of cells derived from the vertebral column of gilthead seabream by targeting the BMP-2 transcript and type II BMP receptor, and in this way, regulating BMP signaling pathway. miR-20a was observed to be upregulated during cell differentiation. Its overexpression produced an inhibition of mineralization and an upregulation of matrix Gla protein (MGP), known to be a physiological inhibitor of calcification that has been found to form a complex with BMP-2 [172].

DNA methylation and histone modifications are some of the main epigenetic mechanisms that are involved in skeletal formation and differentiation, but little is known in fish. Nevertheless, histone deacetylases (Hdac) have been implicated in the development of skeletal tissues. For instance, Hdac4 has been shown to repress the ossification of cartilage and regulate the development of neural crest formation during the embryo development of zebrafish [173], as well as Hdac-1, which is required for the normal formation of craniofacial cartilages and pectoral fins in the zebrafish. Mutant and morphant embryos for *hdac-1* displayed a severe or total absence of cartilage formation in the cranium and pectoral fins [174].

9.5 Mitochondrial Epigenetics

The mitochondrion is an organelle that has its own genome known as mitogenome or mitochondrial DNA (mtDNA) and functions as a cellular "powerhouse," participating in different metabolic processes as well as cell respiration [175].

The number of genes in the mitochondrial genome of eukaryotes (including unicellular organisms) varies among taxa, i.e., the number of protein-coding genes (PCGs) may vary from 3 to 67, and transport RNAs (tRNAs) genes may be absent altogether [176, 177]. Nevertheless, the mitogenome of multicellular animals, with the exception of representatives of the Cnidaria, usually contains 37 genes: 13 PCGs (subunits of the oxidative phosphorylation (OXPHOS) system), 22 tRNAs, and two ribosomal RNA genes (12S rRNA and 16S rRNA), and a noncoding region that contains mitochondrial gene promoters and *D-loop* region. The other genes that are needed for mitochondria function are located in the nuclear genome [178]. Nucleotide sequence diversity of the mitogenome has been extensively explored in functional, human ancestry, population, and medical studies [179–182]. In stark contrast, epigenetic modifications of mtDNA (Figure 9.1) and its functional impact are poorly understood [183]. Moreover, the first studies that were carried out on vertebrates could not find conclusive evidence of mtDNA methylation in humans and other vertebrates [184]. This view changed with the identification of DNA-methyltransferase 1 (DNMT1) in mammalian mitochondria and the description of the dynamics of mtDNA methylation in relation to mitochondrial transcription factors, which stimulated mitoepigenome studies [185, 186].

In mammals, genetic and epigenetic changes in the mitogenome can have a great impact on a variety of physiological and pathophysiological processes [187]. A more detailed and accurate assessment of various epigenetic modifications of mitochondria has been recognized only recently, with the advent of methods for resolving individual cells [188, 189]. In mammals, methylated cytosines in mtDNA are predominantly localized in a non-CpG context, in contrast to the predominantly CpG methylation pattern in the nuclear genome. The functional significance of such non-CpG methylation has been described in skeletal muscle cells, where it was found that methylation of the *PGC-1a* promoter controls mitochondrial density [190]. mtDNA has a lower level of methylation than the nuclear genome [191, 192], but it possibly also varies

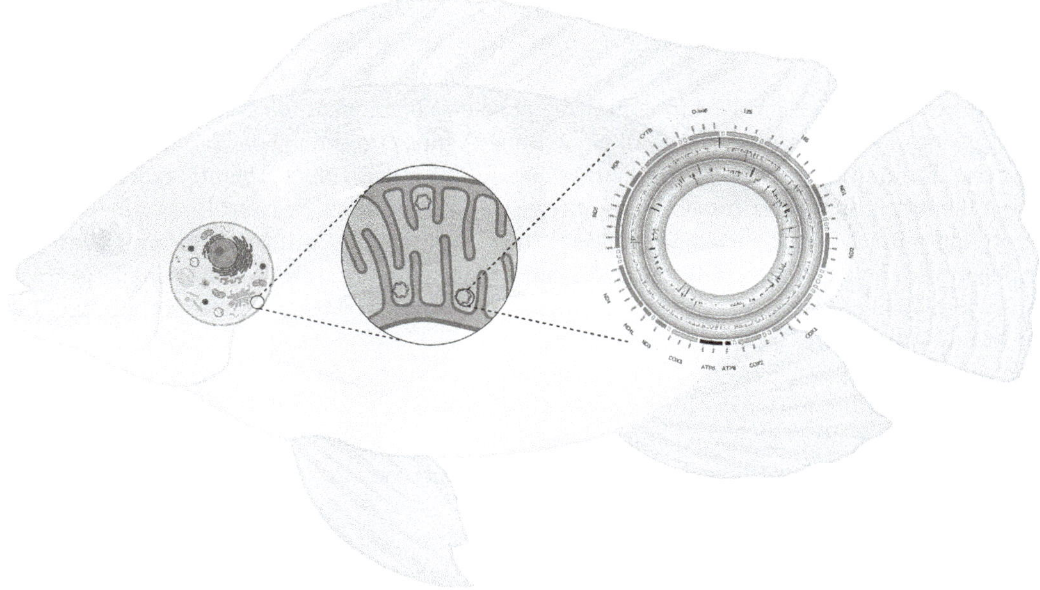

Figure 9.3 Nile tilapia mitochondrial genome and its 5mC methylation map. The double-stranded 16,626 bp mitogenome is preferentially methylated in the light (minus) strand. *Source:* Modified from Nedoluzhko et al. [183]/Elsevier. Created with BioRender.com.

depending on the type of tissue, age, stage of development, environmental conditions, and other factors [193, 194]. mtDNA methylation changes have led to a certain level of debate about the methods of mtDNA methylation analyses, while a certain type of sequencing method and data analysis might result in certain conclusions, e.g., significant differences in CpG-context methylation patterns between primate species – human (*Homo sapiens*) and golden lion Tamarin (*Leontopithecus rosalia*) [195, 196]. Nowadays, epigenetic studies, which include the functional significance of mtDNA methylation, as well as its dependence on environmental factors, form a fundamentally new frontier in genomics – epimitogenomics [197] (Figure 9.3).

9.5.1 Link Between Mitochondrial Function and Muscle Growth

Mitochondrial activity affects all processes in the cell, including myogenesis. Inhibition of mitochondrial translation of myoblast cells completely blocks their further differentiation [198]. Moreover, age-related changes in the number of mitochondria and a decrease in their effectiveness are associated with the accumulation of mutations that leads to the degeneration of muscle tissue [199]. In addition to the actual production of energy, mitochondria are involved in the metabolism of fatty acids, the production of reactive oxygen species, macromolecule synthesis, apoptosis regulation, and epigenetic remodeling [200]. Mitochondria, as part of cellular machinery, are also able to determine the fate of stem cells, including muscle cells [201]. In particular, it is the transition from fatty acid oxidation to glycolysis that is the starting point for the transition of quiescent stem cells (myosatellite cells) to differentiating myoblasts. While most mechanisms of such transition remain unclear, mitochondria participate in dynamic changes during myogenesis and their functionality includes fatty acid oxidation and OXPHOS at the quiescent myosatellite cell stage, the highly glycolytic state during the proliferation of myogenic precursor cells, and significant OXPHOS during the formation of the muscular fibers [201]. Moreover, reactive oxygen species levels in mitochondrion increase during muscle cell differentiation [202].

9.5.2 Introduction to Different Types of DNA Modifications in the Mitoepigenome and Their Implications for Mitochondrial Function

Mitochondrial gene expression seems to be regulated by different epigenetic factors, such as DNA modifications and even noncoding RNAs, just like the nuclear genome. However, the regulation of mitochondrial function and their genome through various epigenetic mechanisms are still poorly understood, even if some results suggest their contribution.

Compared to the nuclear genome, mtDNA is packed by transcription and mtDNA maintenance factor (TFAM)-based nucleoids (nucleoprotein complexes) without using histones [203]. These TFAM complexes regulate the replication and transcription of the mitogenome and pack it as a mitochromosome [204]. During its vital activity, the mitochondria produce ATP and folate, which are used for S-adenosyl-L-methionine synthesis, and methyl group donors are used during DNA methylation in cells. Thus, the mitochondria indirectly regulate the methylation of their own as well as the nuclear genome.

Despite the low level of methylation and hydroxymethylation of the mitochondrial genome, these mechanisms seem to play an important role in the functioning of the cell. It has been suggested that environmental factors, chemicals, and the type of diet play an important role in cell function by influencing the level of DNA hydroxymethylation (5hmC) and classical DNA methylation (5mC) of the mitochondrial genome. Often these changes affect the mitochondrial *D-loop*, which is where translation and transcription initiation sites are located [205].

9.5.3 Mitoepigenome in Fish

The mechanisms of mitogenome functionality (including mtDNA methylation) have not yet been investigated in fish. Nowadays, there are only a few reports in zebrafish and Nile tilapia that describe the methylation of the mitochondrial genome. These studies have shown an almost completely hypomethylated mtDNA status in adult Nile tilapia liver [183] as well as in *D. rerio* sperm and oocytes [206]. Nevertheless, mtDNA methylation levels change during zebrafish embryogenesis [207].

Similar to what is observed in mammals, the highest number of methylated cytosines (more than 80%) are located in a non-CpG context and the noncoding region of the *D-loop* is the most methylated part of the mitogenome [183]. Another feature that is common among fish and other higher vertebrates is the significant difference in the average 5mC methylation level between the heavy and light chains of the strands of mitogenome (Figure 9.3). Recently, 5hmC methylation was first described in wild and early domesticated Nile tilapia. Interestingly, domestication has a significant influence on 5hmC changes in the nuclear genome while mtDNA patterns remain stable [63].

9.5.4 Association Between Growth, Mitochondrial Methylation, and Hydroxymethylation

The involvement of mitochondrial methylation in embryonic vertebrate development has been clearly demonstrated [206, 208], but nothing is known about its plasticity with environmental factors. To date, studies related to the effect of mtDNA epigenetic modifications on growth are very limited. The analysis of hydroxymethylation marks in the mitochondrial genome of the Nile tilapia showed that there is no significant difference between 5hmC methylation patterns in wild and domesticated families of *O. niloticus* selected for improved growth [63]. However, we have preliminary evidence of differential cytosine methylation in the Nile tilapia mitoepigenome between fast- and slow-growing fish.

9.6 Conclusion

Similar to all other living organisms, the teleost epigenomes are extremely dynamic and provide a key link between environmental factors and genomic, physiological, and phenotypic adaptations. However, we are just starting to unveil the complex epigenetic network that modulates growth in teleosts. The evidence discussed in this chapter clearly demonstrates that DNA modifications (methylation and hydroxymethylation), histone modifications, and noncoding RNAs – particularly miRNAs – play a key role in skeletal muscle development and thus in the regulation of somatic growth and phenotypic plasticity. For example, improved growth rate has been linked to differential methylation of key myogenic genes and biological processes, such as the TGF-β1-signaling pathway in Nile tilapia. Also, higher water temperature during metamorphosis promotes growth in Senegalese sole, concomitantly with a decrease in expression of DNA methyltransferases and methylation levels in the *myogenin* promoter. It is imperative to move from association studies and hypothetical inference to functional analysis that can demonstrate the effect of epigenome changes on the phenotype. To this aim, CRISPR-based epigenomic editing – although still in its infancy – is a promising and powerful tool for understanding the biological function of epigenetic mechanisms. There are many other knowledge gaps in the field, especially regarding the epigenetic regulation of the teleost skeleton. Also, virtually nothing is known about sex-specific DNA hydroxymethylation and differences in chromatin accessibility, and epimitogenomics is a new frontier in genomics that will allow us to gain valuable insights into the involvement of the epigenetic modifications of the mitochondrial genome and its relationship with growth in animals. In addition to revealing the fascinating mechanisms underlying muscle and skeleton growth in fish, this knowledge has profound implications for nutritional and thermal programming in fish farming. The rapid development of new DNA sequencing technologies enables the effective determination of epigenetic quantitative trait loci of domesticated animals and their wild ancestors at the level of their epigenomes and epitranscriptomes. The identification of these epigenetic markers will add another dimension beyond the management of genetic diversity for the selection of relevant traits, thus contributing to the sustainability of the aquaculture sector. Nowadays we have an opportunity to create new fish diets, breeding programs, and veterinary treatments using in-depth analysis of molecular pathways in many commercial fish species. Modern molecular techniques of genome, and even epigenome, editing will bring breeding programs to a new level in the near future.

Acknowledgments

We acknowledge the support provided by the European Research Council (ERC) under the European Union's Horizon 2020 research and innovation program (grant agreement no. 683210, 2016) and by the Research Council of Norway under the Toppforsk programme (grant agreement no. 250548/F20, 2016).

References

1 Wright, W.E., Sassoon, D.A., and Lin, V.K. (1989). Myogenin, a factor regulating myogenesis, has a domain homologous to MyoD. *Cell* **56** (4): 607–617.
2 Pinney, D.F., Pearson-White, S.H., Konieczny, S.F. et al. (1988). Myogenic lineage determination and differentiation: evidence for a regulatory gene pathway. *Cell* **53** (5): 781–793.
3 Davis, R.L., Weintraub, H., and Lassar, A.B. (1987). Expression of a single transfected cDNA converts fibroblasts to myoblasts. *Cell* **51** (6): 987–1000.
4 Rudnicki, M.A., Schnegelsberg, P.N., Stead, R.H. et al. (1993). MyoD or Myf-5 is required for the formation of skeletal muscle. *Cell* **75** (7): 1351–1359.
5 Braun, T., Buschhausen-Denker, G., Bober, E. et al. (1989). A novel human muscle factor related to but distinct from MyoD1 induces myogenic conversion in 10T1/2 fibroblasts. *EMBO J.* **8** (3): 701–709.
6 Kassar-Duchossoy, L., Gayraud-Morel, B., Gomes, D. et al. (2004). Mrf4 determines skeletal muscle identity in Myf5:Myod double-mutant mice. *Nature* **431** (7007): 466–471.
7 Rhodes, S.J. and Konieczny, S.F. (1989). Identification of MRF4: a new member of the muscle regulatory factor gene family. *Genes Dev.* **3** (12B): 2050–2061.
8 Edmondson, D.G. and Olson, E.N. (1989). A gene with homology to the myc similarity region of MyoD1 is expressed during myogenesis and is sufficient to activate the muscle differentiation program. *Genes Dev.* **3** (5): 628–640.
9 Weintraub, H., Davis, R., Tapscott, S. et al. (1991). The myoD gene family: nodal point during specification of the muscle cell lineage. *Science* **251** (4995): 761–766.
10 Lassar, A.B., Buskin, J.N., Lockshon, D. et al. (1989). MyoD is a sequence-specific DNA binding protein requiring a region of myc homology to bind to the muscle creatine kinase enhancer. *Cell* **58** (5): 823–831.
11 Weintraub, H., Dwarki, V.J., Verma, I. et al. (1991). Muscle-specific transcriptional activation by MyoD. *Genes Dev.* **5** (8): 1377–1386.
12 Hammond, C.L., Hinits, Y., Osborn, D.P. et al. (2007). Signals and myogenic regulatory factors restrict pax3 and pax7 expression to dermomyotome-like tissue in zebrafish. *Dev. Biol.* **302** (2): 504–521.
13 Groves, J.A., Hammond, C.L., and Hughes, S.M. (2005). Fgf8 drives myogenic progression of a novel lateral fast muscle fibre population in zebrafish. *Development* **132** (19): 4211–4222.
14 Maves, L., Waskiewicz, A.J., Paul, B. et al. (2007). Pbx homeodomain proteins direct Myod activity to promote fast-muscle differentiation. *Development* **134** (18): 3371–3382.
15 Rescan, P.Y. (2001). Regulation and functions of myogenic regulatory factors in lower vertebrates. *Comp. Biochem. Physiol. B Biochem. Mol. Biol.* **130** (1): 1–12.
16 Hirsinger, E., Stellabotte, F., Devoto, S.H. et al. (2004). Hedgehog signaling is required for commitment but not initial induction of slow muscle precursors. *Dev. Biol.* **275** (1): 143–157.

17 Stickney, H.L., Barresi, M.J., and Devoto, S.H. (2000). Somite development in zebrafish. *Dev. Dyn.* **219** (3): 287–303.

18 Nguyen-Chi, M.E., Bryson-Richardson, R., Sonntag, C. et al. (2012). Morphogenesis and cell fate determination within the adaxial cell equivalence group of the zebrafish myotome. *PLoS Genet.* **8** (10): e1003014.

19 Biressi, S., Molinaro, M., and Cossu, G. (2007). Cellular heterogeneity during vertebrate skeletal muscle development. *Dev. Biol.* **308** (2): 281–293.

20 Devoto, S.H., Stoiber, W., Hammond, C.L. et al. (2006). Generality of vertebrate developmental patterns: evidence for a dermomyotome in fish. *Evol. Dev.* **8** (1): 101–110.

21 Devoto, S.H., Melancon, E., Eisen, J.S. et al. (1996). Identification of separate slow and fast muscle precursor cells in vivo, prior to somite formation. *Development* **122** (11): 3371–3380.

22 Stellabotte, F., Dobbs-McAuliffe, B., Fernandez, D.A. et al. (2007). Dynamic somite cell rearrangements lead to distinct waves of myotome growth. *Development* **134** (7): 1253–1257.

23 Daggett, D.F., Domingo, C.R., Currie, P.D. et al. (2007). Control of morphogenetic cell movements in the early zebrafish myotome. *Dev. Biol.* **309** (2): 169–179.

24 Fernandes, J.M., Mackenzie, M.G., Elgar, G. et al. (2005). A genomic approach to reveal novel genes associated with myotube formation in the model teleost, *Takifugu rubripes. Physiol. Genom.* **22** (3): 327–338.

25 Riggs, A.D. (1975). X inactivation, differentiation, and DNA methylation. *Cytogenet. Cell Genet.* **14** (1): 9–25.

26 Holliday, R. and Pugh, J.E. (1975). DNA modification mechanisms and gene activity during development. *Science* **187** (4173): 226–232.

27 Taylor, S.M. and Jones, P.A. (1982). Changes in phenotypic expression in embryonic and adult cells treated with 5-azacytidine. *J. Cell. Physiol.* **111** (2): 187–194.

28 Taylor, S.M. and Jones, P.A. (1979). Multiple new phenotypes induced in 10T1/2 and 3T3 cells treated with 5-azacytidine. *Cell* **17** (4): 771–779.

29 Stein, R., Razin, A., and Cedar, H. (1982). in vitro methylation of the hamster adenine phosphoribosyltransferase gene inhibits its expression in mouse L cells. *Proc. Natl. Acad. Sci. U. S. A.* **79** (11): 3418–3422.

30 Busslinger, M., Hurst, J., and Flavell, R.A. (1983). DNA methylation and the regulation of globin gene expression. *Cell* **34** (1): 197–206.

31 Konieczny, S.F. and Emerson, C.P. Jr. (1984). 5-Azacytidine induction of stable mesodermal stem cell lineages from 10T1/2 cells: evidence for regulatory genes controlling determination. *Cell* **38** (3): 791–800.

32 Brunk, B.P., Goldhamer, D.J., and Emerson, C.P. Jr. (1996). Regulated demethylation of the myoD distal enhancer during skeletal myogenesis. *Dev. Biol.* **177** (2): 490–503.

33 Campos, C., Valente, L.M., Conceicao, L.E. et al. (2013). Temperature affects methylation of the myogenin putative promoter, its expression and muscle cellularity in Senegalese sole larvae. *Epigenetics* **8** (4): 389–397.

34 Campos, C., Valente, L.M., and Fernandes, J.M. (2012). Molecular evolution of zebrafish dnmt3 genes and thermal plasticity of their expression during embryonic development. *Gene* **500** (1): 93–100.

35 Naguibneva, I., Ameyar-Zazoua, M., Polesskaya, A. et al. (2006). The microRNA miR-181 targets the homeobox protein Hox-A11 during mammalian myoblast differentiation. *Nat. Cell Biol.* **8** (3): 278–284.

36 Rao, P.K., Kumar, R.M., Farkhondeh, M. et al. (2006). Myogenic factors that regulate expression of muscle-specific microRNAs. *Proc. Natl. Acad. Sci. U. S. A.* **103** (23): 8721–8726.

37 Campos, C., Sundaram, A.Y., Valente, L.M. et al. (2014). Thermal plasticity of the miRNA transcriptome during Senegalese sole development. *BMC Genomics* **15**: 525.

38 Yan, B., Zhu, C.D., Guo, J.T. et al. (2013). miR-206 regulates the growth of the teleost tilapia (*Oreochromis niloticus*) through the modulation of IGF-1 gene expression. *J. Exp. Biol.* **216** (Pt 7): 1265–1269.

39 Zhou, Y., Zhou, B., Pache, L. et al. (2019). Metascape provides a biologist-oriented resource for the analysis of systems-level datasets. *Nat. Commun.* **10** (1): 1523.

40 Das, A., Das, A., Das, D. et al. (2020). Circular RNAs in myogenesis. *Biochim. Biophys. Acta Gene Regul. Mech.* **1863** (4): 194372.

41 Rbbani, G., Nedoluzhko, A., Galindo-Villegas, J. et al. (2021). Function of circular RNAs in fish and their potential application as biomarkers. *Int. J. Mol. Sci.* **22** (13).

42 Zhu, M., Liu, J., Xiao, J. et al. (2017). Lnc-mg is a long non-coding RNA that promotes myogenesis. *Nat. Commun.* **8**: 14718.

43 Jin, J.J., Lv, W., Xia, P. et al. (2018). Long noncoding RNA SYISL regulates myogenesis by interacting with polycomb repressive complex 2. *Proc. Natl. Acad. Sci. U. S. A.* **115** (42): E9802–E9811.

44 Sui, Y., Han, Y., Zhao, X. et al. (2019). Long non-coding RNA Irm enhances myogenic differentiation by interacting with MEF2D. *Cell Death Dis.* **10** (3): 181.

45 Dong, A., Preusch, C.B., So, W.K. et al. (2020). A long noncoding RNA, LncMyoD, modulates chromatin accessibility to regulate muscle stem cell myogenic lineage progression. *Proc. Natl. Acad. Sci. U. S. A.* **117** (51): 32464–32475.

46 Mathavan, S., Lee, S.G., Mak, A. et al. (2005). Transcriptome analysis of zebrafish embryogenesis using microarrays. *PLoS Genet.* **1** (2): 260–276.

47 Parichy, D.M., Elizondo, M.R., Mills, M.G. et al. (2009). Normal table of postembryonic zebrafish development: staging by externally visible anatomy of the living fish. *Dev. Dyn.* **238** (12): 2975–3015.

48 Hollway, G.E., Bryson-Richardson, R.J., Berger, S. et al. (2007). Whole-somite rotation generates muscle progenitor cell compartments in the developing zebrafish embryo. *Dev. Cell* **12** (2): 207–219.

49 Stellabotte, F. and Devoto, S.H. (2007). The teleost dermomyotome. *Dev. Dyn.* **236** (9): 2432–2443.

50 Rowlerson, A. and Veggetti, A. (2001). Cellular mechanisms of post-embryonic muscle growth in aquaculture species. In: *Muscle Development and Growth* (ed. I.A. Johnston), 103–140. San Diego: Academic Press.

51 Campos, C., Valente, L.M., Borges, P. et al. (2010). Dietary lipid levels have a remarkable impact on the expression of growth-related genes in Senegalese sole (*Solea senegalensis* Kaup). *J. Exp. Biol.* **213** (2): 200–209.

52 Canada, P., Engrola, S., Mira, S. et al. (2016). The supplementation of a microdiet with crystalline indispensable amino-acids affects muscle growth and the expression pattern of related genes in Senegalese sole (*Solea senegalensis*) larvae. *Aquaculture* **458**: 158–159.

53 Hagen, O., Fernandes, J.M., Solberg, C. et al. (2009). Expression of growth-related genes in muscle during fasting and refeeding of juvenile Atlantic halibut, *Hippoglossus hippoglossus* L. *Comp. Biochem. Physiol. B Biochem. Mol. Biol.* **152** (1): 47–53.

54 Skjaerven, K.H., Jakt, L.M., Fernandes, J.M.O. et al. (2018). Parental micronutrient deficiency distorts liver DNA methylation and expression of lipid genes associated with a fatty-liver-like phenotype in offspring. *Sci. Rep.* **8** (1): 3055.

55 Campos, C., Fernandes, J.M.O., Conceição, L.E.C. et al. (2013). Thermal conditions during larval pelagic phase influence subsequent somatic growth of Senegalese sole by modulating gene expression and muscle growth dynamics. *Aquaculture* **414–415**: 46–55.

56 Nagasawa, K., Giannetto, A., and Fernandes, J.M. (2012). Photoperiod influences growth and mll (mixed-lineage leukaemia) expression in Atlantic cod (*Gadus morhua*). *PLoS One* **7** (5): e36908.

57 Giannetto, A., Nagasawa, K., Fasulo, S. et al. (2013). Influence of photoperiod on expression of DNA (cytosine-5) methyltransferases in Atlantic cod (*Gadus morhua*). *Gene* **519** (2): 222–230.

58 Burgerhout, E., Mommens, M., Johnsen, H. et al. (2017). Genetic background and embryonic temperature affect DNA methylation and expression of myogenin and muscle development in Atlantic salmon (*Salmo salar*). *PLoS One* **12** (6): e0179918.

59 Fuentes, E.N., Zuloaga, R., Nardocci, G. et al. (2014). Skeletal muscle plasticity induced by seasonal acclimatization in carp involves differential expression of rRNA and molecules that epigenetically regulate its synthesis. *Comp. Biochem. Physiol. B Biochem. Mol. Biol.* **172–173**: 57–66.

60 Campos, C., Valente, L.M., Conceicao, L.E. et al. (2013). Incubation temperature induces changes in muscle cellularity and gene expression in Senegalese sole (*Solea senegalensis*). *Gene* **516** (2): 209–217.

61 Anastasiadi, D., Diaz, N., and Piferrer, F. (2017). Small ocean temperature increases elicit stage-dependent changes in DNA methylation and gene expression in a fish, the European sea bass. *Sci. Rep.* **7** (1): 12401.

62 Koganti, P.P., Wang, J., Cleveland, B. et al. (2017). 17beta-estradiol increases non-CpG methylation in exon 1 of the rainbow trout (*Oncorhynchus mykiss*) MyoD gene. *Mar. Biotechnol. (NY)* **19** (4): 321–327.

63 Konstantinidis, I., Saetrom, P., Mjelle, R. et al. (2020). Major gene expression changes and epigenetic remodelling in Nile tilapia muscle after just one generation of domestication. *Epigenetics* **15** (10): 1052–1067.

64 Podgorniak, T., Dhanasiri, A., Chen, X. et al. (2022). Early fish domestication affects methylation of key genes involved in the rapid onset of the farmed phenotype. *Epigenetics* **17**: 1281–1298.

65 Podgorniak, T., Brockmann, S., Konstantinidis, I. et al. (2019). Differences in the fast muscle methylome provide insight into sex-specific epigenetic regulation of growth in Nile tilapia during early stages of domestication. *Epigenetics* **14** (8): 818–836.

66 Konstantinidis, I., Anastasiadi, D., Saetrom, P. et al. (2021). Epigenetic mapping of the somatotropic axis in Nile tilapia reveals differential DNA hydroxymethylation marks associated with growth. *Genomics* **113** (5): 2953–2964.

67 Hall, B.K. (2005). *Bones and Cartilage: Developmental and Evolutionary Skeletal Biology*. Academic Press.

68 Lagler, K., Bardach, J., Miller, R., and Passino, D.M. (1987). *Ichthyology*, 2e. USA: Wiley.

69 Davesne, D., Meunier, F.J., Schmitt, A.D. et al. (2019). The phylogenetic origin and evolution of acellular bone in teleost fishes: insights into osteocyte function in bone metabolism. *Biol. Rev.* **94** (4): 1338–1363.

70 Kölliker, A. (1859). On the different types in the microscopic structure of the skeleton of osseous fishes. *Proc. R. Soc. Lond.* **9**: 656–668.

71 Ekanayake, S. and Hall, B.K. (1987). The development of acellularity of the vertebral bone of the Japanese medaka, *Oryzias latipes* (Teleostei; Cyprinidontidae). *J. Morphol.* **193** (3): 253–261.

72 Hall, B.K. and Witten, P.E. (2007). Plasticity of and transitions between skeletal tissues in vertebrate evolution and development. In: *Major Transitions in Vertebrate Evolution Bloomington* (ed. J.S. Anderson and H.D. Sues), 13–56. Indiana University Press.

73 Meunier, F.J. and Huysseune, A. (1992). The concept of bone tissue in Osteichthyes. *Neth. J. Zool.* **42**: 445–458.

74 Moss, M.L. (1961). Osteogenesis of acellular teleost fish bone. *Am. J. Anat.* **108**: 99–110.

75 Weiss, R. and Watabe, N. (1979). Studies on the biology of fish bone. III. Ultrastructure of osteogenesis and resorption in osteocytic (cellular) and anosteocytic (acellular) bones. *Calcif. Tissue Int.* **28** (1): 43–56.

76 Huysseune, A. and Sire, J.Y. (1998). Evolution of patterns and processes in teeth and tooth-related tissues in non-mammalian vertebrates. *Eur. J. Oral Sci.* **106** (S1): 437–481.

77 Sire, J.Y., Donoghue, P.C., and Vickaryous, M.K. (2009). Origin and evolution of the integumentary skeleton in non-tetrapod vertebrates. *J. Anat.* **214** (4): 409–440.

78 Wagner, D.O. and Aspenberg, P. (2011). Where did bone come from? *Acta Orthop.* **82** (4): 393–398.

79 Donoghue, P.C., Sansom, I.J., and Downs, J.P. (2006). Early evolution of vertebrate skeletal tissues and cellular interactions, and the canalization of skeletal development. *J. Exp. Zool. B Mol. Dev. Evol.* **306** (3): 278–294.

80 Haridy, Y., Osenberg, M., Hilger, A. et al. (2021). Bone metabolism and evolutionary origin of osteocytes: novel application of FIB-SEM tomography. *Sci. Adv.* **7** (14): eabb9113.

81 O'Shea, J., Keating, J.N., and Donoghue, P.C.J. (2019). The dermal skeleton of the jawless vertebrate *Tremataspis mammillata* (Osteostraci, stem-Gnathostomata). *J. Morphol.* **280** (7): 999–1025.

82 Moss, M.L. (1961). Studies on the acellular bone of teleost fish. I. Morphological and systematic variations. *Acta Anat.* **46**: 343–362.

83 Hirasawa, T. and Kuratani, S. (2015). Evolution of the vertebrate skeleton: morphology, embryology, and development. *Zool. Lett.* **1** (1): 1–17.

84 Huxley, T.H. (1864). *Lectures on the Elements of Comparative Anatomy: On the Classification of Animals and on the Vertebrate Skull*. John Churchill.

85 Arratia, G., Schultze, H.P., and Casciotta, J. (2001). Vertebral column and associated elements in dipnoans and comparison with other fishes: development and homology. *J. Morphol.* **250** (2): 101–172.

86 Criswell, K.E., Coates, M.I., and Gillis, J.A. (2017). Embryonic origin of the gnathostome vertebral skeleton. *Proc. R. Soc. B Biol. Sci.* **284** (1867): 20172121.

87 Bensimon-Brito, A., Cardeira, J., Cancela, M.L. et al. (2012). Distinct patterns of notochord mineralization in zebrafish coincide with the localization of Osteocalcin isoform 1 during early vertebral centra formation. *BMC Dev. Biol.* **12**: 28.

88 Grotmol, S., Nordvik, K., Kryvi, H. et al. (2005). A segmental pattern of alkaline phosphatase activity within the notochord coincides with the initial formation of the vertebral bodies. *J. Anat.* **206** (5): 427–436.

89 Morin-Kensicki, E.M., Melancon, E., and Eisen, J.S. (2002). Segmental relationship between somites and vertebral column in zebrafish. *Development* **129** (16): 3851–3860.

90 Fleming, A., Keynes, R., and Tannahill, D. (2004). A central role for the notochord in vertebral patterning. *Development* **131** (4): 873–880.

91 Dash, S. and Trainor, P.A. (2020). The development, patterning and evolution of neural crest cell differentiation into cartilage and bone. *Bone* **137**: 115409.

92 Anastasiadi, D. and Piferrer, F. (2019). Epimutations in developmental genes underlie the onset of domestication in farmed European sea bass. *Mol. Biol. Evol.* **36** (10): 2252–2264.

93 Laizé, V., Gavaia, P.J., and Cancela, M.L. (2014). Fish: a suitable system to model human bone disorders and discover drugs with osteogenic or osteotoxic activities. *Drug Discov. Today Dis. Models* **13**: 29–37.

94 Javidan, Y. and Schilling, T.F. (2004). Development of cartilage and bone. *Methods Cell Biol.* **76**: 415–436.

95 Boglione, C., Gavaia, P., Koumoundouros, G. et al. (2013). A review on skeletal anomalies in reared European larvae and juveniles. Part 1: normal and anomalous skeletogenic processes. *Rev. Aquac.* **5**: S99–S120.

96 Giovannone, D., Paul, S., Schindler, S. et al. (2019). Programmed conversion of hypertrophic chondrocytes into osteoblasts and marrow adipocytes within zebrafish bones. *elife* **8**: e42736.

97 Bertin, L. (1958). Tissus squeletiques. In: *Taité de Zoologie – Anatomie, Systématique,Biologie* (ed. P.P. Grassé), 532–550. Paris: Masson et Cie editeurs.

98 Shahar, R. and Dean, M.N. (2013). The enigmas of bone without osteocytes. *Bonekey Rep.* **2**: 343.

99 Witten, P.E. and Huysseune, A. (2009). A comparative view on mechanisms and functions of skeletal remodelling in teleost fish, with special emphasis on osteoclasts and their function. *Biol. Rev. Camb. Philos. Soc.* **84** (2): 315–346.

100 Rosa, J.T., Laize, V., Gavaia, P.J. et al. (2021). Fish models of induced osteoporosis. *Front. Cell. Dev. Biol.* **9**: 672424.

101 Ducy, P. and Karsenty, G. (1998). Genetic control of cell differentiation in the skeleton. *Curr. Opin. Cell Biol.* **10** (5): 614–619.

102 Lian, J.B., Stein, G.S., van Wijnen, A.J. et al. (2012). MicroRNA control of bone formation and homeostasis. *Nat. Rev. Endocrinol.* **8** (4): 212–227.

103 Jadin, K.D., Wong, B.L., Bae, W.C. et al. (2005). Depth-varying density and organization of chondrocytes in immature and mature bovine articular cartilage assessed by 3d imaging and analysis. *J. Histochem. Cytochem.* **53** (9): 1109–1119.

104 Junqueira, L.C., Carneiro, J., and Kelley, R.O. (1995). *Basic Histology*, 8e. USA: Appleton & Lange.

105 Gartner, L.P. and Hiatt, J.L. (1999). *Tratado de Histologia em Cores*, 1e. Rio de Janeiro: Guanabara Koogan.

106 Benjamin, M. (1989). The development of hyaline-cell cartilage in the head of the black molly, *Poecilia sphenops*. Evidence for secondary cartilage in a teleost. *J. Anat.* **164**: 145–154.

107 Benjamin, M. and Ralphs, J.R. (1991). Extracellular matrix of connective tissues in the heads of teleosts. *J. Anat.* **179**: 137–148.

108 Witten, P.E., Huysseune, A., and Hall, B.K. (2010). A practical approach for the identification of the many cartilaginous tissues in teleost fish. *J. Appl. Ichthyol.* **26**: 257–262.

109 Shine, R. (1979). Sexual selection and sexual dimorphism in the amphibia. *Copeia* **2**: 297–306.

110 Owens, I.P.F. and Hartley, I.R. (1998). Sexual dimorphism in birds: why are there so many different forms of dimorphism? *Proc. R. Soc. Lond. B* **265**: 397–407.

111 Colwell, R.K. (2000). Rensch's rule crosses the line: convergent allometry of sexual size dimorphism in hummingbirds and flower mites. *Am. Nat.* **156** (5): 495–510.

112 Kratochvil, L. and Frynta, D. (2002). Body size, male combat and the evolution of sexual dimorphism in eublepharid geckos (Squamata: Eublepharidae). *Biol. J. Linn. Soc.* **76** (2): 303–314.

113 Johansson, F., Crowley, P.H., and Brodin, T. (2005). Sexual size dimorphism and sex ratios in dragonflies (Odonata). *Biol. J. Linn. Soc.* **86**: 507–513.

114 Herler, J., Kerschbaumer, M., Mitteroecker, P. et al. (2010). Sexual dimorphism and population divergence in the Lake Tanganyika cichlid fish genus *Tropheus*. *Front. Zool.* **7**: 4.

115 Santos, E.S. and Machado, G. (2016). Sexual dichromatism in wing pigmentation of New World dragonflies follows Rensch's rule. *J. Evol. Biol.* **29** (7): 1447–1454.

116 Karp, N.A., Mason, J., Beaudet, A.L. et al. (2017). Prevalence of sexual dimorphism in mammalian phenotypic traits. *Nat. Commun.* **8**: 15475.

117 Shine, R. (1989). Ecological causes for the evolution of sexual dimorphism: a review of the evidence. *Q. Rev. Biol.* **64** (4): 419–461.

118 Darwin, C., Murray, J., William, C. et al. (1871). *The Descent of Man: And Selection in Relation to Sex*, 1e, 2 issue ed. Albermarle Street, London: John Murray.

119 Laporte, M., Berrebi, P., Claude, J. et al. (2018). The ecology of sexual dimorphism in size and shape of the freshwater blenny *Salaria fluviatilis*. *Curr. Zool.* **64** (2): 183–191.

120 De Lisle, S.P. (2019). Understanding evolution ecological sex differences: integrating character displacement and the Darwin-Bateman paradigm. *Evol. Lett.* **3** (5): 434–447.

121 Swann, J.B., Holland, S.J., Petersen, M. et al. (2020). The immunogenetics of sexual parasitism. *Science* **369** (6511): 1608–1615.

122 Beardmore, J.A., Mair, G.C., and Lewis, R.I. (2001). Monosex male production in finfish as exemplified by tilapia: applications, problems, and prospects. *Aquaculture* **197** (1): 283–301.

123 Lind, C.E., Safari, A., Agyakwah, S.K. et al. (2015). Differences in sexual size dimorphism among farmed tilapia species and strains undergoing genetic improvement for body weight. *Aquacult. Rep.* **1**: 20–27.

124 Rutten, M.J.M., Komen, H., and Bovenhuis, H. (2005). Longitudinal genetic analysis of Nile tilapia (*Oreochromis niloticus* L.) body weight using a random regression model. *Aquaculture* **246** (1): 101–113.

125 Toguyeni, A., Fauconneau, B.T., Fostier, A. et al. (2002). Influence of sexual phenotype and genotype, and sex ratio on growth performances in tilapia, *Oreochromis niloticus*. *Aquaculture* **207** (3): 249–261.

126 Wang, D., Mao, H.L., Chen, H.X. et al. (2009). Isolation of Y- and X-linked SCAR markers in yellow catfish and application in the production of all-male populations. *Anim. Genet.* **40** (6): 978–981.

127 Ma, W., Wu, J., Zhang, J. et al. (2016). Sex differences in the expression of GH/IGF axis genes underlie sexual size dimorphism in the yellow catfish (*Pelteobagrus fulvidraco*). *Sci. China Life Sci.* **59** (4): 431–433.

128 Yang, Y., Zhou, T., Liu, Y. et al. (2022). Identification of an epigenetically marked locus within the sex determination region of channel catfish. *Int. J. Mol. Sci.* **23** (10): 5471.

129 Mei, J. and Gui, J.F. (2015). Genetic basis and biotechnological manipulation of sexual dimorphism and sex determination in fish. *Sci. China Life Sci.* **58** (2): 124–136.

130 Zhai, G., Shu, T., Chen, K. et al. (2022). Successful production of an all-female common carp (*Cyprinus carpio* L.) population using cyp17a1-deficient neomale carp. *Engineering* **8**: 181–189.

131 Faggion, S., Vandeputte, M., Vergnet, A. et al. (2021). Sex dimorphism in European sea bass (*Dicentrarchus labrax* L.): new insights into sex-related growth patterns during very early life stages. *PLoS One* **16** (4): e0239791.

132 Anastasiadi, D., Vandeputte, M., Sanchez-Baizan, N. et al. (2018). Dynamic epimarks in sex-related genes predict gonad phenotype in the European sea bass, a fish with mixed genetic and environmental sex determination. *Epigenetics* **13** (9): 988–1011.

133 Navarro-Martin, L., Vinas, J., Ribas, L. et al. (2011). DNA methylation of the gonadal aromatase (*cyp19a*) promoter is involved in temperature-dependent sex ratio shifts in the European sea bass. *PLoS Genet.* **7** (12): e1002447.

134 Shao, C., Li, Q., Chen, S. et al. (2014). Epigenetic modification and inheritance in sexual reversal of fish. *Genome Res.* **24** (4): 604–615.

135 Wan, Z.Y., Xia, J.H., Lin, G. et al. (2016). Genome-wide methylation analysis identified sexually dimorphic methylated regions in hybrid tilapia. *Sci. Rep.* **6**: 35903.

136 Nestor, C.E., Ottaviano, R., Reddington, J. et al. (2012). Tissue type is a major modifier of the 5-hydroxymethylcytosine content of human genes. *Genome Res.* **22** (3): 467–477.

137 Lister, R., Mukamel, E.A., Nery, J.R. et al. (2013). Global epigenomic reconfiguration during mammalian brain development. *Science* **341** (6146): 1237905.

138 Kamstra, J.H., Loken, M., Alestrom, P. et al. (2015). Dynamics of DNA hydroxymethylation in zebrafish. *Zebrafish* **12** (3): 230–237.

139 Cisternas, C.D., Cortes, L.R., Bruggeman, E.C. et al. (2020). Developmental changes and sex differences in DNA methylation and demethylation in hypothalamic regions of the mouse brain. *Epigenetics* **15** (1–2): 72–84.

140 Xiao, J., Zhong, H., Zhou, Y. et al. (2014). Identification and characterization of microRNAs in ovary and testis of Nile tilapia (*Oreochromis niloticus*) by using solexa sequencing technology. *PLoS One* **9** (1): e86821.

141 Juanchich, A., Le Cam, A., Montfort, J. et al. (2013). Identification of differentially expressed miRNAs and their potential targets during fish ovarian development. *Biol. Reprod.* **88** (5): 128.

142 Bizuayehu, T.T., Babiak, J., Norberg, B. et al. (2012). Sex-biased miRNA expression in Atlantic halibut (*Hippoglossus hippoglossus*) brain and gonads. *Sex. Dev.* **6** (5): 257–266.

143 Gay, S., Bugeon, J., Bouchareb, A. et al. (2018). MiR-202 controls female fecundity by regulating medaka oogenesis. *PLoS Genet.* **14** (9): e1007593.

144 Davalos, A., Goedeke, L., Smibert, P. et al. (2011). miR-33a/b contribute to the regulation of fatty acid metabolism and insulin signaling. *Proc. Natl. Acad. Sci. U. S. A.* **108** (22): 9232–9237.

145 Hu, Z., Shen, W.J., Kraemer, F.B. et al. (2017). Regulation of adrenal and ovarian steroidogenesis by miR-132. *J. Mol. Endocrinol.* **59** (3): 269–283.

146 Wang, W., Liu, W., Liu, Q. et al. (2016). Coordinated microRNA and messenger RNA expression profiles for understanding sexual dimorphism of gonads and the potential roles of microRNA in the steroidogenesis pathway in Nile tilapia (*Oreochromis niloticus*). *Theriogenology* **85** (5): 970–978.

147 Pallan, P.S., Nagy, L.D., Lei, L. et al. (2015). Structural and kinetic basis of steroid 17alpha,20-lyase activity in teleost fish cytochrome P450 17A1 and its absence in cytochrome P450 17A2. *J. Biol. Chem.* **290** (6): 3248–3268.

148 Tao, W., Sun, L., Shi, H. et al. (2016). Integrated analysis of miRNA and mRNA expression profiles in tilapia gonads at an early stage of sex differentiation. *BMC Genom.* **17**: 328.

149 Lau, E.S., Zhang, Z., Qin, M. et al. (2016). Knockout of zebrafish ovarian aromatase gene (*cyp19a1a*) by TALEN and CRISPR/Cas9 leads to all-male offspring due to failed ovarian differentiation. *Sci. Rep.* **6**: 37357.

150 Eshel, O., Shirak, A., Dor, L. et al. (2014). Identification of male-specific amh duplication, sexually differentially expressed genes and microRNAs at early embryonic development of Nile tilapia (*Oreochromis niloticus*). *BMC Genomics* **15**: 774.

151 Griffiths-Jones, S., Hui, J.H., Marco, A. et al. (2011). MicroRNA evolution by arm switching. *EMBO Rep.* **12** (2): 172–177.

152 Pinhal, D., Bovolenta, L.A., Moxon, S. et al. (2018). Genome-wide microRNA screening in Nile tilapia reveals pervasive isomiRs' transcription, sex-biased arm switching and increasing complexity of expression throughout development. *Sci. Rep.* **8** (1): 8248.

153 Huang, C.W., Li, Y.H., Hu, S.Y. et al. (2012). Differential expression patterns of growth-related microRNAs in the skeletal muscle of Nile tilapia (*Oreochromis niloticus*). *J. Anim. Sci.* **90** (12): 4266–4279.

154 Duran, B.O., Fernandez, G.J., Mareco, E.A. et al. (2015). Differential microRNA expression in fast- and slow-twitch skeletal muscle of *Piaractus mesopotamicus* during growth. *PLoS One* **10** (11): e0141967.

155 Soltanoff, C.S., Yang, S., Chen, W. et al. (2009). Signaling networks that control the lineage commitment and differentiation of bone cells. *Crit. Rev. Eukaryot. Gene Expr.* **19** (1): 1–46.

156 Gaur, T., Hussain, S., Mudhasani, R. et al. (2010). Dicer inactivation in osteoprogenitor cells compromises fetal survival and bone formation, while excision in differentiated osteoblasts increases bone mass in the adult mouse. *Dev. Biol.* **340** (1): 10–21.

157 Roberto, V.P., Gavaia, P., Nunes, M.J. et al. (2018). Evidences for a new role of miR-214 in chondrogenesis. *Sci. Rep.* **8** (1): 3704.

158 Zhao, X., Xu, D., Li, Y. et al. (2014). MicroRNAs regulate bone metabolism. *J. Bone Miner. Metab.* **32** (3): 221–231.

159 Hassan, M.Q., Gordon, J.A., Beloti, M.M. et al. (2010). A network connecting Runx2, SATB2, and the miR-23a~27a~24-2 cluster regulates the osteoblast differentiation program. *Proc. Natl. Acad. Sci. U. S. A.* **107** (46): 19879–19884.

160 Li, Y.P., Wei, X.C., Li, P.C. et al. (2015). The role of miRNAs in cartilage homeostasis. *Curr. Genom.* **16** (6): 393–404.

161 Zhang, Y., Xie, R.L., Gordon, J. et al. (2012). Control of mesenchymal lineage progression by microRNAs targeting skeletal gene regulators Trps1 and Runx2. *J. Biol. Chem.* **287** (26): 21926–21935.

162 Bhushan, R., Grünhagen, J., Becker, J. et al. (2013). miR-181a promotes osteoblastic differentiation through repression of TGF-β signaling molecules. *Int. J. Biochem. Cell Biol.* **45** (3): 696–705.

163 Wan, S.M., Yi, S.K., Zhong, J. et al. (2015). Identification of microRNA for intermuscular bone development in blunt snout bream (*Megalobrama amblycephala*). *Int. J. Mol. Sci.* **16** (5): 10686–10703.

164 Kloosterman, W.P., Steiner, F.A., Berezikov, E. et al. (2006). Cloning and expression of new microRNAs from zebrafish. *Nucleic Acids Res.* **34** (9): 2558–2569.

165 Swingler, T.E., Wheeler, G., Carmont, V. et al. (2012). The expression and function of microRNAs in chondrogenesis and osteoarthritis. *Arthritis Rheum.* **64** (6): 1909–1919.

166 Mei, Y., Bian, C., Li, J. et al. (2013). miR-21 modulates the ERK-MAPK signaling pathway by regulating SPRY2 expression during human mesenchymal stem cell differentiation. *J. Cell. Biochem.* **114** (6): 1374–1384.

167 Roberto, V.P., Tiago, D.M., Gautvik, K. et al. (2015). Evidence for the conservation of miR-223 in zebrafish (*Danio rerio*): implications for function. *Gene* **566** (1): 54–62.

168 Desvignes, T., Contreras, A., and Postlethwait, J.H. (2014). Evolution of the miR199-214 cluster and vertebrate skeletal development. *RNA Biol.* **11** (4): 281–294.

169 Gan, S., Huang, Z., Liu, N. et al. (2016). MicroRNA-140-5p impairs zebrafish embryonic bone development via targeting BMP-2. *FEBS Lett.* **590** (10): 1438–1446.

170 Salazar, V.S., Gamer, L.W., and Rosen, V. (2016). BMP signalling in skeletal development, disease and repair. *Nat. Rev. Endocrinol.* **12** (4): 203–221.

171 Ning, G., Liu, X., Dai, M. et al. (2013). MicroRNA-92a upholds bmp signaling by targeting noggin3 during pharyngeal cartilage formation. *Dev. Cell* **24** (3): 283–295.

172 Tiago, D.M., Marques, C.L., Roberto, V.P. et al. (2014). Mir-20a regulates in vitro mineralization and BMP signaling pathway by targeting BMP-2 transcript in fish. *Arch. Biochem. Biophys.* **543**: 23–30.

173 DeLaurier, A., Nakamura, Y., Braasch, I. et al. (2012). Histone deacetylase-4 is required during early cranial neural crest development for generation of the zebrafish palatal skeleton. *BMC Dev. Biol.* **12**: 16.

174 Pillai, R., Coverdale, L.E., Dubey, G. et al. (2004). Histone deacetylase 1 (HDAC-1) required for the normal formation of craniofacial cartilage and pectoral fins of the zebrafish. *Dev. Dyn.* **231** (3): 647–654.

175 Spinelli, J.B. and Haigis, M.C. (2018). The multifaceted contributions of mitochondria to cellular metabolism. *Nat. Cell Biol.* **20** (7): 745–754.

176 Castellana, S., Vicario, S., and Saccone, C. (2011). Evolutionary patterns of the mitochondrial genome in Metazoa: exploring the role of mutation and selection in mitochondrial protein coding genes. *Genome Biol. Evol.* **3**: 1067–1079.

177 Kayal, E., Bentlage, B., Collins, A.G. et al. (2012). Evolution of linear mitochondrial genomes in medusozoan cnidarians. *Genome Biol. Evol.* **4** (1): 1–12.

178 Boore, J.L. (1999). Animal mitochondrial genomes. *Nucleic Acids Res.* **27** (8): 1767–1780.

179 Macken, W.L., Lucassen, A.M., Hanna, M.G. et al. (2021). Mitochondrial DNA variants in genomic data: diagnostic uplifts and predictive implications. *Nat. Rev. Genet.* **22** (9): 547–548.

180 Pakendorf, B. and Stoneking, M. (2005). Mitochondrial DNA and human evolution. *Annu. Rev. Genom. Hum. Genet.* **6**: 165–183.

181 Shtolz, N. and Mishmar, D. (2019). The mitochondrial genome – on selective constraints and signatures at the organism, cell, and single mitochondrion levels. *Front. Ecol. Evol.* **7**: 342.

182 Stewart, J.B. and Chinnery, P.F. (2021). Extreme heterogeneity of human mitochondrial DNA from organelles to populations. *Nat. Rev. Genet.* **22** (2): 106–118.

183 Nedoluzhko, A., Mjelle, R., Renstrom, M. et al. (2021). The first mitochondrial 5-methylcytosine map in a non-model teleost (*Oreochromis niloticus*) reveals extensive strand-specific and non-CpG methylation. *Genomics* **113** (5): 3050–3057.

184 Groot, G.S. and Kroon, A.M. (1979). Mitochondrial DNA from various organisms does not contain internally methylated cytosine in -CCGG- sequences. *Biochim. Biophys. Acta* **564** (2): 355–357.

185 Rebelo, A.P., Williams, S.L., and Moraes, C.T. (2009). in vivo methylation of mtDNA reveals the dynamics of protein-mtDNA interactions. *Nucleic Acids Res.* **37** (20): 6701–6715.

186 Shock, L.S., Thakkar, P.V., Peterson, E.J. et al. (2011). DNA methyltransferase 1, cytosine methylation, and cytosine hydroxymethylation in mammalian mitochondria. *Proc. Natl. Acad. Sci. U. S. A.* **108** (9): 3630–3635.

187 Gao, D., Zhu, B., Sun, H., and Wang, X. (2017). Mitochondrial DNA methylation and related disease. In: *Mitochondrial DNA and Diseases Advances in Experimental Medicine and Biology* (ed. H. Sun and X. Wang), 117–132. Singapore: Springer.

188 Smallwood, S.A., Lee, H.J., Angermueller, C. et al. (2014). Single-cell genome-wide bisulfite sequencing for assessing epigenetic heterogeneity. *Nat. Methods* **11** (8): 817–820.

189 Yu, B., Dong, X., Gravina, S. et al. (2017). Genome-wide, single-cell DNA methylomics reveals increased non-CpG methylation during human oocyte maturation. *Stem Cell Rep.* **9** (1): 397–407.

190 Barres, R., Osler, M.E., Yan, J. et al. (2009). Non-CpG methylation of the PGC-1alpha promoter through DNMT3B controls mitochondrial density. *Cell Metab.* **10** (3): 189–198.

191 Liu, B., Du, Q., Chen, L. et al. (2016). CpG methylation patterns of human mitochondrial DNA. *Sci. Rep.* **6**: 23421.

192 Pollack, Y., Kasir, J., Shemer, R. et al. (1984). Methylation pattern of mouse mitochondrial DNA. *Nucleic Acids Res.* **12** (12): 4811–4824.

193 Devall, M., Smith, R.G., Jeffries, A. et al. (2017). Regional differences in mitochondrial DNA methylation in human post-mortem brain tissue. *Clin. Epigen.* **9**: 47.

194 Sharma, N., Pasala, M.S., and Prakash, A. (2019). Mitochondrial DNA: epigenetics and environment. *Environ. Mol. Mutagen.* **60** (8): 668–682.

195 Patil, V., Cuenin, C., Chung, F. et al. (2019). Human mitochondrial DNA is extensively methylated in a non-CpG context. *Nucleic Acids Res.* **47** (19): 10072–10085.

196 Wanner, N., Larsen, P.A., McLain, A. et al. (2021). The mitochondrial genome and epigenome of the golden lion tamarin from fecal DNA using nanopore adaptive sequencing. *BMC Genom.* **22** (1): 726.

197 Manev, H. and Dzitoyeva, S. (2013). Progress in mitochondrial epigenetics. *Biomol. Concepts* **4** (4): 381–389.

198 Rochard, P., Rodier, A., Casas, F. et al. (2000). Mitochondrial activity is involved in the regulation of myoblast differentiation through myogenin expression and activity of myogenic factors. *J. Biol. Chem.* **275** (4): 2733–2744.

199 Hartmann, N., Reichwald, K., Wittig, I. et al. (2011). Mitochondrial DNA copy number and function decrease with age in the short-lived fish *Nothobranchius furzeri*. *Aging Cell* **10** (5): 824–831.

200 Chandel, N.S. (2018). Mitochondria: back to the future. *Nat. Rev. Mol. Cell Biol.* **19** (2): 76.

201 Bhattacharya, D. and Scime, A. (2020). Mitochondrial function in muscle stem cell fates. *Front. Cell Dev. Biol.* **8**: 480.

202 Malinska, D., Kudin, A.P., Bejtka, M. et al. (2012). Changes in mitochondrial reactive oxygen species synthesis during differentiation of skeletal muscle cells. *Mitochondrion* **12** (1): 144–148.

203 Farge, G. and Falkenberg, M. (2019). Organization of DNA in mammalian mitochondria. *Int. J. Mol. Sci.* **20** (11): 2770.

204 Cavalcante, G.C., Magalhaes, L., Ribeiro-Dos-Santos, A. et al. (2020). Mitochondrial epigenetics: non-coding RNAs as a novel layer of complexity. *Int. J. Mol. Sci.* **21** (5): 1838.

205 Matilainen, O., Quiros, P.M., and Auwerx, J. (2017). Mitochondria and epigenetics – crosstalk in homeostasis and stress. *Trends Cell Biol.* **27** (6): 453–463.

206 Potok, M.E., Nix, D.A., Parnell, T.J. et al. (2013). Reprogramming the maternal zebrafish genome after fertilization to match the paternal methylation pattern. *Cell* **153** (4): 759–772.

207 Dou, X., Boyd-Kirkup, J.D., McDermott, J. et al. (2019). The strand-biased mitochondrial DNA methylome and its regulation by DNMT3A. *Genome Res.* **29** (10): 1622–1634.

208 Sirard, M.A. (2019). Distribution and dynamics of mitochondrial DNA methylation in oocytes, embryos and granulosa cells. *Sci. Rep.* **9** (1): 11937.

209 Waddington, C.H. (1957). The Strategy of the Genes. Routledge, London, UK. ISBN 9781138998131.

210 Rbbani, G., Nedoluzhko, A., Siriyappagouder, P., Sharko, F., Galindo-Villegas, J., Raeymaekers, J. A. M., Joshi, R., & Fernandes, J. M. O. (2023). The novel circular RNA CircMef2c is positively associated with muscle growth in Nile tilapia. *Genomics*, **115** (3), 110598. https://doi.org/10.1016/j.ygeno.2023.110598

10

Epigenetics in Fish Nutritional Programming

Kaja H. Skjærven, Anne-Catrin Adam, Takaya Saito, Rune Waagbø, and Marit Espe

Feed and Nutrition, Institute of Marine Research (IMR), Bergen, Norway

10.1 Epigenetic Basis of Nutritional Programming

Epigenetic regulation is responsible for controlling heritable changes in gene expression without alterations of the underlying DNA sequence [1]. The epigenetic chromatin landscape responds to various types of environmental factors and stressors and can lead to an acquired phenotype. Depending on the developmental stage, tissue type, or strength of the signals, abiotic factors can trigger a persistent phenotypic change or program the gene expression in such a way that it cannot be reverted at a later timepoint [2, 3]. For instance, adjustments in water temperature affect the sex ratio of populations epigenetically in several fish species [4–6].

Nutrition is one of the key factors that control epigenetic mechanisms [1]. In honeybees, differences in royal jelly intake during early development determine whether genetically identical larvae grow into queens or worker bees via adjusting the epigenetic mechanism [7]. In mice, maternal intake of vitamin C epigenetically affects the development of female fetal germ cells [8], and in fish, nutrients can influence epigenetic inheritance from parental diet to offspring phenotype [9, 10]. Epigenetic inheritance patterns controlled by nutritional programming (for definition, see Section 10.2 and Box 10.1) can potentially play a role at especially sensitive periods called critical windows, like broodstock, egg or larvae stages (Figure 10.1).

The epigenetic regulation, or heritable changes in gene expression, can be transferred either through mitosis or meiosis. For the former, epigenetic alterations can be established by regulating the chromatin within one generation. During mitosis, the cell undergoes cell division and copies the DNA as well as the epigenetic signature to both daughter cells, and thereby keeps the specified linage restrictions for gene expression, depending on their epigenetic profile. The expression of pluripotency genes is epigenetically decreased and the expression of lineage-specific genes is increased as development proceeds from the zygote to a specified somatic cell type [11]. Embryonic development, somatic stem cells, and primordial germ cells (PGCs) tightly control the epigenetic signature through the phases of proliferation to retain the cell-type identity. However, changes in nutrients surrounding these cells can influence the acquired phenotype, which might program the future phenotype of the daughter cells into a persistent phenotype later in life [12]. Three main classes of epigenetic mechanisms are DNA methylation, histone modifications, and noncoding RNAs (Figure 10.1).

10.1.1 DNA Methylation

DNA methylation is the most well-studied epigenetic mechanism [13]. Although both adenine and cytosine can be methylated, DNA methylation dominantly occurs at cytosines within CpG sites, where cytosine (C) is followed by guanine (G) in vertebrates [14, 15]. Cytosine methylation is often symmetric as the cytosines in both strands of DNA are simultaneously methylated in a CpG site. Methylation at the 5th carbon of cytosine (5-methylcytosine, 5mC) is reversible, which requires several steps from 5mC to unmethylated cytosine via 5-hydroxymethylcytosine (5hmC) [16, 17]. The primary function of DNA methylation is silencing [18, 19], but DNA methylation also mediates various types of genetic regulations in different genomic regions [20–22]. Furthermore, not only methylated regions, but also unmethylated regions can be genetic markers of gene regulation. For instance, highly expressed genes tend to have unmethylated clusters of CpGs, called CpG islands, in their promoters.

Epigenetics in Aquaculture, First Edition. Edited by Francesc Piferrer and Han-Ping Wang.

Box 10.1 Glossary Terms

Nutritional epigenetics: The effects of nutrition on epigenetic gene regulation.
Nutritional or metabolic programming: Nutritional or metabolic changes acquired during sensitive periods in early development that induce alterations in animal metabolism and physiology, as a result of changes at the cellular, molecular, and biochemical levels, even in the absence of the stimulus that initiated them [50, 115].
One-carbon (1C) metabolism: Integrated sequence of metabolic reactions including the folate cycle, methionine remethylation, and the trans-sulfuration pathways [96, 98].
Programming: Events early in life that influence long-term outcomes [50].

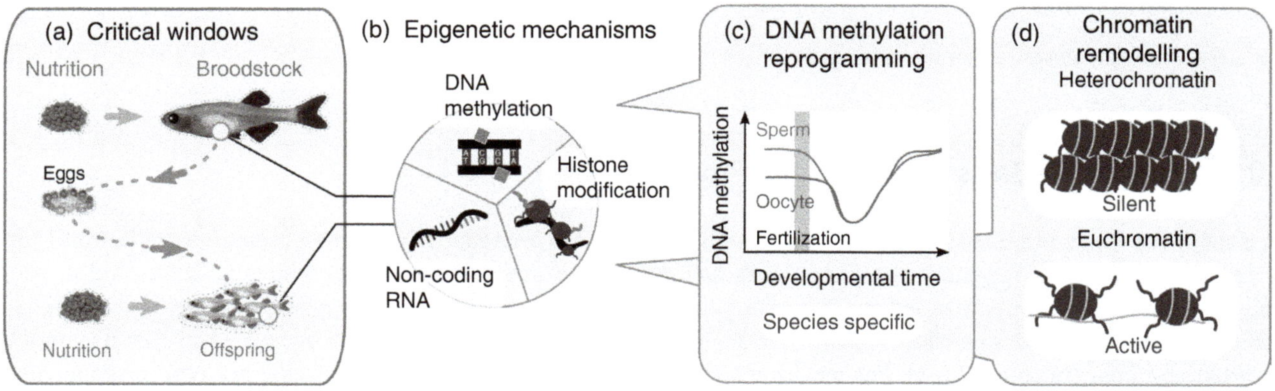

Figure 10.1 Critical windows for nutritional programming and alterations in epigenetic mechanisms. The figure shows the relationship between nutritional programming from broodstock to offspring and epigenetic regulation. (a) A schematic diagram showing critical windows for nutritional programming using an example of the female lineage in fish. (b) Three major classes of epigenetic mechanisms that are potentially controlled by nutritional programming. (c) An example of alterations of methylation rates in the early development stages around fertilization. Alteration patterns can be species-specific in fish. (d) Two types of chromatin structures that are associated with gene expression status (active vs. silent). (*See colour plate section for colour representation of this figure.*)

DNA methyltransferases (DNMTs) are enzymes that transfer a methyl group to DNA and regulate epigenetic inheritance. DNMT1 is responsible for the maintenance of methylation pattern as it transfers DNA methylation status to daughter cells during cell duplication [23]. DNMT3A and DNMT3B account for initiating new methylation as they catalyze de novo methylation [24]. After most 5mCs turn unmethylated around fertilization [25, 26], spatiotemporal patterns of DNA methylation are gradually established as cytosine sites become methylated progressively. This process is referred to as DNA methylation reprogramming, and the regulations can be species-specific, especially in fish. In zebrafish (*Danio rerio*) and Atlantic cod (*Gadus morhua*), the sperm genome is hypermethylated relative to the unfertilized oocyte (Figure 10.1), but the global methylation rates are much higher and more stable compared to mammals, which means many CpG sites remain methylated in zebrafish while most CpG sites become unmethylated in mammals at the beginning of DNA methylation reprogramming [27–29]. Like in zebrafish and cod, sperm is hypermethylated, and oocyte is hypomethylated in medaka fish (*Oryzias latipes*), but most CpG sites turn unmethylated during DNA methylation reprogramming in medaka, which is different from zebrafish but more similar to mammals [28]. Hence, it is important to consider species-specific regulations when studying DNA methylation inheritance in fish.

10.1.2 Histone Modifications and Chromatin Structure

In addition to DNA methylation, histone modifications are also important epigenetic regulators that control gene expression (Figure 10.1). Histones are primary components of chromatin, which packs long linear DNA molecules in a coordinated manner within a chromosome [30]. Histone modifications are covalent modifications that occur at histone tail residues of core histones, which consist of four histone proteins: H2A, H2B, H3, and H4 [31]. Even though all core histone tails are targeted by diverse posttranslational modifications, which span over 200 different types of modifications [32], the tails of H3 are major modification sites that influence gene expression and chromatin structures. H3K27me3 (trimethylation at the 27th lysine of histone H3), for example, associates with gene silencing when it occurs in promoters [33], whereas H3K4me3 on promoters associates with actively transcribed genes or genes that are ready for transcription [34]. Chromatin structures can be lightly packed euchromatin or condensed heterochromatin (Figure 10.1); the former often associates with active gene expression, whereas the latter associates with gene silencing [35]. Heterochromatin marked with H3K9me3, which also acts as a gene-silencing marker, has an epigenetic role in embryonic stem cells [36]. Chromatin structures are dynamically altered during development, and the process is referred to as chromatin remodeling (Figure 10.1).

Histone modifications are also intragenerationally heritable. During DNA replication, all histones are detached from the DNA, but chromatin structures of the daughter DNA are reassembled with identical modifications inherited from the parent DNA molecules [37]. There is some, but little, evidence of the inheritance of histone modifications [38]. An obesity study reported that acute paternal sugar controls the F0 generation and affected phenotypic outcomes in F1 of fruit fly (*Drosophila melanogaster*), which requires several loci with the same status of H3K9me3 and H3K27me3 across the germline and zygote [39]. The study also suggests that humans and mice potentially have a similar system that may affect obesity susceptibility over generations, which is originally controlled by the nutrient composition of F0 diet.

An area in which research is much needed concerns the transmission of histone modifications to the offspring, i.e., whether they are transmitted through meiosis in a similar manner to the mitotic transmission [40]. Although core histones are highly conserved in terms of sequences and structures across a wide range of species [39], how well the inheritance mechanisms are conserved is unknown. Hence, fish species can be useful to examine the common inheritance mechanism of histone modifications among animals as it is generally easier to design experiments that cover multiple generations than mammals.

10.1.3 Noncoding RNAs in Epigenetic Inheritance

Noncoding RNAs (ncRNAs) produce no protein output but play important roles in various cell regulatory mechanisms including epigenetic regulations [41, 42] (Figure 10.1). Long noncoding RNA (lncRNA), small interfering RNA (siRNA), microRNA (miRNA), and piwi-interacting RNA (piRNA) are four classes of ncRNAs that are often linked to epigenetics in literature. Numerous lncRNAs localize in chromatin, where they interact with various chromatin modifiers [41]; therefore, lncRNAs potentially have indirect roles in intragenerational inheritance through chromatin remodeling. Both siRNAs and miRNAs are short ncRNAs that serve as posttranscriptional gene silencers through RNA interference (RNAi) mechanisms [43, 44]. Some endogenous siRNAs associate with chromatin remodeling during development [45]; nonetheless, identified endogenous siRNAs are few in vertebrates. In *Caenorhabditis elegans*, exogenous siRNAs introduced in the F0 generation are transmitted to offspring over multiple generations [46]. Nonetheless, this potential transgenerational epigenetic transfer of siRNAs is mainly observed in *C. elegans* with artificial laboratory conditions [38], and it is unclear whether this form of inheritance is also common in vertebrates. Several miRNAs have been reported to exist in the germline of mammals, and therefore, they may act as mediators of epigenetic inheritance [38]. piRNAs are a class of short ncRNAs that mainly suppress transposable elements in the germline of animal cells [47] but also associate with the maintenance of stem cells and epigenetic regulations during germ cell development [48].

10.2 Nutritional Programming

10.2.1 Definition

Nutritional programming can be defined as the process by which the composition of the feed, or dietary exposure to specific nutrients, during critical windows (or sensitive periods, see below) can affect the development of somatic structures or adjust the metabolism and systemic physiology, resulting in permanent changes even after the nutritional exposure has ended [49, 50]. The programming effect can be at the cellular, molecular, and biochemical level, and is mediated through alterations of active metabolic pathways and/or by epigenetic mechanisms on the DNA [1, 51, 52]. Nutritional programming of desired traits or phenotypes is an important possibility for sustainable growth in aquaculture.

10.2.2 Critical Windows

"Critical windows" represent specific developmental periods or stages during life when the organism's phenotype is particularly sensitive to changing environments. The response to altered environments, like nutrients, temperature, or stress, will lead to a change in the developmental outcome or to a persistent phenotypic change in the whole organism or in specific tissues [2, 53]. Several factors determine the response to the altered environment on the organism's phenotype, and these include which stage of development the organism has reached at the time of exposure, how long the exposure lasts, dose-dependent exposure (e.g., if a nutrient is high, low or absent in a diet affects the response on the epigenome [54]), and the interactions among them [2].

In aquaculture, critical windows can be used to develop specific qualities or desired traits. The possibilities here are numerous. For example, it has been shown that the broodstock can be influenced via the feed so that the offspring have changed phenotype, are adapted to new feed alternatives, or are adapted to a special farming environment, probably by affecting specific tissues or signaling pathways during critical windows at an early age [10, 55–62].

Studying cause-and-effect relationships between environmental cues acting at specific critical windows coinciding with a given developmental stage and the resulting phenotype is not as straightforward as one might assume since the same stimuli might not always lead to the same phenotype due to a phenomenon called phenotypic heterogeneity [63]. Phenotypic heterogeneity explains the broad variation in cell response to the same environment within an isogenic cell population. The cell responses are not always identical, as the genes are not autonomous agents but rather act together with other genes and cell signals to give rise to a unique response [2].

Early developmental periods are especially sensitive to environments. Classic nutritional studies have shown that nutritionally imbalanced diets can cause alterations in metabolic homeostasis, chronic diseases, and immune deficiency, and can increase susceptibility to infectious diseases and poor growth [64–70]. Fish have both species- and stage-specific, as well as conditional, requirements for macro- and micronutrients, including vitamins and minerals [71–75]. In early-life stages, a higher number of nutrients are classified as essential compared to ongrowing stages, as is the case with larvae that still have an underdeveloped and immature gastrointestinal tract [76]. Although it seems to be species-specific, broodstock fish feeding strategies also require increased levels of specific nutrients prior to and during sexual maturation [10, 60, 77–80]. This means that special care and attention to nutritional needs should be taken at all stages of aquaculture production. So, deficiencies are prevented and fish welfare is secured. Indeed, nutrition is focused as an attractive way to enhance the robustness of farmed fish species through functional feeds [80–83].

As fish have external fertilization and embryonic development, the nutritional delivery from the female broodstock to offspring depends on the quantity and quality of nutrients delivered to the yolk [12] (Figure 10.2). The choice of feed ingredients and nutrient composition (feed formulation) will affect the nutritional status in broodstock (F0) and in the eggs (future F1). Early embryonic development is the time when a cell develops from the zygote stage, the two-cell stage, to the

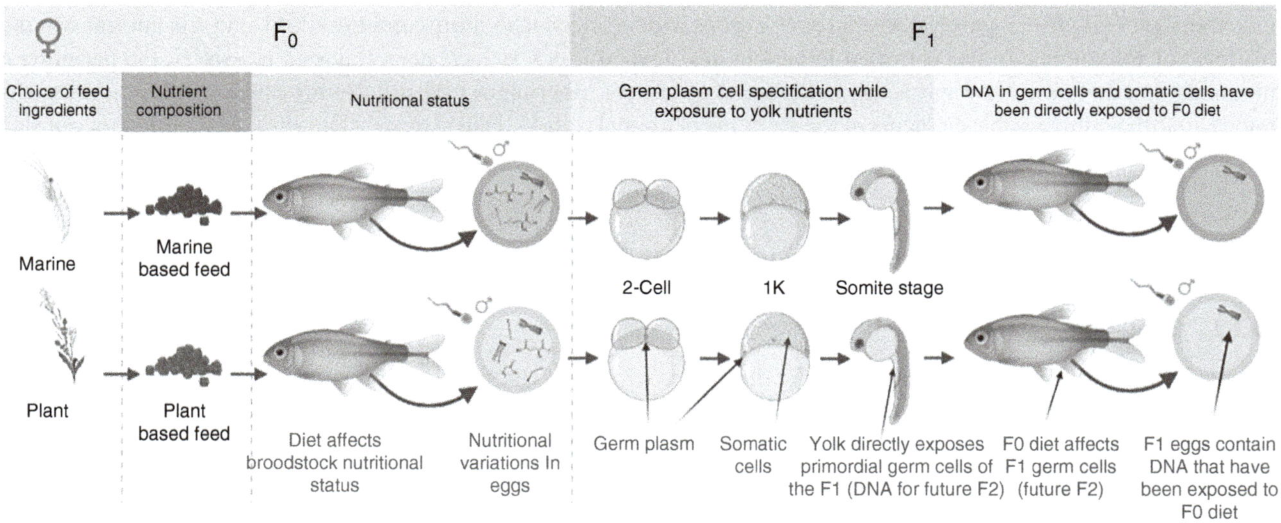

Figure 10.2 Broodstock nutrition and potential long-term effects in offspring generations. The choice of feed ingredients and nutrient composition (feed formulation) will affect the nutritional status in broodstock (F0) and in the eggs (future F1). After fertilization (F1), the cells are in direct contact with the nutrients in the yolk. The germplasm (aggregates of germline determinant RNAs and proteins in the cytoplasm) establishes the germline and separates the PGCs, the precursor cell type of all gametes, from the somatic cells. The germplasm is visualized at the two-cell stage and is incorporated in specific cells at the blastula stage (1K [1000 cell] stage). At the somite stage, the PGCs (future F2-generation) migrate to the genital ridge for proliferation followed by meiotic commencement. During early development and PGC proliferation and specification, the nutritional status of the yolk directly exposes the DNA of F1 gametes. F1 eggs contain DNA that has been exposed to F0 diet through the inclusion of nutrients in the F0 eggs. The figure was created using http://BioRender.com. (*See colour plate section for colour representation of this figure.*)

pluripotent blastula stage (1K stage) to differentiated germ layers, cell types, and organs while still influenced by the yolk [11]. Nutrients induce metabolic signals that through their availability can control cell fate decisions at the molecular and cellular levels and thereby impact the differentiation or proliferation processes, which potentially lead to acquired phenotypes [84, 85]. During this process, the cell fate transitions are tightly regulated by many control layers including changes in gene expression and epigenetic mechanisms [86].

The germplasm (aggregates of germline determinant RNAs and proteins in the cytoplasm) establishes the germline and separates the PGCs, the precursor cell type of all gametes, from the somatic cells. The germplasm is visualized at the two-cell stage and is incorporated in specific cells at the blastula stage (1K [1000 cell] stage) [63]. At the somite stage, the PGCs (future F2-generation) migrate to the genital ridge for proliferation followed by meiotic commencement. The cells that become the PGCs are defined at the early embryonic developmental stage and they migrate to the position for further gonad development to eventually differentiate after spermatogenesis in males and oogenesis in females into sperm and oocytes, respectively [87]. During early development and PGC proliferation and specification, the nutritional status of the yolk directly exposes the DNA of F1 gametes. The PGCs of the F1 generation will give rise to the F2 generation. Even though there are differences in the developmental stage and PGCs migration, settlement, and gonadal formation times between fish species [88–92], the PGCs giving rise to the F2 will be directly exposed to the yolk nutrients and metabolites at the F1 embryo stage [12]. This suggests, also for fish, that the genetic material of the germ cells (that will give rise to the F2 generation) may also be influenced by the parental (F0) nutrient status via the yolk content (Figure 10.2).

Both broodstock and larval feed formulations will provide nutritional-based programming stimuli to improve either gamete or larva quality. Targeted programming can either focus on early critical windows or periods when fish exhibit high sensitivity and plasticity in their development and thereby provide permanent epigenetic programming potential for specific phenotypes [85].

Few studies have addressed the question as to how the nutritional profile of the yolk contributes to the phenotype during the embryonic stages in teleosts. When zebrafish received either a control diet with high 1C nutrients (folate, methionine, choline, vitamin B12, and vitamin B6) or an experimental diet with low 1C nutrients, while both offspring groups received the control diet until maturity, it was shown that the nutritional profile of parents can adjust the gene expression pattern already at the somite stage and lead to altered phenotype later in life [10, 60]. The parental 1C nutrient status lowered the maternal fecundity, but no embryonic phenotypic changes (deformities, fertilization, hatching, survival, and growth rate) were observed, except from the disturbed gene expression of 365 genes in descendant embryos. The altered mRNAs encode proteins associated with overall health and result in inherited aberrations in the 1C cycle and lipid metabolism. When the offspring of the experimental group reached maturity, they had developed a phenotypic pale liver color compared to the control group. Histological examination revealed a fatty-liver-like phenotype with a corresponding downregulation of the mRNA encoding proteins in the functional categories related to lipid biosynthesis and mitochondrial protein synthesis. Differential DNA methylation was found enriched in promoter regions and differed due to parental feed. This points toward parental nutrient status as an important factor for offspring health and that the effect can persist as locus-specific DNA methylation marks in offspring; however, it is not known at which developmental stage there is a critical window toward the phenotypic change observed in the hepatocytes. This highlights the importance of parental fish nutrient status for the health of the offspring.

New knowledge in the field points to micronutrients in the feed as active regulators of the chemical groups and intermediate metabolites like S-adenosylmethionine (SAM) that are needed for epigenetic mechanisms [93]. The concentration of SAM also varies in newly fertilized eggs [94]. By this means, alterations in the metabolome can adjust chromatin accessibility. This implies that nutrition can be a direct link between environmental exposure and chromatin accessibility, which through the concept of nutritional programming regulates long-term metabolism and systemic physiology. In teleosts, this is a rather new and unexplored field. However, a few studies have indicated that the composition of feed, or dietary exposure to specific nutrients, can adjust the epigenetic chromatin landscape that results in altered gene expression even after the nutritional exposure has ended. Nutritional programming in teleosts was recently reviewed by Hou and Fuiman [85].

10.3 Key Nutrients and Metabolites for Epigenetic Mechanisms

Cell metabolism is sensitive to the availability of nutrients and their metabolites, and some of these are directly involved as substrates for the enzymes that modify the epigenetic mechanisms [1]. The cell adapts to its metabolic environment through nutrient catabolization into intermediates, which are important substrates of the epigenetic "writer" and "eraser"

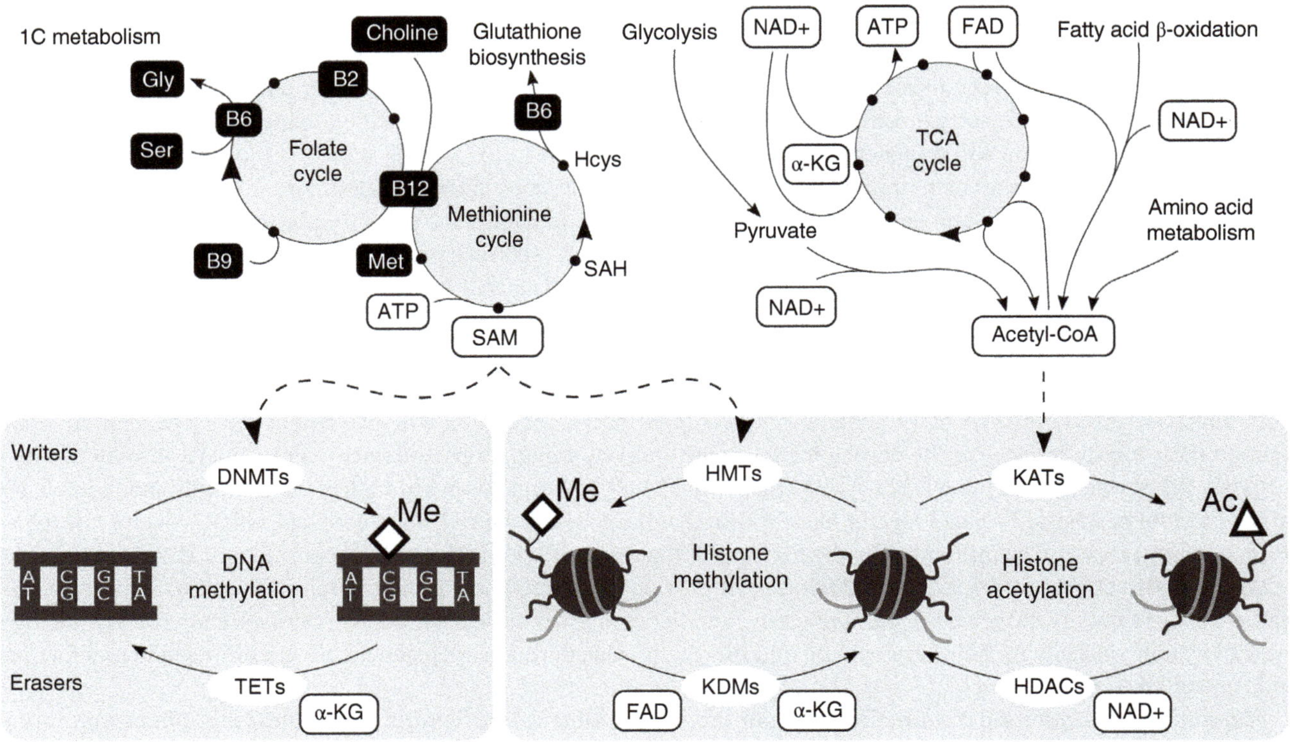

Figure 10.3 Key nutrients and metabolites influencing epigenetic mechanisms. Metabolites (white boxes) and nutrients (black boxes) participate in protein, carbohydrate, lipid, and glutathione pathways, and 1C metabolism (folate and methionine cycle). Fluctuations in those intermediates can result from dietary changes affecting nutrient availability and may thereby impact epigenetic mechanisms, so-called writers and erasers. Those chromatin re-modelling enzymes utilize the metabolites from the intermediary metabolism as co-factors or substrates. Ac, acetylation mark; ATP, adenosine triphosphate; Acetyl-CoA, acetyl coenzyme A; DNMTs, DNA methyltransferases; FAD, flavin adenine dinucleotide; Gly, glycine; Hcys, homocysteine; HDACs, histone deacetylases; HMTs, histone methyltransferases; KATs, lysine acetyltransferases; KDMs, lysine demethylases; Me, methylation mark; Met, methionine; NAD+, nicotinamide adenine dinucleotide; SAH, S-adenosylhomocysteine; SAM, S-adenosylmethionine; Ser, serine; TCA cycle, tricarboxylic acid cycle, also known as Krebs or citric acid cycle; TETs, ten-eleven translocation methylcytosine dioxygenases; α-KG, α-ketoglutarate.

enzymes modifying the chromatin [95]. Among the important key intermediate metabolites are SAM produced through the one-carbon (1C) metabolism, and acetyl coenzyme A (acetyl-CoA) synthesized either through the catabolism of some amino acids or from pyruvate through glycolysis and β-oxidation of fatty acids [96] (Figure 10.3).

Other main substrates are nicotinamide adenine dinucleotide (NADH/NAD+), adenosine triphosphate (ATP), flavin adenine dinucleotide (FAD), and α-ketoglutarate (α-KG) (Figure 10.3) [1, 97]. Together with SAM and Acetyl-CoA, they can be categorized as epigenetic metabolites. The enzymes that are responsible for chromatin remodeling utilize metabolites from intermediary metabolism as substrates or cofactors to write or erase methylation or acetylation marks on the chromatin (Figure 10.3) [95]. Thus, fluctuations in these metabolites, resulting from dietary changes, may thereby affect the epigenetic regulation of genes.

The availability of SAM is essential to the methylation of DNA and histone tails. SAM is produced in an ATP-dependent process from methionine through the methionine cycle that, together with the folate cycle, comprises the 1C metabolism [96, 98] (Figure 10.3). The production of SAM depends on the B vitamins folate (B9), riboflavin (B2), cobalamin (B12), and pyridoxine (B6) as well as choline (betaine) and the amino acids methionine, glycine, and serine, all regulating the 1C metabolism [96]. In the case of lower folate or vitamin B12 levels, homocysteine generated from S-adenosylhomocysteine (SAH) in the methionine cycle gets converted to glutathione. Glutathione is the main cellular redox scavenger, and as such, the homocysteine is eliminated and converted via the transsulfuration pathway, which is a multistep and vitamin B6-dependent pathway [96].

Another epigenetic metabolite is acetyl-CoA that fuels the tricarboxylic acid (TCA) cycle. Acetyl-CoA is a product of degradation from certain amino acids, glycolysis through pyruvate, or fatty acid β-oxidation (Figure 10.3). Both SAM and acetyl-CoA can be utilized directly as substrates for DNA or histone methyltransferases (DNMTs and HMTs), and histone lysine acyltransferases (KATs), respectively (Figure 10.3). Demethylation and deacetylation are also subject to metabolite regulation via ten-eleven translocation (TET) enzymes or histone demethylases (KDMs) that use FAD or α-KG as co-substrate, while histone deacetylases (HDACs) use NAD+ (Figure 10.3). The metabolites SAH and 2-hydroxyglutarate, in the case of methyltransferases and demethylases, and β-hydroxybutyrate for deacetylases, function as inhibitors of the DNA and histone tail-modifying enzymes. Thus, as mentioned earlier, fluctuations in these key metabolites result from dietary changes and may thereby affect epigenetic regulation and possibly heritable patterns of gene expression.

10.4 Case Examples

Most aquaculture species have defined requirement levels for macro- and micronutrients [71]. The chemical nutritional composition, bioavailability of nutrients, and the presence of antinutrients in the feed modulate the absorption in the intestine and together alter the homeostasis in both intra- and intercellular spaces. The feed or diet, along with other environmental factors such as temperature or stress, will set a baseline for cellular functions and interact with the genome by adjusting chromatin accessibility by altering the epigenetic pattern [1].

The location and type of epigenetic pattern can be modified by the cellular milieu via the nutrients in the feed. The epigenetic changes will alter the control of active metabolic pathways, leading to acquired phenotypes [96]. Phenotypes can be transmitted through cell divisions from mother cells to daughter cells (intragenerational), or across generations from parents to offspring (inter- or multigenerational), or persist as epimutations (transgenerational) (see chapter 20 for definitions). As such, the design of these case examples can be categorized into either intragenerational or multigenerational, as no fish nutrition studies have analyzed transgenerational effects.

10.4.1 Programming of Broodstock to Affect the Offspring

The programming effect of changing feed is higher during early developmental stages or during sensitive periods when the organism's phenotype is particularly vulnerable to changing environments [2]. So far, it is still unknown if the feed given to great-grandparents may lead to persistent epigenetic marks transmitted through meiosis in fish as reported for yeast, flies, and mice [40]. However, recent research has shown that zebrafish offspring DNA methylation is altered due to parental nutrition [9, 10], and that Atlantic salmon (*Salmo salar*) male gonad DNA methylation is sensitive to nutritional composition [99]. Epigenetic research is still young in the field of fish nutrition. However, a few studies have provided new perceptions of nutritional epigenetics and/or programming through investigating DNA methylation and histone tail modifications.

The broodstock diet has tremendous effects on fecundity, gene expression pattern, and DNA methylation in the offspring [10, 60]. Defining nutritional requirement levels for broodstock in various fish species has been problematic due to high running costs of broodstock feeding trials. The solution has generally been to produce broodstock feed, which is rich in essential fatty acids, vitamins, and minerals [77]. An enriched broodstock diet is expensive and may not be beneficial, as we know that too high surplus of vitamins might be energetically expensive and thereby reduce growth [73]. Broodstock diets should be re-evaluated to fulfill nutritional requirements for different production regimes as this also influences the yolk composition of the eggs that will give rise to the F1 [94]. The yolk includes a variety of mRNA transcripts that accumulate during oocyte growth. These transcripts have specific functions during oocyte maturation or during embryonic development [100]. Focus on broodstock nutrient composition is important as the maternal and zygotic transcripts in embryos adjust their mRNA level depending on the parental feed composition [60] or broodstock nutritional status [101].

When it comes to intergenerational effects, implications of both the rearing environment and broodstock feed on fecundity, fertilization, deformities, hatching rate, robustness, and growth of the larvae have been reported and reviewed elsewhere [77, 100, 102]. As mentioned earlier, broodstock feed is important for the next generation as the yolk of the oocyte includes a variety of nutrients and transcripts that accumulate during oocyte growth, and have specific functions during oocyte maturation or during embryonic development [100]. Rainbow trout fed a low methionine diet for six months prior to spawning altered the DNA methylation of specific genes in the progeny [103]. For Atlantic salmon, adjustment of the spawning season changed the nutrient allocation into the egg, which had consequences for the expression of genes controlling DNA methylation in newly hatched larvae and for offspring growth modified by nutritional programming [94, 101]. Studies on zebrafish revealed DNA methylation-modulating effects in adult offspring liver from fish given a life cycle plant-based diet, either low in the micronutrients methionine, folate, choline, vitamin B6 and B12, or a diet enriched in arachidonic acid [9, 10]. The parental feed low in the selected micronutrients gave a fatty-liver-like phenotype in the offspring, downregulated mRNA levels of lipid-related genes, and revealed gene-specific changes in DNA methylation in the liver [10]. In rainbow trout, feed that is deficient in methionine, or enriched with two different selenium sources, altered DNA methylation patterns in offspring fry [103, 104]. In addition, a parental diet supplemented with α-linoleic acid-rich oils changed the DNA methylation of selected genes expressed in Gilthead sea bream liver [59, 105].

10.4.2 Programming of Larvae to Affect Later Stages of Development

For intragenerational nutritional epigenetics, the effect of the diet can be measured while the fish are still fed the experimental diet, or after pausing the experimental diet and returning to the control diet. For the former, the diet continuously affects the intermediary metabolism and the epigenetic gene regulation. For the latter, the dietary signal is not active but adjusts later outcomes via nutritional programming of traits. The epigenetic adjustments of the chromatin might lead to permanent epimutations in both cases.

Here, we first give examples of continuous feeding trials. Few studies have investigated how a dietary change can affect epigenetic gene regulation while the experimental diet continuously influences the intermediary metabolism. For example, the transfer from marine to plant-based diets led to a re-evaluation of dietary micronutrients throughout the whole life cycle of farmed Atlantic salmon. DNA methylation and transcription were studied in the liver of Atlantic salmon that were fed graded micronutrient levels provided in packages with double and four times the amounts recommended [71] for optimal growth and fish welfare [54, 99]. The results indicate that micronutrient supplementation suppresses gene expression in liver lipid metabolism in a dose-dependent manner and broadly affects DNA methylation in cell-adhesion and cell-signaling genes. The micronutrient level in the feed increases the DNA methylation levels on the acetyl-CoA carboxylase alpha (*acaca*) promoter in a concentration-dependent manner, which further suggests that *acaca*, which is an upstream regulator controlling its downstream lipid biosynthesis activities, is regulated by epigenetic mechanisms [54]. Moreover, the micronutrient supplementation also affected the DNA methylation of the promoter regions of genes associated with cell signaling, synaptic signaling, and immuneresponse. In addition, the DNA methylation in male gonads also responded to micronutrient levels in the feed [99]. DNA methylation signatures of gonads are of great importance in the field of nutritional epigenetics because the signatures affected by the nutritional profile surrounding the cells could be transferred to the next generation, especially since studies have shown that the DNA methylation signature is more similar to sperm than to eggs for fish [106, 107].

The smoltification period, when Atlantic salmon adapt from freshwater to saltwater, is a challenging period with increased mortality. To investigate if the feed given in the freshwater phase could have a potential programming effect and improve health, growth, and robustness throughout the smoltification period, additional levels of methionine, folate, vitamin B12, and B6 were fed six weeks prior to and three months after smoltification. These four nutrients are important in the 1C metabolism (see Section 10.3, and Figure 10.3 for details). The 1C metabolism consists of the folate cycle, the methionine cycle, and the transsulfuration pathway, which are essential for the generation of SAM that can donate the methyl group to methyltransferases for both DNA methylation and histone tail methylation. It was concluded that moderate surplus levels of 1C nutrients improved growth in the seawater period [73]. Extra amounts adjusted the metabolic profile in muscle, particularly the SAM/SAH ratio that directly impacts the methylation potential of both DNA and histone tails with possible consequences for gene expression [93]. The overall metabolic state and gene expression profiles of muscle showed that the dietary effect before seawater transfer was intensified post-smolt, suggesting metabolic programming, possibly through epigenetic programming [93].

Muscular and hepatic global DNA hypomethylation was associated with increased carbohydrate amounts in the feed for rainbow trout [55, 108]. Varying sources of lipids or dietary fatty acid levels affected liver lipid metabolism associated with hepatic DNA methylation changes observed in Atlantic salmon and Japanese sea bass (*Lateolabrax japonicus*) [109, 110].

Intragenerational metabolic programming of growth was found in Nile tilapia (*Oreochromis niloticus*) alevins that received microinjections of glucose into the yolk reserves, which later displayed muscular DNA hypomethylation in juveniles, suggesting an epigenetic effect [56].

To date, only a few fish nutrition studies explored nutritional programming effects on histone tail modifications intragenerationally on a global level, and no studies have explored them from a multigenerational perspective. For rainbow trout, hepatic global DNA hypomethylation and hypoacetylation of histone H3K9 were displayed when fed carbohydrate-enriched diets [111]. Surplus vitamins at first feeding changed hepatic global DNA methylation, global H3 acetylation, and H3K4 methylation-dependent on the vitamin intake in early life in rainbow trout [112]. Feeding Gilthead Sea bream (*Sparus aurata*) larvae with various plant-based sources such as soybean meal increased global histone H3 acetylation and global hypomethylation in the larvae [113].

10.5 Conclusions and Perspectives for Nutritional Programming in Aquaculture

The aforementioned examples increase our understanding of the importance that nutrients have in altering epigenetic mechanisms during fish nutritional programming. The early nutritional stimuli, long-term feeding, parental feeding, or maternal oocyte contribute to consequences for health, growth, and reproduction in farmed fish. The findings indicate that feed composition or specific nutrient variations adjust the long-term chromatin landscape and gene expression patterns, even after the nutritional exposure has ended.

Still, several questions remain unsolved in the field of nutritional programming. The field is cross-disciplinary, linking reproductive physiology, developmental biology, molecular biology, nutrition, and bioinformatics. Essential principles to consider for nutritional epigenetics studies are to follow the best practice for each of the fields involved.

For the fish nutrition field, it is important to include: (i) feed formulation with details on which nutrients and/or raw materials are altered between diets and that the feeds are energetically balanced; (ii) analyze the nutrient composition of the feeds and tissues; (iii) include minimum triplicate tanks per feed group in the experimental design; and (iv) carefully decide on the sample collected (organism, tissue, or cell) for epigenetic comparisons. For other fields like molecular methods, statistics, and bioinformatics, the methods need to be at a comprehensive level to support the investigated matter and the conclusion drawn. The bioinformatics pipeline should be carefully applied to the generated datasets to uncover solid epigenetic differences and to reduce false positive assumptions between experimental groups [12, 54].

Domestication of wild fish changes the methylome within just one generation of captivity [114]. For aquaculture, several potential applications can be improved by nutritional programming. Attention toward nutritional needs of broodstock is important. The phenotypic plasticity effect from broodstock nutritional profile to offspring's development of beneficial and advantageous traits remains an exciting future topic of research. Nutritional programming implies that environmental exposures, to some extent, can possibly shape future development toward optimal growth and can prevent previous exposures from becoming a metabolic burden for later growth and developmental potential. A beneficial nutrient status surrounding the cells, from the first cell division, might provide nutritional improvement and prohibit nutritional status from becoming a burden in later life stages or even for future generations.

Acknowledgments

This work was supported by the Research Council of Norway through the grant NutrEpi 267787 and by IMRs (Institute of Marine Research, Bergen, Norway) Nutritional programming project 15469.

References

1 Gut, P. and Verdin, E. (2013). The nexus of chromatin regulation and intermediary metabolism. *Nature* **502** (7472): 489–498.

2 Burggren, W.W. (2019). Phenotypic switching resulting from developmental plasticity: fixed or reversible? *Frontiers in Physiology* **10**: 1634.

3 Heard, E. and Martienssen, R.A. (2014). Transgenerational epigenetic inheritance: myths and mechanisms. *Cell* **157** (1): 95–109.

4 Valdivieso, A., Ribas, L., Monleon-Getino, A. et al. (2020). Exposure of zebrafish to elevated temperature induces sex ratio shifts and alterations in the testicular epigenome of unexposed offspring. *Environmental Research* **186**: 109601.

5 Shen, Z.G. and Wang, H.P. (2014). Molecular players involved in temperature-dependent sex determination and sex differentiation in Teleost fish. *Genetics Selection Evolution* **46**: 26.

6 Navarro-Martin, L., Vinas, J., Ribas, L. et al. (2011). DNA methylation of the gonadal aromatase (cyp19a) promoter is involved in temperature-dependent sex ratio shifts in the European sea bass. *PLoS Genetics* **7** (12): e1002447.

7 Kucharski, R., Maleszka, J., Foret, S. et al. (2008). Nutritional control of reproductive status in honeybees via DNA methylation. *Science* **319** (5871): 1827–1830.

8 DiTroi, S.P., Percharde, M., Guerquin, M.J. et al. (2019). Maternal vitamin C regulates reprogramming of DNA methylation and germline development. *Nature* **573** (7773): 271–275.

9 Adam, A.-C., Lie, K.K., Whatmore, P. et al. (2019). Profiling DNA methylation patterns of zebrafish liver associated with parental high dietary arachidonic acid. *PLoS one* **14** (8): e0220934.

10 Skjaerven, K.H., Jakt, L.M., Fernandes, J.M.O. et al. (2018). Parental micronutrient deficiency distorts liver DNA methylation and expression of lipid genes associated with a fatty-liver-like phenotype in offspring. *Scientific Reports* **8** (1): 3055.

11 Berdasco, M. and Esteller, M. (2011). DNA methylation in stem cell renewal and multipotency. *Stem Cell Research & Therapy* **2** (5): 42.

12 Skjærven, K.H., Adam, A.-C., Saito, T. et al. (2022). Nutritional epigenetics. In: *Cellular and Molecular Approaches in Fish Biology* (ed. I.F. Monzón and J.M.O. Fernandes), 161–192. Elsevier Inc.

13 Zemach, A., McDaniel, I.E., Silva, P. et al. (2010). Genome-wide evolutionary analysis of eukaryotic DNA methylation. *Science* **328** (5980): 916–919.

14 Law, J.A. and Jacobsen, S.E. (2010). Establishing, maintaining and modifying DNA methylation patterns in plants and animals. *Nature Review Genetics* **11** (3): 204–220.

15 Ehrlich, M., Gama-Sosa, M.A., Carreira, L.H. et al. (1985). DNA methylation in thermophilic bacteria: N4-methylcytosine, 5-methylcytosine, and N6-methyladenine. *Nucleic Acids Research* **13** (4): 1399–1412.

16 Kriaucionis, S. and Heintz, N. (2009). The nuclear DNA base 5-hydroxymethylcytosine is present in Purkinje neurons and the brain. *Science* **324** (5929): 929–930.

17 Wu, X. and Zhang, Y. (2017). TET-mediated active DNA demethylation: mechanism, function and beyond. *Nature Reviews Genetics* **18** (9): 517–534.

18 Watt, F. and Molloy, P.L. (1988). Cytosine methylation prevents binding to DNA of a HeLa cell transcription factor required for optimal expression of the adenovirus major late promoter. *Genes & Development* **2** (9): 1136–1143.

19 Boyes, J. and Bird, A. (1991). DNA methylation inhibits transcription indirectly via a methyl-CpG binding protein. *Cell* **64** (6): 1123–1134.

20 Slotkin, R.K. and Martienssen, R. (2007). Transposable elements and the epigenetic regulation of the genome. *Nature Reviews Genetics* **8** (4): 272–285.

21 Maunakea, A.K., Nagarajan, R.P., Bilenky, M. et al. (2010). Conserved role of intragenic DNA methylation in regulating alternative promoters. *Nature* **466** (7303): 253–257.

22 Zhu, H., Wang, G., and Qian, J. (2016). Transcription factors as readers and effectors of DNA methylation. *Nature Reviews Genetics* **17** (9): 551–565.

23 Goll, M.G. and Bestor, T.H. (2005). Eukaryotic cytosine methyltransferases. *Annual Review of Biochemistry* **74**: 481–514.

24 Jones, P.A. and Liang, G. (2009). Rethinking how DNA methylation patterns are maintained. *Nature Reviews Genetics* **10** (11): 805–811.

25 Smallwood, S.A. and Kelsey, G. (2012). De novo DNA methylation: a germ cell perspective. *Trends in Genetics* **28** (1): 33–42.

26 Lokk, K., Modhukur, V., Rajashekar, B. et al. (2014). DNA methylome profiling of human tissues identifies global and tissue-specific methylation patterns. *Genome Biology* **15** (4): r54.

27 Mhanni, A.A. and McGowan, R.A. (2004). Global changes in genomic methylation levels during early development of the zebrafish embryo. *Development Genes and Evolution* **214** (8): 412–417.

28 Wang, X. and Bhandari, R.K. (2020). DNA methylation reprogramming in medaka fish, a promising animal model for environmental epigenetics research. *Environmental Epigenetics* **6** (1): dvaa008.

29 Skjaerven, K.H., Hamre, K., Penglase, S. et al. (2014). Thermal stress alters expression of genes involved in one carbon and DNA methylation pathways in Atlantic cod (*Gadus morhua*) embryos. *Comparative Biochemistry and Physiology Part A* **173C**: 17–27.

30 Kornberg, R.D. and Lorch, Y. (1999). Twenty-five years of the nucleosome, fundamental particle of the eukaryote chromosome. *Cell* **98** (3): 285–294.

31 Rothbart, S.B. and Strahl, B.D. (2014). Interpreting the language of histone and DNA modifications. *Biochimica et Biophysica Acta* **1839** (8): 627–643.

32 Zhao, Y. and Garcia, B.A. (2015). Comprehensive catalog of currently documented histone modifications. *Cold Spring Harbor Perspectives in Biology* **7** (9): a025064.

33 Ferrari, K.J., Scelfo, A., Jammula, S. et al. (2014). Polycomb-dependent H3K27me1 and H3K27me2 regulate active transcription and enhancer fidelity. *Molecular Cell* **53** (1): 49–62.

34 Sims, R.J., Nishioka, K., and Reinberg, D. (2003). Histone lysine methylation: a signature for chromatin function. *Trends in Genetics* **19** (11): 629–639.

35 Weiler, K.S. and Wakimoto, B.T. (1995). Heterochromatin and gene expression in *Drosophila*. *Annual Review of Genetics* **29**: 577–605.

36 Nicetto, D., Donahue, G., Jain, T. et al. (2019). H3K9me3-heterochromatin loss at protein-coding genes enables developmental lineage specification. *Science* **363** (6424): 294–297.

37 Budhavarapu, V.N., Chavez, M., and Tyler, J.K. (2013). How is epigenetic information maintained through DNA replication? *Epigenetics & Chromatin* **6** (1): 32.

38 Perez, M.F. and Lehner, B. (2019). Intergenerational and transgenerational epigenetic inheritance in animals. *Nature Cell Biology* **21** (2): 143–151.

39 Ost, A., Lempradl, A., Casas, E. et al. (2014). Paternal diet defines offspring chromatin state and intergenerational obesity. *Cell* **159** (6): 1352–1364.

40 Fitz-James, M.H. and Cavalli, G. (2022). Molecular mechanisms of transgenerational epigenetic inheritance. *Nature Reviews Genetics* **23**: 325–341.

41 Statello, L., Guo, C.J., Chen, L.L. et al. (2021). Gene regulation by long non-coding RNAs and its biological functions. *Nature Reviews Molecular Cell Biology* **22** (2): 96–118.

42 Kaikkonen, M.U., Lam, M.T., and Glass, C.K. (2011). Non-coding RNAs as regulators of gene expression and epigenetics. *Cardiovascular Research Journal* **90** (3): 430–440.

43 Hamilton, A.J. and Baulcombe, D.C. (1999). A species of small antisense RNA in posttranscriptional gene silencing in plants. *Science* **286** (5441): 950–952.

44 Bartel, D.P. (2004). MicroRNAs: genomics, biogenesis, mechanism, and function. *Cell* **116** (2): 281–297.

45 Fukagawa, T., Nogami, M., Yoshikawa, M. et al. (2004). Dicer is essential for formation of the heterochromatin structure in vertebrate cells. *Nature Cell Biology* **6** (8): 784–791.

46 Grishok, A., Tabara, H., and Mello, C.C. (2000). Genetic requirements for inheritance of RNAi in *C. elegans*. *Science* **287** (5462): 2494–2497.

47 Seto, A.G., Kingston, R.E., and Lau, N.C. (2007). The coming of age for Piwi proteins. *Molecular Cell* **26** (5): 603–609.

48 Siomi, M.C., Sato, K., Pezic, D. et al. (2011). PIWI-interacting small RNAs: the vanguard of genome defence. *Nature Reviews Molecular Cell Biology* **12** (4): 246–258.

49 Agosti, M., Tandoi, F., Morlacchi, L. et al. (2017). Nutritional and metabolic programming during the first thousand days of life. *La Pediatria Medica e Chirurgica* **39** (2): 157.

50 Lucas, A. (1998). Programming by early nutrition: an experimental approach. *The Journal of Nutrition* **128**: 401S–406S.

51 Gluckman, P.D., Hanson, M.A., and Low, F.M. (2011). The role of developmental plasticity and epigenetics in human health. *Birth Defects Research Part C: Embryo Today: Reviews* **93** (1): 12–18.

52 Portha, B., Fournier, A., Kioon, M.D. et al. (2014). Early environmental factors, alteration of epigenetic marks and metabolic disease susceptibility. *Biochimie* **97**: 1–15.

53 Monaghan, P. (2008). Early growth conditions, phenotypic development and environmental change. *Philosophical Transactions of the Royal Society B* **363** (1497): 1635–1645.

54 Saito, T., Whatmore, P., Taylor, J.F. et al. (2021). Micronutrient supplementation affects transcriptional and epigenetic regulation of lipid metabolism in a dose-dependent manner. *Epigenetics* **16** (11): 1217–1234.

55 Song, Y., Alami-Durante, H., Skiba-Cassy, S. et al. (2019). Higher glycolytic capacities in muscle of carnivorous rainbow trout juveniles after high dietary carbohydrate stimulus at first feeding. *Nutrition and Metabolism* **16**: 77.

56 Kumkhong, S., Marandel, L., Plagnes-Juan, E. et al. (2020). Glucose injection into yolk positively modulates intermediary metabolism and growth performance in juvenile Nile tilapia (*Oreochromis niloticus*). *Frontiers in Physiology* **11**: 286.

57 Kumkhong, S., Marandel, L., Plagnes-Juan, E. et al. (2020). Early feeding with hyperglucidic diet during fry stage exerts long-term positive effects on nutrient metabolism and growth performance in adult tilapia (*Oreochromis niloticus*). *Journal of Nutritional Science* **9**: e41.

58 Izquierdo, M.S., Turkmen, S., Montero, D. et al. (2015). Nutritional programming through broodstock diets to improve utilization of very low fishmeal and fish oil diets in gilthead sea bream. *Aquaculture* **449**: 18–26.

59 Turkmen, S., Perera, E., Zamorano, M.J. et al. (2019). Effects of dietary lipid composition and fatty acid desaturase 2 expression in broodstock gilthead sea bream on lipid metabolism-related genes and methylation of the fads2 gene promoter in their offspring. *International Journal of Molecular Sciences* **20** (24): 6250.

60 Skjaerven, K.H., Jakt, L.M., Dahl, J.A. et al. (2016). Parental vitamin deficiency affects the embryonic gene expression of immune-, lipid transport- and apolipoprotein genes. *Scientific Reports* **6**: 34535.

61 Rocha, F., Dias, J., Engrola, S. et al. (2014). Glucose overload in yolk has little effect on the long-term modulation of carbohydrate metabolic genes in zebrafish (*Danio rerio*). *Journal of Experimental Biology* **217** (7): 1139–1149.

62 Rocha, F., Dias, J., Geurden, I. et al. (2016). Dietary glucose stimulus at larval stage modifies the carbohydrate metabolic pathway in gilthead seabream (*Sparus aurata*) juveniles: an in vivo approach using (14)C-starch. *Comparative Biochemistry and Physiology Part A* **201**: 189–199.

63 Gilbert, S.F. and Barresi, M.J.F. (ed.) (2016). *Developmental Biology*. Sunderland, MA: Sinauer Associates, Inc., Publishers.

64 Halver, J. and Hardy, R. (2002). *Fish Nutrition*, 3e. Academic Press.

65 Kiron, V. (2012). Fish immune system and its nutritional modulation for preventive health care. *Animal Feed Science and Technology* **173** (1–2): 111–133.

66 Lall, S.P. and Lewis-McCrea, L.M. (2007). Role of nutrients in skeletal metabolism and pathology in fish – an overview. *Aquaculture* **267** (1–4): 3–19.

67 Trichet, V.V. (2010). Nutrition and immunity: an update. *Aquaculture Research* **41** (3): 356–372.

68 Turchini, G.M., Torstensen, B.E., and Ng, W.K. (2009). Fish oil replacement in finfish nutrition. *Reviews in Aquaculture* **1** (1): 10–57.

69 Waagbø, R. (2006). Feeding and disease resistance in fish. In: *Biology of Growing Animal* (ed. R. Mosenthin, J. Zentek and T. Zebrowska), 387–415. Elsevier.

70 Waagbø, R. (2008). Reducing production related diseases in farmed fish. In: *Improving Farmed Fish Quality and Safety* (ed. Ø. Lie), 363–398. Woodhead Publishing.

71 NRC (2011). *Nutrient Requirements of Fish and Shrimp*. Washington, D.C: The National Academic Press.

72 Hemre, G.-I., Lock, E.J., Olsvik, P.A. et al. (2016). Atlantic salmon (*Salmo salar*) require increased dietary levels of B-vitamins when fed diets with high inclusion of plant based ingredients. *PeerJ* **4**: e2493.

73 Espe, M., Vikesa, V., Thomsen, T.H. et al. (2020). Atlantic salmon fed a nutrient package of surplus methionine, vitamin B12, folic acid and vitamin B6 improved growth and reduced the relative liver size, but when in excess growth reduced. *Aquaculture Nutrition* **26** (2): 477–489.

74 Vera, L.M., Metochis, C., Taylor, J.F. et al. (2017). Early nutritional programming affects liver transcriptome in diploid and triploid Atlantic salmon, *Salmo salar*. *BMC Genomics* **18** (1): 886.

75 Waagbø, R., Olsvik, P., and Remø, S. (2020). Nutritional and metabolic disorders. In: *Climate Change and Non-Infectious Fish Disorders (CCNFD)* (ed. P. Woo and G. Iwama), 102–124. CAB International.

76 Moren, M., Waagbø, R., and Hamre, K. (2011). Micronutrients. In: *Larval Fish Nutrition* (ed. C. Holt), 117–149. Wiley.

77 Izquierdo, M.S., Fernandez-Palacios, H., and Tacon, A.G.J. (2001). Effect of broodstock nutrition on reproductive performance of fish. *Aquaculture* **197** (1–4): 25–42.

78 Christiansen, R. and Torrissen, O.J. (1997). Effects of dietary astaxanthin supplementation on fertilization and egg survival in Atlantic salmon (*Salmo salar* L). *Aquaculture* **153** (1–2): 51–62.

79 Mæland, A., Rønnestad, I., and Waagbø, R. (2003). Folate in eggs and developing larvae of Atlantic halibut, *Hippoglossus hippoglossus* L. *Aquaculture Nutrition* **9** (3): 185–188.

80 Waagbo, R., Glette, J., Raa-Nilsen, E. et al. (1993). Dietary vitamin C, immunity and disease resistance in Atlantic salmon (*Salmo salar*). *Fish Physiology and Biochemistry* **12** (1): 61–73.

81 Panserat, S., Marandel, L., Seiliez, I. et al. (2019). New insights on intermediary metabolism for a better understanding of nutrition in teleosts. *Annual Review of Animal Biosciences* **7**: 195–220.

82 Sweetman, J.W., Torrecillas, S., Dimitroglou, A. et al. (2010). Enhancing the natural defences and barrier protection of aquaculture species. *Aquaculture Research* **41** (3): 345–355.

83 Waagbø, R. and Remø, S. (2020). Functional diets in fish health management. In: *Aquaculture Health Management* (ed. F. Kibenge and M. Powell), 187–234. Elsevier.

84 D'Aniello, C., Cermola, F., Patriarca, E.J. et al. (2019). Metabolic-epigenetic axis in pluripotent state transitions. *Epigenomes* **3** (3): 13.

85 Hou, Z.X. and Fuiman, L.A. (2020). Nutritional programming in fishes: insights from mammalian studies. *Reviews in Fish Biology and Fisheries* **30** (1): 67–92.

86 Tatapudy, S., Aloisio, F., Barber, D. et al. (2017). Cell fate decisions: emerging roles for metabolic signals and cell morphology. *EMBO Reports* **12**: 2105–2118.

87 Raz, E. (2003). Primordial germ-cell development: the zebrafish perspective. *Nature Reviews Genetics* **4** (9): 690–700.

88 Fernandez, J.A., Bubner, E.J., Takeuchi, Y. et al. (2015). Primordial germ cell migration in the yellowtail kingfish (*Seriola lalandi*) and identification of stromal cell-derived factor 1. *General and Comparative Endocrinology* **213**: 16–23.

89 Saito, T., Otani, S., Fujimoto, T. et al. (2004). The germ line lineage in ukigori, *Gymnogobius* species (*Teleostei: Gobiidae*) during embryonic development. *The International Journal of Developmental Biology* **48** (10): 1079–1085.

90 Zhao, C., Xu, S., Liu, Y. et al. (2017). Gonadogenesis analysis and sex differentiation in cultured turbot (*Scophthalmus maximus*). *Fish Physiology and Biochemistry* **43** (1): 265–278.

91 Yang, Y., Liu, Q., Xiao, Y. et al. (2018). Germ cell migration, proliferation and differentiation during gonadal morphogenesis in all-female Japanese flounder (*Paralichthys Olivaceus*). *The Anatomical Record* **301** (4): 727–741.

92 Wang, X. and Bhandari, R.K. (2020). The dynamics of DNA methylation during epigenetic reprogramming of primordial germ cells in medaka (*Oryzias latipes*). *Epigenetics* **15** (5): 483–498.

93 Adam, A.C., Saito, T., Espe, M. et al. (2021). Metabolic and molecular signatures of improved growth in Atlantic salmon (*Salmo salar*) fed surplus levels of methionine, folic acid, vitamin B6 and B12 throughout smoltification. *British Journal of Nutrition* **127**: 1–14.

94 Skjaerven, K.H., Mommens, M., Adam, A.-C. et al. (2022). Earlier or delayed seasonal broodstock spawning changes nutritional status and metabolic programming of growth for next-generation Atlantic salmon. *Aquaculture* **554**: 738187.

95 Gillette, T.G. and Hill, J.A. (2015). Readers, writers, and erasers chromatin as the whiteboard of heart disease. *Circulation Research* **116** (7): 1245–1253.

96 Ducker, G.S. and Rabinowitz, J.D. (2017). One-carbon metabolism in health and disease. *Cell Metabolism* **25** (1): 27–42.

97 Wang, Z.J., Long, H., Chang, C. et al. (2018). Crosstalk between metabolism and epigenetic modifications in autoimmune diseases: a comprehensive overview. *Cellular and Molecular Life Sciences* **75** (18): 3353–3369.

98 Xu, J. and Sinclair, K.D. (2015). One-carbon metabolism and epigenetic regulation of embryo development. *Reproduction, Fertility and Development* **27** (4): 667–676.

99 Saito, T., Whatmore, P., Taylor, J.F. et al. (2022). Micronutrient supplementation affects DNA methylation in male gonads with potential intergenerational epigenetic inheritance involving the embryonic development through glutamate receptor-associated genes. *BMC Genomics* **23** (1): 115.

100 Lubzens, E., Bobe, J., Young, G. et al. (2017). Maternal investment in fish oocytes and eggs: the molecular cargo and its contributions to fertility and early development. *Aquaculture* **472**: 107–143.

101 Skjaerven, K.H., Oveland, E., Mommens, M. et al. (2020). Out-of-season spawning affects the nutritional status and gene expression in both Atlantic salmon female broodstock and their offspring. *Comparative Biochemistry and Physiology Part A* **247**: 110717.

102 Lindeman, R.E. and Pelegri, F. (2010). Vertebrate maternal-effect genes: insights into fertilization, early cleavage divisions, and germ cell determinant localization from studies in the zebrafish. *Molecular Reproduction and Development* **77** (4): 299–313.

103 Veron, V., Marandel, L., Liu, J. et al. (2018). DNA methylation of the promoter region of *bnip3* and *bnip3l* genes induced by metabolic programming. *BMC Genomics* **19** (1): 677.

104 Wischhusen, P., Larroquet, L., Durand, T. et al. (2020). Oxidative stress and antioxidant response in rainbow trout fry exposed to acute hypoxia is affected by selenium nutrition of parents and during first exogenous feeding. *Free Radical Biology & Medicine* **155**: 99–113.

105 Perera, E., Turkmen, S., Simo-Mirabet, P. et al. (2020). Stearoyl-CoA desaturase (*scd1a*) is epigenetically regulated by broodstock nutrition in gilthead sea bream (*Sparus aurata*). *Epigenetics* **15** (5): 536–553.

106 Potok, M.E., Nix, D.A., Parnell, T.J. et al. (2013). Reprogramming the maternal zebrafish genome after fertilization to match the paternal methylation pattern. *Cell* **153** (4): 759–772.

107 Jiang, L., Zhang, J., Wang, J.J. et al. (2013). Sperm, but not oocyte, DNA methylome is inherited by zebrafish early embryos. *Cell* **153** (4): 773–784.

108 Craig, P.M. and Moon, T.W. (2013). Methionine restriction affects the phenotypic and transcriptional response of rainbow trout (*Oncorhynchus mykiss*) to carbohydrate-enriched diets. *British Journal of Nutrition* **109** (3): 402–412.

109 Irvine, N.A., Ruyter, B., Ostbye, T.K. et al. (2019). Dietary fish oil alters DNA methylation of genes involved in polyunsaturated fatty acid biosynthesis in muscle and liver of Atlantic salmon (*Salmo salar*). *Lipids* **54** (11–12): 725–739.

110 Xu, H., Dong, X., Ai, Q. et al. (2014). Regulation of tissue LC-PUFA contents, Delta6 fatty acyl desaturase (FADS2) gene expression and the methylation of the putative FADS2 gene promoter by different dietary fatty acid profiles in Japanese sea bass (*Lateolabrax japonicus*). *PLoS one* **9** (1): e87726.

111 Marandel, L., Lepais, O., Arbenoits, E. et al. (2016). Remodelling of the hepatic epigenetic landscape of glucose-intolerant rainbow trout (*Oncorhynchus mykiss*) by nutritional status and dietary carbohydrates. *Scientific Reports* **6**: 32187.

112 Panserat, S., Marandel, L., Geurden, I. et al. (2017). Muscle catabolic capacities and global hepatic epigenome are modified in juvenile rainbow trout fed different vitamin levels at first feeding. *Aquaculture* **468**: 515–523.

113 Perera, E. and Yufera, M. (2017). Effects of soybean meal on digestive enzymes activity, expression of inflammation-related genes, and chromatin modifications in marine fish (*Sparus aurata* L.) larvae. *Fish Physiology and Biochemistry* **43** (2): 563–578.

114 Podgorniak, T., Brockmann, S., Konstantinidis, I. et al. (2019). Differences in the fast muscle methylome provide insight into sex-specific epigenetic regulation of growth in Nile tilapia during early stages of domestication. *Epigenetics* **14** (8): 818–836.

115 Patel, M.S. and Srinivasan, M. (2002). Metabolic programming: causes and consequences. *Journal of Biological Chemistry* **277** (3): 1629–1632.

11

Microbiome, Epigenetics and Fish Health Interactions in Aquaculture

Sofia Consuegra, Tamsyn Uren Webster, and Ishrat Anka

Biosciences Department, Faculty of Science and Engineering, Swansea University, Swansea, UK

11.1 Introduction

Increasing understanding of the critical role of the microbiome (the complex community of endogenous microorganisms present in the gut and other mucosal surfaces, and their genes; Box 11.1) for human health has changed our perspective of health, disease, well-being, nutrition, and even behavior [1]. The Human Microbiome Project, launched in 2007 [2], helped uncover the bidirectional interaction between host and microbiome, and is opening the door to personalized medicine and nutrition [3]. More recently, the capacity of the microbiome to epigenetically regulate host gene expression, and the influence of environmental factors (like diet) on microbiome composition and function has further contributed to the development of precision medicine [4]. Following this trend, research on the microbiome in aquaculture has also increased exponentially for the last 10 years, with a large focus on disease, nutrition, and probiotics (Box 11.1), reflecting the growing importance of the fish-farming industry and the need for alternatives to both antimicrobials and fishmeal [5] to make its intensification sustainable. Epigenetics research in aquaculture has also increased in importance over the same period, although the integration of both epigenetics and microbiome is severely lacking compared to research in mammals and terrestrial model organisms (Figure 11.1), in part due to the much larger diversity of fish species, habitats, and diets. Recent reviews have comprehensively summarized the current knowledge about the fish microbiome and its importance for aquaculture [6–10]. This chapter, therefore, highlights the potential of combining research on microbiome and epigenetics for aquaculture in terms of sustainability and production.

11.2 The Fish Microbiome in Aquaculture

Fish represent 20% of the daily animal protein intake for a growing human population [11] and are key for the food security of millions of people. Thus, the importance of the aquaculture sector has grown continuously for the last 40 years [12]. However, aquaculture-intensive rearing conditions often result in high levels of stress in fish due to confinement, crowding, handling, and environmental mismatch, which can impair immunocompetence and increase disease susceptibility [13]. Increasing awareness of the consequences of antimicrobial resistance, as well as concerns regarding fish welfare, has prompted a search for alternative disease management approaches that allow aquaculture sustainable growth with a lower ecological footprint. As a result, there has been an exponentially increasing interest in the role of the fish microbiome on the health and performance of cultured fish over the last 10 years, particularly in the use of alternatives to fishmeal and fish oil, dietary additives, and pre/probiotics aiming to enhance growth, performance, and disease resistance [8, 14].

Epigenetics in Aquaculture, First Edition. Edited by Francesc Piferrer and Han-Ping Wang.

Box 11.1 Definitions Used in this Chapter

Alpha and beta microbiome diversity: Two measures used to describe organisms' microbiome diversity individually (alpha) or as community heterogeneity between two samples (beta). Alpha diversity measures include richness (number of taxa) or distribution (evenness) while beta diversity quantifies the differences in the overall taxonomic composition between samples. Phylogenetic information can be included in measures of both alpha and beta diversity.
Core microbiome: A group of microbial taxa commonly associated with a host taxon, usually present in a defined proportion of individuals and/or above a given abundance.
DNA methylation: The addition of methyl groups to cytosine bases in DNA, most often at CpG dinucleotides. DNA methylation in regulatory and coding regions alters associated gene transcription. Bacterial metabolites are known to alter the activity and availability of enzymes and cofactors essential for the regulation of host DNA methylation.
Dysbiosis: Alteration of the microbiome homeostasis with modifications in composition, diversity, or function (e.g., a reduction in diversity or an imbalance in the community composition with dominance of particular taxa). Dysbiosis modifies microbiome metabolic activity and results in changes in host disease susceptibility and/or maladaptation.
Fish microbiome: The fish microbiota and their genomes found in a particular environment.
Fish microbiota: The community of microorganisms (bacteria, viruses, eukaryotes, and protists) associated with the gastrointestinal tract, skin, gills, and olfactory mucosal tissues of fish. They can be resident (permanently associated with the organism) or transient (temporarily associated).
Gnotobiotic fish: Fish that are germ-free (axenic) or derived from germ-free eggs and colonized by well-characterized microbial lineages used as models for microbiome experiments. Comparisons between gnotobiotic fish and those conventionally raised (exposed to normal microbiota) are used to understand the mechanisms of host-microbiome interactions.
Histone modification: Posttranslational modifications to histone proteins that collectively alter chromatin state and the accessibility of genes for transcription. Metabolites of microbial origin (such as SCFAs) induce modifications in the chromatin structure of the host animal.
Microbiome engineering: Selection of the microbiome to modify its function and improve host performance. This can be done by indirect selection or by artificial selection applied to the host [91]. It is a very promising, but still underused, avenue to increase resilience to stress and disease in aquaculture.
Microbiome-epigenome crosstalk: A complex series of interactions mediated by microbial and host gene expression and metabolite production. Together, metabolites of both microbial and host origin regulate aspects of the host epigenome and transcriptome and the microbiome.
Noncoding RNA regulation: Collectively various ncRNAs (including micro RNAs and long noncoding RNAs) that regulate gene expression at transcriptional and posttranscriptional levels. Host ncRNAs are differentially expressed in the presence of microbiota and their metabolites.
Prebiotic: Nondigestible food ingredient (dietary fiber) that favors the growth of particular groups of commensal microorganisms that exert beneficial effects on the host's health or well-being.
Probiotic: Feed supplements consisting of live microorganisms that confer nutritional and immunological benefits to the host by manipulating their microbial community. The use of probiotics has increased in aquaculture as substitutes for antimicrobials.
SFCAs: Short-chain (less than six carbon atoms) fatty acids produced by commensal bacteria in the intestine by fermenting carbohydrates and fiber that play a role in the host neuroendocrine and immune regulation, potentially through epigenetic mechanisms.

11.2.1 The Fish Microbiome Diversity and Composition

Animals are associated with a diverse community of microorganisms that colonize all body surfaces exposed to environmental factors, including skin, respiratory, and gastrointestinal systems, influencing their immunity, extending their metabolic repertoire, and affecting their nutritional status, behavior [15], and ultimately fitness [16]. In vertebrates, the microbiome influences both host physiological activity and health through the release of metabolites [17], including neurotransmitters [18], hormones [19], vitamins [20], and short-chain fatty acids (SFCAs). As for other vertebrates, the fish microbiota (Box 11.1) consists of a community of bacteria, viruses, eukaryotes, and protists associated with mucosal tissues in the gastrointestinal tract [21], skin [22], and gills [23]. The diversity and composition of the gut microbiome, and

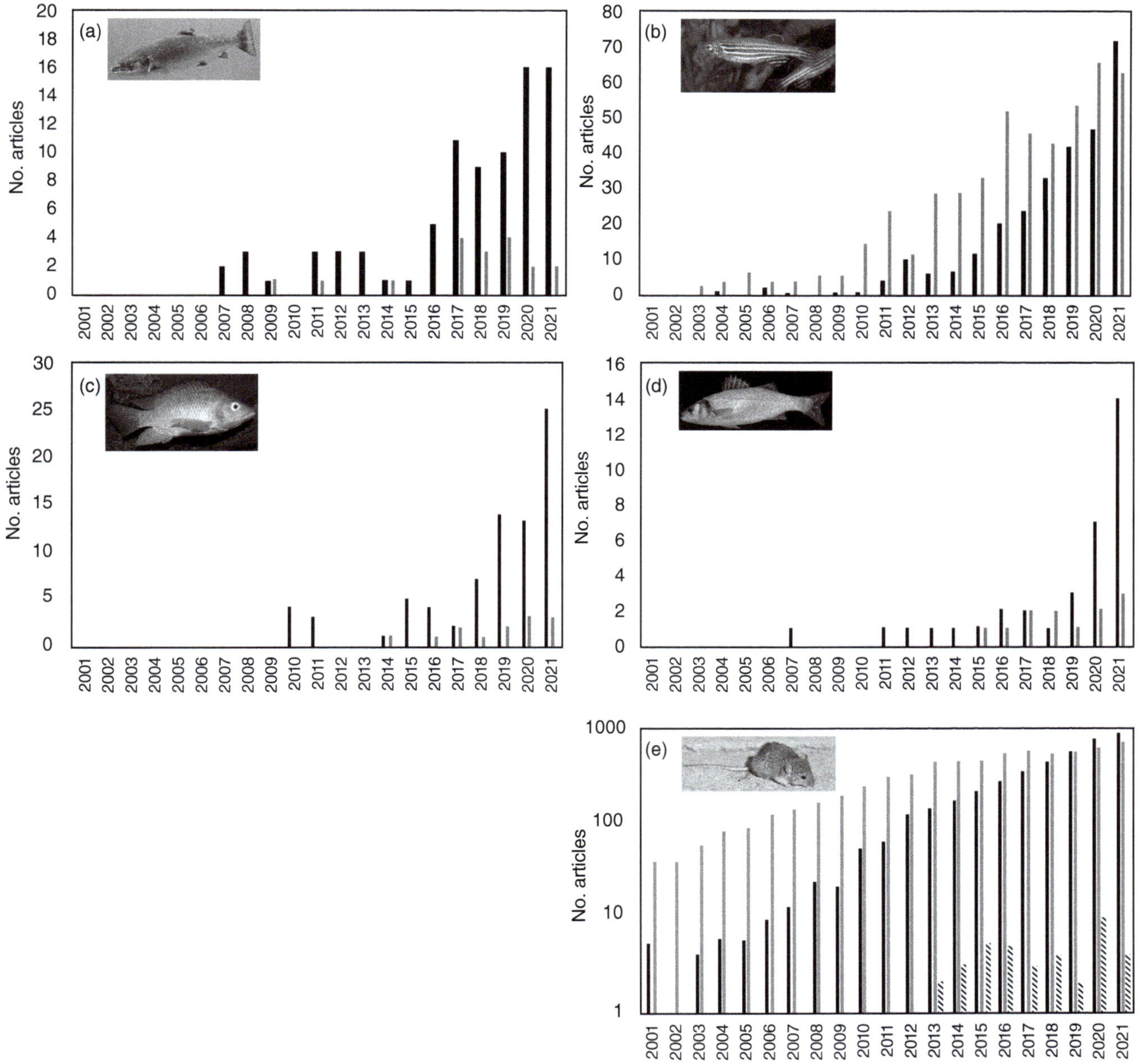

Figure 11.1 Number of articles published on the microbiome or epigenome for select species up until the end of 2021. (a) Atlantic salmon *Source:* Hans-Petter Fjeld/Wikimedia Commons/CC BY SA 2.5, (b) zebrafish *Source:* Azul/Wikimedia Commons/public domain, (c) Nile tilapia *Source:* Bjørn Christian Tørrissen/Wikimedia Commons/CC BY SA 3.0, (d) European sea bass *Source:* Hans Hillewaert/Wikimedia Commons/CC BY SA 4.0, and (e) mouse *Source:* Amirekul/Wikimedia Commons/CC BY SA 4.0. Searches were conducted in PubMed (3 November 2021) for terms in the Title or Abstract as follows: (((Common name) OR (Species name)) AND ((Microbiome) OR (Microbiota))); (((Common name) OR (Species name)) AND ((Epigenome) OR (Epigenetic))); and (((Common name) OR (Species name)) AND ((Microbiome) OR (Microbiota)) AND ((Epigenome) OR (Epigenetic))). Review articles were excluded. Black: microbiome, Gray: epigenome; Hatched: microbiome and epigenome. (*See colour plate section for colour representation of this figure.*)

increasingly that of the skin and gills, has been described in many fish species. An influence on immunity and nutrient absorption is widely assumed, although the mechanisms underneath this bidirectional relationship are largely still to be determined [21].

Compared to terrestrial animals, the outer mucosal surfaces (gills and skin) of fish and their food items are constantly in contact with water, which provides increased opportunities for microbial colonization [21]. Fish are also much more diverse than any other vertebrate group, in a number of species, habitat conditions, and diet adaptations, potentially contributing to microbiome diversity as well. Most of the available information refers to a reduced number of model species with genome

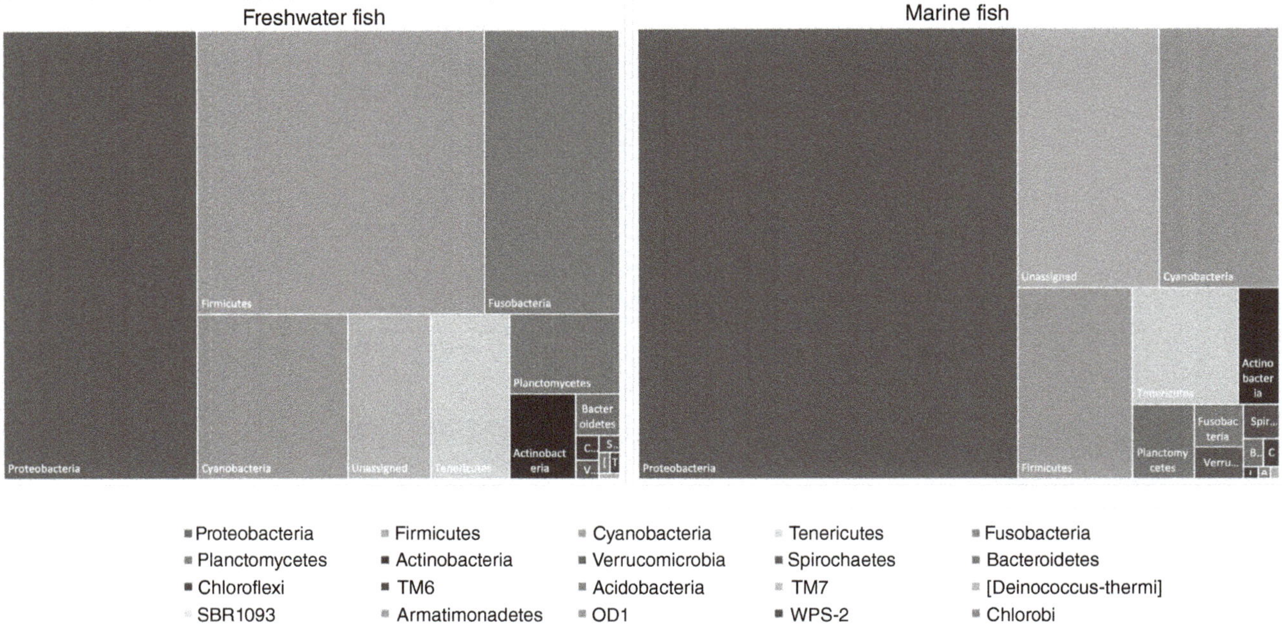

Figure 11.2 Relative abundance of different bacterial taxa in the gut microbiota of fish from freshwater and marine habitats based on data from 14 orders, 42 families, 79 genera, and 85 species. *Source:* Adapted from data in Kim et al. [35]. (*See colour plate section for colour representation of this figure.*)

availability, e.g., zebrafish (*Danio rerio*), three spine stickleback (*Gasterosteus aculeatus*) [24], or species of economic value (mainly salmonids and some cyprinids like carps [25], although the number of studies exploring wild fish populations (e.g., [26–28]) and additional aquaculture species (e.g., [29, 30]) is increasing. Fish microbiome research has been largely focused on marine fish [31], predominantly on their symbiotic bacteria [32]. Current studies on freshwater fish literature refer to less than 1% of freshwater fish species, predominantly on Cyprinidae, followed by Salmonidae and Cichlidae, reflecting the most common and economically important aquaculture species including common carp (*Cyprinus carpio*), Atlantic salmon (*Salmo salar*), rainbow trout (*Oncorhynchus mykiss*), and Nile tilapia (*Oreochromis niloticus*) [25]. Different fish organs (gut, skin, gills) have specialized microbiome compositions, with the skin being more influenced by water bacterioplankton and abiotic factors than the gut, which tends to be mainly determined by diet and life stage.

Despite the diversity of species and environmental conditions, 16S rRNA sequencing has identified that Proteobacteria, Firmicutes, and *Bacteroidetes* represent up to 90% of the gut microbiota in most species [33]. This is also supported by a comparison of the gut microbiome of 12 wild fish and three shark species that identified the dominant phyla including Actinobacteria, Bacteroidetes, Firmicutes, Fusobacteria, Planctomycetes, Proteobacteria, and Tenericutes [34], and by a broad species comparison of freshwater and marine fish [35] (Figure 11.2). In contrast with the mammalian gut, which is dominated by Firmicutes and Bacteriodetes, Proteobacteria predominate in the fish gut, with Firmicutes and Bacteriodetes in much lower abundance [34]. *Terenicutes*, typically *Mycoplasma* sp., have commonly been found to dominate the gut microbiome of adult farmed salmonids, although their role in the microbial community is largely unknown [36]. Also, in contrast to mammals and birds, the dominant indigenous lactic acid-producing bacteria in fish include *Lactobacillus*, *Lactococcus*, *Leuconostoc*, *Enterococcus*, *Streptococcus*, *Carnobacterium*, *Weissella*, and *Pediococcus* [37], but not *Bifidobacterium*. Some of these, such as bacteria from the genus *Carnobacterium* [38], are common in healthy fish and widely used as probiotics because of their beneficial effects on gut health and ability to inhibit the growth of fish pathogens [39, 40], but pathogenic species from the genera *Streptococcus*, *Enterococcus*, *Lactobacillus*, *Carnobacterium*, and *Lactococcus* have been identified to cause mortalities in aquaculture settings and need to be controlled [37].

11.2.2 Extrinsic and Intrinsic Factors that Affect Fish Microbiome Composition

The abundance and composition of the fish microbiome are affected by environmental factors including temperature, season, and diet, as well as by the host genetics [23], although the species-specific core gut microbial communities seem largely resilient to the rearing environment [41]. Here, we summarize some of the main extrinsic and intrinsic factors influencing the composition of the fish microbiome, with particular emphasis on aquaculture conditions.

11.2.2.1 Extrinsic Factors (Habitat, Diet, and Stress)

A meta-analysis of 12 fish species representing different habitats (marine, freshwater, or estuarine) and trophic levels (herbivorous, carnivorous, and omnivorous) identified marked differences in gut microbiota composition determined both by *salinity* and type of diet, and also by the method of sampling, although not by the fish-rearing environment (laboratory, aquaculture, or natural habitat) [26]. A comprehensive study including 14 orders, 42 families, 79 genera, and 85 fish species identified fish habitat as the major determinant of the gut microbiome composition, differing significantly from that of other vertebrates [35]. Habitat-related gut microbiome differences between marine and freshwater fish can also be identified in single species, for example, in gnotobiotic (colonized only by known microorganisms or germ-free; Box 11.1) three-spine stickleback (*Gasterosteus aculeatus*) [42]. In this case, while marine and freshwater sticklebacks shared the core intestinal microbiota (consisting of 12 different genera from the Proteobacteria, Firmicutes, and Actinobacteria phyla), marine fish mounted a stronger immune response to resident microbiota compared to their freshwater counterparts, potentially reflecting the adaptation to the less environmentally variable freshwater environment [42]. Temperature also has an important effect on the diversity of the fish gut microbiota in salmonids [43, 44] and seems to affect alpha diversity (individual) and, particularly, beta diversity (community differentiation) of fish skin communities, which can be also influenced by dissolved oxygen [45].

Diet strongly influences the microbiome; a varied diet is consistent with a more varied gut microbiome, with omnivorous species having the highest richness and phylogenetic diversity [34]. This is also reflected in the metabolic role of the microbiome, as the functional profile of the gut microbiome in the omnivorous freshwater pufferfish (*Tetraodon cutcutia*) was similar in protein metabolism to carnivorous salmon, but to herbivorous grass carp in carbohydrate metabolism [46]. This has important implications for aquaculture. On the one hand, wild individuals seem to have a more diverse microbiome than their captive-bred counterparts, although sharing a core microbiome, as seen both in zebrafish [41] and mumimichogs (*Fundulus heteroclitus*) [34]. On the other hand, in addition to the reduction of disease susceptibility and the search for alternatives to antimicrobials, the sustainability of aquaculture also depends on finding alternatives to the use of fishmeal and fish oil in commercial feeds, particularly for carnivorous species. Plant-derived components, soybean or grains, and more recently microalgae and yeast, have been used as alternatives or supplements to cover the fish's nutritional needs [30] but some of them cause enteropathy and inflammation of the fish intestine [47], as well as potential imbalance in the microbiome [30], which can have unpredictable effects. Soybean, for example, alters the proportion of *Bacteroidetes* in the gut, which can be compensated for by using amino acid supplementation with alanine-glutamine that reduces the relative abundance of *Bacteroides* spp., and the inflammation effects. While Ala-Glu supplementation can alleviate the inflammatory effects of soybean, decrease the expression of the proinflammatory cytokine (TNF-α), and enhance the digestive and absorptive ability, it also reduces the microbiome diversity, which could result in the loss of the barrier effect provided by the microbiome [48]. The diet can also have a synergic effect with other aquaculture conditions, as shown in the yellowtail kingfish (*Seriola lalandi*), where the substitution of fish meal for soya protein concentrate combined with housing at high temperature increased the abundance of *Photobacterium* sp. in the skin, affecting growth and immune defense mechanisms [49].

The *aquaculture environment* results in differences in the fish microbiome compared to the natural environment of the species, despite the maintenance of the core microbiome in many cases. In the yellowtail kingfish, for example, Proteobacteria was the most abundant phylum in the wild while Firmicutes was the most abundant phylum in the aquaculture-reared populations, with differences between both rearing environments related to amino acid and carbohydrate metabolism, strongly linked to the different diets [29]. Marked differences were also found in the skin and mostly the gut of wild and hatchery Atlantic salmon, with the hatchery fish displaying not only different composition, but also lower bacterial diversity, as could be expected from farming conditions consisting of less variable diet and environment [50]. However, a temporal comparison of the water and biofilm of semiclosed containment systems (S-CCS) in the sea and on land recirculating aquaculture systems (RAS) for Atlantic salmon identified changes in the microbiota between both types of systems, depending on salinity, and importantly found reservoirs of potential pathogens that could affect the health and performance of the fish held in those systems [51].

Fish farming intensification is often associated with increased *stress*, not only mainly due to crowding and confinement, but also handling or disruption of social behaviors. Stress impairs immunocompetence, both in wild and farmed fish [52], and can have long-lasting effects on the immune system, potentially increasing susceptibility to disease [53, 54]. Early life stages are particularly sensitive, reflecting developmental plasticity during the maturation of the vertebrate nervous and immune system, as well as periods of epigenetic reprogramming and microbiome assembly [55]. The microbiome–brain–gut axis is essential for maintaining fish homeostasis through the previously mentioned interaction between mutualistic gut microbes and the host physiological processes, modulating immunoregulation and maintaining metabolic equilibrium [56]. The fish microbiome has been shown to be disrupted by a number of different aquaculture-related stressors. Given the importance of the microbiome to many aspects of host health, dysbiosis (i.e., disruption of the microbiome homeostasis; Box 11.1) of the host-microbiome relationship is likely to compound the adverse impacts of stress directly on the animal. An increase in cortisol has been related to changes in the diversity and structure of the fish gut microbiome [57], which may, in turn, contribute to the adverse effects of stress on animal health [56]. Stress disrupts microbial communities associated both with the gut and skin [58, 59], but with different levels of resilience. The skin microbiome seems less impacted by changes in localized cortisol concentration [57], while the gut microbiome recovers more quickly from pH stress [60]. Different types of stress, acute and chronic, also have different responses in the gut and skin microbiome. While acute stress during late embryogenesis has been shown to have a lasting effect on the skin microbiome, likely due to disruption of initial seeding communities, the effect is not so clear on the gut microbiome, while chronic stress seems to have a more pronounced effect in the gut [61]. This suggests that the indigenous microbiota can be a good indicator of physiological stress and potential immunocompromise [62]. Although it is difficult to generalize among species and conditions, changes in the abundance of particular microorganisms could also be used as indicators of stress. For example, in Atlantic salmon, an increase in the levels of fecal cortisol resulting from experimental stress was associated with significant changes in the gut microbiome (although not in the skin), in particular with the pronounced decline of the *Carnobacterium* sp. [57], while in the Eurasian perch (*Perca fluviatilis*), both predator and low food ration stress resulted in an increase in *Fusobacteria*, particularly *Cetobacterium* [59]. In salmon, some microbiome changes in response to acute and chronic stress have common signatures (e.g., increases in representatives from *Acinetobacter* and *Aeromonas*, which include several important opportunistic fish pathogens), which could be useful as general microbial biomarkers of stress [57].

However, the microbiome is also involved in the host's adaptation to stress [63]. For example, temporal changes in the microbiome of the milkfish (*Chanos chanos*) subjected to high temperatures, which correlated to changes in the hepatosomatic index (HSI, as a proxy for fish energy status), were attributed to habituation to high-temperature stress, potentially mediated by microbiome effects [64]. Osmotic stress in the Japanese rice fish (*Oryzias melastigma*) indicated that the resulting changes in the microbiota induced the production of different metabolites and the activation of different metabolic pathways that helped the resident bacteria survive the stress, and potentially maintain the biological function of the gut [65], contributing to the resilience of the host. The gut microbiome has been hypothesized to play a role in the response to crowding stress in the snout bream (*Megalobrama amblycephal*), for which high densities affected not only the growth but also the composition and abundance of the gut microbiome and the metabolites it produced [66]. Effects of the microbiome seem to be generalized even in distant taxa. Thus, the microbiome, together with epigenetic adaptation, has also been proposed to aid in the transgenerational acclimatization of coral reef organisms to environmental stress due to climate change, though changes in abundance that could be selected and vertically transmitted [67].

11.2.2.2 Intrinsic Factors (Host Genetics, and Age)

The influence of host genetics on the microbiome is variable and depends on the host species, the organ analyzed, and the environmental conditions where species live [27]. The individual genotype has been shown to influence the composition of the gut microbiome, and also to some extent the skin and gill microbiome [58, 68]. In brook trout (*Salvelinus fontinalis*), for example, three QTLs (quantitative trait loci) were found to have a major effect on the abundance of three genera in the skin, *Lysobacter*, *Rheinheimera*, and *Methylobacteriumin*, all were involved in the production of antimicrobial compounds. The QTLs explained between 17 and 41% of the variance in the abundance of the three genera, with *Methylobacteriumin* presence being negatively correlated with the pathogen *Flavobacterium psychrophilum*, indicating a role of the host genetics in controlling the microbiome composition in relation to its own fitness [58]. However, there seems to be weak evidence in fish of microbiota phylosymbiosis (higher intraspecific than interspecific similarity in the structuring of the microbial communities) [69, 70]. Thus, while in Amazonian fishes, the gut (but not the skin) microbiome is mainly modulated by species-specific factors [27], the gut microbiome composition of species from the *Gadidae* and *Sparidae* families displays a larger influence of environmental factors (such as diet or temperature) than of their phylogenetic relationship [70, 71].

The microbiome colonization of the gut is species-specific and initially originates from the egg chorion, then is influenced by the surrounding water and, strongly, by first feeding [31, 72]. Afterward, changes in the microbiome composition are observed with *age* and the development of the immune system, although some interact with physiological changes induced by migration, for example. In wild Arctic char (*Salvelinus alpinus*), the main driver of gut microbiota changes seems to be seasonal salinity changes, but an influence of age was also found in fish samples at the same time [73]. Similarly, a comparison of chinook salmon (*Oncorhynchus tshawytscha*) farmed under freshwater and marine conditions revealed higher microbial diversity in freshwater, but also an effect of fish age on the beta diversity in both the marine and freshwater environments [74].

11.2.3 Microbiome Interaction with Fish Health and Immunity

Aquatic species are, by nature, often exposed to pathogenic and opportunistically pathogenic microorganisms in the surrounding water [75]. Interactions between the host and its microbiome play a key role in maintaining long-term fish health [76]. This is because, not only does the composition of the microbiota affect the host immune response and its susceptibility to disease, but environmental conditions, diet, health, and the genetic background of the host also influence their microbiome diversity. In mammals, the normal functioning of the innate and adaptive immune systems has been related to the maintenance of healthy intestinal microbiota, potentially through epigenomic mechanisms involving host epithelial and immune cells, and the microbiome [77]. However, the understanding of the functional interplay between host immunity and the microbiome in fish is more restricted, despite increasing information about fish microbiota composition [78]. The disruption of the host-microbiome functional relationship (dysbiosis) has been associated with human chronic diseases [79], and interactions between innate and adaptive immune responses and the gut microbiome are increasingly well-known in humans and rodents [20]. In mice, for example, certain strains of *Clostridium* that are part of the intestine indigenous microbiota control the number and function of $CD4^+$ regulatory T cells, which are known to be needed for maintaining immune homeostasis in mice and humans [80]. Similarly, in zebrafish, the disruption of the intestinal microbiome by chemical exposure to plastic-derived phthalates (DEHP) suggests a link between the microbiome and the adaptive immune mechanisms in the host intestine. Exposure to DEHP increased Bacteroidetes and Gammaproteobacteria and decreased Verrucomicrobia in both the male and female gastrointestinal system, as well as resulted in other sex-specific microbiome changes. These alterations could modify their metabolite production, potentially mediating the response of the adaptive immune system to DEHP, which includes increasing Th1/Th2 immune response [81]. The microbiome of the skin and gills can also reflect gut health. For example, the dysbiosis produced by enteritis in the yellowtail kingfish (*S. lalandi*) results in a reduction in diversity and an enrichment in Proteobacteria and Actinobacteria, mainly associated with the early stages of the disease [82]. Resident bacteria are essential for maintaining fish individual health, by creating an environment that excludes pathogens through competitive exclusion mechanisms [83, 84], such as the exclusion of *Aeromonas* by lactic acid bacteria in salmonids [85]. Resident microbiota also protect the host against the development of inflammatory diseases. Research in germ-free zebrafish indicates that colonization by commensal microorganisms in newly hatched fish induces proinflammatory and antiviral gene expression, increasing the resistance to viral infections [86]. One of the mechanisms for this interaction is through microbial-derived SCFAs that are produced by commensal bacteria in the intestine by fermenting carbohydrates and fiber, such as acetate and propionate produced by the *Bacteroidetes* phylum. These SCFAs can bind and activate the G-protein-coupled receptor GPR43, which in humans and mice is highly expressed in neutrophils and eosinophils, and whose expression is closely linked to receptors important for innate immunity [87]. This relationship between gut microbiome and immunity can be exploited to develop preventive and therapeutic measures against fish disease [84]. For example, it has been suggested that this SCFA–GPR43 pathway can be related to the effects of diet and pre- and probiotics on the immune response [87].

11.2.4 Microbiome Engineering

The extensive knowledge on the relationship between microbiome and human health has paved the way for potential applications of microbiome engineering in farmed animal systems that have the advantage of diet and genetic diversity being more standardized and susceptible to be manipulated [88]. This can be particularly important to reduce the use of antibiotics in intensive aquaculture, as many antimicrobial drugs currently used have been classified as critically important or highly important for human use by the World Health Organization and there is a strong positive correlation between human clinical antimicrobial resistance and aquaculture-derived multi-antibiotic resistance indices [89]. An increase in awareness about the risks of antimicrobial resistance has played an important role in the development of new measures of disease control and prevention, including microbiome engineering [90, 91]. While vaccination has been effective in reducing antimicrobial use for the Norwegian salmon industry [92], vaccines are not available for many diseases and are unsuitable when animals are infected by multiple opportunistic pathogens [92]. Modulation of the microbiome (and epigenome; see later) is already being recognized as mediating lasting adverse effects of stress on immunity, offering a promising way of promoting animal resilience to stress and disease, but this has yet to be widely exploited in fish farming [93, 94]. Nutritional additives, such as pre- and probiotics, are increasingly used to enhance the immune system, improve growth, and can be used for disease prevention by enhancing animal resilience to disease and stress [95]. Probiotics (live exogenous microorganisms) that enhance the composition of resident communities can be used to maintain host health along with prebiotics (nondigestible food ingredients that favor the growth of particular groups of commensal microorganisms), although in fish, the detailed knowledge about their mechanisms of action is still limited [96]. Probiotics applied to feeds colonize the animal intestinal tract, ensuring varied benefits to the host [97]. For example, the intake of probiotics improves the digestibility of nutrients through enhanced enzymatic activity. Moreover, probiotic bacteria can compete with potentially pathogenic microorganisms for nutrients and adhesion sites within the gut of the host, being able to positively modulate the microbial community of the gut. Besides that, a diverse range of probiotic species can produce antimicrobial molecules, minimizing the growth of pathogenic bacteria, stimulating an immune response, and increasing fish resistance against diseases. Probiotics have proved to be beneficial for the fish immune response by helping modified microbiota to return to their normal status and also by activating the complement system or modulating the secretion and anti-inflammatory cytokines [84].

Probiotics are also effective at minimizing the stress effect of handling [98] and can enhance growth [99], representing a strong and bio-secure alternative to the use of antibiotics for the industry [100]. The growing interest in probiotics within the aquaculture industry has been extensively reviewed (see References [95, 101, 102] for examples). Some common probiotics like *Bacillus* and *Lactobacillus* can stimulate the expression of inflammatory cytokines in the fish gut [103] or increase phagocytic activity and resistance against particular pathogens [104]. Thus, the use of host-associated probiotics has been identified as one of the promising avenues to avoid dependence on those derived from nonaquatic sources (e.g., *Bacillus*, *Lactobacillus*, *Enterococcus*, and *Carnobacterium*) [101].

11.3 Microbiome-Epigenome Interactions

The link between environmentally induced epigenetic modifications, the microbiome composition and function, and how they can influence the host remains an open question in fish [105]. Early-life acute stress has been shown to have stimulatory effects on immunity [54], and modify the epigenome in a developmental-stage-specific manner [54, 106]. There is also growing evidence that epigenetic modifications, either through the nutrition of the parents or during early development, can affect fish growth, muscle development, and meat quality, suggesting that epigenetic programming could be applied to enhance production and disease resistance in aquaculture [107]. Moreover, it is now clear that the farming environment results in rapid epigenetic changes that can affect more than one generation [108]. Thus, the same environmental factors that affect the microbiome (temperature, diet, host genotype, and age) are influencing the epigenome, with consequences for the health and performance of the host, suggesting a common ground for incorporating microbiome and epigenetic engineering in management practices to improve animal fitness as a promising solution to achieve sustainable aquaculture intensification [109] (Figure 11.3).

11.3.1 Mammals and Model Species

Characterizing the interactions between the microbiome and the epigenome in humans and mammalian models is an active area of current research. However, the complexity of these interactions means that much is still unknown, with new links constantly being identified. The predominant focus has been on the impacts of gut microbiota, and their metabolites, on the epigenome and the epitranscriptome of different tissues (including gut, brain, and liver) in the host animal, although epigenetic regulation of microbiome composition and function is also well-documented [110–112], and recent evidence of epigenetic changes within microbiotas themselves is increasing [113]. Together, a highly complex landscape of interactions between the microbiome and the epigenome and transcriptome is being uncovered, with metabolites of both bacterial and host origin regulating the crosstalk between host epigenome/transcriptome and the microbiome [112, 114, 115]. Diet fundamentally influences the nature of these interactions in mammals, and the disruption of this crosstalk is clearly implicated in the etiology of many diseases, particularly metabolic and inflammatory-type conditions [110, 116–118]. Recent reviews cover how the microbiome interacts with the host epigenome via chromatin remodeling, DNA methylation, and noncoding RNAs in detail, and identify key knowledge gaps [111, 119, 120].

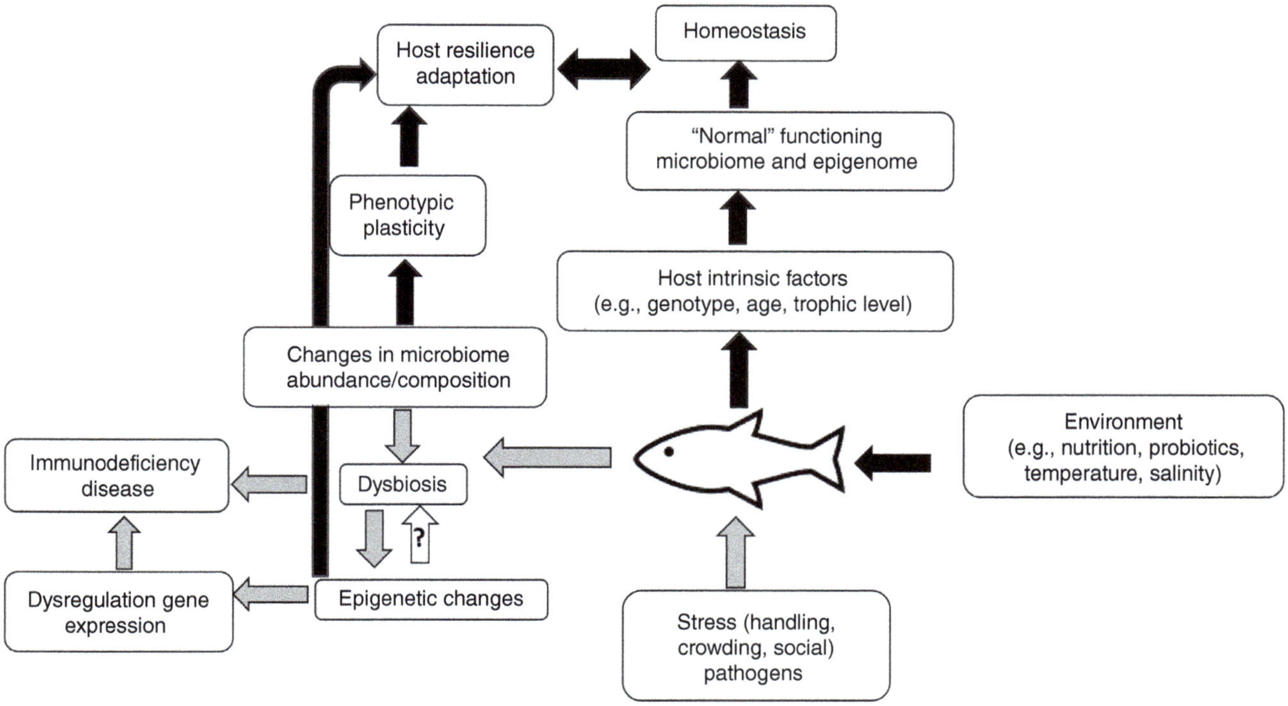

Figure 11.3 Schematic representation of the different host-related extrinsic (environmental) and intrinsic factors affecting the fish microbiome in aquaculture settings, the potential interaction with the epigenome and the health consequences for the host. Gray/black arrows indicate negative and positive impacts on the host's health respectively.

11.3.1.1 Histone Modifications

Perhaps the best understood and most widely characterized interaction is the modification of histones by microbial metabolites [121]. This reflects not only a long history of study, but also more opportunities for interaction with diverse microbial metabolites. In particular, microbial SCFA production has been well documented to alter histone acetylation via the inhibition of histone deacetylases (HDACs). SCFAs, including acetate, propionate, and butyrate, are produced through the breakdown of complex carbohydrates by gut bacteria, including *Bifidobacterium* sp., a common probiotic member of human microbiota [122]. SCFAs have diverse and important impacts on many aspects of host physiology, including energy metabolism, tissue proliferation, and inflammatory responses, with histone modifications thought to be one of the key ways that SCFAs induce these physiological impacts [121]. Using gnotobiotic models, dietary-induced microbial SCFA production has been shown to alter histone acetylation in the gut, and more recently, in the liver and adipose tissues, where it has been associated with considerable changes in the transcription of metabolic genes [123]. Manipulation of microbiota in rodent models has been used to demonstrate the impact of microbial-derived SCFAs on other histone modifications too, with histone crotonylation appearing to be a particularly important, recently discovered mechanism for microbiome-host crosstalk in the gut [119, 121]. Changes in histone methylation also demonstrate how gut microbes can induce epigenetic changes associated with the induction of inflammatory conditions in the intestine [124]. Apart from SCFAs, intestinal microbiota secrete a suite of other metabolites, some of which are also known to directly influence chromatin modification. These include folate, which is synthesized by bacteria and needed for methylation of histones and DNA [125], and some polyphenols that alter host HDAC and histone acetylase (HAT) activity whilst also regulating microbial presence [115]. HDACs also play an important role in regulating the presence of gut bacteria, with HDAC-null mice showing considerably altered microbiomes, as well as intestinal epithelial integrity [126]. The stimulation of histone acetylation of the FoxP3 locus mediated by SCFAs produced by the microbiota of the large intestine seem to be also involved in T-cells' anti-inflammatory effects [77].

11.3.1.2 DNA Methylation

DNA methylation patterns have also been strongly associated, in some cases causally, with the occurrence of various microbiota (pathogenic and commensal) within the intestinal microbiome. For example, in gnotobiotic mice, the presence of commensal microbiota induces genome-wide changes in DNA methylation, including localized changes at TET2/3-dependent regulatory regions that are necessary for the proper expression of a suite of developmental genes crucial for the normal development and maintenance of intestinal homeostasis [127]. In humans, microbiota composition, especially ratios of Bacteroidetes to Firmicutes, has been associated with widespread differential methylation of genes involved in glucose and energy metabolism and obesity [128]. As with histone modifications, it appears that microbial regulation of DNA methylation is due to either alteration in the availability of chemical donors or modulation of enzymes responsible for methylation [111]. S-Adenosylmethioine (SAM) is the main methyl-donating substrate for methylation enzymes, DNA methyltransferases (DNMTs), and histone methyltransferases (HMTs). Its generation is dependent on microbially derived molecules including folate and other B vitamins. Modulation of diet and microbiota alters folate availability for SAM production and, hence, DNA methylation. This has been well associated with obesity and metabolic-type diseases [118, 129]. Recently, stimulation of folate production by *Bacteroides* sp., and subsequent SAM availability, has been shown to alleviate hepatic steatosis in rats [130]. Alteration of DNMT activity has also been shown to be directly associated with microbial metabolism of a number of bioactive dietary compounds, including some flavonoids and polyphenols [131].

11.3.1.3 Noncoding RNAs

Host microbiome interactions mediated by noncoding RNAs (ncRNA) have recently received considerable attention. Microbial regulation of micro-RNAs (miRNAs) and long noncoding RNAs (lncRNAs) has been the most widely documented [110, 120]. Germ-free and conventional rodent models have been used to show that both miRNAs and lncRNAs are differentially expressed in the presence of gut microbiota; however, while some studies identified specific bacteria involved, the exact mechanisms of this microbial regulation are not well known [110, 132–134]. Disruption of the microbiome, and subsequent dysregulation of miRNAs and lncRNAs, have also been implicated in several diseases including obesity, diabetes, and colon cancer [110, 134, 135]. Diet-induced changes in the microbiome are also known to modulate host miRNAs, which are differentially expressed in humans with omnivorous, vegetarian, and vegan diets [136]. The role of circular RNAs (circRNAs), which "sponge" and inhibit the activities of miRNAs, in modulating the interactions between miRNAs and microbiota adds an additional layer of complexity to this crosstalk [120]. In the brain, interactions between circRNA, miRNA, and circulating microbial metabolites have been studied to some extent and implicated in several neurological disorders, but this has been identified as a considerable knowledge gap and priority for future research [120]. In addition, the potential role of other ncRNAs in mediating host-microbiome crosstalk is yet to be widely explored. A recent study, using gnotobiotic models, was the first to demonstrate how tRNA expression profiling is remodeled in a tissue-specific manner in the presence of microbiota [137]. Few studies have examined the impacts of endogenously expressed host ncRNAs on microbiota. However, a landmark study by Liu et al. [138] showed that miRNAs secreted by gut epithelial cells enter and regulate gene expression in gut bacteria, modulating their abundance.

11.3.2 Microbiome and Epigenetic Interactions in Aquaculture

Independently, the role of the microbiome and the epigenome on fish health, nutrition, and development are active foci of research in aquaculture. It is becoming clear that both the host-associated microbiome and the epigenome are sensitive to environmental factors, including diet, stress, and other conditions experienced in aquaculture, and that both are likely to have an extensive impact on animal health and productivity. Emerging research based on mammalian models suggests that these interactions may be extensive and help to explain previously unaccountable complexity.

Studies on diet and nutritional conditioning constitute a major focus for aquaculture research. Both the microbiome and the epigenome are considered key mechanisms by which diet can be used to promote animal health, welfare, and productivity. Perhaps surprisingly, potential interactions between these mechanisms have not been directly studied, and only two recent studies on zebrafish have administered nutritional supplements, designed to alter the gut microbiome, to examine potential epigenetic effects. Cuomo et al. [139] administered a probiotic (*Lactobacillus rhamnosus*) for 28 days and characterized some small changes in promoter DNA methylation and transcription of *BDNF* and *Tph1A* in the brain and gut associated with behavioral changes. Shirani et al. [140] administered a yeast-based prebiotic for 90 days and found some evidence of differential global DNA methylation. However, neither of these studies confirmed that specific probiotic-induced changes in the microbiome were associated with epigenome modulation. While there is a lack of direct evidence for diet-microbiome-epigenome interactions, there are several examples by which similar dietary treatments have been shown to induce both microbiome and epigenetic changes, for example, the inclusion of plant-based products such as soybean in zebrafish diets [141, 142]. As for mammalian systems, there is likely to be a fundamental interaction between diet, gut microbiome, and epigenome, with SCFAs representing a critical link. The beneficial effects of SCFAs on intestinal health, as well as wider fish health, together with the essential role of microbiota in synthesizing them, is well recognized [143, 144]. A considerable amount of recent research has focused on developing diets and nutritional supplements (i.e., prebiotics such as immunosaccharides FOS, GOS, and XOS) that promote SCFA-producing bacteria (such as *Lactobacillus* and *Bifidobactrium*) or directly introduce these probiotic bacteria with the same aim [144]. Increasing SCFA circulation has been shown to reduce intestinal inflammation, promote immune response, and improve feed utilization and growth rates [96, 143]. While the likely interactions between SCFAs and the epigenome have been recognized [143, 144], these interactions have not been investigated to date.

An improved understanding of how environmental conditions, including stressors, experienced in aquaculture affect animal health is another driver for research examining microbiome-epigenome interactions. Intensive aquaculture is often associated with stress, for example, due to crowding, handling, pollutant exposure, or lack of natural stimuli [145, 146]. As with diet, a number of aquaculture-relevant stressors have been shown to induce changes in both the microbiome and the epigenome, but the potential for interactions between them, and how this contributes to animal health effects, has not been considered. Thermal and pollutant-associated stressors have been widely studied and therefore give the clearest indication of likely interactions. For example, exposure to thermal stress has been shown to induce lasting epigenetic changes and, in some cases, were associated with various health effects including thermal tolerance, immunocompetence, and performance in a number of fish species [54, 147–150]. Thermal modulation of the microbiome has also been widely demonstrated in fish and shellfish, in some cases, with lasting disruptive effects on the host-microbiome symbiosis, even after the original stressor has subsided [61, 64, 151, 152]. Early life stages are particularly sensitive to environmental stressors and represent windows of heightened sensitivity for modulation of both the epigenome and microbiome through periods of epigenetic reprogramming and initial microbial colonization and proliferation. Exposure to stressors during early life is known to induce particularly marked and/or lasting effects on the epigenome and microbiome with associated animal health effects [153, 154].

11.4 Gaps in Knowledge and Future Research Avenues

Modulation of both the microbiome and epigenome can contribute to both positive and negative lasting effects in the host, with windows of heightened environmental sensitivity in early life. Crucially, this means that both mechanisms represent promising avenues for nutritional programming or other means of environmental conditioning, with the aim of improving health, welfare, and productivity of farmed aquatic species. But despite these commonalities, in terms of effects and goals, there has been a striking lack of research examining interactions between the microbiome and epigenome in farmed aquatic species. To our knowledge, no studies have examined direct interactions between the microbiome and epigenome in an aquatic-cultured species. For example, a better understanding of how microbially derived SCFAs can benefit fish health, welfare, and ultimately, aquaculture sustainability, via epigenetic interactions represents a unique opportunity for future research. Improved knowledge of interactions between the microbiome and epigenome will enable a better understanding of the impacts of aquaculture-related stressors on animal health, disease susceptibility, and productivity/sustainability (Figure 11.4). On the other hand, highly regulated hatchery-rearing systems may also represent a considerable opportunity for environmental conditioning in early life, the development of which will also be benefited from improved knowledge of epigenetic and microbiome mechanisms and their interactions.

11.5 Conclusions

Research on the importance of epigenetics and microbiome as separate factors in the context of aquaculture has hugely increased over the last 5–10 years. Yet there is a striking lack of knowledge regarding their interactions. In contrast to terrestrial animals, aquatic species are constantly exposed to aquatic microorganisms. As well as posing challenges, like constant exposure to potential pathogens, this also offers unique opportunities to aquaculture, such as conditioning the microbiome via the rearing water. Aquaculture conditions, including diet, water treatment, or stress due to handling and crowding influence both the microbiome and the epigenome, further highlighting the opportunities for likely interactions, conditioning, and selection. Pre- and probiotics have shown beneficial effects on host immunity and nutritional status

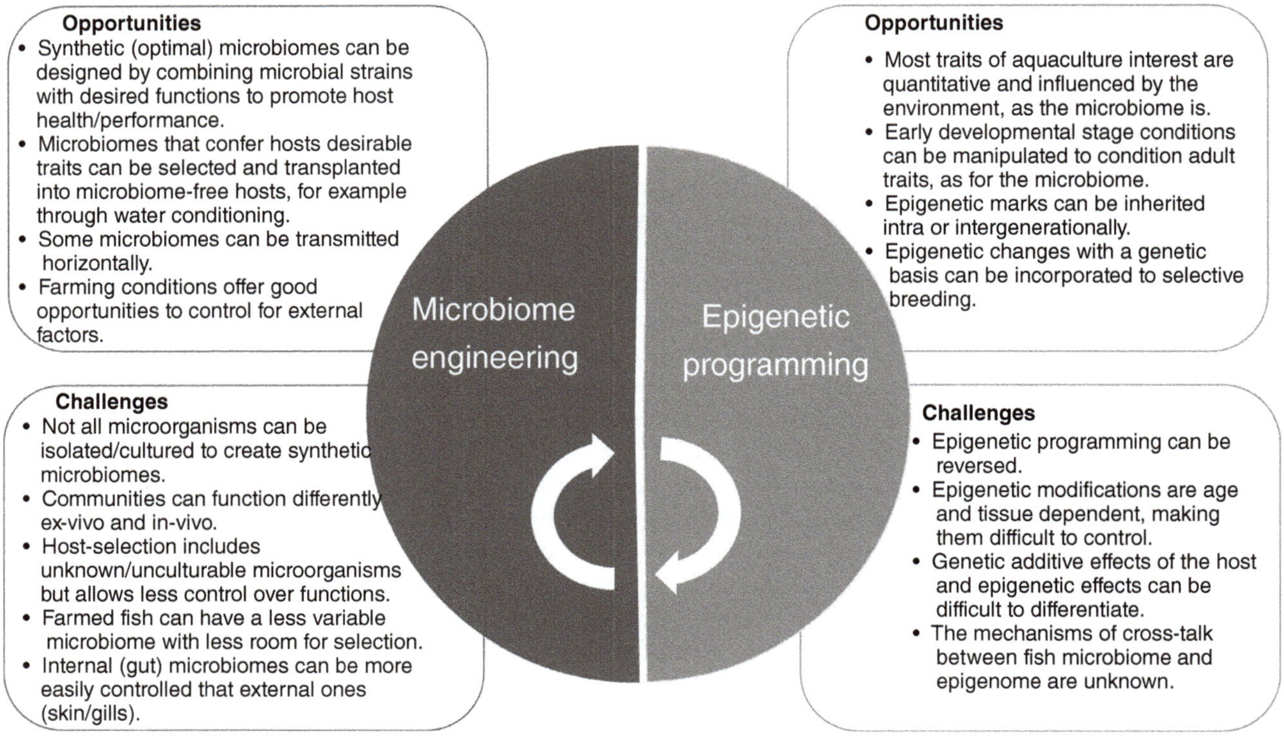

Figure 11.4 Challenges and opportunities for aquaculture of epigenetic programming and microbiome engineering; arrows indicating the epigenetics-microbiome crosstalk.

through the modification of microbiome communities and the metabolites they produce that, in turn, influence host gene expression. Research in mammals indicates that some of these metabolites (e.g., SCFAs) influence the host metabolism, energy status, and immunity through epigenetic regulatory mechanisms. Aquaculture-controlled conditions and the current, rapidly expanding knowledge about fish microbiome and epigenome suggest a great potential for the application of combined epigenetic and microbiome conditioning to improve fish health and performance under farmed conditions. Such an approach is showing promising results in medicine and agriculture. However, there are important differences between fish and mammalian microbiomes, and it is not possible to fully translate results from nonfish model species to aquaculture. Research on mechanisms of microbiome-epigenome crosstalk and their influence on host health and performance is emerging as the next challenge for research on sustainable aquaculture.

References

1 Gilbert, J.A., Blaser, M.J., Caporaso, J.G. et al. (2018). Current understanding of the human microbiome. *Nature Medicine* **24** (4): 392–400.

2 Turnbaugh, P.J., Ley, R.E., Hamady, M. et al. (2007). The human microbiome project. *Nature* **449** (7164): 804–810.

3 Pallister, T. and Spector, T.D. (2016). Food: a new form of personalized (gut microbiome) medicine for chronic diseases? *Journal of the Royal Society of Medicine* **109** (9): 331–336.

4 Riscuta, G., Xi, D., Pierre-Victor, D. et al. (2018). Diet, microbiome, and epigenetics in the era of precision medicine. *Cancer Epigenetics for Precision Medicine* 141–156.

5 Schmidt, V., Amaral-Zettler, L., Davidson, J. et al. (2016). Influence of fishmeal-free diets on microbial communities in Atlantic salmon (*Salmo salar*) recirculation aquaculture systems. *Applied and Environmental Microbiology* **82** (15): 4470–4481.

6 Perry, W.B., Lindsay, E., Payne, C.J. et al. (1926). The role of the gut microbiome in sustainable teleost aquaculture. *Proceedings of the Royal Society B* **2020** (287): 20200184.

7 de Bruijn, I., Liu, Y., Wiegertjes, G.F., and Raaijmakers, J.M. (2018). Exploring fish microbial communities to mitigate emerging diseases in aquaculture. *FEMS Microbiology Ecology* **94** (1): fix161.

8 Yukgehnaish, K., Kumar, P., Sivachandran, P. et al. (2020). Gut microbiota metagenomics in aquaculture: factors influencing gut microbiome and its physiological role in fish. *Reviews in Aquaculture* **12** (3): 1903–1927.

9 Rajeev, R., Adithya, K., Kiran, G.S., and Selvin, J. (2021). Healthy microbiome: a key to successful and sustainable shrimp aquaculture. *Reviews in Aquaculture* **13** (1): 238–258.

10 Infante-Villamil, S., Huerlimann, R., and Jerry, D.R. (2021). Microbiome diversity and dysbiosis in aquaculture. *Reviews in Aquaculture* **13** (2): 1077–1096.

11 Stentiford, G.D., Sritunyalucksana, K., Flegel, T.W. et al. (2017). New paradigms to help solve the global aquaculture disease crisis. *PLoS Pathogens* **13** (2): e1006160.

12 FAO I (2016). *The State of World Fisheries and Aquaculture 2016*, 200. Publications of Food and Agriculture Organization of the United Nations Rome.

13 Vindas, M.A., Madaro, A., Fraser, T.W.K. et al. (2016). Coping with a changing environment: the effects of early life stress. *Royal Society Open Science* **3** (10).

14 Trinh, L.T., Bakke, I., and Vadstein, O. (2017). Correlations of age and growth rate with microbiota composition in Atlantic cod (*Gadus morhua*) larvae. *Scientific Reports* **7** (1): 1–9.

15 Hacquard, S., Garrido-Oter, R., González, A. et al. (2015). Microbiota and host nutrition across plant and animal kingdoms. *Cell Host & Microbe* **17** (5): 603–616.

16 Gould, A.L., Zhang, V., Lamberti, L. et al. (2018). Microbiome interactions shape host fitness. *Proceedings of the National Academy of Sciences* **115** (51): E11951–E11960.

17 Nayak, S.K. (2010). Role of gastrointestinal microbiota in fish. *Aquaculture Research* **41** (11): 1553–1573.

18 Strandwitz, P. (2018). Neurotransmitter modulation by the gut microbiota. *Brain Research* **1693**: 128–133.

19 Antwis, R.E., Edwards, K.L., Unwin, B. et al. (2019). Rare gut microbiota associated with breeding success, hormone metabolites and ovarian cycle phase in the critically endangered eastern black rhino. *Microbiome* **7** (1): 1–12.

20 Kau, A.L., Ahern, P.P., Griffin, N.W. et al. (2011). Human nutrition, the gut microbiome and the immune system. *Nature* **474** (7351): 327–336.

21 Banerjee, G. and Ray, A.K. (2017). Bacterial symbiosis in the fish gut and its role in health and metabolism. *Symbiosis* **72** (1): 1–11.

22 Chiarello, M., Auguet, J.-C., Bettarel, Y. et al. (2018). Skin microbiome of coral reef fish is highly variable and driven by host phylogeny and diet. *Microbiome* **6** (1): 1–14.

23 Merrifield, D.L. and Rodiles, A. (2015). The fish microbiome and its interactions with mucosal tissues. In: *Mucosal Health in Aquaculture*, 273–295. Academic Press.

24 Butt, R.L. and Volkoff, H. (2019). Gut microbiota and energy homeostasis in fish. *Frontiers in Endocrinology* **10** (9): https://www.frontiersin.org/articles/10.3389/fendo.2019.00009/full.

25 Parata, L., Sammut, J., and Egan, S. (2021). Opportunities for microbiome research to enhance farmed freshwater fish quality and production. *Reviews in Aquaculture* **13** (4): 2027–2037.

26 Sullam, K.E., Essinger, S.D., Lozupone, C.A. et al. (2012). Environmental and ecological factors that shape the gut bacterial communities of fish: a meta-analysis. *Molecular Ecology* **21** (13): 3363–3378.

27 Sylvain, F.-É., Holland, A., Bouslama, S. et al. (2020). Fish skin and gut microbiomes show contrasting signatures of host species and habitat. *Applied and Environmental Microbiology* **86** (16): e00789–e00720.

28 Friberg, I.M., Taylor, J.D., and Jackson, J.A. (2019). Diet in the driving seat: natural diet-immunity-microbiome interactions in wild fish. *Frontiers in Immunology* **10**: 243.

29 Ramírez, C. and Romero, J. (2017). The microbiome of *Seriola lalandi* of wild and aquaculture origin reveals differences in composition and potential function. *Frontiers in Microbiology* **8**: 1844.

30 Rimoldi, S., Gini, E., Koch, J. et al. (2020). Effects of hydrolyzed fish protein and autolyzed yeast as substitutes of fishmeal in the gilthead sea bream (*Sparus aurata*) diet, on fish intestinal microbiome. *BMC Veterinary Research* **16** (1): 1–13.

31 Egerton, S., Culloty, S., Whooley, J. et al. (2018). The gut microbiota of marine fish. *Frontiers in Microbiology* **9** (873).

32 Ou, W., Yu, G., Zhang, Y., and Mai, K. (2021). Recent progress in the understanding of the gut microbiota of marine fishes. *Marine Life Science & Technology.* **3** (4): 434–448.

33 Ghanbari, M., Kneifel, W., and Domig, K.J. (2015). A new view of the fish gut microbiome: advances from next-generation sequencing. *Aquaculture* **448**: 464–475.

34 Givens, C.E., Ransom, B., Bano, N., and Hollibaugh, J.T. (2015). Comparison of the gut microbiomes of 12 bony fish and 3 shark species. *Marine Ecology Progress Series* **518**: 209–223.

35 Kim, P.S., Shin, N.-R., Lee, J.-B. et al. (2021). Host habitat is the major determinant of the gut microbiome of fish. *Microbiome* **9** (1): 1–16.

36 Uren Webster, T.M., Rodriguez-Barreto, D., Castaldo, G. et al. (2020). Environmental plasticity and colonization history in the Atlantic salmon microbiome: a translocation experiment. *Molecular Ecology* **29** (5): 886–898.
37 Ringø, E., Hoseinifar, S.H., Ghosh, K. et al. (2018). Lactic acid bacteria in finfish – an update. *Frontiers in Microbiology* **9**: 1818.
38 Balcázar, J.L., De Blas, I., Ruiz-Zarzuela, I. et al. (2006). The role of probiotics in aquaculture. *Veterinary Microbiology* **114** (3–4): 173–186.
39 Ringø, E., Salinas, I., Olsen, R. et al. (2007). Histological changes in intestine of Atlantic salmon (*Salmo salar* L.) following in vitro exposure to pathogenic and probiotic bacterial strains. *Cell and Tissue Research* **328** (1): 109–116.
40 Ringø, E. (2008). The ability of carnobacteria isolated from fish intestine to inhibit growth of fish pathogenic bacteria: a screening study. *Aquaculture Research* **39** (2): 171–180.
41 Roeselers, G., Mittge, E.K., Stephens, W.Z. et al. (2011). Evidence for a core gut microbiota in the zebrafish. *The ISME Journal* **5** (10): 1595–1608.
42 Milligan-Myhre, K., Small, C.M., Mittge, E.K. et al. (2016). Innate immune responses to gut microbiota differ between oceanic and freshwater threespine stickleback populations. *Disease Models & Mechanisms* **9** (2): 187–198.
43 Huyben, D., Sun, L., Moccia, R. et al. (2018). Dietary live yeast and increased water temperature influence the gut microbiota of rainbow trout. *Journal of Applied Microbiology* **124** (6): 1377–1392.
44 Neuman, C., Hatje, E., Zarkasi, K.Z. et al. (2016). The effect of diet and environmental temperature on the faecal microbiota of farmed Tasmanian Atlantic Salmon (*Salmo salar* L.). *Aquaculture Research* **47** (2): 660–672.
45 Krotman, Y., Yergaliyev, T.M., Shani, R.A. et al. (2020). Dissecting the factors shaping fish skin microbiomes in a heterogeneous inland water system. *Microbiome* **8** (1): 1–15.
46 Deb, S., Das, L., and Das, S.K. (2020). Composition and functional characterization of the gut microbiome of freshwater pufferfish (*Tetraodon cutcutia*). *Archives of Microbiology* **202** (10): 2761–2770.
47 Marjara, I.S., Chikwati, E.M., Valen, E.C. et al. (2012). Transcriptional regulation of IL-17A and other inflammatory markers during the development of soybean meal-induced enteropathy in the distal intestine of Atlantic salmon (*Salmo salar* L.). *Cytokine* **60** (1): 186–196.
48 Liu, Y., Chen, Z., Dai, J. et al. (2018). The protective role of glutamine on enteropathy induced by high dose of soybean meal in turbot, Scophthalmus maximus L. *Aquaculture* **497**: 510–519.
49 Horlick, J., Booth, M.A., and Tetu, S.G. (2020). Alternative dietary protein and water temperature influence the skin and gut microbial communities of yellowtail kingfish (*Seriola lalandi*). *PeerJ* **8**: e8705.
50 Uren Webster, T.M., Consuegra, S., Hitchings, M., and Garcia de Leaniz, C. (2018). Interpopulation variation in the Atlantic salmon microbiome reflects environmental and genetic diversity. *Applied and Environmental Microbiology* **84** (16): e00691–e00618.
51 Rud, I., Kolarevic, J., Holan, A.B. et al. (2017). Deep-sequencing of the bacterial microbiota in commercial-scale recirculating and semi-closed aquaculture systems for Atlantic salmon post-smolt production. *Aquacultural Engineering* **78**: 50–62.
52 Tort, L. (2011). Stress and immune modulation in fish. *Developmental and Comparative Immunology* **35** (12): 1366–1375.
53 Ellison, A.R., Uren Webster, T.M., Rey, O. et al. (2018). Transcriptomic response to parasite infection in Nile tilapia (*Oreochromis niloticus*) depends on rearing density. *BMC Genomics* **19** ((1)): 723.
54 Uren Webster, T.M., Rodriguez-Barreto, D., Martin, S.A. et al. (2018). Contrasting effects of acute and chronic stress on the transcriptome, epigenome, and immune response of Atlantic salmon. *Epigenetics* **13** (12): 1191–1207.
55 Foster, J., Rinaman, L., and Cryan, J. (2017). Neurobiology of stress & the gut-brain axis: regulation by the microbiome. *Neurobiology of Stress* **7**: 124–136.
56 Francino, M. (2016). Antibiotics and the human gut microbiome: dysbioses and accumulation of resistances. *Frontiers in Microbiology* **6**: 1543.
57 Uren Webster, T.M., Rodriguez-Barreto, D., Consuegra, S., and Garcia de Leaniz, C. (2020). Cortisol-related signatures of stress in the fish microbiome. *Frontiers in Microbiology* **11**: 1621.
58 Boutin, S., Sauvage, C., Bernatchez, L. et al. (2014). Inter individual variations of the fish skin microbiota: host genetics basis of mutualism? *PLoS One* **9** (7): e102649.
59 Zha, Y., Eiler, A., Johansson, F., and Svanbäck, R. (2018). Effects of predation stress and food ration on perch gut microbiota. *Microbiome* **6** (1): 1–12.
60 Sylvain, F.-É., Cheaib, B., Llewellyn, M. et al. (2016). pH drop impacts differentially skin and gut microbiota of the Amazonian fish tambaqui (*Colossoma macropomum*). *Scientific Reports* **6** (1): 1–10.

61 Webster, T.M.U., Consuegra, S., and de Leaniz, C.G. (2021). Early life stress causes persistent impacts on the microbiome of Atlantic salmon. *Comparative Biochemistry and Physiology Part D: Genomics and Proteomics.* **40**: 100888.

62 Llewellyn, M.S., Boutin, S., Hoseinifar, S.H., and Derome, N. (2014). Teleost microbiomes: the state of the art in their characterization, manipulation and importance in aquaculture and fisheries. *Frontiers in Microbiology* **5**: 207.

63 Rea, K., Dinan, T.G., and Cryan, J.F. (2016). The microbiome: a key regulator of stress and neuroinflammation. *Neurobiology of Stress* **4**: 23–33.

64 Hassenrück, C., Reinwald, H., Kunzmann, A. et al. (2021). Effects of thermal stress on the gut microbiome of juvenile milkfish (*Chanos chanos*). *Microorganisms* **9** (1): 5.

65 Lai, K.P., Lin, X., Tam, N. et al. (2020). Osmotic stress induces gut microbiota community shift in fish. *Environmental Microbiology* **22** (9): 3784–3802.

66 Du, F., Li, Y., Tang, Y. et al. (2019). Response of the gut microbiome of *Megalobrama amblycephala* to crowding stress. *Aquaculture* **500**: 586–596.

67 Webster, N.S. and Reusch, T.B. (2017). Microbial contributions to the persistence of coral reefs. *The ISME Journal* **11** (10): 2167–2174.

68 Pratte, Z.A., Besson, M., Hollman, R.D., and Stewart, F.J. (2018). The gills of reef fish support a distinct microbiome influenced by host-specific factors. *Applied and Environmental Microbiology* **84** (9): e00063–e00018.

69 Sevellec, M., Derome, N., and Bernatchez, L. (2018). Holobionts and ecological speciation: the intestinal microbiota of lake whitefish species pairs. *Microbiome* **6** (1): 47.

70 Escalas, A., Auguet, J.-C., Avouac, A. et al. (2021). Ecological specialization within a carnivorous fish family is supported by a herbivorous microbiome shaped by a combination of gut traits and specific diet. *Frontiers in Marine Science* **8** (91): 622883.

71 Riiser, E.S., Haverkamp, T.H., Varadharajan, S. et al. (2020). Metagenomic shotgun analyses reveal complex patterns of intra-and interspecific variation in the intestinal microbiomes of codfishes. *Applied and Environmental Microbiology* **86** (6): e02788–e02719.

72 Lauzon, H., Gudmundsdottir, S., Petursdottir, S. et al. (2010). Microbiota of Atlantic cod (*Gadus morhua*) rearing systems at pre-and posthatch stages and the effect of different treatments. *Journal of Applied Microbiology* **109** (5): 1775–1789.

73 Element, G., Engel, K., Neufeld, J.D. et al. (2020). Seasonal habitat drives intestinal microbiome composition in anadromous Arctic char (*Salvelinus alpinus*). *Environmental Microbiology* **22** (8): 3112–3125.

74 Zhao, R., Symonds, J.E., Walker, S.P. et al. (2020). Salinity and fish age affect the gut microbiota of farmed Chinook salmon (*Oncorhynchus tshawytscha*). *Aquaculture* **528**: 735539.

75 Ellis, A. (2001). Innate host defense mechanisms of fish against viruses and bacteria. *Developmental and Comparative Immunology* **25** (8–9): 827–839.

76 Gauthier, J., Lavoie, C., Charette, S.J., and Derome, N. (2019). Host-microbiota interactions and their importance in promoting growth and resistance to opportunistic diseases in salmonids. In: *Microbial Communities in Aquaculture Ecosystems*, 21–50. Cham: Springer.

77 Alenghat, T. (2015). Epigenomics and the microbiota. *Toxicologic Pathology* **43** (1): 101–106.

78 Legrand, T.P., Wynne, J.W., Weyrich, L.S., and Oxley, A.P. (2020). A microbial sea of possibilities: current knowledge and prospects for an improved understanding of the fish microbiome. *Reviews in Aquaculture* **12** (2): 1101–1134.

79 Clemente, J.C., Ursell, L.K., Parfrey, L.W., and Knight, R. (2012). The impact of the gut microbiota on human health: an integrative view. *Cell* **148** (6): 1258–1270.

80 Atarashi, K., Tanoue, T., Shima, T. et al. (2011). Induction of colonic regulatory T cells by indigenous clostridium species. *Science* **331** (6015): 337–341.

81 Adamovsky, O., Buerger, A.N., Vespalcova, H. et al. (2020). Evaluation of microbiome-host relationships in the zebrafish gastrointestinal system reveals adaptive immunity is a target of bis(2-ethylhexyl) phthalate (DEHP) exposure. *Environmental Science & Technology* **54** (9): 5719–5728.

82 Legrand, T.P., Catalano, S.R., Wos-Oxley, M.L. et al. (2018). The inner workings of the outer surface: skin and gill microbiota as indicators of changing gut health in yellowtail kingfish. *Frontiers in Microbiology* **8**: 2664.

83 Boutin, S., Bernatchez, L., Audet, C., and Derôme, N. (2012). Antagonistic effect of indigenous skin bacteria of brook charr (*Salvelinus fontinalis*) against *Flavobacterium columnare* and *F. psychrophilum*. *Veterinary Microbiology* **155** (2–4): 355–361.

84 Gómez, G.D. and Balcázar, J.L. (2008). A review on the interactions between gut microbiota and innate immunity of fish. *FEMS Immunology and Medical Microbiology* **52** (2): 145–154.

85 Luis Balcázar, J., De Blas, I., Ruiz-Zarzuela, I. et al. (2006). Growth inhibition of *Aeromonas* species by lactic acid bacteria isolated from salmonids. *Microbial Ecology in Health and Disease* **18** (1): 61–63.

86 Galindo-Villegas, J., García-Moreno, D., De Oliveira, S. et al. (2012). Regulation of immunity and disease resistance by commensal microbes and chromatin modifications during zebrafish development. *Proceedings of the National Academy of Sciences* **109** (39): E2605–E2614.

87 Maslowski, K.M., Vieira, A.T., Ng, A. et al. (2009). Regulation of inflammatory responses by gut microbiota and chemoattractant receptor GPR43. *Nature* **461** (7268): 1282–1286.

88 Jin Song, S., Woodhams, D.C., Martino, C. et al. (2019). Engineering the microbiome for animal health and conservation. *Experimental Biology and Medicine* **244** (6): 494–504.

89 Reverter, M., Sarter, S., Caruso, D. et al. (2020). Aquaculture at the crossroads of global warming and antimicrobial resistance. *Nature Communications* **11** (1): 1–8.

90 Brandt, K.K., Amézquita, A., Backhaus, T. et al. (2015). Ecotoxicological assessment of antibiotics: a call for improved consideration of microorganisms. *Environment International* **85**: 189–205.

91 Mueller, U.G. and Sachs, J.L. (2015). Engineering microbiomes to improve plant and animal health. *Trends in Microbiology* **23** (10): 606–617.

92 Brudeseth, B.E., Wiulsrød, R., Fredriksen, B.N. et al. (2013). Status and future perspectives of vaccines for industrialised fin-fish farming. *Fish & Shellfish Immunology* **35** (6): 1759–1768.

93 Houston, R.D., Bean, T.P., Macqueen, D.J. et al. (2020). Harnessing genomics to fast-track genetic improvement in aquaculture. *Nature Reviews. Genetics* **1-21**.

94 Wiley, N., Dinan, T., Ross, R. et al. (2017). The microbiota-gut-brain axis as a key regulator of neural function and the stress response: implications for human and animal health. *Journal of Animal Science* **95** (7): 3225–3246.

95 Hoseinifar, S.H., Sun, Y.-Z., Wang, A., and Zhou, Z. (2018). Probiotics as means of diseases control in aquaculture, a review of current knowledge and future perspectives. *Frontiers in Microbiology* **9**: 2429.

96 Nawaz, A., Irshad, S., Hoseinifar, S.H., and Xiong, H. (2018). The functionality of prebiotics as immunostimulant: evidences from trials on terrestrial and aquatic animals. *Fish & Shellfish Immunology* **76**: 272–278.

97 Nayak, S.K. (2010). Probiotics and immunity: a fish perspective. *Fish & Shellfish Immunology* **29** (1): 2–14.

98 Tarnecki, A.M., Wafapoor, M., Phillips, R.N., and Rhody, N.R. (2019). Benefits of a Bacillus probiotic to larval fish survival and transport stress resistance. *Scientific Reports* **9** (1): 1–11.

99 Liu, C.-H., Chiu, C.-H., Wang, S.-W., and Cheng, W. (2012). Dietary administration of the probiotic, Bacillus subtilis E20, enhances the growth, innate immune responses, and disease resistance of the grouper, Epinephelus coioides. *Fish & Shellfish Immunology* **33** (4): 699–706.

100 Resistance ICGoA (2019). *No Time to Wait: Securing the Future from Drug-Resistant Infections*. Geneva: World Health Organization.

101 Van Doan, H., Hoseinifar, S.H., Ringø, E. et al. (2020). Host-associated probiotics: a key factor in sustainable aquaculture. *Reviews in Fisheries Science & Aquaculture* **28** (1): 16–42.

102 Kuebutornye, F.K., Abarike, E.D., and Lu, Y. (2019). A review on the application of *Bacillus* as probiotics in aquaculture. *Fish & Shellfish Immunology* **87**: 820–828.

103 He, S., Ran, C., Qin, C. et al. (2017). Anti-infective effect of adhesive probiotic *Lactobacillus* in fish is correlated with their spatial distribution in the intestinal tissue. *Scientific Reports* **7** (1): 1–12.

104 Chen, S.-W., Liu, C.-H., and Hu, S.-Y. (2019). Dietary administration of probiotic *Paenibacillus ehimensis* NPUST1 with bacteriocin-like activity improves growth performance and immunity against *Aeromonas hydrophila* and *Streptococcus iniae* in Nile tilapia (*Oreochromis niloticus*). *Fish & Shellfish Immunology* **84**: 695–703.

105 Eirin-Lopez, J.M. and Putnam, H.M. (2019). Marine environmental epigenetics. *Annual Review of Marine Science* **11**: 335–368.

106 Moghadam, H.K., Johnsen, H., Robinson, N. et al. (2017). Impacts of early life stress on the methylome and transcriptome of Atlantic salmon. *Scientific Reports* **7** (1): 5023.

107 Moghadam, H., Mørkøre, T., and Robinson, N. (2015). Epigenetics – potential for programming fish for aquaculture? *Journal of Marine Science and Engineering* **3** (2): 175–192.

108 Rodriguez-Barreto, D., Garcia de Leaniz, C., Verspoor, E. et al. (2019). DNA methylation changes in the sperm of captive-reared fish: a route to epigenetic introgression in wild populations. *Molecular Biology and Evolution* **36**: 2205–2211.

109 Aubin, J., Callier, M., Rey-Valette, H. et al. (2019). Implementing ecological intensification in fish farming: definition and principles from contrasting experiences. *Reviews in Aquaculture* **11** (1): 149–167.

110 Allen, J. and Sears, C.L. (2019). Impact of the gut microbiome on the genome and epigenome of colon epithelial cells: contributions to colorectal cancer development. *Genome Medicine* **11**: 11.

111 D'Aquila, P., Lynn Carelli, L., De Rango, F. et al. (2020). Gut microbiota as important mediator between diet and DNA methylation and histone modifications in the host. *Nutrients 12* (3): 597.

112 Qin, Y. and Wade, P.A. (2018). Crosstalk between the microbiome and epigenome: messages from bugs. *The Journal of Biochemistry.* **163**: 105–112.

113 Morovic, W. and Budinoff, C.R. (2021). Epigenetics: a new frontier in probiotic research. *Trends in Microbiology* **29**: 117–126.

114 Cortese, R., Lu, L., Yu, Y. et al. (2016). Epigenome-microbiome crosstalk: a potential new paradigm influencing neonatal susceptibility to disease. *Epigenetics* **11**: 205–215.

115 Koudoufio, M., Desjardins, Y., Feldman, F. et al. (2020). Insight into polyphenol and gut microbiota crosstalk: are their metabolites the key to understand protective effects against metabolic disorders? *Antioxidants* **9**: 982.

116 Amatullah, H. and Jeffrey, K.L. (2020). Epigenome-metabolome-microbiome axis in health and IBD. *Current Opinion in Microbiology* **56**: 97–108.

117 Hullar, M.A.J. and Fu, B.C. (2014). Diet, the gut microbiome, and epigenetics. *Cancer journal (Sudbury, Mass).* **20**: 170–175.

118 Sharma, M., Li, Y., Stoll, M.L., and Tollefsbol, T.O. (2020). The epigenetic connection between the gut microbiome in obesity and diabetes. *Frontiers in Genetics* **10**: 1329.

119 Fellows, R., Denizot, J., Stellato, C. et al. (2018). Microbiota derived short chain fatty acids promote histone crotonylation in the colon through histone deacetylases. *Nature Communications* **9**: 105.

120 Malmuthuge, N. and Guan, L.L. (2021). Noncoding RNAs: regulatory molecules of host–microbiome crosstalk. *Trends in Microbiology* **29**: 713–724.

121 Fellows, R. and Varga-Weisz, P. (2020). Chromatin dynamics and histone modifications in intestinal microbiota-host crosstalk. *Molecular Metabolism* **38**: 100925.

122 Koh, A., De Vadder, F., Kovatcheva-Datchary, P., and Bäckhed, F. (2016). From dietary fiber to host physiology: short-chain fatty acids as key bacterial metabolites. *Cell* **165**: 1332–1345.

123 Krautkramer, K.A., Kreznar, J.H., Romano, K.A. et al. (2016). Diet-microbiota interactions mediate global epigenetic programming in multiple host tissues. *Molecular Cell* **64**: 982–992.

124 Kelly, D., Kotliar, M., Woo, V. et al. (2018). Microbiota-sensitive epigenetic signature predicts inflammation in Crohn's disease. *JCI Insight* 3.

125 Kok, D.E., Steegenga, W.T., and McKay, J.A. (2018). Folate and epigenetics: why we should not forget bacterial biosynthesis. *Epigenomics 10* (09): 1147–1150.

126 Alenghat, T., Osborne, L.C., Saenz, S.A. et al. (2013). Histone deacetylase 3 coordinates commensal-bacteria-dependent intestinal homeostasis. *Nature* **504**: 153–157.

127 Ansari, I., Raddatz, G., Gutekunst, J. et al. (2020). The microbiota programs DNA methylation to control intestinal homeostasis and inflammation. *Nature Microbiology* **5**: 610–619.

128 Ramos-Molina, B., Sánchez-Alcoholado, L., Cabrera-Mulero, A. et al. (2019). Gut microbiota composition is associated with the global DNA methylation pattern in obesity. *Frontiers in Genetics* **10**: 613.

129 Lazar, V., Ditu, L.-M., Pircalabioru, G.G. et al. (2019). Gut microbiota, host organism, and diet trialogue in diabetes and obesity. *Frontiers in Nutrition* **6**: 21.

130 Qiao, S., Bao, L., Wang, K. et al. (2020). Activation of a specific gut bacteroides-folate-liver axis benefits for the alleviation of nonalcoholic hepatic steatosis. *Cell Reports* **32**: 108005.

131 Gerhauser, C. (2018). Impact of dietary gut microbial metabolites on the epigenome. *Philosophical Transactions of the Royal Society, B: Biological Sciences* **373**: 20170359.

132 Dempsey, J., Zhang, A., and Cui, J.Y. (2018). Coordinate regulation of long non-coding RNAs and protein-coding genes in germ-free mice. *BMC Genomics* **19**: 1–24.

133 Liang, L., Ai, L., Qian, J. et al. (2015). Long noncoding RNA expression profiles in gut tissues constitute molecular signatures that reflect the types of microbes. *Scientific Reports* **5**: 1–8.

134 Virtue, A.T., McCright, S.J., Wright, J.M. et al. (2019). The gut microbiota regulates white adipose tissue inflammation and obesity via a family of microRNAs. *Science Translational Medicine* 11.

135 Nakata, K., Sugi, Y., Narabayashi, H. et al. (2017). Commensal microbiota-induced microRNA modulates intestinal epithelial permeability through the small GTPase ARF4. *The Journal of Biological Chemistry* **292**: 15426–15433.

136 Tarallo, S., Ferrero, G., De Filippis, F. et al. (2021). Stool microRNA profiles reflect different dietary and gut microbiome patterns in healthy individuals. *Gut* **71** (7): 1302–1314.

137 Huang, J., Chen, W., Zhou, F. et al. (2021). Tissue-specific reprogramming of host tRNA transcriptome by the microbiome. *Genome Research* **31**: 947–957.

138 Liu, S., da Cunha, A.P., Rezende Rafael, M. et al. (2016). The host shapes the gut microbiota via fecal microRNA. *Cell Host & Microbe* **19**: 32–43.

139 Cuomo, M., Borrelli, L., Della Monica, R. et al. (2021). DNA methylation profiles of Tph1A and BDNF in gut and brain of L. Rhamnosus-treated zebrafish. *Biomolecules 11* (2): 142.

140 Shirani, M., Mahboobi Soofiani, N., Farahmand, H., and Jalali, A.H. (2020). Comparative epigenomic profiling and gene expression patterns of zebrafish, *Danio rerio*, administrated by dietary agrimos®. *Journal of Agricultural Science and Technology 22* (6): 1487–1500.

141 Dhanasiri, A., Chen, X., Dahle, D. et al. (2020). Dietary inclusion of plant ingredients induces epigenetic changes in the intestine of zebrafish. *Epigenetics* **15** (10): 1035–1051.

142 Patula, S., Wojno, M., Pinnell, L.J. et al. (2021). Nutritional programming with dietary soybean meal and its effect on gut microbiota in zebrafish (*Danio rerio*). *Zebrafish* **18** (2): 125–138.

143 Abdel-Latif, H.M., Abdel-Tawwab, M., Dawood, M.A. et al. (2020). Benefits of dietary butyric acid, sodium butyrate, and their protected forms in aquafeeds: a review. *Reviews in Fisheries Science & Aquaculture* **28** (4): 421–448.

144 Tran, N.T., Li, Z., Wang, S. et al. (2020). Progress and perspectives of short-chain fatty acids in aquaculture. *Reviews in Aquaculture* **12** (1): 283–298.

145 Conte, F. (2004). Stress and the welfare of cultured fish. *Applied Animal Behaviour Science* **86** (3–4): 205–223.

146 Jennings, S., Stentiford, G.D., Leocadio, A.M. et al. (2016). Aquatic food security: insights into challenges and solutions from an analysis of interactions between fisheries, aquaculture, food safety, human health, fish and human welfare, economy and environment. *Fish and Fisheries* **17** (4): 893–938.

147 Anastasiadi, D., Díaz, N., and Piferrer, F. (2017). Small ocean temperature increases elicit stage-dependent changes in DNA methylation and gene expression in a fish, the European sea bass. *Scientific Reports* **7** (1): 1–12.

148 Beemelmanns, A., Ribas, L., Anastasiadi, D. et al. (2021). DNA methylation dynamics in Atlantic salmon (Salmo salar) challenged with high temperature and moderate hypoxia. *Frontiers in Marine Science* **7**: 604878.

149 Moghadam, H.K., Johnsen, H., Robinson, N. et al. (2017). Impacts of early life stress on the methylome and transcriptome of Atlantic salmon. *Scientific Reports* **7** (1): 1–11.

150 Ryu, T., Veilleux, H.D., Munday, P.L. et al. (2020). An epigenetic signature for within-generational plasticity of a reef fish to ocean warming. *Frontiers in Marine Science* **7**: 284.

151 Lokmer, A. and Wegner, K.M. (2015). Hemolymph microbiome of Pacific oysters in response to temperature, temperature stress and infection. *The ISME Journal* **9** (3): 670–682.

152 Li, Y.-F., Chen, Y.-W., Xu, J.-K. et al. (2019). Temperature elevation and *Vibrio cyclitrophicus* infection reduce the diversity of haemolymph microbiome of the mussel *Mytilus coruscus*. *Scientific Reports* **9** (1): 1–10.

153 Foster, J.A., Rinaman, L., and Cryan, J.F. (2017). Stress & the gut-brain axis: regulation by the microbiome. *Neurobiology of Stress* **7**: 124–136.

154 Burns, S.B., Almeida, D., and Turecki, G. (2018). The epigenetics of early life adversity: current limitations and possible solutions. *Progress in Molecular Biology and Translational Science* **157**: 343–425.

12

Epigenetics of Stress in Farmed Fish: An Appraisal

Bruno Guinand[1] and Athanasios Samaras[2]

[1] *Institut des Sciences de l'Evolution de Montpellier, University of Montpellier, CNRS, IRD, EPHE, Montpellier, France*
[2] *Department of Biology, University of Crete, Heraklion, Crete, Greece*

12.1 Introduction

Most of the time, stress refers to negative outcomes experienced at any time during life that affect the health and well-being of individuals. It may translate into neurobehavioral and neuropsychiatric disorders impacting the individual, but also that individual's sociality and environment. Stress can be divided into "good," "tolerable," and "toxic" [1, 2]. It might be experienced positively to biologically embed an appropriate response that may promote robustness and/or performance of the individuals through short- or long-lasting mechanisms. In aquaculture, managing the stress response of organisms is generally achieved by modifying the hatchery practices and rearing conditions to which organisms are subjected. While certainly unachieved [3], stress monitoring is increasingly devoted to improving animal welfare and sustainability of production [4–9]. However, aquaculture systems invariably impose some level of stress on organisms [10], and should be thoroughly monitored to keep welfare in the "good" to "tolerable" range. Thus, it is necessary to evaluate how chronic exposure to common hatchery stressors affects fish growth, development, immune system, reproduction, and survival [10].

Over the years, a wide range of indicators of fish stress has been proposed, ranging from the cellular to the organismal level [10, 11]. This includes classical monitoring of blood parameters [12, 13], especially cortisol [14]. Monitoring of gene expression using transcriptomics [15, 16], protein production using proteomics (e.g., [17]), behavior, and/or physiological states using biosensing technologies [11, 18, 19] are increasingly used. Granada et al. [20] emphasized epigenetics – defined as changes in the molecular machinery that would induce phenotypic variation without change in the DNA sequence as a "last frontier" in aquaculture, and more generally livestock [21–23]. They put stimuli – i.e., positively defined stressors – at the forefront of induced changes in disease resistance, nutrition and growth, sex ratio management, and stress resilience to build more robust aquaculture species [20]. Epigenetics is thought to shape the stress response and individual performances, with anticipation that epigenetic inputs may fuel aquaculture breeding programs [24–27]. The relationship between timing and stress intensity and epigenetics and performance has traditionally been centered around what is referred to as "early-life stress" (ELS) (review in [28]) (Box 12.1). Thus, stress-induced epigenetic processes could boost the effectiveness of the stress response and impact productivity. However, mostly loose, discursive relationships between hatchery-related stress and epigenetics have been provided thus far [29]. Furthermore, while ELS is recognized to alter the neural architecture to respond to stressors (e.g., [2, 30, 31] and many others), brain epigenetics or neuroepigenetics that might contribute to shaping patterns of gene expression and organismal performance, including production traits, is missing in farmed fish.

After an introduction to stress and the stress response, this chapter will focus on brain epigenetics and knowledge accumulated in mammalian models to illustrate and promote further work in aquaculture. It further expands on how epigenetics might be used for the biomonitoring of the stress response.

Epigenetics in Aquaculture, First Edition. Edited by Francesc Piferrer and Han-Ping Wang.

Box 12.1 Early Life Stress, Epigenetic Programming, and the Epigenetics of Stress

Early-life stress (ELS) or "early life adversity" refers to the critical period(s) during early developmental stages (referred to also as "critical window(s)" or "sensitive period(s)") at which an organism is especially sensitive to priming by environmental stressors. According to Waddington's definition of epigenetics, ELS induces the complex and dynamic interactions between the developmental environment and the genome that led to the production of an individual phenotype without considering a change at the DNA level. ELS thus refers to early intragenerational reprogramming of some organismal functions, with stressors leaving a signature in the whole epigenetic machinery ("epigenetic imprinting" or "memory") that would further drive and regulate genome-wide expression of the organism, including neuronal function and future behaviors. In that sense, stressors experimented on during early-life may favor phenotypic plasticity and adaptability to the surrounding developmental environment, while maintaining canalization to ensure organismal functions. In production settings, ELS needs to be monitored to ensure organisms that, e.g., protective immunity, social behaviors, physical and cognition performances, are maintained in a resting zone favoring health and welfare, and should thus promote eustress rather than distress.

In fish, sensitivity to ELS is classically considered to extend from the fertilized egg stage to hatching, plus a post-hatch period that is certainly species-specific. It could possibly be extended over a larger developmental period, and the late-developmental boundary of the critical period at which ELS could remain active is misunderstood or neglected. Indeed, the critical period may extend over the first years of life in humans, with an impact later in life. Translated to farmed fishes, this suggests that potential amelioration of production traits through epigenetic modulation could extend far beyond larval rearing, which is perhaps not the single critical period. Furthermore, how stressors that differ in strength and/or duration interplay during the critical period to shape a phenotype and promote eustress condition is certainly dependent on the individual genotypes that represent a landscape for epigenetic programming. However, this landscape remains unknown.

ELS is certainly the best entry to deal with the epigenetics of stress. However, ELS should not be considered as a substitute for stress epigenetics, as it only captures the action of a particular stressor or a set of stressors in early life, while those that may arise at other important developmental periods (e.g., sexual differentiation, sexual and gamete maturation) remain poorly assessed. Furthermore, the focus is made on stressor(s) and a phenotype of interest for production (e.g., growth rate), rather than on the stress response, which is a complex set of responses that cannot be equated to a phenotype of interest, but rather to the processes mediated by organs, hormones, and behaviors that participate to build this phenotype while preserving welfare. It is important not to turn ELS into a "buzzword" regarding the epigenetics of stress that has to be fueled by multidisciplinary inputs from, e.g., physiology, neurobiology, and behavioral sciences.

12.2 Stress and Stress Response

12.2.1 Stress

Stress is a concept that is widely used in our everyday lives, although no universally accepted definition of stress exists. Stress was first described by Selye [32] as "the nonspecific response of the body on any demand on it." More complex definitions have been suggested, including "stress is a state in which homeostasis is actually threatened or perceived to be so; homeostasis is re-established by a complex repertoire of behavioral and physiological adaptive responses of the organism" [33]. This definition incorporates the stressor itself, the perception of the stressor by the organism, as well as a set of adaptive responses in order to re-establish homeostasis. It is, therefore, obvious that stress is a multidimensional concept that includes both the actors (stressors) and the responses (stress response) (see [Box 12.2] for definitions).

Stressors are common in aquaculture (Figure 12.1). Irrespective of their nature, these can be divided into many dichotomies, such as internal and environmental, acute and chronic, adaptive and maladaptive, or causing eustress or distress (Box 12.2), to name a few. Of particular interest is the (i) acute and chronic stress division that sets different demands on animals, yet the distinction between them can be fuzzy, and the (ii) eustress and distress dichotomy, which was described by Selye [1]. Incorporating Selye's idea of "not all stress is bad" into advances in neurobiology, a recent theory, called allostasis, has been developed. Allostasis, in contrast to homeostasis, suggests that an organism "achieves constancy through change" [34, 35]. This is performed by adjusting set points of the regulatory mechanism and involves the synthesis

Box 12.2 Definitions of Stressors and Stress Response

Acute stress: The effect of stressors, usually of short duration and severe intensity. The animal's responses are mainly the fight-flight response, through the activation of the sympathetic nervous system and the stimulation of the HPI axis, which result in the release of catecholamines and cortisol into circulation, respectively.

Chronic stress: Less well-defined than acute stress, deals with the fact that stressors can be severe or mild, predictable or unpredictable, and controllable or uncontrollable. Chronic stress can be comprised of various combinations of the above-mentioned parameters. The responses of animals to chronic stress vary between adaptation through habituation and the development of pathological conditions like hypersensitivity, hyposensitivity, or prolonged activation of the stress-response mechanism.

Distress: The physiological alterations caused by a stressor that can induce detrimental effects on the organism.

Eustress: The set of responses of an organism to a stressor that leads to enhanced survival, performance, and adaptation.

Hypothalamus-pituitary-interrenal (HPI) axis: A collection of structures that provide feedforward and feedback signals for the regulation of glucocorticoid release in fish. It is a system homologous to the hypothalamus-pituitary-adrenal (HPA) axis in mammals.

Stressor: A factor that triggers a stress response. Stressors can vary in nature: environmental, biological, physical, psychological, and others.

Stress response: The organism's set of reactions to the stressor. Common examples of stress responses are the "fight-flight" response and the release of cortisol into circulation.

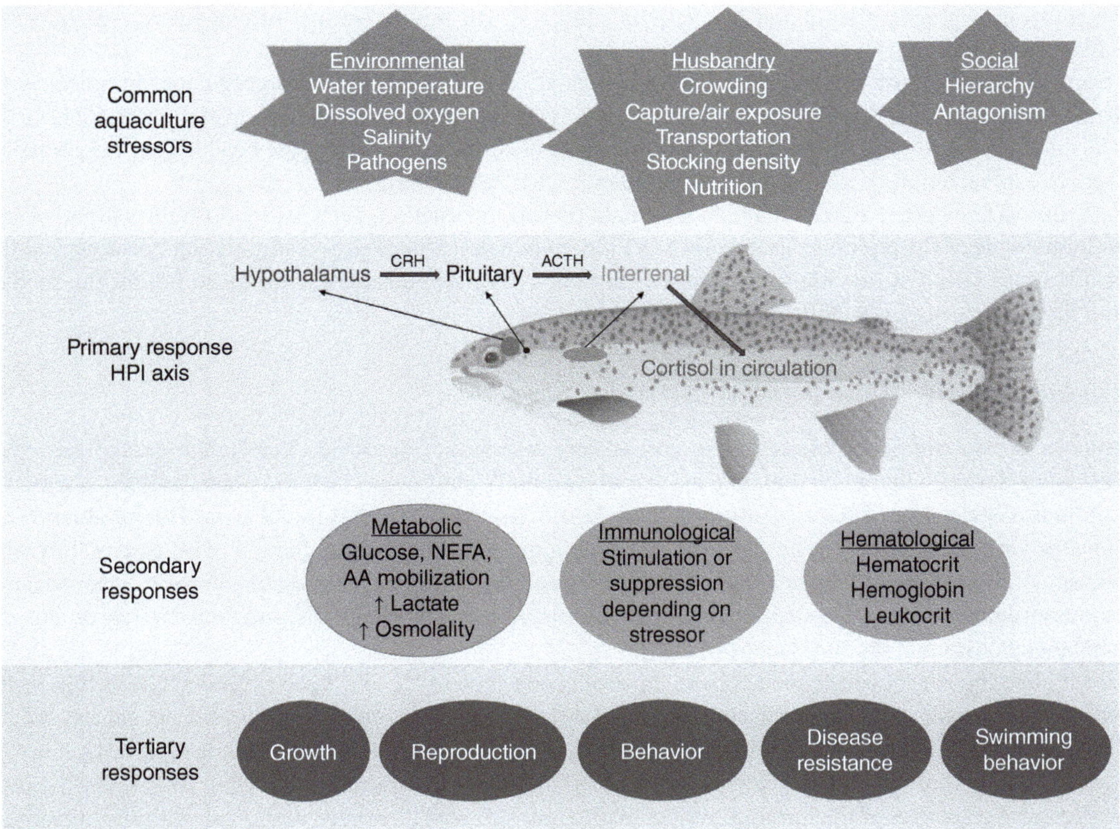

Figure 12.1 Stress responses induced by common aquaculture stressors and associated with the hypothalamic-pituitary-interrenal (HPI) axis. Abbreviations: AA, amino-acids; ACTH, adrenocorticotropin hormone; CRH, Corticotropin-releasing hormone; NEFA, nonesterified fatty acids). See text for details. Rainbow trout image by Tracey Saxby, Integration and Application Network (ian.umces.edu/media-library) under Attribution-ShareAlike 4.0 International (CC BY-SA 4.0) license. (*See colour plate section for colour representation of this figure.*)

of prior knowledge and predicted current needs. When the cumulative cost of allostasis to the body exceeds certain thresholds, allostatic overload occurs, and pathological conditions start to arise. The latter condition is usually observed under chronic stress [36].

12.2.2 HPI/HPA Axis and the Stress Hormones

The mechanisms that regulate the response to stress are complex, involving the actions of many biological systems of the organism. First, the stressful stimuli are perceived by the central nervous system (CNS), which subsequently regulates the response (Figure 12.1). The first level of response is referred to as the primary response and describes the hormonal (mainly corticosteroids and catecholamines) responses. These responses are observed from early larval stages in most teleost fish species [37], depicting therefore the importance of early development in the physiological and epigenetic regulation of subsequent developmental stages.

Upon perception of a stressful stimulus, the hypothalamic-sympathetic axis is activated, followed by the activation of the hypothalamus-pituitary-interrenal (HPI) axis, which is homologous to the hypothalamus-pituitary-adrenal (HPA) axis in mammals. The hypothalamus-sympathetic axis results in the release of epinephrine and norepinephrine to induce hyperglycemia and fuel the fight or flight response [38]. The release of catecholamines after stress is rapid because these monoamines are stored in vesicles in chromaffin cells [39], located in the head kidney, or interrenal, of fish, while the recovery to baseline levels is also fast [40]. The activation of the HPI axis starts with the secretion of corticotrophin-releasing hormone (CRH) from the hypothalamic preoptic area, which stimulates the secretion of the adrenocorticotropin hormone (ACTH) from the pituitary gland [41]. ACTH stimulates the interrenal cells to synthesize and secrete cortisol into the bloodstream [41, 42]. Cortisol actions are regulated by corticosteroid receptors, which in most teleost fish species are one mineralocorticoid (MR) and two glucocorticoid receptors (GRs). Once cortisol is bound, these receptors act as transcription factors and bind to specific DNA sequences (GR- and MR-responsive elements) in target-gene promoters controlling glucocorticoid and MR activities [42, 43]. GR is a major site for epigenetic control of the stress response (e.g., [44]) (see later).

Apart from the feedforward mechanisms described earlier, cortisol synthesis is also regulated by negative feedback mechanisms of the HPI axis. Specifically, cortisol exerts a negative feedback regulation in the production of CRH and ACTH [43], and ultra-short-loop negative feedback on the site of cortisol production, the head kidney [45]. Finally, ACTH can also regulate its secretion by inhibition of hypothalamic CRH release [46].

There are also various other signals that can stimulate cortisol production, such as the hormones arginine vasotocin, AVT [47] (homologue of the arginine vasopressin [AVP] in mammals), α-melanocyte-stimulating hormone (α-MSH [38]), and catecholamines [48]. On the other hand, other hormones (e.g., somatostatin, leptin) also inhibit the production of ACTH, and therefore cortisol [49, 50].

12.2.3 Stress Responses

Stress response is also comprised of secondary and tertiary responses (Figure 12.1). Secondary responses refer to the primary response-derived alterations that take place in the animal's physiology, including metabolic, biochemical, hematological, immunological, and osmoregulatory responses [39]. These responses mainly act toward the production of energy in order to cope with the stressful event. Initially, catecholamines stimulate the production of glucose mainly via hepatic glycogenolysis, while cortisol participates at a later stage of the stress response by stimulating hepatic gluconeogenesis [42]. When the energy demands set by the stressor cannot be met by aerobic metabolism, fish are forced to rely on the utilization of anaerobic metabolism and produce lactate [40, 51].

Apart from affecting carbohydrate metabolism, cortisol exerts proteolytic and lipolytic actions, resulting in increased amino and free fatty acids concentrations in circulation [42, 43] to fuel glucose production via gluconeogenesis. Additionally, through binding to the MR, cortisol plays a crucial role in fish osmoregulation and ionic balance [42, 43].

Stress, acting mainly through cortisol, increases the susceptibility of vertebrates to pathogenic diseases [52]. Both activating and suppressive responses of the immune system to different stressors have been described, with the outcome depending on the timing as well as the nature of the stressor [52, 53]. In this context, acute stress in the form of short-term challenges can potentially enhance the function of the innate immune system by acting as *eustress*. Long-lasting, repeatable and severe stressors, on the other hand, may lead to suppression of the immune system and therefore act as *distress* [52, 53] (Box 12.2).

Finally, tertiary responses involve changes at the whole animal level, like swimming behavior [54], reproduction [55], feeding behavior, and growth [56] (Figure 12.1). Such behaviors are adversely affected by stress, as for instance the reduced appetite and cessation of feeding that follows a stressful event, even for days [57].

12.2.4 Individual Differences in Stress Response and Coping Styles

It is well documented that basal circulating cortisol levels and the stress response pattern can differ between fish species [58]. Variation in cortisol concentrations exists between individuals within the same species as well, as a result of different environmental conditions (temperature [59, 60], photoperiod [61], salinity [62]), but also husbandry [63, 64].

Moreover, genetic origin can affect the stress response within the same species [65, 66]. In this context, divergence in cortisol response has been described in several fish species [65–67], and consistent differences among individuals of the same species have also been described (i.e., identification of low- and high-responding fish). Cortisol responsiveness has also been associated with behavioral differences and coping styles [68, 69], as well as animal performance such as differential growth [70] and metabolic functions [66, 71]. The main driving forces behind these differences are the biosynthetic capacity of the head kidney [72, 73] and its sensitivity to ACTH [72, 74]. Moreover, cortisol response seems to be affected by strong genetic [75, 76] and epigenetic [77, 78] components.

12.2.5 Common Measurements of the Stress Response

Since hormones regulate the physiological response to stress, measuring their concentration is one of the most common and informative techniques used to learn about the stress response. Among the major stress hormones, circulating cortisol is the most commonly used acute stress indicator [14]. However, care should be taken when measuring circulating cortisol since (i) cortisol response is variable among species [39, 51, 58] and (ii) cortisol can be influenced by environmental factors such as temperature, dissolved oxygen, and water quality in general [59, 60, 64]. Catecholamines, on the other hand, display a very fast release and recovery, rendering their measurement unreliable due to the fact that, unless fish are catheterized, it is almost impossible to achieve unstressed levels of both epinephrine and norepinephrine [40, 42]. Other components of the HPI axis, such as ACTH and CRH, cannot be easily and reliably measured in fish, either as a direct measurement of their concentration or through molecular techniques [38, 41].

Common measurements of the stress response also target the secondary responses, and specifically parameters of carbohydrates metabolism such as glucose and lactate [51, 58, 59], and less commonly, lipids such as nonesterified fatty acids, cholesterol and triglycerides, and proteins [50]. Stress effects on hematology are commonly assessed through hematocrit and hemoglobulin [66], while the effects on osmoregulation are assessed through osmolality and ion concentration measurements [51, 60, 71]. It is also common to measure innate immune system bioindicators (lysozyme and complement C) [52, 79]. Finally, tertiary response indicators include aspects of feeding, appetite and growth [56, 57, 80], swimming behavior, and respiration rate. Specifically, increase in the respiration rate (opercular beat rate) and in the time spent at the bottom of tanks (depending on species), as well as induction of "freezing behaviors" (preference for dark areas and loss of schooling consistency), are all cited as characteristic examples of behavioral responses to stress [81, 82].

Overall, "stress is in the brain" as the stress response primarily depends on the encoding of stressful events in the CNS. However, this response is better monitored by targeting the downstream manifestation of stressors on the organisms. Focus on upstream epigenetically driven brain processes, e.g., neurogenesis and neuronal regulation that may modulate the HPI axis, on the stress response and phenotypic or behavioral trait plasticity certainly deserve more attention.

12.3 Is There an Epigenetics of Stress in Cultured Fish?

12.3.1 A Brief State of the Art

Fish in aquaculture are exposed to various stimuli perceived as stressors. The effects of stress on fish performance depend on the (i) developmental stage of the fish; (ii) condition of the fish in terms of nutritional, physiological, and health status; and (iii) type of the stressor, including whether it is acute or chronic, its severity and duration, the time of application, and whether it is repeatable, predictable, controllable, or not. In larval stages, stress factors include husbandry practices, water quality such as temperature, dissolved oxygen concentration and salinity, tank size and water velocity [83], as well as sudden changes of these parameters. This may induce ELS during sensitive periods of development that can affect physiology later in life, depending on stress intensity (e.g., [77, 84, 85]) (Box 12.1). In later developmental stages, fry, juveniles, and adults are subjected to stressful husbandry practices (crowding, capture and air exposure, transportation, sorting, and administration of veterinary treatments), which can be added to the stressors described earlier. Finally, broodstock are usually maintained in separate rearing systems and/or are exposed to different husbandry practices than fish grown to be harvested. It is common that broodstock are tagged, anesthetized for sexual maturation stage analysis involving stripping, ultrasounds, or biopsies, as well as hormonal therapies and alterations in environmental parameters for out-of-season egg production. For example, stressful practices in females during oocyte maturation and spawning can affect the cortisol egg content [86, 87]. Finally, it is common that stressful factors are presented for long periods or in a repeated form, and/or in combination between them and/or in manners that render them uncontrollable and unpredictable. They become chronic stressors with detrimental effects on fish because they may arrest feeding and growth, dysregulate the HPI axis [81, 88], suppress immune functions [52, 53, 89], and disrupt reproductive processes [15].

The impact of rearing and/or manipulation of the rearing environment on the epigenome has been repeatedly reported in fish [77, 90–97, see also 98 for an introduction to mechanisms], and intragenerational epigenetic programming associated with ELS (Box 12.1) is emphasized in order to improve production traits in the near term (e.g., [25, 27, 29]). However, a relationship between ELS, epigenetic changes, and improved fish performance has been demonstrated only once in Atlantic salmon (*Salmo salar*) in a hatchery context [77]. Using whole fry, authors showed (i) consistent differential epigenetic signatures between stressed (submitted to cold-shock and air exposure) and nonstressed (control) individuals, (ii) a correlation between methylation (i.e., a methyl group added to some cytosines in the DNA sequence) and transcriptional differences for a set of genes, and (iii) superior growth performance observed in stressed fish compared to controls after one year. A companion study showed that relationships between gene expression variation and cytosine methylation differences at candidate genes, including genes linked to the HPI axis such as the proopiomelanocortin (*POMC*), *CRH*, and Nuclear Receptor Subfamily 3 Group C Member 1 (*NR3C1*) genes – the last one encoding for GR, could be complex and dependent on the life-stage stressor acted upon fish (egg, post-hatch, or at both stages) [94].

The role of stress-related genes is, however, reinforced by other studies. For example, using whole genome bisulfite sequencing, it was shown that captivity of adult Atlantic salmon induced differential methylation in F1 offspring that was associated with several biological functions, but primarily genes involved in the nervous system (e.g., glutamate receptors), together with genes associated with the immune system [99]. Results are in line with studies reporting epigenetic modifications at stress-susceptibility genes involved in the glutamatergic system in fish [100] and mammals (e.g., [101]), but also illustrate a potential intergenerational inheritance of stress. Putative multigenerational inheritance (i.e., parental influence on the F1) of epigenetic changes have been reported on numerous occasions in fish (e.g., [93, 97, 102–107]), but brain tissues were not targeted. To the best of our knowledge, only Reiser et al. [108] studied brain epigenetic profiles of individuals submitted to a barren vs. an enriched environment (i.e., a stress gradient) across life stages of the rainbow trout (*Oncorhynchus mykiss*). The enriched environment was shown to modulate epigenetic patterns, but phenotypic counterparts were few and nonsignificant in this study. Furthermore, the method used did not allow for the identification of differentially imprinted genes. Concurrently, when investigating some aspects of epigenetic inheritance in seven genes in European sea bass (*Dicentrarchus labrax*), Anastasiadi et al. [107] showed that some marks were related to stress-responsive genes and conserved from parents to offspring. This was, however, observed for sperm and not the brain. Stress-responsive genes included epigenetic variation at *NR3C1*, which is strongly involved in the regulation of the HPI axis, and regularly screened for gene expression variation in fish (e.g., [109–113]).

Furthermore, emphasizing ELS might obscure other important epigenetic contributions to the stress response (Box 12.1), notably in studies aiming to better apprehend the impact of intergenerational inheritance. For example, in brook charr (*Salvelinus alpinus*), it was shown that the thermal regime during parental sexual maturation, but not during the rearing of offspring, modulates DNA methylation [114]. This study is not directly related to the stress response, but it does ask about stressful practices during broodstock management.

12.3.2 Unbalances and Misperceptions

Apart from their number, which is still low, actual limitations in the study of the epigenetics of stress in cultured fish can be summarized as follows:

1) Studies have poor relationships with the stress axis. ELS indicates that individuals were submitted to a stressor that may influence epigenetic outputs (e.g., lists of differentially methylated genes), but brain epigenetics, which represents a fundamental driver of the process, is ignored.

2) Studies were interested in changes in epigenetic profiles, but only a few of them considered the impact of the stressor(s) on traits relevant to the hatchery environment (e.g., growth, weight at age, immune competence).
3) Stress-responsive genes from the HPI axis are not routinely screened for epigenetic modifications although they have been the subject of an impressive number of studies in mammals. At best, these genes, or the pathway they belong to, are detected by chance [99, 107] or proposed as candidates [98, 115], but not systematically studied yet.

In the next section, we focus on results obtained in mammals regarding epigenetically imprinted genes within brain tissues.

12.4 The Neuroepigenetics of Stress: Fishing with Mammalian Models

Mechanisms producing epigenetic alterations in the CNS participate in brain functional plasticity and encode stressful life events and experiences – notably not only in early life, but also adulthood [116, 117] – following stress exposure. These events are encoded in discrete neuronal populations that, once activated, would propagate or promote specific responses to stress across the whole lifespan and influence the health and well-being of organisms. In humans, it was shown that glucocorticoid-induced DNA methylation altered certain gene expression patterns throughout their lifetimes by priming the GR response of neighboring loci, rendering them more responsive (i.e., enhanced transcriptional expression) to a subsequent acute glucocorticoid exposure [118]. Various authors [116, 119–121] reviewed how stressors in early life, and sometimes later in adulthood, induced neuroepigenetic changes at the major central stress axis genes and could mediate not only HPA dysfunctions, but also behavioral changes in humans and mice.

12.4.1 Rationales and Limitations

Since data have rapidly accumulated in mammalian models, a short review of epigenetic variation observed at some key genes of the HPA/HPI axis can be provided for fish, also including the main players of the vasopressin and oxytocin systems (Figure 12.2). The epigenetics of the catecholaminergic system is addressed by Gonzalez [122]. Choices are subjective, as complex molecular processes, including epigenetic modifications, cannot be summarized into a few genes. However, stress-induced variation in the expression of genes presented in Figure 12.2 has been studied in fish, notably in the brain (e.g., [84, 109, 123]). These studies represent an affordable first step for a better understanding of the epigenetic impact of stress in aquaculture, as variation in gene expression and/or observed polymorphisms have been shown to be associated with the stress response, neuropsychiatric disorders, neurodegenerative pathologies, and/or aberrant behavioral phenotypes in mammals [124, 125]. This set of genes includes the hormonal/endocrine (e.g., *CRH*, *CRHR*, and *POMC*) and the neuronal/synaptic (e.g., *C-FOS*, *EGR-1*, and *SLCA4C*) activation of the stress axis to emphasize a potential stressor-specific and dual activation of epigenetic programming (see Figure 12.2 for the list of abbreviations). A stimulus-induced neural activation could primarily initiate modifications in the underlying epigenetic regulatory systems, while parallel induction of the HPA axis activity could lead to hormonal priming of the epigenetic machinery for long-term effects via the release of glucocorticoids.

Nevertheless, even if the mammalian HPA and fish HPI axes are homologous, increased knowledge on the neuroepigenetics of stress has been mostly gained from placental mammalian species with developed social behavior. Transfer of knowledge to oviparous fish species with more limited social behavior and parental care remains presumptive. Although zebrafish (*Danio rerio*) was proposed as a translational model for the study of stress-related disorders, it is not fully established that comparisons with mammals are *always* effective. Evidence is reported for anxiety-like behaviors [126] and for stress disorder symptoms [127]. However, micro-RNAs (i.e., 16–22 base-pair long, noncoding RNA molecules), which are important dynamic actors of epigenetic tuning at the posttranscriptional level major regulators of the brain response to stress [128, 129] and of glucocorticoid actions [130], have undergone extensive rewiring in teleost fish compared to mammalian models [131]. This induces limitations for functional analysis [98]. Furthermore, teleost-specific genome duplication is a common issue in both genetics and epigenetics of fish. It implies the testing of more genes than in mammalian models and assessing how the functions of these additional genes evolved or not [98]. For example, Campos et al. [132] showed that the three DNA methyl transferase 3 (*DNMT3*) genes present in mammals correspond to six paralogs with differential sensitivity to external stimuli in zebrafish. Duplications of the main HPA axis genes are, however, relatively well-known [133]. Furthermore, because the stress response is variable among fish species (e.g., cortisol peak and duration)

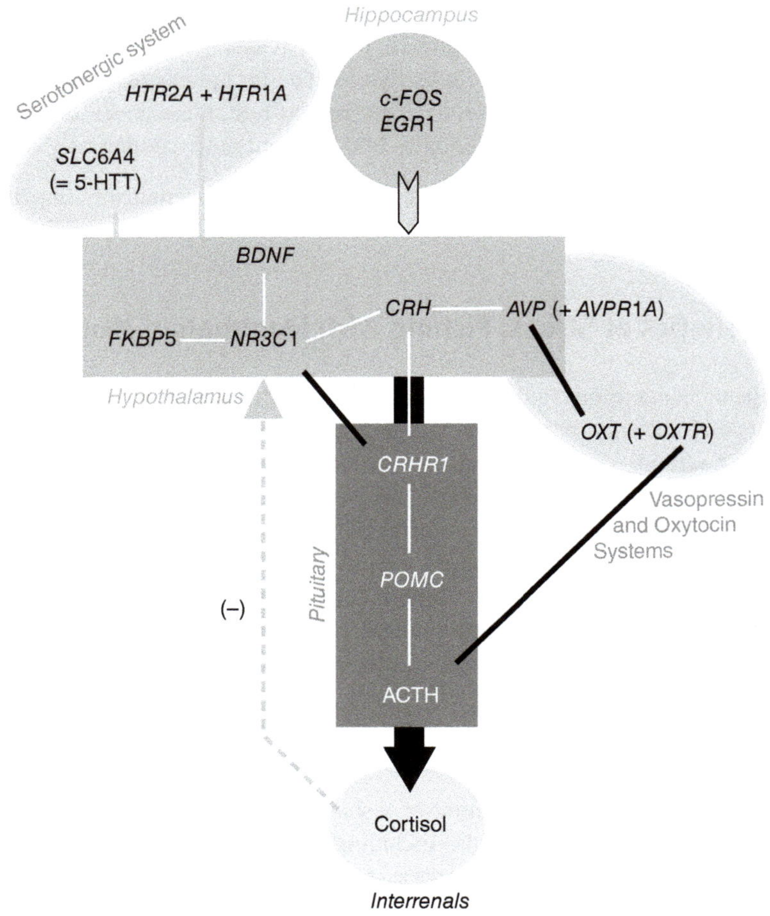

Figure 12.2 Genes considered relevant candidates for epigenetic studies in fish. A focus is made onto the HPI axis, the hippocampus, and other systems for genes in which stress-induced epigenetic modifications, mainly DNA methylation changes, have been observed. These modifications can rewire brain/neuronal molecular pathways and translate into neurobehavioral disorders in humans and rodents, sometimes resulting in differences in cortisol production. Gene names are reported as in mammals. Solid lines refer to epigenetic interactions among genes or the action of genes on specific organs (serotonergic system). (–): negative feedback loop induced by cortisol in the hypothalamus. Abbreviations: ACTH, adrenocorticotropin hormone; AVP, arginine vasopressin (AVT, arginine-vasotocin in fish); AVPR1A, arginine vasopressin receptor 1A; BDNF, brain-derived neurotrophic factor; c-FOS, Fos Proto-Oncogene; CRH, corticotropin-releasing hormone; CRHR1, corticotropin-releasing hormone receptor 1; EGR-1, early growth receptor-1; FKBP5, FK506-binding protein 5; HTR1A, 5-hydroxytryptamine receptor 1A; HTR2A, 5-hydroxytryptamine receptor 2A; POMC, proopiomelanocortin; OXT, oxytocin (IT, isotocin in fish); OXTR, oxytocin receptor. (*See colour plate section for colour representation of this figure.*)

and reflects physiological variation that has a root in the regulation of molecular interactions, a detailed comparison with the zebrafish model should be cautiously considered. Therefore, species-specific epigenetic knowledge should be developed.

12.4.2 Immediate Early Genes in the Hippocampus

The hippocampus is particularly sensitive to the effects of glucocorticoids secreted during the diurnal rhythm and chronic stress that may suppress neurogenesis [134]. The Fos Proto-Oncogene AP-1 Transcription Factor Subunit (C-FOS) and the Early Growth Receptor-1 (EGR-1) are transcription factors, also defined as stress-induced immediate early genes that are transiently induced in the brain, especially in the dentate gyrus of the mammalian hippocampus when submitted to an acute stressor. They are used as functional markers of activity in neurons and neuronal circuitries after a variety of stimuli/stressors (e.g., [135–139]). *C-FOS* can be used as a potent indicator of the neuronal activity of memory retrieval after conditioning to aversive stimuli, and a marker of the stress response in the aquaculture environment [140–142]. ELS and acute restrain stress were found to change the expression of *C-FOS* in brain areas of the gilthead seabream (*Sparus aurata*) [84] and zebrafish [143], respectively. Gene expression variation at *egr-1* predicted glucocorticoid receptor expression in the brain of a cichlid (*Neolamprologus pulcher*) [144]. However, related epigenetic signaling has not been studied so far in fish.

While *C-FOS* and *EGR-1* are not the only genes affected by stress-induced epigenetic modifications in the hippocampus, they are widely studied in fish. The transcriptional control of *C-FOS* and *EGR-1* is traditionally thought to occur predominantly by the activation of the extracellular signal-regulated kinase mitogen-activated protein kinase (ERK-MAPK) signaling cascade that interacts with GRs. This control involves the activation of signaling molecules, which results in specific phosphoacetylation of histone H3, epigenetic chromatin modeling and modification of gene expression. It includes the upregulation of *C-FOS* and *EGR-1*, and combines with changes in DNA methylation patterns of both genes. In rats, methylation-dependent interactions between GR and the EGR-1 protein were also observed [145, 146]. This suggests that the methylation status observed at the *NR3C1* promoter and downstream expression of at least one of the genes could be functionally related. How the epigenetic statuses of *NR3C1* and *ERG-1* precisely interact to shape distinct stress-sensitive phenotypes with different stress-induced memories remains unknown.

12.4.3 Brain-Derived Neurotrophic Factor

Brain-derived neurotrophic factor (BDNF) is a neurotrophic factor, for which stress-related expression is commonly reported, that is widely expressed in the brain [147–149]. BDNF directly regulates the HPA axis [150–152] and has particular functions over dendritic remodeling, neuron survival, and synaptic plasticity in the hippocampus and the basal amygdala at early life stages [153–155]. One of the BDNF receptors, tropomyosin-related kinase receptor B (TrkB; aka NTRK2: Neurotrophic Receptor Kinase 2), and GR are coexpressed in hippocampal neurons, suggesting a strong interplay between the hormonal microenvironment and the neurotrophin regulation for the control of behavioral tasks [156]. BDNF-TrkB signaling pathway has been reviewed [157] and its role in the development and maintenance of aminergic neuronal populations reported for zebrafish [158].

Under stress, methylation or histone acetylation levels of the *BDNF* varies in humans [159, 160] and rodents [161–163]. Epigenetic modifications at *BDNF*, and especially the methylation status of the *BDNF* promoter, impact the binding of the Methyl CpG Binding-Protein 2 (MeCP2), therefore repressing transcription. The MeCP2 protein has been suggested to regulate critical periods [164] and has a central role in activity-dependent neuronal plasticity and transcription during brain development. Early life experiences are thought to imprint *BDNF* methylation and induce changes in the HPA axis regulation and behavior later in life [165, 166]. Epigenetically driven regulation within the BDNF-TrkB signaling pathway also impacts long-term fear memory [167]. In the European sea bass, methylation variation between pre- and poststress challenge fish pointed to genes that regulate BDNF synthesis, but not BDNF itself [78]. This, however, suggested that the epigenetic machinery surrounding the *BDNF* gene itself is fairly complex and misunderstood.

Finally, strain-dependent *BDNF* methylation inducing complex modulation of RNA transcription and expression has been reported in mice [168]. This relationship was dependent on the early social environment and might be of some interest in aquaculture. Furthermore, mice reared in an enriched social environment had increased acetylation in the *BDNF* promoter, translating to increases in *BDNF* expression when responding to a stress challenge later in life [169].

12.4.4 Serotonin Signaling

The brain serotoninergic system including serotonin (5-HT), its transporter (5-HTT; *SLC6A4* gene [Solute Carrier Family 6 Member A4]) and numerous associated receptors (e.g., 5-HT1A or 5-HT2A; *HTR1A* and *HTR2A* genes [5-Hydroxytryptamine Receptor 1A/2A], respectively), is widely conserved across taxa [170, 171]. Serotonin is considered to be involved in more behavioral, physiological, and pathological mechanisms than any other neurotransmitter [172]. Serotonin is an important regulator of neural plasticity, behavior, and stress coping in fish [173–175]. It interacts with corticosteroid metabolism and the HPI axis to regulate stress coping style [173, 176]. Accordingly, studies monitoring the concentration and/or expression of serotonin and its allies to assess stress and stress coping are widespread in fish (e.g., [84, 147, 148]), but epigenetics studies are lacking. However, increased methylation in the proximal promoter region of *SLC6A4* was positively correlated to threat [177] and aberrant compulsive disorders [178]. Methylation landscapes of the *HTR1A* and *HTR2A* receptor genes also respond to stress exposure and could be related to neurobehavioral impairments (e.g., schizophrenia and panic disorders) [179, 180]. The expression of *SLC6A4*, *HTR1A*, and *OXTR* (oxytocin receptor, see later) correlates in stressed zebrafish [181, 182], but epigenetic changes have not yet been investigated. Interestingly, correlated methylation differences in these genes have been suggested to "imprint" the wolf-to-dog domestication process [183].

Additionally, several studies have demonstrated the role of serotonin and BDNF on emotion and memory-related networks [184–186]. For example, Atlantic salmon submitted to uncontrollable chronic stress presented higher growth rates than control fish, with the expression of *BDNF* negatively or positively interacting with serotonergic activity in the controls and the stressed fish, respectively [148]. Nevertheless, the epigenetics of this system needs to be better documented.

12.4.5 Hypothalamic Connections

CRH and AVT in fish, or AVP in mammals, released from the hypothalamus are the essential regulators of the HPA/HPI axis (Figures 12.1 and 12.2). Together with BDNF, CRH and vasotocin interactions with specific receptor subtypes induce responses to anxiogenic stimuli [187]. In mammals, in addition to CRH and AVT, oxytocin (OXT, a mammalian homologue of isotocin, IT in fish) moderates the release of ACTH and cortisol by either altering sensitivity to the feedback mechanism or by decreasing CRH release, hence promoting the return of cortisol levels to a "baseline" level after stress was experienced [188]. AVT (AVP) and IT (OXT) are also involved in the stress axis, with further implications in resulting group behavior and social interactions [189–192].

Stress-induced epigenetic changes have been observed for most of these genes. This includes epigenetic modulation of *OXT* and its receptor *OXTR* [193–195], and *AVP* [196, 197]. The epigenetic regulation of *AVP* through interaction with MeCP2 has been reviewed [198, 199]. Stress-induced epigenetic modulation of *AVPR1A* has been investigated in mice [200] and in prairie voles (*Microtus ochrogaster*), and is associated with changes in social behavior [201, 202].

Nevertheless, because of their interaction and the impact of stressors on HPA axis dysregulation, methylation changes at *NR3C1* [145, 203–205] and its co-chaperone FKBP5 (FK506 Binding Protein 5) [206–208] are the most thoroughly studied areas of stress vulnerability and resilience. *FKBP5* methylation was particularly studied in relation to stress sensitivity and induced pain, both of particular interest for welfare [204]. Methylation at exon 1_7 (rat) or 1_F (human) of *NR3C1* has been the target of many studies (review in [209]), but other methylated sites in the promoter region, or the gene body of *NR3C1*, are also involved in the regulation of the phenotypic outputs [204, 210]. For *FKBP5*, methylation at intron seven of the gene has been exclusively investigated thus far. It has been shown that intronic demethylation increased *FKBP5* expression and induced symptoms of GR resistance [211]. The precise interplay between the two genes and their methylation status in regulating the HPA axis remains complex and necessitates that further genes involved in the epigenetic regulation of stress and coping styles be considered and studied.

Furthermore, no significant relationship between *FKBP5* methylation and cortisol output has been detected so far [212, 213]. Conversely, a relationship between cortisol production under various stress exposure and *NR3C1* methylation was observed [214–216]. Intergenerational programming of *NR3C1* methylation profiles in offspring has been reported in humans [217–219], as well as methylation at the alternative receptor gene (*NR3C2*, encoding for MR) [220]. However, epigenetic data related to *NR3C2* are far less developed, perhaps even neglected.

Apart from methylation, histone modifications, and also noncoding RNA signaling at the posttranscriptional level, are considered important to the stress-induced regulation of the HPA axis and glucocorticoid receptors [120, 145, 221]. For example, breed-dependent epigenetic modulation at the promoter of *NR3C1* by histone deacetylases was found in the liver of piglets (brain not investigated) and associated with differences in weight among individuals [222]. They received far less attention, however.

12.4.6 Other Hypothalamic and Pituitary Candidates

Data regarding the epigenetic regulation of the main hypothalamic hormone CRH and its receptor CRHR1, when submitted to various stressors, have also accumulated. They concern methylation, histone modifications, MeCP2 levels, or the impact of miRNAs [223–225]. Different methylation status of the CRH/CRHR1 duo has also been recorded in the hippocampus of stressed vs. control rats and may translate over at least two generations [226].

POMC is produced in the brain stem and the arcuate nucleus of the hypothalamus, but it is mainly produced by the pituitary and cleaved to ACTH (Figure 12.1). Methylation and other epigenetic modifications of *POMC* induced by ELS have been observed (e.g., [215, 227]). For example, *POMC* methylation level correlated to the methylation of upstream genes – notably *FKBP5* – and impacts ACTH and cortisol production depending on the stressor type, duration, and frequency [215]. Impairments in the regulation of food intake and nutritional programming related to the epigenetic modifications observed at POMC were also evaluated in humans [228, 229], and certainly deserve more attention in aquaculture.

Epigenetic modifications occurring at, or close to, these and other genes appear essential for the regulation of many stress-induced mood or anxiety-like disorders that will affect fish welfare and regulation of the full stress response later in life. How this can be managed to improve traits of interest in a productive environment is still not understood.

12.5 Epigenetic Biomonitoring of Stress

By targeting the main genes of the HPI axis and a few others for which information has been gained over the years, we hereby favored a candidate gene approach over epigenome-wide, high-throughput approaches. High-throughput studies are now achievable in many important aquaculture species that have their genome sequenced. Such studies could provide important functional information on the impact of stress at a/an (epi)genome-wide level and a source of new candidates that may complement patterns investigated in commonly studied genes. On the contrary, although the candidate gene approach targets only a part of the genome, it may participate to translate observations made in humans and rodents to fish, including species with different stress responses. Regardless of the method used, the epigenetic monitoring of stress must satisfy several criteria.

12.5.1 Tracking Changes

Epigenetic profiles have to be analyzed to estimate a cumulative impact of stress over the organism and be compared to other estimators (e.g., cortisol, glucose, and lysozyme). As biomonitoring could be broadly defined as a method for assessing exposure to stressors, it could be important to address this issue for a set of epigenetic markers acting as biomarkers. Epigenetic profiles of genes related to the HPA/HPI axis are an interesting target to address this issue by tracking dynamic epigenetic modifications from early to important developmental stages (including sexual differentiation), and during adulthood under different conditions to estimate the impact of different rearing conditions and/or practices. This longitudinal tracking is not only essential to capture the full complexity of the epigenotype-to-phenotype relationship, but also a difficult task.

Progress has been made mainly for sex determination-related issues [230], and a general framework certainly has to be established for longitudinal stress monitoring of fish cohorts within rearing tanks at different life stages. However, this biomonitoring is already available in fish through blood parameters, including cortisol that establishes a causal link with the stress axis [12, 14]. As blood cells are nucleated in fish, genome-wide epigenetic information is accessible without sacrificing fish, also limiting invasiveness compared to tissue biopsies. Blood cells may provide complementary data to other blood metabolite parameters such as cortisol, glucose, lactate, and other stress indicators. In fish, the use of red blood cells (RBCs) is promising as they are nucleated, providing more DNA for epigenetic evaluation compared to mammals, where only lymphocytes are nucleated. Monitoring of blood epigenetics has already been performed in fish and it is thus technically amenable [78, 91, 231], but it was never correlated to a common proxy of stress such as cortisol. In hens, methylation of RBCs correlated to rearing conditions (open aviaries vs. cages), and thus authors envisioned methylation as a proxy of long-term stress [232]. They initiated the search for epigenetic stress biomarkers in hens [233] and reported differentially methylated candidate genes in the blood that could reflect methylation at stress-responsive brain genes [234]. Similar observations were made for the European sea bass [77]. Results remain presumptive as comparisons with methylation of target genes directly in brain tissues have not yet been performed (see later).

Besides blood, which might be very difficult to sample from small fish, any other cells could potentially integrate a portion of the epigenetic stress response in fish. This includes scales that are also a source of cortisol monitoring [14, 88], but it could also be possible to extend this approach to water by translating environmental DNA (eDNA) protocols on the epigenetic side (epi-eDNA). Epi-eDNA has been proposed for wild populations of aquatic organisms [235, 236], but could be fruitful to monitor an "epigenetic stress index" in the enclosed hatchery environment for welfare assessment at the tank level. This remains a challenging task to obtain well-preserved samples useful for epigenetic analyses that depend on many factors such as water chemistry (basically fresh vs. seawater) or temperature [236]. However, targeting epigenetic profiles of a tissue or of a few candidate genes, as presented in the previous section, or genes traditionally used to monitor differences in stress response and/or coping styles might be achievable. A final advantage of blood – or even water – is the capacity to store samples in biobanks for future assessment [237].

12.5.2 Tissue Dependency

Although blood offers many advantages, drawbacks are present. Perhaps the main one is derived from the fact that epigenetic profiles are tissue dependent. Thus, it is necessary to establish if the epigenetic profiles of a peripheral tissue, such as blood, are a reasonable proxy of stress, and also if they are valid across species and conditions. Most of the literature on humans is based on the use of blood (lymphocytes), buccal cells, and saliva samples, and impressive progress has been made in the field. Studies provided support for the blood–brain comparison or provided tools to evaluate this concordance [238–240]. It is, however, clear that evidence from some of the most studied genes is still partly contentious, as recently shown for BDNF [241]. In humans, it was impossible to translate a recognized signature between kit-ligand (KITLG) methylation, adversity, and cortisol level when using monocytes compared to previous analysis that used a variety of tissues [242].

12.5.3 Time-Dependency

Some epigenetics profiles can also be age-dependent and allow building epigenetic clocks for age estimation. An epigenetic clock is a mathematical procedure that estimates the age of one individual through determining and combining the methylation values of selected loci across the genome. Age-predictive models of epigenetic clocks are tissue-dependent and necessitate rigorous calibration [243, 244]. One has been proposed in European sea bass, however [245]. Methylation values and rate of accumulation might be modified under stress. Therefore, a discrepancy between epigenetic and chronological ages could be predictive of cumulative stress periods over the organism, impaired future events, and maybe a prerequisite for phenotypic prediction [246, 247]. Many epigenetic clocks use blood samples as starting tissue and, in humans, it was shown that "epigenetic aging" measured at methylation sites occurred in glucocorticoid response elements, suggesting that the history of stress exposure can be strongly related to glucorticoid signaling [248]. Epigenetic clocks need to be established for brain and blood tissues in aquaculture fishes and could be indices of fish welfare and well-being in hatcheries. Such data are lacking and certainly challenging as, for example, distinct brain areas may put different imprinting of stress exposures, shaping their own epigenetic clocks.

12.5.4 Critical Period

Time dependency also refers to critical periods at which organisms are sensitive to ELS. This is, however, a very general definition that triggers neural plasticity in early life, but does not stipulate the extent and the main boundaries of a critical period. For example, in humans, most of the ELS literature focuses on pre- and early postnatal critical periods. However, it was also shown that stressors may induce epigenetic alterations until five years of age, possibly until adolescence, with later life effects also [249, 250]. This suggests that some information on "windows of opportunity" for stress to imprint the epigenetic machinery and induce allostatic load-limiting performance, but also affecting well-being later in life, is possibly lacking in fish (Box 12.1). It is clear that early developmental stages (e.g., hatching and mouth-opening) deserve attention, but the dynamics of epigenetic changes over the first months of life have to be investigated in greater detail in order to concentrate improvements of hatchery practices over critical stressful periods that may extend beyond larval rearing. From the epigenetics side, if unstressed individuals are sufficient to investigate how experimental manipulations of stressors affect the epigenome in individual studies, a standardized protocol would improve further comparisons across laboratories, strains, and species. This will facilitate the design of collaborative studies to prime eustress and adequately modify a phenotype.

12.6 Conclusions

While epigenetic modifications have been shown to be linked to numerous neurobehavioral disorders in humans and rodents, the study of epigenetics of stress in fish and aquaculture is still in its infancy. This research area has to be improved by targeting more specific systems or genes that mediate the stress response. The HPI axis represents a starting point as the epigenetic regulation of stress-responsive genes has received considerable attention not only in mammals, but also in gene expression studies in fish. Other systems (e.g., the sympathetic nervous system) may deserve parallel attention. At the scale of a single generation, it is necessary to adopt a more longitudinal approach, to go beyond ELS and to define more precisely the "critical period" at which a stressor may program an epigenome and translate into a significant impact on health, welfare, and production. Such knowledge should open new avenues to epigenetic biomonitoring of stress using peripheral tissues (i.e., blood) to complement more traditional cortisol evaluation of well-being, and create another level of evaluation of coping styles. This necessitates costly programmers and the development of new resources targeting the analyses of epigenome-wide variation. Increased accessibility to (epi)genomic resources for most aquaculture species, as well as the development of models and tools, are promising to make the epigenetics of stress an exciting research area.

Acknowledgments

We greatly thank colleagues who provided support during this work, especially E. Desmarais and S. Ben Chehida. The kindness and patience of editors have been appreciated. Funding by the EU ERA-NET FP7 programme for the Cooperation in Fisheries, Aquaculture and Seafood Processing (COFASP) and the Agence Nationale pour la Recherche (ANR; grant number: ANR-16-0001-COFA) has also to be acknowledged.

References

1 Selye, H. (1950). Stress and the general adaptation syndrome. *British Medical Journal* **1**: 1383–1392.
2 Shonkoff, J.P., Boyce, W.T., and McEwen, B.S. (2009). Neuroscience, molecular biology and the childhood roots of health disparities. *Journal of the American Medical Association* **301**: 2252–2259.
3 Saraiva, J.L., Castanheira, M.F., Arechavala-Lopez, P. et al. (2019). Domestication and welfare in farmed fish. In: *Animal Domestication* (ed. F. Teletchea). London: InTechOpen https://doi.org/10.5772/intechopen.77251.
4 Ellis, T., Yildiz, H.Y., López-Olmeda, J. *et al.* I. (2012) Cortisol and finfish welfare. Fish Physiology and Biochemistry, **38** (1), 163–188.
5 Conte, F.S. (2004). Stress and the welfare of cultured fish. *Applied Animal Behaviour Science* **86**: 205–223.
6 Sneddon, L.U., Wolfenden, D.C., and Thomson, J.S. (2016). Stress management and welfare. In: *Fish Physiology, Biology of Stress in Fish*, vol. **35** (ed. C.B. Schreck, L. Tort, A.P. Farrell and C.J. Brauner), 463–539. New York: Academic Press.
7 Castanheira, M.F., Conceição, L.E., Millot, S. et al. (2017). Coping styles in farmed fish: consequences for aquaculture. *Reviews in Aquaculture* **9** (1): 23–41.
8 Martos-Sitcha, J.A., Mancera, J.M., Prunet, P., and Magnoni, L.J. (2020). Welfare and stressors in fish: challenges facing aquaculture. *Frontiers in Physiology* **11**: 162.
9 Chandararathna, U., Iversen, M.H., Korsnes, K. et al. (2021). Animal welfare issues in capture-based aquaculture. *Animals* **11** (4): 956.
10 Afonso, L.O. (2020) Identifying and managing maladaptive physiological responses to aquaculture stressors, in Fish Physiology, *Vol.* **38** (eds. T. Benfey, A.P. Farrell and C.J. Brauner), Academic Press, New York, pp. 163–191.
11 Barreto, M.O., Rey Planellas, S., Yang, Y. et al. (2021). Emerging indicators of fish welfare in aquaculture. *Reviews in Aquaculture* **14**: 343–361.
12 Rebl, A., Seibel, H., and Baßmann, B. (2021). Blood will tell: what hematological analyses can reveal about fish welfare. *Frontiers in Veterinary Science* **8**: 194.
13 Burgos-Aceves, M.A., Lionetti, L., and Faggio, C. (2019). Multidisciplinary haematology as prognostic device in environmental and xenobiotic stress-induced response in fish. *Science of the Total Environment* **670**: 1170–1183.
14 Sadoul, B. and Geffroy, B. (2019). Measuring cortisol, the major stress hormone in fishes. *Journal of Fish Biology* **94** (4): 540–555.
15 Eissa, N. and Wang, H.P. (2016). Transcriptional stress responses to environmental and husbandry stressors in aquaculture species. *Reviews in Aquaculture* **8** (1): 61–88.
16 Chandhini, S. and Rejish Kumar, V.J. (2019). Transcriptomics in aquaculture: current status and applications. *Reviews in Aquaculture* **11** (4): 1379–1397.
17 de Magalhães, C.R., Schrama, D., Farinha, A.P. et al. (2020). Protein changes as robust signatures of fish chronic stress: a proteomics approach to fish welfare research. *BMC Genomics* **21**: 309.
18 Endo, H. and Wu, H. (2019). Biosensors for the assessment of fish health: a review. *Fisheries Science* **85** (4): 641–654.
19 Brijs, J., Føre, M., Gräns, A. et al. (2021). Bio-sensing technologies in aquaculture: how remote monitoring can bring us closer to our farm animals. *Philosophical Transactions of the Royal Society B* **376** (1830): 20200218.
20 Granada, L., Lemos, M.F., Cabral, H.N. et al. (2018). Epigenetics in aquaculture – the last frontier. *Reviews in Aquaculture* **10** (4): 994–1013.
21 Ibeagha-Awemu, E.M. and Zhao, X. (2015). Epigenetic marks: regulators of livestock phenotypes and conceivable sources of missing variation in livestock improvement programs. *Frontiers in Genetics* **6**: 302.
22 Triantaphyllopoulos, K.A., Ikonomopoulos, I., and Bannister, A.J. (2016). Epigenetics and inheritance of phenotype variation in livestock. *Epigenetics & Chromatin* **9** (1): 31.

23 Wang, M. and Ibeagha-Awemu, E.M. (2020). Impacts of epigenetic processes on the health and productivity of livestock. *Frontiers in Genetics* **11**: 613636.

24 Pittman, K., Yúfera, M., Pavlidis, M. et al. (2013). Fantastically plastic: fish larvae equipped for a new world. *Reviews in Aquaculture* **5**: S224–S267.

25 Gavery, M.R. and Roberts, S.B. (2017). Epigenetic considerations in aquaculture. *PeerJ* **5**: e4147.

26 Moghadam, H., Mørkøre, T., and Robinson, N. (2015). Epigenetics – potential for programming fish for aquaculture? *Journal of Marine Science and Engineering* **3** (2): 175–192.

27 Houston, R.D., Bean, T.P., Macqueen, D.J. et al. (2020). Harnessing genomics to fast-track genetic improvement in aquaculture. *Nature Reviews Genetics* **21** (7): 389–409.

28 Fogelman, N. and Canli, T. (2019). Early life stress, physiology and genetics: a review. *Frontiers in Psychology* **10**: 1668.

29 Li, M. and Leatherland, J.F. (2013). The implications for aquaculture practice of epigenomic programming of components of the endocrine system of teleostean embryos: lessons learned from mammalian studies. *Fish and Fisheries* **14** (4): 528–553.

30 McEwen, B.S., Bowles, N.P., Gray, J.D. et al. (2015). Mechanisms of stress in the brain. *Nature Neuroscience* **18** (10): 1353–1363.

31 Kundakovic, M. and Champagne, F.A. (2015). Early-life experience, epigenetics and the developing brain. *Neuropsychopharmacology* **40** (1): 141–153.

32 Selye, H. (1946). The general adaptation syndrome and the diseases of adaptation. *Journal of Clinical Endocrinology and Metabolism* **6** (2): 117–230.

33 Chrousos, G.P. (2009). Stress and disorders of the stress system. *Nature Reviews Endocrinology* **5** (7): 374–381.

34 Sterling, P. and Eyer, J. (1988). Handbook of life stress, cognition and health. In: *Allostasis: a New Paradigm to Explain Arousal Pathology* (ed. S. Fisher and J. Reason), 629–649. New York: Wiley.

35 McEwen, B.S. and Wingfield, J.C. (2003). The concept of allostasis in biology and biomedicine. *Hormones and Behavior* **43**: 2–15.

36 Korte, S.M., Olivier, B., and Koolhaas, J.M. (2007). A new animal welfare concept based on allostasis. *Physiology and Behavior* **92**: 422–428.

37 Pavlidis, M., Karantzali, E., Fanouraki, E. et al. (2011). Onset of the primary stress in European sea bass *Dicentrarchus labrax*, as indicated by whole body cortisol in relation to glucocorticoid receptor during early development. *Aquaculture* **315** (1–2): 125–130.

38 Gorissen, M. and Flik, G. (2016). The endocrinology of the stress response in fish – an adaptation-physiological view. In: *Fish Physiology, Biology of Stress in Fish*, vol. **35** (ed. C.B. Schreck, L. Tort, A.P. Farrell and C.J. Brauner), 75–111. London: Academic Press.

39 Barton, C. (2002). Stress in fishes: a diversity of responses with particular reference to changes in circulating corticosteroids. *Integrative and Comparative Biology* **42**: 517–525.

40 Milligan, C.L. (1996). Metabolic recovery from exhaustive exercise in rainbow trout. *Science* **113**: 51–60.

41 Flik, G., Klaren, P.H.M., Van Den Burg, E.H. et al. (2006). CRF and stress in fish. *General and Comparative Endocrinology* **146** (1): 36–44.

42 Wendelaar Bonga, S.E. (1997). The stress response in fish. *Physiological Reviews* **77** (3): 591–625.

43 Mommsen, T.P., Vijayan, M.M., and Moon, T.W. (1999). Cortisol in teleosts: dynamics, mechanisms of action and metabolic regulation. *Reviews in Fish Biology and Fisheries* **9**: 211–268.

44 Gray, J.D., Kogan, J.F., Marrocco, J., and McEwen, B.S. (2017). Genomic and epigenomic mechanisms of glucocorticoids in the brain. *Nature Reviews Endocrinology* **13** (11): 661–673.

45 Bradford, S.C., Fitzpatrick, M.S., and Schreck, C.B. (1992). Evidence for ultra-short-loop feedback in ACTH-induced interrenal steroidogenesis in coho salmon: acute self-suppression of cortisol secretion in vitro. *General and Comparative Endocrinology* **87** (2): 292–299.

46 Suda, T., Tomori, N., Yajima, F. et al. (1987). A short negative feedback mechanism regulating corticotropin-releasing hormone release. *Journal of Clinical Endocrinology and Metabolism* **64** (5): 909–913.

47 Balment, R.J., Lu, W., Weybourne, E., and Warne, J.M. (2006). Arginine vasotocin a key hormone in fish physiology and behaviour: a review with insights from mammalian models. *General and Comparative Endocrinology* **147** (1): 9–16.

48 Rotllant, J., Ruane, N.M., Dinis, M.T. et al. (2006). Intra-adrenal interactions in fish: catecholamine stimulated cortisol release in sea bass (*Dicentrarchus labrax* L.). *Comparative Biochemistry and Physiology – Part A Molecular and Integrative Physiology* **143** (3): 375–381.

49 Eigler, T. and Ben-Shlomo, A. (2014). Somatostatin system: molecular mechanisms regulating anterior pituitary hormones. *Journal of Molecular Endocrinology* **53** (1): R1–R19.

50 Gorissen, M., Bernier, N.J., Manuel, R. et al. (2012). Recombinant human leptin attenuates stress axis activity in common carp (*Cyprinus carpio* L.). *General and Comparative Endocrinology* **178** (1): 75–81.

51 Samaras, A., Papandroulakis, N., Costari, M., and Pavlidis, M. (2016). Stress and metabolic indicators in a relatively high (European sea bass, *Dicentrarchus labrax*) and a low (meagre, *Argyrosomus regius*) cortisol responsive species, in different water temperatures. *Aquaculture Research* **47** (11): 3501–3515.

52 Tort, L. (2011). Stress and immune modulation in fish. *Developmental and Comparative Immunology* **35** (12): 1366–1375.

53 Yada, T. and Tort, L. (2016). Stress and disease resistance: immune system and immunoendocrine interactions. In: *Fish Physiology, Biology of Stress in Fish*, vol. **35** (ed. C.B. Schrek, L. Tort, A.P. Farell and C.J. Brauner), 365–403. Amsterdam: Elsevier Inc.

54 Juell, J.-E. and Fosseidengen, J.E. (2004). Use of artificial light to control swimming depth and fish density of Atlantic salmon (*Salmo salar*) in production cages. *Aquaculture* **233** (1–4): 269–282.

55 Schreck, C.B. (2010). Stress and fish reproduction: the roles of allostasis and hormesis. *General and Comparative Endocrinology* **165** (3): 549–556.

56 Leal, E., Fernández-Durán, B., Guillot, R. et al. (2011). Stress-induced effects on feeding behavior and growth performance of the sea bass (*Dicentrarchus labrax*): a self-feeding approach. *Journal of Comparative Physiology – B BiochemIcal, Systems and Environmental Physiology* **181** (8): 1035–1044.

57 Rubio, V.C., Sánchez, E., and Cerdá-Reverter, J.M. (2010). Compensatory feeding in the sea bass after fasting and physical stress. *Aquaculture* **298** (3–4): 332–337.

58 Fanouraki, E., Mylonas, C.C., Papandroulakis, N., and Pavlidis, M. (2011). Species specificity in the magnitude and duration of the acute stress response in Mediterranean marine fish in culture. *General and Comparative Endocrinology* **173**: 313–322.

59 Madaro, A., Folkedal, O., Maiolo, S. et al. (2018). Effects of acclimation temperature on cortisol and oxygen consumption in Atlantic salmon (*Salmo salar*) post-smolt exposed to acute stress. *Aquaculture* **497**: 331–335.

60 Samaras, A., Papandroulakis, N., Lika, K., and Pavlidis, M. (2018). Water temperature modifies the acute stress response of European sea bass, *Dicentrarchus labrax* L. (1758). *Journal of Thermal Biology* **78**: 84–91.

61 Pascoli, F., Lanzano, G.S., Negrato, E. et al. (2011). Seasonal effects on hematological and innate immune parameters in sea bass *Dicentrarchus labrax*. *Fish & Shellfish Immunology* **31** (6): 1081–1087.

62 Tsui, W.C., Chen, J.C., and Cheng, S.Y. (2012). The effects of a sudden salinity change on cortisol, glucose, lactate and osmolality levels in grouper *Epinephelus malabaricus*. *Fish Physiology and Biochemistry* **38** (5): 1323–1329.

63 Samaras, A., Pavlidis, M., Lika, K. et al. (2017). Scale matters: performance of European sea bass, *Dicentrarchus labrax*, L. (1758), reared in cages of different volumes. *Aquaculture Research* **48** (3): 990–1005.

64 Santos, G.A., Schrama, J.W., Mamauag, R.E.P. et al. (2010). Chronic stress impairs performance, energy metabolism and welfare indicators in European sea bass (*Dicentrarchus labrax*): the combined effects of fish crowding and water quality deterioration. *Aquaculture* **299**: 73–80.

65 Pottinger, T.G., Pickering, A.D., and Hurley, M. (1992). Consistency in the stress response of two strains of rainbow trout, *Oncorhynchus mykiss*. *Aquaculture* **103**: 275–289.

66 Samaras, A., Dimitroglou, A., Sarropoulou, E. et al. (2016). Repeatability of cortisol stress response in the European sea bass (*Dicentrarchus labrax*) and transcription differences between individuals with divergent responses. *Scientific Reports* **6**: 34858.

67 Øverli, Ø., Sørensen, C., Pulman, K.G.T. et al. (2007). Evolutionary background for stress-coping styles: relationships between physiological, behavioral and cognitive traits in non-mammalian vertebrates. *Neuroscience and Biobehavioral Reviews* **31** (3): 396–412.

68 Øverli, Ø., Winberg, S., and Pottinger, T.G. (2005). Behavioral and neuroendocrine correlates of selection for stress responsiveness in rainbow trout–a review. *Integrative and Comparative Biology* **45** (3): 463–474.

69 de Ruiz-Gomez, M.L., Huntingford, F.A., Øverli, Ø. et al. (2011). Response to environmental change in rainbow trout selected for divergent stress coping styles. *Physiology and Behavior* **102** (3–4): 317–322.

70 Pottinger, T.G. (2006). Context dependent differences in growth of two rainbow trout (*Oncorhynchus mykiss*) lines selected for divergent stress responsiveness. *Aquaculture* **256** (1–4): 140–147.

71 Tort, L., Montero, D., Robaina, L. et al. (2001). Consistency of stress response to repeated handling in the gilthead sea bream *Sparus aurata* Linnaeus, 1758. *Aquaculture Research* **32**: 593–598.

72 Hori, T.S., Rise, M.L., Johnson, S.C. et al. (2012). The mRNA expression of cortisol axis related genes differs in Atlantic cod (*Gadus morhua*) categorized as high or low responders. *General and Comparative Endocrinology* **175** (2): 311–320.

73 Samaras, A. and Pavlidis, M. (2018). Regulation of divergent cortisol responsiveness in European sea bass, *Dicentrarchus labrax* L. *PLoS One* **13** (8): e0202195.

74 Øverli, Ø., Pottinger, T.G., Carrick, T.R. et al. (2001). Brain monoaminergic activity in rainbow trout selected for high and low stress responsiveness. *Brain, Behaviour and Evolution* **57** (4): 214–224.

75 Chatziplis, D., Oikonomou, S., Loukovitis, D. et al. (2020). QTL for stress and disease resistance in European sea bass, *Dicentrarchus labrax* L. *Animals* **10**: 1668.

76 Fevolden, S.E., Roed, K.H., Fjalestad, K.T., and Stien, J. (1999). Poststress levels of lysozyme and cortisol in adult rainbow trout: heritabilities and genetic correlations. *Journal of Fish Biology* **54** (900): 900–910.

77 Moghadam, H.K., Johnsen, H., Robinson, N. et al. (2017). Impacts of early life stress on the methylome and transcriptome of Atlantic salmon. *Scientific Reports* **7**: 5023.

78 Krick, M.V., Desmarais, E., Samaras, A. et al. (2021). Family-effects in the epigenomic response of red blood cells to a challenge test in the European sea bass (*Dicentrarchus labrax*, L.). *BMC Genomics* **22**: 111.

79 Tort, L., Gómez, E., Montero, D., and Sunyer, J.O. (1996). Serum haemolytic and agglutinating activity as indicators of fish immunocompetence: their suitability in stress and dietary studies. *Aquaculture International* **4**: 31–41.

80 Samaras, A., Santo, C.E., Papandroulakis, N. et al. (2018). Allostatic load and stress physiology in European sea bass (*Dicentrarchus labrax* L.) and gilthead seabream (*Sparus aurata* L.). *Frontiers in Endocrinology* **9**: 451.

81 Millot, S., Cerqueira, M., Castanheira, M.F. et al. (2014). Behavioural stress responses predict environmental perception in European sea bass (*Dicentrarchus labrax*). *PLoS One* **9** (9): e108800.

82 Kleinhappel, T.K., Pike, T.W., and Burman, O.H.P. (2019). Stress-induced changes in group behaviour. *Scientifc Reports* **9** (1): 17200.

83 Lika, K., Pavlidis, M., Mitrizakis, N. et al. (2015). Do experimental units of different scale affect the biological performance of European sea bass *Dicentrarchus labrax* larvae? *Journal of Fish Biology* **86** (4): 1271–1285.

84 Vindas, M.A., Fokos, S., Pavlidis, M. et al. (2018). Early life stress induces long-term changes in limbic areas of a teleost fish: the role of catecholamine systems in stress coping. *Scientific Reports* **8**: 5638.

85 Varsamos, S., Flik, G., Pepin, J.F. et al. (2006). Husbandry stress during early life stages affects the stress response and health status of juvenile sea bass, *Dicentrarchus labrax. Fish & Shellfish Immunology* **20**: 83–96.

86 Faught, E., Best, C., and Vijayan, M.M. (2016). Maternal stress-associated cortisol stimulation may protect embryos from cortisol excess in zebrafish. *Royal Society Open Science* **3**: 160032.

87 Sopinka, N.M., Capelle, P.M., Semeniuk, C.A.D., and Love, O.P. (2017). Glucocorticoids in fish eggs: variation, interactions with the environment and the potential to shape offspring fitness. *Physiological Biochemistry and Zoology* **90** (1): 15–33.

88 Samaras, A., Dimitroglou, A., Kollias, S. et al. (2021). Cortisol concentration in scales is a valid indicator for the assessment of chronic stress in European sea bass, *Dicentrarchus labrax* L. *Aquaculture* **545**: 737257.

89 Balasch, J.C. and Tort, L. (2019). Netting the stress responses in fish. *Frontiers in Endocrinology* **10**: 62.

90 Le Luyer, J., Laporte, M., Beacham, T.D. et al. (2017). Parallel epigenetic modifications induced by hatchery rearing in a Pacific salmon. *Proceedings of the National Academy of Sciences* **114** (49): 12964–12969.

91 Gavery, M.R., Nichols, K.M., Berejikian, B.A. et al. (2019). Temporal dynamics of DNA methylation patterns in response to rearing juvenile steelhead (*Oncorhynchus mykiss*) in a hatchery versus simulated stream environment. *Genes* **10** (5): 356.

92 Nilsson, E., Sadler-Riggleman, I., Beck, D., and Skinner, M.K. (2021). Differential DNA methylation in somatic and sperm cells of hatchery *vs* wild (natural-origin) steelhead trout populations. *Environmental Epigenetics* **7** (1): dvab002.

93 Berbel-Filho, W.M., Berry, N., Rodríguez-Barreto, D. et al. (2020). Environmental enrichment induces intergenerational behavioural and epigenetic effects on fish. *Molecular Ecology* **29** (12): 2288–2299.

94 Robinson, N.A., Johnsen, H., Moghadam, H. et al. (2019). Early developmental stress affects subsequent gene expression response to an acute stress in Atlantic salmon: an approach for creating robust fish for aquaculture? *G3 Genes|Genomes|Genetics* **9** (5): 1597–1611.

95 Kelly, T., Johnsen, H., Burgerhout, E. et al. (2020). Low oxygen stress during early development influences regulation of hypoxia-response genes in farmed Atlantic salmon (*Salmo salar*). *G3: Genes|Genomes|Genetics* **10** (9): 3179–3188.

96 Uren Webster, T.M., Rodriguez-Barreto, D., Martin, S.A. et al. (2018). Contrasting effects of acute and chronic stress on the transcriptome, epigenome and immune response of Atlantic salmon. *Epigenetics* **13** (12): 1191–1207.

97 Venney, C.J., Wellband, K.W., and Heath, D.D. (2021). Rearing environment affects the genetic architecture and plasticity of DNA methylation in Chinook salmon. *Heredity* **126** (1): 38–49.

98 Best, C., Ikert, H., Kostyniuk, D.J. et al. (2018). Epigenetics in teleost fish: from molecular mechanisms to physiological phenotypes. *Comparative Biochemistry and Physiology Part B: Biochemistry and Molecular Biology* **224**: 210–244.

99 Wellband, K., Roth, D., Linnansaari, T. et al. (2021). Environment-driven reprogramming of gamete DNA methylation occurs during maturation and is transmitted intergenerationally in Atlantic salmon. *G3 Genes|Genomes|Genetics* **11**: jkab353.

100 Blanc, M., Antczak, P., Cousin, X. et al. (2021). The insecticide permethrin induces transgenerational behavioral changes linked to transcriptomic and epigenetic alterations in zebrafish (*Danio rerio*). *Science of the Total Environment* **779**: 146404.

101 Marrocco, J., Gray, J.D., Kogan, J.F. et al. (2019). Early life stress restricts translational reactivity in CA3 neurons associated with altered stress responses in adulthood. *Frontiers in Behavioral Neuroscience* **13**: 157.

102 Anastasiadi, D. and Piferrer, F. (2019). Epimutations in developmental genes underlie the onset of domestication in farmed European sea bass. *Molecular Biology and Evolution* **36** (10): 2252–2264.

103 Leitwein, M., Laporte, M., Le Luyer, J. et al. (2021). Epigenomic modifications induced by hatchery rearing persist in germ line cells of adult salmon after their oceanic migration. *Evolutionary Applications* **14** (10): 2402–2413.

104 Christensen, K.A., Le Luyer, J., Chan *et al.* (2021) Assessing the effects of genotype-by-environment interaction on epigenetic, transcriptomic and phenotypic response in a Pacific salmon. G3: Genes|Genomes|Genetics, **11** (2), jkab021.

105 Venney, C.J., Love, O.P., Drown, E.J., and Heath, D.D. (2020). DNA methylation profiles suggest intergenerational transfer of maternal effects. *Molecular Biology and Evolution* **37** (2): 540–548.

106 Valdivieso, A., Ribas, L., Monleon-Getini, A. et al. (2020). Exposure of zebrafish to elevated temperature induces sex ratio shifts and alterations in the testicular epigenome of unexposed offspring. *Environmental Research* **186**: 109601.

107 Anastasiadi, D., Vandeputte, M., Sánchez-Baizán, N. et al. (2018). Dynamic epimarks in sex-related genes predict gonad phenotype in the European sea bass, a fish with mixed genetic and environmental sex determination. *Epigenetics* **13** (9): 988–1011.

108 Reiser, S., Pohlmann, D.M., Blancke, T. et al. (2021). Environmental enrichment during early rearing provokes epigenetic changes in the brain of a salmonid fish. *Comparative Biochemistry and Physiology – Part D: Genomics and Proteomics* **39**: 100838.

109 Pavlidis, M., Theodoridi, A., and Tsalafouta, A. (2015). Neuroendocrine regulation of the stress response in adult zebrafish, *Danio rerio*. *Progress in Neuro-Psychopharmacology and Biological Psychiatry* **60**: 121–131.

110 Malandrakis, E.E., Dadali, O., Golomazou, E. et al. (2016). DNA damage and differential gene expression associated with physical stress in gilthead seabream (*Sparus aurata*). *General and Comparative Endocrinology* **236**: 98–104.

111 Rosengren, M., Thörnqvist, P.O., Winberg, S., and Sundell, K. (2018). The brain- gut axis of fish: rainbow trout with low and high cortisol response show innate differences in intestinal integrity and brain gene expression. *General and Comparative Endocrinology* **257**: 235–245.

112 Alfonso, S., Sadoul, B., Gesto, M. et al. (2019). Coping styles in European sea bass: the link between boldness, stress response and neurogenesis. *Physiology and Behavior* **207**: 76–85.

113 Goikoetxea, A., Sadoul, B., Blondeau-Bidet, E. et al. (2021). Genetic pathways underpinning hormonal stress responses in fish exposed to short- and long-term warm ocean temperatures. *Ecological Indicators* **120**: 106937.

114 Venney, C.J., Wellband, K.W., Normandeau, E. et al. (2022). Thermal regime during parental sexual maturation, but not during offspring rearing, modulates DNA methylation in brook charr (*Salvelinus fontinalis*). *Proceedings of the Biological Sciences* **289**: 20220670.

115 Hattori, R.S., Castañeda-Cortés, D.C., Padilla, L.A. et al. (2020). Activation of stress response axis as a key process in environment-induced sex plasticity in fish. *Cellular and Molecular Life Sciences* **77**: 4223–4236.

116 Van Bodegom, M., Homberg, J.R., and Henckens, M.J.A.G. (2017). Modulation of the hypothalamic-pituitary-adrenal axis by early life stress exposure. *Frontiers in Cellular Neuroscience* **11**: 87.

117 Dirven, B.C.J., Homberg, J.R., Kozicz, T., and Henckens, M.J.A.G. (2017). Epigenetic programming of the neuroendocrine stress response by adult life stress. *Journal of Molecular Endocrinology* **59**: R11–R31.

118 Provençal, N., Arloth, J., Cattaneo, A. et al. (2020). Glucocorticoid exposure during hippocampal neurogenesis primes future stress response by inducing changes in DNA methylation. *Proceedings of the National Academy of Sciences* **117** (38): 23280–23285.

119 Anacker, C., O'Donnell, K.J., and Meaney, M.J. (2014). Early life adversity and the epigenetic programming of hypothalamic-pituitary-adrenal function. *Dialogues in Clinical Neuroscience* **16** (3): 321.

120 Matthews, S.G. and McGowan, P.O. (2019). Developmental programming of the HPA axis and related behaviours: epigenetic mechanisms. *Journal of Endocrinology* **242** (1): T69–T79.

121 Dick, A. and Provencal, N. (2018). Central neuroepigenetic regulation of the hypothalamic–pituitary-adrenal axis. *Progress in Molecular Biology and Translational Science* **158**: 105–127.

122 González, C.R. and Gonzalez, B. (2020). Exploring the stress impact in the paternal germ cells epigenome: can catecholamines induce epigenetic reprogramming? *Frontiers in Endocrinology* **11**: 630948.

123 Moltesen, M., Laursen, D.C., Thörnqvist, P.O. et al. (2016). Effects of acute and chronic stress on telencephalic neurochemistry and gene expression in rainbow trout (*Oncorhynchus mykiss*). *Journal of Experimental Biology* **219** (24): 3907–3914.

124 Argentieri, M.A., Nagarajan, S., Seddighzadeh, B. et al. (2017). Epigenetic pathways in human disease: the impact of DNA methylation on stress-related pathogenesis and current challenges in biomarker development. *eBioMedicine* **18**: 327–350.

125 Park, C., Rosenblat, J.D., Brietzke, E. et al. (2019). Stress, epigenetics and depression: a systematic review. *Neuroscience and Biobehavioral Reviews* **102**: 139–152.

126 Collier, A.D., Kalueff, A.V., and Echevarria, D.J. (2017). Zebrafish models of anxiety-like behaviors. In: *The Rights and Wrongs of Zebrafish: Behavioral Phenotyping of Zebrafish* (ed. A.V. Kalueff), 45–72. Springer.

127 Yang, L., Wang, J., Wang, D. et al. (2020). Delayed behavioral and genomic responses to acute combined stress in zebrafish, potentially relevant to PTSD and other stress-related disorders: focus on neuroglia, neuroinflammation, apoptosis and epigenetic modulation. *Behavioural Brain Research* **389**: 112644.

128 Yapijakis, C. (2020). Regulatory role of microRNAs in brain development and function. *GeNeDis* **2018**: 237–247.

129 Clayton, S.A., Jones, S.W., Kurowska-Stolarska, M., and Clark, A.R. (2018). The role of microRNAs in glucocorticoid action. *Journal of Biological Chemistry* **293** (6): 1865–1874.

130 Lakstygal, A.M., de Abreu, M.S., and Kalueff, A.V. (2018). Zebrafish models of epigenetic regulation of CNS functions. *Brain Research Bulletin* **142**: 344–351.

131 Hollins, S.L. and Cairns, M.J. (2016). MicroRNA: small RNA mediators of the brains genomic response to environmental stress. *Progress in Neurobiology* **143**: 61–81.

132 Campos, C., Valente, L.M., and Fernandes, J.M. (2012). Molecular evolution of zebrafish dnmt3 genes and thermal plasticity of their expression during embryonic development. *Gene* **500** (1): 93–100.

133 Alsop, D. and Vijayan, M. (2009). The zebrafish stress axis: molecular fallout from the teleost-specific genome duplication event. *General and Comparative Endocrinology* **161** (1): 62–66.

134 McEwen, B.S. (2006). Protective and damaging effects of stress mediators: central role of the brain. *Dialogues in Clinical Neuroscience* **8** (4): 367.

135 Cabrera-Álvarez, M.J., Swaney, W.T., and Reader, S.M. (2017). Forebrain activation during social exposure in wild-type guppies. *Physiology and Behavior* **182**: 107–113.

136 Fuss, T. and Schluessel, V. (2018). Immediate early gene expression related to learning and retention of a visual discrimination task in bamboo sharks (*Chiloscyllium griseum*). *Brain Structure and Function* **223** (9): 3975–4003.

137 Kasper, C., Colombo, M., Aubin-Horth, N., and Taborsky, B. (2018). Brain activation patterns following a cooperation opportunity in a highly social cichlid fish. *Physiology and Behavior* **195**: 37–47.

138 Almeida, O., Félix, A.S., Oliveira, G.A. et al. (2019). Fighting assessment triggers rapid changes in activity of the brain social decision-making network of cichlid fish. *Frontiers in Behavioral Neuroscience* **13**: 229.

139 D'Agostino, S., Testa, M., Aliperti, V. et al. (2019). Expression pattern dysregulation of stress-and neuronal activity-related genes in response to prenatal stress paradigm in zebrafish larvae. *Cell Stress & Chaperones* **24** (5): 1005–1012.

140 Samaras, A. and Pavlidis, M. (2020). Behavioural and physiological responses to a conditioning protocol for adult zebrafish, *Danio rerio*, held in groups. *Behavioural Processes* **179**: 104201.

141 Ellison, A.R., Webster, T.M.U., Rodriguez-Barreto, D. et al. (2020). Comparative transcriptomics reveal conserved impacts of rearing density on immune response of two important aquaculture species. *Fish & Shellfish Immunology* **104**: 192–201.

142 Burren, A. and Pietsch, C. (2021). Distress regulates different pathways in the brain of common carp: a preliminary study. *Animals* **11** (2): 585.

143 Assad, N., Luz, W.L., Santos-Silva, M. et al. (2020). Acute restraint stress evokes anxiety-like behavior mediated by telencephalic inactivation and gabaergic dysfunction in zebrafish brains. *Scientific Reports* **10**: 5551.

144 Nyman, C., Fischer, S., Aubin-Horth, N., and Taborsky, B. (2018). Evolutionary conserved neural signature of early life stress affects animal social competence. *Proceedings of the Royal Society B: Biological Sciences* **285** (1871): 20172344.

145 Weaver, I.C., Cervoni, N., Champagne, F.A. et al. (2004). Epigenetic programming by maternal behavior. *Nature Neuroscience* **7** (8): 847–854.

146 Weaver, I.C., Hellstrom, I.C., Brown, S.E. et al. (2014). The methylated-DNA binding protein MBD2 enhances NGFI-A (egr-1)-mediated transcriptional activation of the glucocorticoid receptor. *Philosophical Transactions of the Royal Society, B: Biological Sciences* **369** (1652): 20130513.

147 Alfonso, S., Sadoul, B., Gesto, M. et al. (2019). Coping styles in European sea bass: the link between boldness, stress response and neurogenesis. *Physiology and Behavior* **207**: 76–85.

148 Vindas, M.A., Madaro, A., Fraser, T.W. et al. (2017). Uncontrollable chronic stress reduces growth disparities in farmed Atlantic salmon. *Physiology and Behavior* **179**: 246–252.

149 Tognoli, C., Rossi, F., Di Cola, F. et al. (2010). Acute stress alters transcript expression pattern and reduces processing of proBDNF to mature BDNF in *Dicentrarchus labrax*. *BMC Neuroscience* **11**: 4.

150 Tapia-Arancibia, L., Rage, F., Givalois, L., and Arancibia, S. (2004). Physiology of BDNF: focus on hypothalamic function. *Frontiers in Neuroendocrinology* **25** (2): 77–107.

151 Naert, G., Maurice, T., Tapia-Arancibia, L., and Givalois, L. (2007). Neuroactive steroids modulate HPA axis activity and cerebral brain-derived neurotrophic factor (BDNF) protein levels in adult male rats. *Psychoneuroendocrinology* **32** (8–10): 1062–1078.

152 Jeanneteau, F.D., Lambert, W.M., Ismaili, N. et al. (2012). BDNF and glucocorticoids regulate corticotrophin-releasing hormone (CRH) homeostasis in the hypothalamus. *Proceedings of the National Academy of Sciences* **109** (4): 1305–1310.

153 Huang, E.J. and Reichardt, L.F. (2001). Neurotrophins: roles in neuronal development and function. *Annual Review of Neuroscience* **24** (1): 677–736.

154 Autry, A.E. and Monteggia, L.M. (2012). Brain-derived neurotrophic factor and neuropsychiatric disorders. *Pharmacological Reviews* **64** (2): 238–258.

155 Bennett, M.R. and Lagopoulos, J. (2014). Stress and trauma: BDNF control of dendritic-spine formation and regression. *Progress in Neurobiology* **112**: 80–99.

156 Daskalakis, N.P., De Kloet, E.R., Yehuda, R. et al. (2015). Early life stress effects on glucocorticoid – BDNF interplay in the hippocampus. *Frontiers in Molecular Neuroscience* **8**: 68.

157 Mohammadi, A., Amooeian, V.G., and Rashidi, E. (2018). Dysfunction in brain-derived neurotrophic factor signaling pathway and susceptibility to schizophrenia, Parkinson's and Alzheimer's diseases. *Current Gene Therapy* **18** (1): 45–63.

158 Sahu, M.P., Pazos-Boubeta, Y., Pajanoja, C. et al. (2019). Neurotrophin receptor Ntrk2b function in the maintenance of dopamine and serotonin neurons in zebrafish. *Scientific Reports* **9** (1): 2036.

159 Mitchelmore, C. and Gede, L. (2014). Brain derived neurotrophic factor: epigenetic regulation in psychiatric disorders. *Brain Research* **1586**: 162–172.

160 Bakusic, J., Ghosh, M., Polli, A. et al. (2020). Epigenetic perspective on the role of brain-derived neurotrophic factor in burnout. *Translational Psychiatry* **10** (1): 1–9.

161 Doherty, T.S., Forster, A., and Roth, T.L. (2016). Global and gene-specific DNA methylation alterations in the adolescent amygdala and hippocampus in an animal model of caregiver maltreatment. *Behavioural Brain Research* **298**: 55–61.

162 Niknazar, S., Nahavandi, A., Peyvandi, A.A. et al. (2017). Effect of maternal stress prior to conception on hippocampal BDNF signaling in rat offspring. *Molecular Neurobiology* **54** (8): 6436–6445.

163 Fachim, H.A., Corsi-Zuelli, F., Loureiro, C.M. et al. (2021). Early-life stress effects on BDNF DNA methylation in first-episode psychosis and in rats reared in isolation. *Progress in Neuro-Psychopharmacology and Biological Psychiatry* **108**: 110188.

164 Picard, N. and Fagiolini, M. (2019). MeCP2: an epigenetic regulator of critical periods. *Current Opinion in Neurobiology* **59**: 95–101.

165 Zimmermann, C.A., Hoffmann, A., Raabe, F., and Spengler, D. (2015). Role of mecp2 in experience-dependent epigenetic programming. *Genes* **6** (1): 60–86.

166 Chen, W.G., Chang, Q., Lin, Y. et al. (2003). Derepression of BDNF transcription involves calcium-dependent phosphorylation of MeCP2. *Science* **302** (5646): 885–889.

167 Kim, H.J., Hur, S.W., Park, J.B. et al. (2019). Histone demethylase PHF2 activates CREB and promotes memory consolidation. *EMBO Reports* **20** (9): e45907.

168 Kundakovic, M., Lim, S., Gudsnuk, K., and Champagne, F.A. (2013). Sex-specific and strain-dependent effects of early life adversity on behavioral and epigenetic outcomes. *Frontiers in Psychiatry* **4**: 78.

169 Branchi, I., Karpova, N.N., D'Andrea, I. et al. (2011). Epigenetic modifications induced by early enrichment are associated with changes in timing of induction of BDNF expression. *Neuroscience Letters* **495** (3): 168–172.

170 Puglisi-Allegra, S. and Andolina, D. (2015). Serotonin and stress coping. *Behavioural Brain Research* **277**: 58–67.

171 Edsinger, E. and Dölen, G. (2018). A conserved role for serotonergic neurotransmission in mediating social behavior in octopus. *Current Biology* **28** (19): 3136–3142.

172 Azmitia, E.C. (2007). Serotonin and brain: evolution, neuroplasticity and homeostasis. *International Review of Neurobiology* **77**: 31–56.

173 Lanfumey, L., Mongeau, R., Cohen-Salmon, C., and Hamon, M. (2008). Corticosteroid–serotonin interactions in the neurobiological mechanisms of stress-related disorders. *Neuroscience and Biobehavioral Reviews* **32** (6): 1174–1184.

174 Winberg, S. and Thörnqvist, P.O. (2016). Role of brain serotonin in modulating fish behavior. *Current Zoology* **62** (3): 317–323.

175 Backström, T. and Winberg, S. (2017). Serotonin coordinates responses to social stress – what we can learn from fish. *Frontiers in Neuroscience* **11**: 595.

176 Winberg, S., Nilsson, A., Hylland, P. et al. (1997). Serotonin as a regulator of hypothalamic-pituitary-interrenal activity in teleost fish. *Neuroscience Letters* **230** (2): 113–116.

177 Nikolova, Y.S., Koenen, K.C., Galea, S. et al. (2014). Beyond genotype: serotonin transporter epigenetic modification predicts human brain function. *Nature Neuroscience* **17** (9): 1153–1155.

178 Grünblatt, E., Marinova, Z., Roth, A. et al. (2018). Combining genetic and epigenetic parameters of the serotonin transporter gene in obsessive-compulsive disorder. *Journal of Psychiatric Research* **96**: 209–217.

179 Lesseur, C., Paquette, A.G., and Marsit, C.J. (2014). Epigenetic regulation of infant neurobehavioral outcomes. *Medical Epigenetics* **2** (2): 71–79.

180 Iurato, S., Carrillo-Roa, T., Arloth, J. et al. (2017). DNA methylation signatures in panic disorder. *Translational Psychiatry* **7** (12): 1287.

181 Parker, M.O., Annan, L.V., Kanellopoulos, A.H. et al. (2014). The utility of zebrafish to study the mechanisms by which ethanol affects social behavior and anxiety during early brain development. *Progress in Neuro-Psychopharmacology and Biological Psychiatry* **55**: 94–100.

182 Attaran, A., Salahinejad, A., Crane, A.L. et al. (2019). Chronic exposure to dietary selenomethionine dysregulates the genes involved in serotonergic neurotransmission and alters social and antipredator behaviours in zebrafish (*Danio rerio*). *Environmental Pollution* **246**: 837–844.

183 Banlaki, Z., Cimarelli, G., Viranyi, Z. et al. (2017). DNA methylation patterns of behavior-related gene promoter regions dissect the gray wolf from domestic dog breeds. *Molecular Genetics and Genomics* **292** (3): 685–697.

184 Martinowich, K. and Lu, B. (2008). Interaction between BDNF and serotonin: role in mood disorders. *Neuropsychopharmacology* **33**: 73–83.

185 Homberg, J.R., Molteni, R., Calabrese, F., and Riva, M.A. (2014). The serotonin–BDNF duo: developmental implications for the vulnerability to psychopathology. *Neuroscience and Biobehavioral Reviews* **43**: 35–47.

186 Molteni, R., Cattaneo, A., Calabrese, F. et al. (2010). Reduced function of the serotonin transporter is associated with decreased expression of BDNF in rodents as well as in humans. *Neurobiology of Disease* **37** (3): 747–755.

187 Landgraf, R. (2001). Neuropeptides and anxiety-related behavior. *Endocrine Journal* **48** (5): 517–533.

188 Pierrehumbert, B., Torrisi, R., Ansermet, F. et al. (2012). Adult attachment representations predict cortisol and oxytocin responses to stress. *Attachment & Human Development* **14** (5): 453–476.

189 Culbert, B.M., Gilmour, K.M., and Balshine, S. (2019). Social buffering of stress in a group-living fish. *Proceedings of the Royal Society B* **286** (1910): 20191626.

190 Sokołowska, E., Gozdowska, M., and Kulczykowska, E. (2020). Nonapeptides arginine vasotocin and isotocin in fishes: advantage of bioactive molecules measurement. *Frontiers in Marine Science* **7**: 610.

191 Martos-Sitcha, J.A., Cádiz, L., Gozdowska, M. et al. (2019). Arginine vasotocin and cortisol co-regulate vasotocinergic, isotocinergic, stress and thyroid pathways in the gilthead sea bream (*Sparus aurata*). *Frontiers in Physiology* **10**: 261.

192 Chuang, H.J., Chang, C.Y., Ho, H.P., and Chou, M.Y. (2021). Oxytocin signaling acts as a marker for environmental stressors in zebrafish. *International Journal of Molecular Sciences* **2** (14): 7459.

193 Maud, C., Ryan, J., McIntosh, J.E., and Olsson, C.A. (2018). The role of oxytocin receptor gene (OXTR) DNA methylation (DNAm) in human social and emotional functioning: a systematic narrative review. *BMC Psychiatry* **18** (1): 1–13.

194 Kraaijenvanger, E.J., He, Y., Spencer, H. et al. (2019). Epigenetic variability in the human oxytocin receptor (OXTR) gene: a possible pathway from early life experiences to psychopathologies. *Neuroscience and Biobehavioral Reviews* **96**: 127–142.

195 Sanwald, S., Gahr, M., Widenhorn-Müller, K. et al. (2020). Relation of promoter methylation of the oxytocin gene to stressful life events and depression severity. *Journal of Molecular Neuroscience* **70** (2): 201–211.

196 Jayanthi, S., Gonzalez, B., McCoy, M.T. et al. (2018). Methamphetamine induces TET1-and TET3-dependent DNA hydroxymethylation of Crh and Avp genes in the rat nucleus accumbens. *Molecular Neurobiology* **55** (6): 5154–5166.

197 Murgatroyd, C. and Spengler, D. (2014). Polycomb binding precedes early-life stress responsive DNA methylation at the Avp enhancer. *PLoS One* **9** (3): e90277.

198 Murgatroyd, C., Wu, Y., Bockmühl, Y., and Spengler, D. (2010). Genes learn from stress: how infantile trauma programs us for depression. *Epigenetics* **5** (3): 194–199.

199 Murgatroyd, C. and Spengler, D. (2011). Epigenetic programming of the HPA axis: early life decides. *Stress* **14** (6): 581–589.

200 Bodden, C., van den Hove, D., Lesch, K.P., and Sachser, N. (2017). Impact of varying social experiences during life history on behaviour, gene expression and vasopressin receptor gene methylation in mice. *Scientific Reports* **7**: 8719.

201 Phelps, S.M., Okhovat, M., and Berrio, A. (2017). Individual differences in social behavior and cortical vasopressin receptor: genetics, epigenetics and evolution. *Frontiers in Neuroscience* **11**: 537.

202 Okhovat, M., Berrio, A., Wallace, G. et al. (2015). Sexual fidelity trade-offs promote regulatory variation in the prairie vole brain. *Science* **350** (6266): 1371–1374.

203 Watkeys, O.J., Kremerskothen, K., Quidé, Y. et al. (2018). Glucocorticoid receptor gene (NR3C1) DNA methylation in association with trauma, psychopathology, transcript expression, or genotypic variation: a systematic review. *Neuroscience and Biobehavioral Reviews* **95**: 85–122.

204 Bockmühl, Y., Patchev, A.V., Madejska, A. et al. (2015). Methylation at the CpG island shore region upregulates Nr3c1 promoter activity after early-life stress. *Epigenetics* **10** (3): 247–257.

205 Berretta, E., Guida, E., Forni, D., and Provenzi, L. (2021). Glucocorticoid receptor gene (NR3C1) methylation during the first thousand days: environmental exposures and developmental outcomes. *Neuroscience and Biobehavioral Reviews* **125**: 493–502.

206 Sabbagh, J.J., O'Leary, J.C. III, Blair, L.J. et al. (2014). Age-associated epigenetic upregulation of the FKBP5 gene selectively impairs stress resiliency. *PLoS One* **9** (9): e107241.

207 Géranton, S.M. (2019). Does epigenetic 'memory' of early-life stress predispose to chronic pain in later life? A potential role for the stress regulator FKBP5. *Philosophical Transactions of the Royal Society B* **374** (1785): 20190283.

208 Wiechmann, T., Röh, S., Sauer, S. et al. (2019). Identification of dynamic glucocorticoid-induced methylation changes at the FKBP5 locus. *Clinical Epigenetics* **11** (1): 83.

209 Daskalakis, N.P. and Yehuda, R. (2014). Site-specific methylation changes in the glucocorticoid receptor exon 1F promoter in relation to life adversity: systematic review of contributing factors. *Frontiers in Neuroscience* **8**: 369.

210 Tyrka, A.R., Parade, S.H., Eslinger, N.M. et al. (2015). Methylation of exons 1D, 1F and 1H of the glucocorticoid receptor gene promoter and exposure to adversity in preschool-aged children. *Development and Psychopathology* **27** (2): 577–585.

211 Klengel, T. and Binder, E.B. (2015). FKBP5 allele-specific epigenetic modification in gene by environment interaction. *Neuropsychopharmacology* **40** (1): 244.

212 Farrell, C., Doolin, K., O'Leary, N. et al. (2018). DNA methylation differences at the glucocorticoid receptor gene in depression are related to functional alterations in hypothalamic–pituitary–adrenal axis activity and to early life emotional abuse. *Psychiatry Research* **265**: 341–348.

213 Alexander, N., Kirschbaum, C., Stalder, T. et al. (2020). No association between FKBP5 gene methylation and acute and long-term cortisol output. *Translational Psychiatry* **10** (1): 175.

214 Schür, R.R., van Leeuwen, J.M., Houtepen, L.C. et al. (2018). Glucocorticoid receptor exon 1F methylation and the cortisol stress response in health and disease. *Psychoneuroendocrinology* **97**: 182–189.

215 Lewis, C.R., Breitenstein, R.S., Henderson, A. et al. (2021). Harsh parenting predicts novel HPA receptor gene methylation and NR3C1 methylation predicts cortisol daily slope in middle childhood. *Cellular and Molecular Neurobiology* **41**: 783–793.

216 Jimeno, B., Hau, M., Gómez-Díaz, E., and Verhulst, S. (2019). Developmental conditions modulate DNA methylation at the glucocorticoid receptor gene with cascading effects on expression and corticosterone levels in zebra finches. *Scientific Reports* **9**: 15869.

217 Mulligan, C., D'Errico, N., Stees, J., and Hughes, D. (2012). Methylation changes at NR3C1 in newborns associate with maternal prenatal stress exposure and newborn birth weight. *Epigenetics* **7** (8): 853–857.

218 Braithwaite, E.C., Kundakovic, M., Ramchandani, P.G. et al. (2015). Maternal prenatal depressive symptoms predict infant NR3C1 1F and BDNF IV DNA methylation. *Epigenetics* **10** (5): 408–417.

219 Van Aswegen, T., Bosmans, G., Goossens, L. et al. (2021). Epigenetics in families: covariance between mother and child methylation patterns. *Brain Sciences* **11** (2): 190.

220 Perroud, N., Rutembesa, E., Paoloni-Giacobino, A. et al. (2014). The Tutsi genocide and transgenerational transmission of maternal stress: epigenetics and biology of the HPA axis. *The World Journal of Biological Psychiatry* **15** (4): 334–345.

221 Moisiadis, V.G. and Matthews, S.G. (2014). Glucocorticoids and fetal programming part 2: mechanisms. *Nature Reviews Endocrinology* **10** (7): 403–411.

222 Zou, H., Li, R., Jia, Y. et al. (2012). Breed-dependent transcriptional regulation of 5′-untranslated GR (NR3C1) exon 1 mRNA variants in the liver of newborn piglets. *PLoS One* **7** (7): e40432.

223 Zhu, J., Chen, Z., Tian, J. et al. (2017). miR-34b attenuates trauma-induced anxiety-like behavior by targeting CRHR1. *International Journal of Molecular Medicine* **40** (1): 90–100.

224 Schartner, C., Ziegler, C., Schiele, M.A. et al. (2017). CRHR1 promoter hypomethylation: an epigenetic readout of panic disorder? *European Neuropsychopharmacology* **27** (4): 360–371.

225 Zhou, J.N. and Fang, H. (2018). Transcriptional regulation of corticotropin-releasing hormone gene in stress response. *IBRO Reports* **5**: 137–146.

226 Xu, Y.J., Sheng, H., Wu, T.W. et al. (2018). CRH/CRHR1 mediates prenatal synthetic glucocorticoid programming of depression-like behavior across 2 generations. *The FASEB Journal* **32** (8): 4258–4269.

227 Wu, Y., Patchev, A.V., Daniel, G. et al. (2014). Early-life stress reduces DNA methylation of the Pomc gene in male mice. *Endocrinology* **155** (5): 1751–1762.

228 Marco, A., Weller, A. and Meiri, N. (2016) Epigenetic programming of hypothalamic POMC regulates feeding and obesity, in Epigenetics and Neuroendocrinology. Epigenetics and Human Health (eds. D. Spengler and E. Binder), Springer, Berlin, pp. 135–163.

229 Candler, T., Kühnen, P., Prentice, A.M., and Silver, M. (2019). Epigenetic regulation of POMC: implications for nutritional programming, obesity and metabolic disease. *Frontiers in Neuroendocrinology* **54**: 100773.

230 Piferrer, F. (2018). Epigenetics of sex determination and differentiation in fish. In: *Sex Control in Aquaculture* (ed. H.P. Wang, F. Piferrer, S.L. Chen and Z.G. Shen), 65–83. Oxford: Wiley.

231 Gavery, M.R., Nichols, K.M., Goetz, G.W. et al. (2018). Characterization of genetic and epigenetic variation in sperm and red blood cells from adult hatchery and natural-origin steelhead, *Oncorhynchus mykiss*. *G3: Genes, Genomes, Genetics* **8** (11): 3723–3736.

232 Pértille, F., Brantsæter, M., Nordgreen, J. et al. (2017). DNA methylation profiles in red blood cells of adult hens correlate with their rearing conditions. *Journal of Experimental Biology* **220** (19): 3579–3587.

233 Pértille, F., Ibelli, A.M.G., Sharif, M.E. et al. (2020). Putative epigenetic biomarkers of stress in red blood cells of chickens reared across different biomes. *Frontiers in Genetics* **11**: 1202.

234 Guerrero-Bosagna, C., Pértille, F., Gomez, Y. et al. (2020). DNA methylation variation in the brain of laying hens in relation to differential behavioral patterns. *Comparative Biochemistry and Physiology Part D: Genomics and Proteomics* **35**: 100700.

235 Sigsgaard, E.E., Jensen, M.R., Winkelmann, I.E. et al. (2020). Population-level inferences from environmental DNA – current status and future perspectives. *Evolutionary Applications* **13** (2): 245–262.

236 Schadewell, Y. and Adams, C.I. (2021). Forensics meets ecology – environmental DNA offers new capabilities for marine ecosystem and fisheries research. *Frontiers in Marine Science* **8**: 404.

237 Hebels, D.G., Georgiadis, P., Keun, H.C. et al. (2013). Performance in omics analyses of blood samples in long-term storage: opportunities for the exploitation of existing biobanks in environmental health research. *Environmental Health Perspectives* **121** (4): 480–487.

238 Hannon, E., Lunnon, K., Schalkwyk, L., and Mill, J. (2015). Interindividual methylomic variation across blood, cortex and cerebellum: implications for epigenetic studies of neurological and neuropsychiatric phenotypes. *Epigenetics* **10** (11): 1024–1032.

239 Edgar, R.D., Jones, M.J., Meaney, M.J. et al. (2017). BECon: a tool for interpreting DNA methylation findings from blood in the context of brain. *Translational Psychiatry* **7** (8): e1187–e1187.

240 Vasanthakumar, A., Davis, J.W., Idler, K. et al. (2020). Harnessing peripheral DNA methylation differences in the Alzheimer's disease neuroimaging initiative (ADNI) to reveal novel biomarkers of disease. *Clinical Epigenetics* **12**: 84.

241 Arosio, B., Guerini, F.R., Voshaar, R.C.O., and Aprahamian, I. (2021). Blood brain-derived neurotrophic factor (BDNF) and major depression: do we have a translational perspective? *Frontiers in Behavioral Neuroscience* **15**: 626906.

242 Frach, L., Tierling, S., Schwaiger, M. et al. (2020). The mediating role of KITLG DNA methylation in the association between childhood adversity and cortisol stress reactivity does not replicate in monocytes. *Psychoneuroendocrinology* **116**: 104653.

243 Stubbs, T.M., Bonder, M.J., Stark, A.K. et al. (2017). Multi-tissue DNA methylation age predictor in mouse. *Genome Biology* **18**: 68.

244 Shireby, G.L., Davies, J.P., Francis, P.T. et al. (2020). Recalibrating the epigenetic clock: implications for assessing biological age in the human cortex. *Brain* **143** (12): 3763–3775.

245 Anastasiadi, D. and Piferrer, F. (2020). A clockwork fish: age prediction using DNA methylation-based biomarkers in the European sea bass. *Molecular Ecology Resources* **20** (2): 387–397.

246 Marini, S., Davis, K.A., Soare, T.W. et al. (2020). Adversity exposure during sensitive periods predicts accelerated epigenetic aging in children. *Psychoneuroendocrinology* **113**: 104484.

247 Raddatz, G., Arsenault, R.J., Aylward, B. et al. (2021). A chicken DNA methylation clock for the prediction of broiler health. *Communications in Biology* **4** (1): 1–8.

248 Zannas, A.S., Arloth, J., Carrillo-Roa, T. et al. (2015). Lifetime stress accelerates epigenetic aging in an urban, African American cohort: relevance of glucocorticoid signaling. *Genome Biology* **16** (1): 266.

249 Li, M., Santpere, G., Kawasawa, Y.I. et al. (2018). Integrative functional genomic analysis of human brain development and neuropsychiatric risks. *Science* **362** (6420): eaat7615.

250 Pérez, R.F., Santamarina, P., Tejedor, J.R. et al. (2019). Longitudinal genome-wide DNA methylation analysis uncovers persistent early-life DNA methylation changes. *Journal of Translational Medicine* **17**: 15.

13

Epigenetics in Hybridization and Polyploidization of Aquatic Animals

Li Zhou and Jian-Fang Gui

State Key Laboratory of Freshwater Ecology and Biotechnology, Institute of Hydrobiology, Wuhan, China

13.1 Hybridizing and Hybridization

The term "hybridizing" refers to the crossing of two genetically distinct individuals. The merging of two divergent genomes in the resulting hybrids fuels a series of rapid, complicated, and non-Mendelian genetic and epigenetic alterations to handle the conflict between parental genomes [1, 2]. This process is defined as hybridization. Hybridization that occurs between different lineages, strains, or populations within species is referred to as intraspecific hybridization. This process generally produces fertile autodiploid hybrids without changing chromosome numbers (Figure 13.1a). In contrast, interspecific or distant hybridizing between different species may result in varied hybrid offspring. For fertility, some distant crosses produce sterile hybrids, but the other hybrids can overcome reproductive barriers. Distant hybridizing enables interspecific genome transfer and may lead to genome-level alteration [3–6]. Some hybrids are allodiploid hybrids with the same ploidy as their parents, and some hybrids are defined as doubled interspecific hybrids (allopolyploids) that have a higher ploidy generated by whole genome duplications (WGDs), either in gametogenesis or after fertilization to "rescue" hybrid incompatibility (Figure 13.1b).

Hybridization facilitates speciation and adaptive radiation. It can instantaneously boost genetic variation, especially the traits associated with reproductive isolation or adaptation, which has been demonstrated to link with speciation and adaptive radiation [7, 8]. A famous case is the Lake Victoria region superflock of cichlid fish, encompassing over 700 diverse species. Genomic data have shown that the species superflock may have evolved from hybridization [7].

13.2 Polyploidy and Polyploidization

Polyploidy (Box 13.1) or WGD, defined as the duplication of entire chromosome sets (genomes), is common and widespread in plants [9]. Although polyploidy is much rarer in animals, two rounds of WGD are hypothesized to have occurred during early vertebrate evolution [10], and an additional WGD shared by all teleosts [11] has been universally accepted through genome sequencing analyses [12]. Moreover, many examples of polyploids with different ploidy have been discovered in fish, amphibians, and reptiles [13–18]. For example, Cyprininae contains about 400 polyploid species across more than 30 genera [16, 19, 20]. The global geographical distribution of the earliest occurrence record for 2052 Cyprinidae species, including 1680 diploids and 372 polyploids, also indicates that Cyprinidae polyploid species are more prolific in areas of higher altitude and lower mean annual temperature, which may be associated with the continuous uplift of the Tibetan Plateau [21]. Polyploids are classified into two categories: autopolyploids (Figure 13.1a) and allopolyploids (Figure 13.1b) [22]. The former, multiplying genomes within a species, arise from chromosome set doubling, unreduced gametes, or polyspermy [14], and the latter are produced by doubling genomes between different species [3–5].

Polyploidization (Box 13.1) is an ongoing process. After polyploidy, the newly formed polyploids step into the complex evolutionary trajectory of polyploidization, and most polyploids evolve into diploids through subsequent rediploidization. During the process of polyploidization and rediploidization, complicated non-Mendelian genomic and epigenetic changes may occur: these include chromosomal rearrangements [23–25], gene loss, pseudogenization (to become a pseudogene

Epigenetics in Aquaculture, First Edition. Edited by Francesc Piferrer and Han-Ping Wang.

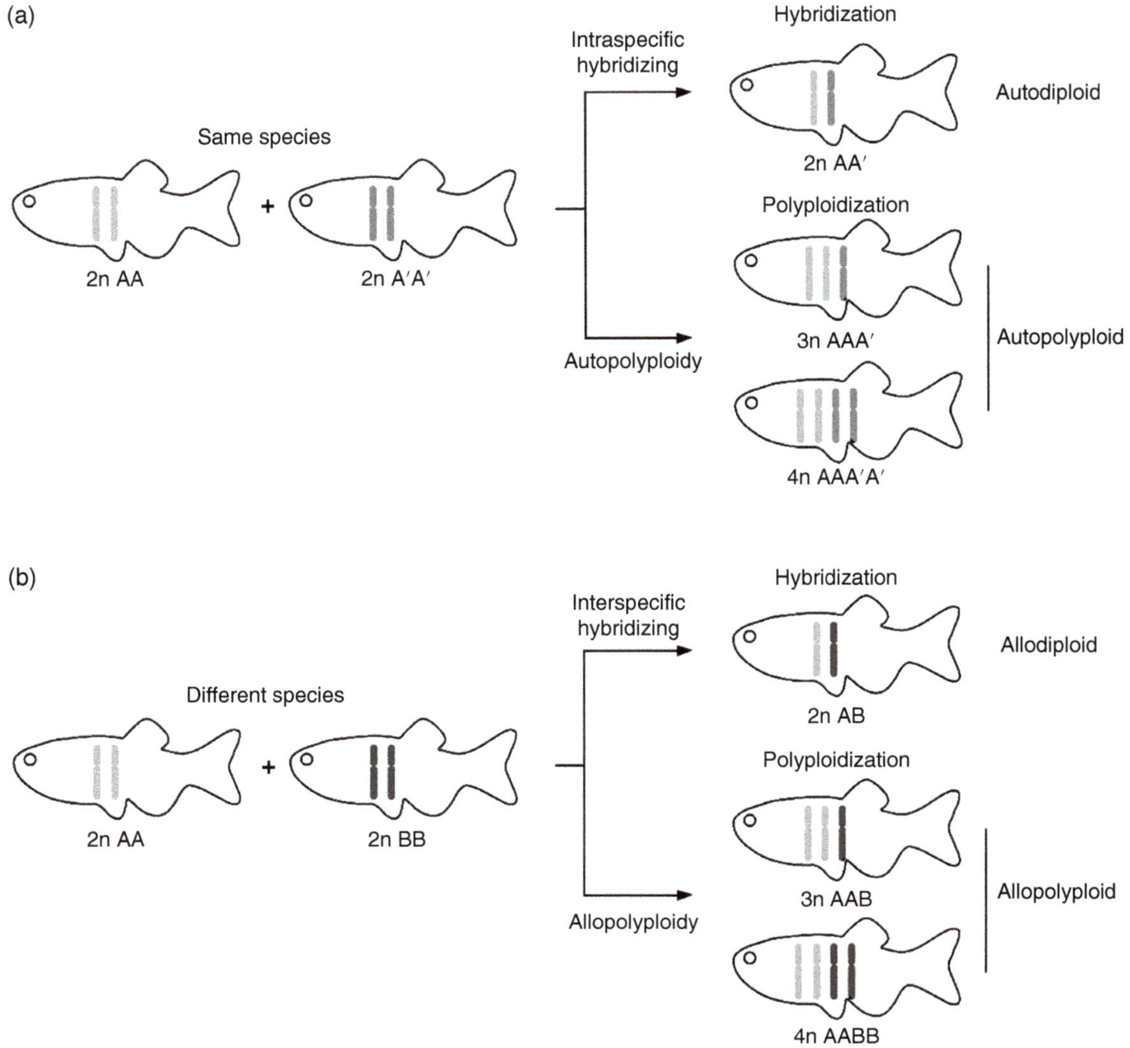

Figure 13.1 Schematic diagram of hybridizing, hybridization, polyploidy, and polyploidization. Generally, intraspecific hybridizing and subsequent hybridization that occur between different lineages, strains, or populations within species produce fertile autodiploid hybrids (2nAA′) without changing chromosome numbers (a). In contrast, interspecific hybridization between different species may result in allodiploid hybrids (2nAB) with the same ploidy as their parents, or allopolyploids (3n AAB or 4n AABB) that have a higher ploidy generated by polyploidization (b). In addition to allopolyploids, autopolyploids (3n AAA′ or 4n AAA′A′) are arisen from multiplying genomes within a species. 2n, 3n, and 4n refer to diploid, triploid, and tetraploid, respectively. A and A′ indicate two genomes with subtle differences within a species, while B indicates a different genome from another species.

Box 13.1 Polyploidy and Polyploidization

The term "polyploidy" refers to an instantaneously occurring event that results in the duplication of entire chromosome sets (genomes). After polyploidy, the newly formed polyploids may undergo "genomic shock" and incur complicated non-Mendelian genomic and epigenetic changes. This ongoing process is defined as polyploidization. Although many changes may lead to instability, some neopolyploids benefit from these changes to pass through a bottleneck of sterility and subsequently evolve into new diploids through rediploidization [16, 22]. Additionally, a few polyploids, such as gibel carp (*Carassius gibelio*), bypass meiotic obstacles and gain the ability to unisexually reproduce following polyploidization [16].

that is a noncoding homolog of protein-coding gene), sub- or neofunctionalization [26], transposon reactivation [27, 28], and epigenetic modifications [29–31]. Recently, the importance and profound impact of polyploidization and rediploidization on evolutionary innovations and ecological adaptation has been widely discussed [7–9, 12, 26, 32–34]. For example, the duplicated genes provide raw materials for functional divergence via gene subfunctionalization or neofunctionalization, and the latter may offer an evolutionary potential to evolve key innovations [9, 26].

It is generally believed that polyploidization and rediploidization may lead to the diversity, variability, and complexity of organisms [9, 26]. All plants have evolved from polyploid ancestors. As a result of the genome sequencing explosion, at least 45 ancient polyploidy events have been discovered in plants [9]. A WGD shared by teleost is believed to be a major driving force behind the dramatic rise and rapid radiation of teleost fish [12, 14, 16]. The subgenome dominance and the fates of duplicated genes have been well documented in paleopolyploids [26]. Recently, we elaborated on the divergent functions of multiple duplicated *foxl2* and *viperin* homeologs and their alleles in a recurrent polyploid fish, auto-allo-hexaploid gibel carp (*Carassius gibelio*) [35, 36]. These studies provide typical examples of different homeolog diversification, retention/loss, biased expression, and sub-/neofunctionalization in the evolution of duplicated genes driven by polyploidization and rediploidization from recurrent polyploid fish. We successfully edited multiple *foxl2* and *viperin* homeologs and/or alleles using CRISPR/Cas9. *Cgfoxl2a*-B deficiency led to ovary developmental arrest or complete sex reversal, while complete disruption of *Cgfoxl2b*-A and *Cgfoxl2b*-B resulted in the depletion of germ cells [35]. These results, in combination with detailed cellular localization information, suggest that *Cgfoxl2a* and *Cgfoxl2b* might have subfunctionalized and cooperated to regulate folliculogenesis and gonad differentiation; further, *Cgfoxl2b* might have evolved a new function in oogenesis [35].

13.3 Epigenetic Changes and Effects During Hybridization and Polyploidization in Aquatic Animals

13.3.1 Epigenetic Reprogramming During Hybridization and Polyploidization

In addition to genetic changes, accumulating evidence in diverse plant taxa suggests that epigenetic reprogramming might occur in a burst following hybridization and polyploidization. This results in remodeling of gene expression networks, which allows hybrids or polyploids to grow vigorously, adapt to environments, and even establish new cytotypes [31, 37, 38]. For example, the combined analyses of transcriptomes, methylomes, and siRNAomes in newly synthesized autotetraploid rice have demonstrated that, despite a lack of dramatic differences between diploid and tetraploid overall methylomes and transcriptomes, polyploidization induces widespread DNA methylation variation of transposable elements (TEs). This variation is found in class II TEs and is accompanied by the alternation of 24-nt siRNA abundance [39]. Similar overall methylome stability and locus-specific DNA methylation repatterning were observed in the nascent segmental rice allotetraploid hybrids generated via the crossing of two subspecies (*indica* and *japonica*) of cultivated Asian rice (*Oryza sativa*) [38]. On the one hand, the multi-'omic data indicate that there is an intrinsic homeostatic mechanism to maintain overall methylation stability. On the other hand, these data imply that DNA methylation repatterning may help autotetraploid rice to adapt genome-dosage effects by suppressing neighboring gene expression [38, 39].

13.3.2 Nonadditive Gene Expression

In plant hybrids, including *Arabidopsis thaliana*, rice, wheat, and maize, a common observation is that many genes involved in hormone signaling (i.e., phytohormone pathways) and stress responses have changed their expression levels [40]. Some changes are nonadditive, meaning that gene expression deviates from the average expression value of the two parents (midparent value, MPV). Nucleolar dominance, a well-known case of nonadditive gene expression in interspecific hybrids, refers to the silencing of rRNA genes from one parent by DNA methylation, histone modifications, or sRNAs [40, 41]. For example, many miRNAs and their target genes showed nonadditive expression in the allotetraploids derived from *A. thaliana* and *A. arenosa* [42]. rRNA genes from *A. thaliana* are silenced [43] and *A. arenosa* genes are dominant over *A. thaliana* genes [44]. Similar nonadditive expressions of duplicated genes are also observed in polyploids via three possible scenarios, including expression-level dominance, transgressive expression, and homeolog expression bias [45]. Increasing evidence from transcriptomic (RNA-seq) and genomic data indicates that nonadditive gene expression is a genome-wide phenomenon in plant allopolyploids [46–48]. Several factors, including maternal-paternal influence, gene dosage balance, cis- and/or transregulation, and epigenetic regulation are separately or in concert attributed to nonadditive gene expression [45]. Many plant polyploid cases showed that transcriptional activation of retrotransposons [27], methylated TE insertions [49, 50], and siRNA-targeted TEs [49, 51] affect the expression of nearby genes, implying an important role of epigenetic changes in allopolyploid nonadditive gene expression [45].

Similar to plant hybrids, the phenomenon of nonadditive gene expression has also been reported in the hybrids of aquatic animals. Given high homology, it is difficult to compare the expression pattern between intraspecific hybrids and their parents. Several studies have attempted to demonstrate divergent expression in intraspecific hybrids generated via the crossing of two parents with significant genetic variation, including in wild and domesticated Atlantic salmon (*Salmo salar*) [52], rainbow trout (*Oncorhynchus mykiss*) [53], brook charr (*Salvelinus fontinalis*) [54], different inbred families of pearl oyster (*Pinctada fucata martensii*) [55], and sympatric species complex [56]. Bougas et al. (2010) have compared the differentially expressed genes (DEGs) among one domestic population, two wild populations (Rupert and Laval), and their crossed hybrids of brook charr. The vast majority (94%) of the DEGs between the parent populations are additive in the hybrid of domestic ♀ × Rupert ♂, whereas more than one-third of DEGs are nonadditive for the hybrids of Laval ♀ × Rupert ♂ (45.7%), and Rupert ♀ × Laval ♂ (37.5%) [54].

In rainbow trout hybrids between wild and domesticated populations, about two-thirds of DEGs show recessive or dominant expression patterns [53]. Similar studies have been performed in two sympatric species pairs of lake whitefish *Coregonus clupeaformis*, referred to as normal (benthic) and dwarf (limnetic). Compared with the F_1-hybrids of normal ♀ × dwarf ♂, nonadditive DEGs are more prevalent in the backcross hybrids of (normal ♀ × dwarf) ♀ × normal ♂ [56].

Interspecific hybrids are more suitable for studying gene expression patterns during hybridization. The proportion of genes showing nonadditive expression varies and seems to be species or cross-specific. For example, about 76.4 and 90.0% of genes in the liver of F_1 and F_2 hybrids of *Megalobrama amblycephala* ♀ × *C. alburnus* ♂, respectively, displayed nonadditive expression, implying the instability of genomes and large-scale nonadditive gene regulation in progenies. Interestingly, the nonadditive pattern could be inherited from F_1 to F_2 [57]. In contrast, only a small portion (1.9 and 8.3%) of all expressed transcripts and proteins, mainly enriched pathways for immune defense, metabolism, digestion, and absorption, were nonadditive in the liver of yellow catfish Huangyou-1 hybrids of *Pelteobagrus fulvidraco* ♀ × *P. vachelli* ♂ [58]. For nucleolar dominance, the inherited and expressed 45S rRNA genes in the F_1 hybrids of *M. amblycephala* ♀ × *C. alburnus* ♂ were derived from both parents. In addition, five patterns of 45S rRNA gene inheritance and expression were observed in the F_2 hybrids [59]. However, nucleolar dominance was established in the F_1 allotetraploid hybrid ($4n = 148$) of *C. auratus* red var. ♀ × *M. amblycephala* ♂ and was inherited by F_2 autotetraploids ($4n = 200$). Therefore, the authors speculated that tetraploidization may facilitate the quick establishment of nucleolar dominance [60]. Although several common findings about nonadditive expression have emerged from studies on plant hybrids [40], there is no consensus on the proportion or category of nonadditive expressed genes in aquatic animals. Further, the causes for such highly variable expression patterns are not yet clear.

Sometimes distant hybridizing and polyploidy occur simultaneously. Allodiploid ($2n = 48$) and allotriploid ($3n = 72$) hybrids were obtained from F_1 generation of the crossing of *M. amblycephala* ♀ × *X. davidi* ♂. Although the number of DEGs (compared to both parents) was asymmetric in the two different ploidy hybrids, both hybrids showed maternal expression level dominance in the testis. Further, the allotriploid hybrids had reduced expression levels relative to allodiploid hybrids, indicating a strong maternal effect on gene expression patterns and the existence of dosage compensation in polyploid hybrid fish [61]. A similar dosage compensation, either by gene-copy silencing or other mechanisms, was also observed in the allotriploid form of *Squalius alburnoides* [62–64] and allotriploid hybrids of *C. idellus* ♀ × *M. amblycephala* ♂ [65]; this dosage compensation may contribute to the success of polyploidization in lower vertebrates [62].

13.3.3 DNA Methylation

Consistent with nonadditive gene expression, variability of DNA methylation in aquatic animals during hybridization and/or polyploidization is a wide-ranging, species-dependent characteristic. Minor methylation differences (12 loci, 2.94%) were detected between diploid and triploid siblings of brown trout (*Salmo trutta*) in six different tissues (brain, gill, heart, kidney, liver, and muscle) [66]. A similar phenomenon, that is, no difference in global methylation state, was observed in the comparison between diploid and triploid Pacific oysters. Only 13 and 8 loci (1.84 and 1.13%) had altered methylation patterns between oysters with different ploidies in adductor muscle or gonad, respectively [67]. In comparison with their parents or diploids, 38.31% of 355 randomly selected CCGG sites were observed with methylation changes in the allotetraploid hybrid of *C. auratus* red var. ♀×*C. carpio* ♂, and the major changes were hypermethylation [68]. Loach (*Misgurnus anguillicaudatus*) populations are mixed diploid-polyploid, with even bisexual and unisexual forms coexisting in the same localities. Most common diploid ($2n = 50$), triploid ($3n = 75$), and tetraploid ($4n = 100$) individuals have been frequently recorded in some localities of China and Japan [69]. Through methylation-sensitive amplified polymorphism (MSAP) technology, a total of 42.21, 38.78, and 37.03% methylation rates were identified in diploid, triploid, and tetraploid loach, respectively, indicating the gradual decrease of methylation levels as ploidy level increases. This is significantly different from the results detected in brown trout [66] where 73.05 and 68.17% of methylation patterns changed in triploid and tetraploid loach, respectively, relative to diploid loach [70].

Gynogenetic diploid offspring with two sets of maternal-derived chromosomes and several paternal-derived microchromosomes were obtained in the first generation of distant hybridization of *C. auratus* red var. ♀×*M. amblycephala* ♂. In comparison with their parents, the DNA methylation level of gynogenetic offspring was lower, and 45.29% of their methylation patterns were altered [71]. Moreover, 60.31% of methylation patterns in the fertile *C. auratus* autotetraploid ($4n = 200$), derived from the same distant hybridization, were altered relative to *C. auratus* red var. [72]. These results imply that rapid epigenetic changes occur in the newly established gynogenetic diploid and autotetraploid fish derived from distant hybridization.

The genome of natural gynogenetic hexaploid gibel carp can integrate the alien genome or subgenomic fragments, and several novel allo-octoploid clones have also been synthesized [16, 73–83]. Generally, these clones still maintain unisexual gynogenesis capability. For example, a novel stable allopolyploid (allo) with 206 chromosomes was established from gibel carp clone D (hexa) activated by heterologous sperm from red common carp (*Cc*) [80]. Among the synthetic allopolyploid, gibel carp clone D, and common carp, we confirmed the whole genome incorporation of common carp and the additive effect in the synthetic allopolyploid of gibel carp. This was done through multiple analyses including genomic in situ hybridization (GISH), sequence comparison and chromosome localization of 45S rDNAs, mitochondrial D-loop sequence comparison, and MSAP analysis. Pooled MSAP analysis revealed that an overwhelming majority (98.82%, 924/935) of cytosine methylation patterns in the synthetic allopolyploid were inherited from its parents: the hexaploid gibel carp clone D and common carp. Compared to those in its parents, these clear and reproducible MSAP bands in the synthetic allopolyploid were grouped into five main patterns, though very few (1.18%, 11/935) variant bands were detected. The variation patterns were categorized into two classes, which may be caused by cytosine methylation changes or mutations in nucleotide sequence. However, we determined that these 11 DNA fragments in the synthetic allopolyploid were caused by interindividual variation, recombination, deletion, and mutation through individual sample-based MSAP and sequencing. Therefore, the epigenetic changes in the newly synthetic gynogenetic allopolyploid are minimal [84], similar to the results detected in nine newly synthesized allotetraploid or allohexaploid cotton (*Gossypium*) [85]. So far, the molecular mechanisms or evolutionary implications responsible for varied methylation changes in different polyploids are unknown. In our case, methylation changes may not be dispensable during the formation and gynogenetic generation transmission of the synthetic allopolyploid.

Captive-rearing in fish can cause significant phenotypic and genetic variations after just one generation, including nonadditive expression and epigenetic (methylation) alterations in immune and stress-related genes [86–88]. For example, 6.4% (298) of all detected genes (16,006) showed significantly different transcription levels (Q-value <0.01) between the Atlantic salmon progeny of Norwegian wild and backcross hybrid of (Farmed×Wild) ♀×Wild ♂, and 82% of this expression was nonadditive [52]. Rodriguez-Barreto et al. [89] compared the genome-wide DNA methylation profiles between the sperm of wild and hatchery-reared Atlantic salmon males and identified 43 DMRs (differentially methylated regions) shared by different from the wild and two hatchery groups. Moreover, some of the DMRs maintained the same methylation pattern in the offspring and escaped the epigenetic reprogramming that occurs in primordial germ cells. This result indicates that captive-rearing may trigger epigenetic modifications in the sperm of hatchery-reared Atlantic salmon, which may explain the phenotypic and genetic changes observed in the intraspecific hybrid between farmed and wild fish [89].

13.3.4 Histone Modification and Other Epigenetic Changes

Besides nonadditive gene expression and DNA methylation, studies of the other epigenetic changes during hybridization and polyploidization in aquatic animals are scarce. Zhou et al. [90] found that four histone H3 modifications, including H3K4me3, H3K9me3, H3K79me, and H3K79me3, differed significantly in the ovaries between fertile *C. auratus* red var. and sterile triploid hybrid fish of tetraploid hybrid of (*C. auratus* red var. ♀×*C. carpio* ♂) ♀×*C. auratus* red var. ♂, while the levels of 17 other histones (i.e., H3K4me1, H3K9me2, and H3K36me2) modifications were similar. Moreover, the authors also observed obvious expression differences of corresponding histone methyltransferase and demethylase genes between the two fish ovaries and thereby speculated that these changes might associate with the pachytene arrest of oocytes in triploid hybrid fish [90]. Tao et al. [91] identified dozens of differentially expressed miRNAs in testes among fertile diploid *C. auratus* red var., sterile triploid hybrid of *C. auratus* red var. ♀×(*C. auratus* red var. ♀×*C. carpio* ♂) ♂, and fertile allotetraploid hybrid of *C. auratus* red var. ♀×*C. carpio* ♂. Their targeted genes were enriched in four pathways involved in spermatogenesis. The germ cells of the triploid hybrid developed into round spermatids with misshapen nuclei and then degenerated. The upregulated expression of several miRNAs in the triploid hybrid, such as miR-199-5p, coincided with the downregulated expression of their target key genes (*dnah3* and *tekt1*), which regulate the formation of sperm flagellum [91].

Hybridogenetic *Squalius alburnoides* complex (Steindachner) is sympatric with their parental species *S. pyrenaicus*. It is composed of diploid hybrids (PA, P genome from *S. pyrenaicus* and A genome from an extinct paternal ancestor related to *Anaecypris hispanica*) and triploid hybrids (PAA) [92]. As described earlier, the deviation for additive expression and dosage compensation in triploid hybrids has been reported [62–64]. To explore the role of microRNAs in nonadditive gene expression, the global miRNA profiles among PAA, PA, PP, and AA were compared. Surprisingly, the triploid hybrid exhibited more (>72%) miRNAs with higher expression than the diploid hybrid. In addition, the majority (93 and 62%) of upregulated miRNAs in PAA/PA also increased their expression in at least one or all of the other triploid/diploid comparisons (PAA/PP or/and PAA/AA). Further expression level analysis confirmed that miRNA expression in the triploid hybrid deviated from additive expression. Therefore, a genome-wide regulatory response by miRNAs might play an important role in the dosage compensation in the triploid hybrid. Since no dramatic differences in the increased or decreased miRNA expression levels were found between the diploid hybrid (AA) and nonhybrid diploids (PP and AA), it is possible that polyploidization, not hybridization, may impact the miRNA profiles of the *S. alburnoides* complex [93].

13.4 Association of Epigenetic Changes with Heterosis

The application of heterosis (Box 13.2) [94] has revolutionized breeding and production in agriculture and aquaculture. Heterosis is defined as the combination of effects of different alleles carried by the hybrid parents; however, the mechanisms underlying heterosis are largely unknown. Several models, such as dominance, overdominance, and epistasis, have been proposed to explain heterosis [40, 95]. Generally, heterosis is positively correlated to genetic distance between the two parents in many hybrid crosses, but the correlation is not absolute [96–98]. For example, two *Arabidopsis* ecotypes (C24 and Ler) have similar genomes (approximately 96% identical DNA sequences) but different epigenomes. Hybrids generated between these ecotypes have 250% increased biomass [99–101]. These examples highlight the important role of epigenetics in heterosis [97, 98, 102]. Small interfering RNAs (siRNAs), approximately 24 nucleotides in length, are components of the epigenome and direct de novo DNA methylation [103]. It is a common phenomenon that 24 nt siRNAs are downregulated in a tissue-specific manner when hybridizing two divergent epigenomes; this correlates with transallelic effects of DNA methylation in genes and gene flanking regions, resulting in altered gene expression, as seen in *Arabidopsis* [42, 104, 105], maize (*Zea mays*) [106, 107], rice (*O. sativa*) [108], and tomato (*Solanum lycopersicum*) [109]. Altered methylation occurs within the DMRs between the parents [110]. In addition to siRNAs and DNA methylation,

Box 13.2 Heterosis

Heterosis, also known as hybrid vigor, refers to the superiority of agronomic or economic traits of the hybrids over their parents. It was systematically characterized by Charles Darwin [94], who documented the growth patterns of more than 60 self- and cross-fertilized plants and concluded that "the crossed plants, when fully grown, were plainly taller and more vigorous than the self-fertilized ones" [94]. Although hybrids have been widely applied for more than a century and have revolutionized breeding and production in agriculture, the mechanisms underlying heterosis are largely unknown. Several models, such as dominance, overdominance, and epistasis, have been proposed to explain heterosis [40, 95]. However, heterosis is very complicated and a general mechanism could not be found. New research suggests that heterosis might result from complex genetic and epigenetic changes during the integration of two different parental genomes; these changes promote the growth, disease resistance, stress tolerance, and other fitness characteristics of hybrids.

histone modifications play a role in heterosis. The lower level of two histone modifications (H3K9ac and H3K4me2) in the promoter regions of *circadian clock associated 1* (*cca1*) and *late elongated hypocotyl* (*lhy*) genes repressed the expression of these two circadian clock genes and thereby altered circadian rhythms in an allotetraploid hybrid of *A. thaliana* × *A. arenosa*, which contributed to growth vigor [111]. Other studies showed that most of the histone modifications in hybrids were relatively stable and additive, except the histone modifications directly associated with siRNA and DNA methylation (i.e., H3K9 dimethylation) [110, 112, 113].

Although the molecular basis of heterosis has been investigated in model plant species (*Arabidopsis*) and important crops (i.e., wheat, rice, barley, and canola) [100], information on the role of epigenetics in aquatic species heterosis is lacking. Nonadditive gene expression is one of the most common contributors to heterosis in hybrids. The earliest attempt of identifying gene-expression patterns underlying growth heterosis in aquatic species was done in Pacific oysters. Generally, bivalve mollusks display high levels of inbreeding depression and heterosis [114]. The comparison of gene-expression patterns among two partially inbred (inbreeding coefficient $f = 0.375$) and two reciprocal hybrid larval populations showed that the overwhelming majority (>96%) of signature sequences detected by MPSS (megacloning and massively parallel signature sequencing) was predominantly nonadditive and approximately half of the categorized signatures showed overexpression in both reciprocal hybrids, which was expected under the classical hypotheses of heterosis [115].

Growth is mainly regulated by GH/IGF (growth hormone/insulin-like growth factor 1) axis. In comparison with its parents, *igf1*, *igfbp-1* (*insulin-like growth factor binding protein 1*), *igfbp-2*, *igfbp-4*, and *igfbp-5a* were significantly upregulated in the liver of Hulong grouper hybrid of *Epinephelus fuscogutatus* × *E. lanceolatus* [116]. The tilapia hybrid of *Oreochromis niloticus* ♀ × *O. aureus* ♂, an important cultured fish in China, showed growth superiority and resistance vigor to cold and the pathogen *Aeromonas sobria* relative to its parents [117]. The expression of both *gh* and *igf-1* were nonadditive and their allelic expression was asymmetric in the tilapia hybrid [118, 119]. A similar expression pattern was also observed in the liver of a hybrid of *M. amblycephala* ♀ × *C. alburnus* ♂ [57], allotriploid ($3n = 72$) hybrid of *Ctenopharyngodon idella* ♀ ($2n = 48$) × *M. amblycephala* ♂ ($2n = 48$) [65], and yellow catfish Huangyou-1 hybrid of *P. fulvidraco* ♀ × *P. vachelli* ♂ [58]. Therefore, the nonadditive expressed gene related to the GH/IGF axis is considered to be in accordance with the growth superiority of the hybrid. Moreover, many nonadditive expressed genes were enriched into the pathways associated with metabolism, digestion and absorption, immune defense, and antioxidant capacity. For example, eight lipid metabolism pathway-related genes, such as *low-density lipoproteins receptor* (*ldlr*) and *fatty acid-binding protein 3* (*fabp3*), showed nonadditive expression in the intestine of tilapia hybrid [120].

Generally, aquatic hybrids are more resistant to disease. The pathways enriched by nonadditive expressed genes included "complement and coagulation cascades," "Chemokine signaling pathway," "B cell receptor signaling pathway," "hematopoietic cell lineage," and so on [58, 121]. In addition, a total of 205, 215, and 184 miRNAs were nonadditive in the livers of F_1 and F_2 hybrid of *M. amblycephala* ♀ × *C. alburnus* ♂ and yellow catfish Huangyou-1 hybrid of *P. fulvidraco* ♀ × *P. vachelli* ♂, respectively [57, 58]. Several miRNAs with well-known functions, such as miR-130, miR-27a, and miR-301, involved in immune responses and cell development, were identified in both studies. Moreover, several miRNA-mRNA-protein pairs identified in the yellow catfish hybrid were involved in the pathways of immune defense, and metabolism, digestion, and absorption, which may be responsible for heterosis [58].

13.5 Conclusions and Future Perspectives

Although many new insights into the epigenetic consequences of hybridization and polyploidization have been obtained from plants, we still know very little about these processes in aquatic animals. Limited data show that epigenetic reprogramming, including DNA methylation, histone modifications, and miRNA, occurs during hybridization and polyploidization in aquatic animals; these epigenetic processes may result in nonadditive gene expression (Figure 13.2). There is still much to study on this topic. First, it is important to understand the regulative mechanisms of merged or doubled genomes from a pluralistic framework, including 'omics analyses (phenome, genome, transcriptome, proteome, metabolome, and epigenome) and a systems biology perspective [122]. Multiple 'omics studies on the newly synthetic alloploids derived from the two parents with clear genetic backgrounds will deepen our understanding of the consequences of hybridization and polyploidization, including heterosis and hybrid incompatibility. Second, the key epigenetic variations associated with phenotypic variations during hybridization and polyploidization need to be explored extensively. Comparative combined analysis of multiple 'omics data would identify genetically dependent epigenetic marks. Finally, more direct and stronger evidence is needed to support the association between epigenetic and phenotypic variations in aquatic animals. It is

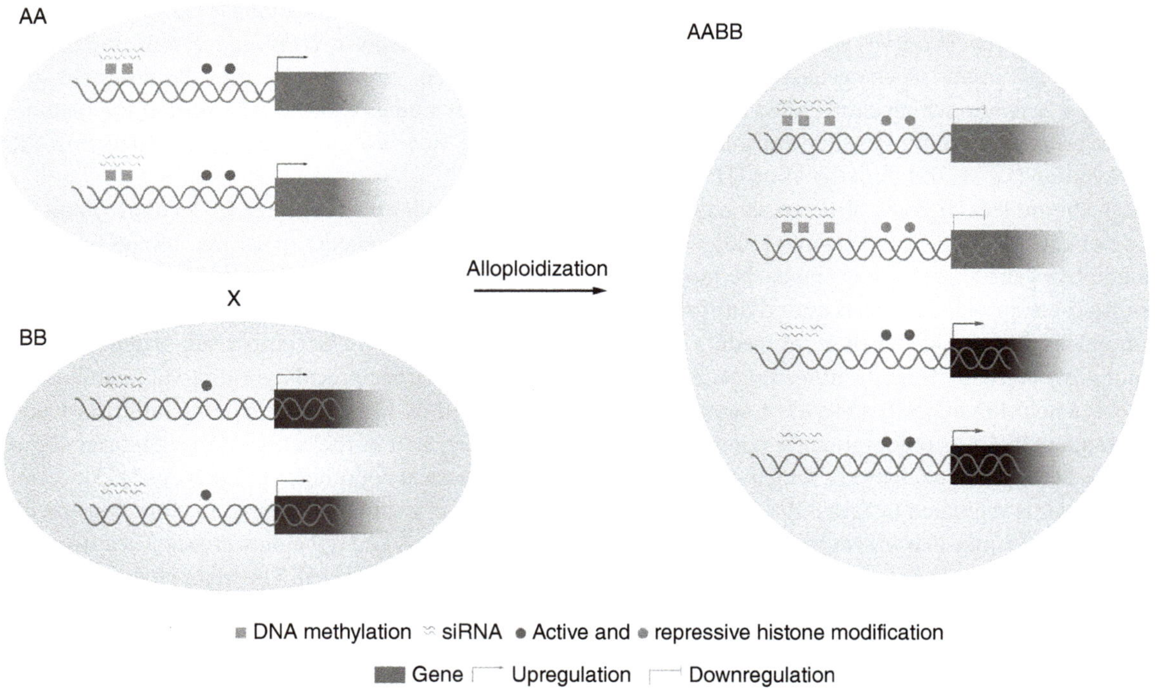

Figure 13.2 Epigenetic changes during hybridization and polyploidization in aquatic animals (allotetraploid as an example). In addition to non-Mendelian genomic changes that occurred during alloploidization, the newly formed allopolyploids may also undergo epigenetic changes, such as DNA methylation, histone modifications, and small RNAs, which affect gene expression. A and B indicate two genomes from different species. (*See colour plate section for colour representation of this figure.*)

necessary to develop gene-editing technologies in aquatic hybrids and polyploids to modify the key epigenetic marks. A better understanding of the changes in epigenetic states contributing to the heterotic phenotype may ultimately help us to improve the economic traits in aquatic species. Thus, these future studies will provide new insights into the epigenetic consequences of hybridization and polyploidization, which can help us understand why aquatic polyploids or hybrids possess rapid growth, extensive adaptability, and disease resistance, and will open a door to develop novel biotechnologies, such as epigenetic selection and key epigenetic mark editing.

Acknowledgments

This work was supported by the Strategic Priority Research Program of the Chinese Academy of Sciences (XDB31000000), the National Natural Science Foundation (31930111), Consulting Research Project of Hubei Institute of Chinese Engineering Development Strategies (HB2019A01), and Academic Divisions of the Chinese Academy of Sciences (2021-SM02-B-010), the Key Program of Frontier Sciences of the Chinese Academy of Sciences (Grant No. QYZDY-SSW-SMC025), and the China Agriculture Research System (CARS-45-07).

References

1 McClintock, B. (1984). The significance of responses of the genome to challenge. *Science* **226** (4676): 792–801.
2 Ng, D.W.K., Lu, J., and Chen, Z.J. (2012). Big roles for small RNAs in polyploidy, hybrid vigor, and hybrid incompatibility. *Current Opinion in Plant Biology* **15** (2): 154–161.
3 Liu, S. (2010). Distant hybridization leads to different ploidy fishes. *Science China Life Sciences* **53** (4): 416–425.
4 Liu, S. (2015). *Fish Distant Hybridizations*. Beijing: Science Press.
5 Wang, S., Tang, C.C., Tao, M. et al. (2019). Establishment and application of distant hybridization technology in fish. *Science China Life Sciences* **62** (1): 22–45.

6 Liu, Q.F., Liu, J.M., Yuan, L.J. et al. (2020). The establishment of the fertile fish lineages derived from distant hybridization by overcoming the reproductive barriers. *Reproduction* **159** (6): R237–R249.

7 Meier, J.I., Marques, D.A., Mwaiko, S. et al. (2017). Ancient hybridization fuels rapid cichlid fish adaptive radiations. *Nature Communications* **8**: 14363.

8 Nieto Feliner, G., Casacuberta, J., and Wendel, J.F. (2020). Genomics of evolutionary novelty in hybrids and polyploids. *Frontiers in Genetics* **11**: 792.

9 Van de Peer, Y., Mizrachi, E., and Marchal, K. (2017). The evolutionary significance of polyploidy. *Nature Reviews Genetics* **18** (7): 411–424.

10 Dehal, P. and Boore, J.L. (2005). Two rounds of whole genome duplication in the ancestral vertebrate. *PLoS Biology* **3** (10): e314.

11 Amores, A., Force, A., Yan, Y.L. et al. (1998). Zebrafish hox clusters and vertebrate genome evolution. *Science* **282** (5394): 1711–1714.

12 Van de Peer, Y., Maere, S., and Meyer, A. (2009). The evolutionary significance of ancient genome duplications. *Nature Reviews Genetics* **10** (10): 725–732.

13 Otto, S.P. and Whitton, J. (2000). Polyploid incidence and evolution. *Annual Review of Genetics* **34**: 401–437.

14 Leggatt, R.A. and Iwama, G.K. (2003). Occurrence of polyploidy in the fishes. *Reviews in Fish Biology and Fisheries* **13** (3): 237–246.

15 Mable, B.K., Alexandrou, M.A., and Taylor, M.I. (2011). Genome duplication in amphibians and fish: an extended synthesis. *Journal of Zoology* **284** (2011): 151–182.

16 Zhou, L. and Gui, J.F. (2017). Natural and artificial polyploids in aquaculture. *Aquaculture and Fisheries* **2**: 103–111.

17 Zhou, L. and Gui, J.F. (2018). Applications of genetic breeding biotechnologies in Chinese aquaculture. In: *Aquaculture in China: Success Stories and Modern Trends* (ed. J.F. Gui, Q.S. Tang, Z.J. Li, et al.), 465–496. Oxford: Wiley.

18 Gui, J.F., Zhou, L., Yin, Z. et al. (2021). *Genetic Breeding in Aquaculture*. Beijing: Science Press.

19 Yu, X.J., Zhou, T., Li, Y.C. et al. (1989). *Chromosomes of Chinese Freshwater Fishes*. Beijing: Science Press.

20 Yang, L., Sado, T., Vincent Hirt, M. et al. (2015). Phylogeny and polyploidy: resolving the classification of cyprinine fishes (Teleostei: Cypriniformes). *Molecular Phylogenetics and Evolution* **85**: 97–116.

21 Li, X. and Guo, B. (2020). Substantially adaptive potential in polyploid cyprinid fishes: evidence from biogeographic, phylogenetic and genomic studies. *Proceedings of the Biological Sciences* **287** (1920): 20193008.

22 Comai, L. (2005). The advantages and disadvantages of being polyploid. *Nature Reviews Genetics* **6** (11): 836–846.

23 Lien, S., Koop, B.F., Sandve, S.R. et al. (2016). The Atlantic salmon genome provides insights into rediploidization. *Nature* **533** (7602): 200–205.

24 He, Z.S., Wang, L.H., Harper, A.L. et al. (2017). Extensive homoeologous genome exchanges in allopolyploid crops revealed by mRNAseq-based visualization. *Plant Biotechnology Journal* **15** (5): 594–604.

25 Xiong, Z.Y., Gaeta, R.T., and Pires, J.C. (2011). Homoeologous shuffling and chromosome compensation maintain genome balance in resynthesized allopolyploid *Brassica napus*. *Proceedings of the National Academy of Sciences of the United States of America* **108** (19): 7908–7913.

26 Cheng, F., Wu, J., Cai, X. et al. (2018). Gene retention, fractionation and subgenome differences in polyploid plants. *Nature Plants* **4** (5): 258–268.

27 Kashkush, K., Feldman, M., and Levy, A.A. (2003). Transcriptional activation of retrotransposons alters the expression of adjacent genes in wheat. *Nature Genetics* **33** (1): 102–106.

28 Zou, J., Gong, H.H., Qian, W. et al. (2011). *De novo* genetic variation associated with retrotransposon activation, genomic rearrangements and trait variation in a recombinant inbred line population of *Brassica napus* derived from interspecific hybridization with *Brassica rapa*. *Plant Journal* **68** (2): 212–224.

29 Guan, X.Y., Song, Q.X., and Chen, Z.J. (2014). Polyploidy and small RNA regulation of cotton fiber development. *Trends in Plant Science* **19** (8): 516–528.

30 Jackson, S.A. (2017). Epigenomics: dissecting hybridization and polyploidization. *Genome Biology* **18** (1): 117.

31 Song, Q.X. and Chen, Z.J. (2015). Epigenetic and developmental regulation in plant polyploids. *Current Opinion in Plant Biology* **24**: 101–109.

32 Clark, J.W. and Donoghue, P. (2018). Whole-genome duplication and plant macroevolution. *Trends in Plant Science* **23** (10): 933–945.

33 Mandakova, T. and Lysak, M.A. (2018). Post-polyploid diploidization and diversification through dysploid changes. *Current Opinion in Plant Biology* **42**: 55–65.

34 Fox, D.T., Soltis, D.E., Soltis, P.S. et al. (2020). Polyploidy: a biological force from cells to ecosystems. *Trends in Cell Biology* **30** (9): 688–694.

35 Gan, R.H., Wang, Y., Li, Z. et al. (2021). Functional divergence of multiple duplicated *foxl2* homeologs and alleles in a recurrent polyploid fish. *Molecular Biology and Evolution* **38** (5): 1995–2013.

36 Mou, C.Y., Li, S., Lu, L.F. et al. (2021). Divergent antiviral mechanisms of two *viperin* homeologs in a recurrent polyploid fish. *Frontiers in Immunology* **12**: 702971.

37 Ding, M. and Chen, Z.J. (2018). Epigenetic perspectives on the evolution and domestication of polyploid plant and crops. *Current Opinion in Plant Biology* **42**: 37–48.

38 Li, N., Xu, C., Zhang, A. et al. (2019). DNA methylation repatterning accompanying hybridization, whole genome doubling and homoeolog exchange in nascent segmental rice allotetraploids. *New Phytologist* **223** (2): 979–992.

39 Zhang, J., Liu, Y., Xia, E.H. et al. (2015). Autotetraploid rice methylome analysis reveals methylation variation of transposable elements and their effects on gene expression. *Proceedings of the National Academy of Sciences of the United States of America* **112** (50): E7022–E7029.

40 Chen, Z.J. (2013). Genomic and epigenetic insights into the molecular bases of heterosis. *Nature Reviews Genetics* **14** (7): 471–482.

41 Pikaard, C.S. (2000). The epigenetics of nucleolar dominance. *Trends in Genetics* **16** (7): 495–500.

42 Ha, M., Lu, J., Tian, L. et al. (2009). Small RNAs serve as a genetic buffer against genomic shock in *Arabidopsis* interspecific hybrids and allopolyploids. *Proceedings of the National Academy of Sciences of the United States of America* **106** (42): 17835–17840.

43 Chen, Z.J., Comai, L., and Pikaard, C.S. (1998). Gene dosage and stochastic effects determine the severity and direction of uniparental ribosomal RNA gene silencing (nucleolar dominance) in *Arabidopsis* allopolyploids. *Proceedings of the National Academy of Sciences of the United States of America* **95** (25): 14891–14896.

44 Wang, J., Tian, L., Lee, H.S. et al. (2006). Genomewide nonadditive gene regulation in *Arabidopsis* allotetraploids. *Genetics* **172** (1): 507–517.

45 Yoo, M.J., Liu, X., Pires, J.C. et al. (2014). Nonadditive gene expression in polyploids. *Annual Review of Genetics* **48**: 485–517.

46 Garsmeur, O., Schnable, J.C., Almeida, A. et al. (2014). Two evolutionarily distinct classes of paleopolyploidy. *Molecular Biology and Evolution* **31** (2): 448–454.

47 Murat, F., Zhang, R., Guizard, S. et al. (2014). Shared subgenome dominance following polyploidization explains grass genome evolutionary plasticity from a seven protochromosome ancestor with 16K protogenes. *Genome Biology and Evolution* **6** (1): 12–33.

48 Woodhouse, M.R., Cheng, F., Pires, J.C. et al. (2014). Origin, inheritance, and gene regulatory consequences of genome dominance in polyploids. *Proceedings of the National Academy of Sciences of the United States of America* **111** (14): 5283–5288.

49 Parisod, C., Salmon, A., Zerjal, T. et al. (2009). Rapid structural and epigenetic reorganization near transposable elements in hybrid and allopolyploid genomes in *Spartina*. *New Phytologist* **184** (4): 1003–1015.

50 Hollister, J.D., Smith, L.M., Guo, Y.L. et al. (2011). Transposable elements and small RNAs contribute to gene expression divergence between *Arabidopsis thaliana* and *Arabidopsis lyrata*. *Proceedings of the National Academy of Sciences of the United States of America* **108** (6): 2322–2327.

51 Wolf, J.B. (2009). Cytonuclear interactions can favor the evolution of genomic imprinting. *Evolution* **63** (5): 1364–1371.

52 Roberge, C., Normandeau, E., Einum, S. et al. (2008). Genetic consequences of interbreeding between farmed and wild Atlantic salmon: insights from the transcriptome. *Molecular Ecology* **17** (1): 314–324.

53 White, S.L., Sakhrani, D., Danzmann, R.G., and Devlin, R.H. (2013). Influence of developmental stage and genotype on liver mRNA levels among wild, domesticated, and hybrid rainbow trout (*Oncorhynchus mykiss*). *BMC Genomics* **14**: 673.

54 Bougas, B., Granier, S., Audet, C., and Bernatchez, L. (2010). The transcriptional landscape of cross-specific hybrids and its possible link with growth in brook charr (*Salvelinus fontinalis* Mitchill). *Genetics* **186** (1): 97–107.

55 Yang, J.M., Luo, S.J., Li, J.H. et al. (2018). Transcriptome analysis of growth heterosis in pearl oyster *Pinctada fucata martensii*. *FEBS Open Bio* **8** (11): 1794–1803.

56 Renaut, S., Nolte, A.W., and Bernatchez, L. (2009). Gene expression divergence and hybrid misexpression between lake whitefish species pairs (*Coregonus* spp. Salmonidae). *Molecular Biology and Evolution* **26** (4): 925–936.

57 Zhou, Y., Ren, L., Xiao, J. et al. (2015). Global transcriptional and miRNA insights into bases of heterosis in hybridization of Cyprinidae. *Scientific Reports* **5**: 13847.

58 Zhang, G., Li, J., Zhang, J. et al. (2019). Integrated analysis of transcriptomic, miRNA and proteomic changes of a novel hybrid yellow catfish uncovers key roles for miRNAs in heterosis. *Molecular & Cellular Proteomics* **18** (7): 1437–1453.

59 Xiao, J., Hu, F., Luo, K. et al. (2016). Unique nucleolar dominance patterns in distant hybrid lineage derived from *Megalobrama amblycephala* × *Culter alburnus*. *BMC Genetics* **17** (1): 150.

60 Cao, L., Qin, Q., Xiao, Q. et al. (2018). Nucleolar dominance in a tetraploidy hybrid lineage derived from *Carassius auratus* red var. (♀) × *Megalobrama amblycephala* (♂). *Frontiers in Genetics* **9**: 386.

61 Li, W., Liu, J., Tan, H. et al. (2018). Asymmetric expression patterns reveal a strong maternal effect and dosage compensation in polyploid hybrid fish. *BMC Genomics* **19** (1): 517.

62 Pala, I., Coelho, M.M., and Schartl, M. (2008). Dosage compensation by gene-copy silencing in a triploid hybrid fish. *Current Biology* **18** (17): 1344–1348.

63 Pala, I., Schartl, M., Brito, M. et al. (2010). Gene expression regulation and lineage evolution: the North and South tale of the hybrid polyploid *Squalius alburnoides* complex. *Proceedings of the Biological Sciences* **277** (1699): 3519–3525.

64 Matos, I., Machado, M.P., Schartl, M., and Coelho, M.M. (2015). Gene expression dosage regulation in an allopolyploid fish. *PLoS One* **10** (3): e0116309.

65 Ren, L., Tang, C., Li, W. et al. (2017). Determination of dosage compensation and comparison of gene expression in a triploid hybrid fish. *BMC Genomics* **18** (1): 38.

66 Covelo-Soto, L., Leunda, P.M., Pérez-Figueroa, A., and Morán, P. (2015). Genome-wide methylation study of diploid and triploid brown trout (*Salmo trutta* L.). *Animal Genetics* **463** (3): 280–288.

67 Jiang, Q., Li, Q., Yu, H., and Kong, L. (2016). Inheritance and variation of genomic DNA methylation in diploid and triploid pacific oyster (*Crassostrea gigas*). *Marine Biotechnology* **18** (1): 124–132.

68 Xiao, J., Song, C., Liu, S.J. et al. (2013). DNA methylation analysis of allotetraploid hybrids of red crucian carp (*Carassius auratus* red var.) and common carp (*Cyprinus carpio* L.). *PLoS One* **8** (2): e56409.

69 Yu, P., Zhou, L., Yang, W.T. et al. (2021). Comparative mitogenome analyses uncover mitogenome features and phylogenetic implications of the subfamily Cobitinae. *BMC Genomics* **22** (1): 50.

70 Zhou, H., Ma, T.Y., Zhang, R. et al. (2016). Analysis of different ploidy and parent-offspring genomic DNA methylation in the loach *Misgurnus anguillicaudatus*. *International Journal of Molecular Sciences* **17** (8): 1299.

71 Qin, Q., Wang, C.Q., Zhou, Y.W. et al. (2020). Rapid genomic and epigenetic alterations in gynogenetic *Carassius auratus* Red Var. derived from distant hybridization. *Marine Biotechnology* **22** (3): 433–442.

72 Wang, C., Zhou, Y., Qin, H. et al. (2021). Genetic and epigenetic changes are rapid responses of the genome to the newly synthesized autotetraploid *Carassius auratus*. *Frontiers in Genetics* **11**: 576260.

73 Gui, J.F., Liang, S.C., Zhu, L.F., and Jiang, Y.G. (1993a). Discovery of multiple tetraploids in artificially propagated populations of allogynogenetic silver crucian carp and their breeding potentialities. *Chinese Science Bulletin* **38** (4): 327–331.

74 Gui, J.F., Liang, S.C., Zhu, L.F., and Jiang, Y.G. (1993b). Preliminary confirmation of gynogenetic reproductive mode in artificial multiple tetraploid allogynogenetic silver crucian carp. *Chinese Science Bulletin* **38** (1): 67–70.

75 Gui, J.F. and Zhou, L. (2010). Genetic basis and breeding application of clonal diversity and dual reproduction modes in polyploid *Carassius auratus gibelio*. *Science China Life Sciences* **53** (4): 409–415.

76 Yi, M.S., Liu, Y.Q., Zhou, L. et al. (2003). Molecular cytogenetic detection of paternal chromosome fragments in allogynogenetic gibel carp, *Carassius auratus gibelio* Bloch. *Chromosome Research* **11** (7): 665–671.

77 Zhu, H.P. and Gui, J.F. (2007). Identification of genome organization in the unusual allotetraploid form of *Carassius auratus gibelio*. *Aquaculture* **265** (1–4): 109–117.

78 Knytl, M., Kalous, L., Symonová, R. et al. (2013). Chromosome studies of european cyprinid fishes: cross-species painting reveals natural allotetraploid origin of a *Carassius* female with 206 chromosomes. *Cytogenetic and Genome Research* **139** (4): 276–283.

79 Avise, J.C. (2015). Evolutionary perspectives on clonal reproduction in vertebrate animals. *Proceedings of the National Academy of Sciences of the United States of America* **112** (29): 8867–8873.

80 Li, Z., Liang, H.W., Wang, Z.W. et al. (2016). A novel allotetraploid gibel carp strain with maternal body type and growth superiority. *Aquaculture* **458**: 55–63.

81 Lu, M., Wang, Z.W., Hu, C.J. et al. (2018). Genetic identification of a newly synthetic allopolyploid strain with 206 chromosomes in polyploid gibel carp. *Aquaculture Research* **491**: 1–10.

82 Lu, M., Li, X.Y., Li, Z. et al. (2021). Regain of sex determination system and sexual reproduction ability in a synthetic octoploid male fish. *Science China Life Sciences* **64** (1): 77–87.

83 Li, Z., Lu, M., Zhou, L. et al. (2020). An effective method for creating Allo-octoploids by integrating exogenous sperm genome into gibel carp eggs. *Acta Hydrobiologica Sinica* **44** (1): 26–35.

84 Shao, G.M., Li, X.Y., Wang, Y. et al. (2018). Whole genome incorporation and epigenetic stability in a newly synthetic allopolyploid of gynogenetic gibel carp. *Genome Biology and Evolution* **10** (9): 2394–2407.

85 Liu, B., Brubaker, C.L., Mergeai, G. et al. (2001). Polyploid formation in cotton is not accompanied by rapid genomic changes. *Genome* **44** (3): 321–330.

86 Araki, H., Cooper, B., and Blouin, M.S. (2007). Genetic effects of captive breeding cause a rapid, cumulative fitness decline in the wild. *Science* **318** (5847): 100–103.

87 Stringwell, R., Lock, A., Stutchbury, C.J. et al. (2014). Maladaptation and phenotypic mismatch in hatchery-reared Atlantic salmon *Salmo salar* released in the wild. *Journal of Fish Biology* **85** (6): 1927–1945.

88 Le Luyer, J., Laporte, M., Beacham, T.D. et al. (2017). Parallel epigenetic modifications induced by hatchery rearing in a Pacific Salmon. *Proceedings of the National Academy of Sciences of the United States of America* **114** (49): 12964–12969.

89 Rodriguez-Barreto, D., de Leaniz, C.G., and Verspoor, E. (2019). DNA methylation changes in the sperm of captive-reared fish: a route to epigenetic untrogression in wild populations. *Molecular Biology and Evolution* **36** (10): 2205–2211.

90 Zhou, R., Shang, R.J., Gong, D.B. et al. (2019). Characterization of H3 methylation in regulating oocyte development in cyprinid fish. *Science China Life Sciences* **62** (6): 829–837.

91 Tao, M., Zhou, Y., Li, S. et al. (2018). MicroRNA alternations in the testes related to the sterility of triploid fish. *Marine Biotechnology* **20** (6): 739–749.

92 Alves, M.J., Coelho, M.M., and Collares-Pereira, M.J. (2001). Evolution in action through hybridisation and polyploidy in an Iberian freshwater fish: a genetic review. *Genetica* **111** (1–3): 375–385.

93 Inácio, A., Pinho, J., Pereira, P.M. et al. (2012). Global analysis of the small RNA transcriptome in different ploidies and genomic combinations of a vertebrate complex–the *Squalius alburnoides*. *PLoS One* **7** (7): e41158.

94 Darwin, C.R. (1876). *The Effects of Cross- and Self-Fertilization in the Vegetable Kingdom*. John Murry.

95 Lippman, Z.B. and Zamir, D. (2007). Heterosis: revisiting the magic. *Trends in Genetics* **23** (2): 60–66.

96 Birchler, J.A., Yao, H., Chudalayandi, S. et al. (2010). Heterosis. *Plant Cell* **22** (7): 2105–2112.

97 Groszmann, M., Greaves, I.K., Fujimoto, R. et al. (2013). The role of epigenetics in hybrid vigour. *Trends in Genetics* **29** (12): 684–690.

98 Greaves, I.K., Gonzalez-Bayon, R., Wang, L. et al. (2015). Epigenetic changes in hybrids. *Plant Physiology* **168** (4): 1197–1205.

99 Schneeberger, K., Ossowski, S., Ott, F. et al. (2011). Reference-guided assembly of four diverse *Arabidopsis thaliana* genomes. *Proceedings of the National Academy of Sciences of the United States of America* **108** (25): 10249–10254.

100 Groszmann, M., Greaves, I.K., Albert, N. et al. (2011a). Epigenetics in plants-vernalisation and hybrid vigour. *Biochimica et Biophysica Acta* **1809** (8): 427–437.

101 Schmitz, R.J., Schultz, M.D., Urich, M.A. et al. (2013). Patterns of population epigenomic diversity. *Nature* **495** (7440): 193–198.

102 Qiu, T., Liu, Z., and Liu, B. (2020). The effects of hybridization and genome doubling in plant evolution via allopolyploidy. *Molecular Biology Reports* **47** (7): 5549–5558.

103 Law, J.A. and Jacobsen, S.E. (2010). Establishing, maintaining and modifying DNA methylation patterns in plants and animals. *Nature Reviews Genetics* **11** (3): 204–220.

104 Groszmann, M., Greaves, I.K., Albertyn, Z.I. et al. (2011b). Changes in 24-nt siRNA levels in *Arabidopsis* hybrids suggest an epigenetic contribution to hybrid vigor. *Proceedings of the National Academy of Sciences of the United States of America* **108** (6): 2617–2622.

105 Shen, H., He, H., Li, J.G. et al. (2012). Genome-wide analysis of DNA methylation and gene expression changes in two *Arabidopsis* ecotypes and their reciprocal hybrids. *Plant Cell* **24** (3): 875–892.

106 Barber, W.T., Zhang, W., Win, H. et al. (2012). Repeat associated small RNAs vary among parents and following hybridization in maize. *Proceedings of the National Academy of Sciences of the United States of America* **109** (26): 10444–10449.

107 He, G.M., Chen, B.B., Wang, X.C. et al. (2013). Conservation and divergence of transcriptomic and epigenomics variation in maize hybrids. *Genome Biology* **14** (6): R57.

108 He, G., Zhu, X.P., Elling, A.A. et al. (2010). Global epigenetic and transcriptional trends among two rice subspecies and their reciprocal hybrids. *Plant Cell* **22** (1): 17–33.

109 Shivaprasad, P.V., Chen, H.M., Patel, K. et al. (2012). A microRNA superfamily regulates nucleotide binding site-leucine-rich repeats and other mRNAs. *Plant Cell* **24** (3): 859–874.

110 Greaves, I.K., Gonzalez-Bayon, R., Ying, H. et al. (2012). Trans chromosomal methylation in *Arabidopsis* hybrids. *Proceedings of the National Academy of Sciences of the United States of America* **109** (9): 3570–3575.

111 Ni, Z., Kim, E.D., Ha, M. et al. (2009). Altered circadian rhythms regulate growth vigour in hybrids and allopolyploids. *Nature* **457** (7227): 327e331.

112 Moghaddam, A.M.B., Roudier, F., Seifert, M. et al. (2011). Additive inheritance of histone modifications in *Arabidopsis thaliana* intra-specific hybrids. *Plant Journal* **67** (4): 691–700.

113 Dong, X., Reimer, J., Göbel, U. et al. (2012). Natural variation of H3K27me3 distribution between two *Arabidopsis* accessions and its association with flanking transposable elements. *Genome Biology* **13** (12): R117.

114 Hedgecock, D., McGoldrick, D.J., and Bayne, B.L. (1995). Hybrid vigor in Pacific oysters: an experimental approach using crosses among inbred lines. *Aquaculture* **137** (1): 285–298.

115 Hedgecock, D., Lin, J.Z., DeCola, S. et al. (2007). Transcriptomic analysis of growth heterosis in larval Pacific oysters (*Crassostrea gigas*). *Proceedings of the National Academy of Sciences of the United States of America* **104** (7): 2313–2318.

116 Sun, Y., Guo, C.Y., Wang, D.D. et al. (2016). Transcriptome analysis reveals the molecular mechanisms underlying growth superiority in a novel grouper hybrid (*Epinephelus fuscogutatus*♀ × *E. lanceolatus*♂). *BMC Genetics* **17**: 24.

117 Cai, W.Q., Li, S.F., and Ma, J.Y. (2004). Diseases resistance of Nile tilapia (*Oreochromis niloticus*), blue tilapia (*Oreochromis aureus*) and their hybrid (female Nile tilapia × male blue tilapia) to *Aeromonas sobria*. *Aquaculture* **229** (1–4): 79–87.

118 Zhou, Y., Zhang, X., Xu, Q. et al. (2019a). Nonadditive and allele-specific expression of insulin-like growth factor 1 in Nile tilapia (*Oreochromis niloticus*, ♀) blue tilapia (*O. aureus*, ♂) hybrids. *Comparative Biochemistry and Physiology – Part B: Biochemistry & Molecular Biology* **232**: 93–100.

119 Zhong, H., Zhang, X., Xu, Q. et al. (2019). Nonadditive and asymmetric allelic expression of growth hormone in hybrid tilapia. *Frontiers in Genetics* **10**: 961.

120 Zhou, Y., Zhang, X., Xu, Q. et al. (2019b). Nonadditive expression of lipid metabolism pathway-related genes in intestine of hybrids of Nile tilapia females (*Oreochromis niloticus*) and blue tilapia males (*Oreochromis aureus*). *Molecular Biology Reports* **46** (1): 425–432.

121 Šimková, A., Vojtek, L., Halačka, K. et al. (2015). The effect of hybridization on fish physiology, immunity and blood biochemistry: a case study in hybridizing *Cyprinus carpio* and *Carassius gibelio* (Cyprinidae). *Aquaculture* **435**: 381–389.

122 Gui, J.F., Zhou, L., and Li, X.Y. (2022). Rethinking fish biology and biotechnologies in the challenge era for burgeoning genome resources and strengthening food security. *Water Biology and Security* **1**: 100002.

14

Epigenetics in Aquatic Toxicology

Sara J. Hutton[1] *and Susanne M. Brander*[2]

[1] *Department of Environmental and Molecular Toxicology, Oregon State University, Corvallis, OR, USA*
[2] *Coastal Oregon Marine Experiment Station, Department of Fisheries, Wildlife, and Conservation Sciences, Oregon State University, Newport, OR, USA*

14.1 Introduction

From the subcellular (i.e., epigenetic) to ecosystem scale, aquatic toxicologists hail from, and thus study, a wide variety of subdisciplines to assess the risk of various environmental stressors to aquatic species and ecosystems. Toxicants, herein referring to synthetic pollutants such as pesticides, pharmaceuticals, industrial waste, oil spills, and microplastics, enter the aquatic environment and can alter epigenetic, molecular, organismal, and ecosystem functions [1–3]. Toxins, herein referring to naturally occurring chemicals like microcystins produced by cyanobacteria, also cause stress and alter epigenetic mechanisms in aquatic organisms [4]. Additionally, the growing threat of climate change adds another layer of complexity to the field as some climate change stressors induce similar effects as some toxins or toxicants, cause increased production of certain types of toxins, such as those from harmful algal blooms, and alter resistance to environmental stressors [5]. Toxicity can occur across a range of molecular mechanisms depending on the toxicant or toxin and exposure route. Over the past decade, as our understanding of molecular biology has advanced, a sizeable focus of research has been on 'omics and their role in mechanisms of toxicity [6]. Epigenetic changes are possible routes by which gene expression can be altered without changing the DNA sequence. A stronger understanding of the role of epigenetic mechanisms in gene regulation, subsequent protein production, and enzyme activation has allowed toxicologists to better study the genes and pathways involved in modulating responses to environmental stress. As science has progressed and our understanding of molecular biology advances, toxicologists can better incorporate epigenetic mechanisms into studies on environmental stress and pollution and improve our understanding of how organisms respond to these stressors. Epigenetics is new to the field of aquatic toxicology and thus, much more information is likely to be gained in the coming years as recent epigenetic work has improved our understanding of toxicity and prediction of risk.

Following exposure to a toxic compound, an organism will likely exhibit a physiological response as a result of molecular changes. Much of our understanding of a compound's mechanism of action (MOA) comes from the study of pharmacology and human health (mammalian) toxicology. While many of the compounds studied in pharmacology are intended to deliver a beneficial outcome, comprehension of basic biology, including epigenetic mechanisms, is needed to understand which compounds activate specific molecular changes in both pharmacology and toxicology. Toxicology is primarily focused on identifying the toxicity of compounds found in the environment, either natural or anthropogenic, and elucidating their mechanisms of toxicity, including their safety or threshold level. Aquatic toxicologists are focused on how toxic compounds affect aquatic organisms, populations, communities, and ecosystems. For example, pyrethroids insecticides bind to sodium transport channels, preventing their closure, and causing continual firing of axons [7]. Pyrethroids are up to 1000 times more toxic to fish than mammals since fish lack carboxylesterase, one of the enzymes involved in pyrethroid metabolism. Thus, they provide a good example of how mammalian studies can provide information on the possible MOA of pyrethroid toxicity, but do not provide a solid understanding of their metabolism or potential toxicity to aquatic organisms due to differences in physiology.

Epigenetics in Aquaculture, First Edition. Edited by Francesc Piferrer and Han-Ping Wang.

Historically, toxicity was measured mainly at the whole organism level, frequently by determining the lethal dose of exposure. In the early days of toxicology, the MOA was not known to scientists due to limitations in technology; thus, toxicologists were often left to assess the concentration of a contaminant that caused an apical adverse effect, often lethality [8]. However, advancements in science have made it clear that lower concentrations of compounds may not result directly in lethality, but can still cause adverse effects on organisms and populations through sublethal toxicity [9]. Epigenetic modification through changes in DNA methylation (DNAm), chromatin structure through histone modification, and noncoding RNAs (ncRNAs) are outcomes that have recently been noted following exposure to toxic compounds [3]. Epigenetic changes due to toxic exposure can act as both potential biomarkers for stress, used to indicate that organisms are exhibiting susceptibility to environmental stress, and as a way to better understand the compound's MOA [10]. Since epigenetic modulation is correlated with gene expression, it is suspected that epigenetic endpoints can help determine which pathways are altered in an organism following toxic exposure and if the epigenetic mechanisms were involved in inducing an adverse effect [11].

The extent to which a compound may induce a toxic effect on an aquatic organism is governed by the quantity of that compound that is biologically available. In toxicology, the amount of a compound that is bioavailable is commonly governed by the concept of ADME: absorption, distribution, metabolism, and excretion [12]. Absorption is determined by how much of the compound is absorbed into the organism, often through either the skin, gastrointestinal tract, or lungs/gills. Distribution describes the route the compound takes throughout the body to reach the target organ, such as the liver, brain, heart, reproductive axis, or gills. Metabolism is the process by which a compound is broken down so that it can be excreted from the body. Generally, larger, hydrophobic compounds must undergo some form of metabolism to become smaller and more hydrophilic so they can be excreted from the body. Lastly, excretion through urine or feces removes the compound from the body. Metabolism can also complicate toxicity. Some toxicants, referred to as *pro-drugs* in pharmacology, have increased toxicity after metabolism, particularly phase I, has occurred. Organophosphates (OPs), a large class of pesticides, are a common example of this phenomenon, where the *-oxon* metabolite is more toxic than the parent compound. Thus, how a compound is absorbed, distributed, metabolized, and excreted dictates how bioavailable it will be to an organism. In terms of epigenetic influences on ADME, epigenetic changes to enzymes involved in metabolism may either increase or decrease toxicity dependent upon whether the change elicits an increase or decrease in metabolizing enzymes.

All organisms have enzymes that are designed to remove toxins or toxicants and mitigate harm. Metabolizing enzymes chemically alter compounds to change their physiochemical properties and reduce toxicity. These types of enzymes generally function within two different phases of metabolism: phase I and II [13]. Phase I is the process by which oxygenation or hydrolysis occurs and an oxygen molecule is added to the compound to increase the hydrophilicity of the molecule. Cytochrome P450s (CYPs) and flavin-containing monooxygenase are two of the main enzyme classes involved in phase I metabolism [14]. Epigenetic mechanisms can influence the transcription of genes involved in the production of phase I enzymes [15, 16]. Histone modification influences CYP expression through its influence on the transcription of nuclear receptors (NRs), which bind to DNA and regulate CYP transcription [17]. For example, cells with the histone 3 lysine 9 methyl transferase enzyme (G9a) knocked out had decreased gene expression of some NRs, which demonstrated a relationship between histone-modifying enzymes and gene expression [17]. Additionally, knockout cells had decreased CYP expression when induced with an increase in fatty acid conditions, linking decreased G9a with altered NR and CYP transcription [17]. Inhibition of DNA methyltransferases, the enzymes involved in transferring the methyl group from the donor compound to the DNA, lead to hypomethylation of CYP promoter sites and increased CYP transcription [18]. ncRNAs are also involved in the expression of CYPs: when ncRNAs involved in CYP production are affected by environmental stress, this causes downstream effects on metabolism which can increase toxic effects [16]. Following phase I, some metabolites then undergo phase II metabolism. Phase II is described as the conjugation phase where endogenous metabolites such as sugars, glutathione, or sulfate are added to the metabolite to increase the polarity, decrease toxicity, and increase the compound's ability to be excreted [19, 20]. In some cases, a third phase may occur and primarily involves either further metabolism of the phase II metabolite or elimination from the cell [21]. The production of these enzymes can be triggered by the introduction of an exogenous compound into the body, and the expression of the genes involved in metabolism is regulated by epigenetic mechanisms at every phase [22, 23]. Therefore, changes at the epigenetic level to the regulation and production of metabolizing enzymes can affect the degree of toxicity an organism experiences. CYPs are well known for their diversity in function; CYP enzyme expression not only impacts detoxification, but also normal physiological functions such as reproduction and development [24]. Disruption or alteration of epigenetic mechanisms and functions related to the production and regulation of CYPs and other metabolizing enzymes have implications for the extent to which metabolism and toxicity occur.

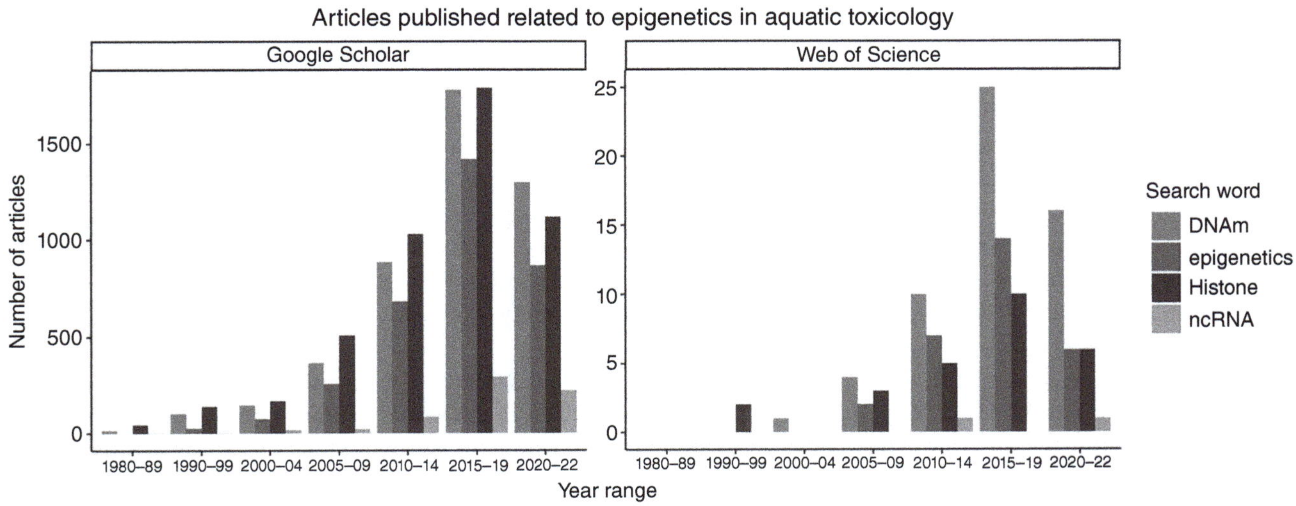

Figure 14.1 Articles published within the last 42 years with search terms related to epigenetics in aquatic toxicology. Google Scholar searches were done with search terms "aquatic" AND "toxicology" AND *search word*. Web of Science searches were done using ALL = ("aquatic" AND "toxicology" AND *search word*). *Search word* was one of the four terms from the legend. Searches conducted on 24 January 2022.

Epigenetic approaches are widely used in human health and disease studies, as well as those related to environmental toxicology and public health. While epigenetic mechanisms are less studied in the context of aquatic toxicology and ecotoxicology, studies in these fields have used epigenetic mechanisms more frequently in recent years to assess toxicity. A drastic increase in epigenetic mechanisms used in aquatic toxicology studies has occurred over the last two to three decades (Figure 14.1). The growing threat of climate change has further added to the list of stressors aquatic toxicologists are using epigenetics to study, as aquatic toxicologists can study a variety of environmental stressors, including climate change, land use changes (including aquaculture), and pollution [25, 26]. Within this chapter, we will explore the role of epigenetics in aquatic toxicology and how the field of aquatic toxicology can benefit from these studies by further incorporating work conducted to date. Discussion of the use of epigenetics in risk assessment and application for aquaculture is considered.

14.2 Epigenetic Endpoints in Aquatic Toxicology Studies

The role of epigenetics in regulating genes responsible for enzyme activation and protein production has led to an interest in how epigenetic dysregulation is involved in adverse, toxic responses. How DNAm, histone modification, and ncRNA interact with environmental pollution and stress; what biomarkers are useful for determining risk; and how epigenetics alterations lead to increased or decreased tolerance of contaminants are all examples of broad questions toxicologists might ask [16]. Much of the toxicology research done to date uses model species such as mice, rats, and zebrafish [27–29]. Zebrafish have become a popular model organism in both traditional toxicology and ecotoxicology as their genome is fully sequenced, they reproduce easily, and have a similar enough physiology and genome to humans to warrant using them as a model for human health [30]. Studies from zebrafish inform on the role of epigenetics in embryogenesis and early development, which has advanced the understanding of the role of epigenetics in both teleost fish and human health [31]. These sorts of data help progress aquatic toxicology studies by introducing new biomarkers and endpoints to study, which improves our ability to screen compounds for toxicity [32]. Advances made with model organisms help toxicologists better understand general epigenetic effects following an environmental stressor.

Less work has been done using alternate models or nonmodel organisms relevant to aquatic toxicology [33]. Much of this is likely due to the high cost of performing these types of experiments, as well as the need for extensively annotated genomes, which are not available for many nonmodel organisms. As the cost of studies comes down, the occurrence of these experiments will continue to increase. For now, there are still many data gaps within the literature on epigenetic mechanisms used in aquatic toxicology and ecotoxicology. DNAm remains the most studied epigenetics mechanism in toxicology studies. While this trend appears to continue with aquatic toxicity studies, only a few have looked at DNAm at an individual gene level [3, 34]. Most aquatic toxicology studies use global genome methylation as an indicator for whether a toxin/toxicant or environmental stressor causes hyper- or hypomethylation [35, 36]. Some studies may not even assess methylation at all, but instead, look at the expression of *dnmt1* and *dnmt3* (genes that maintain and initiate methylation, respectively) to determine whether changes in DNAm may be occurring or not [1]. Since nonmodel species may be more sensitive, and therefore at risk of experiencing toxicity, limited information makes it challenging to relate epigenetic endpoints alone to risk for these species. Studies on model organisms, even those from mammalian and terrestrial invertebrates (i.e., mice and earthworms respectively), can still help inform how epigenetic mechanisms can be incorporated into aquatic toxicology by providing baseline knowledge of epigenetic biology from which aquatic toxicologists can build. In this regard, epigenetic endpoints in aquatic toxicology studies provide a bridge between existing toxicology literature and the epigenetic mechanisms that govern aquatic toxicity. Another possibility is the use of epigenetic dysregulation as a biomarker, demonstrating that toxicity has occurred.

In this section, the roles of the three main epigenetic mechanisms: DNAm, histone modification, and ncRNA are explored, starting with a broad description of their processes, followed by an assessment of their roles in aquatic toxicology studies. All three of these epigenetic mechanisms have shown the potential to be altered following exposure to environmental stress and are suspected to be heritable [37] (Figure 14.2).

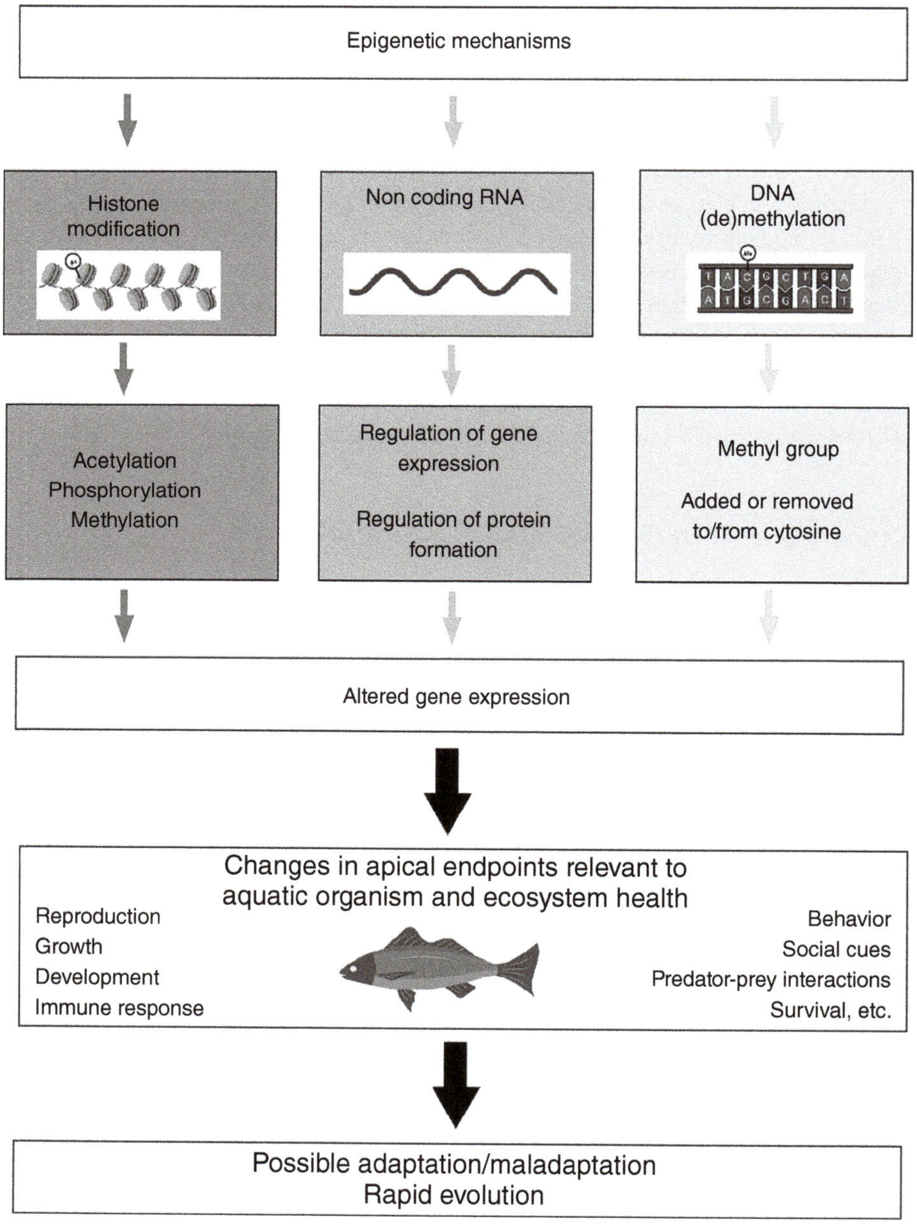

Figure 14.2 Schematic diagram of epigenetic mechanisms involved in aquatic toxicology. *Source:* From © BioRender.

14.2.1 DNA Methylation in Relation to Aquatic Toxicology

DNAm is by far the most common epigenetic mechanism to be used as an endpoint and biomarker in aquatic toxicology studies [38]. DNAm involves the addition of a methyl group to the fifth carbon on a cytosine. Cytosine methylation often occurs next to guanine, known as a CpG dinucleotide site; the grouping of CpG dinucleotide sites makes up CpG islands. Depending on whether or not a CpG island site is within the promoter region or gene body may determine if a gene will be up- or downregulated [39]. The relationship between methylation in the promoter region and gene expression has long been known to be negatively correlated; however, the relationship between gene body methylation and gene expression is less clear [40]. In recent years, positive correlations between gene body methylation and gene expression have been seen in studies with both humans and zebrafish (*Danio rerio*) a commonly used fish in toxicology [41]. However, other studies using the model fish inland silverside (*Menidia beryllina*) did not find a correlation between gene body methylation and gene expression [3, 42]. Differences in cell and species type between studies likely contribute to a range of conclusions. This also highlights the need to focus on a variety of aquatic organisms in future epigenetic studies so that species-specific differences can be better understood.

DNAm is used as an endpoint and biomarker in field-based studies, taking a whole ecosystem or monitoring approach as well as laboratory studies assessing potential relationships between epigenetics and apical endpoints relevant for risk assessment [3, 43, 44]. In toxicology studies, DNAm is often used to determine biomarkers associated with different diseases, for example, helping to detect different cancer types or as a tool to determine potential MOA [45, 46]. As mentioned, DNAm is frequently studied for its correlation to gene expression. For example, exposure to benzo(a)pyrene in larval zebrafish caused an increase in promoter methylation and a decrease in expression of genes related to gonadotropin-releasing hormone, a hormone involved in regulation of other hormones important for reproduction [47]. These changes were correlated with decreased reproduction, which linked changes in DNAm to reproduction, a higher-level endpoint. In another example, fipronil, a chiral insecticide banned in the European Union but commonly used in other countries including the United States, has both an *S* and an *R* form that cause differential toxicity. The neurotoxicity of fipronil was investigated for changes in global DNAm patterns and differential DNAm in zebrafish exposed to the *S*- and *R*-fipronil [38]. Both enantiomers of fipronil caused increased DNAm in the promoter region of genes that are differentially expressed and related to neurological disorders, in conjunction with altered behavior in zebrafish. This study showed enantioselectivity in the hypermethylation of genes related to neuro-activity, which was correlated with similar differences seen in swimming behavior. Here, differential methylation in the promoter region was correlated with the different enantiomers [38]. In both studies, differential methylation in the promoter region was correlated with changes in gene expression; there were also links found between differential promoter methylation and effects on apical endpoints. These types of studies are less common in aquatic toxicology due to the high cost of sequencing, but provide valuable information on the site of differential methylation, which can help determine the mechanisms by which toxicity occurs.

Endocrine-disrupting compounds (EDC) are compounds that, like their name suggests, disrupt the endocrine functions within an organism, causing a hormonal imbalance that can have long term, negative consequences for organism fitness, reproduction, and survival [48, 49]. They include pesticides used in aquaculture, agriculture, and urban areas, as well as in pharmaceuticals that are not removed during wastewater treatment [50, 51]. These compounds are detected in the aquatic environment at concentrations as low as the parts per trillion, or ng/L, range and have effects on behavior, reproduction, and gene expression [3, 52]. Exposure to the EDCs levonorgestrel, trenbolone, and bifenthrin resulted in genes showing greater differential methylation in the gene body than the promoter region in inland silverside [3] (Figure 14.3). DNAm was altered more within the gene body in the F1 and F2s than in the F0s, which demonstrated that EDCs resulted in a higher number of CpGs that were more differentially methylated in the indirectly exposed and unexposed F1 and F2s than directly exposed F0s [3].

This suggests EDCs differentially methylated the indirectly exposed and unexposed F1s and F2s more than the directly exposed F0s [3]. From this study, it was shown that these select EDCs have the potential to alter DNAm in both the promoter and gene body, but changes in the gene body were more distinctive compared to fipronil, which altered methylation only in the promoter region. A select number of genes from this EDC study were selected for comparison between gene expression and DNAm, of which no correlation was found [42]. Whether or not differentially methylated genes (DMGs) in the gene body are correlated with gene expression is not as clear, as studies have found conflicting results; however, it is clear that DMG in the gene body are associated with ancestral methylation patterns as well as gene splicing, which may mean DMG could alter gene splicing [53–55]. DNAm was also found to have a nonmonotonic effect on mangrove rivulus (*Kryptolebias marmoratus*) following exposure to EDCs [56]. This is of note since many EDCs display nonmonotonic, dose-response relationships for higher-level apical endpoints in aquatic model organisms as well. Future studies linking dose-response effects from epigenetic endpoints to apical endpoints will help advance the use of epigenetics in aquatic toxicology since many toxicologists and risk assessors rely on dose-response data for management and risk assessments.

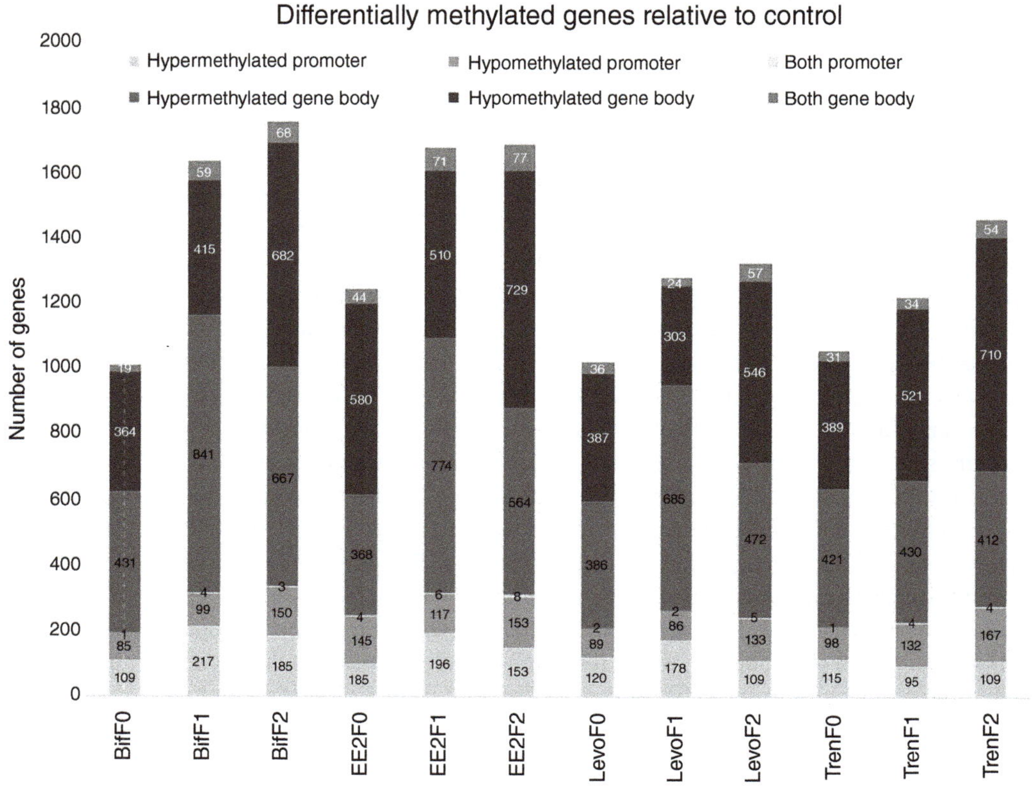

Figure 14.3 Bar graph depicting the differential methylation relative to control by treatment Bifenthrin (Bif), 17α-ethynylestradiol (EE2), levonorgestrel (Levo), trenbolone (Tren), generation (F0, F1, F2), directional methylation change (hyper, hypo) and region (promoter, gene body) for all genes, as informed by the inland silverside transcriptome. Differential methylation was more prominent in the gene body than the promoter regions for all treatments and generations. While most genes only showed a singular directional change in methylation by region relative to the control, a minority showed both hypo- and hypermethylation at different loci in the same gene. *Source:* From Major et al. [3].

Other forms of assessing DNAm are through global DNAm or changes in the expression of genes involved in methylation. To determine if epigenetic modifications are occurring, the genes *dmnt1* and *dmnt3* are frequently analyzed as epigenetic biomarkers. In rats, exposure to glyphosate induced differential methylation in these genes. The DMGs showed promise as biomarkers for transgenerational disease inheritance for diseases such as prostate disease, kidney disease, and/or obesity [57]. Japanese medaka fish (*Oryzias latipes*) exposed to glyphosate during embryonic development had decreased transcription of *dmnt1* in only the males, but not the females [1]. Altered expression of these genes can provide insight into whether there are disruptions in DNAm maintenance or de novo methylation, but are less informative since they do not directly inform on which genes are dysregulated or where dysregulation is occurring. Sex-specific differences have also been noted in other fishes following exposure to toxicants. In the Chinese rare minnow (*Gobiocypris rarus*) global DNAm was increased in sperm cells after exposure to bisphenol-A (BPA), which may be related to decreased sperm mobility and success [58]. Due to the reduced cost, global DNAm is frequently used as an endpoint to help determine whether a compound induces a response through epigenetic-mediated mechanisms. For example, paracetamol and ciprofloxacin are both pharmaceuticals commonly found in wastewater. Global DNAm was increased in zebrafish exposed to paracetamol, but not in ciprofloxacin [59]. Global DNAm in the adult Chinese rare minnow gonads was negatively correlated with expression of the *cyp19a1a* gene under BPA exposure after both 7 and 21 days of exposure. This suggested DNAm was involved in the regulation of *cyp19a1a* [60]. While global DNAm is less informative in terms of specific pathways that are dysregulated, it can provide useful evidence that dysregulation is occurring (hyper or hypo) and the potential for epigenetic dysregulation from exposure to aquatic pollutants. While DNAm can both increase and decrease gene expression, it is clear other epigenetic modifications are involved in gene expression and regulation.

In aquatic invertebrates, the genome is sparsely methylated, and methylation is mostly restricted to gene bodies and silenced repetitive segments [61]. Methylation patterns in invertebrates suggest a complicated role of DNAm in gene expression that varies across species. For example, in the Pacific oyster (*Crassostrea gigas*) global DNAm was not correlated with changes in gene expression following exposure to copper [62]. Following exposure to ocean acidification, eastern oyster (*Crassostrea virginica*) had significantly increased and decreased DMG throughout the gene body, but no association between gene expression and biological processes was found [63]. In reef corals, DNA was hypomethylated following heat stress and ocean acidification. Physiological processes were not investigated, but DNAm was closely associated with gene expression in three out of the six corals studied [64]. The sea cucumber *Apostichopus japonicus* also showed a correlation between methylation in the promoter region and gene expression [65]. DNAm, and all other epigenetic mechanisms, are still largely unexplored in aquatic invertebrates. The variety of the invertebrate species likely lends to a wide diversity in the role of DNAm and epigenetic mechanisms within the genome and gene expression of invertebrates, and much more research is needed to fully understand the diversity of functions DNAm has in toxic response in aquatic invertebrates.

14.2.2 Histone Modification in Relation to Aquatic Toxicology

DNAm does not work alone; histone modification and ncRNAs work in concert with DNAm to induce epigenetic change [33]. Histone modification involves either the acetylation, phosphorylation, methylation, and/or ubiquinition and sumoylation of a histone. Acetylation alters the charge of the histone, which changes the binding and conformation of DNA. In this way, histones help control which DNA is bound up as heterochromatin (inaccessible DNA, less transcription) and which is available for RNA polymerase to bind to, as euchromatin (accessible DNA, more transcription) [66]. Phosphorylation involves either the addition (kinases) or removal (phosphatase) of phosphate groups that alter DNA repair, chromatin compaction during meiosis and mitosis, and regulation of transcription [67]. Methylation alters recruitment and binding of regulatory proteins to chromatin, which have downstream effects on gene expression [68]. Ubiquinition and sumoylation involve the attachment of ubiquitin or ubiquitin-like modifiers, respectively, which influence protein translocation, DNA damage signaling, and transcriptional regulation [69, 70]. Any of these change to histone charge can change chromatin structure, signaling for transcription machinery, and DNA repair enzymes that alter transcription without changes to the DNA sequence.

In aquatic toxicology, study of histone modification is largely lacking; however, the studies that do exist show that environmental stress does affect histone charge, structure, and signaling like it does in human and rodent models. Histone modification acts by either structurally changing the chromatin through induction of either a heterochromatin or euchromatin state, which alter the accessibility of chromatin to RNA polymerase, or by altering signaling for transcription machinery and DNA repair enzymes [33]. Histone H3 at lysine 9 (H3K9) and 27 (H3K27) are involved in gene repression and are some of the best-studied examples of histone modification [71–73]. Thus, the role of histone modification in epigenetic inheritance following exposure to an environmental stressor has become of increased interest to aquatic toxicologists. Recent work has shown that Chinese rare minnow exposed to BP-A had increased expression of genes related to histone methylation and altered the distribution of histones in the testes [58]. In work done with zebrafish, BP-S and permethrin-altered genes related to both histone demethylation and deacetylation, suggesting that these compounds have the potential to initiate epigenetic modification through histone modulation. Early development is a sensitive time for an organism and environmental stress during this period can have lifelong effects [74, 75]. In Pacific oysters, suppression of Jumonji-C domain-containing proteins (which aid in histone demethylation) during development had significantly increased histone methylation associated with developmental deformities (Figure 14.4) [76]. Given the importance of histones in development and sex determination, it is likely that stressors will influence histone structure and modify histone signaling when exposure occurs during this period [77].

The mechanisms that trigger histone modification following contaminant exposure are not yet clear. Following exposure to cadmium, evidence suggests that it may replace zinc in the center of histone demethylase as they share similar chemical properties. The inhibition of histone demethylation results in an increase in H3K4me3 and H3K9me2 [78, 79]. However, mechanisms of histone dysregulation following exposure to other contaminants have largely been unexplored. Evidence does suggest that histone modification is correlated with pesticide metabolism and detoxification in plants, but the exact mechanisms by which histone modification is involved are unknown [80]. In the coming years, research on histone modification and its role in epigenetic inheritance following environmental stress is likely to grow, and these data will greatly help the field understand the specific mechanisms by which environmental stress can induce histone modification.

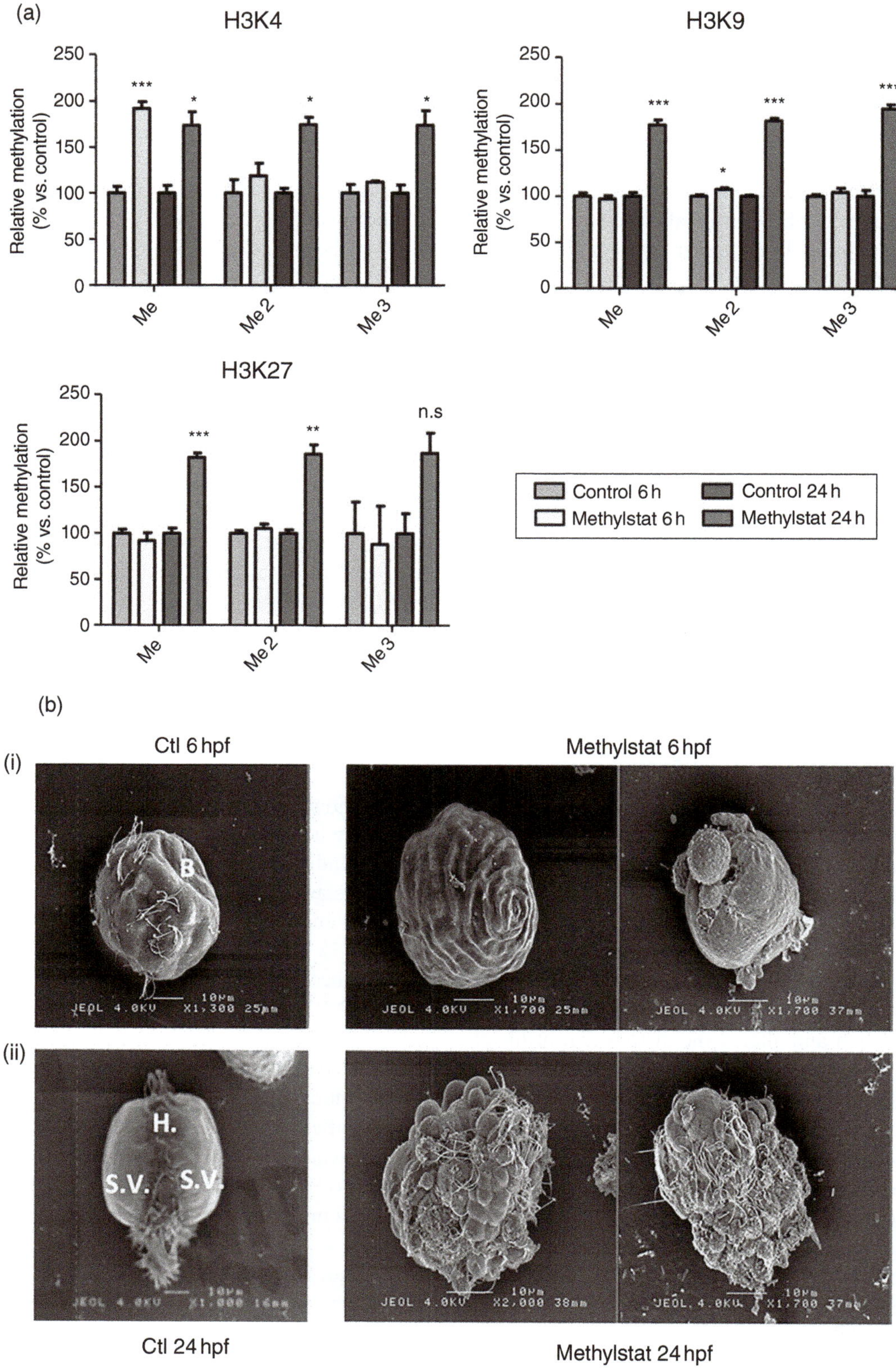

Figure 14.4 (a) Hypermethylation of histone lysine residues in the presence of 10 μM Methylstat. Global H3K4 methylation increases in the presence of methylstat, but only H3K4me1 and H3K4me3 are significant. Global H3K9 methylation increases in the presence of methylstat, but only H3K9me2 is significant. H3K27me and H3K27me2 increase in the presence of methylstat. Asterisk in the legend indicates significant variation of the indicated mark (One-way ANOVA; $p < 0.05$ was considered significant (<0.05 [*], <0.001 [**], <0.0001 [***]). (b) Abnormal development. Different phenotypes observed in control condition and under methylstat treatment at 6 hours after fertilization (i) and 24 hours after fertilization (ii). *Source:* From Fellous et al. [76]/MDPI/public domain/Licensed under CC BY 4.0.

14.2.3 Noncoding RNA in Relation to Aquatic Toxicology

ncRNAs are RNA molecules that do not translate directly into protein, but are still in regulatory and supportive roles. They are becoming obvious essential players in gene regulation and protein formation [81, 82]. We now know that ncRNAs play a role in protein production, target specific genes, and are linked to gene silencing and regulation [83, 84]. MicroRNAs (miRNA), small interfering RNAs (siRNAs), small nucleolar RNAs (snoRNAs), small nuclear RNAs (snRNAs), long-noncoding RNAs (lncRNA), and Piwi-associated RNA (piRNA), ncRNAs that code for Piwi proteins, are examples of ncRNAs that play essential, supportive roles during transcription that affect gene expression and protein production [85]. While the roles of many ncRNAs are still largely unknown. research from human health and traditional toxicology help inform ecotoxicology studies; as such, aquatic toxicology studies are beginning to use ncRNAs, like miRNA, as biomarkers and have important roles in inducing toxicity that are worth understanding [86].

lncRNAs are ncRNAs greater than 200 nucleotides in length that can be transcribed by transcription factors and preform many biologically necessary functions that are still being explored, such as alternative splicing [87, 88]. lncRNAs can serve many different functions including acting as signals for different developmental stages, acting as the transcript for smaller ncRNAs, sequestering miRNAs or DNA-binding proteins, and either guiding or preventing them from reaching their targets [89, 90]. lncRNA may also be important in the transfer of genetic material necessary for development from parents to offspring [91]. Studies in zebrafish have found that the lncRNA, *Cyrano*, is transferred to a zygote following fertilization of the oocyte and sperm, is necessary for neurodevelopment, and interacts with RNA binding proteins; without the transfer of *Cyrano*, offspring development was altered [91]. However, others have found *Cyrano* to not be necessary for embryogenesis in zebrafish, which contradicts earlier study and begs the question of when ncRNAs are necessary and under what conditions they are not [92]. Due to the growth in research showing the wide diversity of roles lncRNA plays in pre- and post-transcription, many have recently become interested in how lncRNAs can be used to better understand and detect toxicity. In invertebrates, nanopolystyrene exposure in the earthworm *Caenorhabditis elegans* caused differential expression of up to 37 lncRNA [93]. In fishes, zebrafish exposed to mercury had increased expression of the lncRNA metastasis-associated lung adenocarcinoma transcript 1 (*malat1*) in the brain, eyes, and notochord [94]. lncRNAs are involved in AHR-mediated toxicity following zebrafish exposure to polycyclic aromatic hydrocarbons (PAHs) [95]. In the large yellow croaker (*Larimichthys crocea*) regulation of lncRNA was altered, along with impaired immune response, after exposure to hypoxia, demonstrating that multiple types of environmental stress can lead to changes in lncRNAs [96, 97] (Figure 14.5). The use of lncRNA as a biological endpoint is worth studying, and its use in potential biomonitoring is still quite new and advancements are still being made.

All other ncRNAs are considered short ncRNAs (sncRNAs). miRNAs (~22 nts) cause decreased gene expression by binding to the mRNA and suppressing translation [98]. miRNAs are active participants in both translation and transcription and have been established as important biomarkers in response to environmental stress [99–101]. miRNAs are actively involved in gene silencing and changes in miRNA can have cascading effects that lead to developmental effects, disease, reduced fitness, and/or death, which are all important endpoints related to aquatic toxicology and risk assessment [102–104]. miRNA expression is altered under exposure to heat, salinity, pesticides, metals, and other environmental stressors [105–107]. It is important to note that to date, there has not been evidence to suggest altered miRNA can have a protective response in offspring and has so far only been found to be maladaptive [108]. miRNAs are well studied as far as ncRNA research goes; other sncRNAs are far less studied, especially in an aquatic toxicology context. One recent study did look at miRNAs and piRNAs in the fathead minnow (*Pimephales promelas*) piRNAs (~29 nt) are involved in the formation of piwi-clade Argonaute proteins, which are a part of the piRNA pathway and necessary for gametogenesis [109–111]. They were previously thought to only have been expressed in germ cells, but are now known to be in somatic cells as well [112]. piRNAs have also been described in process of *de novo* DNAm, further supporting their use as a molecular biomarker [113]. Following exposure to EE2 to environmentally relevant concentrations of EDCs, 2.5 and 10 ng/L, fathead minnows showed differential expression of miRNAs and piRNAs, suggesting that these two sncRNAs may be useful for biomonitoring of EDCs [114]. Other sncRNAs are likely also susceptible to environmental stress, but are far less studied, with no studies to date focused on aquatic organisms [115].

One limitation of ncRNA and aquatic toxicology studies is the lack of data and RNA library for model organisms. Regardless, there is still much to learn about the role of ncRNAs in not only toxicology and environmental response, but in biology as well. Toxicologists see potential in ncRNAs to be useful for biomonitoring and mechanistic studies in both human and aquatic health.

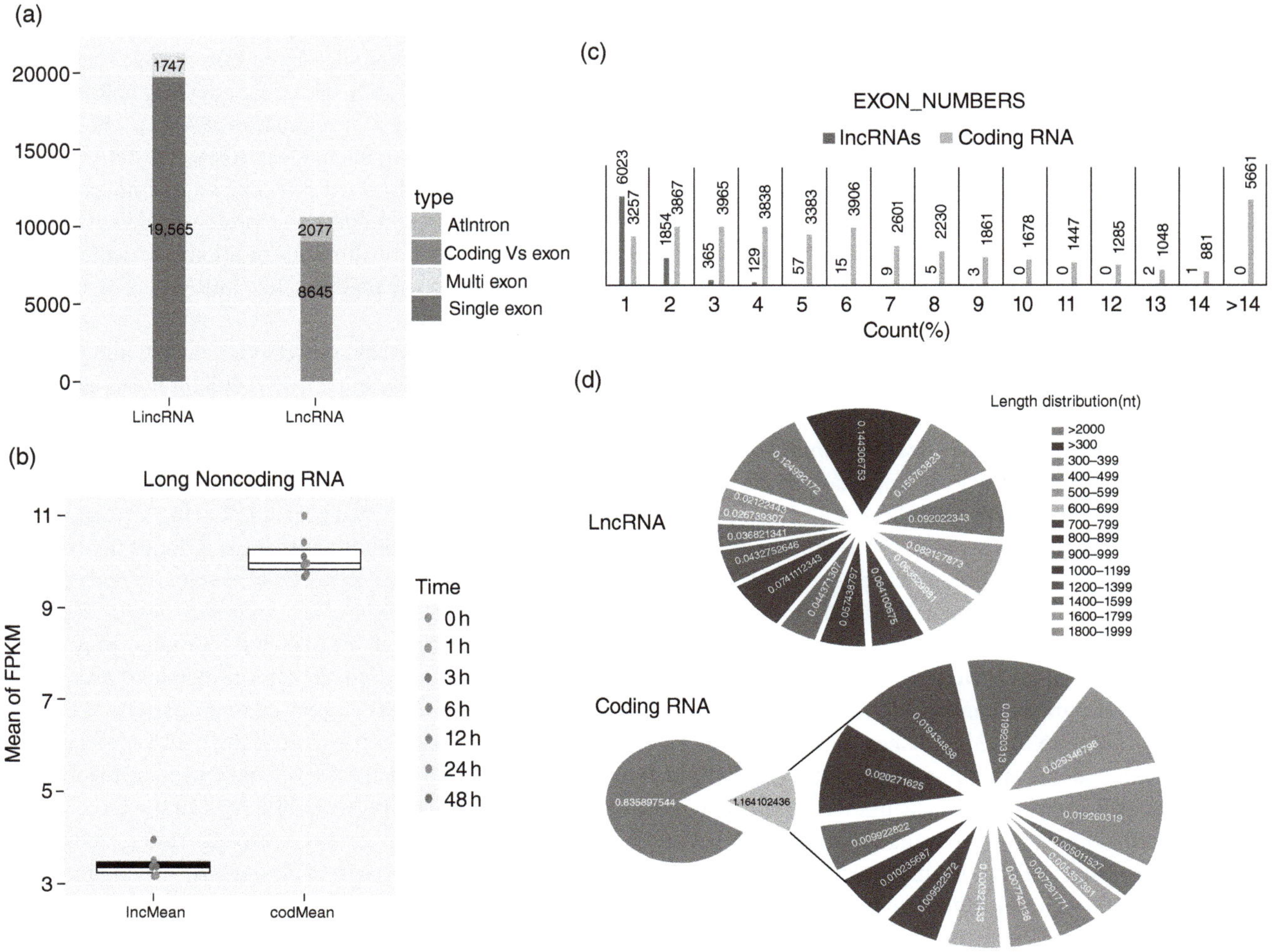

Figure 14.5 Long noncoding RNA characteristics in large yellow croaker. (a) Distribution of lncRNA classes. There were 1747 multi-exon long noncoding RNA; 19565 single-exon long noncoding RNA; 2027 long noncoding RNA generated from intron; 8645 long noncoding RNA overlapping intron. (b) Average expression level of lncRNAs and the protein-coding genes in the brain tissue. (c) Distribution of exon numbers of lncRNAs and coding RNA. (d) Ratio of length of lncRNAs and coding RNA. *Source:* Reproduced with permission from Liu et al. [96]. (*See colour plate section for colour representation of this figure.*)

14.3 Epigenetics During Early Development Related to Toxicology

Development is a critical time in the life history of an organism and biologists are still working to understand the timing and role of epigenetics during this sensitive window [116]. Exposure to toxicants during early life stages can have adverse effects (such as diabetes, cancer, decreased reproduction, and increased risk of other diseases) later in life; therefore, epigenetics is a large area of research in developmental toxicology [117, 118]. Some factors attributed to increased sensitivity in early life stages are underdeveloped organs, for example, the liver, which is involved in detoxification as well as immature immune systems and homeostasis response. Another reason is the timing of DNAm reprogramming, which plays a role in organ development and cell differentiation, making it a particularly sensitive time during embryogenesis [10]. During this time, the embryo undergoes a rapid change in global DNAm status as methylation patterns shift throughout different stages of development.

In fish, DNAm reprogramming differs depending on the species [119]. For example, Japanese medaka and zebrafish, two commonly used species in lab studies, exhibit different methylation erasure patterns and timing. Japanese medaka undergo methylation reprogramming similar to mammals, in contrast to zebrafish, which differ from most mammals, including humans. In Japanese medaka, both the paternal and maternal genomes are demethylated immediately following fertilization; in most mammals, this occurs over several cycles of cell division, then global DNAm levels increase into the gastrula stages, like in mammals [119]. In zebrafish, the methylation patterns of the early-developing embryo are similar to the sperm. Here, the maternal genome is reprogrammed to mimic the patterns of the paternal genome, which is not the case in mammals.

It appears that invertebrates exhibit altogether different methylation patterns, where the global methylation levels are much lower than those in mammals and fish. Most invertebrates have either none or very little methylation throughout their genomes [61]. Of the methylation that does occur, it is concentrated in active transcription regions throughout the gene body instead of the promoter region, like in vertebrates. This implies that in invertebrates, DNAm is not fully related to transcription and instead may have applications in invertebrate physiology yet to be fully understood [120].

It is unclear how different reprogramming patterns affect toxicity or whether one type of methylation pattern makes an organism susceptible to adverse effects later in life, but it is possible that the different patterns of methylation reprogramming play a role in determining life history traits. Methylation dysregulation during sensitive life stages may influence the degree of epigenetic inheritance due to environmental stress. The degree to which epigenetic dysregulation influences future generations is a recent interest in aquatic toxicology studies. Maternal epigenetic changes can be passed down to offspring and remain during early development. After epigenetic reprogramming in chinook salmon, it was found that maternal epigenetic changes subsided during the fry stage, but evidence of maternal effects remained throughout the duration of the second generation's life [121]. The implications for this for risk assessment are still unknown, but knowledge of the role of DNAm reprogramming will be important in understanding how early life exposures influence adult and later generation's health.

14.4 Multigenerational and Transgenerational Toxicology

Multigenerational and transgenerational studies describe those generations, other than the parent, that are either indirectly exposed or not exposed at all, respectively (Figure 14.6) [33, 122]. In oviparous organisms, the F2 generation will be the first to be unexposed to a stressor, while viviparous organisms will still be indirectly exposed in the F2 generation as germ cells from the F0s (Figure 14.6). Transgenerational toxicity occurs when environmental stress alters epigenetic factors that are inherited through the germ line and passed down to unexposed generations [123]. For example, in earthworms, DNA replication was disrupted during embryogenesis, causing changes to chromatin structure for up to five generations before it was restored [124]. Similar types of molecular disruption can occur following environmental stress. When an

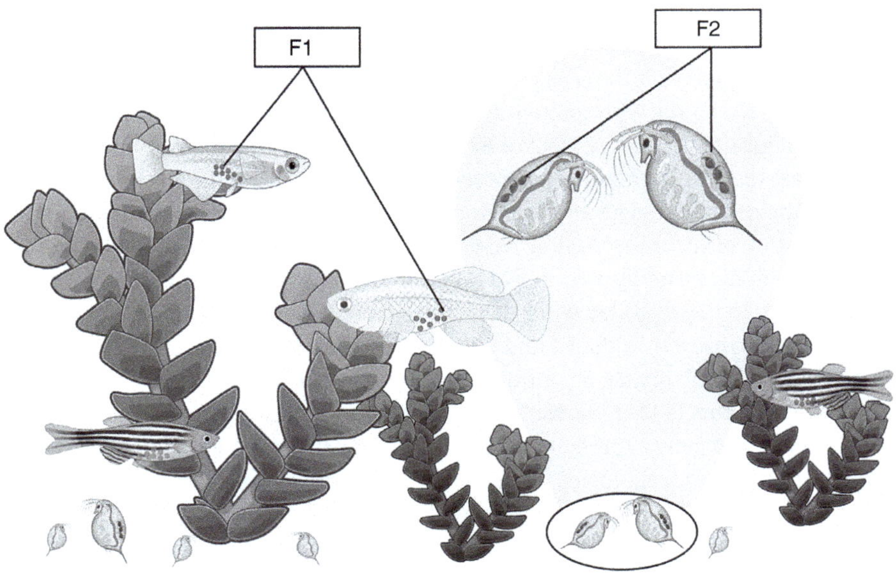

Figure 14.6 True transgenerational exposure occurs in the F2 and F3 generations of oviparous and viviparous species. Most fish are oviparous, meaning they produce eggs that develop outside of the parent. In oviparous species, the F1 generation is indirectly exposed as germ cells. Cladoceran species (e.g., *Ceriodaphnia dubia* and *Daphnia magna*), mainly reproduced through parthenogenesis, are viviparous. In viviparous organisms, the F2 generation germ cells are also indirectly exposed during F1 generation development in the brood pouch. *Source:* From © BioRender. (*See colour plate section for colour representation of this figure.*)

organism is exposed to environmental stress, its germ cells can be affected, leading to changes in epigenetic inheritance and mechanisms that impact future generations [125, 126]. Multigenerational studies are more common than transgenerational studies in aquatic toxicology, particularly in nonmodel organisms [127–131]. Model organisms make multigenerational and transgenerational studies more feasible since we have a better understanding of their life history, their ease of culturing and reproducing, and generally greater availability of their genomes, which is particularly important in studies on epigenetics (i.e., DNAm and ncRNA) [3, 132–136].

Understanding how epigenetic mechanisms drive transgenerational toxicity is critical to better comprehending the mechanisms contributing to transgenerational effects. Molecular mechanisms (i.e., DNAm, histone modification, and ncRNA) that are involved in epigenetic inheritance are used in aquatic toxicology studies to help assess and better understand the mechanisms behind multi- and transgenerational effects of environmental stress [3, 42, 136–138]. Transgenerational, epigenetic experiments are particularly relevant to aquatic toxicology because longer-term experiments are often the most environmentally realistic and can provide insight into the potential implications to aquatic populations and communities. For example, methyl mercury (MeHg) is a naturally occurring toxicant that results from the anerobic metabolism of mercury and is readily bioaccumulated up the food chain. Mercury not only occurs in the environment naturally, but has also been a byproduct of gold mining and coal burning that pollute waterways around the globe, making it a pollutant of concern. MeHg is well known for its long-term neurotoxic effects. In zebrafish, MeHg caused transgenerational toxicity associated with epigenetic changes in sperm; the sperm of F2 individuals had increased DMG associated with neurological functions as well as hypoactive swimming, showing that paternal epigenetic modifications following exposure to MeHg can be passed down to future generations [139]. The neurotoxic effects of MeHg can persist past the initially exposed generation, and the prolonged effects of MeHg may be in part due to changes in DNAm patterns caused by the direct exposure in past generations [139]. In another example, chlorpyrifos, an OP used in aquaculture and agriculture, is a known neurotoxin that inhibits acetylcholinesterase (AChE) activity. In zebrafish, it was found that exposure to chlorpyrifos toxic metabolite, chlorpyrifos-oxon, in developing larvae induced hypomethylation and hypermethylation in the F1 and F2 progeny larvae, respectively [140]. Additionally, the F2, unexposed generation showed increased AChE activity and hyperactivity relative to F2 controls, whereas the F0s showed no changes in AChE and hypoactivity relative to F0 controls. A study of other OPs in zebrafish found tris(1,3-dichloro-2-propyl) phosphate caused hypermethylation of genes involved in thyroid hormone transportation in F1 indirectly exposed larvae [141]. This shows a link between epigenetic changes and changes in hormone levels that contribute to neurotoxicity following indirect OP exposure. To better understand the role of epigenetics in modulating neuroactivity, more in-depth studies are required.

Exposure to EDCs, which were introduced earlier, influences apical endpoints such as reproduction, sex ratio, and embryo hatching in fish, and has the potential to cause population-level decline [42, 142]. In Japanese medaka, reproduction was decreased up through the F4 generation following exposure to atrazine [143]. Additionally, EDCs can induce changes in DNAm in unexposed grad-offspring (F2) [3, 42]. In laboratory studies, EDCs have been found to alter DNAm in the unexposed generation in correlation with altered apical endpoints like reproduction, development, and other life-history traits [144]. Transgenerational effects of DNAm in inland silverside, after exposure to EDCs, showed that DNAm can be passed down to at least the F2 generation (unexposed), following exposure in F0s for 21 days [42]. Different genes have shown differential methylation across generations as well and even sometimes skip generations. For example, *tgfb*, which transcribes the production of the transcription growth factor protein, was hypomethylated in F0s exposed to EDCs, not changed in the F1s, but then hypermethylated in F2s [42]. Interestingly, gene expression for *tgfb* was not dysregulated despite changes in methylation. In an expanded study of inland silversides exposed to EDCs in the early life stages, it was found that differential DNAm was greater in the F1 and F2 generations compared to F0 [3]. In another example, zebrafish were exposed to both hypoxia and venlafaxine, an antidepressant commonly found in wastewater [145]. Hypoxia and venlafaxine are both endocrine disruptors [146–148]. Here, miRNA expression was altered through the F1 generation, but not the F2. This is of note as changes in gene expression of heat shock proteins were found through to the F2. It was unclear from the study what other epigenetic mechanisms were involved in carrying the changes in gene expression down to the unexposed generations. However, it was clear that altered miRNA expression was passed down to at least the F1, and other mechanisms must be involved for gene expression to still be altered in the F2. Outside of the laboratory, it was not shown until 2007 that EDCs, such as estrogen 17α-ethynylestradiol (EE2) used in birth control, could alter the structure of wild populations. Use of the Experimental Lakes Area in Canada allowed for the controlled addition of EE2 to a natural aquatic system, where the fathead minnow populations severely declined and reproduction almost halted, leading to a near extinction event [142]. Clearly, EDCs have transgenerational effects, but we are just beginning to understand what role epigenetic modification plays in inheritance of the effects.

Several invertebrates are also used in aquatic toxicology studies and are invaluable model organisms for studying multi- and transgenerational toxicity. As mentioned earlier, studies using earthworms can provide useful pieces of the puzzle to better understand epigenetic inheritance and are emerging as an important model organism in epigenetic studies [149, 150]. In oysters, exposure to diuron caused hypermethylation in both the F0 and F1 generations [151]. The water flea *Daphnia magna* is another invertebrate species commonly used in both regulatory and research studies [135, 152, 153]. Early life stress to high salinity in water flea F0s showed evidence of hypomethylation in the F0 exposed organisms, through to the unexposed F3 offspring. This was correlated with a significant increase in reproduction in subsequent generations [154]. Water fleas exposed to radium had the greatest level of DMGs in the F2 generation (indirectly exposed as germ cells) and the least in the F3s (unexposed) [155]. Many invertebrates used are either asexual and/or reproduce reliably as a culture, making them ideal for long-term, multi-, and transgenerational studies. Invertebrates are used extensively as biomarkers in both freshwater and marine environments and are essential to the food chain, making them ecologically relevant as well.

A range of chemical classes have been found to induce multi- and transgenerational toxicity, including PAHs, metals, pesticides, and pharmaceuticals among others [156, 157]. The extent that changes in epigenetic mechanisms induced by environmental stress is adaptive or maladaptive remains to be understood. Future transgenerational studies should consider investigating the adaptive potential of organisms in unexposed generations to the original stressor to better understand how epigenetic modification is involved in adaptation and tolerance [158]. Multi- and transgenerational studies help scientists understand the role of epigenetic mechanisms in evolutionary responses to pollution and other environmental stressors since we are not yet able to predict toxicity in later generations.

14.5 Epigenetics in Ecological Risk Assessment

The use of epigenetics in toxicology has become common in human clinical disease and public health studies where research areas such as personalized medicine and individual exposure risk have become popular [159]. Aquatic toxicologists are often interested in the overall health of a community and whether it can support human and wildlife activity. Epigenetics play a pivotal role in helping aquatic toxicologists understand the mechanisms of toxicity following exposure to pollution that may be putting organisms and ecosystems at risk [33, 160]. “Footprints” from past environmental exposure to chemicals are considered a potential route of interest in epigenetics and ecotoxicology [25]. A current challenge in ecotoxicology is making these connections with epigenetic data due to the limited research in this field. Recent work using laboratory and field data to predict population success aims to address these gaps and challenges [161]. Establishing models that can link data from molecular (i.e., epigenetic mechanisms) to apical endpoints (i.e., reproduction, growth, and survival) will be important to move the field forward. Relating molecular endpoints to entire populations remains a topic of exploration as the ecotoxicology discipline moves through the twenty-first century and more efforts are focused on increasing in silico and in vitro work to decrease reliance on whole animal testing [162, 163]. Challenges remain in relating epigenetics to aquatic and ecotoxicology as many data gaps exist, but efforts to incorporate epigenetics into risk assessment are growing.

Much of the work done in ecotoxicological risk assessment is focused on the effects on nontarget organisms and unintended toxicity. Epigenetic endpoints are promising tools for biomonitoring and risk assessment and have key roles in determining MOA [164]. The use of molecular endpoints in biomonitoring and risk assessment is still in its infancy and many unknowns exist that make it challenging to incorporate epigenetic biomarkers into risk assessment. For example, while it is known that there is a correlation between DNAm and increased risk of adverse apical outcomes, it is not known how severe or to what degree DMGs contribute to endpoints relevant to risk assessment. Weak and inconsistent correlations between molecular changes and adverse outcomes (AOs) make it challenging to use epigenetic data alone to predict more hierarchical endpoints, such as reproduction and survival. Additionally, risk assessors frequently rely on high-quality data spanning multiple concentrations of exposure to establish a dose–response relationship between the contaminant and exposed population. Due to the high cost of molecular sequencing, it has historically been challenging to produce high-quality epigenetic data to study toxicity in a dose–response manner. Nonetheless, it is promising to see interest in incorporating epigenetic molecular markers in risk assessment grow in recent years, and it is important to remember that even responses to single concentrations can be used in the assessment of risk via approaches such as species sensitivity distributions if NOECs or LOECs are incorporated to determine thresholds (e.g., Belanger et al. [165] and Adam et al. [166]).

First proposed in 2010, an AO pathway (AOP) follows the biological reactions induced by exposure to environmental stress. The AOP explains the cascade of reactions and effects that traverse the biological hierarchy following environmental stress. AOPs connect causal events to establish a biological-based framework for connecting molecular-initiating events (MIEs) and an apical, AO. Evidence show that those epigenetic modifications may be a part of key events (KEs) that link the MIE and AO. In this framework, exposure to a toxin can affect all levels of the biological hierarchy from the molecular (i.e., DNAm) to the population (i.e., reproduction and/or immune response). One example of epigenetic mechanisms being used in an AOP-type framework is through arsenic toxicity. Arsenic metabolism shares the same methyl donor as DNAm, *S-adenosyl methionine* (SAM), which may lead to a prolonged decrease in SAM availability if the exposure is chronic. Decreased availability of SAM may lead to hypomethylation due to fewer available methyl groups [167]. Arsenic exposure may lead to cancer, diabetes, other diseases, or even death. Epigenetic mechanisms show promise in helping to better understand molecular pathways involved AOPs (Figure 14.2). With the AOP framework, we can see a clear example through arsenic of how new knowledge of epigenetic mechanisms can fit into pre-existing data and help inform risk of specific compounds.

14.6 Rapid Evolution

Epigenetic inheritance allows for the passage of genetic information from parent to offspring without direct changes in DNA sequence. While this does not fall under the umbrella of traditionally described evolutionary mechanisms, this phenomenon has often been referred to as "rapid evolution," as it increases the resistance of an organism's offspring and subsequent generations. It is hypothesized that epigenetic mechanisms play a role in rapid evolution. A unified theory of evolution describes the role of the environment in genotypic variation through epigenetic modifications. Inheritance of epigenetic modifications can alter both genotypic and phenotypic variation of future generations and has even been shown to play a role in speciation [122, 168]. The Atlantic killifish (*Fundulus heteroclitus*) is one of the first species to show an evolutionary adaptive response to pollution that has enabled it to persist even in highly polluted waters [169]. Killifish living in water polluted with polychlorinated biphenyl (PCB) exhibit decreased global DNAm and expression of *dnmt1*, as well as increased expression of Ten-Eleven Translocation (TET) genes, which are involved in demethylation, compared to populations living in unpolluted waters. The relationship between gene expression, epigenetic modification, and inheritance has been proposed as a route for rapid evolution, while literature is still sparse in terms of understanding the specific epigenetic mechanism that led to the type of adaptive inheritance required for rapid evolution, there is evidence to support its role [33, 145].

Recent studies suggest epigenetic mechanisms are involved in inherited changes in gene expression, for example, DNAm can lead to increased resistance in future generations [170]. Zebrafish exposed to crude oil showed adaptive responses as soon as the F1 generation. Interestingly, it was found that F1 performed better in water contaminated with crude oil than in clean water, potentially hinting at the likelihood of tradeoffs in adaptive responses [171]. Another example of rapid evolution is in seagrasses previously exposed to heat stress. Those exposed to heat were more tolerant than those not pre-exposed or "primed" during a second heat wave simulation [172]. An analysis of gene expression related to methylation, stress, and photosynthesis found that methylation-related genes were more closely related to heat stress than the others [172]. Although the study on seagrass was not multigenerational, the high correlation between methylation-related genes and thermal tolerance suggests the role of epigenetic modification in environmental stress memory for seagrasses. Future studies of assessing how these modifications manifest in later generations and contribute to generational tolerance to heat and other stressors will be informative to better understand the role of epigenetics in rapid evolution.

Climate change poses the question of whether adaptive epigenetic modification and phenotypic plasticity will be enough to prevent extinction for some species [5]. Early evidence does suggest that epigenetic mechanisms, like DNAm, can increase phenotypic plasticity and rapid evolution following climate-related stressors in grasslands [173]. For example, Pacific oysters from separate tidal areas showed differential responses to temperature correlated with DMGs, where it was suggested that phenotypic plasticity from DMG allowed for rapid adaptation to heat stress [11]. Under climate change scenarios, invasive species are more likely to do well than native species due to their increased propensity for adaptation and rapid evolution [174]. Invasive species show similar effects following prolonged exposure to pesticides used in agriculture. It is worth mentioning that species that have demonstrated rapid evolution and tolerance to pesticides can help drive increased use of pesticides [3], which is expected to increase under climate change scenarios. Additionally, we should note that organisms found to have resistance to different stressors may be making a tradeoff for survival by reducing resources and energy normally used for things such as reproduction and immune response [175]. Climate change will put pressure on species to quickly evolve to survive; this will likely lead to the spread of persistence of invasive, non-native species, while more sensitive, native species that have evolved to only survive in a particular niche will be less successful and risk extinction.

14.7 Epigenetics in Aquaculture

The growing challenge to produce enough food to feed the world's population has expanded the use of aquaculture farms [176]. Sea lice are ectoparasites that have been a common problem in salmon aquaculture for years [177]. Pyrethroids and OPs, two compounds discussed throughout this chapter, and avermectins and hydrogen peroxide are routinely used to treat sea lice [178]. One study found a significant correlation between gene expression and phenotypic variability in sea lice following exposure to azamethiphos, cypermethrin, and deltamethrin [179]. lncRNA expression was also linked to the insecticide exposures and was associated with different susceptibility to azamethiphos in different sea-lice populations. This study demonstrates the use of epigenetic markers to improve understanding of both sea-lice resistance and insecticide toxicity [180].

A risk assessment of black tiger shrimp (*Penaeus monodon*) farmed off the coast of Queensland, Australia, found they were sensitive to fipronil, bifenthrin, and imidacloprid. These insecticides were also found in the intake waters of the aquaculture farm [181]. The shrimp were especially sensitive to fipronil and bifenthrin, which were discussed in terms of DNAm effects on fishes in Section 14.2.1. Limited research on epigenetics and invertebrates makes it challenging to compare across these species. However, in the future, data on what epigenetic changes lead to endpoints relevant to risk assessment will help improve the use of these markers. This further highlights the discrepancy in studies on epigenetic modification in invertebrate species, which will be necessary if aquaculture and risk assessment are to incorporate epigenetics into their framework.

Biomonitoring for signs of biochemical and physiological changes helps determine early signs of stress, but epigenetic approaches in aquaculture species are not as common compared to environmental studies [182–184]. Using epigenetics as a biomonitoring endpoint for aquaculture will require a greater understanding of how commonly used compounds impact aquaculture species. Similar to the trend of increased use of epigenetic endpoints in aquatic toxicology studies, the use of epigenetics in studying and monitoring aquaculture health is likely to increase as costs become less prohibitive. Studying a wide range of organisms can provide more data points to commonly used risk assessment tools, such as species sensitivity distributions.

Box 14.1 Highlights of Epigenetics in Aquatic Toxicology

- Epigenetic studies provide insight into implications for toxic exposure past the initially exposed generation, which is important for risk assessment [44].
- Epigenetic mechanisms can be used in the development of AOP [167].
- Epigenetic mechanisms may lead to rapid evolution and increased tolerance to some contaminants [33].
- Epigenetic approaches in aquaculture species are not as common compared to environmental studies [183, 184].
- Early efforts to incorporate epigenetic endpoints into aquatic toxicology demonstrate that it may be a useful tool in aquaculture biomonitoring [114, 183, 184].

14.8 Conclusion and Perspectives

Epigenetics is a growing field of research within the aquatic toxicology realm (Box 14.1). While the consideration of epigenetic endpoints in aquatic toxicology studies is growing, data are still quite limited relative to human health studies, and more research is needed to better connect changes at the molecular, epigenetic level to higher level apical endpoints such as reproduction, growth, and survival. However, several commonly used aquatic toxicology model species such as zebrafish *Danio rerio* and inland silversides are becoming more prominent in epigenetic literature. Epigenetic dysregulation, because of environmental stress and toxicity, provides insight into the mechanisms of toxicity and the potential heritability of AOs past the F0 exposed generation. These studies can provide insight into the potential implications for toxic exposure past the initially exposed generation, and are important for assessing risk to a population and ecosystem. A stronger understanding of how epigenetic modifications drive adverse effects is needed. As the exact mechanisms involved in epigenetic-related toxicity in human studies is still limited, the work in ecotoxicology is still in its infancy. Understanding how toxins interact with epigenetics and how epigenetic mechanisms influence response to toxic injury will help biomonitoring, and field and laboratory studies deliver on more high-throughput studies. Epigenetics will continue to develop as an important part of aquatic and ecotoxicology studies as knowledge regarding the utility of epigenetic changes as biomarkers expands and our understanding of epigenetic mechanisms in response to stressors continues to grow [163].

References

1 Smith, C.M., Vera, M.K., and Bhandari, R.K. (2019). Developmental and epigenetic effects of roundup and glyphosate exposure on Japanese medaka (*Oryzias latipes*). *Aquatic Toxicology* **210**: 215–226.

2 Portnoy, D.S., Fields, A.T., Greer, J.B., and Schlenk, D. (2020). Genetics and oil: transcriptomics, epigenetics, and population genomics as tools to understand animal responses to exposure across different time scales. In: *Deep Oil Spills*, 515–532.

3 Major, K.M., DeCourten, B.M., Li, J. et al. (2020). Early life exposure to environmentally relevant levels of endocrine disruptors drive multigenerational and transgenerational epigenetic changes in a fish model. *Frontiers in Marine Science* 7. https://doi.org/10.3389/fmars.2020.00471.

4 Asselman, J., De Coninck, D.I., Beert, E. et al. (2017). Bisulfite sequencing with *Daphnia* highlights a role for epigenetics in regulating stress response to *Microcystis* through preferential differential methylation of serine and threonine amino acids. *Environmental Science and Technology* **51** (2): 924–931.

5 McGuigan, K., Hoffmann, A.A., and Sgrò, C.M. (1826). How is epigenetics predicted to contribute to climate change adaptation? What evidence do we need? *Philosophical Transactions of the Royal Society, B: Biological Sciences* **2021** (376): 20200119.

6 Brockmeier, E.K., Hodges, G., Hutchinson, T.H. et al. (2017). The role of omics in the application of adverse outcome pathways for chemical risk assessment. *Toxicological Sciences* **158** (2): 252–262.

7 Ullah, S., Li, Z., Zuberi, A. et al. (2019). Biomarkers of pyrethroid toxicity in fish. *Environmental Chemistry Letters* **17** (2): 945–973.

8 Zbinden, G. and Flury-Roversi, M. (1981). Significance of the LD 50-test for the toxicological evaluation of chemical substances. *Archives of Toxicology* **47** (2): 77–99.

9 Brander, S.M., Connon, R.E., He, G. et al. (2013). From 'omics to otoliths: responses of an estuarine fish to endocrine disrupting compounds across biological scales. *PLoS One* **8** (9): e74251.

10 Barouki, R., Melén, E., Herceg, Z. et al. (2018). Epigenetics as a mechanism linking developmental exposures to long-term toxicity. *Environment International* **114**: 77–86.

11 Wang, X., Li, A., Wang, W. et al. (2021). DNA methylation mediates differentiation in thermal responses of Pacific oyster (*Crassostrea gigas*) derived from different tidal levels. *Heredity* **126** (1): 10–22.

12 Hutchinson, T.H. (2007). Small is useful in endocrine disrupter assessment – four key recommendations for aquatic invertebrate research. *Ecotoxicology* **16** (1): 231–238.

13 Crosby, D.G. (1998). *Environmental Toxicology and Chemistry*. New York: Oxiford University.

14 Pá, D. (2002). The cytochrome P450 superfamily: biochemistry, evolution and drug metabolism in humans. *Current Drug Metabolism* **3** (6): 561–597.

15 Park, H.-J., Choi, Y.-J., Kim, J.W. et al. (2015). Differences in the epigenetic regulation of cytochrome P450 genes between human embryonic stem cell-derived hepatocytes and primary hepatocytes. *PLoS One* **10** (7): e0132992.

16 Li, D., Tolleson, W.H., Yu, D. et al. (2019). Regulation of cytochrome P450 expression by microRNAs and long noncoding RNAs: epigenetic mechanisms in environmental toxicology and carcinogenesis. *Journal of Environmental Science and Health, Part C* **37** (3): 180–214.
17 Pande, P., Zhong, X.-b., and Ku, W.W. (2020). Histone methyltransferase G9a regulates expression of nuclear receptors and cytochrome P450 enzymes in HepaRG cells at basal level and in fatty acid induced steatosis. *Drug Metabolism and Disposition* **48** (12): 1321–1329.
18 Taneja, G., Maity, S., Jiang, W. et al. (2019). Transcriptomic profiling identifies novel mechanisms of transcriptional regulation of the cytochrome P450 (Cyp) 3a11 gene. *Scientific Reports* **9** (1): 1–13.
19 Nebert, D.W. and Vasiliou, V. (2004). Analysis of the glutathione S-transferase (GST) gene family. *Human Genomics* **1** (6): 1–5.
20 Landi, S., Gemignani, F., Moreno, V. et al. (2005). A comprehensive analysis of phase I and phase II metabolism gene polymorphisms and risk of colorectal cancer. *Pharmacogenetics and Genomics* **15** (8): 535–546.
21 Xu, C., Li, C.Y.-T., and Kong, A.-N.T. (2005). Induction of phase I, II and III drug metabolism/transport by xenobiotics. *Archives of Pharmacal Research* **28** (3): 249–268.
22 Chen, L., Hong, C., Chen, E.C. et al. (2013). Genetic and epigenetic regulation of the organic cation transporter 3, SLC22A3. *The Pharmacogenomics Journal* **13** (2): 110–120.
23 Peng, L. and Zhong, X. (2015). Epigenetic regulation of drug metabolism and transport. *Acta Pharmaceutica Sinica B* **5** (2): 106–112.
24 Nebert, D.W. and Dalton, T.P. (2006). The role of cytochrome P450 enzymes in endogenous signalling pathways and environmental carcinogenesis. *Nature Reviews Cancer* **6** (12): 947–960.
25 Jeremias, G., Barbosa, J., Marques, S.M. et al. (2018). Synthesizing the role of epigenetics in the response and adaptation of species to climate change in freshwater ecosystems. *Molecular Ecology* **27** (13): 2790–2806.
26 Nobre, R.L.G., Caliman, A., Cabral, C.R. et al. (2020). Precipitation, landscape properties and land use interactively affect water quality of tropical freshwaters. *Science of the Total Environment* **716**: 137044.
27 Kamstra, J.H., Aleström, P., Kooter, J.M., and Legler, J. (2015). Zebrafish as a model to study the role of DNA methylation in environmental toxicology. *Environmental Science and Pollution Research* **22** (21): 16262–16276.
28 Bhattacharya, S., Fontaine, A., MacCallum, P.E. et al. (2019). Stress across generations: DNA methylation as a potential mechanism underlying intergenerational effects of stress in both post-traumatic stress disorder and pre-clinical predator stress rodent models. *Frontiers in Behavioral Neuroscience* **13**: 113.
29 Olsvik, P.A., Whatmore, P., Penglase, S.J. et al. (2019). Associations between behavioral effects of bisphenol A and DNA methylation in zebrafish embryos. *Frontiers in Genetics* **10**: 184.
30 Lakstygal, A.M., de Abreu, M.S., and Kalueff, A.V. (2018). Zebrafish models of epigenetic regulation of CNS functions. *Brain Research Bulletin* **142**: 344–351.
31 Balasubramanian, S., Raghunath, A., and Perumal, E. (2019). Role of epigenetics in zebrafish development. *Gene* **718**: 144049.
32 Blanc, M., Rüegg, J., Scherbak, N., and Keiter, S.H. (2019). Environmental chemicals differentially affect epigenetic-related mechanisms in the zebrafish liver (ZF-L) cell line and in zebrafish embryos. *Aquatic Toxicology* **215**: 105272.
33 Brander, S.M., Biales, A.D., and Connon, R.E. (2017). The role of epigenomics in aquatic toxicology. *Environmental Toxicology and Chemistry* **36** (10): 2565–2573.
34 Shenker, N. and Flanagan, J. (2012). Intragenic DNA methylation: implications of this epigenetic mechanism for cancer research. *British Journal of Cancer* **106** (2): 248–253.
35 Bordoni, L., Nasuti, C., Mirto, M. et al. (2015). Intergenerational effect of early life exposure to permethrin: changes in global DNA methylation and in Nurr1 gene expression. *Toxics* **3** (4): 451–461.
36 Cocci, P., Mosconi, G., Bracchetti, L. et al. (2018). Investigating the potential impact of polycyclic aromatic hydrocarbons (PAHs) and polychlorinated biphenyls (PCBs) on gene biomarker expression and global DNA methylation in loggerhead sea turtles (*Caretta caretta*) from the Adriatic Sea. *Science of the Total Environment* **619**: 49–57.
37 Skinner, M.K., Maamar, M.B., Sadler-Riggleman, I. et al. (2018). Alterations in sperm DNA methylation, non-coding RNA and histone retention associate with DDT-induced epigenetic transgenerational inheritance of disease. *Epigenetics & Chromatin* **11** (1): 1–24.
38 Qian, Y., Ji, C., Yue, S., and Zhao, M. (2019). Exposure of low-dose fipronil enantioselectively induced anxiety-like behavior associated with DNA methylation changes in embryonic and larval zebrafish. *Environmental Pollution* **249**: 362–371.
39 Wan, Z.Y., Xia, J.H., Lin, G. et al. (2016). Genome-wide methylation analysis identified sexually dimorphic methylated regions in hybrid tilapia. *Scientific Reports* **6** (1): 1–14.

40 Ball, M.P., Li, J.B., Gao, Y. et al. (2009). Targeted and genome-scale strategies reveal gene-body methylation signatures in human cells. *Nature Biotechnology* **27** (4): 361–368.

41 McGaughey, D.M., Abaan, H.O., Miller, R.M. et al. (2014). Genomics of CpG methylation in developing and developed zebrafish. *G3: Genes, Genomes, Genetics* **4** (5): 861–869.

42 DeCourten, B.M., Forbes, J.P., Roark, H.K. et al. (2020). Multigenerational and transgenerational effects of environmentally relevant concentrations of endocrine disruptors in an estuarine fish model. *Environmental Science & Technology* **54** (21): 13849–13860.

43 Jung, J.-H., Kim, M., Yim, U.H. et al. (2011). Biomarker responses in pelagic and benthic fish over 1 year following the Hebei Spirit oil spill (Taean, Korea). *Marine Pollution Bulletin* **62** (8): 1859–1866.

44 Beal, A.P., Hackerott, S., Franks, B. et al. (2021). Epigenetic responses in juvenile Lemon sharks (*Negaprion brevirostris*) during a coastal dredging episode in Bimini, Bahamas. *Ecological Indicators* **127**: 107793.

45 Perera, B.P., Svoboda, L.K., and Dolinoy, D.C. (2019). Genomic tools for environmental epigenetics and implications for public health. *Current Opinion in Toxicology* **18**: 27–33.

46 Mahna, D., Puri, S., and Sharma, S. (2021). DNA methylation modifications: mediation to stipulate pesticide toxicity. *International Journal of Environmental Science and Technology* **18** (2): 531–544.

47 Gao, D., Lin, J., Ou, K. et al. (2018). Embryonic exposure to benzo (a) pyrene inhibits reproductive capability in adult female zebrafish and correlation with DNA methylation. *Environmental Pollution* **240**: 403–411.

48 Colborn, T., Vom Saal, F.S., and Soto, A.M. (1993). Developmental effects of endocrine-disrupting chemicals in wildlife and humans. *Environmental Health Perspectives* **101** (5): 378–384.

49 Hayes, T.B., Collins, A., Lee, M. et al. (2002). Hermaphroditic, demasculinized frogs after exposure to the herbicide atrazine at low ecologically relevant doses. *Proceedings of the National Academy of Sciences* **99** (8): 5476–5480.

50 Weston, D.P. and Lydy, M.J. (2010). Urban and agricultural sources of pyrethroid insecticides to the Sacramento-San Joaquin delta of California. *Environmental Science & Technology* **44** (5): 1833–1840.

51 Horak, I., Horn, S., and Pieters, R. (2021). Agrochemicals in freshwater systems and their potential as endocrine disrupting chemicals: a South African context. *Environmental Pollution* **268**: 115718.

52 Kuivila, K. and Hladik, M. (2008). Understanding the occurrence and transport of current-use pesticides in the San Francisco estuary watershed. *San Francisco Estuary and Watershed Science* **6** (3): 1–19.

53 Laurent, L., Wong, E., Li, G. et al. (2010). Dynamic changes in the human methylome during differentiation. *Genome Research* **20** (3): 320–331.

54 Sarda, S., Zeng, J., Hunt, B.G., and Yi, S.V. (2012). The evolution of invertebrate gene body methylation. *Molecular Biology and Evolution* **29** (8): 1907–1916.

55 Bewick, A.J., Zhang, Y., Wendte, J.M. et al. (2019). Evolutionary and experimental loss of gene body methylation and its consequence to gene expression. *G3: Genes, Genomes, Genetics* **9** (8): 2441–2445.

56 Voisin, A.-S., Suarez Ulloa, V., Stockwell, P. et al. (2021). Genome-wide DNA methylation of the liver reveals delayed effects of early-life exposure to 17-α-ethinylestradiol in the self-fertilizing mangrove rivulus. *Epigenetics* **17** (5): 1–25.

57 Ben Maamar, M., Beck, D., Nilsson, E.E. et al. (2021). Epigenome-wide association study for glyphosate induced transgenerational sperm DNA methylation and histone retention epigenetic biomarkers for disease. *Epigenetics* **16** (10): 1–18.

58 Zhu, L., Yuan, C., Wang, M. et al. (2020). Bisphenol A-associated alterations in DNA and histone methylation affects semen quality in rare minnow *Gobiocypris rarus*. *Aquatic Toxicology* **226**: 105580.

59 Nogueira, A.F., Pinto, G., Correia, B., and Nunes, B. (2019). Embryonic development, locomotor behavior, biochemical, and epigenetic effects of the pharmaceutical drugs paracetamol and ciprofloxacin in larvae and embryos of *Danio rerio* when exposed to environmental realistic levels of both drugs. *Environmental Toxicology* **34** (11): 1177–1190.

60 Liu, Y., Yuan, C., Chen, S. et al. (2014). Global and cyp19a1a gene specific DNA methylation in gonads of adult rare minnow *Gobiocypris rarus* under bisphenol A exposure. *Aquatic Toxicology* **156**: 10–16.

61 Xu, X., Li, G., Li, C. et al. (2019). Evolutionary transition between invertebrates and vertebrates via methylation reprogramming in embryogenesis. *National Science Review* **6** (5): 993–1003.

62 Sussarellu, R., Lebreton, M., Rouxel, J. et al. (2018). Copper induces expression and methylation changes of early development genes in *Crassostrea gigas* embryos. *Aquatic Toxicology* **196**: 70–78.

63 Venkataraman, Y.R., Downey-Wall, A.M., Ries, J. et al. (2020). General DNA methylation patterns and environmentally-induced differential methylation in the eastern oyster (*Crassostrea virginica*). *Frontiers in Marine Science* **7**: 225.

64 Dimond, J.L. and Roberts, S.B. (2016). Germline DNA methylation in reef corals: patterns and potential roles in response to environmental change. *Molecular Ecology* **25** (8): 1895–1904.

65 Han L, Sun Y, Cao Y, Gao P, Quan Z, Chang Y, et al. Analysis of the gene transcription patterns and DNA methylation characteristics of triploid sea cucumbers (*Apostichopus japonicus*). *Scientific Reports* 2021;**11**(1):1-15.

66 Bannister, A.J. and Kouzarides, T. (2011). Regulation of chromatin by histone modifications. *Cell Research* **21** (3): 381–395.

67 Oki, M., Aihara, H., and Ito, T. (2007). Role of histone phosphorylation in chromatin dynamics and its implications in diseases. *Chromatin and Disease* **41**: 323–340.

68 Hyun, K., Jeon, J., Park, K., and Kim, J. (2017). Writing, erasing and reading histone lysine methylations. *Experimental & Molecular Medicine* **49** (4): e324.

69 Seeler, J.-S. and Dejean, A. (2003). Nuclear and unclear functions of SUMO. *Nature Reviews Molecular Cell Biology* **4** (9): 690–699.

70 Aquila, L. and Atanassov, B.S. (2020). Regulation of histone ubiquitination in response to DNA double strand breaks. *Cell* **9** (7): 1699.

71 Nakayama, J.-i., Rice, J.C., Strahl, B.D. et al. (2001). Role of histone H3 lysine 9 methylation in epigenetic control of heterochromatin assembly. *Science* **292** (5514): 110–113.

72 Peters, A.H., Mermoud, J.E., O'Carroll, D. et al. (2002). Histone H3 lysine 9 methylation is an epigenetic imprint of facultative heterochromatin. *Nature Genetics* **30** (1): 77–80.

73 Ho, J.W., Jung, Y.L., Liu, T. et al. (2014). Comparative analysis of metazoan chromatin organization. *Nature* **512** (7515): 449–452.

74 Diotel, N., Le Page, Y., Mouriec, K. et al. (2010). Aromatase in the brain of teleost fish: expression, regulation and putative functions. *Frontiers in Neuroendocrinology* **31** (2): 172–192.

75 Suderman, M., McGowan, P.O., Sasaki, A. et al. (2012). Conserved epigenetic sensitivity to early life experience in the rat and human hippocampus. *Proceedings of the National Academy of Sciences* **109** (Supplement 2): 17266–17272.

76 Fellous, A., Le Franc, L., Jouaux, A. et al. (2019). Histone methylation participates in gene expression control during the early development of the pacific oyster *Crassostrea gigas*. *Genes* **10** (9): 695.

77 Jonsson, B. and Jonsson, N. (2019). Phenotypic plasticity and epigenetics of fish: embryo temperature affects later-developing lift-history traits. *Aquatic Biology* **28**: 21–32.

78 Das, P., Samantaray, S., and Rout, G. (1997). Studies on cadmium toxicity in plants: a review. *Environmental Pollution* **98** (1): 29–36.

79 Xiao, C., Liu, Y., Xie, C. et al. (2015). Cadmium induces histone H3 lysine methylation by inhibiting histone demethylase activity. *Toxicological Sciences* **145** (1): 80–89.

80 Zhang, J.J. and Yang, H. (2021). Metabolism and detoxification of pesticides in plants. *Science of the Total Environment* **790**: 148034.

81 Nikinmaa, M. and Rytkönen, K.T. (2011). Functional genomics in aquatic toxicology – do not forget the function. *Aquatic Toxicology* **105** (3-4): 16–24.

82 Mattick, J.S. and Makunin, I.V. (2006). Non-coding RNA. *Human Molecular Genetics* **15** (suppl_1): R17–R29.

83 Frías-Lasserre, D. and Villagra, C.A. (2017). The importance of ncRNAs as epigenetic mechanisms in phenotypic variation and organic evolution. *Frontiers in Microbiology* **8**: 2483.

84 Qu, Z. and Adelson, D.L. (2012). Evolutionary conservation and functional roles of ncRNA. *Frontiers in Genetics* **3**: 205.

85 Cech, T.R. and Steitz, J.A. (2014). The noncoding RNA revolution – trashing old rules to forge new ones. *Cell* **157** (1): 77–94.

86 Guo, S., Zhong, Y., Zhang, Y. et al. (2021). Transcriptome analysis provides insights into long noncoding RNAs in medaka gonads. *Comparative Biochemistry and Physiology Part D: Genomics and Proteomics* **39**: 100842.

87 Guttman, M., Amit, I., Garber, M. et al. (2009). Chromatin signature reveals over a thousand highly conserved large non-coding RNAs in mammals. *Nature* **458** (7235): 223–227.

88 Gonzalez, I., Munita, R., Agirre, E. et al. (2015). A lncRNA regulates alternative splicing via establishment of a splicing-specific chromatin signature. *Nature Structural & Molecular Biology* **22** (5): 370–376.

89 Kaikkonen, M.U., Lam, M.T., and Glass, C.K. (2011). Non-coding RNAs as regulators of gene expression and epigenetics. *Cardiovascular Research* **90** (3): 430–440.

90 Cao, J. (2014). The functional role of long non-coding RNAs and epigenetics. *Biological Procedures Online* **16** (1): 1–13.

91 Sarangdhar, M.A., Chaubey, D., Srikakulam, N., and Pillai, B. (2018). Parentally inherited long non-coding RNA Cyrano is involved in zebrafish neurodevelopment. *Nucleic Acids Research* **46** (18): 9726–9735.

92 Goudarzi, M., Berg, K., Pieper, L.M., and Schier, A.F. (2019). Individual long non-coding RNAs have no overt functions in zebrafish embryogenesis, viability and fertility. *eLife* **8**: e40815.

93 Qu, M., Zhao, Y., Zhao, Y. et al. (2019). Identification of long non-coding RNAs in response to nanopolystyrene in *Caenorhabditis elegans* after long-term and low-dose exposure. *Environmental Pollution* **255**: 113137.

94 Cao, M., Song, F., Yang, X. et al. (2019). Identification of potential long noncoding RNA biomarker of mercury compounds in zebrafish embryos. *Chemical Research in Toxicology* **32** (5): 878–886.

95 Garcia, G.R., Shankar, P., Dunham, C.L. et al. (2018). Signaling events downstream of AHR activation that contribute to toxic responses: the functional role of an AHR-dependent long noncoding RNA (slincR) using the zebrafish model. *Environmental Health Perspectives* **126** (11): 117002.

96 Liu, W., Liu, X., Wu, C., and Jiang, L. (2018). Transcriptome analysis demonstrates that long noncoding RNA is involved in the hypoxic response in *Larimichthys crocea*. *Fish Physiology and Biochemistry* **44** (5): 1333–1347.

97 Liu, X., Li, W., Jiang, L. et al. (2019). Immunity-associated long non-coding RNA and expression in response to bacterial infection in large yellow croaker (*Larimichthys crocea*). *Fish & Shellfish Immunology* **94**: 634–642.

98 Bartel, D.P. (2004). MicroRNAs: genomics, biogenesis, mechanism, and function. *Cell* **116** (2): 281–297.

99 Bushati, N. and Cohen, S.M. (2007). MicroRNA functions. *Annual Review of Cell and Developmental Biology* **23**: 175–205.

100 Lema, C. and Cunningham, M.J. (2010). MicroRNAs and their implications in toxicological research. *Toxicology Letters* **198** (2): 100–105.

101 Fay, M.J., Alt, L.A., Ryba, D. et al. (2018). Cadmium nephrotoxicity is associated with altered microRNA expression in the rat renal cortex. *Toxics* **6** (1): 16.

102 Hong, W.Y. and Cho, W.C. (2015). The role of microRNAs in toxicology. *Archives of Toxicology* **89** (3): 319–325.

103 Gonçalves, A.T., Valenzuela-Muñoz, V., and Gallardo-Escarate, C. (2020). Brain microRNAs in rainbow trout are modulated by functional additives and fish density. *Aquaculture* **519**: 734754.

104 Raza, S.H.A., Abdelnour, S.A., Alotaibi, M.A. et al. (2022). MicroRNAs mediated environmental stress responses and toxicity signs in teleost fish species. *Aquaculture* **546**: 737310.

105 Zhao, X., Yu, H., Kong, L. et al. (2016). High throughput sequencing of small RNAs transcriptomes in two *Crassostrea* oysters identifies microRNAs involved in osmotic stress response. *Scientific Reports* **6** (1): 1–11.

106 Huang, J., Li, Y., Ma, F. et al. (2018). Identification and characterization of microRNAs in the liver of rainbow trout in response to heat stress by high-throughput sequencing. *Gene* **679**: 274–281.

107 Ma, J., Li, Y., Wu, M. et al. (2018). Serum immune responses in common carp (*Cyprinus carpio* L.) to paraquat exposure: the traditional parameters and circulating microRNAs. *Fish & Shellfish Immunology* **76**: 133–142.

108 Nemoto, T. and Kakinuma, Y. (2020). Fetal malnutrition-induced catch up failure is caused by elevated levels of miR-322 in rats. *Scientific Reports* **10** (1): 1–14.

109 Houwing, S., Kamminga, L.M., Berezikov, E. et al. (2007). A role for Piwi and piRNAs in germ cell maintenance and transposon silencing in zebrafish. *Cell* **129** (1): 69–82.

110 Aravin, A.A., Hannon, G.J., and Brennecke, J. (2007). The Piwi-piRNA pathway provides an adaptive defense in the transposon arms race. *Science* **318** (5851): 761–764.

111 Esteller, M. (2011). Non-coding RNAs in human disease. *Nature Reviews Genetics* **12** (12): 861–874.

112 Yan, Z., Hu, H.Y., Jiang, X. et al. (2011). Widespread expression of piRNA-like molecules in somatic tissues. *Nucleic Acids Research* **39** (15): 6596–6607.

113 Aravin, A.A., Sachidanandam, R., Bourc'his, D. et al. (2008). A piRNA pathway primed by individual transposons is linked to de novo DNA methylation in mice. *Molecular Cell* **31** (6): 785–799.

114 Toth, G.P., Bencic, D.C., Martinson, J.W. et al. (2021). Development of omics biomarkers for estrogen exposure using mRNA, miRNA and piRNAs. *Aquatic Toxicology* **235**: 105807.

115 Choudhuri, S. (2010). Small noncoding RNAs: biogenesis, function, and emerging significance in toxicology. *Journal of Biochemical and Molecular Toxicology* **24** (3): 195–216.

116 Labbé, C., Robles, V., and Herraez, M.P. (2017). Epigenetics in fish gametes and early embryo. *Aquaculture* **472**: 93–106.

117 Jin, Y., Liu, Z., Peng, T., and Fu, Z. (2015). The toxicity of chlorpyrifos on the early life stage of zebrafish: a survey on the endpoints at development, locomotor behavior, oxidative stress and immunotoxicity. *Fish & Shellfish Immunology* **43** (2): 405–414.

118 Chen, Q., An, J., Xie, D. et al. (2021). Suppression and recovery of reproductive behavior induced by early life exposure to mercury in zebrafish. *Comparative Biochemistry and Physiology Part C: Toxicology & Pharmacology* **239**: 108876.

119 Wang, X. and Bhandari, R.K. (2019). DNA methylation dynamics during epigenetic reprogramming of medaka embryo. *Epigenetics* **14** (6): 611–622.

120 Ying, H., Hayward, D.C., Klimovich, A. et al. (2022). The role of DNA methylation in genome defense in Cnidaria and other invertebrates. *Molecular Biology and Evolution* **39** (2): msac018.

121 Venney, C.J., Love, O.P., Drown, E.J., and Heath, D.D. (2020). DNA methylation profiles suggest intergenerational transfer of maternal effects. *Molecular Biology and Evolution* **37** (2): 540–548.

122 Skinner, M.K. (2015). Environmental epigenetics and a unified theory of the molecular aspects of evolution: a neo-Lamarckian concept that facilitates neo-Darwinian evolution. *Genome Biology and Evolution* **7** (5): 1296–1302.

123 Rissman, E.F. and Adli, M. (2014). Minireview: transgenerational epigenetic inheritance: focus on endocrine disrupting compounds. *Endocrinology* **155** (8): 2770–2780.

124 Klosin, A., Reis, K., Hidalgo-Carcedo, C. et al. (2017). Impaired DNA replication derepresses chromatin and generates a transgenerationally inherited epigenetic memory. *Science Advances* **3** (8): e1701143.

125 Heard, E. and Martienssen, R.A. (2014). Transgenerational epigenetic inheritance: myths and mechanisms. *Cell* **157** (1): 95–109.

126 Klosin, A., Casas, E., Hidalgo-Carcedo, C. et al. (2017). Transgenerational transmission of environmental information in *C. elegans*. *Science* **356** (6335): 320–323.

127 Chen, L., Li, X., Hong, H., and Shi, D. (2018). Multigenerational effects of 4-methylbenzylidene camphor (4-MBC) on the survival, development and reproduction of the marine copepod Tigriopus japonicus. *Aquatic Toxicology* **194**: 94–102.

128 Henry, J., Brand, J.A., Bai, Y. et al. (2021). Multi-generational impacts of exposure to antidepressant fluoxetine on behaviour, reproduction, and morphology of freshwater snail *Physa acuta*. *Science of the Total Environment* **814**: 152731.

129 Jeong, C.-B., Kang, H.-M., Byeon, E. et al. (2021). Phenotypic and transcriptomic responses of the rotifer *Brachionus koreanus* by single and combined exposures to nano-sized microplastics and water-accommodated fractions of crude oil. *Journal of Hazardous Materials* **416**: 125703.

130 Kim, J. and Rhee, J.-S. (2021). Biochemical and physiological responses of the water flea *Moina macrocopa* to microplastics: a multigenerational study. *Molecular & Cellular Toxicology* **17** (4): 523–532.

131 Le Du-Carrée, J., Saliou, F., Cachot, J. et al. (2021). Developmental effect of parental or direct chronic exposure to environmental concentration of glyphosate on the larvae of rainbow trout, *Oncorhynchus mykiss*. *Aquatic Toxicology* **237**: 105894.

132 Vandegehuchte, M.B., Lemière, F., Vanhaecke, L. et al. (2010). Direct and transgenerational impact on *Daphnia magna* of chemicals with a known effect on DNA methylation. *Comparative Biochemistry and Physiology Part C: Toxicology & Pharmacology.* **151** (3): 278–285.

133 Bhandari, R.K. (2016). Medaka as a model for studying environmentally induced epigenetic transgenerational inheritance of phenotypes. *Environmental Epigenetics* **2** (1): 1–9.

134 Cavalieri, V. and Spinelli, G. (2017). Environmental epigenetics in zebrafish. *Epigenetics & Chromatin* **10** (1): 1–11.

135 Hearn, J., Chow, F.W.N., Barton, H. et al. (2018). Daphnia magna microRNAs respond to nutritional stress and ageing but are not transgenerational. *Molecular Ecology* **27** (6): 1402–1412.

136 Valles, S., Hernández-Sánchez, J., Dipp, V.R. et al. (2020). Exposure to low doses of inorganic arsenic induces transgenerational changes on behavioral and epigenetic markers in zebrafish (*Danio rerio*). *Toxicology and Applied Pharmacology* **396**: 115002.

137 Bautista, N.M., Crespel, A., Crossley, J. et al. (2020). Parental transgenerational epigenetic inheritance related to dietary crude oil exposure in *Danio rerio*. *The Journal of Experimental Biology* **223** (16): jeb222224.

138 McCaw, B.A., Stevenson, T.J., and Lancaster, L.T. (2020). Epigenetic responses to temperature and climate. *Integrative and Comparative Biology* **60** (6): 1469–1480.

139 Carvan, M.J. III, Kalluvila, T.A., Klingler, R.H. et al. (2017). Mercury-induced epigenetic transgenerational inheritance of abnormal neurobehavior is correlated with sperm epimutations in zebrafish. *PLoS One* **12** (5): e0176155.

140 Schmitt, C., Peterson, E., Willis, A. et al. (2020). Transgenerational effects of developmental exposure to chlorpyrifos-oxon in zebrafish (*DANIO RERIO*). *Toxicology and Applied Pharmacology* **408**: 115275.

141 Li, R., Yang, L., Han, J. et al. (2021). Early-life exposure to tris (1,3-dichloro-2-propyl) phosphate caused multigenerational neurodevelopmental toxicity in zebrafish via altering maternal thyroid hormones transfer and epigenetic modifications. *Environmental Pollution* **285**: 117471.

142 Kidd, K.A., Blanchfield, P.J., Mills, K.H. et al. (2007). Collapse of a fish population after exposure to a synthetic estrogen. *Proceedings of the National Academy of Sciences* **104** (21): 8897–8901.

143 Bhandari, R.K., Vom Saal, F.S., and Tillitt, D.E. (2015). Transgenerational effects from early developmental exposures to bisphenol A or 17α-ethinylestradiol in medaka, *Oryzias latipes*. *Scientific Reports* **5** (1): 1–5.

144 Schwindt, A.R. (2015). Parental effects of endocrine disrupting compounds in aquatic wildlife: is there evidence of transgenerational inheritance? *General and Comparative Endocrinology* **219**: 152–164.

145 Luu, I., Ikert, H., and Craig, P.M. (2021). Chronic exposure to anthropogenic and climate related stressors alters transcriptional responses in the liver of zebrafish (*Danio rerio*) across multiple generations. *Comparative Biochemistry and Physiology Part C: Toxicology & Pharmacology* **240**: 108918.

146 Wu, R.S., Zhou, B.S., Randall, D.J. et al. (2003). Aquatic hypoxia is an endocrine disruptor and impairs fish reproduction. *Environmental Science & Technology* **37** (6): 1137–1141.

147 Vera-Chang, M.N., Moon, T.W., and Trudeau, V.L. (2019). Ancestral fluoxetine exposure sensitizes zebrafish to venlafaxine-induced reductions in cortisol and spawning. *Endocrinology* **160** (9): 2137–2142.

148 Thompson, W.A. and Vijayan, M.M. (2021). Zygotic exposure to venlafaxine disrupts cortisol stress axis activity in multiple generations of zebrafish. *Environmental Pollution* **274**: 116535.

149 Weinhouse, C., Truong, L., Meyer, J.N., and Allard, P. (2018). *Caenorhabditis elegans* as an emerging model system in environmental epigenetics. *Environmental and Molecular Mutagenesis* **59** (7): 560–575.

150 Wang, Z., Dai, S., Wang, J. et al. (2021). Assessment on chronic and transgenerational toxicity of methamphetamine to *Caenorhabditis elegans* and associated aquatic risk through toxicity indicator sensitivity distribution (TISD) analysis. *Environmental Pollution* **288**: 117696.

151 Bachère, E., Barranger, A., Bruno, R. et al. (2017). Parental diuron-exposure alters offspring transcriptome and fitness in Pacific oyster *Crassostrea gigas*. *Ecotoxicology and Environmental Safety* **142**: 51–58.

152 Ellis, L.J.A., Kissane, S., Hoffman, E. et al. (2020). Multigenerational exposures of *Daphnia magna* to pristine and aged silver nanoparticles: epigenetic changes and phenotypical ageing related effects. *Small* **16** (21): 2000301.

153 Frost, P.C., Ebert, D., Larson, J.H. et al. (2010). Transgenerational effects of poor elemental food quality on *Daphnia magna*. *Oecologia* **162** (4): 865–872.

154 Jeremias, G., Barbosa, J., Marques, S.M. et al. (2018). Transgenerational inheritance of DNA hypomethylation in *Daphnia magna* in response to salinity stress. *Environmental Science & Technology* **52** (17): 10114–10123.

155 Trijau, M., Asselman, J., Armant, O. et al. (2018). Transgenerational DNA methylation changes in *Daphnia magna* exposed to chronic γ irradiation. *Environmental Science & Technology* **52** (7): 4331–4339.

156 Knecht, A.L., Truong, L., Marvel, S.W. et al. (2017). Transgenerational inheritance of neurobehavioral and physiological deficits from developmental exposure to benzo(a) pyrene in zebrafish. *Toxicology and Applied Pharmacology* **329**: 148–157.

157 Zheng, J.-L., Zhu, Q.-L., Hu, X.-C. et al. (2022). Transgenerational effects of zinc in zebrafish following early life stage exposure. *Science of The Total Environment* **828**: 154443.

158 Brady, S.P., Monosson, E., Matson, C.W., and Bickham, J.W. (2017). Evolutionary toxicology: toward a unified understanding of lifes response to toxic chemicals. *Evolutionary Applications* **10** (8): 745.

159 Egger, G., Liang, G., Aparicio, A., and Jones, P.A. (2004). Epigenetics in human disease and prospects for epigenetic therapy. *Nature* **429** (6990): 457–463.

160 Head, J.A. (2014). Patterns of DNA methylation in animals: an ecotoxicological perspective. *Integrative and Comparative Biology* **54** (1): 77–86.

161 White, J.W., Cole, B.J., Cherr, G.N. et al. (2017). Scaling up endocrine disruption effects from individuals to populations: outcomes depend on how many males a population needs. *Environmental Science & Technology* **51** (3): 1802–1810.

162 National Research Council (2007). *Toxicity Testing in the 21st Century: a Vision and a Strategy*. National Academies Press.

163 Jeremias, G., Gonçalves, F.J., Pereira, J.L., and Asselman, J. (2020). Prospects for incorporation of epigenetic biomarkers in human health and environmental risk assessment of chemicals. *Biological Reviews* **95** (3): 822–846.

164 LaRocca, J., Johnson, K.J., LeBaron, M.J., and Rasoulpour, R.J. (2017). The interface of epigenetics and toxicology in product safety assessment. *Current Opinion in Toxicology* **6**: 87–92.

165 Belanger, S., Barron, M., Craig, P. et al. (2017). Future needs and recommendations in the development of species sensitivity distributions: estimating toxicity thresholds for aquatic ecological communities and assessing impacts of chemical exposures. *Integrated Environmental Assessment and Management* **13** (4): 664–674.

166 Adam, V., von Wyl, A., and Nowack, B. (2021). Probabilistic environmental risk assessment of microplastics in marine habitats. *Aquatic Toxicology* **230**: 105689.

167 Sharavanan, V.J., Sivaramakrishnan, M., Sivarajasekar, N. et al. (2020). Pollutants inducing epigenetic changes and diseases. *Environmental Chemistry Letters* **18** (2): 325–343.

168 Skinner, M.K., Gurerrero-Bosagna, C., Haque, M.M. et al. (2014). Epigenetics and the evolution of darwins finches. *Genome Biology and Evolution* **6** (8): 1972–1989.

169 Whitehead, A., Clark, B.W., Reid, N.M. et al. (2017). When evolution is the solution to pollution: key principles, and lessons from rapid repeated adaptation of killifish (*Fundulus heteroclitus*) populations. *Evolutionary Applications* **10** (8): 762–783.

170 Bal, N., Kumar, A., and Nugegoda, D. (2017). Assessing multigenerational effects of prednisolone to the freshwater snail, *Physa acuta* (Gastropoda: Physidae). *Journal of Hazardous Materials* **339**: 281–291.

171 Bautista, N.M. and Burggren, W.W. (2019). Parental stressor exposure simultaneously conveys both adaptive and maladaptive larval phenotypes through epigenetic inheritance in the zebrafish (*Danio rerio*). *The Journal of Experimental Biology* **222** (17): jeb208918.

172 Nguyen, H.M., Kim, M., Ralph, P.J. et al. (2020). Stress memory in seagrasses: first insight into the effects of thermal priming and the role of epigenetic modifications. *Frontiers in Plant Science* **11**: 494.

173 van Moorsel, S.J., Schmid, M.W., Wagemaker, N.C. et al. (2019). Evidence for rapid evolution in a grassland biodiversity experiment. *Molecular Ecology* **28** (17): 4097–4117.

174 Clements, D.R. and Jones, V.L. (2021). Rapid evolution of invasive weeds under climate change: present evidence and future research needs. *Frontiers in Agronomy* **3**: 10.

175 Heim, J.R., Weston, D.P., Major, K. et al. (2018). Are there fitness costs of adaptive pyrethroid resistance in the amphipod, *Hyalella azteca*? *Environmental Pollution* **235**: 39–46.

176 Zhang, Y., Yu, J., Su, Y. et al. (2019). Long-term changes of water quality in aquaculture-dominated lakes as revealed by sediment geochemical records in Lake Taibai (Eastern China). *Chemosphere* **235**: 297–307.

177 Roth, M., Richards, R., and Sommerville, C. (1993). Current practices in the chemotherapeutic control of sea lice infestations in aquaculture: a review. *Journal of Fish Diseases* **16** (1): 1–26.

178 Burridge, L., Weis, J.S., Cabello, F. et al. (2010). Chemical use in salmon aquaculture: a review of current practices and possible environmental effects. *Aquaculture* **306** (1–4): 7–23.

179 Ernst, W., Doe, K., Cook, A. et al. (2014). Dispersion and toxicity to non-target crustaceans of azamethiphos and deltamethrin after sea lice treatments on farmed salmon, *Salmo salar*. *Aquaculture* **424**: 104–112.

180 Xing, H., Wang, C., Wu, H. et al. (2015). Effects of atrazine and chlorpyrifos on DNA methylation in the brain and gonad of the common carp. *Comparative Biochemistry and Physiology Part C: Toxicology & Pharmacology* **168**: 11–19.

181 Hook, S.E., Doan, H., Gonzago, D. et al. (2018). The impacts of modern-use pesticides on shrimp aquaculture: an assessment for North eastern Australia. *Ecotoxicology and Environmental Safety* **148**: 770–780.

182 Barisic, J., Cannon, S., and Quinn, B. (2019). Cumulative impact of anti-sea lice treatment (azamethiphos) on health status of Rainbow trout (*Oncorhynchus mykiss*, Walbaum 1792) in aquaculture. *Scientific Reports* **9** (1): 1–13.

183 Varó, I., Navarro, J., Nunes, B., and Guilhermino, L. (2007). Effects of dichlorvos aquaculture treatments on selected biomarkers of gilthead sea bream (*Sparus aurata* L.) fingerlings. *Aquaculture* **266** (1–4): 87–96.

184 Stoeck, T., Frühe, L., Forster, D. et al. (2018). Environmental DNA metabarcoding of benthic bacterial communities indicates the benthic footprint of salmon aquaculture. *Marine Pollution Bulletin* **127**: 139–149.

15

Epigenetics in Mollusks

Manon Fallet

Man-Technology-Environment Research Centre (MTM), School of Science and Technology, Örebro University, Örebro, Sweden

15.1 Introduction

15.1.1 Definition and Presentation of Mollusks

Mollusks represent a large phylum of invertebrate organisms with more than 130,000 described species [1]. They are the second most diverse animal phylum after arthropods. They are characterized by a soft body comprised of three main parts: a head, a muscular foot, and a visceral mass [2]. The foot is large, muscular, and originally used for locomotion and anchorage. The visceral mass contains all the organs responsible for digestive, reproductive, respiratory, and excretory systems. The gills serve as a means of gas exchange with cilia that beat and create a water flow carrying oxygen. In some species like bivalves, gills are enlarged and used in filter-feeding. The circulatory system is open with large cavities filled by hemolymph. The body of mollusks is covered by a specific dorsal epidermis called the mantle. In many mollusks, the mantle secretes a calcium carbonate shell. In some other species, the mantle has other functions like respiration. Most mollusks possess a radula, an organ with chitinous tooth-like appendages that are used for scraping or shredding food. It has also been transformed into a drill or a poison dart for some species [2].

Despite these common features, mollusks are highly diverse in size, morphology, and habitats. They range in size from a millimeter to 18 m. Most mollusks are marine organisms, and the phylum Mollusca is the predominant one in marine environments with 23% of all the named marine organisms being mollusks [3]. Nevertheless, mollusks can also be found in other environments. For instance, gastropods are predominant in freshwater or terrestrial habitats. The phylum Mollusca is divided into eight classes [4]. The Caudofoveata are deep sea-living mollusks, usually very small (only a few centimeters). Around 130 species have been categorized as Caudofoveata [5]. They exhibit a worm-like shape and have no shell. Instead, their body is covered by a cuticula and protected by calcareous spicules. They also lack a foot, but possess an anterior muscular structure called an oral shield [5]. The other worm-like and shell-less mollusks belong to the Solenogastres, which are comprised of 280 species [5]. They are distinguished from Caudofoveata by their narrow and midventral foot and the absence of an oral shield [5]. Monoplacophores are small organisms with a body enclosed in a single cap-like shell. Their head is not well developed, and the mouth is found on the ventral surface, surrounded by an anterior lip and tentacles. The digestive system is looped, and some organs (heart, kidneys, and gills) are serially repeated [6]. The class Polyplacophora, also called chitons, encompasses animals that bear an armor-like shell with 7 to 17 overlapping plates [6]; they are all marine species and can be found worldwide. Their nervous system is simple with a unique buccal ganglion. The mantle is covered by calcareous spines and extends beyond the shell to form a girdle. The polyplacophoran foot is ventral, broad, and adapted to suction on rocks and substrates. Their radula is specialized for substrate scraping. Approximately, 7500 species of mollusks are Bivalvia, which includes clams, oysters, mussels, scallops, and geoducks. Bivalves can be found both in marine and freshwater habitats. The body of bivalves is protected by a two-valve shell [7]. Their head is poorly developed without radula. They feed by filtrating water and suspending food through their branchias. Respiration is achieved through a pair of ctenidia, while excretion and osmoregulation are made possible by a pair of nephridia. Some bivalves can create a calcareous pearl around foreign particles that enter the mantle cavity. Scaphopods, also called tusk shells, live inside a tubular shell produced by the mantle. The shell is open at the two longitudinal ends. About 800 species are known to be

Epigenetics in Aquaculture, First Edition. Edited by Francesc Piferrer and Han-Ping Wang.

Box 15.1 Glossary of Mollusk Biology

Abiotic: Due to physical and nonbiological (not derived from living organisms) parameters.
Biomarkers: A natural biological cue (molecule, gene, or characteristic) allowing for the identification of a specific pathological or physiological process.
Biotoxins: Poisonous substances excreted by a living organism.
Ectothermic: Qualification of an organism that regulates its internal temperature by exchanging heat with its environment.
Epizooties: Disease outbreak affecting many animals of one kind at the same time.
Eutrophication: Excessive gradual increase in nutrients in an aquatic ecosystem that is responsible for the dense growth of plants.
Shadow reflex avoidance: Defensive withdrawal behavior that allows some organisms to protect themselves from predation.
Trochophore: Specific planktonic larval stage of some invertebrates with a spherical body, cilia, and spinning motion.
Veliger: Final larval stage of some mollusks using two cilia for feeding and swimming.

Scaphopoda and all are essentially marine. They can move using their muscular foot that extends outside the shell, but mostly live burrowed in sand or mud, from which they feed on foraminifera using their head tentacles [8]. They have no gills and respiration is achieved via water circulation in the mantle thanks to cilia. The Gastropoda include snails, slugs, conchs, sea hares, and sea butterflies. They are the largest group of mollusks, comprising more than 62,000 described species. The typical characteristics of gastropods are their asymmetrical morph, the torsion of their visceral mass, and the specialization of their foot for crawling. Gastropods can be found in all kinds of habitats (marine, freshwater, and terrestrial) [8]. Usually, gastropods have separate sexes, but some species are hermaphrodites. Most marine gastropod's larval development includes a trochophore and a veliger (Box 15.1) phase before metamorphosis to juvenile, while freshwater and terrestrial taxa have direct development. Cephalopods include octopuses, squids, cuttlefish, nautilus, and ammonites. They are all marine predators with vivid coloration useful for camouflage and modified beak-like radula used for defeating prey. Cephalopods possess eight tentacles derived from the ancestral foot and they are the most mobile among mollusks thanks to their modified mantle that allows locomotion via jet-propulsion (ejection of water through siphon). The reproduction of cephalopods is distinct from other mollusks, as the larvae are already like young adults and do not undergo metamorphosis [9].

15.1.2 Mollusks' Place in Aquaculture

The use of mollusks in aquaculture dates back to antiquity [10]. Mollusks, after fish and aquatic plants, represent the third main resource for aquaculture with overall mollusk production reaching 18 million tons in 2018, accounting for 10% of the global and mondial aquaculture (FAO, 2019). Moreover, since 2020, the production of mollusks has been growing at an average rate of 3.46% per year [11]. Today, aquaculture encompasses approximately 65 different mollusk species, among which most are bivalves [11, 12]. Bivalves are particularly attractive for sustainable aquaculture, as most of them do not require feed inputs [12]. Among the 30 principal groups or species listed by the Aquatic Sciences and Fisheries Information System (AFSIS) for aquacultural production in 2019, four were mollusks: cup oysters *Crassostrea* spp. with 5,271,844 tons produced in 2019; the Japanese carpet shell (*Ruditapes philippinarum*) (4,059,993 tons); scallops (1,838,427 tons); and the jumbo flying squid *Dosidicus gigas* (526,902 tons). The overall production of oysters altogether (13 ASFIS species) in 2020 ranked sixth on the list of species groups regarding quantity in world aquaculture production with a 2.13% production increase compared to 2018 (FAO, 2021). Clams, cockles, and ark shells (26 ASFIS species) are ranked eighth on the same list with a production decrease of 1.08% since 2018. Clams and the Pacific cupped oysters (*Crassostrea gigas*) represent two-thirds of the mollusks' global production [12]. Most of the produced mollusks come from Asia, where China is the top consumer and producer of mollusks with 84% of the global cultivated volume in 2017 [12] and 250,963 tons of shellfish produced in 2018. Finally, mollusk products coming from aquaculture are also used in industry as fertilizers, poultry grit, construction material, jewelry, pharmaceuticals, and nutraceuticals. For instance, French Polynesia is the top producer of pearls.

Extensive mollusk aquaculture, however, can also negatively impact ecosystems. The introduction of many filter-feeding mollusks in an environment can lead to the depletion of phytoplankton and an increase in sedimentation rates. A high density of bivalves in localized areas presents a high risk of disease spread. Overstocked systems and poor aquaculture management can have a serious impact on ecosystem balance [12]. Thus, mollusk aquaculture history is marked by massive species' extinctions due to epizooties (Box 15.1), especially of bivalve species (bonamiosis, iridovirus, Pacific oyster mortality syndrome (POMS), etc.).

In conclusion, mollusks represent a high economic interest in aquaculture, and they provide significant benefits for humans. They are highly sensitive to environmental conditions and susceptible to infectious diseases that can cause massive economic loss.

15.1.3 Sensitivity of Mollusks to Environmental Changes and Anthropogenic Pressures

Nowadays, the high impact human activities have on ecosystems has led to profound modifications of the environment: greenhouse gas emission, climate change, depletion of fossil resources, release of toxic pollutants, decrease in natural habitats, overexploitation and impoverishment of soils, as well as the import of invasive species [13, 14]. These changes weaken established ecosystems and lead to the emergence of new diseases and threats affecting natural populations. Organisms must adapt rapidly, or they will vanish. In such context, epigenetic mechanisms are of major interest, as they have been linked to adaptation to fast-changing environments. Shifts in ocean temperature and chemistry, variations in precipitation, more common and stronger heatwave phenomenon, and modifications of sea surface salinity are the major identified threats due to climate change that may have serious consequences on marine environments (Intergovernmental Panel on Climate Change, 2018). Mollusks, and particularly marine bivalves, as ectothermic (Box 15.1) sedentary calcifying organisms, are particularly sensitive to environmental changes. For instance, among abiotic (Box 15.1) changes, increase in water temperature, summer heatwaves, modification of sea surface salinity, and water acidification can have multiple deleterious consequences on bivalves with alteration of spawning, modification of soft tissue and lipid composition, change in metabolism, immunosuppression, delay in development, reduced growth, reduction of byssus attachment, inability to develop protective layers, and increased mortalities [15–25]. Loss of habitats due to extreme fluctuations in precipitation and anthropic actions, like river drainage, reduce mollusk species' richness. It has been demonstrated for unionacean mussels that they are sensitive to hydrological variation and changes [26–28]. Moreover, the relationships between mollusks and microorganisms, as well as the phytoplankton and nutrient balance in water [29], are also dependent on climatic conditions. For instance, the disease POMS that kills up to 80% of *C. gigas* spats is caused by OsHV-1, a microvariant-like virus that induces 84 and 77% mortality at 26 and 22 °C, respectively, while only 23% at 18 °C and 0% at 14 °C, respectively [30]. In addition, a loss in phytoplankton diversity may lead to an overall loss of species diversity [31] and a decrease in available food for bivalves.

Furthermore, mollusks are also sensitive to climate change. Modifications of wind, waves, and natural water currents, together with nutrient balance disturbances and species transportation, can be responsible for harmful algal bloom increases. The biotoxins released from these blooms (Box 15.1) can decimate mollusk populations [32]. On the other hand, urbanization, increased aquaculture, leaching, fertilizer runoff, and climate change are responsible for increased pollutants concentration, especially in coastal areas and estuaries, and also lead to oxygen depletion and eutrophication (Box 15.1) [33].

The sensitivity of gastropods to changes in natural photoperiods has also been investigated in 11 species [34]. Ten of the 11 species studied have displayed physiological and/or behavioral alterations when the photoperiod has been modified or when they were exposed to red light. In addition, natural populations can become invasive or, on the contrary, vanish because of changes in predation risk and/or a reduction of shadow reflex avoidance (Box 15.1) [34].

As filter-feeding organisms, a lot of mollusks absorb pollutants, chemicals, and microorganisms from the environment [11]. Among pollutants, oestrogens have been studied in gastropods and were shown to be responsible for sexual abnormalities [35–39]. Various other pollutants are reducing mollusk fitness by affecting hatching success, shell structure, behavior, and causing morphological deformities [40–43].

Moreover, the sensitivity of marine mollusks to microplastic pollution has been investigated in 10 species [44]. Results have highlighted the adverse effects of microplastics on physiology with a reduction in shell length, shell height, and soft tissue weight, all correlated to the microplastic abundance.

Mollusks are thus highly sensitive to environmental stressors and all these changes and increased disease outbreaks can lead to severe mortalities, extinctions, and massive economic losses. More precisely, for aquaculture, strong changes in local environments may limit production and force aquaculturists to relocate their livestock to more suitable areas [45]. Hatcheries can also be affected by climatic changes via seawater (SW) supply [45]. Restrictions in spat production and commercialization may be needed to avoid the spread of diseases. Also, weaker and thinner shells will facilitate their destruction during transportation. Therefore, there is an imminent need for research to detect threats on mollusks before they impact livestock. In this regard, epigenetic biomarkers (Box 15.1) could be a promising tool for future aquaculture. Moreover, among all studies of environmental and anthropogenic threats on mollusks, several have highlighted the sensitivity and vulnerability of precocious and larval stages to environmental fluctuations [22]. As epigenetic processes are particularly sensitive to environmental cues during early developmental stages, such results raise the question of how epigenetic processes are involved in the response of mollusks to environmental changes and anthropogenic pressures, and how they can be used to prevent mollusk extinctions.

15.2 DNA Modifications in Mollusk Species

The most well-known DNA modification is DNA methylation, which in animals mostly occurs in cytosines in a CpG nucleotide context (5mC). 5mC is the epigenetic mechanism that has been the most studied in mollusks. The DNA methylation machinery has been identified in various mollusk species with homologous sequences to vertebrate's DNA methyltransferases (DNMTs), ten-eleven translocation (TETs) methylcytosine dioxygenases, and proteins of the methyl-CpG binding domain family (MBDs) [46–50]. However, some species have been proven to miss DNMT3 [47, 51]. The DNA methylation pattern in mollusks is mosaic with wide areas fully methylated (housekeeping functions) separated by wide nonmethylated areas (inducible functions) [52–54]. Methylation of DNA principally arises on cytosine, and this modification has been identified in gastropods, bivalves, and cephalopods [55]. Methylation is predominantly found in a CpG context, where 5–15% of cytosines are methylated [55]. Methylated areas are essentially found inside genes where methylation levels (MLs) have been linked to different types of exons [56], while repetitive sequences and intergenic regions are less methylated [52]. This pattern, however, is not so clear because a recent study on *C. virginica* has shown that DNA methylation can also be present in promoters and transposable elements [57]. DNA methylation has been mostly studied in *C. gigas*, for which gene length and expression are correlated with methylation rate [50, 58, 59]. Moreover, genes linked to housekeeping functions are hypermethylated while genes linked to inducible functions are hypomethylated [54]. For some genes involved in conserved mechanisms, such as developmental pathways, a correlation between expression differential and change in methylation rate at promoters has been identified [60]. A positive correlation between promoter methylation and associated gene expression has been reported in *Pinctada fucata* [61]. Despite these features, promoter methylation is not always correlated to differential gene expression [59, 60, 62–64].

DNA methylation in mollusks has been shown to be essential for the development of organisms. The development of gametes and reproduction success are dependent on DNA MLs in *C. gigas* [65] (Figure 15.1b) and *B. glabrata* [68]. In gonadal tissue, genes encoding for the DNA methylation machinery are overexpressed compared to somatic tissue. Studies using DNA demethylating treatments in *B. glabrata* [68, 69] alter egg production and embryo development, further highlighting this link between DNA methylation and gametes development.

Interestingly, whilst in most animals, mitochondrial DNA (mtDNA) is strictly inherited via the mother [70], at least 105 bivalve species possess a specific mitochondrial inheritance system called "doubly uniparental inheritance" and can also inherit the mtDNA of the father [71]. The distribution and expression of the two mtDNA strands have been suggested to be regulated by epigenetic mechanisms. DNA methylation and hydroxymethylation in mtDNA cytosines have been identified in mammals [72–74] and changes in mtDNA methylation patterns have been involved in some diseases [75, 76]. The mtDNA methylation patterns have been shown to be tissue-specific [74, 77]. In male gonads of mollusks, where the two inherited mtDNA are present, the mtDNA ML is higher than in other tissues [78]. Moreover, transcriptomic analysis has highlighted a male-biased overexpression of DNA methyltransferases in gonads (including the DNMT1 involved in mitochondrial methylation) [79, 80]. Thus, mtDNA methylation could be an important mechanism involved in mitochondrial inheritance and/or in sexual determination in bivalves.

An impact of DNA methylation has also been reported during the early development of mollusks. In *C. gigas* larval development, the expression of genes encoding DNA methylation machinery proteins varies according to tissue and developmental stage [50], leading to global DNA methylation increase in exons and decrease in other genomic regions [67] (Figure 15.1c–e). Such ML changes have also been observed in other mollusk species. In the snail *P. acuta* [81] and the scallop *Chlamys farreri* [82], a decrease of 5mC levels as juveniles aged was observed. In the same way, a global demethylation process occurs during the development of *O. vulgaris* [83, 84]. For *P. fucata*, changes in DNA ML also occur in early development and among tissues, and the level of DNA methylation increases with development [85]. Finally, in *Aplysia*, synaptic plasticity has been extensively studied and DNA methylation variations have been reported to be primordial for long-lasting modifications in neurons' functional state [62].

Hydroxy-methyl cytosine has also been identified in *B. glabrata* with a rate of 0.0009% [86] and more studies are required to learn more about the significance of this DNA modification.

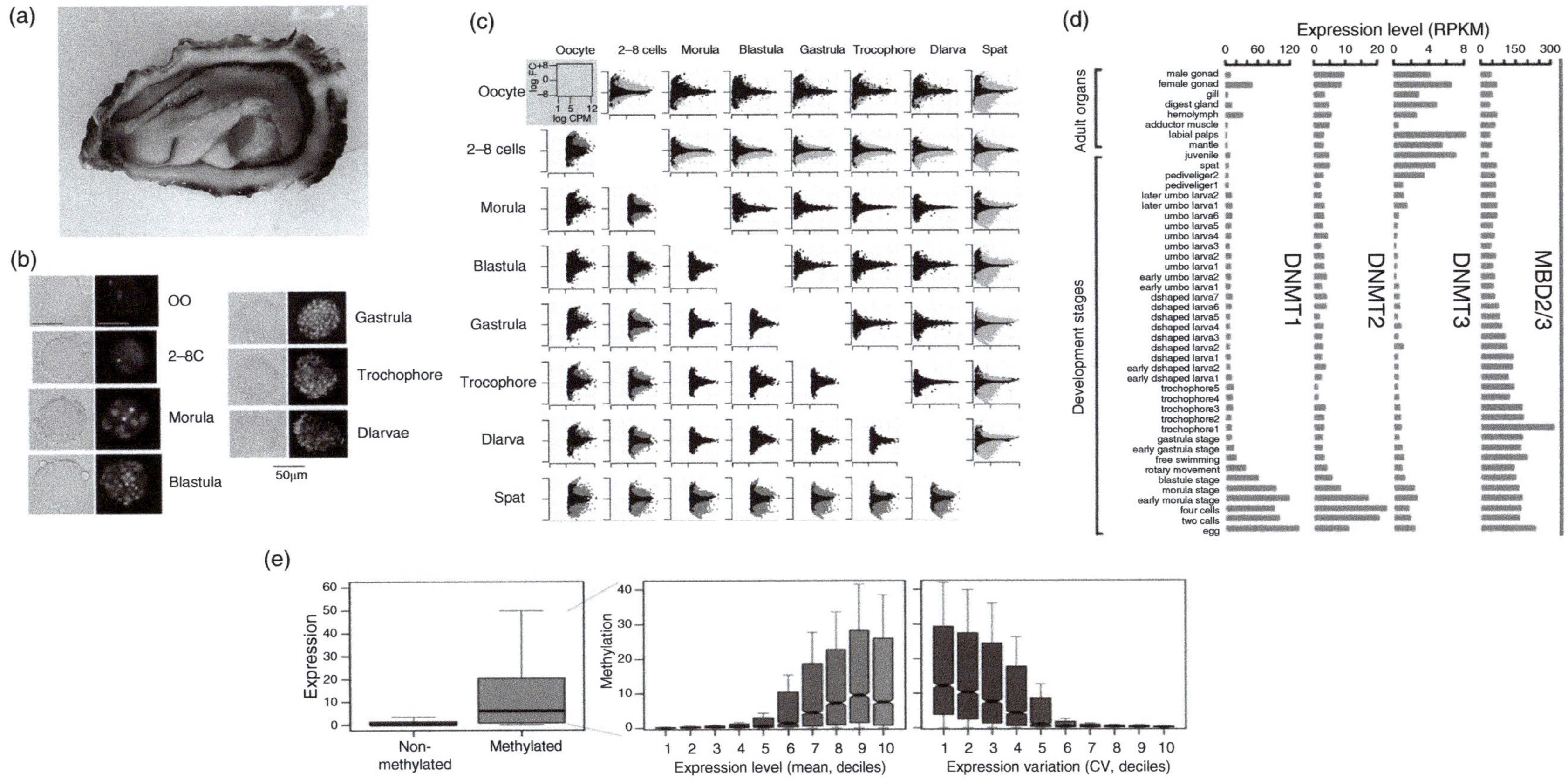

Figure 15.1 Impact of DNA methylation during the development of the cupped oyster *C. gigas*. (a) Opened *Crassostrea gigas* oyster with minor (ventral) valve removed, exposing mantle; (b) Localization of methylated DNA within *C. gigas* embryos using a sheep anti-5-mC antibody and an anti-sheep secondary antibody; left, transmitted bright field photonic microscopy; right, transmitted UV light epifluorescence (Box 15.2) microscopy using a 395–420 nm filter; the embryonic stage is indicated; 2–8C, 2–8 cells; magnification, ×400; Nb, bubbles at embryo edges are mounting artifacts; oo, oocytes. Scale bar 50 μm. (c) Oyster developmental methylomes exhibit feature and stage-dependent landscapes highlighting cleavage and metamorphosis. Pairwise comparisons of the methylation of coding sequences (CDS, upper-right) and transposable elements (TE, lower-left) among development stages are listed in chronological order. Colored points are significantly differentially methylated CDS (top-right) and TEs (bottom-left) between stages (Student's T test, $p < 0.01$). The ML (log counts per million, *x* axis) and its variation (log fold change, *y* axis) are shown. (d) Expression levels of the oyster DNMTs and MBD2/3 genes in different developmental stages and adult organs, denoted by reads per kilobase of transcript per million mapped reads (RPKM). Numbers after sample names indicate the orders in development, with smaller numbers representing earlier stages in development. (e) The expression level of nonmethylated and methylated genes (CDS methylation is shown, left). Relation between CDS methylation and mRNA levels (mean across development stages, deciles) of the methylated genes (middle). Relation between CDS methylation and expression variability of the methylated genes (coefficient of variation [CV] across development stages, deciles) (right). *Source:* (a) From Leverett and Thain [66]/International Council for the Exploration of the Sea; (b) From Riviere et al. [65]/Springer Nature; (c, e) Riviere et al. [67]/PLOS/CC BY 4.0; (d) Wang et al. [50]/Springer Nature/CC BY 4.0. (*See colour plate section for colour representation of this figure.*)

15.3 Chromatin Conformation and Histone Modifications/Variants in Mollusks

The chromatin in animals is highly structured and chromosome territories have been identified in the snail *B. glabrata* [87]. Such territorial organization of chromosomes leads to spaces where genes can be mobile [88, 89] and RNA can be transcribed [90, 91]. This spatial organization is therefore involved in the regulation of gene expression. Moreover, histone proteins are key constituents of chromatin and DNA is wrapped around histones to form nucleosomes. The histones possess N-terminal tails that can carry different chemical modifications. Such modifications act in chromatin structural changes and may have consequences on gene expression.

Histone modifications have been shown to be crucial as keys for mollusk development. In *C. gigas* and the scallop *Patinopecten yessoensis*, the expression of histone-demethylases varies according to tissue and developmental stage, leading to a clear role of histone-demethylases for both gametogenesis and embryogenesis [92, 93]. Moreover, the expression of histone deacetylases and histone acetyltransferases is biased toward males in the gonads of bivalves, suggesting a role of these histone modifications in gamete formation and/or sex differentiation [79]. In *C. gigas*, for instance, the histone methylation profiles of H3K9 change during the development [94] (Figure 15.2a). The testes of various mollusks possess

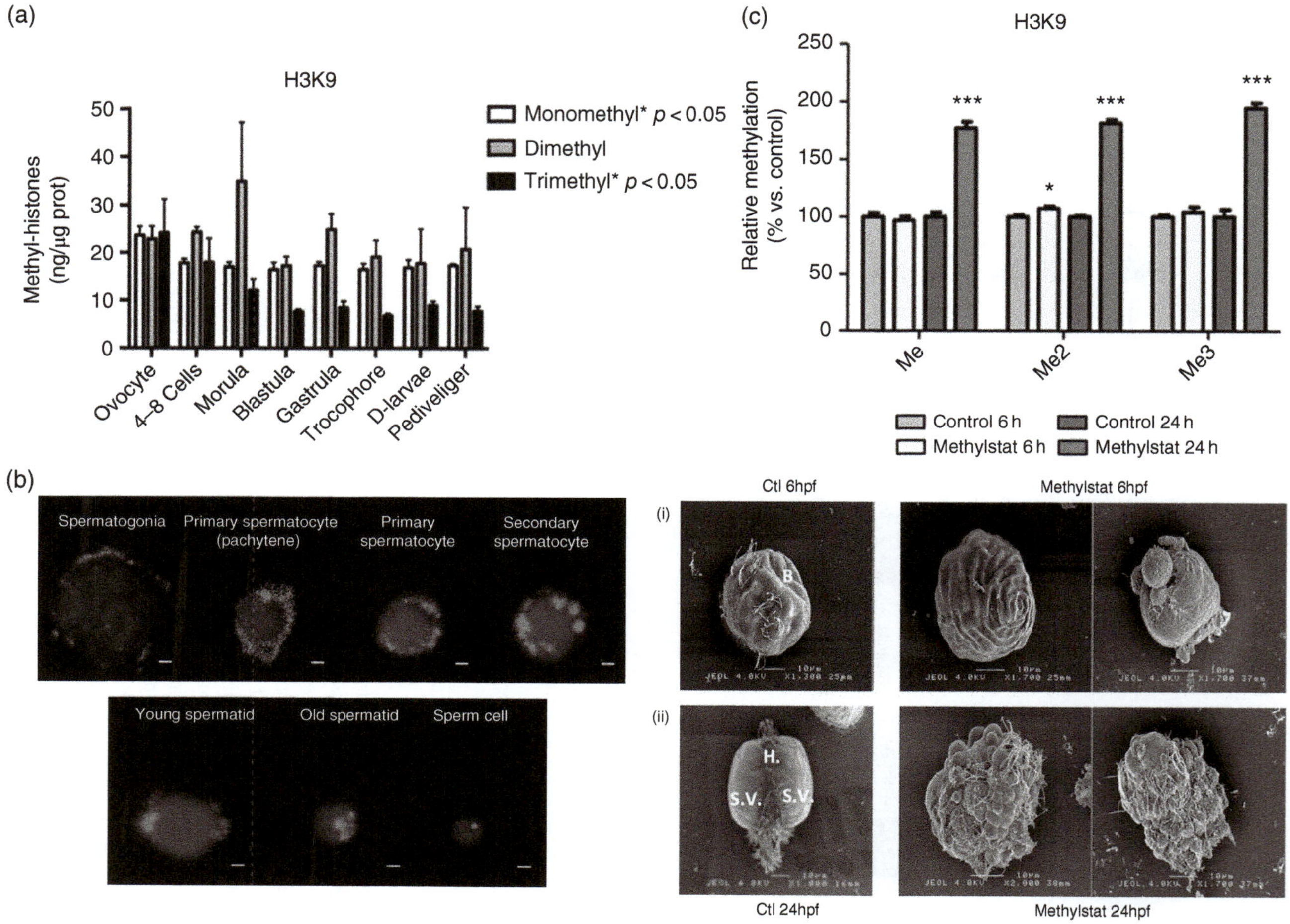

Figure 15.2 Histone modifications during the development of the cupped oyster *C. gigas*, examples of H3K9 methylation and Cg-CenH3 variant. (a) Histone methylation profiles of H3K9 during the early development of *C. gigas*. Asterisk in the legend indicates significant variation of the mark during development. A one-way ANOVA was used to determine whether MLs were statistically and significantly different between development stages for each residue, and a $p<0.05$ was considered significant. (b) Distribution of Cg-CenH3 in male *C. gigas* gonadal cells in different stages of maturation. Scale bar = 1 µm. (c) Hypermethylation of histone lysine residues in the presence of 10 µM methylstat. Global H3K9 methylation increases in the presence of methylstat. Asterisk in the legend indicates a significant variation of the indicated mark (One-way ANOVA; $p<0.05$ was considered significant (<0.05 (*), <0.001 (**), <0.0001 (***)) (top). Different phenotypes observed in control conditions and under methylstat (Jmj inhibitor) treatment at 6 hours after fertilization (i) and 24 hours after fertilization (ii) (bottom). *Source:* (a, c) Fellous et al. [94]/MDPI/CC BY 4.0; (b) From Cvitanić et al. [95]/MDPI/CC BY 4.0. (*See colour plate section for colour representation of this figure.*)

Box 15.2 Glossary of Mollusk Physiology

Centromere: Region of a chromosome that plays a key role in cell division, and to which the microtubules of the spindle attach via the kinetochore.
Chorion peroxidase: Enzyme catalyzing a protein involved in the formation of the chorion (most external foetal membrane around the embryo in amniotes).
Epifluorescence: The fluorescence of an object in an optical microscope when irradiated from the viewing side.
Long-term potentiation: Process that consists of a persistent synaptic strength increase after a chemical synapse stimulation leading to a long-lasting increase in signal transmission between neurons.
Luciferase reporter assay: Technique used commonly in molecular biology to study gene regulation and expression by using luciferase enzyme activity and a luminometer.
Miracidiae: The free-swimming ciliated larvae of a digenetic trematode.
Nucleus grafting: Operation done on oysters or mussels to induce nacre secretion and culture pearl growth by introducing a nucleus and a piece of donor mantle tissue into the oyster/mussel.
Protamine: Basic simple proteins rich in arginine and associated with DNA in place of histone in the sperm cells of various animals.
Short- and long-term sensitization: Short-term sensitization lasts seconds to minutes while long-term sensitization lasts from days to weeks.
Synapsin: Gene coding for proteins found in nearly all synapses in the nervous system. They regulate the neurotransmitter releases at synapses.

protamine-type (Box 15.2), protamine-like-type, or histone H1-type sperm that replace canonical histones [96–98]. In *C. gigas*, a specific histone variant, Cg-CenH3, has been identified. This variant is present in centromeres (Box 15.2) (crucial regions for the segregation of genetic material) and not distributed uniformly in *C. gigas* gonad cells, suggesting a possible role of this variant in sex differentiation and gametogenesis [95] (Figure 15.2b). In addition, variations in the JmJ gene expression for *C. gigas* are responsible for abnormal development and are correlated to modification in the levels of H3K4, H3K9, H3K27, and H3K36 [94, 99] (Figure 15.2c).

The formation of mature sperm in the cephalopod *Sepia officinalis* depends on a progressive reorganization of the chromatin, which relies on H4K12me1 followed by a huge hyperacetylation of H4 [100].

Like DNA methylation, histone modifications play a role in learning and long-term memory formation. In the gastropod *Helix lucorum*, stage-specific increase of H3 acetylation happens during food aversion learning in the right parietal ganglion [101, 102]. This change in histone acetylation is followed by an increase in H3K4me3 and H3K9me2 after learning [103]. Complementary results have been obtained using the histone deacetylase inhibitor sodium butyrate that increases the amplitude of long-term potentiation (Box 15.2) of synaptic response in command neurons, demonstrating the involvement of histone modifications in synaptic plasticity [104]. Another example of the importance of histone modifications for mollusks in learning comes from studies on *Aplysia* sp. The expression of genes involved in the formation of new synapses is dependent on histone acetylation that regulates chromatin conformation [105, 106]. This finding is highlighted using the histone deacetylase inhibitor trichostatin A, which allows for the reestablishment of long-term memory in the case of dysfunction [107].

15.4 Noncoding RNAs in Mollusks

Various noncoding RNAs (NcRNAs) have been identified in mollusks and they are involved in various biological processes [108]. In *C. gigas*, 11,668 long noncoding RNAs (lncRNAs) have been identified and their overall expression is lower than for protein-coding genes [108]. Among these *C. gigas* lncRNAs, 1175 are differentially expressed according to the studied tissue [109]. miRNAs have also been identified in other mollusks, and their biogenesis is due to products of the following genes: DROSHA, DGCR8, EXP5, RAN, DICER TARBP2, AGO, and PIWI [110]. A last class of ncRNAs, the piRNAs, are also present in mollusks, but they seem to have differential roles inside the phylum. Indeed, as in *L. stagnalis* and *C. gigas*, piRNA clusters displayed lineage-specific adaptations in their transposon composition, highlighting differential transposon activity between gastropods and bivalves [111].

NcRNAs are key molecules in differentiation, development, and growth of mollusks. In *A. californica*, the biogenesis of ovotestes is regulated by neuromodulators and linked to a neuronal piwi protein. piRNAs may be involved in this biogenesis, as they are massively found in neurons and highly expressed outside of the ovotestes [62]. In *Crassostrea hongkongensis*, the expression of miRNAs is different between ovaries and testes and miRNAs have thus been highlighted as key players for gonad differentiation and function maintenance [112]. The same pattern is found in *P. fucata* with significant differential expression of piRNAs between somatic and gonad tissues [113] (Figure 15.3b). In *C. gigas*, 32.5% of lncRNAs are expressed preferentially in the gonads of males. Such long noncoding RNAs expression varies with time and co-expression network analysis of five lncRNAs has shown that their expression is linked to key genes in development with an overexpression of some lncRNAs during juvenile stages and metamorphosis [109]. Moreover, several piRNAs with unique transposon sequences are differentially expressed during oyster development [111]. Additionally, during *C. gigas* development, the expression of various miRNAs is changing across different developmental stages [118]. Research on *Haliotis discus hannai* has identified 10 miRNAs linked to muscle growth and development [119]. In *P. fucata*, one miRNA has been associated with regulation of abductor muscle cell proliferation [117] (Figure 15.3e).

NcRNAs are also involved in memory storage and formation. In *Aplysia californica*, the miRNA miR-124 may be important in the mature nervous system by playing a role in plasticity of synapses and participating in the switch from short- to long-term memory [120]. In the same snail, piRNAs may mediate transient external stimuli action on long-term memory storage by inducing specific DNA methylation. Indeed, piRNAs can induce the methylation of CREB2, a gene involved in plasticity and memory transcriptional repression [121]. Such piRNAs control of CREB2 DNA methylation may be a mechanism that allows long-lasting gene expression changes to participate in long-term memory storage [62].

A specific lncRNA in *C. gigas* regulates chorion peroxidase (Box 15.2) and may be involved in the melanin synthetic pathway [122]. In *P. fucata*, some lncRNAs have been suggested to play a role in nucleus grafting (Box 15.2) [123] and in nacre formation [115, 124, 125] (Figure 15.3c). For instance, results show that the decreased LncMSEN1 expression in mantle can cause the disordered growth of crystals on the inner surface of nacre in the shells, as well as the decreased expression of N–U8, suggesting an involvement of LncMSEN1 in the formation of nacre and innate immune response in pearl oyster. Many miRNAs are involved in mollusk biomineralization [116, 126], shell formation [127], and shell regeneration [116]. In *P. fucata*, novel_miR_1 regulates biomineralization by downregulating biomineralization-related genes (Figure 15.3d).

In *N. lapillus*, piRNAs are involved in genome defense, as piRNA-like small RNAs expressed in the soma can match virus and transposon sequences [128].

The different epigenetic mechanisms are highly interconnected and can work together in the production of a new phenotype. One example of such interconnexion has been given by studies on the memory formation in the gastropod *Aplysia*. The formation of the memory relies on the formation of new synaptic connections, leading to a switch from short- to long-term facilitation (LTF) (lasting minutes to hours) [121]. The expression of genes involved in LTF is dependent on

Figure 15.3 (Continued) relative to the internal control is expressed as the mean ± S.D. ($n = 3$). Significant differences from the control are indicated with an asterisk ($p < 0.05$) or with two asterisks ($p < 0.01$). The expression levels of the lncRNA: LncMSEN1 and the nacre matrix protein: N–U8 after RNAi (i). LncMSEN1 and N–U8 expression levels in the mantel are positively correlated (results not shown). LncMSEN1-dsRNA: experimental group, DEPC: blank control group, RFP (red fluorescent protein)-dsRNA: negative control group, and * indicates significant difference ($p < 0.05$). Influence of RNAi on shell nacre ultrastructure (ii). (A, D) are the blank control groups; (B, E) are the negative control groups; (C, F) are the experiment groups; (D–F) are the enlarged versions of (A–C), respectively. (d) Relative expression levels of the miRNA: novel_miR_1 in different tissues by qRT-PCR (left). The expression level of novel_miR_1 in the mantle pallial was used as the control. Interactions between novel_miR_1 and target genes in vitro (right). (i) Putative binding sites of novel_ miR_1 on the CDS region of the protein Prisilkin-39, involved in shell formation, predicted by miRanda. (ii) The relative luminescence ratio in Prisilkin-39 CDS luciferase reporter assay (Box 15.2). (iii) Putative binding sites of novel_miR_1 on the CDS region of ACCBP (Amorphous calcium carbonate binding protein 1), also involved in shell formation, predicted by miRanda. (iv) The relative luminescence ratio in ACCBP CDS luciferase reporter assay. Blank groups were used as controls. (e) Expression pattern of the mature miRNA pm-miR-133 in different tissues of *P. fucata* suggesting it could be involved in adductor muscles functioning. Different letters indicate significant differences ($p < 0.05$) determined through one-way ANOVA and Duncan's multiple comparisons. The bar represents standard deviation (SD). *Source:* (a) From Du et al. [114]/Oxford University Press/CC BY 4.0; (b) Huang et al. [113]/Springer Nature/CC BY 4.0; (c) Modified from Zheng et al. [115]/Elsevier; (d) Zhu et al. [116]/Springer Nature/CC BY 4.0; (e) Zheng et al. [117]/ with permission of Elsevier.

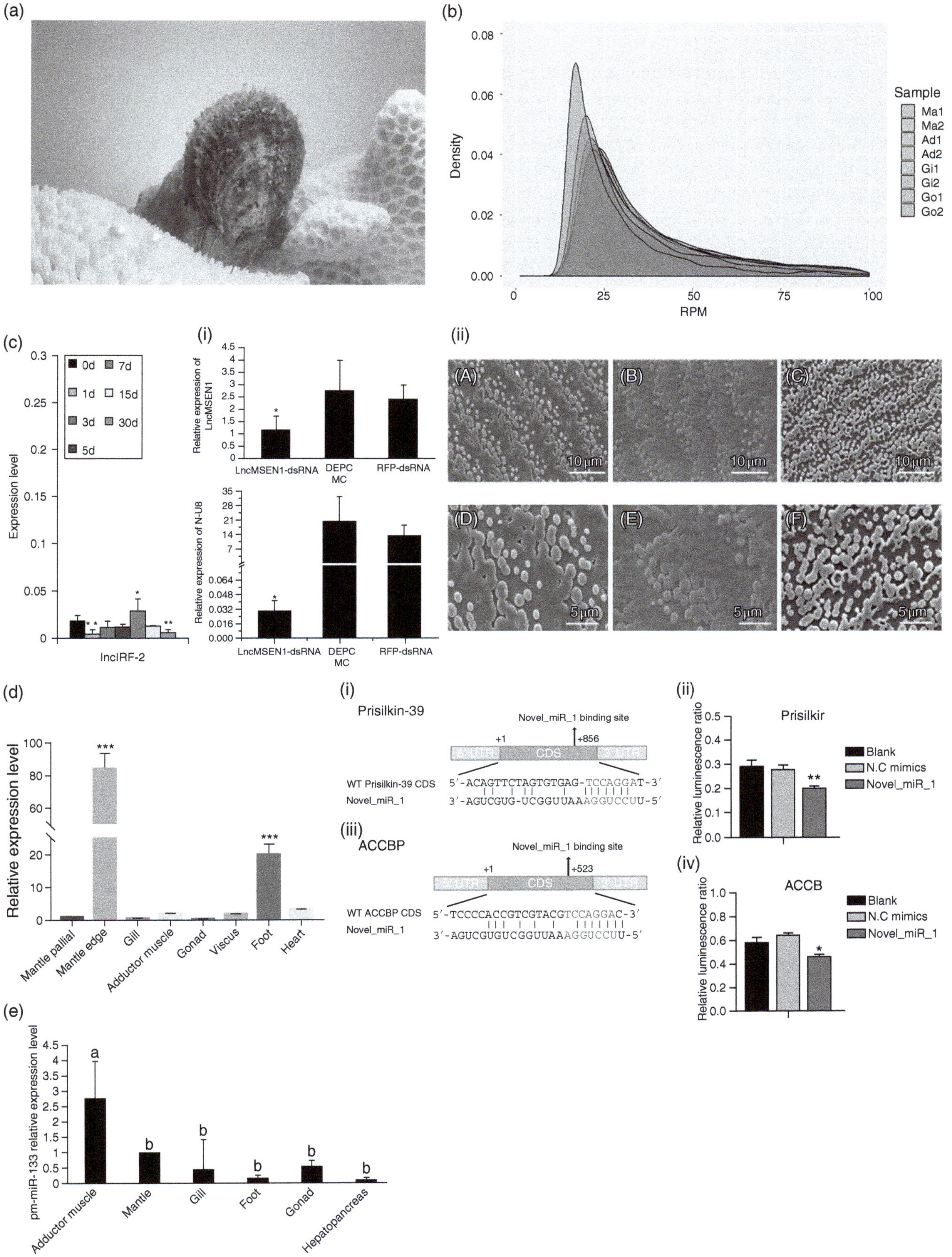

Figure 15.3 Examples of NcRNAs mechanisms in the pearl oyster *P. fucata*. (a) The pearl oyster *Pinctada fucata martensii*. (b) Expression density of putative piRNAs from the somatic and gonadal tissues of pearl oysters, suggesting the important functions of piRNAs in the germline and beyond. Ad, adductor muscle; Gi, Gill tissue; Go, Gonad tissue; RPM, Reads per million reads; Ma, Mantle tissues.
(c) Expression level of two lncRNAs in *P. fucata*. lncIRF-2 expression level in the haemolymph after nucleus grafting operation showing the lncRNA involvement in the immune response to the graft (left). The expression levels were determined by qPCR. The expression

a cascade of events. The neurotransmitter serotonin (5-HT) is released and leads to covalent modification of pre-existing proteins. The serotonin acts on a transmembrane serotonin receptor to activate the enzyme adenylyl cyclase (AC), which converts ATP to the second messenger cyclic AMP (cAMP), an intracellular signal transductor. In turn, cAMP recruits the cAMP-dependent protein kinase A (PKA) by binding to the regulatory subunits (spindles), causing them to dissociate from and free the catalytic subunits (ovals). These subunits can then phosphorylate substrates (channels and exocytosis machinery) in the presynaptic terminals, leading to enhanced transmitter availability and release. cAMP increase will also lead to an increase of cAMP-induced transcriptional regulation through action of CREB1 (cAMP responsive element binding protein 1). CREB1 will bind to the cAMP response element (CRE) sequence in the promoter region of target genes like *synapsin* (Box 15.2) or CAAT box enhancer binding protein (C/EBP) that leads to the synthesis of the proteins needed for the growth of new synapses [129]. In long-term sensitization, repeated stimulation causes the level of cAMP to rise and persist for several minutes. The catalytic subunits can then translocate to the nucleus and recruit the mitogen-activated protein kinase (MAPK). In the nucleus, PKA and MAPK phosphorylate and activate the CRE-binding (CREB) protein and remove the repressive action of CREB-2, an inhibitor of CREB-1. CREB-1, in turn, activates several immediate-response genes including a ubiquitin hydrolase necessary for regulated proteolysis of the regulatory subunit of PKA. Cleavage of the (inhibitory) regulatory subunit results in persistent activity of PKA, leading to persistent phosphorylation of the substrate proteins of PKA. A second immediate-response gene activated by CREB-1 is C/EBP, which acts both as a homodimer and as a heterodimer with an activating factor (AF) to activate downstream genes (Figure 15.4b). All three epigenetic information carriers: DNA methylation, histones marks, and noncoding RNAs are involved in this synaptic cascade of events and in the memory formation in *Aplysia*. Indeed, after 5-HT treatment, binding of CREB1 to *synapsin* promoter increases *synapsin* transcription and is linked to an increase of histone H3 and H4 acetylation together with a decrease in histone deacetylase 5 activity near the CRE site [106]. FMRFa (F) treatment resulted in the recruitment of CREB2 and the removal of CREB1a from the C/EBP promoter. HDAC5 was also recruited to the promoter, presumably causing the deacetylation of histone H4. In FMRFa and 5-HT cotreatment (FS), CREB2 and HDAC5 were recruited and histone H4 deacetylated. After the 5-HT treatment, CREB1a was recruited and CREB2 was removed from the promoter. In addition, specific histone lysine residues (H4-K8 and H3-K14) were acetylated after the 5-HT treatment. As controls, all the samples were also analyzed by the primers specific to the promoter of Aplysia histone H4 and no change was observed [106] (Figure 15.4c). In a same way, 5-HT treatment leads to binding of CREB1 on C/EBP promoter and histone acetylation. Summary data indicated that treatment with 5-HT led to a significant increase in acetylation of histones H3 and H4, and decreased association of HDAC5 surrounding the CREB1 binding site of the *synapsin* promoter region (compared with vehicle-treated ganglia)

Figure 15.4 (Continued) promoter is accompanied by histone acetylation after 5-HT treatment. Representative DNA gel from a ChIP assay assessing the state of histone acetylation of the *synapsin* promoter region containing the CRE (left). Ganglia were treated with 5-HT or vehicle. Densitometry measurements of bands from PCR-amplified DNA precipitated with antibodies against acetylated histone H3 (anti-acetyl-H3), acetylated histone H4 (anti-acetyl-H4), and anti-histone deacetylase 5 (anti-HDAC5) were normalized to input control. Basal acetylation of the *synapsin* promoter was observed with vehicle treatment (left lane) as well as basal association of HDAC5 (left lane). After treatment with 5-HT, association of histones H3 and H4 (right lane) and of HDAC5 (right lane) was modified compared to vehicle. (e) Real-time PCR experiments show that knockdown of Piwi produces a significant increase in CREB2 RNA levels when normalized to GAPDH levels (effects quantified as a mean of four independent trials ±SD) (left). Real-time pyrosequencing of the CREB2 (CRE-binding protein 2 inhibiting long-term memory formation) promoter region shows a significant baseline level of methylation (gray) at individual CpG sites that is robustly upregulated with exposure to serotonin (5HT) (middle bars), and this effect is abolished in the presence of the DNMT inhibitor RG108 (third bars) (right). Effects were quantified as the mean of four independent trials and SDs were calculated, but were so low that they are not shown on the graphs for clarity of the figure. (f) miR-124 directly regulates CREB and facilitates the switch that converts short-term into long-term synaptic facilitation with 1. (left) a northern blot with 10 mg total RNA from *Aplysia* pleural ganglia loaded in each lane, after treatment with either 20-*O*-methyl oligonucleotides antisense to miR-124 conjugated to penetratin or with 20-*O*-methyl control oligonucleotides antisense to miR-194 or with vehicle alone. Blots are probed for miR-124 to show efficient and specific knockdown of miR-124 by penetratin conjugates. Blots were reprobed to detect tRNA, without stripping, to verify equal loading of all lanes. Level of knockdown is quantified by taking the mean percentage reduction of antisense miR-124 as compared to antisense miR-194 over four independent trials ±SD. 2. (right) A western blot loaded with 15 mg total protein in each lane, after treatment with either 20-*O*-methyl oligonucleotides antisense to miR-124 conjugated to penetratin or with 20-*O*-methyl control oligonucleotides antisense to miR-194 or with vehicle alone. Changes in protein levels were quantified as a ratio of band intensity between anti-miR-124 and control oligo treatment after each was normalized to the loading control. Data are shown as a mean of 10 independent trials for CREB1 and at least five for all others ±SD. *Source:* (a) From Robertson and Walter [130]/Cambridge University Press; (b) From Kandel [129]; (c) Modified from Guan et al. [105]/Elsevier; (d) Modified from Hart et al. [106]/Society for Neuroscience/CC BY 4.0; (e) Rajasethupathy et al. [62]/with permission of Elsevier; (f) Modified from Rajasethupathy et al. [120]/Elsevier.

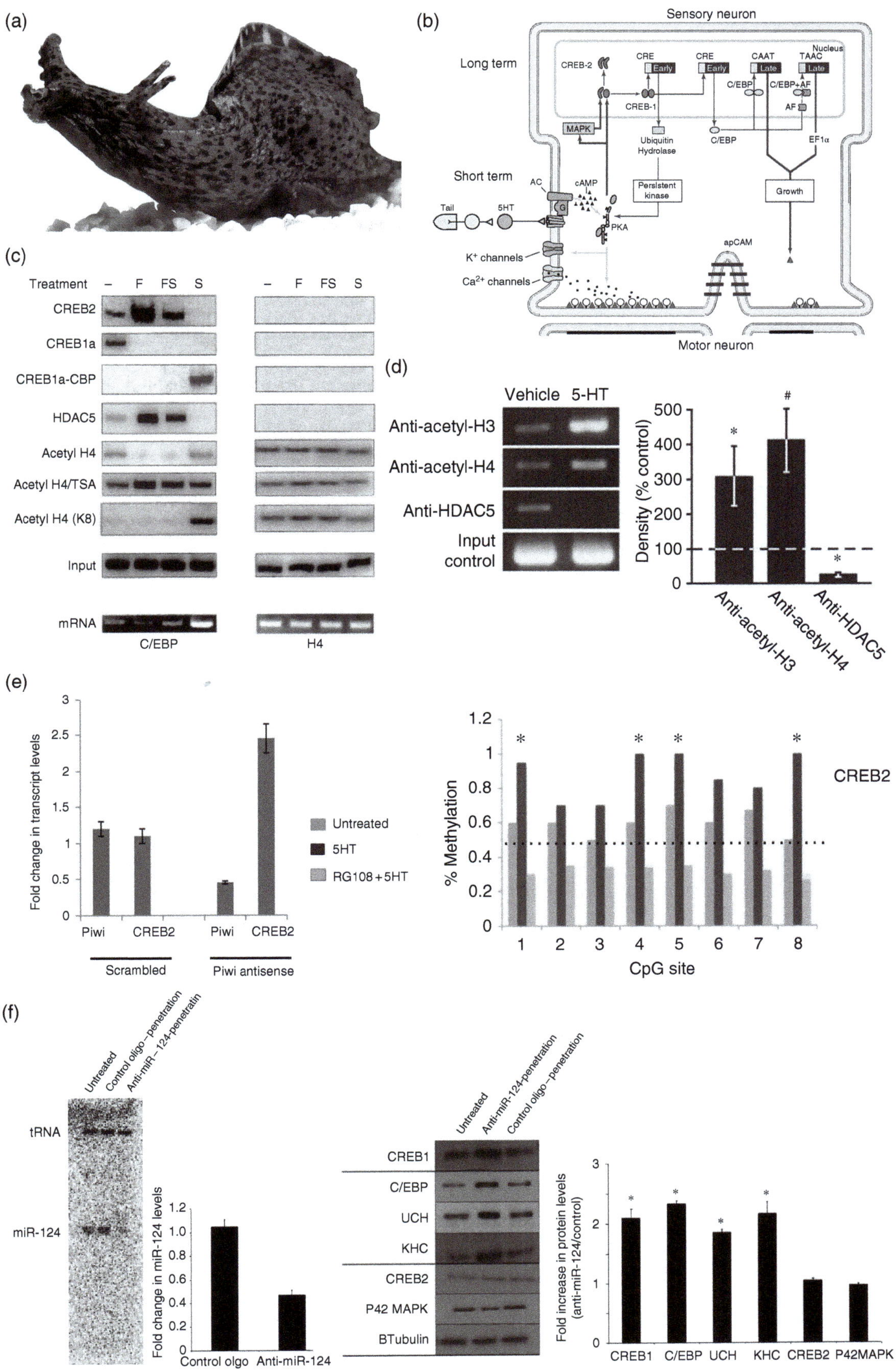

Figure 15.4 Epigenetic mechanisms' interaction in the memory formation of *Aplysia*. (a) The sea slug *Aplysia californica*. (b) Effects of short- and long-term sensitization (Box 15.2) on the monosynaptic component of the gill-withdrawal reflex of *Aplysia*. (c) Soluble chromatin was prepared and immunoprecipitated with antibodies separately against CREB2, CREB1a, HDAC5, acetylated histone H4 (acetyl H4) or histone H4 acetylated at lysine 8 (acetyl H4 [K8]). As controls, chromatin samples were also analyzed before immunoprecipitation (Input). The transcription of C/EBP is shown at the bottom of the figure. (d) Binding of CREB1 to Aplysia *synapsin*

(*$p < 0.05$, $p < 0.01$) (right). No change in input control was observed [105] (Figure 15.4d). However, miR-124, a miRNA present in sensory-motor synapses, constrained LTF by regulation of CREB. Indeed, CREB1 and three of its downstream targets (kinesin heavy chain [KHC], CAAT enhancer binding protein [C/EBP], and ubiquitin C-terminal hydrolase [UCH]) are upregulated after miR-124 inhibition. CREB2 and MAPK are not altered by miR-124 (Figure 15.4f) [120]. Moreover, while CREB1 is involved in memory formation, the displacement of CREB1 by CREB2 block the LTF by histone deacetylation through HDAC5 recruitment [105]. The silencing of CREB2, necessary for LTF, occurs thanks to a PIWI/piRNA complex that facilitates 5-HT-dependent methylation of a CpG island in CREB2 promoter (Figure 15.4e). miR-124 can also be regulated by the action of serotonin [120].

15.5 Epigenetic Responses to Environmental Fluctuations in Mollusks

Environmental stimuli have been shown to be able to alter the phenotype of organisms through epigenetic mechanisms, and the same holds for mollusks.

First, abiotic factors may affect epigenetic marks. Temporal DNA methylation pattern varies according to the seasonality in oyster *Isognomon alatus* [131] (Figure 15.5a) and with both temperature and salinity in Pacific oyster *Haliotis discus hannai* [132] (Figure 15.5b). Variations in water acidity also induce DNA methylation modifications in *Crassotrea virginica* [57] (Figure 15.5e). Changes in response to current speed variations in *Potamopyrgus antipodarum* [134] (Figure 15.5d) and to water depth in the pearl oyster *Pinctada margaritifera var. cumingii.* have been highlighted [133] (Figure 15.5c). For the pearl oyster, an investigation of the whole methylome at various depths has allowed for the identification of key genes involved in pigmentations and biomineralization processes that displayed DNA methylation changes in CpGs. Such methylation changes are related to the color of the shell and the nacre [133].

The transcriptome and the DNA methylation landscape of two subpopulations of the Pacific abalone *H. discus hannai* have been investigated. These two populations exhibited two different DNA methylation patterns and various differentially expressed genes linked to growth, development, transduction, and apoptosis. Particularly, a genomic hypomethylation was observed for the southern population that could be explained by long-term acclimatization to heat [135] (Figure 15.6a, b).

Young invasives populations of *Xenostrobus securis* that manage to settle in a new areas are less methylated than the older, well-established population, suggesting a role of DNA methylation loss in phenotypic plasticity [136] (Figure 15.6c).

Mollusks have been widely used in risk assessment studies and the impact of environmental chemicals on epigenetic information has been investigated in few studies. An acute benzoalphapyrene (Bap) exposure on the blood clam *Tegillarca granosa* cause gill damage, neurotoxicity (inhibition of choline enzymes), and oxidative stress (overexpression of antioxidants). Moreover, the global DNA ML is reduced, and DNA methylation is negatively correlated to antioxidant activities while positively linked to activities of choline enzymes. These results suggest a role of a methylation switch to increase and reduce the transcriptional activity of the antioxidants and the choline enzymes, respectively [42] (Figure 15.7a).

Figure 15.5 (Continued) comparison; 22 and 32; and 42 and 32, for salinity comparison, respectively. Total/no. tested is the total number of differentially methylated loci (DMLs) over the total number tested. (c) Methylation direction changes in *P. margaritifera var. cumingii* in candidate genes linked to inner shell color darkening in the depth treatment individuals at three successive sampling times (T-S1, T-S2, and T-S3). The first four genes are involved in pigmentation processes, and the last two in biomineralization processes. "+" illustrates a significant hypermethylation of at least one CpG and "−" illustrates a significant hypomethylation of at least one CpG. +/T-S(a) and −/T-S(b) mean that the methylation in the present sampling time is more methylated than in the T-S(a) and less than in the T-S(b). ABCC1, multidrug resistance-associated protein 1; GART, trifunctional purine biosynthetic protein adenosine-3; GRL101, G-protein coupled receptor GRL101; MAPKAP1, target of rapamycin complex 2 subunit MAPKAP1; MGAT1, alpha-1,3-mannosyl-glycoprotein-2-beta-*n*-acetylglucosaminyltransferase. Pictures showed are illustrative. (d) Number of differentially methylated regions (DMR) associated genes in the snail *P. antipodarum* between habitats with current speed variations correlated to specific gene categories using DMR at $p < 10-3$ threshold. (e) Heat map of DMLs in *C. virginica* reproductive tissue created using a Euclidean distance matrix. Samples in control pCO_2 conditions are represented by gray, and samples in elevated pCO_2 conditions are represented by a black bar. Loci with higher percent methylation are represented by darker colors. A logistic regression identified 598 DMLs, defined as individual CpG dinucleotide with at least a 50% methylation change between treatment and control groups, and a *q*-value <0.01 based on correction for false discovery rate with the SLIM method. The density of DML at each percent methylation value is represented in the heat map legend. *Source:* (a) Adapted from Suarez-Ulloa et al. [131]; (b) Kong et al. [132]/Springer Nature; (c) Modified from Stenger et al. [133]/Frontiers Media S.A/CC BY 4.0; (d) Thorson et al. [134]/Springer Nature/CC BY 4.0; (e) Venkataraman et al. [57]/Frontiers Media S.A/CC BY 4.0.

(a)

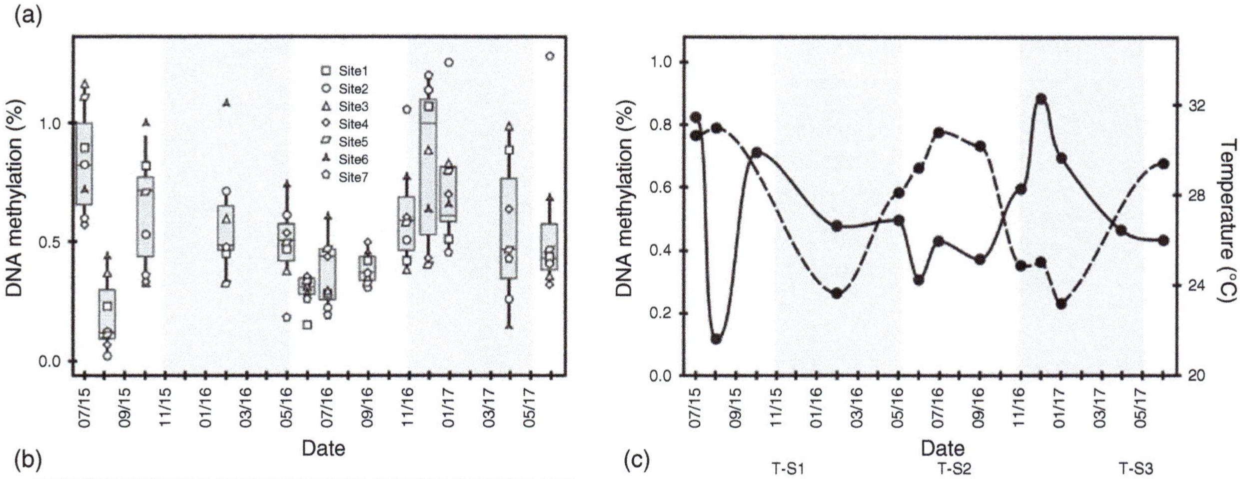

(b)

Primer set	No. significant		Total/no. tested	No. significant		Total/no. tested
	20 °C[a]	30 °C[b]		22[c]	42[d]	
Set 1	2	2	4/25	2	0	2/20
Set 2	6	3	8/46	1	2	3/39
Set 3	5	4	9/58	4	4	7/74
Set 4	4	3	7/45	4	9	12/53
Set 5	8	2	10/89	14	7	18/94
Set 6	7	2	9/66	5	0	5/58
Set 7	1	0	1/36	0	1	1/35
Set 8	7	2	9/58	3	2	3/39
Set 9	5	1	6/49	5	3	7/69
Set 10	3	1	4/26	2	5	5/41
Total	48	20	67/498 (13.5%)	40	33	63/542 (11.6%)

(c)

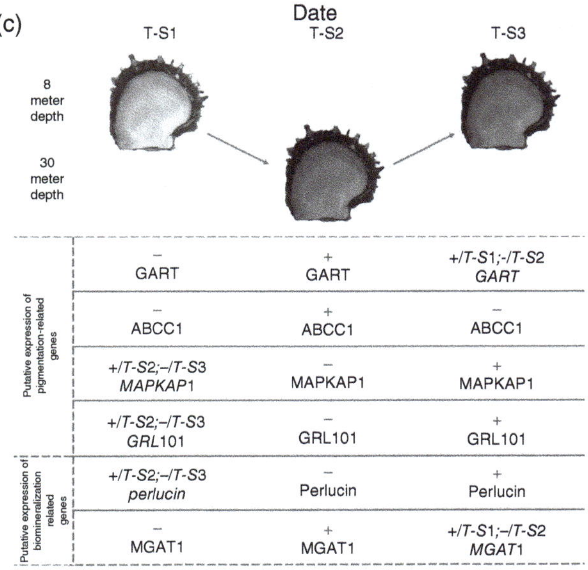

(d)

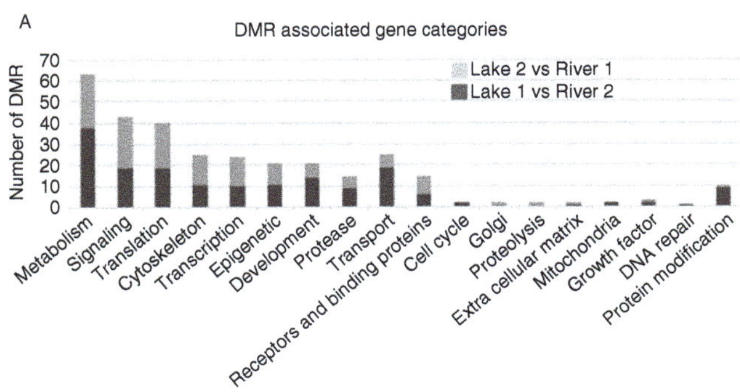

(e)

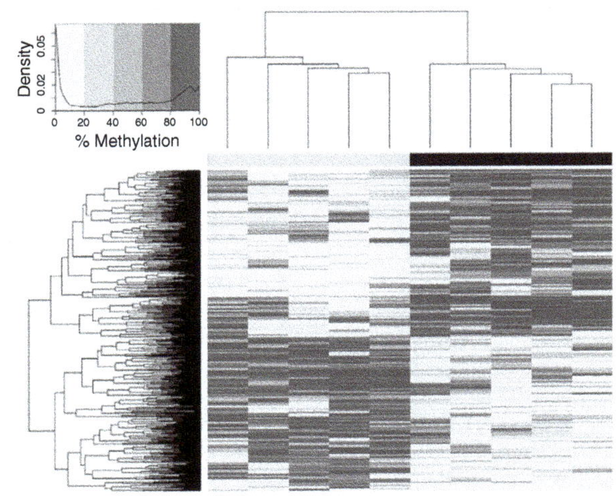

Figure 15.5 Epigenetic response to climate change and habitat features in mollusks. (a) Seasonal variation in global genome-wide DNA methylation in *I. Alatus* from North Biscayne Bay, Florida, displaying significant differences between averages in dry and wet seasons (*P* value = 2E-3, t-test) (left). Each site is represented by a biological sample of pooled gill tissue from five oyster individuals. Months corresponding to dry seasons are highlighted by gray boxes, as opposed to wet season months, over the two years' experiment. DNA methylation data (solid line) are compared with temperature data (dashed line, medians across sites) showing opposite seasonal trends (right). (b) Exact tests for difference in methylation frequency among the methylation-susceptible polymorphism locus in *H. discus hannai* in selected temperature and salinity treatments. [a], [b], [c], and [d] are the number of loci showing a significant difference in methylation frequency between treatments of 20 and 24 °C; 30 and 24 °C for temperature

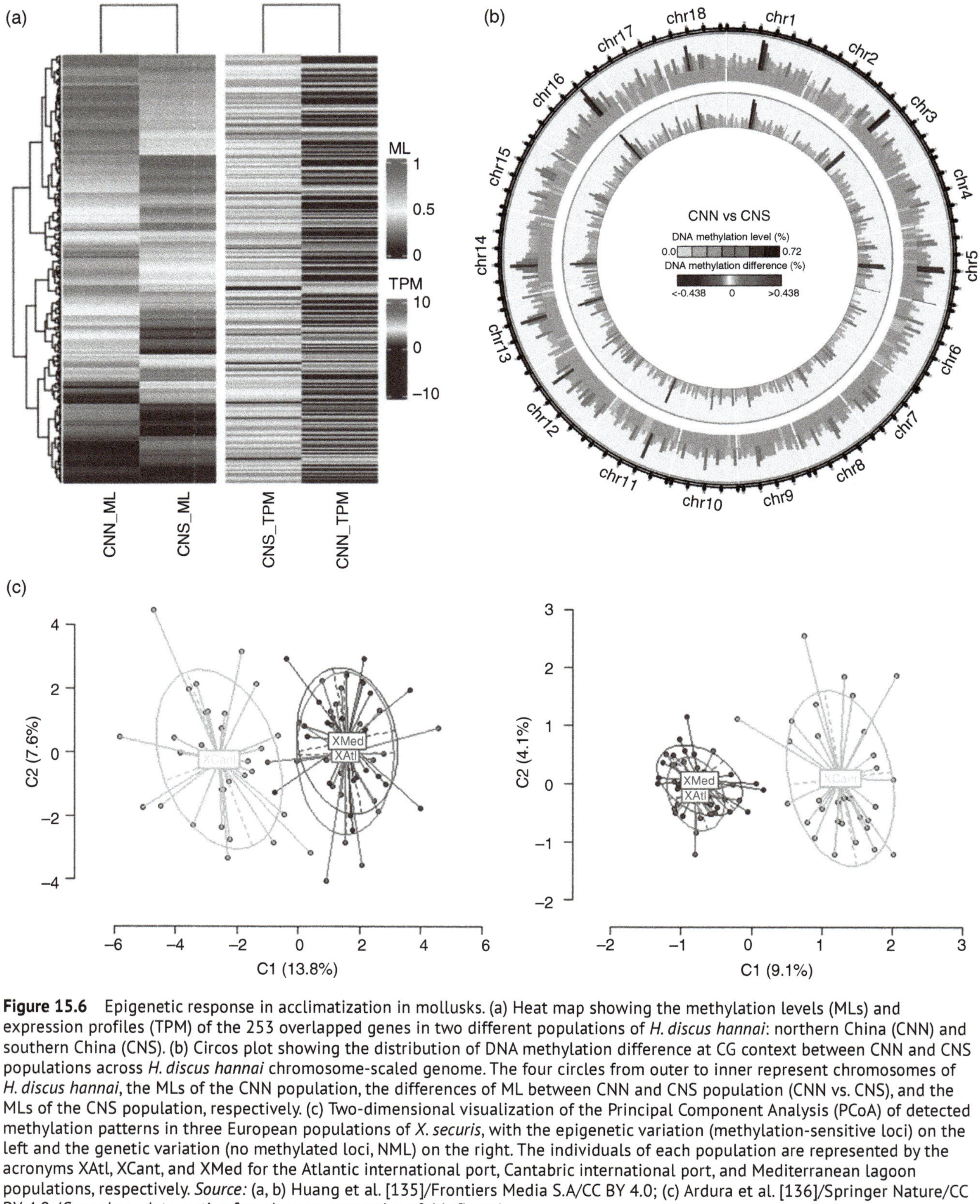

Figure 15.6 Epigenetic response in acclimatization in mollusks. (a) Heat map showing the methylation levels (MLs) and expression profiles (TPM) of the 253 overlapped genes in two different populations of *H. discus hannai*: northern China (CNN) and southern China (CNS). (b) Circos plot showing the distribution of DNA methylation difference at CG context between CNN and CNS populations across *H. discus hannai* chromosome-scaled genome. The four circles from outer to inner represent chromosomes of *H. discus hannai*, the MLs of the CNN population, the differences of ML between CNN and CNS population (CNN vs. CNS), and the MLs of the CNS population, respectively. (c) Two-dimensional visualization of the Principal Component Analysis (PCoA) of detected methylation patterns in three European populations of *X. securis*, with the epigenetic variation (methylation-sensitive loci) on the left and the genetic variation (no methylated loci, NML) on the right. The individuals of each population are represented by the acronyms XAtl, XCant, and XMed for the Atlantic international port, Cantabric international port, and Mediterranean lagoon populations, respectively. *Source:* (a, b) Huang et al. [135]/Frontiers Media S.A/CC BY 4.0; (c) Ardura et al. [136]/Springer Nature/CC BY 4.0. (*See colour plate section for colour representation of this figure.*)

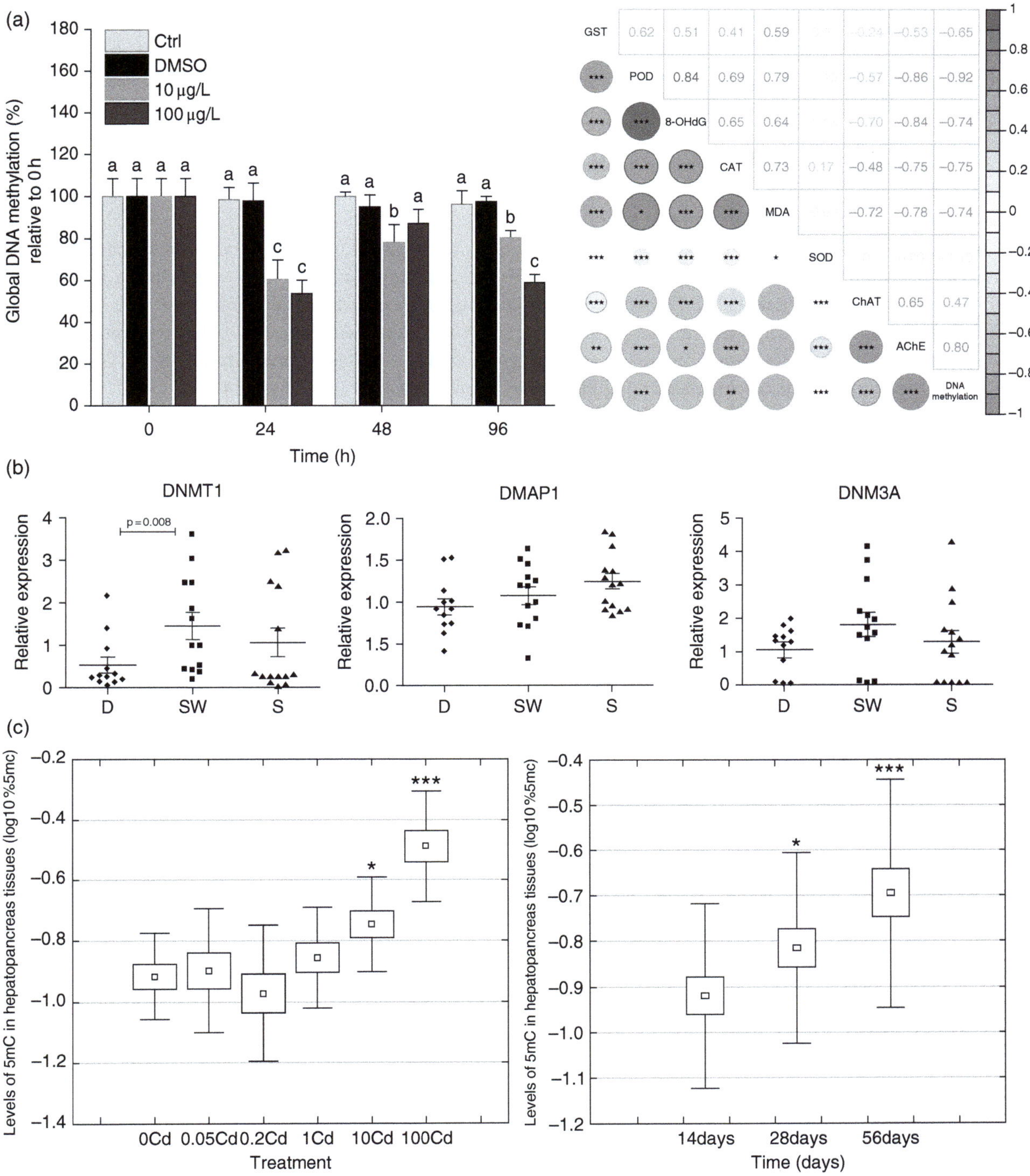

Figure 15.7 Examples of epigenetic response to chemical pollution in mollusks. (a) Effects of acute Bap exposure on global DNA ML in gills of blood clams *T. granosa*. The vertical bars represent the mean ± SD ($n = 3$). Different letters indicate significant differences ($p < 0.05$) (left) and Spearman rank correlation results after Bap exposure for different indicators: (i) oxidative stress with antioxidant activities of superoxide dismutase (SOD), catalase (CAT), peroxidase (POD) and glutathione-s-transferase (GST), lipid peroxidation (LPO) level and 8-hydroxy-2′-deoxyguanosine (8-OhdG) content; (ii) neurotoxicity with acetylcholinesterase (AChE) and choline acetyltransferase (ChAT) activities; and (iii) global DNA ML. The circle size is proportional to the correlation value (Spearman rank correlation coefficient), which represents correlation level. Positive values represent positive correlation and negative values represents negative correlation. *p* values: $^{*}p \leqslant 0.05$, $^{**}p \leqslant 0.01$, $^{***}p \leqslant 0.001$. (b) Expression levels of genes relevant to DNA methylation in *C. gigas* oyster genitors exposed respectively to Diuron (D) in comparison with seawater (SW) and solvent (S) control groups. Relative expression data were normalized with the CT geometric mean value of three reference genes (*Cg-ef1α*, *Cg- rpl40*, and *Cg-rps6*) according to the $2^{-\Delta\Delta CT}$ method. Significant differences in relative gene expressions between oyster groups (D, SW, and S) were determined by Mann–Whitney U tests and *p* values are given if significant at the $p < 0.05$ threshold. (c) Effect of dietary cadmium on genome-wide 5mC levels in DNA of the hepatopancreas of *C. aspersus* adults. Estimated marginal means for dietary dose (as the main effect) (left) and estimated marginal means for time (as the main effect) (right). Four snails were sampled for each treatment group at each time point (14, 28, and 56 days). Data are shown on a log10 scale as mean (point) with one standard error (box) and one standard deviation (error bar). Marked boxes (*) indicate significant differences as compared to the reference group (Newman-Keuls test, $^{***}p < 0.001$, $^{*}p < 0.05$). *Source:* (a) Guo et al. [42]/Springer Nature/CC BY 4.0; (b) Akcha et al. [137]/Springer Nature; (c) Nica et al. [138]/PLOS One/CC BY 4.0.

Phthalates are plastic additives widely found in the environment. An exposure of the snail *Physella acuta* to the butyl benzyl phthalate (BBP) is responsible for an overexpression of genes linked to epigenetic regulation (DNMT1, KAT6B and HDAC1) [139].

Diuron exposure on *C. gigas* during gametogenesis decreases DNMT1 activity [137] (Figure 15.7b). In addition, a parental exposure of oysters to diuron has shown no change in global methylation rate of offspring, but changes in the DNA methylation pattern touching specific genomic regions [59].

Both the effects of the fungicide vinclozolin and the glucocorticoid prednisolone (PDS) have been tested on *P. acuta*. A long-term exposure (45 days) to vinclozolin altered the snails' survival, together with DNA methylation patterns [140], while no effect of PDS has been detected. F0-exposed snails survived but had reduced fecundity, and both shell and DNA methylation changed in F1 and F2 generations [41]. For the gastropod *Cantareus aspersus*, cadmium exposure led to an increase in both 5-methylcystosine and 5-hydroxymethylcytosine (5hmC) in the hepatopancreas [138, 141] (Figure 15.7c). The global level of 5hmC has also been affected (decreased) by copper exposure in *C. gigas* [142].

Epigenetic changes in response to infectious agents and parasite challenges have been investigated in oysters. Haemocytes of *C. gigas* challenged using lipopolysaccharides (LPS) in vitro, or *Vibrio splendidus* displayed an increase in DNMT3 mRNA together with an overexpression of cytokines involved in oyster immune response. Such reactions were inhibited when the oysters were pretreated with a DNA methyltransferase inhibitor [143]. After having been exposed to microbiota-enriched water during development, *C. gigas* adults displayed improved survival to the POMS diseases together with specific DNA methylation changes [144]. The oyster *C. gasar*, when infected by the protozoan parasite *Perkinsus marinus* or *P. olseni*, display gills and gastrointestinal hypomethylation [145]. A similar hypomethylation event was detected in *C. virginica* in response to infection with toxic algae [146].

Other than in oysters, DNA methylation response to infectious stress has been studied in the snail *B. glabrata*. The parasite *Schistosoma mansoni* can manipulate the genome of *B. glabrata* by relocating the gene responsible for the heat-shock protein hsp70 to a nuclear compartment and thus lead to the expression of hsp70, which has been shown to enable the parasite to establish in the host [147]. In the susceptible snails, the *Hsp70* gene was repositioned after infection from an intermediate position to an internal location within interphase nuclei. This repositioning was temporally correlated with upregulation of gene expression two hours post infection. No repositioning or any significant change in expression is observed in the resistant strain snails. Further, there was no nuclear relocation of the *Hsp70* gene loci and no real *Hsp70* expression was detected by quantitative PCR when the two snail lines were infected with irradiated miracidia. (Figure 15.8a). These results are the only ones available showing an implication of nuclear structure in stress response in mollusks. Both an infection to the parasite *S. mansoni* and heat-shock stress led to DNA hypomethylation. In the case of the parasite infection, the hypomethylation was tissue-specific and presumably linked to differential expression of genes involved in stress response and differences in the DNA methylation machinery [68, 148] (Figure 15.8b, c). Surprisingly, in the snail *Zeacumantus subcarinatus*, after a trematode infection (*Philophthalmus* sp. or *Maritrema novaezealandensis*), no modifications in the level of 5mC or in global histones post-translational tails were observed [149].

To conclude, mollusks are sensitive to a wide array of environmental factors that can affect their epigenetic machinery. Nevertheless, such epigenetic modifications seem to be time-, stress-, tissue-, and species-specific. Moreover, the molecular mechanisms linking the environmental cues to the epimutations remain undescribed. The involvement of the cellular metabolism in such information transfer has been suggested [55]. Indeed, both mitochondria and microbiota can produce intermediate metabolites (e.g., acetyl coA, S-adenosylmethionine) known to act on epigenetic marks in the nucleus [150–152]. Corroborating to this hypothesis, the mitochondrion has been shown as a specific target of environmental stress for mollusks: in response to cadmium in clams [153], to titanium particles in mussels [154], and to nematode infection in snail *B. glabrata* [155].

15.6 Mechanisms of Meiotic Epigenetic Inheritance in Mollusks and Their Impact in Evolution

Intergenerational effects in mollusks have been demonstrated in 14 mollusk species representing eight mollusk families [55]. Most of them have focused on the effect of parental exposure (F0) on the direct offspring (F1). Only four among them have studied more than one generation of descendants [41, 81, 156, 157]. As in mollusks, germ cells are produced early in development [158–161]; transgenerational inheritance can be observed only in studies conducted for a minimum

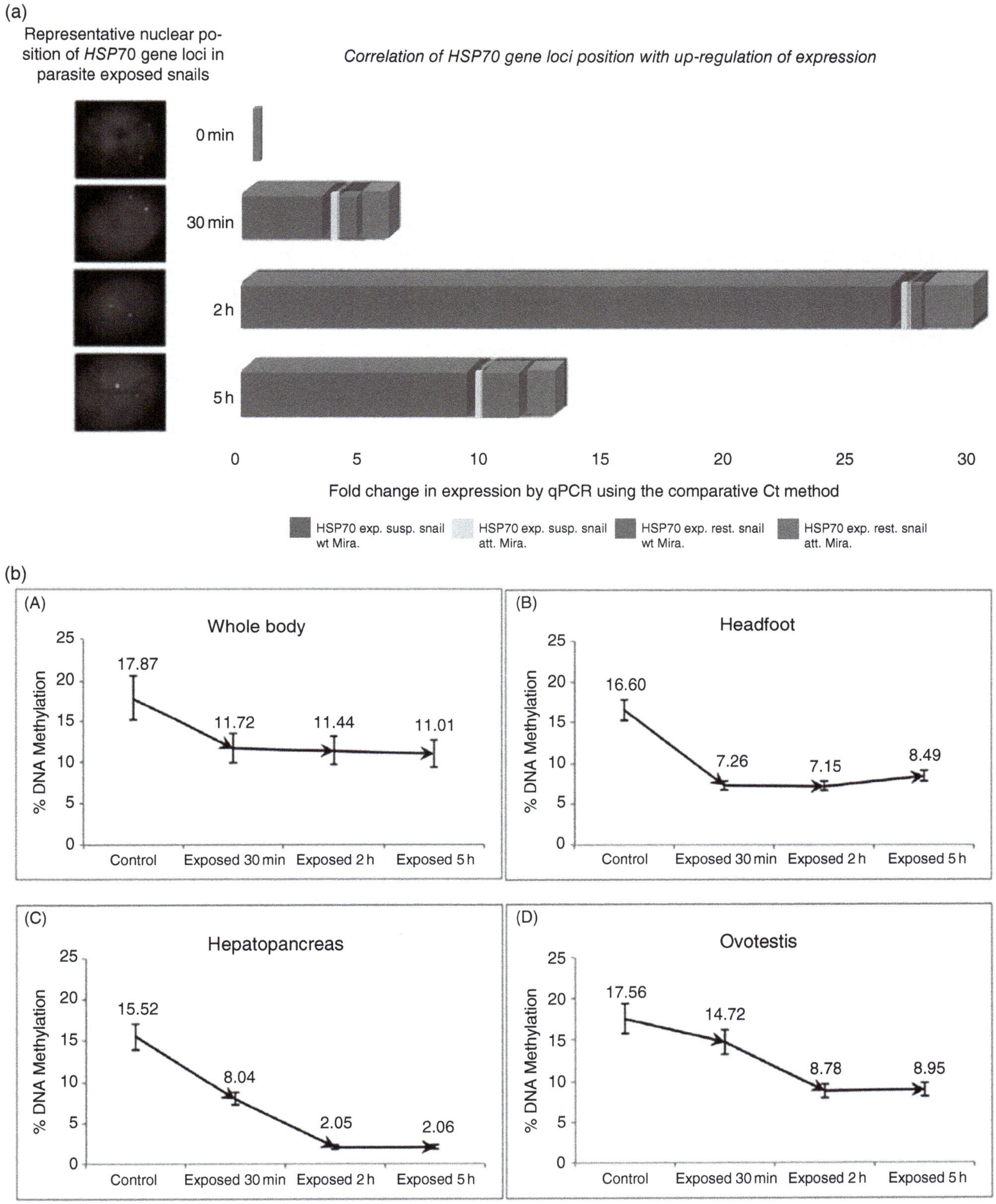

Figure 15.8 Epigenetic modifications in response to *Schistosoma mansoni* infection in *B. glabrata*. (a) Chart displaying the fold change in the expression of the *B. glabrata Hsp70* gene in the interphase nuclei of cells derived from the susceptible and resistant snail strains, pre- and post exposure to *S. mansoni* miracidia, as measured by quantitative PCR. Representative images of these positions are shown in columns. (b) Cytosine DNA MLs of the *Bg-hsp70* gene in *B. glabrata* whole body tissue, head foot, hepatopancreas, and ovotestis after *S. mansoni* miracidiae (Box 15.2) exposure (10 miracidia/snail) and heat shock treatment, respectively. DNA methylation percentages were calculated from individual samples that were run in triplicates, and from four independent experiments.

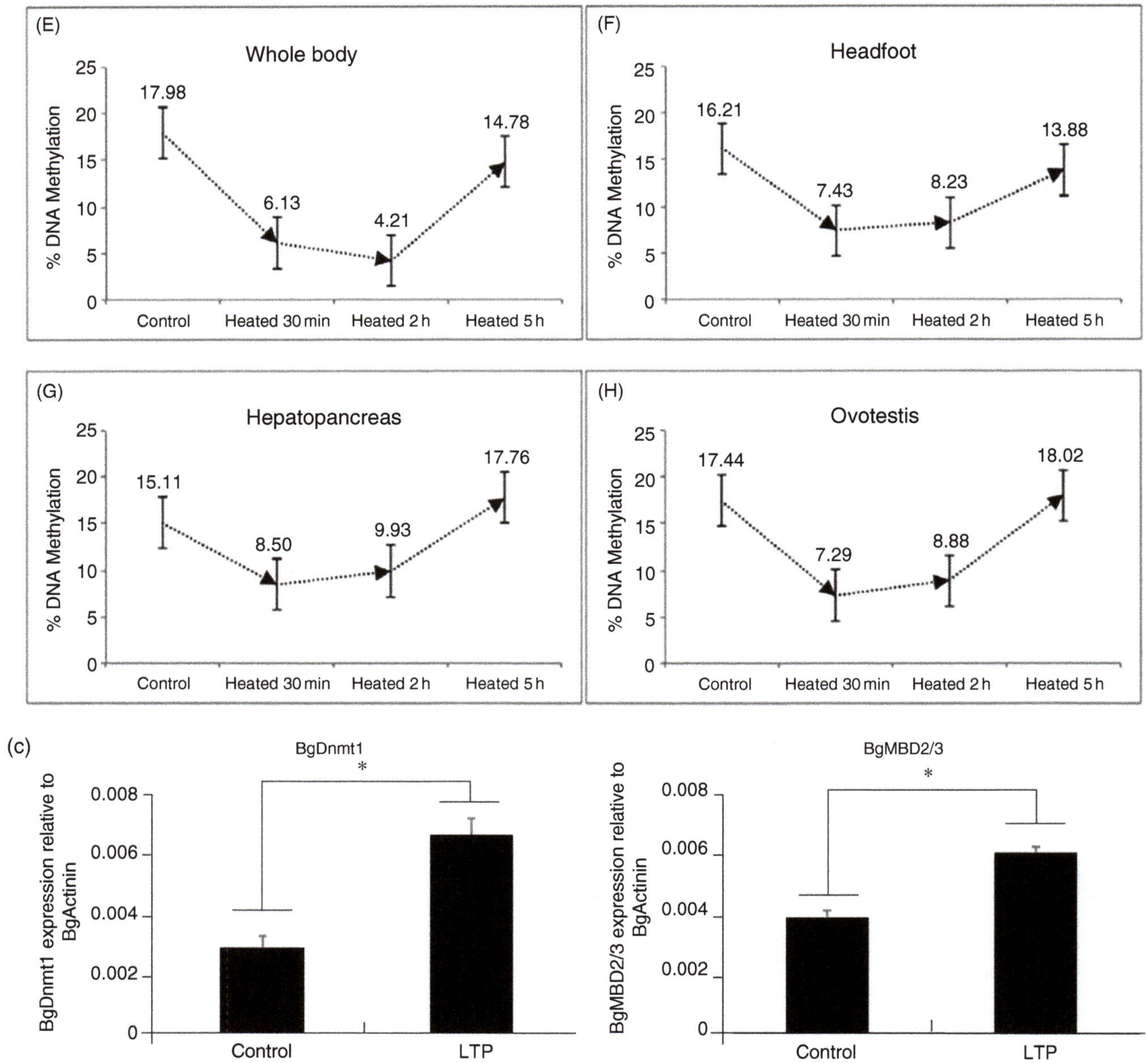

Figure 15.8 (Continued) The average of delta Ct by qRT-PCR with SYTO-9 was used for ML calculation. Percent values are expressed as means +/− SD. (c) qRT-PCR expression analysis of the DNA methyltransferases *Bgmbd2/3* and *Bgdnmt1* in untreated *B. glabrata* embryonic (Bge) cells (control) versus cells treated with larval transformation products (LTPs) derived from *S. mansoni* miracidia to sporocyst (Box 15.2) transformation. Data is derived from duplicate biological samples and qRT-PCR reactions were performed in technical duplicates. The Ct-values of target genes were normalized to *Bgactin*. Error bars represent SD of the normalized means and a student's t-test was performed to identify expression differences between LTP-treated and untreated (Control) Bge cells (*$p < 0.05$). *Source:* (a) Redrawn and modified from Knight et al. [147]/Elsevier; (b) Ittiprasert et al. [148]/Academic Journals/CC BY 4.0; (c) Geyer et al. [68]/PLOS/CC BY 4.0.

of three generations (F0 = directly exposed generation; F1 = indirectly exposed generation via germ cells and F2: unexposed generation). However, studies on intergenerational effects in mollusks have shown that various environmental factors, both biotic and abiotic, may have an impact on the phenotype of the offspring.

Predator cues like crab odors, used for parental exposure in *N. lapillus*, affected the offspring by increasing its performances via enhancement of offspring size at birth and during growth. Moreover, it also led to an increase of the anti-predator behavior of the offspring. Such results are nevertheless observed only when the offspring face the similar cues encountered by the parents [162–164]. The same effect has been observed in *P. acuta*, in which parental exposure to crayfish also induced an increase in anti-predator behavior, shell thickness, and shell size compared to the control [165, 166]. Such an impact has also been identified in *B. glabrata*: parental exposure to predator cues led to an increased cadmium tolerance in the offspring [167]. Finally, when *Crassostrea gasar* mothers have been exposed to a virus, their offspring survive better to the same virus exposure [168].

Such phenomenon is also observed in response to abiotic factors. In response to a water acidification event endured by parents, both *R. philippinarum* and *Saccostrea glomerata* offspring displayed an increase in shell growth and a modified metabolism, even if they were also exposed to acid pH [21, 169]. Acidification had negative effects on the parents (reduced growth, alteration of developmental rate, and survival). Conversely, clams and scallops seem unable to acclimate to ocean acidification because parental exposure to acid water led to more vulnerable offspring [170]. Interestingly, Parker et al. have also shown that a parental exposure to elevated CO_2 in *S. glomerata* improved offspring survival if exposed solely to the same stressor, but is maladaptive if the offspring faced several stressors [171]. Also surprisingly, when looking at the effect of parental exposure to salinity, reverse effects have been observed on the hatching success of gastropods *Helisoma trivolvis* (increase of hatching success) and *Physa. pomilia* (decrease of hatching success) [172].

The effect of cadmium exposure of F0 embryos or juveniles on the hatching success of F1 snails of *Physella pomilia* has been investigated. When parents were exposed as embryos, the hatching success of their offspring was reduced, while juvenile exposure of the parents did not influence the offspring phenotype [173]. An exposure to cadmium at the adult stage led to an increased tolerance of the offspring to cadmium [167]. Moreover, Reategui-Zirena et al. have exposed adult *L. stagnalis* to cadmium and studied the consequences on both the embryonic (hatching success and time to hatch) and postembryonic (juvenile survival) stages [174]. The parental exposure did not induce any effects on the hatching success while it increased cadmium tolerance and survival of juveniles.

The adverse effects due to a direct exposure of eggs from *L. stagnalis* to the fungicide pyraclostrobin are amplified if the parents have been exposed, leading to an increase in the eggs' hatching time, a decrease in the eggs' hatching success, and developmental defects in the larvae. Such results have been suggested to be linked to a change in adult metabolism, leading to modification of the macronutrient content of the eggs [40].

Intergenerational effects have thus been highlighted in response to various environmental stressors. Such effects constitute interesting and promising cues for further epigenetic investigations. Nevertheless, and unfortunately, only a few studies have addressed the involvement of epigenetic mechanisms in intergenerational effects.

Among studies investigating epigenetic mechanisms in intergenerational effects, Bal, Kumar, and Nugegoda [41] have exposed three generations of *P. acuta* snails to PDS (a glucocorticoid). This continued exposure has resulted in shell abnormalities and underdeveloped F1 and F2 snails with a tendency of adverse effects in additional generations. Moreover, the global methylation pattern (% of 5-methyl cytosine) in parents has not been affected by the exposure, but resulted in a decrease in the F1 at the higher concentrations tested; the 5mC levels of *P. acuta* did not significantly change in PDS-treated adult snails of F0 generation in comparison to the adult snails of the control group. The multigenerational PDS exposure resulted in a linear decrease of %5mC content in adult snails of the F1 generation to <0.4% at 32–64 μg/L, while control snails had %5mC content in the range of $0.68 \pm 0.03\%$ ($p < 0.001$). (Figure 15.9a). In *C. gigas*, parental exposure to diuron (an herbicide) has also revealed a consecutive change in the progeny DNA methylation [59] (Figure 15.9b). In another study, on *C. gigas*, a parental exposure to a nonpathogenic natural microbiota has also induced DMR in the progeny [144] (Figure 15.9c). A final interesting discovery provided by literature is in the chromatin of oyster germs cells: most of the histones are replaced by protamine-like proteins, but some histones persist. Such persisting histones could be involved in transgenerational epigenetic inheritance (TEI) [96, 175].

The published studies on intergenerational and transgenerational effects in mollusks allow the identification of major avenues for the study of epigenetic inheritance [55]. Results on intergenerational effects have shown that the inheritance of phenotypes depends on the environmental conditions experienced by the offspring; some phenotypes will be visible only if offspring meet the same exact cue as the parents. Furthermore, the developmental stage at which the ancestors are exposed, as well as the developmental stage studied in the offspring, are crucial for experimental designs because different phenotypes can be observed when these variables are modified. Regarding epigenetic inheritance per se, relationships between phenotypical changes, inheritance patterns, and molecular aspects of epigenetic information have been barely addressed in mollusks and several challenges remain in this field [55].

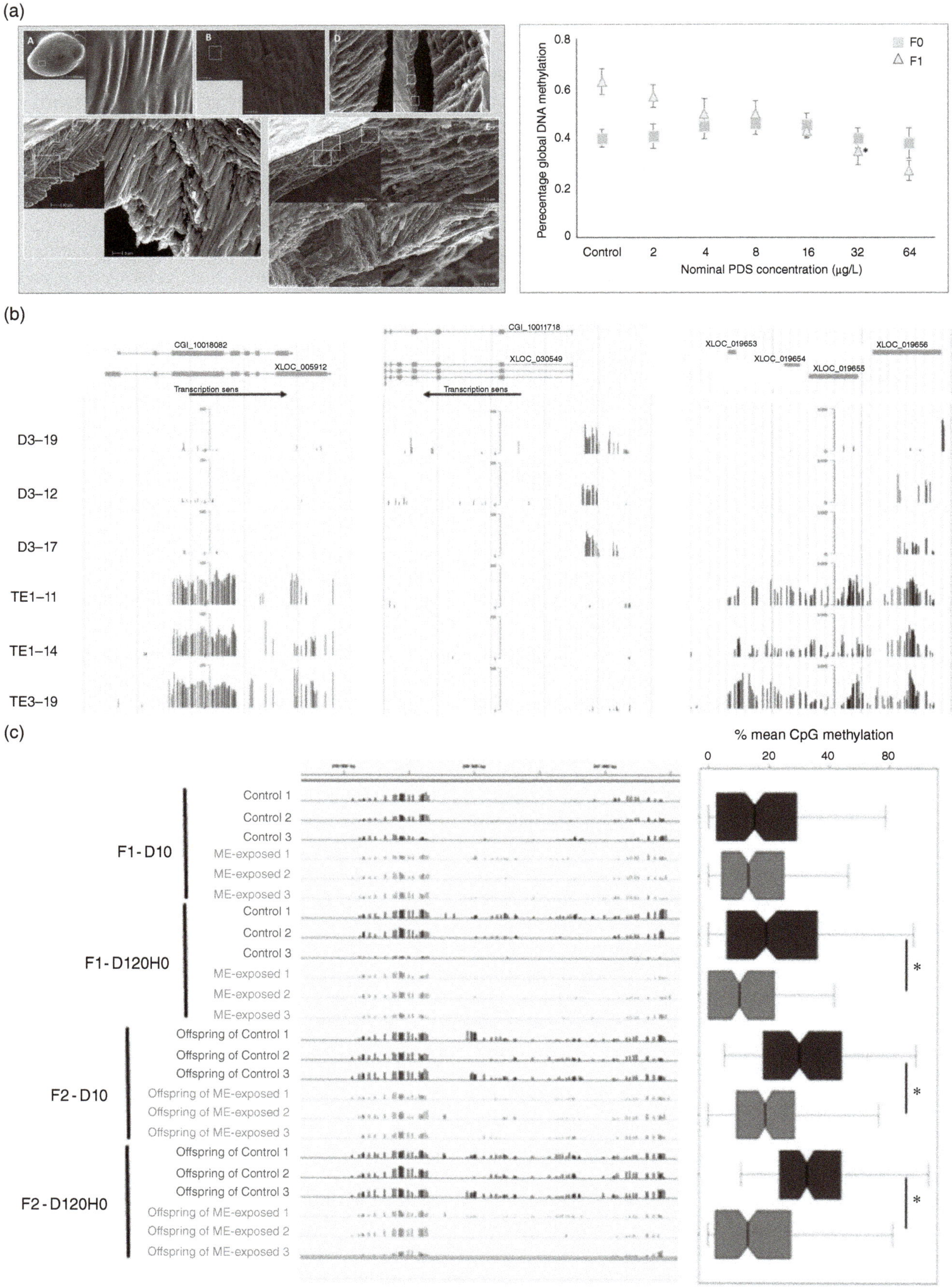

Figure 15.9 Epigenetic mechanisms in multigenerational effects in mollusks. (a) SEM micrographs *P. acuta* snail shell structures of adult F1 generation (left). Pictures showing normal shell structures of the control group (A–C) and abnormally formed shell structure of PDS exposed snails (D, E) at 32–64 μg/L. INL, Inner nacreous layer; MPL, Middle prismatic layer; OPL, outer periostracum layer. Percentage of global DNA methylation in snails of F0 (exposed to PDS) and F1 generations (mean ± SD) (right). Error bars indicate SDs. (b) Extract of the *C. gigas* jbrowse interface, which allows visualization of transcriptomic and DNA methylation data for three regions

15.7 Perspectives

15.7.1 Remaining Challenges in Mollusk Epigenetics

Epigenetic modifications in response to a wide array of environmental changes have been highlighted in various mollusk species. Such epigenetic changes have also been identified in a few intergenerational responses in mollusks. Unfortunately, in this context, only DNA methylation has been studied extensively and a lack of information about the other epigenetic mechanisms remains. Moreover, the mechanisms by which the environment impacts epigenetic marks are also still unknown. Regarding answers about epigenetic mechanisms and their role in evolutionary processes, more studies are needed. Major approaches for studying inter- and TEI have been highlighted by Fallet et al. [55]:

Investigating how and to what extent the epigenetic processes are related to phenotypic changes over generations. For this purpose, it is necessary to study (i) whether specific environmental factors lead to both phenotypic and epigenetic responses; (ii) if the reversibility of both the phenotypic response and the epigenetic changes are the same, and the possible asymmetry in response for both, between male and female. To realize this objective it is necessary to define the developmental windows for the exposure of the organisms, considering that the precocious stages are the most sensitive to environmental factors [176]. Another prerequisite is to detect and focus on the organs or tissues that are the most appropriate to study the selected phenotype. A whole genome study will be the most suitable to detect epigenetic changes. In general, validation studies are lacking in this field, but the Crispr-cas9 tool is promising to study the cause/effect relationship of an epimutation for a specific phenotype. Crispr-cas9 contains an inactivated cas9 protein that can be combined with a specific enzyme, for instance the DNMT3 enzyme [177], to study the impact of methylation. This system would allow researchers to methylate a specific locus and to induce targeted epimutations whose functional impact may be investigated.

Estimating the phenotypic outcome (adverse or adaptative) of epigenetic processes and identifying the relevant drivers for TEI among the collateral, stochastic, or silent epimutations. In this regard, it is crucial to examine the fitness (survival and fecundity) in several generations and assess it in both controlled and exposed conditions. If epigenetic changes are identified as being linked to phenotypic changes affecting fitness, then such epigenetics processes are promising as evolutionary players.

Tracking down epigenetic marks/modifications from the somatic to the germ cells over generations is a good approach to determine whether such changes are vehicles of heritable information or merely phenotypic byproducts. Here, the choice of the studied tissues is crucial. To be transmitted over generations, such epigenetic markers need to be transmitted to the germ cells and, therefore, to escape to the chromatin reprogramming that occurs during germ cells production and after fertilization. Such reprogramming has been identified in mammals [178] and is suspected to occur also in mollusks [46]. The transmission of epigenetic information over generations would thus require transient support, allowing for re-writing later in development. In mollusks, some histones in germ cells are not replaced by protamines and persist and could then participate in the information transfer. Moreover, noncoding RNAs are interesting targets for information transfer identified in other organisms [179, 180], and should also be studied in mollusks. Furthermore, some marks (DNA methylation and histone modifications) escape reprograming and are transmitted to the next generation [181]. Understanding if these "resistant" bearers have been environmentally induced and why/how they do not undergo this erasure is a crucial issue.

Figure 15.9 (Continued) strongly impacted by a parental exposure to Diuron: (left) Scaffold 1255 from 364,500 to 370,000 containing XLOC_005912 (CGI_10018082), (middle) Scaffold 43,664 from 222,500 to 235,000 containing XLOC_030549 (CGI_10011718), and (right) Scaffold 309 from 130,000 to 137,500 containing four unique exon genes (XLOC_019653, XLOC_019654, XLOC_019655, XLOC_019656). Upper panel represents the annotation of genes in the current genome assembly V9 (CGI number), and the CDS. Bottom panel represents the methylated cytosine position. *X*-axis represents the position of each methylated cytosine. *Y* axis (bars) represents the percentage of methylation observed for each methylated cytosine (scale is 0–100). The six tracks represent the three biological replicates of individual spats from genitors treated with diuron or control (from top to bottom). (c) Snapshot of Integrative Genomics Viewer windows (left), which display CpG methylation data from scaffold 1792:239,000–242,000 for oyster parents (F0) exposed to a nonpathogenic natural microbiota (Microorganisms Enriched SW-exposed: ME-exposed) and oyster parents raised in filtered and UV-treated SW (control) and their respective F1 offspring. D10 = day 10 after fecundation, D120 = 120 days after fecundation. Each bar indicates a methylated CpG position on a scale from 0 to 100%. The hypomethylation displayed in the F1 juvenile oysters (F1 – D120H0) in ME-exposed (gray) compared to control oysters (black) was detected with DMRseq software and was inherited from F1 to F2 generation in larvae (D10) and juveniles (D120H0); and box plots (right) representing the distribution of the level of methylation (mean of three biological replicates) for each CpG position present in the selected region. * indicates a p-val < 0.05 of a Mann–Whitney analysis comparing the ME-exposed and control oysters for each studied condition. *Source:* (a) Modified from Bal et al. [41]/with permission of Elsevier; (b) Rondon et al. [59]/Oxford University Press; (c) Fallet et al. [144]/Springer Nature/CC BY 4.0.

Distinguishing between genetic/epigenetic and other processes of information inheritance between organisms and generations. Epigenetics and genetics are known to affect each other and there is no reason to study them separately. Therefore, all studies aiming at investigating epigenetic effects should also investigate genetic mutations, SNPs, and copy number variations to be sure that the observed phenotype is not due to genetic changes.

Discovering the complex interplay between the different epigenetic processes and how they can influence each other to draw an overall picture of epigenetic processes and interactions. Epigenetic processes and epigenetic bearers of information are highly interconnected [182]. Therefore, it is recommended to study them all together and not look solely at one of them. Unfortunately, in mollusks, the focus has been mainly on DNA methylation, and, therefore, integrative studies are needed.

Epigenetic studies in response to environmental changes have increased in the last 20 years. A lot of different analytical study designs and analytical tools have emerged, making comparison between studies tricky. Moreover, some teams have become specialists regarding one epigenetic mark and/or one specific organism. Therefore, the use of collaborative platforms and the creation of multidisciplinary projects are recommended to harmonize experimental design and facilitate future research.

15.7.2 Application of Epigenetic Markers in Aquaculture and Aquatic Species Conservation

Various mollusk species have been used as bioindicators because they (i) are effective sentinel organisms for the study of pollution [141, 183–188] and oxidative stress [189], as they are nonmigratory, ubiquitous, abundant, and easy to sample; (ii) can stand high concentrations of several toxic substances and environmental alterations; (iii) can bioconcentrate contaminants and live long enough to allow comparisons between ages; and (iv) have an important place in the food chain and dose-effect relationships can be observed with them [190]. Moreover, environmental stressors like temperature changes [191], water parameter variations [169], pollutant exposures [59, 192], pharmaceuticals [192] nutritional deficiencies [191], pathogens [193], or predator presence [194, 195] are likely to affect the epigenome. As mentioned above, such epigenetic changes can persist after the disappearance of the initial stressors and may be transmitted to the next generations, making epigenetic modifications promising tools as predictive biomarkers for the sensitivity toward environmental factors [196, 197]. Therefore, combining epigenetic studies in mollusks will be an asset for mollusk preservation. The reactivity of epigenetic marks to environmental changes in mollusk species could then be used to detect threats to mollusk's aquaculture. An example has been provided by Rondon et al. [59], who have identified two predictive differentiated DNA methylated regions in response to parental Diuron exposure in *C. gigas*. Such regions could be used as biomarkers for risk assessment for aquaculture. Identifying relevant epigenetic modifications in the context of ecotoxicology and further developing fast and accurate tools for their detection in various organisms could help identify deleterious environments affecting both mollusks and other organisms' phenotype for the aquaculture industry. Such fast identification could allow rapid actions to protect and conserve threatened aquaculture species and fisheries.

15.8 General Conclusions

Approximately, 130,000 mollusks species currently have been identified and they share some specific features like a soft body with three main parts: the head, the foot, and the visceral mass. Nevertheless, mollusks species differ in their nutrition mode, reproductive system, and habitat. They can be found in all types of environments (terrestrial, freshwater, and marine). Many mollusk species are of economic interest and used in aquaculture, or as fertilizers, poultry grit, construction material, jewelry, pharmaceuticals, and nutraceuticals. However, mollusks are also very sensitive to environmental changes and anthropic pollutions. Moreover, mollusks' aquaculture history has been marked by numerous epidemics that have led to major economic loss. The sensitivity of mollusks to various stressors has led to the study of epigenetic changes. In this regard, the epigenetic machinery has been identified in mollusk species that indeed, possess DNA methylation, DNA hydroxymethylation, histones marks and variants, as well as noncoding RNAs. Some studies have allowed researchers to highlight the link between environmental changes and epigenetic modifications, together with the connection between environmental stress and intergenerational effects, but only a few studies have investigated epigenetic inheritance and more efforts need to be focused on this field. Identifying relevant epigenetic modifications in ecotoxicology and further developing fast and accurate tools for their detection in various organisms using mollusks could help identify deleterious environments that affect aquaculture species and industry.

References

1 Parkhaev, P.Y. (2017). Origin and the early evolution of the phylum Mollusca. *Paleontol. J.* **51** (6): 663–686.

2 Purves, W.K., Sadava, D., Orians, G.H., and Heller, H.C. (ed.) (2003). *Life: The Science of Biology: Plants and Animals*, 6e, vol. **III**, 450pp. Sinauer Associates, INC: Sunderland, MA.

3 Joseph, A. (2017). Oceans: abode of nutraceuticals, pharmaceuticals, and biotoxins. In: *Investigating Seafloors and Oceans* (ed. A. Joseph), 493–554. Elsevier.

4 Gökoglu, N. (2021). Molluscan shellfish. In: *Shellfish Processing and Preservation* (ed. N. Gökoglu), 129–250. Springer.

5 Kocot, K.M., Todt, C., Mikkelsen, N.T., and Halanych, K.M. (2019). Phylogenomics of *Aplacophora* (Mollusca, Aculifera) and a solenogaster without a foot. *Proc. R. Soc. B Biol. Sci.* **286** (1902): 01–15.

6 Lindberg, D.R. (2009). Monoplacophorans and the origin and relationships of mollusks. *Evol. Educ. Outreach.* **2** (2): 191–203.

7 Gosling, E. (2008). *Bivalve Molluscs: Biology, Ecology and Culture*. Oxford: Blackwell Publishing.

8 Haszprunar, G. and Wanninger, A. (2012). Molluscs. *Curr. Biol.* **22** (13): 510–514.

9 Lindberg, D.R. (2001). Molluscs. In: *Encyclopedia of Biodiversity*, 2e (ed. S.A. Levin), 373–383. Amsterdam: Elsevier, Inc.

10 Stickney, R.R. and Treece, G.D. (2012). History of aquaculture. In: *Aquaculture Production Systems* (ed. J.H. Tidwell), 15–50. https://doi.org/10.1002/9781118250105.ch2.

11 Tacon, A.G.J. (2020). Trends in global aquaculture and aquafeed production: 2000–2017. *Rev. Fish. Sci. Aquac.* **28** (1): 43–56.

12 Naylor, R.L., Hardy, R.W., Buschmann, A.H. et al. (2021). A 20-year retrospective review of global aquaculture. *Nature* **591** (7851): 551–563.

13 Western, D. (2001). Human-modified ecosystems and future evolution. *Proc. Natl. Acad. Sci.* **98** (10): 5458–5465.

14 Ceballos, G., García, A., and Ehrlich, P.R. (2010). The sixth extinction crisis loss of animal populations and species. *J. Cosmol.* **8**: 1821–1831.

15 Rodrigues, L.C., Van Den Bergh, J.C.J.M., Massa, F. et al. (2015). Sensitivity of Mediterranean bivalve mollusc aquaculture to climate change, ocean acidification, and other environmental pressures: findings from a producer survey. *J. Shellfish Res.* **34** (3): 1161–1176.

16 Clements, J.C., Hicks, C., Tremblay, R., and Comeau, L.A. (2018). Elevated seawater temperature, not pCO2, negatively affects post-spawning adult mussels (*Mytilus edulis*) under food limitation. *Conserv. Physiol.* **6** (1): 1–14.

17 Talmage, S.C. and Gobler, C.J. (2011). Effects of elevated temperature and carbon dioxide on the growth and survival of larvae and juveniles of three species of Northwest Atlantic bivalves. *PLoS One* **6** (10).

18 Gazeau, F., Alliouane, S., Bock, C. et al. (2014). Impact of ocean acidification and warming on the Mediterranean mussel (*Mytilus galloprovincialis*). *Front. Mar. Sci.* **1** (11): 1–12.

19 Gazeau, F., Quiblier, C., Jansen, J.M. et al. (2007). Impact of elevated CO2 on shellfish calcification. *Geophys. Res. Lett.* **34** (7): 1–5.

20 Kroeker, K.J., Sanford, E., Jellison, B.M., and Gaylord, B. (2014). Predicting the effects of ocean acidification on predator-prey interactions: a conceptual framework based on coastal molluscs. *Biol. Bull.* **226** (3): 211–222.

21 Coleman, D.W., Byrne, M., and Davis, A.R. (2014). Molluscs on acid: gastropod shell repair and strength in acidifying oceans. *Mar. Ecol. Prog. Ser.* **509**: 203–211.

22 Zhao, L., Schöne, B.R., Mertz-kraus, R., and Yang, F. (2017). Sodium provides unique insights into transgenerational effects of ocean acidification on bivalve shell formation. *Sci. Total Environ.* **577**: 360–366.

23 Gazeau, F., Parker, L.M., Comeau, S. et al. (2013). Impacts of ocean acidification on marine shelled molluscs. *Mar. Biol.* **160** (8): 2207–2245.

24 López-Landavery, E.A., Carpizo-Ituarte, E.J., Pérez-Carrasco, L. et al. (2021). Acidification stress effect on umbonate veliger larval development in *Panopea globosa*. *Mar. Pollut. Bull.* **163** (7): 111945.

25 Mackenzie, C.L., Lynch, S.A., Culloty, S.C., and Malham, S.K. (2014). Future oceanic warming and acidification alter immune response and disease status in a commercial shellfish species, *Mytilus edulis* L. *PLoS One* **9** (6): e99712.

26 Morris, T.J. and Corkum, L.D. (1996). Assemblage structure of freshwater mussels (Bivalvia: Unionidae) in rivers with grassy and forested riparian zones. *J. North Am. Benthol. Soc.* **15** (4): 576–586.

27 Hastie, L.C., Boon, P.J., Young, M.R., and Way, S. (2001). The effects of a major flood on an endangered freshwater mussel population. *Biol. Conserv.* **98** (1): 107–115.

28 Di Maio, J. and Corkum, L.D. (1995). Relationship between the spatial distribution of freshwater mussels (Bivalvia: Unionidae) and the hydrological variability of rivers. *Can. J. Zool.* **73** (4): 663–671.

29 Käse, L. and Geuer, J.K. (2018). Phytoplankton responses to marine climate change – an introduction. In: *YOUMARES 8 – Ocean Across Boundaries Learn from Each Other* (ed. S. Jungblut, V. Liebich and M. Bode) Proceedings of the 2017 Conference for YOUng MARine RESearchers, Kiel, Germany, 55–71. Cham: Springer Nature.

30 de Kantzow, M., Hick, P., Becker, J., and Whittington, R. (2016). Effect of water temperature on mortality of Pacific oysters *Crassostrea gigas* associated with microvariant ostreid herpesvirus 1 (OsHV-1 μVar). *Aquac. Environ. Interact.* **8** (2010): 419–428.

31 Worm, B., Barbier, E.B., Beaumont, N. et al. (2006). Impacts of biodiversity loss on ocean ecosystem services. *Science* **314** (11): 787–790.

32 Hallegraeff, G., Anderson, D.M., Hole, W., and Cembella, A. (ed.) (2004). *Manual on Harmful Marine Microalgae*, 2e. Paris: UNESCO Publishing.

33 Callaway, R., Shinn, A.P., Grenfell, S.E. et al. (2012). Review of climate change impacts on marine aquaculture in the UK and Ireland. *Aquat. Conserv. Mar. Freshwat. Ecosyst.* **22** (3): 389–421.

34 Hussein, A.A.A., Bloem, E., Fodor, I. et al. (2021). Slowly seeing the light: an integrative review on ecological light pollution as a potential threat for mollusks. *Environ. Sci. Pollut. Res.* **28** (5): 5036–5048.

35 Oehlmann, J., Schulte-Oehlmann, U., Tillmann, M., and Markert, B. (2000). Effects of endocrine disruptors on prosobranch snails (mollusca: Gastropoda) in the laboratory. Part I: bisphenol A and octylphenol as xeno-estrogens. *Ecotoxicology* **9** (6): 383–397.

36 Duft, M., Schulte-Oehlmann, U., Weltje, L. et al. (2003). Stimulated embryo production as a parameter of estrogenic exposure via sediments in the freshwater mudsnail *Potamopyrgus antipodarum*. *Aquat. Toxicol.* **64** (4): 437–449.

37 Jobling, S., Casey, D., Rodgers-Gray, T. et al. (2003). Comparative responses of molluscs and fish to environmental estrogens and an estrogenic effluent. *Aquat. Toxicol.* **65** (2): 205–220.

38 Oehlmann, J., Schulte-Oehlmann, U., Bachmann, J. et al. (2006). Bisphenol A induces superfeminization in the ramshorn snail *Marisa cornuarietis* (Gastropoda: Prosobranchia) at environmentally relevant concentrations. *Environ. Health Perspect.* **114** (SUPPL.1): 127–133.

39 Schulte-Oehlmann, U., Tillmann, M., Casey, D. et al. (2001). Xeno-estrogenic effects of bisphenol A in prosobranchs (Mollusca: Gastropoda: Prosobranchia). *Originalarbeiten* **13** (6): 319–333.

40 Fidder, B.N., Reategui-Zirena, E.G., Olson, A.D., and Salice, C.J. (2016). Energetic endpoints provide early indicators of life history effects in a freshwater gastropod exposed to the fungicide, pyraclostrobin. *Environ. Pollut.* **211**: 183–190.

41 Bal, N., Kumar, A., and Nugegoda, D. (2017). Assessing multigenerational effects of prednisolone to the freshwater snail, *Physa acuta* (Gastropoda: Physidae). *J. Hazard. Mater.* **339**: 281–291.

42 Guo, B., Feng, D., Xu, Z. et al. (2021). Acute benzo[a]pyrene exposure induced oxidative stress, neurotoxicity and epigenetic change in blood clam *Tegillarca granosa*. *Sci. Rep.* **11** (1): 1–13.

43 Piscopo, M., Trifuoggi, M., Notariale, R. et al. (2018). Protamine-like proteins' analysis as an emerging biotechnique for cadmium impact assessment on male mollusk *Mytilus galloprovincialis* (Lamarck 1819). *Acta Biochim. Pol.* **65** (2): 259–267.

44 Liu, J., Zhu, X., Teng, J. et al. (2021). Pollution characteristics of microplastics in mollusks from the coastal area of Yantai, China. *Bull. Environ. Contam. Toxicol.* **107**: 693–699.

45 Nascimento-Schulze, J.C., Bean, T.P., Houston, R.D. et al. (2021). Optimizing hatchery practices for genetic improvement of marine bivalves. *Rev. Aquac.* **13** (4): 2289–2304.

46 Li, Y., Zhang, L., Li, Y. et al. (2019). Dynamics of DNA methylation and DNMT expression during gametogenesis and early development of scallop *Patinopecten yessoensis*. *Mar. Biotechnol.* **21** (2): 196–205.

47 Männer, L., Schell, T., Provataris, P. et al. (2021). Inference of DNA methylation patterns in molluscs. *Philos. Trans. R. Soc. B* **376** (1825): 20200166.

48 Zhang, J., Luo, S., Gu, Z. et al. (2020). Genome-wide DNA methylation analysis of mantle edge and mantle central from pearl oyster *Pinctada fucata martensii*. *Mar. Biotechnol.* **22** (3): 380–390.

49 Gerdol, M., La Vecchia, C., Strazzullo, M. et al. (2021). Evolutionary history of DNA methylation related genes in bivalvia: new insights from *Mytilus galloprovincialis*. *Front. Ecol. Evol.* **9** (July): 1–15.

50 Wang, X., Li, Q., Lian, J. et al. (2014). Genome-wide and single-base resolution DNA methylomes of the Pacific oyster *Crassostrea gigas* provide insight into the evolution of invertebrate CpG methylation. *BMC Genomics* **15** (1): 1119.

51 Dabe, E.C., Sanford, R.S., Kohn, A.B. et al. (2015). DNA methylation in basal metazoans: insights from ctenophores. *Integr. Comp. Biol.* **55** (6): 1096–1110.

52 Sarda, S., Zeng, J., Hunt, B.G., and Yi, S.V. (2012). The evolution of invertebrate gene body methylation. *Mol. Biol. Evol.* **29** (8): 1907–1916.

53 Aliaga, B., Bulla, I., Mouahid, G. et al. (2019). Universality of the DNA methylation codes in eucaryotes. *Sci. Rep.* **9** (1): 1–11.
54 Gavery, M.R. and Roberts, S.B. (2010). DNA methylation patterns provide insight into epigenetic regulation in the Pacific oyster (*Crassostrea gigas*). *BMC Genomics* **11** (1): 483.
55 Fallet, M., Luquet, E., David, P., and Cosseau, C. (2019). Epigenetic inheritance and intergenerational effects in mollusks. *Gene* **729** (June): 144–166.
56 Song, K., Li, L., and Zhang, G. (2017). The association between DNA methylation and exon expression in the Pacific oyster *Crassostrea gigas*. *PLoS One* **12** (9): 1–12.
57 Venkataraman, Y.R., Downey-Wall, A.M., Ries, J. et al. (2020). General DNA methylation patterns and environmentally-induced differential methylation in the eastern oyster (*Crassostrea virginica*). *Front. Mar. Sci.* **7** (April): 1–14.
58 Olson, C.E. and Roberts, S.B. (2014). Genome-wide profiling of DNA methylation and gene expression in *Crassostrea gigas* male gametes. *Front. Physiol.* **17** (5): 224.
59 Rondon, R., Grunau, C., Fallet, M. et al. (2017). Effects of a parental exposure to diuron on Pacific oyster spat methylome. *Environ. Epigenet.* **3** (1): 1–13.
60 Tran, T.K.A., MacFarlane, G.R., Kong, R.Y.C. et al. (2016). Potential mechanisms underlying estrogen-induced expression of the molluscan estrogen receptor (ER) gene. *Aquat. Toxicol.* **179**: 82–94.
61 Li, Y., Huang, X., Guan, Y. et al. (2015). DNA methylation is associated with expression level changes of galectin gene in mantle wound healing process of pearl oyster, *Pinctada fucata*. *Fish Shellfish Immunol.* **45** (2): 912–918.
62 Rajasethupathy, P., Antonov, I., Sheridan, R. et al. (2012). A role for neuronal piRNAs in the epigenetic control of memory-related synaptic plasticity. *Cell* **149** (3): 693–707.
63 Saint-Carlier, E. and Riviere, G. (2015). Regulation of Hox orthologues in the oyster *Crassostrea gigas* evidences a functional role for promoter DNA methylation in an invertebrate. *FEBS Lett.* **589** (13): 1459–1466.
64 Wei, L., Xu, F., Wang, Y. et al. (2018). The molecular differentiation of anatomically paired left and right mantles of the Pacific oyster *Crassostrea gigas*. *Mar. Biotechnol.* **20** (4): 425–435.
65 Riviere, G., Wu, G.C., Fellous, A. et al. (2013). DNA methylation is crucial for the early development in the oyster *C. gigas*. *Mar. Biotechnol.* **15** (6): 739–753.
66 Leverett, D., Thain, J. (2013). Oyster embryo-larval bioassay (revised). ICES Techniques in Marine Environmental Sciences, No. 54.
67 Riviere, G., He, Y., Tecchio, S. et al. (2017). Dynamics of DNA methylomes underlie oyster development. *PLoS Genet.* **13** (6): 1–16.
68 Geyer, K.K., Niazi, U.H., Duval, D. et al. (2017). The *Biomphalaria glabrata* DNA methylation machinery displays spatial tissue expression, is differentially active in distinct snail populations and is modulated by interactions with *Schistosoma mansoni*. *PLoS Negl. Trop. Dis.* **11** (5): 1–29.
69 Aparicio, N.L. (2021). The DNA methylation of the snail *Biomphalaria glabrata*, role and impact on the generation of phenotypic plasticity. PhD thesis. Université Perpignan Via Domitia, Perpignan.
70 Hayashi, J.-I., Yonekawa, H., Gotoh, O. et al. (1978). Strictly maternal inheritance of rat mitochondrial DNA. *Biochem. Biophys. Res. Commun.* **83** (3): 1032–1038.
71 Gusman, A., Lecomte, S., Stewart, D.T. et al. (2016). Pursuing the quest for better understanding the taxonomic distribution of the system of doubly uniparental inheritance of mtDNA. *PeerJ* **4**: e2760.
72 Shock, L.S., Thakkar, P.V., Peterson, E.J. et al. (2011). DNA methyltransferase 1, cytosine methylation, and cytosine hydroxymethylation in mammalian mitochondria. *Proc. Natl. Acad. Sci. U. S. A.* **8** (9): 3630–3635.
73 Bianchessi, V., Vinci, M.C., Nigro, P. et al. (2016). Methylation profiling by bisulfite sequencing analysis of the mtDNA non-coding region in replicative and senescent endothelial cells. *Mitochondrion* **27**: 40–47.
74 Bellizzi, D., D'aquila, P., Scafone, T. et al. (2013). The control region of mitochondrial DNA shows an unusual CpG and non-CpG methylation pattern. *DNA Res.* **20** (6): 537–547.
75 Blanch, M., Mosquera, J.L., Ansoleaga, B. et al. (2016). Altered mitochondrial DNA methylation pattern in Alzheimer disease-related pathology and in Parkinson disease. *Am. J. Pathol.* **186** (2): 385–397.
76 Ghosh, S., Singh, K.K., Sengupta, S., and Scaria, V. (2015). Mitoepigenetics: the different shades of grey. *Mitochondrion* **25**: 60–66.
77 Devall, M., Smith, R.G., Jeffries, A. et al. (2017). Regional differences in mitochondrial DNA methylation in human post-mortem brain tissue. *Clin. Epigenetics* **9** (1): 1–15.
78 Bouvet, P.K. (2019). Épigénétique mitochondriale chez des espèces avec DUI. Master thesis. Université de Montréal, Montréal.

79 Capt, C., Renaut, S., Ghiselli, F. et al. (2018). Deciphering the link between doubly uniparental inheritance of mtDNA and sex determination in bivalves: clues from comparative transcriptomics. *Genome Biol. Evol.* **10** (2): 577–590.

80 Capt, C., Renaut, S., Stewart, D.T. et al. (2019). Putative mitochondrial sex determination in the bivalvia: insights from a hybrid transcriptome assembly in freshwater mussels. *Front. Genet.* **10** (September): 1–13.

81 Müller, R., Charaf, S., Scherer, C. et al. (2016). Phenotypic and epigenetic effects of vinclozolin in the gastropod *Physella acuta*. *J. Molluscan Stud.* **82** (2): 320–327.

82 Lian, S., He, Y., Li, X. et al. (2015). Changes in global DNA methylation intensity and DNMT1 transcription during the aging process of scallop *Chlamys farreri*. *J. Ocean Univ. China* **14** (4): 685–690.

83 Díaz-Freije, E., Gestal, C., Castellanos-Martínez, S., and Morán, P. (2014). The role of DNA methylation on *Octopus vulgaris* development and their perspectives. *Front. Physiol.* **5** (February): 1–7.

84 García-Fernández, P., García-Souto, D., Almansa, E. et al. (2017). Epigenetic DNA methylation mediating *Octopus vulgaris* early development: effect of essential fatty acids enriched diet. *Front. Physiol.* **8** (May): 1–9.

85 Li, Y., Guan, Y., Li, Q., and He, M. (2014). Analysis of DNA methylation in tissues and development stages of pearl oyster *Pinctada fucata*. *Genes Genom.* **37**: 263–270.

86 Fneich, S., Dheilly, N., Adema, C. et al. (2013). 5-methyl-cytosine and 5-hydroxy-methyl-cytosine in the genome of *Biomphalaria glabrata*, a snail intermediate host of *Schistosoma mansoni*. *Parasit. Vectors* **6**: 167.

87 Odoemelam, E., Raghavan, N., Miller, A. et al. (2009). Revised karyotyping and gene mapping of the *Biomphalaria glabrata* embryonic (Bge) cell line. *Int. J. Parasitol.* **39** (6): 675–681.

88 Kosak, S.T. and Groudine, M. (2004). Gene order and dynamic domains. *Science* **306** (5696): 644–647.

89 Kosak, S.T. and Groudine, M. (2004). Form follows function: the genomic organization of cellular differentiation. *Genes Dev.* **18** (12): 1371–1384.

90 Lampel, S., Bridger, J.M., Zirbel, R.M. et al. (1997). Nuclear RNA accumulations contain released transcripts and exhibit specific distributions with respect to Sm antigen foci. *DNA Cell Biol.* **16** (10): 1133–1142.

91 Bridger, J.M., Kalla, C., Wodrich, H. et al. (2005). Nuclear RNAs confined to a reticular compartment between chromosome territories. *Exp. Cell Res.* **302** (2): 180–193.

92 Fellous, A., Favrel, P., Guo, X., and Riviere, G. (2014). The Jumonji gene family in *Crassostrea gigas* suggests evolutionary conservation of Jmj-C histone demethylases orthologues in the oyster gametogenesis and development. *Gene* **538** (1): 164–175.

93 Guo, Z., Zhang, L., Li, Y. et al. (2020). Expression profiling of the Kdm genes in scallop *Patinopecten yessoensis* suggests involvement of histone demethylation in regulation of early development and gametogenesis. *Comp. Biochem. Physiol. Part – B Biochem. Mol. Biol.* **243–244** (December 2019): 110434.

94 Fellous, A., Lefranc, L., Jouaux, A. et al. (2019). Histone methylation participates in gene expression control during the early development of the Pacific oyster *Crassostrea gigas*. *Genes* **10** (9): 695.

95 Cvitanić, M.T., Zeljko, T.V., Pasantes, J.J. et al. (2020). Sequence composition underlying centromeric and heterochromatic genome compartments of the pacific oyster *Crassostrea gigas*. *Genes* **11** (6): 1–20.

96 Eirín-López, J.M. and Ausio, J. (2009). Origin and evolution of chromosomal sperm proteins. *BioEssays* **31** (10): 1062–1070.

97 Mennella, M.R.F., Farina, B., Irace, M.V. et al. (2002). Histone H1-like protein and a testis-specific variant in the reproductive tracts of *Octopus vulgaris*. *Mol. Reprod. Dev.* **63** (3): 355–365.

98 Martinez-Soler, F., Kurtz, K., Ausio, J., and Chiva, M. (2007). Transition of nuclear proteins and chromatin structure in spermiogenesis of *Sepia officinalis*. *Mol. Reprod. Dev.* **74** (3): 360–370.

99 Fellous, A., Favrel, P., and Riviere, G. (2015). Temperature influences histone methylation and mRNA expression of the Jmj-C histone-demethylase orthologues during the early development of the oyster *Crassostrea gigas*. *Mar. Genomics* **19**: 23–30.

100 Kurtz, K., Martínez-Soler, F., Ausió, J. et al. (2007). Acetylation of histone H4 in complex structural transitions of spermiogenic chromatin. *J. Cell. Biochem.* **102** (6): 1432–1441.

101 Danilova, A.B., Kharchenko, O.A., Shevchenko, K.G., and Grinkevich, L.N. (2010). Histone H3 acetylation is asymmetrically induced upon learning in identified neurons of the food aversion network in the mollusk *Helix lucorum*. *Front. Behav. Neurosci.* **4** (November): 1–7.

102 Danilova, A.B. and Grinkevich, L.N. (2012). Failure of long-term memory formation in juvenile snails is determined by acetylation status of histone H3 and can be improved by NaB treatment. *PLoS One* **7** (7): e41828.

103 Grinkevich, L.N. (2014). Studies of histone H3 methylation during the formation of long-term memory. *Neurosci. Behav. Physiol.* **44** (5): 571–575.

104 Kolotova, D.E., Malyshev, A.Y., and Balaban, P.M. (2021). Histone deacytylase inhibitor enhances long-term synaptic potentiation in neurons of a grape snail. *J. Evol. Biochem. Physiol.* **57** (3): 704–708.

105 Guan, Z., Giustetto, M., Lomvardas, S. et al. (2002). Integration of long-term-memory-related synaptic plasticity involves bidirectional regulation of gene expression and chromatin structure. *Cell* **111** (4): 483–493.

106 Hart, A.K., Fioravante, D., Liu, R.-Y. et al. (2011). Serotonin-mediated synapsin expression is necessary for long-term facilitation of the *Aplysia* sensorimotor synapse. *J. Neurosci.* **31** (50): 18401–18411.

107 Chen, S., Cai, D., Pearce, K. et al. (2014). Reinstatement of long-term memory following erasure of its behavioral and synaptic expression in *Aplysia. elife* **3** (November): 1–21.

108 Hongkuan, Z., Karsoon, T., Shengkang, L. et al. (2021). The functional roles of the non-coding RNAs in molluscs. *Gene* **768** (July): 145300.

109 Yu, H., Zhao, X., and Li, Q. (2016). Genome-wide identification and characterization of long intergenic noncoding RNAs and their potential association with larval development in the Pacific oyster. *Sci. Rep.* **6** (February): 20796.

110 Rosani, U., Pallavicini, A., and Venier, P. (2016). The miRNA biogenesis in marine bivalves. *PeerJ* **4**: e1763.

111 Jehn, J., Gebert, D., Pipilescu, F. et al. (2018). PIWI genes and piRNAs are ubiquitously expressed in mollusks and show patterns of lineage-specific adaptation. *Commun. Biol.* **1** (137).

112 Wei, P., He, P., Zhang, X. et al. (2019). Identification and characterization of microRNAs in the gonads of *Crassostrea hongkongensis* using high-throughput sequencing. *Comp. Biochem. Physiol. – Part D Genomics Proteom.* **31** (June): 100606.

113 Huang, S., Ichikawa, Y., Igarashi, Y. et al. (2019). Piwi-interacting RNA (piRNA) expression patterns in pearl oyster (*Pinctada fucata*) somatic tissues. *Sci. Rep.* **9** (1): 1–10.

114 Du, X., Fan, G., Jiao, Y. et al. (2017). The pearl oyster *Pinctada fucata martensii* genome and multi-omic analyses provide insights into biomineralization. *Gigascience* **6** (8): 1–12.

115 Zheng, Z., Li, W., Xu, J. et al. (2020). LncMSEN1, a mantle-specific LncRNA participating in nacre formation and response to polyI:C stimulation in pearl oyster *Pinctada fucata martensii. Fish Shellfish Immunol.* **96** (October 2019): 330–335.

116 Zhu, X., Chen, Y., Zhang, Z. et al. (2020). A species-specific miRNA participates in biomineralization by targeting CDS regions of Prisilkin-39 and ACCBP in *Pinctada fucata. Sci. Rep.* **10** (1): 1–12.

117 Zheng, Z., Huang, R., Tian, R. et al. (2016). Pm-miR-133 hosting in one potential lncRNA regulates RhoA expression in pearl oyster *Pinctada martensii. Gene* **591** (2): 484–489.

118 Xu, F., Wang, X., Feng, Y. et al. (2014). Identification of conserved and novel microRNAs in the pacific oyster *Crassostrea gigas* by deep sequencing. *PLoS One* **9** (8): e104371.

119 Huang, J., Luo, X., Huang, M. et al. (2018). Identification and characteristics of muscle growth-related microRNA in the Pacific abalone, Haliotis discus hannai. *BMC Genomics* **19** (915): 1–11.

120 Rajasethupathy, P., Fiumara, F., Sheridan, R. et al. (2009). Characterization of small RNAs in *Aplysia* reveals a role for miR-124 in constraining synaptic plasticity through CREB. *Neuron* **63** (6): 803–817.

121 Bartsch, D., Ghirardi, M., Skehel, P.A. et al. (1995). *Aplysia* CREB2 represses long-term facilitation: relief of repression converts transient facilitation into long-term functional and structural change. *Cell* **83** (6): 979–992.

122 Feng, D., Li, Q., Yu, H. et al. (2018). Transcriptional profiling of long non-coding RNAs in mantle of *Crassostrea gigas* and their association with shell pigmentation. *Sci. Rep.* **8** (1436): 1–10.

123 Huang, X., Dai, J., Lin, K. et al. (2018). Regulation of IL-17 by lncRNA of IRF-2 in the pearl oyster. *Fish Shellfish Immunol.* **81** (July): 108–112.

124 Zheng, Z., Xiong, X., Zhang, J. et al. (2019). The global effects of PmRunt co-located and co-expressed with a lincRNA lncRunt in pearl oyster *Pinctada fucata martensii. Fish Shellfish Immunol.* **91** (February): 209–215.

125 Bahn, S.Y., Jo, B.H., Hwang, B.H. et al. (2015). Role of Pif97 in nacre biomineralization: in vitro characterization of recombinant Pif97 as a framework protein for the association of organic–inorganic layers in nacre. *Cryst. Growth Des.* **15** (8): 3666–3673.

126 Huang, S., Yoshitake, K., Asaduzzaman, M. et al. (2021). Discovery and functional understanding of MiRNAs in molluscs: a genome-wide profiling approach. *RNA Biol.* **18** (11): 1702–1715.

127 Xiong, X., Xie, B., Zheng, Z. et al. (2019). PfmPif97-like regulated by Pfm-miR-9b-5p participates in shell formation in *Pinctada fucata martensii. PLoS One* **14** (12): 1–14.

128 Waldron, F.M., Stone, G.N., and Obbard, D.J. (2020). Metagenomic sequencing suggests a diversity of RNA interference-like responses to viruses across multicellular eukaryotes. *PLoS Genet.* **16** (3): 1–37.

129 Kandel, E.R. (2015). The molecular biology of memory storage: a dialogue between genes and synapses. *Science* **294** (November 2001): 1030–1039.

130 Robertson, M. and Walter, G. (2010). Eric Kandel and *Aplysia californica*: their role in the elucidation of mechanisms of memory and the study of psychotherapy. *Acta Neuropsychiatr.* **22** (4): 195–196.
131 Suarez-Ulloa, V., Rivera-Casas, C., and Michel, M. (2019). Seasonal DNA methylation variation in the flat tree oyster *Isognomon alatus* from a mangrove ecosystem in North Biscayne Bay, Florida. *J. Shellfish Res.* **38** (1): 79.
132 Kong, N., Liu, X., Li, J. et al. (2017). Effects of temperature and salinity on survival, growth and DNA methylation of juvenile Pacific abalone, *Haliotis discus hannai* Ino. *Chin. J. Oceanol. Limnol.* **35** (5): 1248–1258.
133 Stenger, P.L., Ky, C.L., Reisser, C.M.O. et al. (2021). Environmentally driven color variation in the pearl oyster *Pinctada margaritifera* var. *cumingii* (Linnaeus, 1758) is associated with differential methylation of CpGs in pigment- and biomineralization-related genes. *Front. Genet.* **12** (March): 1–18.
134 Thorson, J.L.M., Smithson, M., Beck, D. et al. (2017). Epigenetics and adaptive phenotypic variation between habitats in an asexual snail. *Sci. Rep.* **7** (1): 1–11.
135 Huang, Z., Xiao, Q., Yu, F. et al. (2021). Comparative transcriptome and DNA methylation analysis of phenotypic plasticity in the Pacific abalone (*Haliotis discus hannai*). *Front. Physiol.* **12** (June): 1–11.
136 Ardura, A., Zaiko, A., Morán, P. et al. (2017). Epigenetic signatures of invasive status in populations of marine invertebrates. *Sci. Rep.* **7** (January): 42193.
137 Akcha, F., Barranger, A., Bachère, E. et al. (2016). Effects of an environmentally relevant concentration of diuron on oyster genitors during gametogenesis: responses of early molecular and cellular markers and physiological impacts. *Environ. Sci. Pollut. Res.* **23** (8): 8008–8020.
138 Nica, D.V., Popescu, C., Draghici, G.A. et al. (2017). High-level dietary cadmium exposure is associated with global DNA hypermethylation in the gastropod hepatopancreas. *PLoS One* **12** (9): 1–15.
139 Prieto-Amador, M., Caballero, P., and Martínez-Guitarte, J. (2021). Analysis of the impact of three phthalates on the freshwater gastropod *Physella acuta* at the transcriptional level. *Sci. Rep.* **11**: 11411.
140 Sánchez-Argüello, P., Aparicio, N., Guevara, M.A. et al. (2016). Effects on reproduction, genotoxicity and DNA methylation pattern after chronic exposure of the freshwater snail *Physa acuta* (Gastropoda, Pulmonata) to vinclozolin. *Austin. J. Environ. Toxicol.* **2** (1): 1008.
141 Nica, D., Popescu, C., Draghici, G. et al. (2017). Effect of cadmium on cytosine hydroxymethylation in gastropod hepatopancreas. *Environ. Sci. Pollut. Res.* **24** (17): 15187–15195.
142 Sussarellu, R., Lebreton, M., Rouxel, J. et al. (2018). Copper induces expression and methylation changes of early development genes in *Crassostrea gigas* embryos. *Aquat. Toxicol.* **196** (January): 70–78.
143 Zhao, Q., Wang, W., Li, J.X. et al. (2021). The DNA cytosine-5-methyltransferase 3 (DNMT3) involved in regulation of CgIL-17 expression in the immune response of oyster *Crassostrea gigas*. *Dev. Comp. Immunol.* **123** (January).
144 Fallet, M., Montagnani, C., Petton, B. et al. (2022). Early life microbial exposures shape the *Crassostrea gigas* immune system for lifelong and intergenerational disease protection. *Microbiome* **10**: 85.
145 Farias, N.D., de Oliveira, N.F.P., and da Silva, P.M. (2017). Perkinsus infection is associated with alterations in the level of global DNA methylation of gills and gastrointestinal tract of the oyster *Crassostrea gasar*. *J. Invertebr. Pathol.* **149** (August): 76–81.
146 Gonzalez-Romero, R., Suarez-Ulloa, V., Rodriguez-Casariego, J. et al. (2017). Effects of Florida red tides on histone variant expression and DNA methylation in the eastern oyster *Crassostrea virginica*. *Aquat. Toxicol.* **186**: 196–204.
147 Knight, M., Ittiprasert, W., Arican-Goktas, H.D., and Bridger, J.M. (2016). Epigenetic modulation, stress and plasticity in susceptibility of the snail host, *Biomphalaria glabrata*, to *Schistosoma mansoni* infection. *Int. J. Parasitol.* **46** (7): 389–394.
148 Ittiprasert, W., Miller, A., Knight, M. et al. (2015). Evaluation of cytosine DNA methylation of the *Biomphalaria glabrata* at shock protein 70 locus after biological and physiological stresses. *J. Parasitol. Vector Biol.* **7** (10): 182–193.
149 Joe, M. (2013). The epigenetic effect of trematode infection on the snail host *Zeacumantus subcarinatus*. Master thesis. University of Otago, New Zealand.
150 Matilainen, O., Quirós, P.M., and Auwerx, J. (2017). Mitochondria and epigenetics – crosstalk in homeostasis and stress. *Trends Cell Biol.* **27** (6): 453–463.
151 Weinhouse, C. (2017). Mitochondrial-epigenetic crosstalk in environmental toxicology. *Toxicology* **391**: 5–17.
152 Penkov, S., Mitroulis, I., Hajishengallis, G., and Chavakis, T. (2019). Immunometabolic crosstalk: an ancestral principle of trained immunity? *Trends Immunol.* **40** (1): 1–11.
153 Ji, C., Lu, Z., Xu, L. et al. (2019). Evaluation of mitochondrial toxicity of cadmium in clam *Ruditapes philippinarum* using iTRAQ-based proteomics. *Environ. Pollut.* **251**: 802–810.

154 Wang, T., Huang, X., Jiang, X. et al. (2019). Differential in vivo hemocyte responses to nano titanium dioxide in mussels: effects of particle size. *Aquat. Toxicol.* **212**: 28–36.

155 Tunholi-Alves, V.M., Tunholi, V.M., Amaral, L.S. et al. (2019). Alterations in the mitochondrial physiology of *Biomphalaria glabrata* (Mollusca: Gastropoda) after experimental infection by *Angiostrongylus cantonensis* (Nematoda: Metastrongylidae). *Acta Parasitol.* **64**: 693–699.

156 Tariel, J., Plénet, S., and Luquet, É. (2019). Transgenerational plasticity of inducible defenses: combined effects of grand-parental, parental and current environments. *Ecol. Evol.* **10**: 2367–2376.

157 Parker, L.M., O'Connor, W.A., Raftos, D.A. et al. (2015). Persistence of positive carryover effects in the oyster, *Saccostrea glomerata*, following transgenerational exposure to ocean acidification. *PLoS One* **10** (7): 1–19.

158 Luchtel, D. (1972). Gonadal development and sex determination in pulmonate molluscs. *Z. Zellforsch.* **301** (5): 279–280.

159 Extavour, C.G. (2003). Mechanisms of germ cell specification across the metazoans: epigenesis and preformation. *Development* **130** (24): 5869–5884.

160 Fabioux, C., Huvet, A., Lelong, C. et al. (2004). Oyster vasa-like gene as a marker of the germline cell development in *Crassostrea gigas*. *Biochem. Biophys. Res. Commun.* **320**: 592–598.

161 Fabioux, C. (2004). Origine et développement des cellules germinales chez l'huître *Crassostrea gigas*: intérêt pour le contrôle de la reproduction en écloserie. PhD thesis. Université de Bretagne occidentale, Brest.

162 Donelan, S.C. and Trussell, G.C. (2015). Parental effects enhance risk tolerance and performance in offspring. *Ecology* **96** (8): 2049–2055.

163 Donelan, S.C. and Trussell, G.C. (2018). Parental and embryonic experiences with predation risk affect prey offspring behaviour and performance. *Proc. R. Soc. B Biol. Sci.* **285** (1874): 20180034.

164 Donelan, S.C. and Trussell, G.C. (2018). Synergistic effects of parental and embryonic exposure to predation risk on prey offspring size at emergence. *Ecology* **99**: 68–78.

165 Luquet, E. and Tariel, J. (2016). Offspring reaction norms shaped by parental environment: interaction between within- and trans-generational plasticity of inducible defenses. *BMC Evol. Biol.* **16**: 209.

166 Beaty, L.E., Wormington, J.D., Kensinger, B.J. et al. (2016). Shaped by the past, acting in the present: transgenerational plasticity of anti-predatory traits. *Oikos* **125** (11): 1570–1576.

167 Plautz, S.C. and Salice, C.J. (2013). Plasticity in offspring contaminant tolerance traits: developmental cadmium exposure trumps parental effects. *Ecotoxicology* **22** (5): 847–853.

168 Lafont, M., Goncalves, P., Guo, X. et al. (2019). Transgenerational plasticity and antiviral immunity in the Pacific oyster (*Crassostrea gigas*) against Ostreid herpesvirus 1 (OsHV-1). *Dev. Comp. Immunol.* **91**: 17–25.

169 Parker, L.M., Ross, P.M., O'Connor, W.A. et al. (2012). Adult exposure influences offspring response to ocean acidification in oysters. *Glob. Chang. Biol.* **18** (1): 82–92.

170 Griffith, A.W. and Gobler, C.J. (2017). Transgenerational exposure of North Atlantic bivalves to ocean acidification renders offspring more vulnerable to low pH and additional stressors. *Sci. Rep.* **7**: 11394.

171 Parker, L.M., Coleman, R.A., O'Connor, W.A. et al. (2017). Adult exposure to ocean acidification is maladaptive for larvae of the Sydney rock oyster *Saccostrea glomerata* in the presence of multiple stressors. *Biol. Lett.* **13** (2): 20160798.

172 Suski, J.G., Salice, C.J., and Patinos, R. (2012). Species-specific and transgenerational responses to increasing salinity in sympatric freshwater gastropods. *Environ. Toxicol. Chem.* **31** (11): 2517–2524.

173 Kimberly, D.A. and Salice, C.J. (2014). If you could turn back time: understanding transgenerational latent effects of developmental exposure to contaminants. *Environ. Pollut.* **184**: 419–425.

174 Reategui-Zirena, E., Fidder, B.N., Olson, A.D. et al. (2017). Transgenerational endpoints provide increased sensitivity and insight into multigenerational responses of *Lymnaea stagnalis* exposed to cadmium. *Environ. Pollut.* **224**: 572–580.

175 Galindo, M., Rodriguez, H., and Olivares, C. (1992). Sperm basic nuclear proteins in the bivalve mollusc *Mesodesma donacium*: characterization and comparison with histone-like and protamine-like proteins of other molluscs. *Comp. Biochem. Physiol. – Part B Biochem.* **102** (4): 947–952.

176 Faulk, C. and Dolinoy, D.C. (2011). Timing is everything the when and how of environmentally induced changes in the epigenome of animals. *Epigenetics* **6** (7): 791–797.

177 McDonald, J.I., Celik, H., Rois, L.E. et al. (2016). Reprogrammable CRISPR/Cas9-based system for inducing site-specific DNA methylation. *Biol. Open.* **5** (6): 866–874.

178 Morgan, H.D., Santos, F., Green, K. et al. (2005). Epigenetic reprogramming in mammals. *Hum. Mol. Genet.* **14** (SPEC. ISS. 1): 47–58.

179 De Vanssay, A., Bougé, A.L., Boivin, A. et al. (2012). Paramutation in *Drosophila* linked to emergence of a piRNA-producing locus. *Nature* **490** (7418): 112–115.

180 Chen, Q.Q., Yan, M., Cao, Z. et al. (2015). Sperm tsRNAs contribute to intergenerational inheritance of an acquired metabolic disorder. *Science (80-)* **351**: 397–400.

181 Migicovsky, Z. and Kovalchuk, I. (2011). Epigenetic memory in mammals. *Front. Genet.* **2** (JUNE): 1–7.

182 Eirin-Lopez, J.M. and Putnam, H.M. (2018). Marine environmental epigenetics. *Annu. Rev. Mar. Sci.* **11** (1): 335–368.

183 Collin, H., Meistertzheim, A.-L., David, E. et al. (2010). Response of the Pacific oyster *Crassostrea gigas*, Thunberg 1793, to pesticide exposure under experimental conditions. *J. Exp. Biol.* **213** (Pt 23): 4010–4017.

184 Campos, A., Tedesco, S., Vasconcelos, V., and Cristobal, S. (2012). Proteomic research in bivalves. Towards the identification of molecular markers of aquatic pollution. *J. Proteom.* **75** (14): 434643–434659.

185 Suárez-Ulloa, V., Fernández-Tajes, J., Aguiar-Pulido, V. et al. (2013). The CHROMEVALOA database: a resource for the evaluation of okadaic acid contamination in the marine environment based on the chromatin-associated transcriptome of the mussel *Mytilus galloprovincialis*. *Mar. Drugs.* **11** (3): 830–841.

186 Fernández-Tajes, J., Flórez, F., Pereira, S. et al. (2011). Use of three bivalve species for biomonitoring a polluted estuarine environment. *Environ. Monit. Assess.* **177** (1–4): 289–300.

187 Milan, M., Pauletto, M., Patarnello, T. et al. (2013G). Ene transcription and biomarker responses in the clam *Ruditapes philippinarum* after exposure to ibuprofen. *Aquat. Toxicol.* **126**: 17–29.

188 Lüchmann, K.H., Mattos, J.J., Siebert, M.N. et al. (2012). Suppressive subtractive hybridization libraries prepared from the digestive gland of the oyster *Crassostrea brasiliana* exposed to a diesel fuel water-accommodated fraction. *Environ. Toxicol. Chem.* **31** (6): 1249–1253.

189 Klimova, Y.S., Chuiko, G.M., Pesnya, D.S., and Ivanova, E.S. (2020). Biomarkers of oxidative stress in freshwater bivalve mollusks (review). *Inl. Water Biol.* **13** (4): 674–683.

190 Oehlmann, J. and Schulte-Oehlmann, U. (2003). Molluscs as bioindicators. In: *Trace Metals and Other Contaminants in the Envionment* (ed. B.A. Markert, A.M. Breure and H.G. Zechmeister). Amsterdam: Elsevier.

191 Weyrich, A., Lenz, D., and Fickel, J. (2019). Environmental change-dependent inherited epigenetic response. *Genes* **10** (1).

192 Baccarelli, A. and Bollati, V. (2011). Epigenetics and environmental chemicals. *Curr. Opin. Pediatr.* **21** (2): 243–251.

193 St-Cyr, S. and McGowan, P.O. (2015). Programming of stress-related behavior and epigenetic neural gene regulation in mice offspring through maternal exposure to predator odor. *Front. Behav. Neurosci.* **9** (JUNE): 1–10.

194 Tariel, J., Plénet, S., and Luquet, É. (2020). Transgenerational plasticity in the context of predator–prey interactions. *Front. Ecol. Evol.* **8** (September): 548660.

195 Suarez-ulloa, V., Gonzalez-romero, R., and Eirin-Lopez, J.M. (2015). Environmental epigenetics: a promising venue for developing next-generation pollution biomonitoring tools in marine invertebrates. *Mar. Pollut. Bull.* **98** (1–2): 5–13.

196 Vandegehuchte, M.B. and Janssen, C.R. (2014). Epigenetics in an ecotoxicological context. *Mutat. Res. Genet. Toxicol. Environ. Mutagen.* **64** (765): 36–45.

197 Mirbahai, L. and Chipman, J.K. (2014). Epigenetic memory of environmental organisms: a reflection of lifetime stressor exposures. *Mutat. Res. Genet. Toxicol. Environ. Mutagen.* **764–765**: 10–17.

16

Epigenetics in Crustaceans

Günter Vogt

Faculty of Biosciences, University of Heidelberg, Heidelberg, Germany

16.1 Introduction

Epigenetic marks on the DNA and chromatin such as DNA methylation and histone methylation, acetylation, phosphorylation, and ubiquitination, as well as noncoding RNA pathways, can significantly affect gene expression and modify morphological, physiological, behavioral, and life history traits [1–3]. The epigenetic signatures can persist throughout life and across multiple generations, and occasionally, they may even be genetically assimilated in the long range [4–6]. Epigenetic marks can be changed stochastically or in response to environmental stimuli and are accessible for experimental manipulation [7–10]. In fish aquaculture, epigenetic mechanisms were already shown to be capable of inducing morphological and behavioral changes during domestication [11, 12]. Due to these reasons, epigenetics should be considered a promising future tool for improving crustacean aquaculture.

Epigenetics research in crustaceans is a relatively young discipline that has gained momentum in the last few years. Regev et al. [13] and Wang et al. [14] were the first to report on epigenetic mechanisms in crustaceans, namely DNA methylation in the green tiger prawn (*Penaeus semisulcatus*) and the land crab *Gecarcinus lateralis*, in the years 1998 and 1999, respectively. DNA methylation data for freshwater crayfish were first published in 2007 [15], for water fleas in 2009 [16], and for brine shrimps in 2014 [17]. The application of advanced analytical techniques like whole genome bisulfite sequencing (WGBS), which enable the determination of DNA methylation genome-wide at single-nucleotide resolution, considerably improved the detection and measurement of DNA methylation in crustaceans. So far, WGBS has been applied for several crustaceans including water fleas *Daphnia magna* and *Daphnia pulex* [18–20] and crayfish *Procambarus virginalis* [21].

Histone modifications and noncoding RNAs, the other well-known epigenetic mechanisms, are only sparsely investigated in crustaceans. Histone acetylation and deacetylation were studied in brine shrimp *Artemia franciscana* [17, 22] and miRNAs were characterized in *D. magna* [23] and swimming crab *Portunus trituberculatus* [24].

Parthenogenetic strains of *Artemia* and *Daphnia* and the obligatory parthenogenetic *P. virginalis* are particularly suitable for investigations into crustacean epigenetics because confounding genetic variation is minimal [25–27]. Clonal laboratory lineages for experimentation can easily be produced from single females of these species. Such clonal crustaceans have been used to address basic issues of epigenetics like the involvement of epigenetic mechanisms in gene regulation and their potential to change phenotypic traits and the inheritance of epigenetic signatures across generations, which will be detailed later.

Brine shrimps and daphnids and several decapod species are cultured on an industrial scale. The small entomostracans are produced as live feed for fish and shrimp hatcheries and nurseries [28–31]. Copepods and amphipods are also considered live food but are not yet cultured on an industrial scale [32, 33]. The Decapoda, which include the most commercially valuable shrimps, crayfish, crabs, and lobsters, are cultured for human consumption [34]. About 3000–4000 tons of *Artemia* cysts with an estimated value of $100–200 million are consumed annually by aquarists and the aquaculture industry [31], and aquaculture of decapods yields 9.4 million tons/year (Table 16.1) with a value of $69.3 billion [34].

The following sections will describe the epigenetics work that has been done so far with crustaceans and outline how epigenetic interventions could help improve the production of entomostracan and decapod crustaceans.

Epigenetics in Aquaculture, First Edition. Edited by Francesc Piferrer and Han-Ping Wang.

Table 16.1 Crustaceans produced in world aquaculture for human consumption.

Common name	Scientific name	Production 2018 (thousand tons)	Percentage
Whiteleg shrimp	*Penaeus vannamei*	4966.2	52.9
Red swamp crayfish	*Procambarus clarkii*	1711.3	18.2
Chinese mitten crab	*Eriocheir sinensis*	757.0	8.1
Giant tiger prawn	*Penaeus monodon*	750.6	8.0
Oriental river prawn	*Macrobrachium nipponense*	237.1	2.5
Giant river prawn	*Macrobrachium rosenbergii*	234.4	2.5
Other crustaceans		729.9	7.8
Crustaceans total		9386.5	100

Source: Compiled from FAO [34].

16.2 Epigenetics Research with Brine Shrimps and Copepods

Brine shrimps are small entomostracan crustaceans of the genus *Artemia* (Branchiopoda: Anostraca) occuring worldwide in tropical, subtropical, and temperate regions. They have been found in more than 500 inland saltwater lakes and coastal lagoons of salinities up to 25% [25, 35]. *Artemia* is a complex of morphologically very similar sibling species defined by reproductive isolation. There are sexually reproducing and asexual populations in the same species. The obligatory parthenogenetic strains, which have a broad range of ploidy levels ($2n$–$5n$) [35], are of particular interest for epigenetics research. *Artemia* produces dormant embryos (cysts) that can be stored for years and hatched on-demand to provide a convenient form of live feed for larval fish and shellfish [25, 29]. *Artemia* is also used as a model organism for toxicity testing of chemicals [36].

Brine shrimps have body lengths of about 10 mm. Under optimal conditions, they can grow from the hatching nauplius stage to adulthood in only eight days, produce up to 300 offspring every four days, and live for several months. Females can reproduce either ovoviviparously, releasing free-living nauplii, or oviparously, releasing encysted dormant embryos. Ovoviviparity is predominant in stable environments and oviparity is triggered by extreme conditions like high salinity and high temperature. *Artemia* is a nonselective filter feeder of organic detritus, microscopic algae, and bacteria [25, 35]. A draft genome is already established for *Artemia*, which is a good prerequisite for epigenetics research. The haploid genome size is 849 MB [37].

Artemia cysts and nauplii have a size of about 0.4 mm and constitute the most widely used food item in the larviculture of fish and shellfish. Therefore, they are in high demand. The cysts are collected either in natural saline waters like the Great Salt Lake (United States) or are produced in ponds and tanks [29]. *Artemia* shows a large diversity in various quantitative traits, some of them changing from year to year or even season to season [35]. Such differences within genetically uniform strains are probably caused by epigenetic variation that is produced stochastically or in response to environmental cues. Phenotypic traits like cyst size, hatching rate, caloric content, and fatty acid composition determine whether a particular cyst product is suitable for use in aquaculture and how much it costs. *Artemia* nauplii provide not only basic nutrients, but also enzymes and other crucial dietary elements to the early life stages of fishes and decapods [35].

Norouzitallab et al. exposed a population of parthenogenetic *Artemia* sp. raised from a single female at 28 °C to two heat shocks (35 °C for 30 minutes and 38 °C for 30 minutes) [17]. These heat shocks induced higher levels of the 70-kDa heat shock protein (Hsp70) and increased tolerance to lethal heat stress and resistance against the pathogenic bacterium *Vibrio campbellii* when compared to the control. Hsp70 is a ubiquitous molecular chaperone that acts in cellular protein folding and remodeling processes and is thus crucial for maintaining protein homeostasis [38]. Interestingly, the induced tolerance to heat stress and bacterial infection was transmitted to the three successive generations investigated, none of which was exposed to the parental stressor. Moreover, trait alteration in the heat-shocked group was associated with alterations of global DNA methylation and histone acetylation, which were transgenerationally inherited as well. These results suggest an involvement of epigenetic mechanisms in the heritability of acquired phenotypic traits across generations, although no correlation between specific traits and specific epigenetic marks were established yet.

The innate immune system of invertebrates can be adaptive to some extent, which is termed "trained immunity." The memory characteristics and mechanisms behind this phenomenon remain unclear, but epigenetic mechanisms are prime candidates. To study trained immunity and its inheritance in detail, Norouzitallab et al. exposed *A. franciscana* larvae to sublethal concentrations of pathogenic *V. campbellii* bacteria for four days and monitored the effects in their filial generations exposed to different *Vibrio* strains using unexposed parents and their *Vibrio*-exposed progeny as controls [39]. The results demonstrated an increased resistance of the *Vibrio campbellii*-trained group to *V. campbellii* (Figure 16.1a) but not to other *Vibrio* strains for three successive generations, indicating the existence and inheritance of trained immunity in crustaceans. The resistant phenotype showed increased reproduction (Figure 16.1b) and elevated levels of Hsp70 (Figure 16.1c) and displayed alterations in the expression of key genes related to innate immunity. The filial generations F1–F3 also showed altered levels of acetylation and methylation of histones H3 and H4 (Figure 16.1d). Although these epigenetic changes were inconsistent in the three filial generations, the authors assumed that trained immune responses in *Artemia* may be based on epigenetic reprogramming of the innate immune effectors.

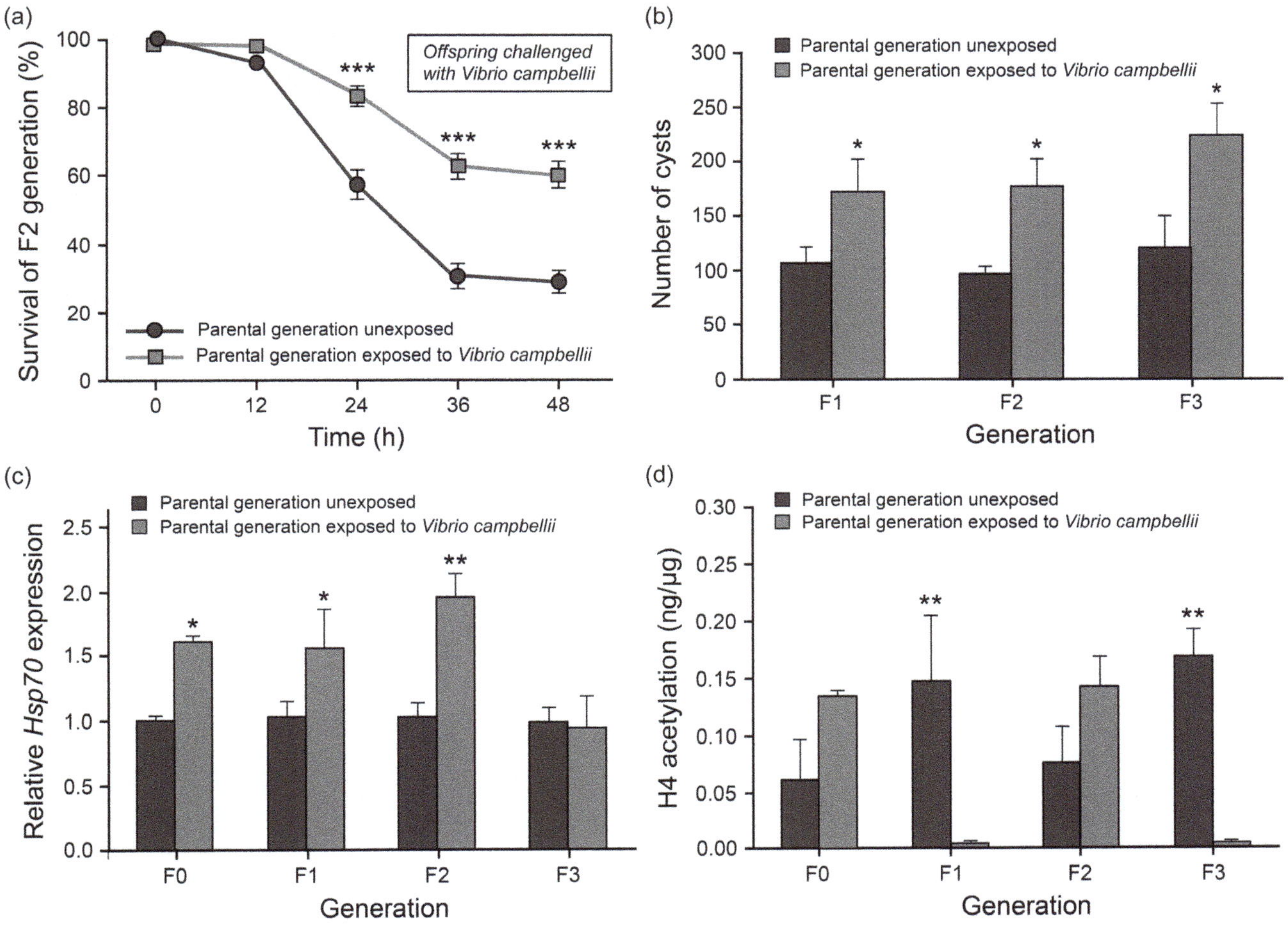

Figure 16.1 Stimulation and transgenerational inheritance of trained immunity in brine shrimp *Artemia franciscana*. (a) Survival of F2 progeny of *Vibrio campbellii*-exposed and unexposed parents after challenge with the same bacterium. Survival is significantly higher in the offspring of the exposed parents. (b) Increase of cyst production in the F1–F3 progeny of *V. campbellii*-exposed parents in comparison to the progeny of unexposed parents. Values represent means ± SD of eight specimens per group. Asterisks indicate significant difference at $p < 0.05$. (c) Comparison of relative expression of heat shock protein gene *Hsp70* in unexposed and *V. campbellii*-exposed parental generation F0 and their unexposed filial generations F1–F3. RNA expression was analyzed by qPCR. The expression value of the unexposed group was set at 1.0. *Hsp70* expression is enhanced in the exposed parental generation and their unexposed filial generations F1 and F2. $n = 3$ per group; *, significantly different at $p < 0.05$; **, significantly different at $p < 0.01$. (d) Enhancement of histone H4 acetylation in *V. campbellii*-exposed parental generation compared to the unexposed parental generation and fluctuation of the acetylation level in the unexposed generations F1–F3. $n = 4$ per group; **, significantly different at $p < 0.01$. *Source:* Adapted from Norouzitallab et al. [39].

Roy et al. tested the efficiency of the plant-based phenolic compound phloroglucinol in preventing *Vibrio* disease outbreaks in crustaceans [40]. Using *A. franciscana* as a test species, the authors showed that treatment of the parental generation with 2 μM phloroglucinol at 1–16 days post-hatching resulted in increased survival of the F1–F3 progeny when challenged with pathogenic *Vibrio parahaemolyticus* bacteria (Figure 16.2a). Studying the expression profiles of immune genes showed that phloroglucinol treatment of the parents significantly enhanced the expression of Hsp70, Hsp90 (heat-shock protein 90), proPO (prophenoloxidase), and LGBP (lipopolysaccharide and beta-1,3-glucan-binding protein) in three subsequent generations, suggesting transgenerational inheritance of the protecting effects. An example is shown in Figure 16.2b for the *proPO* gene that encodes the key molecule of the prophenoloxidase activating system, which is the most powerful immune defense system of crustaceans. The proPO-system consists of an enzymatic cascade that leads to melanization and encapsulation of pathogens and infected tissue areas [41].

Epigenetic mechanisms such as DNA methylation (Figure 16.2c) and histone acetylation and methylation (Figure 16.2d) responded to phloroglucinol treatment as well, suggesting that they may have been involved in the observed changes in gene expression and the transgenerational inheritance of increased resistance against the pathogenic bacteria. The study shows that a robust phenotype and associated epigenetic signatures can be induced in brine shrimp by a chemical compound and that these features can be transgenerationally inherited.

Copepods are considered a good alternative live feed for fish hatcheries [42]. These entomostracan crustaceans include about 13,000 species and are abundant in the plankton of limnic and marine waters. In the ocean, they are the major constituent of the zooplankton and are important prey for larval fish. Their larval and adult stages have wide size ranges within and between species and biochemical profiles well-suitable for fish larvae, including high levels of polyunsaturated fatty acids (PUFAs). Due to their small sizes, the early-stage nauplii (~50 μm) and copepodites are particularly suitable as starter

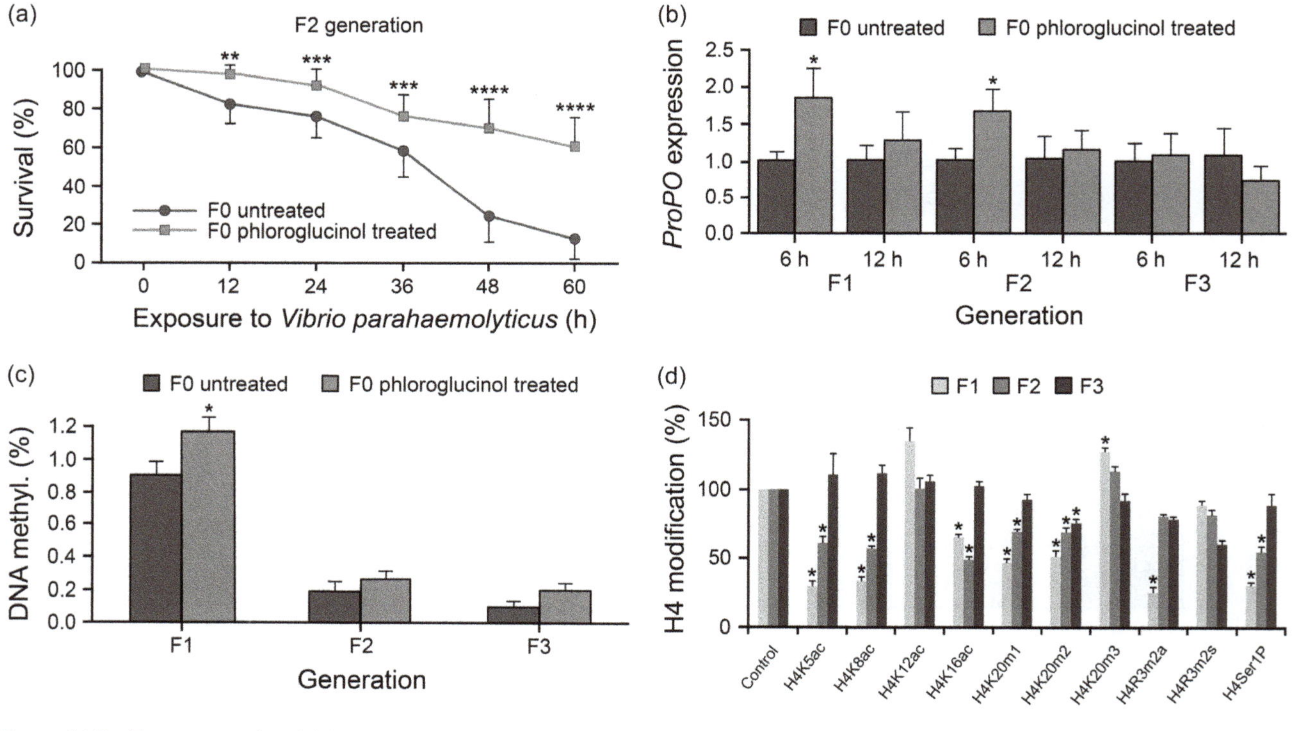

Figure 16.2 Transgenerational inheritance of phloroglucinol-induced resistance to bacterial infection in *Artemia franciscana*. (a) Survival of filial generation F2 of phloroglucinol-treated and untreated parents after infection with the pathogenic bacterium *Vibrio parahaemolyticus*, showing increased resistance after phloroglucinol treatment. $n = 7$ per group; asterisks indicate significant differences: *, $p < 0.05$; **, $p < 0.01$; ***, $p < 0.001$; ****, $p < 0.0001$; *****, $p < 0.00001$. (b) Relative expression of immune gene *proPO* in *Vibrio harveyi*-challenged F1–F3 offspring of phloroglucinol-treated and untreated parents, showing increased expression levels in F1 and F2 of treated parents. The larvae were exposed to the bacteria for 6 and 12 hours. $n = 3$ per group. (c) Global DNA methylation in F1–F3 offspring of phloroglucinol-treated and untreated parents after 12 hours of infection with *V. harveyi*. Methylation is significantly higher in the F1 offspring of the treated parents, but only slightly higher in the F2 and F3 generations. $n = 3$ per group. (d) H4 histone modifications in *V. harveyi*-challenged F1–F3 offspring of phloroglucinol-treated and untreated parents (control was set at 100%), showing significant alterations of several histone modifications after treatment. $n = 2$ per group. *Source:* Adapted from Roy et al. [40].

feeds for very small fish larvae. So far, copepods have been successfully grown in extensive systems but are not yet cultured on an industrial scale.

To study the potential effects of future ocean acidification on zooplankton organisms, Aluru et al. [42] exposed the copepod *Acartia clausii* to 400 and 900 ppm pCO_2 for 24 hours and compared transcriptional and DNA methylation patterns using RNA sequencing and reduced representation bisulfite sequencing (RRBS), respectively. They revealed differential expression of genes associated with energy metabolism, cytoskeletal and extracellular matrix functions, hypoxia, and one-carbon metabolism. Principal component analysis of differentially methylated regions (DMRs) between the control and 900 ppm pCO_2 showed clear clustering of samples based on conditions. The authors found a total of 34 DMRs, 25 of them being hypomethylated and nine being hypermethylated. The observed DNA methylation changes were not functionally linked to particular gene expression changes.

16.3 Epigenetics Research with Water Fleas

Water fleas (Branchiopoda: Cladocera) are small and shortlived entomostracan crustaceans that are abundant in zooplankton and therefore ecologically very important. They are well investigated and established as model organisms in ecology and ecotoxicology [26, 43, 44]. *Daphnia* and *Moina* species are frequently used food sources in freshwater fish larviculture and in the ornamental fish industry [28, 30]. They are produced in ponds and tanks [28, 45].

Daphnids usually reproduce by cyclic parthenogenesis, the alternation of sexual and asexual reproduction. Parthenogenetic reproduction is used in favorable conditions and sexual reproduction in unfavorable conditions [46]. Monoclonal asexual lineages can easily be produced and raised in the laboratory for experimentation, but there are also numerous wild asexual lineages [47, 48]. Obligatory parthenogenetic lineages of *Daphnia* offer a variety of benefits for the study of epigenetics such as short generation time, adaptation to a wide range of environmental conditions, and plastic phenotypic responses to predators (cyclomorphosis) [49, 50]. *Daphnia* species can produce normal embryos and encysted dormant embryos resembling *Artemia* cysts. The genomes of *Daphnia pulex* and *Daphnia magna* are fully sequenced [51–53], providing a good basis for epigenetics research.

D. pulex is a very abundant cosmopolitan species, and *D. magna* occurs in the Northern Hemisphere and South Africa. Both are key species in a wide range of lentic habitats and are prey for many invertebrate and vertebrate predators. Most of the time, they stay in the water column and filter-feed on phytoplankton, bacteria, and detritus. The adults measure 0.2–5 mm depending on species and sex. Longevity in the laboratory is about three months. The females produce one clutch with several eggs every three–four days. The offspring develop directly in the brood chamber of the mother until the hatching of a miniature adult. Development from the laid egg to maturation lasts about 11 days at 10 °C and only 2 days at 25 °C [28]. Alternatively, the eggs can be enclosed in a cyst (ephippium) that is formed in the brood chamber and then released into the water. These cysts can remain viable for up to 700 years [48, 54].

Vandegehuchte et al. were the first to investigate DNA methylation in *Daphnia* [16]. They found homologs of the human DNA methyltransferases Dnmt1 and Dnmt3A and the RNA methyltransferase Dnmt2 in the partially available *D. magna* genome. Using an optimized amplification of intermethylated sites (AIMS) technique, two methylated DNA fragments were discovered. No sequence homologs were found in databases for these fragments. In another paper, the authors determined global cytosine methylation of the genome by liquid chromatography and mass spectrometry and revealed 5mC/total C ratios between 0.13 and 0.81% [55].

Daphnids possess both highly methylated genes and genes devoid of methylation in a background of relatively low global methylation levels of 0.70% in *D. pulex* and 0.52% in *D. magna* [18]. These species showed differences in the DNA methylation pattern (Figure 16.3a), which may be a consequence of sequence divergence and differences in the number of CpGs. Methylation differences were primarily found in small gene families, whereas large gene families tended to lack methylation variation, suggesting that gene body methylation may help regulate gene family expansion and functional diversification, leading to increased phenotypic variation.

Methylation in *Daphnia* is largely restricted to CpG dinucleotides. Methylated cytosines are most prevalent in coding exons of gene bodies (Figure 16.3b), particularly in exons 2–5, indicating a directed, nonrandom distribution in the genome [20]. This distribution pattern differs from that of vertebrates, which show particularly high methylation in exon 1 [56]. Genes with high levels of CpG methylation were highly transcribed in *Daphnia* and apparently evolutionarily conserved.

Daphnia was also shown to possess the chemical modification 5-hydroxymethylcytosine (5hmC) [57]. 5hmC is twice as frequent in exons when compared to introns. Functional analysis indicated that 5hmC is abundant in genes that are

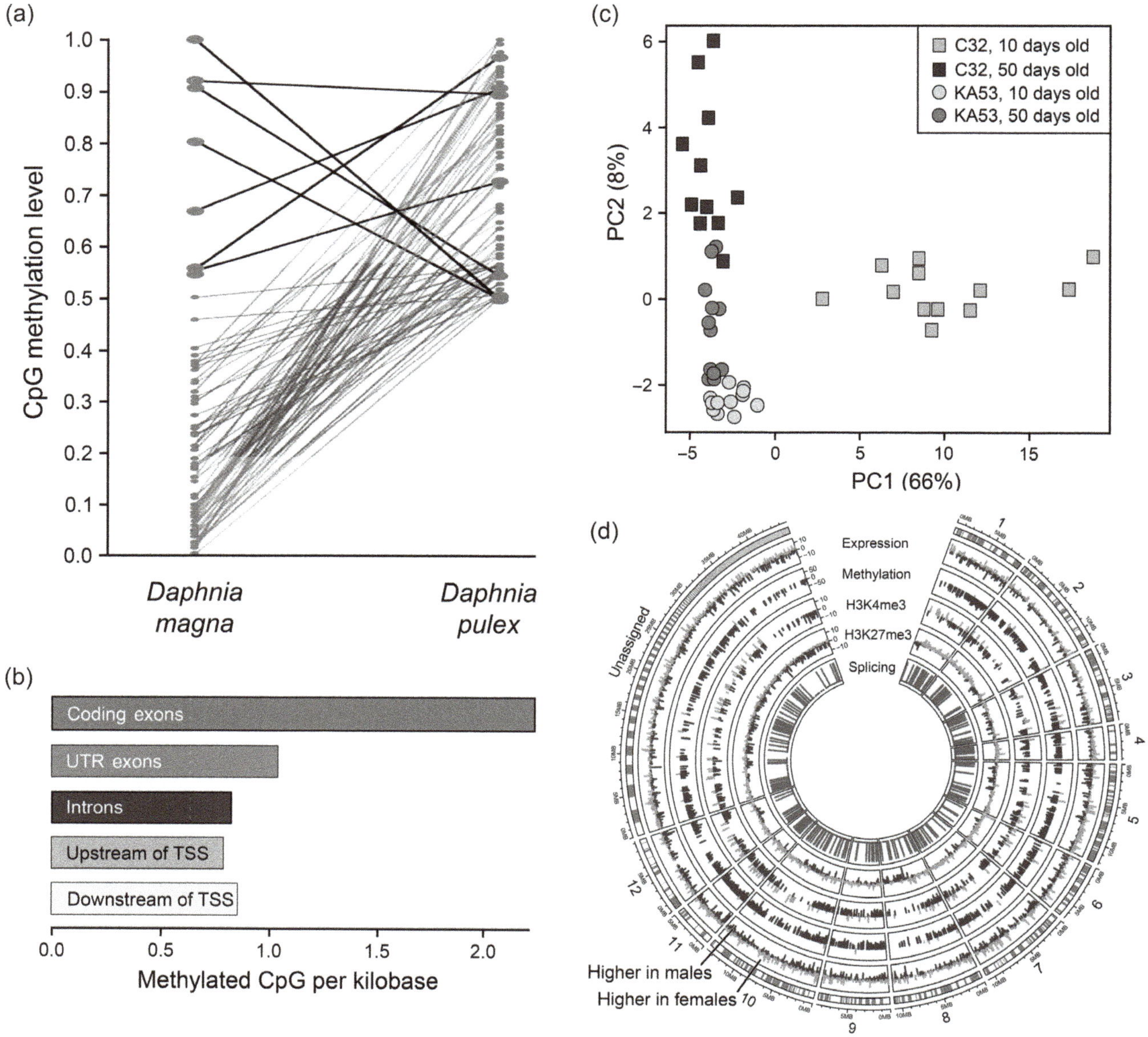

Figure 16.3 DNA methylation in *Daphnia* water fleas and age- and sex-related differences. (a) CpG methylation levels of highly methylated genes in *Daphnia magna* and *Daphnia pulex*, showing considerable differences between species. Bold lines highlight genes that are highly methylated in both species. (b) Frequency of methylated CpGs in different parts of the genome of *D. magna*, showing predominance in coding exons. (c) Principal component analysis for age-significant CpGs in two strains of *D. magna*, showing marked differences between 10-day- and 50-day-old groups in strain C32. (d) Circos-plot showing sex-specific differences for *D. pulex* in multiple omics datasets, distributed across the genome. Inner bars in circles indicate higher values in males and outer bars indicate higher values in females. There are marked differences between males and females in gene expression, DNA methylation, histone modifications, and gene splicing. *Source:* (a) Adapted from Asselman et al. [18]. (b, c) Adapted from Hearn et al. [20]. (d) Adapted from Kvist et al. [19].

involved in the adenylate cyclase-activating G-protein-coupled receptor signaling pathway, morphogenesis, cell fate determination, and molting cycle. Genes involved in mRNA-related processes usually lacked 5hmC.

The DNA methylation patterns in water fleas can change in different life history phases including aging and sex determination, but can also be modified by environmental cues. An example of age-related changes is published for *D. magna*. Using WGBS, Hearn et al. investigated the dynamics of DNA methylation during aging in two genotypically different strains (KA53 and C32) and found 61 CpGs that were hypermethylated in 50-day-old individuals when compared to

10-day-old individuals. Principal component analysis showed a prominent separation of the young and old groups in strain C32 and a less clear separation in strain KA53 (Figure 16.3c). The authors identified a subset of CpGs that was highly correlated with chronological age, suggesting their potential use as an epigenetic clock.

Males and females from a given water flea strain are genetically identical but display large differences in morphology, behavior, lifespan, and metabolic activity. To achieve a better understanding of the epigenetic factors that underlie the phenotypic differences between sexes, Kvist et al. integrated gene expression, DNA methylation, and histone modification analyses from female and male *D. pulex* raised under controlled laboratory settings (Figure 16.3d) [19]. They revealed a positive correlation between gene expression and DNA methylation levels. Gene expression was also positively correlated with histone H3 trimethylation at lysine 4 (H3K4me3) at predicted promoter regions but negatively correlated with histone H3 trimethylation at lysine 27 (H3K27me3) distributed across the entire transcript length. Epigenetic modifications that globally promote elevated gene expression were predominant in males, while epigenetic modifications that globally reduce gene expression were more frequent in females. These data demonstrate that there are vast epigenetic differences between males and females of *D. pulex* despite genetic identity, which might explain the prominent morphological and life history differences between sexes.

Gomez et al. studied the dynamics of the two posttranslational histone modifications H3S10ph and H3K9m3 during spermatogenesis of *D. magna* [58]. Phosphorylation of histone H3 at serine 10 is involved in many animals in regulation of the condensation and relaxation of chromosomes and chromatid cohesion during meiosis. H3S10ph was not detected in premeiotic cells. It first appeared over the entire chromatin in prometaphase I and disappeared in late telophase I. H3S10ph was again detected in metaphase II chromosomes and disappeared again in late telophase II. In contrast, methylation of lysine 9 of histone 3 was consistently maintained throughout all stages of meiosis, suggesting that this modification is not directly involved in regulation of the various meiotic chromatin states.

The impacts of environmental cues on DNA methylation and chromatin structure in water fleas were studied for a variety of factors. For example, *D. magna* exposed to the DNA methylation inhibitor 5-azacytidine, hypoxia, hyperoxia, and arsenic showed numerous differentially methylated CpGs with relatively little overlap between conditions (Figure 16.4a) [56]. In another paper, 5-azacytidine was shown to cause a consistent reduction in DNA methylation, and this effect was transmitted to the F1 and F2 generations [61].

Lindeman et al. exposed egg-carrying *D. magna* to a sublethal concentration (10 mg/L) of 5-azacytidine for seven days and investigated phenotypic, transcriptional, and epigenetic responses and their causal linkages [59]. Exposure significantly reduced fecundity in the F0 generation when compared to the control and resulted in differential expression of genes in the one-carbon-cycle metabolic pathway (Figure 16.4b). These features were paralleled by a decrease in global DNA methylation but not by changes in methylated histones H3K4me3 and H3K27me3. In the F1 generation, changes in gene expression relative to the controls occurred as well, but these were accompanied by the reduction of global DNA methylation plus changes in histone modifications. The study suggests that DNA methylation is apparently an important player in the epigenetic control of gene expression and that DNA methylation inhibitors can cause more pronounced effects in the unexposed offspring than in the exposed parental generation.

Hu et al. investigated the toxic effects of the juvenile hormone analogue (JHA) insecticides methoprene and fenoxycarb on freshwater water flea *Moina macrocopa* [62]. JHA insecticides are endocrine disrupters that interfere with the hormonal system of arthropods by mimicking juvenile hormones. Chronic exposure of the water flea to the JHAs caused a series of toxic effects, including shortening of life expectancy, repression of body growth, and reduction in fecundity. Furthermore, they disturbed the expression of genes involved in the juvenile hormone signaling pathway, cuticle development, and carbohydrate, amino acid, and ATP metabolisms. Moreover, JHA exposure of the parents impaired growth and reproduction of the unexposed offspring. These responses were associated with changes in the expression of genes related to histone methylation, suggesting that histone modifications may play a role in the transgenerational inheritance of the effects. Likewise, the endocrine disruptor vinclozolin caused a consistent reduction in DNA methylation in *D. magna* and this effect was transmitted to the F1 and F2 generations as well [61].

Aluru et al. exposed the marine water flea *Evadne nordmanii* to expected future ocean acidification conditions (400 and 900 ppm pCO_2) for 24 hours [42]. These treatments changed DNA methylation patterns and induced differential expression of genes associated with energy metabolism, cytoskeletal and extracellular matrix functions, hypoxia, and one-carbon metabolism.

Ionizing radiation is another environmental factor capable of inducing epigenetic changes in aquatic animals. Trijau et al. investigated epigenetic effects of radiation in *D. magna* and their offspring by a 25-day chronic γ-irradiation of the parental generation to 6.5 μGy or 41.3 mGy/h [63]. The F1 was exposed together with the parents as embryos in the brood

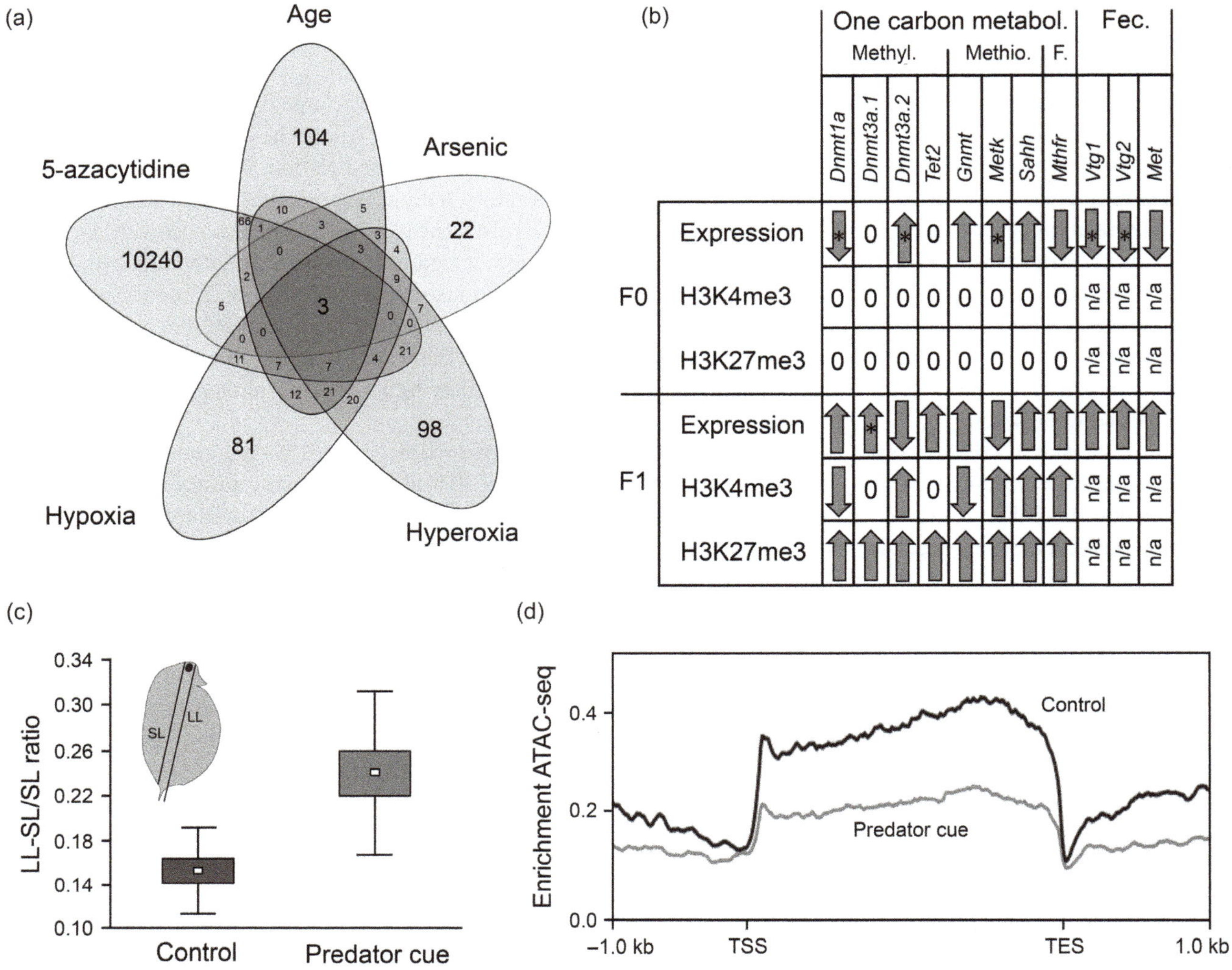

Figure 16.4 Changes of epigenetic signatures and chromatin structure in *Daphnia* in response to environmental cues. (a) Venn diagram of differentially methylated CpGs in *Daphnia magna* exposed to different stress conditions and the normal aging process (5 versus 14 days), showing relatively little overlap between conditions. (b) Changes of gene expression and histone modifications in the one-carbon-metabolism of *D. magna* exposed to 10 mg/L 5-azacytidine for seven days. Exposure of the egg-carrying females induced gene expression changes in the exposed F0 and unexposed F1 generation, but histone modification changes occurred only in the F1 generation. Arrows indicate upregulation and downregulation of gene expression and changes in histone modifications relative to the unexposed controls. Asterisks, significant difference at $p < 0.05$; 0, no changes; n/a, no data available. Dnmt 1, Dnmt3A.1 and Dnmt3A.2 are DNA methyltransferases; F., folate cycle; Fec., fecundity; Gnmt, glycine *N*-methyltransferase; Methio., methionine cycle; Met, methoprene-tolerant protein; Methyl., methylation cycle; Metk, methionine adenosyltransferase; Mthfr, methylenetetrahydrofolate reductase; Sahh, S-adenosylhomocysteine hydrolase; Tet, methylcytosine dioxygenase; Vtg1 and Vtg2, vitellogenins. (c) Body shape of *Daphnia pulex* in the absence and presence of predator fish, showing environmental induction of anti-predator morphs. LL, long length; SL, short length. (d) Metagene ATACseq profiles of predator-exposed and unexposed *D. pulex* populations. Predator-exposed groups had on average fewer reads over genes than control samples indicating induction of major changes in chromatin structure by the predator cues. TES, transcription end site; TSS, transcription start site. *Source:* (a) Adapted from Kvist et al. [56]. (b) Adapted from Lindeman et al. [59]. (c, d) Adapted from Augusto et al. [60].

chamber and the F2 was exposed as germ line cells in the F1 embryos. Thus, the F3 was the first truly unexposed generation. In F0, a significant reduction in fecundity was observed at the highest tested dose rate only, but there were no effects on fecundity in subsequent generations. No radiation effect was observed on survival and growth at any dose rate in any generation. In contrast, bisulfite sequencing revealed significant methylation changes at specific CpG positions in every generation with a majority of hypomethylation. Comparison of consistent methylation patterns between the indirectly exposed F2 and unexposed F3 generation implies that some epigenetic modifications may have been transgenerationally inherited.

Thaulow et al. chronically exposed adult *D. magna* for seven days to external ^{60}Co gamma radiation levels of 0.4, 1.4, 10, and 40 mGy/h [64]. They observed an increase in global DNA methylation associated with loci-specific alterations of histone H3K9 methylation and acetylation. These effects were accompanied by the downregulation of genes involved in DNA methylation, one-carbon metabolism, antioxidant defense, DNA repair, apoptosis, calcium signaling, and endocrine regulation of development and reproduction. Fecundity was not significantly changed by the gamma exposure. Based on their data, the authors proposed a model for the role of epigenetic mechanisms in stress response to low-dose ionizing radiation.

Water fleas can develop defense structures such as large helmets or abdominal spines in response to predator cues (cyclomorphosis) (Figure 16.4c). To better understand this adaptive mechanism on the molecular level, Augusto et al. developed an ATAC-seq assay (Assay for Transposase-Accessible Chromatin using sequencing) for *D. pulex* [60]. ATAC-seq is a relatively new technique for assaying chromatin accessibility genome-wide. The authors found that the presence of predator stimuli induced profound reorganization of the chromatin (Figure 16.4d) parallel to the morphological changes, suggesting that both are functionally linked. Since chromatin remodeling usually requires histone modifications [65], epigenetic mechanisms are assumed to be involved in the expression of cyclomorphosis.

Noncoding RNAs are only sparsely investigated in water fleas. Coucheron et al. identified 66 conserved miRNAs and 13 novel miRNAs in juvenile, subadult, and adult life stages of *D. magna* [23], providing a first basis for future work on miRNA-dependent epigenetics in water fleas. These miRNAs showed variation in expression levels between life stages. Forty-one of these miRNAs genes were concentrated in 13 genomic clusters, which probably reflects the formation of new miRNAs close to preexisting ones.

16.4 Epigenetics Research with Amphipods

Amphipods of the genus *Gammarus* are used as a food source for aquarium fish and other pet animals and as models in aquatic ecotoxicology and water quality assessment [33]. There are approximately 200 *Gammarus* species worldwide, mainly in freshwater habitats. They have adult body sizes of approximately 1–2 cm. These malacostracan crustaceans can be used as live feed and as a partial replacement for fish meal, which is among the most expensive feed ingredients in aquaculture. *Gammarus* extract was shown to increase innate immune reaction, stress resistance, food consumption, and weight gain in fish. *Gammarus* products (live, dried flakes, or powdered) are traded in international markets. The specimens used are mainly collected from the wild. Culture in semi-intensive or intensive conditions has not yet been developed. There are no statistics available on worldwide *Gammarus* production [33].

The level of global cytosine methylation in the freshwater amphipod *Gammarus fossarum* ranges from 0.4 to 1.4% [66]. DNA methylation varied with sampling site and age, but not with sex. Decreased water temperature and starvation caused hypermethylation after one month of exposure. Environmentally relevant cadmium concentrations (0.05–5 μg/L) induced significant hypomethylation after 14 days followed by a trend for hypermethylation after 1 month. Cribiu et al. concluded from their work that epigenetic marks could serve as valuable biomarkers in water flea ecotoxicology. Alves et al. [67] investigated the effect of the cholesterin-lowering drug simvastatin, one of the most prescribed human pharmaceuticals, on DNA methylation of the marine *Gammarus locusta*. Exposure to 0.32, 1.6, and 8 μg/L simvastatin induced significant downregulation of Dnmt1 and a reduction of global DNA methylation in all concentrations. These effects were paralleled by an increase in the expression of housekeeping genes and growth.

16.5 Epigenetics Research with Freshwater Crayfish

The parthenogenetic marbled crayfish (*Procambarus virginalis*) is one of the most intensely studied crustaceans with respect to epigenetics. It is used in the following to illustrate the spectrum of phenotypes that can be produced from the same genome by epigenetic mechanisms, either stochastically or by environmental induction. It is further used to illustrate associations of phenotypic traits and DNA methylation patterns during development, environmental adaptation, and evolution. The highly invasive marbled crayfish itself is not very suitable for aquaculture because of its relatively small body size and the high ecological risk of escapes in extensive pond culture [68], the only economically feasible culture method, but it is well suited as an epigenetics model for decapods.

Marbled crayfish was detected in 1995 in the German aquarium trade [69], and since then, it has been intensely investigated [27, 70]. In 2008, I proposed to use this species as a research model in epigenetics [71]. Marbled crayfish is an autotriploid, apomictic parthenogenetic descendant of the Floridian slough crayfish (*Procambarus fallax*) [72–74]. It produces offspring genetically identical among each other and to the mother [75–79]. Marbled crayfish is very robust and can be exposed to a wide range of environmental conditions, which is an important property for research on environmental epigenetics. Moreover, its genome is fully sequenced, comprising 21,772 predicted genes [78]. The haploid genome size was estimated to be 3.7 GB, which is a bit larger than the human genome size [78, 79], and the haploid chromosome number is 92 [73].

Marbled crayfish have a maximum total length (TL) (tip of cephalothorax to end of pleon) of ca. 12 cm [80]. They are directly developing without larval stages and mostly reproduce twice a year [81]. The clutches consist of ~50–730 offspring, depending on female size [74], providing an extraordinary source of genetically identical clutchmates for experimentation. The eggs and first three juvenile stages are brooded on the pleopods. Maturity is reached after about 6 months and the maximum age is 4.5 years [82].

As a result of releases, marbled crayfish have invaded numerous highly diverse habitats in tropical to cold-temperate biomes. There are thriving populations in 17 European and four Asian countries and in Madagascar. In Europe, marbled crayfish occur in ponds, lakes, and rivers from the Netherlands to Ukraine and from Estonia to Malta [27]. In Madagascar, the species has spread from an initial introduction near the capital Antananarivo in about 2005 to over more than 100,000 km^2, mostly by human dispersal, inhabiting rivers, lakes, rice fields, and thermal and polluted waters [78, 83]. By combining laboratory experiments with stable isotope analyses of field samples, Linzmaier et al. [84] established that marbled crayfish are highly plastic with respect to food utilization, trophic position and niche breadth, depending on habitat, the availability of food and shelter, and the presence of competitors and predators.

Global DNA methylation in marbled crayfish is about 2.4% as shown by mass spectrometry [77]. This value is in the order of magnitude of mammals, but an order of magnitude higher than in many insects [85]. A reference methylome, which demonstrates the distribution of methylation marks over the entire genome, was established by WGBS [21, 86, 87]. Analysis of the methylome revealed that 41% of genes are heavily methylated, 33% are moderately methylated, and 26% are unmethylated. About 25% of the 20 longest scaffolds were ubiquitously methylated, 70% were mosaically methylated (Figure 16.5a), and 5% were sporadically methylated [86]. The DNA methylation toolkit of marbled crayfish consists of single homologs of the DNA methyltransferases Dnmt1 and Dnmt3 and the ten-eleven-translocation enzyme Tet [21]. Dnmt1 serves for the maintenance of methylation patterns after DNA replication, Dnmt3 for de novo methylation of the DNA, and Tet for demethylation of the DNA.

DNA methylation in marbled crayfish is CpG-specific as in other animal genomes and present in coding genes, intergenic regions, and repeats (Figure 16.5b) [21]. Gene body methylation is highest in evolutionarily old housekeeping genes and moderately expressed genes. Repeats are generally hypomethylated. Housekeeping genes are good examples of constitutive, almost permanently transcribed genes. They are suitable to investigate the involvement of DNA methylation in gene function. Analysis of housekeeping genes in marbled crayfish revealed low methylation of gene bodies in unexpressed genes and higher methylation levels in moderately expressed genes (Figure 16.5c) [86], suggesting that gene body methylation may fine-tune gene activity.

The analysis of global DNA methylation and methylomes from different marbled crayfish and its parent species *P. fallax* revealed differences between individuals, tissues, physiologically different life periods (growth and reproduction phases), and species (Figure 16.5d) [86–88]. Methylation differences were also observed among clutchmates raised in the same narrowly controlled environment [76]. Comparative analysis of 697 variably methylated genes demonstrated that inter-individual differences are highest in the hepatopancreas [21, 87], which is the central metabolic organ of crayfish [89].

The integrative analysis of DNA methylation, chromatin accessibility, and mRNA expression patterns revealed that high gene body methylation is correlated with limited chromatin accessibility and stable gene expression (Figure 16.5e). In contrast, lowly methylated genes often reside in chromatin with higher accessibility and show increased expression variation [21].

To investigate the association of phenotypes and DNA methylation during the life cycle, various embryonic stages, juveniles, and adults of laboratory-raised marbled crayfish were analyzed. The embryonic period starts with spawning and ends with the hatching of the first juvenile stage (Figure 16.6a). It lasts 17–28 days, depending on conditions. The juvenile period lasts about five months and includes approximately 15 stages that are separated by molts. The adult life period begins with the first reproduction and is characterized by alternating reproduction and growth phases, which are separated by molts. Typically, it lasts 1.5–2.5 years, but in some cases, up to 4 years [82, 93–95].

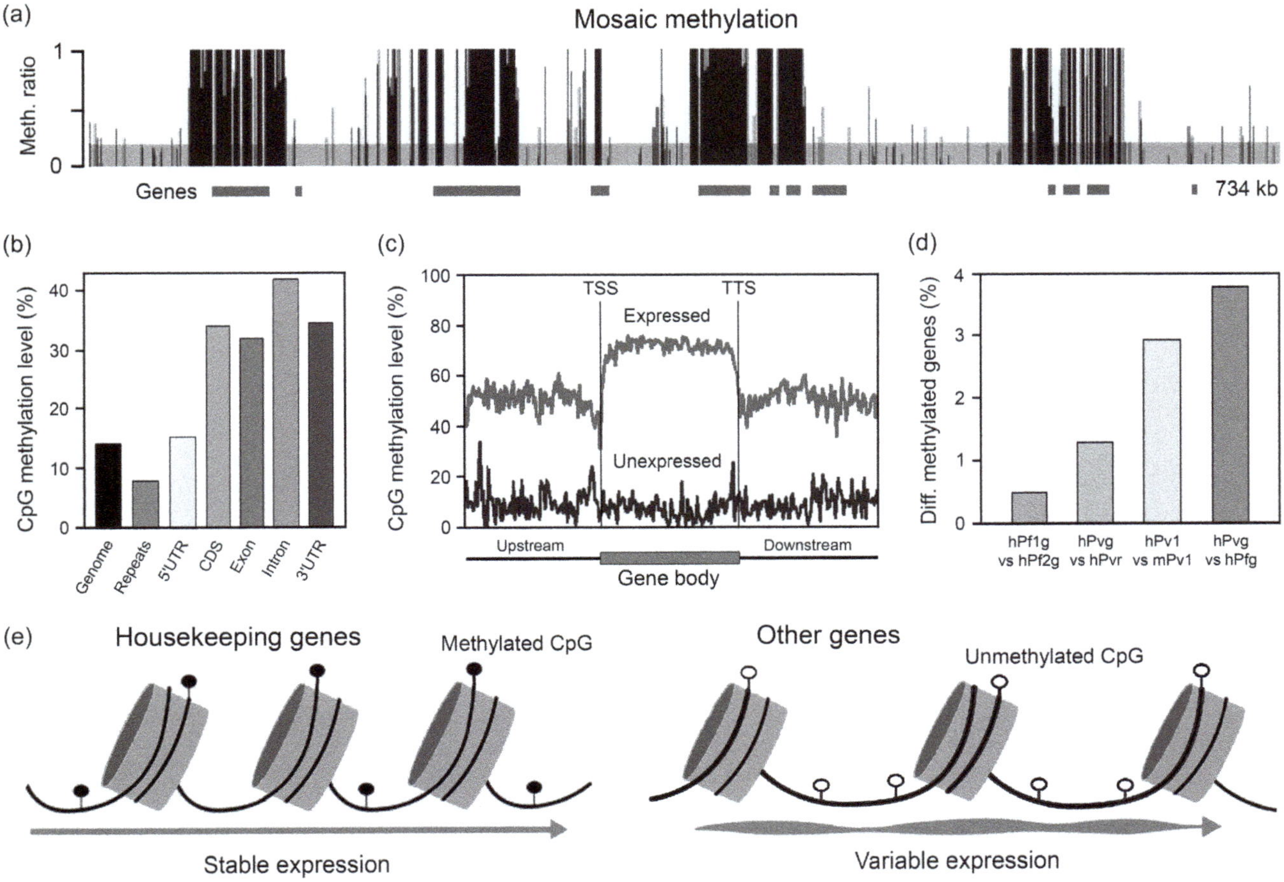

Figure 16.5 DNA methylation in marbled crayfish (*Procambarus virginalis*). (a) Example of mosaically methylated genomic scaffold analyzed by WGBS. Methylation ratios of CpGs are indicated by vertical bars. Methylation ratios below 0.2 are considered bisulfite conversion artifacts. Methylation is particularly prominent in genic regions (marked by horizontal bars). (b) Methylation levels of different genomic regions showing the highest values in introns and the lowest values in repeats. 5′UTR and 3′UTR, untranslated regions; CDS, protein-coding sequences. (c) Comparison of average methylation along unexpressed and moderately expressed housekeeping genes, starting 4 kb upstream of the transcription start site (TSS) and ending 4 kb downstream of the transcription termination site (TTS). (d) Differences in methylation of gene bodies between individuals, tissues, different physiological states, and species. Given is the percentage of genes with a methylation difference >0.2. Compared are the hepatopancreases of two *Procambarus fallax* (parent species of marbled crayfish) in growth phase (hPf1g/hPf2g), the hepatopancreases of *P. virginalis* in growth phase and reproduction phase (hPvg/hPvr), the hepatopancreas and abdominal musculature of the same *P. virginalis* specimen (hPv1/mPv1), and the hepatopancreases of size-matched *P. virginalis* and *P. fallax* females in growth phase (hPvg/hPfg). The numbers of genes analyzed were 6347, 6333, 5145, and 6303, respectively. (e) Functional scheme of gene body methylation. Highly methylated genes like housekeeping genes are commonly found in chromatin with limited accessibility. Methylation stabilizes the expression of such poorly accessible genes. Lowly methylated genes are localized in more accessible chromatin and are more variably expressed. *Source:* (a–d) Adapted from Vogt [88], previously compiled with data by Falckenhayn [86]. (e) Adapted from Gatzmann et al. [21].

DNA methylation changes were expected to be particularly prominent during embryonic development because this life period is characterized by the most dramatic phenotypic changes. Based on morphological criteria, the embryonic development of marbled crayfish was divided into 10 stages [90]. Stage 1, which lasts from the spawned oocyte to the completion of the blastoderm, was further divided into eight substages [96]. The first cleavages of the nucleus occur in the center of the spawned oocyte and are not externally visible (Figure 16.6a, embryo 1.1). In the 128-nucleus stage, the nuclei migrate to the egg surface (Figure 16.6a, embryo 1.4) and further development of the embryo is now well visible. Cell membranes between the nuclei are only present from the 256-cell stage (Figure 16.6a, embryo 1.5). Further important developmental events like gastrulation, segmentation, and formation of the analgen of the appendages occur in embryonic stages 2–5. The embryonic stages 6–10 are characterized by differentiation of the pleomeres, pereiopods, and internal tissues and organs. The heart is beating from stage 6 and pigmentation of the eyes is visible from stage 8. The hatchlings or stage 1 juveniles (Figure 16.6a)

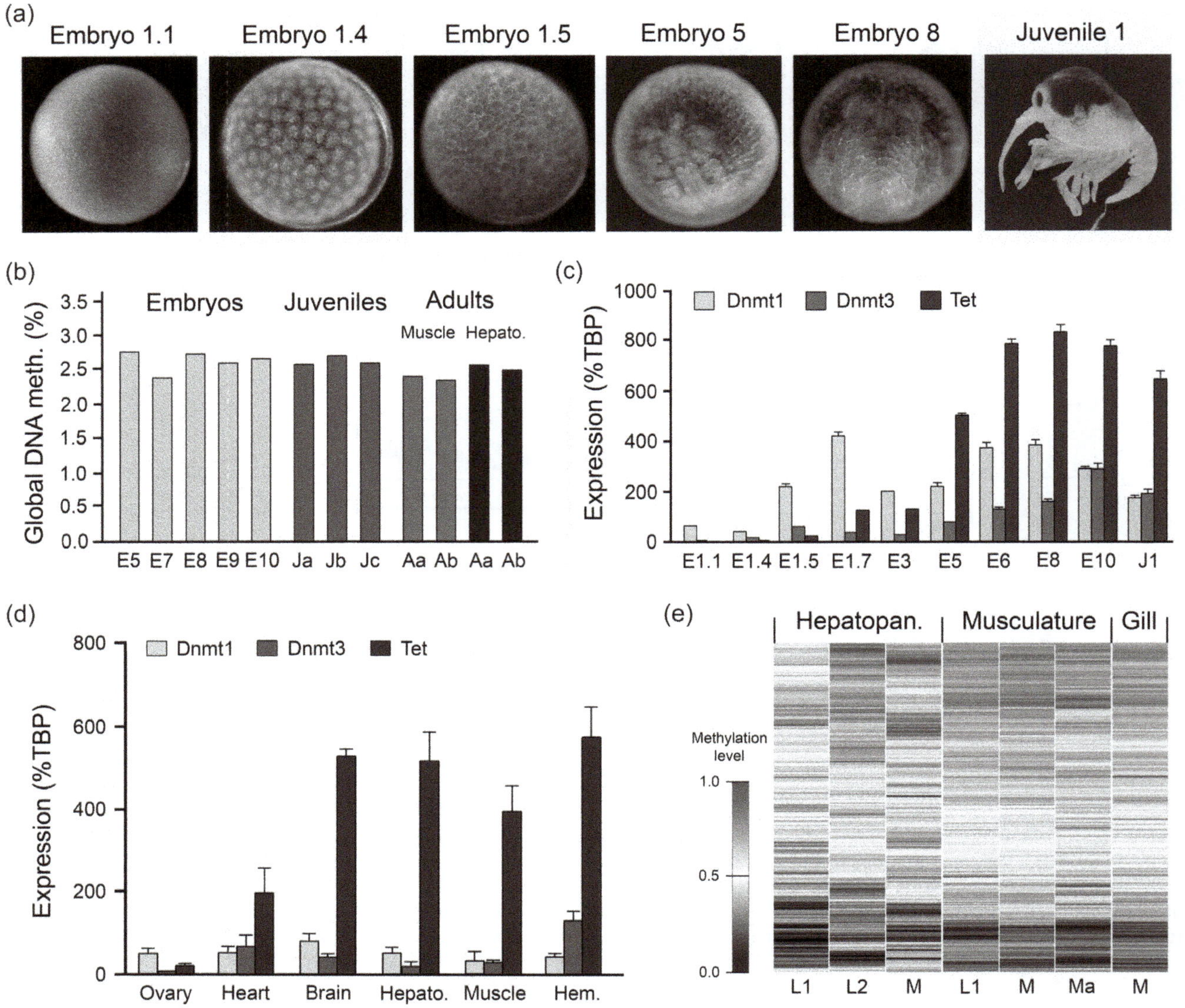

Figure 16.6 Dynamics of DNA methylation during the life cycle of marbled crayfish and methylation differences between tissues. (a) Phenotypic changes between spawned oocyte and first juvenile stage (hatchling). (b) Global DNA methylation levels in embryonic stages 7–10, three 154-day-old juveniles (Ja–Jc) and two 2-year-old adults (Aa and Ab). The DNA methylation level seems to decrease slightly with age. (c) Expression of DNA methylation enzymes Dnmt1 and Dnmt3 and demethylation enzyme Tet during embryonic development. Expression levels are given relative to the expression of the TATA-box binding protein (TBP). The Dnmt1, Dnmt3, and Tet genes show very low expression until embryo 1.4 and quite different expression dynamics thereafter. Embryo 5 corresponds to about 50% of embryonic development, in which tissues are not yet differentiated. Bars indicate means ± SD of three measurements. (d) Expression of Dnmt1, Dnmt3, and Tet in various tissues and organs relative to TBP expression. The graph shows marked differences between tissues for all enzymes. Bars indicate means ± SD of three measurements. (e) Comparative analysis of 697 variably methylated genes between hepatopancreas, abdominal musculature, and gills from two laboratory-reared specimens (L1 and L2), one specimen from Lake Moosweiher (M), and one specimen from a rice field in Moramanga, Madagascar (Ma). The heat map shows differences in methylation patterns between individuals and tissues. *Source:* (a) Left image from Alwes and Scholtz [90], right image from Vogt and Tolley [91], pictures in between from Vogt [88]. (b–d) Adapted from Gatzmann et al. [21]. (e) Adapted from Tönges et al. [92]. (*See colour plate section for colour representation of this figure.*)

and the following two juvenile stages are brooded on the pleopods until all appendages and sense organs are fully developed in juvenile stage 3, which resembles miniature adults [91].

The DNA is well methylated from embryonic stage 5, the earliest stage where we could reliably determine global DNA methylation by mass spectrometry. 5mC/total C was 2.78% in this stage (Figure 16.6b), which corresponds to about 50% embryonic development. Measurement of global DNA methylation revealed a slight reduction of the methylation level with age. In an experiment analyzed by mass spectrometry, DNA methylation was, on average, 2.62% in the second half of embryonic development, 2.58% in 154 days' old juveniles, and 2.41% in adults of about 2 years (Figure 16.6b), corresponding to a reduction of ~8% during lifetime (most marbled crayfish that reach adulthood die at an age of 2–3 years). In a previous experiment, in which capillary electrophoresis with a laser-induced fluorescence detector was used for DNA methylation measurement, adults of 626 days had ~10% lower 5mC/total C levels when compared to 188 days old juveniles [76].

Expression of the methylation enzymes Dnmt1 and Dnmt3 and the demethylation enzyme Tet was generally low until the 124 nucleus stage (embryo 1.4) (Figure 16.6c) [21]. Dnmt1 was strongly upregulated between embryonic stage 1.5 (256–512 cell stage) and embryo 1.7 and fluctuated then on a relatively high level, whilst Dnmt3 expression increased more continuously from stage 1.5 to stage 10. Tet mRNA levels increased strongly during mid-embryogenesis and remained high until hatching (Figure 16.6c). The dynamics of the methylation and demethylation enzymes suggest that DNA methylation of the genome is intensely remodeled in the time in which the appendages and tissues are formed. This dynamic is not shown by the global DNA methylation values, which reflect the sum of methylation and demethylation events.

Determination of global DNA methylation levels, WGBS, and measurement of the expression of methylation and demethylation enzymes also revealed differences in DNA methylation between tissues and organs of adults (Figure 16.6d, e). Dnmt1 showed the smallest difference in the methylation and demethylation toolkit between tissues, and Dnmt3 showed the largest difference (Figure 16.6d). Tet expression was high in the hemocytes, brain, hepatopancreas, and abdominal musculature, moderate in the heart and very low in the ovary (Figure 16.6d). Thus, Dnmt3 that serves for de novo methylation of the DNA represents the most tissue-specific enzyme of the methylation machinery [21]. Comparative analysis of 697 variably methylated genes from different organs revealed marked intra-individual differences between hepatopancreas, abdominal musculature and gills. The highest inter-individual variability was observed for the hepatopancreas (Figure 16.6e) [87, 92].

There are also considerable phenotypic differences between individuals of the same life stage despite their genetic identity. These differences are caused by developmental stochasticity and environmental induction and are most probably mediated by epigenetic mechanisms. The stochastic proportion became evident by rearing clutchmates in identical laboratory settings [76]. Such experiments revealed considerable variation in all traits investigated. Variation was lowest for morphological characters such as carapace and body lengths or numbers of olfactory and gustatory sense organs, higher for life history traits including growth, reproduction, and longevity, and particularly high for behavioral traits. The marbled coloration pattern was extremely variable. So, each specimen was individually identified. Coloration differed markedly between mother and offspring and between clutchmates and was not inherited [76]. There were also differences in global DNA methylation between clutchmates, but these differences were not yet mechanistically linked to the variation of particular morphological, behavioral, and life-history traits.

The impact of environmental cues on phenotypic traits and DNA methylation in marbled crayfish became evident by comparison of specimens from differently adapted, but genetically identical populations. Near genetic identity was demonstrated by whole genome sequencing (WGS) of individuals from different sources [78, 79]. For example, WGS of 19 representatives from 15 wild populations in Europe and Madagascar with the reference genome revealed extremely low genetic variation of a total of 16,564 single nucleotide variants (SNVs) in the 3.7 GB genome (Figure 16.7a) [79]. The reference genome was established from a descendant of my laboratory population, which was founded in 2003 with a single specimen from the oldest known aquarium lineage started in 1995. Since the generation time of marbled crayfish is about seven months, the various populations examined are separated from the laboratory reference for no more than 40 generations. About 74% of the SNVs were located in intergenic regions and 4% were located in coding regions, among them very few nonsynonymous variants that change amino acids in proteins [78, 79]. These data suggest that any phenotypic differences between specimens from different sources must be caused by epigenetic variation rather than genetic variation.

A comparison of marbled crayfish from the laboratory and German lakes revealed considerable differences in both phenotypic traits and DNA methylation [80, 92, 97]. Specimens raised under stringent laboratory conditions for many generations reproduced well and grew old (4.5 years), but reached maximum TLs of only 9 cm and weights of 18 g, whereas their relatives from Lake Moosweiher, Germany, and other lakes grew to ~12 cm and 40 g (Figure 16.7b). The wild specimens had a uniform dark greenish-brown tanning on their dorsal sites, whereas the laboratory-reared specimens displayed a broad spectrum of greenish-brown, reddish, brown, and bluish dorsal colorations (Figure 16.7b). Furthermore, wild specimens from different German lakes had prominent sharp spines on their carapace and chelipeds, but laboratory-raised specimens lacked these spines and had only small blunt knobs instead [80]. Specimens transferred from Lake Moosweiher to the laboratory maintained their spines through the following molts, but in the F1 generation they were considerably reduced, resembling specimens of the laboratory colony.

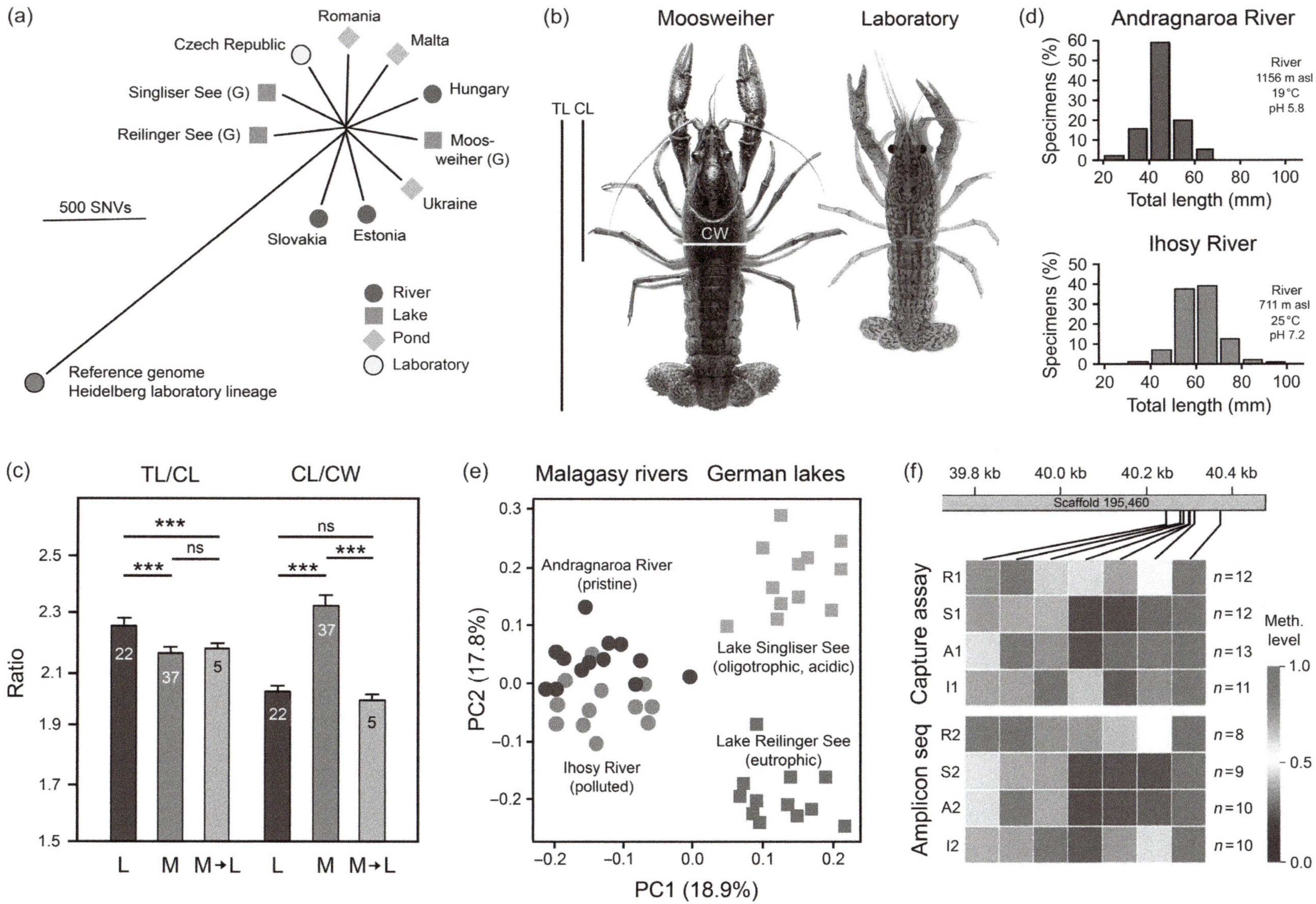

Figure 16.7 Genetic uniformity and phenotypic and epigenetic diversity in marbled crayfish adapted to different conditions. (a) Genetic difference on the whole genome scale between representatives from different European populations. A descendant of the oldest known marbled crayfish laboratory population was used as a reference. G, Germany. (b) Comparison of maximum body size and coloration between populations from the laboratory and wild. TL of the largest specimen from Lake Moosweiher (Germany) exceeded that of the largest specimen from my laboratory colony by about one-third. Coloration of the dorsal side was uniformly dark greenish-brown in the lake specimens, but quite variable in laboratory specimens. CL, carapace length; CW, carapace width. (c) Body proportions of specimens from my laboratory colony (L) and Lake Moosweiher (M), showing significant differences in total length (TL) to carapace length (CL) ratio and CL to carapace width (CW) ratio. The adult offspring of a specimen transferred from Lake Moosweiher to the laboratory (M→L) had a TL-to-CL ratio similar to the wild population, but a CL-to-CW ratio similar to the laboratory population. Figures in columns give numbers of specimens investigated; ***, significantly different (p<0.001); ns, not significantly different. (d) Significant differences in population structure between the pristine and montane Andragnaroa River and heavily polluted Ihosy River in Madagascar. p<0.05. (e) Principal component analysis of methylation of 122 genes separating four populations from rivers and lakes in Germany and Madagascar. (f) Persistent DNA methylation fingerprints of populations from Andragnaroa River (A), Ihosy River (I), Lake Reilinger See (R) and Lake Singliser See (S) in consecutive years (1 and 2), exemplified for a small genic region of the hepatopancreatic DNA. The samples were collected at intervals of 12–21 months and analyzed with two different methods. *Source:* (a) Adapted from Maiakovska et al. [79]. (b) Adapted from Vogt [97]. (c) Adapted from Vogt [10]. (d) Adapted from Andriantsoa et al. [83]. (e, f) Adapted from Tönges et al. [92]. (*See colour plate section for colour representation of this figure.*)

The laboratory-raised crayfish also had significantly longer pleons and broader carapaces when compared to equal-sized specimens from Lake Moosweiher (Figure 16.7c) [10]. The offspring of a specimen that was transferred from Lake Moosweiher to the laboratory and reproduced there several months later had a total length-to-carapace length ratio similar to the wild population and the mother, but a carapace-length-to-carapace-width ratio similar to the laboratory population (Figure 16.7c), suggesting partial adaptation to the new environment [10].

Laboratory-raised and wild specimens from Lake Moosweiher showed no marked difference in global DNA methylation. However, a comparative analysis of 697 variably methylated genes of specimens from the same sources revealed methylation differences in numerous genes (Figure 16.6e). These differences were particularly prominent in the hepatopancreas, the central metabolic organ of decapods, and less prominent in the abdominal musculature.

Phenotypic differences in marbled crayfish were also found with respect to population structure. For example, Andriantsoa et al. [83] revealed marked differences in size-frequency distribution among five populations from ponds, lakes, rivers, and rice fields in Madagascar (Figure 16.7d), despite genetic identity. Tönges et al. [92] investigated the association between environmental adaptation and epigenetic signatures in more detail on the example of two Malagasy and two German populations from strikingly different habitats. One of the Malagasy populations lives in a clear, slowly flowing mountain river (Andragnaroa River) with relatively low pH and temperature and is characterized by a low population density of 20 CPUE (catch per unit effort; number of specimens caught by two persons per hours) and a bias toward small-sized animals (Figure 16.7d). The second Malagasy population inhabits a slowly flowing, turbid and polluted city river (Ihosy River) with higher pH and temperature and high contents of iron and aluminum. It is further characterized by a high population density (152 CPUE) and comparably large animals (Figure 16.7d). The German lake populations inhabit acidic, oligotrophic Lake Singliser See (former lignite mining site, no fishes present) and slightly basic, eutrophic Lake Reilinger See (predatory fishes present). Principal component analysis of DNA methylation of a subset of 122 genes of the WGBS data revealed a clear separation of the four populations (Figure 16.7e).

Interestingly, the specific DNA methylation fingerprints of these populations remained stable over consecutive years in both the hepatopancreas and abdominal musculature (Figure 16.7f) [92]. Gene ontology analysis of the variably methylated genes revealed a significant enrichment of GTP-binding proteins, which transmit signals from the external environment into the cells. Enrichment was also recorded for proteins involved in regulation of transcription and translation, RNA metabolism, response to stress, and immune response. Since no SNVs were found in the coding regions of the differently methylated genes, the study provides conclusive evidence for location-specific epigenetic variation that is independent of genetic variation [92]. It also argues for the existence of epigenetic ecotypes.

Comparison of the triploid marbled crayfish with its diploid parent species *Procambarus fallax* demonstrated the degree of phenotypic and epigenetic changes that can result from polyploid speciation. The generation of polyploids is used in fish and crustacean aquaculture to improve performance [98, 99]. In marbled crayfish, triploidization was accompanied or followed by a transition from gonochorism to parthenogenesis and associated with reproductive isolation and marked alterations in life history traits, DNA content, and DNA methylation [21, 74, 77]. However, their morphological habitus and marbled coloration remained very similar to the parent species.

In laboratory experiments, marbled crayfish females readily copulated with males of *P. fallax* but the offspring was always pure marbled crayfish as demonstrated by microsatellite analysis [77], indicating that reproductive isolation was caused by cytological or genetic rather than behavioral incompatibility. The life history alterations concerned body size, longevity, and fecundity. The neo-species marbled crayfish grows to a maximum body length of ~12 cm corresponding to a body weight of 40 g, whereas the parent species *P. fallax* grows only to a maximum of ~9 cm and 18 g [74]. Marbled crayfish have a maximum life span of 1620 days in the laboratory, but *P. fallax* reaches only 882 days under the same conditions [82, 88]. Egg number per clutch is on average much higher in marbled crayfish due to bigger body size, but it is also considerably higher in marbled crayfish of equal sizes (Figure 16.8a) [74]. The enhancement of growth, fertility, and longevity indicates superior fitness of marbled crayfish when compared to its parent species.

The triploid marbled crayfish has a 1.4-fold instead of an expected 1.5-fold DNA content when compared to its diploid parent species (Figure 16.8b), suggesting loss of some DNA after polyploidization [77]. DNA reduction after polyploidization is quite common in animals and plants. Global DNA methylation is about 20% lower in marbled crayfish (Figure 16.8c) [77] and there are differences in gene body methylation in hundreds of genes between both species (Figure 16.8d) [21], arguing for a considerable rearrangement of the DNA methylation pattern and gene expression during polyploid speciation.

Marçal et al. studied the influence of the herbicide penoxsulam (Px, 23 μg/L) and the genotoxicant ethyl methanesulfonate (EMS, 5 mg/L) on global DNA methylation in the crayfish *Procambarus clarkii* using a global 5-mC ELISA

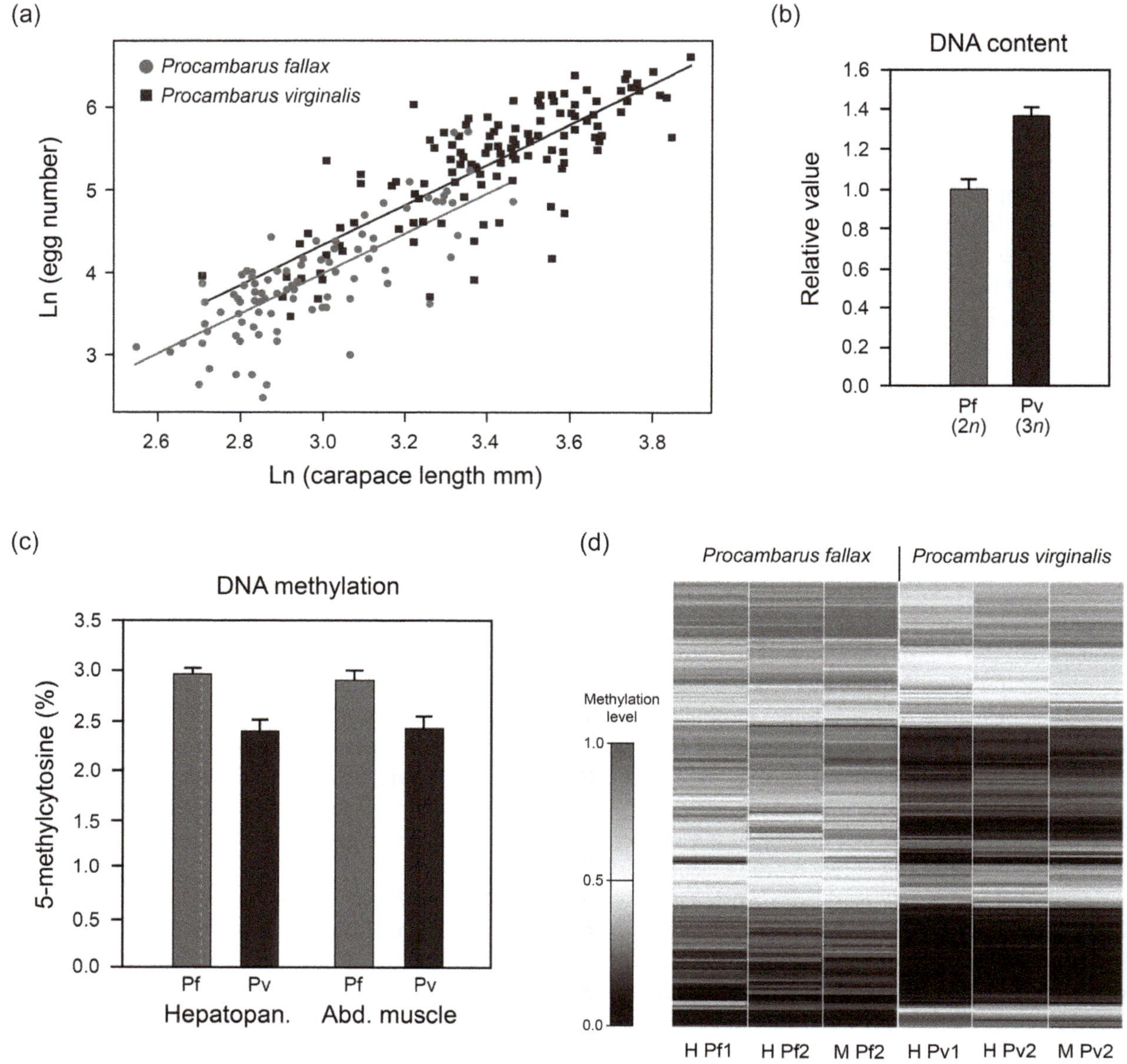

Figure 16.8 Comparison of fecundity, DNA content, and DNA methylation between triploid marbled crayfish (*Procambarus virginalis*) and its diploid parent species *Procambarus fallax*. (a) Pleopodal egg numbers per female and clutch. Fecundity is, on average, much higher in *P. virginalis* due to bigger body size, but it is also ca. 40% higher in specimens of equal size, as indicated by the linear model prediction lines. (b) DNA content in hemocytes demonstrating a ~1.4-fold higher value in *P. virginalis*. Bars indicate mean ± SD of two biological and three technical replicates. (c) Global DNA methylation in hepatopancreas and abdominal musculature. Mass spectrometry revealed a ca. 20% lower DNA methylation level in *P. virginalis*. Bars indicate mean ± SD of three laboratory-raised females. (d) Gene body methylation of 2357 genes in two *P. fallax* (Pf 1 and Pf2) and two *P. virginalis* (Pv1 and Pv2). The heat map of the hepatopancreas (H) and abdominal musculature (M) samples shows numerous differentially methylated genes between both species, most of them being hypomethylated in *P. virginalis*. *Source:* (a) Adapted from Vogt et al. [74]. (b, c) Adapted from Vogt et al. [77]. (d) Adapted from Gatzmann et al. [21].

colorimetric kit [100]. They also determined transmission of the effects to the offspring. Global DNA methylation of the musculature varied between sexes, being higher in females (4.3% 5mC) than in males (1.7% 5mC). Px did not induce significant changes in DNA methylation in the exposed F0 generation, but hypomethylation occurred in the unexposed F1 juveniles, which was more pronounced in males than in females. EMS exposure of the F0 generation induced hypomethylation, particularly in males. Hypomethylation was also observed in the unexposed F1 generation, indicating a transgenerational epigenetic effect. The results demonstrate that exposure to chemicals can have epigenetic effects in the unexposed offspring even if no effects are visible in the exposed parental generation.

16.6 Epigenetics Research with Shrimps and Crabs

Apart from the crayfish *P. virginalis* and *P. clarkii*, there are only a few epigenetics studies on decapod crustaceans. Full genome sequences, which facilitate epigenetics research, are available for the penaeid shrimps *Penaeus monodon* and *Penaeus vannamei* [101, 102], the crabs *Portunus trituberculatus* [103] and *Eriocheir sinensis* [104], and the American lobster *Homarus americanus* [105].

Shrimps are the leading products of crustacean aquaculture followed by freshwater crayfish and crabs (Table 16.1) [34]. The cultured shrimps, often also called prawns, are marine Dendrobranchiata of the genus *Penaeus* and freshwater Caridea of the genus *Macrobrachium*. Among the cultured crabs, the Chinese midden crab (*Eriocheir sinensis*) and species of the mud crabs genus *Scylla* are predominant [106, 107]. Commercially traded clawed lobsters, spiny lobsters, and slipper lobsters are mainly fished from the wild. Vietnam is the only country where farming of lobsters is fully developed and commercially successful [108]. It currently produces around 1600 tons valued at $120 million, primarily of spiny lobster *Panulirus ornatus*.

Wang et al. [109] investigated the responses of *Penaeus japonicus* to air exposure, a common stress phenomenon during transportation of shrimps. They measured time-dependent changes in survival and heart rate (Figure 16.9a) and total antioxidant capacity, as well as significant apoptosis in the hepatopancreas and gills. The expression of apoptosis genes

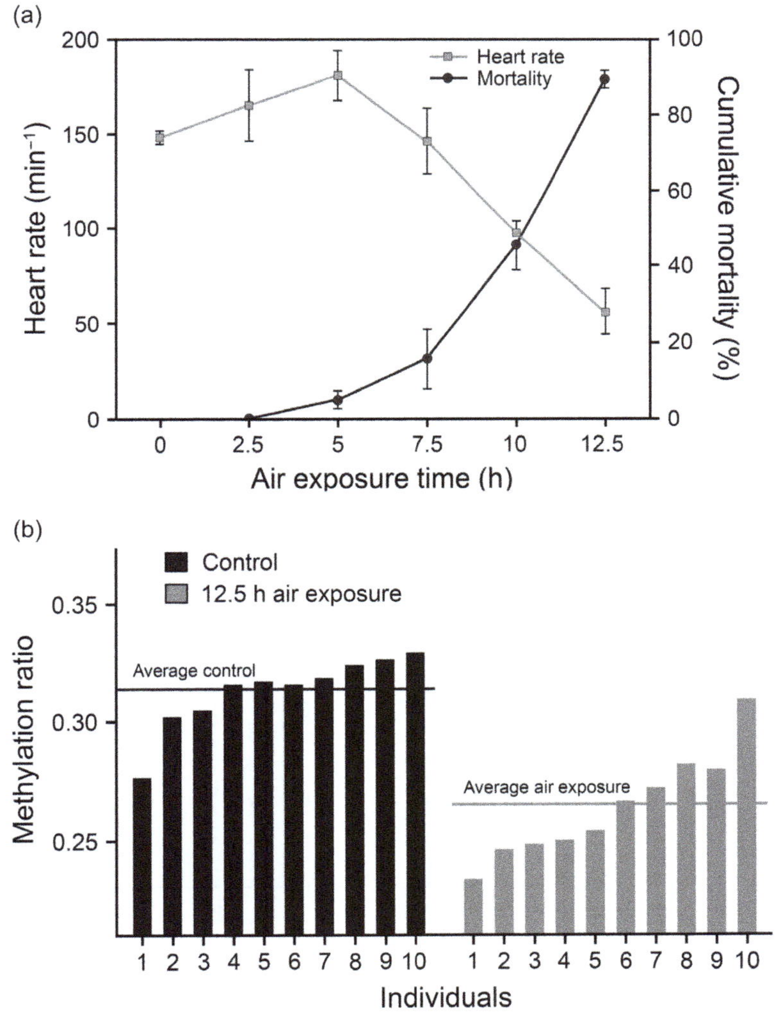

Figure 16.9 Survival and alterations of heart rate and DNA methylation in shrimp *Penaeus japonicus* after air exposure. (a) Increase in mortality and decrease in heart rate after prolonged air exposure. Bars in the heart rate curve represent mean ± SD of six specimens. The mortality curve is based on 200 specimens. (b) DNA methylation levels in the hepatopancreas of air-exposed and unexposed shrimps, showing lower average methylation values after air exposure. *Source:* Adapted from Wang et al. [109].

(*IAP*, *TXNIP*, *caspase*, and *caspase-3*) and hypoxia-related genes (*Hsp70*, *hif-1a* and *HcY*) were all significantly enhanced in the hepatopancreas and gills of air-exposed specimens. Moreover, shrimps exposed to air for 12.5 hours had, on average, lower methylation levels in the hepatopancreas than their nonexposed counterparts (Figure 16.9b), as calculated from MSAP (methylation-sensitive amplification polymorphism) data obtained with different DNA digestion combinations that were used to calculate the methylation ratio. This study shows that stressors can change DNA methylation in tissues within a few hours.

Wang et al. investigated the distribution, transcription, and methylation status of satellite DNA in the land crab *Gecarcinus lateralis* [14]. This satellite DNA is rich in C and G, displays a repeat unit of 2.1 kb, and makes up approximately 3% of the genome. The authors found that some satellites were transcribed because of their spatial linkage to functional genes, whilst those outside of genes were not transcribed. The latter satellites were methylated, whereas those interspersed in genes were unmethylated. This fits well with the literature data on the silencing of repeats by methylation [110]. The authors also found tissue-specific DNA methylation patterns: the gills showed higher methylation levels than the musculature, hepatopancreas, testis, and ovary.

Jiang et al. investigated the role of DNA methylation in the expression of the crustacean female sex hormone (CFSH), a key regulator of crustacean sex differentiation, in the mud crab *Scylla paramamosain* [107]. *CFSH* showed a tissue- and gender-specific expression pattern (Figure 16.10a) as expected. To explore the role of DNA methylation in *CFSH* expression and inhibition, the 5′-flanking region of the gene was cloned and a CpG island containing 12 CpG sites was identified by MethPrimer (Figure 16.10b). Bisulfite sequencing and methylated DNA immunoprecipitation showed that CpG island methylation of the *CFSH* gene was significantly lower in the eyestalk ganglion of females than in the eyestalk ganglion of males (Figure 16.10c). The eyestalk ganglion includes the X-organ that synthesizes the CFSH [111]. Moreover, in females, CpG island methylation of the *CFSH* gene was significantly lower in the eyestalk ganglion when compared to the

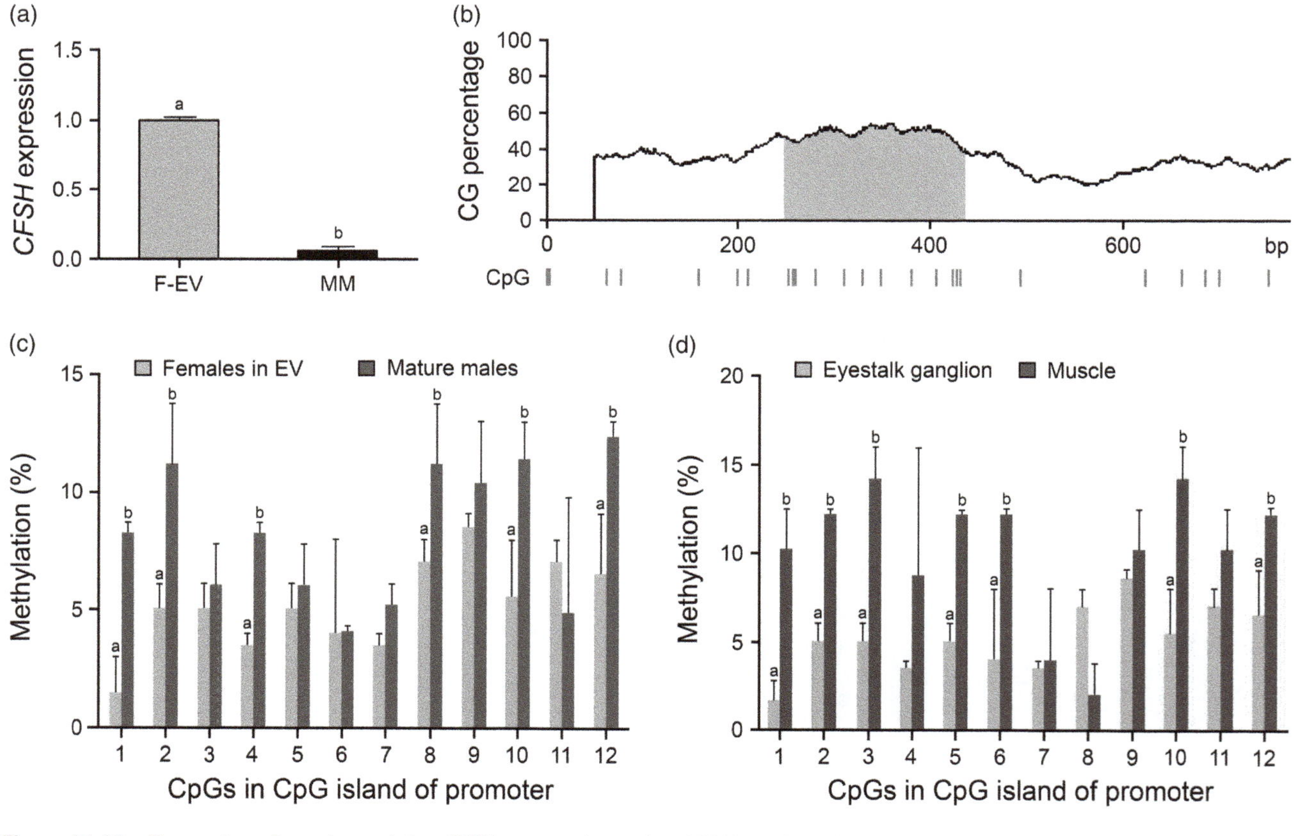

Figure 16.10 Expression of sex-determining *CFSH* gene and associated DNA methylation in mud crab *Scylla paramamosain*. (a) *CFSH* gene expression in eyestalk ganglion of early-vitellogenic females (FEVs), the female stage with the lowest *CFSH* expression, and mature males (MMs), showing the pronounced difference between sexes. $n = 6$ per group; $p < 0.05$. (b) CpG island (shaded) resembling vertebrate promoter region in 5′-flanking sequence of *CFSH* gene containing 12 CpGs (vertical bars). (c) Comparison of CpG methylation in *CFSH* promoter region between eyestalk ganglia of FEVs and MMs, showing significantly higher values in males. $n = 3$ per group; $p < 0.05$. (d) Comparison of CpG methylation in *CFSH* promoter region between eyestalk ganglion and musculature of early vitellogenic females, showing significantly higher values in the musculature. $n = 3$ per group; $p < 0.05$. *Source:* Adapted from Jiang et al. [107].

musculature (Figure 16.10d). These findings suggest that hypermethylation of the promoter CpGs suppresses *CFSH* expression in mud crab, contributing to its regulation in a gender- and tissue-specific manner. Analysis of the *CFSH* promoter activity further revealed that methylation inhibited expression by blocking the binding of transcription factor Sp1 to the DNA.

Alterations of the ovarian miRNA profile and proteome after eyestalk ablation were investigated by Meng et al. in the marine crab *Portunus trituberculatus*, the most widely fished crab species in the world [24]. Eyestalk ablation is a common means to induce ovarian maturation in cultured shrimps [111]. The authors recorded differential expression of 163 protein-coding genes and 31 miRNA genes between the ablated and unablated groups. Integration of miRNA and proteome expression data with miRNA target prediction algorithms generated a potential regulatory network consisting of 26 miRNAs and 30 proteins linked by multiple functional associations. This knowledge could perhaps be used to epigenetically manipulate ovarian maturation in cultured shrimps.

16.7 State of the Art of Epigenetics Research in Crustaceans

Epigenetics in crustaceans is a relatively young field of research. Although it is far behind epigenetics research in mammals and insects, it has significantly increased in the last several years. Scientific articles on crustacean epigenetics are available for brine shrimps, copepods, water fleas, amphipods, freshwater crayfish, penaeid shrimps, and crabs (Box 16.1). Most of these papers date from the last few years. The brine shrimp *Artemia franciscana*, the water fleas *Daphnia pulex* and *Daphnia magna*, and the freshwater crayfish *Procambarus virginalis* are presently being developed as models for basic and applied epigenetics.

The available studies demonstrate that all taxa of interest for crustacean aquaculture possess methylated DNA, histone modifications, and ncRNA pathways. This cannot be taken for granted because several arthropod groups, including dipteran insects and mites, have unmethylated DNA [112–114]. The global DNA methylation values of 0.1–0.4% in brine shrimps [17], 0.5–0.7% in daphnids [18], and 0.4–1.4% in amphipods [66] are small and moderate when compared with other animal taxa [85], but the values of 2.4% in *P. virginalis* [77] and 4.5% in *Penaeus semisulcatus* [13] are relatively high, being in the range of vertebrates. Full genome sequences, which are an important prerequisite to investigate epigenetic marks genome-wide, are meanwhile available for the brine shrimp *Artemia* [37], the water fleas *D. pulex* and *D. magna* [51–53], the crayfish *P. virginalis* [78], the shrimps *Penaeus vannamei* and *Penaeus monodon* [101, 102], the crabs *Portunus trituberculatus* and *Eriocheir sinensis* [103, 104], and the American lobster *Homarus americanus* [105].

Box 16.1 Overview of Epigenetics Research in Crustaceans and Perspectives for Aquaculture

Crustacean groups used in aquaculture: Brine shrimps and water fleas are cultured as food for fish and shellfish larviculture, copepods and amphipods are used as food ingredients, and shrimps, crabs, and spiny lobsters are cultured for human consumption.

Crustaceans used so far for epigenetics research: Brine shrimp *Artemia franciscana*, copepod *Acartia clausii*, water fleas *Daphnia magna*, *Daphnia pulex*, *Evadne nordmanni* and *Moina macrocopa*, amphipods *Gammarus locusta* and *Gammarus fossarum*, penaeid shrimps *Penaeus japonicus*, *Penaeus semisulcatus* and *Penaeus vannamei*, freshwater crayfish *Procambarus virginalis*, *Procambarus fallax* and *Procambarus clarkii*, mud crab *Scylla paramamosain*, and land crab *Gecarcinus lateralis*.

Epigenetic mechanisms investigated: Primarily DNA methylation, to a lesser extent histone modifications and least often noncoding RNAs.

Epigenetics research targeted on crustacean aquaculture: There is one paper on alterations of DNA methylation in response to air exposure in shrimp *Penaeus japonicus* and another paper on production system specific epigenetic fingerprints in *Penaeus vannamei*.

Functions of epigenetic mechanisms in crustaceans: Involvement in gene regulation, tissue differentiation, phenotypic variation, development, growth, reproduction, aging, sex determination, response to adverse environmental conditions and toxicants, trained immunity, transgenerational inheritance of acquired characters, and polyploid speciation.

Future potential of epigenetics in crustacean aquaculture: Improvement of growth and reproduction, sex manipulation to optimize yield and economic revenue, enhancement of resistance against adverse environmental conditions, toxicants and diseases, and development of epigenetic biomarkers for animal welfare and origin tracing of aquaculture products.

Epigenetics work with crustaceans has shown that DNA methylation, DNA hydroxymethylation, and histone acetylation and methylation marks can dynamically change, as in other animals [21, 39, 56, 57]. DNA methylation is generally confined to CpG dinucleotides [18, 21, 56]. It was found in exons and introns of genes, intergenic regions, promoter regions, and transposons [14, 21, 107]. Methylation of promoters and repetitive elements apparently promotes silencing [14, 107], whereas methylation of gene bodies seems to stabilize gene expression, particularly in housekeeping genes [21].

DNA methylation in crustaceans can considerably differ between tissues as shown for global DNA methylation values, methylation of individual genes, and expression of DNA methylating and demethylating enzymes [21]. In decapods, the hepatopancreas is probably the most dynamic organ with respect to DNA methylation. This is plausible because it is the central organ of metabolism serving many functions such as synthesis of digestive enzymes, absorption and processing of the digestive products, provisioning of energy to other organs, storage of glycogen and lipid reserves, detoxification of xenobiotics, and synthesis of compounds involved in vitellogenesis and immune defense [89]. Moreover, it contributes considerably to adaptation to harsh environmental conditions. The air exposure experiment with shrimps has shown that some methylation marks in this organ can change within a few hours. DNA methylation fingerprints of the hepatopancreas could provide a valuable biomarker for health and environmental adaptation of decapods. It is likely that there are short-term epigenetic responses to rather rapidly fluctuating factors and long-term responses to rather stable and persisting conditions. This issue must be investigated in more detail to find the appropriate sets of genes and epigenetic fingerprints that can be used as biomarkers for either purpose. For comparison of long-term adaptation to different stable conditions, the use of organs with lower methylation and demethylation dynamics like the musculature may be better suitable than the hepatopancreas.

Functionally, DNA methylation in crustaceans was shown to be involved in development and aging [20, 21, 56, 97], compensation of environmental stress [17, 56], sex determination [19], immune defense [39], and polyploid speciation [10, 21, 77]. Epigenetic marks can change stochastically [9, 76] or in response to strong environmental signals such as high and low temperatures, hypoxia, air exposure, toxicants, and γ-irradiation [17, 56, 59, 61, 63, 109]. Reduction of global DNA methylation in a given species or genus seems to promote phenotypic variation. For example, the globally hypomethylated *P. virginalis* is a highly invasive crayfish that has conquered diverse ecosystems in different bioclimatic regions, whereas its higher methylated parent species *Procambarus fallax* is less invasive and confined to the subtropics of Florida and southern Georgia [27].

The question of the involvement of DNA methylation in the transgenerational inheritance of acquired characters has repeatedly been addressed in crustaceans. There are examples of the inheritance of induced phenotypes and associated epigenetic marks up to the third filial generation [39, 40, 59, 61], but cause-effect relationships are not yet established.

Histone modifications were shown to respond to heat shock, bacterial infection, and phloroglucinol treatment in *Artemia* [17, 39, 40]. They were also involved in sex differentiation of *Daphnia* [19] and probably in the transgenerational inheritance of acquired phenotypes in brine shrimps and water fleas [17, 39, 40, 63]. Noncoding RNA pathways are currently the least investigated epigenetic mechanisms in crustaceans.

Genomic and epigenetic resources such as fully sequenced genomes, transcriptomes, genome-wide methylomes, histone modification patterns, and chromatin structures are increasingly available for crustaceans [19, 21, 60, 78, 115]. They facilitate investigations on the relationships between epigenetic features, gene expression, and phenotypic traits. Moreover, the techniques for DNA sequencing and determination of DNA methylation, histone modifications, and chromatin structure have been greatly improved in the last several years [60, 115–117]. The same holds true for bioinformatics interpretation tools [118, 119]. The new analytical techniques are fast and affordable and require only small sample sizes. Therefore, future genomic and epigenetic studies with crustaceans can be done by single laboratories in a reasonable time and at reasonable costs.

16.8 Potential Application of Epigenetics in Crustacean Aquaculture

Studying the epigenetic mechanisms behind different biological aspects of aquaculture species may offer great opportunities to improve their culture [120]. The knowledge of epigenetics in crustaceans has not yet been applied to aquaculture, but the foundations for possible epigenetic improvements in growth, reproduction, and health of cultured species are presently being laid. It is well documented that environmental cues such as nutrients, stressful water conditions, toxicants, and pathogens can change epigenetic marks in animals, resulting in transcriptional changes, which in turn lead to phenotypic changes [121–123]. Understanding of these mechanisms and relationships is required for developing epigenetic interventions suitable to improve the performance of cultured species and increase the economical revenue of crustacean aquaculture enterprises. Epigenetic interventions should principally be possible in crustaceans because they possess DNA methylation, histone modifications, and ncRNAs that are responsive to environmental signals and experimental manipulation.

The influence of nutrition on the epigenome is rather well investigated in mammals [121, 124, 125]. Methyl-supplemented food or feeding of DNA methyltransferase-inhibiting substances significantly changed the DNA methylation pattern and associated phenotypic traits [125]. Feeding of DNA methyltransferase-inhibiting substances also changed the DNA methylation pattern in *Daphnia* [56, 59], suggesting that nutritional epigenetic interventions could be used to manipulate growth in cultured crustaceans. A particularly impressive example of the epigenetic power of nutrition is the castes of honeybees. The same bee larva can develop either into a queen or a worker with strikingly different size, morphology, physiology, behavior, reproduction, and longevity, depending on whether it is fed with royal jelly or pollen. Such genetically identical but phenotypically diverse honeybee females exhibit extensive genome-wide differences in DNA methylation and histone modifications, many of which correlate with caste-specific transcription [126, 127].

Sex manipulation aimed at generating all-male or all-female populations is regarded as a beneficial technique for achieving higher yields in crustacean aquaculture. In crayfish and freshwater prawns, males grow larger and faster, and as such, all-male rearing is desired [128]. For example, all-male monosex culture of the freshwater prawn *Macrobrachium rosenbergii* was 63 and 60% more profitable than mixed and all-female cultures, respectively [129]. In contrast, females grow larger and faster than males in cultured marine shrimps and crabs [130]. For this reason, understanding of sex differentiation and its manipulability is crucial. DNA hypermethylation was demonstrated to promote silencing of the female-determining *CFSH* gene in decapod males [107], and this knowledge could be used to specifically produce females and males.

Environmental stressors such as toxicants and high and low salinities, temperature, and oxygen were shown to affect the epigenetic profiles of diverse animals [131, 132]. A crustacean example is *Penaeus japonicus*, in which air exposure induced changes in the expression of stress resistance genes and their methylation profiles [109]. The artificial modification of the epigenetic signatures of such genes could be used to enhance tolerance of cultured crustaceans to unfavorable environmental conditions. The induced resistance may last for some generations due to transgenerational inheritance of the underlying epigenetic marks as shown for *Artemia* and *Daphnia*.

Particularly promising is the application of epigenetics in the prevention of diseases. Approximately, 40% of shrimp production for human consumption is lost annually by diseases [133]. About 60% of these diseases are caused by viruses, and a further 20% by bacteria. Experiments with brine shrimp revealed that the innate immune system of crustaceans can be trained for increased resistance against pathogens, that epigenetic mechanisms are involved in this process, and that the enhanced resistance can last for several generations in the absence of the stimulating agent [39, 40]. It is likely that this mechanism also works in higher crustaceans inclusive of the cultured decapods, and that epigenetic conditioning of the innate immune system of the broodstock could be a powerful novel strategy for prophylaxis against infectious diseases in crustacean aquaculture.

If the epigenetic signatures of desired phenotypes in cultured crustaceans were known, then these epigenetic marks could be experimentally manipulated in a targeted manner. Particularly suitable for this purpose is the CRISPR-Cas technique, which could be used to manipulate genes that encode DNA methyltransferases and histone-modifying enzymes or, even better, to add and remove individual methylation marks on the DNA [134–136]. CRISPR-Cas has already been applied to crustaceans. It proved highly effective in producing heritable genetic knockouts and knockins in *D. pulex* [137, 138]. CRISPR-Cas interventions could now be developed for specifically customizing the epigenome of crustaceans to the cultural requirements.

References

1 Jaenisch, R. and Bird, A. (2003). Epigenetic regulation of gene expression: how the genome integrates intrinsic and environmental signals. *Nature Genetics* **33**: 245–254.

2 Lennartsson, A. and Ekwall, K. (2009). Histone modification patterns and epigenetic codes. *Biochimica et Biophysica Acta* **1790** (9): 863–868.

3 Moutinho, C. and Esteller, M. (2017). MicroRNAs and epigenetics. *Advances in Cancer Research* **135**: 189–220.

4 Jablonka, E. and Raz, G. (2009). Transgenerational epigenetic inheritance: prevalence, mechanisms, and implications for the study of heredity and evolution. *Quarterly Review of Biology* **84** (2): 131–176.

5 Burggren, W. (2016). Epigenetic inheritance and its role in evolutionary biology: re-evaluation and new perspectives. *Biology* **5** (2): 24.

6 Chen, Q., Yan, W., and Duan, E. (2016). Epigenetic inheritance of acquired traits through sperm RNAs and sperm RNA modifications. *Nature Reviews Genetics* **17**: 733–743.

7 Leung, C., Breton, S., and Angers, B. (2016). Facing environmental predictability with different sources of epigenetic variation. *Ecology and Evolution* **6** (15): 5234–5245.

8 Angers, B., Perez, M., Menicucci, T. et al. (2020). Sources of epigenetic variation and their applications in natural populations. *Evolutionary Applications* **13** (6): 1262–1278.

9 Vogt, G. (2020). Disentangling the environmentally induced and stochastic developmental components of phenotypic variation. In: *Phenotypic Switching: Implications in Biology and Medicine* (ed. H. Levine, M.K. Jolly, P. Kulkarni and V. Nanjundiah), 207–251. San Diego, CA: Academic Press.

10 Vogt, G. (2021). Epigenetic variation in animal populations: sources, extent, phenotypic implications and ecological and evolutionary relevance. *Journal of Biosciences* **46**: 24.

11 Anastasiadi, D. and Piferrer, F. (2019). Epimutations in developmental genes underlie the onset of domestication in farmed European sea bass. *Molecular Biology and Evolution* **36** (10): 2252–2264.

12 Konstantinidis, J., Sætrom, P., Mjelle, R. et al. (2020). Major gene expression changes and epigenetic remodelling in Nile tilapia muscle after just one generation of domestication. *Epigenetics* **15** (10): 1052–1067.

13 Regev, A., Lamb, M.J., and Jablonka, E. (1998). The role of DNA methylation in invertebrates: developmental regulation or genome defense? *Molecular Biology and Evolution* **15** (7): 880–891.

14 Wang, S.Y., Biesiot, P.M., and Skinner, D.M. (1999). Toward an understanding of satellite DNA function in Crustacea. *American Zoologist* **39** (3): 471–486.

15 Schiewek, R., Wirtz, M., Thiemann, M. et al. (2007). Determination of the DNA methylation level of the marbled crayfish: an increase in sample throughput by an optimised sample preparation. *Journal of Chromatography B* **850** (1–2): 548–552.

16 Vandegehuchte, M.B., Kyndt, T., Vanholme, B. et al. (2009). Occurrence of DNA methylation in *Daphnia magna* and influence of multigeneration Cd exposure. *Environment International* **35** (4): 700–706.

17 Norouzitallab, P., Baruah, K., Vandegehuchte, M. et al. (2014). Environmental heat stress induces epigenetic transgenerational inheritance of robustness in parthenogenetic *Artemia* model. *FASEB Journal* **28** (8): 3552–3563.

18 Asselman, J., De Coninck, D.I.M., Pfrender, M.E. et al. (2016). Gene body methylation patterns in *Daphnia* are associated with gene family size. *Genome Biology and Evolution* **8** (4): 1185–1196.

19 Kvist, J., Gonçalves Athanàsio, C., Pfrender, M.E. et al. (2020). A comprehensive epigenomic analysis of phenotypically distinguishable, genetically identical female and male *Daphnia pulex*. *BMC Genomics* **21**: 17.

20 Hearn, J., Plenderleith, F., and Little, T.J. (2021). DNA methylation differs extensively between strains of the same geographical origin and changes with age in *Daphnia magna*. *Epigenetics & Chromatin* **14**: 4.

21 Gatzmann, F., Falckenhayn, C., Gutekunst, J. et al. (2018). The methylome of the marbled crayfish links gene body methylation to stable expression of poorly accessible genes. *Epigenetics & Chromatin* **11**: 57.

22 Norouzitallab, P. (2015). Use of Artemia as model organism to study epigenetic control of phenotypes relevant for aquaculture species. PhD thesis. University of Ghent.

23 Coucheron, D.H., Wojewodzic, M.W., and Bøhn, T. (2019). MicroRNAs in *Daphnia magna* identified and characterized by deep sequencing, genome mapping and manual curation. *Scientific Reports* **9**: 15945.

24 Meng, X., Zhang, M., Gao, B. et al. (2020). Integrative proteomic and microRNA analysis: insights into mechanisms of eyestalk ablation-induced ovarian maturation in the swimming crab *Portunus trituberculatus*. *Frontiers in Endocrinology* **11**: 533.

25 Abatzopoulos, T.J., Beardmore, J., Clegg, J.S., and Sorgeloos, P. (ed.) (2002). *Artemia: Basic and Applied Biology*. Dordrecht: Springer.

26 Stollewerk, A. (2010). The water flea *Daphnia* – a 'new' model system for ecology and evolution? *Journal of Biology* **9** (2): 21.

27 Vogt, G. (2020). Biology, ecology, evolution, systematics and utilization of the parthenogenetic marbled crayfish, *Procambarus virginalis*. In: *Crayfish: Evolution, Habitat and Conservation Strategies* (ed. F.B. Ribeiro), 137–227. Nova Science Publishers, Hauppauge.

28 Delbare, D. and Dhert, P. (1996). Cladocerans, nematodes and trochophora larvae. In: *Manual on the Production and Use of Live Food for Aquaculture* (ed. P. Lavens and P. Sorgeloos). FAO Fisheries Technical Paper 361, 283–295. Rome: Food and Agriculture Organization of the United Nations.

29 Merchie, G. (2016). *Artemia*: use of nauplii and meta-nauplii. In: *Manual on the Production and Use of Live Food for Aquaculture* (ed. P. Lavens and P. Sorgeloos). FAO Fisheries Technical Paper 361, 137–163. Rome: Food and Agriculture Organization of the United Nations.

30 Radhakrishnan, D.K., Ali, I.A., Schmidt, B.V. et al. (2020). Improvement of nutritional quality of live feed for aquaculture: an overview. *Aquaculture Research* **51**: 1–17.

31 Camara, M.R. (2020). After the gold rush: a review of *Artemia* cyst production in northeastern Brazil. *Aquaculture Reports* **17**: 100359.

32 Hansen, B.W. (2017). Advances using copepods in aquaculture. *Journal of Plankton Research* **39** (6): 972–974.

33 Harlıoğlu, M.M. and Farhadi, A. (2018). Importance of *Gammarus* in aquaculture. *Aquaculture International* **26**: 1327–1338.

34 FAO (2020). *The State of World Fisheries and Aquaculture 2020. Sustainability in Action*. Rome: Food and Agriculture Organization of the United Nations.

35 Van Stappen, G. (1996). *Artemia*: introduction, biology and ecology of *Artemia*. In: *Manual on the Production and Use of Live Food for Aquaculture* (ed. P. Lavens and P. Sorgeloos). FAO Fisheries Technical Paper 361, 79–106. Rome: Food and Agriculture Organization of the United Nations.

36 Libralato, G., Prato, E., Migliore, L. et al. (2016). A review of toxicity testing protocols and endpoints with *Artemia* spp. *Ecological Indicators* **69**: 35–49.

37 De Vos, S., Rombauts, S., Coussement, L. et al. (2021). The genome of the extremophile *Artemia* provides insight into strategies to cope with extreme environments. *BMC Genomics* **22**: 635.

38 Rosenzweig, R., Nillegoda, N.B., Mayer, M.P., and Bukau, B. (2019). The Hsp70 chaperone network. *Nature Reviews Molecular Cell Biology* **20**: 665–680.

39 Norouzitallab, P., Baruah, K., Biswas, P. et al. (2016). Probing the phenomenon of trained immunity in invertebrates during a transgenerational study, using brine shrimp *Artemia* as a model system. *Scientific Reports* **6**: 21166.

40 Roy, S., Kumar, V., Bossier, P. et al. (2019). Phloroglucinol treatment induce transgeneration genetic inherited resistance against *Vibrio* infections and thermal stress in a brine shrimp (*Artemia franciscana*) model. *Frontiers in Immunology* **10**: 2745.

41 Cerenius, L. and Söderhäll, K. (2004). The prophenoloxidase activating system in invertebrates. *Immunological Reviews* **198** (1): 116–126.

42 Aluru, N., Fields, D.M., Shema, S. et al. (2021). Gene expression and epigenetic responses of the marine cladoceran, *Evadne nordmanni*, and the copepod, *Acartia clausi*, to elevated CO_2. *Ecology and Evolution* **11** (23): 16776–16785.

43 Altshuler, I., Demiri, B., Xu, S. et al. (2011). An integrated multi-disciplinary approach for studying multiple stressors in freshwater ecosystems: *Daphnia* as a model organism. *Integrative and Comparative Biology* **51** (4): 623–633.

44 Teixidó, E., Leuthold, D., de Crozé, N. et al. (2020). Comparative assessment of the sensitivity of fish early-life stage, *Daphnia*, and algae tests to the chronic ecotoxicity of xenobiotics: perspectives for alternatives to animal testing. *Environmental Toxicology and Chemistry* **39** (1): 30–41.

45 Monowar, A.K., Mohibul, H.K., Kizar, A.S., and Harunur, R. (2020). Culture of freshwater zooplankton *Daphnia magna* fed with different feed combination. *Bangladesh Journal of Fisheries* **32** (1): 55–59.

46 Dufresne, F. (2011). The history of the *Daphnia pulex* complex: asexuality, hybridization, and polyploidy. In: *Phylogeography and Population Genetics in Crustacea* (ed. C. Held, S. Koenemann and C.D. Schubart). Crustacean Issues **19**, 217–232. Boca Raton, FL: CRC Press.

47 Hebert, P.D.N., Beaton, M.J., Schwartz, S.S., and Stanton, D.J. (1989). Polyphyletic origins of asexuality in *Daphnia pulex*. I. breeding-system variation and levels of clonal diversity. *Evolution* **43** (5): 1004–1015.

48 Frisch, D., Morton, P.K., Chowdhury, P.R. et al. (2014). A millennial-scale chronicle of evolutionary responses to cultural eutrophication in *Daphnia*. *Ecology Letters* **17** (3): 360–368.

49 Tollrian, R. (1995). Predator-induced morphological defenses: costs, life history shifts, and maternal effects in *Daphnia pulex*. *Ecology* **76** (6): 1691–1705.

50 Harris, K.D.M., Bartlett, N.J., and Lloyd, V.K. (2012). *Daphnia* as an emerging epigenetic model organism. *Genetics Research International* **2012**: 147892.

51 Colbourne, J.K., Pfrender, M.E., Gilbert, D. et al. (2011). The ecoresponsive genome of *Daphnia pulex*. *Science* **331** (6017): 555–561.

52 Ye, Z., Xu, S., Spitze, K. et al. (2017). A new reference genome assembly for the microcrustacean *Daphnia pulex*. *G3 Genes Genomes Genetics* **7** (5): 1405–1416.

53 Lee, B.-Y., Choi, B.-S., Kim, M.-S. et al. (2019). The genome of the freshwater water flea *Daphnia magna*: a potential use for freshwater molecular ecotoxicology. *Aquatic Toxicology* **210**: 69–84.

54 Ellegaard, M., Clokie, M.R.C., Czypionka, T. et al. (2020). Dead or alive: sediment DNA archives as tools for tracking aquatic evolution and adaptation. *Communications Biology* **3**: 169.

55 Vandegehuchte, M.B., Lemière, F., and Janssen, C.R. (2009). Quantitative DNA-methylation in *Daphnia magna* and effects of multigeneration Zn exposure. *Comparative Biochemistry and Physiology Part C: Toxicology & Pharmacology* **150** (3): 343–348.

56 Kvist, J., Gonçalves Athanàsio, C., Solari, O.S. et al. (2018). Pattern of DNA methylation in *Daphnia*: evolutionary perspective. *Genome Biology and Evolution* **10** (8): 1988–2007.

57 Strepetkaitė, D., Alzbutas, G., Astromskas, E. et al. (2016). Analysis of DNA methylation and hydroxymethylation in the genome of crustacean *Daphnia pulex*. *Genes* **7** (1): 1.

58 Gómez, R., Van Damme, K., Gosálvez, J. et al. (2016). Male meiosis in Crustacea: synapsis, recombination, epigenetics and fertility in *Daphnia magna*. *Chromosoma* **125** (4): 769–787.

59 Lindeman, L.C., Thaulow, J., Song, Y. et al. (2019). Epigenetic, transcriptional and phenotypic responses in two generations of *Daphnia magna* exposed to the DNA methylation inhibitor 5-azacytidine. *Environmental Epigenetics* **5** (3): dvz016.

60 Augusto, R.C., Minod, A., Rey, O. et al. (2021). A simple ATAC-seq protocol for population epigenetics, version 2. *Wellcome Open Research* **5**: 121.

61 Vandegehuchte, M.B., Lemière, F., Vanhaecke, L. et al. (2010). Direct and transgenerational impact on *Daphnia magna* of chemicals with a known effect on DNA methylation. *Comparative Biochemistry and Physiology Part C: Toxicology & Pharmacology* **151** (3): 278–285.

62 Hu, H.L., Tang, Y.Y., Kwok, M.L. et al. (2020). Impact of juvenile hormone analogue insecticides on the water flea *Moina macrocopa*: growth, reproduction and transgenerational effect. *Aquatic Toxicology* **220**: 105402.

63 Trijau, M., Asselman, J., Armant, O. et al. (2018). Transgenerational DNA methylation changes in *Daphnia magna* exposed to chronic γ irradiation. *Environmental Science & Technology* **52** (7): 4331–4339.

64 Thaulow, J., Song, Y., Lindeman, L.C. et al. (2020). Epigenetic, transcriptional and phenotypic responses in *Daphnia magna* exposed to low-level ionizing radiation. *Environmental Research* **190**: 109930.

65 Becker, P.B. and Workman, J.L. (2013). Nucleosome remodeling and epigenetics. *Cold Spring Harbor Perspectives in Biology* **5**: a017905.

66 Cribiu, P., Chaumot, A., Geffard, O. et al. (2018). Natural variability and modulation by environmental stressors of global genomic cytosine methylation levels in a freshwater crustacean, *Gammarus fossarum*. *Aquatic Toxicology* **205**: 11–18.

67 Alves, N., Neuparth, T., Barros, S., and Santos, M.M. (2021). The anti-lipidemic drug simvastatin modifies epigenetic biomarkers in the amphipod *Gammarus locusta*. *Ecotoxicology and Environmental Safety* **209**: 111849.

68 Vogt, G. (2021). Evaluation of the suitability of the parthenogenetic marbled crayfish for aquaculture: potential benefits versus conservation concerns. *Hydrobiologia* **848**: 285–298.

69 Scholtz, G., Braband, A., Tolley, L. et al. (2003). Parthenogenesis in an outsider crayfish. *Nature* **421**: 806.

70 Vogt, G. (2018). Annotated bibliography of the parthenogenetic marbled crayfish, *Procambarus virginalis*, a new research model, potent invader and popular pet. *Zootaxa* **4418** (4): 301–352.

71 Vogt, G. (2008). The marbled crayfish: a new model organism for research on development, epigenetics and evolutionary biology. *Journal of Zoology* **276** (1): 1–13.

72 Martin, P., Dorn, N.J., Kawai, T. et al. (2010). The enigmatic Marmorkrebs (marbled crayfish) is the parthenogenetic form of *Procambarus fallax* (Hagen, 1870). *Contributions to Zoology* **79** (3): 107–118.

73 Martin, P., Thonagel, S., and Scholtz, G. (2016). The parthenogenetic Marmorkrebs (Malacostraca: Decapoda: Cambaridae) is a triploid organism. *Journal of Zoological Systematics and Evolutionary Research* **54** (1): 13–21.

74 Vogt, G., Dorn, N.J., Pfeiffer, M. et al. (2019). The dimension of biological change caused by autotriploidy: a meta-analysis with the triploid *Procambarus virginalis* and its diploid parent *Procambarus fallax*. *Zoologischer Anzeiger* **281**: 53–67.

75 Martin, P., Kohlmann, K., and Scholtz, G. (2007). The parthenogenetic Marmorkrebs (marbled crayfish) produces genetically uniform offspring. *Naturwissenschaften* **94** (10): 843–846.

76 Vogt, G., Huber, M., Thiemann, M. et al. (2008). Production of different phenotypes from the same genotype in the same environment by developmental variation. *Journal of Experimental Biology* **211** (4): 510–523.

77 Vogt, G., Falckenhayn, C., Schrimpf, A. et al. (2015). The marbled crayfish as a paradigm for saltational speciation by autopolyploidy and parthenogenesis in animals. *Biology Open* **4** (11): 1583–1594.

78 Gutekunst, J., Andriantsoa, R., Falckenhayn, C. et al. (2018). Clonal genome evolution and rapid invasive spread of the marbled crayfish. *Nature Ecology and Evolution* **2**: 567–573.

79 Maiakovska, O., Andriantsoa, R., Tönges, S. et al. (2021). Genome analysis of the monoclonal marbled crayfish reveals genetic separation over a short evolutionary timescale. *Communications Biology* **4**: 74.

80 Vogt, G., Lukhaup, C., Pfeiffer, M. et al. (2018). Morphological and genetic characterization of the marbled crayfish, including a determination key. *Zootaxa* **4524** (3): 329–350.

81 Vogt, G. (2015). Bimodal annual reproduction pattern in laboratory-reared marbled crayfish. *Invertebrate Reproduction and Development* **59** (4): 218–223.

82 Vogt, G. (2010). Suitability of the clonal marbled crayfish for biogerontological research: a review and perspective, with remarks on some further crustaceans. *Biogerontology* **11**: 643–669.

83 Andriantsoa, R., Tönges, S., Panteleit, J. et al. (2019). Ecological plasticity and commercial impact of invasive marbled crayfish populations in Madagascar. *BMC Ecology* **19**: 8.

84 Linzmaier, S.M., Musseau, C., Matern, S., and Jeschke, J.M. (2020). Trophic ecology of invasive marbled and spiny-cheek crayfish populations. *Biological Invasions* **22**: 3339–3356.

85 Vogt, G. (2023). Evolution and functions of epigenetic mechanisms in animals. In: *Handbook of Epigenetics*, 3rd edition (ed. T. Tollefsbol), 521–549. New York: Academic Press.

86 Falckenhayn, C. (2016). The methylome of the marbled crayfish Procambarus virginalis. Doctoral thesis. University of Heidelberg.

87 Gatzmann, F. (2019). DNA methylation in the marbled crayfish Procambarus virginalis. Doctoral thesis. University of Heidelberg.

88 Vogt, G. (2018). Investigating the genetic and epigenetic basis of big biological questions with the parthenogenetic marbled crayfish: a review and perspectives. *Journal of Biosciences* **43** (1): 189–223.

89 Vogt, G. (2019). Functional cytology of the hepatopancreas of decapod crustaceans. *Journal of Morphology* **280** (9): 1405–1444.

90 Alwes, F. and Scholtz, G. (2006). Stages and other aspects of the embryology of the parthenogenetic Marmorkrebs (Decapoda, Reptantia, Astacida). *Development Genes and Evolution* **216**: 169–184.

91 Vogt, G. and Tolley, L. (2004). Brood care in freshwater crayfish and relationship with the offspring's sensory deficiencies. *Journal of Morphology* **262** (2): 566–582.

92 Tönges, S., Venkatesh, G., Andriantsoa, R. et al. (2021). Location-dependent DNA methylation signatures in a clonal invasive crayfish. *Frontiers in Cell and Development Biology* **9**: 794506.

93 Vogt, G., Tolley, L., and Scholtz, G. (2004). Life stages and reproductive components of the marmorkrebs (marbled crayfish), the first parthenogenetic decapod crustacean. *Journal of Morphology* **261** (3): 286–311.

94 Vogt, G. (2008). Investigation of hatching and early post-embryonic life of freshwater crayfish by in vitro culture, behavioral analysis, and light and electron microscopy. *Journal of Morphology* **269** (7): 790–811.

95 Seitz, R., Vilpoux, K., Hopp, U. et al. (2005). Ontogeny of the Marmorkrebs (marbled crayfish): a parthenogenetic crayfish with unknown origin and phylogenetic position. *Journal of Experimental Zoology A* **303** (5): 393–405.

96 Grimmer, A.U. (2015). Analysis of DNA methylation dynamics during marmorkrebs (*Procambarus fallax forma virginalis*) development. Master's thesis. University of Heidelberg.

97 Vogt, G. (2022). Studying phenotypic variation and DNA methylation across development, ecology and evolution in the clonal marbled crayfish: a paradigm for investigating epigenotype-phenotype relationships in macro-invertebrates. *The Science of Nature* **109**: 16.

98 Sellars, M.J., Degnan, B.M., and Preston, N.P. (2006). Production of triploid Kuruma shrimp, *Marsupenaeus* (*Penaeus*) *japonicus* (Bate) nauplii through inhibition of polar body I, or polar body I and II extrusion using 6-dimethylaminopurine. *Aquaculture* **256** (1–4): 337–345.

99 Piferrer, F., Beaumont, A., Falguière, J.-C. et al. (2009). Polyploid fish and shellfish: production, biology and applications to aquaculture for performance improvement and genetic containment. *Aquaculture* **293** (3–4): 125–156.

100 Marçal, R., Llorente, L., Herrero, O. et al. (2021). Intergenerational patterns of DNA methylation in *Procambarus clarkii* following exposure to genotoxicants: a conjugation in past simple or past continuous? *Toxics* **9** (11): 271.

101 Zhang, X., Yuan, J., Sun, Y. et al. (2019). Penaeid shrimp genome provides insights into benthic adaptation and frequent molting. *Nature Communications* **10**: 356.

102 Uengwetwanit, T., Pootakham, W., Nookaew, I. et al. (2021). A chromosome-level assembly of the black tiger shrimp (*Penaeus monodon*) genome facilitates the identification of growth-associated genes. *Molecular Ecology Resources* **21** (5): 1620–1640.

103 Tang, B., Zhang, D., Li, H. et al. (2020). Chromosome-level genome assembly reveals the unique genome evolution of the swimming crab (*Portunus trituberculatus*). *GigaScience* **9** (1): giz161.

104 Cui, Z., Liu, Y., Yuan, J. et al. (2021). The Chinese mitten crab genome provides insights into adaptive plasticity and developmental regulation. *Nature Communications* **12**: 2395.

105 Polinski, J., Zimin, A.V., Clark, K.F. et al. (2021). The American lobster genome reveals insights on longevity, neural, and immune adaptations. *Science Advances* **7**: eabe8290.

106 Wang, Q., Liu, J., Zhang, S. et al. (2016). Sustainable farming practices of the Chinese mitten crab (*Eriocheir sinensis)* around Hongze Lake, lower Yangtze river basin China. *Ambio* **45** (3): 361–373.

107 Jiang, Q., Lin, D., Huang, H. et al. (2020). DNA methylation inhibits the expression of CFSH in mud crab. *Frontiers in Endocrinology* **11**: 163.
108 Jones, C.M., Le Anh, T., and Priyambodo, B. (2019). Lobster aquaculture development in Vietnam and Indonesia. In: *Lobsters: Biology, Fisheries and Aquaculture* (ed. E. Radhakrishnan, B. Phillips and G. Achamveetil), 541–570. Singapore: Springer.
109 Wang, P., Wang, J., Su, Y. et al. (2020). Air exposure affects physiological responses, innate immunity, apoptosis and DNA methylation of Kuruma shrimp, *Marsupenaeus japonicus*. *Frontiers in Physiology* **11**: 223.
110 Mitsuda, S. and Shimizu, N. (2016). Epigenetic repeat-induced gene silencing in the chromosomal and extrachromosomal contexts in human cells. *PLoS One* **11** (8): e0161288.
111 Sittikankaew, K., Pootakham, W., Sonthirod, C. et al. (2020). Transcriptome analyses reveal the synergistic effects of feeding and eyestalk ablation on ovarian maturation in black tiger shrimp. *Scientific Reports* **10**: 3239.
112 Raddatz, G., Guzzardo, P.M., Olova, N. et al. (2013). Dnmt2-dependent methylomes lack defined DNA methylation patterns. *Proceedings of the National Academy of Sciences U.S.A.* **110** (21): 8627–8631.
113 Provataris, P., Meusemann, K., Niehuis, O. et al. (2018). Signatures of DNA methylation across insects suggest reduced DNA methylation levels in *Holometabola*. *Genome Biology and Evolution* **10** (4): 1185–1197.
114 Liu, S., Aagaard, A., Bechsgaard, J., and Bilde, T. (2019). DNA methylation patterns in the social spider, *Stegodyphus dumicola*. *Genes* **10** (2): 137.
115 O'Geen, H., Echipare, L., and Farnham, P.J. (2011). Using ChIP-seq technology to generate high-resolution profiles of histone modifications. In: *Epigenetics Protocols*, Methods in Molecular Biology, vol. **791** (ed. T. Tollefsbol), 265–286. Totowa, NJ: Humana Press.
116 Simpson, J.T., Workman, R.E., Zuzarte, P.C. et al. (2017). Detecting DNA cytosine methylation using nanopore sequencing. *Nature Methods* **14**: 407–410.
117 Van Dijk, E.L., Jaszczyszyn, Y., Naquin, D., and Thermes, C. (2018). The third revolution in sequencing technology. *Trends in Genetics* **34** (9): 666–681.
118 Angarica, V.E. and Del Sol, A. (2017). Bioinformatics tools for genome-wide epigenetic research, in *Neuroepigenomics*. In: *Aging and Disease*, Advances in Experimental Medicine and Biology, vol. **978** (ed. R. Delgado-Morales), 489–512. Cham: Springer.
119 Yin, Q., Wu, M., Liu, Q. et al. (2019). DeepHistone: a deep learning approach to predicting histone modifications. *BMC Genomics* **20** (Suppl. 2): 193.
120 Granada, L., Lemos, M.F.L., Cabral, H.N. et al. (2017). Epigenetics in aquaculture – the last frontier. *Reviews in Aquaculture* **10** (4): 994–1013.
121 Zhang, Y. and Kutateladze, T.G. (2018). Diet and the epigenome. *Nature Communications* **9**: 3375.
122 Cavalli, G. and Heard, E. (2019). Advances in epigenetics link genetics to the environment and disease. *Nature* **571**: 489–499.
123 Chung, F.F.-L. and Herceg, Z. (2020). The promises and challenges of toxico-epigenomics: environmental chemicals and their impacts on the epigenome. *Environmental Health Perspectives* **128** (1): 1–20.
124 Ideraabdullah, F.Y. and Zeisel, S.H. (2018). Dietary modulation of the epigenome. *Physiological Reviews* **98** (2): 667–695.
125 Pizzorusso, T. and Tognini, P. (2020). Interplay between metabolism, nutrition and epigenetics in shaping brain DNA methylation, neural function and behavior. *Genes* **11** (7): 742.
126 Foret, S., Kucharski, R., Pellegrini, M. et al. (2012). DNA methylation dynamics, metabolic fluxes, gene splicing, and alternative phenotypes in honey bees. *Proceedings of the National Academy of Sciences U.S.A.* **109** (13): 4968–4973.
127 Wojciechowski, M., Lowe, R., Maleszka, J. et al. (2018). Phenotypically distinct female castes in honey bees are defined by alternative chromatin states during larval development. *Genome Research* **28** (10): 1–11.
128 Harlıoğlu, M.M. and Farhadi, A. (2017). Feminization strategies in crustacean aquaculture. *Aquaculture International* **25**: 1453–1468.
129 Mohanakumaran Nair, C., Salin, K.R., Raju, M.S., and Sebastian, M. (2006). Economic analysis of monosex culture of giant freshwater prawn (*Macrobrachium rosenbergii* De Man): a case study. *Aquaculture Research* **37** (9): 949–954.
130 Gopal, C., Gopikrishna, G., Krishna, G. et al. (2010). Weight and time of onset of female-superior sexual dimorphism in pond reared *Penaeus monodon*. *Aquaculture* **300** (1–4): 237–239.
131 Chatterjee, N., Gim, J., and Choi, J. (2018). Epigenetic profiling to environmental stressors in model and non-model organisms: ecotoxicology perspective. *Environmental Health and Toxicology* **33** (3): e2018015.
132 Uren Webster, T.M., Rodriguez-Barreto, D., Martin, S.A.M. et al. (2018). Contrasting effects of acute and chronic stress on the transcriptome, epigenome, and immune response of Atlantic salmon. *Epigenetics* **13** (12): 1191–1207.

133 Stentiford, G.D., Neil, D.M., Peeler, E.J. et al. (2012). Disease will limit future food supply from the global crustacean fishery and aquaculture sectors. *Journal of Invertebrate Pathology* **110** (2): 141–157.

134 Liu, X.S., Wu, H., Ji, X. et al. (2016). Editing DNA methylation in the mammalian genome. *Cell* **167** (1): 233–247.

135 Xu, X., Tao, Y., Gao, X. et al. (2016). CRISPR-based approach for targeted DNA demethylation. *Cell Discovery* **2**: 16009.

136 Nakamura, M., Gao, Y., Dominguez, A.A., and Lei, S. (2021). CRISPR technologies for precise epigenome editing. *Nature Cell Biology* **23**: 11–22.

137 Nakanishi, T., Kato, Y., Matsuura, T., and Watanabe, H. (2014). CRISPR/Cas-mediated targeted mutagenesis in *Daphnia magna*. *PLoS One* **9** (5): e98363.

138 Xu, S., Pham, T., and Neupane, S. (2020). Delivery methods for CRISPR/Cas9 gene editing in crustaceans. *Marine Life Science & Technology* **2**: 1–5.

139 Whelan, R., Tönges, S., Böhl, F. et al. (2023). Epigenetic biomarkers for animal welfare monitoring. *Frontiers in Veterinary Science* **9**, 1107843.

140 Venkatesh, G., Tönges, S., Hanna, K. et al. (2023). Context-dependent DNA methylation signatures in animal livestock. *Environmental Epigenetics* **9**, dvad001.

17

Epigenetics in Algae

Christina R. Steadman

Earth & Environmental Sciences Division, Climate, Ecology & Environment Group, Los Alamos National Laboratory, Los Alamos, NM, USA

17.1 Introduction: What Are Algae

The term “algae” encompasses a wide range and number of organisms from microscopic algae, such as pond scum found in small bodies of water, to large seaweeds, such as kelp that span hundreds of feet in the ocean. Specifically, algae are nonflowering organisms that are morphologically simple: they can perform photosynthesis like plants but lack traditional plant traits including roots, stems, and leaves. Further, while algae contain copious amounts of chlorophyll *a* (responsible for the green color of plants, leaves, and some algae), many species express other pigments, resulting in a rainbow of colors from golden browns to deep reds. Algae can be classified based on size, typically micro versus macro. In this chapter, we will focus on microalgae, the type of algae that are unicellular and thrive in a variety of aquatic systems, including freshwater and seawater (Figure 17.1).

Most people are familiar with certain forms of microalgae: green algae in lakes, phytoplankton floating in the oceans, and vast algal blooms that spread across miles of open water. Microalgae are classified into two major categories: prokaryotic cyanobacteria (such as blue-green algae), which are actually bacteria that are able to perform photosynthesis, and eukaryotic protists. Both classifications include unicellular organisms with the capacity to fix carbon dioxide, perform photosynthesis, and release oxygen. The algae community estimates there are anywhere from 40,000–200,000 species of microalgae [1], suggesting extraordinary genetic diversity, particularly given their diverse phenotypes. Microalgae are found in almost every ecological niche around the world, reflecting a wide range of environmental adaptations. Further, microalgae naturally thrive in environments with scarce resources, and, as such, are able to colonize extreme environments.

Taxonomic classification of algae is an ever-evolving field and quite contentious. Algae are polyphyletic, that is, phylogenetically, one common ancestor has not been identified for this group of organisms. It is therefore not appropriate to place algae within the same taxon given this lack of commonality. Genomics analyses suggest that algae evolved through endosymbiosis of other organisms (and organelles); thus, they may be categorized into eight to nine phyla over the course of evolution. Other reports suggest that algae can be distributed into at least a dozen major groups primarily based on pigment and aesthetic (or morphological) composition. While some algae may be aesthetically similar, their bioproduct storage profiles and ultrastructural features can be dramatically different. Thus, algal phyla are continually changing as taxonomic profiles are updated and new genomic information is released. Algae are spread across the tree of life and are phylogenetically linked to bacteria, archaea, and plants (Figure 17.2).

The phylogenetic and genomic diversity of algae is reflected in the various types, locations, and functions of epigenetic modifications in various species (Box 17.1). This distribution of algae is beneficial for understanding mechanisms of epigenetic evolution. Several thousands of species coevolved in diverse environments simultaneously during various epochs. This resulted in highly divergent genomes, and potential variability in regulatory mechanisms governing the architecture and expression of these genomes. Indeed, initial efforts into characterizing algal epigenetics have found different epigenetic strategies are employed by different species of algae, likely based on genetic and environmental influence.

Epigenetics in Aquaculture, First Edition. Edited by Francesc Piferrer and Han-Ping Wang.

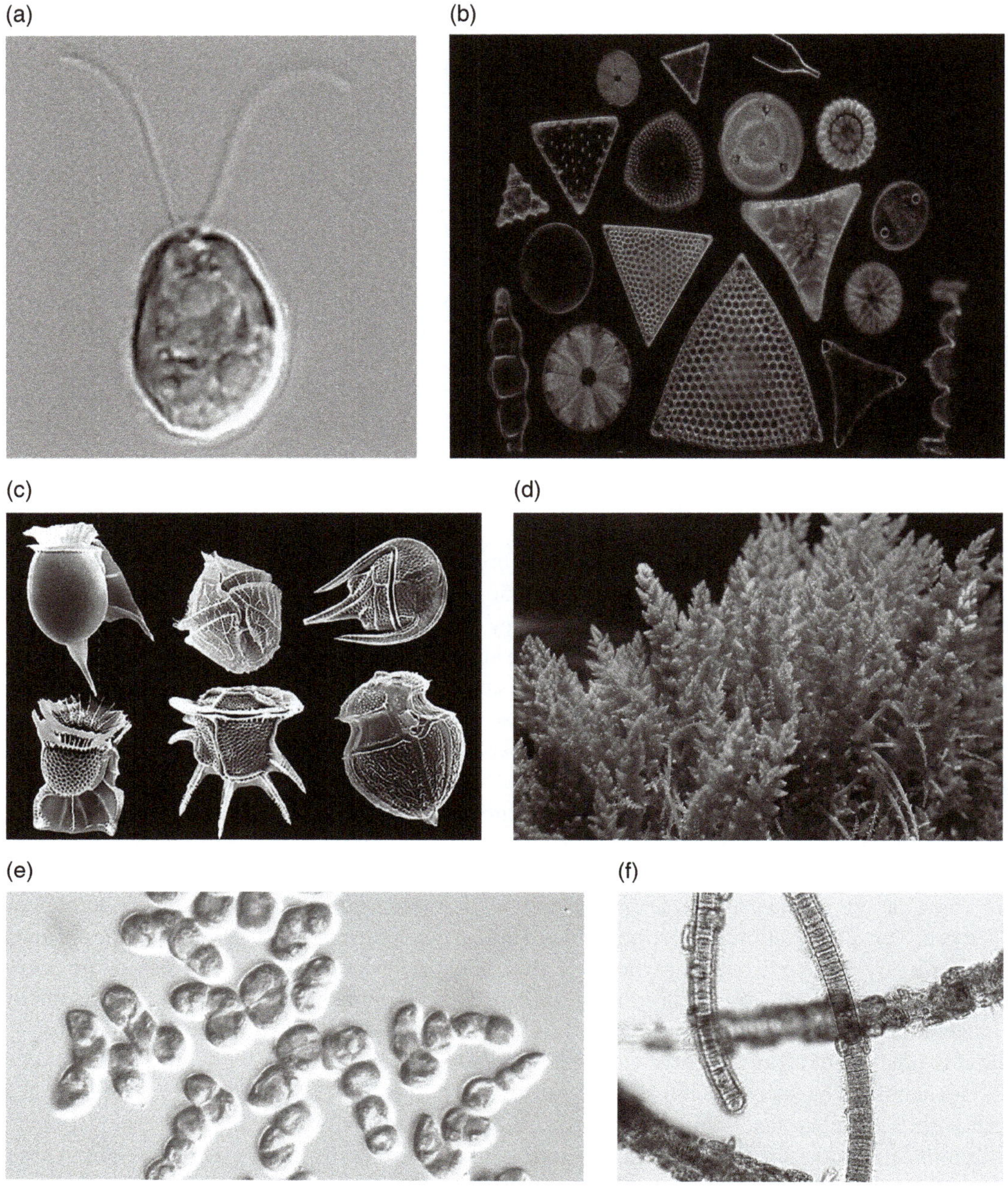

Figure 17.1 Images of model algae species. (a) *Chlamydomonas reinhardtii*, *Source:* J. L. Carson, Ph.D/Science Source, (b) a collection of diatom algae, *Source:* Dr. N. Lange/Adobe Stock, (c) a collection of dinoflagellate algae, *Source:* Fickleandfreckled/Flickr/CC BY 2.0, (d) representative of macro red algae, *Source:* Damedias/Adobe Stock, (e) representative of chlorophytes (green algae), *Source:* Lebendkulturen.de/Shutterstock, and (f) blue-green algae or cyanobacteria, *Source:* Elif/Adobe Stock. (*See colour plate section for colour representation of this figure.*)

17.1.1 Algae's Prominent Role in the Environment

Microalgae are more efficient at fixing carbon than plants, resulting in greater biomass generation (based on size). They play a paramount role in the global carbon cycle and thereby impact climate change. Microalgae that inhabit the oceans are referred to as phytoplankton; these include prokaryotic cyanobacteria (blue-green algae) and eukaryotic green algae, silica-encased diatoms, dinoflagellates, and coccolithophores. All these taxa of microalgae consume and store carbon similar to plants. Algae are the base of the oceanic food web, feeding krill, macro-zooplankton, shrimp, and small fish and squid, all of which feed animals ranging from penguins and seals to larger fish and whales. Upon consumption, fixed carbon is transferred from algae to the oceanic ecosystem and even into the terrestrial food web. When algae die, they carry fixed carbon to the ocean floor. After a millennia, this carbon can be harvested as fossil fuel for energy. Essentially, the gasoline you put into your vehicle comes from ancient dead algae. Approximately 10 gigatons of carbon are transferred from the atmosphere to the deep ocean every year; thus, even small changes in algae growth in the open ocean can impact carbon dioxide concentrations. Anthropogenic and natural variations in carbon dioxide concentrations alter ocean surface temperatures, resulting in a circular effect on microalgae growth. The role of epigenetic regulation in carbon fixation and the resulting changes in algal physiology are relatively unexplored.

Given microalgae's prominent role in ocean biology and atmospheric composition, altered algal productivity can have significant influence on climate, carbon dioxide concentrations, biodiversity, fisheries, and even the human food supply. Productivity (growth) is dependent upon several factors including sunlight, nutrient availability, water temperature, salinity, depth, wind disturbances, and predators that graze on algae. Phytoplankton predominately live in the upper layers of

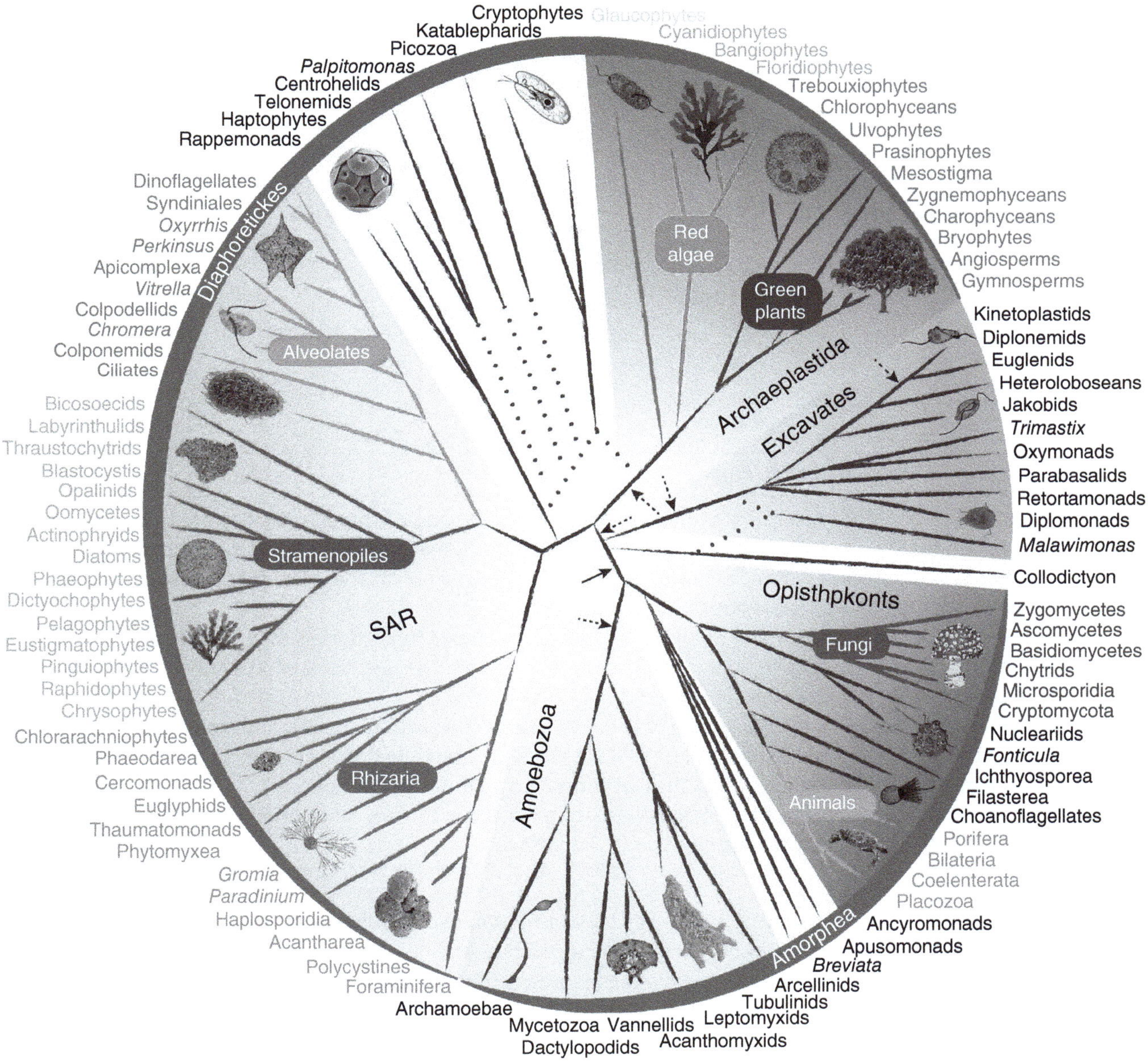

Figure 17.2 The tree of life can be modeled in several different ways, but in all models, algae species are spread across the tree. The red stars indicate the different families and phyla to which various algae belong across the tree of life. Epigenetics analysis has been performed in less than 25 total species, but these include algae from several different families and groups including stramenopiles (which include diatoms and brown algae), alveolates (which include dinoflagellates), red algae, and green algae (chlorophytes, prasinophytes, etc.). *Source:* Burki (2014)/Cold Spring Harbor Laboratory Press, and Delsuc et al. (2005)/Springer Nature. (*See colour plate section for colour representation of this figure.*)

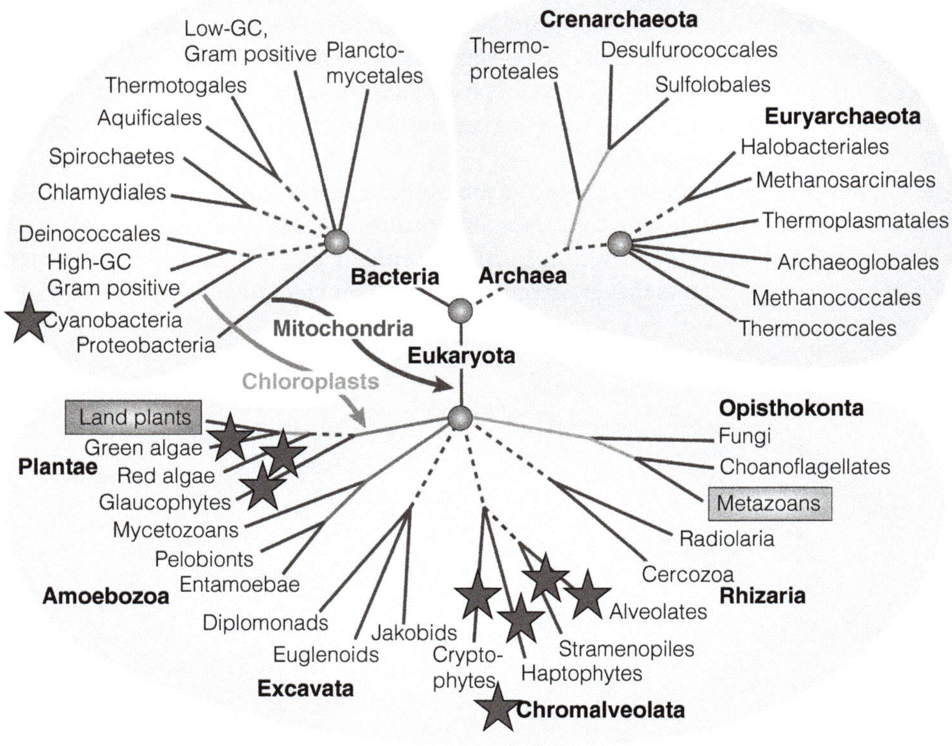

Figure 17.2 (Continued)

Box 17.1 Glossary of Terms Related to Algae

Algae: Nonflowering photosynthetic uni- or multicellular eukaryotic aquatic organisms that are morphologically simple, and lack traditional plant features, but contain chlorophyll. (Some "algae" like blue-green algae are actually bacteria that can perform photosynthesis; these are instead called cyanobacteria. Microalgae are invisible to the naked eye, while macroalgae are visible and large (like sea kelp).

Carbon fixation: The act of taking up and sequestering carbon in various forms; for example, algae (and plants) use carbon dioxide from the atmosphere to produce glucose for themselves. The process of photosynthesis also produces water and releases oxygen into the atmosphere.

Phylogenetics: The study of relationships among or between groups of algae. Morphology and molecular assessments (like genomic information) are used to group algae that are closely related into different categories (phyla). The closer the algae are genetically and morphologically, the closer they are on the phylogenetic tree.

Productivity: The act of algae proliferating in aquatic ecosystems and producing biomass and releasing oxygen from carbon fixation processes like photosynthesis.

the water column toward the ocean surface where they have maximum photosynthetic efficiency. Given that microalgae require specific amounts of nitrogen, phosphorus, silica, and calcium, the flux of these nutrients in aquatic ecosystems directly influences the balance of algae species, promoting the growth of one species while hindering the growth of another. Vertical mixing of the water column induces nutrient flux and feedback, and is thus contingent upon water conditions like temperature and salinity. As surface waters warm due to increased greenhouse gases, the water column becomes more stratified with less nutrient flux from the deep ocean to surface water, resulting in decreased algal productivity. Certain species, particularly larger microalgae such as diatoms, will likely not survive this warming, in contrast to smaller algae like cyanobacteria. This is likely to induce a shift in oceanic algae species composition, which may result in a cascade of consequences throughout the marine food web that will impact carbon dioxide concentrations and climate. We, and others, have demonstrated that perturbations in these factors alter the epigenomes of microalgae, resulting in changes in phenotype and productivity. Therefore, it is paramount that we develop a deeper understanding of gene-by-environment interactions, specifically epigenetic regulation of algal productivity, to determine the effects of climate change on algal physiology and the resulting reciprocal impact on climate.

Importantly, high concentrations of nitrogen and phosphorus induce rapid proliferation of algae species over short periods of time, resulting in the formation of harmful algal blooms. These blooms contaminate and destroy the vitality of ecosystems by depleting resources and, in some cases, producing toxins like microcystin. Blooms can occur over hundreds of miles in multiple aquatic ecosystems (visible from space) and can last for several days to weeks, dependent upon the species (Figure 17.3). Climate change and anthropogenic activities have contributed to increased bloom frequency and expanded blooms to areas where they were previously uncommon, such as Arctic ecosystems [2, 3]. The mechanisms that induce bloom formation and the underlying regulation of bloom physiology are relatively unexplored. We, and others, have demonstrated that bloom species have epigenetic modifications, but a deeper understanding of the function of these modifications and their impact on bloom dynamics is still needed.

The diversity of microalgae and their high nutrient value provide the potential for the production of various bioproducts ranging from monomers for bioplastics to lipids for biofuels. Microalgae can thrive in brackish water, seawater, and wastewater, all of which are unsuitable for agriculture. Algae species can therefore be used for purification and treatment of water. They consume excess carbon dioxide for autotrophic growth, though some species can consume sugar and starch for

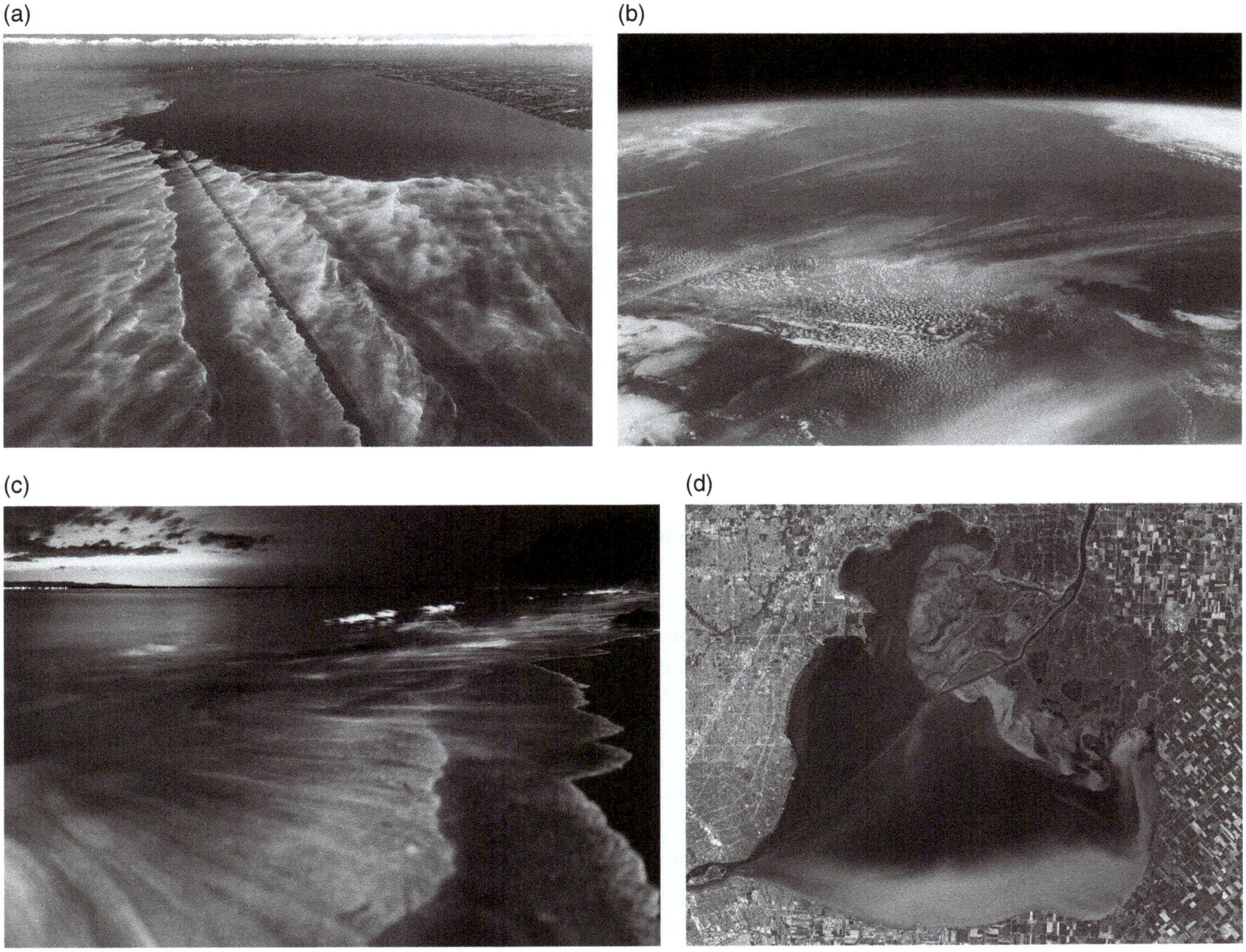

Figure 17.3 Harmful algal blooms occur when algae rapidly proliferate, consume the resources of an aquatic ecosystem, and release harmful substances. Numerous species can form blooms ranging from blue-green cyanobacteria ((a): Lake Erie, United States), *Source:* NOAA Great Lakes Environmental Research Laboratory/Flickr. Algal blooms can be observed from space ((b): Bering Sea, NASA/Public Domain.) (c) Dinoflagellates, *Source:* Gary/Adobe Stock and via satellite imagery ((d): Lake St. Clair, United States, *Source:* Joshua Stevens/NASA/Public Domain). (*See colour plate section for colour representation of this figure.*)

heterotrophic growth (or mixotrophic growth if they consume carbon in multiple forms). It is unclear how growth in these types of environments impacts algal physiology or genetic programs; potentially, we could utilize algae to clean up our water and simultaneously produce biomolecules for biofuel. Research has focused on synthesis and formation of bioproducts, with particular effort on bioengineering genetic programs to increase production of specific biomolecules. While the field has discovered that certain environmental conditions promote the formation of these products, significant advances, particularly in bioengineering, have been limited. We hypothesize that investigation of epigenetic modifications and chromatin architecture will provide great value to the field for genetic engineering efforts. Further, characterization of differential epigenetic responses, based on nutrient availability and water composition, which directly impact metabolism, growth, and bioproduct formation, would be particularly helpful to advance the state of knowledge for the algae bioproduct field.

17.1.2 Algae in Aquaculture

In addition to their prominent role in the environment, microalgae play a critical role in the aquaculture industry. Taking a cue from nature, the aquaculture industry utilizes microalgae as fish feed and as a source of biologically active compounds. Various algae species have replaced traditional fish meal and are utilized as live feeds for the cultivation of fish from various genera and for bivalve mollusks, including oysters, scallops, and mussels [4]. Microalgae species produce natural pigments that contribute to the color of organisms that feed upon them. Thus, the culturing of prawns, shrimp, and salmon includes the use of microalgae that contain specific pigments and high quantities of lipids and proteins [5]. Algae are also used to treat wastewater derived from aquaculture; a beneficial side effect of this process includes the uptake of high quantities of nitrogen and phosphorus, which induces the synthesis of molecules of interest to the biofuel industry. However, some species of algae are invasive to open aquaculture settings in lake and seawater environments. These species can cause serious damage, particularly under high temperatures or nutrient conditions, and some induce harmful algae blooms. Protection against these species is a major concern in the industry. Microalgae's nutrient, biomolecular, and pigment compositions are inherently predicated by the species' characteristics, as some microalgae are more nutritious or colorful than others. However, nutritive value and biomolecular composition are dynamic in response to environmental conditions, and thus could be sensitive to aquaculture conditions. Thus, a deep understanding of the phylogenetic and epigenetic characteristics of microalgae species and how these impact phenotypic traits are important when considering the use of algae in aquaculture and fisheries.

17.2 Algae Epigenetics

Microalgae employ a wide range of evolutionary strategies for survival and adaptation. Many groups have focused on characterizing both phenotypic and genomic algal profiles for such adaptation to a myriad of stresses. Dynamic, efficient adaptation to turbulent environmental conditions likely involves the modification and regulation of existing genes rather than mutation in underlying DNA sequences. Epigenetic processes are ideal candidate mechanisms for this regulation; further, they may have directly contributed to algae genomic evolution, ultimately influencing phenotype, including growth and biomass yield. While epigenetic mechanisms were discovered in plant species, the occurrence of DNA methylation, histone modifications, and small noncoding RNAs suggests that epigenetic mechanisms themselves are ancient, highly conserved, and likely contributed to the evolution of higher eukaryotes. Thus, deciphering the epigenetic code in algae species will further our understanding of the age of the epigenetic code and how it evolved [6].

Several techniques have been employed to study algae epigenetics. Techniques for characterizing the quantity and presence of epigenetic modifications include mass spectrometry, antibody-based immunohistochemistry, restriction enzyme digestion, and whole genome bisulfite sequencing (WGBS). However, epigenetic sequencing technologies to obtain site-specific information require well-annotated genomes for alignment and base calling. While there are potentially hundreds of thousands of species of microalgae, we have little genomic (and even less epigenomic) information. A conservative estimate in 2012 suggested there were approximately 40,000 algae in AlgaeBase; as of 2022, there are 166,000 species and infraspecific names in AlgaeBase [7]. A recent survey of algae genomes that primarily focused on eukaryotic algae reported 222 de novo assemblies for 149 species from 97 genera, which only represent 32 of the 50 algae classes [8]. Many of these assemblies are for different strains of the same species, with each strain having appreciable phenotypic differences due to environmental variation (though very little epigenetic data is available). The quality of these genomes is highly variable given the lack of a comprehensive standard of practice for algae genome assembly, though significant efforts at creating such standards are underway. While tremendous effort has been made to sequence algae genomes, particularly for those species that have potential importance for the bio-based economy, relatively little effort has been focused on sequencing epigenomes. To date, we have information on the epigenomes, either DNA methylation or histone modifications, for less than 25 algae species, including our unpublished data on two species. Potentially, the dearth of epigenomic information is simply due to the small field of algae biology itself and the lack of significant epigenetics expertise.

17.2.1 DNA Methylation and Associated Chromatin Modifying Enzymes in Other Organisms

Algae are ancient organisms that are closely associated with bacteria, archaea, and plants; thus, we anticipate algae utilize epigenetic processes similar to these organisms, including DNA methylation. This particular epigenetic modification is ubiquitous and found in almost every organism except for a select number of species, including *Drosophila* spp. and *Caenorhabditis elegans*, though recent reports have contested lack of DNA methylation in these two species [9, 10]. In the broadest sense, DNA methylation signifies covalent addition of a methyl group on adenosine (N^6-methyladenosine; 6 mA) or the fourth or fifth carbon of cytosine (N^4-cytosine, 4mC and C^5-cytosine, 5mC, respectively) [11]. Typically, the phrase "DNA methylation" conveys 5mC methylation in the CpG dinucleotide context unless otherwise specified. The presence of a particular type of DNA (or histone) modification is contingent upon the expression of chromatin-modifying enzymes (CMEs). These enzymes are responsible for writing, erasing, and reading covalent additions to amino acids in histones and nucleic acids in DNA [12]. Investigation of the presence and expression of these enzymes in algae genomes can provide clues as to the types of potential DNA and histone modifications and their genomic contexts in algae.

Mammals have three conserved DNA methyltransferases (DNMT1, DNMT3a, and DNMT3b) responsible for the maintenance and de novo methylation of the 5′ carbon of cytosine in CpG dinucleotides. The presence of the 5mC modification in promoter regions in CpG islands recruits CMEs for specific histone modifications resulting in heterochromatin formation and subsequent repression of gene expression [13]. Ours and others' work suggests that algae do not have direct sequence homology to mammalian CMEs for DNA methylation [14, 15].

In plants, de novo 5mC methylation occurs via RNA-directed DNA methylation pathway (RdDM) and recruits the methyltransferase DRM2 (domains rearranged methyltransferase 2). The 5mC modification is not limited to CpG islands and occurs in multiple dinucleotide contexts, including CpG, CHH, and CHG (where H can be A, C, or T). The CMEs responsible for these modifications include DNA methyltransferase enzyme MET1 (methyltransferase 1), CMT2 (chromomethylase 2), and CMT3 (chromomethylase 3). However, a recent report suggests de novo 5mC methylation potential for CMT3 [16]. See the study by Niederhuth and Schmitz for a detailed review of plant DNA methylation [17]. Localization and enrichment of DNA methylation in transposable elements (TEs), genomic repeats, and regulatory regions of plants induces silencing of transcriptional expression and influences mRNA-processing events. However, some DNA methylation promotes gene expression mediated by CME DNA methylation "readers," including the SU(VAR)3–9 protein homologs SUVH1 and SUVH3, but this is context-dependent [18]. Copious 5mC CpG methylation is localized to gene bodies in plants and is thought to be permissive.

Bacteria and archaea also contain low levels of 5mC methylation and associated methyltransferases, including DCM, which recognizes CCAGG and CCTGG sequences, and JHP1050, which recognizes GCGC [19]. However, most DNA methylation occurs as 4mC and 6mA in bacteria. All three DNA modifications play a role in the restriction-modification (R-M) system, which is important for defense against foreign DNA; as such, much of this methylation represses gene expression. Advances in single molecule real-time (SMRT) sequencing have allowed for extensive characterization of 4mC and 6mA in many prokaryotic species. However, the occurrence of 4mC in eukaryotes is widely debated given that detection of this modification is below the false positive rate for SMRT sequencing [20]. Over the past 10 years, extensive research has characterized the presence and localization of 6mA in multiple species, including algae [21]. Understanding the types of DNA modifications, their localization and function, and the responsible CMEs provides a roadmap for what epigenetic processes to investigate in algae. Given the phylogenic relationships of algae with bacteria and plants, we anticipate that microalgae likely have the same CMEs (or enzymes with similar sequences) and employ the same types of DNA modifications.

17.2.2 DNA Methylation in the Model Algae Species, *Chlamydomonas reinhardtii*

A handful of research groups have done extensive work on epigenetic regulation in *C. reinhardtii*. This unicellular green alga is a reference organism for studying eukaryotic processes including cellular function and metabolism, photosynthetic biology, nutrient use and homoeostasis, and both sex and cell cycles [22]. As sequencing technology and epigenetic assessments rapidly evolve, our knowledge of the amount and localization of DNA methylation in *C. reinhardtii* changes. In 2010, it was thought that this species contained ~5% CpG and ~2.5% CHH and CHG 5mC DNA methylation [13]. In 2015, the DNA methylome across the *C. reinhardtii* lifecycle was characterized, and this study found less than 0.75% CpG DNA methylation in the nuclear genome, corroborating earlier studies [23–25]. Hypermethylation (greater than 80% methylation) was found in only 23 genomic loci that contained low protein-coding density [22]. Interestingly, *C. reinhardtii* chloroplast DNA methylation undergoes dynamic changes throughout the lifecycle, though this is still less than 2% methylation globally. Chloroplast CpG methylation increases after gametes mate and during zygote development – this is unique not only for this alga, but in comparison to gamete fusion and zygote formation in mammalian species, where extensive DNA demethylation (except on imprinted genes) occurs. Thus, for the 120 MB haploid genome of *C. reinhardtii*, there is less than 1% CpG methylation [26]. The functions of *C. reinhardtii* methylation likely include gametogenesis, sexual reproduction, and transgene repression, while heterochromatin formation may be responsible for gene expression silencing [23, 25, 27, 28]. While extensive work has focused on *C. reinhardtii*, understanding epigenetic processes in this particular species has not provided a roadmap for epigenetics in other species. Indeed, what we can glean from the majority of algae DNA methylation assessments is that the quantity, distribution, localization, and perhaps function of this modification are highly dependent upon the species.

DNA methylation has been assessed in 17 species of algae (including micro and macro). In general, a common DNA methylation pattern is lacking, though most algae have DNA methylation in all three (or at least two) contexts (CpG, CHH, and/or CHG), similar to plants. Table 17.1 provides a summary of the amount of global 5mC methylation on all three contexts for all microalgae that have been assessed to date. DNA methylation content, specifically in the CpG context, is not predicated on the size of the genome as depicted in Figure 17.4. The amount of 5mC methylation and its role is described later.

Table 17.1 DNA methylation in microalgae species.

		Percent methylation in genomic context		
Microalgae species	**Nuclear genome size (MB)**	**CpG**	**CHG**	**CHH**
Ostreococcus lucimarinus	13.2	22	0.13	0.13
Picochlorum soloecismus	15	12.1	0.8	0.9
Micromonas pusilla	15	17.7	0.14	0.13
Bathycoccus prasinos	15	20.9	0.21	0.19
Cyanidioschyzon merolae	16	0.2	0.18	0.19
Phaeodactylum tricornutum	27.4	1	0.19	0.2
Thalassiosira pseudonana	34.5	2.57	0.13	0.12
Chlorella variabilis NC64A	46	80	0.08[a]	N/A
Aureococcus anophagefferens	57	24	0.19	0.18
Fragilariopsis cylindrus	61	8.63	0.13	0.13
Chlamydomonas reinhardtii	120	0.75	0	0
Volvox carteri	138	1.1	0	0
Emiliania huxleyi	141	16.36	5.63	0.38
Ectocarpus siliculosus	214	0	0	0
Saccharina japonica	537	25	19	57
Fugacium (Symbiodinium) kawagutii	1180	78	1.19	0.99
Breviolum (Symbiodinium) minutum	1500	68	1.55	1.15

N/A denotes data not available.
[a]within false discovery rate.

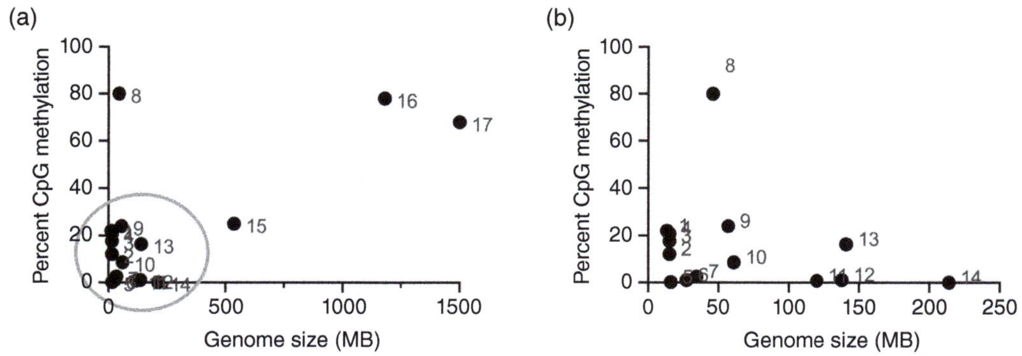

Figure 17.4 DNA methylation in the CpG context for all algae species measured to date. (a) No particular relationship between 5mC methylation in the CpG context and genome size was found in algae. (b) Most algae species that have genomes smaller than 250 MB have less than 25% CpG methylation, though there are some outliers. However, picoeukaryotic algae species that have very small genomes still have significant quantities of 5mC CpG methylation. The orange circle in (a) indicates the species that are graphed in (b).

17.2.3 Chlorophyte DNA Methylation

As the number of genomic sequences for algae increases, so does the evolution of their classification. Broadly, microalgae can be distributed into the "red" lineage (primarily Stramenopiles, discussed later) that dominate marine waters, and the "green" lineage, which are closely related to plants and primarily occupy terrestrial ecosystems, though they are also found in some oceanic ecosystems [29]. The chloroplast origins of these algae are vastly different: green algae chloroplasts have two membranes derived from a single endosymbiotic event with chlorophyll *a* and *b* pigments, whereas red algae contain chloroplasts without membranes, unstacked thylakoids, and phycobiliprotein pigments [30]. This is obviously a gross oversimplification of algae classification, but is useful for discussing DNA methylation. Green algae belonging to the Chlorophyta phylum will be discussed here, and red algae belonging to Stramenopiles (Heterokonts) will be discussed later.

Volvox carteri is a multicellular green alga in the Chlorophyta phylum and is closely related to *C. reinhardtii*. It contains approximately 2000 somatic cells and 16 germ cells within a spheroid. This alga typically uses asexual haploid reproduction, but can undergo sexual differentiation. *V. carteri* is a useful model organism for the study of the evolution of multicellularity. The entire *V. carteri* genome, comprised of nuclear and organelle (chloroplast and mitochondria) nucleic acid content, is 138 MB; however, the nuclear genome is 19.6 MB [31]. This species contains 1.1% 5mC methylation only in CpG contexts primarily within TEs in the nuclear genome; methylation is markedly absent in promoter regions. CpG methylation is responsible for the silencing of TEs and is inversely correlated with transcriptional activation [32–34]. Compared to *C. reinhardtii*, most DNA methylation studies in microalgae focus on nuclear (not organelle) DNA methylation.

Chlorella variabilis NC64A is a freshwater, unicellular green alga in the Chlorophyta phylum. Chlorophytes are distributed into two classes: Chlorophyceae and Trebouxiophyceae, the latter of which includes *C. variabilis* NC64A. While most Chlorophytes exist as standalone species, NC64A is an endosymbiont with photosynthetic capacity (and thus a photobiont) of the protozoan *Paramecium bursaria* [35]. This relationship is circumstantial and not essential, as *in vitro* laboratory experiments are able to separately cultivate these two species. Interestingly, *C. variabilis* NC64A hosts several large dsDNA viruses (named *Chlorella* viruses); thus, this species is used to study the relationship between viruses and algae [36, 37]. The *C. variabilis* NC64A 46 MB nuclear genome contains methylation in all three cytosine contexts with ~80% CpG methylation in almost every gene; promoter methylation negatively correlates with gene expression [34]. It is likely that this high CpG methylation content is due to the presence of viral DNA [35, 38]. *C. variabilis* NC64A encodes a homolog for a chromomethyltransferase (CMT) enzyme and has substantial CHG methylation found within repetitive elements, but not genes. CHH methylation was reported at 0.08%, but this falls within the false discovery rate of the reported sequencing metrics [34]. Of the 100 *Chlorella* species to date, only NC64A has DNA methylation sequencing analysis completed. We have determined that three strains of *Chlorella sorokiniana* (1412, 1228, and 1230), a Chlorophyte related to NC64A, have homologous sequences for DNA methyltransferase enzymes, particularly the MET1 sequence from *A. thaliana*, though no similarity to CMTs [14]. This *in silico* data suggests that this species likely uses 5mC DNA methylation, though this remains to be determined.

Picochlorum soloecismus is a unicellular green alga belonging to the Trebouxiophyceae Chlorophytes. This species was isolated from a culture of *Nannochloropsis salina* CCMP1776 after it outcompeted the native *Nannochloropsis* under heat stress [39]. *P. soloecismus* is relatively small, approximately 2.2 μm, though its size almost doubles under nitrogen stress [40]. Its 15 MB nuclear genome contains 12.1% 5mC CpG methylation with 48,667 CpG sites containing greater than 80% methylation [41]. *P. soloecismus* has low, but detectable, 5mC methylation in CHG (0.8%) and CHH (0.9%) contexts, and a bimodal distribution of 5mC methylation in all contexts, as observed in other eukaryotic species. Interestingly, DNA methylation is evenly distributed in the genome among promoters, intergenic regions, coding regions, and UTRs. The function of DNA methylation in *P. soloecismus* is not currently known, as transcriptomic evaluation has yet to be performed. During fatty acid accumulation, however, there is markedly decreased DNA methylation on genes responsible for fatty acid synthesis, suggesting a repressive function for DNA methylation in this species [41].

The next three algae species are treated collectively as picoeukaryotes, as they have similar size and genomic composition with relatively simple cellular organization (one chloroplast and one mitochondrion) [42]. Further, these three green algae are ubiquitously distributed worldwide and play a paramount role in the cycling of carbon and oxygen in coastal regions. They also belong to the Mamiellales order (previously Prasinophyte), which hosts common Chlorophyte algae. *Micromonas pusilla* of *Micromonas* is less than 2 μm in diameter. Its 15 MB genome contains 5mC methylation in all three contexts with 17.7% CpG, 0.14% CHG, and 0.13% CHH methylation [15]. *Ostreococcus lucimarinus* of *Ostreococcus* is the smallest known eukaryote, approximately 1 μm in diameter [43]. *O. lucimarinus*' 13.2 MB genome has 5mC DNA methylation in 22% CpG, 0.13% CHG, and 0.13% CHH contexts [15, 43]. *Bathycoccus prasinos* categorized within Bathycoccaceae is a coccoid covered in patterned scales and is nonmotile. *O. lucimarinus* is relatively naked (no covering), but also nonmotile, while *M. pusilla* is naked, but has one long flagellum [42]. The 15 MB *B. prasinos* genome contains 5mC methylation in all three contexts: 20.9% CpG, 0.21% CHG, and 0.19% CHH methylation [15]. Thus, all three algae have similar levels of DNA methylation in all three contexts: ~17–22% CpG methylation, ~0.13–0.21% CHG, and ~0.13–0.19% CHH methylation within genomes approximately 15 MB in size. These species also lack the *Dnmt1* gene, but have similar methyltransferase homologs, namely *Dnmt5* that could be responsible for methylation. *Ostreococcus* encodes two bacterial methyltransferases fused to a chromatin domain and a plant DNA glycosylase sequence [43]. Huff and Zilberman report that methylation in these species occurs in regularly spaced intervals with 160 and 206 bp periodicity, which is highly unusual [15]. Elevated 5mC methylation occurs, particularly in gene bodies, regardless of transcription levels; this pattern is also observed in *Emiliania huxleyi* and *A. anophagefferens* discussed in Section 17.2.4. All these genomes are relatively small with high quantities of DNA methylation considering their size. These methylation concentrations are equivalent to those observed in mammalian cells. However, mammals have significant quantities of heterochromatin that are not transcriptionally active due to highly differentiated tissues. Algae genomes are very transcriptionally active, particularly given their high frequency of mitosis, and typically lack collections of differentiated cells. Thus, these CG-rich genomes with dense gene body methylation may utilize this methylation to stabilize their chromatin structures [15].

17.2.4 Stramenopile (Heterokont) DNA Methylation

As the tree of life evolves as more genomes are sequenced, categorizing algae will continually change. However, it is within reason to state that Stramenopiles (also known as Heterokonts) are not closely associated with terrestrial plants nor green algae (Chlorophytes) and are phylogenetically distinct. Stramenopiles have a characteristic flagellate stage at some point during their life cycle Members are unicellular while others are multicellular and some have photosynthetic capacity, while others do not. Three main categories of Stramenopiles will be discussed; these include diatoms, dinoflagellates, and both micro and macro brown algae, of which there are harmful algal bloom species.

Diatoms are unicellular phytoplankton that inhabit both fresh and marine waters. They contribute approximately 40% of the oceanic primary production and 20% of the carbon fixation worldwide. Their defining feature is their elaborate and intricate cell walls, also called frustules, which are composed of amorphous silica. Diatoms are thought to have acquired a chloroplast from a secondary endosymbiotic event of a photosynthetic eukaryotic species (likely a red algae endosymbiont), contrasting with the primary endosymbiotic event of green algae engulfing a prokaryotic cyanobacteria containing chlorophyll [44]. Diatom genomes are relatively compact, dominated by TEs, particularly from the LTR retrotransposon family, with short genic and intergenic regions. There are seven groups of diatom-specific TEs called CoDi (Copia-like elements from diatoms) [45], and these are believed to have significantly contributed to adaptation and diversification of diatoms over the course of evolution. Given methylation-induced repression of TEs in plants, we would anticipate diatoms containing at least some form of DNA methylation. Diatom 5mC methylation is predominantly localized to CpGs with negligible methylation in CHG and CHH contexts; methylation percentages linearly correlate with genome size and TE genomic quantity.

Phaeodactylum tricornutum is a marine pennate diatom less than 10 μm in diameter. It is highly responsive to environmental conditions, resulting in various phenotypes. The most comprehensive profiling of the *P. tricornutum* 27.4 MB genome suggests that it contains 5% mosaic genomic DNA methylation in all three cytosine contexts with approximately 3.3% of its genes methylated [44, 46]. Mosaic methylation refers to small islands of highly methylated regions surrounded by regions with no methylation. This finding contrasts with earlier reports of 2.5% genomic DNA methylation for diatoms including *P. tricornutum*; however, previous techniques for DNA methylation assessment were of lower resolution [47]. *P. tricornutum* housekeeping genes have mid-to-low methylation with constitutive expression. Genes with high methylation are typically those with signaling and metabolic function. These genes are under tight regulatory control and have differential methylation status and gene expression under specific conditions. 5mC methylation likely has a suppressive role in gene regulation for this species, though methylation in promoter regions does not impact gene expression. Larger TEs have markedly high methylation with surrounding regions containing spreading methylation, suggesting that insertion of TEs promotes heterochromatin formation as observed in terrestrial plant species. The authors used a DNA methylation tiling array comprised of 2.1 million probes for the entire positive strand of the nuclear genome. Array methylation data was validated with WGBS of 28 randomly selected loci of genes and TEs distributed throughout the genome. Thus, specific CpG, CHG, and CHH methylation percentages were not available in this study. Another study using WGBS of the entire *P. tricornutum* genome reported 1% CpG, 0.19% CHG, and 0.2% CHH methylation; thus, *P. tricornutum* has CHH and CHG methylation, but not in appreciable quantities. 5mC methylation was predominantly localized in TEs, supporting the 2013 study; this species lacks periodic methylation observed in some green algae species [15].

Diatom research has mainly been focused on *P. tricornutum* and *Thalassiosira pseudonana*, the first marine phytoplankton species with a complete genome [48]. *T. pseudonana* is approximately 4 μm, though its size is predicated on nutrient availability [49]. Huff and Zilberman report 2.57% CpG, 0.13% CHG, and 0.12% CHH 5mC methylation in the *T. pseudonana* 34.5 MB genome, on par with methylation quantities reported for *P. tricornutum*. Indeed, similar to *P. tricornutum*, *T. pseudonana* lacks periodic methylation and 5mC is localized to TE elements likely to silence expression [15]. These two diatoms share approximately 6200 genes, 461 of which are methylated only in *T. pseudonana* and 408 of which are only methylated in *P. tricornutum* [50]. Differential methylation of these genes is reflected in gene expression and gene ontology (GO) category enrichment; thus, further investigation will illuminate how these two diatoms diverged from one another.

Fragilariopsis cylindrus is a pennate diatom that has adapted to colder marine temperatures and is used as an indicator species for polar water [51]. This diatom becomes trapped at the bottom of sea ice during the winter and is then released into open water as the ice melts [52]; thus, it likely evolved adaptive mechanisms, perhaps epigenetic in nature, to cope with this environmental variability. The 61 MB *F. cylindrus* genome has 8.63% 5mC CpG methylation and negligible CHG and CHH methylation (0.13% for both) [15]. As Huff and Zilberman discuss for other diatoms, *F. cylindrus* is largely methylated in TEs, where methylation functions to silence expression.

An initial report in 2010 suggested that brown algae genomes did not have detectable levels of genomic cytosine methylation [53]. Brown algae belong to the Heterkont family but are one of the only algae lineages comprised of unicellular and multicellular members. The brown alga, *Ectocarpus siliculosus,* is a giant kelp (macroalgae) that can extend more than 40 m. This organism is used to study brown algae developmental biology, as it has a haploid-diploid life cycle comprised of a gametophyte and a sporophyte, respectively [54]. Its 214 MB genome contains ~16,000 protein-coding genes with introns and long 3′ UTRs, and approximately 23% of the genome encodes for TEs. The authors reported a marked absence of DNA methylation based on high-performance liquid chromatography (HPLC) analysis [53]. This is highly unusual for a eukaryotic organism of such complexity with such a large genome and abundance of TEs. This finding has not been validated using traditional DNA methylation analysis techniques.

A recent study used a combination of MeDIP (methylated DNA immunoprecipitation) and WGBS sequencing of the multicellular (macro) brown algae *Saccharina japonica* [55]. The authors report approximately 1.4% of all cytosines in the 537 MB genome to have 5mC methylation, predominately in the CHH contexts (57%), with 25% CpG and 19% CHG sites methylated [56]. Interestingly, loci encoding for circRNAs and lncRNA were the most highly methylated genomic elements with 81% and 66% 5mC methylation, respectively. However, *S. japonica* does not have high levels of methylation in TEs. The authors proposed the CME protein DNMT2 is responsible for this methylation, given that the *S. japonica* genome does not contain sequences for any other methyltransferase enzyme. The overall methylation level in this larger genome is lower than typically observed for multicellular organisms. However, this methylation contributes to the regulation of gene expression, particularly noncoding RNAs, and likely plays a role in life cycle stages of *S. japonica*. While chloroplast methylation is not widely analyzed in microalgae, a recent 2021 paper reported 5mC methylation in 17.1% CpG, 58.1% CHG, and 58.8% CHH contexts in *S. japonica* chloroplasts, though methylation at any one position was relatively low [57]. As observed in *C. reinhardtii*, methylation levels in gametophytes are substantially higher than levels in sporophytes, suggesting that even chloroplast methylation is important for lifecycles of this alga.

Aureococcus anophagefferens is a unicellular brown alga within the Pelagophyceae class of Heterokonts. This species is small (2–3 μm), coccoid, and nonmotile; excessive proliferation causes the formation of harmful blooms in shallow marine waters with elevated concentrations of carbon and nitrogen, typically from anthropogenic contamination [58]. *A. anophagefferens* has approximately 24% CpG methylation, 0.19% CHG, and 0.18% CHH methylation in its 57 MB genome [15]. This methylation occurs in regularly spaced intervals with elevated methylation in gene bodies regardless of transcription levels, suggesting that 5mC methylation may have a structurally stabilizing effect in this genome. The mitochondrial genome of *A. anophagefferens* has methylation in two palindromic sequences, 5′gCnnGc3′ and 5′gGCc3′ (where the capitalized nucleotide represents where methylation can occur on both strands), which has not been reported for any other eukaryote to date [15]. The nuclear genome encodes predicted DNA methyltransferases that are not homologous to other known eukaryotes but are similar to bacterial cytosine methyltransferases. Interestingly, the *A. anophagefferens* mitochondrial genome encodes for adenine and cytosine methyltransferases (Dam/Dcm) likely derived from a viral or prokaryotic lateral gene transfer [59].

To date, DNA methylation has been assessed in one other species of harmful algae, *Emiliania huxleyi*. This marine haplotype belongs to the Coccolithophore class, a diverse collection of algae in which all have coccolith coverings. These coverings are small oval-shaped rings composed of calcite ($CaCO_3$). Thus, *E. huxleyi*, and other coccolithophores, play a paramount role in the cycling of carbon through the synthesis and decomposition of their coccoliths, contributing ~50% of the open ocean $CaCO_3$ sediments [60]. *E. huxleyi* is small (4–5 μm), unicellular, nonmotile, and each is covered in 10–15 coccoliths. It forms massive blooms covering hundreds of thousands of square kilometers of ocean, and the calcite shells of these algae make blooms easily detectable via satellites. These massive blooms appear seasonally over several weeks, and then suddenly collapse due to infection with viruses or bacteria [61, 62]. Interestingly, there is significant variation in genome size among *E. huxleyi* isolates from different regions: genomes range from 99 to 133 MB, though the haploid genome is estimated to be 141.7 MB [63]. This genome contains 16.36% CpG, 5.63% CHG, and 0.38% CHH 5mC DNA methylation [15]. *E. huxleyi* encodes for a CMT, likely responsible for CHG methylation, the quantity of which seems to compensate for lower CpG methylation. Both CpG and CHG methylation appear every 182 bp; this episodic methylation, like that of *A. anophagefferens*, results in densely methylated clusters (both CpG and CHG) that are independent of sequence context. Significant quantities of nucleosome linker methylation in *E. huxleyi* and other microalgae species (*A. anophagefferens* and three algae of the Mamiellales order, discussed in Section 17.2.3) suggest that the function of this methylation is evolutionarily conserved and important.

Cyanidioschyzon merolae is a thermoacidophilic, haploid, unicellular, red microalgae approximately 2 μm in diameter that inhabits hot springs rich in sulfur (pH 1.5, 45 °C) [64]. Its 16 MB genome has ~5300 genes, 80% of which are expressed at any given time. This red alga lacks genes for DNMT1 and DNMT5 and does not have appreciable CpG-specific methylation: 0.20% CpG, 0.18% CHG, and 0.19% CHH [15]. This is a considerable finding given that most eukaryotic organisms have some amount of CpG methylation; however, *C. merolae* is by far the simplest photosynthetic alga. It contains one mitochondrion that replicates once per cell cycle, a minimum number of organelles, ~4700 protein-encoding genes, and lacks a rigid cell wall. These features allow for efficient chromosome editing and bioengineering; as such, *C. merolae* has become a new model organism for investigation [65, 66].

Dinoflagellates are a class of microalgae unto themselves. They are unicellular and characterized by a pair of flagella. Most are photosynthetic, and collectively they are the second largest primary producers (after the diatoms) in the world's oceans [67]. They occupy marine, brackish, and freshwater ecosystems. Several members form harmful algal blooms, including the "Red Tide" in coastal marine waters. Further discussion of dinoflagellate's interesting nuclear composition, and striking lack of histones, will be discussed in Section 17.2.8. Information on DNA methylation has only been acquired for two dinoflagellate species, *Fugacium (Symbiodinium) kawagutii* and *Breviolum (Symbiodinium) minutum*, both of which are coral endosymbionts. WGBS revealed significant 5mC CG methylation in these species. *S. kawagutii* has 78% CpG, 1.19% CHG, and 0.99% CHH methylation with total genomic methylation of 0.27 (mC/C) [68]. *S. minitum* has 68% CpG, 1.55% CHG, and 1.15% CHH methylation with total genomic methylation of 0.32 (mC/C). High levels of CG methylation are potentially due to genome size: *Fugacium kawagutii* has a 1180 MB genome, while *Breviolum minutum* has a 1500 MB genome [69, 70]. Indeed, dinoflagellates have unusually large genomes given their size and complexity, likely not the result of increased gene composition, but rather tandem repeats [71]. Methylation quantity, localization, and function are so similar between *F. kawagutii* and *Branchioglossum minutum* that the authors reported their findings collectively for both species. They found that extensive (~70%) CG methylation occurred in a unimodal distribution, which is highly unusual – only 0.1% of cytosines in the CG context were unmethylated (compared to 3% in mammalian species) [68]. Both species lacked symmetric CpG methylation, likely due to a lack of DNMT1 homologs for maintenance methylation. Further, environmental perturbation, including heat stress, did not significantly alter methylation levels. The authors conclude that the default state of both species is hypermethylation that appears homogeneous across the genome (i.e., no depletion in promoters and within constitutively expressed genes and TEs alike). However, CHG methylation was enriched in TEs. Both species lack significant differentially methylated regions; methylation is not environmentally responsive, nor does it correlate with gene expression. Thus, no definitive role of methylation has been determined for these species.

17.2.5 Other Types of DNA Methylation

There has been very little assessment of 4mC and 6 mA DNA methylation in microalgae species. However, many of the genomes that we have queried (unpublished observations, Steadman) suggest that some algae species have the CMEs necessary for both 4mC and 6mA DNA methylation. Indeed, a series of papers have reported identifying 6mA in multiple species including the green algae, *C. reinhardtii* [21]. The authors reported the presence of 6mA in more than 14K genes, typically clustered around transcriptional start sites. Further, 6mA is found in linker regions between nucleosomes. Potentially, the 6mA modification promotes gene expression in *C. reinhardtii*, but this has yet to be assessed. To our knowledge, the 4mC modification has yet to be identified in eukaryotic algae to date.

17.2.6 Histone Modifications and Associated Chromatin-Modifying Enzymes in Other Organisms

In addition to DNA methylation, modifications on histone proteins are a form of epigenetic regulation. Histone proteins themselves are the most conserved protein in the tree of life and many algae species have exactly the same histone protein sequences as humans. In the nucleus, 147 base pairs of DNA are tightly coiled around several histone proteins and are collectively referred to as the nucleosome. For most of eukaryotic biology, nucleosomes are the essential organizational unit of the genome. They are comprised of eight core histone proteins: two H2A/H2B dimers complexed with a tetramer of H3 and H4 proteins, with the linker histone, H1, that binds the proteins together. The histone proteins have an abundance of lysine (K) and arginine (R) residues (over 60 sites) located on their N-terminal domains or "tails." These residues are readily available for posttranslational modification (PTM) via covalent addition of at least eight modifications, including methylation (me), acetylation (ac), and phosphorylation (ph). To further complicate matters, lysine residues can become mono-, di-, or trimethylated (me1, me2, and me3, respectively), while arginine residues can be mono- or dimethylated symmetrically or asymmetrically. Each histone tail can have multiple modifications, and these modifications can change in response to environmental factors, cellular signaling, and the state of cellular differentiation. Modifications on histone proteins alter the interactions of the tails with the DNA and the result of this interaction is epigenetic regulation of the genome. Researchers in the biomedical and plant fields have identified a continually evolving "histone code" of the types of modifications and enzymes that are present at different stages of development, under different environmental conditions, and in different tissue types. The range of modifications among complex eukaryotes seems to be the same, though studies in humans and mice have shown that both have the same types of modifications, expression of CMEs, and similar responses. Naturally, as we have previously learned with DNA modifications, this pattern of commonality is not necessarily true for the diversity of algae species.

As with DNA modifications, the types of histone PTMs that are employed by an organism are dependent upon the CMEs that are available to write, erase, or read PTMs. Unlike DNA modifications, where a handful of CMEs are responsible for imparting modifications, there are hundreds of CMEs for histone PTMs. For example, in humans, there are at least nine histone acetyltransferase enzyme families, each with several members. CME groups include histone methyltransferases (HMT, sometimes KMT), demethylases (HDM, sometimes KDM), acetyltransferases (HAT), deacetyltransferases (HDAC), phosphatases, dephosphatases, and the CMEs responsible for ubiquitination and SUMOylation. CME expression profiles are dependent upon tissue type, stage of development, cell-to-cell interactions and signaling, and environmental conditions. Many groups interested in algae epigenetics have first identified histone sequences, which are highly conserved, and the CMEs encoded in a particular algae genome before experimentally assessing the quantity and function of a histone PTM. Our work has shown that sequence homology among CMEs in algae genomes is rather low, suggesting that simply using a CME sequence found in a plant or other algae species will not necessarily constitute positive identification of a

CME in a new algae species (unpublished observations, Steadman [41]). There is much to be learned about the evolution of histone modifications by examining which PTMs (and their corresponding CMEs) are found in algae species across the tree of life. The sheer number and complexity of CMEs and histone PTMs make this research in algae one of the most daunting, yet exciting, areas of epigenetics research. For this particular section, we will describe histone genes, CMEs, or their associated histone modifications in seven algae species.

17.2.7 Histone Modifications in the Model Alga Species, *C. reinhardtii*

The first extraction of chromatin from *C. reinhardtii* was reported in 1975. It was suggested that while this microalga contained nuclear proteins, it lacked similarity to higher-ordered eukaryotic histones [72]. In 1990, *C. reinhardtii's* five histone proteins were isolated and mapped using micrococcal nuclease digestion, nuclear protein isolation, and electrophoresis. The authors reported that *C. reinhardtii* had the full complement of histone proteins found in more complex eukaryotes with a nucleosome repeat length of 189 base pairs [73]. Over the next several years, research into histone proteins and modifications in *C. reinhardtii* burgeoned. Major findings included the discovery of 15 gene clusters, three genes for each of the four core histone proteins and one for the H1 linker histone in the *C. reinhardtii* genome [74]. New methods for histone extraction and protein purification provided deeper insight into PTMs. Using reverse phase HPLC, acetylation of H3 proteins was reported, with at least four acetylated lysine residues per histone 3 molecule, including K9, K14, K18, and K23 [75]. The authors also reported methylation, to the exclusion of acetylation, on K4, K9, and K27 of histone 3. The histone H3 sequence was confirmed by protein sequencing, though a threonine residue was found at position 28 rather than a serine residue. The study also demonstrated that certain isoforms of histones were present, dependent upon culture conditions: photosynthetic cultures with greater mitotic activity were dominated by the H1.2 linker and H2A.1 histone isoforms, while slower-growing mixotrophic cultures possessed the H1.1 linker and H2A.2 histone isoforms [75]. The first impacts of environmental conditions on histone PTMs in *C. reinhardtii* were also reported: a dramatic decrease in global histone acetylation levels occurred during the dark phase of the light cycle for photosynthetic cultures [75]. Further investigation revealed approximately 20% of the *C. reinhardtii* genome contained acetylation, which had a high turnover rate (~two minutes), particularly for tri- and pennate acetylation of H3 [76].

A follow-up study in 1999 focused on lysine-rich histone proteins, reporting that *C. reinhardtii* has two isoforms of the H1 linker histone, which appeared as a doublet of two distinct isoforms, H1A and H1B [77]. Sequencing of the *C. reinhardtii* genome revealed that histone genes are arranged as pairs, with associations between H2A and H2B and H3 with H4 [26, 78]. Each major histone class is present as a four-gene cluster (H2A, H2B; H3, H4), corresponding to 84 (out of 117) histone genes [78]. Histone gene expression is tightly coordinated with the cell cycle, where shortly before mitosis (S/M phase), expression is the highest, though a small number of histone genes are constitutively expressed during the diurnal cycle [79, 80].

Deeper characterization of histone modifications, including their identity, quantity, localization, and function, followed over the next decade from a key group of researchers. Studies from the Cerutti lab demonstrated that histone modifications play an important role in transgene silencing in *C. reinhardtii*. The group discovered and characterized three CMEs responsible for methylation and phosphorylation of histone proteins in *C. reinhardtii*. They found MUT11p interacted with components of the Trithorax complex, known for inducing methylation of H3K4 in other eukaryotic systems. Mut11p is capable of methylating H2A, H3, and H4 histones. Loss of Mut11p induced a global reduction in H3K4 monomethylation, deficits in trimethylation of H3K4, and a concomitant increase in dimethylation of H3K4. These changes resulted in derepression of transcriptionally silenced loci, including transgenes and retrotransposons [81]. Thus, the H3K4me1 modification is likely important for repressing euchromatin states, while the H3K4me2 modification serves to mark active chromatin conformations in *C. reinhardtii*. The Cerutti group further characterized two other CME proteins, MUT9p and SET3p, important for imparting phosphorylation and methylation on specific histone proteins. The kinase MUT9p phosphorylates H3T3 and H2A, resulting in long-term heritable gene silencing in *C. reinhardtii* [82]. A homolog of the CME SU(VAR)3–9, SET3p, was found to be responsible for monomethylation of K9 on histone 3 [83]. This particular modification localized to heterochromatin regions and transcriptionally repressed genes, suggesting a suppressive role, as observed in other eukaryotes including *Arabidopsis thaliana* and *Zea mays*. Interestingly, SET3p was not responsible for trimethylation of K9 on histone 3, which was markedly absent from silenced transgenes and associated with a transcriptionally active gene (RPS3). However, this reported localization of the H3K9me3 modification to active genes has been observed in *Arabidopsis* euchromatin regions, suggesting a divergent role from that of repressor in mammals. A 2015 study confirmed the localization of the H3K9me3 modification in *C. reinhardtii*, further reporting that the modification colocalized with trimethylation of K4 and acetylation of K27 on histone 3 to the exclusion of the H3K27me3 modification; this novel finding may be exclusive to green algae [84]. The Cerutti group was unable to detect dimethylation of K9 on histone 3 in whole cell extracts, nor from partially purified histone extractions, suggesting that the H3K9me2 modification is either absent or present at low levels [81, 85]. This highly contrasts with the nearly ubiquitous presence of the H3K9me2 modification in plant chromatin and the intricate role it plays in establishing DNA methylation in higher-ordered plants. In a follow-up study, the E(z) homology (EZH) subunit from the Polycomb Repressive Complex 2 (PRC2) was found in *C. reinhardtii*; the PRC2 complex is responsible for methylation of K27 on histone 3 [85]. The authors reported the widespread presence of both mono- and dimethylated H3K27 and acetylated H3K27. RNA interference (RNAi) of EZH resulted in massive demethylation of H3K27, defective transgene and retrotransposon silencing, and a concomitant increase in global levels of H3K4 trimethylation and H4 acetylation [85]. The Cerutti group also reported the presence of the EZH subunit and other components of the PRC2 complex responsible for H3K27 methylation in several algae species, including *A. anophagefferens*, *E. huxleyi*, *T. pseudonana*, *C. merolae*, *O. tauri*, *V. carteri*, and *Coccomyxa* sp. C16; to date, the presence of K27 methylation on histone 3 has not been confirmed in any of these species [85].

Further evidence suggested conserved epigenetic silencing mechanisms in *C. reinhardtii*. A recent paper reported a putative Sir2-type histone deacetylase domain named SRTA [86]. Sir NAD+-dependent deacetylases have important regulatory roles in mammalian development and are also found in *Arabidopsis* and yeast. The *C. reinhardtii* transgene mutants, UVM4 and UVM11, which are well characterized, have high levels of transgene expression; both carry point mutations in the Cre10.g462200 gene in the Sir2 domain of SRTA [86]. The authors reported loss of the STRA protein did not impact endogenous gene expression, but rather resulted in activation of silent transgenes. The authors suggested that this HDAC induces epigenetic transgene suppression, largely mediated on foreign DNA lacking significant chromatin modifications [86]. Given that invading foreign DNA is prevalent in aquatic environments, and that unicellular organisms lack a protected germline, it is possible that *C. reinhardtii* evolved this epigenetic tool as a protective mechanism.

Mass spectrometry methods have been employed to assess histone variants and their potential modifications in *C. reinhardtii*; however, these methods have not fully characterized the litany of modifications that were reported in prior publications. In 2018, the first use of top-down mass spectrometry (TDMS) was reported for characterizing histone proteins and modifications in *C. reinhardtii* [87]. While challenges in protein oxidation required alternative histone extraction and purification methods, the authors reported minimal modification on histones. Only acetylation of H2A, H2B, and H4, and monomethylation of H3K4 and H3K27 were found, though a novel, previously unreported modification was also found on H4K79 (methylation) [87]. Improved methods for TDMS resulted in the identification of 86 proteoforms of all histone proteins, including the noncanonical variant H2A.Z with high levels of acetylation [88]. Table 17.2 shows a table of these proteoforms in *C. reinhardtii*. The authors identified high acetylation of H2A and H2B, and a bimodal distribution of modified H3 where the H3K4me1 modification correlated with H3 hypoacetylation and the H3K4m3 modification associated with H3 hyperacetylation [88]. Mass spectrometry methods, while useful for identifying the potential histone PTMs employed by a species, do not provide information about the potential function of these modifications.

Chromatin immunoprecipitation followed by sequencing (ChIP-seq) provides information on the quantity and genomic context of particular histone PTMs. Using this technique, a recent report characterized the chromatin landscape of *C. reinhardtii* and discovered localizations of several histone PTMs and their potential functions. The authors reported that nearly all promoters in *C. reinhardtii* contained trimethylation of K4, K9, and K36, and acetylation of K27 on histone 3 and RNA polymerase II [84]. The H3K36me3 modification had distinctive sharp, narrow promoter-centric peaks in these promoter regions. In other species, this modification is primarily found in broad regions of actively transcribed genes. Potentially, the H3K36me3 modification has a divergent role in *C. reinhardtii* from that observed in other eukaryotes. The H3K36me3 modification also colocalized with the repressive H3K27me3 modification. Transcripts from these regions had low expression [84]. Bivalent domains such as these are often found in mammalian genomes, but not in terrestrial plants. Dimethylated H3K4 and acetylated H3K27 were found in distal regulatory elements, including enhancers. The authors assessed changes in histone PTMs under nitrogen and sulfur starvation culture conditions, and results of this are provided in Section 17.3. Further investigation of several microalgae species is needed to acquire deep knowledge of the chromatin landscape and histone code of algae, particularly in response to environmental perturbation.

In summary, *C. reinhardtii* contains H3K27me3, H3K4me1, and H3K9me1, all of which have been associated with transcriptionally silent regions, particularly TEs. Localization of the modifications H3K4me2, H3K27ac, and H3K9me3 in transcriptionally compliant regions suggests contributions to euchromatin and active gene states. *C. reinhardtii* promoters contain several sequential histone PTMs, including methylation of H3K4, H3K9, and H3K36, the latter of which may have a novel function in algae. *C. reinhardtii* seems to lack the capacity to methylate H4K20 and dimethylate H3K9. Acetylation

Table 17.2 Top-Down Mass. Spectrometry characterization of *C. reinhardtii* histone proteins and modifications shows all 86 proteoforms of histone proteins and potential PTMs. Nα- denotes amino modifications, while the other histone posttranslational modifications (PTMs) follow standardized nomenclature.

H1	H1.1Nα-ac	acH1.2Nα-ac			
H2A	H2A.0Nα-ac	H2A.1	H2A.1Nα-ac	H2A.1Nα-acK5ac	H2A.1Nα-acK188ac
	H2A. 2	H2A.2Nα-ac	H2A.2Nα-acK5ac	H2A.2Nα-acK188ac	H2A.3
	H2A.3Nα-ac	H2A.3Nα-acK5ac	H2A.4Nα-ac	H2A.ZNα-ac	H2A.ZNα-acK6acK14ac
	H2A.ZNα-acΔ143	H2A.ZNα-acΔ142-143			
H2B	H2BNα-me3.1[c]	H2BNα-me3.2[c]	H2BNα-me3.3[c]	H2BNα-me3.5[c]	H2BNα-me3.8[c]
	H2BNα-me3.9[c]	H2BNα-me3.11[c]	H2BNα-me3.12[c]	H2BNα-me3.13[c]	H2BNα-me3.14[c]
	H2BNα-me3.15[c]	H2BNα-me3.v1	H2BNα-me3.K7ac	H2BNα-me3.v1K11ac	H2BNα-me3.v1K7acK11ac
	H2BNα-me3.v1K7ac K11acK12ac	H2BNα-me3.v1K7acK11acK12acK16ac			
H3	H3	H3K4[b]/9me1	H3K4me1 + 14 Da (K35-Q86)	H3K4/9me1 + 28 Da (K27-M119)	H3K4me1 + 42 Da (T28-Y40)
	H3K4me1 + 56 Da (P37–F57)	H3K4me1 + 71 Da (L20-I118)	H3K4me1 + 84 Da (T28-T117)	H3K4me1 + 98 Da (K23-L99)	H3K4me1 + 112 Da (A23-R115)
	H3K4me1 + 125 Da (Q54-F77)	H3K4me1 + 140 Da (K23-T44)	H3K4me1 + 155 Da (K23-D105)	H3K4me1 + 167 Da (A25-F83)	H3K4me1 + K18ac + 140 Da (Q19-L91)
	H3K9me1 + 197 Da (K23-I118)	H3K4/9me1 + 211 Da (G32-F103)	H3K4me3K9acK14acK18ac + 70 Da (Q19-I118)	H3K4me3K9acK14ac + 128 Da (Q23-T79)	H3K4/9me1 + 255 Da unlocalized
	H3K4me3K9acK14acK18ac + 115 Da (Q19-F77)	H3K4me3 + 174 Da (Q19-I118) or H3K4me1 + 283 Da (Q19-I118)			
H4	H4Nα-ac	H4Nα-acK79me1	H4Nα-acR3me1K79me1	H4Nα-acK5acK79me1	H4Nα-acK8acK79me1
	ac4K12acK79me1	H4Nα-acK16acK79me1	H4Nα-acK5acK8acK79me1	H4Nα-acK5[b]/8acK12ac K79me1	H4Nα-acK12acK16ac K79me1
	H4Nα-acK5[b]/8acK16ac K79me1	H4Nα-acK5acK8ac K12acK79me1	H4Nα-acK5acK8ac K16acK79me1	H4Nα-acK5acK12ac K16acK79me1	H4Nα-acK8acK12ac K16acK79me1
	H4Nα-acK5acK8ac K12acK16acK79me1	H4Nα-acK5acK12acK16acK20acK79me1	H4Nα-acK5acK8acK12ac K16acK20acK79me1		

Superscript a ([a]) denotes that proteoforms that differ in methionine or cysteine oxidation were not considered. Superscript b ([b]) indicates an ambiguous isomer assignment, while superscript c ([c]) indicates evidence of monoubiquitylation.
Source: Adapted from Rommelfanger, *Journal of the American Society for Mass Spectrometry 2021, 32, 1671–1688.*

occurs in approximately 20% of the *C. reinhardtii* genome and its turnover rate is high. The *C. reinhardtii* genome has potential CME sequences that are homologs to CMEs and protein domains of epigenetic complexes found in higher-ordered eukaryotes. Histone PTMs in *C. reinhardtii* can be dynamically turned over at a relatively high rate, particularly during synchronized light–dark phase cultures [89], which suggests that at least the histone component of the *C. reinhardtii* epigenome is highly responsive to environmental stimulation.

17.2.8 Histone Modifications in Other Microalgae

Assessment of histone PTMs and potential CME sequences has been limited to a handful of species. In this section, we will discuss histone PTMs in *V. carteri*, two diatoms, and the brown algae *Ectocarpus*. We will begin with studies investigating the presence of CMEs using genomic search tools, which are more broadly applicable and encompass several other microalgae species.

To determine the potential for histone methylation, an appropriate starting point is to query an algae genome for the dominant methyltransferase enzymes. Most methylation of lysine residues is conducted by CMEs that contain SET domains, specificity for which lysine residue on a particular histone protein is typically determined by adjoining protein domains, flanking sequencing, and cofactors. The Cerutti group surveyed 14 algae genomes for SET polypeptides from *Arabidopsis salina* and *Homo sapiens* protein domains using BLAST or PSI-BLAST [90]. They identified potential homologs containing sequence similarity for alignment and phylogenetic tree construction. Of particular interest were the KMT1 and KMT6 homologs, both of which are histone methyltransferases responsible for H3K9 methylation and H3K27 methylation, respectively. KMT1 homologs were limited to certain species including *C. reinhardtii*, *Chlorella sorokiniana*, *Cocomyxa subellipsoidea*, *C. variabilis* NC64A, and *V. carteri*. KMT1 homologs had high sequence homology to terrestrial plant KMT1 polypeptides within the SET domain and the pre- and post-SET motifs; further, KMT1 homologs contained the SET and RING-associated (SRA) domain, which recognizes DNA methylation of CpG and CHH contexts. KMT6 proteins are putative components of the PRC2 complex for trimethylation of H3K27 and were nearly ubiquitous in the species examined. KMT6 homology was found in *B. prasinos*, *C. merolae*, *C. reinhardtii*, *C. sorokiniana*, *C. subellipsoidea*, *C. variabilis* NC64A, *Galdieria sulphuraria*, *M.* sp. RC299, *M. pusilla* CCMP1545, *O. lucimarinus*, *O. tauri*, and *V. carteri* [90]. The virus that infects *C. variabilis* NC64A, named *Paramecium bursaria* chlorella virus 1 (PBCV-1), encodes a KMT6 protein, named vSET: PBCV-1 induces trimethylation of H3K27 on *Chlorella* genes and suppresses host transcription after infection [91]. Except for *C. reinhardtii*, to date, the presence of methylation of H3K9 and H3K27 has yet to be confirmed in many of these species.

Another group surveyed more components of the PRC2 complex (in addition to the EZH subunit containing the SET methyltransferase domain). They found several species of algae contain a broad distribution of the PRC2 complex components, namely E(z), ESC, and p55. These algae include Prasinophytes and red algae, *C. reinhardtii* (published before), *Cyanophora paradoxa*, *C. merolae*, *M. pusilla*, and *O. tauri* [92]. As mentioned earlier, the PRC2 complex is heavily involved in establishing trimethylation of K27 on histone 3. The authors biochemically verified the presence of the H3K27me3 modification in *C. merolae*, *M. pusilla*, and *O. tauri*. Deeper, genome-wide characterization of H3K27me3 modification was performed using ChIP-seq on *C. merolae*, the model unicellular red alga occupying highly acidic environments with high temperatures. The authors found that this modification covered 14% of the total nuclear genome with 1300 peaks per replicate and an average peak width of 1792 bp [92]. The *C. merolae* genome contains relatively few repetitive elements, only 0.7% TEs and 5% repeats corresponding to a truncated open reading frame from a virus. The H3K27me modification was localized to 172 repetitive elements or about 50% of their total number. Potentially, this modification is necessary for protecting the *C. merolae* genome, given its abundance of repetitive elements. Enrichment was found on 4% of genes and negatively correlated with gene expression from RNA-seq studies, suggesting a repressive role. Consistent with findings from other studies, the authors reported occupancy of H3K27me3 and decreased expression of genes involved development, protein maturation, and splicing in *C. merolae*. The H3K27me3 modification is recognized and read by the Polycomb Repressive Complex 1 (PRC1), which initiates mono-ubiquitination of H2A to further suppress H3K27me3-occupied genomic contexts. There is a marked absence of PRC1 components, including Psc/BMI1, Pc/LHP1, EMF1, and Scm in several algae species, including *M. pusilla*, *O. tauri*, *C. merolae*, and *C. paradoxa*. However, *C. reinhardtii* encodes for a Psc/BMI1 homolog and thus may have PRC1 functionality. Interestingly, the RING1 subunit of the PRC1 complex, which has mono-ubiquitination activity, was found in the genomes of *C. reinhardtii*, *M. pusilla*, and *C. paradoxa*, though its function may not be related to PRC1.

In addition to studies of *C. reinhardtii*, deep characterization of histone modifications has been performed by the Tirichine group for diatoms (Stramenopiles/Heterokonts). These microalgae are some of the most diverse species on the planet with approximately 100,000 members, all of which have frustules, or silica-containing cell walls. Diatoms play a paramount role in global carbon cycling, accounting for 20% of carbon fixation and 40% of primary production. *P. tricornutum* and *Thalassiosira pseudonana* have emerged as model diatoms for investigation, as both are relatively simple to grow under laboratory conditions and have fully sequenced genomes with interesting composition, demonstrating acquisition of genes from several origins. They also contain the full spectrum of CMEs for modification of DNA (discussed earlier) and histone PTMs. *P. tricornutum* has 14 histone genes encoding for nine histone proteins including H1, H2A, H2B, H3, and H4 located on five of its chromosome scaffolds [44]. Histone genes are clustered as doublets of H2A-H2B and H3-H4, similar to *C. reinhardtii*, though *P. tricornutum* encodes for variants of H1, H2A, and H3. The presence of histone proteins in *P. tricornutum* and *T. pseudonana* was verified via Western blot and mass spectrometry [93]. In silico analysis of both the *P. tricornutum* and *T. pseudonana* genomes revealed several CMEs for histone modifications including all histone methyltransferases, demethylates, acetyltransferases, and deacetylases [47]. Table 17.3 lists all the CMEs found for these organisms, including those for the brown alga (Heterokont) *Ectocarpus siliculosus*.

Using high-resolution mass spectrometry (HR-MS), 62 histone PTMs on the core and histone variants of *P. tricornutum* were identified, while eight novel histone PTMs were found [94]. As observed in other studies using HR-MS, this method revealed a plethora of histone PTMs but missed several important modifications, including di- and trimethylation of K3K9, both of which had been previously verified. Further, methylation of arginine was markedly absent from the HR-MS data, though arginine methylases are encoded in genome. Novel modifications include acetylation of K2, K6, K34, K48, and K107 residues on H2A; acetylation of K31 and K59 residues on H4; ubiquitination of K11 on H2A; and mono-, di-, and trimethylation of K79 on H4. Interestingly, the multivalency methylation of H4K79 is found in metazoans, but not plants. The full complement of histone PTMs from this analysis is shown in Figure 17.5.

Genome-wide mapping using ChIP-seq for five histone PTMs demonstrated that ~40% of the *P. tricornutum* genome, and likely more, is occupied. These modifications included H3K4me2 (~29%), H3K9/14Ac (~25%), H3K9me2 (~25%), H3K9me3 (~8%), and H3K27me3 (~14%), where percentages signify genomic occupancy [94]. Approximately, 119,000 regions including genes, intergenic regions, and TEs (total of 1.3 M base pairs) are shared between all five modifications. TEs were enriched in H3K27me3 (3493 TEs), though H3K9me2 and H3K9me3 occupied 1281 and 1510 TEs, respectively. The function of these modifications, particularly H3K27me3 and H3K9me3, is predominantly repressive, which for the latter, is distinctly different from *A. thaliana* where H3K9me3 is found in euchromatic regions and positively correlates with gene expression. The H3K27me3 modification peak was broad and covered large domains. This is counter to the types of peaks observed in *A. thaliana*, but analogous to those found in mammalian species. The H3K4me2 and H3K9/14 Ac modifications occupied very few TEs, 605 and 74, respectively. As expected, most genes were occupied by acetylation of H3K9/K14 (~63%) and dimethylation of H3K4 (~84%), where enrichment was predominately in the 5′ region of coding sequences with sharp acetylation peaks. Occupancy of the H3K4me2 modification in genes did not appear to correlate with gene expression and thus may not play a direct role in transcriptional activation. Conversely, as observed in other species, acetylation is positively correlated with transcriptional activation and is enriched in gene promoters. The H327me3 modification was also found on some genes (~7%) and exhibited a broad distribution dissimilar from the sharp peaks observed in terrestrial plants, suggesting this modification is analogous to that observed in mammalian epigenomes [45]. The H3K27me3 modification covers approximately 14% of the *P. tricornutum* genome, more than several other species including *A. thaliana*, *D. melanogaster*, and even mammals, which only have 6% of their genomes with H3K27me3 occupancy [45]. The H3K9me2 modification was also found on a substantial number of repressed genes. Nucleosome occupancy assessment revealed that most of the genome (~60%) contains nucleosomes in exons, TEs, and intergenic regions; nucleosome-depleted regions were found upstream of transcriptional start sites for highly expressed genes [94]. The most recent experimental work from the Tirichine group showed that the H3K27me3 modification selectively regulates the *P. tricornutum* phenotype [95]. Using gene knockouts of the E(z) component of the PRC2 complex, the group demonstrated that E(z) is directly responsible for di- and trimethylation of H3K27. Removal of the H3K27me2 and H3K27me3 modifications in the knockout mutants abolished the *P. tricornutum* phenotype, specifically preventing its differentiation into different morphotypes. This work is the first piece of evidence clearly delineating the role of PRC2 and H3K27me3 modification for cellular differentiation of a unicellular eukaryote. Interestingly, *P. tricornutum* seems to combine types and locations, and presumably function, of histone PTMs from both mammals and plants, further promoting the notion of a multiorigin genome [45].

Ectocarpus siliculosus is a brown macroalgae species belonging to the Stramenopile (Heterokont) family and is closely related to diatoms. In the previous section, it was noted that this particular multicellular algae species lacks DNA

Table 17.3 Reverse BLAST was used to identify chromatin-modifying enzymes for histone modifications in Stramenopiles: *P. tricornutum* (diatom), *Thalassiosira pseudonana* (diatom), and *Ectocarpus siliculosus* (macro brown algae).

Histone modifiers	Residues modified	Homologs in *P. tricornutum*	Homologs in *T. pseudonana*	Homologs in *E. siliculosus*
Lysine acetyltransferases (KATs)				
HAT1 (KAT1)	H4 (K5, K12)	54,343	1397, 22,580	Esi0002_0151
GCN5 (KAT2)	H3 (K9, K14, K18, K23, K36)	46,915	15,161	Esi0090_0053
Nejire (KAT3); CBP/p300 (KAT3A/B)	H3 (K14, K18, K56) H4 (K5, K8); H2A (K5)H2B (K12, K15)	45,703, 45,764, 54,505	24,331, 269,496, 263,785	Esi0053_0101
MYST1 (KAT8)	H4 (K16)	24,733, 24,393	37,928, 36,275	Esi0090_0053, Esi0084_0032
ELP3 (KAT9)	H3	50,848	9040	Esi0264_0014
Lysine deacetylases (HDACs)				
RPD3 (Class I HDACS)	H2, H3, H4	51,026, 49,800	41,025, 32,098, 261,393	Esi0147_0031, Esi0181_0051, Esi0092_0077
HDA1 (Class II HDACS)	H2, H3, H4	45,906, 50,482, 35,869	268,655, 269,060, 3235, 15,819	Esi0168_0016, Esi0157_0055, Esi0157_0057, Esi0040_0047
NAD+ dependant (Class III HDACS)	H4 (K16)	52,135, 45,850, 24,866, 45,909, 52,718, 21,543, 39,523	269,475, 264,809, 16,405, 35,693, 264,494, 16,384, 35,956	Esi0054_0036, Esi0014_0023, Esi0026_0040
Lysine methyltransferases				
MLL	H3 (K4)	40,183, 54,436, 42,693, 47,328, 49,473, 49,476, 44,935	35,182, 35,531, 22,757	Esi0069_0089
ASH1/WHSC1	H3 (K4)	43,275	264,323	Esi0016_0129, Esi0000_0259
SETD1	H3 (K36), H4 (K20)	Not found	Not found	Esi0070_0045, Esi0020_0054, Esi0043_0074
SETD2	H3 (K36)	50,375	35,510	Esi0028_0153
SETDB1	H3 (K9)	Not found	Not found	Esi0162_0064
SETMAR	H3 (K4, K36)	Not found	Not found	Esi0100_0024
SMYD	H3 (K4)	bd1647, 43,708	23,831, 24,988	Esi0013_0081, Esi0189_0013, Esi0189_0015, Esi0286_0007, Esi0015_0158
TRX-related		Not found	Not found	Esi0115_0076, Esi0076_0098, Esi0043_0074, Esi0094_0082, Esi0455_0010
E(Z)	H3 (K9, K27)	32,817	268,872	Not found
EHMT2	H3 (K9, K27)	Not found	Not found	Esi0453_0004
SET+JmjC	Unknown	bd1647	Not found	Not found
Lysine demethylases (KDM)				
LSD1 (KDM1)	H3 (K4, K9)	51,708, 44,106, 48,603	Not found	Esi0073_0073, Esi0060_0020, Esi0014_0196
FBXL (KDM2)	H3 (K36)	42,595	Not found	Esi0025_0159
JMJD2 (KDM4)/ JARID	H3 (K9, K36)	48,747	2137	Esi0145_0049, Esi0014_0020, Esi0084_0076
JMJ-MBT	Unknown	48,109	22,122	Not found
JMJ-CHROMO	Unknown	Pt1-40322	1863	Not found
Arginine methyltransferases				
CARM1 (PRMT4)	H3 (R17)	17,184	3690, 28,185	Esi0137_0007, Esi0000_0236, Esi0165_0056
PRMT5	H3 (R8), H4 (R3)	49,565	24,429	Esi0006_0115
PRMT6	H3 (R2)	54,710	bd1828	Esi0153_0004, Esi0000_0300
PRMT7	H4 (R3)	49,245	25,810	Esi0133_0084, Esi0122_0102
Arginine demethylases				
JMJD6	H3 (R2), H4 (R3)	3251, 35,989, 46,234	24,024	Esi0055_0014, Esi0131_0061, Esi0189_0080, Esi0055_0061

Source: Maumus [47]/Bentham Science Publishers.

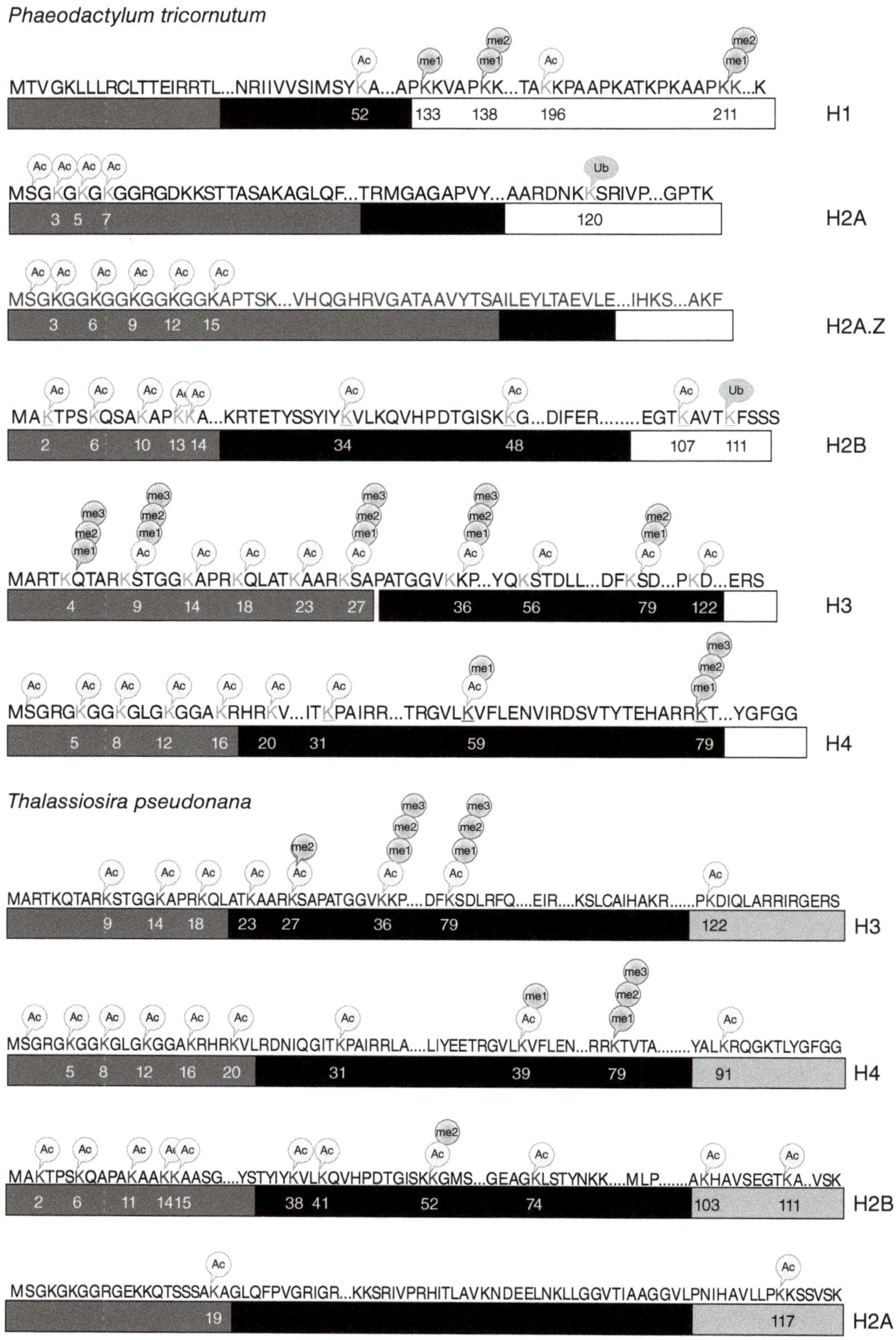

Figure 17.5 Histone PTMs identified in *P. tricornutum* (upper) and *Thalassiosira pseudonana* (lower) via high-resolution mass spectrometry. The amino acid residue number is indicated below the protein sequence. Dark gray indicates the N-terminal domain, black indicates the globular histone core, and light gray indicates the C-terminal domain. Acetylation is shown in green indicated by Ac. Methylation is shown in red indicated by me1, me2, and me3. Ubiquitination is shown in yellow indicated by Ub. Novel histone PTMs found in these species are underlined. For *P. tricornutum*, novel modifications identified include H2BKac, H2BK34Ac, H2BK48Ac, H2BK107Ac, H2BK111Ub, H4K59me1, and H4K79me1, me2, me3. For *T. pseudonana*, novel modifications identified include H4K31, H4K59Ac, H4K59me, H4K79me, me2; H4K20Ac, H4K20me, H3K79me, me2; H2BK107Ac. *Source:* Adapted from Veluchamy *Genome Biology* 2015, 16: 102 and Ragosti *AIMS Genetics* 2015, 2 [3]: 173–191.

methylation. A 2021 study identified histone and CME genes and histone PTMs [54]. The authors reported 34 core histones and nine H1 genes with possible variant forms. Histone genes appeared as clusters of H4, H1, H3, H2B, and H2A, suggesting duplication, rearrangement, and potential fragmentation. The authors identified putative histone acetyltransferases and methyltransferases, and homologs for deacetylase and demethylase CMEs. Surprisingly, this algae species lacked components of both PRC1 and PRC2 complexes. Using mass spectrometry, the authors identified 47 PTMs on core and variant histones. One novel modification was found: monomethylation of R38 on H2A.Z. Similar to reports in *P. tricornutum*, the H3K9me2 and H3K9me3 modifications were not found. Most of the other histone PTMs have been reported in other species, though some only overlap with the diatoms or were only found in *E. siliculosus*.

Genome-wide assessment of eight histone PTMs was determined using ChIP-seq for *E. siliculosus*. Transcriptional start sites of genes (though not gene bodies) were preferentially occupied with the H3K4me2, H3K4me3, H3K9ac, H3K14ac, and H3K27ac modifications [54]. Interestingly, the H3K4me2 modification was found on each side of the transcriptional start site. These modifications were found in ~13,500 genes (out of 17,000) and positively correlated with gene expression, suggesting a permissive function in *E. siliculosus*. Similar to mammalian epigenomes, most gene bodies (~74% of the genome) were occupied with trimethylation of H3K36, which positively correlated with gene expression. Repetitive sequences, introns, and transposons were occupied by the H4K20me3 modification, which correlated with lower gene expression. The distribution and function of the H4K20me3 modification in *E. siliculosus* is more closely aligned with that of mammals than plants. H3K79 dimethylation peaks were broad, found in large regions spanning multiple genes with some periodicity, and occupied ~37% of the genome. The H3K79me2 and H4K20me3 modifications were typically colocalized for regions with very low gene expression, suggesting a repressive function. These two modifications also occupied a large insertion of viral DNA, similar to the *Ectocarpus* phaeovirus EsV-1 virus, located on chromosome 6. It is possible that the H3K79me2 and H4K20me3 modifications serve repressive functions attributed to the H3K9me2 and H3K9me3 modifications, which were not detectable through ChIP-seq profiling. The assayed histone PTMs were all relatively stable during the lifecycle states of the algae except for a handful of generational-biased genes that showed differential occupancy dependent upon the lifecycle stage. It is still unclear what role histone PTMs play in facilitating or inducing gene expression; further, the authors conclude that it is possible that histone PTMs are actually a result of gene expression changes, rather than the cause, in this species [54].

A handful of other reports on histone gene sequences and PTMs in other species suggest that while most microalgae contain some subset of histone PTMs, the function of these modifications is highly dependent on the lineage of the species. The model species, *V. carteri*, contains half as many histone genes as *C. reinhardtii* with two H1 genes, and a pair of H3–H4 genes [96–98]. Its histone proteins were isolated in 1974, yet, to date, there has not been a follow-up study on histone PTMs in this particular species [99]. The species *O. tauri* and *O. lucimarinus* contain histone gene sequences, but a definitive analysis of how many genes and their sequence clusters has yet to be performed [43]. The genomes were queried for 104 chromatin proteins, of which 76 were found; both have histone chaperones and CMEs [43]. A more comprehensive characterization of histone genes, CME genes, and histone PTMs has yet to be done for these species. In 2018, a methods paper detailing the protocol for ChIP-seq in *Nannochloropsis oceanica* was published [100]. This oleaginous microalgae species produces high quantities of triacylglycerols and polyunsaturated fatty acids; it is therefore an important model species for the biofuel and bioproduct industries. *N. oceanica* has a relatively small, compact 29.3 MB genome with approximately 7330 genes [101]. A previous publication on transcriptomic changes during specific photoperiods for *N. oceanica* reported that CMEs, including histone acetyltransferases, deacetylases, and methyltransferases had dynamic expression during the diurnal cycle [102]. In the study, the authors biochemically verified the presence of acetylation and trimethylation of K9 and K27 on histone 3. Using ChIP-qPCR, the authors analyzed the occupancy of these modifications at various time points during the growth cycle on carbon fixation genes including carbonic anhydrase (CA), bicarbonate transporter (BCT), pyruvate orthophosphate dikinase (PPDK), phosphoenolpyruvate carboxylase (PEPC), phosphoenolpyruvate carboxylase kinase (PEPCK), and malate dehydrogenase (MDH). Differences in occupancy among the time points were found for the H3K9me3, H3K27me3, and H3K27ac modifications on CA, the H3K9me3 modification on PPDK and PEPCK, and the H3K27me3 modification on PEPC. Further characterization of genome-wide occupancy of acetylation of H3K9 via ChIP-seq revealed enrichment in genic regions, particularly at transcription start sites. Peaks were not readily detected in exon, intron, or intergenic regions, though the number of peaks was dynamic and dependent upon timing in the growth cycle [100]. Further characterization of this ChIP-seq data with transcriptomics would provide information to determine the epigenetic regulation of gene expression in *N. oceanica*.

Finally, we must include a note on epigenetic mechanisms and histone-like proteins in dinoflagellates. These microalgae form symbiotic relationships with corals and help maintain tropical reef ecosystems, though some species form harmful algal blooms. Until recently, it was widely accepted that, unlike most eukaryotes, dinoflagellates lack histone proteins. Indeed, the dinoflagellate nucleus is unlike that of other organisms. It contains high quantities of DNA content (~200 pg, whereas the human nucleus contains ~3 pg DNA), and DNA can be modified to form 5-hdyroxymethyluracil, replacing almost 70% of thymidine [103]. Dinoflagellates have a skewed DNA-to-protein ratio of 10 : 1, versus 1 : 1 ratio observed in other eukaryotes. Further, they have permanently condensed chromosomes of liquid crystalline DNA. Until recently, it was thought that dinoflagellate nuclei completely lacked histone proteins. Early reports in the late 1970s and early 1980s suggested that some dinoflagellate species contain two distinct and dissimilar nuclei, one derived from a diatom, and have histone-like, though not canonical, proteins [104, 105]. Subsequent studies demonstrated that other dinoflagellate species lacked canonical histone proteins entirely. This is likely due to the types of techniques employed to assess the presence of histone proteins. Given that dinoflagellates have low protein content in the nucleus, there were no observations of histone proteins using chromatin spreading, microscopy, or chromatin digestion [106]. Advancements in next-generation sequencing helped aid in the "rediscovery" of histone proteins in dinoflagellates. Reports identified histone-like proteins in genomic and transcriptomics data, suggesting that some species encode and express H2A, H2B, H3, and H4 [107–110]. However, the sequences for these histone-like proteins are highly divergent and in particular, for histone 3, they lack key residues that are modified in other eukaryotic species. Further, their expression levels are very low, even 25 times lower in some species compared to eukaryotes [109]; expression is also independent of the cell cycle, calling the function of these histone-like proteins into question. In addition, it is still quite clear that dinoflagellates lack nucleosomes [111–113]. Potentially, H3 mediates some transcriptional elongation, though further investigation is warranted. It should be noted that in silico analysis showed several homologs for CME-like proteins and DNA-binding domains in several dinoflagellate species, though the function of these enzymes is unknown [114]. A deeper investigation demonstrating the isolation of histone proteins, their potential occupancy of DNA, and their functional significance is required for dinoflagellate species. For further reading, refer to the study by Marinov and Lynch and the work of Gornik et al. [106, 114].

17.3 Environmental Stress Alters Microalgae Epigenomes

Algae are highly adaptive to microenvironments, including responding to changes in temperature, pH, light intensities, mechanical forces, nutrients, and the presence of other organisms, pathogens, and toxins. Several groups have assessed the impact of nutrient deprivation, particularly nitrogen, on algae phenotypes, and measured subsequent changes in the epigenome. We reported that nitrogen starvation resulted in genome-wide demethylation of 5mC in the biofuel platform species *P. soloecismus*, resulting in decreased methylation on genes involved in lipid synthesis with a concomitant increase in lipid production. This was further corroborated by the treatment of algae cultivars with 5-aza-2′deoxycytidine, which imparted the same epigenomic and phenotypic effects as nitrogen depletion, suggesting an alternative approach for early induction of lipid synthesis [41]. Nitrogen starvation in culture conditions also imparted significant changes in DNA methylation and histone modifications in the diatom *P. tricornutum*. This environmental change induced hypomethylation (of 5mC) on genes involved in nitrogen metabolism, including a carbamoyl phosphate synthase and a nitrite reductase, and activation of the retrotransposon *Blackbeard* [115, 116]. The authors concluded that diatoms may be highly responsive to alterations in the environment, leading to loss of gene and TE methylation and inducing reversal of transcriptional repression. Indeed, reduced growth rates and chlorotic phenotypes in response to nitrate limitation were correlated with significant differential histone PTMs, including a loss of H3K9 trimethylation (~31%), loss of H3K4 dimethylation (~20%), and gain of H3K9/K14 acetylation (~16%) [94]. Genes with increased acetylation and decreased methylation had corresponding increased gene expression and included genes for nitrate and nitrite transport and nitrate metabolism. Increased occupancy of H3K4 dimethylation occurred on genes for desaturase enzymes, heme and chlorophyll synthesis, and light harvesting under low nitrate conditions [94]. Nitrogen starvation altered the histone modification landscape of the model green algae *C. reinhardtii* as well. Genes involved in triacylglycerol (fatty acid) accumulation and stress responses, including DGAT (diacylglycerol acyltransferase), had altered chromatin modifications, including transitions from Type II to Type IV promoters associated with increased transcription. Of particular interest was the increase of permissive histone modifications on H3, including K27ac, K4me3, K9me3, and K36me3 on the promoter of the novel transcription factor PSR1, which may mediate the transition to lipid accumulation in *C. reinhardtii* [84]. Comprehensive analysis of chromatin landscapes in these species, particularly under nutrient deprivation, is useful for identifying new regulators of metabolic pathways.

Further insight into other species, particularly marine phytoplankton, is needed to elucidate the role epigenetic modifications play in responding to nutrient availability. This is even more urgent for understanding the impact of climate change on global carbon cycling. The increase in carbon dioxide availability in the atmosphere may alter the productivity of the world's primary producers, thereby altering the amount of carbon fixed, nitrogen utilized, and oxygen produced. A recent study demonstrated that increased partial pressure of carbon dioxide (pCO_2) did not alter the physiological performance of *P. tricornutum* after 15 generations of exposure [117]. However, transcription of histone genes and genes involved in chromatin remodeling and structure were downregulated; this was concomitant with an increase in TE expression and genes for histone acetyltransferases, suggesting differential responsiveness to pCO_2. A series of long noncoding RNAs (lncRNAs) that were specifically responsive to increased CO_2 was identified: this finding underscores a potential regulatory role for a relatively uncharacterized component of the *P. tricornutum* genome [117]. Further epigenomic characterization, rather than simply transcriptomic analysis, is needed to understand the impact of altered carbon availability on this species and others.

Rising global temperatures is another component of climate change in addition to an increase in atmospheric CO_2. A 2011 assessment of the impact of heat shock suggested that chromatin remodeling occurs in response to this environmental change [80]. The heat shock response is regulated by heat shock proteins (HSP) that are typically activated via phosphorylation and bind to heat shock elements in the genome. The authors used ChIP and FAIRE techniques to determine how the binding of transcription factors impacted the chromatin landscape and regulated gene expression. They reported that two particular transcription factors, HSF1 and CRR1, which mediate heat shock and copper starvation responses, respectively, regulated gene expression by altering histone acetylation, histone methylation (particularly monomethylation of H3K4), nucleosome eviction, and loading or activation of polymerase. The mechanism of chromatin remodeling was unique and dependent upon target genes. Inactive target promoters embedded in closed chromatin received different chromatin remodeling than those that were weakly active within partially open chromatin. Histone deacetylation independent of transcription factors also occurred at particular promoters after heat shock, suggesting that there is an interplay between TFs and more general adaptive responses to changing environmental conditions in *C. reinhardtii*. More research on the effects of increased temperatures on species that occupy the global oceans, particularly diatoms, is needed to understand the impending impact of climate change on these organisms.

A recent study used an experimental evolution model of *C. reinhardtii* to determine how epigenetic processes contribute to adaptation. Cultures were exposed to high salt, low phosphorus, or high CO_2 environments and treated with a cocktail of chemicals to induce demethylation, including 5-aza-2′deoxycytidine and zebularine, both of which are cytosine analogs that cannot be methylated and L-ethionine, which outcompetes methionine for DNA methylation. Some cultures were also treated with 5-aza-2′deoxycytidine, or L-ethionine and a histone deacetylase inhibitor, Trichostatin A. The authors reported that inhibition of epigenetic modifications reduced adaptation to the three different environments. Changes in DNA methylation patterns occurred in evolved populations in response to environmental perturbation. These findings suggest that adaptation to environmental stress (whether positive or negative) is dependent upon the availability of epigenetic modifications to provide a means for rapid response and evolved phenotype [118].

These studies demonstrate that several species of microalgae are responsive to environmental perturbations through epigenetic modifications. Greater detail and genome-wide profiling are needed to understand the regulatory nature of epigenetic modifications on processes controlling physiology, particularly in light of the changes in nutrients, temperature, pH, pathogens, and toxins as climate change progressively impacts aquatic ecosystems.

17.4 Conclusions and Perspectives

Given algae's importance in biological systems across the world, we have fairly limited knowledge of what controls their genome organization, structure, and function. While analysis of epigenetic processes in a select number of microalgae has revealed interesting findings, further investigation is warranted, particularly for determining how these microorganisms will respond to changing global environmental conditions. From the studies that have been performed, we have learned that epigenetic processes are ancient, and their functionality is highly dependent upon the needs of the organism.

A global consensus on DNA methylation quantity and localization is likely not appropriate for all microalgae species. The methylation context, location, and percentage (so far) all vary significantly. This is likely due to the highly divergent nature of algae genomes. The quantity and distribution of this methylation is highly variable, with no one DNA methylation pattern from one alga predictive to the next. Some groups of algae have distinctive patterns of DNA methylation, including more 5mC methylation in CpG contexts (save for one species), potential periodicity in methylation regardless of genomic context, and high quantities of methylation compared to their small genome sizes. Other algae display methylation patterns similar to plants, showing robust methylation on TEs, while others demonstrate methylation patterns similar to mammals. Further, DNA methylation in algae is likely not dependent on genome size. *E. siliculosus*, one of the largest macroalgae with a 214 MB genome, completely lacks DNA methylation, while *Ostreococcus lucimarinus*, the smallest known picoeukaryote, contains a whopping 22% CpG methylation in its 13.2 MB genome. Despite this interesting finding, there may be a correlation between the amount of transposable element content in an algae genome and the amount of DNA methylation. Many microalgae species seem to have DNA methylation located within TEs and other repetitive regions; this localization is correlated with lower gene expression of these genomic elements, suggesting that DNA methylation represses transcriptional activation of TEs. Further, DNA methylation may provide genomic protection: *Chlorella variabilis* NC64A has approximately 80% CpG methylation likely due to the high quantity of viral DNA in its genome. DNA methylation is found in multiple genome contexts, including gene bodies, and the pattern of periodicity found by Huff and Zilberman suggests a potentially stabilizing role of this epigenetic modification [15]. Further, while DNA methylation is responsive to extreme environmental fluctuations, ranging from deliberate inhibition via drugs or severe nutrient deprivation, it is relatively stable throughout the life cycle of most algae assessed. While the quantity of DNA methylation is highly variable, when it occurs, it is likely found in CpG contexts, within TEs (and potentially other genomic regions), and potentially functions to either stabilize the genome or repress foreign DNA. Determining dynamic DNA methylation, particularly in response to changing environmental conditions like temperature, increased carbon dioxide, and nutrient availability would help elucidate the role of this modification in controlling physiological responses to climate change.

Histone PTMs play a paramount and dynamic role in the life cycles of the algae assessed. Deep profiling using mass spectrometry, immunoblotting, or ChIP-seq has been performed in seven microalgae species, including *C. reinhardtii* [84], *C. merolae* [92], *E. siliculosus* [54], *N. oceanica* [102], *P. tricornutum* [94, 95], *T. pseudonana* [93], and *V. carteri* [97]. *In silico* analysis of the presence of CMEs and histone genes has been performed in an additional 10 species including *B. prasinos*, *C. paradoxa*, *Chlorella sorokiniana*, *C. subellipsoidea*, *C. variabilis* NC64A, *Galdieria sulphuraria*, *M.* sp. RC299, *M. pusilla* CCMP1545, *O. lucimarinus*, and *O. tauri* [15, 47, 92]. Thus, our collective knowledge of histone proteins and their PTMs in microalgae is fairly limited. From these studies, we can presume that most algae, except for dinoflagellates, have conserved canonical histone proteins with basic amino acid residues readily available for PTMs. Some of these histone PTMs are common to the microalgae assessed, in particular, H3K27me3, though the presence of a specific histone PTM is highly dependent on the presence and expression of associated CMEs. The localization and potential function of these histone PTMs are also highly dependent on the organism, with some expressing histone PTMs similar to plants (*C. reinhardtii*), while others have functionality more akin to mammals (*E. siliculosus*). Further, novel histone PTMs that have yet to be identified in mammalian and plant species have been found in microalgae [94]. Thus, the algal lineage of histone PTMs is useful for understanding the emergence of particular modifications at certain periods of evolution, which may be correlated to environment and location. For further reading, please see Vigneau and Borg [119]. The functionality of histone modifications, particularly methylation of histone 3, is variable; however, histone acetylation, for its part, seems to have a conserved permissive function, always correlated with higher gene expression for the regions it occupies. We can glean that most histone PTMs assessed have some functionality in either protecting the genome or responding to environmental variability for altering phenotype and adaptation, particularly for controlling morphotype and responses to the photoperiod and nutrient availability. Further investigation on the types and functions of histone PTMs will likely depend upon the research question posed, particularly for those interested in identifying epigenetic regulation of specific microalgae functions. For example, deeper characterization of histone PTMs in microalgae species prominent in the bioproduct and biofuel industries may elucidate epigenetic mechanisms that can be manipulated to drive phenotypes toward industry goals.

Investigation of epigenetic modifications in species that may be used for various purposes in aquaculture would similarly provide useful information about the regulation of these organisms' functions. Potential strategies for epigenetic bioengineering of algae for either the bioeconomy (bioproduct) or aquaculture industries range from small molecule additions in larger cultures to CRISPR-based technologies [120, 121]. These could include transient increases in certain histone PTMs or removal of methylation on certain CpG sites that impact the physiology of algae, thereby engineering more suitable feedstocks for fisheries, aquaculture, and bioproduct synthesis. Interestingly, a common functionality of histone PTMs is their control of adaptation. Microalgae that lack the ability to perform epigenetic modifications have difficulty adapting to environmental variability. This finding underscores the importance of epigenetic modifications in the evolution and adaptation of species; thus, epigenetics assessment is essential for a deeper understanding of how these organisms will respond to the rapidly changing global environment.

References

1 Guiry, M.D. (2012). How many species of algae are there? *J. Phycol.* **48**: 1057–1063.
2 Anderson, D.M., Cembella, A.D., and Hallegraeff, G.M. (2012). Progress in understanding harmful algal blooms: paradigm shifts and new technologies for research, monitoring, and management. *Ann. Rev. Mar. Sci.* **4**: 143–176. https://doi.org/10.1146/annurev-marine-120308-081121.
3 Anderson, D., Richlen, M., and Lefebvre, K. (2018). Harmful Algal Blooms in the Arctic. Annual Artic Report Card, NOAA.
4 Ahmad, A., Hassan, S.W., and Banat, F. (2022). An overview of microalgae biomass as a sustainable aquaculture feed ingredient: food security and circular economy. *Bioengineered* **13**: 9521–9547.
5 Nagappan, S., Das, P., AbdulQuadir, M. et al. (2021). Potential of microalgae as a sustainable feed ingredient for aquaculture. *J. Biotechnol.* **341**: 1–20.
6 Tirichine, L. and Bowler, C. (2011). Decoding algal genomes: tracing back the history of photosynthetic life on earth. *Plant J.* **66**: 45–57.
7 AlgaeBase. World-wide electronic publication, National University of Ireland, Galway. www.algaebase.org (accessed March 2022).
8 Hanschen, E.R. and Starkenburg, S.R. (2020). The state of algal genome quality and diversity. *Algal Research* **50** (5): 101968. https://doi.org/10.1016/j.algal.2020.101968.
9 Capuano, F., Mülleder, M., Kok, R. et al. (2014). Cytosine DNA methylation is found in *Drosophila melanogaster* but absent in *Saccharomyces cerevisiae*, *Schizosaccharomyces pombe*, and other yeast species. *Anal. Chem.* **86**: 3697–3702.
10 Tang, Y., Gao, X.D., Wang, Y. et al. (2012). Widespread existence of cytosine methylation in yeast DNA measured by gas chromatography/mass spectrometry. *Anal. Chem.* **84**: 7249–7255.
11 Otto, G. (2019). 6mA in stress tolerance across generations. *Nat. Rev. Mol. Cell Biol.* **20**: 66–67.
12 Kouzarides, T. (2007). Chromatin modifications and their function. *Cell* **128** (4): 693–705. https://doi.org/10.1016/j.cell.2007.02.005.
13 Feng, S., Cokus, S., Zhang, X. et al. (2010). Conservation and divergence of methylation patterning in plants and animals. *PNAS* **107** (19): 8689–8694.
14 Hovde, B.T., Hanschen, E., Steadman Tyler, C.R. et al. (2018). Genomic characterization reveals significant divergence within *Chlorella sorokiniana* (Chlorellales, Trebouxiophyceae). *Algal Research* **35**: 449–461.
15 Huff, J.T. and Zilberman, D. (2014). Dnmt1-independent CG methylation contributes to nucleosome positioning in diverse eukaryotes. *Cell* **156**: 1286–1297.
16 Wendte, J.M., Zhang, Y., Ji, L. et al. (2019). Epimutations are associated with chromomethylase-3-induced *de novo* DNA methylation. *Elife* **8**: e47891.
17 Niederhuth, C.E. and Schmitz, R.J. (2017). Putting DNA methylation in context: from genomes to gene expression in plants. *Biochim. Biophys. Acta – Gene Regul. Mech.* **1860**: 149–156.
18 Gallego-Bartolomé, J. (2020). DNA methylation in plants: mechanisms and tools for targeted manipulation. *New Phytol.* **227**: 38–44.
19 Sánchez-Romero, M.A. and Casadesús, J. (2020). The bacterial epigenome. *Nat. Rev. Microbiol.* **18**: 7–20.
20 O'Brown, Z.K., Boulias, K., Wang, J. et al. (2019). Sources of artifact in measurements of 6mA and 4mC abundance in eukaryotic genomic DNA. *BMC Genomics* **20**: 445.
21 Fu, Y., Luo, G.Z., Chen, K. et al. (2015). N6-methyldeoxyadenosine marks active transcription start sites in *Chlamydomonas*. *Cell* **161**: 879–892.
22 Lopez, D.A., Hamaji, T., Kropat, J. et al. (2015). Dynamic changes in the transcriptome and methylome of *Chlamydomonas reinhardtii* throughout its life cycle. *Plant Physiol.* **169**: 2730–2743.
23 Hattman, S., Kenny, C., Berger, L., and Pratt, K. (1978). Comparative study of DNA methylation in three unicellular eucaryotes. *J. Bacteriol.* **135**: 1156–1157.
24 Feng, T.Y. and Chiang, K.S. (1984). The persistence of maternal inheritance in *Chlamydomonas* despite hypomethylation of chloroplast DNA induced by inhibitors. *PNAS* **81**: 3438–3442.
25 Cerutti, H., Johnson, A.M., Gilham, N.W., and Boynton, J.E. (1997). Epigenetic silencing of a foreign gene in nuclear transformants of chlamydomonas. *Plant Cell* **9**: 925–945.
26 Merchant, S.S., Prochnik, S.E., Vallon, O. et al. (2007). The *Chlamydomonas* genome reveals the evolution of key animal and plant functions. *Science* **318**: 245–250.

27 Wu-Scharf, D., Jeong, B., Zhang, C., and Cerutti, H. (2000). Transgene and transposon silencing in *Chlamydomonas reinhardtii* by a DEAH-box RNA helicase. *Science* **290**: 1159–1162.

28 Jeong Br, B., Wu-Scharf, D., Zhang, C., and Cerutti, H. (2002). Suppressors of transcriptional transgenic silencing in *Chlamydomonas* are sensitive to DNA-damaging agents and reactivate transposable elements. *PNAS* **99**: 1076–1081.

29 Falkowski, P.G., Schofield, O., Katz, M.E. et al. (2004). Why is the land green and the ocean red? In: *Coccolithophores* (ed. H.R. Thierstein and J.R. Young). Berlin, Heidelberg: Springer https://doi.org/10.1007/978-3-662-06278-4_16.

30 Wehr, J.D. and Sheath, R.G. (2003). Freshwater habitats of algae. In: *Freshwater Algae of North America: Ecology and Classification*. Elsevier Aquatic Ecology. https://doi.org/10.1016/B978-0-12-741550-5.X5000-4.

31 Prochnik, S.E., Umen, J., Nedelcu, A.M. et al. (2010). Genomic analysis of organismal complexity in the multicellular green alga *Volvox carteri*. *Science* **329**: 223–226.

32 Babinger, P., Kobl, I., Mages, W. et al. (2001). A link between DNA methylation and epigenetic silencing in transgenic *Volvox carteri*. *Nucleic Acids Res.* **29**: 1261–1271.

33 Babinger, P., Völkl, R., Cakstina, I. et al. (2007). Maintenance DNA methyltransferase (Met1) and silencing of CpG-methylated foreign DNA in *Volvox carteri*. *Plant Mol. Biol.* **63**: 325–336.

34 Zemach, A., McDaniel, I.E., Silva, P., and Zilberman, D. (2010). Genome-wide evolutionary analysis of eukaryotic DNA methylation. *Science* **328**: 916–919.

35 Blanc, G., Duncan, G., Agarkova, I. et al. (2010). The *Chlorella variabilis* NC64A genome reveals adaptation to photosymbiosis, coevolution with viruses, and cryptic sex. *Plant Cell* **22**: 2943–2955.

36 Quispe, C.F., Esmael, A., Sonderman, O. et al. (2017). Characterization of a new chlorovirus type with permissive and non-permissive features on phylogenetically related algal strains. *Virology* **500**: 103–113.

37 Van Etten, J.L., Lane, L.C., and Dunigan, D.D. (2010). DNA viruses: the really big ones (giruses). *Annu. Rev. Microbiol.* **64**: 83–99.

38 Rowe, J.M., Jeanniard, A., Gurnon, J.R. et al. (2014). Global analysis of *Chlorella variabilis* NC64A mRNA profiles during the early phase of *Paramecium bursaria* chlorella virus-1 infection. *PLoS One* **9**: e90988–e90988.

39 Gonzalez-Esquer, C.R., Twary, S.N., Hovde, B.T., and Starkenburg, S.R. (2018). Nuclear, chloroplast, and mitochondrial genome sequences of the prospective microalgal biofuel strain *Picochlorum soloecismus*. *Genome Announc.* **6**: 1–2.

40 Gonzalez-Esquer, C.R., Wright, K., Sudasinghe, N. et al. (2019). Demonstration of the potential of *Picochlorum soloecismus* as a microalgal platform for the production of renewable fuels. *Algal Res.* **43**: 101658.

41 Steadman, C.R., Banerjee, S., Sanders, C.K. et al. (2020). Epigenetic manipulation of DNA methylation in *Picochlorum soloecismus* alters productivity. *Front. Genet. Epigenomics* **11**: 560444.

42 Moreau, H., Verhelst, B., Couloux, A. et al. (2012). Gene functionalities and genome structure in *Bathycoccus prasinos* reflect cellular specializations at the base of the green lineage. *Genome Biol.* **13**: R74.

43 Palenik, B., Grimwood, J., Aerts, A. et al. (2007). The tiny eukaryote *Ostreococcus* provides genomic insights into the paradox of plankton speciation. *PNAS* **104**: 7705–7710.

44 Bowler, C., Allen, A.E., Badger, J.H. et al. (2008). The *Phaeodactylum* genome reveals the evolutionary history of diatom genomes. *Nature* **456**: 239–244.

45 Tirichine, L., Rastogi, A., and Bowler, C. (2017). Recent progress in diatom genomics and epigenomics. *Curr. Opin. Plant Biol.* **36**: 46–55.

46 Veluchamy, A., Lin, X., Maumus, F. et al. (2014). Insights into the role of DNA methylation in diatoms by genome-wide profiling in *Phaeodactylum tricornutum*. *Nat. Commun.* **4**: 2091. https://doi.org/10.1038/ncomms3091.

47 Maumus, F., Rabinowicz, P., Bowler, C., and Rivarola, M. (2011). Stemming epigenetics in marine *Stramenopiles*. *Curr. Genomics* **12**: 357–370.

48 Armbrust, E.V., Berges, J., Bowler, C. et al. (2004). The genome of the diatom *Thalassiosira pseudonana*: ecology, evolution, and metabolism. *Science* **306**: 79–86.

49 Wohlrab, S., Tillmann, U., Cembella, A., and John, U. (2016). Trait changes induced by species interactions in two phenotypically distinct strains of a marine dinoflagellate. *ISME J.* **10**: 2658–2668.

50 Rastogi, A., Lin, X., Lombard, B. et al. (2015). Probing the evolutionary history of epigenetic mechanisms: what can we learn from marine diatoms. *AIMS Genet.* **2** (3): 173–191.

51 Mock, T., Otillar, R., Strauss, J. et al. (2017). Evolutionary genomics of the cold-adapted diatom *Fragilariopsis cylindrus*. *Nature* **541**: 536–540.

52 Kang, S.H. and Fryxell, G.A. (1992). *Fragilariopsis cylindrus* (Grunow) Krieger: the most abundant diatom in water column assemblages of Antarctic marginal ice-edge zones. *Polar Biol.* **12**: 609–627.

53 Cock, J.M., Sterck, L., Rouze, P. et al. (2010). The *Ectocarpus* genome and the independent evolution of multicellularity in brown algae. *Nature* **465**: 617–621.

54 Bourdareau, S., Tirichine, L., Lombard, B. et al. (2021). Histone modifications during the life cycle of the brown alga *Ectocarpus*. *Genome Biol.* **22**: 12.

55 Fan, X., Han, W., Teng, L. et al. (2020). Single-base methylome profiling of the giant kelp *Saccharina japonica* reveals significant differences in DNA methylation to microalgae and plants. *New Phytol.* **225**: 234–249.

56 Ye, N., Zhang, X., Miao, M. et al. (2015). *Saccharina* genomes provide novel insight into kelp biology. *Nat. Commun.* **6**: 6986.

57 Teng, L., Han, W., Fan, X. et al. (2021). Integrative analysis of chloroplast DNA methylation in a marine alga – *Saccharina japonica*. *Plant Mol. Biol.* **105**: 611–623.

58 Gobler, C.J., Berry, D., Dyhrman, S. et al. (2011). Niche of harmful alga *Aureococcus anophagefferens* revealed through ecogenomics. *PNAS* **108**: 4352–4357.

59 Sibbald, S.J., Lawton, M., and Archibald, J.M. (2021). Mitochondrial genome evolution in Pelagophyte algae. *Genome Biol. Evol.* **13**: evab018.

60 Broecker, W. and Clark, E. (2009). Ratio of coccolith $CaCO_3$ to foraminifera $CaCO_3$ in late Holocene deep sea sediments. *Paleoceanography* **24**: PA3205. https://doi.org/10.1029/2009PA001731.

61 Segev, E., Wyche, T., Kim, K. et al. (2016). Dynamic metabolic exchange governs a marine algal-bacterial interaction. *Elife* **5**: e17473.

62 Haunost, M., Riebesell, U., and Bach, L.T. (2013). The calcium carbonate shell of *Emiliania huxleyi* provides limited protection against viral infection. *Front. Mar. Sci.* **7**: 735.

63 Read, B.A., Kegel, J., Klute, M. et al. Pan genome of the phytoplankton *Emiliania* underpins its global distribution. *Nature* **499**: 209–213.

64 Matsuzaki, M., Misumi, O., Shin-i, T. et al. (2004). Genome sequence of the ultrasmall unicellular red alga *Cyanidioschyzon merolae* 10D. *Nature* **428**: 653–657.

65 Miyagishima, S.Y. and Tanaka, K. (2021). The unicellular red alga *Cyanidioschyzon merolae* – the simplest model of a photosynthetic eukaryote. *Plant Cell Physiol.* **62**: 926–941.

66 Kato, S., Okamura, E., Matsunaga, T. et al. (2019). *Cyanidioschyzon merolae* aurora kinase phosphorylates evolutionarily conserved sites on its target to regulate mitochondrial division. *Commun. Biol.* **2**: 477.

67 Bercovici, A. and Vellekoop, J. (2017). Methods in paleopalynology and palynostratigraphy: an application to the K-Pg boundary. In: *Terrestrial Depositional Systems: Deciphering Complexities Through Multiple Stratigraphic Methods* (ed. K.E. Zeigler and W.G. Parker), 127–164. Elsevier https://doi.org/10.10d16/B978-0-12-803243-5.00003-0.

68 Mendoza, A., Bonnet, A., Vargas-Landin, D. et al. (2018). Recurrent acquisition of cytosine methyltransferases into eukaryotic retrotransposons. *Nat. Commun.* **9**: 1341.

69 Shoguchi, E., Shinzato, C., Kawashima, T. et al. (2013). Draft assembly of the *Symbiodinium minutum* nuclear genome reveals dinoflagellate gene structure. *Curr. Biol.* **23**: 1399–1408.

70 Lin, S., Cheng, S., Song, B. et al. (2015). The *Symbiodinium kawagutii* genome illuminates dinoflagellate gene expression and coral symbiosis. *Science* **350**: 691–694.

71 Hou, Y. and Lin, S. (2009). Distinct gene number-genome size relationships for eukaryotes and non-eukaryotes: gene content estimation for dinoflagellate genomes. *PLoS One* **4**: e6978.

72 Robreau, G. and Le Gal, Y. (1975). Isolation and partial characterization of *Chlamydomonas reinhardtii* chromatin. *Biochimie* **57**: 703–710.

73 Morris, R.L., Keller, L.R., Zweidler, A., and Rizzo, P.J. (1990). Analysis of *Chlamydomonas reinhardtii* histones and chromatin. *J. Protozool.* **37**: 117–123.

74 Fabry, S., Muller, K., Lindauer, A. et al. (1995). The organization structure and regulatory elements of *Chlamydomonas* histone genes reveal features linking plant and animal genes. *Curr. Genet.* **28**: 333–345.

75 Waterborg, J.H., Robertson, A.J., Tatar, D.L. et al. (1995). Histones of *Chlamydomonas reinhardtii*: synthesis, acetylation, and methylation. *Plant Physiol.* **109**: 393–407.

76 Waterborg, J.H. (1998). Dynamics of histone acetylation in *Chlamydomonas reinhardtii*. *J. Biol. Chem.* **273**: 27602–27609.

77 Morris, R.L., Salinger, A.P., and Rizzo, P.J. (1999). Analysis of lysine-rich histones from the unicellular green alga *Chlamydomonas reinhardtii*. *J. Eukaryot. Microbiol.* **46**: 648–654.

78 Salomé, P.A. and Merchant, S.S. (2021). Co-expression networks in *Chlamydomonas* reveal significant rhythmicity in batch cultures and empower gene function discovery. *Plant Cell* **33**: 1058–1082.

79 Zones, J.M., Blaby, I.K., Merchant, S.S., and Umen, J.G. (2015). High-resolution profiling of a synchronized diurnal transcriptome from *Chlamydomonas reinhardtii* reveals continuous cell and metabolic differentiation. *Plant Cell* **27**: 2743–2769.

80 Strenkert, D., Schmollinger, S., Sommer, F. et al. (2011). Transcription factor-dependent chromatin remodeling at heat shock and copper-responsive promoters in *Chlamydomonas reinhardtii*. *Plant Cell* **23**: 2285–2301.

81 van Dijk, K., Marley, K.E., Jeong, B. et al. (2005). Monomethyl histone 3 lysine 4 as an epigenetic mark for silenced euchromatin in *Chlamydomonas*. *Plant Cell* **17**: 2439–2453.

82 Casas-Mollano, J.A., Jeong, B.-R., Xu, J. et al. (2008). The MUT9p kinase phosphorylates histone H3 threonine 3 and is necessary for heritable epigenetic silencing in *Chlamydomonas*. *PNAS* **105**: 6486–6491.

83 Casas-Mollano, J.A., van Dijk, K., Eisenhart, J., and Cerutti, H. (2007). SET3p monomethylates histone H3 on lysine 9 and is required for the silencing of tandemly repeated transgenes in *Chlamydomonas*. *Nucleic Acids Res.* **35**: 939–950.

84 Ngan, C.Y., Wong, C., Choi, C. et al. (2015). Lineage-specific chromatin signatures reveal a regulator of lipid metabolism in microalgae. *Nat. Plants* **1**: 1–11.

85 Shaver, S., Casas-Mollano, J.A., Cerny, R.L., and Cerutti, H. (2010). Origin of the polycomb repressive complex 2 and gene silencing by an E(z) homolog in the unicellular alga *Chlamydomonas*. *Epigenetics* **5**: 301–312.

86 Neupert, J., Gallaher, S., Lu, Y. et al. (2020). An epigenetic gene silencing pathway selectively acting on transgenic DNA in the green alga *Chlamydomonas*. *Nat. Commun.* **11**: 6269.

87 Khan, A., Eikani, C.K., Khan, H. et al. (2018). Characterization of *Chlamydomonas reinhardtii* core histones by top-down mass spectrometry reveals unique algae-specific variants and post-translational modifications. *J. Proteome Res.* **17**: 23–32.

88 Rommelfanger, S.R., Zhou, M., Shaghasi, H. et al. (2021). An improved top-down mass spectrometry characterization of *Chlamydomonas reinhardtii* histones and their post-translational modifications. *J. Am. Soc. Mass Spectrom.* **32**: 1671–1688.

89 Strenkert, D., Schmollinger, S., Gallaher, S. et al. (2019). Multiomics resolution of molecular events during a day in the life of *Chlamydomonas*. *PNAS* **116**: 2374–2383.

90 Kim, E.-J., Ma, X., and Cerutti, H. (2015). Gene silencing in microalgae: mechanisms and biological roles. *Bioresour. Technol.* **184**: 23–32.

91 Mujtaba, S., Manzur, K., Gurnon, J. et al. (2008). Epigenetic transcriptional repression of cellular genes by a viral SET protein. *Nat. Cell Biol.* **10**: 1114–1122.

92 Mikulski, P., Komarynets, O., Fachinelli, F. et al. (2017). Characterization of the polycomb-group mark H3K27me3 in unicellular algae. *Front. Plant Sci.* **8**: 607.

93 Tirichine, L., Lin, X., Thomas, Y. et al. (2014). Histone extraction protocol from the two model diatoms *Phaeodactylum tricornutum* and *Thalassiosira pseudonana*. *Mar. Genomics* **13**: 21–25.

94 Veluchamy, A., Ragosti, A., Lin, X. et al. (2015). An integrative analysis of post-translational histone modifications in the marine diatom *Phaeodactylum tricornutum*. *Genome Biol.* **16**: 1–18. https://doi.org/10.1186/s13059-015-0671-8.

95 Zhao, X., Ragosti, A., Deton Cabanillas, A. et al. (2021). Genome wide natural variation of H3K27me3 selectively marks genes predicted to be important for cell differentiation in *Phaeodactylum tricornutum*. *New Phytol.* **229**: 3208–3220.

96 Müller, K., Lindauer, A., Brüderlein, M., and Schmitt, R. (1990). Organization and transcription of *Volvox* histone-encoding genes: similarities between algal and animal genes. *Gene* **93**: 167–168.

97 Lindauer, A., Müller, K., and Schmitt, R. (1993). Two histone H1-encoding genes of the green alga *Volvox carteri* with features intermediate between plant and animal genes. *Gene* **129**: 59–68.

98 Müller, K. and Schmitt, R. (1988). Histone genes of *Volvox carteri*: DNA sequence and organization of two H3-H4 gene loci. *Nucleic Acids Res.* **16**: 4121–4136.

99 Bradley, D.M. (1974). Histone analysis in *Volvox*. *FEBS Lett.* **41**: 219–222.

100 Wei, L. and Xu, J. (2018). Optimized methods of chromatin immunoprecipitation for profiling histone modifications in industrial microalgae *Nannochloropsis* spp. *J. Phycol.* **54** (3): 358–367.

101 Guo, L., Liang, S., Zhang, Z. et al. (2019). Genome assembly of *Nannochloropsis oceanica* provides evidence of host nucleus overthrow by the symbiont nucleus during speciation. *Commun. Biol.* **2**: 249.

102 Poliner, E., Panchy, N., Newton, L. et al. (2015). Transcriptional coordination of physiological responses in *Nannochloropsis oceanica* CCMP1779 under light/dark cycles. *Plant J.* **83**: 1097–1113.

103 Rae, P.M. (1973). 5-Hydroxymethyluracil in the DNA of a dinoflagellate. *PNAS* **70**: 1141–1145.

104 Rizzo, P.J. and Cox, E.R. (1977). Histone occurrence in chromatin from *Peridinium balticum*, a binucleate dinoflagellate. *Science* **198**: 1258–1260.

105 Rizzo, P.J. and Burghardt, R.C. (1982). Histone-like protein and chromatin structure in the wall-less dinoflagellate *Gymnodinium nelsoni*. *Biosystems* **15**: 27–34.

106 Gornik, S.G., Hu, I., Lassadi, I., and Waller, R.F. (2019). The biochemistry and evolution of the dinoflagellate nucleus. *Microorganisms* **7**: 245.

107 Okamoto, O.K. and Hastings, J.W. (2003). Genome-wide analysis of redox-regulated genes in a dinoflagellate. *Gene* **321**: 73–81.

108 Hackett, J.D., Scheetz, T., Yoon, H. et al. (2005). Insights into a dinoflagellate genome through expressed sequence tag analysis. *BMC Genomics* **6**: 80.

109 Roy, S. and Morse, D. (2012). A full suite of histone and histone modifying genes are transcribed in the dinoflagellate lingulodinium. *PLoS One* **7** (4): e34340. https://doi.org/10.1371/journal.pone.0034340.

110 Riaz, S. and Sui, Z. (2018). Molecular cloning, transcriptome profiling, and characterization of histone genes in the dinoflagellate *Alexandrium pacificum*. *J. Microbiol. Biotechnol.* **28**: 1185–1198.

111 Bodansky, S., Mintz, L.B., and Holmes, D.S. (1979). The mesokaryote *Gyrodinium cohnii* lacks nucleosomes. *Biochem. Biophys. Res. Commun.* **88**: 1329–1336.

112 Herzog, M. and Soyer, M.O. (1981). Distinctive features of dinoflagellate chromatin: absence of nucleosomes in a primitive species *Prorocentrum micans E*. *Eur. J. Cell Biol.* **23**: 295–302.

113 Díaz, M., de la Espina, S., Alverca, E. et al. (2005). Organization of the genome and gene expression in a nuclear environment lacking histones and nucleosomes: the amazing dinoflagellates. *Eur. J. Cell Biol.* **84**: 137–149.

114 Marinov, G.K. and Lynch, M. (2016). Diversity and divergence of dinoflagellate histone proteins. *G3 Genes|Genomes|Genetics* **6** (2): 397–422.

115 Maumus, F., Allen, A., Mhiri, C. et al. (2009). Potential impact of stress activated retrotransposons on genome evolution in a marine diatom. *BMC Genomics* **10**: 624.

116 Veluchamy, A., Lin, X., Maumus, F. et al. (2013). Insights into the role of DNA methylation in diatoms by genome-wide profiling in *Phaeodactylum tricornutum*. *Nat. Commun.* **4**: 2091.

117 Huang, R., Ding, J., Gao, K. et al. (2019). A potential role for epigenetic processes in the acclimation response to elevated pCO_2 in the model diatom *Phaeodactylum tricornutum*. *Front. Microbiol.* **9**: 3342.

118 Kronholm, I., Bassett, A., and Baulcombe, D. (2017). Epigenetic and genetic contributions to adaptation in *Chlamydomonas*. *Mol. Bio. Evol.* **34**: 2285–2306.

119 Vigneau, J. and Borg, M. (2021). The epigenetic origin of life history transitions in plants and algae. *Plant Reprod.* **34**: 267–285.

120 Patel, V.K., Soni, N., Prasad, V. et al. (2019). CRISPR-Cas9 system for genome engineering of photosynthetic microalgae. *Mol. Biotechnol.* **61** (8): 541–561.

121 Naduthodi, M.I.S., Südfeld, C., Avitzigiannis, E.K. et al. (2021). Comprehensive genome engineering toolbox for microalgae *Nannochloropsis oceanica* based on CRISPR-Cas systems. *ACS Synth. Biol.* **10** (12): 3369–3378. https://doi.org/10.1021/acssynbio.1c00329.

Part III

Implementation of Epigenetics in Aquaculture

18

Development of Epigenetic Biomarkers in Aquatic Organisms

Dafni Anastasiadi[1] *and Anne Beemelmanns*[2]

[1] *The New Zealand Institute for Plant and Food Research Limited, Nelson Research Centre, Nelson, New Zealand*
[2] *Institut de Biologie Intégrative et des Systèmes (IBIS), Université Laval, Québec City, Quebec, Canada*

18.1 Biomarkers

18.1.1 General Concepts and Definitions of Biomarkers

A biomarker is defined by the World Health Organization (WHO) as "any substance, structure, or process that can be measured in the body or its products, and which influences or predicts the incidence of the outcome of diseases" [1]. Biomarkers have been developed predominantly for human health purposes and are measurable (accurately and reproducibly) characteristics that serve as indicators of an individual's health by quantifying normal biologic processes, pathogenic processes, as well as allergic, stress and toxification responses, or pharmacologic responses to therapeutics [2]. Examples of biomarkers range from basic clinical measurements (e.g., blood pressure) to more complex molecular tests (e.g., gene and protein expression) using tissues or body fluids (e.g., blood, plasma, and urine) [2]. Herein, we adapt the definitions of clinical *diagnostic*, *prognostic*, *predictive*, *monitoring*, and *safety* biomarkers according to the US Food and Drug Administration (FDA) to aquatic organisms (see Box 18.1).

18.1.2 Biomarkers in Aquatic Organisms

In aquatic organisms, molecular, histological, biochemical, or physiological biomarkers have been used as indicators of animal health [3] and for the detection of environmental stress and pollution [4]. In aquaculture, diagnostic biomarkers for quantifying the impact of environmental stressors on an organism's physiology and health have received attention [5]. Levels of mRNA transcripts, proteins, or hormones in specific tissues and body fluids as well as alterations of the DNA sequence have been traditionally used as biomarkers [2, 3]. For common aquaculture species, biomarkers are primarily developed to detect diseases or infections [6], past exposures to pollutants [7], to assess life history characteristics [8], and to determine the physiological tolerance range to environmental stressors [9–11].

18.2 Epigenetic Biomarkers

The sensibility of epigenetic modifications to environmental signals and their relative stability over time suggests that they are very powerful biomarkers. Among animals, it was found that nutrition [12], exposure to toxins [13], pathogens [14], salinity [15, 16], temperature [11, 17–19], hypoxia [20, 21], and ocean acidification [22, 23], as well as domestication such as aquaculture farming [24–27] are associated with epigenetic changes within and between generations.

Given that epigenetic changes upon environmental influences can be stably associated with a particular phenotypic response throughout an individual's lifetime or even transgenerationally [28], epigenetic biomarkers (EBs) have the potential to serve as tools to predict the phenotypic outcome and facilitate accurate future assessments according to all types of

Epigenetics in Aquaculture, First Edition. Edited by Francesc Piferrer and Han-Ping Wang.

Box 18.1 Definitions

Bias: Difference between the expected predicted values and the correct values.
Biomarker: Any substance, structure, or process that can be measured in the body or its products that predict the incidence and/or the outcome of a specific condition (e.g., disease, disorder, infection, stress).
Diagnostic biomarker: Indicates the presence of a disease, stress condition, or a phenotypic trait of interest that can also provide information on the past or the present status.
DNA methylation: The enzymatic addition of a methyl group (CH_3) at 5′ carbon of cytosines (5mC) that may be of importance for the regulation of gene expression.
EBDP: Epigenetic biomarker development procedure.
Epigenetic biomarker (EB): Any induced epigenetic modification that can be used to detect phenotypic traits, environmental stressors, pollutants, external treatments, or diseases.
Histone modifications: Covalent posttranslational modifications (PTMs) to histones (e.g., methylation, phosphorylation, acetylation) that can affect gene expression through alterations of the chromatin structure.
Metastable epialleles: Environmentally responsive loci showing similar trends in epigenetic changes across tissues of the same individual, stable within lifetime and different between genetically identical individuals.
Monitoring biomarker: Provides information on the exposure to, or the effect of, a product or an environmental agent and reflects an ongoing status over time through repeated measures.
Noncoding RNAs (ncRNAs): Noncoding RNAs (ncRNAs), including microRNAs (miRNAs) that can regulate posttranscriptional gene expression responses.
Predictive biomarker: Identifies individuals with more probabilities than others for future development and progression of a disease or stress condition when exposed to a specific environment.
Prognostic biomarker: Predicts the outcome of a disease, stress condition, or phenotypic trait in exposed individuals with the condition or trait.
Safety biomarker: Indicates the likelihood, presence, or extent of toxicity as a harmful effect measured before or after exposure to a medical product or an environmental agent.
Variance: Variability of a model prediction for a given data point.

biomarkers (see Box 18.1). Pioneer studies in a clinical context have developed EBs for the prognosis of cancer, metabolic syndrome, infertility, pregnancy, allergy reactions, respiratory diseases, and neurodegenerative disorders [29].

EBs present several important properties and advantages over traditional molecular biomarkers (adapted from [29]):

1) EBs are highly sensitive to environmental influences and natural history.
2) EBs show cell-type specificity that can be used as a diagnostic tool to distinguish cell types.
3) EBs are generally stable in time in different cells, tissues, and body fluids.
4) EBs are detectable during the early stages of a particular condition.
5) EBs are measurable in all genomic contexts on individual and population levels.

In brief, EBs are useful as (i) highly sensitive diagnostic biomarkers (e.g., environmental stress exposures, diseases, age); (ii) prognostic biomarkers capable of estimating the likelihood of the future outcome (e.g., severity of disease, survival, sex, and aging); (iii) predictive biomarkers able to predict risks of future exposures or disease development; (iv) monitoring biomarkers to assess the effect of an exposure on an organism or an ecosystem; and (v) safety biomarkers to evaluate harmful effects. However, specific properties of EBs that need to be taken properly into account during *Epigenetic Biomarker Development Procedure (EBDP)* are that they (i) have dependencies on multiple levels (e.g., cell, tissue, genetics, sex, and age), which can cause biases, and (ii) can be reversible and change during tissue development and upon external stimuli. Consequently, the (iii) heterogeneous cellular composition of tissues makes their application challenging (i.e., single-cell sequencing technology could overcome this problem); and (iv) may rely on costly technologies (e.g., when antibody usage is involved) [29, 30].

18.2.1 DNA Methylation

DNA methylation is the enzymatic addition of a methyl group (CH_3) at the 5′ carbon of cytosines (5mC), in animals, generally relevant when followed by a guanine (CpG dinucleotide), and is one key epigenetic mark in the regulation of gene expression [31]. DNA methylation biomarkers have been successfully developed in clinical research to target disease conditions (e.g., cancer, disorders, and infections) and are currently used as noninvasive tools for the prognosis of cancer and other diseases [30, 32]. Disease susceptibility can be revealed when removal of specific epigenetic marks in progenitor cells leads to transcriptional changes that cause a disease or facilitate the progression of carcinogenesis [33].

DNA methylomes of aquatic organisms important for aquaculture such as rainbow trout (*Oncorhynchus mykiss*) [34], coho salmon (*Oncorhynchus kisutch*) [24, 27], Atlantic salmon (*Salmo salar*) [20, 35], half-smooth tongue sole (*Cynoglossus semilaevis*) [36], and the Pacific oyster (*Crassostrea gigas*) [37] are increasingly being assessed. DNA methylation changes have been used as diagnostic biomarkers for age in vertebrates including fish and lobsters [38–41], for sex and past environmental exposures [42–44]. DNA methylation changes upon climate change-related challenges in aquaculture were identified in Atlantic salmon and may guide future research to develop EBs for diagnosis and prediction of stress tolerance [20].

18.2.2 Histone Modifications

Histone modifications are covalent posttranslational modifications (PTMs) to histones (e.g., methylation, acetylation) that affect gene expression by alterations of the chromatin structure [45]. The quantitative detection of these histone PTMs, as well as the enzymes that catalyze them, have been explored for their potential as EBs and for the development of therapeutic medication in humans [46]. Histone PTMs are utilized as diagnostic biomarkers for cancer (e.g., pancreatic or colorectal cancer) [46] and diseases (e.g., Alzheimer or systematic lupus erythematosus) [30]. Thus, targeting histone modifications in liquid biopsies (e.g., blood and plasma) is a promising approach for the development of early diagnostic biomarkers for diseases or toxin exposures in aquatic organisms [47, 48]. However, in aquatic organisms, research for the development of histone PTMs-based EBs is still rare and more challenging as it relies on costly techniques using antibodies, which hamper their practical application as biomarkers.

18.2.3 Noncoding RNAs

Transcriptional responses can be regulated through small noncoding RNAs (ncRNAs) including microRNAs (miRNAs) [49]. Understanding the regulatory mechanism mediated by ncRNAs and the identification of new miRNA-based targets supports cancer research and the design of novel therapeutics and treatment monitoring [29, 50]. Interestingly, Keller et al. [51] developed a mathematical algorithm to accurately predict diseases based on dysregulated miRNA data measured in the blood of cancer patients. The human miRNA-associated disease database (HMDD) (http://www.cuilab.cn/hmdd) and Tools4miRs (https://tools4mirs.org) are fundamental resources to perform screenings of miRNA profiles.

In aquatic organisms, a number of studies have detected miRNAs indicative of reproductive status [52, 53], sensitive to pathogens [54, 55], immune-stimulants [55], as well as temperature fluctuations [56], hypoxia [57], nutrition and immunity [58], and are providing the foundation for the development of ncRNA-based biomarkers. In addition, circular RNAs (circRNAs) have been suggested as strong candidate EBs for muscle growth and immune responses in aquatic organisms [59]. For example, miRNAs involved in regulating pigmentation or immune responses in oysters (*C. gigas*) may prove to be valuable biomarkers to control coloration or prevent diseases [60].

18.3 Development of Epigenetic Biomarkers

To develop EBs in aquatic organisms, some considerations need to be accounted for. Advances in methodologies for the detection of epigenetic marks and the availability and accessibility of statistical methods, including machine learning (ML), as well as knowledge from more advanced fields (e.g., clinical), provide a good starting point to design EBDP in aquatic organisms, a field still in its infancy. In the following sections, we will discuss these aspects.

18.3.1 Considerations for the Design of Experiments for Biomarker Discovery

There are five key considerations to be applied in broad contexts: genetic variation, developmental stage, reproductive stage, sex, and tissue (cell type). Genetic variation must be considered since epigenetic modifications are interdependent of the underlying DNA sequence [61]. Even when epigenetic variants are inducible by the environment, the genome may define the range of epigenomic responses. Indeed, a strong family component has been found in all species tested [62, 63]. Thus, including samples from distinct families to encompass as much genetic variation as possible should be considered to avoid bias. Depending on the reproductive mode of each taxon [64], erasure and resetting of epigenetic patterns may occur during early development [65], while epigenetic changes occur during development and aging in adulthood. Furthermore, epigenetic patterns in somatic tissues may differ according to sex [26, 43]. Identifying the target life history stage and testing biomarkers across stages and sexes will allow for the detection of age- and sex-independent biomarkers. Lastly, epigenetic patterns are associated with specific tissues since they contribute to cell commitment and differentiation. To identify tissue-independent EBs, several representative tissues need to be investigated simultaneously, including tissues that can be sampled nonlethally and in a minimally invasive manner. The screening could target *metastable epialleles*: environmentally responsive loci showing similar trends in epigenetic changes across tissues of the same individual, stable within its lifetime, and regardless of its genotype. Metastable epialleles have been described in mammals [66–68] and fish [69], and are considered important to deduce ancient experiences [66]. For example, in humans, epigenetic patterns in six metastable epialleles may help deduce whether children were born during the rainy or the dry season in Gambia [67, 68], and in the European sea bass (*Dicentrarchus labrax*), DNA methylation of nine metastable epialleles can be used to infer whether fish experienced a marine heatwave during juvenile stages [69].

18.3.2 Epigenetic Biomarker Discovery Procedure in the Era of Omics

In recent years, the decreasing cost of sequencing has led to increased availability of molecular "omics" datasets. These data concern several molecular layers, such as genomics, transcriptomics, epigenomics, proteomics, and metabolomics. In humans, complete omics datasets are utilized to screen EBs for diseases or genetic disorders using ML approaches in large datasets of often thousands of patients, including longitudinal data [70]. In nonmodel aquatic species, the integration of several omics datasets is still rarely achieved, while obtaining large datasets of samples from a variety of families and populations remains a challenge. However, state-of-the-art methods for epigenetic data generation and analysis can be adapted to design assays for EBDP in aquatic organisms. Here, we suggest what we consider as a solid procedure for EBDP by combining epigenomics high-throughput sequencing (HTS) data and ML modeling. We provide guidelines for each step while considering variations of the timeline or essential steps based on resources available and objectives to attain.

18.3.3 General Systematic Approach

A systematic approach using epigenomics data for biomarker development in clinical applications was outlined by Bock [71] and adapted to test the model of the "Conserved Epigenetic Regulation of Sex" (CERS) in fish [72]. The approach described herein is an updated version after a decade of research. Our current approach focuses on the EBDP for aquatic organisms and includes a wide range of phenotypes and epigenetic data. The specific parameters may be adjusted per study depending on final goals and resources.

The EBDP involves tradeoffs between a number of samples and a number of epigenetic marks during each step to maximize resources (Figure 18.1). Each step represents a separate experimental event in terms of methods, both lab and statistical, and eventually samples. The starting point includes a low n_s, where n_s = number of samples, in combination with a whole-genome or genome-wide sequencing protocol that interrogates a high n_f, where n_f = number of features or epigenetic marks. ML and related techniques (e.g., feature selection) are not recommended in Steps 1 and 2 because the n_s is expected to be low and, thus, the statistical power reduced. Several representative accessible tissues should be included, although, in some cases, only one tissue might be enough, e.g., fin that is routinely sampled for genetics. After each step, informative epigenetic marks are filtered, while the n_s increases with the final goal to reach the lowest n_f to validate across the highest possible n_s. To achieve this workflow, whole-genome HTS approaches are appropriate in the first steps with decreasing importance toward the last steps, whereby the methods can be progressively replaced by locus-specific (targeted) approaches.

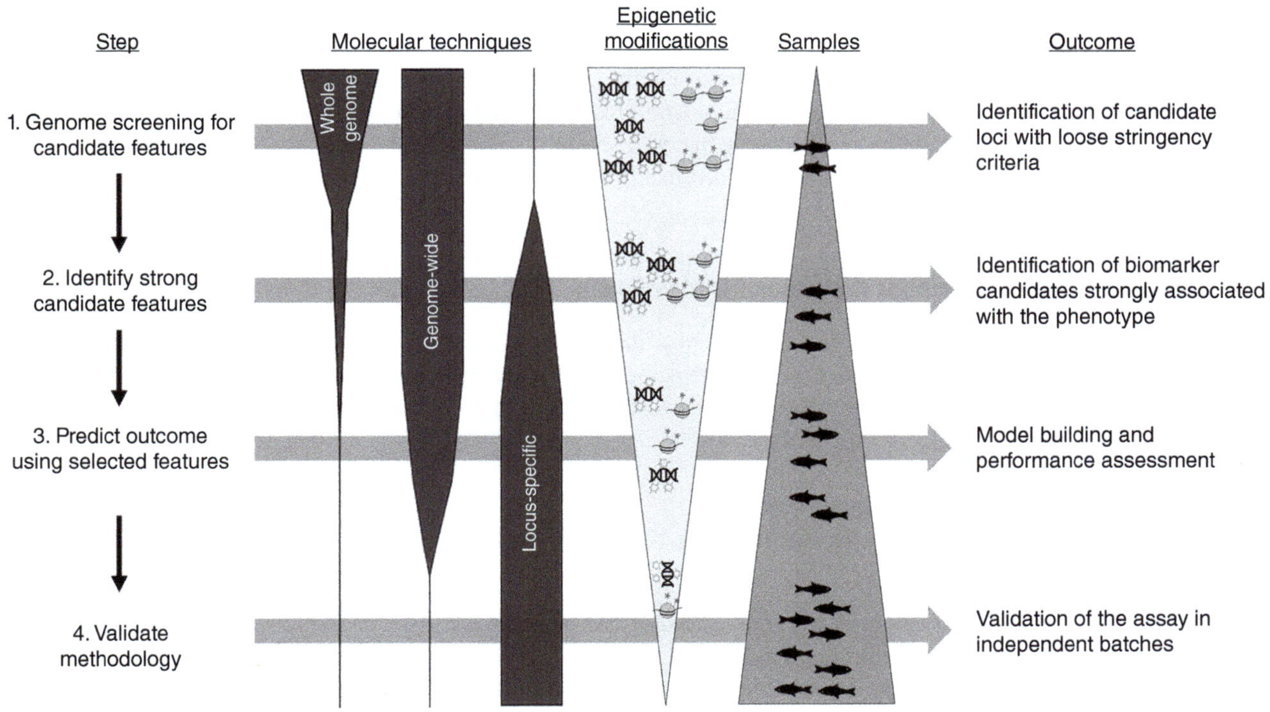

Figure 18.1 General systematic procedure for epigenetic biomarker (EB) development (EBDP). The procedure occurs vertically from Step 1 to Step 4. Depending on the specific context, steps can be skipped or adjusted. During each step, molecular techniques are selected to cover different parts of the genome, which gives information on several epigenetic modifications. From Step 1 towards Step 4, resources optimization can be achieved by using more samples and targeting a smaller number of epigenetic loci. The outcome expected from each step is described on the right pane.

18.3.3.1 Procedure Step-by-Step

Our proposed EBDP is divided into four general steps (Figure 18.1):

Step 1: Genome Screening for Candidate Features The procedure starts with hundreds of thousands of available features that need to be filtered for their informative value determined by statistical tests as explained later. The expected result is a list of candidates to be determined with loose stringency criteria, ideally in the order of n_f = 10^4–10^6 (Figure 18.1). If the outcome variable is qualitative or categorical, we are addressing the question of whether epigenetic features differ between two groups (e.g., healthy vs. disease, male vs. female, and stress vs. control). Then, statistical methods can be employed to detect differences between two groups (e.g., Student's *t*-tests, Mann–Whitney, Wilcoxon rank sum, or Wald tests). If the outcome variable is quantitative (e.g., growth or age), we are addressing the question of which epigenetic features show a higher association (correlation) with the outcome variable. In both cases, the candidate features can be identified using less stringent filter criteria such as raw *p*-values (e.g., $p < 1$) without any further multiple testing correction (e.g., Bonferroni and False Discovery Rate). In addition, they can be based on small differences between the two groups or in lower than standard (adjusted) R^2. These considerations aim to retain the maximum candidate biomarker loci.

Step 2: Identify Strong Candidate Features The selected features (n_f = 10^4–10^6) identified in a few samples are evaluated in a larger n_s (Figure 18.1). The objective is to identify features that strongly associate with the outcome variable and the final list is in the order of n_f = 10^1–10^4. The recommended statistical methods are similar to Step 1. However, during this step, the higher n_s and the previously filtered n_f allow for the identification of more robust associations. Thus, stringent criteria (e.g., $p < 0.001$ or after multiple testing corrections or high R^2) may be needed depending on the procedure and setup.

Step 3: Predict Outcome This is the core step for applying ML methods for EBDP and will be described in Section 18.3.4.

Step 4: Validate Methodology In previous steps, the minimum n_f capable of predicting the outcome variable with the highest accuracy and precision was identified. Here, validation of the final biomarker panel ($n_f = 1–10^2$) and prediction in independent batches of a large n_s need to be performed (Figure 18.1). The ML methods corroborated for technical validity in Step 3 (see Section 18.3.4) are used for prediction. The greatest breadth of biological variability possible should be sampled since the EB model should be generalizable in a wide range of genetic backgrounds, ages, developmental stages, tissues, geographic origins, sexes, populations, and, ultimately, species. This approach will allow the identification of any biological limitations of the method.

18.3.4 Machine Learning Methods for Epigenetic Biomarkers

The application of ML methods for EBDP holds great potential. In aquatic organisms, ML methods with epigenetic data have been applied only in limited studies, e.g., with DNA methylation markers to predict sex, age, and past temperature experiences. The application of ML approaches is expected to increase in the future. Thus, we describe herein the basic procedures for EBs (see: [73, 74]).

18.3.4.1 Classification of Machine Learning Methods

The first level of classification of ML methods relates to the amount and type of supervision that is required for model training (Figure 18.2).

1) **Unsupervised methods.** These methods build models that describe the data in the absence of a target outcome variable and of a priori knowledge of the presence of groups [73, 74]. Typically, they include clustering (e.g., k-means or hierarchical clustering) and Principal Component Analysis (PCA). In EBDP, a priori knowledge of groups should usually lead the procedure. However, unsupervised methods can be used at the initial steps to perform exploratory data analysis, an important part of the procedure to understand the dataset before ML development. These methods have been utilized for the discrimination of mixed cell populations from heterogeneous cancers [75] and could be similarly envisaged to be applied to discriminate heterogeneous cell types or even groups of conditions, e.g., different reproductive or nutritional statuses in aquatic organisms. Furthermore, unsupervised methods can be used during feature selection as part of building a predictive model (see Section 18.3.4.2).
2) **Supervised methods.** These methods construct models that are predictive [73, 74]. The dataset is used to construct or train a model that can predict the outcome variable of unseen data. More specifically, supervised methods learn a target function (f) that best maps input variables (X) to an output variable (Y) based on example input–output pairs: $Y = f(X)$. Supervised methods are divided into (i) classification techniques to predict a categorical outcome and (ii) regression techniques to predict a quantitative outcome.

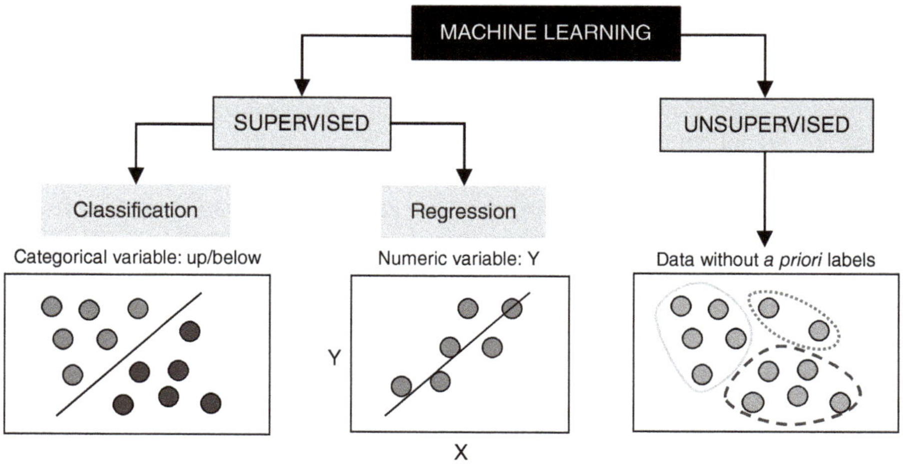

Figure 18.2 Main categories of machine learning (ML) methods. For all three subcategories, vector diagrams represent the task accomplished by each method. In classification, the algorithm searches for the line (threshold) that separates dots, where upper/lower dot groups can be, for example, male/female. In regression, the algorithm searches for the regression line that best fits the points, where Y can be age, growth, or any other continuous variable. In unsupervised methods, the algorithm searches for the optimal grouping (solid/dashed/dotted lines around data points). (*See colour plate section for colour representation of this figure.*)

In a classification problem, the goal of the algorithm is to learn the best boundary that divides the classes (Figure 18.2). The ways to learn this boundary (i.e., the separation line between the two groups of data points) may include linear classifiers if the dataset can be linearly separable, such as logistic regression or linear discriminant analysis (Figure 18.2). However, more complex datasets may require nonlinear boundaries and algorithms, such as kernel support vector machine, naïve Bayes classifiers, or random forests, which may provide a better estimation. In a regression problem, the goal is to predict a continuous outcome variable based on the values of one or more independent variables by fitting a curve that explains the majority of the input data. Typical regression algorithms include linear regression, regularized (also known as penalized) regression, neural networks regression, or decision trees. There are algorithms specific for classification or regression, such as logistic regression (classification) or linear regression (regression) and other algorithms that can be applied to both types of problems with modifications in their loss functions, such as decision trees and artificial neural networks. For EBDP in aquatic organisms, our outcome variable may be categorical or quantitative. In fact, the question can be formulated in different ways so that the outcome variable is categorical or continuous and variable prediction may be either a classification or a regression problem. This depends on whether one aims to develop biomarkers for traits or conditions that have occurred in the past (e.g., previous stress exposure), are present at the moment, or are predicted to be expressed in the future (Figure 18.3). For example, during EBDP for temperature stress in a fish, one may ask two distinct questions and each one will require different ML approaches to be answered:

1) Is the fish stressed? The answer is "yes" or "no," a binary decision, and represents a classification problem. The training dataset should contain features or columns of epigenetic markers values, while rows should include the phenotype category "stressed/yes" and "non-stressed/no." Then the ML task is to find the boundary between the two classes.
2) How stressed is the fish? The answer is numeric (e.g., on a scale of 0–10 with 0 = not stressed and 10 = very stressed) and the outcome can be any given number. This would represent a regression problem, where the features are used to fit a curve to predict the numeric outcome. Notice, however, that many times a regression problem can be reconverted into a classification problem by binarizing the target variable.

	Past	Present	Future
Hatchery-rearing	C		
Development	C		
Maturation		C	R
Growth		R	R
Nutrition	C	C	C
Sex		C	C
Age	R	R	R
Health	C	C	C/R
Environment	C/R	C/R	C/R
Contaminants	C/R	C/R	
Morphology			C

Figure 18.3 Supervised ML methods for key traits and processes of aquatic organisms. EBs can be used for building classification (C) models for categorical outcomes or regression (R) models for numeric outcomes for each key trait and process. The type of the expected outcome may change over time: for example, the question can be at present whether an organism is mature (categorical), but in the future, how fast an organism will mature (numeric outcome). In another example, we may aim to know whether an environmental stressor has occurred (yes/no) or how much it has affected the organisms (numeric) in the past. The list presented here aims to cover the most common objectives for EB development, but is not exhaustive. (*See colour plate section for colour representation of this figure.*)

18.3.4.2 Sample Size Determination

The determination of the sample size per se is a very important factor to take into consideration during biomarker development. Depending on the natural complexity of the dataset, the minimum sample size required can range from a few tens to several thousands of samples. Thus, the optimal n_s required to build a model that accurately predicts the outcome variable of interest should be estimated prior to further analysis. Standardized sample size determination reporting for ML prediction models is still scarce in the medical field [76]. The sample size required for training an ML model (see Section 18.3.4.3) needs to be addressed specifically in each case since it depends on the complexity of the problem, the learning algorithm and in classification, and how well the classes are separated [77]. For example, the optimal n_s differs for the recognition of red blood cells vs. leukocytes vs. breast tumor cells [77]. One valid approach is to rely on previous studies and recommendations. For calibrating age epigenetic clocks in wild animals, Mayne et al. [78] conducted simulations using human and zebrafish (*Danio rerio*) data to determine the optimal sample size and concluded that ideally 134 individuals are required for accurate predictions, a recommendation that can be followed for every new epigenetic clock.

An easy classification problem with a clear separation of classes may require 10 training samples and 58 test samples to reach >95% sensitivity [77]. Another approach is to use methods that allow researchers to predict the performance of an ML model for given variable sample sizes through the creation of learning curves, such as the method by Figueroa et al. [79], which was specifically designed to help researchers decide sample sizes for supervised ML. During model training, the sample size will be crucial to determine data splitting strategies and resampling techniques that need to be used with limited sample sizes. Thus, the definition of small or large datasets will depend on the specific problem developed and researchers will need to evaluate the required level of accuracy. For example, being able to place animals in classes of stressed/not stressed using a slightly inaccurate model might be enough for the purposes of a specific study, while targeting the level of stress with more details will require more accurate models. Researchers need to evaluate whether investing resources to improve the accuracy of their model is worthwhile.

18.3.4.3 Machine Learning Workflow

A typical ML workflow includes some basic steps to build an accurate and generalizable model that performs well in additional (unseen in ML terminology) data [73, 74]. This workflow can be built up in the *R* environment using the specialized *caret* package (*C*lassification *A*nd *RE*gression *T*raining) [80]. The workflow includes (Figure 18.4a):

1) **Data splitting.** Data splitting allows the evaluation of the model performance. Ideally, when a large n_s is available (see Section 18.3.4.2 for an explanation about sample sizes), data can be split between training, validation (internal validation), and test (external validation) datasets. However, data splitting between training (70–80%), test (20–30%), and resampling techniques instead of a validation dataset are typically used, the so-called cross-validation (CV) methods. Training is the core dataset used throughout the workflow to train the algorithms and tune the parameters. The test is used when the optimal model has been built, to give an unbiased assessment of the model's performance. During training, performance of the models needs to be evaluated to reassure that the model can generalize in unseen data (test). A validation approach, where training is further split into training and validation sets, must be applied to assess the performance of models trained with the training sets. Instead, when n_s are limited, resampling techniques can be used. In resampling, iterative cycles of splitting between training and validation sets and model fitting allow for the averaging of prediction errors across each training/validation dataset (Figure 18.4b). Typical resampling techniques suitable for this procedure are k-fold CV, leave-one-out-cross-validation (LOOCV), repeated k-fold CV, random subsampling, and bootstrapping to mention a few (see examples in Figure 18.4b). Bootstrapping is considered the most accurate method to estimate performance in validation procedures. The double or nested CV (further nested splitting within training into training/validation) can be used to estimate the quality of trained models (training/test data), as well as to select the best model and its parameters.
2) **Data preparation and preprocessing.** This step includes: (i) replacing categorical with binary (0–1) variables; (ii) excluding features with zero or near-zero variance among groups in the training dataset; (iii) identifying correlated features; (iv) centering and scaling; (v) normalization and nonlinear transformations to improve gaussianity; and (vi) imputing missing values if necessary and performing proper transformations of the data [80].
3) **Feature selection.** In the case that n_f remains high (i.e., this needs to be evaluated specifically for each project), feature selection techniques can be used to further reduce it to improve computational effort and resources, and eventually model performance. Feature selection techniques are divided again into unsupervised and supervised, with the latter ones further divided into wrapper, filter, and intrinsic [70, 81, 82]. One example of an unsupervised feature selection is included in the previous step where features with near-zero variance or correlated features are eliminated. Supervised feature selection techniques are intended to quantify the importance of each predictor [70, 81, 82]. Some supervised ML algorithms, such as decision trees or regularized regressions, intrinsically perform feature selection [82]. Filter-based supervised feature selection relies on the relationship of the features with the outcome variable that can be selected based on statistical methods, e.g., ANOVAs, Pearson's or Spearman correlation, or χ^2. Wrapper methods build and evaluate many models using subsets of features and select the best one based on performance metrics. Examples include forward, backward, stepwise selection or recursive feature elimination [82].
4) **Model tuning.** The parameters of the algorithm (e.g., linear discriminant analysis or penalized regression) are tuned.
5) **Model evaluation.** The model is evaluated using resampling techniques (e.g., k-fold CV, LOOCV, or bootstrapping). Repeated runs lead to error minimization and optimal model selection.
6) **Assembling the predictions.** The performance of the final model is evaluated using the 20–30% unseen test set. If the performance against the test dataset is not satisfactory, we may need to restart from Steps 3–5 and select a different

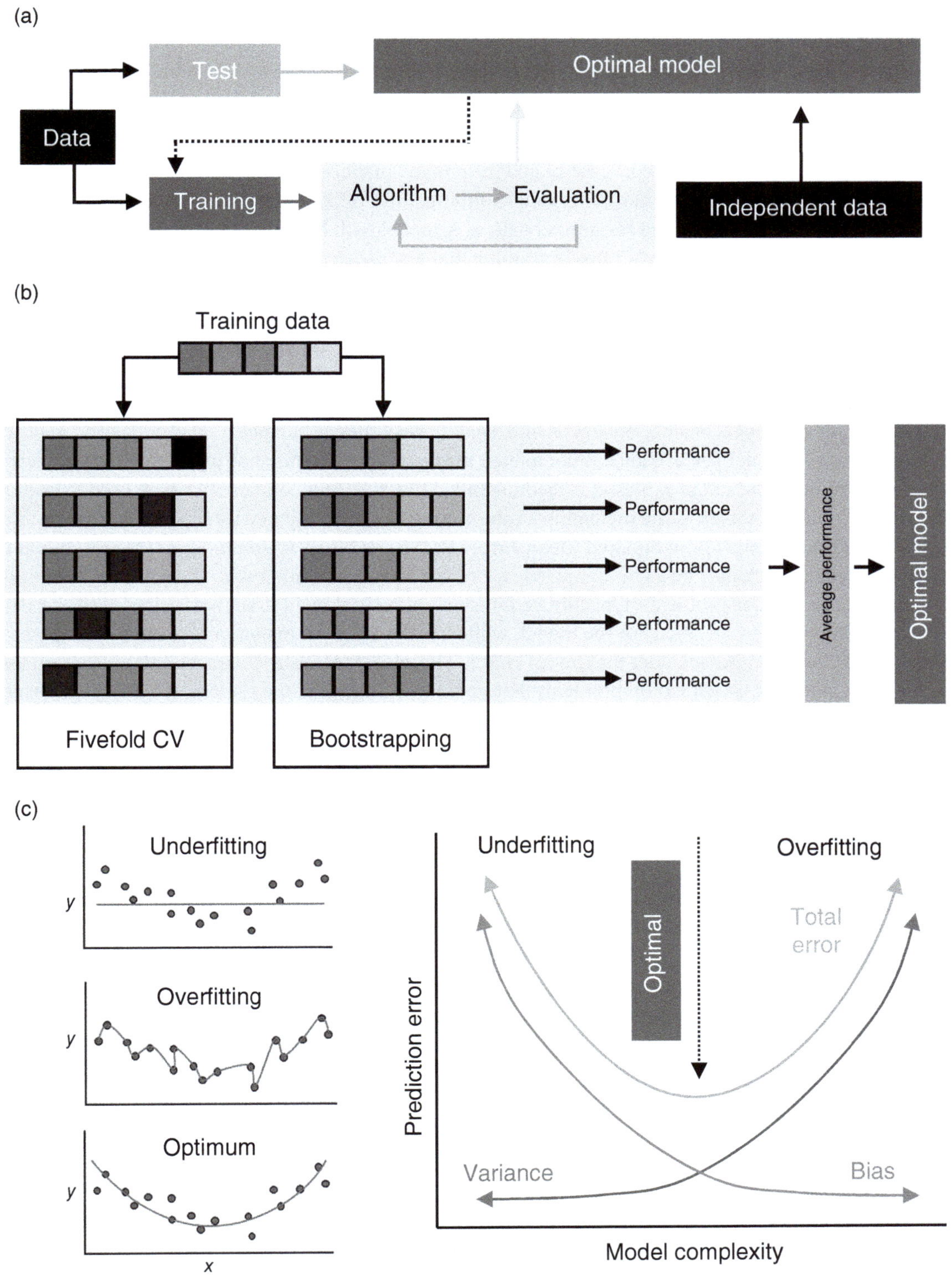

Figure 18.4 Basic concepts for ML. (a) Workflow for building the optimal model: data divided into test and training, the optimal model is chosen through repeated evaluation of the parameters using resampling methods (see c) and evaluated using the test data. The dotted line shows that the procedure might reiterate if the prediction of test data fails to perform well. Independent data can be predicted at the end of the procedure. (b) The two most common resampling techniques that allow the evaluation of the performance using training data: in k-fold CV, fivefold in this example, the training data are divided into fivefolds. Onefold is left aside during each round, the remaining fourfolds are used to build the model and the fifth is used to evaluate its performance. This occurs using all folds. A similar process is used in bootstrapping, where data are taken with replacement, allowing for duplicates to be present. The original observations not contained in each bootstrapping are then used to evaluate the performance of the model. Resampling methods lead to the estimation of an average model performance. (c) The optimal model aims to minimize the error resulting from tradeoff between bias and variance. Low variance/high bias result in underfitting, while high variance/low bias result in overfitting. The more terms are added to the model to avoid underfitting, the more complex a model becomes. (*See colour plate section for colour representation of this figure.*)

algorithm and/or reiterate the parameter tuning process. Two main strategies are used to build the final models. First, the chosen ML procedure or final model is applied to the total training dataset. Second, ensemble methods can be used that allow the combination of multiple models into a single best model. These methods include BAGGing (Bootstrap AGGregating) or random forests [82].

Model performance is evaluated by assessing errors that are formed by bias and variance (Figure 18.4c) [74]. To visualize the bias-variance tradeoff, a hypothetical curve estimated from many models has been plotted in Figure 18.4c for illustration purposes, even though in realistic scenarios of model building, this might not always be possible. Bias is the difference between the expected predicted values and the correct values. A model with high bias fails to capture well the underlying structure of the data and leads to high error rates in training and test dataset. Error due to variance is the variability of a model prediction for a given data point and provides information regarding the spread of data. A model with high variance fits extremely well with the training data and leads to low error rates in training datasets, but high errors in test datasets, thus, it fails to be generalizable. Therefore, there is a tradeoff between bias and variance that relates to the complexity of the model, and which is regulated by tuning model parameters. A simple model with few parameters may have high bias/low variance (underfitting), while a very complex model with a lot of parameters may have low bias/high variance (overfitting). Parameter tuning and model evaluation aim to find a good bias-variance balance.

A modern evaluation of model performance is completed through the assessment of predictive accuracy via loss functions that compare predicted vs. actual values. It is recommended that the same loss functions are used to evaluate performance during model training, as well as the final model. Interpreting loss function metrics depends on the specific problem. Briefly, regression metrics such as Root Squared Mean Error (RMSE) or Mean Absolute Error (MAE) should be used in combination with the R^2, which may be a popular metric for researchers not familiar with the ML field but has limitations (see [74]). In binary classification problems, typically a confusion matrix is applied to support calculations for the accuracy, precision, recall, sensitivity, and specificity of the model, while a multi-label confusion matrix can be computed for multi-class problems. Often, the AUC (Area Under the Curve) or ROC (Receiver Operating Characteristic) Curve are used instead to evaluate the performance of classification problems. For imbalanced datasets, the F-score, which is the harmonic mean of precision and recall, or the Balanced Accuracy are recommended. Since this list is only indicative, for detailed and clear further information, we recommend to see [74, 82].

18.3.5 Examples of Machine Learning Methods for Epigenetic Biomarker Discovery in Aquatic Organisms

18.3.5.1 Sex: A Classification Problem

Sex is a binary outcome that can generally take only one value for a given individual: female or male (hermaphrodite species are not considered). Building an ML model capable to diagnose sex in the present or to predict sex in the future is a classification problem (Figure 18.3). EBs for sex prediction have been developed in the European sea bass [44] and more recently in zebrafish [83]. Previous knowledge of DNA methylation levels of sex-related genes allowed researchers to start from Step 3 of the EBDP (Figure 18.1) using a targeted approach (Multiplex Bisulfite Sequencing; MBS) [44] to interrogate the DNA methylation levels of 10 genes in a total of 70 zebrafish samples (females = 33, males = 37). A discriminant analysis was used, which predicts the probability of belonging to a certain class (female/male). Linear discriminant analysis (LDA) finds linear combinations of features that separate the two classes. Flexible discriminant analysis (FDA) was considered more appropriate in this case, as a flexible extension of LDA for nonlinear combinations. The performance of FDA models was evaluated using a tenfold CV repeated five times. The accuracy of the model was calculated from the confusion matrix and was 88%. Subsequently, in Step 4 (Figure 18.1), the final FDA model was used to predict sex in a biologically independent batch of samples using the DNA methylation values of *cyp19a1a*. The results were correct in 100% of the cases.

18.3.5.2 Age: A Regression Problem

Age is a continuous numeric outcome variable that can take any value above zero. Since the first epigenetic clock was developed in humans [84], the ML algorithm that has dominated the development of epigenetic clocks in all species is penalized (or regularized) regression, which fits well with the structure of the datasets and the purposes of EBDP. In most cases, the structure consists of a large multivariate dataset with high n_f (CpGs) in the order of 10^2–10^5 and a comparatively low n_s in the order of 10^2. The purpose is to obtain a final n_f in the order of 10^{-1}–10^2. Penalized regressions allow the creation of a linear regression with penalized coefficients, or a weight associated with each feature (each CpG). This shrinks some coefficients to zero (shrinkage or regularization) and allows the elimination of features with zero contribution [73]. Tuning of the α (άλφα/alpha) parameter defines which type of penalized regression is used:

1) Ridge (α = 0). Coefficients are shrinking close to zero, but all features are kept.
2) LASSO (Least Absolute Shrinkage and Selection Operator; α = 1). Coefficients are forced to zero and the model is of reduced complexity.
3) Elastic net (α = anywhere between 0 and 1). Some coefficients are shrinking as in ridge and others are forced to zero, as in LASSO.

Another tuning parameter is λ (λάμδα/lambda), which adjusts the amount of shrinkage of the coefficients. The choice between types of penalized regressions depends on the specific dataset. Ridge should be an uncommon choice for epigenetic clock construction since it keeps all features. However, if data are obtained using a targeted approach (MBS) for specific CpGs, ridge might be a good option like in Australian lungfish (*Neoceratodus forsteri*), Murray cod (*Maccullochella peelii*), and Mary River cod (*Maccullochella mariensis*) [85]. LASSO is expected to perform better when a few features have large coefficients and a lot of features have small coefficients, while the elastic net is usually selected since it combines the best characteristics of each model. In the European sea bass [86], the European lobster (*Homarus gammarus*) [40], the northern red snapper (*Lutjanus campechanus*), and the red grouper (*Epinephelus morio*) [87], all three types of penalized regressions were evaluated and the best tuning parameters of the elastic net models were automatically chosen: this approach allowed choosing the best models based on a combination of relevant metrics (i.e., R^2, RMSE, and MAE), as well as the n_f in the final model. The n_f is an important characteristic of an epigenetic clock in terms of applicability in aging hundreds or thousands of individuals and should be minimized (step 4, Figure 18.1).

18.4 Epigenetic Biomarkers in Aquatic Organisms and their Applications in Aquaculture

Aquaculture practices may persistently affect the epigenome of cultured species and include alterations of abiotic (e.g., temperature, hypoxia, and salinity) or biotic (e.g., density and diet) factors. Even though environmental epigenetics in aquatic animals is still in the early stages, a growing number of studies suggest great potential for the application of EBs in aquaculture together with several recent reviews on the potential of epigenetics in aquaculture [47, 48, 88–90]. Understanding epigenetic changes in conjunction with phenotypic changes supports the development of EBs predictive of traits such as morphology, maturation, growth, reproduction, sex, aging, and nutrition, but also for monitoring animal welfare, population dynamics [91], and ecosystem health, and for safety assessment of environmental impacts (Figure 18.5).

Within the current aquaculture framework, research focuses on approaches to modulate specific phenotypic traits, such as sex, enhanced growth, environmental stress resistance, and immune competence for more efficient productivity. However, animal welfare, health, and fitness, as well as the environmental and ecosystemic impacts of farming aquatic animals should be an equally important research focus since they concern a wider audience and are still mostly neglected. Monitoring fish welfare should focus on key domains: resource deprivation, environmental challenges, diseases and injuries, behavioral restrictions, and mental and physical suffering [92]. Particular traits can be "programmed" through exposure to environmental factors during early development to obtain specific adult phenotypes through the so-called "epigenetic priming effect" [47, 48, 93]. This highlights the possibility for EBDP with implications on the management of aquaculture operations to reflect mitigating strategies for improving animal welfare, as well as monitoring, controlling, and reporting of environmental impact.

Finally, environmental epigenetics can contribute to the understanding of population adaptation and speciation processes. Epigenomic changes should be monitored in symphony with the genome to better understand adaptation to new environments, especially to rapidly changing conditions. Aquaculture, as a domestication process of aquatic organisms, is likely to result in species locally adapted, often to habitats of low complexity with high population density known to associate with morphological changes in fish [94] and likely to result in speciation. EBs may provide a tool to predict transgenerational future trajectories of aquaculture species [47, 73, 93]. Furthermore, EBs can be employed to monitor undergoing adaptation and speciation processes due to rearing, as a means to report and control negative ecosystem impacts as artificial selection occurs. Later, we discuss epigenetic studies relevant to applying EBs in aquatic organisms and their future perspectives. These studies can be used as the first step toward EBDP for similar traits and conditions since the epigenomic responses reported could form the basis for downstream discovery within our proposed framework.

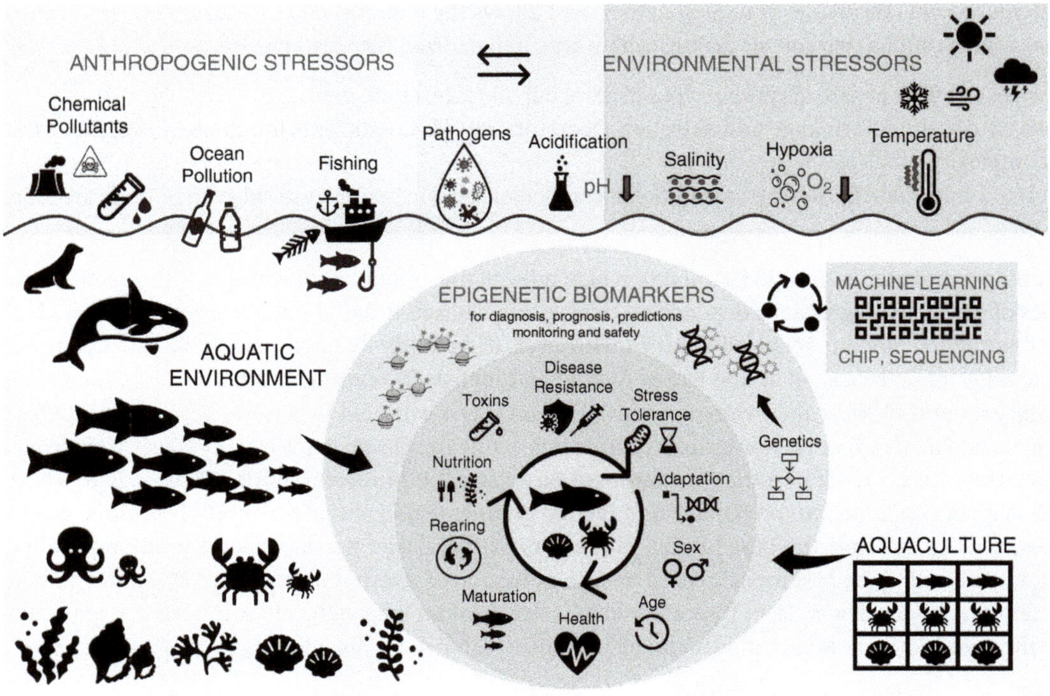

Figure 18.5 Development and application of epigenetic biomarkers in aquatic organisms. Illustration of the main anthropogenic and environmental stressors that are challenging for living organisms in an aquatic environment, aquaculture conditions, and the application of EBs developed based on machine learning (ML) and advanced chip or sequencing technologies. EBs can be applied for the diagnosis, prognosis, prediction, monitoring, and safety of an organism's disease resistance, environmental and anthropogenic stress tolerance, nutritional status, rearing and early development, maturation, general health condition, age, sex, adaptation/ speciation processes, and the exposure of aquatic contaminants.

18.4.1 Hatchery Rearing and Fitness

EBs predictive of population fitness (e.g., survival and growth) may help improve rearing conditions in aquaculture, restocking programs, and biomonitoring. Hatchery rearing can result in lower fitness and survival while inducing genome-wide DNA methylation and hydroxyl-methylation changes [24, 25, 27, 64, 95, 96].

In salmonids, hatchery rearing-induced persistent DNA methylation changes in the germline of adults after ocean migration [27] that may have been passed on to their offspring. Interestingly, between two independent hatchery-reared coho salmon populations, there was shared epigenetic variation induced without genetic differentiation, and this could affect their overall fitness [24, 27]. Nile tilapia (*Oreochromis niloticus*) showed differential hydroxyl-methylation in 126 genes related to immune, growth, and neuronal-related pathways between wild and domesticated populations after one generation [96]. In the European sea bass, 57 genes were differentially methylated in red blood cells after density-associated stress, suggesting a minimally invasive EB approach [63]. In marbled crayfish (*Procambarus virginalis*), genome-wide DNA methylation differences were also found between wild and lab-reared individuals [97].

The impacts of hatchery rearing on the overall fitness of cultured organisms are important to consider. Farmed organisms that escape from sea cages or are released into the wild, if able to survive and reproduce, may possess DNA methylation patterns induced by artificial rearing that may be passed on to wild populations. This process may lead to long-term ecological consequences in terms of fitness. Therefore, classification algorithms can be applied to develop generic diagnostic EBs to detect past rearing under captivity. CpG methylation marks identified in the most phylogenetically close species can be used to start EBDP from Steps 2 or 3. These EBs will be beneficial for improved management of restocking programs, biomonitoring, and animal conservation.

18.4.2 Developmental Processes, Maturation, and Growth

In aquatic animals, early developmental processes and morphology are regulated by epigenetic mechanisms [56, 98]. In clonal individuals of dace hybrids (*Chrosomus eos-neogaeus*), differences in epigenetic marks between the left and right sides of symmetrical animals were suggested as a measure of developmental instability and called "epigenetic fluctuating asymmetry" [99]. In the sea lamprey (*Petromyzon marinus*), during metamorphosis from a sessile filter feeder to a migrating tissue-feeding adult, significant changes in whole-genome methylation levels, including developmentally important *hox* genes, were associated with muscle development [100]. Similarly, DNA methylation changes are associated with early maturation and smoltification in salmonids [101]. In rainbow trout, 57 differentially methylated regions (DMRs) functionally associated with smoltification were identified between migratory and nonmigratory F2 juveniles [34]. In half-smooth tongue sole, an association of genome (single-nucleotide polymorphism; SNP) and epigenome in the *pacap* gene was linked to growth traits of different genotypes [102].

Furthermore, ncRNAs are crucial regulators of yolk sac developmental processes in fish. Rainbow trout showed strong correlations between miRNAs and mRNA transcripts of metabolic genes during the feeding stage transition [103]. Moreover, miRNAs contribute to skeletal muscle growth in Nile tilapia and are differently expressed according to fast- and slow-growing strains [104]. Classification algorithms can be applied to detect normal or abnormal development and whether an individual is mature or not (diagnostic EBs), while regression algorithms can be applied to identify how well animals are growing or to predict when they will mature and how they will be growing (prognostic, predictive, and monitoring EBs). The epigenetic marks associated with development in dace hybrids, maturation in sea lamprey and rainbow trout, and growth in Nile tilapia can be used as starting points to further develop EBs in other species to monitor past developmental conditions and predict future tissue development, maturation, and growth rates.

18.4.3 Sex Ratios

Since epigenetic patterns differ between two sexes and mediate sex determination [72, 105], sex EBs allow the prediction of the sexual phenotype (see Section 18.3.5.1 for a detailed example of sex EBDP). In poikilothermic vertebrates such as fish, a combination of genetic and environmental factors, including temperature, determine the phenotypic sex [106]. Often in aquaculture settings, abnormally high temperatures during sensitive developmental stages result in genotypic females developing as males (masculinization) and skewed sex ratios [36]. Temperature effects are mediated by DNA methylation changes of genes involved in sexual development including *cyp19a1a* and *dmrt1* [105, 107]. Interestingly, transgenerational epigenetic inheritance of environmentally induced masculinization can occur [36]. Together these results led to the model of CERS, a roadmap for developing sex EBs and strong candidate genes involved in sexual development [43, 72]. This framework should help further sex EBDP since the initial steps of the procedure can be avoided and few candidate loci (in the order of 10^1–10^2) can be directly tested in hundreds of samples using classification algorithms. EBs that predict sex ratios of a population (see Section 18.3.5.1) can help to determine whether farmed populations have been epigenetically trapped, leading to sex-biased broodstocks, and oversee sex control programs in aquaculture [108, 109].

18.4.4 Nutrition and Nutritional Programming

EBs may help uncover malnutrition and diet-requirement restrictions and support the management of nutritional modulation and programming. The supplemental usage of fatty acids and lipids with the feed is an important aspect of the efficient provision of nutrients to aquatic animals, such as farmed fish. Indeed, Geay et al. [110] discovered in European sea bass that diets supplemented with high unsaturated fatty acids (HUFA) in their composition differentially affected the methylation of *fads2* gene promoter in liver, important for fatty acid metabolism. In gilthead sea bream (*Sparus aurata*), the replacement of fish oil with vegetal oils in parents resulted in differential methylation and expression of *scd1a* in the liver of the offspring, likely associated with lower fat accumulation [111]. In rainbow trout, glucose intolerance was linked to DNA methylation changes [112] and microRNAs regulated nutritional processes such as postprandial lipogenesis [113], diet-induced hyperglycemia [114], and the postprandial regulation of hepatic insulin pathway [115] in the liver. Thus, the connection between epigenetic changes and dietary disorders, as well as diet compositions (e.g., fatty acids), provides a promising background for the development of EBs. Classification algorithms can be used based on promoter methylation of genes related to metabolic processes and microRNAs prevalent in the liver to monitor nutritive conditions and to improve nutrition in aquaculture operations (all five types of EBs) see Figure 18.4.

18.4.5 Aging

Age is crucial to comprehend an individual's fitness and key population parameters linked to fish phenotype (see Section 18.3.5.2 for details on age EBDP). An acceleration or deceleration of the aging process occurs and can be detected in mammals [84, 116, 117] and fish [118]. Epigenetic clocks were developed using DNA methylation as an extremely reliable method for age estimation in mammals, birds, fish, and even an invertebrate species [38–40, 85–87]. Epigenetic clocks are calibrated based on the correlation between the methylation levels of specific CpG sites and an individual's age. Recently, a universal mammalian clock was published, paving the way for similar approaches in other taxonomic groups such as fish or crustaceans [119]. Information on the aging process contributes to evaluating the lifestyle and overall welfare of aquatic animals [120, 121]. Further, epigenetic clocks would modernize fisheries and conservation assessments and management that require age estimation as a crucial population parameter. Nowadays, the estimation of age is fundamentally relying on the reading of otoliths and other hard structures (e.g., scales), which requires lethal sampling and a high level of expertise for processing and reading, while it is impossible in the absence of seasonal variation (e.g., tropical species) or on length-at-age growth modeling, which is often hindered by poor quality data from biased sampling [122]. Since epigenetic clocks represent a noninvasive method for age determination and hold great potential for the evaluation of age acceleration, they could be implemented to monitor lifestyle conditions in aquaculture. Hence, predictive epigenetic age clocks would be useful to identify individuals with the slowest age acceleration to select "less aged" or the most vital and youngest individuals with high reproductive output, and consequently, to evaluate the overall lifestyle and rearing conditions.

18.4.6 Health and Disease Resistance

The framework for developing clinical biomarkers can easily be adapted to aquaculture, since EBs capable of diagnosing diseases and disease susceptibility and predicting disease progression and survival may prove valuable for risk management. To date, only a few studies have investigated epigenomics related to diseases, infections, or parasite load in aquatic organisms [14, 33, 42, 123]. In the common flatfish dab (*Limanda limanda*), 20% occurrence of liver tumors were diagnosed and genes related to cancer and apoptosis showed altered DNA methylation and transcription [33]. Exposure of zebrafish to bacteria lipopolysaccharides was associated with DNA methylation of two genes with immune functions, *casp9* and *IL1β* [123]. Three-spined sticklebacks (*Gasterosteus aculeatus*) infected with their nematode parasite (*Camallanus lacustris*) revealed genome-wide DNA methylation modifications, likely linked to parasite-mediated immune suppression [14].

miRNAs in biofluids could also be used for early disease diagnosis [124]. Atlantic salmon injected with salmonid alphavirus (SAV) [125] or infected with *Piscirickettsia salmonis* bacteria [126] displayed differentially expressed miRNAs targeting mRNAs important for inflammatory and immune responses. A simulation of immune challenge and parallel feeding with an immunostimulant (unmethylated CpG) revealed further miRNA biomarkers [55]. These miRNA biomarkers can be utilized for diets containing immune stimulants to prevent infectious diseases in aquaculture and to trace their impacts.

DNA methylation marks of immune genes or miRNAs detected in these studies can serve as a starting point for developing diagnostic EBs with the ability to detect past immune stress that could help toidentify early exposures to pathogens or the start of detrimental infections in aquatic animals [47, 48, 93]. Classification algorithms can be used to detect healthy or diseased states in the past, present, and future, while regression algorithms can help to predict responses to treatments. They would contribute to a time-efficient disease assessment and help to reduce the risk of challenges such as disease outbreaks, parasite infections, and mass mortality events.

18.4.7 Stress Due to Environmental Factors

18.4.7.1 Temperature

For aquatic poikilotherms, temperature is crucial for most biological processes including development, growth, and reproduction [127]. It is now well-established that environmental temperature affects the DNA methylation of aquatic animals [18–20, 62, 128] with effects during early development being long-lasting [127].

Temperature is able to affect the expression of DNA methyltransferases in muscle tissues [17] and liver [129] as well as the DNA methylation of genes related to muscle growth [17, 19]. DNA methylation in fish responds to temperature stress during early larvae development [17], with even three months early exposure resulting in lasting effects in adulthood [69]. Thermal stresses in adulthood associated with acclimation responses also affect genome-wide DNA methylation in muscle tissue of sticklebacks [128], and dynamic gene-specific DNA methylation changes were associated with long-term thermal and hypoxia exposure in the liver of Atlantic salmon [20].

Histone modifications and miRNAs also regulate temperature-driven responses. In zebrafish, cold temperature exposure showed differentially methylated genes related to chromatin modifications, indicating the interplay of epigenetic mechanisms to regulate cold-acclimation pathways [130]. Subtypes of the histone variant H2A.Z, known for mediating environmental responsiveness, were seasonally enriched in common carp fish (*Cyprinus carpio*) [131]. Temperature-induced phenotypic plasticity of growth was associated with altered miRNAs in Senegalese sole (*Solea senegalensis*) [132].

Metastable epialleles indicative of marine heatwaves experienced early during development were identified in adult European sea bass, paving the way for conserved inference of past life history events across tissues [69]. ML approaches were used in zebrafish to determine past thermal history in a sex-dependent manner using DNA methylation biomarkers in single genes (a similar approach to Section 18.3.5.1) [83]. Together, these studies demonstrate the sensitivity of the epigenome to temperature stimuli during early and adult life and highlight loci as valuable candidate diagnostic EBs to detect past temperature exposures and suggest the possibility of developing metastable epialleles in single loci conserved across species to infer past life history.

18.4.7.2 Hypoxia

Aquatic organisms need to cope with varying oxygen availability. Steep hypoxic events in open-water sea cages caused by seasonal temperature increases, tidal fluctuations, algae blooms, and high fish density are a concerning problem in aquaculture [133]. The role of epigenetics in aquatic animals exposed to hypoxia is mostly still unexplored. In rainbow trout fry after hypoxia, *bnip3a* and *bnip3lb1* genes were differentially methylated and expressed [134] in whole bodies. Wang et al. [135] unraveled, for the first time, the transgenerational epigenetic basis of hypoxia on the reproduction in marine medaka (*Oryzias melastigma*). Hypoxia induced differential DNA methylation of histone-modifying genes in sperm, together with transcriptomic and proteomic alterations of F1 and F2 males [135]. Further, acute hypoxic stimulation and reoxygenation resulted in decreased DNA methylation of two immune-related genes (*stat3* and *vegf*) in the Japanese flounder (*Paralichthys olivaceus*) [21]. In the water flea (*Daphnia magna*), both hypoxia and hyperoxia conditions induced DNA methylation changes [136]. Hypoxia exposure in aquatic organisms could result in cell death and overexpressed miRNAs were linked to apoptosis [57]. The epigenetic changes in these genes could pave the way for the future development of EBs for hypoxia stress using either classification (stress/not stressed) or regression (how much stressed) algorithms in aquatic organisms.

18.4.7.3 Temperature and Hypoxia Combined

Seasonal fluctuations in temperature and hypoxia are particularly challenging for aquaculture species confined in open-water net pens [137]. EBs able to detect simultaneous environmental effects will enhance mitigating strategies for real-world conditions. Beemelmanns et al. [20] exposed Atlantic salmon to an incremental temperature increase (up to 20 °C) alone or combined with moderate hypoxia (70% air saturation), and these extreme conditions were maintained for four weeks and showed changes in DNA methylation of thermal- and hypoxia-responsive genes that correlated with their expression in liver tissue [20]. In the downregulated gene *ucp2* (mitochondrial uncoupling protein 2), CpG methylation changes were maintained after four weeks of stress exposure [20]. By regulating the coupling of mitochondrial respiration to ATP synthesis during chronic temperature exposure, the lower *ucp2* expression may have contributed to a reduction in cellular reactive oxygen species (ROS), suggesting a mechanism to facilitate long-term acclimatization [20]. Theory predicts that if the epigenetic pattern is inherited (Figure 18.6), the F1 generation may be able to cope more efficiently with the same environmental challenges that the parental generation had already experienced through a "transgenerational priming effect." Thus, future EBDP for combined effects of temperature and hypoxia should focus on directly screening F1 offspring for DNA methylation changes using targeted detection approaches to determine multigenerational effects (Figure 18.6).

Taken together, we propose that these genes can be targeted to develop EBs to (i) detect past exposure to acute or chronic temperature and hypoxia stress; (ii) detect early responses to sublethal stressors; and (iii) predict multigenerational thermal and hypoxia tolerance if epigenetic fixation occurs (Figure 18.6). Prognostic EBs may help predict fitness and performance of individuals that have experienced environmental stress, while predictive EBs may identify individuals more likely to respond to future environmental stresses. These EBs would support mitigating measures upon climate-related challenges in aquaculture and for ecological conservation programs of wild populations.

18.4.7.4 Salinity

Fluctuations in salinity impose strong physiological stress on aquatic animals to maintain osmotic homeostasis. Recent evidence revealed associations with changes in DNA methylation and miRNA abundance. In half-smooth tongue sole, salinity stress exposure resulted in reduced growth and was linked to high DNA methylation and lower expression of the

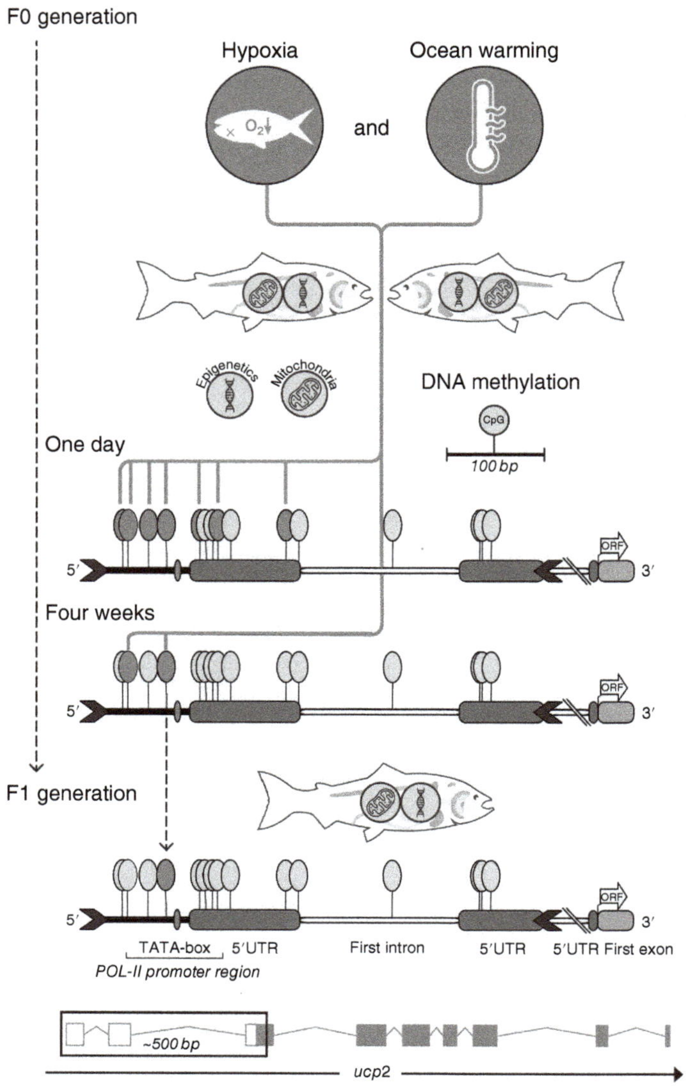

Figure 18.6 Candidate transgenerational epigenetic biomarkers (EBs) for high temperature in combination with hypoxia. Schematic illustration of induced DNA methylation marks in CpGs of *ucp2* (mitochondrial uncoupling protein 2) upon high temperature and moderate hypoxia (70% air saturation) exposure in Atlantic salmon for one day and four weeks, and the hypothetical transgenerational transfer from the F0 to the F1 generation indicated as dotted arrows. CpG methylation marks within different genomic elements around the transcription start of *ucp2* are indicated as lollipops, with stress-induced CpG methylation marks, represented as dark shaded lollipops which can be considered as strong candidate EBs. The targeted genomic region around the transcription start site (5′upstream, 5′UTR, first coding exon, and first intron), including CpG positions, is based on the *Salmo salar* genome (ssa) version ICSASG_v2 with the in silico predicted promoter region. The open reading frame (ORF) is indicated by an arrow. Exons (boxes) and introns (lines) are shown in the transcript diagram at the bottom (modified from Ensembl). The targeted genomic area (~500 bp) used for Multiplex Bisulfite Sequencing (MBS) is indicated by a black frame. *Source:* Modified from Beemelmanns et al. [20].

insulin growth factor (*igf*) [15]. In brown trout (*Salmo trutta*), differences in global DNA methylation were found in the gills of fish held in freshwater compared to fish acclimated to seawater [138], while diet-induced DNA methylation modifications may be important for salinity acclimation [138]. Similarly, sticklebacks from a salinity gradient with different salinity tolerances showed genome-wide DNA methylation changes associated with osmoregulatory processes, implying that methylation may be a complementary mechanism for adaptation [16]. In addition, osmotic stress impacts gill and kidney miRNAs that regulate processes to maintain homeostasis and salt tolerance [139, 140]. Together, these results could be used for future development of diagnostic EBs applying classification methods and prognostic EBs utilized regression methods EBD to evaluate the tolerance and adaptability of aquatic organisms to salinity.

18.4.8 Aquatic Contaminants

Traditional molecular biomarkers (DNA, RNA, and protein) have been used for environmental monitoring and risk assessment since they can identify the presence of pollutants [141]. Even in low concentrations, pollutants can induce sublethal metabolic effects before they cause noticeable damage as a consequence of their accumulation. A sublethal effect could be an activation, inactivation, overexpression, or interference effect of different metabolic routes in response to the presence of a xenobiotic compound that is acting on cellular receptor levels [7]. Variability in these biomarkers occurs due to life history, taxon, and seasons, thus efforts to standardize them as robust indicators of ecosystem health are underway [141].

Aquatic contaminants including metals, pesticides, or Bisphenol A (BPA), some of them known as Endocrine Disrupting Chemicals (EDC), affect epigenetic modifications such as histone modifications in zebrafish [142], miRNAs, and genome-wide DNA methylation in several species [48], and the reported transgenerational effects [143]. Thus, diagnostic and prognostic EBs within organisms can contribute to evaluate the health status of their surrounding ecosystem. Further, monitoring and safety EBs with unique properties of stability over time and environmental responsiveness can help to identify past or present contamination events if evaluated in sessile or with low dispersed organisms, and the past environmental experiences of contamination in migrating organisms.

18.4.9 Adaptation and Speciation

Epigenetic variation can increase the phenotypic range encoded by a genome and, thus, may help populations cope with rapid environmental changes and eventually contribute to adaptation and speciation. This is especially relevant in populations with low genetic diversity [144], a high possibility in farmed populations. In a rapidly changing environment, such as open ocean sea cages during frequently occurring marine heatwaves [145], epigenetic changes may confer an advantage and, if stable, epigenetic inheritance may accelerate adaptation [146]. This can become an epigenetic trap if maladaptive epigenetically mediated phenotypes are stably inherited [108, 109].

For biological conservation and aquaculture operations, verifying DNA methylation in transposable elements can be important to monitor reproductive isolation between stocked and wild-type fish. Laporte et al. [147] uncovered that DNA methylation of transposable elements can cause genetic incompatibilities between two different whitefish (*Coregonus clupeaformis*) species. Consequently, EBs on transposable elements should be developed to evaluate genetic incompatibilities and reproductive isolation that may lead to speciation due to domestication, and to monitor the anthropogenic effects associated with rearing.

18.5 Future Perspectives

Challenges remain for the development of EBs in aquatic organisms, nevertheless, technological advances are available and hold great potential to be adapted to the field in the near future. Among the herein discussed five major dependencies that should be carefully considered in the experimental design of studies for EBDP (i.e., genetic variation, developmental and reproductive stage, sex, and tissue), the bias of tissue heterogeneity due to the cell specificity of epigenetic patterns can be resolved via single-cell sequencing technologies in future research. These new methods allow for the description of the epigenome at single-cell resolution and are already successfully developed in medical research [148, 149]. For example, recent advances in HTS and single-cell technologies allowed the exploration of cell heterogeneity in cancer and the discovery of different methylation profiles of tumor single cells [149]. Hence, new protocols can be established for aquatic organisms to uncover epigenomic profiles of cell populations and disentangle major cell types within a tissue, and, thus, to determine biomarkers based on single cells, although the costs of these technologies still limit their accessibility for aquatic animals.

For the routine application of future EB at a large scale in aquatic organisms, cost-effective epigenetic tools that interrogate multiple epigenetic marks in a high number of individuals need to be implemented. One key factor to be considered for future research is to advance research on the blood/hemolymph of aquatic organisms in order to monitor epigenetic changes over time in a nonlethal way.

Future EB studies should focus on the construction of high-throughput multiplex PCR assays similar to successfully established Genotyping-in-Thousands by sequencing (GTseq) approaches of targeted SNPs for thousands of individuals in a single Illumina HiSeq lane [148]. The adaptation of similar protocols would allow for the assessment of multiple

EBs (e.g., 100–500 CpG sites) for thousands of individuals, thereby enabling rapid and cost-effective measurements (e.g., <$10 per sample). For example, in humans and mice, ultra-cheap and targeted methylation sequencing technologies such as the Tagmentation-based Indexing for Methylation Sequencing (TIMEseq) were recently established for epigenetic clocks and DNA methylation biomarkers from hundreds of individuals [150]. Methylation arrays measuring thousands of conserved CpGs across species based on probes (e.g., 36K CpGs probes) would provide deep coverage of conserved cytosines and facilitate the development of cross-species EB microarrays in aquatic taxa, similarly to the mammalian methylation array [151]. Consequently, TIMEseq, CpG microarrays and single-cell sequencing technologies will open up ways for the establishment of low-cost EB tools applicable to aquatic organisms.

18.6 Concluding Remarks

EBs hold great potential for assessing the impacts of artificial rearing on aquatic organisms' wellness and fitness, as well as the effects of anthropogenic pollution of the environment. This is because of their unique biological properties compared to other traditionally used molecular biomarkers, and especially their stability over time combined with their responsiveness to environmental changes. In the era of low-cost sequencing and omics data, ML approaches can contribute to maximizing information extraction from an abundant and rich number of datasets. We suggested a flexible general systematic approach for nonlethal EBs development in aquatic organisms that can be adapted to the specific needs of the phenotypic trait to be predicted, as well as the organism or environment under question. Furthermore, this approach can be adjusted to balance resource availability. Knowledge from previous studies that reported epigenetic changes in phenotypic traits relevant for an aquaculture context (e.g., growth, sex, aging, health, and disease resistance), but also as responses to environmental stress, should be considered to target strong candidates for EBs. Adopting such a strategy paves the way for developing conserved EBs that would accurately predict the desired outcome across species.

Acknowledgments

We are thankful to Drs. Santiago Marco, Alejandro Valdivieso, Núria Sánchez-Baizán, and the editors for providing constructive criticism of the manuscript. We thank Susanne Landis @scienstration for her support with graphical artwork.

References

1 World Health Organization & International Programme on Chemical Safety (2001). *Biomarkers in Risk Assessment: Validity and Validation*. WHO Environmental health criteria; No. 222.

2 Strimbu, K. and Tavel, J.A. (2010). What are biomarkers? *Curr. Opin. HIV AIDS* **5** (6): 463–466.

3 Warne, R.W., Proudfoot, G.A., and Crespi, E.J. (2015). Biomarkers of animal health: integrating nutritional ecology, endocrine ecophysiology, ecoimmunology, and geospatial ecology. *Ecol. Evol.* **5** (3): 557–566.

4 Monserrat, J.M., Martínez, P.E., Geracitano, L.A. et al. (2007). Pollution biomarkers in estuarine animals: critical review and new perspectives. *Comp. Biochem. Physiol. Part C Toxicol. Pharmacol.* **146** (1–2): 221–234.

5 Eissa, N. and Wang, H. (2016). Transcriptional stress responses to environmental and husbandry stressors in aquaculture species. *Rev. Aquac.* **8** (1): 61–88.

6 Eslamloo, K., Caballero-Solares, A., Inkpen, S.M. et al. (2020). Transcriptomic profiling of the adaptive and innate immune responses of Atlantic salmon to *Renibacterium salmoninarum* infection. *Front. Immunol.* **11**: 2487.

7 Valdivieso Quintero, W. and Zafra, G. (2016). Use of molecular biomarkers in studies of aquatic environmental impact. *J. Ind. Pollut. Control.* **32** (1): 381–389.

8 Houde, A.L.S., Günther, O.P., Strohm, J. et al. (2019). Discovery and validation of candidate smoltification gene expression biomarkers across multiple species and ecotypes of Pacific salmonids. *Conserv. Physiol.* **7** (1): coz051.

9 Akbarzadeh, A., Günther, O.P., Houde, A.L. et al. (2018). Developing specific molecular biomarkers for thermal stress in salmonids. *BMC Genomics* **19** (1): 749.

10 Akbarzadeh, A., Houde, A.L.S., Sutherland, B.J.G. et al. (2020). Identification of hypoxia-specific biomarkers in salmonids using RNA-sequencing and validation using high-throughput qPCR. *G3 Genes|Genomes|Genetics* **10** (9): 3321–3336.

11 Beemelmanns, A., Zanuzzo, F.S., Xue, X. et al. (2021). The transcriptomic responses of Atlantic salmon (*Salmo salar*) to high temperature stress alone, and in combination with moderate hypoxia. *BMC Genomics* **22** (1): 1–33.

12 Panserat, S., Marandel, L., Geurden, I. et al. (2017). Muscle catabolic capacities and global hepatic epigenome are modified in juvenile rainbow trout fed different vitamin levels at first feeding. *Aquaculture* **468**: 515–523.

13 Bowman, J.D., Rahman, S.M., and Choudhury, M. (2017). Endocrine disruptors and epigenetics. In: *Gene Regulation, Epigenetics and Hormone Signaling* (ed. S.S. Mandal), 577–606. Wiley-VCH Verlag GmbH & Co. KGaA.

14 Sagonas, K., Meyer, B.S., Kaufmann, J. et al. (2020). Experimental parasite infection causes genome-wide changes in DNA methylation. *Mol. Biol. Evol.* **37** (8): 2287–2299.

15 Li, S., He, F., Wen, H. et al. (2017). Low salinity affects cellularity, DNA methylation, and mRNA expression of igf1 in the liver of half smooth tongue sole (*Cynoglossus semilaevis*). *Fish Physiol. Biochem.* **43** (6): 1587–1602.

16 Heckwolf, M.J., Meyer, B.S., Häsler, R. et al. (2020). Two different epigenetic information channels in wild three-spined sticklebacks are involved in salinity adaptation. *Sci. Adv.* **6** (12): eaaz1138.

17 Campos, C., Valente, L., Conceição, L. et al. (2013). Temperature affects methylation of the myogenin putative promoter, its expression and muscle cellularity in Senegalese sole larvae. *Epigenetics* **8** (4): 389–397.

18 Anastasiadi, D., Díaz, N., and Piferrer, F. (2017). Small ocean temperature increases elicit stage-dependent changes in DNA methylation and gene expression in a fish, the European sea bass. *Sci. Rep.* **7** (1): 1–12.

19 Burgerhout, E., Mommens, M., Johnsen, H. et al. (2017). Genetic background and embryonic temperature affect DNA methylation and expression of myogenin and muscle development in Atlantic salmon (*Salmo salar*). *PLoS one* **12** (6): e0179918.

20 Beemelmanns, A., Ribas, L., Anastasiadi, D. et al. (2021). DNA methylation dynamics in Atlantic salmon (*Salmo salar*) challenged with high temperature and moderate hypoxia. *Front. Mar. Sci.* **7**: 1076.

21 Li, X., Liu, B., Yang, J. et al. (2021). DNA methylation in promoter region of immune related genes STAT3 and VEGF and biochemical parameters change in muscle of Japanese flounder under acute hypoxia. *Dev. Comp. Immunol.*, https://doi.org/10.1016/j.dci.2021.104295 104295.

22 Lim, Y.-K., Cheung, K., Dang, X. et al. (2021). DNA methylation changes in response to ocean acidification at the time of larval metamorphosis in the edible oyster, *Crassostrea hongkongensis*. *Mar. Environ. Res.* **163**: 105214.

23 Venkataraman, Y.R., Downey-Wall, A.M., Ries, J. et al. (2020). General DNA methylation patterns and environmentally-induced differential methylation in the eastern oyster (*Crassostrea virginica*). *Front. Mar. Sci.* **7**: 225.

24 Le Luyer, J., Laporte, M., Beacham, T.D. et al. (2017). Parallel epigenetic modifications induced by hatchery rearing in a Pacific salmon. *Proc. Natl. Acad. Sci.* **114** (49): 12964–12969.

25 Anastasiadi, D. and Piferrer, F. (2019). Epimutations in developmental genes underlie the onset of domestication in farmed European sea bass. *Mol. Biol. Evol.* **36** (10): 2252–2264.

26 Podgorniak, T., Brockmann, S., Konstantinidis, I. et al. (2019). Differences in the fast muscle methylome provide insight into sex-specific epigenetic regulation of growth in Nile tilapia during early stages of domestication. *Epigenetics* **14** (8): 818–836.

27 Leitwein, M., Laporte, M., Le Luyer, J. et al. (2021). Epigenomic modifications induced by hatchery rearing persist in germ line cells of adult salmon after their oceanic migration. *Evol. Appl.* **14** (10): 2402–2413.

28 Mirbahai, L. and Chipman, J.K. (2014). Epigenetic memory of environmental organisms: a reflection of lifetime stressor exposures. *Mutat. Res. – Genet. Toxicol. Environ. Mutagen.* **764–765**: 10–17.

29 García-Giménez, J.L. (2016). *Epigenetic Biomarkers and Diagnostics*. Academic Press.

30 Berdasco, M. and Esteller, M. (2019). Clinical epigenetics: seizing opportunities for translation. *Nat. Rev. Genet.* **20** (2): 109–127.

31 Bird, A. (2002). DNA methylation patterns and epigenetic memory. *Genes Dev.* **16** (1): 6–21.

32 Leygo, C., Williams, M., Jin, H.C. et al. (2017). DNA methylation as a noninvasive epigenetic biomarker for the detection of cancer. *Dis. Markers* **2017**: 1–13.

33 Mirbahai, L., Yin, G., Bignell, J.P. et al. (2011). DNA methylation in liver tumorigenesis in fish from the environment. *Epigenetics* **6** (11): 1319–1333.

34 Baerwald, M.R., Meek, M.H., Stephens, M.R. et al. (2016). Migration-related phenotypic divergence is associated with epigenetic modifications in rainbow trout. *Mol. Ecol.* **25** (8): 1785–1800.

35 Moghadam, H.K., Johnsen, H., Robinson, N. et al. (2017). Impacts of early life stress on the methylome and transcriptome of Atlantic salmon. *Sci. Rep.* **7** (1): 1–11.

36 Shao, C., Li, Q., Chen, S. et al. (2014). Epigenetic modification and inheritance in sexual reversal of fish. *Genome Res.* **24** (4): 604–615.

37 Gavery, M.R. and Roberts, S.B. (2010). DNA methylation patterns provide insight into epigenetic regulation in the Pacific oyster (*Crassostrea gigas*). *BMC Genomics* **11**: 483.
38 De Paoli-Iseppi, R., Deagle, B.E., Polanowski, A.M. et al. (2019). Age estimation in a long-lived seabird (*Ardenna tenuirostris*) using DNA methylation-based biomarkers. *Mol. Ecol. Resour.* **19** (2): 411–425.
39 Polanowski, A.M., Robbins, J., Chandler, D. et al. (2014). Epigenetic estimation of age in humpback whales. *Mol. Ecol. Resour.* **14** (5): 976–987.
40 Fairfield, E.A., Richardson, D.S., Daniels, C.L. et al. (2021). Ageing European lobsters (*Homarus gammarus*) using DNA methylation of evolutionarily conserved ribosomal DNA. *Evol. Appl.* **14** (9): 2305–2318.
41 Fairfield, E. (2020). Ageing the unageable: investigating novel methods of ageing crustaceans. Doctoral dissertation. University of East Anglia.
42 Moraleda-Prados, J., Caballero-Huertas, M., Valdivieso, A. et al. (2021). Epigenetic differences in the innate response after immune stimulation during zebrafish sex differentiation. *Dev. Comp. Immunol.* **114**: 103848.
43 Piferrer, F. (2019). Epigenetics of sex determination and differentiation in fish. In: *Sex Control in Aquaculture* (ed. H.-P. Wang, F. Piferrer, S.-L. Chen, and Z.-G. Shen), 65–83. Wiley.
44 Anastasiadi, D., Vandeputte, M., Sánchez-Baizán, N. et al. (2018). Dynamic epimarks in sex-related genes predict gonad phenotype in the European sea bass, a fish with mixed genetic and environmental sex determination. *Epigenetics* **13** (9): 988–1011.
45 Wang, L., Zhang, J., Duan, J. et al. (2014). Programming and inheritance of parental DNA methylomes in mammals. *Cell* **157** (4): 979–991.
46 Gezer, U., Yörüker, E.E., Keskin, M. et al. (2015). Histone methylation marks on circulating nucleosomes as novel blood-based biomarker in colorectal cancer. *Int. J. Mol. Sci.* **16** (12): 29654–29662.
47 Gavery, M.R. and Roberts, S.B. (2017). Epigenetic considerations in aquaculture. *PeerJ* **5**: e4147.
48 Best, C., Ikert, H., Kostyniuk, D.J. et al. (2018). Epigenetics in teleost fish: from molecular mechanisms to physiological phenotypes. *Comp. Biochem. Physiol. Part B Biochem. Mol. Biol.* **224**: 210–244.
49 Holoch, D. and Moazed, D. (2015). RNA-mediated epigenetic regulation of gene expression. *Nat. Rev. Genet.* **16** (2): 71–84.
50 Hayes, J., Peruzzi, P.P., and Lawler, S. (2014). MicroRNAs in cancer: biomarkers, functions and therapy. *Trends Mol. Med.* **20** (8): 460–469.
51 Keller, A., Leidinger, P., Bauer, A. et al. (2011). Toward the blood-borne miRNome of human diseases. *Nat. Methods* **8** (10): 841–843.
52 Cardona, E., Guyomar, C., Desvignes, T. et al. (2021). Circulating miRNA repertoire as a biomarker of metabolic and reproductive states in rainbow trout. *BMC Biol.* **19** (1): 235.
53 Papadaki, M., Kaitetzidou, E., Mylonas, C.C. et al. (2020). Non-coding RNA expression patterns of two different teleost gonad maturation stages. *Mar. Biotechnol.* **22** (5): 683–695.
54 Andreassen, R. and Høyheim, B. (2017). miRNAs associated with immune response in teleost fish. *Dev. Comp. Immunol.* **75**: 77–85.
55 Xue, X., Woldemariam, N.T., Caballero-Solares, A. et al. (2019). Dietary immunostimulant CpG modulates microRNA biomarkers associated with immune responses in Atlantic salmon (*Salmo salar*). *Cells* **8** (12): 1592.
56 Bizuayehu, T.T., Johansen, S.D., Puvanendran, V. et al. (2015). Temperature during early development has long-term effects on microRNA expression in Atlantic cod (*Gadus morhua*). *BMC Genomics* **16** (1): 305.
57 Huang, C., Chen, N., Wu, X.-J. et al. (2015). The zebrafish miR-462/miR-731 cluster is induced under hypoxic stress via hypoxia-inducible factor 1α and functions in cellular adaptations. *FASEB J.* **29** (12): 4901–4913.
58 Zhou, W., Xie, Y., Li, Y. et al. (2021). Research progress on the regulation of nutrition and immunity by microRNAs in fish. *Fish Shellfish Immunol.* **113**: 1–8.
59 Rbbani, G., Nedoluzhko, A., Galindo-Villegas, J. et al. (2021). Function of circular RNAs in fish and their potential application as biomarkers. *Int. J. Mol. Sci.* **22** (13): 7119.
60 Feng, D., Li, Q., Yu, H. et al. (2020). Integrated analysis of microRNA and mRNA expression profiles in *Crassostrea gigas* to reveal functional miRNA and miRNA-targets regulating shell pigmentation. *Sci. Rep.* **10** (1): 20238.
61 Adrian-Kalchhauser, I., Sultan, S.E., Shama, L. et al. (2021). Inherited gene regulation unifies molecular approaches to nongenetic inheritance: response to Edelaar et al. *Trends Ecol. Evol.* **36** (6): 477.
62 Valdivieso, A., Ribas, L., Monleón-Getino, A. et al. (2020). Exposure of zebrafish to elevated temperature induces sex ratio shifts and alterations in the testicular epigenome of unexposed offspring. *Environ. Res.* **186**: 109601.

63 Krick, M.V., Desmarais, E., Samaras, A. et al. (2021). Family-effects in the epigenomic response of red blood cells to a challenge test in the European sea bass (*Dicentrarchus labrax*, L.). *BMC Genomics* **22** (1): 111.

64 Anastasiadi, D., Venney, C.J., Bernatchez, L. et al. (2021). Epigenetic inheritance and reproductive mode in plants and animals. *Trends Ecol. Evol.* **36** (12): 1124–1140.

65 Ortega-Recalde, O. and Hore, T.A. (2019). DNA methylation in the vertebrate germline: balancing memory and erasure. *Essays Biochem.* **63** (6): 649–661.

66 Gokhman, D., Malul, A., and Carmel, L. (2017). Inferring past environments from ancient epigenomes. *Mol. Biol. Evol.* **34** (10): 2429–2438.

67 Waterland, R.A., Kellermayer, R., Laritsky, E. et al. (2010). Season of conception in rural Gambia affects DNA methylation at putative human metastable epialleles. *PLoS Genet.* **6** (12): e1001252.

68 Dominguez-Salas, P., Moore, S.E., Baker, M.S. et al. (2014). Maternal nutrition at conception modulates DNA methylation of human metastable epialleles. *Nat. Commun.* **5** (1): 3746.

69 Anastasiadi, D., Shao, C., Chen, S. et al. (2021). Footprints of global change in marine life: inferring past environment based on DNA methylation and gene expression marks. *Mol. Ecol.* **30** (3): 747–760.

70 Shi, K., Lin, W., and Zhao, X. (2020). Identifying molecular biomarkers for diseases with machine learning based on integrative omics. *IEEE/ACM Trans. Comput. Biol. Bioinforma.* **18**: 2514–2525.

71 Bock, C. (2009). Epigenetic biomarker development. *Epigenomics* **1** (1): 99–110.

72 Piferrer, F., Anastasiadi, D., Valdivieso, A. et al. (2019). The model of the conserved epigenetic regulation of sex. *Front. Genet.* **10**: 857.

73 Kassambara, A. (2018). *Machine Learning Essentials: Practical Guide in R*. Sthda.

74 Boehmke, B. and Greenwell, B. (2019). *Hands-on Machine Learning with R*. Boca Raton, FL: Chapman and Hall/CRC.

75 Crowgey, E.L., Marsh, A.G., Robinson, K.G. et al. (2018). Epigenetic machine learning: utilizing DNA methylation patterns to predict spastic cerebral palsy. *BMC Bioinform.* **19** (1): 225.

76 Balki, I., Amirabadi, A., Levman, J. et al. (2019). Sample-size determination methodologies for machine learning in medical imaging research: a systematic review. *Can. Assoc. Radiol. J.* **70** (4): 344–353.

77 Beleites, C., Neugebauer, U., Bocklitz, T. et al. (2013). Sample size planning for classification models. *Anal. Chim. Acta* **760**: 25–33.

78 Mayne, B., Berry, O., and Jarman, S. (2021). Optimal sample size for calibrating DNA methylation age estimators. *Mol. Ecol. Resour.* **21** (7): 2316–2323.

79 Figueroa, R.L., Zeng-Treitler, Q., Kandula, S. et al. (2012). Predicting sample size required for classification performance. *BMC Med. Inform. Decis. Mak.* **12** (1): 8.

80 Kuhn, M. (2008). Building predictive models in R using the caret package. *J. Stat. Softw.* **28** (5): 1–26.

81 Rauschert, S., Raubenheimer, K., Melton, P.E. et al. (2020). Machine learning and clinical epigenetics: a review of challenges for diagnosis and classification. *Clin. Epigenetics* **12** (1): 51.

82 Kuhn, M. and Johnson, K. (2013). *Applied Predictive Modeling*. Springer.

83 Valdivieso, A., Anastasiadi, D., Ribas, L. et al. (2022). Development of epigenetic biomarkers for the identification of sex and thermal stress in fish using DNA methylation analysis and machine learning procedures. *Mol. Ecol. Resour.* **23**: 453–470.

84 Horvath, S. (2013). DNA methylation age of human tissues and cell types. *Genome Biol.* **14** (10): R115.

85 Mayne, B., Espinoza, T., Roberts, D. et al. (2021). Nonlethal age estimation of three threatened fish species using DNA methylation: Australian lungfish, Murray cod and Mary River cod. *Mol. Ecol. Resour.* **21** (7): 2324–2332.

86 Anastasiadi, D. and Piferrer, F. (2020). A clockwork fish: age prediction using DNA methylation-based biomarkers in the European sea bass. *Mol. Ecol. Resour.* **20** (2): 387–397.

87 Weber, D.N., Fields, A.T., Patterson, W.F. et al. (2021). Novel epigenetic age estimation in wild-caught Gulf of Mexico reef fishes. *Can. J. Fish. Aquat. Sci.* **99** (999): 1–5.

88 Li, M. and Leatherland, J.F. (2013). The implications for aquaculture practice of epigenomic programming of components of the endocrine system of teleostean embryos: lessons learned from mammalian studies. *Fish* **14** (4): 528–553.

89 Moghadam, H., Mørkøre, T., and Robinson, N. (2015). Epigenetics – potential for programming fish for aquaculture? *J. Mar. Sci. Eng.* **3** (2): 175–192.

90 Granada, L., Lemos, M.F.L., Cabral, H.N. et al. (2018). Epigenetics in aquaculture – the last frontier. *Rev. Aquac.* **10** (4): 994–1013.

91 Heydenrych, M.J., Saunders, B.J., Bunce, M. et al. (2021). Epigenetic measurement of key vertebrate population biology parameters. *Front. Ecol. Evol.* **9**: 835.

92 de Raposo, Magalhães, C.S.F., Cerqueira, M.A.C., Schrama, D. et al. (2020). A proteomics and other omics approach in the context of farmed fish welfare and biomarker discovery. *Rev. Aquac.* **12** (1): 122–144.

93 Eirin-Lopez, J.M. and Putnam, H.M. (2019). Marine environmental epigenetics. *Annu. Rev. Mar. Sci.* **11** (1): 335–368.

94 Wringe, B.F., Purchase, C.F., and Fleming, I.A. (2016). In search of a "cultured fish phenotype": a systematic review, meta-analysis and vote-counting analysis. *Rev. Fish Biol. Fish.* **26** (3): 351–373.

95 Venney, C.J., Wellband, K.W., and Heath, D.D. (2021). Rearing environment affects the genetic architecture and plasticity of DNA methylation in Chinook salmon. *Heredity (Edinb).* **126** (1): 38–49.

96 Konstantinidis, I., Sætrom, P., Mjelle, R. et al. (2020). Major gene expression changes and epigenetic remodelling in Nile tilapia muscle after just one generation of domestication. *Epigenetics* **15** (10): 1052–1067.

97 Gatzmann, F., Falckenhayn, C., Gutekunst, J. et al. (2018). The methylome of the marbled crayfish links gene body methylation to stable expression of poorly accessible genes. *Epigenetics Chromatin* **11** (1): 57.

98 Metzger, D.C.H. and Schulte, P.M. (2016). Epigenomics in marine fishes. *Mar. Genomics* **30**: 43–54.

99 Angers, B., Perez, M., Menicucci, T. et al. (2020). Sources of epigenetic variation and their applications in natural populations. *Evol. Appl.* **13** (6): 1262–1278.

100 Covelo-Soto, L., Saura, M., and Morán, P. (2015). Does DNA methylation regulate metamorphosis? The case of the sea lamprey (*Petromyzon marinus*) as an example. *Comp. Biochem. Physiol. Part B Biochem. Mol. Biol.* **185**: 42–46.

101 Morán, P. and Pérez-Figueroa, A. (2011). Methylation changes associated with early maturation stages in the Atlantic salmon. *BMC Genet.* **12** (1): 86.

102 Si, Y., He, F., Wen, H. et al. (2016). Genetic polymorphisms and DNA methylation in exon 1 CpG-rich regions of PACAP gene and its effect on mRNA expression and growth traits in half smooth tongue sole (*Cynoglossus semilaevis*). *Fish Physiol. Biochem.* **42** (2): 407–421.

103 Mennigen, J.A., Skiba-Cassy, S., and Panserat, S. (2013). Ontogenesis of expression of metabolic genes and microRNAs in rainbow trout alevins during the transition from the endogenous to the exogenous feeding period. *J. Exp. Biol.* **16**: 1597–1608.

104 Huang, C.W., Li, Y.H., Hu, S.Y. et al. (2012). Differential expression patterns of growth-related microRNAs in the skeletal muscle of Nile tilapia (*Oreochromis niloticus*). *J. Anim. Sci.* **90** (12): 4266–4279.

105 Navarro-Martín, L., Viñas, J., Ribas, L. et al. (2011). DNA methylation of the gonadal aromatase (*cyp19a*) promoter is involved in temperature-dependent sex ratio shifts in the European sea bass. *PLoS Genet.* **7** (12): e1002447.

106 Devlin, R.H. and Nagahama, Y. (2002). Sex determination and sex differentiation in fish: an overview of genetic, physiological, and environmental influences. *Aquaculture* **208** (3–4): 191–364.

107 Domingos, J.A., Budd, A.M., Banh, Q.Q. et al. (2018). Sex-specific *dmrt1* and *cyp19a1* methylation and alternative splicing in gonads of the protandrous hermaphrodite barramundi. *PLoS one* **13** (9): 1–19.

108 Consuegra, S. and Rodríguez López, C.M. (2016). Epigenetic-induced alterations in sex-ratios in response to climate change: an epigenetic trap? *BioEssays* **38** (10): 950–958.

109 Piferrer, F. (2016). Altered sex ratios in response to climate change-who will fall into the (epigenetic) trap? *BioEssays* **38** (10): 939–939.

110 Geay, F., Zambonino-Infante, J., Reinhardt, R. et al. (2012). Characteristics of *fads2* gene expression and putative promoter in European sea bass (*Dicentrarchus labrax*): comparison with salmonid species and analysis of CpG methylation. *Mar. Genomics* **5**: 7–13.

111 Perera, E., Turkmen, S., Simó-Mirabet, P. et al. (2020). Stearoyl-CoA desaturase (scd1a) is epigenetically regulated by broodstock nutrition in gilthead sea bream (*Sparus aurata*). *Epigenetics* **15** (5): 536–553.

112 Marandel, L., Lepais, O., Arbenoits, E. et al. (2016). Remodelling of the hepatic epigenetic landscape of glucose-intolerant rainbow trout (*Oncorhynchus mykiss*) by nutritional status and dietary carbohydrates. *Sci. Rep.* **6** (1): 32187.

113 Mennigen, J.A., Plagnes-Juan, E., Figueredo-Silva, C.A. et al. (2014). Acute endocrine and nutritional co-regulation of the hepatic omy-miRNA-122b and the lipogenic gene fas in rainbow trout, *Oncorhynchus mykiss*. *Comp. Biochem. Physiol. Part B Biochem. Mol. Biol.* **169**: 16–24.

114 Kostyniuk, D.J., Marandel, L., Jubouri, M. et al. (2019). Profiling the rainbow trout hepatic miRNAome under diet-induced hyperglycemia. *Physiol. Genomics* **51** (9): 411–431.

115 Mennigen, J.A., Panserat, S., Larquier, M. et al. (2012). Postprandial regulation of hepatic microRNAs predicted to target the insulin pathway in rainbow trout. *PLoS one* **7** (6): e38604.

116 Simpson, D.J. and Chandra, T. (2021). Epigenetic age prediction. *Aging Cell* **20** (9): e13452.

117 Bell, C.G., Lowe, R., Adams, P.D. et al. (2019). DNA methylation aging clocks: challenges and recommendations. *Genome Biol.* **20** (1): 249.

118 Bertucci, E.M., Mason, M.W., Rhodes, O.E. et al. (2021). Exposure to ionizing radiation disrupts normal epigenetic aging in Japanese medaka. *Aging (Albany NY)* **13** (19): 22752–22771.

119 Consortium, M.M., Lu, A.T., Fei, Z. et al. (2021). Universal DNA methylation age across mammalian tissues. *BioRxiv* **5**: 410.

120 Parrott, B.B. and Bertucci, E.M. (2019). Epigenetic aging clocks in ecology and evolution. *Trends Ecol. Evol.* **34** (9): 767–770.

121 Field, A.E., Robertson, N.A., Wang, T. et al. (2018). DNA methylation clocks in aging: categories, causes, and consequences. *Mol. Cell* **71** (6): 882–895.

122 Smart, J.J. and Grammer, G.L. (2021). Modernising fish and shark growth curves with Bayesian length-at-age models. *PLoS one* **16** (2): e0246734.

123 Caballero-Huertas, M., Moraleda-Prados, J., Joly, S. et al. (2020). Immune genes, IL1β and Casp9, show sexual dimorphic methylation patterns in zebrafish gonads. *Fish Shellfish Immunol.* **97**: 648–655.

124 Mitchell, P.S., Parkin, R.K., Kroh, E.M. et al. (2008). Circulating microRNAs as stable blood-based markers for cancer detection. *Proc. Natl. Acad. Sci.* **105** (30): 10513–10518.

125 Andreassen, R., Woldemariam, N.T., Egeland, I.Ø. et al. (2017). Identification of differentially expressed Atlantic salmon miRNAs responding to *Salmonid alphavirus* (SAV) infection. *BMC Genomics* **18** (1): 349.

126 Valenzuela-Miranda, D., Valenzuela-Muñoz, V., Farlora, R. et al. (2017). MicroRNA-based transcriptomic responses of Atlantic salmon during infection by the intracellular bacterium *Piscirickettsia salmonis*. *Dev. Comp. Immunol.* **77**: 287–296.

127 Jonsson, B. and Jonsson, N. (2019). Phenotypic plasticity and epigenetics of fish: embryo temperature affects later-developing lift-history traits. *Aquat. Biol.* **28**: 21–32.

128 Metzger, D.C.H. and Schulte, P.M. (2017). Persistent and plastic effects of temperature on DNA methylation across the genome of threespine stickleback (*Gasterosteus aculeatus*). *Proc. R. Soc. B Biol. Sci.* **284** (1864): 20171667.

129 Beemelmanns, A., Zanuzzo, F.S., Sandrelli, R.M. et al. (2021). The Atlantic salmon's stress- and immune-related transcriptional responses to moderate hypoxia, an incremental temperature increase, and these challenges combined. *G3 Genes|Genomes|Genetics* **11** (7): jkab102.

130 Han, B., Li, W., Chen, Z. et al. (2016). Variation of DNA methylome of zebrafish cells under cold pressure. *PLoS one* **11** (8): e0160358.

131 Simonet, N.G., Reyes, M., Nardocci, G. et al. (2013). Epigenetic regulation of the ribosomal cistron seasonally modulates enrichment of H2A.Z and H2A.Zub in response to different environmental inputs in carp (*Cyprinus carpio*). *Epigenetics Chromatin* **6** (1): 22.

132 Campos, C., Sundaram, A.Y., Valente, L.M. et al. (2014). Thermal plasticity of the miRNA transcriptome during *Senegalese sole* development. *BMC Genomics* **15** (1): 525.

133 Burt, K., Hamoutene, D., Mabrouk, G. et al. (2012). Environmental conditions and occurrence of hypoxia within production cages of Atlantic salmon on the south coast of Newfoundland. *Aquac. Res.* **43** (4): 607–620.

134 Veron, V., Marandel, L., Liu, J. et al. (2018). DNA methylation of the promoter region of *bnip3* and *bnip3l* genes induced by metabolic programming. *BMC Genomics* **19** (1): 677.

135 Wang, S.Y., Lau, K., Lai, K.-P. et al. (2016). Hypoxia causes transgenerational impairments in reproduction of fish. *Nat. Commun.* **7** (1): 12114.

136 Kvist, J., Gonçalves Athanàsio, C., Shams Solari, O. et al. (2018). Pattern of DNA methylation in *Daphnia*: evolutionary perspective. *Genome Biol. Evol.* **10** (8): 1988–2007.

137 Burke, H., Gardner, I., and Farrell, A.P. (2020). A Review of the 2019 Newfoundland and Labrador South Coast Cultured Atlantic Salmon Mortality Event. *Dep. Fish. L. Resour. Gov. Newfoundl. Labrador, Spec. Stud. Reports*.

138 Morán, P., Marco-Rius, F., Megías, M. et al. (2013). Environmental induced methylation changes associated with seawater adaptation in brown trout. *Aquaculture* **392–395**: 77–83.

139 Yan, B., Zhao, L.-H., Guo, J.-T. et al. (2012). miR-429 regulation of osmotic stress transcription factor 1 (OSTF1) in tilapia during osmotic stress. *Biochem. Biophys. Res. Commun.* **426** (3): 294–298.

140 Yan, B., Guo, J.-T., Zhao, L.-H. et al. (2012). MiR-30c: a novel regulator of salt tolerance in tilapia. *Biochem. Biophys. Res. Commun.* **425** (2): 315–320.

141 Madeira, D., Mendonça, V., Madeira, C. et al. (2019). Molecular assessment of wild populations in the marine realm: importance of taxonomic, seasonal and habitat patterns in environmental monitoring. *Sci. Total Environ.* **654**: 250–263.

142 Santangeli, S., Maradonna, F., Gioacchini, G. et al. (2016). BPA-induced deregulation of epigenetic patterns: effects on female zebrafish reproduction. *Sci. Rep.* **6** (1): 21982.

143 Kamstra, J.H., Sales, L.B., Aleström, P. et al. (2017). Differential DNA methylation at conserved non-genic elements and evidence for transgenerational inheritance following developmental exposure to mono(2-ethylhexyl) phthalate and 5-azacytidine in zebrafish. *Epigenetics Chromatin* **10** (1): 20.

144 Vogt, G. (2021). Epigenetic variation in animal populations: sources, extent, phenotypic implications, and ecological and evolutionary relevance. *J. Biosci.* **46** (1): 24.

145 Smale, D.A., Wernberg, T., Oliver, E.C.J. et al. (2019). Marine heatwaves threaten global biodiversity and the provision of ecosystem services. *Nat. Clim. Chang.* **9** (4): 306–312.

146 Stajic, D. and Jansen, L.E.T. (2021). Empirical evidence for epigenetic inheritance driving evolutionary adaptation. *Philos. Trans. R. Soc. B Biol. Sci.* **376** (1826): 20200121.

147 Laporte, M., Le Luyer, J., Rougeux, C. et al. (2019). DNA methylation reprogramming, TE derepression, and postzygotic isolation of nascent animal species. *Sci. Adv.* **5** (10): eaaw1644.

148 Kelsey, G., Stegle, O., and Reik, W. (2017). Single-cell epigenomics: recording the past and predicting the future. *Science(80-)* **358** (6359): 69–75.

149 Chen, Q.-F., Gao, H., Pan, Q.-Y. et al. (2021). Analysis at the single-cell level indicates an important role of heterogeneous global DNA methylation status on the progression of lung adenocarcinoma. *Sci. Rep.* **11** (1): 23337.

150 Griffin, P.T., Kane, A.E., Trapp, A. et al. (2021). Ultra-cheap and scalable epigenetic age predictions with TIME-Seq. *BioRxiv* https://doi.org/10.1101/2021.10.25.465725.

151 Arneson, A., Haghani, A., Thompson, M.J. et al. (2022). A mammalian methylation array for profiling methylation levels at conserved sequences. *Nat. Commun.* **13** (1): 783.

19

Genetics and Epigenetics in Aquaculture Breeding

Shokouoh Makvandi-Nejad[1] *and Hooman Moghadam*[2]

[1] *Department of Immunology and Virology, Norwegian Veterinary Institute, Ås, Norway*
[2] *Breeding and Genetics, Benchmark Genetics Norway AS, Bergen, Norway*

19.1 Overview

The Neolithic revolution, which began around 10,000 to 15,000 years ago, fundamentally altered the relationship between *Homo sapiens* and their environment. The transition from a hunter–gatherer lifestyle to agriculture-based production and the domestication of various plant and animal species provided a plethora of knowledge and technological advancements. Domestication was a gradual process followed by selecting plants or animals with a more desirable appearance or superior performance, such as coat color, high yield, faster growth, or pathogen resistance. Breeding and husbandry principles have been applied for thousands of years, even though much of its theoretical concepts and molecular mechanisms are new knowledge that we do not fully understand. In fact, over the past century, genetic inheritance theories have fueled much of the research in biological sciences. The idea that the expression of specific phenotypes is mainly due to variations in genes and genetic materials provided the core approach to our understanding of the molecular mechanism of trait regulation. Over the past few decades, the significant advances in the production of farmed plants and animals or the created diversity in the look, size, behavior, and health of the various domesticated dog and chicken breeds are a testament to the power of genetic approaches and genomic information in selecting for improved phenotypes.

Today, with the availability of complete genome sequence information from many species of economic importance and various genotyping platforms, the efficiency of our selective breeding programs has advanced significantly. We have developed new genomic tools and devised new strategies such as genomic selection or GWAS (Box 19.1) to assess the genetic values of individuals for various traits based on signatures obtained from the entirety of their genomes. These new tools have allowed us to capture a sizable proportion of sequence-based heritability, explaining part of the phenotypic variation. However, polymorphisms in sequence data alone cannot fully explain the observed diversity in phenotypes, and at least for some traits, there remains a significant portion of unaccounted heritable components [1]. The non-Mendelian inheritance patterns, phenomena such as epistasis, extranuclear inheritance, and gene conversion, offer some additional modes of molecular modification that can potentially contribute to the observed differences between individuals [2–4]. In recent years, however, based on the findings from various model and nonmodel organisms, there has been considerable attention paid to the potential of epigenetic information in explaining parts of the heritable variation [5]. We now know that epigenetic marks play vital roles in regulating different aspects of cell-specific gene expression, gene silencing, chromatin remodeling, imprinting, and the spatiotemporal patterns of cell differentiation [6, 7]. In the face of ever-increasing challenges such as limited resources, increasing population, and environmental constraints due to the changing climate, we need new strategies and a deeper understanding of the sources contributing to phenotypic diversity. In this regard, this chapter aims to provide the reader with a brief overview of advances in selective breeding in aquaculture along with the evolution and application of genomic tools suitable for breeding programs in these species, and then introduces the concept of epigenetics in breeding, some of the critical current knowledge, existing gaps, challenges, and potentials for future applications in selective breeding.

Epigenetics in Aquaculture, First Edition. Edited by Francesc Piferrer and Han-Ping Wang.

Box 19.1 Glossary of Terms

Epigenetic Marks: Molecular modifications to DNA, such as methylation of cytosine residues, that are often associated with transcriptional activities of genes.
Epigenome: The profile of epigenetic marks throughout the genome for a given tissue or cell at a given time.
Epigenome-Wide Association Studies (EWAS): Studies that aim to investigate the association between epigenetic modifications and a particular trait.
Genome-Wide Association Studies (GWAS): Studies that involve genotyping large numbers of individuals with the aim to test the association or correlation between genetic variations, throughout the genome, with a phenotype of interest.
Genomic Selection (GS): A selection method based on the predicted breeding values computed from the estimated effect size of the genetic markers covering the entire genome.
Heritability: The proportion of phenotypic variation in a trait due to genetic variation between individuals. Heritability is a standardized measure, with values ranging from zero (no genetic variation contributing to phenotypic differences) to one (all of the variations in the population are due to heritable genetic factors).
Linkage Disequilibrium (LD): A measure of nonrandom association and segregation of alleles at distinct loci due to limited or lack of recombination.
Missing Heritability: The observation for some complex traits is that the total phenotypic variance explained by all individually significant genomic loci identified in a genome-wide association is less than the heritability estimates obtained from traditional quantitative genetic methods.
Pedigree Information: Animals within a specific population or breed whose parents are known.
Quantitative Trait Loci (QTL): Regions in the genome that harbor the causative mutations that contribute to the observed phenotypic variation.
Single Nucleotide Polymorphism (SNP): Single base pair differences in the DNA sequence at any given position.
Transgenerational Epigenetic Inheritance: The transmission of epigenetic marks from the parents to the offspring and across generations.

19.2 Breeding in Aquaculture and Evolution of Genetic Markers

Food production from various aquaculture sources is growing faster than any other farmed animal sector [8] (Figure 19.1). This success is primarily due to global demand for high-quality protein food and the improvements in husbandry and farming conditions, advancements in feed and nutrition, and the successful implementation of breeding programs. Breeding can substantially accelerate genetic gains and improve economic returns. The structure of a breeding program is mainly influenced by the available technology and the biology of the species. In its simplest form, mass selection has historically been applied in many aquaculture programs (e.g., [9, 10]). Even today, for some species, particularly those with high fecundity and in some geographical regions, this approach might still be the method of choice. For instance, data from the Pacific

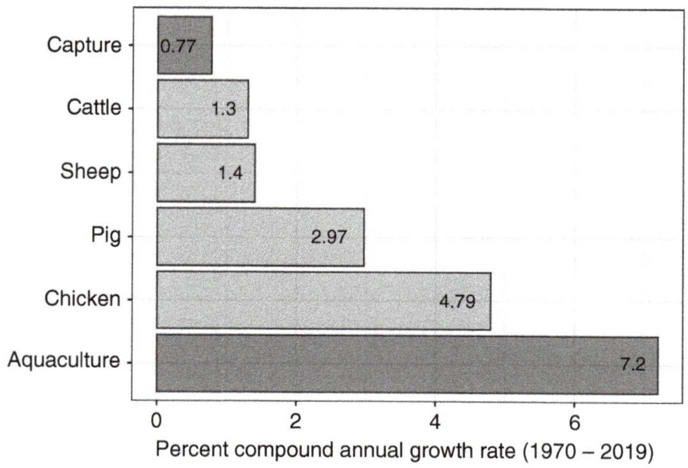

Figure 19.1 The compound annual growth rate for aquaculture, poultry, pig, sheep, cattle, and fisheries between 1970 and 2019. Values are expressed in percent. *Sources:* https://www.fao.org/fishery/statistics-query and https://www.fao.org/faostat.

oyster (*Crassostrea gigas*) breeding programs in China have shown that mass selection might be an effective way to improve the shell height or growth of this species [11, 12]. However, one of the main concerns with this selection method is the risk associated with the potential increase in inbreeding depression within the population. This is mainly the result of a significantly higher contribution of only a few individuals to the next generation. Such a biased representation from a limited number of animals can, as a result, mask the potential genetic impacts that could have been manifested if all the breeding candidates had the opportunity to make a higher contribution. The other limitation of mass selection is that it is restricted only to phenotypes that can be recorded on the breeding candidates. As a result, this method cannot be applicable to many economic traits of interest, such as carcass traits or disease resistance.

The family-based selection addresses these concerns, and nowadays, it constitutes the primary selection strategy for many aquaculture species with a more advanced breeding program. The approach is highly efficient in genetically improving the traits targeted for selection [13]. Under this method, the parental broodstock forms genetically diverse nucleus families, and the breeding scheme takes advantage of many progenies resulting from different crosses. The performance of the full or half-siblings against various pathogens or the slaughter and quality traits will then be evaluated to obtain average family data. Animals can be identified using different tagging strategies [14] or by utilization of genetic markers such as microsatellites or single nucleotide polymorphism (SNP) (Box 19.1). These methods will allow the tracing of the animals back to their family of origin, facilitating the selection process. Nevertheless, maintaining such an operation routinely is an endeavor that requires knowledge, expertise, and financial resources. In addition, the genetic variation within families cannot be fully exploited using this approach alone.

In fact, over the past few decades, significant efforts have been made to develop genomic information and genetic markers for many aquaculture species of economic importance. During the early years, the primary sources of genetic variations used for different applications in aquatic species (e.g., conservation, population genetics, evolutionary studies, and breeding) were molecular markers such as allozymes (polymorphic enzymes), DNA restriction sites, minisatellites, and microsatellites [15]. Notably, during the late 1990s and the early 2000s, with the popularity of PCR technology, microsatellites started to emerge as the dominant genetic marker. Also known as short tandem repeats (STR) or simple sequence repeats (SSR), they are tandemly repeated short DNA motifs with high mutation rates [16]. As a result, microsatellites are highly polymorphic, abundant in the genome with relatively even distribution, and are easy to screen. They proved to be vastly informative for various genetic applications, particularly for the parentage assignment, sex identification, and quantitative trait loci (QTL) (Box 19.1) studies and construction of genetic maps. In fact, by facilitating the construction of linkage maps, microsatellites provided us with the first glimpse into the genome architecture of many aquaculture species such as Atlantic salmon (*Salmo salar*), rainbow trout (*Oncorhynchus mykiss*), and Arctic char (*Salvelinus alpinus*) (e.g., [17, 18]) with a higher resolution than what was previously known. Due to the highly polymorphic nature in size variation of the microsatellite markers, these also played an essential role in providing the basis for mapping the duplicated regions of the genomes in species such as salmonid fishes [18] that have originated from ancestors that had experienced polyploidization events in recent evolutionary times [19].

However, nowadays, the use of microsatellites is mainly restricted to applications such as parentage assignments that do not require a high-density panel of genetic markers. In fact, in species with more advanced genomic data, such as Atlantic salmon and rainbow trout, the use of microsatellites has now been primarily replaced by single nucleotide polymorphic markers or SNPs. In addition to being the most abundant genetic marker throughout the genome, SNPs also have the advantage of being suitable for automation and high-throughput processing [15]. Although the initial identification and development of any high-density SNP panel can be costly, time-consuming, and needs a priori knowledge regarding the allelic variations in the population or species under selection, once established, it can conveniently be used to screen thousands of samples over tens of thousands of genetic markers at a relatively low cost. Considering the density of the SNP panels, which are usually in the range of tens of thousands but occasionally can be up to hundreds of thousands, SNP arrays provide a higher resolution platform to narrow down the genomic regions of interest.

Selection based on such genetic markers has proven to be highly efficient, particularly for traits that are under the control of mainly one or a few major loci. Perhaps one of the most notable examples is the case of selection for higher resistance against the infectious pancreatic necrosis virus (IPNV) in Atlantic salmon. The virus can affect fish during various life cycles, leading to significant economic losses and deteriorating animal welfare. A series of works across salmon populations identified microsatellite genetic markers with a significant association for higher resistance against IPNV [20, 21]. The IPNV resistance was found to have high heritability (Box 19.1), where most of the genetic variation could be explained by a single QTL located on chromosome 26 [20–22]. A follow-up investigation identified mutations in the epithelial cadherin gene (*cdh1*) as the putative causative genetic variation for IPNV resistance [23]. Following the inclusion of this QTL into breeding programs, the industry has witnessed a sharp decline in IPN outbreaks [24]. However, the role of *cdh1* in conferring genetic resistance against this virus has recently been challenged [25].

Unlike resistance to IPNV, however, most economically important traits have a polygenic nature, in which the products of many genes regulate their expressions. Modern aquaculture breeding programs have started implementing advanced methods such as genomic selection (GS) (Box 19.1) to select such traits. The concept of GS was initially proposed by Meuwissen et al. [26], and later its applications in breeding programs became possible through the availability of mid-to-high-density SNP arrays. In this method, the information from all the genetic markers is used to estimate the genomic breeding value for a given trait. In order to estimate the breeding values in the breeding candidates or the test group, the trait of interest is first measured in a reference or a training population. Then, all the animals from both the reference and the breeding candidates are genotyped on the SNP panel. The marker effect is estimated based on the measurements from the training dataset. A key advantage of GS is that it is not essential to record the trait on the breeding candidates. Also, unlike the more traditional family-based selection, no pedigree information (Box 19.1) is usually required. Utilizing genomic markers can lead to a more accurate estimation of genomic relationships between animals. It can also result in higher accuracy in predicting breeding values than pedigree-based assessments [27]. Of course, the GS predictions are most accurate if the test and the training groups are closely related and might be of limited value for distantly related animals [28].

19.3 Epigenetics and Missing Heritability

The main mechanisms by which epigenetic modifications can alter gene expression patterns can broadly be divided into four domains: DNA methylation, histone modification, chromatin structure, and noncoding RNA (e.g., [29–32]). Molecular mechanisms of these processes have been described in previous chapters and elsewhere (e.g., [33]). Indeed, physiological and biological signatures of environmental exposures during critical developmental stages of an individual can potentially have profound, perhaps life-long or even multi-generational impact on the organism (e.g., [34]). In other words, changes in the epigenome (Box 19.1) can potentially provide the molecular means necessary for a phenotypic inheritance that is not entirely dependent on DNA sequence. This brings us to the concept of missing heritability (Box 19.1), which has been one of the mysteries when comparing GWAS estimates to those of classical genetic studies [1]. For some traits, at least, it has been found that the SNP-based heritability estimates from GWAS are lower than estimates from the traditional genetic analysis. Perhaps one of the best-known examples is the case of missing heritability for human height. According to classical studies, this is a highly heritable trait with h^2 about 0.8 [35]. However, the estimates from different GWAS studies have set the h^2 to only 0.05–0.45 (e.g., [1, 36]). Several explanations have been proposed as possible reasons to account for the differences between the two estimates. First, most estimates from GWAS studies rely on common variants, i.e., polymorphisms with frequencies greater than 0.01–0.05, disregarding the contributions of the rare variants. However, there are examples of rare variants with potentially significant phenotypic effects (e.g., [37]). Second, for some traits, the phenotypic expression can be attributed to the contribution of many variants with minor effects that are difficult to capture using current methods. Third, one can expect that in specific settings and production systems, estimations from the classical-based methods to be inflated due to the shared environmental effects, as they can cause the phenotypes to cluster and aggregate based on the experienced environment. However, a recent paper has suggested that part of the discrepancy might be due to how the word genetic has been defined [38]. While in the GWAS, the word "genetic" exclusively refers to DNA, in classical studies, this word can encompass any genetic or nongenetic factor (e.g., epigenetic) that can be transmitted from the parents to the offspring and act in an additive manner. As a result, GWAS remains uninformative to non-DNA elements.

As one of the non-DNA elements, variations in the epigenetic marks have been suggested to at least partially account for some aspects of missing heritability when it comes to complex traits. For instance, it has been shown that a stable inheritance of differentially methylated loci can act as epigenetic quantitative trait loci (*epiQTL*), independent of the differences in the DNA sequences, and further can be subjected to artificial selection [39]. For this to happen, however, some epigenetic marks should be able to bypass the reset process during the gametogenesis or early developmental reprogramming stages, and hence be transmitted to the next generation. The concept of trait heritability based on transmissible and non-transmissible epigenetic and environmental elements has been the subject of various theoretical studies. In a recent work, Banta and Richards [40] have shown the profound effects of epigenetic mechanisms on various evolutionary processes and demonstrated that epigenetics could contribute to all different components of phenotypic variance (V_P), i.e., the genetic variance (V_G), environmental variance (V_E), genotype by environment interactions (V_{GxE}), the genotype-environment covariance ($2COV_{GE}$), and residual variance ($V\epsilon$). Slatkin [41] suggested that, under a single environmental model, the contribution of epigenetics to the similarities between siblings depends on the likelihood of the epigenetic marks being induced or reset between generations. In other words, no epigenetic-based heritable resemblance is expected between siblings in the absence of transgenerational memory, even though the epigenetic state might significantly contribute to the onset of the trait. In another work, Furrow et al. [42] investigated the theoretical concepts of missing heritability under the assumption of correlation between the epigenetic and environmental effects. In their model, the rate of epigenetic change is dependent on the environment that an individual experiences. Hence, a correlation between the environmental conditions experienced by the parents and the offspring can be reflected in the state of their epigenetic marks. They have shown that the estimated heritability of a trait can be highly dependent on the transmission rates of epigenetic factors, environmental conditions, and their interactions. Based on this study, even in the absence of efficient transmission of epigenetic marks, a high correlation with environmental factors might lead to high heritability estimates and result in familial covariances comparable to those expected from genetic data. These findings suggest that epigenetic variations, either through parental inheritance or correlations with environmental conditions, can play critical roles in the heritability estimate of a trait and help to untangle elements of the missing heritability.

To test if, and to what degree, the non-DNA elements contribute to the phenotypic resemblance between relatives, Tal et al. [43] proposed a method that essentially can be considered an extension of the classical quantitative genetic models of family studies. Their model considers the epigenetic transmissibility (i.e., the probability of transmission of the epigenetic state) and environmental induction (i.e., the probability that an epigenetic state is induced), with the assumption that during the transmission of the epigenetic marks from the parents to the offspring, an unknown proportion of these marks are lost during the resetting stages, the gametogenesis and embryogenesis. This unknown probability is referred to as the reset coefficient or v and can vary between the parents (v_{sire} and v_{dam}). The complement of the reset coefficient, i.e., $1-v$, is known as the coefficient of epigenetic transmissibility (Figure 19.2). They introduced a new term into the standard model, the heritable epigenetic variance (V_C), which is used to estimate epigenetic heritability (γ^2), the proportion of phenotypic variation that is determined by heritable epigenetic variance. Estimates of V_C can be obtained by measuring the covariance between parents and offspring, between siblings, and between uncles and nephews. By comparing the covariances between these relatives, one can infer the heritable epigenetic variance and epigenetic transmissibility using information about the number of opportunities for epigenetic reset following meiosis or egg fertilization. In cases of a frequent reset of the epigenetic state, the similarity of parent-offspring due to γ^2 tends to be low. On the other hand, with a stable, transgenerational transfer of the epigenetic marks, the parent-offspring similarity resulting from epigenetic contributions should be high.

However, it should be noted that so far the models remain quite general concerning the processes of epigenetic inheritance and their knock-on effects on the phenotypic expression in the next generation [42, 44, 45]. As will be discussed in the next section, the efficiency of epigenetic inheritance can vastly vary between species and among parents. Unlike monogenic traits, the expression of many phenotypes can be influenced by the interaction of genetic and nongenetic factors. Genotype by environment interaction, for instance, can cause challenges for breeding programs as the performance of the selected animals can substantially vary in different environments [46]. The dynamics of such interactions can be very complex and not trivial to model, especially considering our limited understanding of the nature of such biological processes. For instance, many of these models do not include the contributions of the genetic elements to the onset of phenotypic differences, but rather assume independence between the genetic and epigenetic variations [42, 47]. While this might be true in some circumstances [47, 48], it should not be considered a rule. Also, confirming the epigenetic inheritance over generations, separating it from modifications due to environmental exposures, and demonstrating that the epigenetic variants are independent of the cis- or trans-acting DNA elements and can modulate a phenotypic expression is a major challenge in genetic studies.

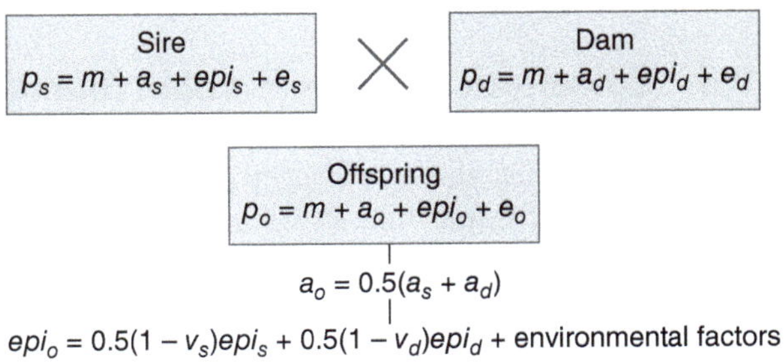

Figure 19.2 Diagram depicting the transmission of various inherited elements from the parents to offspring. a, genetic; d, dam; e, residual (includes other nonepigenetic environmental effects and error); epi, epigenetic; m, mean; o, offspring; p, phenotype; s, sire; v, reset coefficient. *Source:* Figure adapted from David and Ricard [45].

19.4 Transgenerational Inheritance of Epigenetic Marks

There exists complex machinery in multicellular organisms responsible for clearing epigenetic states between generations. However, there is now a growing interest in the idea that some epigenetic marks might have the ability to escape the reprogramming step of the primordial germ cells (PGCs) and be passed from one generation to the next. Studies in plants provided the first line of evidence of such epigenetic inheritance. A classic example is a study by Cubas et al. [49], investigating a naturally occurring mutation causing the morphology of the toadflax flower (*Linaria vulgaris*) to change from bilateral to radial. The mutation results from the transcriptional silencing of a gene known as *Lcyc* due to promoter hypermethylation. The hypermethylated state was then shown to be heritable, indicating that epigenetic mutations can make essential contributions to phenotypic evolution. However, such transgenerational epigenetic inheritance (Box 19.1) does not seem to be a universal phenomenon. In mammals, for instance, different studies have indicated that an almost complete erasure of the epigenetic marks happens between generations [50, 51]. The resetting process happens during two stages, gametogenesis and embryogenesis. Our understanding of the epigenetic reset and inheritance in aquatic species during gametogenesis or embryogenesis is minimal. There is suggestive evidence of multi-generational transmission of epigenetic marks in a few species such as Atlantic salmon [52], Chinook salmon (*Oncorhynchus tshawytscha*) [53], European sea bass (*Dicentrarchus labrax*) [54], Pacific oyster [50] and half-smooth tongue sole (*Cynoglossus semilaevis*) [55]. However, high-resolution studies that, with much higher certainty and accuracy, can trace changes in the epigenetic signatures following each cell division are limited to only a few model species, including zebrafish (*Danio rerio*) and medaka (*Oryzias latipes*) [56–58], showing that each exhibit a different mechanism of epigenetic fate in subsequent generations. In zebrafish, the sperm is hypermethylated, and it is the paternal methylome that seems to be stably inherited during early development. On the other hand, the oocyte undergoes demethylation before reprogramming to a profile like that of the sperm [56]. This suggests an epigenetic state similar to sperm is more likely to be passed to the next generation [56]. Medaka, unlike zebrafish, more resembles the mammalian reprogramming model, where methylation marks are erased from the embryo within the first cell cycle followed by another round of demethylation and reprogramming during the PGC step [57, 58]. There have been suggestions that this might be linked to the sex-determination mechanism of the species [57, 59], as medaka, like mammals, has a male heterogametic system (XX/XY) [60], whereas zebrafish lacks the genetic sex-determination machinery [61].

To our knowledge, for all the important aquaculture species, similar detailed information and understanding of the epigenetic inheritance mechanisms are currently lacking. Whether the epigenetic marks can pass from one generation to the next and to what extent such a mechanism is essential to the organism's biology are subjects for extensive future research studies. For each aquaculture species, we further need to understand how the maternal and paternal lineages might contribute to the epigenetic profile in the next generation. At the same time, we also need to better understand the molecular basis of quantitative traits and how various molecular and cellular components contribute to the onset of a phenotype. In this regard, it is quite likely that some of the variation in quantitative traits under selection in our breeding programs is due to variations in the epigenetic rather than genetic sequences. Combining this data with the inheritance profiles of the epigenetic marks would be a piece of essential knowledge if we are to incorporate such information into our breeding programs. For highly stable and heritable epigenetic marks that follow the Mendelian inheritance pattern, we expect to find them in linkage disequilibrium (LD) (Box 19.1) with other genetic variations such as SNPs. In such cases, utilizing only the information from the genetic markers or pedigree data will suffice for trait improvement. Similarly, some epigenetic marks might also be controlled by genetic variations [62, 63] and can be expected to be heritable in an additive manner. On the other hand, markers that are less stably inherited or have inheritance patterns that are not similar to the expectations from Mendelian segregation, or are easier to be triggered by perturbation in the environment, most likely do not show LD with any genetic marker. As a result, no information from these epigenetic marks will be captured on the SNP arrays or genomic assessments. This will add more noise and uncertainty to our ability to predict the phenotype of the offspring.

19.5 Epigenetic Marks – Possible Biomarkers to Improve Breeding

As mentioned earlier, the state of an epigenetic mark can be a function of direct or indirect environmental cues. Parameters such as temperature, nutrition, stress, and infection can directly impact the epigenome, particularly if those exposures happen during the early stages of life (e.g., [5]). The environmental conditions that the parents experience might also be reflected and play a role in shaping the state of some epigenomic targets [57]. Such interindividual differences in the epigenetic signatures can be translated into variations in gene expression or patterns of alternative splicing [64] and subsequently contribute to the observed phenotypic variations, both within and between populations. This is one of the characteristics that makes using epigenetic marks, as biomarkers, a possibly attractive option for assessing and improving animal health, welfare, management, and breeding. Further, due to the dynamic nature of the epigenetic marks, it is also possible to use the epigenetic state as a gauge to monitor the changing conditions in a production facility or to evaluate how the implementation of new management decisions and strategies affects the health and welfare of the animals. The term biomarker, or biological marker, has been defined as a measurable indicator suitable for assessing a biological state or a phenotypic condition [65]. Such conditions can include parameters such as animal welfare, predicting an infection outcome and interaction with pathogens, monitoring overall health, sex determination, or response to a vaccine. A suitable biomarker should be phenotype-sensitive and be specific to the conditions of interest. In other words, the ideal biomarker should mainly be attributed to a specific state rather than a variety of circumstances that an individual might experience. It should also be highly reproducible, easy to assess, and inexpensive to analyse. These characteristics can make the use of specific biomarkers attractive, mainly if many samples are processed routinely.

Biomarkers can broadly be classified into three main domains: prognostic, predictive, and diagnostic. In the context of management, health, and breeding, prognostic and predictive biomarkers can mainly be used to evaluate the outcome of a process, e.g., response to infection with a specific pathogen or to treatment [66]. Prognostic biomarkers can provide information on the likely course of a disease in untreated animals, while predictive biomarkers are helpful for assessing the effectiveness of a treatment [66]. On the other hand, diagnostic biomarkers are mostly served to indicate the state of a condition, for instance, to measure the expression or the state of a biomarker to find out if animals are stressed or infected. Although utilizing epigenetic signatures as biomarkers currently has limited applications in farmed animals, it is conceivable to anticipate that these marks show all the necessary characteristics that can make them suitable for assessing various conditions. Probably one of the best examples of the use of epigenetic biomarkers in an aquaculture setting is the discovery that the methylation state of individual CpG sites from a few candidate genes such as *cyp19a1a* and *dmrt1* in the European sea bass can be used as a tool to predict the sex of an animal [54]. As research in this area progresses, we can expect to find specific changes in the epigenetic state in a particular tissue to be associated with performance (e.g., growth or response to treatment) or to indicate a particular condition (e.g., whether the animal is infected). Alternatively, one might also consider inferring the epigenetic state of an animal and its information as a biomarker in an indirect manner in cases where there is an established link between the epigenetic marks and the gene activities. Since epigenetic signatures can alter the regulatory patterns of the genes they control, we should be able to accurately predict the state of an animal for a particular condition under investigation by measuring the level of expression or splicing profile of the candidate genes. This approach has the advantage of having already well-established protocols for efficiently collecting, storing, extracting, and quantifying RNA. Developing real-time PCR assays is easy, as they can be highly specific, inexpensive to process, and can be performed in any standard molecular biology laboratory.

19.6 Association Analysis and Search for Epigenetic Biomarkers

Various works have now started to use massively parallel sequencing technologies to investigate the association of the epigenetic factors and to identify suitable biomarkers through methods like whole genome bisulfite sequencing (WGBS) [67], chromatin immunoprecipitation sequencing (CHIP-Seq) [68], or reduced representation bisulfite sequencing (RRBS) [69]. Such efforts are valuable in revealing how different environmental parameters can modify the epigenome in a specific tissue or cell. These methods are high-resolution at the DNA base-pair level, making them beneficial, particularly at the discovery stage, by profiling the epigenome. However, their applications for breeding purposes are generally limited due to the high cost, extensive computational power necessary for analysis, storage space requirements, and competence in handling large datasets. In studying complex traits at the population level, one of the general approaches to identifying epigenetic marks associated with the phenotype of interest is through EWAS (Box 19.1). Notably, in cases where environmental conditions induce epigenetic modifications, association studies based on the variations of the epigenetic marks can be an efficient way to narrow down the genomic regions and identify the responsible candidate genes or genetic elements. Similar to GWAS, in EWAS, one will collect target tissues from a large number of individuals that express a particular phenotype (e.g., fillets with darker pigmentation or animals with higher disease resistance) and compare their epigenome profile with those of the control group. The general idea is to find epigenetic marks with significantly different frequencies between the test and the control datasets. Association mainly relies on the LD between markers and the trait under investigation. It is of particular importance to consider the relatedness between the animals and correct for possible population stratification [70, 71].

In addition to the applications in the case–control studies, where the aim is to identify the association between the epigenetic marks and the phenotype of interest, EWAS might also be applied in the parent-offspring analysis to investigate and infer possible transgenerational transmission of the epigenetic signatures. Through genomic–epigenomic integrated screening of the parent-offspring trios, it should be possible to compare the epigenome of an offspring to the parents. The study setup should allow one to accurately identify the parents' allele of origin. Usually, a large number of samples are needed, so the study has enough statistical power for identifying the targeted loci. The popularity of GWAS follows various technological advancements that allowed the affordable calling of many SNPs. For large-scale, epigenome-wide association assessments, comparable platforms are only available for profiling methylation signatures, chiefly in humans or a few model organisms (e.g., through Illumina EPIC Microarray). While other epigenetic mechanisms are similarly important in regulating genes and various cellular activities, the nature of these marks makes them more challenging for high-throughput processing. Commercial tools for assessing many samples over thousands of methylation marks are currently lacking for any aquaculture species. Alternative methods such as epi-genotyping by sequencing (epiGBS) [72] or bisulfite Restriction-Associated-Digest sequencing (bsRADseq) [73] offer some opportunities for more cost-effective screening of a large number of samples, even in the absence of a reference genome. These methods are variations of the RRBS [69] application that mainly rely on the choice of restriction enzymes and DNA fragment size selection. Such steps will help to reduce the size of the studied portion of the genome. Barcoded adaptors can be added to each sample to allow pooling and construction of multiplexed libraries, followed by high-throughput sequencing. Such methods have already been used to study the genome-wide methylation profiles in farmed fish [74]. However, these methods also suffer from certain limitations, including difficulty in targeting mainly the promoter regions, as the most functionally essential segments of the genes in the methylation context [51], and being limited to sampling only a tiny portion of the genome (usually less than 1%).

A more common method to investigate the association of epigenetic markers with the traits of interest is through a targeted approach where one typically screens a small number of candidate genes or loci that are believed to be involved in the biological processes of regulating the phenotype under study. The choice of selecting for such candidate genes or genomic regions can be based on the literature (e.g., findings from other species) and/or according to the findings from sequencing projects, such as RNASeq, RRBS, or WGBS, where a few replicates from extreme phenotypes are sequenced to very high depths [75]. Of course, identification of the causative epimutations to the traits of interest will not be an easy task. Currently, we have little idea of the importance of epimutations in the estimates of genetic variations. Simultaneous assessment of phenotypic variation within a population through GWAS and EWAS over several generations will accelerate our understanding of the epigenome (in particular the methylome), its connection with genetic variations, and its contribution in subsequent generations. We can expect significant improvement in the accuracy of the prediction of breeding values by accounting for these mechanisms.

19.7 Concluding Remarks

Our understanding of the importance of epigenetic mechanisms among aquaculture species is rapidly growing, although research is still in its infancy in this field. The methodological tools necessary for studying and population-wide screening of the epigenetic marks are expensive or inaccessible, and their potential applications to advance animal health, improve management strategies, and design for more efficient breeding programs are still unappreciated. A quick PubMed search (of the title or abstract) using the keywords "methylation," "human," "mouse," and "salmon" in 2020 and 2021 resulted in 3451, 1055, three, and 12 publications, respectively. However, it is reasonable to expect a paradigm shift in the near future where the importance of epigenetic regulation and variations in aquaculture settings are better recognized. We are now aware of the potential impacts of the environment in shaping the epimutation, which along with the genotype of an animal, can jointly drive characteristic expressions of a phenotype. It is reasonable to expect that, considering how various environmental parameters are controlled in a modern aquaculture facility, epigenetic variations might not significantly shape the phenotypic diversity between the animals. Rather, such epigenetic mechanisms might play much more important roles when we compare variations between populations or across different year classes. Hence, in practical settings, if we are to implement epigenetic contributions more efficiently in improving the performance of our animals, it is inevitable to better understand the role of management strategies and various environmental conditions in how they might help to manifest the favorable epigenetic marks. Such manifestations, particularly those imprinted during the early life stages, can critically impact the epigenome, potentially persist during the animal's life, and might even be passed to the next generation.

References

1 Maher, B. (2008). Personal genomes: the case of the missing heritability. *Nature* **456**: 18–21.

2 Le Gros, P.-A., Nagard, H., and Tenaillon, O. (2009). The evolution of epistasis and its links with genetic robustness, complexity and drift in a phenotypic model of adaptation. *Genetics* **182**: 277–293.

3 Birky, C.W. Jr. (1994). Relaxed and stringent genomes: why cytoplasmic genes don't obey Mendel's laws. *Journal of Heredity* **85**: 355–365.

4 Lassalle, F., P'erian, S., Bataillon, T. et al. (2015). GC-content evolution in bacterial genomes: the biased gene conversion hypothesis expands. *PLoS Genetics* **11**: e1004941.

5 Wang, Y., Liu, H., and Sun, Z. (2017). Lamarck rises from his grave: parental environment-induced epigenetic inheritance in model organisms and humans. *Biological Reviews of the Cambridge Philosophical Society* **92**: 2084–2111.

6 Diez, C.M., Roessler, K., and Gaut, B.S. (2014). Epigenetics and plant genome evolution. *Current Opinion in Plant Biology* **18**: 1–8.

7 Zhang, Y.-Y., Fischer, M., Colot, V., and Bossdorf, O. (2013). Epigenetic variation creates potential for evolution of plant phenotypic plasticity. *The New Phytologist* **197**: 314–322.

8 Anderson, J.L., Asche, F., Garlock, T., and Chu, J. (2017). Aquaculture: its role in the future of food. In: *World Agricultural Resources and Food Security*, Frontiers of Economics and Globalisation, 17, 159–173. Emerald Publishing Limited.

9 Bentsen, H.B. and Olesen, I. (2002). Designing aquaculture mass selection programs to avoid high inbreeding rates. *Aquaculture* **204**: 349–359.

10 Gjedrem, T. and Rye, M. (2018). Selection response in fish and shellfish: a review. *Reviews in Aquaculture* **10**: 168–179.

11 Xu, C., Li, Q., Chong, J. et al. (2019). Mass selection for growth improvement in black shell line of Pacific oyster *Crassostrea gigas*. *Journal of Ocean University of China* **18**: 1411–1416.

12 Zhang, J., Li, Q., Xu, C., and Han, Z. (2019). Response to selection for growth in three selected strains of the pacific oyster *Crassostrea gigas*. *Aquaculture* **503**: 34–39.

13 Boudry, P., Allal, F., Aslam, M.L. et al. (2021). Current status and potential of genomic selection to improve selective breeding in the main aquaculture species of international council for the exploration of the sea (ICES) member countries. *Aquaculture Reports* **20**: 100700.

14 Sandford, M., Castillo, G., and Hung, T.-C. (2020). A review of fish identification methods applied on small fish. *Reviews in Aquaculture* **12**: 542–554.

15 Schlötterer, C. (2004). The evolution of molecular markers – just a matter of fashion? *Nature Reviews Genetics* **5**: 63–69.

16 Brinkmann, B., Klintschar, M., Neuhuber, F. et al. (1998). Mutation rate in human microsatellites: influence of the structure and length of the tandem repeat. *American Journal of Human Genetics* **62**: 1408–1415.

17 Reid, D.P., Szanto, A., Glebe, B. et al. (2005). QTL for body weight and condition factor in Atlantic salmon (*Salmo salar*): comparative analysis with rainbow trout (*Oncorhynchus mykiss*) and arctic charr (*Salvelinus alpinus*). *Heredity* **94**: 166–172.

18 Sakamoto, T., Danzmann, R.G., Gharbi, K. et al. (2000). A microsatellite linkage map of rainbow trout (*Oncorhynchus mykiss*) characterised by large sex-specific differences in recombination rates. *Genetics* **155**: 1331–1345.

19 Danzmann, R.G., Davidson, E.A., Ferguson, M.M. et al. (2008). Distribution of ancestral proto-actinopterygian chromosome arms within the genomes of 4R-derivative salmonid fishes (rainbow trout and Atlantic salmon). *BMC Genomics* **9**: 557.

20 Houston, R.D., Haley, C.S., Hamilton, A. et al. (2010). The susceptibility of Atlantic salmon fry to freshwater infectious pancreatic necrosis is largely explained by a major QTL. *Heredity* **105**: 318–327.

21 Moen, T., Baranski, M., Sonesson, A.K., and Kjøglum, S. (2009). Confirmation and fine-mapping of a major QTL for resistance to infectious pancreatic necrosis in Atlantic salmon (*Salmo salar*): population-level associations between markers and trait. *BMC Genomics* **10**: 368.

22 Houston, R.D., Gheyas, A., Hamilton, A. et al. (2008). Detection and confirmation of a major QTL affecting resistance to infectious pancreatic necrosis (IPN) in Atlantic salmon (*Salmo salar*). *Developments in Biologicals* **132**: 199–204.

23 Moen, T., Torgersen, J., Santi, N. et al. (2015). Epithelial cadherin determines resistance to infectious pancreatic necrosis virus in Atlantic salmon. *Genetics* **200**: 1313–1326.

24 Sommerset, I., Walde, C.S.B., Jensen, B. et al. (2020). Fiskehelserapporten 2019, Norway.

25 Pavelin, J., Jin, Y.H., Gratacap, R.L. et al. (2021). The nedd-8 activating enzyme gene underlies genetic resistance to infectious pancreatic necrosis virus in Atlantic salmon. *Genomics* **113**: 3842–3850.

26 Meuwissen, T.H., Hayes, B.J., and Goddard, M.E. (2001). Prediction of total genetic value using genome-wide dense marker maps. *Genetics* **157**: 1819–1829.

27 Meuwissen, T., Hayes, B., and Goddard, M. (2016). Genomic selection: a paradigm shift in animal breeding. *Animal Frontiers* **6**: 6–14.

28 Meuwissen, T.H.E., Odegard, J., Andersen-Ranberg, I., and Grindflek, E. (2014). On the distance of genetic relationships and the accuracy of genomic prediction in pig breeding. *Genetics, Selection, Evolution, GSE* **46**: 49.

29 Aravin, A.A., Lagos-Quintana, M., Yalcin, A. et al. (2003). The small RNA profile during *Drosophila melanogaster* development. *Developmental Cell* **5**: 337–350.

30 Bannister, A.J. and Kouzarides, T. (2011). Regulation of chromatin by histone modifications. *Cell Research* **21**: 381–395.

31 Bird, A. (2002). DNA methylation patterns and epigenetic memory. *Genes & Development* **16**: 6–21.

32 Kouzarides, T. (2007). Chromatin modifications and their function. *Cell* **128**: 693–705.

33 Cavalli, G. and Heard, E. (2019). Advances in epigenetics link genetics to the environment and disease. *Nature* **571**: 489–499.

34 Skinner, M.K. (2011). Environmental epigenetic transgenerational inheritance and somatic epigenetic mitotic stability. *Epigenetics* **6**: 838–842.

35 Visscher, P.M. (2008). Sizing up human height variation. *Nature Genetics* **40**: 489–490.

36 Yang, J., Benyamin, B., McEvoy, B.P. et al. (2010). Common SNPs explain a large proportion of the heritability for human height. *Nature Genetics* **42**: 565–569.

37 Yang, J., Bakshi, A., Zhu, Z. et al. (2015). Genetic variance estimation with imputed variants finds negligible missing heritability for human height and body mass index. *Nature Genetics* **47**: 1114–1120.

38 Bourrat, P., Lu, Q., and Jablonka, E. (2017). Why the missing heritability might not be in the DNA. *BioEssays: News and Reviews in Molecular, Cellular and Developmental Biology* **39**: 1–2.

39 Cortijo, S., Wardenaar, R., Colom'e-Tatch'e, M. et al. (2014). Mapping the epigenetic basis of complex traits. *Science (New York, N.Y.)* **343**: 1145–1148.

40 Banta, J.A. and Richards, C.L. (2018). Quantitative epigenetics and evolution. *Heredity* **121**: 210–224.

41 Slatkin, M. (2009). Epigenetic inheritance and the missing heritability problem. *Genetics* **182**: 845–850.

42 Furrow, R.E., Christiansen, F.B., and Feldman, M.W. (2011). Environment-sensitive epigenetics and the heritability of complex diseases. *Genetics* **189**: 1377–1387.

43 Tal, O., Kisdi, E., and Jablonka, E. (2010). Epigenetic contribution to covariance between relatives. *Genetics* **184**: 1037–1050.

44 Day, T. and Bonduriansky, R. (2011). A unified approach to the evolutionary consequences of genetic and non-genetic inheritance. *The American Naturalist* **178**: E18–E36.

45 David, I. and Ricard, A. (2019). A unified model for inclusive inheritance in livestock species. *Genetics* **212**: 1075–1099.

46 Sae-Lim, P., Gjerde, B., Nielsen, H.M. et al. (2016). A review of genotype-by-environment interaction and micro-environmental sensitivity in aquaculture species. *Reviews in Aquaculture* **8**: 369–393.

47 Herrera, C.M. and Bazaga, P. (2011). Untangling individual variation in natural populations: ecological, genetic and epigenetic correlates of long-term inequality in herbivory. *Molecular Ecology* **20**: 1675–1688.

48 Verhoeven, K.J.F., van Jansen, J.J., Dijk, P.J., and Biere, A. (2010). Stress-induced DNA methylation changes and their heritability in asexual dandelions. *The New Phytologist* **185**: 1108–1118.

49 Cubas, P., Vincent, C., and Coen, E. (1999). An epigenetic mutation responsible for natural variation in floral symmetry. *Nature* **401**: 157–161.

50 Skvortsova, K., Iovino, N., and Bogdanovi'c, O. (2018). Functions and mechanisms of epigenetic inheritance in animals. *Nature Reviews Molecular Cell Biology* **19**: 774–790.

51 Guo, H., Zhu, P., Yan, L. et al. (2014). The DNA methylation landscape of human early embryos. *Nature* **511**: 606–610.

52 de Barreto, D.R., Leaniz, C.G., Verspoor, E. et al. (2019). DNA methylation changes in the sperm of captive-reared fish: a route to epigenetic introgression in wild populations. *Molecular Biology and Evolution* **36**: 2205–2211.

53 Venney, C.J., Love, O.P., Drown, E.J., and Heath, D.D. (2020). DNA methylation profiles suggest intergenerational transfer of maternal effects. *Molecular Biology and Evolution* **37**: 540–548.

54 Anastasiadi, D., Vandeputte, M., Sánchez-Baizán, N. et al. (2018). Dynamic epimarks in sex-related genes predict gonad phenotype in the European sea bass, a fish with mixed genetic and environmental sex determination. *Epigenetics* **13**: 988–1011.

55 Shao, C., Li, Q., Chen, S. et al. (2014). Epigenetic modification and inheritance in sexual reversal of fish. *Genome Research* **24**: 604–615.

56 Jiang, L., Zhang, J., Wang, J.-J. et al. (2013). Sperm, but not oocyte, DNA methylome is inherited by zebrafish early embryos. *Cell* **153**: 773–784.

57 Wang, X. and Bhandari, R.K. (2020). DNA methylation reprogramming in medaka fish, a promising animal model for environmental epigenetics research. *Environmental Epigenetics* **6**: dvaa008.

58 Wang, X. and Bhandari, R.K. (2020). The dynamics of DNA methylation during epigenetic reprogramming of primordial germ cells in medaka (*Oryzias latipes*). *Epigenetics* **15**: 483–498.

59 Koopman, P., Münsterberg, A., Capel, B. et al. (1990). Expression of a candidate sex-determining gene during mouse testis differentiation. *Nature* **348**: 450–452.

60 Matsuda, M., Nagahama, Y., Shinomiya, A. et al. (2002). DMY is a y-specific DM-domain gene required for male development in the medaka fish. *Nature* **417**: 559–563.

61 Howe, K., Clark, M.D., Torroja, C.F. et al. (2013). The zebrafish reference genome sequence and its relationship to the human genome. *Nature* **496**: 498–503.

62 Coolen, M.W., Statham, A.L., Qu, W. et al. (2011). Impact of the genome on the epigenome is manifested in DNA methylation patterns of imprinted regions in monozygotic and dizygotic twins. *PLoS One* **6**: e25590.

63 Liu, Y., Li, X., Aryee, M.J. et al. (2014). GeMes, clusters of DNA methylation under genetic control, can inform genetic and epigenetic analysis of disease. *American Journal of Human Genetics* **94**: 485–495.

64 Lyko, F., Foret, S., Kucharski, R. et al. (2010). The honeybee epigenomes: differential methylation of brain DNA in queens and workers. *PLoS Biology* **8**: e1000506.

65 Biomarkers Definitions Working Group (2001). Biomarkers and surrogate endpoints: preferred definitions and conceptual framework. *Clinical Pharmacology and Therapeutics* **69**: 89–95.

66 Ballman, K.V. (2015). Biomarker: predictive or prognostic? *Journal of Clinical Oncology: Official Journal of the American Society of Clinical Oncology* **33**: 3968–3971.

67 Frommer, M., McDonald, L.E., Millar, D.S. et al. (1992). A genomic sequencing protocol that yields a positive display of 5-methylcytosine residues in individual DNA strands. *Proceedings of the National Academy of Sciences of the United States of America* **89**: 1827–1831.

68 Park, P.J. (2009). ChIP-seq: advantages and challenges of a maturing technology. *Nature Reviews Genetics* **10**: 669–680.

69 Meissner, A., Gnirke, A., Bell, G.W. et al. (2005). Reduced representation bisulfite sequencing for comparative high-resolution DNA methylation analysis. *Nucleic Acids Research* **33**: 5868–5877.

70 Kang, H.M., Sul, J.H., Service, S.K. et al. (2010). Variance component model to account for sample structure in genome-wide association studies. *Nature Genetics* **42**: 348–354.

71 Sul, J.H., Martin, L.S., and Eskin, E. (2018). Population structure in genetic studies: confounding factors and mixed models. *PLoS Genetics* **14**: e1007309.

72 van Gurp, T.P., Wagemaker, N.C.A.M., Wouters, B. et al. (2016). epiGBS: reference-free reduced representation bisulfite sequencing. *Nature Methods* **13**: 322–324.

73 Trucchi, E., Mazzarella, A.B., Gilfillan, G.D. et al. (2016). BsRADseq: screening DNA methylation in natural populations of non-model species. *Molecular Ecology* **25**: 1697–1713.

74 Krick, M.V., Desmarais, E., Samaras, A. et al. (2021). Family-effects in the epigenomic response of red blood cells to a challenge test in the European sea bass (*Dicentrarchus labrax*, L.). *BMC Genomics* **22**: 111.

75 Moghadam, H.K., Johnsen, H., Robinson, N. et al. (2017). Impacts of early life stress on the methylome and transcriptome of Atlantic salmon. *Scientific Reports* **7**: 5023.

20

Epigenetics in Aquaculture: Knowledge Gaps, Challenges, and Future Prospects

Francesc Piferrer

Institute of Marine Sciences, Spanish National Research Council (CSIC), Barcelona, Spain

20.1 Introduction

The field of epigenetics in aquaculture is very young, but promising. Several papers have previously reviewed its potential from different points of view. Some have focused on the epigenetic regulation of endocrine systems in fish [1], on the different epigenetic regulatory mechanisms in aquatic animals [2] or, more generally, on the actual application of epigenetics in aquaculture [3, 4]. Some reviews focus on more specific aspects such as the epigenetics of gametes [5] and how epigenetics can be integrated into current breeding programs [6]. Conversely, other reviews focus on the prospects of epigenetics to address different questions regarding how aquatic organisms adapt to new environments [7].

Epigenetics in aquaculture has its parallel in terrestrial animal production. Thus, epigenetics has also been regarded as a nonpharmacological approach to induce desired phenotypes by environmental manipulation in the post-genomic era [8, 9]. It has been regarded as a promising way to improve phenotypic prediction because epigenetics can explain what has been referred to as the "missing heritability" [10–12].

In the preceding chapters of this book, authors have provided a comprehensive and compelling argument on the different state-of-the-art aspects of the current application of epigenetics in aquaculture and of its study in aquatic organisms. Here, I have attempted not to simply repeat what authors have already written in their conclusion sections. Instead, the idea is to provide some additional thoughts, point out some knowledge gaps that need to be addressed, attempt to identify overall challenges that must be overcome, and outline some prospects in the years to come. These knowledge gaps, challenges, and prospects can be considered of a general nature and apply to several, if not most, chapters because they touch upon fundamental aspects of epigenetics and its application to aquaculture.

There are several reasons that one could find to justify why more research on the contribution to epigenetics to the farming of aquatic animals is needed. Here, some will be briefly discussed.

First, in aquaculture, like in any other type of animal farming, desired phenotypic traits are of paramount importance since they heavily influence the costs of production and hence profitability, in addition to the need of a more efficient production. These phenotypic traits mainly include aspects related to reproduction, growth, disease resistance and product quality (Chapter 1). Thus, their improvement has been the subject of classic genetic selection and the focus of selective breeding programs (Chapter 19). Nowadays, the contribution of epigenetic regulatory mechanisms in integrating genomic and different environmental cues (Figure 20.1) to shape the phenotype is firmly established [11] (Chapter 5). In this context, it should be remembered an important asset: that aquatic organisms have an advantage because of having both external fertilization and external early development and thus are much more prone to environmental influences. Therefore, in the context of new advances in epigenetics, quantifying how epigenetic mechanisms integrate genomic and environmental information and influence the phenotype of farmed aquatic organisms is of crucial relevance, also in an ever more evident scenario of climate change.

A second reason is that, as we have seen, the still fragmentary information we have nowadays seems to suggest that, at least within vertebrates, epigenetic information can be more heritable in fish compared to mammals [13]. Taking into account that epigenetics convey environmental information, if this superior transmission of environmentally-induced epigenetic changes is confirmed, then this can be exploited to pass epigenetic marks known to correlate with desired traits

Epigenetics in Aquaculture, First Edition. Edited by Francesc Piferrer and Han-Ping Wang.

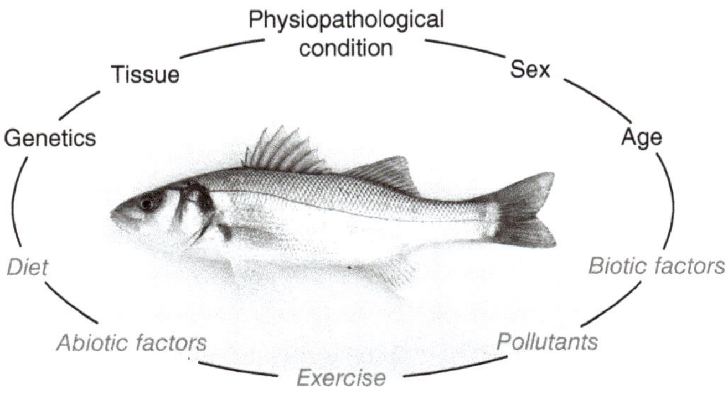

Figure 20.1 Some of the internal factors (above the fish) and external factors (in italics, below the fish) influencing the epigenome. *Source:* Photograph of the European sea bass (*Dicentrarchus labrax*) by Peter van der Sluijs/Wikimedia Commons/CC BY-SA 4.0.

to the offspring [10]. This paints an even more promising situation for the application of epigenetics to the farming of aquatic organisms when compared to terrestrial livestock.

A third reason comes from the fact that animal farming, in any of its inceptions, constitutes a process of domestication. Understanding the changes occurring during domestication is a topic that caught the interest of Darwin himself [14]. Understanding what domestication actually is, and its underlying molecular mechanisms is a hot research topic nowadays [15, 16]. Most studies on domestication have compared the genomes of current domesticates with those of their wild ancestors to identify genes under positive selection [17]. In many classical domesticates, the process started thousands of years ago and, in fact, comparisons such as those between dog vs. wolf, or chicken vs. jungle fowl, are then made between what actually have become two differentiated species. Thus, knowing what are the underlying processes that act at the initial stages of domestication and, in particular, the contribution of epigenetics, is hampered or influenced by the aforementioned genetic differences. Fortunately, this is not the case in most fish and other aquatic farmed animals since typically their domestication started only recently [18] and thus the number of generations of selective breeding is much reduced compared to nonaquatic farmed animals. Epigenetic changes are now known to underly the initial stages of domestication [19]. Further, farm-raised fish exhibit a characteristic set of epigenetic marks that can be used as a telltale sign of the influence of the early environment on these animals [20]. Thus, fish can be used as models to address broader questions related to domestication, with implications for the evolution of humans and the appearance of some cognitive disorders [21]. Thus, taking the layer of epigenetic information into account allows for a more nuanced description and understanding of the process of domestication.

20.2 Knowledge Gaps

What are the pressing questions on which we need to focus the research, and what are the most important aspects given the current state of knowledge, or lack thereof? Without pretending to be exhaustive, I hereby outline some questions that, in my opinion, constitute important knowledge gaps.

20.2.1 Phylogenetic Considerations

In which aquatic organism types can it be rewarding to study epigenetics? Important genomic and physiological differences do exist among different taxa. When compared to fish, invertebrates are clearly less studied, although nowadays the field of marine epigenetics, focusing on corals, mollusks, and algae is experimenting with rapid growth [7]. However, as vertebrates, fish have the lead because they also benefit from research conducted in model organisms such as the zebrafish (*Danio rerio*) and medaka (*Oryzias latipes*) and for their close phylogenetic relationship with mammals, where epigenetics research is more advanced for obvious reasons connected to human health. However, of merit is the research on epigenetics done in salmonids, particularly in rainbow trout (*Oncorhynchus mykiss*) and Atlantic salmon (*Salmo salar*), Nile tilapia (*Oreochromis niloticus*), tongue sole (*Cynoglossus semilaevis*), and the European sea bass (*Dicentrarchus labrax*). Nevertheless, important advances have been made in invertebrates, particularly in mollusks (Chapter 15) and crustaceans (Chapter 16). However, a better picture of the peculiarities/differences with respect to the situation in vertebrates is still needed. In addition, more work is also needed in microalgae: some do not have epigenetic mechanisms, and some resemble the situation in plants, while others that in mammals (Chapter 17). Thus, there is ample variation that needs to be better understood. In this regard, there is application potential of epigenetics in the bioengineering of algae.

20.2.2 Epigenetic Mechanisms

What type of epigenetic modifications need to be studied? There are three main recognized epigenetic mechanisms that regulate gene expression, namely DNA methylation, histone modifications (Chapter 2), and noncoding RNAs (ncRNAs) (Chapter 3). Roughly speaking, it could be said that the research focus is on DNA methylation first, ncRNAs second, and histone modifications last. However, it should be remembered that epigenetic mechanisms work together and a clear picture of their regulatory gene function will only be obtained when we integrate their study. While this is true, it is also true that from an applied point of view, there is no added advantage in interrogating all layers of epigenetic modification if focusing on just one allows us to gain the information pursued. Similarly, emphasis is made on 5-methyl-cytosine (5mC) function and the role of the presence and actions of the different DNA methyltransferases (DNMTs) but, conversely, the role of the enzymes involved in active demethylation, such as ten-eleven translocases (TET), is less studied. To study the effects of the DNMTs and DNA methylation, compounds such as 5-aza-2′-deoxycytidine (5-aza-dC) [22] have been available. More recently, however, newly developed compounds such as Bobcat339, a selective cytosine-based TET enzyme inhibitor, can reduce DNA 5-hydroxy-methyl-cytosine abundance by inhibiting TET enzyme function in living cells [23]. This facilitates the study of 5-hydroxy-methyl-cytosine (5hmC) function and thus of 5-hydroxymethylation, present not only in fish [24] like in the rest of vertebrates, but also in mollusks according to recent reports. Likewise, the role of the mitochondrial epigenome, or mitoepigenome, is only beginning to be deciphered [25]. In any case, studies targeting DNA methylation after bisulfite conversion of DNA prevail because, coupled with next generation sequencing (NGS)-based approaches, they provide single nucleotide resolution of the genome-wide or whole genome methylome at a reasonable cost. Recently, the importance of DNA methylation in regulating the 3D chromatin structure has been demonstrated [26], contributing to a better understanding of the interaction between epimarks and chromatin structure.

On the other hand, many miRNAs are conserved across species and their study benefits from tools such as specific databases like FishmiRNA, dedicated to piscine miRNAs, and thus are relatively easy to study. In contrast, a given miRNA may have many target genes, and thus deciphering their exact regulatory pathway may not be as straightforward as one thinks. For ncRNAs other than miRNAs, target prediction has not progressed as much and can be rather difficult to define. Nevertheless, miRNAs offer great potential as biomarkers associated with given phenotypes, also being considered as potential sentinels of environmental exposure (Chapter 3). Furthermore, miRNAs and other ncRNAs can be measured in blood, urine, and seminal plasma as noninvasive methods. Therefore, ncRNAs can be integrated as biomarkers of product quality and monitor disease outbreaks, development, growth, and reproduction.

20.2.3 Methodological Aspects

The study of gene expression has progressed enormously. The most used approach nowadays is RNA-seq, which provides a complete picture of the genes that are being actively expressed or modified in a cell or tissue at a given time. This allows the description of differentially expressed genes (DEGs) among two or more conditions, followed by the description of GO terms and KEGG pathways. More recent methods include the study of gene networks with approaches such as weighted gene coexpression network analysis (WGCNA) alone and/or in combination with DEG analysis [27]. However, the study of "epigenetic networks," understood as the study of the methylation levels of gene networks and how they influence their expression is clearly lacking and behind the study of gene expression and gene networks. Progress has started to be made in some areas, for example, in sex differentiation by the model of the conserved epigenetic regulation of sex (CERS) [28]. However, mathematical models and bioinformatic pipelines are still needed. Thus, there is no software or packages available to integrate this kind of data. Other areas where similar approaches or developments can be applied are in the study of the regulation of the skeletal muscle development (Chapter 9) or the stress response (Chapter 12) within the hypothalamus-pituitary-interrenal (HPI) axis of fish.

20.2.4 Epigenetic Inheritance

One aspect that can be recurrently observed in the literature on epigenetics is the lack of a uniform clarification regarding how many generations need to be studied to conclude the existence of epigenetic inheritance. Thus, an agreement on the use of terminology is lacking. Since the field of epigenetics in aquaculture is still very young, it is necessary to agree on the terminology used to describe key expressions related to epigenetics. In this regard, here a nomenclature for the transmission of external effects within the same (through mitotic cell division) or subsequent (through meiosis) generations is suggested.

Importantly, this definition is in line and in accordance to what has been proposed for the situation in mammals [29]. Thus, according to the number of generations where the effects of a stimulus can be observed, we have the following types of effects:

1) **Intragenerational**. Effects induced by a stimulus are observed within the same exposed generation, i.e., only the F0 generation. Can be epigenetic or not. Epigenetic marks are passed from mother to daughter cells during mitosis and can lead to long-lasting phenotypic consequences. However, long-lasting effects can be due to, for example, the irreversible alteration of a particular function through effects on cell differentiation and hormone synthesis dysregulation. Thus, not all intragenerational effects have necessarily an underlying epigenetic nature.
2) **Intergenerational**. Effects induced by a stimulus are observed across two or more generations, including the F0 generation. Can be epigenetic or not. Intergenerational effects can in turn be subdivided into two types:
   2.1) **Multigenerational** (also known as parental effects). Effects induced by a stimulus are observed in all the directly exposed generations: the F0, F1, and F2 in a pregnant mammal; the F0 and F1 in a nonpregnant female mammal or in a male mammal; and the F0 and F1 in a fish. Can be epigenetic or not.
   2.2) **Transgenerational**. Effects induced by a stimulus are observed in all directly exposed generations: F0, F1, F2, as well as the F3 (unexposed generation) in a pregnant mammal; the F0, F1 (directly exposed), as well as the F2 (unexposed) in a nonpregnant female or male mammal; the F0 and F1 (directly exposed) as well as the F2 (unexposed) in a fish. Effects can also be observed in additional generations, also those never exposed. Thus, proof of epigenetic inheritance comes from seeing effects at the F3 (and maybe also in the subsequent generations) in a pregnant mammal, and in the F2 (and maybe also in the subsequent generations) in a nonpregnant mammal, a male mammal, or in a fish. Transgenerational effects can only be epigenetic.

Here, two cautionary notes are appropriate in order to clarify some aspects and avoid confusion. First, the first directly exposed generation is always referred to as the F0 (or parental, P) generation by convention. Second, and this is important, while there seems to be consensus in the literature in the meaning of "intragenerational" and "transgenerational" effects, in contrast, the terms "multigenerational" and "intergenerational" are sometimes used interchangeably by different authors. Here, it was adopted as arranged earlier because "intergenerational," rather than "multigenerational," is the opposite of "intragenerational."

One important knowledge gap in fish and other aquatic organisms is to determine the existence of parental imprinting and monoallelic gene expression, as it is well described in mammals and one of the classic epigenetic phenomena [30]. In fish little information exists except in the case of the situation of hybrids or polyploids such as diploid and triploid hybrids of *Squalius alburnoides* [31] (Chapter 13).

Another important aspect, particularly in aquatic toxicology, is to take advantage of species such as *Daphnia magna*, which is an aquatic keystone species where epigenetic transgenerational inheritance has been recently suggested [32]. Species such as those of the genus *Daphnia* can be very useful to study how epigenetic modifications can drive adverse effects. In this regard, the concept of epigenetic traps (Box 20.1), whereby evolutionarily novel factors (e.g., endocrine disruptors) hack into the existing epigenetic machinery, should perhaps receive more attention [33]. However, we need to better define epigenetic endpoints (Chapters 14 and 16).

Box 20.1 Definitions of Some Terms Used in the Text

Environmentally responsible loci (ERL): Loci in which epigenetic modifications such as methylation levels change in response to environmental influences.
Epigenetic trap: Any epigenetic change that arises in response to novel environmental cues that produce maladaptive phenotypes, with no increase in phenotypic variance
Metastable epiallele (ME): Epiallele the methylation of which can vary even in genetically identical individuals, but responds to environmental cues in the same way in all tissues within a given individual. This means that environmental cues influencing MEs occur during very early stages of development.
Methylation-sensitive quantitative trait loci (meQTL): Loci the methylation of which is dependent of the environment, and that are linked to a quantitative phenotypic trait.
Pure epiallele (PE): Epialleles the methylation of which changes in the same direction in response to a given environmental cue in individuals that are not genetically related. PEs are thus independent of the genotype but fully dependent of the environment.

20.2.5 Environmental Influences

Environmental factors affect the epigenome and this can set certain gene expression programs and result in long-lasting phenotypic influences, especially if the environmental input is applied during one of the sensitive windows (Figure 20.2). This figure is meant for fish, but most likely the principles can hold across different taxa.

An important knowledge gap is to better delineate the windows during the life cycle in which organisms are more sensitive to environmental influences and capable of eliciting long-term changes in the epigenome. Thus, in what part of the life cycle is worth focusing our attention in each species? There are three main sensitive windows during the life cycle when environmental influences can induce long-lasting epigenetic changes. Of note, it is important to recall that from an epigenetics point of view, the "environment" is not only the external environment of an organism but anything acting "above" (epi) or outside the chromosome. The first and most sensitive of the three windows begins right after fertilization, when reprogramming takes place, peaking shortly thereafter and lasts throughout embryogenesis, particularly until the time of tissue differentiation. During this sensitive time window the epigenetic marks inherited by the parents are erased to allow the totipotency of the zygote and then the development of the new individual. Thus, any aberrant modification of this pathway and the ability of some genomic regions to escape this reprogramming could allow the inheritance of specific epimarks (Chapter 4). It is very important to improve knowledge of the reprogramming strategies adopted by the species already studied, but also to increase the number of species under study.

The second most important window includes gametogenesis, particularly after reaching adulthood, when animals enter the reproductive season and when the future gametes are being produced (Chapter 7). The third and final sensitive window takes place when fish are still larvae and when there is tissue reorganization during metamorphosis. An important aspect to note is that from a production point of view, most of these sensitive periods affect either broodstock, eggs, or larvae and thus take place when animals are in the hatchery. Once animals leave the hatchery or nursery, there is less chance to manipulate the epigenome. One can do little or nothing when animals are in floating net pens at sea where environmental conditions are much more difficult to manipulate. Thus, induction of epigenetic modifications to achieve a desired phenotype can be applied under the typical controlled conditions already present in hatcheries. This is important because it frees producers of having to deal with environmental manipulation to modify the epigenome during the nursery or grow-out stages, which take place in outdoor tanks or floating net pens in the sea. Thus, locating with precision the limits of the critical windows for epigenome programming is a fundamental knowledge gap. Also, one must ask whether these boundaries are the same for different environmental factors and how these interact in modulating the epigenetic response.

The influences of the diet on epigenetic profiles and nutritional programming is perhaps one of the most promising areas for the practical application of epigenetics in aquaculture, but more research is needed. Nutritional programming can be defined "as the process through which intended manipulation of the nutrients at a given time can result in permanent effects on future growth and health performance within the same generation or in the offspring" [34]. Thus, it can be applied to broodstock during gametogenesis to improve the gametes and thus the chances of good survival and development of the eggs and larvae; or applied to larvae and juveniles to adapt or "program" the young fish to a particular diet

Figure 20.2 Major stages (in uppercase) and processes (in lowercase) of the life cycle and indication of the main periods (shaded areas) where epigenetic marks can be established and thus potentially manipulated: 1) right after fertilization and early embryogenesis, 2) tissue reorganization during metamnorphosis, and 3) gametogenesis. The width of the shaded area is proportional to the sensitivity of each period (early development and embryogenesis > gametogenesis > metamorphosis).

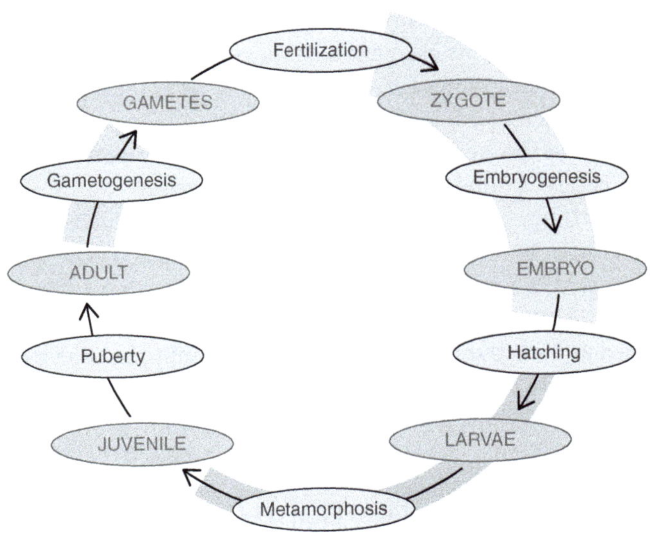

composition in order to become better performing adults. In nutritional programming, one knowledge gap or challenge is to disentangle epigenetic from nonepigenetic effects (parental or, more specifically, maternal effects) (Chapter 10). Of note, nutritional programming holds promise not only in fish but perhaps also in farmed crustaceans, supported by the observation that it is also possible to nutritionally program *Daphnia magna* [35] (Chapter 16).

20.3 Challenges

20.3.1 Overall Challenge

Undoubtedly, there are many challenges related to the application of epigenetics in aquaculture that must be overcome. Perhaps the most important and most obvious one is that, after recognizing the potential of epigenetics, it needs to be convincingly shown that it is useful. In other words, that the application of epigenetics provides an added value to aquaculture, either because it contributes to make it more profitable, more efficient and sustainable, or both. However, it is anticipated that before this becomes a reality, which is not going to be imminent, challenges of a more methodological and technical nature must first be solved. Some of them are briefly discussed below.

20.3.2 Methodological and Technical Challenges

In contrast to the genome, which is the same in all somatic cells of an organism, the epigenome, with its modifications or epigenetic marks, is tissue-specific. Thus, when studying the effect of a factor, be it internal or external (Figure 20.1), a major challenge is to know in which tissue is best to analyze epigenetic changes induced by that factor. The answer to that apparently trivial question is not straightforward. Thus, for example, when analyzing the effect of an environmental factor that may affect reproduction, one has the option to target different tissues along the hypothalamus-pituitary-gonad (HPG) axis. The same thing can be applied when studying factors related to growth or nutritional programming, as well as in other cases such as stress resistance or immune competence. In this context, nonlethal or noninvasive samplings become of interest and, of course, one of the most obvious tissues one can think of is the blood in fishes and the hemolymph of crustaceans and mollusks. In this regard, epigenetic studies have already used the blood, e.g., [36] or the hemolymph, e.g., [37], but whether epigenetic information extracted from blood can be used as an endpoint for different applications is still far from clear. Thus, more research needs to be carried out focusing on blood or hemolymph to better gauge the possibilities that they may offer.

Perhaps one possible way to circumvent the problem of the tissue specificity of epigenetic marks is through the identification and use of metastable epialleles (MEs) as biomarkers. MEs are alleles whose expression can be quite different even in genetically identical individuals due to environmentally-induced epigenetic modifications [38]. A classic example of a ME in mammals is the viable yellow agouti (A^{vy}) allele in mouse, the expression of which is dependent on DNA methylation, in turn dependent on an environmental factor; in this case, the presence of a methyl-donor such as folate or vitamin B12 in the diet [39]. However, while the methylation levels of a given ME vary even in genetically close or identical individuals, it remains constant regardless of tissue in a given individual. This indicates that the methylation must be established very early in development, before or at the beginning of tissue differentiation, and that MEs are very sensitive to environmental influences (see also [40]).

MEs can be identified by finding differentially methylated cytosines (DMCs) or differentially methylated regions (DMRs) in tissues of different embryonic origin between control and exposed animals. The first identification of MEs in fish was recently provided by Anastasiadi et al. [41], who used the methylation level of MEs common in muscle, brain, testis, and liver to infer whether adult European sea bass had been exposed to a marine heatwave when they were larvae. Thus, once robust MEs are identified, one does not need to worry anymore about the tissue specificity of epigenetic marks because ME provide information on the effects of environmental influences regardless of tissue and genetic background. MEs could be used as biomarkers for exposure to an environmental cue during early development. Conserved ME across species would constitute a useful way to monitor the effects of environmental cues at a large scale [41].

A fast-developing area that can also solve the problem of tissue specificity and the issue of cell heterogeneity between different samples is represented by the single-cell (sc)-based approaches, notably scRNAseq. Nowadays, these approaches are also being developed to interrogate the epigenome including scDNA methylation profiling [42] (Chapter 6). Of note, methods are also being developed that interrogate more than one layer of information such as sci-CAR, which is an assay based on combinatorial indexing that jointly profiles chromatin accessibility and mRNA in each of thousands of single cells [43]. Single-cell technology-biomarker is currently expensive, but with refinements and further development they eventually can become affordable for small labs.

Another promising method is multi-omics factor analysis (MOFA), a computational method for discovering the principal sources of variation in multi-omics data sets [44]. MOFA is part of a series of new strategies to integrate genomic information, taking into account epigenomic information [45]. MOFA can be used for what it was designed – to combine, for example, genomic, epigenomic, transcriptomic, proteomic, and metabolomic data. However, with minimal modifications, MOFA can also be applied to consider different layers of epigenetic information such as DNA methylation, histone modifications, and ncRNAs. One can even further subdivide information with one of the three mentioned layers; for example, within DNA methylation to investigate the contribution of the promoter vs. exon vs. intron vs. gene body methylation, in order to obtain a more nuanced picture about in which parts of the genome DNA methylation actually controls gene expression. An application of the latter is that MOFA has been recently used in our lab to discriminate between regular European sea bass males and neomales (sex-reversed) females, which do not exhibit transcriptomic differences, in a system with polygenic sex determination where there are also no genetic differences between females and neomales (Sánchez-Baizán and Piferrer, unpublished observations). Thus, these integrative studies will likely shed light on the complexities of epigenetic regulation of gene expression and will facilitate also the studies considering gene networks [27], not only from a transcriptomic point of view, but also from an epigenetics point of view. This will paint a clearer picture of how organisms regulate developmental processes and respond to environmental change. In this regard, theoretical models such as the calcium and redox state (CaRE) [46] and the CERS model [28] provide testable hypothesis on how the environment, through different signaling pathways including epigenetic regulation contribute to the determine the phenotype – the sexual phenotype in this case (Chapter 8). Advances along these lines will contribute to moving from purely association studies and distinguishing whether a given epigenetic change is a cause or the consequence of a given transcriptomic change, a generalized challenge of epigenetic studies in the broadest sense.

Understanding how individual environmental factors affect the phenotype can be challenging and should be studied under carefully controlled conditions, and it may make sense to study one factor at a time. However, it is important to keep in mind that in farming, organisms are exposed to the combined effect of many different factors at once. This means that the effect resulting of the modification of different factors also needs to be taken into account if we are supposed to get meaningful information.

Another important challenge is to learn how we can condition the microbiome through the rearing water. More efforts are needed at elucidating microbiome-epigenome crosstalk and to study the influence of host health and performance (Chapter 11). Here, discoveries in mammals can show the way, but one should bear in mind that there are important differences in fish vs. mammalian microbiomes. Epigenetic conditioning is a broader term, meaning exploiting the epigenetic machinery to make organisms respond in a particular way to an environmental stimulus. Thus, nutritional programming, microbiome programming, high temperature-resistance programming, and innate immune response programming can be considered under the umbrella of epigenetic programming. Understanding how these different types of epigenetic programming can be best set up is a major challenge.

Methods to interrogate the epigenome at an affordable cost need to be developed similarly to, for example, the microarray available to assess DNA methylation in human samples. In the aquaculture application, similar developments can be envisaged. For example, high-throughput multiplex PCR assays could be developed to determine the methylation level of several CpG sites as done, for example, in the development of epigenetic clocks to determine age in endangered species [47].

20.3.3 Challenges Related to Data Analysis

Last but not least, genetics deals with genomic sequences or genetic markers that can be described precisely. In contrast, epigenetic modifications have a "softer" nature. These, however, can be influenced by the underlying genotype due to associations between certain SNPs and methylation levels of nearby CpG, the environment and, in addition, by a certain degree of stochasticity, which should not be forgotten. Thus, an important challenge is adding or integrating epigenetic information, as a necessary step, if we want to deal with the missing heritability problem, i.e., that phenotypic variability cannot be accounted for only by genetic variability [48]. In the same way that there are methods to calculate genetic heritability (even specifying maternal and paternal contribution), we need methods to calculate the contribution of epigenetic information to the inheritance of traits. In this regard, developments such as the "transmissibility model" of David and

Ricard [49], which takes into account not only genetic but also epigenetic and other type of effects (e.g., microbiota), are very much needed and their development constitutes one of the major challenges.

20.4 Prospects

As we have seen in the preceding lines, epigenetics can aid in understanding the underlying molecular mechanism that explains how the environment influences the genotype, helps to explain the genotype-phenotype gap or the missing heritability problem, and improves accuracy and precision in the identification of the desired trait. Thus, it is envisaged that the integration of epigenetics in aquaculture will pass through the selection of epigenetic biomarkers linked to a desired phenotype (Chapter 18). Several types of epigenetic markers can be envisaged and, in addition to what has been discussed in different chapters of the book, further considerations are briefly outlined below.

20.4.1 Epigenetic Markers

Pure epialleles (PEs) may offer promise because their methylation status is, in contrast to that of MEs, independent of the underlying genotype and, importantly, reactive in the same manner to a given environmental change. For example, a given PE, say a DMC or DMR, could change its methylation level in response to hypoxia and thus facilitate coping with low oxygen concentration. Selecting animals with a desired PE can constitute a sort of "epigenetic selection" that is performed in addition to, or as a complement of, genetic selection. Then, external manipulations (abiotic factors, diet, etc.) are applied to broodstock to influence offspring, or during the early stages of development, to influence subsequent stages (Figure 20.3). Last but not least, the production of desired phenotypes is of a nonpharmacological nature and thus in line with the principles of sustainable aquaculture.

One interesting aspect of epigenetic markers is that they can be not only of diagnostic but also of prognostic or predictive value. This means that epigenetic changes can appear before other types of changes such as transcriptomic changes in

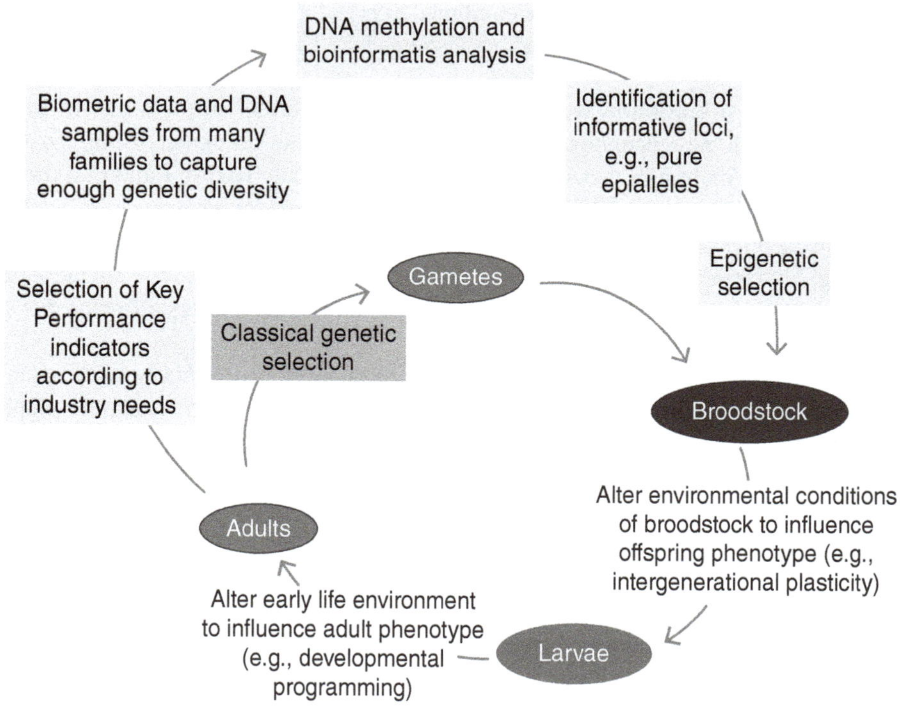

Figure 20.3 Epigenetic integration in aquaculture. First, selection for epigenetic biomarkers (epimutations) through the identification of, e.g., pure epialleles. This epigenetic selection is done in addition, or as a complement to, genetic selection. Then external manipulations (abiotic factors, diet, etc.) are applied to broodstock to influence offspring, or during the early stages of development to influence subsequent stages. The production of desired phenotypes is of a nonpharmacological nature and thus in line with the principles of sustainable aquaculture.

response to a certain stimulus. This characteristic is currently being exploited nowadays to develop prognostic biomarkers in the fight against cancer [50]. The idea, then, is to identify epigenetic markers such as DNA methylation changes or the presence/abundance of ncRNAs that respond to an environmental cue. These epigenetic markers could then be used to screen batches of larvae to determine the quality of their environment and to anticipate (prognosis) their performance when juveniles and adults. An example of such type of marker is the methylation levels of keratin-associated protein 10-4 gene (*krtap10-4*) in the European sea bass. Analysis of the DNA methylation levels of *krtap10-4* allows for the anticipation of long-lasting phenotypic consequences of temperature changes in newly hatched larvae [51]. An illustration of the criteria used to identify these types of epigenetic markers with both diagnostic and prognostic values is shown in Figure 20.4.

One could also anticipate the identification of epialleles that are pure, heritable, and associated with a particular phenotype of interest. For example, linked to resistance to an infection or, in some species, resistance to the masculinizing effect of elevated temperature. Analysis of the epigenome to search these epialleles in parents, gametes, or offspring can then be used to select animals that will, among a cohort of already selected individuals in a breeding program, allow to further select them based on that phenotype of interest. For example, imagine that a breeding company has selected animals for fast growth and good product yield. Then, among the selected animals, one could further select those that, in addition, present a desirable trait such as resistance to a particular type of disease. With epialleles that are pure, heritable, and linked to that trait of interest, one can be confident that they will be passed from parents to offspring (because they are heritable), that will work regardless of genetic background (because they are pure) and will be linked to phenotype. This allows, for example, the classification of broodstock or their offspring as either "desired fish" or "nondesired fish." An actual example, combining genome-wide epigenomic analysis and machine learning procedures in the European sea bass, is exemplified in the identification, among the progeny of broodstock selected for fast growth, of animals that exhibit high survival, a high proportion of females (the desired sex due to their superior growth), and the resistance to the masculinizing effects of elevated temperature (Figure 20.5) (Sánchez-Baizán and Piferrer, unpublished observations).

Expansion of large breeding programs involves the management of large amounts of gametes. This opens the door to the possibility of taking a subset of them and analyzing the gamete's epigenome either as a sort of "quality check," by ensuring

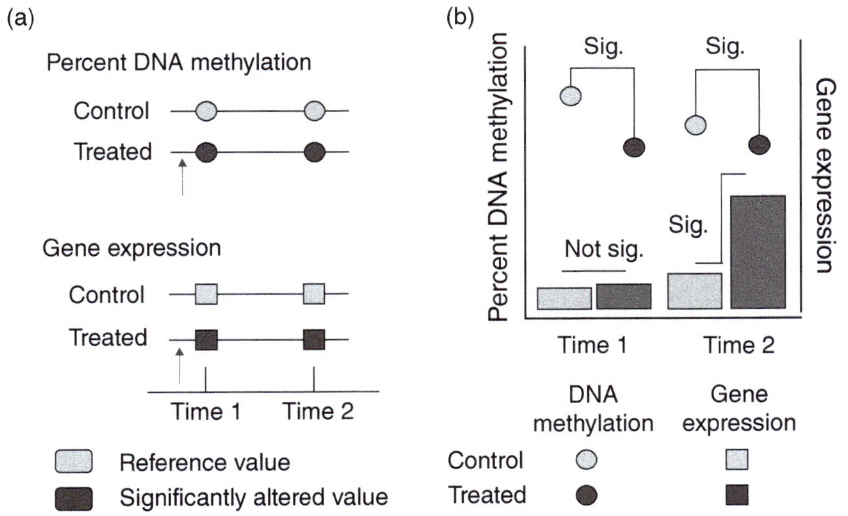

Figure 20.4 Identification of an epigenetic marker with both prognostic and diagnostic value, able to predict a certain phenotype in response to an environmental stimulus. (a) DNA methylation and gene expression are measured in a control and a treated group in at least two different time points: Time 1, shortly after the exposure to the environmental cue (dark arrow), and Time 2, sometime later, which can be days, weeks, months or even years. Differentially methylated cytosines (DMC) or regions (DMR) are identified among the differentially expressed genes (DEG) linked to the phenotype of interest induced by the environmental factor and apparent at time 2. (b) Example of a situation in which a gene becomes DEG in response to the environmental cue only at Time 2, but contains a DMC or a DMR already at Time 1 and is maintained at Time 2. These DMC or DMR constitute epigenetic markers with prognostic and diagnostic value since gene expression data provides no information at Time 1 while DNA methylation levels already anticipate the situation in Time 2. In this example, DNA methylation and gene expression levels are inversely related. Notice that in the control group DNA methylation levels do not need to be constant between Time 1 and Time 2, here undergoing a slight decrease (light dots in b). *Source:* Adapted from Valdivieso et al. [51], where *krtap10–4* is shown to be a gene the methylation levels of which have both diagnostic and prognostic value.

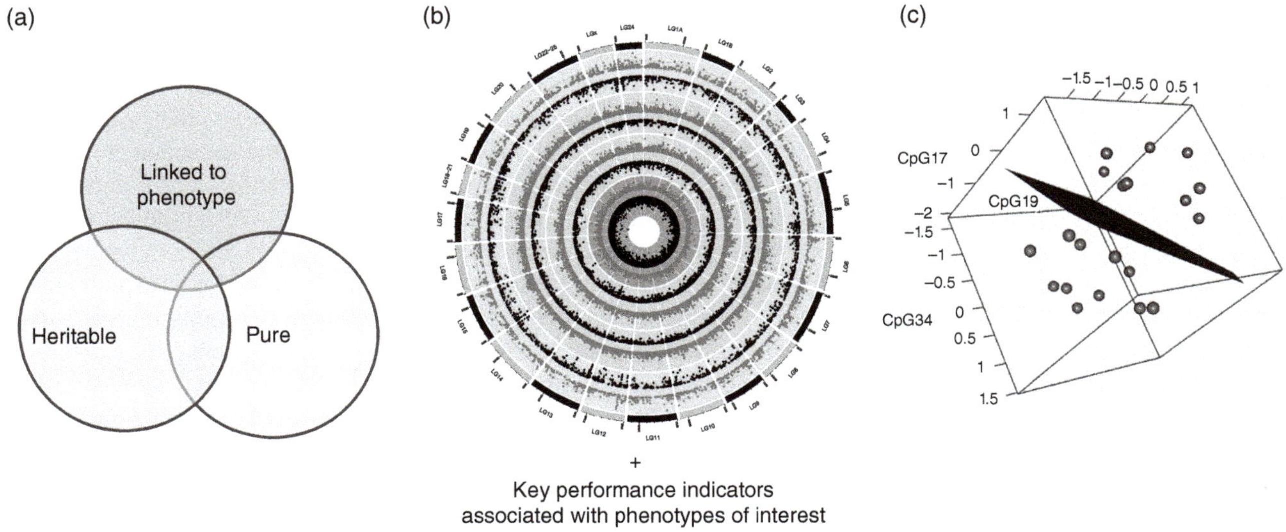

Figure 20.5 Development of epigenetic markers to aid in broodstock selection in aquaculture. (a) Epialleles to be selected should fulfil a series of criteria in order to get the most out of epigenetics. For example, they should be at least associated with a certain phenotype of interest, if possible heritable from parents to offspring, and, ideally, be PEs, meaning that their response to a given environmental cue should be in the same direction regardless of the underlying genetic background. (b) Epigenome-wide analysis and the measure of key performance indicators are conducted in the target population. (c) PCA-based plot showing the classification of fish based on a combination of desired traits such as survival, growth, and resistance to stress, done on a line of genetically selected fish. (*See colour plate section for colour representation of this figure.*)

that the marks indicative of good quality are present while the ones indicative of bad quality are not present in a given sperm sample, or to ensure that the marks indicative of a desired phenotype are being passed to the next generation.

The epigenome has three main types of influences: genetic, environmental, and stochastic. Let us start by dealing with the influences we do not want to focus on here. The latter, because of its random nature, is difficult to account for – or to characterize properly – and represents a sort of "noise" with which we have to live with. As explained earlier, from the point of view of epigenetics, the "environment" is anything happening above (epi) the genetics, i.e., the chromosome, or chromatin, if one wishes. Thus, for the epigenome, the environment may refer to the nuclear environment, the cytoplastic environment, the environment outside the cell, outside the tissue, organ, or even the environment outside the whole individual. As shown in Figure 20.1, there are different internal or external (to the organism) influences on the epigenome. What about the genetic influence on the epigenome? Epigenetic regulation and, in particular, DNA methylation have been linked to the underlying genetic sequence. The individual genotypes at specific loci, e.g., single nucleotide polymorphism (SNP), can influence the pattern of DNA methylation in the same or surrounding loci. Loci that are responsive to environmental influences, known as epigenetically responsive loci (ERLs), and that are linked to a phenotypic trait are referred to as methylation quantitative trait loci (meQTLs). Thus, meQTLs can have a strong influence on the epigenome, even in distant regions, and are believed to account for direct SNP associations or gene-environment interactions [52].

The association between the underlying genetic code and the epigenome in different pathophysiological contexts has allowed for the identification of meQTLs. It has been suggested that determining the relationships between DNA methylation and polymorphic alleles results in a significant improvement in our understanding of key alleles identified in GWAS that lack a satisfactory underlying genetic mechanism [53].

20.4.2 Technical Advances and Training

Significant advances in our ability to select and control desired phenotypes that cannot be accounted for only through genetics will be made with techniques that allow the manipulation of the epigenome in a manner similar to those gene-editing technologies such as CRISPR/Cas allow the manipulation of the genome. In this regard, the newly developed technique CRISPRoff to edit the methylome and introduce repressive histone modifications holds much promise. It allows to program heritable epigenetic memory states, being able in this way to heritable silence most genes, including genes without CpG islands [54].

One last consideration before finishing: The application of epigenetics in aquaculture can benefit from the increasing number of species for which the sequenced and annotated genome is available, something that allows exploiting the power of NGS-based and multi-omics based-approaches. Thus, epigenetic studies will largely benefit from techniques such as tagmentation-based indexing for methylation sequencing (TIME-Seq). This technique allows ultra-cheap and scalable targeted methylation sequencing [55] and hence the characterization of the epigenome and epigenetic marks at a very affordable cost, thus facilitating and democratizing the use of epigenetic information in aquaculture research and production. NGS-based and multi-omics based-approaches require the collection of a considerable number of samples, especially if studies are carried out in close collaboration with breeding companies in order to capture sufficient genetic variability and ensure statistical power. Therefore, one of the most important challenges is the management of technical and biological variation. In this regard, it will become increasingly necessary the establishment of international consortia to tackle the big problems. Only with cooperation can complex problems be solved. In this stimulating task, it is necessary to add the need for the training of personnel specialized in the use of these tools, with a multidisciplinary vision of the problems. Epigenetics is a sort of "bridging" discipline. Thus, a new generation of scientists with good background knowledge in genetics, physiology, immunology, epigenetics, and data analysis with an eye to solving problems in aquaculture is needed. I hope that the information presented in this book will contribute to guiding practitioners, expanding epigenetics research in aquaculture, and aiding in the training of these much-needed new professionals. They will work in an exciting and challenging field with many new discoveries and applications awaiting to be made.

Acknowledgments

Thanks are due to Drs. Paulino Martínez and Han-Ping Wang, as well as the PhD young researchers Núria Sánchez-Baizán and Silvia Beato, for helpful comments and suggestions for improvement. Research on epigenetics at the author's lab has been and is being made possible by continued support from the Ministry of Science and Innovation of the Spanish Government (projects "Sexgene," "Epifarm," "Epimark," and "Epipure"), grants, contracts, networks, and research infrastructures from the European Union ("Performfish," "Cure4Aqua," "Fishgenome," "Epiconcept," and "Aquaexcel"), CSIC grants (i-COOP), as well as contracts with private aquaculture companies (Culmarex, S.A.U.). Also with funding from the "Severo Ochoa Centre of Excellence" accreditation (CEX2019-000928-S).

References

1 Li, M. and Leatherland, J.F. (2013). The implications for aquaculture practice of epigenomic programming of components of the endocrine system of teleostean embryos: lessons learned from mammalian studies. *Fish and Fisheries* **14** (4): 528–553.
2 Best, C., Ikert, H., Kostyniuk, D.J. et al. (2018). Epigenetics in teleost fish: from molecular mechanisms to physiological phenotypes. *Comparative Biochemistry and Physiology Part B: Biochemistry and Molecular Biology* **224**: 210–244.
3 Gavery, M.R. and Roberts, S.B. (2017). Epigenetic considerations in aquaculture. *PeerJ* **5**: e4147.
4 Granada, L., Lemos, M.F.L., Cabral, H.N. et al. (2018). Epigenetics in aquaculture – the last frontier. *Reviews in Aquaculture* **10** (4): 994–1013.
5 Labbé, C., Robles, V., and Herraez, M.P. (2017). Epigenetics in fish gametes and early embryo. *Aquaculture* **472**: 93–106.
6 Moghadam, H., Mørkøre, T., and Robinson, N. (2015). Epigenetics – potential for programming fish for aquaculture? *Journal of Marine Science and Engineering* **3** (2): 175–192.
7 Eirin-Lopez, J.M. and Putnam, H.M. (2019). Marine environmental epigenetics. *Annual Review of Marine Science* **11** (1): 335–368.
8 González-Recio, O. (2012). Epigenetics: a new challenge in the post-genomic era of livestock. *Frontiers in Genetics* https://doi.org/10.3389/fgene.2011.00106.
9 González-Recio, O., Toro, M.A., and Bach, A. (2015). Past, present, and future of epigenetics applied to livestock breeding. *Frontiers in Genetics* https://doi.org/10.3389/fgene.2015.00305.
10 Thompson, R.P., Nilsson, E., and Skinner, M.K. (2020). Environmental epigenetics and epigenetic inheritance in domestic farm animals. *Animal Reproduction Science* **220**: 106316.
11 Pacht, E., Murdoch, B., and McKay, S. (2021). Examining the extent of environmental contributions toward DNA methylation and phenotypic variation. *Animal Frontiers* **11** (6): 83–89.

12 Shah, S., Bonder, M.J., Marioni, R.E. et al. (2015). Improving phenotypic prediction by combining genetic and epigenetic associations. *The American Journal of Human Genetics* **97** (1): 75–85.

13 Wang, X. and Bhandari, R.K. (2020). DNA methylation reprogramming in medaka fish, a promising animal model for environmental epigenetics research. *Environmental Epigenetics* **6** (1): dvaa008.

14 Darwin, C.R. (1871). *The Descent of Man, and Selection in Relation to Sex*. London: John Murray.

15 Purugganan, M.D. (2022). What is domestication? *Trends in Ecology & Evolution* **37** (8): 663–671.

16 Zeder, M.A. (2015). Core questions in domestication research. *Proceedings of the National Academy of Sciences of the United States of America* **112** (11): 3191–3198.

17 Wilkins, A.S. (2017). Revisiting two hypotheses on the "domestication syndrome" in light of genomic data. *Vavilov Journal of Genetics and Breeding* **21** (4): 435–442.

18 Duarte, C.M., Marbá, N., and Holmer, M. (2007). Rapid domestication of marine species. *Science* **316** (5823): 382–383.

19 Anastasiadi, D. and Piferrer, F. (2019). Epimutations in developmental genes underlie the onset of domestication in farmed European sea bass. *Molecular Biology and Evolution* **36** (10): 2252–2264.

20 Le Luyer, J., Laporte, M., Beacham, T.D. et al. (2017). Parallel epigenetic modifications induced by hatchery rearing in a Pacific salmon. *Proceedings of the National Academy of Sciences of the United States of America* **114** (49): 12964–12969.

21 Anastasiadi, D., Piferrer, F., Wellenreuther, M., and Benítez-Burraco, A. (2022). Fish as model systems to study epigenetic drivers in human self-domestication and neurodevelopmental cognitive disorders. *Genes* **13** (6): 987.

22 Ribas, L., Vanezis, K., Imués, M.A., and Piferrer, F. (2017). Treatment with a DNA methyltransferase inhibitor feminizes zebrafish and induces long-term expression changes in the gonads. *Epigenetics & Chromatin* **10** (1): 59.

23 Weirath, N.A., Hurben, A.K., Chao, C. et al. (2022). Small molecule inhibitors of TET dioxygenases: bobcat339 activity is mediated by contaminating copper(II). *ACS Medicinal Chemistry Letters* **13** (5): 792–798.

24 Konstantinidis, I., Anastasiadi, D., Sætrom, P. et al. (2021). Epigenetic mapping of the somatotropic axis in Nile tilapia reveals differential DNA hydroxymethylation marks associated with growth. *Genomics* **113** (5): 2953–2964.

25 Nedoluzhko, A., Mjelle, R., Renström, M. et al. (2021). The first mitochondrial 5-methylcytosine map in a non-model teleost (Oreochromis niloticus) reveals extensive strand-specific and non-CpG methylation. *Genomics* **113** (5): 3050–3057.

26 Buitrago, D., Labrador, M., Arcon, J.P. et al. (2021). Impact of DNA methylation on 3D genome structure. *Nature Communications* **12** (1): 3243.

27 Sánchez-Baizán, N., Ribas, L., and Piferrer, F. (2022). Improved biomarker discovery through a plot twist in transcriptomic data analysis. *BMC Biology* **20** (1): 208.

28 Piferrer, F., Anastasiadi, D., Valdivieso, A. et al. (2019). The model of the conserved epigenetic regulation of sex. *Frontiers in Genetics* **10**: 857.

29 Skinner, M.K. (2011). Environmental epigenetic transgenerational inheritance and somatic epigenetic mitotic stability. *Epigenetics* **6** (7): 838–842.

30 Onyango, P., Jiang, S., Uejima, H. et al. (2002). Monoallelic expression and methylation of imprinted genes in human and mouse embryonic germ cell lineages. *Proceedings of the National Academy of Sciences of the United States of America* **99** (16): 10599–10604.

31 Matos, I., Machado, M.P., Schartl, M., and Coelho, M.M. (2019). Allele-specific expression variation at different ploidy levels in *Squalius alburnoides*. *Scientific Reports* **9** (1): 3688.

32 Feiner, N., Radersma, R., Vasquez, L. et al. (2022). Environmentally induced DNA methylation is inherited across generations in an aquatic keystone species. *Iscience* **25** (5): 104303.

33 O'Dea, R.E., Noble, D.W.A., Johnson, S.L. et al. (2016). The role of non-genetic inheritance in evolutionary rescue: epigenetic buffering, heritable bet hedging and epigenetic traps. *Environmental Epigenetics* **2** (1): dvv014.

34 Piferrer, F. and Ribas, L. (2020). The use of the zebrafish as a model in fish aquaculture research. In: *Fish Physiology*, vol. **38** (ed. T.J. Benfey, A.P. Farrell and C.J. Brauner), 273–313. Academic Press. https://doi.org/10.1016/bs.fp.2020.10.003.

35 Hearn, J., Clark, J., Wilson, P.J., and Little, T.J. (2020). *Daphnia magna* modifies its gene expression extensively in response to caloric restriction revealing a novel effect on haemoglobin isoform preference. *Molecular Ecology* **29** (17): 3261–3276.

36 Krick, M.V., Desmarais, E., Samaras, A. et al. (2021). Family-effects in the epigenomic response of red blood cells to a challenge test in the European sea bass (Dicentrarchus labrax, L.). *BMC Genomics* **22** (1): 111.

37 Jin, L.H., Choi, J.K., Kim, B. et al. (2009). Requirement of split ends for epigenetic regulation of notch signal-dependent genes during infection-induced hemocyte differentiation. *Molecular and Cellular Biology* **29** (6): 1515–1525.

38 Dolinoy, D.C., Das, R., Weidman, J.R., and Jirtle, R.L. (2007). Metastable epialleles, imprinting, and the fetal origins of adult diseases. *Pediatric Research* **61** (5 Part 2): 30R–37R.

39 Bertozzi, T.M. and Ferguson-Smith, A.C. (2020). Metastable epialleles and their contribution to epigenetic inheritance in mammals. *Seminars in Cell & Developmental Biology* **97**: 93–105.

40 Bourc'his, D. (2021). Metastable epialleles are stable in their instability. *Nature Genetics* **53** (8): 1121–1123.

41 Anastasiadi, D., Shao, C., Chen, S., and Piferrer, F. (2021). Footprints of global change in marine life: inferring past environment based on DNA methylation and gene expression marks. *Molecular Ecology* **30** (3): 747–760.

42 Karemaker, I.D. and Vermeulen, M. (2018). Single-cell DNA methylation profiling: technologies and biological applications. *Trends in Biotechnology* **36** (9): 952–965.

43 Cao, J., Cusanovich, D.A., Ramani, V. et al. (2018). Joint profiling of chromatin accessibility and gene expression in thousands of single cells. *Science* **361** (6409): 1380–1385.

44 Argelaguet, R., Velten, B., Arnol, D. et al. (2018). Multi-omics factor analysis – a framework for unsupervised integration of multi-omics data sets. *Molecular Systems Biology* **14** (6): https://doi.org/10.15252/msb.20178124.

45 Kim, D.Y. and Kim, J.M. (2021). Multi-omics integration strategies for animal epigenetic studies – a review. *Animal Bioscience* **34** (8): 1271–1282.

46 Castelli, M.A., Whiteley, S.L., Georges, A., and Holleley, C.E. (2020). Cellular calcium and redox regulation: the mediator of vertebrate environmental sex determination? *Biological Reviews* **95** (3): 680–695.

47 Mayne, B., Espinoza, T., Roberts, D. et al. (2021). Nonlethal age estimation of three threatened fish species using DNA methylation: *Australian lungfish, Murray cod and Mary River cod. Molecular Ecology Resources* **21** (7): 2324–2332.

48 Danchin, É., Charmantier, A., Champagne, F.A. et al. (2011). Beyond DNA: integrating inclusive inheritance into an extended theory of evolution. *Nature Reviews Genetics* **12** (7): 475–486.

49 David, I. and Ricard, A. (2019). A unified model for inclusive inheritance in livestock species. *Genetics* **212** (4): 1075–1099.

50 Costa-Pinheiro, P., Montezuma, D., Henrique, R., and Jerónimo, C. (2015). Diagnostic and prognostic epigenetic biomarkers in cancer. *Epigenomics* **7** (6): 1003–1015.

51 Valdivieso, A., Sánchez-Baizán, N., Mitrizakis, N. et al. (2023). Development of epigenetic biomarkers with diagnostic and prognostic value to assess the lasting effects of early temperature changes in farmed fish. *Aquaculture* **563**: 738918.

52 Smith, A.K., Kilaru, V., Kocak, M. et al. (2014). Methylation quantitative trait loci (meQTLs) are consistently detected across ancestry, developmental stage, and tissue type. *BMC Genomics* **15** (1): 145.

53 Heyn, H., Sayols, S., Moutinho, C. et al. (2014). Linkage of DNA methylation quantitative trait loci to human cancer risk. *Cell Reports* **7** (2): 331–338.

54 Nuñez, J.K., Chen, J., Pommier, G.C. et al. (2021). Genome-wide programmable transcriptional memory by CRISPR-based epigenome editing. *Cell* **184** (9): 2503–2519.e17.

55 Griffin, P.T., Kane, A.E., Trapp, A. et al. (2021). Ultra-cheap and scalable epigenetic age predictions with TIME-Seq. *bioRxiv* 2021.10.25.465725; doi: https://doi.org/10.1101/2021.10.25.465725.

Index-Species

Epigenetics in Aquaculture, First Edition. Edited by Francesc Piferrer and Han-Ping Wang.

Index-Subjects

Epigenetics in Aquaculture, First Edition. Edited by Francesc Piferrer and Han-Ping Wang.

d

e

t

u

v

w

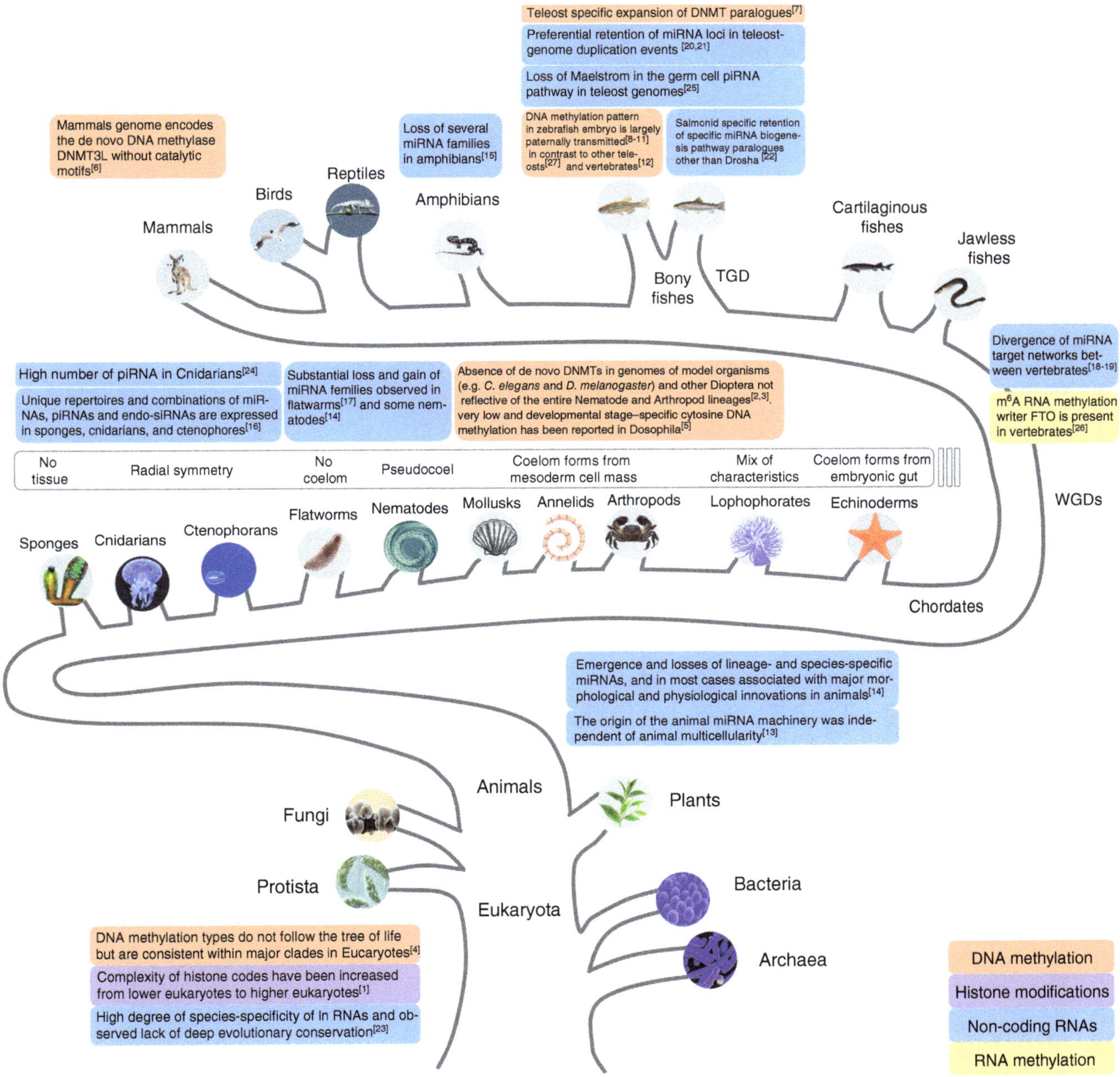

Figure 1.1 Phylogenetic epigenetics showing the common and divergent epigenetic molecular mechanisms in animal taxa. *Source:* Modified from Navarro-Martín et al. [169]. (a) Bradleyblackburn/Adobe Stock, (b) Vadim_petrakov/Adobe Stock, (c) RooM The Agency/Adobe Stock, (d) Brian Gratwicke/Wikimedia Commons/CC BY 2.0, (e) Mirko Rosenau/Adobe Stock, (f) U.S. Fish & Wildlife Service/Wikimedia Commons/Public Domain, (g) U.S. Department of Commerce, (h) Andrei Nekrassov/Adobe Stock, (i) Thanunchakorn/Adobe Stock, (j) Devin/Adobe Stock, (k) Sinhyu/Adobe Stock, (l) Uckyo/Adobe Stock, (m) Siloto/Adobe Stock, (n) Peteri/Adobe Stock, (q) Artphotoclub/Adobe Stock, (s) National Institute of Allergy and Infectious Diseases (NIAID)/CDC Public Health Image Library. References: [1][170]; [2][171]; [3][172]; [4][173]; [5][174]; [6][175]; [7][176]; [8][177]; [9][178]; [10][179]; [11][180]; [12][181]; [13][182]; [14][183]; [15][184]; [16][185]; [17][186]; [18][187]; [19][188]; [20][189]; [21][190]; [22][191]; [23][192]; [24][193]; [25][194]; [26][195]; [27][196].

Epigenetics in Aquaculture, First Edition. Edited by Francesc Piferrer and Han-Ping Wang.

Figure 1.2 Environment (internal vs. external) where fertilization and early development events take place in mammals, birds, and aquatic animals. Because both events take place externally, aquatic animals in principle offer more opportunities for influencing the epigenetic reprogramming events that take place during these early stages. *Source:* F. Piferrer.

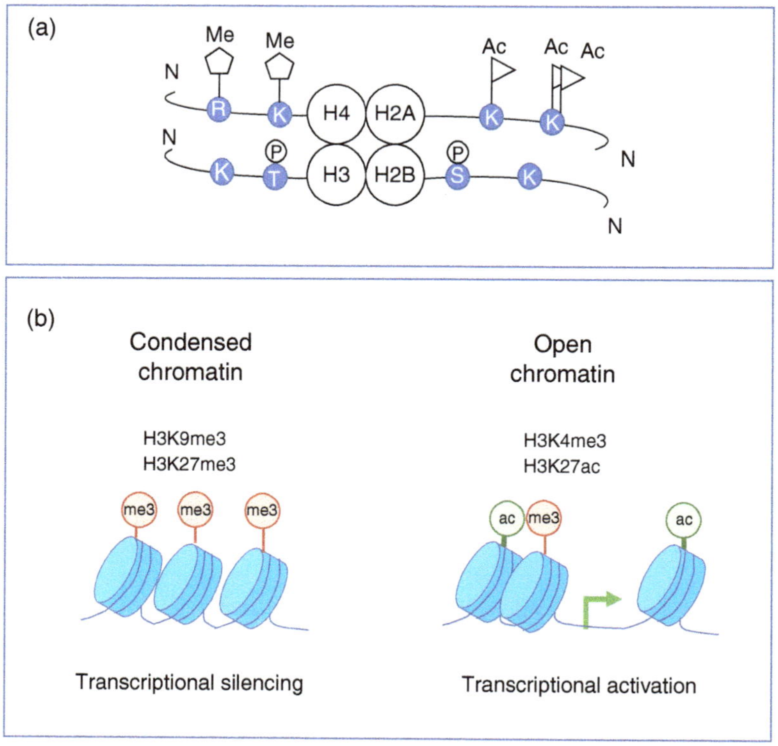

Figure 2.2 Histone modifications (a) and simplified diagram of chromatin state and DNA accessibility linked to transcription (b).

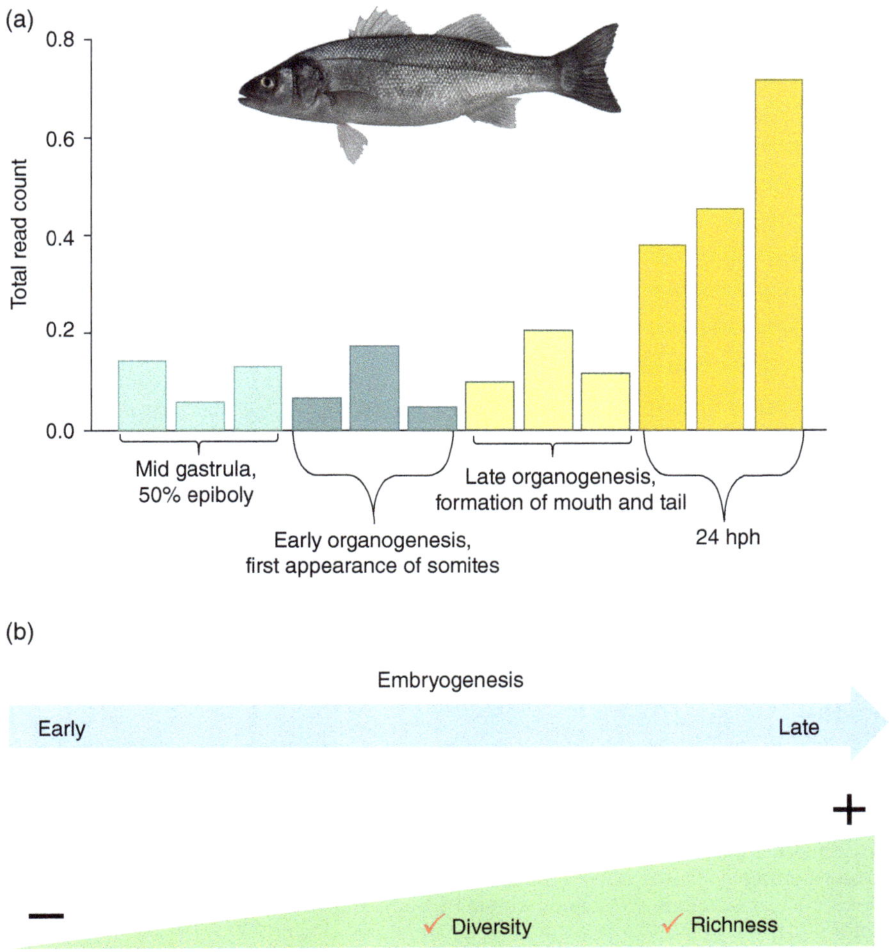

Figure 3.9 miRNA abundance during early teleost development. (a) Total read counts per sample extracted out of the European sea bass miRNA RNA-Seq results, (b) increased miRNA diversity and richness along later developmental stages reported in four different species, i.e. European sea bass. *Source:* Modified from Sarropoulou et al. [58], Kaitetzidou [12], Bizuayehu et al. [59], Robledo et al. [60].

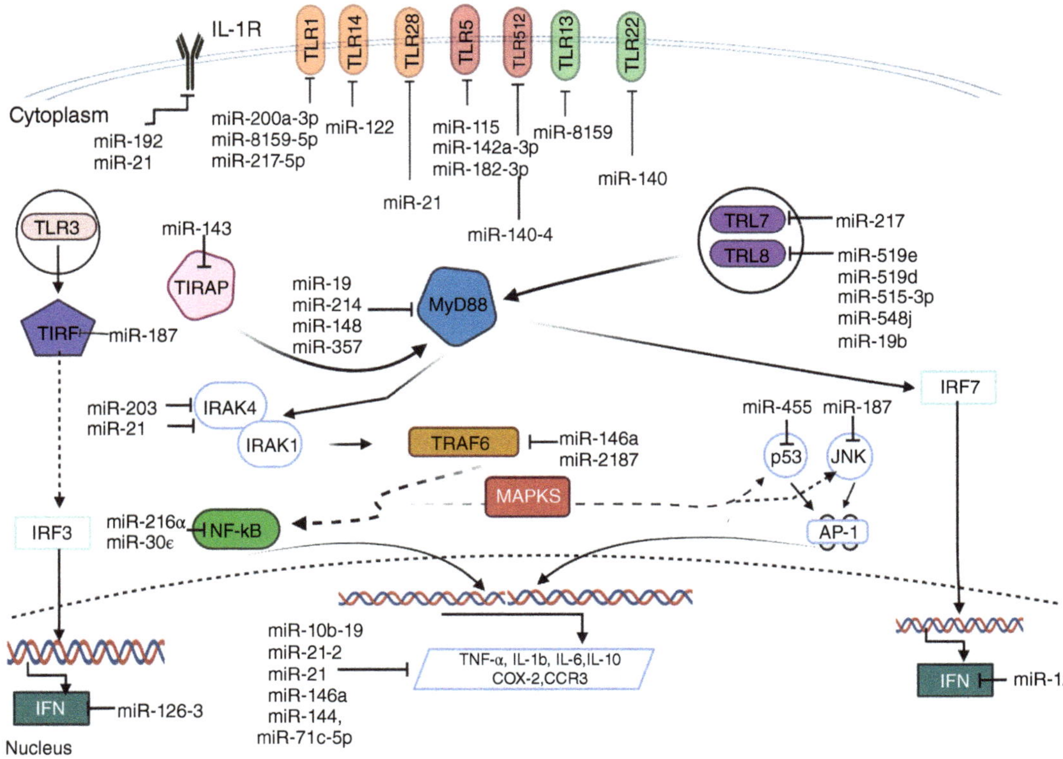

Figure 3.14 Schematic overview of miRs acting in toll-like receptor (TLR) signaling pathway. miRs are set according to published work on the miiuy croaker and orange-spotted grouper as described within the text. AP-1, activator protein; CCR3, CC chemokine receptor 3; COX2, cyclooxygenase-2 enzyme; IFN, interferon; IRAK, IL-1 receptor-associated protein kinase; IRF, activate interferon regulatory factor; JNK, c-Jun N-terminal kinases; MAPKs, mitogen-activated protein kinases; MyD88, myeloid differentiation primary response protein 88; NF-κB, nuclear factor κB; P53, tumor protein P53, also known as p53, cellular tumor antigen p53; TIRAP, TIR domain-containing adaptor protein; TLR, Toll-like receptor; TNF, tumor necrosis factor; TRAF, TNFR-associated factor; TRIF, TIR domain-containing adaptor molecule 1. *Source:* Modified form Chu and Xu [204], Cui et al. [206].

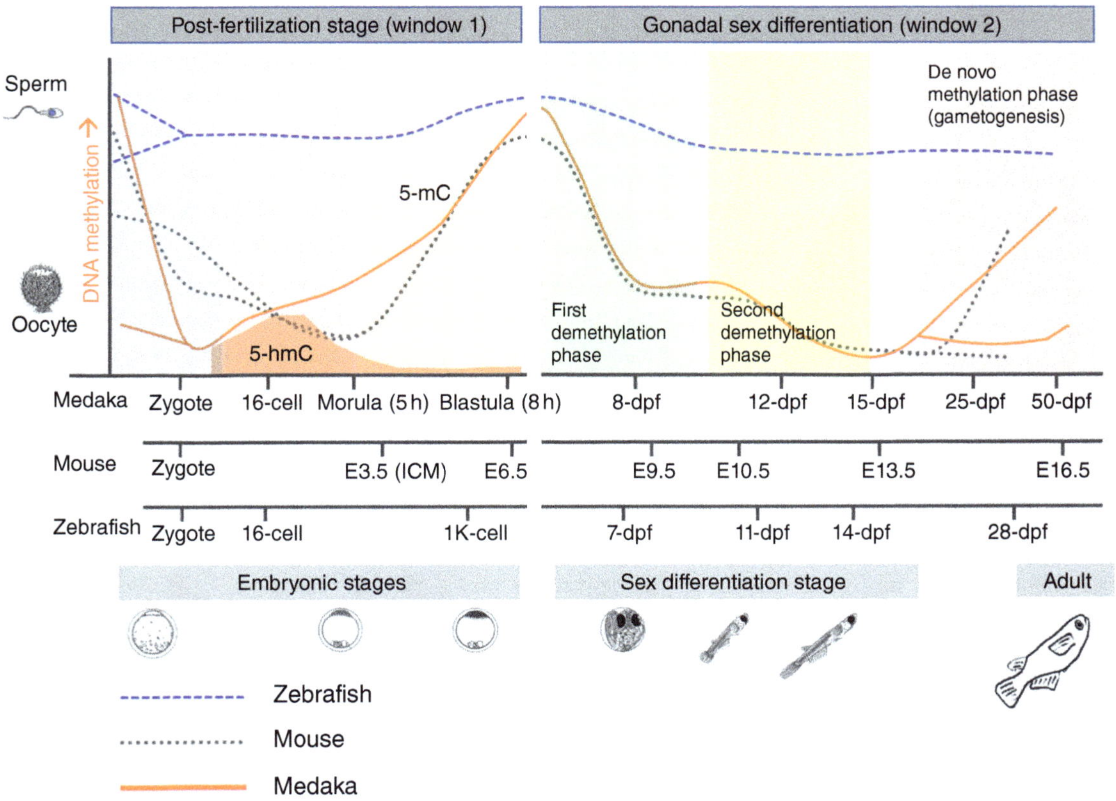

Figure 4.3 DNA methylation reprogramming model during embryogenesis (left panel) and PGC specification (right panel) in medaka. Left panel: Sperm is hypermethylated, and oocyte is hypomethylated. Paternal genomic methylation is erased in the zygote. Embryos maintain genome hypomethylation during the first several cell cycles. Global DNA methylation levels increase from the 16-cell stage to the Blastula stage. Right panel: In PGCs, the demethylation occurs in two phases: the first phase occurs before the 8-dpf stage (first demethylation phase), and the second phase occurs between the 10-dpf and the 12-dpf stages (second demethylation phase). Global DNA methylation level increases from 15-dpf to 25-dpf in male PGCs, whereas female PGCs remain hypomethylated. Data set sources: medaka embryogenesis, medaka PGCs, mouse embryogenesis, mouse PGCs, zebrafish embryogenesis, and zebrafish PGCs. *Source:* Wang and Bhandari [139]/Oxford University Press/public domain CC BY 4.0.

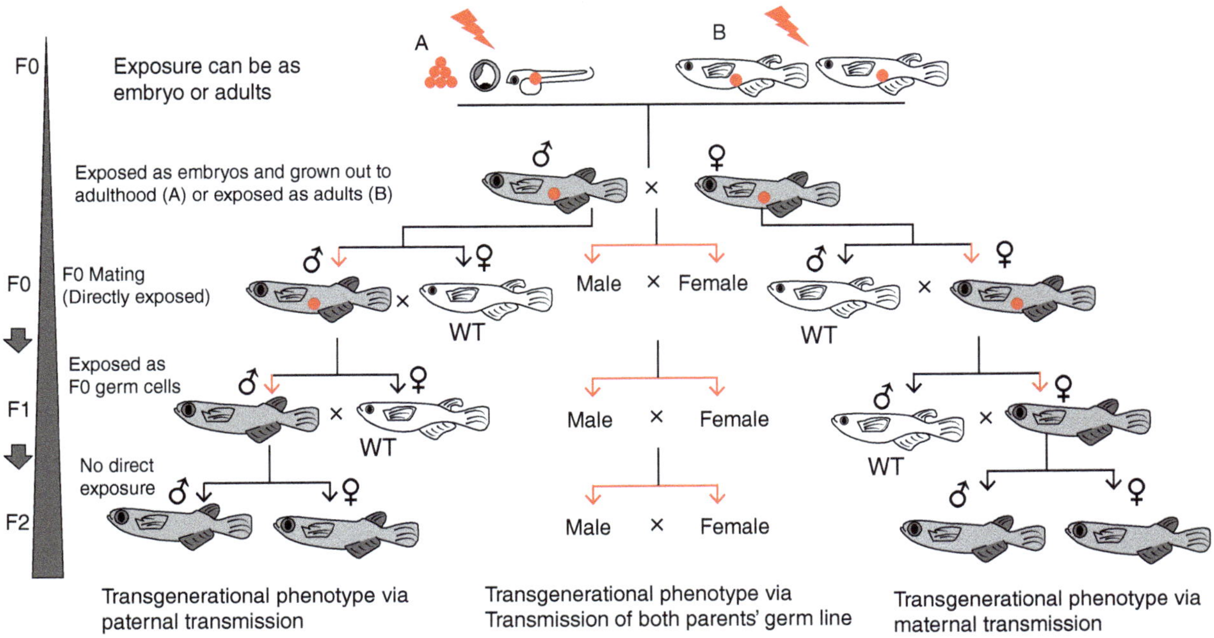

Figure 4.7 Design of transgenerational laboratory experiments in fish. Exposure to F0 generation can be carried out at embryonic stages or in adulthood. Parent-of-origin effects can be studied by mating exposed individuals with the unexposed across generations. To obtain a transgenerational phenotype, the F2 or subsequent generations must be screened. For subsequent generations beyond the F2, the same strategy of breeding should be applied, as shown for the breeding of the F0 and F1 generation fish. The shaded fish shows the lineage of the fish exposed to the stressor at F0 generation (as embryo or adult). The arrows on the top of shaded fish indicate the parent of origin, × indicates crossbreeding, and the line immediately below the fish indicates the founder fish (parent) for the next generation. Dark spots on the abdomen of the fish indicate direct exposure effects on gonad.

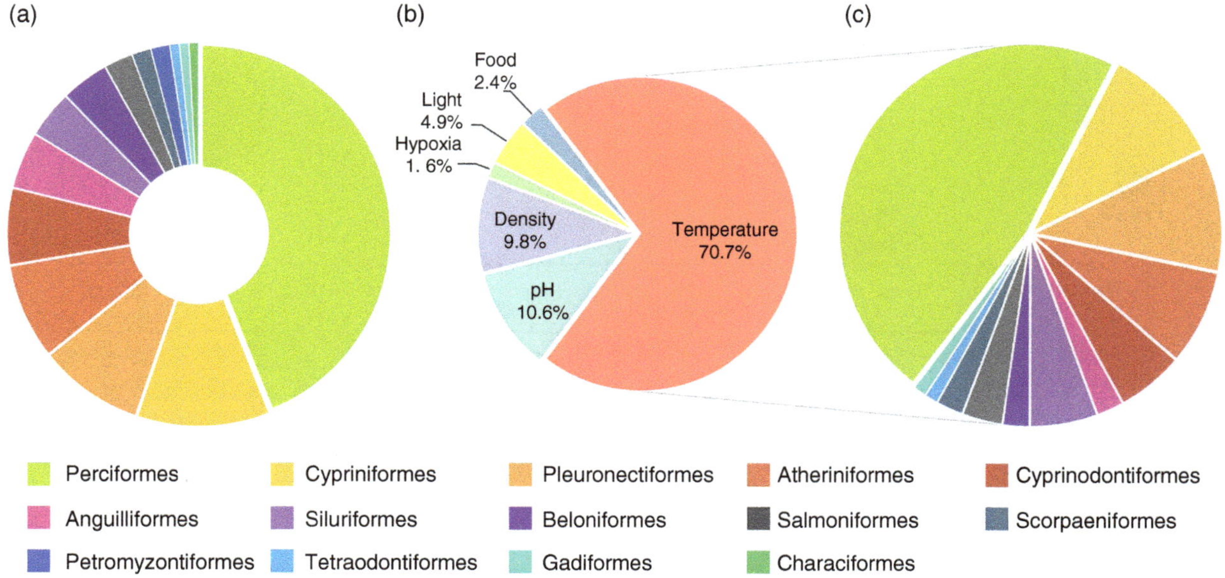

Figure 5.1 A summary of environmental stress on sex differentiation in fish. This survey covers 100 fish species and 123 cases belonging to 14 orders, which is from 65 published papers. A case represents the corresponding relationship between a fish species and a stressor. (a) Distribution of the 123 cases according to the order in which the target species belongs. (b) Percentage of type of environmental stress in the entire 123 cases. (c) Percentage of 14 orders in the 87 cases in temperature-dependent sex differentiation. Temperature, high or low temperature, or stress temperature; pH, acidic or alkali water; density, including stress density and high levels of hierarchy; hypoxia, low dissolved oxygen; light, stress illumination condition, such as a long light period or continuous green light; food, including low food availability and low food quality. References of each order: Petromyzontiformes, *Lampetra aepyptera* [95], *Petromyzon marinus* [96]; Anguilliformes, *Anguilla anguilla* [97], *Anguilla rostrata* [97], *Anguilla japonica* [97], *Anguilla dieffenbachii* [97]; Cypriniformes, *Carassius auratus* [98], *Carassius Carassius* [99, 100], *Carassius gibelio* [101], *Cyprinus carpio* [102], *Danio rerio* [103], *Gnathopogon caerulescens* [104], *Misgurnus anguillicaudatus* [105], *Paramisgurnus dabryanus* [105], *Sabanejewia balcanica* [106]; Siluriformes, *Ictalurus punctatus* [107], *Tachysurus fulvidraco* [108], *Tachysurus vachellii* [109], *Hoplosternum littorale* [110], *Clarias gariepinus* [31, 111]; Characiformes, *Nematobrycon palmeri* [112]; Salmoniformes, *Oncorhynchus nerka* [113], *Oncorhynchus mykiss* [114], *Thymallus thymallus* [115]; Gadiformes, *Gadus chalcogrammus* [116]; Perciformes, *Apistogramma* spp. [117], *Apistogramma caetei* [117], *Pelvicachromis pulcher* [118], *Pelvicachromis subocellatus* [117], *Pelvicachromis taeniatus* [117], *Pelmatochromis* spp. [119], *Oreochromis niloticus* [120], *Oreochromis aureus* [120], *Oreochromis mosambicus* [120], *Dicentrarchus labrax* [121], *Macropodus opercularis* [122], *Trichogaster lalius* [123], *Betta splendens* [124, 125], *Lepomis macrochirus* [126], *Acanthochromis polyacanthus* [127]; Scorpaeniformes, *Sebastes schlegeli* [128, 129], *Anoplopoma fimbria* [130]; Atheriniformes, *Menidia menidia* [131], *Menidia peninsulae* [132], *Odontesthes argentinensis* [133], *Odontesthes bonariensis* [134], *Odontesthes hatcheri* [135], *Chirostoma estor* [136], *Leuresthes tenuis* [137], *Hypoatherina tsurugae* [138]; Tetraodontiformes, *Takifugu rubripes* [139, 140]; Cyprinodontiformes, *Limia melanogaster* [141], *Poeciliopsis lucida* [141], *Poecilia sphenops* [142], *Poecilia reticulate* [143], *Xiphophorus helleri* [144]; Beloniformes, *Oryzias latipes* [145–147], *Oryzias sakaizumii* [148]; Pleuronectiformes, *Paralichthys olivaceus* [149], *Paralichthys dentatus* [150], *Paralichthys lethostigma* [151], *Pseudopleuronectes yokohamae* [152], *Pseudopleuronectes americanus* [153], *Verasper moseri* [154], *Verasper variegatus* [155], *Hippoglossus hippoglossus* [156], *Scophthalmus maximus* [157], *Cynoglossus semilaevis* [158].

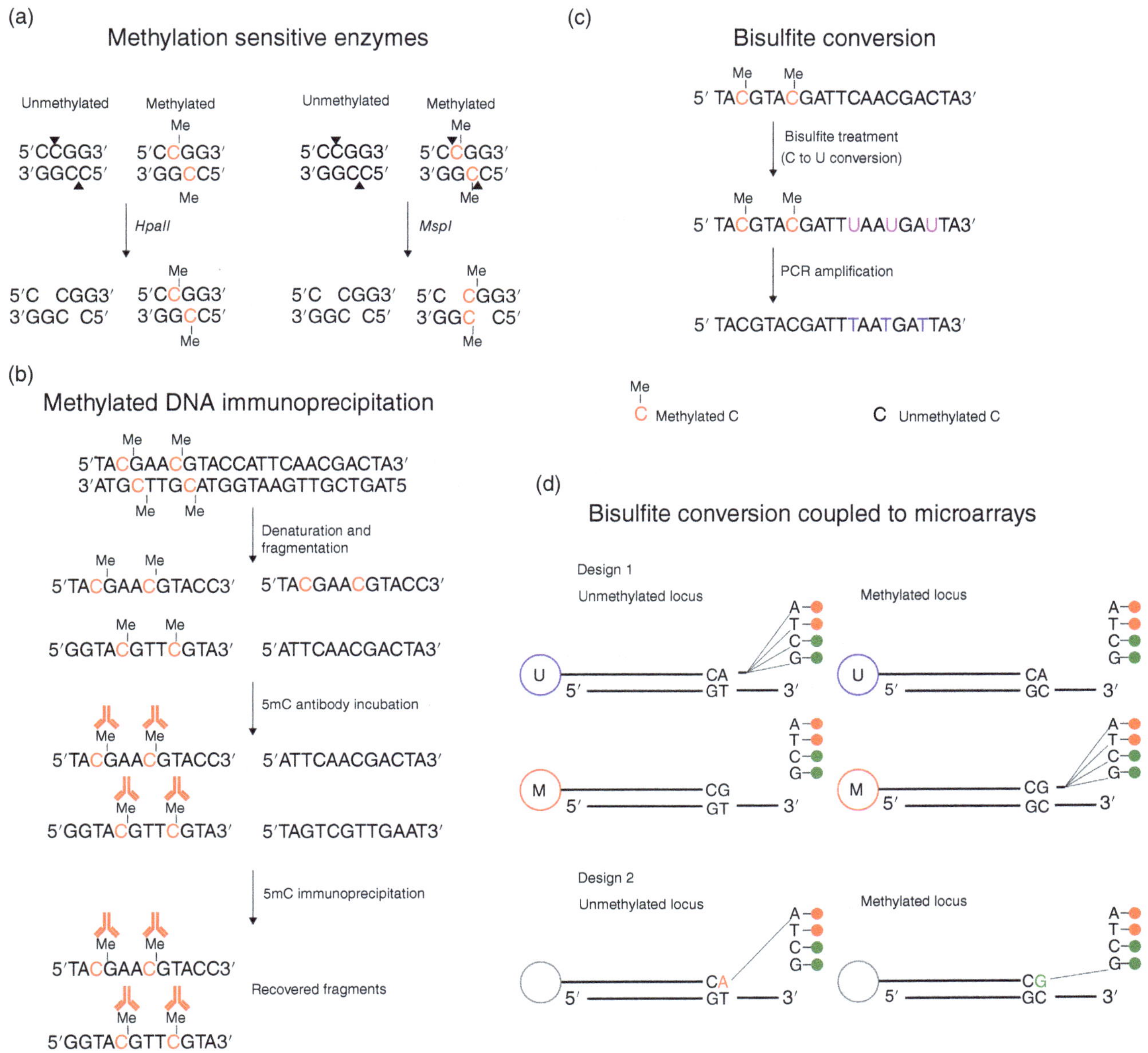

Figure 6.1 Common detection methods used to assess cytosine methylation. (a) Methylation-sensitive enzymes. (b) Bisulfite conversion. (c) Methylated DNA immunoprecipitation. (d) Bisulfite coupled to microarrays. In the Infinium assay design 2, two color channels are used simultaneously, specific incorporation of labeled nucleotides allows methylation status discrimination. Arrows indicate subsequent steps in each strategy.

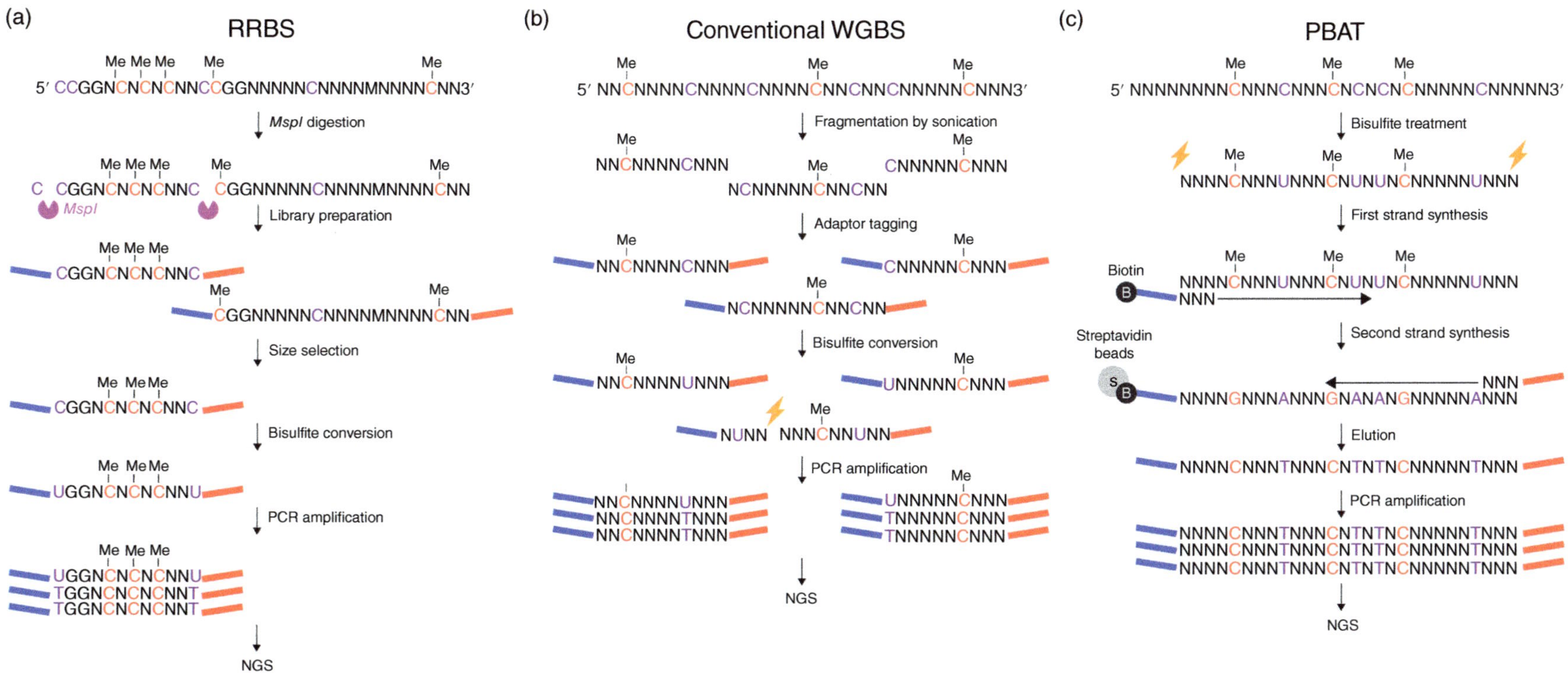

Figure 6.2 Common bisulfite-based techniques coupled to NGS. (a) Reduced representation bisulfite sequencing (RRBS). (b) Conventional Whole Genome Bisulfite Sequencing (WGBS). (c) Post-bisulfite Adaptor Tagging (PBAT).

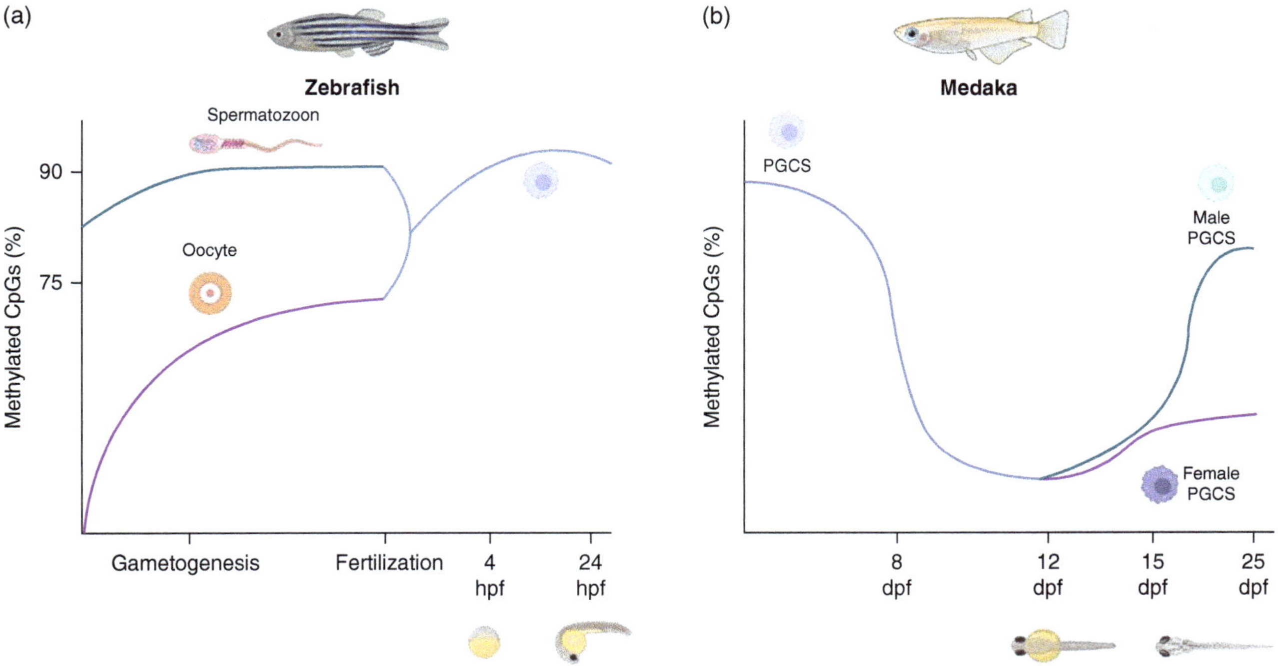

Figure 7.1 DNA methylation dynamics in fish PGCs and gametes. (a) Percentage of CpG methylation in zebrafish gametes and PGCs. The sperm cells have around 91–95% of methylated CpGs, whereas, in the oocyte, the percentage of CpG methylation is 75%. After fertilization, the DNA methylome of PGCs from 4 to 24 hpf is similar to that of spermatozoa. (b) Percentage of CpG methylation in embryonic cells at 4 hpf and in PGCs of medaka. A first round of demethylation would take place before or at 8 dpf (the first PGCs sampling point), reaching the epigenetic ground state in a second round occurring at 12 dpf. At 25 dpf, the male PGCs suffer a process of *de novo* methylation but the female PGCs retain the same pattern as at 15 dpf.

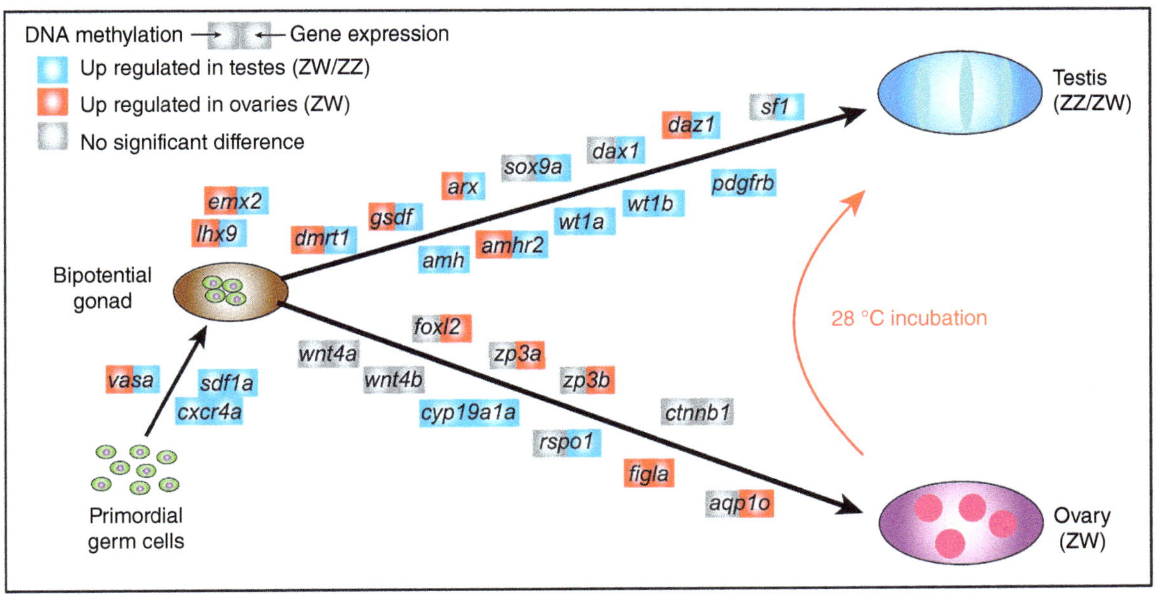

Figure 8.3 Differentially methylated and differentially expressed genes in the putative sex determination pathway of Chinese tongue sole. The compared methylation level (left square) and expression level (right square) between testes and ovaries of each gene are shown by different colors. Dark gray indicates upregulated in testes, light gray indicates upregulated in ovaries, and medium gray indicates no significant difference between testes and ovaries. *Source:* Shao et al. [29]/Cold Spring Harbor Laboratory Press.

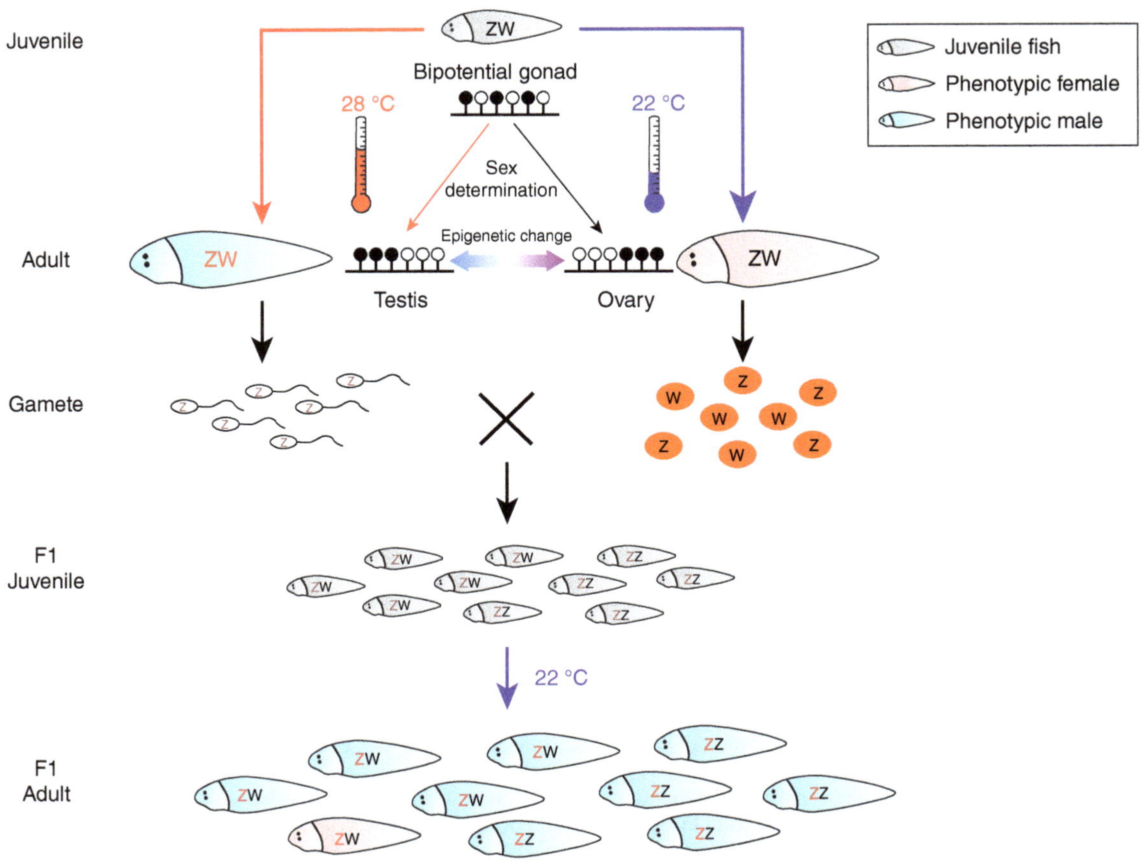

Figure 8.4 Transgenerational epigenetic inheritance of sexual reversal in Chinese tongue sole. High temperature treatment of juvenile ZW fish can change the epigenetic modification of the Z chromosome to develop into neomales. The neomales can only produce Z-type sperm, and some of these alterations are transferred to the next generation. Thus, most of the genetic female offspring form into neomales spontaneously. *Source:* Adapted from Shao et al. [29]/Cold Spring Harbor Laboratory Press.

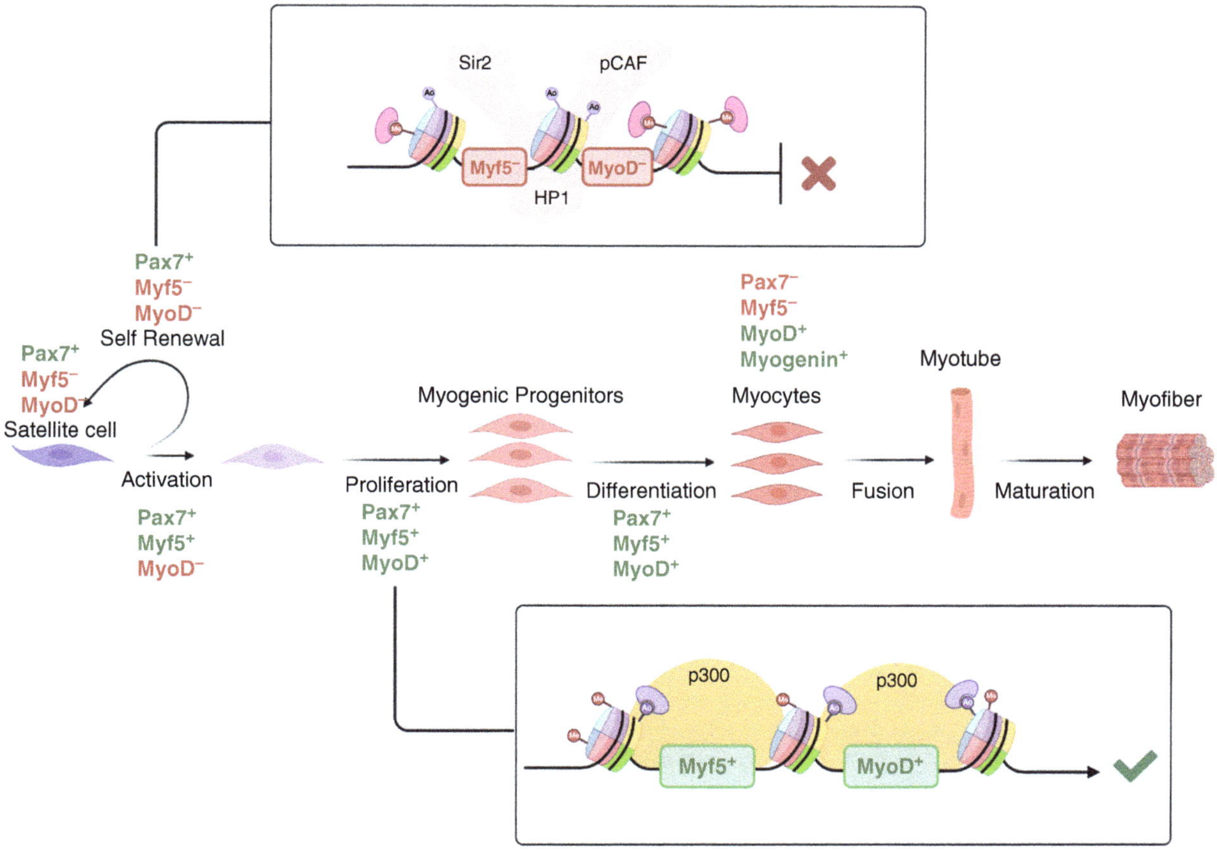

Figure 9.1 The process of myofiber formation in vertebrates. The muscle regulatory factors MyoD, Myf5, and Myogenin are expressed at different stages. The paired box protein Pax 7 plays a key role in myogenesis and is a marker of quiescent satellite cells. Repressive methylation signatures associated with the action of *Sir2*, *pCAF*, and HP1 lead to repression of MyoD and satellite cell self-renewal, whereas phosphorylation and acetylation signals associated with histone acetyltransferase p300 cause differentiation of myogenic progenitor cells. Gene name abbreviations: HP1, Heterochromatin Protein 1-Alpha; Myf5, Myogenic Factor 5; MyoD, Myoblast Determination Protein; MyoG, Myogenin; Pax7, Paired Box Protein 7; p300, E1A Binding Protein P300; pCAF, P300/CBP-Associated Factor; and Sir2, Sirtuin 2. *Source:* Created with BioRender.com.

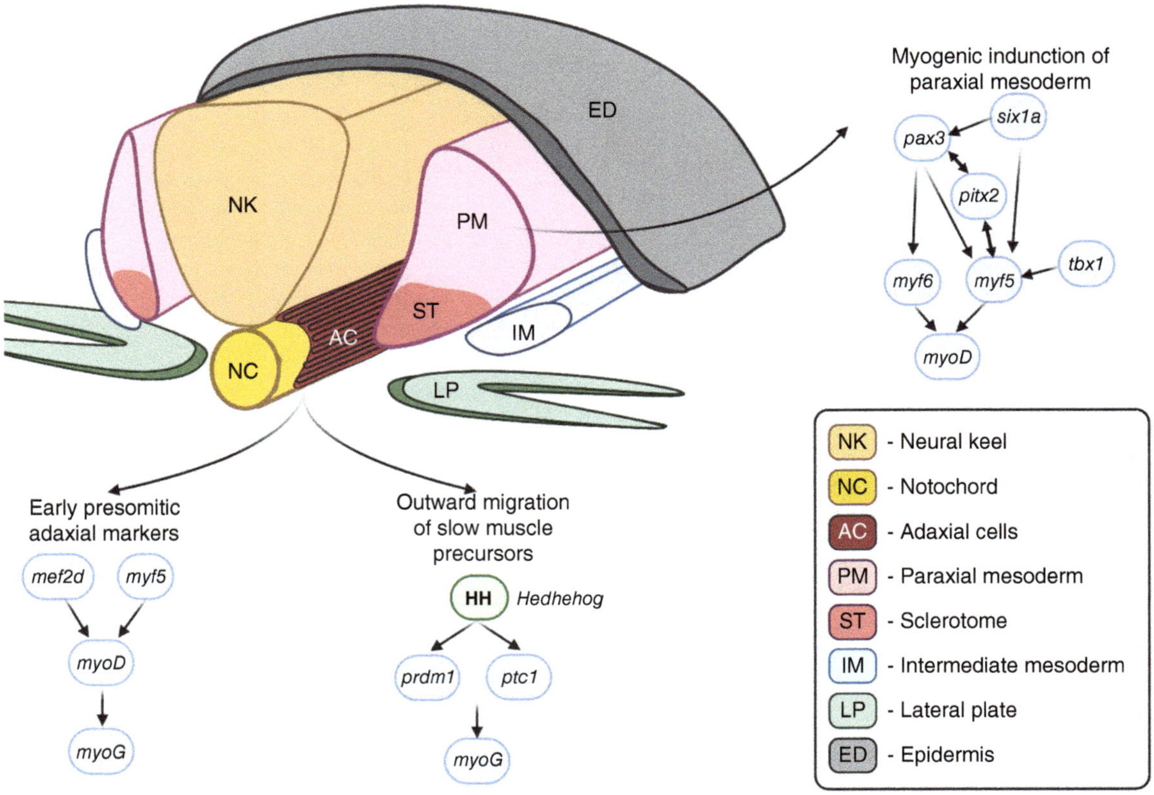

Figure 9.2 Transverse 3D section of the mid trunk during somitogenesis in fish, depicting an overview of the mesodermal structures: (i) the early presomitic phase in which mef2d is an adaxial marker; (ii) the outward migration of slow muscle precursors through the activation of HH signaling and action of prdm1 and ptc1; and (iii) the myogenic induction of the paraxial mesoderm, where pax3, six1a, pitx2, and tbx1 promote the expression of myf5 and myf6. Gene name abbreviations: *Mef2d*, Myocyte Enhancer Factor 2D; *Myf5*, Myogenic Factor 5; *Myf6*, Myogenic Factor 6; *MyoD*, Myoblast Determination Protein; *MyoG*, Myogenin; *Pax3*, Paired Box Protein 3; *Pitx2*, Paired Like Homeodomain 2; *Prdm1*, PR/SET Domain 1; *Ptc1*, Patched 1; *Six1a*, SIX Homeobox 1a; *Tbx1*, T-Box Transcription Factor 1. *Source:* Created with BioRender.com.

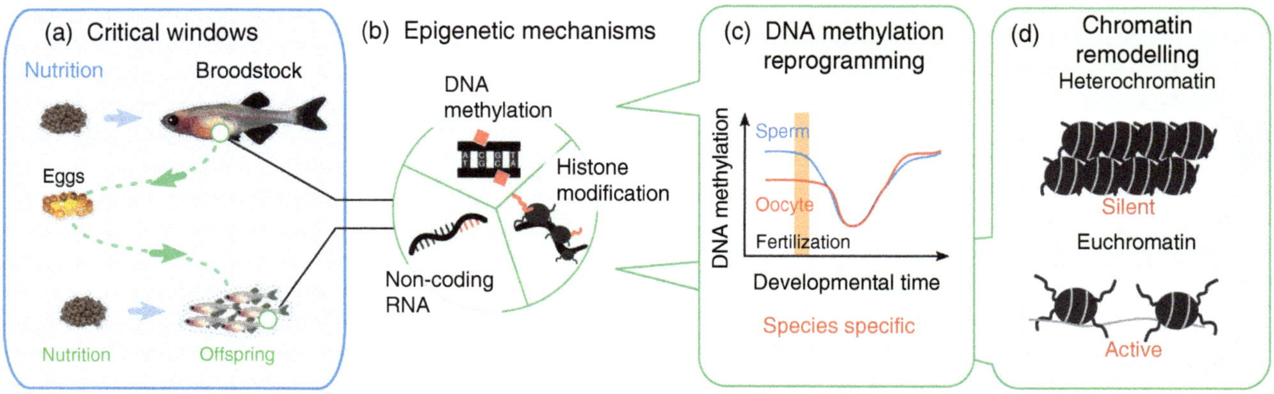

Figure 10.1 Critical windows for nutritional programming and alterations in epigenetic mechanisms. The figure shows the relationship between nutritional programming from broodstock to offspring and epigenetic regulation. (a) A schematic diagram showing critical windows for nutritional programming using an example of the female linage in fish. (b) Three major classes of epigenetic mechanisms that are potentially controlled by nutritional programming. (c) An example of alterations of methylation rates in the early development stages around fertilization. Alteration patterns can be species-specific in fish. (d) Two types of chromatin structures that are associated with gene expression status (active vs. silent).

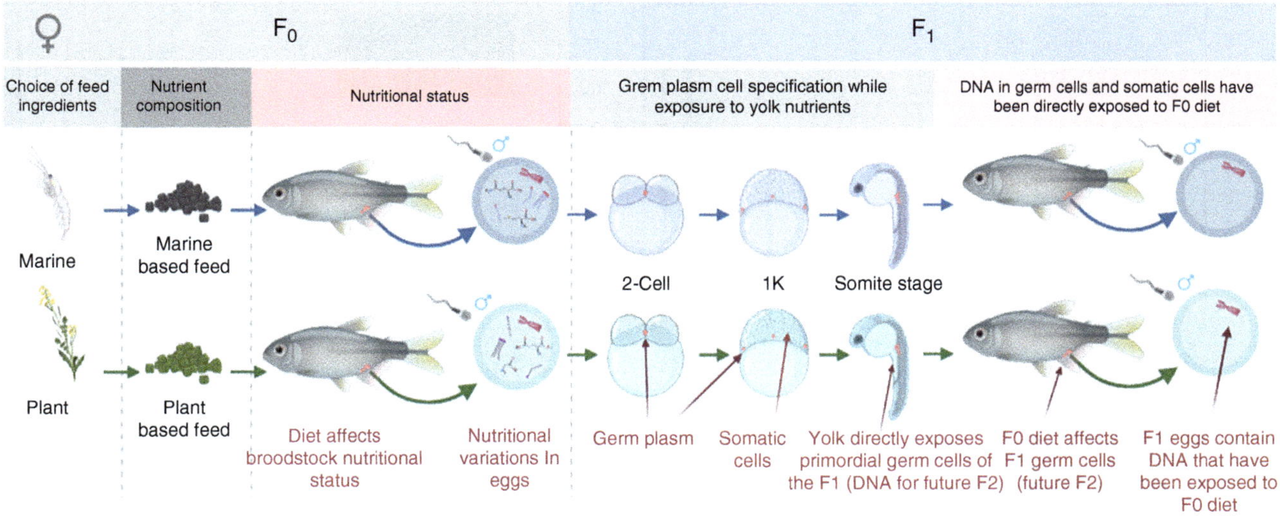

Figure 10.2 Broodstock nutrition and potential long-term effects in offspring generations. The choice of feed ingredients and nutrient composition (feed formulation) will affect the nutritional status in broodstock (F0) and in the eggs (future F1). After fertilization (F1), the cells are in direct contact with the nutrients in the yolk. The germplasm (aggregates of germline determinant RNAs and proteins in the cytoplasm) establishes the germline and separates the PGCs, the precursor cell type of all gametes, from the somatic cells. The germplasm is visualized at the two-cell stage and is incorporated in specific cells at the blastula stage (1K [1000 cell] stage). At the somite stage, the PGCs (future F_2-generation) migrate to the genital ridge for proliferation followed by meiotic commencement. During early development and PGC proliferation and specification, the nutritional status of the yolk directly exposes the DNA of F1 gametes. F1 eggs contain DNA that has been exposed to F0 diet through the inclusion of nutrients in the F0 eggs. *Source:* The figure was created using http://BioRender.com.

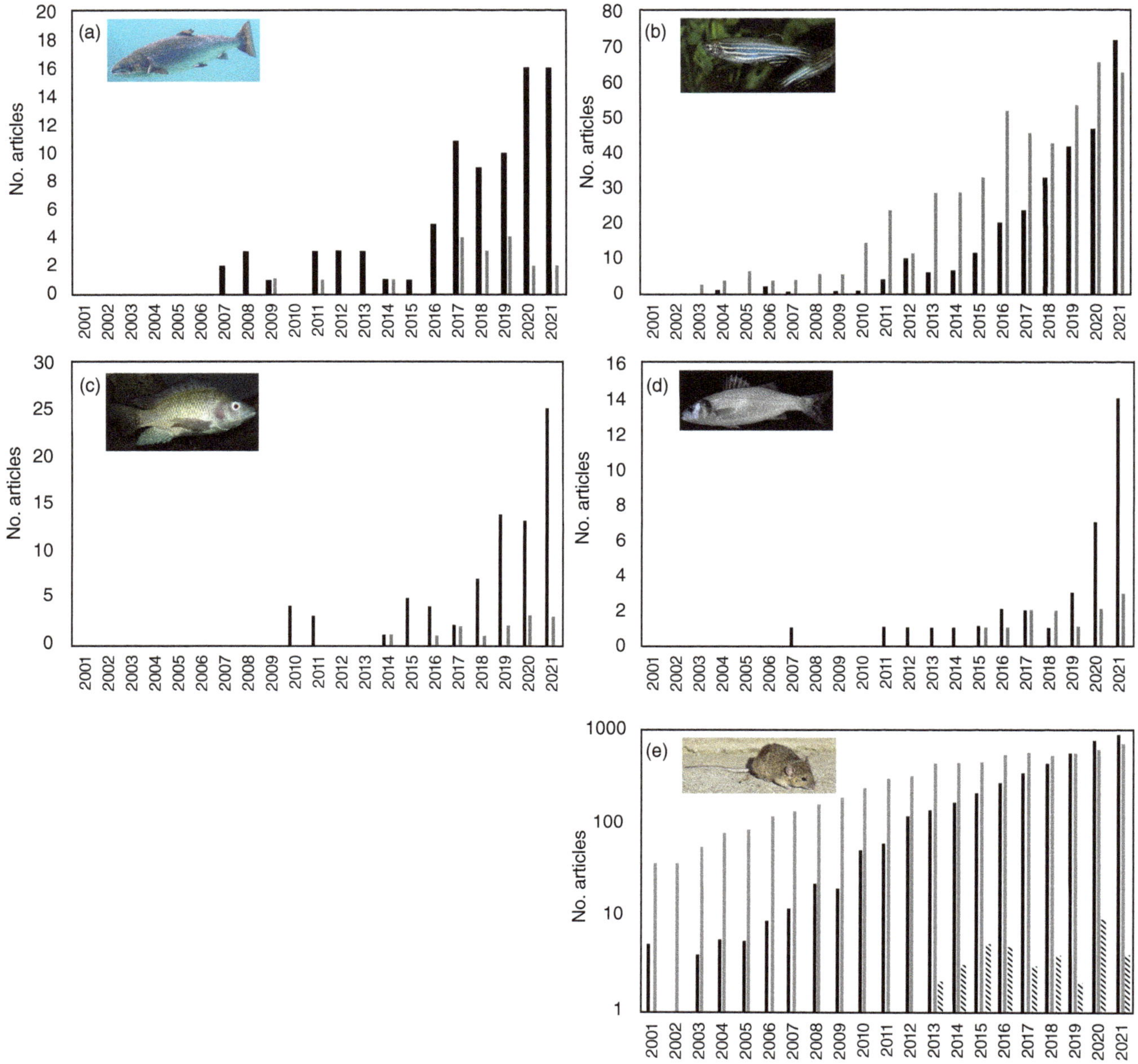

Figure 11.1 Number of articles published on the microbiome or epigenome for select species up until the end of 2021. (a) Atlantic salmon *Source:* Hans-Petter Fjeld/Wikimedia Commons/CC BY SA 2.5, (b) zebrafish *Source:* Azul/Wikimedia Commons/public domain, (c) Nile tilapia *Source:* Bjørn Christian Tørrissen/Wikimedia Commons/CC BY SA 3.0, (d) European seabass *Source:* Hans Hillewaert/ Wikimedia Commons/CC BY SA 4.0, and (e) mouse *Source:* Amirekul/Wikimedia Commons/CC BY SA 4.0. Searches were conducted in PubMed (3 November 2021) for terms in the Title or Abstract as follows: (((Common name) OR (Species name)) AND ((Microbiome) OR (Microbiota))); (((Common name) OR (Species name)) AND ((Epigenome) OR (Epigenetic))); and (((Common name) OR (Species name)) AND ((Microbiome) OR (Microbiota)) AND ((Epigenome) OR (Epigenetic))). Review articles were excluded. Black: microbiome, Gray: epigenome; Hatched: microbiome and epigenome.

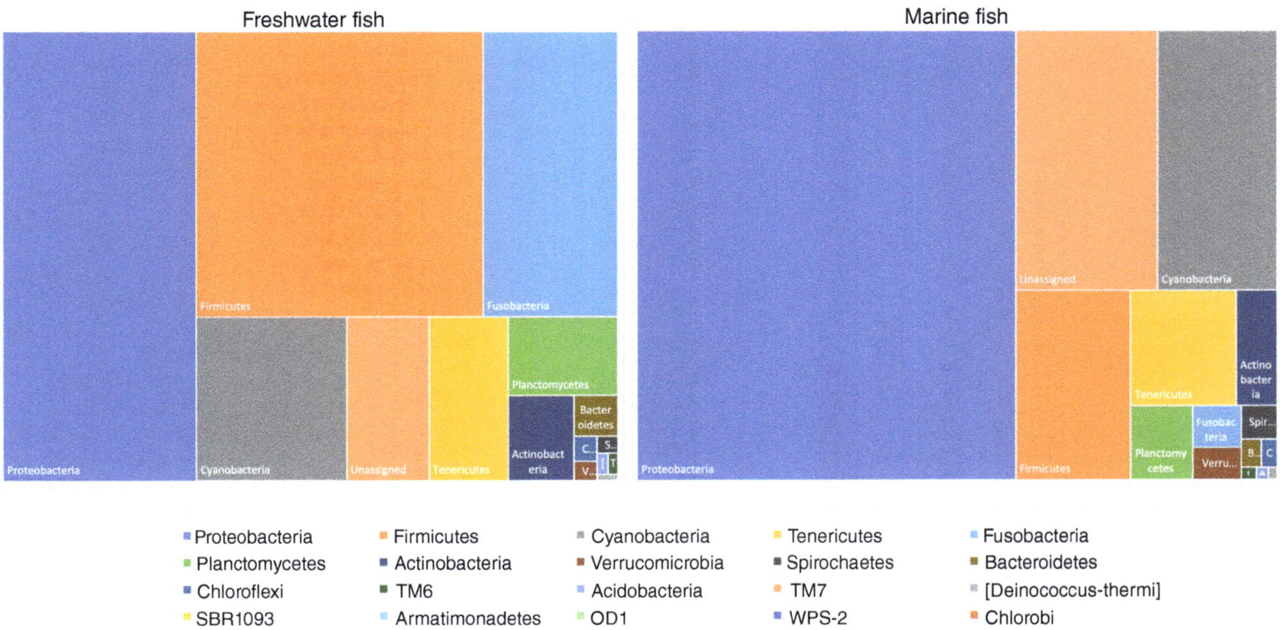

Figure 11.2 Relative abundance of different bacterial taxa in the gut microbiota of fish from freshwater and marine habitats based on data from 14 orders, 42 families, 79 genera, and 85 species. *Source:* Adapted from data in Kim et al. [35].

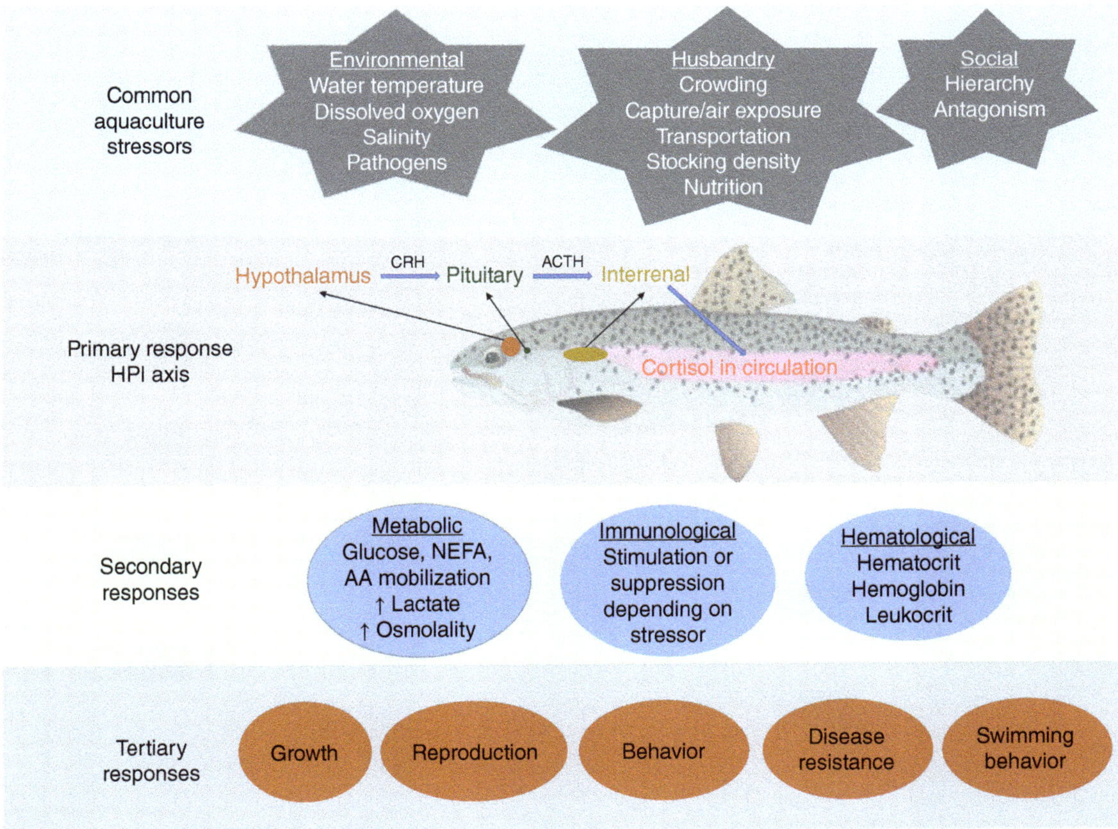

Figure 12.1 Stress responses induced by common aquaculture stressors and associated with the hypothalamic-pituitary-interrenal (HPI) axis. Abbreviations: AA, amino-acids; ACTH, adrenocorticotropin hormone; CRH, Corticotropin-releasing hormone; NEFA, nonesterified fatty acids). See text for details. Rainbow trout image by Tracey Saxby, Integration and Application Network (ian.umces.edu/media-library) under Attribution-ShareAlike 4.0 International (CC BY-SA 4.0) license.

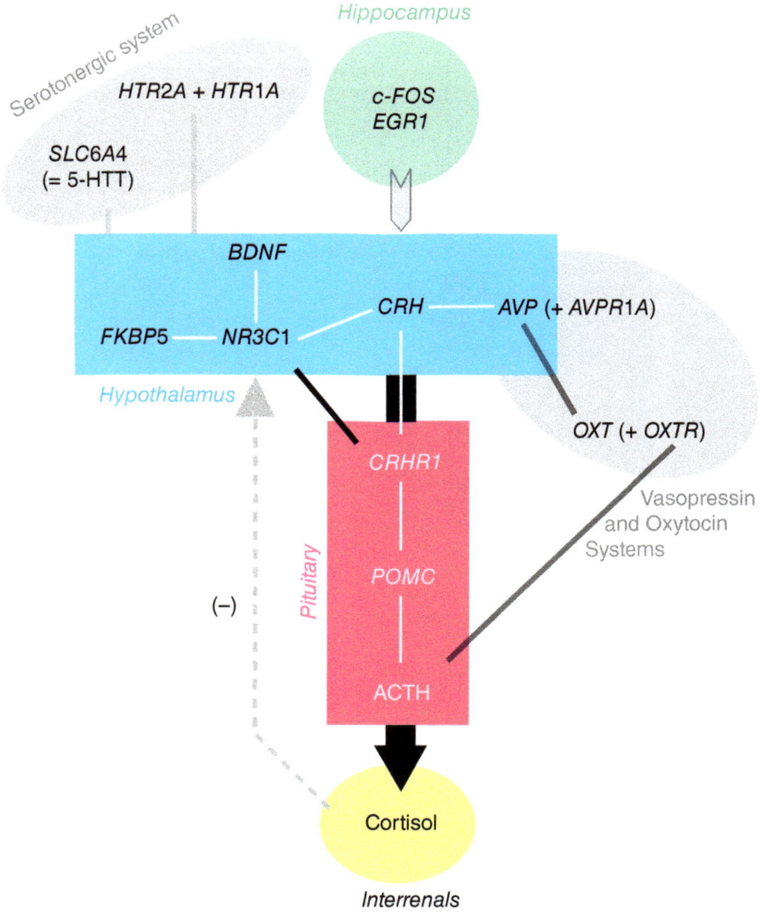

Figure 12.2 Genes considered relevant candidates for epigenetic studies in fish. A focus is made onto the HPI axis, the hippocampus, and other systems for genes in which stress-induced epigenetic modifications, mainly DNA methylation changes, have been observed. These modifications can rewire brain/neuronal molecular pathways and translate into neurobehavioral disorders in humans and rodents, sometimes resulting in differences in cortisol production. Gene names are reported as in mammals. Solid lines refer to epigenetic interactions among genes or the action of genes on specific organs (serotonergic system). (–): negative feedback loop induced by cortisol in the hypothalamus. Abbreviations: ACTH, adrenocorticotropin hormone; AVP, arginine vasopressin (AVT, arginine-vasotocin in fish); AVPR1A, arginine vasopressin receptor 1A; BDNF, brain-derived neurotrophic factor; c-FOS, Fos Proto-Oncogene; CRH, corticotropin-releasing hormone; CRHR1, corticotropin-releasing hormone receptor 1; EGR-1, early growth receptor-1; FKBP5, FK506-binding protein 5; HTR1A, 5-hydroxytryptamine receptor 1A; HTR2A, 5-hydroxytryptamine receptor 2A; POMC, proopiomelanocortin; OXT, oxytocin (IT, isotocin in fish); OXTR, oxytocin receptor.

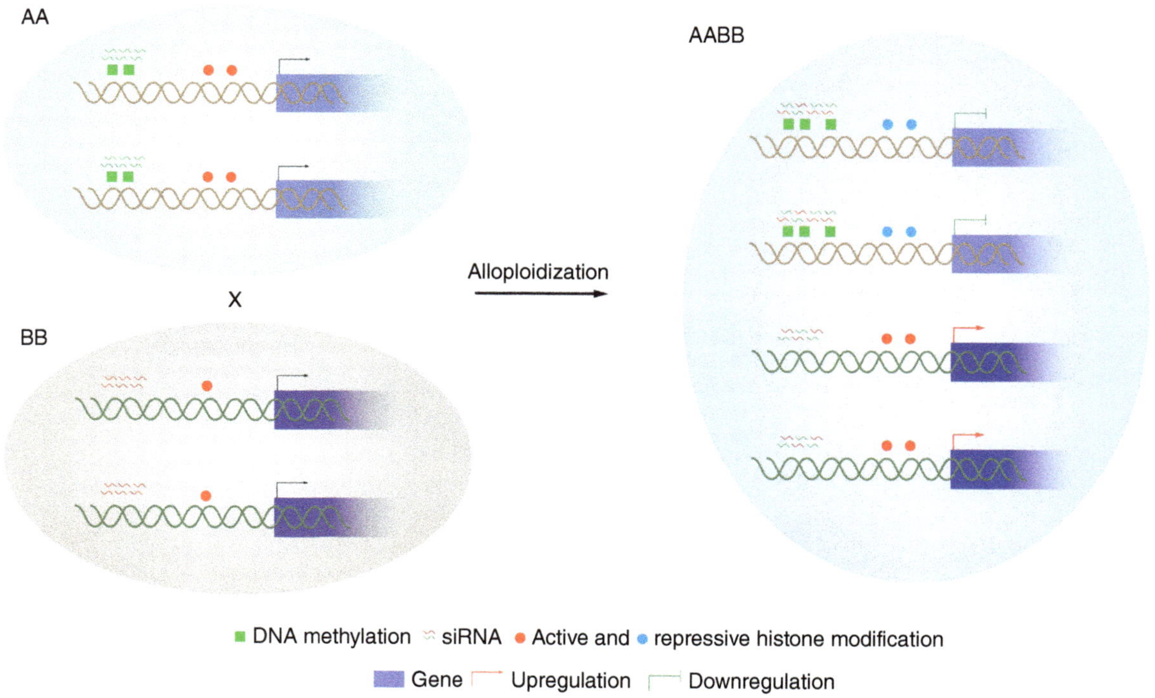

Figure 13.2 Epigenetic changes during hybridization and polyploidization in aquatic animals (allotetraploid as an example). In addition to non-Mendelian genomic changes that occurred during alloploidization, the newly formed allopolyploids may also undergo epigenetic changes, such as DNA methylation, histone modifications, and small RNAs, which affect gene expression. A and B indicate two genomes from different species.

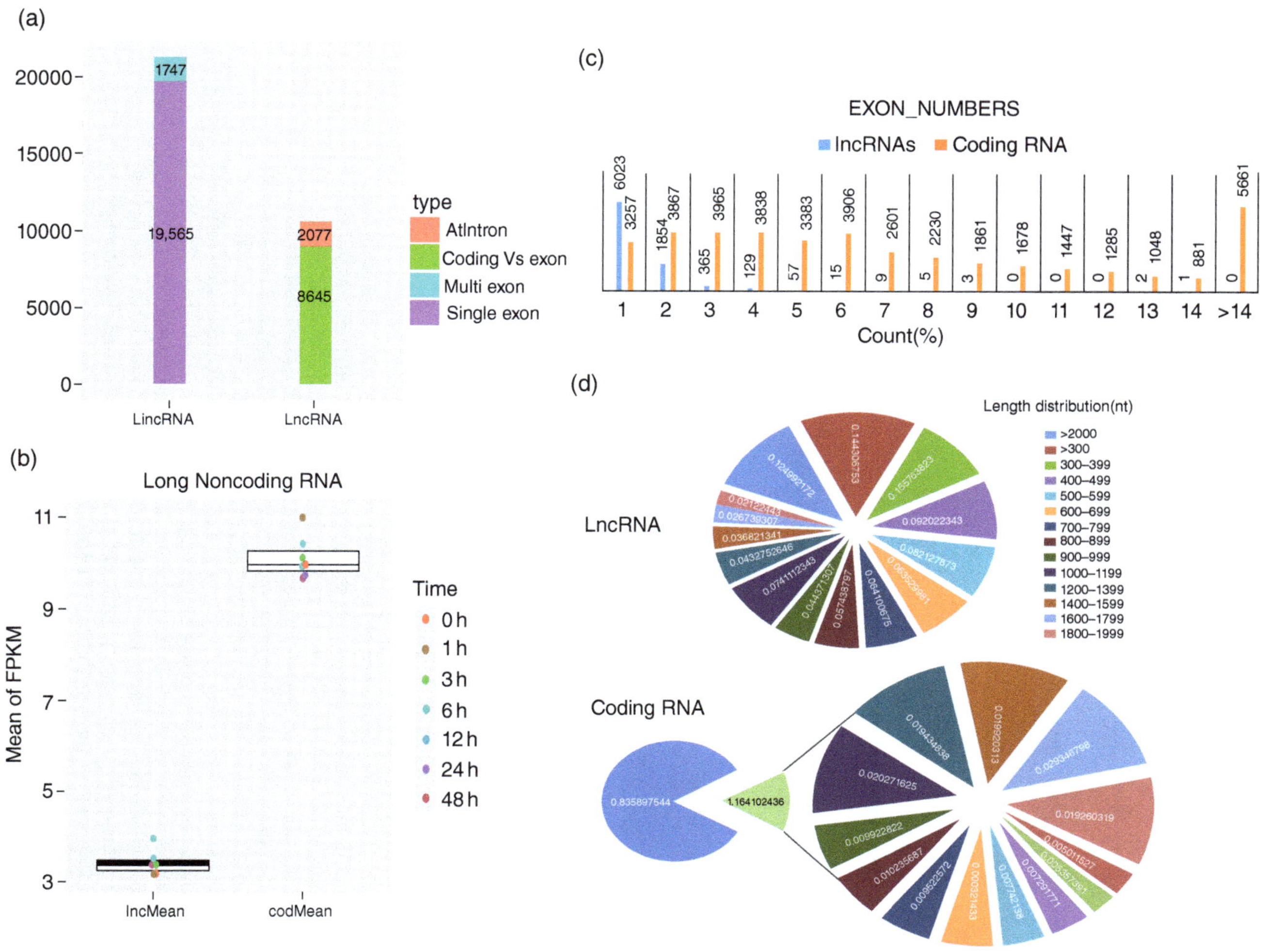

Figure 14.5 Long noncoding RNA characteristics in large yellow croaker. (a) Distribution of lncRNA classes. There were 1747 multi-exon long noncoding RNA; 19565 single-exon long noncoding RNA; 2027 long noncoding RNA generated from intron; 8645 long noncoding RNA overlapping intron. (b) Average expression level of lncRNAs and the protein-coding genes in the brain tissue. (c) Distribution of exon numbers of lncRNAs and coding RNA. (d) Ratio of length of lncRNAs and coding RNA. *Source:* Reproduced with permission from Liu et al. [96].

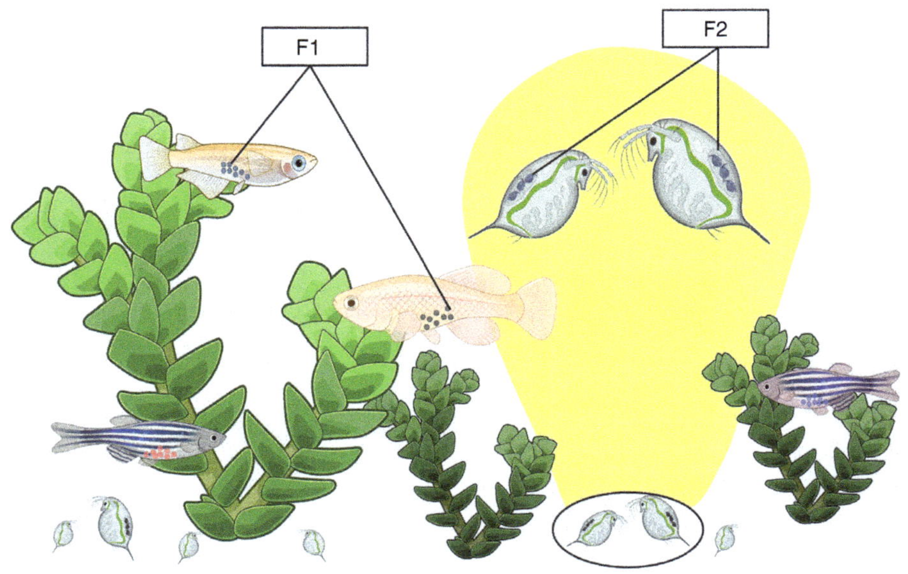

Figure 14.6 True transgenerational exposure occurs in the F2 and F3 generations of oviparous and viviparous species. Most fish are oviparous, meaning they produce eggs that develop outside of the parent. In oviparous species, the F1 generation is indirectly exposed as germ cells. Cladoceran species (e.g. *Ceriodaphnia dubia* and *Daphnia magna*), mainly reproduced through parthenogenesis, are viviparous. In viviparous organisms, the F2 generation germ cells are also indirectly exposed during F1 generation development in the brood pouch. *Source:* From © BioRender.

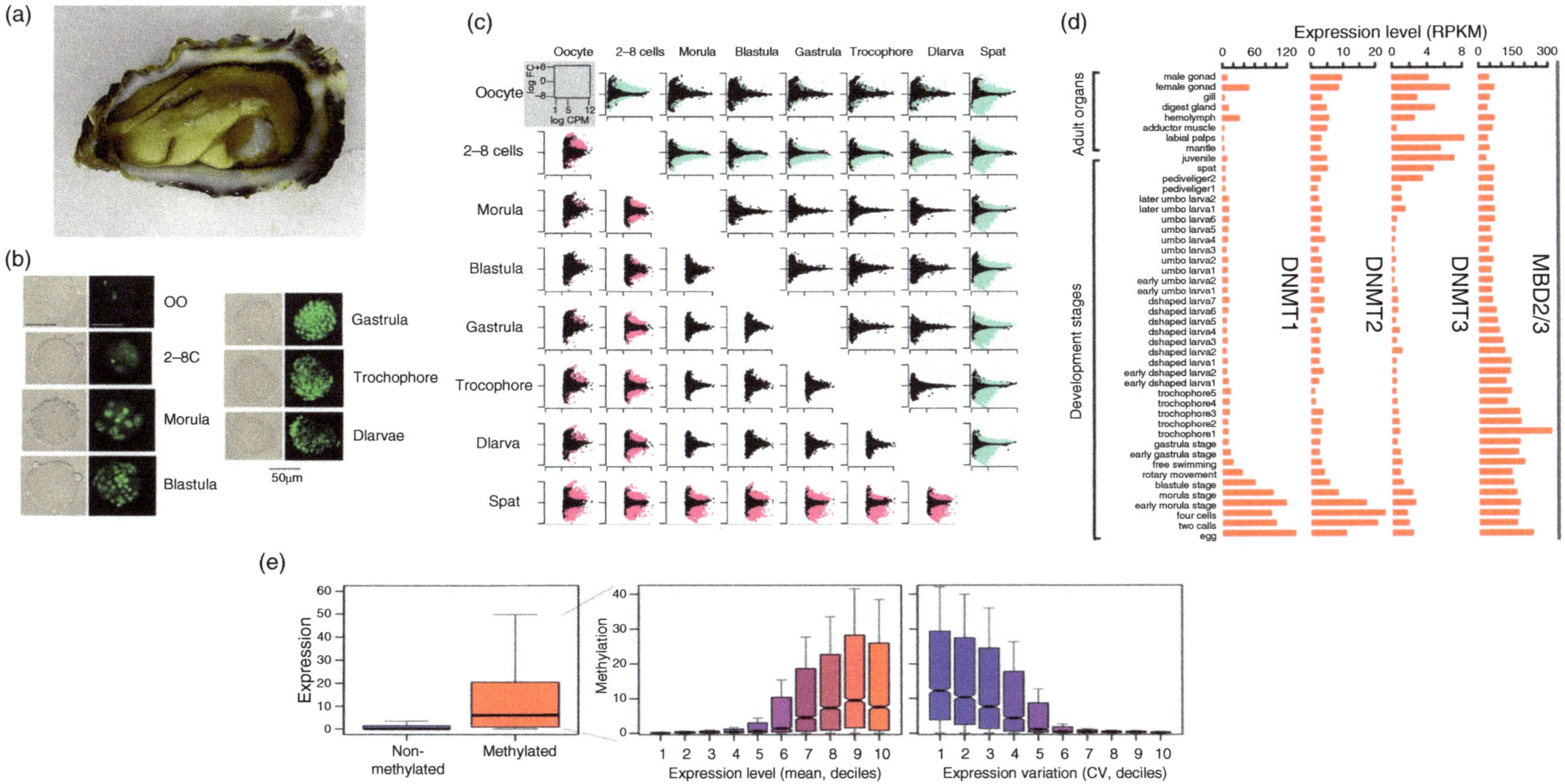

Figure 15.1 Impact of DNA methylation during the development of the cupped oyster *C. gigas*. (a) Opened *Crassostrea gigas* oyster with minor (ventral) valve removed, exposing mantle; (b) Localization of methylated DNA within *C. gigas* embryos using a sheep anti-5-mC antibody and an anti-sheep secondary antibody; left, transmitted bright field photonic microscopy; right, transmitted UV light epifluorescence (Box 15.2) microscopy using a 395–420 nm filter; the embryonic stage is indicated; 2–8C, 2–8 cells; magnification, ×400; Nb, bubbles at embryo edges are mounting artifacts; oo, oocytes. Scale bar 50 μm. (c) Oyster developmental methylomes exhibit feature and stage-dependent landscapes highlighting cleavage and metamorphosis. Pairwise comparisons of the methylation of coding sequences (CDS, upper-right) and transposable elements (TE, lower-left) among development stages are listed in chronological order. Colored points are significantly differentially methylated CDS (top-right) and TEs (bottom-left) between stages (Student's T test, $p < 0.01$). The ML (log counts per million, *x* axis) and its variation (log fold change, *y* axis) are shown. (d) Expression levels of the oyster DNMTs and MBD2/3 genes in different developmental stages and adult organs, denoted by reads per kilobase of transcript per million mapped reads (RPKM). Numbers after sample names indicate the orders in development, with smaller numbers representing earlier stages in development. (e) The expression level of nonmethylated and methylated genes (CDS methylation is shown, left). Relation between CDS methylation and mRNA levels (mean across development stages, deciles) of the methylated genes (middle). Relation between CDS methylation and expression variability of the methylated genes (coefficient of variation [CV] across development stages, deciles) (right). *Source:* (a) From Leverett and Thain [66]/International Council for the Exploration of the Sea; (b) From Riviere et al. [65]/Springer Nature; (c, e) Riviere et al. [67]/PLOS/CC BY 4.0; (d) Wang et al. [50]/Springer Nature/CC BY 4.0.

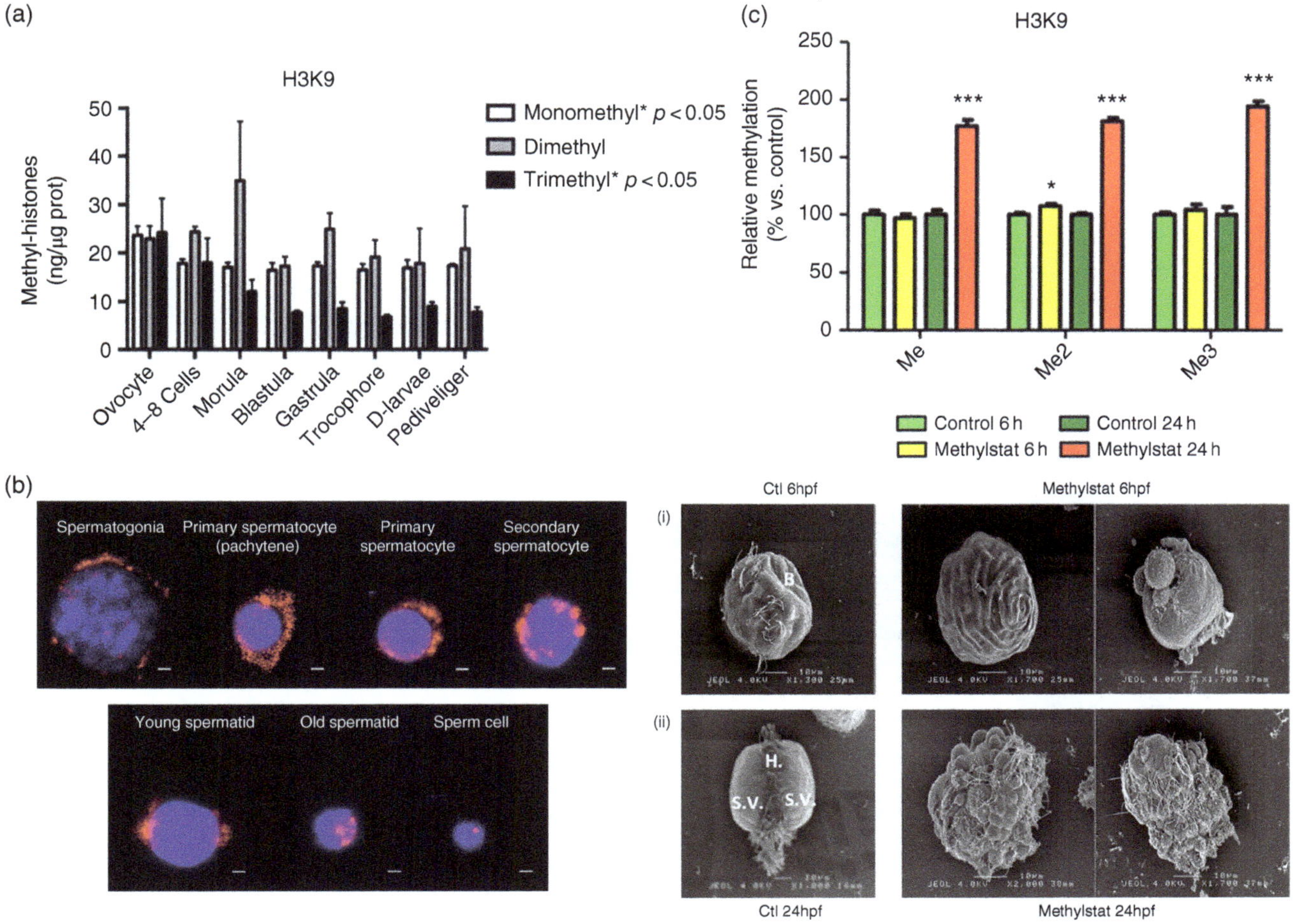

Figure 15.2 Histone modifications during the development of the cupped oyster *C. gigas*, examples of H3K9 methylation and Cg-CenH3 variant. (a) Histone methylation profiles of H3K9 during the early development of *Crassostrea gigas*. Asterisk in the legend indicates significant variation of the mark during development. A one-way ANOVA was used to determine whether MLs were statistically and significantly different between development stages for each residue, and a $p < 0.05$ was considered significant. (b) Distribution of Cg-CenH3 in male *C. gigas* gonadal cells in different stages of maturation. Scale bar = 1 µm. (c) Hypermethylation of histone lysine residues in the presence of 10 µM methylstat. Global H3K9 methylation increases in the presence of methylstat. Asterisk in the legend indicates a significant variation of the indicated mark (One-way ANOVA; $p < 0.05$ was considered significant (<0.05 (*), <0.001 (**), <0.0001 (***)) (top). Different phenotypes observed in control conditions and under methylstat (Jmj inhibitor) treatment at 6 hours after fertilization (i) and 24 hours after fertilization (ii) (bottom). *Source:* (a, c) Fellous et al. [94]/MDPI/CC BY 4.0; (b) From Cvitanić et al. [95]/MDPI/CC BY 4.0.

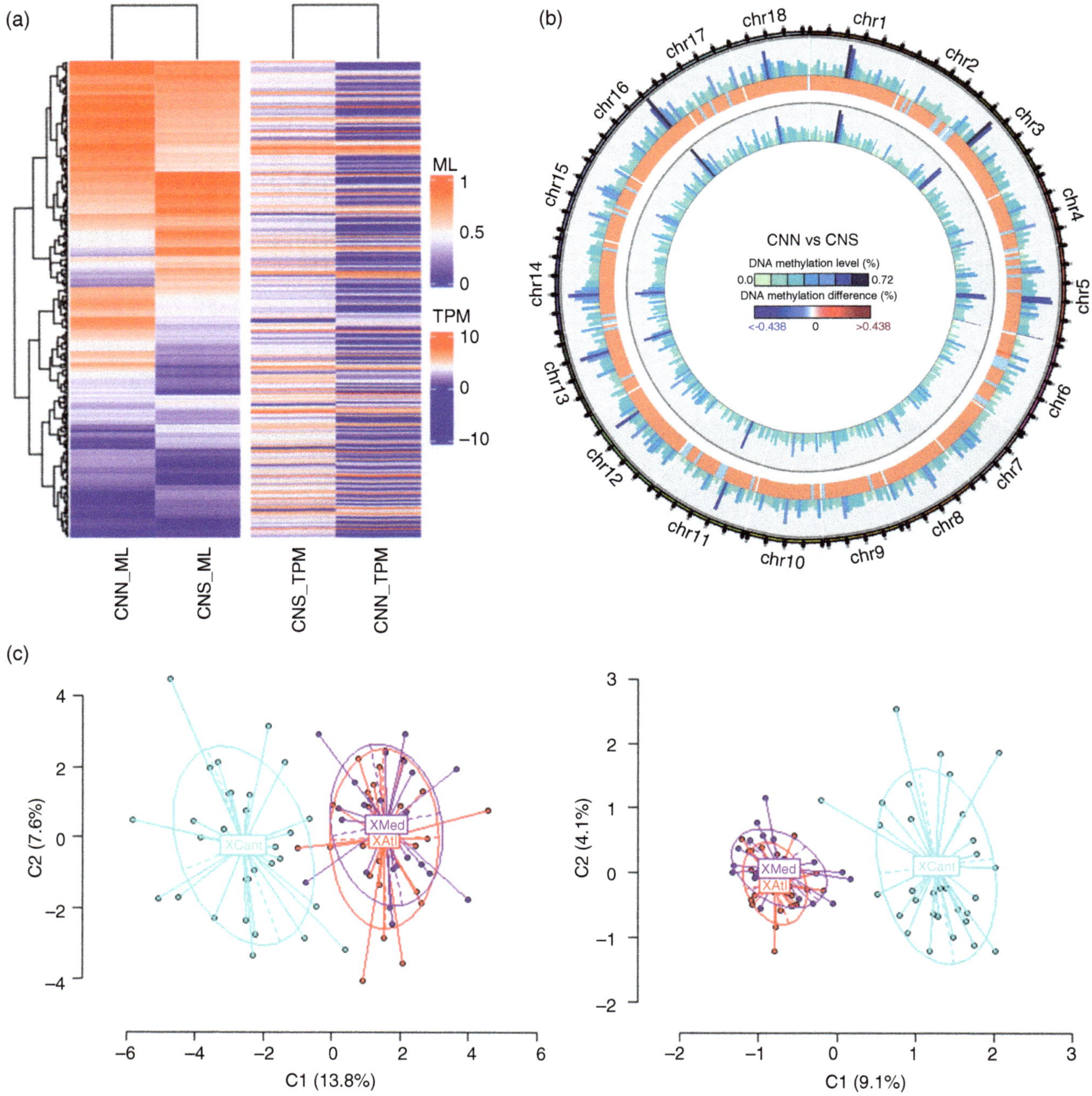

Figure 15.6 Epigenetic response in acclimatization in mollusks. (a) Heat map showing the methylation levels (MLs) and expression profiles (TPM) of the 253 overlapped genes in two different populations of *Haliotis discus hannai*: northern China (CNN) and southern China (CNS). (b) Circos plot showing the distribution of DNA methylation difference at CG context between CNN and CNS populations across *H. discus hannai* chromosome-scaled genome. The four circles from outer to inner represent chromosomes of *H. discus hannai*, the MLs of the CNN population, the differences of ML between CNN and CNS population (CNN vs. CNS), and the MLs of the CNS population, respectively. (c) Two-dimensional visualization of the Principal Component Analysis (PCoA) of detected methylation patterns in three European populations of *Xenostrobus securis*, with the epigenetic variation (methylation-sensitive loci) on the left and the genetic variation (no methylated loci, NML) on the right. The individuals of each population are represented by the acronyms XAtl, XCant, and XMed for the Atlantic international port, Cantabric international port, and Mediterranean lagoon populations, respectively. *Source:* (a, b) Huang et al. [135]/Frontiers Media S.A/CC BY 4.0; (c) Ardura et al. [136]/Springer Nature/CC BY 4.0.

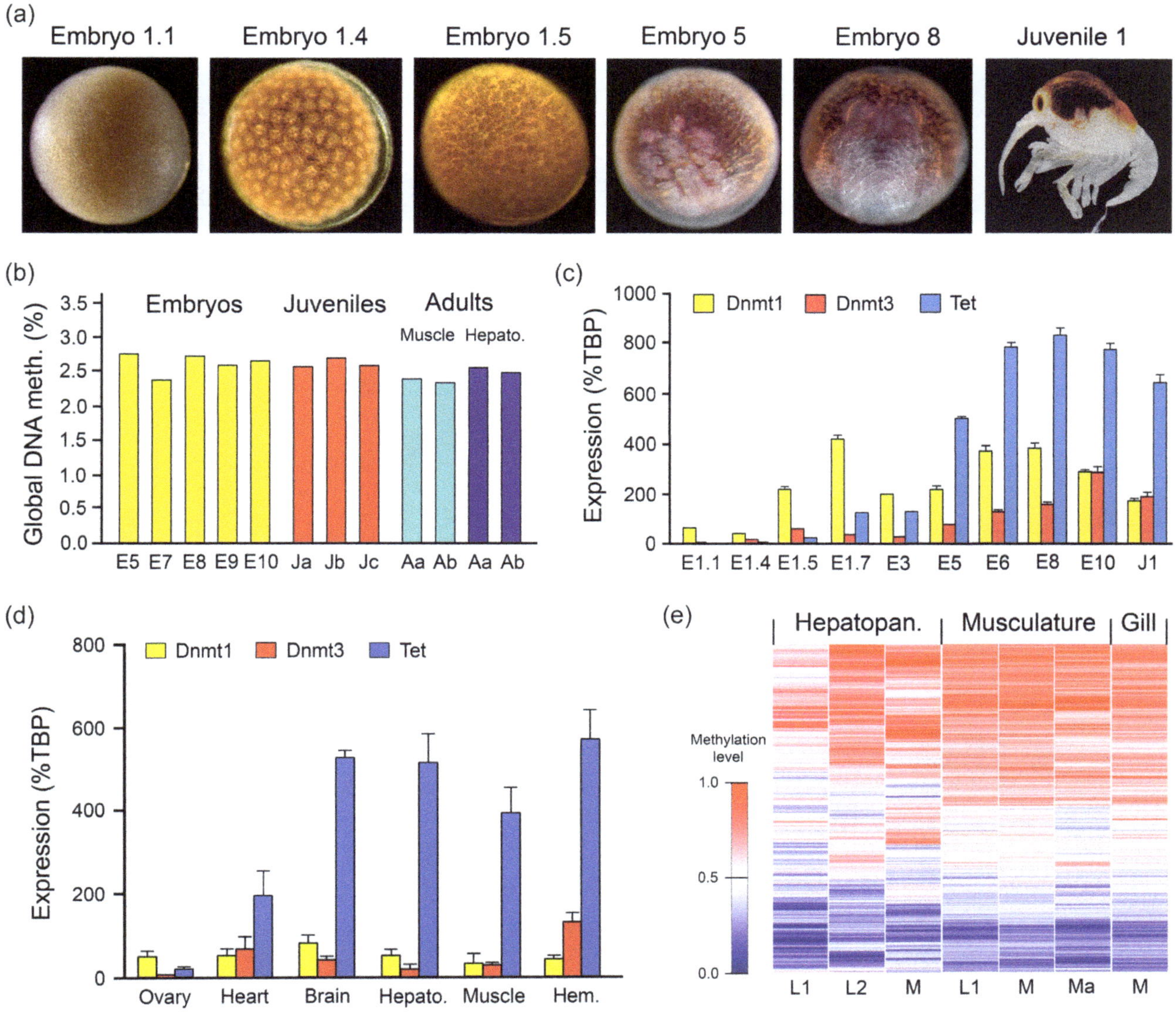

Figure 16.6 Dynamics of DNA methylation during the life cycle of marbled crayfish and methylation differences between tissues. (a) Phenotypic changes between spawned oocyte and first juvenile stage (hatchling). (b) Global DNA methylation levels in embryonic stages 7–10, three 154-day-old juveniles (Ja–Jc) and two 2-year-old adults (Aa and Ab). The DNA methylation level seems to decrease slightly with age. (c) Expression of DNA methylation enzymes Dnmt1 and Dnmt3 and demethylation enzyme Tet during embryonic development. Expression levels are given relative to the expression of the TATA-box binding protein (TBP). The Dnmt1, Dnmt3, and Tet genes show very low expression until embryo 1.4 and quite different expression dynamics thereafter. Embryo 5 corresponds to about 50% of embryonic development, in which tissues are not yet differentiated. Bars indicate means ± SD of three measurements. (d) Expression of Dnmt1, Dnmt3, and Tet in various tissues and organs relative to TBP expression. The graph shows marked differences between tissues for all enzymes. Bars indicate means ± SD of three measurements. (e) Comparative analysis of 697 variably methylated genes between hepatopancreas, abdominal musculature, and gills from two laboratory-reared specimens (L1 and L2), one specimen from Lake Moosweiher (M), and one specimen from a rice field in Moramanga, Madagascar (Ma). The heat map shows differences in methylation patterns between individuals and tissues. *Source:* (a) Left image from Alwes and Scholtz [90]/from Springer Nature; right image from Vogt and Tolley [91]/from John Wiley & Sons, pictures in between from Vogt [88]/from Springer Nature. (b–d) Adapted from Gatzmann et al. [21]. (e) Adapted from Tönges et al. [92].

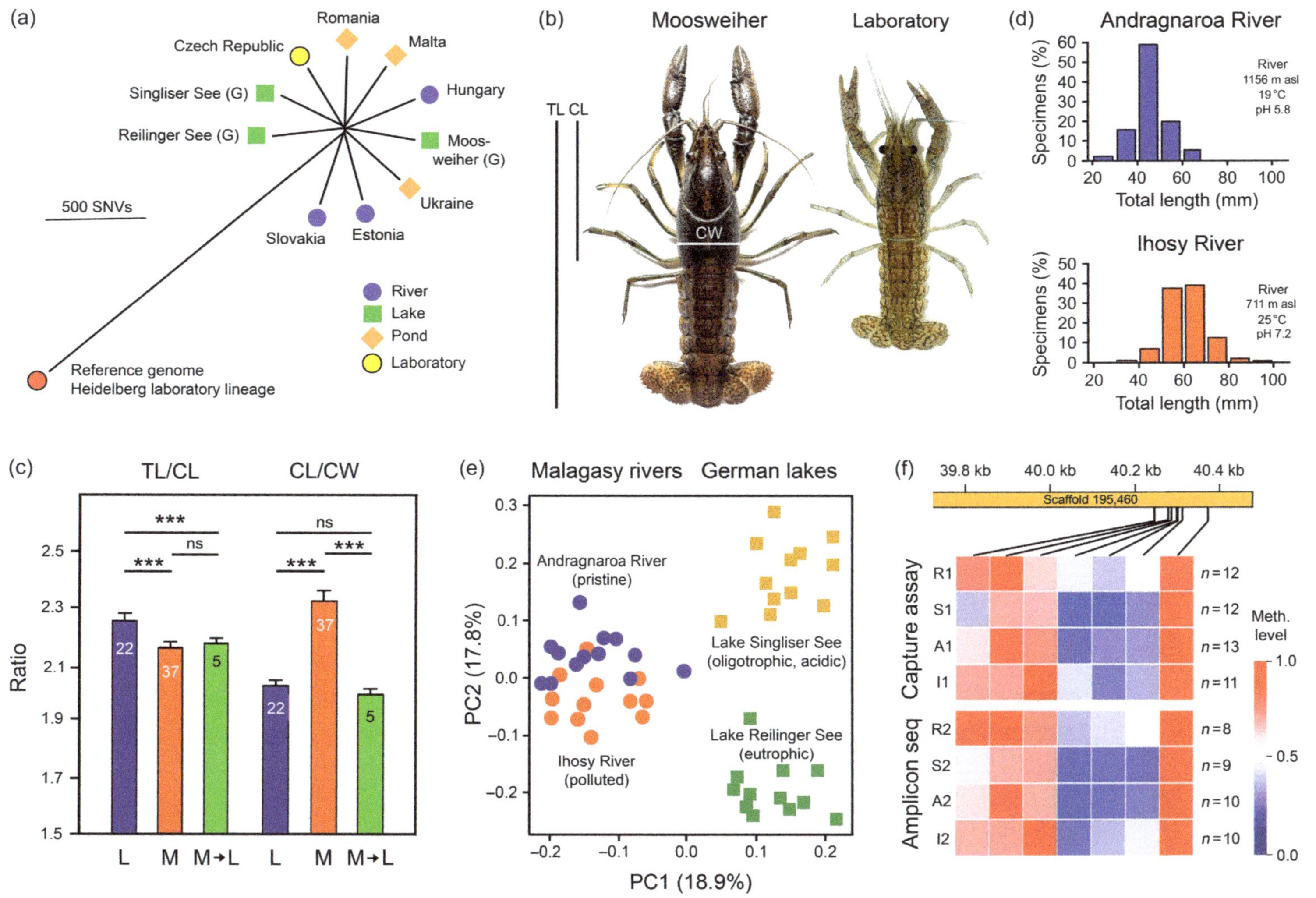

Figure 16.7 Genetic uniformity and phenotypic and epigenetic diversity in marbled crayfish adapted to different conditions. (a) Genetic difference on the whole genome scale between representatives from different European populations. A descendant of the oldest known marbled crayfish laboratory population was used as a reference. G, Germany. (b) Comparison of maximum body size and coloration between populations from the laboratory and wild. TL of the largest specimen from Lake Moosweiher (Germany) exceeded that of the largest specimen from my laboratory colony by about one-third. Coloration of the dorsal side was uniformly dark greenish-brown in the lake specimens, but quite variable in laboratory specimens. CL, carapace length; CW, carapace width. (c) Body proportions of specimens from my laboratory colony (L) and Lake Moosweiher (M), showing significant differences in total length (TL) to carapace length (CL) ratio and CL to carapace width (CW) ratio. The adult offspring of a specimen transferred from Lake Moosweiher to the laboratory (M → L) had a TL-to-CL ratio similar to the wild population, but a CL-to-CW ratio similar to the laboratory population. Figures in columns give numbers of specimens investigated; ***, significantly different ($p < 0.001$); ns, not significantly different. (d) Significant differences in population structure between the pristine and montane Andragnaroa River and heavily polluted Ihosy River in Madagascar. $p < 0.05$. (e) Principal component analysis of methylation of 122 genes separating four populations from rivers and lakes in Germany and Madagascar. (f) Persistent DNA methylation fingerprints of populations from Andragnaroa River (A), Ihosy River (I), Lake Reilinger See (R) and Lake Singliser See (S) in consecutive years (1 and 2), exemplified for a small genic region of the hepatopancreatic DNA. The samples were collected at intervals of 12–21 months and analyzed with two different methods. *Source:* (a) Adapted from Maiakovska et al. [79]. (b) Based on Vogt [97]/from Springer Nature. (c) Vogt [10]/Springer Nature. (d) Adapted from Andriantsoa et al. [83]. (e, f) Adapted from Tönges et al. [92].

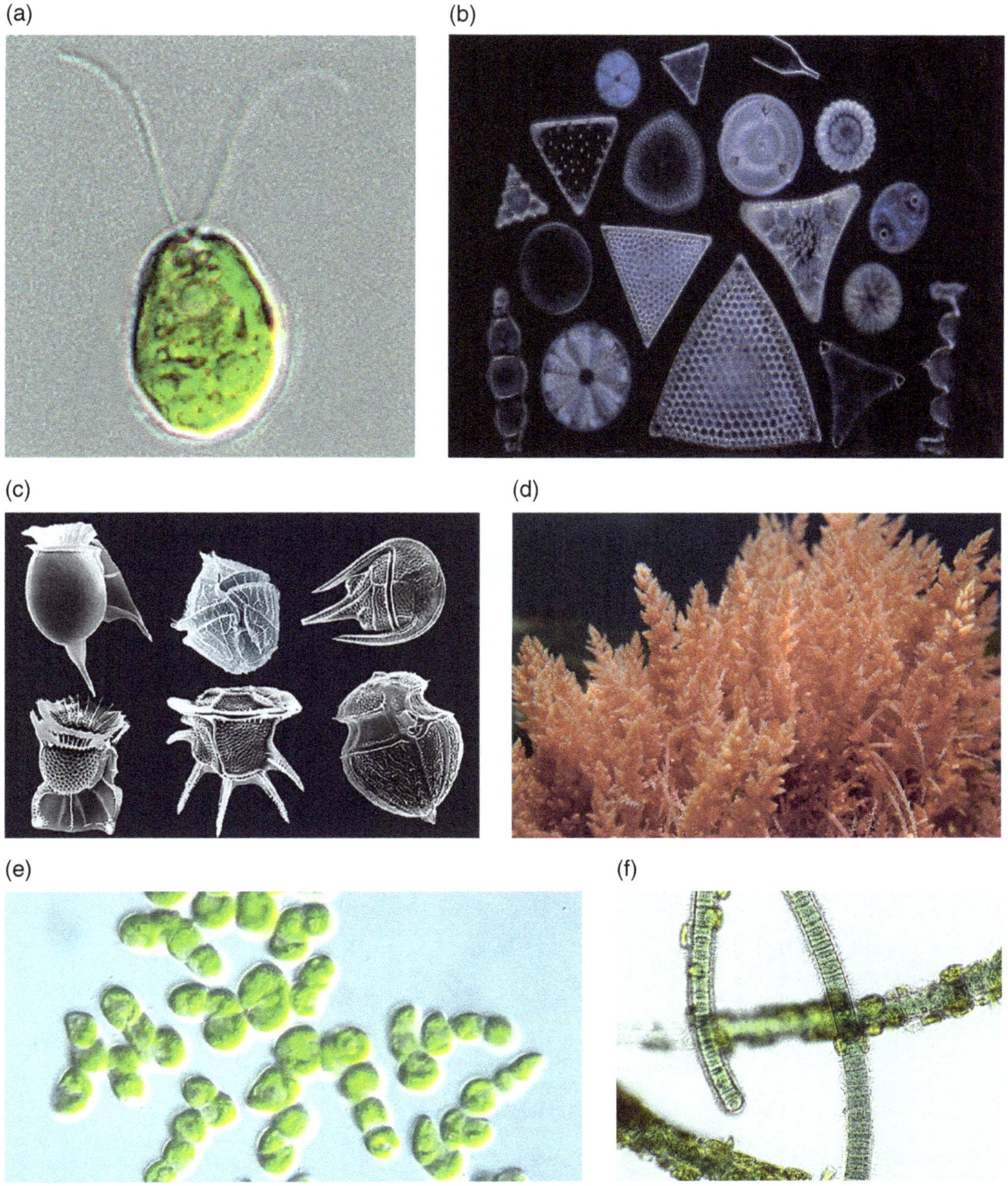

Figure 17.1 Images of the model algae species. (a) *Chlamydomonas reinhardtii*, *Source:* J. L. Carson, Ph.D/Science Source, (b) a collection of diatom algae, *Source:* Dr. N. Lange/Adobe Stock, (c) a collection of dinoflagellate algae, *Source:* Fickleandfreckled/Flickr/ CC BY 2.0, (d) Representative macro red algae, (e) representative of chlorophytes (green algae), *Source:* Lebendkulturen.de/Shutterstock, and (f) blue-green algae or cyanobacteria, *Source:* Elif/Adobe Stock.

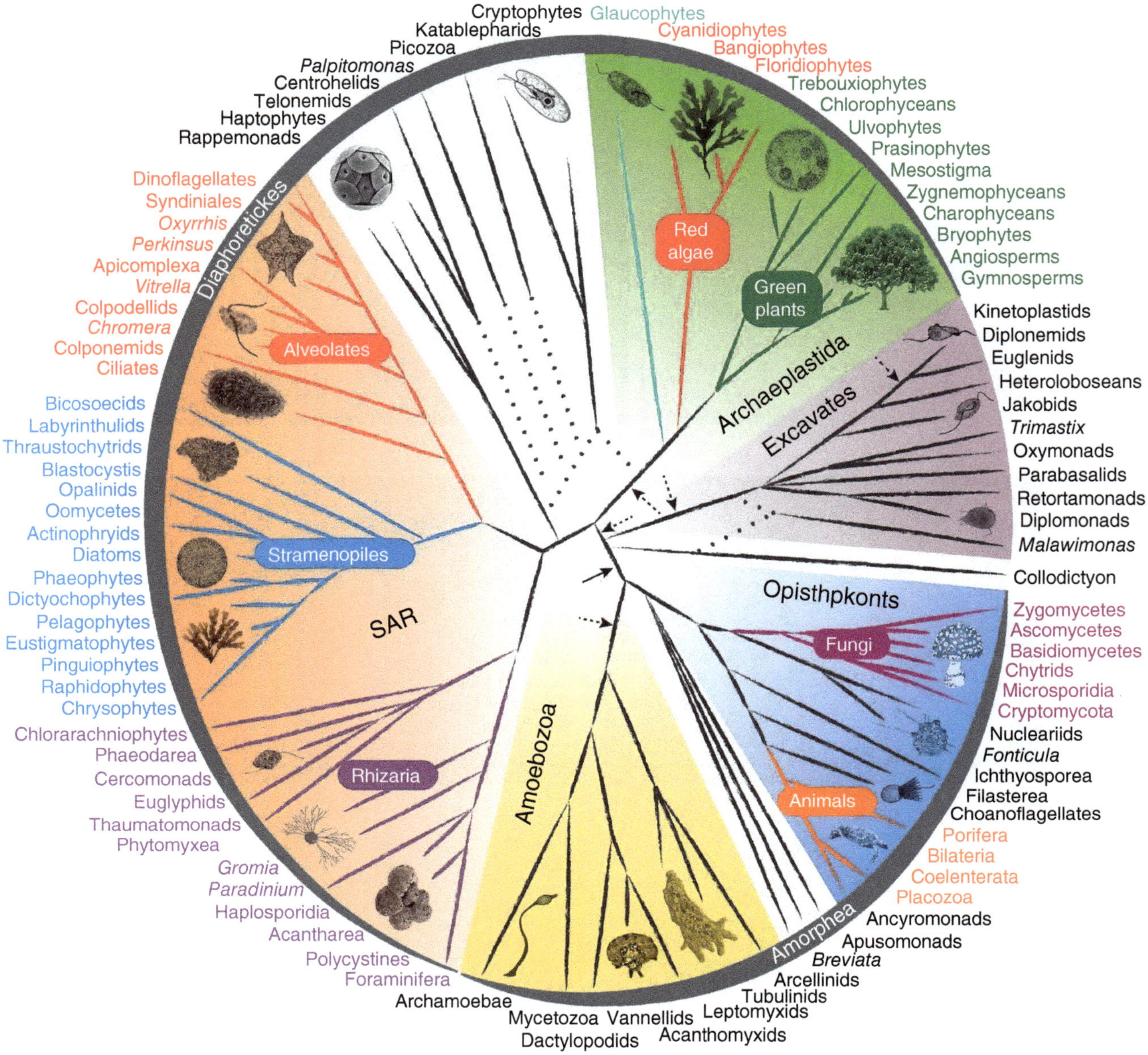

Figure 17.2 The tree of life can be modeled in several different ways, but in all models, algae species are spread across the tree. The red stars indicate the different families and phyla to which various algae belong across the tree of life. Epigenetics analysis has been performed in less than 25 total species, but these include algae from several different families and groups including stramenopiles (which include diatoms and brown algae), alveolates (which include dinoflagellates), red algae, and green algae (chlorophytes, prasinophytes, etc.). *Source:* Burki (2014)/Cold Spring Harbor Laboratory Press, and Delsuc et al. (2005)/Springer Nature.

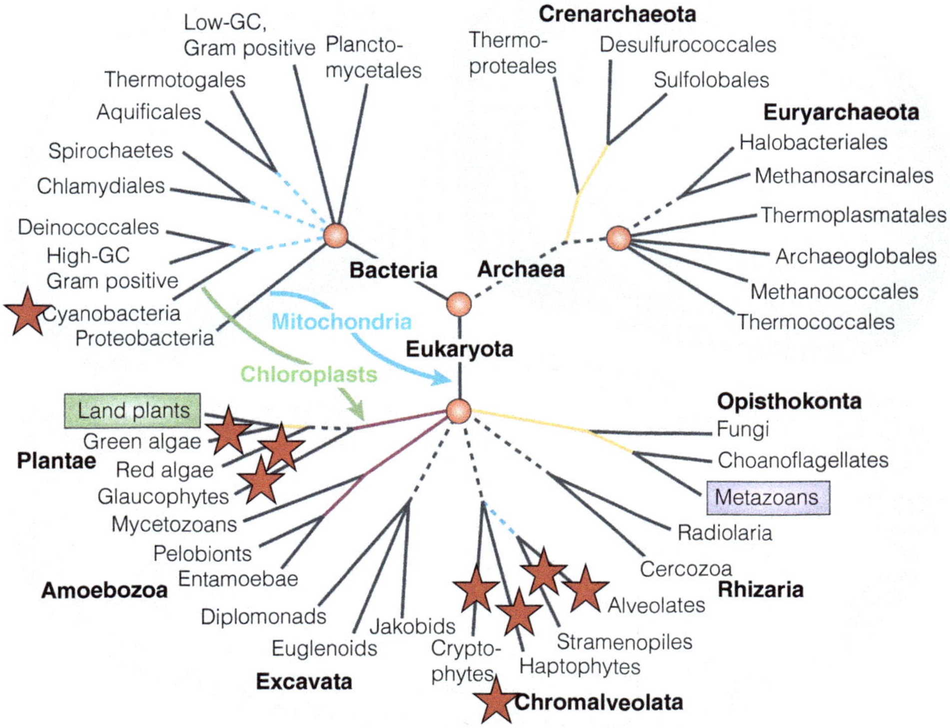

Figure 17.2 (Continued)

(a)

(b)

(c)

(d)

Figure 17.3 Harmful algal blooms occur when algae rapidly proliferate, consume the resources of an aquatic ecosystem, and release harmful substances. Numerous species can form blooms ranging from blue-green cyanobacteria ((a): Lake Erie, United States), *Source:* NOAA Great Lakes Environmental Research Laboratory/Flickr. Algal blooms can be observed from space ((b): Bering Sea, NASA/Public Domain.) (c) Dinoflagellates, *Source:* Gary/Adobe Stock and via satellite imagery ((d): Lake St. Clair, United States, *Source:* Joshua Stevens/NASA/Public Domain).

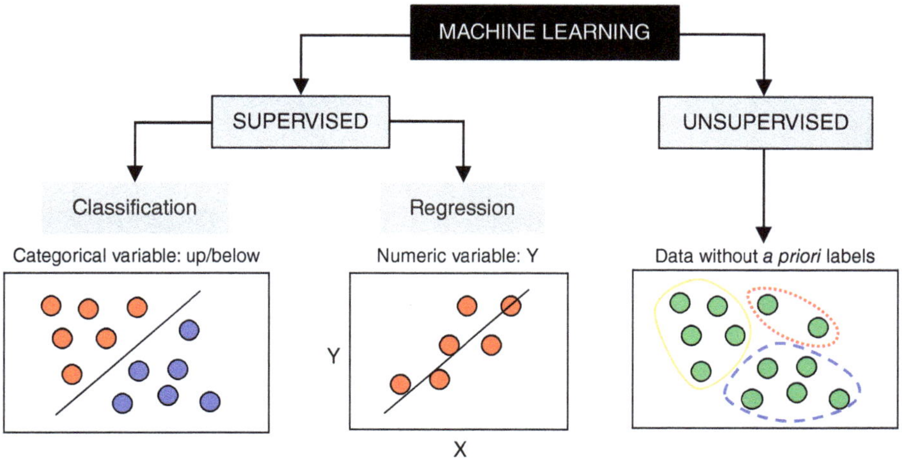

Figure 18.2 Main categories of machine learning (ML) methods. For all three subcategories, vector diagrams represent the task accomplished by each method. In classification, the algorithm searches for the line (threshold) that separates dots, where upper/lower dot groups can be, for example, male/female. In regression, the algorithm searches for the regression line that best fits the points, where *Y* can be age, growth, or any other continuous variable. In unsupervised methods, the algorithm searches for the optimal grouping (solid/dashed/dotted lines around data points).

	Past	Present	Future
Hatchery-rearing	C		
Development	C		
Maturation		C	R
Growth		R	R
Nutrition	C	C	C
Sex		C	C
Age	R	R	R
Health	C	C	C/R
Environment	C/R	C/R	C/R
Contaminants	C/R	C/R	
Morphology			C

Figure 18.3 Supervised ML methods for key traits and processes of aquatic organisms. EBs can be used for building classification (C) models for categorical outcomes or regression (R) models for numeric outcomes for each key trait and process. The type of the expected outcome may change over time: for example, the question can be at present whether an organism is mature (categorical), but in the future, how fast an organism will mature (numeric outcome). In another example, we may aim to know whether an environmental stressor has occurred (yes/no) or how much it has affected the organisms (numeric) in the past. The list presented here aims to cover the most common objectives for EB development, but is not exhaustive.

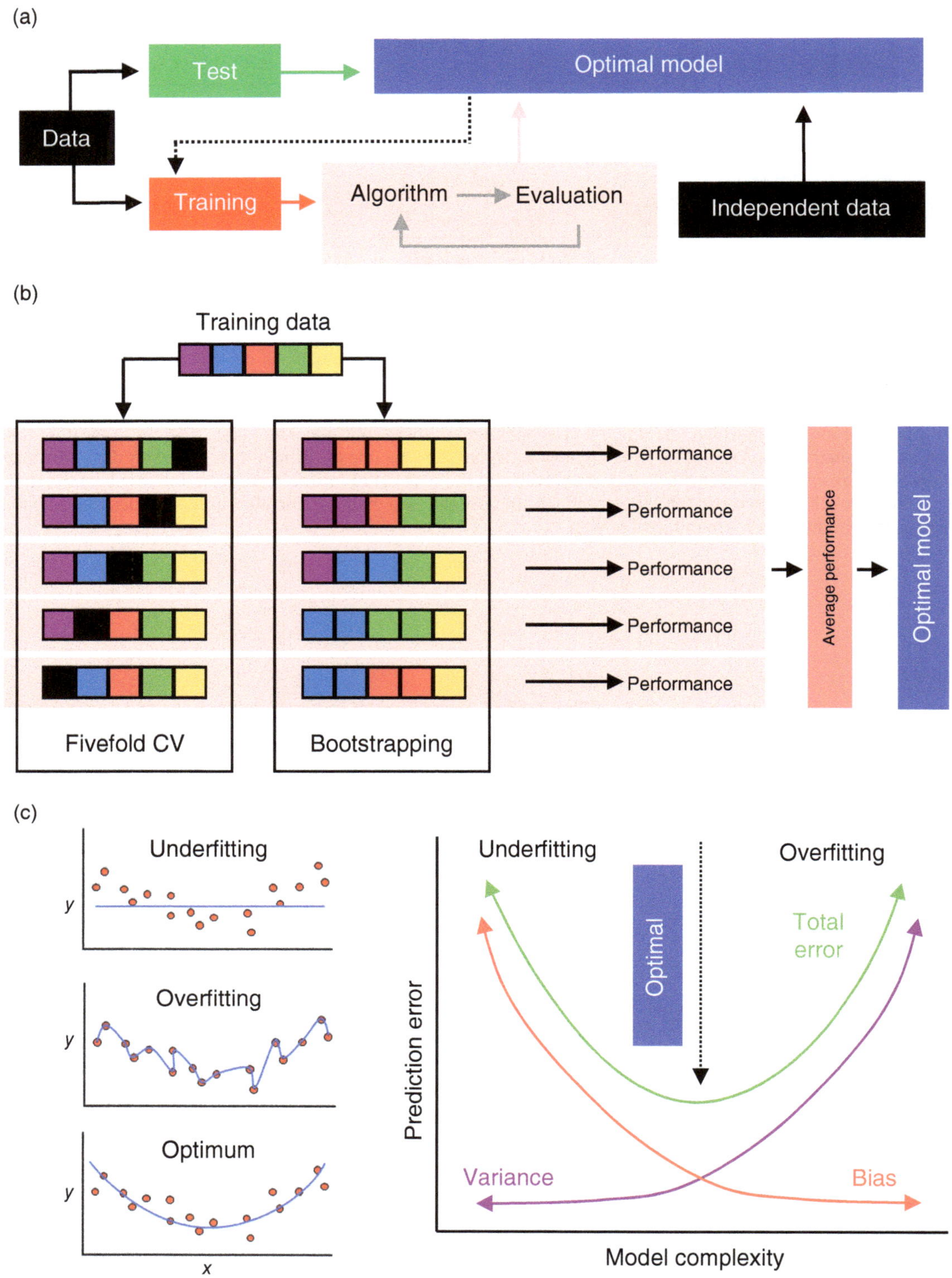

Figure 18.4 Basic concepts for ML. (a) Workflow for building the optimal model: data divided into test and training, the optimal model is chosen through repeated evaluation of the parameters using resampling methods (see c) and evaluated using the test data. The dotted line shows that the procedure might reiterate if the prediction of test data fails to perform well. Independent data can be predicted at the end of the procedure. (b) The two most common resampling techniques that allow the evaluation of the performance using training data: in k-fold CV, fivefold in this example, the training data are divided into fivefolds. Onefold is left aside during each round, the remaining fourfolds are used to build the model and the fifth is used to evaluate its performance. This occurs using all folds. A similar process is used in bootstrapping, where data are taken with replacement, allowing for duplicates to be present. The original observations not contained in each bootstrapping are then used to evaluate the performance of the model. Resampling methods lead to the estimation of an average model performance. (c) The optimal model aims to minimize the error resulting from tradeoff between bias and variance. Low variance/high bias result in underfitting, while high variance/low bias result in overfitting. The more terms are added to the model to avoid underfitting, the more complex a model becomes.

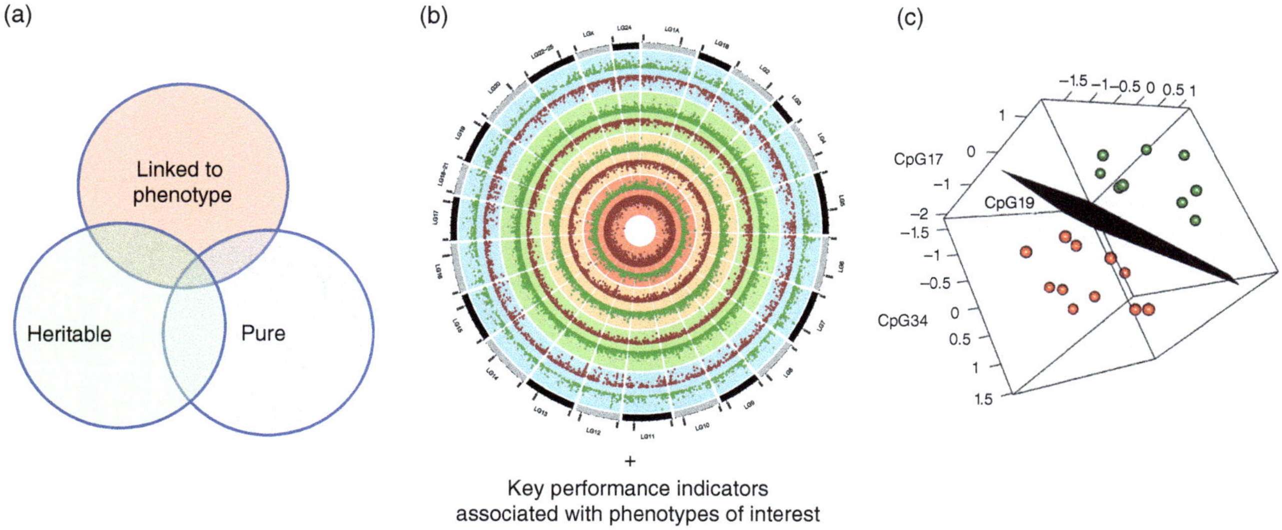

Figure 20.5 Development of epigenetic markers to aid in broodstock selection in aquaculture. (a) Epialleles to be selected should fulfil a series of criteria in order to get the most out of epigenetics. For example, they should be at least associated with a certain phenotype of interest, if possible heritable from parents to offspring, and, ideally, be PEs, meaning that their response to a given environmental cue should be in the same direction regardless of the underlying genetic background. (b) Epigenome-wide analysis and the measure of key performance indicators are conducted in the target population. (c) PCA-based plot showing the classification of fish based on a combination of desired traits such as survival, growth, and resistance to stress, done on a line of genetically selected fish.

www.ingramcontent.com/pod-product-compliance
Ingram Content Group UK Ltd.
Pitfield, Milton Keynes, MK11 3LW, UK
UKHW050845101026
15624UKWH00018B/580

9 781119 821915